VOLUME IX

In Preparation: CHEMICAL EXPERIMENTATION UNDER EXTREME CONDITIONS
Edited by Bryant W. Rossiter

VOLUME X

APPLICATIONS OF BIOCHEMICAL SYSTEMS IN ORGANIC CHEMISTRY, in Two Parts
Edited by J. Bryan Jones, Charles J. Sih, and D. Perlman

VOLUME XI

CONTEMPORARY LIQUID CHROMATOGRAPHY
R. P. W. Scott

VOLUME XII

SEPARATION AND PURIFICATION, 3rd Edition
Edited by Edmond S. Perry and Arnold Weissberger

VOLUME XIII

LABORATORY ENGINEERING AND MANIPULATIONS, 3rd Edition
Edited by Edmond S. Perry and Arnold Weissberger

VOLUME XIV

THIN LAYER CHROMATOGRAPHY, 2nd Edition
Justus G. Kirchner

TECHNIQUES OF CHEMISTRY

ARNOLD WEISSBERGER, *Editor*

VOLUME XIV

THIN-LAYER CHROMATOGRAPHY

Second Edition

TECHNIQUES OF CHEMISTRY

VOLUME XIV

THIN-LAYER CHROMATOGRAPHY

Second Edition

BY

JUSTUS G. KIRCHNER, Retired

Senior Scientist
The Coca-Cola Company

EDMOND S. PERRY, *Editor*

A WILEY-INTERSCIENCE PUBLICATION

JOHN WILEY & SONS
New York • Chichester • Brisbane • Toronto

Library of Congress Cataloging in Publication Data

Kirchner, Justus G., 1911–
Thin-Layer chromatography.

(Techniques of chemistry; v. 14)
"A Wiley-Interscience publication."
Includes bibliographies and index.
1. Thin layer chromatography. I. Title.
QD61.T4 vol. 14 [QD79.C8] 542'.08s [547'.34'92]
ISBN 0-471-93264-7 78-9163

Printed in the United States of America

10 9 8 7 6 5 4 3 2 1

To my daughter Dr. Grace L. Kirchner

INTRODUCTION TO THE SERIES

Techniques of Chemistry is the successor to the Technique of Organic Chemistry Series and its companion—Technique of Inorganic Chemistry. Because many of the methods are employed in all branches of chemical science, the division into techniques for organic and inorganic chemistry has become increasingly artificial. Accordingly, the new series reflects the wider application of techniques, and the component volumes for the most part provide complete treatments of the methods covered. Volumes in which limited areas of application are discussed can be easily recognized by their titles.

Like its predecessors, the series is devoted to a comprehensive presentation of the respective techniques. The authors give the theoretical background for an understanding of the various methods and operations and describe the techniques and tools, their modifications, their merits and limitations, and their handling. It is hoped that the series will contribute to a better understanding and a more rational and effective application of the respective techniques.

My special thanks are due to Dr. Edmond S. Perry for assistance in the editorial work on the present volume and other volumes dealing with techniques of Separation and Purification.

Authors and editors hope that readers will find the volumes in this series useful and will communicate to them any criticisms and suggestions for improvements.

ARNOLD WEISSBERGER

Research Laboratories
Eastman Kodak Company
Rochester, New York

FOREWORD TO THE FIRST EDITION

During the past fifteen years thin-layer chromatography has been established as one of the most powerful, exacting, and useful tools for the chemist in the laboratory. The technique is relatively simple, the equipment required inexpensive. Applicable to both volatile and nonvolatile substances, thin-layer chromatography is useful in many fields—vitamins, steroids, pharmaceuticals in general, synthetic organic materials, dyes, essential oils, resins, pesticides, etc. In many cases it offers the only practical solution of a perplexing problem. Combinations of thin-layer chromatography with other techniques, vapor-phase chromatography among them, are being explored at present by researchers throughout the world.

The present volume by Dr. Justus Kirchner presents an up-to-date and comprehensive treatment of the subject. The first section describing techniques shows the familiarity of the author with the subject from the practical point of view and the thoroughness with which he has studied existing literature. The second section brings together in one place the many and varied applications that have been made of the technique.

We heartily welcome Dr. Kirchner's *Thin-Layer Chromatography*, and are convinced that this work will be indispensable on the book shelves of laboratories in the scientific world as well as in industry everywhere.

ERNEST GUENTHER, PH.D.

Senior Vice President
Fritzsche Brothers, Inc.
New York

FOREWORD TO THE SECOND EDITION

It was in 1951 that the most important paper for thin-layer chromatography was published. In this journal article, Dr. Justus G. Kirchner reported separating terpenes on open layers of silica gel, which he referred to as "chromatostrips." From this auspicious beginning, thin-layer chromatography has become one of the most useful and practical chemical separation techniques. This is most evident by the recent tremendous proliferation of papers published in which thin-layer chromatography is a key method.

The vast quantity of papers published on thin-layer chromatography makes this book a prerequisite not only for routine laboratory work but also for obtaining a thorough theoretical background. Though everyone has different needs and problems, this well-organized book will continually serve the reader as an invaluable ready reference.

RICHARD K. VITEK

New Berlin, Wisconsin

PREFACE TO THE FIRST EDITION

Although the basic principle of using thin layers of adsorbent for chromatographic separations was used by Izmailov and Schraiber in 1938, and by Meinhard and Hall in 1948, it was not until the author dispensed with the idea of drop chromatography, employed by the earlier workers, and originated the use of uniformly coated glass strips and plates for development in a manner analogous to paper chromatography that the possibilities of the method were uncovered.

Since the present system of thin-layer chromatography was introduced by the author in 1951 the method has found wide application in the various fields of chemistry. The past two years have seen a tremendous increase in the number of papers on the specific subject of thin-layer chromatography and also in the number of papers where it has been used as a tool to help in solving a given problem.

The book is divided into two sections, the first presenting the techniques of the method and the second covering the many applications that have been made from its inception through the year 1964, including partial coverage of 1965. In order to make the book complete, a chapter on inorganic ions has been included.

It is with pleasure that I acknowledge my indebtedness to Dr. C. A. Shillinglaw for his interest in this work. I express my appreciation to the following people: to my wife Mildred M. Kirchner for constant encouragement, to Louise M. Blood for typing the manuscript, to Ludwig Renner for preparation of the drawings, to my daughter, Grace L. Kirchner, for assistance in translating, and to Dr. James M. Bobbitt for the use of his reprints.

JUSTUS G. KIRCHNER

Scotch Plains, New Jersey

PREFACE TO THE SECOND EDITION

Recently uncovered work has shown that the basic principle of thin-layer chromatography was demonstrated by Beyerinck in 1889 and again in 1898 by Wijsman. Beyerinck's work predated that of Ismailov and Schraiber in 1938 and indeed even that of Reed on column chromatography in 1893.

Since the first edition of this book in 1967 the Kirchner technique of thin-layer chromatography that I introduced in 1951 has continued to expand, as evidenced by the increase in publications and introduction of new techniques and modifications. The 5000 to 6000 publications available up to 1967 have grown to an estimated 15,000 to 20,000, of which more than 6000 have been cited in this work. The literature has been thoroughly covered through 1975, with some work throughout 1976 and a few from 1977.

Some of the chapters, as for example that on quantitative work, have been completely revised, and all have been modified and updated. The general outline of the first edition has been retained. New techniques and special notations to be considered in using thin-layer chromatography in many fields have been brought together from many sources to assist the worker in solving his or her problems and obtaining the best results. The chapter on detection reagents has been revised and expanded, and a cross index for locating detection reagents for types of compounds or for specific compounds has been added in Appendix A. Rather than include an author index, which takes up valuable space and serves only to show that a given author's work has been cited, I have devoted the space to a thorough subject index where subjects of interest may be readily located. A separate compound index is included so as to make the location of a specific subject easier without searching through columns of compound names.

I acknowledge my indebtedness to Dr. Darshan Bhatia, Director of Corporate Research and Development of the Coca-Cola Company, for making available the facilities of the Technical Information Services and to Bernard Prudhomme, Mary Jane Montesinos, Brenda Pierce, and Cindy Blair of that department for locating and obtaining sources of information. Recognition is extended to Dr. Victor Krample for assistance with Russian, Hungarian, and Czech translations and to Ellen Villanueva for Spanish translations. Thanks are also due Nathaniel Davis for assistance in supplying copies of references.

JUSTUS G. KIRCHNER

Dunwoody, Georgia
May 1978

CONTENTS

PART I TECHNIQUES OF THIN-LAYER CHROMATOGRAPHY

TECHNIQUES OF CHEMISTRY

ARNOLD WEISSBERGER, *Editor*

VOLUME XIV

THIN-LAYER CHROMATOGRAPHY

Second Edition

Part I

TECHNIQUES OF THIN-LAYER CHROMATOGRAPHY

Chapter I

INTRODUCTION, HISTORY, AND GENERAL DESCRIPTION

1 INTRODUCTION

Farradane [1] has pointed out that the credit for the first recorded work on column chromatography should go to Reed [2], who published in 1892 on the use of tubes of kaolin for the separation of potassium chromate from eosin and of ferric chloride from copper sulfate. Prior to this, the idea of paper chromatography appears to have originated with Schoenbein [3] in 1861, with further developments by his pupil Goeppelsroeder [4–7], who called the technique capillary analysis.

Tswett's [8] great contribution to chromatography was his use of pure solvents to develop the chromatogram. Tswett's first experiments (published in 1903) were concerned with the separation of the pigments from a leaf extract by passing a petroleum ether solution through a column of calcium carbonate. The yellow and green pigments separated, and after developing the column with pure solvent, Tswett cut up the column and eluted the various pigments that remained on the column with alcohol, the carotene having passed through the column with the petroleum ether solvent. In this example the bands were readily visible, but some method was needed in order to be able to see what was happening in chromatographing the many colorless compounds which are invisible on the column.

Since that time many methods have been devised to detect the chromatographic bands. Tswett himself proposed the addition of a pigment to a colorless solution in order to relate the colorless zones to a visible standard. The majority of zone-detecting techniques for columns are applied to the eluate as it leaves the column and range all the way from measurement of the refractive index to specific chemical tests for individual compounds. On the column itself techniques such as arbitrary cutting of the column, detection by means of strongly fluorescent columns [9–11], and extrusion from the glass envelope with the brushing on of an indicating reagent [12] have been used to advantage. Miller and Kirchner [13], in 1951, applied the principles of thin-layer chromatography to a self-contained adsorbent column without the usual confining glass tube, so that the column could be sprayed with indicating reagents.

Column chromatography as used by Tswett is concerned with the distribution of compounds between the solid adsorbent and the solvent. In the simplest case, as the developing solvent moves down the column of previously adsorbed material, the latter is desorbed to the extent that an equilibrium is set up between adsorbed material and the material in solution. Then as this solution travels further down the column it comes in contact with fresh adsorbent, and dissolved material is again picked up by the adsorbent. In this way the adsorbed material progresses down the column. By this process, a compound for which the adsorbent has a strong affinity will displace compounds which are less strongly adsorbed. This results in a series of bands or layers with the most strongly adsorbed on top followed in order by less and less strongly adsorbed materials. If the adsorption affinities are close together the bands may appear adjacent to one another, whereas with more widely differing adsorption affinities the bands will appear separated by empty bands of adsorbent. Theoretically, given a long enough column, any two compounds which have different adsorption affinities under a given set of conditions can be separated by chromatography.

A variation of ordinary column chromatography was introduced in 1941 by Martin and Synge [14] with their concept of partition chromatography. Here the column becomes a support for the liquid aqueous phase with which it is treated. Through this is passed a nonmiscible solvent so that a partitioning of the substances to be separated occurs between the two solvents. By placing the hydrophobic solvent on the column and using the hydrophilic solvent as the moving phase one obtains a reverse-phase partition chromatogram.

The paper chromatography introduced by Schoenbein [3] and Goeppelsroeder [4–7] as capillary analysis in 1881 lay dormant until reintroduced by Consden et al. [15]. This technique has played an important role in the analysis of amino acids but the two main drawbacks are its inherent slowness and its very small capacity. In contrast to this, thin-layer chromatography has a high capacity with rapid development which does not detract from its sensitivity. In direct comparison between paper and thin-layer the latter has been found to be more sensitive in detecting small amounts. Fahmy et al. [16] have shown that thin-layer chromatography of amino acids on silica gel G is more than 10 times as sensitive as paper chromatography. Even considering paper chromatography itself it has been shown that cellulose layers on glass plates are faster and give sharper separations than the corresponding work with paper chromatography [17, 18].

Some attention should be given to the term thin-layer chromatography. Although the terms "chromatostrip" and "chromatoplate" were first used in 1951 [19] and 1954 [20], for this present-day method of analysis, the designation thin-layer chromatography (TLC) has become so widely accepted in referring to the general method that terms such as thin film and open column should be discarded.

2 HISTORY OF THIN-LAYER CHROMATOGRAPHY

Thin-layer chromatography actually began with the Dutch biologist Beyerinck, in 1889 [21], when he allowed a drop of a mixture of hydrochloric and sulfuric acids to diffuse through a thin layer of gelatin. The hydrochloric acid traveled faster than the sulfuric acid and formed a ring around the sulfuric acid. The hydrochloric acid zone was made visible by brushing on a solution of silver nitrate, and the sulfuric acid was made visible with barium chloride. Nine years later Wijsman [22], using the same techniques, showed the presence of two enzymes in malt diastase, and also showed that only one of them split off maltose from soluble starch. Wijsman was also the first to use a fluorescent phenomenon for detecting a zone on a thin layer. He incorporated fluorescent bacteria from seawater in a gelatin layer containing starch and allowed the amylase mixture to diffuse in the layer. A fluorescent band appeared only where the β-amylase reacted with the starch. This also happened to be one of the more sensitive visualizing agents encountered in thin-layer work. He was able to detect 1/28,000,000 of a milligram of maltose. This places it at about 40 pg. This bit of history was brought to light by Van Klinkenberg [23].

Izmailov and Schraiber [24], in 1938, discussed the use of a thin layer of aluminum oxide spread on a glass plate. It did not contain a binder and was used for circular chromatography by placing a drop of the solution on the adsorbent and developing into concentric zones with solvent drops. They also pointed out its usefulness for testing adsorbents and solvents for column chromatography.

Lapp and Erali [25], in 1940, published a method of loose-layer chromatography. This was accomplished by spreading a layer 8 cm long on a glass slide. This was then placed on an inclined aluminum sheet which was cooled at its upper end and heated at the lower end. The mixture to be separated was placed at the top of the adsorbent and gradually washed down with a developing solvent.

Crowe [26], in 1941, used a similar technique for selecting solvent and adsorbent combinations. This was accomplished by placing the adsorbent in the cups of spot plates. After the selection was made a thin, wedge-shaped layer of adsorbent was formed in a petri dish by tilting the dish. A drop of the solution flowed onto the adsorbent and was then developed dropwise with solvent.

Békésy [27], in 1942, used a layer of adsorbent between glass plates held apart by cork gaskets. This was filled with an adsorbent slurry and used in the manner of column chromatography. A microcolumn was formed from a glass plate with a shallow channel covered by another glass plate. The two were held together with agar and the shallow column was filled with an adsorbent slurry.

Earlier, in 1939, Brown [28] demonstrated the use of paper chromatography by placing filter paper between two glass plates, the upper one of which contained a small hole for application of the sample and the developing solvent; this resulted in a circular chromatogram. To compensate for the mild adsorption strength of the paper, he proposed the use of a thin layer of alumina between the two sheets of paper. Williams [29], using the same idea, eliminated the paper and used only the adsorbent between the glass plates.

Meinhard and Hall [30], working with inorganic ions, used a starch binder to hold the adsorbent (in this case a mixture of aluminum oxide and a filter aid, Celite) on microscope slides. The Celite was used to prevent checking of the surface of the finished slides. Radial chromatography was employed using a special developing pipet to slowly apply the solvent.

During the period 1945–1954 my associates and I were engaged in isolating and identifying the flavoring components present in citrus juices. Because of the minute quantities of material present in the fruit it was desired to have a microchromatographic method for the purification and identification of the terpenes. Paper chromatography was tried, but it was soon obvious that this was not suitable because of the limited adsorption strength of the paper. Impregnation of the paper was then tried in order to increase the adsorbing strength, and Kirchner and Keller [31] at this time originated the use of silica-impregnated paper. Although this treated paper was promising, there were drawbacks to its use, particularly the tedious preparation of the paper and its still limited capacity. About this time the paper of Meinhard and Hall [30] appeared, and it occurred to me that by suitable modifications a technique could be evolved which would combine the advantages of column and paper chromatography. These modifications were as follows: (1) elimination of the filter aid to obtain a stronger adsorbent and a firmer surface for writing, made possible by careful selection of a starch binder which would not crack; (2) use of other adsorbents and especially silicic acid, that is, silica gel; (3) classification of the adsorbent using only that material which passed through a 100-mesh sieve (149 μ); (4) use of gypsum (plaster of Paris) as an inorganic binder in place of starch whenever the latter would interfere with the color developing agent; (5) use of larger strips and plates to allow for greater development and, consequently, more effective separation; (6) development of these adsorbent-coated glass sheets in a closed tank by the ascending technique analogous to paper chromatography; (7) use of adsorbent-coated plates for two-dimensional chromatography; and (8) development of visualizing agents which would not only locate the compounds but would indicate the type of compounds present. The first paper on the results of this work was published by Kirchner et al. in 1951 [19]. Additional work on modifications and applications of the method was published in a series of papers [13, 32–38] during the years 1952–1957.

Although quite a number of workers used this method successfully during the period 1951–1958, the method attracted little attention at that time. This can be likened to the original column chromatography of Tswett, which also went relatively unnoticed until 1931 (Zechmeister and Cholnoky [39]). It was only after the wide publicity given the thin-layer method by Desaga and Merck in advertising the availability of equipment and adsorbents that the method became popular. It was also during this period that *Chemical Abstracts* recognized thin-layer chromatography as a separate method and listed it as such in the indexes, thus making it easier to locate information on the method.

Larger plates ($5\frac{1}{4} \times 5\frac{1}{4}$ in.) were used by Kirchner et al. [19] for two-dimensional chromatography. Later, Reitsema [20] used these larger plates and originated the term "chromatoplate."

Mottier and Potterat [40, 41], using nonbound alumina layers, developed a procedure in 1952 for separating food dyes by circular chromatography. The solvent in this case was added dropwise.

In 1956 Stahl [42] published his first paper on thin-layer chromatography and asserted that he had improved the method by eliminating the difficulties of "complicated paste production with binders and the troublesome preparation of plaster strips." This was done by using an extremely fine grain (0.5 to 5 μ) silicic acid which he maintained adhered to the glass sufficiently to eliminate the need for binders. Later [43], he returned to the uniform gypsum-bound layers (the so-called silica gel G) originated by Kirchner et al. [19].

Although extremely fine grain silica gels are commercially available so that thin layers can be made to adhere to glass without a binder, these layers are rather soft and research workers have preferred the bound materials.

The advantages of thin-layer chromatography as pointed out by Kirchner et al. [19] may be reiterated here. It combines the best features of paper and column chromatography, that is, ease of locating compounds by spraying with various reagents plus the wide range of adsorbents available in column chromatography. It is very rapid since in some solvents a chromatogram may be run in 30 min. The method can be used for rapidly checking solvents and adsorbents. More drastic reagents can be applied to locate compounds than can be used with paper chromatography. It is a microchromatographic method and can be used when available material is at a minimum.

Any adsorbent used in column work can be applied to thin-layer work, although in some cases the particle size has to be reduced in order to produce a satisfactory thin-layer surface. Kirchner et al. [19] examined 15 different adsorbents and selected silicic acid as the best all-around adsorbent for essential oils. The more commonly used adsorbents have been silica gel, alumina, cellulose, polyamide, and kieselguhr (diatomaceous earth). Numerous other adsorbents have been tried and the commercial market is continually offering new adsorbents especially prepared for this work.

Numerous reviews have been written on thin-layer chromatography and some of these are referred to in the specific sections where they apply. The bibliographies available to the worker in this field include the following: those that appear regularly in various issues of *The Journal of Chromatography* and that are periodically published cumulatively as supplements of that journal by Macek et al. [44–46]; the annotated bibliography (replete with errors) of Haywood [47]; and Jaenchen's bibliography (48), which is published quarterly by CAMAG and which contains in addition a listing of the adsorbents, the solvents, the detection agents, and the quantitative methods used as well as other pertinent facts occasionally. The latter bibliography is also published periodically in cumulative form [49–50b]. There have been two abstracts published by Scott [51, 52] as well as the current abstract series by Saxby [53].

3 GENERAL DESCRIPTION OF THE METHOD

In thin-layer chromatography an adsorbent is applied to a supporting plate in a thin layer. Generally, a binding agent is used to adhere the adsorbent to the support, although some work is done without a binder using very finely divided adsorbent which clings to the support and forms a rather soft layer. This is to be distinguished from the "loose-layer" chromatograms in which the adsorbent does not adhere to the supporting plate and must therefore be developed in a horizontal or near horizontal position. The mixture of adsorbent and binder is applied as a thin slurry and the excess moisture is removed under varying conditions depending on the adsorbent, the binder, and the desired degree of activity. After the starting point is marked about 1.5 cm from the bottom of the plate, the finish line is marked a convenient distance from the starting point. This is done with a very soft lead pencil and care is taken not to disturb the adsorbent layer at the point of sample application since this leads to deformed spots. (It is convenient in marking the finish line to cut through the layer to the support so that a definite break occurs in the adsorbent film. In this way when the solvent reaches the line, development is automatically stopped.) The solution of the compound is deposited at the starting line by means of a micropipet and the plates are then placed in a closed container containing a layer of solvent about 0.5 cm deep. The solvent ascends the plate by capillary attraction until it reaches the finish line, at which time the plate is removed and the solvent allowed to evaporate. The locations of the various substances are then determined by some method which will make the colorless compounds visible, and many different spraying agents have been developed for this purpose. In some cases these are specific in nature, such as *o*-dianisidine for the detection of aldehydes; in other cases they are more general in nature, such as sulfuric acid with an oxidizing agent for detecting most compounds [19]. The R_f value is then determined by measuring from the center of the spot to the origin. The R_f value has the same meaning as in paper chromatography and is defined as the ratio of the distance traveled by the compound to the distance traveled by the solvent. The latter is measured from the origin to the solvent front. Sometimes the R_f value is referred to that of a substance used as a standard, in which case

$$R_{st} = \frac{R_f \text{ of the compound}}{R_f \text{ of the standard}}$$

There are many variations to this general method such as descending development, multiple development, stepwise development, and two-dimensional chromatography; these are all discussed in detail later.

4 THE CHROMATOGRAPHIC PROCESS

The theoretical aspects of column adsorption chromatography are applicable to thin-layer chromatography which, after all, is a microcolumn. No attempt is made here to repeat the mathematical development, but for the reader interested in the theoretical treatment reference is made to the following papers: Wilson [54], DeVault [55], Weiss [56], Offord and Weiss [57], Glueckauf [58–60], Sillén [61], Baylé and Klinkenberg [62–64], Langvad [65], Smit and van den Hoek [66, 67], Klamer and van Krevelen [68], Dixon [69], McQuarrie [70], Tudge [71], and Snyder [72, 73]. Of special interest also is the work of Pataki [74], Brenner and Pataki [75], Belenky et al. [76], Dixon [77], Pollak [78], Rachinskii [79], Rachinskii and Davidova [80], Snyder and Saunders [81], Soczewiński [82, 83], Soczewiński and Golkiewicz [84, 85], Soczewiński et al. [86], Stewart [87], and Turina et al. [88], which deal directly with thin-layer chromatography.

Briefly, the adsorption of a compound from a solution onto a solid surface takes place at the interface of the solid and the liquid. At equilibrium when the number of molecules being adsorbed is equal to the number being desorbed in a given unit of time, the situation can be expressed by the equation

$$a = \frac{c_a}{c_s}$$

where a is the distribution coefficient and c_a and c_s are the amounts adsorbed and in solution, respectively. The equilibrium is affected by the temperature and concentration of the solution, but because chromatographic separations are carried out under essentially adiabatic conditions, the concentration is the only factor with which we are really concerned.

In general, the adsorption coefficient is not a constant but varies with the concentration of the solution. If the adsorption increases rapidly with the dilution of the solution, a graph of the amount adsorbed versus the concentration of the solution shows a parabolic curve with the adsorption coefficient essentially linear at the lower concentrations. The steeper the curve, the more strongly the substance is adsorbed and, consequently, the slower is the rate of travel across the adsorbent with the flow of solvent. Thus the strength of the adsorption as well as the rate of flow of the solvent affects the rate at which the adsorbed component (the adsorbate) is transported across the thin-layer plate.

After the sample is spotted on the thin-layer plate, the chromatogram is developed by allowing a solvent to travel across the plate by capillary attraction. As the solvent moves over the adsorbed spot the equilibrium is shifted and compounds are desorbed, the more tightly adsorbed material to a lesser extent than the loosely held material. This redissolved material is then carried to the edge of the spot where fresh adsorbent is contacted and adsorption now takes place under the new equilibrium conditions. In this way the composition of the moving solution is continuously changing by the interchange of material between the adsorbent and the solvent.

Given two compounds which will adsorb at the same site on the adsorbent, the more strongly adsorbed material will displace the less strongly adsorbed material [89], thus causing the displaced material to form a zone or spot further away from the origin. The closer the adsorption coefficients of two compounds, the more difficult is the separation until the point is reached where a separation cannot be achieved under the given conditions. (For a discussion of the principles involved in mixed solvents see Chapter III, Section 5.)

The spots on a chromatogram are not always sharp, well defined, and circular. They may be diffused and spread out uniformly in all directions or they may be elongated to various degrees ranging from a slightly elongated spot to heavy tailing of the spot toward the rear. This can be caused by a nonlinear adsorption isotherm, most frequently where the adsorption increases strongly with the dilution of the substances. However, tailing can occur even under conditions of linear adsorption, and Giddings and Eyring [90] have proposed a kinetic mechanism for this phenomenon. According to this theory, adsorption in these cases occurs at two different types of sites, one of which is less abundant but of stronger adsorption energy. In the normal process of development the major portion of the zone would have passed on before the more tightly held material on the high energy sites would be desorbed, creating a tailing effect. Recently Giddings [91] developed this theory further. Tailing also occurs when the chromatoplate is overloaded.

Tailing is to be avoided, if possible, because of the danger of overlapping or masking of other zones. It can be diminished by using gradient elution in which the polarity of the solvent mixture is gradually increased. With this procedure the more polar solvent picks up the tailing part and moves it more rapidly toward the front of the zone. Care must be taken not to increase the polarity so much that other more tightly adsorbed compounds could be detached and pushed ahead into another zone.

References

1. J. Farradane, *Nature,* **167,** 120 (1951).
2. L. Reed, *Proc. Chem. Soc.,* **9,** 123 (1893).
3. C. F. Schoenbein, *Verhl. Naturforsch. Ges. Basel,* **3,** 249 (1861).
4. F. Goeppelsroeder, *Anregung zum Studium der auf Capillaritaets und Adsorptionserscheinungen beruhanden Capillaranlyse,* Basel, 1906.
5. F. Goeppelsroeder, *Mitt. k. k. Tech Gewerbemuseums Wien, N.S.,* **3,** 14 (1889).
6. F. Goeppelsroeder, *Mitt. k. k. Tech. Gewerbemuseums Wien, N.S.,* **2,** 86 (1888).
7. F. Goeppelsroeder, *Verhl. Naturforsch. Ges. Basel,* **3,** 268 (1861).
8. M. Tswett, *Proc. Warsaw Soc. Nat. Sci., Biol. Sect.,* **14,** minute #6 (1903).
9. H. Brockmann and F. Volpers, *Chem. Ber.,* **80,** 77 (1947).
10. J. W. Sease, *J. Am. Chem. Soc.,* **69,** 2242 (1947).
11. *Ibid.,* **70,** 3630 (1948).
12. L. Zechmeister, L. Cholnoky, and E. Ujhelyi, *Bull. Soc. Chim. Biol.,* **18,** 1885 (1936).
13. J. M. Miller and J. G. Kirchner, *Anal. Chem.,* **23,** 428 (1951).
14. A. J. Martin and R. L. M. Synge, *Biochem. J.,* **35,** 1358 (1941).
15. R. Consden, A. H. Gordon, and A. J. P. Martin, *Biochem. J.,* **38,** 225 (1944).
16. A. R. Fahmy, A. Niederwieser, G. Pataki, and M. Brenner, *Helv. Chim. Acta,* **44,** 2022 (1961).
17. K. Randerath, *Biochem. Biophys. Res. Commun.,* **6,** 452 (1962).
18. K. Randerath, *Angew. Chem. Int. Ed. Engl.,* **1,** 435 (1962).
19. J. G. Kirchner, J. M. Miller, and G. J. Keller, *Anal. Chem.,* **23,** 420 (1951).
20. R. H. Reitsema, *Anal. Chem.,* **26,** 960 (1954).
21. M. W. Beyerinck, *Z. Phys. Chem.,* **3,** 110 (1889).
22. H. P. Wijsman, *De Diastase, beschouwd als mengsel van Mattase en Dextrinase,* Amsterdam, 1898.
23. G. A. Van Klinkenberg, *Chem. Weekbl.,* **63,** 66 (1967).
24. N. A. Izmailov and M. S. Schraiber, *Farmatsiya (Sofia),* **1938,** 1.
25. C. Lapp and K. Erali, *Bull. Sci. Pharmacol.,* **47,** 49 (1940).
26. M. O'l. Crowe, *Anal. Chem.,* **13,** 845 (1941).
27. N. V. Békésy, *Biochem. Z.,* **312,** 100 (1942).
28. W. G. Brown, *Nature,* **143,** 377 (1939).
29. T. L. Williams, *Introduction to Chromatography,* Blackie, Glasgow, 1947, p. 3.
30. J. E. Meinhard and N. F. Hall, *Anal. Chem.,* **21,** 185 (1949).
31. J. G. Kirchner and G. J. Keller, *J. Am. Chem. Soc.,* **72,** 1867 (1950).
32. J. G. Kirchner and J. M. Miller, *Ind. Eng. Chem.,* **44,** 318 (1952).
33. J. M. Miller and J. G. Kirchner, *Anal. Chem.,* **24,** 1480 (1952).
34. J. G. Kirchner and J. M. Miller, *J. Agric. Food Chem.,* **1,** 512 (1953).
35. J. M. Miller and J. G. Kirchner, *Anal. Chem.,* **25,** 1107 (1953).
36. J. G. Kirchner, J. M. Miller, and R. G. Rice, *J. Agric. Food Chem.,* **2,** 1031 (1954).

37. J. M. Miller and J. G. Kirchner, *Anal. Chem.*, **26**, 2002 (1954).
38. J. G. Kirchner and J. M. Miller, *J. Agric. Food Chem.*, **5**, 283 (1957).
39. L. Zechmeister and L. Cholnoky, *Principles and Practice of Chromatography,* transl. from 2nd German ed. by A. L. Bacharach and F. A. Robinson, Wiley, New York, 1941, pp. 12, 13.
40. M. Mottier and M. Potterat, *Mitt. Geb. Lebensm. Hyg.*, **43**, 118 (1952).
41. *Ibid.*, p. 123.
42. E. Stahl, G. Schroeter, G. Kraft, and R. Renz, *Pharmazie*, **11**, 633 (1956).
43. E. Stahl, *Chem. Ztg.*, **82**, 323 (1958).
44. K. Macek, I. M. Hais, J. Kopecký, and J. Gasparič, Eds., *Bibliography of Paper and Thin-Layer Chromatography 1961–1965,* Elsevier, Amsterdam, 1968.
45. K. Macek, I. M. Hais, J. Kopecký, J. Gasparič, V. Rabek, and J. Churáček, Eds., *Bibliography of Paper and Thin-Layer Chromatography 1966–1969,* Elsevier, Amsterdam, 1972.
46. K. Macek, I. M. Hais, J. Kopecký, S. Schwarz, J. Gasparič, and J. Churáček, Eds., *Bibliography of Paper and Thin-Layer Chromatography 1970–1973,* Elsevier, Amsterdam, 1976.
47. B. J. Haywood, *Thin-Layer Chromatography: An Anotated Bibliography 1964–1968,* Ann Arbor Science, Ann Arbor, Mich., 1968.
48. D. Jaenchen, Ed., *CAMAG Bibliography Service,* CAMAG, Muttenz, Switzerland (published quarterly).
49. D. Jaenchen, Ed., *Thin-Layer Chromatography. Cumulative Bibliography* I *1964–1967,* CAMAG, Muttenz, Switzerland, 1967.
50. D. Jaenchen, Ed., *Thin-Layer Chromatography. Cumulative Bibliography* II *1967–1969,* CAMAG, Muttenz, Switzerland, 1970.
50a. D. Jaenchen, Ed., *Thin-Layer Chromatography. Cumulative Bibliography* III *1969–1973,* CAMAG, Muttenz, Switzerland, 1974.
50b. D. Jaenchen, Ed., *Thin-Layer Chromatography. Cumulative Bibliography* IV *1973–1977,* CAMAG, Muttenz, Switzerland, 1977.
51. R. M. Scott, *Thin-Layer Chromatography Abstracts 1968–1971,* Ann Arbor Science, Ann Arbor, Mich., 1972.
52. R. M. Scott and M. Lundeen, *Thin-Layer Chromatography Abstracts 1971–1973,* Ann Arbor Science, Ann Arbor, Mich., 1974.
53. M. J. Saxby, Ed., *Thin-Layer Chromatography Abstracts,* D. H. Masek, England.
54. J. N. Wilson, *J. Am. Chem. Soc.*, **62**, 1583 (1940).
55. D. DeVault, *J. Am. Chem. Soc.*, **65**, 532 (1943).
56. J. Weiss, *J. Chem. Soc.*, **1943**, 297.
57. A. C. Offord and J. Weiss, *Dis. Faraday Soc.*, **7**, 26, 45 (1949).
58. E. Glueckauf, *Proc. Roy. Soc. London Ser. A,* **186**, 35 (1946).
59. E. Glueckauf, *J. Chem. Soc.*, **1947**, 1302.
60. E. Glueckauf, *Disc. Faraday Soc.*, **7**, 12, 45 (1949).
61. L. G. Sillén, *Ark. Kemi,* **2**, 477 (1950).
62. G. G. Baylé and A. Klinkenberg, *Rec. Trav. Chim.* **73**, 1073 (1954).
63. *Ibid.*, **76**, 593 (1957).

64. *Ibid.*, p. 607.
65. T. Langvad, *Acta Chem. Scand.*, **10,** 1649 (1956).
66. W. M. Smit and A. van den Hoek, *Rec. Trav. Chim.*, **76,** 561 (1957).
67. *Ibid.*, p. 577.
68. K. Klamer and D. W. van Krevelen, *Chem. Eng. Sci.*, **7,** 197 (1958).
69. H. B. F. Dixon, *J. Chromatogr.*, **7,** 467 (1962).
70. D. A. McQuarrie, *J. Chem. Phys.*, **38,** 437 (1963).
71. A. P. Tudge, *Can. J. Phys.*, **39,** 1600 (1961).
72. L. R. Snyder, *J. Chromatogr.*, **23,** 388 (1966).
73. *Ibid.*, **25,** 274 (1966).
74. G. P. Pataki, Ph.D. dissertation, University of Basel, 1962.
75. M. Brenner and G. Pataki, *Helv. Chim. Acta,* **44,** 1420 (1961).
76. B. G. Belenky, V. V. Nesterov, E. S. Gankina, and M. M. Smirnov, *J. Chromatogr.*, **31,** 360 (1967).
77. H. B. F. Dixon, *J. Chromatogr.*, **7,** 467 (1962).
78. V. Pollak, *J. Chromatogr.*, **77,** 245 (1973).
79. V. V. Rachinskii, "Theory of the Longitudinal Distribution of the Mobile Phase in Paper and Thin-Layer Chromatography," in *Stationary Phase in Paper and Thin-Layer Chromatography,* K. Macek and T. M. Hais, Eds., Elsevier, Amsterdam, 1965, p. 284.
80. V. V. Rachinskii and E. G. Davidova, "Sorption Properties of Ion-Exchange Celluloses," in *Stationary Phase in Paper and Thin-Layer Chromatography,* K. Maček and I. M. Hais, Eds., Elsevier, Amsterdam, 1965, p. 111.
81. L. R. Snyder and D. L. Saunders, *J. Chromatogr.* **44,** 1 (1969).
82. E. Soczewiński, *Adv. Chromatogr.*, **5,** 3 (1968).
83. E. Soczewiński, *Anal. Chem.*, **41,** 179 (1969).
84. E. Soczewiński and W. Golkiewicz, *Chem. Anal. (Warsaw),* **14,** 465 (1969).
85. E. Soczewiński and W. Golkiewicz, *Chromatographia,* **6,** 269 (1973).
86. E. Soczewiński, W. Golkiewicz, and H. Szumilo, *J. Chromatogr.*, **45,** 1 (1969).
87. G. H. Stewart, *Sep. Sci.*, **1,** 747 (1966).
88. S. Turina, L. Horvath, and V. Marjanović, *J. Chromatogr.*, **37,** 234 (1968).
89. A. Tiselius, *Ark. Kemi., Miner. Geol.,* **16A** (18) (1943); *Chem. Abstr.*, **38,** 2895 (1944).
90. J. C. Giddings and H. Eyring, *J. Phys. Chem.,* **59,** 416 (1955).
91. J. C. Giddings, *Anal. Chem.*, **35,** 1999 (1963).

Chapter **II**

COMMERCIALLY PREPARED ADSORBENTS FOR THIN-LAYER CHROMATOGRAPHY

1 SILICA GEL OR SILICIC ACID

Silica gel is by far the most widely used adsorbent in thin-layer chromatography; it was introduced in 1951 by Kirchner et al. [1] for thin-layer work for the separation of terpenes. After investigating a series of 15 different adsorbents, they found silicic acid gave the best separation.

There are a number of firms that supply adsorbents for thin-layer chromatography. These supplies are offered in various forms, with and without binders and fluorescent agents. E. Merck and EM Laboratories supply a series of silica gel preparations; the suffix 60, 90, or 150 indicates the mean pore diameter in angstroms, G indicates a calcium sulfate (gypsum) binder and H a silicon dioxide/aluminum oxide binder, the letter F means a fluorescent indicator has been added and a subscript number indicates the excitation wavelength, P is the code designation for preparative thin-layer chromatography, R indicates a specially purified adsorbent, and RP indicates a silanized gel for reverse-phase work.

Silica gel D5 is produced by CAMAG. It is a very fine-grain gel mixed with 5% calcium sulfate as a binder. The same adsorbent with an inorganic fluorescent indicator for observing substances which absorb in ultraviolet regions is listed as silica gel DF-5. For special work where the calcium sulfate cannot be used, this adsorbent is available as silica gel D0 or DF-0, depending on whether or not a fluorescent agent is needed. This will adhere to a glass plate without a binder but produces a rather soft layer. As with other finely divided adsorbents without binder, its ability to adhere to glass gradually diminishes with time so that it cannot be kept indefinitely. CAMAG also has a silica gel line, what they term their DS series. These correspond to the D series just mentioned except that the DS series has an increased running speed owing to somewhat larger particle sizes.

M. Woelm supplies a silica gel of less than 50-μ particle size and approximately neutral pH. This is available with and without calcium sulfate binder and a fluorescent agent.

Macherey, Nagel and Co. supplies two grades of silica gel, one a standard quality and the other a more highly purified silica gel. These are supplied in a series of combinations. The designation G indicates that it has a calcium sulfate binder. N indicates that no binder is used and S indicates that a starch binder is used. The designation HR indicates that the mixture contains higher-purity silica gel. The letters UV added to the designation indicate that the mixture contains an ultraviolet fluorescing material. Thus MN-silicagel G-HR/UV indicates that this Macherey, Nagel silica gel mixture is composed of the higher-purity silica gel containing calcium sulfate binder and an ultraviolet fluorescing material.

In the United States, Applied Science Laboratories supplies a standardized silica gel containing 10% calcium sulfate as a binder. The particle size of this silica gel is in the range of 10 to 20 μ and has acidic properties. This is supplied under the name Adsorbosil-1. They also supply the same silica gel without a binder under the name of Adsorbosil-2. Adsorbosil-S-1 is listed as an acidic powder of 300 to 400 mesh with strong silicic acid properties. It contains a binder but is obtainable without the binder as Adsorbosil-S-2. Adsorbosil-P-1 is a silica gel containing an inorganic phosphor and a starch binder. It is obtainable without the starch binder as Adsorobil-P-2. Their grades ADP-1 and ADP-2 contain an inorganic phosphor and fluorescein indicator; the ADP-1 contains a starch binder, whereas ADP-2 is binder free.

Adsorbosil-3 contains 10% magnesium silicate and Adsorbosil-4 is an extra high-purity silica gel, acid treated, neutralized, solvent extracted, and compounded with 10% magnesium silicate. Adsorbosil-5 is a gel of pH 6.8 to 7 without a binder. Adsorbosil-ADN is impregnated with 25% silver nitrate, and the suffix 1 or 2 designates with 10% calcium sulfate or without binder, respectively. The same company markets a silica gel impregnated with dimethyldichlorosilane to make it nonwettable to polar solvents so that it can be applied to reverse-phase partition chromatography. It is designated as Reversil-3. Slurries of this are prepared with a relatively high-boiling, nonpolar solvent.

Schleicher and Schuell supplies a silica gel designated as silica gel 150 which may also be obtained with 15% gypsum or starch binder and with added fluorescent material.

J. T. Baker offers a neutral silica gel which may also be obtained with a calcium sulfate binder and with a fluorescent agent.

Mallinckrodt offers two series of silica gel-based adsorbents: SilicAR TLC-4 (Code 7097) is an acidic product (10% slurry pH 4 to 5.5) with specifications for a maximum iron content of 0.003% and a heavy metal content maximum of 0.004% (as Pb); SilicAR TLC-7 (Code 7102) has a pH of 6.5 to 7.2 with identical heavy metal specification and a maximum of 0.004% on iron. These two products may also be obtained with added 15% calcium sulfate binder or with 6% inorganic phosphor or with both, in which case the letter(s) G, F, and GF, respectively, are added to the designations. The phosphor used is an inorganic haloapatite which exhibits a brilliant white fluorescence. The average particle size range is 2 to 15 μ.

Bio-Rad Laboratories supplies a series of silica gels in three narrow particle-size ranges of 1 to 10, 10 to 30, and 30 to 60 μ, all available with or without a 5% calcium sulfate binder. All grades are available with a fluorescent zinc silicate.

Whatman provides a silica gel (SG41) for thin-layer work. Average particle size is stated to be 5 to 20 μ with a maximum size of 60 μ. The iron content is 140 ppm and the maximum chloride content 0.01%. A 10% aqueous suspension has a pH of 7.

Analabs provides an adsorbent containing silicon dioxide which is available with calcium sulfate (Anasil B) and without a binder (Anasil S).

2 ALUMINA

Aluminum oxide has been used for a long time in column chromatography and has been the next most widely used adsorbent in thin-layer work after silica gel. Most of the loose-layer chromatography has been on alumina. Special grades prepared for thin-layer chromatography are available commercially. CAMAG produces a series of aluminum oxides called "aluminum oxide for thin-layer chromatography." These grades carry the same numeral and letter designation as the corresponding silica gel mixtures, that is, the D series made with very fine-grain aluminum oxide and the DS series made with a somewhat larger grain but which has an increased running speed. The letter F in the designation indicates that it contains an ultraviolet indicator. The numeral 0 indicates that it does not contain a binder and the numeral 5 indicates that it contains 5% calcium sulfate (plaster of Paris) as a binder. This adsorbent is basic in character with a pH of around 9.5.

Merck and the associated EM Laboratories supply aluminum oxide in grades corresponding to those of their silica gel adsorbents including neutral, acid, and basic aluminum oxide in some grades.

Woelm puts out a basic, a neutral, and an acidic alumina for thin-layer chromatography that is a fine-grain material and does not contain any binding agent. The approximate pH values of these materials are 9, 7.5, and 4, respectively. An aluminum oxide containing 10% calcium sulfate as a binder is also available.

Bio-Rad supplies three grades of alumina designated as acid alumina AG 4, neutral alumina AG 7, and basic alumina AG 10. All grades are obtainable with or without 5% calcium sulfate binder.

Mangold [2] has used Alcoa activated alumina 20 mesh by combining it with 5% its weight of plaster of Paris.

3 DIATOMACEOUS EARTH

E. Merck supplies a diatomaceous earth (kieselguhr G) for thin-layer chromatography. This has an average particle size of 10 μ and contains a calcium sulfate binder. It is used mainly for partition separations where it is impregnated with either a hydrophilic or a hydrophobic substance.

Johns Manville markets a diatomaceous earth as Celite No. 545, which has been used successfully as a thin-layer material.

4 MAGNESIUM SILICATE

Woelm markets a magnesium silicate for thin-layer chromatography that has a pH of about 10. The slurry proportions are about 15 g of magnesium silicate to 45 ml of water. As with all their air-dried inorganic layers, they recommend activation at about 130°C.

Applied Science has a magnesium silicate especially prepared for TLC work designated as Adsorbosil-M-1 (ADM-1) and Adsorbosil-M-2 (ADM-

2). The ADM-1 grade contains a calcium sulfate binder whereas the ADM-2 is without binder.

Bio-Rad has a magnesium silicate M-1 available for TLC work with a particle size of 2 to 44 μ. This contains a calcium sulfate binder but may also be obtained without the binder.

Floridin manufactures an activated magnesium silicate in a number of mesh sizes. The sizes most suitable for thin-layer chromatography are 100 to 200 mesh, finer than 100 mesh, and finer than 200 mesh.

5 MISCELLANEOUS INORGANIC ADSORBENTS

As Kirchner et al. [1] first demonstrated, practically any adsorbent used in column work can be used for thin layers providing the material is ground and/or screened to a suitable particle size. Thin layers have been made of talc, calcium hydroxide, calcium phosphate, calcium sulfate, and carbonates of zinc, calcium, and magnesium, all from commercial sources. Most of these, of course, may be obtained as analytical grade reagents.

Powdered porous glass both with and without plaster of Paris binder is offered by Applied Science Laboratories as Adsorbosil-G-1 and Adsorbosil-G-2, respectively. Bio-Rad Laboratories produces a porous glass powder in five grades, all in a 200-mesh size for use in thin-layer work.

Hydroxyapatite, used by Hofmann for thin-layer work [3, 4], is now available commercially as a suspension in phosphate buffer under the trade name Hypatite (Clarkson Chemical Co.). According to Hofmann [5], this may be washed with water, alcohol, and then acetone and dried to obtain a satisfactory product. It must of course be screened to the proper particle size.

Bio-Rad Laboratories offers a hydroxyapatite powder for TLC work under the designation Bio-Gel HTP.

6 CELLULOSE AND MODIFIED CELLULOSE

Because of the important role that paper has played in the chromatography of amino acids and other compounds, it was only a question of time before thin layers of cellulose would be tried for the same task. The results have been quite successful since the chromatography on thin layers of cellulose has proved to be more rapid and to give sharper separations than the corresponding work on paper. Because of the fibrous nature of cellulose it is possible to work without a binder, although preparations are available that do make use of the latter. Table 2.1 lists

Table 2.1 Cellulose and Treated Cellulose Especially Prepared for Thin-Layer Chromatography and Their Sources[a]

Supplier	*Designation*	*Description*
CAMAG	Cellulose powder, CAMAG, Type D-0	Plain cellulose, no binder
	Cellulose powder, CAMAG, Type DF-0	Cellulose with fluorescent indicator, no binder
Whatman	Whatman, crystalline cellulose powder CC41	Cellulose powder, no binder, of mean particle size passing 200 mesh. High purity: max. ash 0.01%; max. iron 5 ppm; max. copper 5 ppm
Schleicher and Schuell	Cellulose powder #65	Pure cellulose
	DEAE cellulose #66	Diethylaminoethyl cellulose
	ECTEOLA cellulose #67	Anion exchanger from reaction of epichlorohydrin and triethanolamine with cellulose
	Selectacel CM #68	Cation exchanger based on carboxymethyl cellulose
	Selectacel P #69	Cation exchanger of phosphorylated cellulose
	Cellulose 144 dg	Cellulose powder, double acid washed
	Cellulose 144	Very pure and fine cellulose powder
	Cellulose 144 LS 254	Cellulose with 3% fluorescent agent
	Cellulose 144/ac	Acetylated cellulose available as 6, 21, and 41% acetylated
Macherey, Nagel and Co.	MN 300	Cellulose, plain
	MN 300 UV	Cellulose with fluorescent agent (activated zinc silicate)
	MN 300 HR	Cellulose, specially purified

Table 2.1 (Continued)

Supplier	*Designation*	*Description*
	MN 300 Ac	Acetylated cellulose (supplied 10, 20, 30, or 40% acetylated)
	MN300CM	Carboxymethyl cellulose
	MN300P	Phosphorylated cellulose
	MN300 DEAE	Anion exchanger, diethylaminoethyl cellulose
	MN300 ECTEOLA	Anion exchanger
	MN300 PEI	Anion exchanger, polyethylenimine cellulose
	MN300 Poly-P	Cation exchanger, cellulose polyphosphate
Bio-Rad Laboratories	Celex D (DEAE)	Diethylamino cellulose
	Celex E (ECTEOLA)	ECTEOLA cellulose
	Celex PEI	Polyethylenimine cellulose
	Celex CM	Carboxymethyl cellulose
	Celex P	Phosphorylated cellulose
	Celex N-1	Cellulose powder
	Celex MX	Microcrystalline cellulose
Serva-Entwicklungslabor Co.	Serva CM-TLC Cellulose	Carboxymethyl cellulose
	Serva Cellulose TLC	Cellulose
	Serva DEAE-TLC	DEAE cellulose
	Serva ECTEOLA-TLC	ECTEOLA cellulose
	Serva PEI-TLC	Polyethylenimine cellulose
Avicel Sales Division, American Viscose, Div. FMC	Avicel	Microcrystalline cellulose

[a] Addresses of manufacturers and suppliers are listed in Appendix B.

the various cellulose preparations available for thin-layer chromatography with their supplier.

7 ION-EXCHANGE RESINS

Bio-Rad Laboratories offers a number of ion-exchange materials specifically for TLC work. AG50W-X8 and AG1-X8 are cation- and anion-exchange resins, respectively, and are supplied in a 200 to 400-mesh spherical particle with calcium sulfate binder and as a 2- to 44-μ granular material without a binder.

They also offer zirconium phosphate (Bio-Rad ZP-1), zirconium tungstate (Bio-Rad ZT-1), zirconium molybdate (Bio-Rad ZM-1), hydrous zirconium oxide (Bio-Rad HZO-1), and ammonium molybdophosphate (Bio-Rad AMP-1). The latter is 2 to 10 μ in particle size and the remainder are classed as 2 to 44 μ. These inorganic compounds have been used as ion-exchange materials in layers bound with 3% starch [6].

8 MISCELLANEOUS ORGANIC ADSORBENTS

There are several organic materials which have been used to only a limited extent in thin-layer chromatography. However, these materials may be expected to play a more important role in the future.

Polyamides

The following three firms supply a polyamide powder for thin-layer chromatography: Macherey, Nagel & Co., M. Woelm Co., and E. Merck, A. G. These powders do not contain any binder and the thin layers are prepared from a slurry of 5 g of the powder in 45 ml of ethanol.

A number of firms supply polyamides for TLC work. Macherey, Nagel & Co. supply polyamide 6, 66, and 11 each in three grades: standard grade, with fluorescent additive, and as the acetylated derivative. M. Woelm, E. Merck, J. T. Baker, and Wako Pure Chemicals Ltd. all supply a polyamide product for TLC.

Gel Filtration Media

Pharmacia markets a cross-linked dextran gel under the name of Sephadex in various grades and modifications. These have been used in column work for the differentiation of materials based on the difference in molecular weights. Sephadex products prepared specifically for TLC work are available in grades of G-25, G-50, G-75, G-100, G-150, and G-200, all superfine in particle size 10 to 40 μ. These give a range of gels for separating molecules of various sizes. Bio-Rad Laboratories offer polyacrylamide gel beads in grades of P-2, P-4, P-6, P-10, P-30, P-60, P-100, P-150, P-200, and P 300, all in a 400-mesh size for thin-layer application.

Polyethylene

Farbwerke Hoechst produces a polyethylene powder under the name Hostalen S which has been used by Mangold [2] for thin-layer chromatography of fatty acids and their methyl esters.

9 COMMERCIALLY PREPARED PLATES

Analtech introduced adsorbent-coated glass plates ready for use under the name "Uniplates." These are offered in a number of different types of coatings both with and without binders. A scored, precoated plate is also available, so that after developing a series of individual chromatograms they may be separated by snapping off strips of the plate at the appropriate score marks. Also available is a custom coating service. Most of the manufacturers of TLC adsorbents now offer precoated plates and/or foils so that these may be obtained with all the more common adsorbents produced by these sources. In addition to the common adsorbents, Eastman Kodak offers polycarbonate coated on polyester foil. Pharmacia produces polyacrylamide gradient gel slabs 2.7 mm thick.

Macek and Bečvarová [7] have reviewed the sources of precoated plates and flexible sheets as well as papers available for chromatography. The dimensions of the plates are cataloged along with various other features such as the binder or special characteristics.

Another type of prepared product that has been on the market for some time (1962) is the glass fiber sheet impregnated with potassium silicate or silicic acid, manufactured by Applied Science Laboratories. Gelman Instrument Co. also markets a similar product designated Gelman Instant Thin Layer Chromatography" (ITLC), which consists of "sheets of potassium silicate-siliceous fibers." Mallinckrodt offers a silicic acid-impregnated glass fiber sheet in thicknesses of 500 and 1000 μ in 8 × 80 in. rolls.

References

1. J. G. Kirchner, J. M. Miller, and G. J. Keller, *Anal. Chem.*, **23**, 420 (1951).
2. H. K. Mangold, *J. Am. Oil Chem. Soc.*, **38**, 708 (1961).
3. A. F. Hofmann, *J. Lipid Res.*, **3**, 391 (1962).
4. A. F. Hofmann, *Biochim. Biophys. Acta*, **60**, 458 (1962).
5. A. F. Hofmann, "Hydroxyapatite as an Adsorbent for Thin-layer Chromatography; Separations of Lipids and Proteins," in *New Biochemical Separations*, A. T. James and L. J. Morris, Eds., Van Nostrand, New York, 1964, p. 283.
6. B. A. Zabin and C. B. Rollins, *J. Chromatogr.*, **14**, 534 (1964).
7. K. Macek and H. Bečvarová, *Chromatogr. Rev.*, **15**, 1 (1971).

Chapter **III**

PREPARATION OF THE PLATES

1 THE SUPPORT FOR THE LAYER

Because of its inertness to chemicals, glass is of course the universal support for thin-layer chromatography, and glass plates of different sizes, shapes, and thicknesses have been and still are being used to suit the user's particular needs. Meinhard and Hall [1] used microscope slides. Kirchner et al. [2] standardized on $\frac{1}{2} \times 5\frac{1}{4}$ in. strips which fit conveniently in standard test tubes. This keeps the developing chamber to a minimum size so that rapid equilibration of the solvent vapors may be achieved. For two-dimensional work they standardized on a $5\frac{1}{4} \times 5\frac{1}{4}$ in. plate. The latter fits a standard museum jar and both the strips and the plate are of sufficient size to allow for a 10-cm development. This distance was selected as an optimum development distance because it gives sufficient distance for resolving mixtures in a reasonable development time. This distance also allows an ordinary millimeter rule to be used directly for reading the R_f values. The supply houses for thin-layer equipment supply glass plates in 200 × 200, 200 × 400, 200 × 100, and 200 × 50 mm sizes. (Using the 200-mm plate size for 10-cm development wastes one-third of the adsorbent.) Pyrex brand glass plates are available for cases where high-temperature heating of the plate is necessary for making compounds visible; however, because of manufacturing problems they are not as uniform in thickness as are the regular glass plates.

In 1962, three independent workers—Hofmann [3], Peifer [4], and Wasicky [5]—reintroduced the use of microscope slides as supports for thin layers. In contrast to Meinhard and Hall's work with radial development, these workers adopted the ascending method established by Kirchner et al. [2]. Both Peifer and Hofmann also used small square plates; Peifer used lantern slides (10.3 × 8.3 cm) and Hofmann used a square plate 66 × 66 mm. Peifer in addition used microscope cover slips of 4.2 × 2.5 cm for routine testing. The short plates are useful because of the rapidity of development and have been used by a number of workers [6–17], but their use is somewhat limited by their short developing space. At the other extreme in size are the large plates that have been used for preparative thin-layer chromatography. Korzun et al. [18] have used 12 × 15 in. plates and Halpaap [19] has used glass plates 20 cm wide × 1 m long. Shandon Scientific Co. offers 100 × 200 cm plates for preparative work.

It is advantageous, especially where water-based solvents are being used, to apply the adsorbents to a ground-glass or a sandblasted surface. Rigby and Bethune in 1955 [20] were the first to report on the use of sandblasted glass plates as a support for thin layers, and since that time a number of workers have used this type of glass support surface [21–32]. These surfaces can be prepared quickly by grinding with a slurry of

Carborundum and water. An additional advantage of a ground plate is that it helps to even out surface irregularities, thus providing a flat surface. The rough surface not only provides better adhesion for the layer to the plate but if the adsorbent does slip off the plate where it dips beneath the surface of the solvent, the rough surface provides enough capillary attraction to carry the solvent up to the remainder of the layer. When making spots visible by charring with sulfuric acid and heat there is less tendency for bubbles to form in the layer if a ground-glass plate is used [23, 33]. Clapp and Jeter [34] employed a different means to prevent the adsorbent from sliding off the plate. They used a CAMAG "sandwich plate," removing 8 mm of the adsorbent on all four edges instead of just three edges as is normally done. The bottom of the plate was then covered with three or four strips of washed and dried blotting paper which overlapped the adsorbent. With the cover plate clamped in place, the paper held the adsorbent from slipping and served as a wick to feed solvent.

Gamp et al. [35] initiated the use of ribbed, decorative glass. This is convenient since it permits the preparation of thin strips for thin-layer chromatography without using specialized spreading equipment. A simple spreader is used to distribute the slurry over the ribbed glass, and then after it sets for 1 to 2 min, a spatula is used to scrape off the excess material. After drying, the spatula is once more run along the glass ribs in order to get rid of extraneous material. This produces a series of narrow strips of adsorbent separated by glass boundaries; because of the nature of this glass the layers are not as uniform as can be produced by other methods. Naturally this plate cannot be used for two-dimensional work. Hansbury et al. [36] have made their own grooved plates. This is done by masking the area that is to hold the adsorbent with electrical tape. The remainder of the plate is coated with Glyptal varnish (General Electric Glyptal Varnish No. 1202). After the varnish is dried, the tape is removed and the plate is then immersed with agitation in a solution of 114 g of ammonium bifluoride in 1 liter of water for $7\frac{1}{4}$ hr at room temperature. After rinsing and removal of the Glyptal resin a plate is obtained which has a 0.3-mm uniform groove in it. This procedure also provides an etched surface for the binding of the thin layer. Square plates can also be made for two-dimensional work by this method.

Plates containing grooves 10 mm wide × 2 mm deep × 8 in. long spaced $\frac{1}{4}$ in. apart were used by Matherne and Bathalter [37] as a cleanup procedure for pesticide residue analysis by gas chromatography. Collins [38] used a plate with a groove running across it at the origin. This provides a thicker layer of adsorbent at the origin thus incorporating a cleanup procedure into the normal chromatographic technique. Up to tenfold quantities of extract can be applied to this origin, and using a volatile solvent, mobile constituents are moved to a new origin above the groove leaving the proteinaceous material behind. (These plates are available from May and Baker.)

Boyd and Hutton [39] examined the use of quartz plates as a support for silicic acid layers in order to view them by transmitted ultraviolet light. They found that the silicic acid layer absorbed the incident light so that there was a low contrast between the spots and the background. Therefore they returned to the use of glass plates. Weiner and Zak [40] used quartz supports for UV densitometry of thin layers of agar used in electrophoresis. Arreguín [41] used glass plates (Corning Glass filters No. 7910, CS No. 9.54) as supports for DEAE cellulose in the separation of nucleotides. These plates have good light transmission between 250 and 400 mμ.

Squibb [42] applied a thin layer of silica gel to the matted side of an otherwise clear plastic plate. After the development and the location of the spots was completed, the layer was sprayed with Tuffilm Spray No. 543 (M. Grumbacher). This plastic spray penetrated the silica gel and bound it to the supporting plate. The plates were then cut into strips so that they could be run through standard scanning equipment to measure the radioactivity of the spots. The plastic used for the supporting plates was Union Carbide # VCA 3310-C1. Marsh et al. [43] have used plastic trays for electrophoresis.

Thin sheets of plastic for use as a support for thin-layer chromatography were first mentioned in 1964 [44] and since then ready-prepared layers of this nature have become widely available from supply houses. These sheets may be readily cut to any desired size with a pair of scissors, but in general they will accept slightly less sample than glass-supported layers. Rolls of this material (20 cm × 500 cm) are available [45] and Halpaap and Bausch [46] have carried out developments to a length of 1 m on rolled-up foils. Roessler et al. [47] patented a method of coating plastic sheets with TiO_2 or ZrO_2 before coating with adsorbent. This prevents the migration of plasticizers or other material onto the adsorbent. Thin plastic sheeting has been used as a support for the gel in thin-layer electrophoresis to eliminate (1) the zonal profile distortion which occurs in thick layers and (2) the need for slicing the latter [48–51]. Teflon-covered glass paper [40] has also been used for this purpose.

Schweda [52] compared the performance of silica gel-coated glass plates and polyester sheets. In some cases the plates performed better than the flexible layers and in other cases the reverse was true.

Snyder [53] has reported on the use of 4-mm aluminum sheets as a support for thin-layer adsorbents. In using these, it was found necessary to polish them thoroughly with Brillo soap pads before the final washing; the slurry for coating these aluminum plates was made up with a 47%

ethyl alcohol solution, the alcohol serving as a wetting agent. (In preparing the layers a strip of Scotch tape was applied to the edge of the applicator to prevent scratching of the aluminum plate.) These plates can be placed directly on a hot plate for visualizing the components without the danger of breakage that is present with glass plates. Korzun and Brody [54] and Rosmus et al. [55] used aluminum supports for centrifugal thin-layer chromatography and Rabenort [56] and Koss and Jerchel [57] used aluminum foil in descending chromatography. Thin stainless steel strips have also been used as a support material [58] as well as nickel–brass [59] and heavily chrome-plated brass [60]. The stainless steel supports are cut from 0.005-in.-thick shim stock and must be roughened with coarse emery paper to facilitate the adhesion of the thin layer.

Lie and Nye [61] coated the insides of test tubes with thin layers of adsorbent. This was accomplished by filling the tubes with the slurry and then inverting and letting the tubes drain. The coated test tube then became its own developing chamber. Spots were visualized with iodine vapor; of course with this method difficulty would be encountered in using sprays. Also, because the concentration of the compounds is greatest at the surface of the layers [62] detection by observing through the glass at the back side of the layer would be less sensitive than observation of the opposite surface. Similar to inner-coated test tubes are cylindrical tubes coated on the inside [63–66], and these suffer from the same disadvantages although they are open at both ends and can be made larger in diameter. Ikan and Rapaport [67] placed the layer on the outside of the test tube and thus made it readily accessible for spraying, observation, and removal of spots.

Glass rods of various diameters have been used as supports [67–70]. These are coated by dipping in a slurry and can be developed in test tubes. They are convenient for checking solvent systems.

Sen [71, 72] has prepared sticks of calcium sulfate without a support by molding 6- to 8-mm-diameter rods from a mixture of precipitated calcium sulfate ($CaSO_4 \cdot 2H_2O$) and plaster of Paris (55 : 44) moistened with water. After careful drying and activating, the rods were developed as microcolumns in test tubes.

Glass frustums have also been introduced as supports for the layer [73, 74]. These in effect are a form of circular chromatography, but are less convenient to coat and at the same time it is more difficult to apply samples to them.

2 THE ADSORBENT

In Section 4 commercial adsorbents for thin-layer chromatography are listed, particularly those that have been specifically prepared for this purpose. In this section directions are given for those who wish to prepare their own adsorbents or use other sources of supply. This will include preparations for specific purposes as well as a discussion of some of the desirable characteristics of the adsorbents. Treatment which is applied during the preparation of the slurry, for example, buffering, is discussed under slurry preparations.

Silicic Acid or Silica Gel

The terms silica gel and silicic acid refer to somewhat different modifications of the same material; Wren [75] has a good discussion on this subject. Through a comparison of separations on columns by various workers it has been shown that silicic acid is the more active adsorbent [76–84].

Kirchner et al. [2] used a Merck reagent-grade silicic acid which had been sieved to remove all particles larger than 100 mesh. The specifications for this material include a maximum of not over 0.001% iron and 0.003% heavy metals (as Pb). Applewhite et al. [85, 86], Battaile et al. [87], Demole [88], Kuroiwa and Hashimoto [89], Hayward et al. [90], and Onishi et al. [91] used Mallinckrodt silicic acid 100 mesh, No. 2847 for chromatographic analysis. Here again, the maximum iron content is specified as not greater than 0.001% and the heavy metals (as Pb) not over 0.002%. Vogel et al. [92], Doizaki and Zieve [93], and Avigan et al. [94] all used this same brand of silicic acid, but Vogel et al. and Doizaki and Zieve screened the material to take everything at 200 mesh and smaller, and Avigan et al. took everything finer than 325-mesh size. All three groups of workers found that the resolution of the Mallinckrodt silicic acid was better than that of silica gel G. Reitsema [95] and Allentoff and Wright [21] used Fisher reagent-grade precipitated silicic acid.

For those who wish to prepare their own silica gel Adamec et al. [96] give the following directions: water glass is diluted 1:2 with water and then the gel is precipitated with concentrated hydrochloric acid until a pH of 5 is reached. This mixture is dried for 3 days at 80°C in order to give the gel a granular structure. It is then washed in a Büchner funnel until free of chlorides and extracted in a Soxhlet extractor for 8 hr with chloroform. Drying, grinding in a ball mill, and sieving to correct particle size complete the process. In drying silica gel, temperatures above 170°C should not be used, because above this point irreversible loss of water occurs and the adsorption characteristics are permanently impaired.

Shukla et al. [97] prepared a silica gel by pouring a hot 19% solution of sodium silicate into an equal volume of 10% HCl. While this solution was maintained at 50 to 55°C and a pH of 1 to 1.5 with stirring, it was treated with 0.25 ml of hydrofluoric acid per liter of solution. They found that the use of hydrofluoric acid greatly reduced the gelation time while retaining a surface area of 811 m^2/g and increasing the adsorption capacity.

It is not surprising that different commercial samples or even different batches of the same manufacturer show different adsorption characteristics. Separation efficiency depends on grain size, specific surface area, pore size, and the amount of moisture in the silica gel as can be shown by a few examples. Petrova [98] obtained separation of certain antibiotics only when the grain size was less than 20 μ, and Waksmundzki and Różylo [99] showed that silica gel with specific surface areas smaller than 565 m^2/g would not separate isomeric hydroxynaphthalenes.

Gaertner and Griessbach [100] found that the concentration of the starting sol and the pH affected the surface area and the pore structure. Maximum surfaces occurred at medium concentrations and pH values. Tarutani [101], investigating the behavior of silicic acid on Sephadex columns, found that the rate of polymerization of monosilicic acid varied with the pH of the solution. In strongly alkaline solution monosilicic acid is stable, but polysilicic acid forms rapidly in neutral solution. The polymerization is least rapid in an aqueous solution at pH 2, and below this the rate of polymerization increases again. Girgis [102] investigated the effect of temperature and pH on the structure and surface area. An increase in pH caused a decrease in the specific surface area and an increase in both the mean pore radius and pore volume of the xerogel. El Rassi et al. [102a] have studied the influence of water and thermal treatment on the activity of silica gel.

Hinz et al. [103] give directions for preparation of a lepidoid silica. This is made by freezing silica sols or dialyzed water glass solutions. This gel is scaly in particulate structure and can be made with widely varying properties by changing the pH of the solutions, the time of aging, and the temperature at which the material is precipitated out. The chemical purity is very high; only spectroscopic traces of iron, aluminum, calcium, and magnesium are present. Among other uses, the authors propose this material for thin-layer chromatography.

For the preparation of silica gel, Adamec et al. [104] purified the sodium silicate by passing it through an ion exchanger, as proposed by Pitra [105]. To accomplish this the ion-exchange resin (Wofatit KPS-200) was converted to the H^+ cycle with 4% hydrochloric acid. Water glass was then diluted to a density of 1.070 and 20 ml of concentrated ammonium hydroxide added for each liter of solution. By passing this through the cation-exchange column a sol of silicic acid was formed. Gelatinization occurred on adding, with vigorous stirring, 25% by volume of a 10%

ammonium carbonate solution. After standing 24 hr, the gel was dried at 120°C before grinding and classifying for particle size.

Reichelt and Pitra [106] describe a method for altering the adsorption characteristics of a silica gel for loose-layer work by partial deactivation of the gel. This is accomplished by the addition of 25% water or 43% dilute acetic acid to the adsorbent which is allowed to stand for 2 hr in a closed flask. This is shaken occasionally to help distribute the moisture uniformly. Klein [107] recommends the use of controlled atmospheres of known humidity for the deactivation of silica gels to produce more uniform results. Table 3.1 from Stokes and Robinson [108] gives the concentrations of solutions for giving specified vapor pressures and Table 3.2 from the same source gives the vapor pressures of saturated solutions.

Table 3.1 Concentrations of Solutions Giving Specified Vapor Pressures at 25°C[a,b]

	H_2SO_4		$NaOH$		$CaCl_2$	
a_w	*m*	%	*m*	%	*m*	%
0.95	1.263	11.02	1.465	5.54	0.927	9.33
0.90	2.224	17.91	2.726	9.83	1.584	14.95
0.85	3.025	22.88	3.840	13.32	2.118	19.03
0.80	3.730	26.79	4.798	16.10	2.579	22.25
0.75	4.398	30.14	5.710	18.60	2.995	24.95
0.70	5.042	33.09	6.565	20.80	3.400	27.40
0.65	5.686	35.80	7.384	22.80	3.796	29.64
0.60	6.341	38.35	8.183	24.66	4.188	31.73
0.55	7.013	40.75	8.974	26.42	4.581	33.71
0.50	7.722	43.10	9.792	28.15	4.990	35.64
0.45	8.482	45.41	10.64	29.86	5.431	37.61
0.40	9.304	47.71	11.54	31.58	5.912	39.62
0.35	10.21	50.04	12.53	33.38	6.478	41.83
0.30	11.25	52.45	13.63	35.29	7.183	44.36
0.25	12.47	55.01	14.96	37.45	...	...
0.20	13.94	57.76	14.67	40.00	...	...
0.15	15.81	60.80	19.10	43.32	...	...
0.10	18.48	64.45	23.05	47.97	...	...
0.05	23.17	69.44	...	...	...	...

[a] From R. H. Stokes and R. A. Robinson [108]; reproduced with permission of the authors and the American Chemical Society.

[b] a_w = water activity = p/p_0, where p = vapor pressure of solution, and p_0 = vapor pressure of pure water; m = molality = moles of (anhydrous) solute per 1000 g of water; % = percentage of (anhydrous) solute by weight.

Table 3.2 Vapor Pressures of Saturated Solutions at 25°C[a,b]

Solid Phase	a_w	*Solid Phase*	a_w
$K_2Cr_2O_7$	0.9800	$NaBr \cdot 2H_2O$	0.577
KNO_3	0.9248	$Mg(NO_3)_2 \cdot 6H_2O$	0.5286
$BaCl_2 \cdot 2H_2O$	0.9019	$LiNO_3 \cdot 2H_2O$	0.4706
KCl	0.8426	$K_2CO_3 \cdot 2H_2O$	0.4276
KBr	0.8071	$MgCl_2 \cdot 6H_2O$	0.3300
NaCl	0.7528	$K(C_2H_3O_2 \cdot 1.5H_2O$	0.2245
$NaNO_3$	0.7379	$LiCl \cdot H_2O$	0.1105
$SrCl_2 \cdot 6H_2O$	0.7083	$NaOH \cdot H_2O$	0.0703

[a] From R. H. Stokes and R. A. Robinson [108]; reproduced with permission of the authors and the American Chemical Society.

[b] Formula is that of solid, stable in contact with saturated solution at 25°C. a_w = water activity = p/p_0, where p = vapor pressure of solution, and p_0 = vapor pressure of pure water.

Special Gels

In 1949 Dickey [109] first prepared specific silica gels by forming them in the presence of methyl, ethyl, propyl, and butyl orange dyes. The adsorbents were prepared by the addition of 30 ml aqueous sodium silicate (d_{20} = 1.401), 275 ml of water, and 30 ml glacial acetic acid to 0.5 g of finely divided dye. After the mixture was dried at room temperature it was ground and sieved. The correct particle size was then extracted with methyl alcohol to remove the dye. Maximum adsorption for the dye used in the preparation of the adsorbent was shown to range from 4 to 20 times as effective as a control gel. The prepared gels showed decreasing effectiveness for dyes with decreasing similarity of molecular structure. Erlenmeyer and Bartels in 1964 [110] applied the same principle in the preparation of specific gels for thin-layer chromatography. In this case the gels were prepared in the presence of methylphenylamine or ethylphenylamine. The same adsorption selectivity was observed. Majors and Rogers [111, 112] and Reed and Rogers [113] have studied the variables in the preparation of specific gels and also the effects of structural changes of the pretreating agent on the adsorption characteristics of the gels. Bartels [114] has studied the structure of specific gels prepared with 1,10-phenanthroline. Bartels and Prijs [114a] have reviewed the subject of specific silica gels.

The change in the adsorption properties of silica gel by treating it with silver ions was demonstrated by Dimov [115] in 1961. The retention time of ethylene was increased by 100%, that of propylene by 50%; for ethane and butane the retention time remained unchanged. This change is based on the well-known property of silver nitrate to complex with unsaturated compounds. In 1961 Goering et al. [116], and de Vries [117] in 1962, used a silver nitrate-impregnated silica gel for column chromatography and Morris [118] and Barrett et al. [23] used the same principle for thin-layer chromatography of lipids. In general, the silver nitrate is added either to the slurry as it is being prepared or to the prepared plates. These preparations are therefore discussed under the appropriate headings. However, Gupta and Dev [119] give directions for preparing a silver nitrate-impregnated silica gel in a dry powder form. To 125 ml of alcohol is added 7.5 g of silver nitrate in 7.5 ml of water. Fifty grams of silica gel is gradually stirred into this solution and the stirring is continued for 15 min, after which the excess solvent is removed by drying on the water bath, with constant stirring. The drying is completed to constant weight in a vacuum. This material will keep for several months if stored in the dark.

Markl and Hecht [120] prepared a silica gel impregnated with triisooctylamine in order to complex metal ions by treating 16.8 g of silica gel with 2 ml of the base in 25 ml of ether. The ether was removed by placing in a drying oven at 100°C for 2 hr. Tuerler and Hoegl [120a] have used disodium ethylenediaminetetraacetate to tie up interfering metal ions. For this purpose, they made a slurry of 30 g of silica gel G with 50 to 60 ml of water containing 0.1 to 0.2 g of the complexing agent.

Vidrine and Nicholas [121] silanized silica gel by slurrying silica gel in anhydrous toluene with the addition of 1-ml portions of dimethyldichlorosilane with constant stirring until hydrogen chloride stopped bubbling off. After the reaction mixture stood overnight, it was filtered and rinsed, first with 3 volumes of anhydrous benzene and then with 3 volumes of anhydrous methanol. After being vacuum dried for 4 hr, it was dried at 70°C. Mixing 43% of this product with 50% untreated silica gel and 7% gypsum as a binder gave layers that could be used for adsorption chromatography in one direction and reverse-phase partition chromatography in another direction.

Yasuda [122] prepared a trimethylsilylated silica gel by refluxing for 1 hr a mixture of 50 g of silica gel, 160 ml of cyclohexane, and 34 ml of hexamethyldisilazane. After 2 ml of isopropanol was added, the refluxing was continued for 3 hr. The adsorbent was then filtered off, washed with isopropanol, dried at 130°C for 3 days, and stored in a desiccator. Because of the hydrophobic nature of the gel a methanol–water mixture (1 : 1) was used to prepare the layers.

Gilpin and Sisco [122a] compared the adsorption characteristics of silica gels treated with methyl, ethyl, hexyl, dodecyl, and octadecyl alkyltrichlorosilanes.

At times it is necessary to have an especially pure silica gel. For example, in inorganic thin-layer work the adsorbent must be free of trace amounts of cations, and again in certain cases with organic materials it is desirable to keep away from traces of metal ions which may act as catalysts. Seiler and Rothweiler [123] give directions for purifying silica gel. They allow 500 g of silica gel to stand in 1 liter of a 1:1 mixture of concentrated hydrochloric acid and water for a period of time. The yellow iron-containing solution is decanted and the washing process with acid is repeated twice. The silica gel is then washed three times, each time with 1 liter of distilled water by decantation. Then it is filtered and washed with distilled water until the filtrate is neutral or only weakly acidic. It is then washed with 250 ml of ethanol followed by 250 ml of benzene, after which it is dried for 24 hr at 120°C. Lipina [124] treated a silica gel with both concentrated hydrochloric acid and concentrated nitric acid. Passera et al. [125] washed their silica gel with hydrochloric acid (1:1), then with water, and finally with 0.1% (wt/vol) aqueous solution of EDTA to remove interfering ions.

Likewise, there are times when organic materials that have been adsorbed on the silicic acid must be removed before use. This can be done after preparation of the thin layer, as shown by Kirchner et al. [126], by developing in a suitable solvent with subsequent drying. On the other hand, the adsorbent may be washed with a suitable solvent prior to preparation of the layers [127]. Struck [128] kept the silica gel under methyl alcohol overnight. It was then filtered and washed twice with methyl alcohol followed by drying at 100°C for $\frac{1}{2}$ hr. Bowyer et al. [129] used a mixture of chloroform–methyl alcohol (2:1) for extracting their silicic acid. Privett and Blank [130] washed first with chloroform and then with ether. Miller and Kirchner [127] showed that sometimes special precautions must be taken in order to avoid picking up impurities from the air while drying the washed adsorbent. They used a modified, commercial, forced-draft oven with a special packing gland around the fan shaft. Air for the drying was drawn through a filter containing silicic acid. Summers and Mefferd [131] purified their silica gel for lipid work by extracting in a Soxhlet successively with heptane, chloroform, and 95% ethanol, each extraction being carried out overnight. Carreau et al. [132] treated 100 g of silica gel for 30 min at 60 to 70°C with 400 ml of 0.2% sodium ethylate. This was centrifuged and treated under the same conditions with 200 ml of 20% acetic acid in ethanol. The final product was filtered and washed repeatedly with ethanol. Ma [133] has shown that impurities can be adsorbed by storing silica gel in plastic bottles, plastic bags, or in glass bottles closed with rubber stoppers. He recommends storing purified silica gel in a screw-cap glass bottle with a Teflon liner.

Aluminum Oxide

Most of the aluminum oxide used in any form of chromatography is obtained from commercial sources, although according to Reichstein and Shoppee [134], it can be made by heating aluminum hydroxide for about 3 hr with stirring at 380 to 400°C. Aluminum oxide made in this manner always contains free alkali. The properties and chromatographic uses of a fibrous alumina prepared by hydrolyzing an amalgamated alumina has been described by Wislicenus [135], and Huneck [136] has described a similar preparation for use in thin-layer chromatography. This method, a slight modification of the Wislicenus method, consists of treating 50 g of aluminum grits (approximately 0.5 mm in size) with 100 ml of 10% sodium hydroxide until hydrogen is evolved vigorously. The alkali is decanted, the treatment is repeated, and then the aluminum is washed free of alkali. Ten milliliters of saturated mercuric chloride is added and allowed to react with agitation. The gray slime is removed by decanting and the particles are washed with water allowing the excess water to drain off. A vigorous reaction takes place with the evolution of steam. Then 80 to 100 ml of water is added with stirring to produce a dry aluminum oxide powder. The powder is washed with ethyl alcohol and decanted from the unreacted aluminum. After filtering, it is dried, ignited, and sieved through an 0.066-mm sieve. This material was used without a binder.

Baitsholts and Ardell [136a] have prepared neutral aluminum oxide layers by dipping commercial layers in a mixture of 5 ml of 10% acetic acid in 345 ml of ethanol. Acidic layers were prepared by dipping in solution made by dissolving 23.6 g of sodium acetate in 70 ml of water, adding 10.5 ml of acetic acid, and then diluting to 350 ml with ethanol. This gives a pH of 5 to 6.

Halpaap and Reich [137] have studied the structure and adsorption characteristics of aluminum oxides prepared by heating hydrargillite at varying temperatures up to 1150°C.

Hydroxyapatite

For the separation of one and two monoglycerides and of proteins, Hofmann [138, 139] used hydroxyapatite. The latter is the partially hydrolyzed calcium phosphate prepared by the alkaline hydrolysis of dibasic calcium phosphate. It is a weaker adsorbent than silicic acid and can be used for the separation of nonionizing or neutral compounds. The preparation of hydroxyapatite according to Anacker and Stoy [140] is as follows: 250 g of hydrous dibasic calcium phosphate is suspended with stirring in 2.5 liter of 0.05 *M* NaOH at 40°C. When the pH drops to a value of 8 to 9, the liquid is decanted and fresh 0.05 *M* NaOH solution is added at 40°C and allowed to stand for 24 hr. This procedure is repeated three or four times until the pH does not change appreciably. The precipitate is then filtered and washed with 0.005 *M* NaH_2PO_4 solution until the pH value of the wash water is between 6 and 7. It is then washed with alcohol and finally with acetone before drying. The product is sieved through a 170-mesh sieve. The coarse material may be powdered in a ball mill and again sieved. Anacker and Stoy [140] prepared their own diabasic calcium phosphate from a monobasic calcium phosphate.

Magnesium Silicate

Wolfrom et al. [141] found that Magnesol, a synthetic magnesium silicate obtained from the Waverly Chemical Co., had a pH of 9.8 in aqueous slurry and had different chromatographic properties than the Magnesol previously obtainable from Westvaco Chloralkali Division of the Food, Machinery & Chemical Corp. They therefore modified the new Magnesol as follows: 500 g of Magnesol (Waverly) was suspended with stirring in a solution of 100 ml of glacial acetic acid and 2000 ml of water for 1.5 hr. It was then filtered and washed with 2000 ml of water. It was suspended overnight in 2000 ml of water, after which it was filtered and washed with 1 liter of water. This product having a pH of 7.5 was then dried for 10 hr at 100°C. For thin-layer work it was sieved through a 200 mesh screen. Schwarz [142] likewise treated Florisil (Floridin) with acetic acid in order to bring the pH down to 6.5 because it normally has an alkaline reaction of around pH 8.5.

Talc is a magnesium silicate mineral of composition $Mg_3[Si_4O_{10}](OH)_2$ which has been used for a number of separations including antibiotics [143], alkaloids [144], porphyrins [145, 146], and cardeneloids [147].

Magnesium Oxide

Bomhoff [148] has shown that the properties of magnesium oxide depend on the method of preparation. To obtain the maximum surface area of 190 m^2/g, magnesium hydroxide was heated for 12 hr at 400°C. Thin layers were prepared by mixing 7 g each of the oxide and kieselguhr G with 70 ml of chloroform. The plates were air-dried and activated at 120°C. Tewari and Ram [149] made an aqueous slurry of magnesium oxide and starch (4 : 1) for their layers.

Calcium Sulfate

Matis et al. [150] prepared their own calcium sulfate by adding the stoichiometric amount of sulfuric acid to an aqueous solution of calcium chloride with stirring at a temperature of 70 to 80°C. The precipitated calcium sulfate was filtered and washed thoroughly with distilled water until neutral. It was then ground and allowed to dry for 40 hr at 115 to 120°C.

Plaster of Paris ($CaSO_4 \cdot \frac{1}{2}H_2O$) has been used to prepare thin layers [151, 152]. Mitchell [153] prepared his own plaster of Paris by heating $CaSO_4 \cdot 2H_2O$ (Fisher Certified Reagent, Precipitate Calcium Sulfate) at 175°C overnight. Dobici and Grassini [154] used strips of plaster of Paris for electrophoresis work.

Cellulose

Hammerschmidt and Mueller [155] have published a method for treating cellulose powder in order to remove interfering cations. This consists in treating cellulose powder with 1.5% nitric acid with agitation at 50°C for 2 hr. It is then thoroughly washed with distilled water.

Wolfrom et al. [156] stirred 50 g of microcrystalline cellulose (Avicel-Technical Grade) with 1 liter of 0.1 *N* sodium borohydride for 10 hr at 25°C with constant stirring. The solid was allowed to settle before decanting, and then the process was repeated with fresh borohydride. It was then filtered, washed until neutral, and dried over phosphorus pentoxide under vacuum. The resulting cake was then ground in a mortar.

For chromatography of nucleo derivatives, Pataki [157] purified cellulose by suspending 60 g of powder in 500 ml of *n*-propanol–25% ammonia water–water (6:3:1). This was shaken vigorously for 30 min and again filtered. After a final washing with *n*-propanol it was vacuum dried at 60°C.

Haworth and Heathcote [157a] purified cellulose for amino acid work by slurrying 50 g with 200 ml of 80% methanol. This was filtered and washed successively with 300 ml of 2-propanol–acetic acid–water (3:1:1), 200 ml of methanol–water (1:3), 200 ml of methanol–1 *N* hydrochloric acid (3:2), 200 ml of water, and 200 ml of methanol. Drying was completed in a vacuum overnight.

Cellulose Acetate

Wieland et al. [158] give directions for preparing an acetylated cellulose. The cellulose powder (30 g) was dried for 15 to 30 min at 110°C and then over concentrated sulfuric acid in a desiccator for a few hours. This powder was stirred at 70°C in a mixture of 225 ml of acetic anhydride,

675 ml of benzene, and 0.9 ml of concentrated sulfuric acid for 9 hr. The mixture was filtered, washed with methanol, allowed to stand under methanol for a few hours, and then filtered and washed with ether. The final product was dried in a vacuum cabinet at 90°C. Badger et al. [159] acetylated cellulose powder using Spotswood's method [160]. For this purpose 200 g of cellulose powder is stirred frequently for 24 hr at 18°C in a mixture of 1700 ml of thiophene-free benzene, 800 ml redistilled acetic anhydride, 4 g of 92% sulfuric acid, and 4 g of 72% perchloric acid. The liquid is filtered off and the acetylated cellulose is allowed to stand in ethyl alcohol for 24 hr with occasional stirring. It is washed thoroughly with more alcohol and water and finally allowed to stand in distilled water for 2 hr. It is then removed and air-dried.

Keratin

Brady and Hoskins [161] give the following directions for the preparation of keratin for use in thin-layer chromatography. Sixty grams of commercially scoured wool is digested at 65°C for 3 hr in a solution of 20 g of sodium bisulfite and 20 g of papain in 2 liters of water. The pH of the solution is adjusted to 6.5 with the addition of 1 *M* sodium hydroxide. The cortical cells are filtered and washed with distilled water (3 × 500 ml). They are then resuspended in 300 ml of water and the pH is adjusted to 3.0 with 6 *M* HCl. After heating at 75°C for 30 min, they are filtered, washed, and stored in distilled water with a little chloroform to inhibit the growth of microorganisms. For use the cells (14 g) are suspended in 100 ml of a 1:1 mixture of ethanol–water. They also give directions for methylating or deaminating the cells in order to modify the adsorbent characteristics.

Miscellaneous Adsorbents

Although most of the thin-layer work has been done on the more common adsorbents, there are a number of adsorbents that have been used only infrequently or perhaps only once or twice. Ackermann and Frey [162] have compared the use of quartz powder with that of silica gel. Magnesium fluoride was used to separate 11 metal ions [163]. Zinc sulfide served as an adsorbent for the separation of geometric isomers [164]. Zirconium hypophosphate has been applied to the separation of inorganic ions [165]. Maddrell salt gave good separations of seven sugars, nine amino acids, and the lower members of the dicarboxylic acids [166]. Zinc ferrocyanide gave a good separation of sodium, potassium, rubidium, and cesium [167] and has been applied to the separation of some sulfonamides [168]. Barium sulfate has been used to separate food dyes [169] and for the separation of metal dyes [170]. Crystalline titanium and zirconium phosphates [171], cerium phosphate [172, 173], and zirconium arsenate [174] have been examined for their ion exchange properties in separating ions. A mixed layer of strontium sulfate and silica [175] has been used for the separation of ^{90}Sr and ^{90}Y. Titanium oxide was tested as an adsorbent by separating *o*-, *m*-, and *p*-aminophenols [176].

3 BINDERS

Starch

Although there has been a great deal of discussion in the literature concerning the use of finely powdered adsorbents without a binder, by far the most popular adsorbent that has been used is silica gel G which makes use of the plaster of Paris binder originated by Kirchner et al. [2]. My preference is for starch-bound adsorbents wherever they do not interfere by reacting with the compound-locating reagents. This limitation is not as serious as one would imagine; for example, Kirchner et al. [2] found that a concentrated sulfuric acid–nitric acid mixture may be used if the plates are not heated. Smith and Foell [177] found (among other reagents) antimony trichloride, phosphoric acid, and trichloroacetic to be applicable to starch-bound layers. With any given solvent the starch-bound layer is faster than the gypsum-bound layer and can be made with a much firmer surface without losing speed of development. This firm surface allows one to mark the surface with a soft lead pencil and provides for easier spotting of the plates without disturbing the surface of the adsorbent. There are also times when gypsum cannot be used. For example, Seiler [178] found that in separating phosphates, insoluble calcium phosphates were formed with gypsum-bound layers. He therefore used a starch binder.

Kirchner and Flanagan [179, 180] have determined the effect of the binder on the rate of development (Table 3.3). As can be seen, in each

Table 3.3 Effect of Binder on Rate of Development [180]

	Solvent (time in minutes for 10 cm solvent travel)				
Adsorbent	*Hexane*	*Benzene*	*Ethyl Acetate*	*95% Ethanol*	*Water*
Silicic acid plus 2.5% starch	14	19	15	72	20
Silica G	32	39	43	97	49
Silicic acid plus 20% plaster of Paris	46	75	60	223	72

solvent used, the starch-bound layers are considerably shorter in developing time. Even the silicic acid with 20% gypsum does not provide quite as firm a layer as the starch-bound material, and the silica gel G with a lower gypsum content provides a rather fragile layer which is difficult, if not impossible, to write on. Not all starches are satisfactory as a binding agent; some starches are subject to fissuring on drying. After investigating a number of starches, Miller and Kirchner [181] found that either Clinco-15 (a modified starch from Clinton Corn Processing Co.) or a 2:1 mixture of ordinary corn starch and Superior AA tapioca flour (Stein-Hall Co.) were good binding agents. Other starches (besides Amioca and those just mentioned) that have been used by various workers are rice, cornstarch, potato starch, and wheat starch. Kirchner and Flanagan [179] have also found that by decreasing the amount of starch from 5 to 2½% the time–temperature factor required in the previous formulation could be eliminated; with 2½% starch it is only necessary to heat the mixture thoroughly on a boiling water bath.

An additional advantage of the starch binder is that slurries can be kept for months in the refrigerator, and in fact have been kept in my laboratory at room temperature for periods as long as 2 months in subdued light. Thus large batches can be prepared in advance, ready for use.

Gypsum (Plaster of Paris)

This binder was developed by Kirchner et al. [2] for use where the detecting agent would react with a starch binder. The main disadvantages of this binder are the increased time of development required and the decreased stability of the layers as compared to a starch-bound layer. Since the plaster sets rather quickly, large batches of material cannot be prepared ahead of time and care should be taken to mix only that quantity that can be conveniently used in a relatively short period of time before the plaster begins to set.

Although for most purposes the ordinary grade of plaster of Paris obtained from most chemical supply houses is satisfactory as a binder, Battaile et al. [87] and Peifer [4] have found it convenient to prepare their own plaster of Paris by heating reagent-grade $CaSO_4 \cdot 2H_2O$ (gypsum), in the first case at 110°C overnight, and in the second case at 180°C for 24 to 48 hr. The resulting plaster of Paris may be ground to pass through a 200-mesh sieve. As a binder, plaster of Paris has been used in quantities from 5 to 20%. The 5% level gives a very soft layer whereas the 20% gives a firmer layer (but still less firm than the starch-bound layers). The larger the quantity of plaster of Paris binder, the slower the speed of development.

The softness of the gypsum-bound layers is inherent in the materials.

Although gypsum that has been regenerated from plaster of Paris loses water of crystallization at 100°C to form a soluble anhydrite the process takes place slowly. Plates which are dried at 80°C for 24 hr are no softer at the end of this period than are plates dried at 110°C for 30 min. However, on layers that have been dried for 2 hr at 110°C, the gypsum loses its binding power. At 130°C the gypsum loses its binding capacity within 45 min.

Carboxymethyl Cellulose

This is a binder that looks very promising and was introduced by Obreiter and Stowe [182] in 1964. They used 5% of a #70 premium, low-viscosity grade of carboxymethyl cellulose (Hercules Powder Co.) as a binder for silicic acid. I have found that it is a little more sensitive to checking than the starch-bound layers; however, decreasing the amount of binder to 2.5% helps in this respect. Satisfactory layers up to 2 mm thick can be prepared. Uniform layers must be prepared, as a slight nonuniformity in thickness induces checking. The surfaces of the layers are hard and can be readily marked with a lead pencil.

Miscellaneous Binders

Onoe [183] used polyvinyl alcohol as a binding agent by mixing 30 ml of 2.5% polyvinyl alcohol solution with 23 g of silica gel. In preparing this mixture a few drops of alcohol was added to prevent foaming. Randerath [184] has used collodion as a binding agent with ECTEOLA cellulose, and Huettenrauch et al. [185] used the same binder for ion-exchange resins. Hofmann [138, 139] has used Zytel 61 (DuPont), an alcohol-soluble polyamide, as a binding agent with hydroxyapatite. For this preparation 40 mg of Zytel is dissolved in 60 ml of 70% (vol/vol) methyl alcohol by heating in a covered vessel with stirring. After cooling, 15 g of hydroxyapatite is smoothly homogenized into the mixture. Concentrations of 0.5, 1, and 4% of this binder have been tried [138]. Polyethylene [186], epoxy [187], polyacrylamide [188], and polyvinyl acetate [189] have also been used as binders. Sodium silicate has been used as a binder [190]. A slurry of 40 g of silicon dioxide and 115 g of 6° Baumé sodium silicate solution was applied to glass plates. Althaus and Neuhoff [191] dipped glass plates into a slurry of 30 g of silica gel in 130 ml of chloroform–methanol (2:1), and then after drying, the layers were developed in water glass–water (1:20). Birkofer et al. [192] used gelatin as a binder for a mixture of silica gel and Perlon powder, and Gauthier and Mangency [193] made use of Senegal gum for binding formamide-impregnated kieselguhr. Maini [194] proposed using agar-agar as a binder for silica gel and alumina layers. This increased the stability of the layer and the adherence of agar disks when used as a means of applying samples. Jellinek [195] confirmed the use of agar-agar as a useful binder and showed that development speeds were greater than with the gypsum bound layer. Furthermore, the layers prepared with agar-agar have less tendency to slide off of the plate when immersed in the solvents than do the gypsum-bound layers. Okumura et al. [196–198] have used fused powdered glass as a binder for silica gel and alumina layers. These layers are mechanically stable, heat stable, and acid resistant. They can be used repeatedly by immersing in cleaning solution prior to reactivation.

4 PREPARATION OF THE SLURRY

By far the greatest portion of thin-layer work has been accomplished on layers prepared from water-based slurries of the adsorbents. Even with the same amount and type of binder, the amount of water which is employed for a given slurry varies with different brands of adsorbent. If the slurry is to be spread by hand, the ratio of water to adsorbent is not critical. However, if it is to be used in an apparatus for the preparation of thin layers, then the amount of water must be carefully controlled. If the slurry is too thick it will not flow through the spreader, and if it is too thin it will flow too rapidly.

Water-Based Slurries of Silica Gel with Plaster of Paris Binder

For the preparation of silica gel slurries from commercial adsorbent preparations specifically for thin-layer chromatography work, the mixing proportions recommended by the manufacturers can be followed.

For spreading 2- to 5-mm layers for preparative work Honnegger [199] used a slurry of silica gel G–water in a 1 : 1.57 ratio.

In blending the adsorbent with the water, some workers place the ingredients in a flask for mixing; others triturate in a mortar. Svennerholm and Svennerholm [200] shake their slurries vigorously for 1 min in a flask attached to a water pump in order to deaerate the material.

For chromatography of materials which adsorb in the ultraviolet, it is convenient to add ultraviolet fluorescing materials to the slurry during the preparation. This aids in the detection of this type of compound which appears as a dark spot on a fluorescent background under ultraviolet light. Although the introduction of fluorescent layers has been ascribed by Stahl [201] to Gaenshirt, these were first introduced by Kirchner et al. [2] in 1951.

The inorganic fluorescent compounds used by Sease [202, 203] for column chromatography can be used. This is best accomplished by adding 1.8% zinc cadmium sulfide (phosphor #1502, Du Pont) and 1.8% zinc silicate (phosphor #609, Du Pont) to the combined weight of the adsorbent and binder. According to Sease [203], the use of both zinc silicate and zinc sulfide gives a mixture whose excitation range is continuous from 390 to 230 mμ. For illuminating thin layers made with these fluorescent materials a hydrogen lamp may be used, but for most purposes it is convenient to use two lights, one for the short-wave and one for the longer-wave ultraviolet light. Reitsema [95] used Rhodamine 6G as a fluorescing agent and for this purpose employed 0.0011 g of the fluorescent agent (0.0037%) in 30 g total of adsorbent and binder. Stahl [204] used sodium fluorescein as a fluorescing agent by preparing the slurry with a 0.04% aqueous sodium fluorescein solution instead of water. Brown and Johnston [205] used a 0.02% solution of 2′,7′-dichlorofluorescein for making up their slurries. Tschesche et al. [206] used either sodium 3,5-dihydroxypyrene-8,10-disulfonate or sodium 3-hydroxypyrene-5,8,10-trisulfonate. These were incorporated in the slurries, 0.25 mg of the dihydroxy compound and 0.33 mg of the monohydroxy compound per gram of silica gel G. These latter reagents have their limitations, however, because with polar solvents stronger than methanol–chloroform (7:3), they are eluted from the adsorbent.

Water-Based Slurries of Silica Gel with Starch Binder

Kirchner and Flanagan [179] modified the original formulation of Kirchner et al. [2] in order to get away from the time–temperature factor required in the earlier formulation. To prepare a starch slurry, 19 g of silicic acid or silica gel is combined with 0.5 g of starch (Clinco-15 starch, Clinton Corn Processing Co., or a 2:1 mixture of ordinary cornstarch and superior AA tapioca flour, Stein-Hall Co.) and 38 ml of distilled water; 0.37 g each of zinc cadmium sulfide and zinc silicate may be added to produce fluorescent layers. This mixture is heated on a boiling water bath with stirring until the starch has completely gelatinized. More water may be added after heating if a thinner slurry is desired. This slurry is stable for many months if kept in an ordinary refrigerator and may even be kept at room temperature for periods of up to 2 months if kept under subdued lighting conditions.

Water-Based Slurries of Modified Silica Gels

Acidified Silica Gel

Stahl [207], using the idea of Brockmann and Groene [208] for acidified silica in column work, prepared acidic silica gel layers by using 0.5 *N* oxalic acid solution for preparing the slurry instead of distilled water. Deters [209] used a 0.05 *N* oxalic acid solution in preparing his layers, and Seher [210] has shown that there is a 20 to 25% loss of oxalic acid

Table 3.4 Effect of Acidifying the Adsorbent (Silica Gel G) on the R_B Value of Some Dinitrophenylhydrazones of Keto Acids[a,b]

Layer Thickness (mm)	*Adsorbent*	R_B *Value*[c]							
		1	2	3	4	5	6	7	8
0.10	Neutral	0.58	0.56	0.59	0.48	0.28	0.56	0.05	0.02
0.10	Acid	0.48	0.48	0.39	0.29	0.20	0.49	0.02	0.02
0.16	Neutral	0.50	0.51	0.43	0.35	0.24	0.55	0.04	0.01
0.17	Acid	0.45	0.42	0.37	0.30	0.20	0.52	0.02	0.00

[a] From P. Ronkainen [211]; reproduced with permission of the author and the Elsevier Publishing Co.

[b] R_B = R_f of DNP-spot of keto acid/R_f of dinitrophenylhydrazine. Solvent = petroleum ether (60 to 80°C)–ethylformate (13:7) with 0.0104 moles of propionic acid per 100 ml solvent mixture.

[c] (1) α-keto-β-methylvaleric acid; (2) α-ketoisocaproic acid; (3) α-ketoisovaleric acid; (4) α-ketobutyric acid; (5) pyroracemic acid; (6) levulinic acid; (7) α-ketoglutaric acid; (8) oxalacetic acid.

from the plates when the plates are dried at 105°C. Ronkainen [211] acidified silica gel G by using 30 g of silica gel G with 60 ml of water and 5 ml of propionic acid. The effect of this treatment on the R_B value is shown in Table 3.4. Petrowitz [212] compared the effects of various acids as acidifying media on the R_f values of some chlorinated compounds (Table 3.5).

Neutral Silica Gel

Smith and Foell [177] prepared a neutral silica gel by mixing 30 g of finely divided silica gel (Fisher #S-158), 50 ml of water containing 16 ml of 0.1 *N* sodium hydroxide solution, and 1.5 g of rice starch (Matheson, Coleman and Bell). This mixture was heated thoroughly on a steam bath until it thickened. It was further diluted with water to proper consistency. This provided a silica gel with a pH of 6.4.

Alkaline Silica Gel

Stahl [207] likewise prepared an alkaline silica gel by using 0.5 *N* base instead of water for making a slurry. Teichert et al. [213] prepared an alkaline silica gel by mixing 22 g of silica gel G with 45 ml of 0.5 *N* or 0.1 *N* potassium hydroxide, depending on the desired basicity of the resulting layers. Skipski et al. [214] used 0.01 *N* solutions of sodium carbonate or sodium acetate in their preparation of a basic silica gel. Groeger et al. [215] prepared an alkaline silica gel with a starch binder. For this purpose they used Baker and Adamson (Allied Chemical Corporation) silicic acid

Table 3.5 Effect on R_f Values by Acidifying Silica Gel with Various Acids ($R_f \times 100$)[a]

Compound	Acid Solution Used for the Preparation of the Layers																	
	Boric Acid 1%			*Oxalic Acid 1%*			*Tartaric Acid 1%*			*Citric Acid 1%*			*Salicylic Acid Saturated*			*Phthalic Acid 0.5%*		
	n[b]	c[b]	h[b]	n	c	h	n	c	h	n	c	h	n	c	h	n	c	h
Pentachlorophenol	64	70	05	61	66	07	66	72	07	53	74	05	53	60	04	62	65	10
DDT	96	97	50	94	96	45	96	96	48	96	96	46	96	97	46	92	95	48
α-Hexachlorocyclohexane	92	93	27	88	93	26	89	91	26	91	88	24	93	94	27	90	92	22
β-Hexachlorocyclohexane	92	93	04	88	93	03	89	91	06	91	88	04	93	94	04	90	92	04
γ-Hexachlorocyclohexane	92	93	18	88	93	14	89	91	17	91	88	15	93	94	17	90	92	15
δ-Hexachlorocyclohexane	92	87	11	88	88	10	89	91	11	91	88	10	93	83	10	90	92	10

[a] From H.-J. Petrowitz [212]; reproduced with permission of the author and Alfred Huethig Verlag.

[b] Developing solvents, n = benzene, c = chloroform, h = hexane.

#1169 which had been sieved to pass through a #100 sieve. Twenty-five grams of this material was stirred with 80 ml of 1% potassium hydroxide, 1.3 g of Argo corn starch (Best Foods Division, CPC, International) in 10 ml of 1% potassium hydroxide solution was added, and the mixture was heated at 70°C for 15 min. Additional 1% potassium hydroxide solution was then added to produce a mixture that would pour readily.

Buffered Silica Gel Layers

Borke and Kirch [216] in 1953 applied the use of buffered adsorbent in thin-layer chromatography. In this case, they used a mixture of silicic acid and magnesium oxide. Ten grams each of these adsorbents were mixed with 4 g of plaster of Paris and 0.250 g each of zinc silicate and zinc cadmium sulfide as fluorescing agents. These were thoroughly mixed in a mortar, and 38 ml of a phosphate buffer (pH 6.6) was added to the mixture and stirred for 2 min to form a creamy suspension. This was then coated on glass strips. Buffers below 6.6 could not be used with the fluorescent agents because the fluorescent properties were lost below this point. Honegger [217] prepared a sodium citrate buffered plate by mixing 25 g silica gel G with 50 ml of 0.1 *M* sodium citrate buffer (pH 3.8). As with these examples any desired buffer can be applied by simply slurrying the adsorbent with the appropriate buffer in place of water.

Silica Gel Impregnated with Complexing Agents

Meinhard and Hall [1] were the first to propose using complexing agents in thin-layer chromatography; however, their intention was for use as a visualizing agent rather than for the purpose of assisting in the separation. They incorporated 8-hydroxyquinoline in some of their layers but had difficulty in obtaining reproducible results. In 1961 Dimov [115] applied the principle of modifying the surface of the silica gel with silver ions for use in gas chromatography, and in the same year Goering et al. [116] treated silica gel with silver nitrate solution for use in column chromatography. Morris [118] and Barrett et al. [23] in 1962 were the first to use silica gel impregnated with silver nitrate for thin-layer chromatography. The usefulness of this adsorbent is based on the fact that olefins selectively complex with silver cations. This adsorbent is good for the separation of molecules which differ in number of double bonds and in their configuration (cis–trans). Barrett et al. [23] prepared their slurry by mixing 30 g of silica gel G with 60 ml of a 12.5% aqueous silver nitrate solution. Although Morris [19] first applied the silver nitrate by spraying an already prepared plate, he later [218] used a 5% impregnation of the slurry by mixing 23.7 g of silica gel G with 50 ml of a solution containing 1.25 g of silver nitrate. He reported that a 2% impregnation gave as good a separation as a 20 to 30% impregnation. Plates prepared

Table 3.6 Values of 100 × R_f in Thin-Layer Chromatography of Various Phenolic Carboxylic Acids on Kieselgel G Layers Untreated and Treated With Sodium Salts of Chelate-Forming Anions[a]

Solvent System[b]	*Layer Treated with*	*Vanillic Acid*	*Proto-catechuic Acid*	*Guaiacyl-acetic Acid*	*Homo-protocate-chuic Acid*	*Guaiacyl-propionic Acid*	*Dihydro-caffeic Acid*	*Ferulic Acid*	*Caffeic Acid*	*Iso-ferulic Acid*	*Chloro-genic Acid*
1	Nil	92	95	93	91	92	95	96	95	93	50
	Na_2MoO_4	95	76	89	63	87	69	87	61	85	10
	Na_2WO_4	88	59	80	61	84	91	90	85	89	06
	Borax	82	57	78	51	77	64	82	61	82	22
2	Nil	73	38	60	36	57	39	61	42	64	02
	Na_2MoO_4	91	32[c]	84	37[c]	84	42[c]	81	47[c]	82	02
	Na_2WO_4	50	11	34	10	34	19	39	12	41	00
	Borax	42	15	43	13	42	16	49	16	53	00
3	Nil	72	31	54	22	50	27	53	25	53	00
	Na_2MoO_4	80	31[c]	71	31[c]	68	39	73	34	65	00
	Na_2WO_4	47	04	38	03	37	11	41	02	39	00
	Borax	56	02	47	03	43	10	49	05	45	00
4	Nil	45	38	29	23	34	27	35	31	30	01
	Na_2MoO_4	44	10	22	06	29	14	34	14	25	00
	Na_2WO_4	39	24	24	13	32	19	36	22	31	00
	Borax	34	10	18	05	25	09	28	08	25	00
5	Nil	82	81	59	55	56	57	63	65	60	08
	Na_2MoO_4	93	71	69	66	84	80	87	85	86	14
	Na_2WO_4	38	15	19	13	29	23	37	31[c]	33	10
	Borax	53	04	21[c]	03	40	03	50	02	45	00

[a] From J. Halmekoski [222], reproduced with permission of the author and publisher.

[b] The following solvent systems were employed for development: (1) organic layer of *n*-butanol–acetic acid–water (4:1:5), (2) benzene–methanol–acetic acid (45:8:4), (3) benzene–dioxane–acetic acid (90:25:4), (4) *n*-butyl ether (satd. with water)–acetic acid (10:1), and (5) ethyl acetate–isopropanol–water (65:24:11.

[c] Elongated spot.

with the silver nitrate slurry gave better reproducibility than those made by spraying an already prepared silica gel plate. Przybylowicz et al. [219] impregnated preformed silica gel layers on plastic supports by dipping in silver nitrate solution.

Gupta and Dev [119] found that although the R_f value of a compound varied with the amount of silver nitrate impregnation, the relation of the R_f's of two compounds remained fairly constant, as can be seen from Fig. 3.1.

Because some of the solvents used in separating nitrogen hetercyclics on silver nitrate-impregnated layers leached the complexing agent out of the layers, Tabak and Verzola [220] investigated the use of silver oxide as a complexing agent. The method was also applied to some carboxylic acids and amines [221]. These plates can be useful because they can be used with more polar solvents than can the silver nitrate layers. The method of preparation was as follows: 5% silver nitrate solution was added to silica gel to give 5% silver in the total solids. To this was slowly added the stoichiometric amount of 5% sodium hydroxide solution with thorough mixing. After coating, the plates were dried for 45 min at 105 to 110°C.

Various other complexing agents have been used. Halmekoski [222, 223] has shown the effect of a number of complexing agents on the separation of phenol carboxylic acids and adrenalin derivatives (Tables

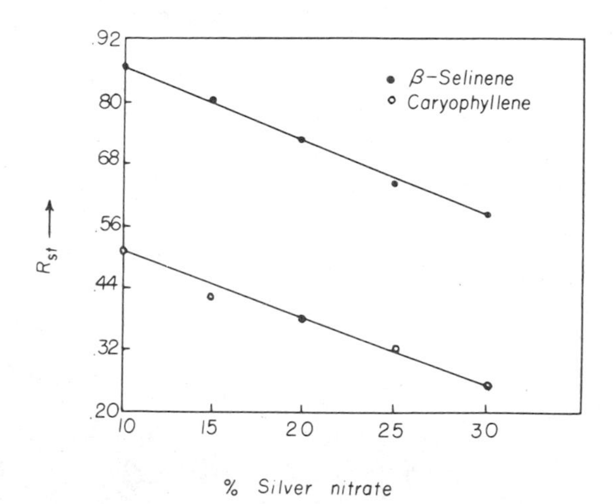

Fig. 3.1 Variation of R_{st} with the silver nitrate content of impregnated silica gel: R_{st} = (R_f of compound/R_f of standard dye). From Gupta and Dev [119]; reproduced with permission of the authors and The Elsevier Publishing Co.

Table 3.7 Values of 100 × R_f in Thin-Layer Chromatography of Various Adrenaline Derivatives on Kieselgel G Layers Buffered to pH 4 and on Kieselgel G Layers to which the Sodium Salts of Different Chelate-Forming Anions were Added[a]

Solvent System[b]	*Layer Treated with*	*Adrenaline*	*Oxedrine*	*Noradrenaline*	*Adrenone*	*Corbadrine*	*Oxamphetamine*	*Isoprenaline*	*Metaoxedrin*	*3-Hydroxytyramine*
1	Nil	27	35	41	38	57	69	45	44	53
	Na_2MoO_4	06	40	12	06	14	71	11	47	17
	Na_2WO_4	05	48	06	05	12	72	14	53	12
	Borax	15	38	14	07	18	51	20	40	28
2	Nil	33	42	52	41	56	64	48	48	54
	Na_2MoO_4	10	40	13	10	21	66	21	50	25
	Na_2WO_4	18	45	17	08	27	71	34	53	29
3	Borax	27	56	30	17	37	73	40	61	46
	Nil	27	30	34	27	38	48	31	34	37
	Na_2MoO_4	05	35	14	06	19	56	17	39	19
	Na_2WO_4	09	36	17	08	22	60	20	40	21
4	Borax	13	34	20	16	22	60	20	36	23
	Nil	42	45	52	45	60	67	53	51	56
	Na_2MoO_4	22	47	25	16	31	64	24	47	24
	Na_2WO_4	24	50	30	19	36	64	31	52	31
	Borax	28	55	36	24	42	73	46	61	42

[a] From J. Halmekoski [223]; reproduced with permission of the author and publishers.

[b] The organic layers of the following solvent mixtures were used for development: (1) *n*-butanol saturated with aqueous sulfur dioxide solution (H_2SO_3); (2) *n*-butanol–acetic acid–H_2SO_3 (4 : 1 : 5); (3) *n*-amyl alcohol–acetic acid–ethanol–H_2SO_3 (4 : 1 : 1 : 5).

Table 3.8 Some Complexing Agents Used for Impregnating Thin Layers

Impregnating Agent	*Base Layer*	*Type of Compounds Separated*	*Reference*
$Pb(NO_3)_2$	Silica gel	Polyols	224
Basic lead acetate	Silica gel	Sugars, anthocyanins, anthocyanidins	225
Manganese acetate	Silica gel	Aromatic amines	226
		Phospholipids	227, 228
Manganese formate	Silica gel	Aromatic amines	226
$Mn \cdot Na_2EDTA \cdot 2H_2O$	Silica gel	Aromatic amines	226
Thallous nitrate	Silica gel	Monoterpene hydrocarbons	229
Cadmium sulfate	Silica gel	Aromatic amines	230, 231
Cadmium acetate	Silica gel	Aromatic amines	231
Cadmium phosphate	Silica gel	Aromatic amines	231
Zinc chloride	Silica gel	Chlorinated anilines	122
Zinc nitrate	Silica gel	Chlorinated anilines	122
	Silanized	Chlorinated anilines	122
Ferric chloride	Silica gel	Steroids	232
		Oxine derivatives	233
Copper sulfate	Silica gel	Hexosamines	234
$CuCO_3$ $Cu(OH)_2$	Cellulose	Amino acids	235
Molybdic acid	Silica gel	Carbohydrates	236, 237
Tungstic acid	Silica gel	Oligosaccharides	237
	Alumina	Oligosaccharides	237
Boric acid	Silica gel	Lipids	238, 239, 240 241, 242
Boric acid	Silica gel	Carbohydrates	236, 237, 243 244, 245, 246, 247
Boric acid	Silica gel	Triterpenoids	248
Boric acid	Silica gel	Polyhydroxy acids	249

Table 3.8 (Continued)

Impregnating Agent	*Base Layer*	*Type of Compounds Separated*	*Reference*
Oxalic acid	Silica gel	Azulenes	250
Oxalic acid	Silica gel	Insecticides	251
Oxalic acid	Silica gel	Nonionic detergents	252
Oxalic acid	Silica gel	Free fatty acids	253
Picric acid	Silica gel	Polycyclic hydrocarbons	254, 255
Sodium molybdate	Silica gel	Phenolic carboxylic acids	222
Sodium molybdate	Silica gel	Adrenaline derivatives	223
Sodium molybdate	Silica gel	Flavonol glycosides	256
Sodium tungstate	Silica gel	Phenolic carboxylic acids	222
Sodium tungstate	Silica gel	Adrenaline derivatives	223
Sodium tungstate	Silica gel	Flavonol glycosides	256
Sodium tetraborate	Silica gel	Nucleotides	257, 258
Sodium tetraborate	Silica gel	Lipids	259, 260, 261 262, 263, 264
Sodium tetraborate	Silica gel	Carbohydrates	236, 237, 265
EDTA	Silica gel	Antibiotics	266, 267
EDTA	Silica gel	8-Hydroxyquinoline derivatives	268
Sodium arsenite	Silica gel	Polyhydroxy compounds	269, 270, 271, 249
Urea	Silica gel	Hydrocarbon waxes	272
	Alumina	Plasticizers	273
Picramide	Silica gel	Amines	274
	Silica + Mg silicate	Amines	274
2,4,6-Trinitroanisole	Silica gel	Amines	274
m-Dinitrobenzene	Silica gel	Amines	275
2,4,6-Trinitrotoluene	Silica gel	Amines	275
Trinitrobenzene	Silica gel	Polycyclic hydrocarbons	276, 277

2,4,7-Trinitrofluoren-9-one	Silica gel	Polycyclic hydrocarbons	255, 276
Pyromellitic dianhydride	Silica gel	Polycyclic hydrocarbons	278, 279
Chloranil	Silica gel	Polycyclic hydrocarbons	279
Tetramethyluric acid	Silica gel	Polycyclic hydrocarbons	279
Caffeine	Silica gel	Polycyclic hydrocarbons	255, 279
Benzoquinone	Silica gel	Polycyclic hydrocarbons	280
Digitonin	Silica gel	Steroids	281
o-Toluidine	Silica + Mg silicate	Aromatic nitro compounds	282
Dimethylaniline	Silica + Mg silicate	Aromatic nitro compounds	282
m-Chloroaniline	Silica + Mg silicate	Aromatic nitro compounds	282
2,4,6,2′,4′,6′-hexanitrodiphenyl sulfide	Silica gel	Conjugated compounds	283
	Alumina	Conjugated compounds	283
2,4,2′,4′-Tetranitrodiphenyl sulfide	Alumina	Conjugated compounds	283
$NiSO_4$ or $CaCl_2$	Alumina	Aliphatic and heterocyclic alcohols	284
Diethyl or dipropyl amine	Silica gel	Cannabinoids	285
Strychnine	Silica gel	Alkyl gallates	286
	Silica + kieselguhr	Alkyl gallates	286
Cinchonine	Silica + kieselguhr	Alkyl gallates	286

3.6 and 3.7). Table 3.8 lists some examples of additional complexing agents and the types of compounds for which they were used.

Water-Based Slurries of Alumina with Plaster of Paris Binder

For the preparation of alumina slurries from commercial adsorbent preparations specifically for thin-layer chromatography work, the mixing proportions recommended by the manufacturers can be followed. In mixing slurries of aluminum oxide, procedures used for silica gel may be observed; that is, they may be triturated in a mortar or mixed in a flask by shaking.

For preparing 1-mm-thick layers for preparative chromatography, Korzun et al. [18] used a slurry of 110 ml of water with 120 g of alumina G, and Honegger [199] in preparing 2- to 5-mm-thick layers for preparative thin-layer work used an alumina G/water ratio of 1:0.9.

For fluorescent aluminum oxide layers the procedure introduced by Brockman and Volpers [287, 288] for column work may be used. They adsorbed morin (a pentamethoxyflavone) on alumina in a ratio of 300 mg to 500 g of the adsorbent. Cěrný et al. [289] used this fluorescent agent impregnated on alumina for thin-layer chromatography. Matthews et al. [290] used G. S. 115 green emission phosphor (U.S. Radium) in the ratio of 100 mg of the phosphor as a fluorescent agent with 30 g of aluminum oxide G. They washed both the adsorbent and the phosphor with 375-ml portions of boiling methyl alcohol to remove interfering materials. The mixture was then dried before preparation of the slurry.

Water-Based Slurries of Alumina with Starch Binder

Very little use has been made of a starch binder with alumina. It was first introduced by Meinhard and Hall [1] and was checked for the separation of terpenes by Kirchner et al. [2]. Petschik and Steger [291] prepared a slurry of 28.5 g aluminum oxide with 1.5 g wheat starch by slurrying in a mortar with distilled water. This mixture was not heated to gelatinize the starch prior to preparation of the layers but instead was applied directly to the plates which were then placed in a 120°C oven for drying. Attempts to gelatinize the starch in this manner in the author's laboratory were unsuccessful and the layers remained soft.

Water-Based Slurries of Alumina without a Binder

According to the firm of CAMAG, the very fine-grain aluminum oxide (aluminum oxide, FDO) which is made for use without a binder can be stored for only limited periods of time. After 6 to 9 months' storage, the ability to adhere to the glass plate diminishes.

For the preparation of an aluminum oxide slurry (acid, basic, or neutral) Woelm recommends using 35 g aluminum oxide with 40 ml of water for use with spreading equipment. However, for preparation of a slurry for pouring directly onto the plate without a spreading apparatus, they recommend 6 g of adsorbent in 15 ml of an ethanol–water mixture (9:1).

Water-Based Slurries of Modified Alumina

Acidified Alumina

For an acidic aluminum oxide slurry, one may either use a commercial acidic adsorbent, preparing the slurry as above, or aluminum oxide G or other suitable adsorbent, slurrying it with 0.2 *N* HCl in the ratio of 1 g of aluminum oxide to 2 ml of acid [292]. Vacíková et al. [293] prepared an acidic aluminum oxide slurry of Brockman activity of grade IV by adjusting a slurry of aluminum oxide grade I or II to a pH of 4 by means of hydrochloric acid.

Alkaline Alumina

Aluminum oxide is usually basic owing to its method of preparation, and if a basic adsorbent is not desired the pH of the material to be used should be checked. Alumina (CAMAG) for thin-layer chromatography is a basic adsorbent with a pH of about 9. Slurries can be prepared by mixing 20 g of the adsorbent with 50 ml of distilled water. Woelm basic aluminum oxide for thin-layer chromatography does not have a binding agent and slurries may be prepared as described previously.

Groeger and Erge [294] prepared an alkaline slurry by mixing 18 g of acidic aluminum oxide with 2 g of plaster of Paris and 40 ml of a 1% potassium hydroxide solution.

Alumina Impregnated with Complexing agents

Not much information is available on the use of complexing agents with aluminum oxide. Morris [249] mentions the use of sodium borate impregnated alumina G which gave a similar degree of separation as did a borate-impregnated silica gel G. A slurry is prepared by the standard procedure of using an aqueous solution of the complexing agent, instead of water, for preparing the slurry. Zinkel and Rowe [295] prepared a silver nitrate aluminum oxide slurry by dissolving 12 g of silver nitrate in 20 ml of water and diluting this further with 40 ml of methyl alcohol. To this was added 30 g of aluminum oxide. This was all thoroughly mixed by shaking. Urbach [296] prepared his slurry by mixing 30 g of aluminum oxide G with a solution of 7.5 g of silver nitrate in 50 ml of water.

Solvent-Based Slurries of Silica Gel

In contrast to most of the work on thin layers which has been done with layers prepared from water-based slurries, a number of workers have used slurries made with solvents. Mueller and Honerlagen [297] were the first to use this type of slurry. Their slurry was based on using 16 g silica gel G with 30 ml acetone. Layers prepared from this mixture were smoother and gave sharper spots. For applying the slurry to the glass plate by means of a spray gun, Joska (reported by Prochazka [298] from a private communication) used a slurry of 25 g silica gel–calcium sulfate mixture in 90 ml of 60% acetone. Duncan [299] used a slurry composed of 1 part silica gel G to 2 parts of a methyl alcohol–water mixture (1:1). Sinclair and Lehrfeld [300] used a 10 to 50% (vol/vol) of ethanol in the preparation of slurries to offset the water-repellent effect of glass contaminated with films from silicone grease.

For the preparation of layers by just pouring the slurry onto the glass support, Hoerhammer et al. [301] prepared a slurry of 1 part silica gel (Woelm) by weight to 3 vol of either ethyl acetate or acetone. They reported that chloroform, benzene, petroleum ether, methanol, or isopropanol were not good for the preparation of layers either because of the low volatility or because these solvents did not form a well bound layer.* In the preparation of solvent slurries, if too much solvent is used, an adsorbent-free zone will form near the edge of the plate. Bhandari et al. [302], of the firm of Woelm, recommend 6 g of silica gel (Woelm) in 15 ml of an ethanol–water mixture (9:1).

For applying a coating by dipping microscope or lantern slides into a slurry, Peifer [4] recommends dispersing 35 g of silica gel G in 100 ml of chloroform or chloroform–methanol (2:1). When he incorporated concentrated sulfuric acid directly into the slurry, he prepared this by blending 45 g of 100-mesh silicic acid and 5 g plaster of Paris in 102.5 ml of chloroform–methanol–concentrated sulfuric acid (70:30:2.5). In order to provide enough water for setting the plaster of Paris, the plates are exposed to steam. Lie and Nyc [61] used 1 g of silicic acid to $2\frac{1}{2}$ ml of chloroform for coating their test tubes.

Solvent-Based Slurries of Alumina

Peifer [4] prepared a solvent-based slurry of alumina by triturating 45 g of activated alumina powder and 15 g of plaster of Paris with a small volume of chloroform–methanol (70:30); after thorough mixing, the solvent was increased to a total 100 ml. This slurry was in preparation for the dipping of the plates and after the solvent evaporated, the plates were exposed to steam in order to provide moisture for the setting of the plaster of Paris binder.

* Woelm silica gel does not contain a binder.

Slurries of Miscellaneous Inorganic Adsorbents

Calcium Sulfate

Matis et al. [150] prepared a slurry of 20 g of calcium sulfate without a binder in 100 ml of water by vigorous shaking. This was poured onto the plate. The calcium sulfate was specially prepared by precipitating calcium chloride with sulfuric acid. After washing the precipitate thoroughly, it was dried for 48 hr at 115 to 120°C and was subsequently ground into a fine powder. They used this adsorbent for the separation of corticoids.

Magnesium Oxide and Hydroxide

Schwartz et al [303] and Schwartz and Parks [304] prepared a slurry by mixing 15 g of partially deactivated magnesia and 6 g of Celite 545 in 50 ml of 95% ethyl alcohol in a 125-ml glass-stoppered Erlenmeyer flask. This was shaken vigorously for 5 min. The magnesia, Seasorb 43 (Fisher), was conditioned by heating in a muffle furnace at 525°C for 16 hr. Plates made from this material were used for the separation of the 2,4-dinitrophenylhydrazones of carbonyl compounds.

Of several magnesias tested, Nicolaides [305] found MX-66 catalytic grade 200-mesh adsorptive powder (Matheson, Coleman and Bell) gave satisfactory plates. A slurry of 20 g of this adsorbent or of adsorbent + $CaSO_4$ binder (9:1) with 40 ml of water was used. Drying and activating was for 2 hr at 180°C after preliminary air-drying. The binder prevented flaking when the plates were subjected to charring procedures.

Snyder [306] has investigated the chromatographic behavior of water-deactivated magnesia. He points out that the major difference of magnesia and alumina is the stronger relative adsorption of the carbon–carbon double bond on magnesia. Magnesia tends to adsorb olefins and aromatics more strongly than other adsorbents.

Keefer [307] and Keefer and Johnson [308] have proposed the use of magnesium hydroxide as an adsorbent for thin layers. A slurry was prepared from 100 g of magnesium hydroxide (Fisher) in 150 ml of water. This adsorbent appears to have a high capacity.

Calcium, Magnesium, and Aluminum Silicates

Tore [309, 310] used calcium silicate for separating sugars and for phenylosazones. Plates were prepared from a slurry of Silene EF (hydrated Ca silicate, Pittsburgh Plate Glass) or from a mixture of Silene EF and acid-washed Celite 535. The mixture used was 11 g of Silene EF, 3 g of Celite 535, and 700 mg of sodium acetate. The exact amount of water was not specified. Wolfrom et al. [141] prepared a slurry of magnesium silicate by mixing 20 g of specially prepared Magnesol (see Section 2 for preparation), and plaster of Paris (13%) with 55 ml of water. This material was used in the separation of sugar acetates and methyl ethers.

For a nonaqueous slurry, Peifer [4] triturated 45 g of Florisil (Floridin) of a 60 to 100 mesh and 10 g of plaster of Paris with 1 ml of glacial acetic acid and a small amount of chloroform-methanol (70:30). This was then diluted so that the suspension contained a total of 101 ml of solvent. This slurry was applied by dipping plates into it, and then exposing the coated plates to an atmosphere of steam. Heacock and Mahon [10] prepared a slurry of 10 g of aluminum silicate (containing 3% plaster of Paris) in 20 ml of water. This material was used for the chromatography of the hydroxyskatoles.

Diatomaceous Earth

The manufacturers of kieselguhr G (Merck) recommend slurrying 30 g of the adsorbent in 60 ml of water. This brand of diatomaceous earth contains plaster of Paris as a binding agent. Vaedtke and Gajewska [311] used Celite 545 (Johns Manville). The diatomaceous earth as well as the plaster of Paris were first sieved through a DIN-1171 sieve (0.07 mm mesh) before slurrying 7 g of Celite and 0.4 g plaster of Paris with 40 ml of water. Because of the weak adsorptive power of diatomaceous earth, it is used mostly as a support for partition chromatography. Thus Knappe and Peteri [312] prepared their impregnated kieselguhr layers by incorporating the impregnating agent directly in the slurry. They used 30 g of kieselguhr G, 0.3 g of sodium diethyldithiocarbaminate, 50 ml of water, and 10 g of polyethylene glycol M 4000. Slurries for the preparation of buffered layers of diatomaceous earth are prepared by slurrying the 30 g of diatomaceous earth with 60 ml of the appropriate buffer solution instead of water. Prey et al. [313] prepared slurries using a mixed silica gel–kieselguhr (1:4) adsorbent slurried with a 0.02 *M* sodium acetate buffer solution. This mixed adsorbent gave better resolution of carbohydrates than buffered kieselguhr. Badings and Wassink [314] prepared silver nitrate-impregnated kieselguhr plates by slurrying 25 g kieselguhr G in 58 ml of water containing 7.5 g of silver nitrate. This material was used for the separation, into classes, of the 2,4-dinitrophenylhydrazones of aliphatic aldehydes and ketones.

Calcium Phosphate (Hydroxyapatite)

Hofmann [11, 138, 139, 315] has fostered the use of hydroxyapatite for the separation of lipids and proteins. He used two types of binders, plaster of Paris and Zytel-61 (soluble nylon, Du Pont). For the nylon binder he dissolved 40 mg of nylon in 60 ml of 70% ethyl alcohol with heating and stirring. After cooling the solution, 15 g of hydroxyapatite is added and thoroughly mixed by homogenizing to get a uniform suspension. Since plaster of Paris binder improves the resolution of the material, 15 g of hydroxyapatite is mixed with 1.2 g of plaster of Paris and 60 ml of water. This is shaken thoroughly in a closed flask or it may be homogenized with a high-speed machine. Seven percent of plaster of Paris binder seems to give the optimum resolution. Hydroxyapatite still remains inferior to silica gel for the separation of lipids. A silver nitrate-impregnated hydroxyapatite may be prepared by slurrying with 40% silver nitrate solution instead of water (239).

Zinc Carbonate

Badings [316] and Badings and Wassink [314] have used basic zinc carbonate layers with 5% starch binder for the separation of 2,4-dinitrophenylhydrazones. For the preparation of a slurry one may take 19 g of basic carbonate with 0.5 g ($\frac{1}{2}$%) starch and 50 ml of water. The mixture is then heated thoroughly on a water bath to gelatinize the starch.

Eisenberg et al. [317] slurried 25 g of zinc carbonate containing 5% soluble starch with 70 ml of water. They also prepared an eluotropic series of solvents for zinc carbonate.

Carbon

For a carbon slurry, Brodasky [318] slurried 30 g of Nuchar-C-190-N (a vegetable carbon black prepared by West Virginia Pulp & Paper Co.) and 1.5 g plaster of Paris with 220 ml of distilled water or of distilled water adjusted to pH 2 with sulfuric acid depending on whether or not an acidified carbon is desired. The acidified carbon must be allowed to stand a minimum of 16 hr prior to preparation of the coated plates. Using starch as a binder, Kirchner and Flanagan [179] prepared a slurry of 19 g of Nuchar C-N charcoal with 2 g of Clinco-15 starch (Clinton) in 38 ml of water. Prior to preparing the slurry, it may be necessary to wash the charcoal with acetone in order to remove adsorbed materials. Hesse and Alexander [319] have prepared layers from slurries of graphite. Procházka [320] suspended 20 g of charcoal with 2 g of polyethylene as a binder in 100 to 150 ml of methylene chloride.

Powdered Glass

In 1962 Rahandraha [321] published on the use of powdered glass for thin layers. These layers did not contain a binder and Rahandraha et al. [322] reported the method of treating the glass powder and preparing the thin layers. One kilogram of powdered glass that passed through a 0.05-mm-mesh screen was stirred in 2 liters of distilled water and allowed to stand for 1 hr. After the supernatant liquid was decanted the powder was next washed in 2 liters of 5% acetic acid and after standing 1 hr was again decanted. The final washing was again with 2 liters of water and after the supernatant liquid was decanted the powder was dried at 140°C. For application of the layer, a thick slurry may be prepared with water or with a water–propanol mixture (4:1). Brud [323] has demonstrated the use of powdered Jena glass.

Kramer et al. [324] used a powdered porous glass which because of its porous nature has a greater surface area than ordinary powdered glass. This was prepared by grinding porous glass plates #7930 (Corning Glass) to a 200- to 250-mesh size. Thirteen percent plaster of Paris was used as a binder and the coated plates were dried at 130°C for 3 hr. In the examples cited of three waxes, the porous glass gave better resolution than silica gel G or aluminum oxide, as evidenced by more distinct spots and in some cases additional spots. Wolf et al. [325] used 5% gypsum as a binder and activated at 200°C. They commented that the efficiency depends on the activation temperature. Marino [326] reviewed the literature on porous glass thin-layer and chromatography and then ran some comparisons with silica gel. In general silica gel was superior.

Slurries of Organic Polymers

Polyamides

Recommendations from the firms that provide a polyamide for thin-layer chromatography are for the preparation of a slurry of 5 g of the powder in 45 ml of ethanol or methanol. These powders, of course, do not contain any binder. Where the slurry is not to be used in a spreading apparatus, but rather is to be poured onto the plate, Woelm recommends a slurry of 1 g of polyamide in 13.5 g of ethanol. Wang [327], Egger [328], and Davídek and co-workers [329, 330] were the first to use polyamides for thin-layer chromatography. The latter workers used the loose-layer method of Mottier and Potterat [331]. The modified method of Wang et al. [332] is to dissolve 20 g of polyamide resin in 100 ml of formic acid (75%). Fifteen milliliters of this viscous solution is poured evenly on each of four glass plates (15 × 15 cm). These are kept in an atmosphere saturated with water vapor for 2 days at 25 ± 2°C. They are then dried at 100°C for 15 min to remove traces of formic acid. Propylene or ethylene glycol may also be used as a solvent for preparing polyamide layers [333]. In this case nylon is dissolved above 100°C to make a 10% (by wt) solution. Glass plates are then dipped into this solution. Urion et al. [334, 335] used a starch-bound slurry of polyamide. To prepare this they heated 1.5 g of rice starch in 54 ml of water for 2 to 3 min. This was then diluted with an additional 30 ml of water and heated to 80°C. After cooling to 50 to 60°C they added 9.5 ml of polyamide, homogenizing the mixture for 30 sec before putting it in a spreader. To prepare an even firmer layer,

Nordby et al. [336] added silica gel to the mixture. With occasional stirring, a mixture of 0.8 g of rice starch, 0.4 g of silica gel (Fisher No. 1 impalpable powder), and 9 ml of water was heated on a steam bath for 40 min. Addition of 1 to 2 ml of water to the covered beaker prevented caking on the sides. It was then rinsed with 3 ml of water into 5.5 g of Woelm polyamide powder in 35 to 40 ml of methanol. Blending in a Waring Blendor for 3 min ultimately gave a smoother surfaced layer. The layers prepared from this were dried for 2 hr at room temperature. Wagner et al. [337] used cellulose as a binder by blending 12 g of polyamide, 2.4 g of cellulose, and 40 ml of methanol in a blendor.

Roesler et al. [338] gave a procedure for preparing a standardized fine-grain polyamide by controlled hydrolysis of polyamide pellets or powder. The final slurry preparation may be kept by addition of a few milliliters of chloroform to prevent fungal growth or it may be dried for storage.

Polyvinylpyrrolidone

Insoluble polyvinylpyrrolidone (PVP) also forms hydrogen bonds with carboxyl and phenolic hydroxyl groups, and therefore can form a very useful adsorbent. Quarmby [339] prepared a slurry by shaking 15 g of PVP (100 to 300 mesh) with 80 ml of water. To prevent settling, the slurry was applied to the plates as soon as it was poured into the spreader set at 500 μ. Moreno Dalmau et al. [340] ground and sieved Polyclar AT (General Aniline and Film) to obtain particles smaller than 40 μ. A slurry was prepared from 8 g of powder in 49 ml of isopropanol. Later [341] they used 6 g of a 37- to 53-μ fraction in 37.5 ml of anhydrous isopropanol. Loomis and Battaile [341a] purified PVP (Polyclar AT) by boiling for 10 min in 10% hydrochloric acid followed by washing with glass-distilled water and then with acetone.

In all cases without a binder the prepared plates were dried overnight at room temperature.

Clifford [342[used a binder by combining 2.4 g of plaster of Paris with 12 g of commercial PVP that passed a 100 mesh screen. This was mixed dry and then homogenized with 96 ml of water. Layers prepared from this slurry were dried 15 min at room temperature and then 30 min at 105°C.

Miscellaneous Organic Polymers

Poropak is the trade name given to a series of polymers produced by Waters Associates for use in gas chromatography. Janák and coworkers [343–345] have examined some of these for use in thin-layer work. Poropak Q (100 to 120 mesh), a polymerized ethylvinylbenzene cross-linked with divinylbenzene, was cold-pressed to form a layer on a glass support [344]. Poropak N and P were also examined. Poropak T, an ethylvinyl–

divinylbenzene modified with polar monomers, was applied to plates as a suspension in propanol (2:5 wt/wt) [345].

Pevikon C-870 (Mercer Chemical Corp.), a copolymer of polyvinyl chloride and polyvinyl acetate, has been used for electrophoresis because there is very little electroosmosis and no interaction with proteins. To prepare a slurry of this [346] 50 g is mixed with 100 ml of buffer prepared from 0.0365 *M* tris(hydroxymethyl)aminomethane, 0.0365 *M* maleic anhydride, and 0.05 *M* sodium hydroxide. After a homogeneous suspension is obtained, the slurry is allowed to settle for 4 min and then 50 ml of the supernatant liquid is discarded. The thick slurry is then applied to a 11 × 20 glass plate.

Raaen [347, 348] has used polytetrafluorethylene (Fluoroglide 200 Grade TWO 218, Chemplast) as a thin-layer adsorbent. Slurries can be prepared in numerous organic liquids in a ratio of polytetrafluoroethylene to liquid of 1:3 wt/vol. Water slurries can be prepared if a wetting agent such as the fluorochemical surfactant FX-173 (3 M Company, Atlanta Branch, Industrial Chemicals Division) is used. A mixed layer was also prepared [349] by slurrying 20 g of Aviamide-6 powder, a microcrystalline nylon (Chemical Research & Development Center, FMC Corp.), with 5 g of the Fluoroglide in 75 ml of *n*-propanol.

Polyvinyl acetate has been used as an adsorbent layer [350]. Varying amounts of polyvinyl acetate (0.5 to 4 g) were dissolved in 25 ml of methanol and 15 ml of ethyl acetate. After dilution to 60 ml with methanol, 25 g of kieselguhr G was stirred into the solution. After the layers were spread they were dried for 3 hr at room temperature. (Equipment should be cleaned with methanol immediately after use.)

Hesse et al. [351] have compared polyacrylonitrile, polyacrylamide, and *N*-acetylpolyacrylamide with polyamide as adsorbents. These are suitable for the separation of water-soluble substances because they form hydrogen bonds.

Epton et al. [352] have synthesized and evaluated a range of cross-linked poly(acryloylmorpholine) gel networks as supports for gel permeation chromatography.

Polyacrylonitrile–Perlon Mixture

Birkofer et al. [192] have prepared thin layers from slurries of a mixture of polyacrylonitrile and Perlon (a polycaprolactam) in a ratio of 7:2 which was slurried with 40 ml of 0.05 *M* primary calcium phosphate solution. This material was used for the separation of anthocyanins.

Urea–Formaldehyde Resins

Ardelt and Lange [353] have used thin layers of urea–formaldehyde resin prepared by making a slurry of 7 g of the resin with 15 ml of a methyl alcohol–water solution (1:6). No binder was used with this material which was applied to the separation of thiourea and related compounds.

Sephadex

Determann [354], Wieland and Determann [355], and Johansson and Rymo [356] introduced the use of the cross-linked polysaccharide, Sephadex, in thin-layer chromatography. In the same year Dose and Krause [357] introduced the use of the same material for thin-layer electrophoresis. Determann, after washing the Sephadex G-25, (200 to 400 mesh) fine (Pharmacia), suspended 7 g of the gel in water to form a pourable paste. Wieland and Determann [355] used slurries of DEAE, Sephadex A-25 (200 to 400 mesh) fine, which had been treated in succession with 0.5 *N* hydrochloric acid, water, 0.5 *N* sodium hydroxide, and water. The pH of the slurry was then adjusted with the acid used in the buffer solution. Johansson and Rymo slurried 10 g of Sephadex G-25 gel with 50 ml of 0.1 *M* sodium chloride or with an appropriate buffer for 1 hr before spreading the slurry on the plates. With Sephadex G-75 fine, and Sephadex G-75 (particle size, 400 mesh and finer) 5 g of the gel was placed in 70 ml of buffer solution for 5 hr with occasional stirring. The buffer generally used was 0.02 *M* sodium phosphate buffer at a pH of 7.0, containing 0.2 *M* sodium chloride. Johansson and Rymo [358] used Sephadex G-50 and G-75 for separating low-molecular-weight proteins and G-100 and G-200 for high-molecular-weight proteins. In this case they prepared the Sephadex plates and then used them in a descending chromatogram. The Sephadex was conditioned by allowing the buffer solution to flow through the plate at least 1 hr before applying the material to be separated.

The same authors [359] also used Sephadex G-100 and G-200 in a two-dimensional separation on human serum by employing gel filtration in one direction followed by electrophoresis perpendicular to the filtration direction. Morris [360, 361] used Sephadex G-100 and G-200 for the separation of proteins having a molecular weight of up to 180,000. He conditioned his material for 48 hr and then allowed 18 hr for equilibration in the developing chamber. Using Sephadex Dose and Krause [357], in 1962, applied thin-layer electrophoresis to the separation of proteins. Sephadex G-50 (fine) was mixed with an excess of buffer solution (1:7.5) for 24 hr. The excess was then filtered off and the resulting gel, which was plastic but not fluid, was poured between guides placed on the glass plates. Fasella et al. [362] used Sephadex G-25, G-100, and G-200 for the separation of proteins. The Sephadex was prepared by stirring it in the appropriate buffer for 30 min and then allowed to settle before decanting the supernatant liquid. This was repeated five or six times so that the total buffer contact time was at least 48 hr for Sephadex G-25 and 72 hr for G-100 and G-200. Vendrely et al. [363] gave greater rigidity to their Sephadex layers for electrophoresis by the addition of agarose. For this purpose, 1.25 g of agarose in 65 ml of buffer was thoroughly blended with 4 g of Sephadex G-200, 5 g of Sephadex G-100, or 6.5 g of Sephadex G-75 after the Sephadex had been equilibrated with buffer solution.

Tortolani and Colosi [364] found that mixed layers of Sephadex and silica gel or cellulose could be used so that ascending development could be used. Slurries were prepared by mixing 16 g fine Sephadex G-10 with 4 g of silica gel GF or microcrystalline cellulose. These were thoroughly washed with distilled water and then equilibrated with 45 ml of phosphate buffer (0.05 M, pH 7.0). The layers were spread and left in a horizontal position for 1 hr, then dried with hot air until almost dry. Development times were much shorter than for untreated Sephadex with comparable R_f values.

Morgan et al. [365] called attention to the fact that Sephadex contains from 2 to 6 μg of zinc per gram of dry Sephadex.

Slagt and Sipman [365a] used a methylated Sephadex prepared by methylating Sephadex G superfine with dimethyl sulfate in sodium hydroxide solution.

Hashizume and Sasaki [366] have used the diethylaminoethyl Sephadex for the separation of nucleotides.

Thin-layer chromatography on Sephadex provides a convenient means of quickly obtaining an approximate molecular weight of proteins and peptides [354, 360, 363] because their motility can be plotted against the molecular weight or, better, against the log of the molecular weight [367, 368]. As illustrated in Fig. 3.2, the degree of separation on Sephadex can be correlated with the Stokes radii [369, 370].

Polyethylene

Mangold [374] has used polyethylene powder for the preparation of chromatoplates to be used for the separation of fatty acids or their methyl esters. The slurries for these plates were prepared by mixing 20 g of Hostalene S (Farbwerke Hoechst) of 250-mesh size and 50 ml of water–ethanol (96%) in a 1:1 ratio.

Urea

In order to apply the principle of clathrate formation to thin-layer chromatography, layers of urea have been used. This material needs a binder, and Hradec and Menšík [375] made a slurry of urea–Celite (3:1). This was homogenized with urea-saturated methanol and poured onto a glass plate using a form to contain the layer until it was dry. Calcium

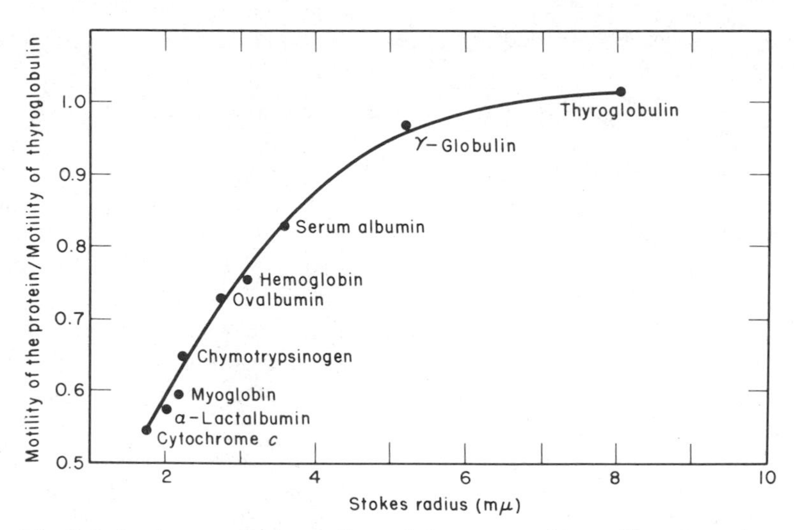

Fig. 3.2 Relation between Stokes radius and chromatographic motility of proteins on thin layers of Sephadex G-100. Motility is expressed as the ratio between the displacement of the protein and the displacement of thyroglobulin which is excluded from the gel. The Stokes radii were calculated from values reported in the literature [369, 371]. Data are included from P. Andrews [367] as well as data obtained by P. Fasella, A. Giartosio, and C. Turano. It is important to note that proteins undergoing reversible dissociation behave abnormally on Sephadex. This abnormality of behavior can be used to obtain information about the equilibrium constant for the dissociation [372, 373]. (Figure specially prepared by P. Fasella.)

sulfate binder can also be used, and a slurry can be made of 25 g of calcium sulfate mixed with 60 ml of a methanolic urea solution [376]. Good layers can be prepared containing 20 to 40% urea.

Sugars

Sugar has proved useful for the separation of pigments. A slurry can be prepared by blending 100 g of confectioner's sucrose (containing 3% cornstarch) in 130 ml of reagent-grade acetone in a low-speed electric blender [377]. Plates are air-dried for 2 hr before use. Mannitol may also be used [378] by blending 65 g of mannitol with 100 ml of acetone for 1 min, adding 1 ml of a 50% cornstarch solution, and blending for another minute. The layers are ready for use after drying 20 to 30 min in the air.

Starch

Thin layers of rice starch may be prepared by suspending 18 g of rice starch and 2 g of plaster of Paris in 20 ml of 96% ethanol and then in 40 ml of distilled water [379]. Cornstarch layers can be prepared, but the starch should be washed several times with water containing a little chloroform or carbon tetrachloride [380]. The two starches have different adsorption characteristics.

Slurries of Ion-Exchange Resins

The first thin layers prepared from ion-exchange resins were prepared by Kuroiwa and Hashimoto [89], although the Zeolite which they used in their preparations was used as an adsorbent rather than an ion-exchange material. These authors slurried 150 g of 90-mesh Zeolite (Dow Chemical) with 10 g of starch and 500 ml of distilled water. This mixture was heated at 80 to 85°C until the starch gelled. Berger and coworkers [381] at first used a slurry of equal parts of cellulose MN 300 G and Dowex 1 (for anions) or Dowex 50 (for cations); later they used a mixture of 5 g of cellulose MN (Macherey, Nagel) with 30 g of the Dowex resin and 60 ml of water. In order to get a uniform distribution it is necessary to triturate the material in a mortar, Greenland et al. [382] found that Dowex 50 cation-exchange resin as well as other polystyrene–divinylbenzene sulfonated cation-exchange resins contained a ninhydrin-positive contaminant which reappeared after washing with ammonium hydroxide if the resin was allowed to stand overnight.

Lepri et al. [383] prepared layers of sodium carboxymethyl cellulose as an ion exchanger from 4.5 g of CMCNa with an exchange capacity of 0.89 meq./g in 40 ml of water as well as layers of alginic acid by slurrying 6 g of the latter and 1.5 g of cellulose in 40 ml of water [384]. Numerous other resins may be used, and Decker and Hoeller [385] give a method of fractionating ion-exchange resins according to particle size by a sedimentation procedure.

There have been a number of papers published on the use of adsorbents impregnated with liquid ion exchangers; those by Brinkman et al. [386] and Graham and Carr [387] can be cited as examples of this work.

Zabin and Rollins [388] have used the inorganic ion exchangers zirconium phosphate and hydrous zirconium oxide. Both of these are cation exchangers, but the latter exhibits anion-exchange properties in acidic mediums. These exchangers can be used without a binder if the layers are not placed directly in the solvent; otherwise they should be mixed with 3% starch. (Note: the preparation of cellulose ion-exchange slurries is discussed under modified celluloses.)

Slurries of Cellulose Powder

For various grades of commercial cellulose powders that are available for thin-layer chromatography, the amount of powder to be mixed with water varies depending on the supplier. For example, Serva recommends using 10 g of cellulose powder in 60 to 80 ml of water, CAMAG recommends 10 g in 65 ml of water, and Whatman recommends 30 g in 75 ml of water. These slurries may be prepared by shaking in a stoppered flask; however, more uniform suspensions are prepared by homogenizing for 30 sec with a mechanical mixer. Roessel [389] has used a starch-bound cellulose layer. He prepared the slurry for this by heating 0.3 g of cornstarch in 90 ml of water to gelatinize the starch; on cooling, 15 g of Schleicher & Schuell (140 DG or 142 DG) cellulose powder was added and thoroughly homogenized with a mixer for 30 sec. He compared layers made of this cellulose with those made from other powders in a corresponding fashion and found that the Schleicher & Schuell cellulose gave plates which developed within 50 min compared to other powders which took from $2\frac{1}{2}$ to 6 hr to develop. Knappe et al. [390] prepared impregnated layers of cellulose by preparing a slurry of 15 g of cellulose powder MN 300 G, 6.2 g of polyester solution (adipic acid triethyleneglycol polyester 80 to 82% in methylglycol; Glasurit-Werk polyester IK 123), and 0.05 g of sodium diethyldithiourea with 35 ml of water and 35 ml of alcohol. Randerath [391] and Randerath and Struck [392] prepared cellulose solvent slurries by mixing 10 g of cellulose powder with 50 to 60 ml of acetone (in this case no binder was used in the mixture). Peifer [4], using the dipping technique for coating slides, prepared a cellulose slurry by triturating 35 g of cellulose and 15 g of plaster of Paris in a minimum volume of methyl alcohol. Schweiger [393] prepared a slurry using cellulose powder that had been pretreated with 0.5% solution of ethylenediaminetetraacetic acid in order to remove interfering cations. Hammerschmidt and Mueller [155] treated their cellulose at 50°C for 2 hr under reflux with a 1.5% nitric acid solution in order to remove interfering ions. The material was filtered and washed thoroughly with water before preparing their slurries. Madaeva and Ryzhkova [394] pretreated their cellulose powder with 2% hydrochloric acid. Wolfrom et al. [394a] prepared a reduced cellulose by mechanically stirring 50 g of cellulose for 10 hr at 25°C with 1 liter of an aqueous solution of 0.1 *N* sodium borohydride. After settling, the supernatant was decanted and the procedure repeated with fresh borohydride. The resulting cellulose was then properly washed and dried over phosphorus pentoxide. It was then ground to a powder in a mortar.

Slurries of Modified Cellulose

DEAE Cellulose

Slurries of diethylaminoethyl cellulose (DEAE) can be prepared by mixing 10 g of the cellulose exchanger with 60 to 80 ml of distilled water either by shaking in a closed Erlenmeyer flask or preferably by homogenizing with an electric mixer. Coffey and Newburgh [395] prepared slurries of DEAE cellulose containing 5% calcium sulfate as a binder by mixing equal portions of MN-cellulose 300-DEAE and MN-cellulose 300 G/DEAE, the latter containing 10% calcium sulfate. For certain nucleic acid degradation products, the calcium sulfate-bound cellulose gave better results than the unbound material. Boerning and Reinicke [396] found that DEAE cellulose contained impurities that adsorbed ultraviolet light in the same region as the nucleotides. They therefore purified this material by stirring 10 g of the exchanger for 30 min with 250 ml of 1 *N* hydrochloric acid. This material was filtered through a sintered glass filter and the process was repeated four times, after which the material was washed with water until the washings were neutral.

ECTEOLA Cellulose

This material is a weak anion-exchange material formed by linking cellulose with triethanolamine (TEOLA) by means of epichlorhydrin (EC). Randerath [184, 397] used this material for the separation of nucleotides. For the preparation of a slurry of this material he mixed 10 g of the powder with 60 to 70 ml of distilled water.

PEI Cellulose

Randerath [398] has proposed the use of PEI cellulose (polyethylenimine cellulose) for the separation of nucleotides. This material is an anion-exchange material and is made by Serva and by Macherey, Nagel; however Weimann and Randerath [399] give the following directions for the preparation of a slurry starting with cellulose powder. Three grams of a 50% polyethylenimine solution (Badische Aniline und Sodafabrik) is dissolved in 6 ml of water. After neutralization with concentrated hydrochloric acid the solution is diluted to a final volume of 15 ml and dialyzed in Visking 36/100 ft dialysis tubing against 1 liter of distilled water. The dialyzing water is changed after 4 and 8 hr. After 20 hr of dialysis the solution is diluted to 150 ml. Ninety milliliters of the dialyzed material is then slurried with 15 g cellulose powder. The dried plates prepared from this material are developed with distilled water and then dried again in order to remove further impurities. Dialysis of the polyethylenimine solution is not necessary if the prepared layers are developed for 5 cm with 10% sodium chloride and then without drying to the top edge with distilled water. After drying in cold air, the water development is repeated [399a].

Acetylated Cellulose

A slurry of this material may be prepared by mixing 10 g with 55 ml of 95% alcohol, preferably with an electric mixer.

The commercial cellulose acetate is not completely acetylated, and in order to prepare a cellulose triacetate, Hesse and Hagel [400] stirred 200 g of cellulose powder (Schleicher & Schuell #123 or 144) with 4 liters of benzene, 800 ml of acetic acid, 6 ml of 60% perchloric acid, and 800 ml of acetic anhydride for 3 days at 35°C. The mixture was then centrifuged, reslurried with methanol, again centrifuged, and dried at 30°C.

Phosphorylated Cellulose

Ten grams of a commercial phosphorylated cellulose (MN 300 P, Macherey, Nagel) may be slurried with 50 ml of water. Randerath and Randerath [401] have given directions for preparation of a polyphosphate cellulose. Twenty grams of powdered cellulose which has been treated for 5 min with 120 ml of a 3% polyethylenimine solution is further diluted with 200 ml of water. This entire mixture is then filtered through a sintered glass funnel and washed twice with 120-ml portions of distilled water. The resulting product is then stirred for 5 min with 120 ml of a 20% water solution of sodium metaphosphate. It is again filtered and washed with 120 ml of 0.25% hydrochloric acid solution followed by three washings of 120 ml of distilled water. The moist product is then shaken with 80 ml of distilled water to produce a good suspension.

5 APPLICATION OF THE ADSORBENT TO THE SUPPORTING PLATE

Pouring the Layers

In general this method is not used to a great extent although it is one of the simpler methods of preparing plates. In order to obtain layers of equal thickness, a measured amount of the slurry is put on a given size plate which is placed on a level surface. The plate is then tipped back and forth to spread the slurry uniformly over the surface. Breccia and Spalletti [402], after pipetting a slurry suspension onto the glass plate, have applied mechanical vibration for 5 min to even out the layer.

Dipping

Peifer [4] has introduced the technique of preparing plates by dipping them two at a time, back to back, in chloroform or chloroform–methanol slurries of the adsorbent. (For preparation of slurries for this purpose, see Section 4, Solvent-Based Slurries.)

Spraying the Adsorbent

This technique for the preparation of thin-layer plates was first proposed by Reitsema [95], who used a small paint sprayer for distribution of the slurry on the glass plate. In this procedure it is necessary to dilute the slurry further in order for the sprayer to operate. With this technique it is difficult to get uniform layers on a single plate, and also there may be variation from plate to plate. Bekersky [403], Morita and Haruta [12], and Metche et al. [335] used a regulation laboratory glass spray apparatus for spraying on slurries. Druding [404] has also used a dentist's oral atomizer. Bekersky [403] has shown a variation of $\pm 40\ \mu$ in the thickness of sprayed layers, and Metche et al. [335] claimed a variation in thickness of less than $20\ \mu$ for sprayed plates.

Use of Guide Strips

Although the statement has been made that uniform layers cannot be obtained by manual methods, this is not true. If uniform glass plates are selected and placed between glass or metal guides which are thicker than the glass plate by the amount that is desired for the layer, very uniform layers of adsorbent may be produced by spreading the slurry with a spatula or a glass rod. This method was introduced by Kirchner et al. [2] in 1951. This method or modifications of it are usually used for the preparation of loose layers. A glass rod with rubber or plastic sleeves or with wire or tape wound around the ends to serve as depth guides may be used as a means of spreading the adsorbent on the plate [405, 289, 330, 96, 407].

Mistryukov [28] used a metal rod thickened on the ends for spreading loose layers of alumina. A commercial design of this same principle is offered by Serva-Entwicklungslabor Co. This is a four-sided plastic rod shaped so that the collar at the end serves to guide the movement along the side edge of the plate (Fig. 3.3). The stepped design allows the two ends of the rod to rest on the top edges of the plate with a narrow gap between for the slurry coating. Three of the sides are for preparing 100-mm-wide layers 0.3, 0.5, and 1.0 mm thick, respectively. The fourth side is for the preparation of 50 mm layers, 0.3 or 1 mm thick.

As guides for spreading a thin layer of adsorbent, Lees and DeMuria [408] have used narrow strips of adhesive tape along the edges of the

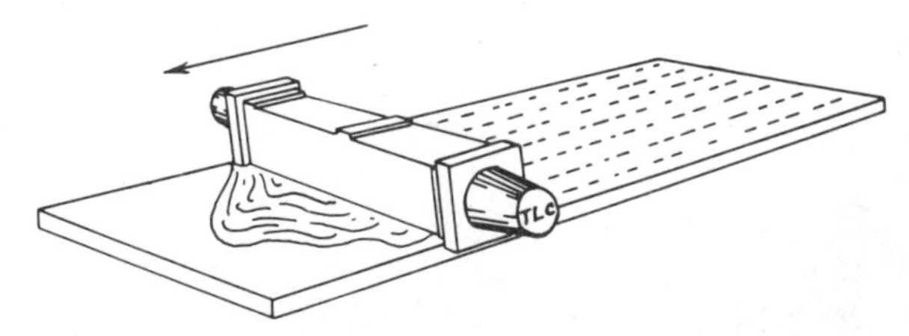

Fig. 3.3 Serva-Entwicklungslabor Co. thin-layer applicator. Courtesy of the Gallard-Schlesinger Chemical Mfg. Corp.

glass plate. Duncan [299] used a pair of stainless steel metal channels which fitted over the edges of the plate. Of course, along with those using guide strips should be included Gamp et al. [35] with their use of ribbed glass and Hansbury et al. [36] with their specially prepared grooved plates (see Section 1).

Specially Designed Equipment

Included here is equipment which has been designed either to allow plates to be pushed through during the coating or to have spreaders which are themselves moved over a series of plates. The first type was designed by Miller and Kirchner [181] in 1954 for the coating of narrow strips of glass. This is shown in Fig. 3.4, and Fig. 3.5 is an exploded view to show the assembly of the parts. In a modification of this apparatus the front gate is also pressed down by a spring, and the thickness of the layers is regulated by an adjustable wire foot which rests upon the glass slide and follows any unevenness in the thickness of the glass. In this apparatus and others of similar design, a plate is first pushed into the apparatus and the hopper is filled with the slurry; then each strip or glass plate is pushed through by the following one, and uniform layers of adsorbent are deposited on the plates.

Fig. 3.4 First apparatus for producing uniform, standard thin-layer coating on glass strips. From Miller and Kirchner [181]; reproduced with permission of the American Chemical Society.

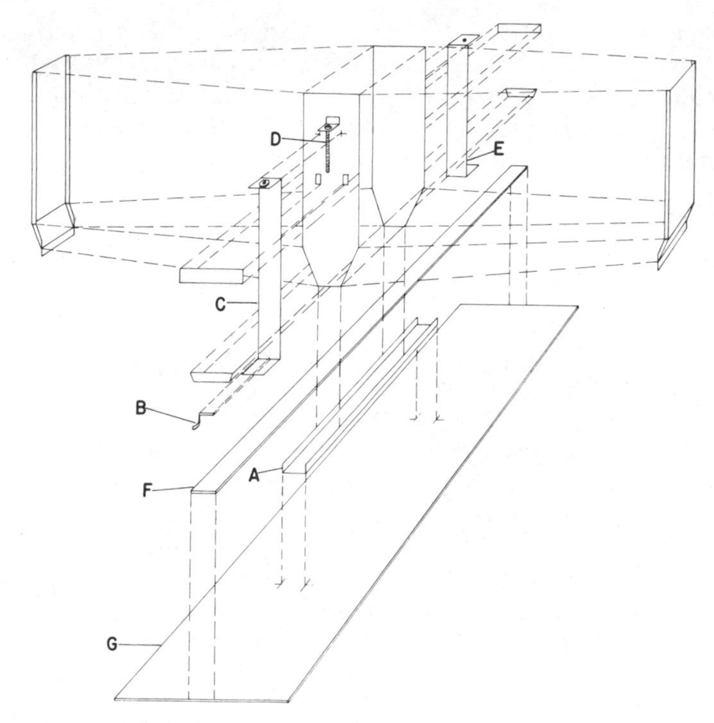

Fig. 3.5 Exploded view of apparatus for preparing uniform, standard thin-layer coatings on glass strips. From Miller and Kirchner [181]; reproduced with permission of the American Chemical Society.

Commercial equipment based on the same principle of operation is available. This equipment was designed by K. Mutter and J. F. Hoffstetter of Hoffmann-La Roche. It is produced and sold by CAMAG in Switzerland. This equipment is illustrated in Fig. 3.6 and comes in two models, one for 100 × 200 mm plates; the other is more versatile and can be used for either 100- or 200-mm-wide plates. Marcucci and Mussini [409] published details on the construction of an electrically driven plate coater. This equipment propels the glass plates under the coating reservoir at a uniform rate. Electrically driven coaters operating on this same general principle have been placed on the market by CAMAG.

All hand-operated applicators which depend for their coating action on the flow of the slurry through a slot are subject to variations in thickness of the deposited layer, regardless of whether the coater is moved over the plate or the latter is pushed through the coater. This shortcoming is inherent in the design since the slurry can continue to flow if the relative motion of the plate and coater is changed. (Obviously this does not apply

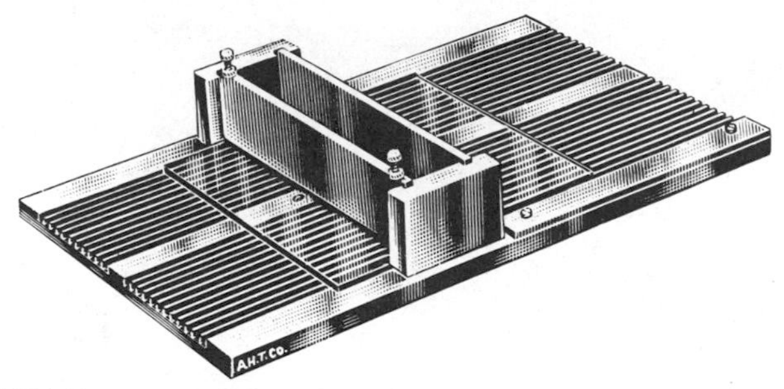

Fig. 3.6 CAMAG apparatus for coating thin-layer plates. Courtesy of Arthur H. Thomas Co.

to the preparation of layers by using a scraper to spread the slurry between fixed guides.) Speeding up the motion can cause less slurry to be deposited because of its viscosity. The thickness can also vary from batch to batch because of slight differences in composition or other variables. To offset these difficulties Moye [410] has modified an apparatus of the Mutter–Hoffstetter type by attaching micrometer depth gauges to each end of the applicator blade. After the coated plates have been dried the applicator blade is properly adjusted so as to shave the top surface of the chromatoplate as it is pushed through the apparatus. On uniform glass plates the uniformity of the layer can thus be improved.

In contrast to the above equipment where the plates are moved through the applicator is equipment where the plates are held stationary while the applicator is moved over the surface of the plates. This type was first introduced by Stahl in 1958 [204]. It is available commercially and is manufactured by Desaga in Germany. The spreader is supplied in two models; one is not adjustable and produces a layer thickness of 275 μ. The other is adjustable (Fig. 3.7) so that the thickness of the layer can be varied from 0 to 2000 μ. In using this apparatus, the glass plates are placed side by side on a smooth surface so that the spreader may be moved smoothly and continuously from start to finish. A little water may be used to hold the plates firmly to the support, but care must be taken not to get the moisture on the upper surface of the glass plate since this will tend to dilute the slurry resulting in a decrease in the density of the gel layer [411]. The applicator is then placed on the plates, filled with slurry, and moved rapidly and uniformly along. Research Specialties Co. and Kensington Scientific Corp. market similar devices.

A modification of the Desaga coater has been described by Stahl [412]. This consists of an adapter so that gradient layers composed of two different adsorbent materials (A and B) may be prepared ranging from

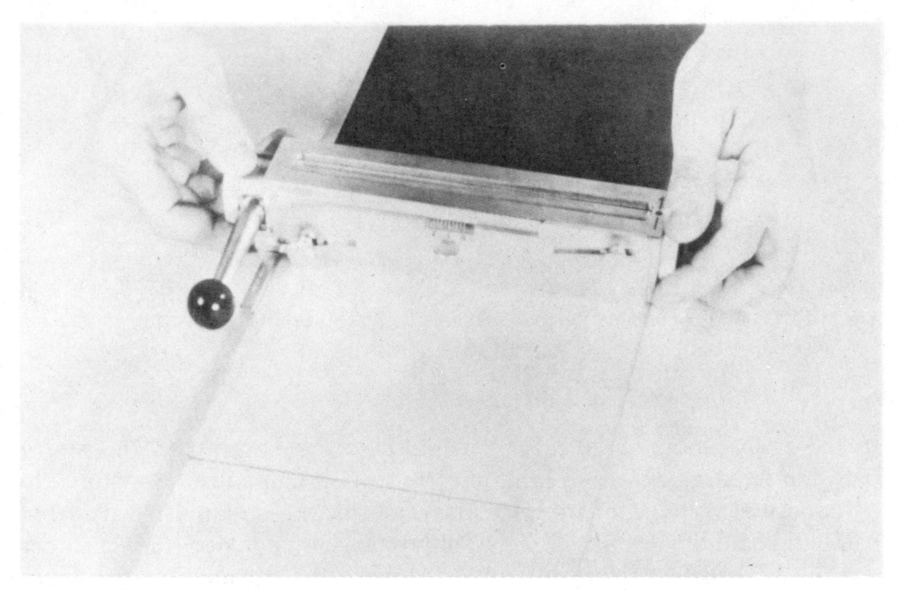

Fig. 3.7 Desaga equipment for coating glass plates. Courtesy of Brinkmann Instruments.

100% A at one end to 100% B at the other. Naturally, the variation can be of any type desired such as pH, buffering, or two different adsorbents (e.g., silica and alumina). The greatest value of this apparatus is in determining the optimum adsorbent for a given separation. An idea of the applicability of this type of plate (although the results were not obtained with this equipment) can be obtained from Fig. 3.8. The equipment is manufactured by Desaga. Warren [413] has published on the construction of a simpler device adapted to the Shandon spreader for the preparation of gradient layers.

Another interesting modification of the movable-chamber coater has been made by Badings [414]. Besides the adjustable gate for varying the thickness of the coating a second slide on the bottom of this apparatus is adjustable to coat various widths of plates.

The main problem with this type of movable-hopper spreader is caused by an uneven rate of moving the applicator and nonuniformity in thickness of the glass plates. Chemetron in Italy and Baird and Tatloch in London market electrically driven coaters of this type which eliminate the problem of uneven rate of coating.

The Shandon Scientific Co. offers a somewhat different approach to this type of spreader. The glass plates are placed on rollers under two guide rails. After the equipment is filled with plates an air bag beneath the rollers is inflated, thus squeezing the plates against the underside of the guide rails. In this way, all the upper surfaces of the plates are placed at the same level even though there might be some difference in the thickness of the plates. The adjustable spreader is then moved rapidly over the plate surfaces. The details of this equipment are shown in Fig. 3.9.

There have been numerous modifications published, especially of the movable-hopper type ranging from simple plastic hoppers [415] to more complex equipment [416–418].

Wasicky [5] has used this type of equipment for coating microscope slides. Berger et al. [419] modified commercial equipment so that adjacent layers of two different adsorbents may be laid down at one time. Abbott and Thomson [420] modified the Desaga apparatus for the preparation of wedge layers and Beckstead et al. [421] used a mold to prepare thicker wedge layers. Kontes offers a grooved wedge-type plate with a graded depth decreasing from 1000 to 125 microns with 2 guiding edges.

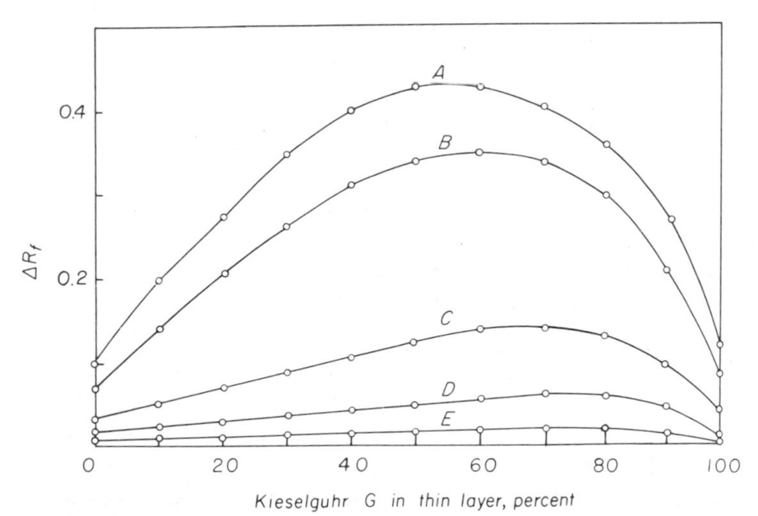

Fig. 3.8 Variation of R_f values with reference to delapon with the composition of the thin layer. Curve *A*, 4-(-chloro-2-methylphenoxy)butyric acid (MCPB); curve *B*, 4-(2,3-dichlorophenoxy)butyric acid (2,4-DB); curve *C*, 4-chloro-2-methylphenoxyacetic acid (MCPA); curve *D*, 2,4-trichlorophenoxyacetic acid (2,4,5-T); curve *E*, 2,4-dichlorophenoxyacetic acid. Reproduced with permission of the authors and The Society for Analytical Chemistry from D. C. Abbott, H. Egan, E. W. Hammond, and J. Thomson, *Analyst (London),* 89, 480 (1964).

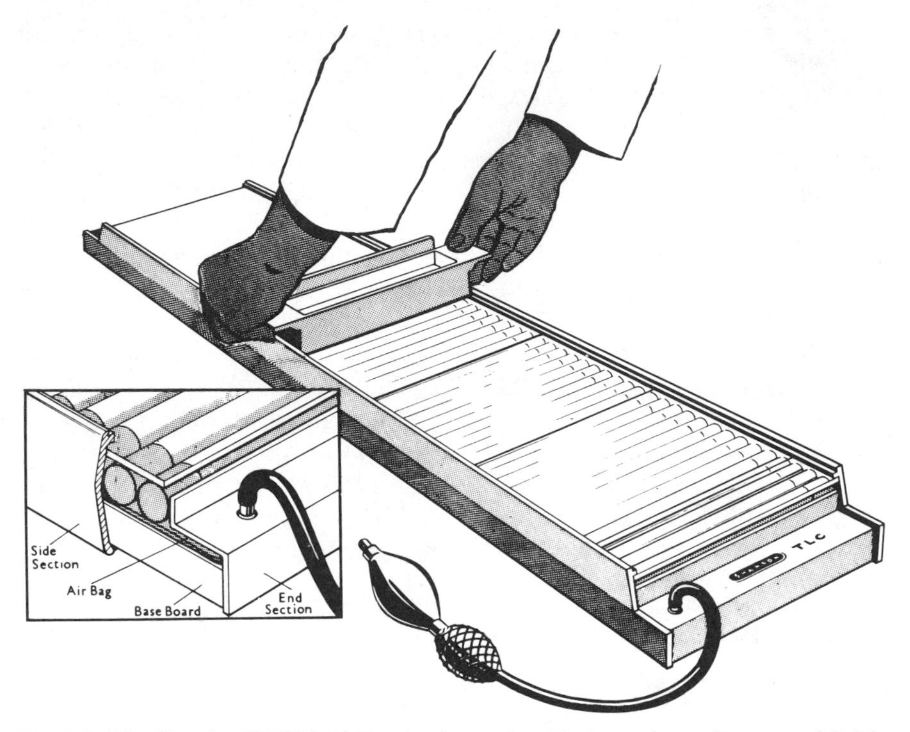

Fig. 3.9 The Shandon UNOPLAN leveler for coating thin-layer plates. Courtesy of Colab Laboratories.

Application of a Dry Adsorbent

Dry adsorbents for loose-layer chromatography may be applied to the supporting plates by any of the simpler methods listed above which use some type of guide to regulate the thickness of the layer.

6 DRYING AND ACTIVATION OF THE LAYERS

The degree of dryness of the adsorbent depends of course on the particular situation, for example, the compounds to be separated or the solvents used. Barbier [422] found that silicic acid-coated plates dried for 1 hr at 95 to 105°C did not give a good separation of *p*-benzoquinones because they were too strongly adsorbed; however, by allowing the plates to stand in the air for 48 hr before using, a good separation could be obtained. Another alternative to deactivating the plates is "multiple development," which is explained later.

Layers that have been prepared from a slurry containing a starch binder may be placed directly in the oven as soon as the layers are formed. However, in the case of plaster of Paris which depends for its binding power on the conversion to gypsum by crystallizing with water of crystallization, time must be allowed for this to take place before the adsorbent can be dried out. Usually 30 min is sufficient for setting to take place. Drying of the adsorbent is usually carried out at 110°C for 30 min to 1 hr. This is the normal drying time for average layers of approximately 250 μ thick. For thicker layers, used for preparative chromatography, longer drying times are necessary. Halpaap [19] allowed the silica gel layers 1.5 to 2 mm thick to dry in the atmosphere until they had become white. They were then activated by 3 hr of heating at 120°C. Korzun et al. [18] allowed their 1-mm-thick layers to set overnight approximately 16 hr before drying in an oven at 80°C for 2 to 4 hr. If the plates are dried too rapidly fissures appear in the surface. (See also Chapter III, Section 3).

Layers made from cellulose powder or modified cellulose can generally be dried by allowing them to stand overnight at room temperature or by drying at 50°C for 40 min. Of course layers that are going to be used for partition chromatography do not need to be thoroughly dried.

Sephadex layers should not be overdried because the solvent migrates more slowly and irregularly [354]. On drying in air these layers lose their moist sheen, and they should be dried only until the gel grains are visible. If dried too much, they can still be used by spraying with water and redrying with a warm-air blower.

Plates that have been activated above room temperature should be stored in a desiccator or other container to prevent access of moisture; a desiccant will help maintain the plates in an active condition. For plates that are to be used with the fluorescein–bromine indicator developed by Kirchner et al. [2] an acidic desiccant should not be used, because the plates will absorb enough acid vapors to interfere with the production of the red eosin color.

Although loose layers have the disadvantage of being fragile, so that they require great care in handling and expecially in spraying on detecting reagents, they have a distinct advantage in preparing aluminum oxide layers of high adsorptive activity. Aluminum oxide should be heated to 200°C in order to obtain an active adsorbent.* Activation temperatures of 200°C are of course out of the question for starch and plaster of Paris binders, since their binding power would be destroyed.

* Milligan [424] has shown that aluminum oxide dried at this temperature still contains around 87% moisture and removal of the remaining moisture can be achieved only at much higher temperatures, causing a loss of activity of the adsorbent.

Urbach [423] dried aluminum oxide G plates which had been treated with silver nitrate in an oven for 20 min with the temperature rising gradually from 115 to 135°C, after the preliminary period for allowing the binder to set. Compounds which form fairly stable complexes can be separated satisfactorily on a plate of low activity. Compounds which form less stable complexes are more easily separated by using a drier plate.

7 SPECIAL TREATMENT OF THE LAYERS

For special purposes there are a number of treatments that can be applied to the dried plate prior to application of the sample. Kirchner et al. [126] in 1954 showed that it was necessary to wash the silicic acid layers prior to using them for the quantitative determination of biphenyl in citrus fruits and fruit products. This was accomplished by carrying out a predevelopment with 95% ethyl alcohol. The layers were then dried at 85°C for 4 min in a mechanical convection oven. Stanley et al. [425] designed a rack which fits into a standard museum jar for washing the layers (Fig. 3.10). In this case, solvent from a trough is fed to the layers by means of a wick of heavy filter paper and the solvent moves down the strips by capillary descent. Kovacs [426] removed interfering chlo-

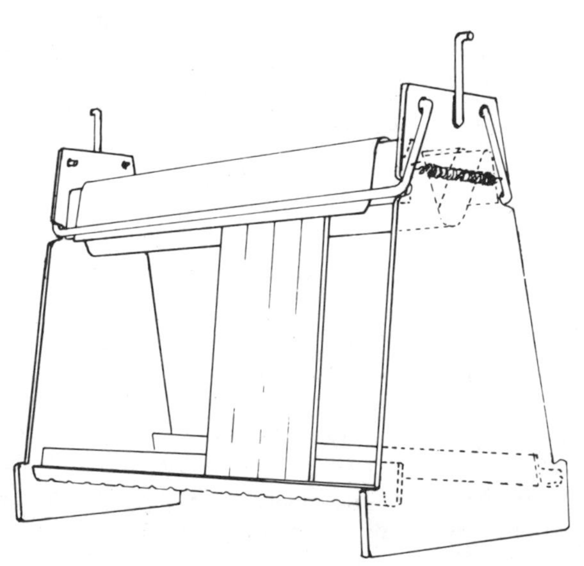

Fig. 3.10 Rack for washing chromatostrips. Overall dimensions to fit 6 × 10 × 10 in. museum jar. From W. L. Stanley, S. H. Vannier, and B. Gentili [425]; reproduced with permission of the authors and the Association of Official Analytical Chemists.

rides by using a prewash of water applied in the same manner as a developing solvent, and Blinn [427] applied a double wash of acetone for removing infrared interfering components in the analysis of pesticides. Brown and Benjamin [428] removed interfering adsorbed material by pouring a methyl alcohol–ether mixture (80:20) across the plate at right angles to the development direction. The plates were dried at 110°C for 25 min. Schweiger [393] used a prewash of 0.5% ethylenediaminetetraacetic acid to remove cations from cellulose layers prior to the separation of uronic acids.

The amount of moisture in the layer can be determined by the Karl Fischer method as used by Miller and Kirchner [127]. For this determination, 0.5 g of the adsorbent is added to 5 ml of formamide in a ground-glass stoppered flask. This is then heated to 120°C on a hot plate for 1 min followed by cooling. Excess Karl Fischer reagent is then added and a back-titration is made with standard water–alcohol mixture. Heřmánek et al. [423] determined the amount of moisture in alumina on loose layers by measuring the loss in weight on heating for 4 hr at 800 to 850°C.

At this state of the operation the plates may be impregnated for partition chromatography or for reverse-phase chromatography, if the impregnating material has not already been included in the preparation of the slurry. There are a number of ways of doing this. The plates may be dipped into a solution of the impregnating agent. Undecane and tetradecane have been used in 5 to 15% solutions in petroleum ether for impregnating by immersion [27, 430, 431]. Impregnation by immersion must be done very slowly and carefully so as not to disturb the thin layer. This method of impregnation has also been used for the application of decalin [432], silicones [433], Carbowax-400 [432], paraffin oil [434, 435], formamide [436], 2-phenoxyethanol and 2-methoxyethanol [434], nitromethane [432], peanut oil [437], and castor oil [438].

Another method of applying the impregnating agent is to allow a solution of the material to ascend or descend the plate in the normal manner of development. This method is less apt to cause damage to the thin layer and has been used for the application of paraffin oil [439], *n*-decane [440], vegetable oils [411], *N,N*-dimethylformamide [442], and propylene glycol [443].

The impregnating agent can also be sprayed onto the plates, although there is evidence that this produces a less uniform dispersion than by immersion or development. It has been less widely used than the two previously mentioned methods. Paraffin oil, formamide, and propylene glycol [311] have been applied in this manner. Besides reagents for partition chromatography, buffering agents have been sprayed on the plates [217] as well as solutions of complexing agents [118, 444, 445, 446, 447].

Morris [218] found that slurry impregnation with silver nitrate gave more uniform and reproducible results than impregnation by spraying the dry plates with a silver nitrate solution.

Lees et al. [448] found that by dipping silica gel plates in a 5% solution of silver nitrate in acetonitrile and then drying in the air to remove the solvent they obtained layers which gave much less background discoloration than those prepared by the usual methods. Perron and Auffret [449] immersed silica gel layers in a saturated (at 30°C) methanolic solution of silver nitrate prepared in anhydrous, aldehyde-free methanol. The layers were dried at room temperature for 30 min. Layers prepared in this manner showed no deterioration after exposure to daylight for 1 month.

Another method of impregnating layers is by exposing the layers to the vapor of the impregnating agent. This has been done with nitromethane and with 5% methanol–water solution [432]. For partition chromatography on kieselguhr G layers, Bennett and Heftmann [450] impregnated their kieselguhr G plate, after spotting with the mixture to be separated, by placing it over a beaker of boiling water until the plate was thoroughly wet. It was then placed in the hood until the water began to recede from the corners of the plate at which time it was placed in the development chamber.

After the impregnating agents are sprayed on, they are usually dried in air for approximately 15 min before use. Plates which have been sprayed with aqueous solutions of complexing agents may be dried in the oven for the normal time. The amount of impregnating agent that has been added to the layer may be determined by weighing [423, 311].

One of the advantages of impregnation of the plate after its formation is realized in two-dimensional chromatography. By impregnating the balance of the plate which has not been used for the separation of compounds in the first development, the adsorbent may be modified so as to take advantage of certain separating characteristics. For example, Kaufmann et al. [451] performed a normal adsorption chromatographic separation in one direction, and then impregnated the balance of the plate with paraffin oil or undecane [452] for reverse-phase partition chromatography in the second direction. Urbach [423] applied this technique by chromatographing on aluminum oxide in one direction, then impregnating the adsorbent with 2-phenoxyethanol before chromatographing in the second direction. As another example of this technique, Bergel'son et al. [445] chromatographed monounsaturated fatty acids in one direction on silica gel impregnated with dodecane. The plates were then impregnated with silver nitrate as a complexing agent before developing in the second direction.

Wang et al. [453] immersed polyamide layers in a mixture of acetic anhydride–anhydrous pyridine (20:7) for 5 min in order to acetylate the layer. The layer was then hung at room temperature for one day and finally dried at 70°C for 30 min. Completeness of acetylation was checked by spraying with ninhydrin solution; completely acetylated layers showed no color change.

References

1. J. E. Meinhard and N. F. Hall, *Anal. Chem.*, **21,** 185 (1949).
2. J. G. Kirchner, J. M. Miller, and G. J. Keller, *Anal. Chem.*, **23,** 420 (1951).
3. A. F. Hofmann, *Anal. Biochem.*, **3,** 145 (1962).
4. J. J. Peifer, *Mikrochim. Acta,* **1962,** 529.
5. R. Wasicky, *Anal. Chem.*, **34,** 1346 (1962).
6. E. Bancher, H. Scherz, and V. Prey, *Mikrochim. Acta,* **1963,** 712.
7. L. L. M. van Deenen and G. H. de Haas, *Biochim. Biophys. Acta,* **70,** 538 (1963).
8. M. Dobiasova, *J. Lipid Res.*, **4,** 481 (1963).
9. J. Hansson, *Explosivstoffe,* **10,** 73 (1963).
10. R. A. Heacock and M. E. Mahon, *Can. J. Biochem. Physiol.*, **41,** 487 (1963).
11. A. F. Hofmann, "Thin-Layer Adsorption Chromatography of Lipids," in *Biochemical Problems of Lipids* (B.B.A. Library Vol. 1), A. C. Frazer, Ed., Elsevier, Amsterdam, 1963, p. 1.
12. K. Morita and F. Haruta, *J. Chromatogr.*, **12,** 412 (1963).
13. M. B. Naff and A. S. Naff, *J. Chem. Educ.*, **40,** 534 (1963).
14. J. J. Peifer, F. Janssen, R. Muesing, and W. O. Lundberg, *J. Am. Oil Chem. Soc.*, **39,** 292 (1962).
15. J. J. Peifer, R. Muesing, and F. Janssen, American Oil Chemists' Society Meeting, Minneapolis, September 30, 1963.
16. K.-T. Wang and Y.-T. Lin, *J. Chinese Chem. Soc. (Taiwan),* **10,** 146 (1963).
17. N. Wiedenhof, *J. Chromatogr.*, **15,** 100 (1964).
18. B. P. Korzun, L. Dorfman, and S. M. Brody, *Anal. Chem.*, **35,** 950 (1963).
19. H. Halpaap, *Chem.-Ing.-Tech.*, **35,** 488 (1963).
20. F. L. Rigby and J. L. Bethune, *Am. Soc. Brewing Chem. Proc.*, **1955,** 174.
21. N. Allentoff and G. F. Wright, *Can. J. Chem.*, **35,** 900 (1957).
22. C. B. Barrett, M. S. J. Dallas, and F. B. Padley, *J. Am. Oil Chem. Soc.*, **40,** 580 (1963).
23. C. B. Barrett, M. S. J. Dallas, and F. B. Padley, *Chem. Ind. (London),* **1962,** 1050.
24. G. Bottura, A. Breccia, F. Marchetti, and F. Spalletti, *Ric. Sci. Rend. Ser. A,* **6,** 373 (1964).
25. A. Breccia, F. Spalletti, G. Bottura, and F. Marchetti, private communication.

26. J. J. Kabara, G. C. Kabara, and R. S. Wojtalik, *J. Chromatogr.*, **15**, 267 (1964).
27. H. P. Kaufmann and T. H. Khoe, *Fette, Seifen, Anstrichm.*, **64**, 81 (1962).
28. E. A. Mistryukov, *Collect. Czech. Chem. Commun.*, **26**, 2071 (1961).
29. S. M. Rybicka, *Chem. Ind. (London)*, **1962**, 308.
30. Z. A. Shevchenko and I. A. Favorskaya, *Vestn. Leningr. Univ. Ser. Fiz. Khim.*, **19**, 107 (1964); through *Chem. Abstr.*, **61**, 8874 (1964).
31. J. J. Wohnlich, *Chromatogr., Symp., 2nd, Brussels*, **1962**, 255.
32. J. J. Wohnlich, *Bull. Soc. Chim. Biol.*, **46**, 729 (1964).
33. J. J. Wohnlich, *J. Pharm. Belg.*, **19**, 53 (1964).
34. M. P. Clapp and J. Jeter, *J. Chromatogr.*, **17**, 578 (1965).
35. A. Gamp, P. Studer, H. Linde, and K. Meyer, *Experientia*, **18**, 292 (1962).
36. E. Hansbury, D. G. Ott, and J. D. Perrings, *J. Chem. Educ.*, **40**, 31 (1963).
37. M. J. Matherne, Jr., and W. H. Bathalter, *J. Assoc. Off. Anal. Chem.*, **49**, 1012 (1966).
38. R. F. Collins, *Chem. Ind. (London)*, **1969**, 614.
39. G. S. Boyd and H. R. B. Hutton, *Biochim. Biophys. Acta*, **69**, 419 (1963).
40. L. M. Weiner and B. Zak, *Clin. Chim. Acta*, **9**, 407 (1964).
41. B. Arreguín, *J. Chromatogr.*, **26**, 527 (1967).
42. R. L. Squibb, *Nature*, **198**, 317 (1963).
43. C. L. Marsh, C. R. Jolliff, and L. C. Payne, *Tech. Bull Regist. Med. Technol.*, **34**, 1 (1964).
44. A. Lestienne, Fr. Pat. 1,370,780 (August 28, 1964).
45. E. Merck, Darmstadt, Germany.
46. H. Halpaap and H. Bausch, *J. Chromatogr.*, **48**, 144 (1970).
47. H. Roessler, H. Halpaap, and K. Klatyk, Ger. Pat. 1,908,695 (October 15, 1970).
48. E. W. Baur, private communication.
49. E. W. Baur, *J. Lab. Clin. Med.*, **61**, 166 (1963).
50. E. Correni, *Naturwissenschaften*, **51**, 40 (1964).
51. W. G. Dangerfield, *Nature*, **202**, 520 (1964).
52. P. Schweda, *J. Chromatogr.*, **63**, 67 (1971).
53. F. Synder, *Anal. Chem.*, **35**, 599 (1963).
54. B. P. Korzun and S. Brody, *J. Pharm. Sci.*, **53**, 454 (1964).
55. J. Rosmus, M. Pavlicek, and Z. Deyl, "Centrifugal Chromatography, XII. Centrifugal Thin-Layer Chromatography," in *Thin-Layer Chromatography*, G. B. Marini-Bettòlo, Ed., Elsevier, Amsterdam, 1964, p. 119.
56. B. Rabenort, *J. Chromatogr.*, **17**, 594 (1965).
57. F. W. Koss and D. Jerchel, *Naturwissenschaften*, **51**, 382 (1964).
58. W. M. Connors and W. K. Boak, *J. Chromatogr.*, **16**, 243 (1964).
59. J. Janák, *J. Chromatogr.*, **15**, 15 (1964).
60. M. Covello and O. Schettino, "The Application of Thin-Layer Chromatography to Investigations of Antifermentatives in Foodstuffs," in *Thin-Layer Chromatography*, G. B. Marini-Bettòlo, Ed., Elsevier, Amsterdam, 1964, p. 215.
61. K. B. Lie and J. F. Nyc, *J. Chromatogr.*, **8**, 75 (1962).

62. J. G. Rimmer, *Chromatographia,* **1,** 219 (1968).
63. P. Pogacar and H. Klein, *Dtsch. Z. Gesamte Gerichtl. Med.,* **60,** 1 (1967); through *Chem.'Abstr.,* **67,** 78639t (1967).
64. P. Pogacar, B. Kienle, P. Krapp, and K. Lueehrsen, *J. Chromatogr.,* **29,** 287 (1967).
65. H. P. Lenk and W. Gleich, *J. Chromatogr.,* **43,** 350 (1969).
66. H. P. Lenk and H. Gruber, *J. Chromatogr.,* **43,** 355 (1969).
67. R. Ikan and E. Rapaport, *J. Chem. Educ.,* **44,** 297 (1967).
68. H. Feltkamp, *Dtsch. Apoth.-Ztg.,* **102,** 1269 (1962).
69. R. Stupnicki and E. Stupnicka, *J. Chromatogr.,* **21,** 150 (1966).
70. V. B. Ramos, *Chem. Quart.,* **8,** 30 (1967).
71. B. N. Sen, *Anal. Chim. Acta,* **23,** 152 (1960).
72. *Ibid.,* **12,** 154 (1955).
73. K. Brendel, R. S. Steele, and E. A. Davidson, *J. Chromatogr.,* **30,** 232 (1967).
74. P. E. Shaw, R. E. Carter, and R. E. Berry, *J. Chromatogr.,* **47,** 507 (1970).
75. J. J. Wren, *J. Chromatogr.,* **4,** 173 (1960).
76. J. Asselineau and E. Lederer, *Biochim. Biophys. Acta,* **17,** 161 (1951).
77. J. A. Lovern, *Biochem. J.,* **63,** 373 (1956).
78. J. A. Lovern, J. Olley, E. F. Hartree, and T. Mann, *Biochem. J.,* **67,** 630 (1957).
79. J. A. Lovern, J. Olley, and H. A. Watson, *J. Sci. Food Agric.,* **10,** 327 (1959).
80. J. A. Lovern, in *Essential Fatty Acids,* H. M. Sinclair, Ed., Butterworths, London, 1958, p. 47.
81. H. Noll and H. Bloch, *J. Biol. Chem.,* **214,** 251 (1955).
82. H. Noll, H. Bloch, J. Asselineau, and E. Lederer, *Biochem. Biophys. Acta,* **20,** 299 (1956).
83. J. Polonsky, G. Ferreol, R. Toubiana, and E. Lederer, *Bul. Soc. Chim. Fr.,* **1956,** 1471.
84. C. Riley and R. F. Nunn, *Biochem. J.,* **74,** 56 (1960).
85. T. H. Applewhite, M. J. Diamond, and L. A. Goldblatt, *J. Am. Oil Chem. Soc.,* **38,** 609 (1961).
86. T. H. Applewhite, J. S. Nelson, and L. A. Goldblatt, *J. Am. Oil Chem. Soc.,* **45,** 101 (1963).
87. J. Battaile, R. L. Dunning, and W. D. Loomis, *Biochim. Biophys. Acta,* **51,** 538 (1961).
88. E. Demole, *J. Chromatogr.,* **1,** 24 (1958).
89. Y. Kuroiwa and H. Hashimoto, *J. Inst. Brewing,* **67,** 347 (1961).
90. L. D. Hayward, R. A. Kitchen, and D. J. Livingstone, *Can. J. Chem.,* **40,** 434 (1962).
91. I. Onishi, H. Tomita, and T. Fukuzumi, *Bull. Agric. Chem. Soc. Japan,* **20,** 61 (1956).
92. W. C. Vogel, W. M. Doizaki, and L. Zieve, *J. Lipid Res.,* **3,** 138 (1962).
93. W. M. Doizaki and L. Zieve, *Proc. Soc. Exp. Biol. Med.,* **113,** 91 (1963).
94. J. Avigan, D. S. Goodman, and D. Steinberg, *J. Lipid Res.,* **4,** 100 (1963).

95. R. H. Reitsema, *Anal. Chem.*, **26,** 960 (1954).
96. O. Adamec, J. Matis, and M. Galvanek, *Lancet,* **1962-T,** 81.
97. R. N. Shukla, A. R. Sharma, P. K. Mukherjee, and A. Sinha, Indian Pat. 113,944 (August 22, 1970); through *Chem. Abstr.*, **76,** 50498h (1972).
98. L. Y. Petrova, *Antibiotiki (Moscow),* **15,** 395 (1970); through *Chem. Abstr.*, **73,** 84453t (1970).
99. A. Waksmundzki and J. Różylo, *Chem. Anal. (Warsaw),* **14,** 1217 (1969).
100. K. Gaertner and R. Griessbach, *Kolloid-Z.*, **160,** 21 (1958).
101. T. Tarutani, *J. Chromatogr.*, **50,** 523 (1970).
102. B. S. Girgis, *J. Appl. Chem. Biotechnol.*, **22,** 905 (1972).
102a. Z. El Rassi, C. Gonnet, and J. L. Rocca, *J. Chromatogr.*, **125,** 179 (1976).
103. W. Hinz, H. Ruttloff, and A. Taeufel, *Silikat Tech.*, **13,** 378 (1962).
104. O. Adamec, J. Matis, and M. Galvanek, *Steroids,* **1,** 495 (1963).
105. J. Pitra, *Chem. Listy,* **56,** 495 (1962).
106. J. Reichelt and J. Pitra, *Collect. Czech. Chem. Commun.*, **27,** 1709 (1962).
107. P. D. Klein, *Anal. Chem.*, **34,** 733 (1962).
108. R. H. Stokes and R. A. Robinson, *Ind. Eng. Chem.*, **41,** 2013 (1949).
109. F. H. Dickey, *Proc. Natl. Acad. Sci. U.S.*, **35,** 227 (1949).
110. H. Erlenmeyer and H. Bartels, *Helv. Chim. Acta,* **47,** 46 (1964).
111. R. E. Majors and L. B. Rogers, *Anal. Chem.*, **41,** 1052 (1969).
112. *Ibid.*, p. 1058.
113. G. H. Reed and L. B. Rogers, *Anal. Chem.*, **37,** 861 (1965).
114. H. Bartels, *J. Chromatogr.*, **30,** 113 (1967).
114a. H. Bartels and B. Prijs, "Specifically Adsorbing Silica Gels," in *Advances in Chromatography,* Vol. 11, J. C. Giddings and R. A. Keller, Eds., Marcel Dekker, New York, p. 115.
115. N. Dimov, *God. Nauchnoizsled. Inst. Goriva Toplotekhn (Sofia),* **7,** 137 (1961); through *Chem. Abstr.*, **58,** 8428 (1963).
116. H. L. Goering, W. D. Closson, and A. C. Olson, *J. Am. Chem. Soc.*, **83,** 3507 (1961).
117. D. de Vries, *Chem. Ind. (London),* **1962,** 1049.
118. L. J. Morris, *Chem. Ind. (London),* **1962,** 1238.
119. A. S. Gupta and S. Dev, *J. Chromatogr.*, **12,** 189 (1963).
120. P. Markl and F. Hecht, *Mikrochim. Acta,* **1963,** 970.
120a. M. Tuerler and D. Hoegl, *Mitt. Geb. Lebensm.-Hyg.*, **52,** 123 (1961).
121. D. W. Vidrine and H. J. Nicholas, *J. Chromatogr.*, **89,** 92 (1974).
122. K. Yasuda, *J. Chromatogr.*, **74,** 142 (1972).
122a. R. K. Gilpin and W. R. Sisco, *J. Chromatogr.*, **124,** 257 (1976).
123. H. Seiler and W. Rothweiler, *Helv. Chim. Acta,* **44,** 941 (1961).
124. T. G. Lipina, *Tr. Khim Khim Tekhnol.*, **1962,** 424.
125. C. Passera, A. Pedrotti, and G. Ferrari, *J. Chromatogr.*, **14,** 289 (1964).
126. J. G. Kirchner, J. M. Miller, and R. G. Rice, *J. Agric. Food Chem.*, **2,** 1031 (1954).
127. J. M. Miller and J. G. Kirchner, *Anal. Chem.*, **24,** 1480 (1952).
128. H. Struck, *Mikrochim. Acta,* **1961,** 634.

129. D. E. Bowyer, W. M. F. Leat, A. N. Howard, and G. A. Gresham, *Biochem. J.*, **89,** 24P (1963).
130. O. S. Privett and M. L. Blank, *J. Am. Oil Chem. Soc.*, **39,** 465 (1962).
131. R. M. Summers and R. B. Mefferd, Jr., *J. Chromatogr.*, **32,** 587 (1968).
132. J. P. Carreau, D. Lapous, and J. Raulin, *J. Chromatogr.*, **42,** 422 (1969).
133. J. C. N. Ma, *J. Chromatogr.*, **21,** 151 (1966).
134. T. Reichstein and C. W. Shoppee, *Disc. Faraday Soc.*, **7,** 305 (1949).
135. H. Wislicenus, *Kolloid-Z.*, **100,** 66 (1942).
136. S. Huneck, *J. Chromatogr.*, **7,** 561 (1962).
136a. A. D. Baitsholts and R. E. Ardell, *J. Chromatogr.*, **30,** 493 (1967).
137. H. Halpaap and W. Reich, *J. Chromatogr.*, **33,** 70 (1968).
138. A. F. Hofmann, *J. Lipid Res.*, **3,** 391 (1962).
139. A. F. Hofmann, *Biochim. Biophys. Acta*, **60,** 458 (1962).
140. W. F. Anacker and V. Stoy, *Biochem. Z.*, **330,** 141 (1958).
141. M. L. Wolfrom, R. M. de Lederkremer, and L. E. Anderson, *Anal. Chem.*, **35,** 1357 (1963).
142. V. Schwarz, *Pharmazie*, **18,** 122 (1963).
143. B. Vasileva and I. Tseslyak, *Antibiotiki*, **10,** 877 (1965); through *Chem. Abstr.*, **64,** 6406b (1966).
144. E. Grigorescu and A. Verbuta, *Rev. Med.*, **13,** 349 (1967); through *Chem. Abstr.*, **68,** 107914g (1968).
145. T. K. With, *Ugeskr. Laeger*, **130,** 641 (1968); through *Chem. Abstr.*, **69,** 33433w (1968).
146. T. K. With, *J. Chromatogr.*, **42,** 389 (1969).
147. B. Pekic, *Acta Pharm. Jugosl.*, **18,** 141 (1968); through *Chem. Abstr.*, **73,** 48568a (1970).
148. G. H. Bomhoff, *Tijdschr. Chem. Instrum.*, **15,** 407 (1968); through *Anal. Abstr.*, **17,** 1853 (1969).
149. S. N. Tewari and L. Ram., *Mikrochim. Acta*, **1970,** 58.
150. J. Matis, O. Adamec, and M. Galvánek, *Nature*, **194,** 477 (1962).
151. H. P. Kaufmann and T. H. Khoe, *Fette, Seifen, Anstrichm.*, **64,** 81 (1962).
152. A. Affonso, *J. Chromatogr.*, **22,** 452 (1966).
153. L. C. Mitchell, *J. Chromatogr.*, **30,** 269 (1967).
154. F. Dobisi and G. Grassini, *J. Chromatogr.*, **10,** 98 (1963).
155. H. Hammerschmidt and M. Mueller, *Papier*, **17,** 448 (1963).
156. M. L. Wolfrom, R. M. Lederkremer, and G. Schwab, *J. Chromatogr.*, **22,** 476 (1966).
157. G. Pataki, *J. Chromatogr.*, **29,** 126 (1967).
157a. C. Haworth and J. G. Heathcote, *J. Chromatogr.*, **41,** 380 (1969).
158. T. Wieland, G. Lueben, and H. Determann, *Experientia*, **18,** 432 (1962).
159. G. M. Badger, J. K. Donnelly, and T. M. Spotswood, *J. Chromatogr.*, **10,** 397 (1963).
160. T. M. Spotswood, *J. Chromatogr.*, **3,** 101 (1960).
161. P. R. Brady and R. M. Hoskinson, *J. Chromatogr.*, **54,** 55 (1971).
162. G. Ackermann and H.-P. Frey, *J. Chromatogr.*, **33,** 53 (1968).

163. H. Kasuya, *Tokyo Gakugei Daigaku Kiyo,* **19,** 127 (1968); through *Chem. Abstr.,* **70,** 53630g (1969).
164. Y. N. Kukushkin, E. I. Karpeiskaya, and V. A. Trofimov, *Zh. Prikl. Khim. (Leningr.),* **44,** 662 (1971); through *Chem. Abstr.,* **75,** 10652b (1971).
165. K.-H. Koenig and K. Demel, *J. Chromatogr.,* **39,** 101 (1969).
166. G. Hesse, H. Engelhardt, and D. Klotz, *Z. Anal. Chem.,* **215,** 182 (1965).
167. S. Kawamure, K. Kurotaki, H. Kuraku, and M. Izawa, *J. Chromatogr.,* **26,** 557 (1967).
168. A. G. Fogg and R. Wood, *J. Chromatogr.,* **20,** 613 (1965).
169. D. Corbi, *G. Med. Mil.,* **114,** 168 (1964); through *Chem. Abstr.,* **61,** 10986a (1964).
170. R. Y. Churi, V. A. Shenai, and R. Parameswaran, *Text. Res. J.,* **42,** 628 (1972).
171. G. Alberti, G. Giammari, and G. Grassini-Strazza, *J. Chromatogr.,* **28,** 118 (1967).
172. G. Alberti, M. A. Massucci, and E. Torracca, *J. Chromatogr.,* **30,** 579 (1967).
173. K.-H. Koenig and H. Graf, *J. Chromatogr.,* **67,** 200 (1972).
174. E. Torracca, U. Costantino, and M. A. Massucci, *J. Chromatogr.,* **30,** 584 (1967).
175. R. Kuroda and K. Oguma, *Anal. Chem.,* **39,** 1003 (1967).
176. W. Grace and Co., Brit. Pat. 1,181,089 (February 11, 1970).
177. L. L. Smith and T. Foell, *J. Chromatogr.,* **9,** 339 (1962).
178. H. Seiler, *Helv. Chim. Acta,* **44,** 1753 (1961).
179. J. G. Kirchner and V. P. Flanagan, Gordon Research Conference, Colby Junior College, New London, N.H., August, 1962.
180. J. G. Kirchner and V. P. Flanagan, 147th Meeting of the American Chemical Society, Philadelphia, Pa., April, 1964.
181. J. M. Miller and J. G. Kirchner, *Anal. Chem.,* **26,** 2002 (1954).
182. J. B. Obreiter and B. B. Stowe, *J. Chromatogr.,* **16,** 226 (1964).
183. K. Onoe, *J. Chem. Soc. Jap., Pure Chem. Sect.,* **73,** 337 (1952).
184. K. Randerath, *Angew. Chem.,* **73,** 436 (1961).
185. R. Huettenrauch, L. Klotz, and W. Mueller, *Z. Chem.,* **3,** 193 (1963).
186. R. Segura Cardona, Sp. Pat., 363,086 (March 1, 1971).
187. W. Page, W. R. Schevey, Fr. Pat., 1,581,874 (September 19, 1969).
188. E. Merck A.-G., Fr. Pat., 1,543,724 (October 25, 1968).
189. J. W. Copius Peerboom, "Separation of Antioxidants on Polyamide Layers," in *Stationary Phase in Paper and Thin-Layer Chromatography,* K. Macek and J. M. Hais, Eds., Elsevier, Amsterdam, 1965, p. 134.
190. E. K. Seybert and P. W. Link, U.S. Pat. 3,594,217 (July 20, 1971).
191. H. H. Althaus and V. Neuhoff, *Z. Physiol. Chem.,* **354,** 1073 (1973).
192. L. Birkofer, C. Kaiser, H. A. Meyer-Stoll, and F. Suppan, *Z. Naturforsch.,* **17B,** 352 (1962).
193. H. Gauthier and G. Mangency, *J. Chromatogr.,* **14,** 209 (1964).
194. P. Maini, *J. Chromatogr.,* **42,** 266 (1969).
195. M. Jellinek, *J. Chromatogr.,* **69,** 402 (1972).

196. T. Okumura and T. Kadono, Ger. Pat., 2,106,689 (August 19, 1971).
197. T. Okumura, T. Kadono, and M. Nakatani, *J. Chromatogr.*, **74**, 73 (1972).
198. T. Okumura and T. Kadono, *J. Chromatogr.*, **86**, 57 (1973).
199. C. G. Honegger, *Helv. Chim. Acta,* **45**, 1409 (1962).
200. E. Svennerholm and L. Svennerholm, *Biochim. Biophys. Acta,* **70**, 432 (1963).
201. E. Stahl, "Development and Application of Thin-Layer Chromatography," in G. B. Marini-Bettòlo, Ed., *Thin-Layer Chromatography,* Elsevier, Amsterdam, 1964, p. 11.
202. J. W. Sease, *J. Am. Chem. Soc.*, **69**, 2242 (1947).
203. *Ibid.*, **70**, 3630 (1948).
204. E. Stahl, *Chem.-Ztg.*, **82**, 323 (1958).
205. J. L. Brown and J. M. Johnston, *J. Lipid Res.*, **3**, 480 (1962).
206. R. Tschesche, G. Biernoth, and G. Wulff, *J. Chromatogr.*, **12**, 342 (1963).
207. E. Stahl, *Arch. Pharm.*, **292/64**, 41 (1959).
208. H. Brockmann and M. Groene, *Chem. Ber.*, **91**, 773 (1958).
209. R. Deters, *Chem.-Ztg.*, **86**, 388 (1962).
210. A. Seher, *Nahrung,* **4**, 466 (1960).
211. P. Ronkainen, *J. Chromatogr.*, **11**, 228 (1963).
212. H.-J. Petrowitz, *Chem.-Ztg.*, **86**, 815 (1962).
213. K. Teichert, E. Mutschler, and H. Rochelmeyer, *Dtsch. Apoth.-Ztg.*, **100**, 477 (1960).
214. V. P. Skipski, R. F. Peterson, and M. Barclay, *J. Lipid Res.*, **3**, 467 (1962).
215. D. Groeger, V. E. Tyler, Jr., and J. E. Dusenberry, *Lloydia,* **24**, 97 (1961).
216. M. L. Borke and E. R. Kirch, *J. Am. Pharm. Assoc. Sci. Ed.*, **42**, 627 (1953).
217. C. G. Honegger, *Helv. Chim. Acta,* **44**, 173 (1961).
218. L. J. Morris, *J. Lipid Res.*, **4**, 357 (1963).
219. E. P. Przybylowicz, W. J. Staudenmayer, E. S. Perry, A. D. Baitsholts, and T. N. Tischer, Pittsburgh Conference on Analytical Chemistry and Applied Spectroscopy, Pittsburgh, Pa., March 1–5, 1965.
220. S. Tabak and M. R. M. Verzola, *J. Chromatogr.*, **51**, 334 (1970).
221. S. Tabak, A. E. Mauro, and A. Del'Acqua, *J. Chromatogr.*, **52**, 500 (1970).
222. J. Halmekoski, *Suom. Kemistil.*, **35B**, 39 (1962).
223. *Ibid.*, **36B**, 58 (1963).
224. V. De Simone and M. Vicedomini, *J. Chromatogr.*, **37**, 538 (1968).
225. P. G. Pifferi, *J. Chromatogr.*, **43**, 530 (1969).
226. K. Yasuda, *J. Chromatogr.*, **87**, 565 (1973).
227. A. Yamamoto, S. Adachi, Z. Ishibe, Y. Shinji, Y. Kaki-Uchi, K. I. Seki, and T. Kitani, *Lipids,* **5**, 566 (1970).
228. G. Rouser, S. Fleischer, and A. Yamamoto, *Lipids,* **5**, 494 (1970).
229. D. A. Baines and R. A. Jones, *J. Chromatogr.*, **47**, 130 (1970).
230. K. Yasuda, *J. Chromatogr.*, **60**, 144 (1971).
231. *Ibid.*, **72**, 413 (1972).
232. D. R. Shapiro and S. S. Kuwahara, *Clin. Chem.*, **19**, 1305 (1973).
233. M. A. Cawthorne, *J. Chromatogr.*, **25**, 164 (1966).

234. M. D. Martz and A. F. Krivis, *Anal. Chem.*, **43,** 790 (1971).
235. F. Jursik and B. Hajek, *Sb. Vys. Sk. Chem.-Technol. Praze, Anorg. Chem. Technol., B14,* **45** (1972); through *Chem. Abstr.*, **78,** 168307q (1973).
236. T. Mezzetti, M. Lato, S. Rufini, and G. Ciuffini, *J. Chromatogr.*, **63,** 329 (1971).
237. T. Mezzetti, M. Ghebregziabhier, S. Rufini, G. Ciuffini, and M. Lato, *J. Chromatogr.*, **74,** 273 (1972).
238. R. J. Warren and J. E. Zarembo, *J. Pharm. Sci.*, **60,** 307 (1971).
239. D. L. Turner, M. J. Silver, E. Baczynski, R. R. Holburn, S. F. Herb, and F. E. Luddy, *Lipids,* **5,** 650 (1970).
240. W. W. Christie and J. H. Moore, *Biochem. Biophys. Acta,* **176,** 445 (1969).
241. R. G. Powell, R. Kleinman, and C. R. Smith, Jr., *Lipids,* **4,** 450 (1969).
242. H. Brockerhoff, *J. Lipid Res.*, **8,** 167 (1967).
243. V. Prey, H. Berbalk, and M. Kausz, *Mikrochim. Acta,* **1962,** 449.
244. D. C. Jeffrey, J. Arditti, and R. Ermst, *J. Chromatogr.*, **41,** 475 (1969).
245. M. Lato, B. Brunelli, G. Ciuffini, and T. Mezzetti, *J. Chromatogr.*, **34,** 26 (1968).
246. *Ibid.*, **36,** 191 (1968).
247. E. J. Shellard and G. H. Jolliffe, *J. Chromatogr.*, **40,** 458 (1969).
248. G. Orzalesi, T. Messetti, C. Rossi, and Bellavita, *Planta Med.*, **19,** 30 (1970).
249. L. J. Morris, *J. Chromatogr.*, **12,** 321 (1963).
250. E. C. Kirby, *J. Chromatogr.*, **80,** 271 (1973).
251. F. Grimmer, W. Dedek, and E. Leibnitz, *Z. Naturforsch.*, **23B,** 10 (1968).
252. H. Koenig, *Z. Anal. Chem.*, **251,** 167 (1970).
253. M. S. J. Dallas, *Nature,* **207,** 1388 (1965).
254. H. Kessler and E. Mueller, *J. Chromatogr.*, **24,** 469 (1966).
255. A. Berg and J. Lam, *J. Chromatogr.*, **16,** 157 (1964).
256. D. B. Harper and H. Smith, *J. Chromatogr.*, **41,** 138 (1969).
257. J. D. Upton, *J. Chromatogr.*, **52,** 171 (1970).
258. S. Hynie, *J. Chromatogr.*, **76,** 270 (1973).
259. S. G. Karlander, K. A. Karlsson, H. Leffler, B. E. Samuelsson, and G. O. Steen, *Biochim. Biophys. Acta,* **270,** 117 (1972).
260. Y. Fujino and M. Nakano, *Biochim. Biophys. Acta,* **239,** 273 (1971).
261. W. R. Morrison and J. D. Hay, *Biochim. Biophys. Acta,* **202,** 460 (1970).
262. J. N. Kanfer, *Chem. Phys. Lipids,* **5,** 159 (1970).
263. K. Karlsson, B. E. Samuelsson, and G. O. Steen, *Acta Chem. Scand.*, **22,** 1361 (1968).
264. K. A. Karlsson, *Chem. Phys. Lipids,* **5,** 6 (1970).
265. M. Rink and S. Herrmann, *J. Chromatogr.*, **12,** 415 (1963).
266. A. Alverez Fernandez, V. Torre Noceda, and E. Sanchez Carrera, *J. Pharm. Sci.*, **58,** 443 (1969).
267. C. Radecka and W. L. Wilson, *J. Chromatogr.*, **57,** 297 (1971).
268. Y. Nishimoto, E. Tsuchida, and S. Toyoshima, *Yakugaku Zasshi,* **86,** 199 (1966); through *Chem. Abstr.*, **64,** 18381b (1966).
269. R. Wood, E. L. Bever, and F. Snyder, *Lipids,* **1,** 399 (1966).

270. R. Wood, C. Piantadosi, and F. Snyder, *J. Lipid Res.*, **10,** 370 (1969).
271. L. J. Morris, *Lab. Pract.*, **15,** 8 (1966).
272. W. Dietsche, *Fette, Seifen, Anstrichm.*, **72,** 778 (1970).
273. E. Hagen, *Plaste Kautsch.*, **15,** 557 (1968).
274. D. B. Parihar, S. P. Sharma, and K. K. Verma, *J. Forensic Sci. Soc.*, **10,** 77 (1970).
275. A. K. Dwivedy, D. B. Parihar, S. P. Sharma, and K. K. Verma, *J. Chromatogr.*, **29,** 120 (1967).
276. R. G. Harvey and M. Halonen, *J. Chromatogr.*, **25,** 294 (1966).
277. M. Franck-Neumann and P. Joessang, *J. Chromatogr.*, **14,** 280 (1964).
278. G. D. Short and R. Young, *Analyst (London)*, **94,** 259 (1969).
279. V. Libíčková, M. Stuchlík, and Ľ. Krasnec, *J. Chromatogr.*, **45,** 278 (1969).
280. G. Schenk, G. L. Sullivan, and P. A. Fryer, *J. Chromatogr.*, **89,** 49 (1974).
281. T. Taylor, *Anal. Biochem.*, **41,** 435 (1971).
282. D. B. Parihar, S. P. Sharma, and K. K. Verma, *J. Forensic Sci.*, **13,** 246 (1968).
283. D. B. Parihar, O. Prakash, I. Bajaj, R. P. Tripathi, and K. K. Verma, *J. Chromatogr.*, **59,** 457 (1971).
284. E. O. Turgel', E. V. Kuznetsova, S. A. Rudoi, and S. L. Skop, *Zh. Anal. Khim.*, **27,** 1194 (1972); through *Thin-Layer Chromatogr. Abstr.*, **2,** 374 (1972).
285. L. Grlic, *J. Chromatogr.*, **48,** 562 (1970).
286. I. Bajaj, K. K. Verma, O. Prakash, and D. B. Parihar, *J. Chromatogr.*, **46,** 261 (1970).
287. H. Brockmann and F. Volpers, *Chem. Ber.*, **80,** 77 (1947).
288. H. Brockmann and F. Volpers, *Naturwissenschaften*, **33,** 58 (1946).
289. V. Cěrný, J. Joska, and L. Lábler, *Collect. Czech. Chem. Commun.*, **26,** 1658 (1961).
290. J. S. Matthews, A. L. Pereda-V., and A. Auguilera-P., *J. Chromatogr.*, **9,** 331 (1962).
291. H. Petschik, and E. Steger, *J. Chromatogr.*, **9,** 307 (1962).
292. R. Tschesche, K. Kometani, F. Kowitz, and G. Snatzke, *Chem. Ber.*, **94,** 3327 (1961).
293. A. Vacíková, V. Felt, and J. Malíková, *J. Chromatogr.*, **9,** 301 (1962).
294. D. Groeger and D. Erge, *Pharmazie*, **18,** 346 (1963).
295. D. F. Zinkel and J. W. Rowe, *J. Chromatogr.*, **13,** 74 (1964).
296. G. Urbach, *J. Chromatogr.*, **12,** 196 (1963).
297. K. H. Mueller and H. Honerlagen, *Arch. Pharm.*, **293/65,** 202 (1960).
298. Z. Prochazka, *Chem. Listy*, **55,** 974 (1961).
299. G. R. Duncan, *J. Chromatogr.*, **8,** 37 (1962).
300. H. B. Sinclair and J. Lehrfeld, *Chemist-Analyst*, **55,** 117 (1966).
301. L. Hoerhammer, H. Wagner, and G. Bittner, *Dtsch. Apoth.*, **14,** 148 (1962).
302. P. R. Bhandari, B. Lerch, and G. Wohlleben, *Pharm. Ztg., Ver. Apoth.-Ztg.*, **107,** 1618 (1962).
303. D. P. Schwartz, M. Keeney, and O. W. Parks, *Michrochem. J.*, **8,** 176 (1964).

304. D. P. Schwartz and O. W. Parks, *Microchem. J.*, **7**, 403 (1963).
305. N. Nicolaides, *J. Chromatogr.*, **8**, 717 (1970).
306. L. R. Snyder, *J. Chromatogr.*, **28**, 300 (1967).
307. L. K. Keefer, *J. Chromatogr.*, **31**, 390 (1967).
308. L. K. Keefer and D. E. Johnson, *J. Chromatogr.*, **69**, 215 (1972).
309. J. P. Tore, *Anal. Biochem.*, **7**, 123 (1964).
310. J. P. Tore, *J. Chromatogr.*, **12**, 413 (1963).
311. J. Vaedtke and A. Gajewska, *J. Chromatogr.*, **9**, 345 (1962).
312. E. Knappe and D. Peteri, *Z. Anal. Chem.*, **190**, 380 (1962).
313. V. Prey, H. Scherz, and E. Bancher, *Mikrochim. Acta*, **1963**, 567.
314. H. T. Badings and J. G. Wassink, *Neth. Milk Dairy J.*, **17**, 132 (1963).
315. A. F. Hofmann, "Thin-Layer Chromatography of Bile Acids and Their Derivatives," in *New Biochemical Separations*, A. T. James and L. J. Morris, Eds., Van Nostrand, London, 1964, p. 261.
316. H. T. Badings, *J. Am. Oil Chem. Soc.*, **36**, 648 (1959).
317. W. C. Eisenberg, R. E. Gilman, and K. T. Finley, *J. Chromatogr.*, **44**, 569 (1969).
318. T. F. Brodasky, *Anal. Chem.*, **35**, 343 (1963).
319. G. Hesse and M. Alexander, *Journees Intern. Etude Methodes Separation Immediate Chromatog., Paris 1961*, 1962, p. 229.
320. Z. Procházka, *J. Chromatogr.*, **48**, 113 (1970).
321. T. Rahandraha, *Chromatogr. Symp., 2nd Brussels*, **1962**, p. 261.
322. T. Rahandraha, M. Chanez, P. Boiteau, and S. Jaquard, *Ann. Pharm. Fr.*, **21**, 561 (1963).
323. W. S. Brud, *J. Chromatogr.*, **18**, 591 (1965).
324. J. K. G. Kramer, E. O. Schiller, H. D. Gesser, and A. D. Robinson, *Anal. Chem.*, **36**, 2379 (1964).
325. F. Wolf, G. Kotte, and J. Hannemann, *Chem. Tech. (Berlin)*, **23**, 550 (1971).
326. V. S. Marino, *J. Chromatogr.*, **46**, 125 (1970).
327. K.-T. Wang, *J. Chinese Chem. Soc. (Taiwan)*, **8**, 241 (1961).
328. K. Egger, *Z. Anal. Chem.*, **182**, 161 (1961).
329. J. Davídek and E. Davídková, *Pharmazie*, **16**, 352 (1961).
330. J. Davídek and Z. Procházka, *Collect. Czech. Chem. Commun.*, **26**, 2947 (1961).
331. M. Mottier and M. Potterat, *Mitt. Geb. Lebensm. Hyg.*, **43**, 118 (1952).
332. K.-T. Wang, J. M. K. Huang, and I. S. Y. Wang, *J. Chromatogr.*, **22**, 362 (1966).
333. Toyo Chemical Industry Co., Ltd., Fr. Pat., 2,112,775 (July 28, 1972).
334. E. Urion, M. Metche, and J. P. Haluk, *Brauwissenschaft*, **16**, 211 (1963).
335. M. Metche, J.-P. Haluk, Q.-H. Nguyen, and E. Urion, *Bull. Soc. Chim. Fr.*, **1963**, 1080.
336. H. E. Nordby, T. J. Kew, and J. F. Fisher, *J. Chromatogr.*, **24**, 257 (1966).
337. H. Wagner, L. Hoerhammer, and K. Macek, *J. Chromatogr.*, **31**, 455 (1967).
338. H. Roesler, W. Heinrich, and T. J. Mabry, *J. Chromatogr.*, **78**, 432 (1973).
339. C. Quarmby, *J. Chromatogr.*, **34**, 52 (1968).

340. J. Moreno Dalmau, J. M. Pla Delfina, and A. Del Pozo Ojeda, *J. Chromatogr.*, **48**, 118 (1970).
341. *Ibid.*, **78**, 165 (1973).
341a. W. D. Loomis and J. Battaile, *Phytochemistry*, **5**, 423 (1966).
342. M. N. Clifford, *J. Chromatogr.*, **94**, 261 (1974).
343. J. Janák, *Chem. Ind. (London)*, **1967**, 1137.
344. J. Janák and V. Kubecova, *J. Chromatogr.*, **33**, 132 (1968).
345. V. Martinu and J. Janák, *J. Chromatogr.*, **65**, 477 (1972).
346. S. Hamada and S. H. Ingbar, *J. Chromatogr.*, **61**, 352 (1971).
347. H. P. Raaen, *J. Chromatogr.*, **44**, 522 (1969).
348. *Ibid.*, **53**, 605 (1970).
349. *Ibid.*, p. 600.
350. J. H. Dohnt, G. J. C. Mulders-Dijkman, J. C. De Beauveser, and G. G. Kuijpers, *J. Chromatogr.*, **52**, 429 (1970).
351. G. Hesse, H. Engelhardt, and R. Kaltwasser, *Chromatographia*, **1**, 302 (1968).
352. E. Epton, S. R. Holden, and J. V. McLaren, *J. Chromatogr.*, **110**, 327 (1975).
353. H. W. Ardelt and P. Lange, *Z. Chem.*, **3**, 266 (1963).
354. H. Determann, *Experientia*, **18**, 430 (1962).
355. T. Wieland and H. Determann, *Experientia*, **18**, 431 (1962).
356. B. G. Johansson and L. Rymo, *Acta Chem. Scand.*, **16**, 2067 (1962).
357. K. Dose and G. Krause, *Naturwissenschaften*, **49**, 349 (1962).
358. B. G. Johansson and L. Rymo, *Acta Chem. Scand.*, **18**, 217 (1964).
359. B. G. Johansson and L. Rymo, *Biochem. J.*, **92**, 5P (1964).
360. C. J. O. R. Morris, *J. Chromatogr.*, **16**, 167 (1964).
361. C. J. O. R. Morris, *Biochem. J.*, **92**, 6P (1964).
362. P. Fasella, A. Giartosio, and C. Turano, "Applications of Thin-Layer Chromatography on Sephadex to the Study of Proteins," in *Thin-Layer Chromatography*, G. B. Marini-Bettòlo, Ed., Elsevier, Amsterdam, 1964, p. 205.
363. R. Vendrely, Y. Coirault, and A. Vanderplancke, *Compt. Rend.*, **258**, 6399 (1964).
364. G. Tortolani and M. E. Colosi, *J. Chromatogr.*, **70**, 182 (1972).
365. R. S. Morgan, N. H. Morgan, and R. A. Guinavan, *Anal. Biochem.*, **45**, 668 (1972).
365a. C. Slagt and W. A. Sipman, *J. Chromatogr.*, **74**, 352 (1972).
366. T. Hashizume and Y. Sasaki, *Agric. Biol. Chem. (Tokyo)*, **27**, 881 (1963); through *Chem. Abstr.*, **60**, 12347 (1964).
367. P. Andrews, *Biochem. J.*, **91**, 222 (1964).
368. E. Nieschlag and K. Otto, *Z. Physiol. Chem.*, **340**, 46 (1965).
369. G. K. Ackers, *Biochem.*, **3**, 723 (1964).
370. L. M. Siegel and K. J. Monty, *Biochim. Biophys. Res. Commun.*, **19**, 494 (1965).
371. K. T. Edsall, in *The Proteins*, Vol. 1, Pt. B, H. Neurath and K. Bailey, Eds., Academic Press, New York, 1953, p. 634.

372. G. K. Ackers and T. E. Thompson, *Proc. Natl. Acad. Sci. U.S.*, **53**, 342 (1965).
373. D. J. Winzor and H. A. Scheraga, *J. Phys. Chem.*, **68**, 338 (1964).
374. H. K. Mangold, *J. Am. Oil Chem. Soc.*, **38**, 708 (1961).
375. J. Hradec and P. Menšík, *J. Chromatogr.*, **32**, 502 (1968).
376. V. M. Bhatnagar and A. Liberii, *J. Chromatogr.*, **18**, 177 (1965).
377. A. S. K. Chan, R. K. Ellsworth, H. J. Perkins, and S. E. Snow, *J. Chromatogr.*, **47**, 395 (1970).
378. L. W. Smith, R. W. Breidenbach, and D. Rubenstein, *Science*, **148**, 508 (1965).
379. S. M. Petrović and S. E. Petrović, *J. Chromatogr.*, **21**, 313 (1966).
380. V. D. Canic, M. N. Turic, and N. U. Perisić, *Z. Anal. Chem.*, **228**, 258 (1967).
381. J. A. Berger, G. Meyniel, J. Petit, and P. Blanquet, *Bull. Soc. Chim. Fr.*, **1963**, 2662.
382. R. D. Greenland, W. Roth, T. Gieske, and I. A. Michaelson, *Anal. Biochem.*, **62**, 305 (1974).
383. L. Lepri, P. G. Desideri, and R. Mascherini, *J. Chromatogr.*, **70**, 212 (1972).
384. L. Lepri, P. G. Desideri, and V. Coas, *J. Chromatogr.*, **52**, 421 (1970).
385. P. Decker and H. Hoeller, *J. Chromatogr.*, **7**, 392 (1962).
386. U. A. T. Brinkman, G. De Vries, and H. R. Leene, *J. Chromatogr.*, **69**, 181 (1972).
387. R. J. T. Graham and A. Carr, *J. Chromatogr.*, **46**, 301 (1970).
388. B. A. Zabin and C. B. Rollins, *J. Chromatogr.*, **14**, 534 (1964).
389. T. Roessel, *Z. Anal. Chem.*, **197**, 333 (1963).
390. E. Knappe, D. Peteri, and I. Rohdewald, *Z. Anal. Chem.*, **197**, 364 (1963).
391. K. Randerath, *Nature*, **205**, 908 (1965).
392. K. Randerath and H. Struck, *J. Chromatogr.*, **6**, 365 (1961).
393. A. Schweiger, *J. Chromatogr.*, **9**, 374 (1962).
394. O. S. Madaeva and V. K. Ryzhkova, *Med. Prom. SSSR*, **17**, 44 (1963).
394a. M. L. Wolfrom, R. M. de Lederkremer, and G. Schwab, *J. Chromatogr.*, **22**, 474 (1966).
395. R. G. Coffey and R. W. Newburgh, *J. Chromatogr.*, **11**, 376 (1963).
396. H. Boernig and C. Reinicke, *Acta Biol. Med. Ger.*, **11**, 600 (1963).
397. K. Randerath, *Nature*, **194**, 768 (1962).
398. K. Randerath, *Angew. Chem. Int. Ed. Engl.*, **1**, 553 (1962).
399. G. Weimann and K. Randerath, *Experientia*, **19**, 49 (1963).
399a. K. Randerath and E. Randerath, *J. Chromatogr.*, **22**, 110 (1966).
400. H. Hesse and R. Hagel, *Chromatographia*, **6**, 277 (1973).
401. E. Randerath and K. Randerath, *J. Chromatogr.*, **10**, 509 (1963).
402. A. Breccia and F. Spalletti, *Nature*, **198**, 756 (1963).
403. I. Bekersky, *Anal. Chem.*, **35**, 261 (1963).
404. L. F. Druding, *J. Chem. Educ.*, **40**, 536 (1963).
405. J. Davídek and J. Pokorný, *Z. Lebensm.-Untersuch.-Forsch.* **115**, 113 (1961).
406. S. Heřmánek, V. Schwarz, and Z. Cekan, *Pharmazie*, **16**, 566 (1961).

407. T. A. Dyer, *J. Chromatogr.*, **11,** 414 (1963).
408. T. M. Lees and P. J. DeMuria, *J. Chromatogr.*, **8,** 108 (1962).
409. F. Marcucci and E. Mussini, *J. Chromatogr.*, **11,** 270 (1963).
410. C. J. Moye, *J. Chromatogr.*, **13,** 56 (1964).
411. S. Hara, H. Tanaka, and M. Takeuchi, *Chem. Pharm. Bull. (Tokyo),* **12,** 626 (1964).
412. E. Stahl, *Angew. Chem. Int. Ed. Engl.*, **3,** 784 (1964).
413. B. Warren, *J. Chromatogr.*, **20,** 603 (1965).
414. H. T. Badings, *J. Chromatogr.*, **14,** 265 (1964).
415. M. Barbier, H. Jaeger, H. Tobias, and E. Wyss, *Helv. Chim. Acta,* **42,** 2440 (1959).
416. W. E. Moore and M. J. Effland, *U.S. Dep. Agric. Forest Serv. Forest Prod. Lab. Res. Note,* **FPL-0119.**
417. R. Wood and F. Snyder, *J. Chromatogr.*, **21,** 318 (1966).
418. E. Von Arx and R. Neher, *J. Chromatogr.*, **25,** 109 (1966).
419. J. A. Berger, G. Meyniel, P. Blanquet, and J. Petit, *Compt. Rend.*, **257,** 1534 (1963).
420. D. C. Abbott and J. Thomson, *Chem. Ind. (London),* **1964,** 481.
421. H. D. Beckstead, W. N. French, and S. J. Smith, *J. Chromatogr.*, **31,** 226 (1967).
422. M. Barbier, *J. Chromatogr.*, **2,** 649 (1959).
423. G. Urbach, *J. Chromatogr.*, **12,** 196 (1963).
424. L. H. Milligan, *J. Phys. Chem.*, **26,** 247 (1922).
425. W. L. Stanley, S. H. Vannier, and B. Gentili, *J. Assoc. Off. Agric. Chem.*, **40,** 282 (1957).
426. M. F. Kovacs, Jr., *J. Assoc. Off. Agric. Chem.*, **46,** 884 (1963).
427. R. C. Blinn, *J. Assoc. Off. Agric. Chem.*, **46,** 952 (1963).
428. T. L. Brown and J. Benjamin, *Anal. Chem.*, **36,** 446 (1964).
429. S. Heřmánek, V. Schwarz, and Z. Čekan, *Collect. Czech. Chem. Commun.*, **26,** 3170 (1961).
430. R. Marcuse, U. Mobech-Hanssen, and P.-O. Goethe, *Fette, Seifen, Anstrichm.*, **66,** 192 (1964).
431. J. W. Copius-Peereboom and H. W. Beekes, *J. Chromatogr.*, **9,** 316 (1962).
432. H. T. Badings and J. G. Wassink, *Neth. Milk Dairy J.*, **17,** 132 (1963).
433. D. Firestone, *J. Am. Oil Chem. Soc.*, **40,** 247 (1963).
434. D. I. Cargill, *Analyst (London),* **87,** 865 (1962).
435. Č. Michalec, M. Šulc, and J. Měšťan, *Nature,* **193,** 63 (1962).
436. K. Teichert, E. Mutschler, and H. Rochelmeyer, *Z. Anal. Chem.*, **181,** 325 (1961).
437. I. D. Jones, L. S. Butler, E. Gibbs, and R. C. White, *J. Chromatogr.*, **70,** 87 (1972).
438. D. Ropte, and J. U. Gu, *Pharmazie,* **27,** 544 (1972).
439. L. Anker and D. Sonanini, *Pharm. Acta Helv.*, **37,** 360 (1962).
440. S. J. Purdy and E. V. Truter, *Analyst (London),* **87,** 802 (1962).
441. K. Egger, *Planta,* **58,** 664 (1962).
442. F. Korte and H. Sieper, *J. Chromatogr.*, **13,** 90 (1964).

443. H. P. Kaufmann and A. K. Sen Gupta, *Fette, Seifen, Anstrichm.*, **65,** 529 (1963).
444. K. Schreiber, O. Aurich, and G. Osske, *J. Chromatogr.*, **12,** 63 (1963).
445. L. D. Bergel'son, E. V. Dyatlovitskaya, and V. V. Voronkova, *Izv. Akad. Nauk SSSR, Otd. Khim. Nauk,* **1963,** 954.
446. J. A. Cornelius and G. Shone, *Chem. Ind. (London),* **1963,** 1246.
447. L. J. Morris, *J. Chromatogr.*, **12,** 321 (1963).
448. T. M. Lees, M. J. Lynch, and F. R. Mosher, *J. Chromatogr.*, **18,** 595 (1965).
449. R. Perron and M. Auffret, *Oleagineux,* **20,** 379 (1965).
450. R. D. Bennett and E. Heftmann, *J. Chromatogr.*, **9,** 353 (1962).
451. H. P. Kaufmann, Z. Makus, and F. Deicke, *Fette, Seifen, Anstrichm.*, **63,** 235 (1961).
452. H. P. Kaufmann and Z. Makus, *Fette, Seifen, Anstrichm.*, **64,** 1 (1960).
453. K.-T. Wang, P.-H. Wu, and T.-B. Shih, *J. Chromatogr.*, **44,** 635 (1969).

Chapter **IV**

SAMPLE APPLICATION

1 PRETREATMENT

Sometimes it is necessary to pretreat a sample before applying it to the chromatographic layer. In some cases the sample is much too dilute and must be concentrated first. Hashmi et al. [1] designed a simple electrolytic cell for concentrating metal ions within the sampling capillary pipet. Zeineh et al. [2] used a single hollow acrylic fiber for concentrating body fluids. A 1-ml sample of protein solution could be concentrated to 0.02 ml with 85% recovery within 4 min. Biomed Instruments and Bio-Rad Laboratories offer equipment for concentrating solutions using the hollow fiber technique. Dechary et al. [3] constructed a reversible filtration apparatus for concentrating protein solutions on dry Sephadex G-25 (coarse). Water and low-molecular-weight substances were absorbed and the concentrated solution was centrifuged through a porous plate. The process could be repeated as often as desired without removing the concentrate from the apparatus. Awdeh and Abu-Samara [4] devised a method for the concentration of micro protein samples using Aquacide No. 2 (a sodium salt of carboxymethyl cellulose, Calbiochem). Ten grams of this compound was spread on a 10 × 20 cm glass plate and covered with a cellophane sheet that had been dipped in distilled water to moisten it. After wrapping the sheet around the plate, it was dried at room temperature for 1 hr. A drop of the dilute solution was then placed on the cellophane sheet. Water and electrolytes were absorbed through the cellophane by the Aquacide. A 200-μl sample concentrated fivefold yielded an 80% recovery; however with an eightfold concentration the recovery dropped to one-half.

Of course evaporation can be used for the concentration of dilute samples, but in some cases a problem arises with increased salt concentrations. In this case a desalting procedure (see Chapter XVII, Section 1 and Chapter XXIV, Section 8) may be applied.

Vacuum sublimation has been used to help cleanup pesticide residue samples [5].

2 APPLICATION TECHNIQUES

Prior to the sample application, the starting point (15 to 20 mm from the bottom edge) is marked and the finish line, usually 10 cm from the application point [6], is also marked. Great care must be taken in marking the starting line not to disturb the surface of the adsorbent because this causes distortion of the spots. This is very difficult, if not impossible, with nonbound layers and with those adsorbents that contain a very low level of plaster of Paris (5–10%). At higher levels of plaster of Paris binder (15 to 20%), the layers may be marked with care. Starch-bound layers are much more stable and can be readily marked with a lead pencil. This is also true of some of the precoated plates and sheets where an organic polymer is used as a binder.

In applying the sample to the thin-layer plate, there are certain techniques which must be practiced in order to obtain the optimum resolution. The sample is applied as a solution in as nonpolar a solvent as possible, since the use of a polar solvent tends to cause the starting spot to spread out and also may affect the R_f value of the compounds, especially when less polar solvents are used for the development [6]. Polar solvents for application of the sample also tend to cause streaking. The solvent used for dissolving the sample should be a relatively volatile one so that it may be removed easily from the plate before the development is started. The area of application should be kept as small as possible, because the smaller the area of application, the sharper the resolution. Usually, in order to keep the size of the spot small, a series of applications is made by allowing the solvent to evaporate after each application. To assist in this evaporation of the solvent, the plate may be warmed prior to the application or a stream of air may be directed at the sample spot. In the case of preparative work where the sample is applied in a narrow band, the width of the band should be kept as narrow as possible.

Calibrated micropipets are particularly convenient for applying the samples, although simple capillary tubes may be used (precision-bore capillaries in various microliter sizes are available from Drummond Scientific Co. Small wire loops may be used for the application of the sample [7–9]. For quantitative work, a micrometer syringe or a micrometer buret can be used conveniently.

There are a number of spotting templates commercially available which, in general, provide a means for marking the starting line and the finish line and for uniformly spacing the spots along the starting line. One form or another of these templates is usually available from the various chemical supply houses.

The Desaga template which is made from transparent plastic, illustrated in Fig. 4.1, also contains a series of circles which permits the estimation of the areas of various size spots.

The CAMAG spotting guide is illustrated in Fig. 4.2. It is sold in two versions, one for 100-mm plates and the other for either a 200-mm plate or two 100-mm plates. This guide contains a series of notches for receiving the spotting pipet.

Wieme [10] and Wieme and Rabaey [11] have introduced the technique of applying the sample in thin-layer electrophoresis of biological material by inserting tissue sections directly into the thin layer, and Curri and coworkers [12] have applied samples as tissue sections in the thin-layer chromatography of lipids. In the latter case the tissue slices were applied

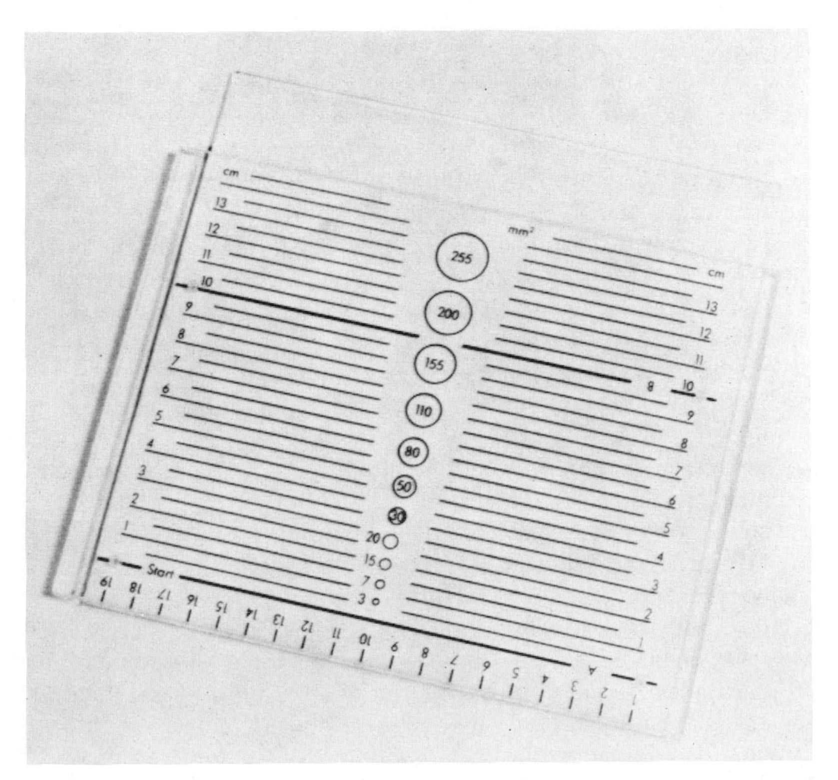

Fig. 4.1 Desaga spotting guide for applying sample. Courtesy of Brinkmann Instruments.

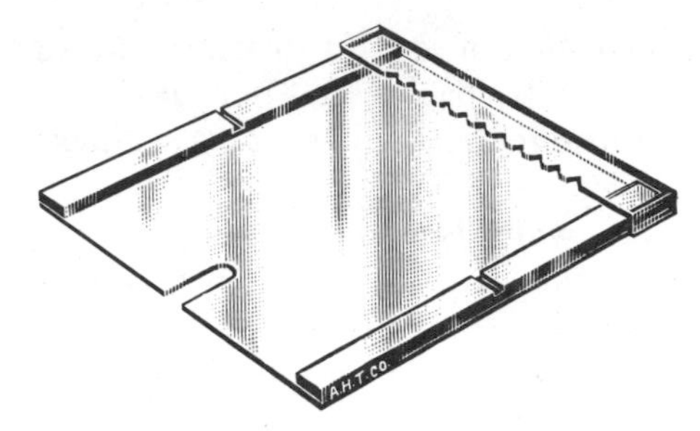

Fig. 4.2 CAMAG spotting guide. Courtesy of Arthur H. Thomas Co.

to 20 × 20 mm cover slides which were then glued with gum arabic to a chromatographic plate. The plate and the tissue section were covered with a 250-μ layer of silica gel. The tissue may also be pressed directly onto a coated thin-layer plate [13, 14]. Culley [15] has applied whole blood samples by placing dried, blood-saturated paper discs on the layer.

Dilute samples may be applied from a capillary while evaporating the solvent with a stream of air or inert gas [16–18].

Kirchner et al. [6] showed that undue exposure of the layer to the atmosphere caused the activity to decrease to a point where it affected the R_f values. Brenner et al. [19] plotted the log of the R_f versus the log of the exposure to air prior to running the chromatogram and found a linear increase in the R_f after 2 min of exposure. To offset this, they covered the upper portion of the layers with a glass plate while spotting the samples. Another way of offsetting this effect and also of protecting oxygen-sensitive compounds is by use of the application box sold by Desaga (Fig. 4.3). This box has a plexiglass cover with a sliding bracket which can be moved across the box. The sample is spotted by means of a micropipet through a hole in the sliding bracket and the box is equipped with connections for the introduction of an inert gas.

It is advantageous to have equipment available in order to make multiple applications either of the same solution or of different solutions; this is especially true if one wishes to build up the concentration of the material from a dilute solution. Equipment of this nature is available commercially. One of these, designed by Morgan [20], is available from Arthur H. Thomas. In this apparatus the sample material is contained in a series of capillaries which are suspended from a bar. By depressing the bar, the capillaries simultaneously contact the thin-layer plate. This equipment is illustrated in Fig. 4.4. A somewhat similar device has been constructed by Medvedev and Smolyaninov [21]. A semi-automatic applicator [22] makes use of 25 Hamilton syringes so that numerous samples

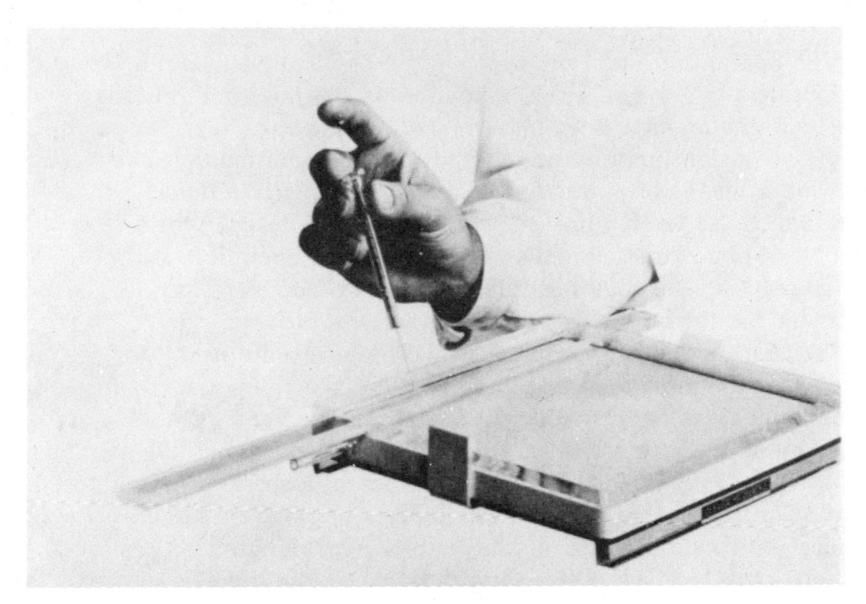

Fig. 4.3 Desaga application box for protection of chromatoplate from moisture during application of the sample. Courtesy of Brinkmann Instruments.

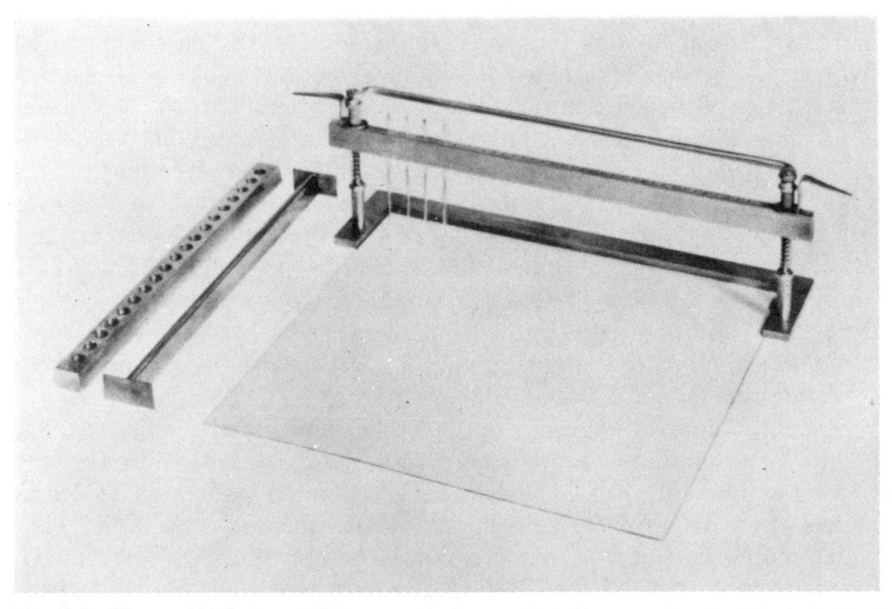

Fig. 4.4 Thomas–Morgan multiple spot applicator. Samples are contained in individual capillaries which contact the chromatoplate simultaneously when the supporting crossbar is depressed. Courtesy of Arthur H. Thomas Co.

can be applied at one time; for multiple applications, hot air is used to evaporate the solvent. In fact, numerous designs, both automatic and semi-automatic, have been published and some have been marketed [23–31]. Two designs permit not only the application of multiple samples, but provide a means of extraction of the sample prior to application of the solution to the layer. Flinn's [32] rather simple design allows 28 powder samples to be extracted and spotted in a single operation. It is useful for drug analysis. The equipment designed by Fosslin and Musil [33, 34] was intended for the lipid extraction of serum samples.

Another application technique which is especially useful for preparative chromatography is that of applying the sample in a streak instead of as individual spots. This technique may also be used for normal separations by applying short streaks instead of spots. Numerous instruments have been designed for streak applications ranging from simple to complicated. Of the simpler ones are those in which (*a*) the sample is held by capillary attraction between two parallel plates or threads separated by a short space [35–37], (*b*) the sample drains from a capillary or is expelled from a syringe as the plate is pushed along by hand [38–45], or (*c*) a simple guide is used to apply the sample with a brush [46].

A line applicator which is designed according to different principles has been constructed by McKibbens et al. [47] for paper chromatography and has been applied to thin-layer work [48, 49]. In this equipment, a hypodermic syringe is fastened to a slide which moves along a fixed bar. The plunger of the syringe slides along an adjustable bar which is placed at an angle to the first bar. As the slide with the syringe is moved along, the plunger is uniformly depressed by the angular bar. By adjusting the angle between the two bars, the rate of expulsion of the liquid can be controlled. There are two versions of this equipment on the market, one the Radin–Pelick sample streaker obtainable from Applied Science Laboratories (Fig. 4.5) and the other by CAMAG. In the apparatus of Contractor [50] the knurled head of a micrometer syringe is rolled along a bar, thus expelling the sample as the sliding carrier is moved across the chromatographic plate.

A slightly different approach to this problem of applying a uniform line of sample to the thin-layer plate has been used by Ritter and Meyer [51]. In their equipment, the sample is expelled from a hypodermic syringe by means of air or gas pressure. The syringe is mounted on a sliding plate which moves between two parallel bars. The apparatus of Coleman [52] is similar in design, but the flow of sample is automatically started and stopped by valves controlling the air pressure as the carrier is moved across the plate. Other similar designs have been published [53, 54]. Waessle and Sandhoff [55] used an artists' spray painting unit for applying

Fig. 4.5 Radin-Pelick thin-layer chromatography streaker. Courtesy of Applied Science Laboratories.

their samples, and Labadie [56] used an automatically driven syringe in his equipment. The CAMAG Linomat II and the Desaga Autoliner use air pressure to apply the samples as the chromatoplate is automatically moved back and forth.

Techniques for obtaining a very thin band of material at the starting line have been used by Truter [57], Fessler and Galley [58], and Musgrave [59]. These techniques although similar in principle are somewhat different in application. In Fessler and Galley's technique, the sample is applied in a continuous streak at the starting region of the thin layer. After the solvent is evaporated, the plate is immersed in a solvent in which all of the components in the mixture have an R_f value approaching unity. The solvent is then allowed to ascend the plate until it has risen above the region where the sample was applied. The plate is then removed and quickly dried with an air dryer. A repeat development is then made with the same solvent to the same point to make sure that no material has been left behind. After the final removal of the solvent, the plate is ready for development by the usual methods. This technique concentrates the sample into a thin starting line.

Truter's technique employs a special slotted lid through which the thin-layer plate projects above the developing chamber. As the solvent front reaches the slit, it evaporates, leaving the dissolved material concentrated in a thin line. In operation, the sample is applied as a line across the plate and then placed in a solvent which will carry the entire sample to the top of the plate where the solvent evaporates. Hydroxylic solvents alone are not satisfactory because they evaporate too slowly, thus giving an irregular line about the slot. The adsorbent must be reactivated by heating if a solvent such as methanol has been used in the production of the hairline origin. Once the hairline has been established and the excess solvent removed, the plate is turned over and developed in the opposite direction with the desired solvent.

Musgrave used a multiple layer consisting of a narrow band (4.5 cm) of kieselguhr that had been treated with 18% hydrochloric acid to remove iron and the calcium sulfate binder. The remainder of the plate (15.5 cm) was coated with silica gel with calcium sulfate binder. The sample applied to the kieselguhr layer was formed into a narrow streak as it reached the silica gel layer on development. Prepared plates of this type are available from Kontes.

As mentioned in the original work by Kirchner et al. [1], it is desirable to spot a control sample to be run at the same time as the experimental spot(s): "On each set of five chromatograms (individual spots), a control chromatogram of limonene was run with hexane in order to make sure the strips were dried properly and *to afford a reference for comparing R_f values.*" Stahl [60] adopted the method of Brockmann and Schodder [61] in using dyes for checking the activity of adsorbents. He used a mixture of three dyes: butter yellow, indophenol, and Sudan III.

References

1. M. H. Hashmi, M. A. Shahid, and F. R. Chughtai, *Mikrochim. Acta,* **1968,** 309.
2. R. A. Zeineh, B. J. Fiorella, E. P. Nijm, and G. Dunea, *Anal. Chem.,* **46,** 477 (1974).
3. J. M. Dechary, A. C. F. Mason, and W. B. Carney, *Anal. Biochem.,* **30,** 142 (1969).
4. Z. L. Awdeh and S. Abu-Samra, *Anal. Biochem.,* **62,** 601 (1974).
5. R. P. Farrow, E. R. Elkins, Jr., and L. M. Beacham, *J. Assoc. Off. Agric. Chem.,* **48,** 738 (1965).
6. J. G. Kirchner, J. M. Miller, and G. J. Keller, *Anal. Chem.,* **23,** 420 (1951).
7. F. Leudy-Tenger, *Pharm. Acta. Helv.,* **37,** 770 (1962).
8. M. E. Tate and C. T. Bishop, *Can. J. Chem.,* **41,** 1801 (1963).

9. S. Samuels and C. Fisher, *J. Chromatogr.*, **71**, 291 (1972).
10. R. J. Wieme, *Behringwerk-Mitt.*, **34**, 27 (1958).
11. R. J. Wieme and M. Rabaey, *Naturwissenschaften*, **44**, 112 (1957).
12. S. B. Curri and Y. Levanon, *Rev. Roum. Biochim.*, **3**, 159 (1966); through *Chem. Abstr.*, **65**, 19106f (1966).
13. G. D. Cheryil and K. S. Scaria, *J. Lipid Res.*, **11**, 378 (1970).
14. B. Kuennert and H. Krug, *Acta Histochem.*, **37**, 194 (1970).
15. W. J. Culley, *Clin. Chem.*, **15**, 902 (1969).
16. A. K. Munson, J. R. Mueller, and M. E. Yannone, *Microchem. J.*, **15**, 95 (1970).
17. A. N. Crabtree, *Lab. Pract.*, **15**, 311 (1966).
18. M. Beroza, M. E. Getz., and C. W. Collier, *Environ. Contam. Toxicol.*, **3**, 18 (1968).
19. M. Brenner, A. Niederwieser, and G. Pataki, *Experientia*, **17**, 145 (1961).
20. M. E. Morgan, *J. Chromatogr.*, **9**, 379 (1962).
21. B. I. Medvedev and V. V. Smolyaninov, *Zh. Fiz. Khim.*, **45**, 1855 (1971); through *Chem. Abstr.*, **75**, 122333e (1971).
22. D. Shapcott, B. Lemieux, and A. Sahapoglu, *J. Chromatogr.*, **70**, 174 (1972).
23. L. S. Bark, R. J. T. Graham, and D. McCormick, *Talanta*, **12**, 122 (1965).
24. G. C. Meadows and R. J. T. Graham, *Int. Symp. Chromatogr. Electrophor., Lect. Pap., 6th 1970*, Ann Arbor Science, Ann Arbor, Mich., 1971, p. 94.
25. J. W. Boag, P. S. Bond, E. M. Fielden, H. Hodt, and Z. Tramer-Zarebska, *J. Chromatogr.*, **73**, 265 (1972).
26. S. Samuels, *J. Chromatogr.*, **32**, 751 (1968).
27. C. Bernhardt, *Z. Med. Labortech.*, **11**, 212 (1970).
28. L. F. Vas'kovskaya, *Gig. Primen. Polim. Mater. Izdelii Nikh*, **1969**, 474; through *Chem. Abstr.*, **75**, 47748p (1971).
29. P. J. Curtis, *Chem. Ind. (London)*, **1966**, 247.
30. R. A. De Zeeuw and G. G. Dull, *J. Chromatogr.*, **110**, 279 (1975).
31. G. I. Zhurbin, V. A. Kokunin, and V. I. Yatsenko, *Lab. Delo*, **1968**, 440; through *Chem. Abstr.*, **69**, 73601f (1968).
32. P. E. Flinn, *J. Chromatogr.*, **82**, 117 (1973).
33. F. Musil and E. Fosslien, *J. Chromatogr.*, **47**, 116 (1970).
34. E. Fosslien, *J. Chromatogr.*, **63**, 59 (1971).
35. R. D. Bennett and E. Heftmann, *J. Chromatogr.*, **12**, 245 (1963).
36. F. K. Klein and H. Rapoport, *J. Chromatogr.*, **47**, 505 (1970).
37. C. R. Turner, *J. Chromatogr.*, **22**, 471 (1966).
38. P. J. Curtis, *Chem. Ind. (London)*, **1966**, 1680.
39. L. J. Altman and J. R. Trudell, *J. Chem. Educ.*, **47**, 404 (1970).
40. H. J. Monteiro, *J. Chromatogr.*, **18**, 594 (1965).
41. K. O. Abraham and L. V. L. Sastry, *Lab. Pract.*, **19**, 1038 (1970).
42. T. Darocha, C. H. Gray, and R. V. Quincey, *J. Chromatogr.*, **27**, 497 (1967).
43. I. R. Shimi and G. M. Imam, *Analyst (London)*, **94**, 62 (1969).
44. F. A. Vandenheuvel, *J. Chromatogr.*, **25**, 102 (1966).
45. G. P. Arsenault, *J. Chromatogr.*, **21**, 155 (1966).
46. Z. Tamura, *J. Chromatogr.*, **19**, 431 (1965).

47. S. W. McKibbens, J. F. Harris, and J. F. Saeman, *J. Chromatogr.*, **5,** 207 (1961).
48. M. A. Millett, W. E. Moore, and J. F. Saeman, *Anal. Chem.*, **36,** 491 (1964).
49. E. Von Arx and R. Neher, *J. Chromatogr.*, **25,** 109 (1966).
50. S. F. Contractor, *J. Chromatogr.*, **20,** 182 (1965).
51. F. J. Ritter and G. M. Meyer, *Nature,* **193,** 941 (1962).
52. M. H. Coleman, *Lab. Pract.*, **13,** 1200 (1964).
53. Z. Tuba, F. Soti, S. Dombi, and J. Magyar, Hung. Pat. 7651 (January 28, 1974).
54. G. Kasang and H. Rembold, *J. Chromatogr.*, **71,** 101 (1972).
55. W. Waessle and K. Sandhoff, *J. Chromatogr.*, **34,** 357 (1968).
56. R. P. Labadie, *Pharm. Weekbl.*, **107,** 421 (1972).
57. E. V. Truter, *J. Chromatogr.*, **14,** 57 (1964).
58. J. H. Fessler and H. Galley, *Nature,* **201,** 1056 (1964).
59. A. Musgrave, *J. Chromatogr.*, **41,** 470 (1969).
60. E. Stahl, G. Schroeter, G. Kraft, and R. Renz, *Pharmazie,* **11,** 633 (1956).
61. H. Brockmann and H. Schodder, *Chem. Ber.*, **74,** 73 (1941).

Chapter **V**

DEVELOPING THE CHROMATOGRAM

1 SELECTION OF THE SOLVENT

In selecting the proper adsorbent and solvent for a particular thin-layer separation, the experimenter will be able to use his previous experience from column chromatography and paper chromatography and also from modifications of these, such as partition chromatography. An example of this is the use of the so-called eluotropic series of solvents which has been established for column chromatography. These are tables of solvents (Table 5.1) arranged in order of effectiveness in removing an adsorbed material from a given adsorbent. The most familiar of these is, of course, the one set up by Trappe [1] and variations of this have been published from time to time. These may be used as guides in picking a solvent or in setting up a solvent mixture to be used as a developing agent. Actually a series like one of these will vary somewhat with different adsorbents and with the compound being adsorbed. Kaufmann and Makus [4] give the following series (listed in order of increasing eluting power) for lipid work: xylene, toluene, benzene, trichloroethylene, ethylene dichloride, methylene chloride, chloroform, isoamyl ether, isopropyl ether, diethyl ether, acetone, and dioxane.

Bulenkov [5], using loose layers of alumina and a selection of dyes, established an eluotropic series of solvents (Table 5.2) which he has divided into three groups, each group made up of solvents having similar eluting properties. Activity coefficients were determined for both polar and nonpolar compounds, and these activity coefficients are the basis for the three groups. The first group, the nonpolar solvents, do not or only barely elute compounds having an R_f of less than 0.3 in the solvents in

Table 5.1 Some Eluotropic Series of Solvents

Trappe [1]	*Wren* [2]	*Strain* [3]
Light petroleum	Ligroin	Light petroleum, 30–50°C
Cyclohexane	Cyclohexane	Light petroleum, 50–70°C
Carbon tetrachloride	Carbon tetrachloride	Light petroleum, 70–100°C
Trichloroethylene	Trichloroethylene	Carbon tetrachloride
Toluene	Chloroform	Cyclohexane
Benzene	1,1,2,2-Tetrachloroethane	Carbon disulide
Dichloromethane	1,1-Dichloroethane	Anhydrous ether
Chloroform	Toluene	Anhydrous acetone
Ether	Benzene	Benzene
Ethyl acetate	Dichloromethane	Toluene
Acetone	Ether	Esters of organic acids
n-Propanol	Ethyl acetate	1,2-Dichloroethane
Ethanol	Methyl acetate	Alcohols
Methanol	Acetone	Water
	1-Propanol	Pyridine
	Ethanol	Organic acids
	Methanol	Mixtures of acids or bases, water, alcohols, or pyridine

Table 5.2 Eluotropic Series with the Activity Coefficients for the Solvents[a]

Solvent	R_f for Test Material			Activity Coefficient for Test Material	
	Nonpolar	Polar	ΔR_f	Nonpolar	Polar
	First Group				
1. Petroleum ether	0.038			0.05	
2. *n*-Heptane	0.040			0.05	
3. *n*-Hexane	0.042			0.05	
4. *n*-Undecane	0.050			0.06	
5. Decalin	0.052			0.07	
6. Carbon disulfide	0.155			0.19	
7. Carbon tetrachloride	0.246			0.31	
	Second Group				
8. Xylene	0.561			0.70	0.68
9. Chlorobenzene	0.615			0.77	0.74
10. Toluene	0.616			0.77	0.80
11. Benzene	0.717			0.90	0.95
12. Chloroform	0.798	0.000	0.798	1.00	1.00
13. Ethyl bromide	0.778	0.010	0.768	0.97	1.01
14. Trichloroethane	0.790	0.015	0.775	0.99	1.02
15. Dichloroethane	0.810	0.040	0.770	1.01	1.05
16. Anisole	0.984	0.040	0.944	1.23	1.04
17. Diethyl ether	0.973	0.160	0.813	1.22	1.20
18. Methyl ethyl ketone	0.878	0.218	0.660	1.10	1.33
19. Malonic ester	1.000	0.405	0.595	1.25	1.68
	Third Group				
20. Dioxane	0.882	0.610	0.272	1.11	3.24
21. Isoamyl alcohol	0.990	0.645	0.345	1.24	2.87
22. *n*-Amyl alcohol	0.980	0.650	0.330	1.23	2.97
23. Ethyl acetate	0.855	0.555	0.300	1.07	2.85
24. *n*-Butanol	0.950	0.690	0.260	1.19	3.65
25. *n*-Propanol	0.880	0.680	0.200	1.10	4.40
26. Isopropanol	0.850	0.685	0.165	1.07	5.15
27. Ethyl acetoacetate	0.000	0.820	0.180	1.25	5.56
28. Acetone	0.846	0.696	0.150	1.06	5.64
29. Absolute ethanol	0.840	0.718	0.122	1.05	6.89
30. Dimethylformamide	0.000	0.880	1.120	1.25	8.33
31. Methanol	0.870	0.785	0.085	1.09	10.02
32. Water	0.930	0.880	0.050	1.16	18.60

[a] From T. I. Bulenkov [5]; reproduced with permission of the author and the Plenum Publishing Corporation.

the other two groups. The second group of solvents show similar eluting power for both polar and nonpolar compounds so that the differences in activity coefficients for these two groups is never greater than 0.43 (malonic ester). The third solvent group is characterized by having a significantly greater activity coefficient for polar compounds than for nonpolar compounds; in other words the rate of movement of the chromatographic zone is greater for polar compounds than for nonpolar compounds.

To the eluotropic series should be added perfluorokerosene (Columbia Chemicals Co.) perfluoroalkane (b.p. 70–80°C) and perfluoro-*n*-hexane (Peninsular ChemResearch) in that order at the top of the nonpolar solvents, for Attaway et al. [6, 7] have shown that the fluorocarbons by reason of their eluting properties are less polar than the hydrocarbons.

Thin-layer chromatography lends itself admirably to the selection of the proper solvent. Izmailov and Schraiber [8] and Crowe [9] were the first to use this technique for selecting solvents for column work. They spotted their samples on loose layers of material and then by adding solvent to the spot from a capillary, they allowed the solvent to spread outward from the center. By trying various solvents they could determine which solvent appeared to give the best separation. Miller and Kirchner [10] used bound layers for this purpose. In this case, narrow plates or strips are spotted with the mixture and then placed in individual test tubes with different solvents. Development is allowed to take place in the normal manner and after evaporation of the excess solvent, the strip is sprayed with the detecting agent so as to disclose the location of the various compounds. If a pure solvent does not give a satisfactory separation, mixed solvents are made by combining solvents that do not move the compounds with solvents that do move the compounds along the layers. These of course are generalizations because no hard and fast rules can be laid down. To give a specific example, Knight and Groenning [11] cite the case of the dye Luxol Fast Red B, which is moved by neither acetone nor water on silica gel even though it is soluble in both solvents; it is, however moved by a mixture of the two solvents.

Numerical analysis has been applied by Massart and De Clercq [12–13b] to the problem of selecting the optimum sets of solvents where these are not readily apparent by simple observation of the results of a large number of available chromatographic systems. Moffat and co-workers [13c–13d] have applied a similar system to the problem of separating and identifying a large group of drugs by determining which solvent systems to use. On the other hand, Turina et al. [14] used numerical analysis to show that it may be used to correct solvent composition ratios so as to obtain optimum resolution.

Baker et al. [15] illustrated the use of experimental design to explore the use of various systems for obtaining a chromatographic separation. They showed that a program of this type results in one of three possibilities: (1) a successful separation is achieved, (2) the scope and means of improving the separation are indicated, or (3) the variables that have been studied are unimportant, and a new approach to the problem must be considered. In working with organic sulfonates they found that by dissolving the mixture in a solvent such as methanol or *N,N*-dimethylformamide and then gradually adding chloroform (in which the sulfonates are insoluble) until a precipitate formed, they obtain a solvent composition that was close to the region of optimum resolution.

Panova et al. [16] showed how to determine the optimum conditions in a chromatographic separation in a three-component system by introducing an optimality criterion with the use of a simplex lattice design.

Rouser [16a] discussed the classification of solvents according to the nature of their hydrogen-bonding, acidic, and basic groups with examples of the effects of various types of solvents. He divided solvents into six classes as follows: "(1) No hydrogen bonding (hydrocarbons), (2) Both proton donor and acceptor in hydrogen bonding (water, alcohols, amines, carboxylic acids, amides, etc., (3) Proton donor only in hydrogen bonding (chloroform), (4) Proton acceptor only in hydrogen bonding (aldehydes, ketones, ethers, esters, nitro compounds, etc.), (5) Acids (hydrogen ion transfer donors), (6) Bases (hydrogen ion transfer acceptors)."

Lofts et al. [17] systematically examined the formulation of solvents for resolving mixtures of amino acids on cellulose layers. From their experiments they decided that there were six factors involved in a satisfactory multicomponent solvent for amino acids although a single component of the mixture could contribute more than one factor. These six factors are as follows:

1. Movers: in general these solvents have too high an eluting power by themselves.
2. Restrainers: these possess too low an eluting power.
3. Homogenizer: this is added in case the mixture of the mover and restrainer are immiscible.
4. pH controller: because of the presence of both acidic and basic amino acids an acidic and a basic solvent are required, and hence a two-dimensional chromatogram is needed.
5. Sharpener: because acidic and basic solvent systems produce more compact spots these are termed sharpeners.
6. Viscosity reducers: when the development time is too long a component may be added to the mixture to decrease the viscosity and increase the rate of flow.

In examining their initial experiments, they established the dependence of the eluting power of the various solvents on the hydroxyl content and the optimum molar hydroxyl fraction (the M value) between 0.45 and 0.7. Based on these facts a first-dimensional solvent of cyclohexanol—acetone—diethylamine—water (10:5:2:5) was developed and also a second-dimensional solvent of t-butanol–acetic acid–water (5:1:1). In the first mixture, according to the authors, the various components performed the following roles:

1. Mover: water assisted by cyclohexanol.
2. Restrainer: cyclohexanol assisted by acetone and diethylamine.
3. Homogenizer: acetone assisted by diethylamine.
4. pH Controller: diethylamine.
5. Sharpener: diethylamine.
6. Viscosity reducer: acetone.

Ościk and Różylo [18] have studied the theoretical calculation of the relation between the R_M values and the composition of two-component solvent mixtures, and its application to the calculation of the optimum separation conditions. Snyder has also studied the theoretical aspects of solvents in relation to both alumina [19] and silica [20].

For partition chromatography, Thoma [21] and Thoma and Perisho [22] examined the theoretical aspects of separation and concluded that the maximum separation between two similar solutes occurs when the zones have migrated one-fourth of the length of the support. On this basis they recommend selecting solvent proportions to give average R_f values of 0.25. On the other hand, if a change in solvent components is desired they suggest choosing a component that will selectively alter the structure of one of the solutes. An example of this would be the addition of a complexing molecule. In lieu of a better solvent system, recourse may be had to multiple development or to continuous chromatography to increase the effective travel distance and the resolution. Thoma [23] recommends using finely divided supports with minimum spot loading and developing with rapidly migrating solvents.

In working with three-component solvent systems, the establishment of the optimum composition is even more difficult than with the two-component systems, and Waksmundzki and Różylo [24] and Różylo [25] have used the Gibbs triangle for interpreting the results, as well as for choosing the three-component mobile phase, if its electron donor–acceptor properties are known in advance.

In preparing mixed solvents, it is essential to measure the components carefully because small variations in composition can alter the R_f values and also the resolution of a given mixture. As an example of the latter,

Nichols [26] found that in the separation of lipids with solvents based on a mixture of diisobutylketone and acetic acid he seldom obtained a good resolution if more than 4 parts water were present. On the other hand, a solvent containing 3.6 to 3.8 parts water provided the best general purpose solvent. (Throughout this work mixtures are given in terms of vol/vol unless otherwise specified or unless a weight relationship clearly exists, for example, mixtures of adsorbents.)

An example of the complexity that may exist in a given sample and equal complexity of a solvent can be illustrated by the work of Kunz and Kosin [27] on the separation of plasma phospholipids. A solvent was formulated for the separation of 12 compounds using 22 components in the solvent system including numerous inorganic ions (see Chapter XXIII, Section 7). In addition, to obtain the desired results a mixture of two silica gels was used for the adsorbent layer.

Care must also be taken in reusing a solvent for more than one development because the composition of the solvent can change due to evaporation. Roeder [28] explored the use of azeotropic mixtures to eliminate this effect. Twenty-one azeotropic mixtures have been listed for the separation of estrogens, gestogens, alkaloids, sulfonamides, psychopharmaceuticals, narcotics, and local anesthetics. In this way solvents can be used repeatedly. Azeotropic solvents which contain a polar component at a low concentration cannot be used for more than 10 repetitive runs because the concentration of this component is altered so that the separation is affected. Manthey and Amundson [29] investigated the use of alcoholic water solution of inorganic salts as a means of avoiding the evaporation effect.

Another factor to be taken into account when working with the separation of volatile mixtures is the change in composition of the solvent owing to the vaporization of the chromatographed materials from the plate with consequent buildup in the solvent itself.

To much emphasis cannot be placed on the purity of the solvents employed in thin-layer chromatography, for a number of artifacts have been reported as due to solvent impurities. A diester of sebacic acid was found as a contaminant of reagent-grade chloroform used in lipid analysis [30]. Dibutyl phthalate has been found in ether [31] and in commercial first-grade benzene [32]. Bucke [33] isolated a fluorescent impurity from acetone, and Crosby and Aharonson [34] examined a number of solvents to find fluorescent contaminants. Of these examined the worst sample was reagent-grade ethyl acetate with 10 mg/liter of impurity. They found that percolation of the solvent through activated carbon, followed by distillation, yielded a solvent almost free of fluorescent substances. Not only are these impurities apt to be harmful directly in thin-layer chromatography work, but also indirectly where large volumes of solvent are used for the extraction of natural products with the impurities eventually being concentrated in the extracts.

2 ASCENDING DEVELOPMENT

To date, by far the greatest amount of thin-layer work has been accomplished by ascending development. This is probably because it is the simplest procedure and it requires the least complicated apparatus. In general any closed container which will hold the coated plate in a vertical or near-vertical position is satisfactory. It is desirable that the volume of the container be as small as possible; all the supply houses handling thin-layer equipment stock various sizes and shapes of developing chambers.

Abbott et al. [35] have observed that the angle at which the plate is supported in the tank affects the shape of the spot as well as the rate of development. The speed of development increases as the plate is tipped backward but spots become more diffuse. As the plate is tipped forward beyond the vertical position, speed again increases but the spots are more compact. The authors recommend an optimum angle of 45° between the adsorbent layer and the solvent surface.

There are a number of ways in which the ascending development can be carried out. A layer of solvent may be introduced into the developing tank to a depth of 5 to 10 mm. If the chromatograms are to be run in a saturated atmosphere, the developing chamber may be lined with filter paper which dips into the solvent to assist in saturating the chamber. The plate with the sample applied is then placed in the chamber with the lower edge dipping into the solvent.

Since only two plates can be run in the average chamber by leaning one plate against each side wall, Rosi and Hamilton [36] increased the capacity of their chamber by coating both sides of the chromatoplates. One side is coated and dried after which the second side can be readily coated. In applying the samples they found it was necessary to support the plate by the edges on a simple framework to avoid damaging the lower layer. Holders have also been devised so that more plates may be run at one time. Two designs are shown in Fig. 5.1. Nybom [39] stacks the plates by placing four plastic pellets on the corners between each set of plates. A plain plate is used as a final cover and the whole assembly is held together with rubber bands. The stack is then placed in a tank for development. Space between the plates is 2 to 3 mm; this minimizes the vapor space. Actually, this is a variation of the sandwich plate, which is described later. CAMAG has made available accessory equipment for running multiples of five plates in a sandwich arrangement. Sloane-Stan-

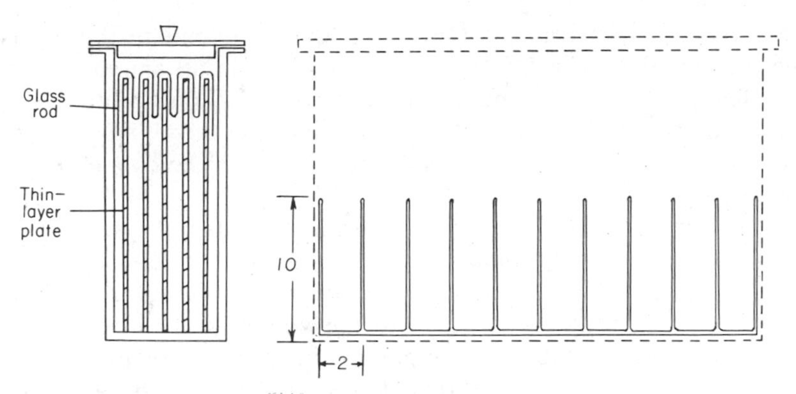

Fig. 5.1 Arrangements for development of multiple plates. Left, according to Brenner and Niederwieser [37]; right, according to Urion et al. [38].

ley and Bowler [40] constructed a holder from glass rod and polyethylene sheet, and Wollish et al. [41] constructed a holder made of glass rod and stainless steep strip. Ford [42] inserted two compression springs in the tank so that the springs were held by their own tension. A number of flexible sheets or plates could then be inserted. A number of plate holders are available commercially and differ somewhat in design and size, ranging from holders for the 5 × 20 cm plates to those for 1 m × 20 cm plates put out by Shandon Scientific Co. Kensington Scientific Corp. has developed a holder which supports two plates and CAMAG and Analtech supply similar equipment. The holders are so designed that the plates may be suspended above the liquid surface of the chromatographic chamber in order to allow the chamber atmosphere and the plate to come to equilibrium with the solvent. After this is accomplished the plates may then be lowered without opening the chromatographic chamber. The same idea may be accomplished in another manner by supporting the plates on glass rods above the surface of a layer of liquid in the chamber until chamber saturation has been reached. Then additional solvent is added through a hole in the cover in order to raise the level of solvent, so that the chromatogram will develop. A somewhat similar procedure is to scrape off the lower portion of the adsorbent from the plate, [43] so that the plates can rest on the bottom of the chamber without the solvent touching the adsorbent; after saturation of the chamber has been reached, additional solvent is poured into the chamber to contact the adsorbent.

Sankoff and Sourkes [44] have gotten around the chamber saturation problem by always keeping solvent in the chambers. Prior to running the chromatographic plate they throw out the old solvent and replace it with fresh solvent. By this means they were able to eliminate the need for lining the chamber with filter paper. In working in an ammonia atmosphere, Jensen [45] found it necessary to place a separate container filled with the ammonium hydroxide solution within the developing chamber, since attempts to add the ammonia directly to the developing solution were not successful.

If plates are developed in an unsaturated chamber the solvent front toward the center of the plate may advance faster than that near the edges, resulting in a concave-shaped arrangement of a group of spots of the same compound. This varies with the volatility of the solvent, the temperature, and the rate of development and is due to evaporation from the edge of the plate. To avoid this effect the chamber can be saturated with solvent vapor as previously mentioned. Another way to solve this problem is by trimming the adsorbent along the edges of the plate [43]. Karpitschka [46], who had no success with paper-lined chambers, trimmed the sides of the layer in a wedge shape so that the layer was narrower at the bottom than at the top.

Of course minimum-volume chambers, such as the sandwich chamber, and chromatostrips developed in test tubes do not present this problem of edge effect.

A special type of ascending-development equipment is the so-called "sandwich plate" [47–50]. In this method a thin-layer plate is cleaned of adsorbent along three edges so as to provide a clear space for a separating rod or gasket. This may be of glass, metal, or even plastic. After depositing the sample on the layer along that side of the plate from which the adsorbent has not been removed, the U-shaped separator is put in place

Fig. 5.2 CAMAG "sandwich type" holder for thin-layer chromatography. Courtesy of Arthur H. Thomas Co.

followed by another glass plate. The whole assembly is held together with spring clips and is placed with the open end dipping into a trough of solvent. This of course provides a very narrow chamber so that the time required for saturation of the free space is very short. When a solvent mixture containing heavy and light components is used, such as a mixture of carbon tetrachloride and hexane, the vapors have a tendency to separate in this type of chamber. The heavier carbon tetrachloride vapor settles to the bottom of the chamber and lighter hexane vapor rises to the top so that the chamber is not homogeneously saturated with vapor [48]. The spots on "sandwich plates" are smaller than on regular plates and the edge effect is eliminated. Wasicky [50] in his arrangement used two coated plates placed face to face with one another and thus doubled the capacity of the chromatogram. Commercial equipment is available from CAMAG (Fig. 5.2) and from Desaga (Fig. 5.3). In the Desaga

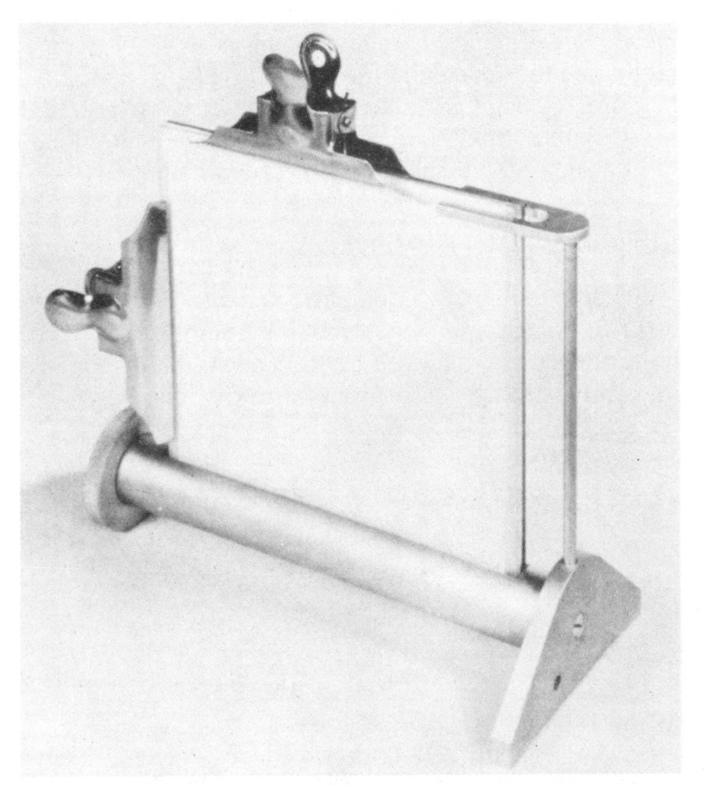

Fig. 5.3 Desaga "sandwich type" chromatographic plate. Courtesy of Brinkmann Instruments.

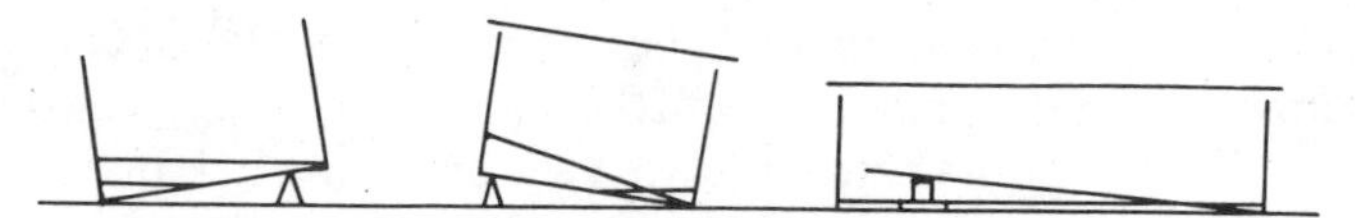

Fig. 5.4 Chromatographic tank for the development of loose-layer plates. The first figure shows the position of the tank during insertion of the plate, the remaining two during development. From Procházka [51]; reproduced with permission of the author and the Publishing House of the Czechoslovak Academy of Science.

equipment the cover plates possesses a fused glass rim so that two thin-layer plates cannot be placed face-to-face with this type. Eastman Kodak supplies a sandwich chamber for thin sheets, and Gelman one constructed of plastic for 65 × 100 mm sheets.

Loose-layer plates cannot be developed in a vertical position and are usually developed at an angle of about 20°. (Loose layers, rather than unbound layers, is the preferred term for this type of chromatography, since there are certain types of adsorbents which do not contain a binder but which can nevertheless be developed in a vertical position.) The general technique for developing the loose-layer chromatogram is shown in the diagrams in Fig. 5.4 taken from Procházka [51]. Davídek and Davídková [52] used the same technique except that the solvent was contained in a separate vessel within the main chamber.

3 DESCENDING DEVELOPMENT

In general, descending development is used to carry out a continuous development for separating slowly moving components. In many cases, better separation can be achieved by increasing the development time in this manner rather than by resorting to a more polar solvent.

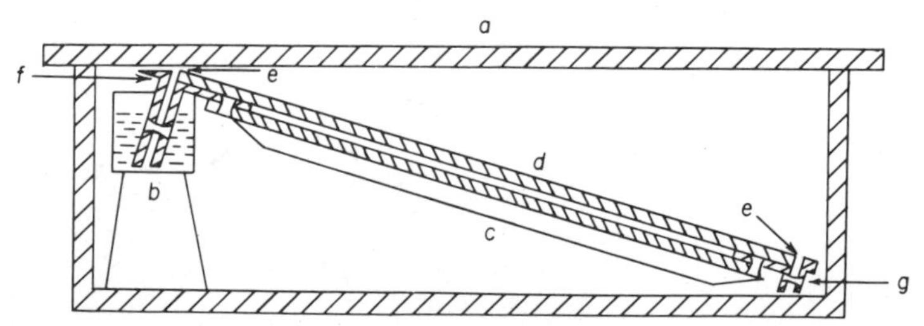

Fig. 5.5 Apparatus for the descending development of loose-layer plates. (*a*) Container 34 × 10 cm with ground cover; (*b*) Solvent tank 24 × 3.5 × 3 cm; (*c*) Supporting frame; (*d*) Glass plate 24 × 24 cm; (*e*) Slits for capillary transport of solvent or for paper wicks; (*f*, *g*) Guides for spread adsorbent layer. From Mistryukov [54]; reproduced with permission of the author and the Elsevier Publishing Co.

Lapp and Erali [53] were the first to run a descending thin-layer chromatograph and this was on a loose-layer basis. Mistryukov [54] developed the apparatus shown in Fig. 5.5 for the descending development of loose layers. The solvent is carried up to and away from the thin layer either by capillary attraction through the narrow slits or by means of strips of filter paper. Johansson and Rymo [55] used an electrophoresis unit for holding loose-layer plates of Sephadex. Pharmacia offers equipment for thin-layer gel chromatography with Sephadex, but since this equipment is made of plastic, the type of solvent required for other applications of descending thin-layer chromatography cannot be used. Stanley and Vannier [56] in 1957 were the first to use descending development with bound layers. Their equipment (Fig. 5.6) was developed for narrow "chroma-

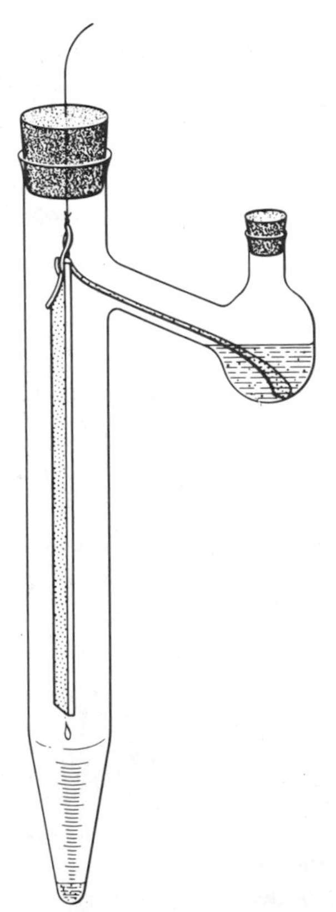

Fig. 5.6 Silicic acid-coated chromatostrip equipped with filter paper solvent wick in side-arm test tube for descending development. Reproduced from Stanley et al., *Food Technol.*, **15**, 382 (1962).

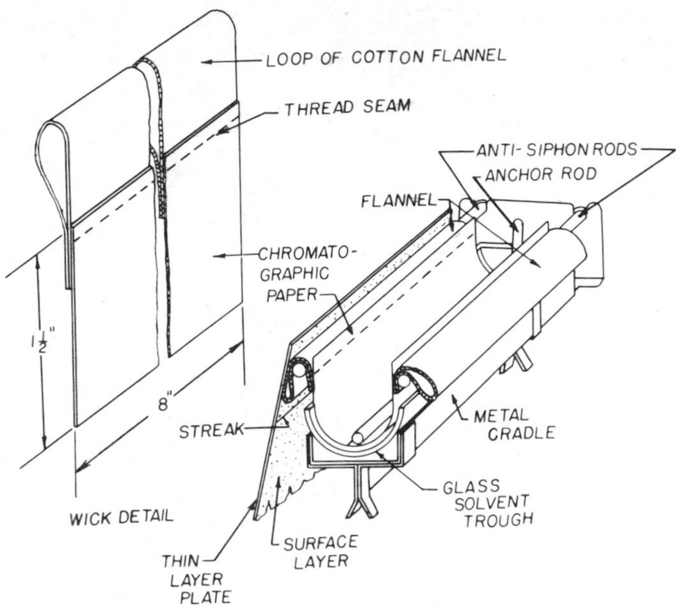

Fig. 5.7 Arrangement of apparatus for continuous downward development of chromatoplates. From Seikel et al. [59]; reproduced with permission of the authors and The Elsevier Publishing Co.

tostrips'' [57]. The coated strip is suspended from the stopper and solvent is carried to the surface of the layer by means of filter paper. Research Specialties Co. offers this device with a modification which consists of a replaceable collection tube for the eluate and a stopcock on the solvent chamber so that the solvent may be drained and replaced. The rack which Stanley et al. [58] developed for the prewashing of the thin layers by a descending technique has been adapted with slight modifications by Seikel et al. [59] for continuous descending chromatography of thin-layer plates (Fig. 5.7). Birkofer et al. [60], Goeldel et al. [61], Reisert and Schumacher [62], and Zoellner and Wolfram [63] have all published their versions of equipment for descending thin-layer chromatography. To remove solvent from the bottom of the plate, Seikel et al. [59] pressed the plate on a pad of filter paper, and Goeldel et al. [61] used a trough filled with diatomaceous earth. Reisert and Schumacher [62] claimed that too steep a slope causes irregular development and therefore the plate was inclined at an angle of 10° from the horizontal.

4 HORIZONTAL DEVELOPMENT*

Linear

For the horizontal development of loose-layer plates, Mistryukov [64] used a shallow dish with a ground-glass cover. The plate was supported on a T-shaped glass piece and the end of the thin-layer plate was pressed against a filter paper held by another glass strip. This arrangement allowed the solvent in the bottom of the dish to be transported up to the thin-layer film. Hesse and Alexandria [65] used bound layers which were developed horizontally with the coated layer facing a tray containing the developing solvent. The solvent was applied to the layer by means of contacts with a felt strip.

Radial and Circular

In principle these two are identical. Circular chromatography forms concentric rings, because of the initial placement of the sample at the center. In radial chromatography the sample is placed to one side of the center so that the resulting chromatogram forms arcs instead of complete circular bands. These methods are less versatile than the various forms of linear chromatography.

Radial chromatography takes us back into the history of thin layers. The Izmailov and Shraiber [8] drop-chromatographic method was a microcircular method on loose layers. Crowe [9] used a semicircular development in his work and Meinhard and Hall [66] used a special pipet for the application of solvent for the radial chromatography of inorganic ions on bound layers. In 1955 Bryant [67] carried out radial chromatography by using the chromatoplate, adsorbent side down, over a 5-in. petri dish containing the solvent. The solvent was transferred to the center of the plate by means of a cotton wick held by a metal support. The rate of development was affected by the diameter of the wick and the degree of contact with the plate. Sherwood [68] and Litt and Johl [69] designed equipment using rectangular sheets or plates so that commercially prepared plates could be utilized. Van Ooij [70] published the details of equipment for circular electrophoresis.

A new technique designated HPTLC (high-performance thin-layer chromatography) has been developed [71]. The method uses a type of silica gel developed by Halpaap [72] which has better resolving power and optical properties than the normal commercial product. This is based on a very fine (60 Å) and a very uniform particle size. Because the flow

* See also Section 5 on Continuous Development.

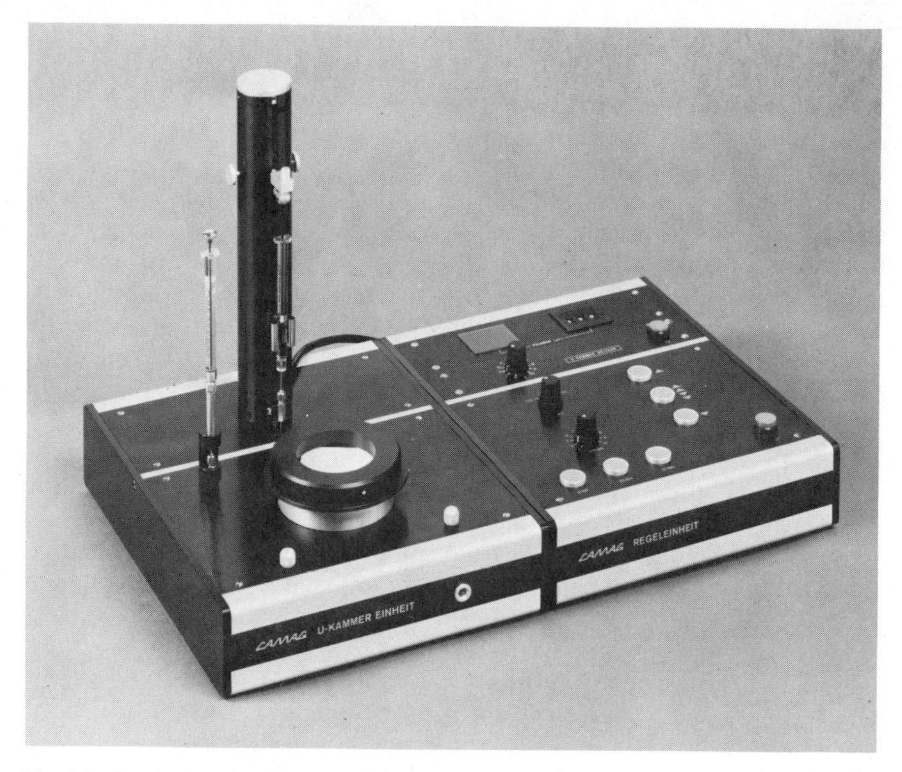

Fig. 5.8 Equipment for HPTLC (high-performance thin-layer chromatography). Courtesy of CAMAG Inc.

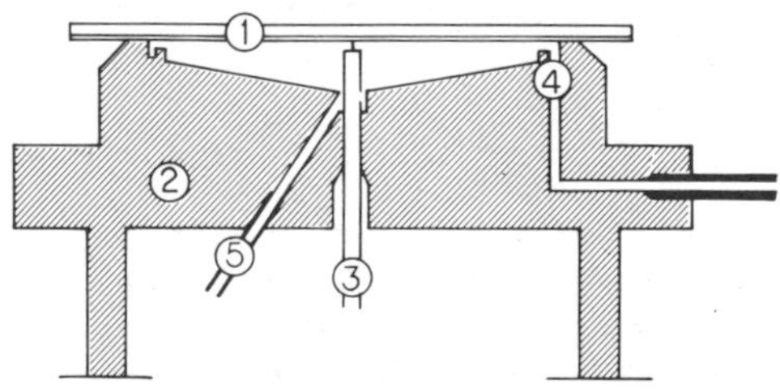

Fig. 5.9 Schematic cross-sectional diagram of U-chamber for HTPLC. A 50 × 50 mm HPTLC plate (1) rests adsorbent side down on the body (2) of the U-chamber. Developing solvent is fed to the center of the plate via a platinum–iridium capillary (3), 0.2 mm i.d. A vapor phase to control layer conditions may be fed through the circular channel (4) and exits via the tube (5). Courtesy of CAMAG Inc.

rates on layers made of this silica gel are lower, smaller plates are used. Equipment making use of this procedure for radial and circular high-performance thin-layer chromatography (HPRC) is available from CAMAG (Fig. 5.8). A cross-section of the instrument (U-chamber) is shown in Fig. 5.9. Solvent is fed to the plate at a constant rate from a syringe operated by a stepper motor with electronic control. Samples may be applied to the dry layer in a circular pattern either 3, 4.5, or 6 mm in diameter around the plate for radial chromatography. Sample may also be introduced via the solvent feed capillary to the center of the plate for circular development. In the latter case on a prewetted layer the results are directly transferable to high-performance liquid chromatography. Precoated plates may be obtained from E. Merck or CAMAG as Silica Gel 60 F-254 for Nano TLC.

5 SPECIAL TECHNIQUES

Continuous Development

This subject has been mentioned briefly under descending development, since this is the main reason for using descending development. However, other forms of continuous chromatography as applied to thin layers have been developed. One method for keeping the solvent moving over the plate is to allow the solvent to evaporate after it has reached the end of the run. This continued flow through evaporation was first applied by Mottier [73] and Mottier and Potterat [74] to the ascending chromatography on loose layers of aluminum oxide. Zoellner and Wolfram [63] applied the evaporative technique to ascending bound layers. In this case, the evaporation was accomplished by simply raising the cover of the chamber on one edge. Truter [75], using a special slotted lid, allowed his plate to project through this slot. As the solvent rose above the chamber it evaporated continuously. Libbey and Day [76] used a piece of Saran for their slotted lid. Anwar [77] applied the same principle to chromatostrips; a paper wick fastened to the end of the strip carried the solvent out beyond the stopper where it could evaporate, and the same idea has been applied to 20 cm × 20 cm plates [78, 79]. Stewart et al. [80, 81] have discussed the theoretical aspects of evaporative thin-layer chromatography.

Using this evaporative technique, Brenner and Niederwieser [82, 83] designed equipment (Fig. 5.10) for continuous chromatography on horizontal, thin-layer plates. With this equipment, the sample is first applied to the plate after which the equipment is assembled. Solvent is fed to the

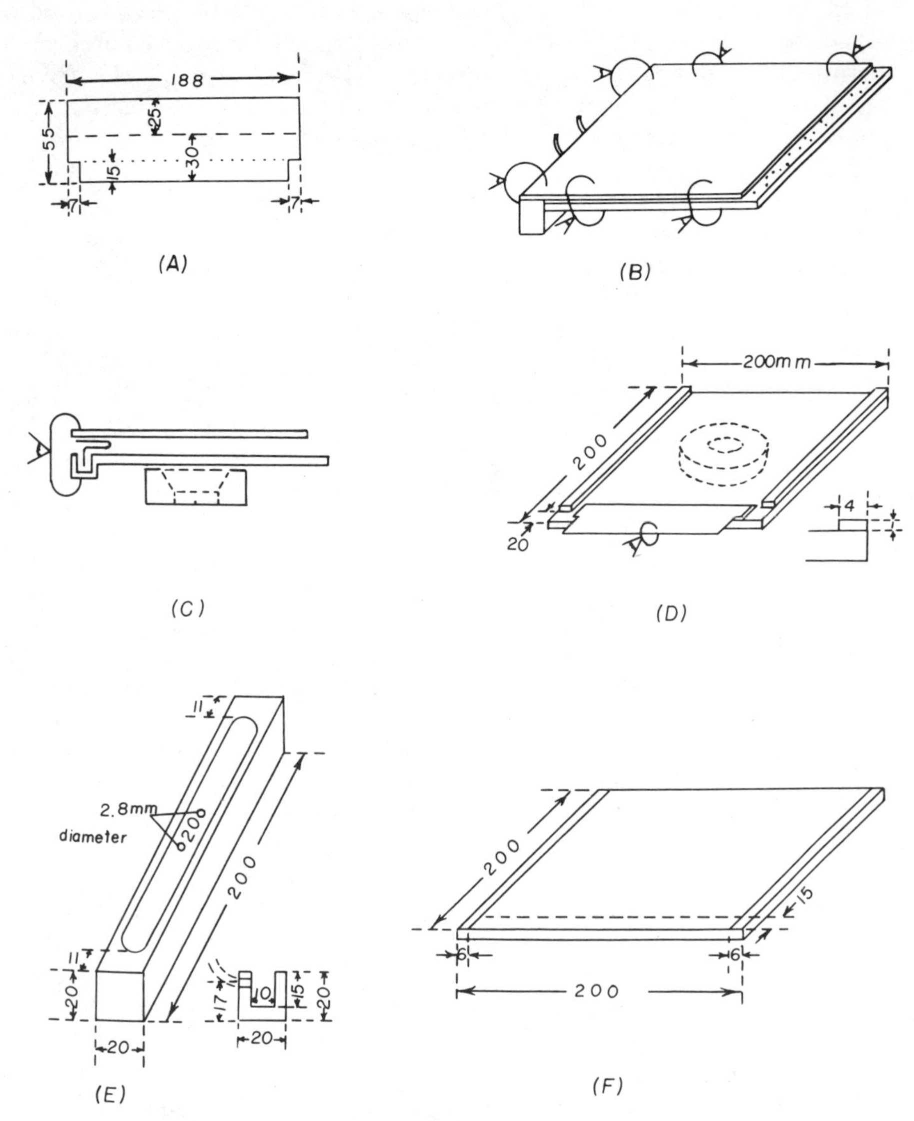

Fig. 5.10 Apparatus for continuous horizontal chromatography. (*A*) Filter paper wick for transferring solvent from the solvent tank (*E*) to the thin-layer plate (*F*), which is shown with the adsorbent removed from two sides preparatory to placing the cover-supporting side strips as in (*D*), where it rests on a cork-supporting ring. (*E*) Stainless steel solvent trough containing two holes for the insertion of polyethylene tubing which provides a means for adding solvent. (*B*) Assembled apparatus. (*C*) Side view with the side supports removed. All measurements are in millimeters. From Brenner and Niederwiesser [83]; reproduced with permission of the authors and Birkhaeuser Verlag.

plate by means of a paper strip from the solvent trough. The thin layer is covered by a second glass plate which rests on narrow strips of glass placed at the sides of the coated plate. One end of the thin layer is exposed to the atmosphere so that as the solvent reaches this point it can evaporate. The assembly is held together with clamps and can be considered to be the forerunner of the present-day "sandwich layers." A modification of this equipment is commercially available from Desaga and the Vario-KS Chamber of CAMAG may also be used for continuous chromatography in the same manner. In the commercial version, provision is made for applying heat to the end of the plate to assist in evaporation of the solvent, and the main portion of the plate may be cooled to prevent condensation of vapors on the cover plate. This arrangement also allows low-temperature chromatography to be carried out. Lees et al. [84] describe a simple setup for carrying out continuous horizontal chromatography. Hara et al. [85, 86] applied the technique of Brenner and Niederwieser to vertical "sandwich layers."

Another method of obtaining continuous flow was used by Procházka [51]. In using thin-layer chromatography without a binder, he placed a mound of adsorbent at the top of the sloping plate, so that as the solvent reached the top it was absorbed. This same principle was applied by Bennett and Heftmann [87] to the ascending chromatography of bound plates. In this case the loose adsorbent was contained in an aluminum pouch which was fastened to the upper part of the plate. A modification of this has been suggested by V. Schwarz (reported by Lábler [88]) in order to increase the capacity of the absorbing system. A thin-layer plate inclined at an angle of approximately 20° makes contact with a descending plate by means of a small mound of adsorbent. The second plate dips into a petri dish filled with more loose adsorbent which can absorb a considerable quantity of solvent.

Johansson and Kashemsanta [89] used a descending preparative method by allowing the eluate to drip from the plate after collecting on the tip of a pointed collector. Samples were applied as streaks.

Van Dijk [90] constructed an apparatus for continuous elution in which the thin layer was coated on a 200-mm length of 6-mm glass capillary. The upper end of the capillary was closed by a piece of fritted glass filter. In operation the sample was placed near the bottom of the layer and development was by the ascending technique. As the solvent reached the top of the layer, it was forced into the capillary by air pressure from a pressure reservoir connected to the developing chamber. The fractions were led to a detecting device or to a microfraction collector. Later [91] the idea was modified by coating the adsorbent on a 230-mm flat disc. The sample was applied as a ring and development took place from the

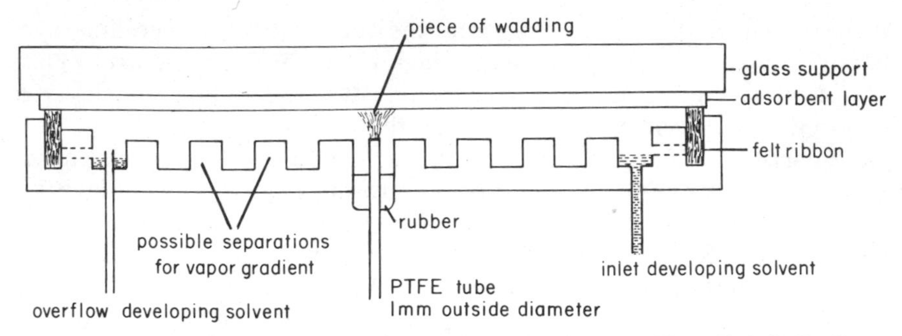

Fig. 5.11 Cross section of apparatus for centripetal development. From V. J. R. De Deyne and A. F. Vetters [92]; reproduced with permission of the authors and the Elsevier Publishing Co.

circumference toward the center (centripetal development). The developing chamber was grooved to permit vapor programming. DeDeyne and Vetters [92] modified the technique so that commercial square plates could be used, and with a different solvent distribution and collecting system the eluent was removed by capillary action and gravity. A cross section of the modified apparatus is shown in Fig. 5.11.

Turina et al. [93] applied two different solvents simultaneously to the sides of a triangular thin-layer plate. The mixture to be separated was placed near the apex of the triangle, and the separated fractions flowed from the base of the plate through paper strips. Sample and solvent feed to the plate were from reservoirs by a similar arrangement. A solution containing ferric chloride and cobalt chloride was separated by this method. Turina and Marjanović-Krajovan [94] also developed a continuous method using a slowly rotating plate covered with adsorbent. The compounds take a spiral path across the plate. Visser [95] published on a rather complicated set up for continuous preparative work. Adsorbent was fed to a continuous belt, the layer was dried, sample was applied continuously, and the developing solvent was applied at one edge of the belt. Development was across the width of the belt. After passing through a drying area, the adsorbent with the separated compounds was scraped from the belt.

Dietrich [96] developed equipment for the continuous fractionation of proteins and mucopolysaccharides by simultaneous two-dimensional gel filtration and electrophoresis.

Multiple Development

The principle and application of multiple development was first propounded by Jeans et al. [97] for application to paper chromatography.

Table 5.3 Number of Solvent Passes Required to Separate Two Solutes 0.1 × Length of Support[a,b]

R_f of Faster-Moving Solute × 100	*R_f of Slower-Moving Solute × 100*								
	Number of Solvent Passes for Required Separation								
	2	3	4	5	6	7	8	9–14	*Impossible*
30	23–21								24–29
29	22–20								23–28
28	21–19	22							23–27
27	20–18	21							22–26
26	19–17	20							21–25
25	18–16	19							20–24
24	17–15	18							19–23
23	16–14	17	18						19–22
22	15–13	16	17						18–21
21	14–12	15	16						17–20
20	13–11	15–14							16–19
19	13–10	14							15–18
18	12–9	13			14				15–17
17	11–8	12		13					14–16
16	10–7	11	12						13–15
15	9–6	10	11						12–14
14	8–5	9	10						11–13
13	7–4	8	9			10			11–12
12	6–3	7	8		9				10–11
11	5–2	6	7	8					9–10
10	4–1	6–5		7					8–9
9	3–1	5–4		6					7–8
8	2–1	4–3		5				6	7
7	1	3–2	4				5		6
6		2–1	3			4			5
5		1	2			3			4
4			1		2			3	
3					1			2	
2								1	

[a] From J. A. Thoma [99]; reproduced with permission of the author and Elsevier Publishing Co.

[b] To determine the number of passes for separation, locate the R_f of the faster-moving component in the left-hand column. In the same row to the right of this figure locate the R_f of the slower-moving component; the number of passes to separate these two components is then found at the head of the column of the slower-moving component.

Mottier and Potterat [74] were the first to apply this technique to thin-layer chromatography.

This technique is carried out by repeated development in the same solvent. That is, the chromatogram is developed in a given solvent, the plate is removed from the chamber, and the solvent is allowed to evaporate; it is then returned to the same solvent and developed a second time. This can be repeated a number of times depending on the separation to be achieved. In effect it lengthens the distance over which the substance must travel. Owing to the difference in R_f values of two given components, and therefore the rates at which they travel with a given solvent, multiple development will tend to separate these components more and more up to a given point, after which they will tend to approach one another. This can be seen from the fact that the solvent reaches the compound of lower R_f value first and starts to move it before the solvent has a chance to reach the second compound. As this process is repeated, the distance over which the lower-R_f-value compound travels before the solvent reaches the higher-R_f component keeps increasing, whereas the distance that the solvent travels after it reaches the second component is continually decreasing. Jeanes et al. [97], Thoma [98, 99], Stárka and Hampl [100], and Lenk [101] have discussed the theoretical aspects of multiple development. Tables 5.3–5.6 taken from Thoma [99] may be used to determine the number of passes required to separate by various degrees two components of given R_f values. To use these tables the degree of separation desired is first ascertained. Then the R_f of the faster-moving component is located in the left-hand column of the appropriate table. In the same row to the right of this figure is then located the R_f of the slower-moving component; the number of passes required to separate these two components will be found at the head of the column in which the slower-moving component is located. Halpaap [102] applied this technique to preparative thin-layer chromatography, and for this purpose gives tables in which the separating distances are greater than those given by Thoma. Petrowitz [102a] also presents tables for this purpose and Goldstein [102b] uses a graph.

For automated multiple development see this chapter under automated thin-layer chromatography.

Stepwise Development

In this technique the chromatogram is developed to different heights with different solvents. In other words, in place of repeated development with the same solvent, different solvents are used and they are developed to different heights. This technique (previously used in paper chromatography) was first applied in 1951 by Miller and Kirchner [103] to the

Table 5.4 Number of Solvent Passes Required to Separate Two Solutes 0.08 × Length of Support[a]

R	R_f *of Slower-Moving Solute* × 100								
R_f *of Faster-Moving Solute* × 100	*Number of Solvent Passes for Required Separation*								
	2	3	4	5	6	7	8	9–14	*Impossible*
30	24–23								25–29
29	23–22	24							25–28
28	22–21	23							24–27
27	21–20	22							23–26
26	20–19	21							22–25
25	19–18	20							21–24
24	18–17	19							20–23
23	17–16	18							19–22
22	17–15		18						19–21
21	16–14		17						18–20
20	15–13	16							17–19
19	14–12	15							16–18
18	13–11	14							15–17
17	12–10	13							14–16
16	11–9	12		13					14–15
15	10–8	11		12					13–14
14	10–9	11				12			13
13	9–8	10			11				12
12	8–7	9		10					11
11	7–6	8		9					10
10	6–5	7	8						9
9	5–4	6	7						8
8	4–3	5	6						7
7	3–2	4	5						6
6	2–1	3	4					5	
5	1	2	3					4	
4		1	2				3		
3			1			2			
2						1			

[a] See footnote *b* to Table 5.3 for directions.

separation of compounds on a "chromatobar." The latter is a modification of the thin-layer principles applied to a column without a containing envelope (a further description of the "chromatobar" is given in Chapter X). In the example cited for the use of the stepwise method of development, the problem was to separate the three terpenes limonene, terpenyl

Table 5.5 Number of Solvent Passes Required to Separate Two Solutes 0.06 × Length of Support[a] [99]

R_f *of Faster-Moving Solute* × 100	*R_f of Slower-Moving Solute* × 100								
	Number of Solvent Passes for Required Separation								
	2	3	4	5	6	7	8	9–14	*Impossible*
30	25	26							27–29
29	24	25							26–28
28	23	24							25–27
27	23–22								24–26
26	22–21								23–25
25	21–20								22–24
24	20–19								21–23
23	19–18								20–22
22	18–17		19						20–21
21	17–16		18						19–20
20	16–15		17						18 19
19	15–14	16							17–18
18	14–13	15							16–17
17	13–12	14							15–16
16	12–11	13							14–15
15	11–10	12							14–13
14	10–9	11				12			13
13	9–8	10			11				12
12	8–7	9		10					11
11	7–6	8		9					10
10	6–5	7	8						9
9	5–4	6	7						8
8	4–3	5	6						7
7	3–2	4	5						6
6	2–1	3	4					5	
5	1	2	3					4	
4		1	2				3		
3			1			2			
2						1			

[a] See footnote *b* to Table 5.3 for directions.

acetate, and α-terpineol. The "chromatobar" was first developed in hexane a distance of two-fifths of the column. The bar was then transferred to the solvent mixture consisting of 15% ethyl acetate in hexane. Development in this solvent was carried out for the full length of the column with the resultant separation of the three components. A good separation was not achievable by a simple development with 15% ethyl acetate in

Table 5.6 Number of Solvent Passes Required to Separate Two Solutes 0.04 × Length of Support[a] [99]

R_f of Faster-Moving Solute × 100	*R_f of Slower-Moving Solute × 100*								
	Number of Solvent Passes for Required Separation								
	2	3	4	5	6	7	8	9–14	*Impossible*
30	27								28–29
29	26								27–28
28	25								26–27
27	24								25–26
26	23								24–25
25	22								23–24
24	21								22–23
23	20								21–22
22	19								20–21
21	18		19						20
20	17		18						19
19	16	17							18
18	15	16							17
17	14	15							16
16	13	14							15
15	12	13							14
14	11	12							13
13	10	11							12
12	9	10							11
11	8	9							10
10	7	8						9	
9	6	7				8			
8	5	6			7				
7	4	5			6				
6	3	4			5				
5	2	3		4					
4	1	2		3					
3		1		2					
2				1					

[a] See footnote *b* to Table 5.3 for directions.

hexane. Later in 1959, Weicker [104] and Stahl [105] applied the stepwise technique to thin-layer plates.

The technique may be applied in various manners. The less polar solvent may be applied first, followed by the more polar solvent, or vice versa. Naturally the stepwise technique is not limited to the use of only two solvents. Kaunitz et al. [106] applied a series of mixtures of methyl alcohol in benzene ranging from 0 to 5%. In this case each succeeding solvent, representing an increase in the percent of methyl alcohol, was allowed to develop a shorter distance than the preceding solvent. The reverse procedure can also be true; that is, the first step may be developed a short distance and the succeeding developments a greater distance [105, 107]. In some cases the technique may be applied where different solvents are allowed to develop the same distance [100, 108, 109].

Weicker [104], Zoellner and Wolfram [110], and Baumann [111] have applied the stepwise technique in a slightly different manner. In this case, the final development was carried out at an angle of 180° to the first development so that the solvent traveled in a direction directly opposite to the previous development. In an example of the application of this method [104] to lipids, the first development was for 30 mm with propanol–ammonia (2:1), and the second development in the same direction was for 100 mm with a chloroform–benzene mixture (3:1). The final development was for 40 mm in carbon tetrachloride in the opposite direction. As with multiple development, the solvent is allowed to evaporate after each run and before placing in the next solvent. In Zoellner and Wolfram's [110] application to the separation of lipids, they spotted the sample in the middle of the plate and then chromatographed the mixture in an ascending fashion twice with a mixture of petroleum ether–benzene–ethyl alcohol–ether mixture (100:20:10:4). The chromatoplate was then rotated through 180° and developed with a mixture of petroleum ether–butyl alcohol–glacial acetic acid (90:6:2).

In both multiple and stepwise development, the procedure involves evaporating the solvent from the plate. Unless this is undertaken in a closed chamber by passing a dry atmosphere over the plate, some deactivation of the plate will occur because of moisture in the air. Some deactivation may also take place because of the type of solvent used. Loose-layer chromatography has an advantage in this respect; the unused portion of the plate after an initial development may easily be removed and replaced with fresh adsorbent [112, 113]. Conversely, sections of the loose-layer chromatogram which contain certain components may easily be removed and placed upon a new plate, the balance of the plate then being coated with fresh adsorbent. However, the adsorbent containing the desired components may also be removed from bound layers and applied to new coated plates [114–116]. The adsorbent containing the spot is removed from the first plate and pressed into a cleared area on the fresh plate. Iacono and Ishikawa [117] applied the material from the first plate as a slurry to form a smooth low mound on the surface of the second layer.

Polyzonal Development

Furukawa [118] first demonstrated the existence of a second solvent front while developing thin layers in a two-component solvent system. This occurs because the adsorption affinity is greater for one of the solvents and the adsorbent selectively removes the more polar solvent. Furukawa observed that compounds whose R_f values occurred in the first or upper solvent zone began to move on the layer as soon as the solvent front reached the sample spot. On the other hand, the compounds whose R_f values lay in the second or lower zone did not move until the second solvent front reached the sample spot. With multicomponent systems, additional solvent fronts are formed. This principle was utilized by Tiselius [119] in his frontal analysis in column chromatography.

Several workers [120–123] have investigated this phenomenon. Frache and Dadone [122] used thin-layer plates in which the glass support was composed of 16 pieces, 1 × 15 cm, held together on a stainless steel plate. After completion of a chromatogram, the individual strips were quickly transferred to glass-stoppered test tubes and the solvent in the layer subsequently analyzed by gas–chromatography. Viricel et al. [121] investigated the interplay of saturation time, unsaturation, demixing, and type of tank on the resolution of a group of several organophosphorus compounds.

Niederwieser and Brenner [120] have incorporated this "demixing" phenomenon into a separation technique designated polyzonal development. Basically, this consists in applying the sample in a series of spots on a diagonal line across the thin-layer plate. In a multicomponent solvent the spots are consecutively treated to a series of different solvent fronts, and the number and duration of each solvent zone depends on the location of the spot with respect to the solvent immersion point. Essentially this gives the effect on a single plate of trying many different solvent systems.

Using the two-component system as the simplest case of "demixing" as illustrated by Niederwieser and Brenner, a number of situations can occur. In the first schematic example [Fig. 5.12] all the compounds move in the first or α zone. Here a better separation could be achieved in the entirely nonpolar solvent; the polar solvent contributes nothing to the separation. In the second case [Fig. 5.13] all the compounds travel in the second or β zone and in addition the component which travels with the β front also travels in the α zone. Increased separation of this latter component could be obtained by a stepwise development, first in the nonpolar solvent and then in the solvent mixture. Figure 5.14 shows the case where compounds are present which travel in both the α and β

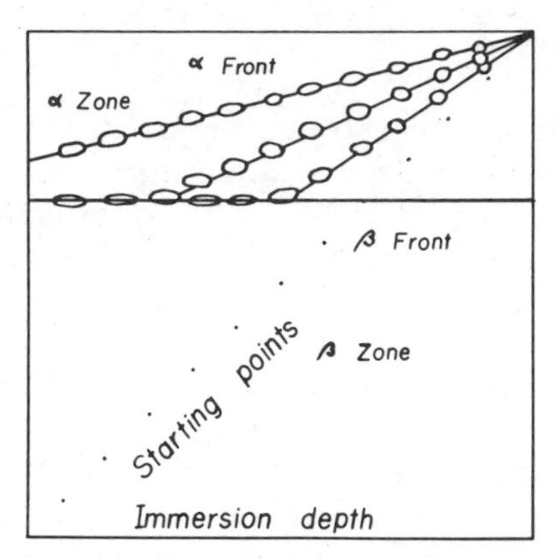

Fig. 5.12 Polyzonal development. Situation in which all substances travel in the α zone. The more polar solvent contributes nothing to the separation. The nonpolar solvent alone will produce a better chromatogram. From Niederwieser and Brenner [120]; reproduced with permission of the authors and Birkhaeuser Verlag.

zones and Fig. 5.15 diagrams the situation where all of the three described separation possibilities exist at one time. Here the optimum separation point (designated by an arrow) is a function of φ which the authors define as

$$\varphi = \frac{\text{Distance from immersion point to spot application point}}{\text{Distance from immersion point to } \alpha \text{ front}}$$

When compound mixtures travel with the β solvent front the separations

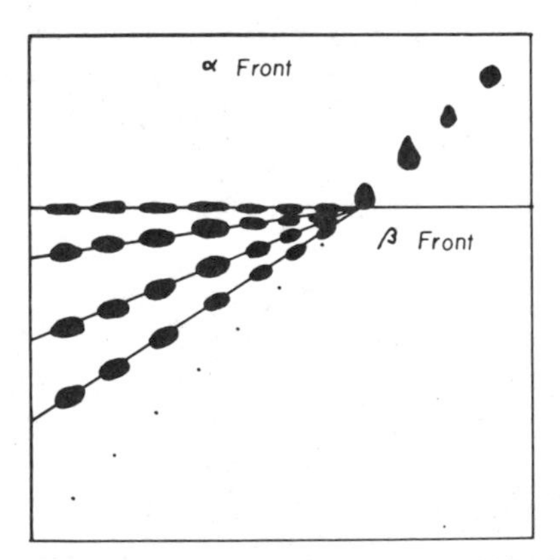

Fig. 5.13 Polyzonal development where all substances travel in the β zone and not in the α zone. They are more polar than the components in Fig. 5.12. From Niederwieser and Brenner [120]; reproduced with permission of the authors and Birkhaeuser Verlag.

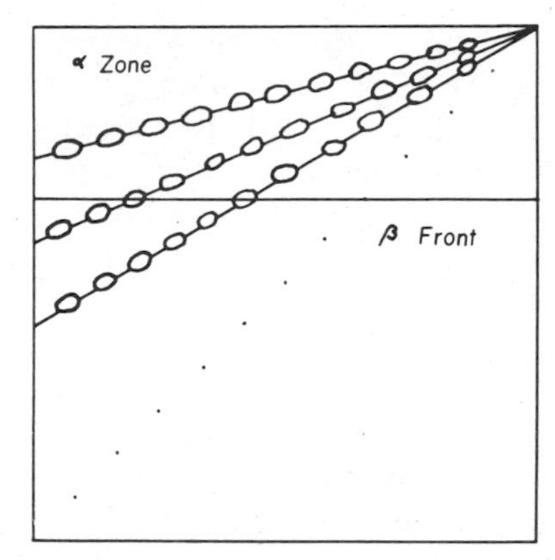

Fig. 5.14 Polyzonal development where the compounds move in both the α and β zones. From Niederwieser and Brenner [120]; reproduced with permission of the authors and Birkhaeuser Verlag.

can be achieved by using less polar solvents and also by using more volatile but polar components (e.g., ether, methylene chloride) in a saturated, large-volume chamber which tends to counteract demixing.

Two-Dimensional Chromatography

This variation in development, which has been used a great deal in paper chromatography, is essentially the application of multiple development or stepwise development in a two-dimensional field. However,

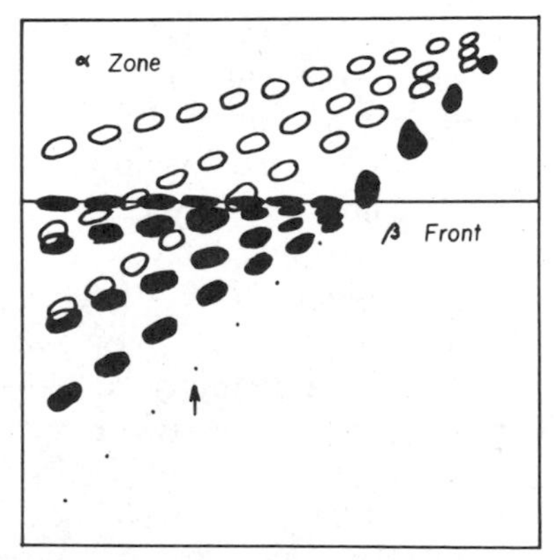

Fig. 5.15 Polyzonal development showing a combination of the situation in Figs. 5.13 and 5.14. The starting point for optimum separation is indicated by an arrow. From Niederwieser and Brenner [120]; reproduced with permission of the authors and Birkhaeuser Verlag.

the two-dimensional technique is somewhat more versatile than either of the other two methods. The first application of the method to thin-layer work was made by Kirchner et al. [57]. This technique is carried out by spotting the sample in one corner of a square plate and developing in the usual manner. The chromatogram is removed from the developing chamber so that the solvent can be evaporated from the layer, and it is then placed in a second solvent so that the development proceeds in a direction perpendicular to that of the first development. Care must be taken in spotting the sample so that the line of developed spots, after the first separation, will not be below the solvent level when the plate is turned at right angles. The versatility of this system lies in the ability to modify the layer prior to the second development. In working with isomeric monounsaturated fatty acids, Bergel'son et al. [124] ran the first development on silica gel impregnated with dodecane and then impregnated the plates with silver nitrate before developing in the second direction. Johansson and Rymo [125] combined thin-layer filtration on Sephadex in one direction and thin-layer electrophoresis in the other direction. Vergani et al. [126] using this technique found that some large-molecule proteins were delayed in the electrophoretic migration. Therefore, after running the gel filtration on Sephadex G 200, they transferred by imprinting on a cellulose acetate membrane, and then ran the electrophoresis on the membrane perpendicular to the gel filtration. Where gel filtration is used in combination with adsorption chromatography, Calderon and Baumann [127] recommend running the adsorption chromatography prior to the gel filtration, because of the higher capacity of the former. Kaufmann and Makus [4, 128] have used a combination of chromatography and reverse-phase partition in the field of lipids. Normal chromatography was carried out in the first dimension on silica gel, and then after evaporation of the solvent the remainder of the plate was impregnated with paraffin oil or undecane dissolved in petroleum ether. Care was exercised to prevent the impregnating solution from washing over the spots along the one edge of the plate. After removal of the excess solvent the plates were then developed across the impregnated layer. Honegger [129] applied electrophoresis in one direction on buffered adsorbents followed by chromatography in the second dimension. Raymond and co-workers [130, 131] worked out a two-dimensional thin-layer electrophoresis method in which the gel concentration differs in the two directions. First the electrophoresis is run in one concentration, then a narrow slice of the gel is removed and placed on a new plate, after which the balance of the plate is filled with a gel of a different concentration. When the gel solidifies the electrophoresis can be run in the second direction. Jimeno de Osso

[132] demonstrated a three-dimensional development and Pollack et al. [133] a four-dimensional development on a single plate (see Fig. 23.2*a*, *b*, and *c*). Zappi et al. [134] spotted extracts of thyroid hormones, which also contained interfering lipids, 4 cm from the top of a cellulose layer. A chloroform development then carried the lipids to the top of the plate without moving the iodoamino acids. The impurities were scraped off and the plate turned 180° and developed in acetone–0.1 *N* acetic acid (2:8). Anet [135] ran two-dimensional separations with one adsorbent in one direction and a different adsorbent in a second direction by first coating the plate with silica gel. After development in the first direction the excess adsorbent was scraped off the plate, leaving a strip on one side which contained the spots of the first separation; the remainder of the plate was then coated with aluminum oxide. The second layer was air-dried prior to developing in the second direction. Kirchner and Flanagan [136, 137] also demonstrated the use of multiple layers for two-dimensional chromatography. In this case both adsorbents were applied to the plate before spotting the sample. Using depth guides, a strip of adsorbent was laid along one edge of the plate, and after drying, the balance of the plate was covered with the second adsorbent. A portion of the second adsorbent was scraped away along one edge after drying so that it would not dip into the first solvent. This was done in order to avoid diffusion effects because the solvent traveled faster in one adsorbent than in the other. A combination charcoal and silicic acid plate demonstrates the usefulness of this procedure. Compounds were spotted on the narrow strip of charcoal and developed with a mixture of benzene–ether–acetic acid (82:9:9) along the charcoal layer. After air-drying to remove solvents, the plate was developed in 15% ether in benzene in order to move the compounds from the charcoal out onto the silicic acid, so that they could be seen (Fig. 5.16). The method was also demonstrated with a combination magnesium silicate and silicic acid layer in chromatographing bergamot oil (Fig. 5.17). Berger and co-workers [138–140] have published a description of an apparatus for simultaneously coating a plate with two adjacent layers of different adsorbents. This was accomplished by placing a plastic insert into a commercial spreader, thus forming two independent chambers. In their case they used multiple layers for one-dimensional chromatography to remove an interfering component, as well as for two-dimensional work. When using multiple layers for two-dimensional work, it is advisable to scratch a line between the two layers prior to the first development. This prevents pulling spots to one side, if the rate of solvent movement is different in the two layers or if solvent has been prevented from entering the second layer during the

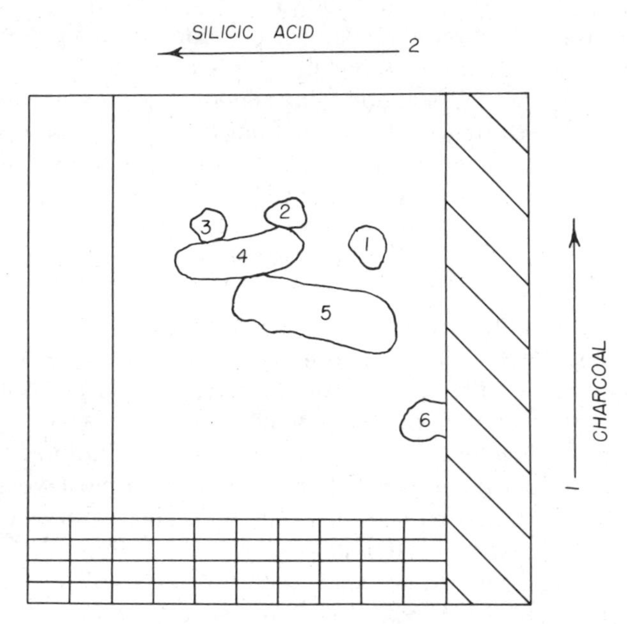

Fig. 5.16 Two-dimensional separation of some ketones on a dual-layer chromatoplate. First development with benzene–ether–acetic acid (82:9:9). Second development with benzene–ether (85:15). Visualizing agent, 2,4-dinitrophenylhydrazine in 2 *N* hydrochloric acid; diagonal hatching, charcoal with 10% starch binder; crosshatching, adsorbent-free area; balance of plate, silicic acid with 2½% starch binder; development distance, 10 cm. (1) Angelica lactone; (2) acetophenone; (3) 7-tridecanone; (4) bromoacetophenone; (5) 2-methylcyclohexanone; (6) 2-hydroxyacetophenone. Data from Kirchner and Flanagan [136].

first development by removing adsorbent from the bottom of the plate [136]. After the first development the scratch may be filled by pressing dry adsorbent into it.

Multiple layers may also be used in one-dimensional chromatography. In this way they have been used for pesticide cleanup work [141–143]. Váradi [144] has used a multiple layer by converting a strongly acidic cation-exchange resin plate partly to the Na^+ and partly to the Li^+ form.

Of course stepwise and multiple development may also be combined with two-dimensional development. In this way Sable-Amplis et al. [145] eliminated noncorticoid substances from plasma extracts by spotting the samples across the center of the plate. Two successive developments with (*a*) *n*-hexane–ethyl acetate (4:1) and (*b*) ethyl acetate–cyclohexane–toluene (10:10:1) eliminated the noncorticoid material which was then scraped from the plate. After 180° rotation, a paper wick carried the developing solvent across the scraped area.

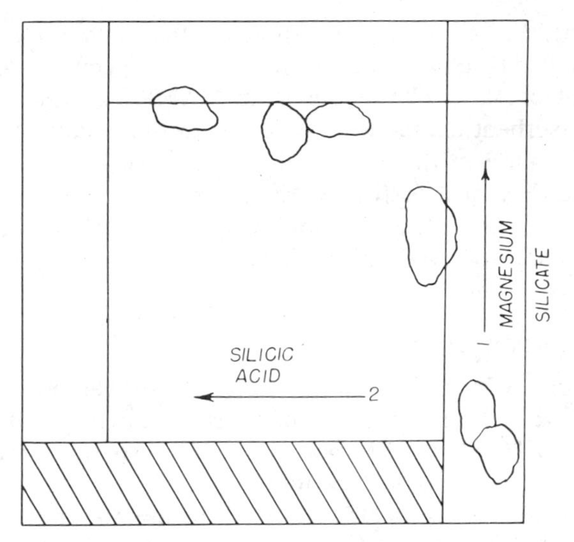

Fig. 5.17 Two-dimensional separation of bergamot oil on a dual-layer plate of magnesium silicate and silicic acid, both adsorbents bound with $2\frac{1}{2}\%$ starch. Solvent, benzene in both directions; crosshatched area, plate cleared of adsorbent. From Kirchner and Flanagan [136].

Two-dimensional development may also be carried out by transferring the spots from one plate to another as described under stepwise development. Another method has been demonstrated by Pumphrey [146]. In this case the development in the first direction was carried out on a narrow 2-cm strip. After drying it was clipped face down onto a 20 × 20 cm plate for development in the second direction. Close contact must be maintained between the two layers for proper development. Bond [147] used the same principle with plastic-backed sheets.

High-Performance Thin-Layer Chromatography (HPTLC)

This technique is based on the use of a very fine grain (60 Å) and a very uniform particle size of silica gel. This has better resolving and optical properties than the silica gels that have been in current use [72], but thin layers made from this product have a lower flow rate, and as a consequence smaller plates are used so as to decrease the total development time. The smaller development distance is more than compensated for by the increased resolution which is obtained. In this technique smaller sample quantities must be used, because the plate becomes overloaded and no separation is achieved with more than 10 μg per sample spot. In order to keep the diameter of the applied spot to a minimum (<2 mm), the sample may be applied either with a one μl Hamilton syringe controlled by a micrometer screw or by a special platinum-iridium capillary (Antech; also obtainable from CAMAG). Other than use of the special adsorbent and careful attention to the details just mentioned the other procedures of thin-layer chromatographer may be applied with this technique. For quantitative work where the compounds are visualized with a spray reagent, an automatic sprayer is used (Anton Paar) to eliminate variations obtained in hand sprayed applications [147a]. A book has been published on this technique [71].

Centrifugal Chromatography

This technique, which involves the use of centrifugal force to accelerate the flow of solvent through the chromatogram, was first applied to paper chromatography [148, 149]. Herndon et al. [150] first mentioned its use with adsorbents, such as silicic acid, starch, and alumina, on paper supports. Korzun and Brody [151] applied the method to thin-layer chromatography, where the layers of plaster-of-Paris-bound aluminum oxide or silica gel were applied to circular glass or aluminum plates having a hole in the center to fit on the centrifuge. The samples were spotted 2.5 cm from the center hole and the solvent was set to permit a constant flow without overloading, with the centrifuge rotating at 500 to 700 rpm. The acceleration decreased the developing time from 35 min to approximately 10 min. Rosmus et al. [152] have also demonstrated the method with dyes and 2,4-dinitrophenylhydrazones. Affonso [153] used plaster of Paris plates and Lepoivre [154] using various adsorbents increased the size of the layers to 40 cm in diameter. Deyl et al. [155] reviewed centrifugal chromatography.

Wedge-Shaped Chromatography

Marchal and Mittwer [156, 157] conceived the idea of modifying paper chromatography so that the sample first traveled along a narrow strip of paper and then was allowed to expand into a wider section. By doing this the spots were transformed into narrow bands, thus enhancing the separation. This effect is similar to that produced by radial chromatography.

The first application of this principle to thin-layer chromatography was by Mottier [73, 158]. Furukawa [160] examined numerous shapes of strips to determine their effect on separation. Hausser [161], Prey et al. [162], and Bayzer [163] used a simple wedge-shaped strip to obtain the same type of results. These strips are diagramed in Fig. 5.18.

Abbott and Thomson [164, 165] have prepared a somewhat different wedge strip or layer. They prepared plates with a layer of varying thick-

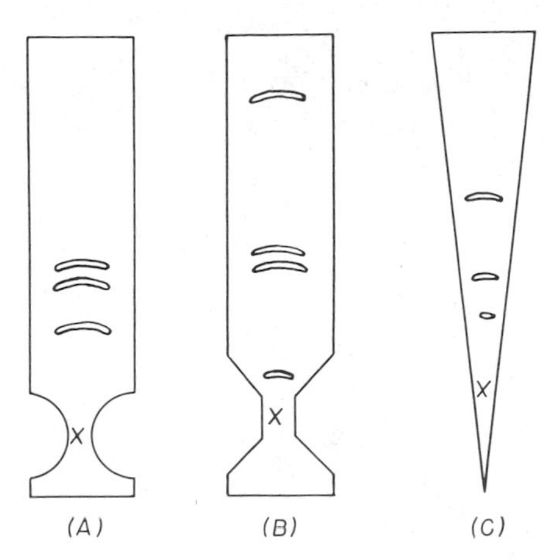

Fig. 5.18 Various forms of "wedge-strip" chromatography. (*A*) According to Mottier [158]; (*B*) according to Peereboom and Beekes [159]; (*C*) according to Hausser [161].

ness so that they have a wedge shape in thickness instead of breadth. The thick end of the wedge is used for the application of the sample, and in their case it was useful for the "cleanup" of pesticide residues. Bazán and Joel [166] have used the same technique for lipid work.

Gradient Techniques

There are a number of gradient techniques that may be employed in thin-layer work. Perhaps the best known is that of gradient elution, because of the carry-over from column work.

In gradient elution the composition of the solvent may be changed either continuously or in steps. Wieland and Determann [167] have designed a special tank for gradient elution work. This chamber, which permits a continual change in the solvent makeup, is diagramed in Fig. 5.19. A magnetic stirrer keeps the solution homogeneous, and a filter paper strip fastened by a rubber band to the lower end of the immersed plate prevents a mechanical dissolution of the layer during mixing. Solvent is added to the chamber by means of a buret or a measuring pump; an overflow outlet keeps the volume constant. In this way various effects can be obtained, such as increasing the strength of a buffer solution [167, 168], increasing the polarity of the solvent [169–171], varying the pH of the solution, or changing any other system in which it is desired to introduce a gradient in the solvent.

Luzzatto and Okoye [172] and Strickland [173] used a wick to feed the changing gradient solvent to the thin-layer plate arranged as for descend-

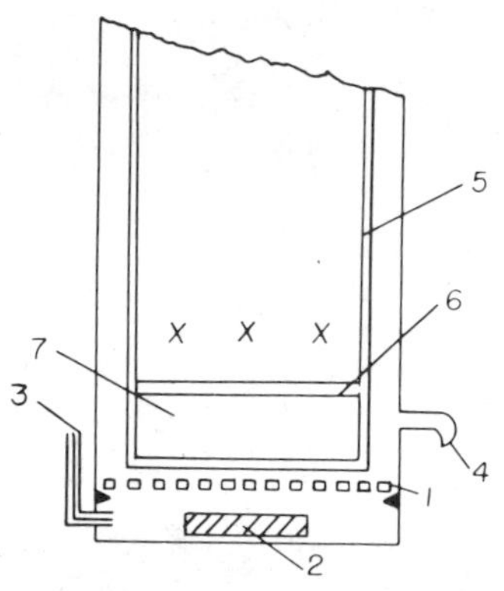

Fig. 5.19 Cross section of lower part of gradient elution tank for thin-layer chromatography: (1) Filter plate, (2) stirring magnet, (3) solvent inlet tube, (4) overflow tube, (5) thin-layer plate 5 × 20 cm, (6) rubber band, (7) paper strip 1.5 × 5 cm to prevent disintegration of lower part of layer. From Wieland and Determann [167]; reproduced with permission of the authors and Birkhaeuser Verlag.

ing or horizontal development, respectively. The gradient solvent was mixed in the solvent supply vessel using magnetic stirrers. A number of devices have been conceived for generating solvent gradients, those of Hegenauer et al. [174], Peterson and Rowland [175], Peterson and Sober [176], and Niederwieser and Honegger [177], to mention a few. Niederwieser and Honegger [123, 177] show several devices for feeding the gradient solvent to the thin-layer plate, especially when using the Brenner-Niederwieser continuous developing apparatus (B–N chamber) or a sandwich layer.

Randerath and Randerath [178] carried out gradient elution by transferring the thin-layer plate from tank to tank containing different solvent proportions without intermediate drying of the plate.

Tarr [179] has developed a simple and unique gradient elution method although it has certain limits as to flexibility. The method consists in placing a trough filled with a relatively polar solvent in a regular tank which is saturated with the vapor of a volatile nonpolar solvent. The tank is also lined with filter paper also saturated with the nonpolar solvent. The TLC plate is placed in the polar solvent trough and allowed to develop. The polar solvent evaporates as it moves up the plate and is gradually replaced by the nonpolar solvent, thus creating a solvent gradient. Tarr tabulates a series of polar solvents suitable for the method. This is similar to the gradient flux system mentioned by Niederwieser [180] at about the same time. In the latter case the mixed solvent is placed in the trough of a vertical sandwich layer system. The back of the thin-layer plate is heated by a heating plate and the cover plate condenses the evaporated solvent, which runs down and returns to the trough. The cover plate may also be cooled with a cooling plate. The compounds concentrate in horizontal lines and their location is independent of the location and area of the sample spot. Here again the more polar compound should have the highest vapor pressure and the nonpolar compound the lowest vapor pressure. Turina et al. [181] also applied heat to the thin-layer plate in their hot plate chromatographic method. In this case a normal tank was used. In cases where a higher resolution was needed for trace components (where the ratio of trace components to the other components was below $1:10^{-5}$) a special plate was interposed between the hot plate and the chromatographic plate, and the temperature of the hot plate was increased [182]. The special plate was composed of an asbestos sheet adjacent to a metal sheet. The high-resolution zone was in the region where the metal and the asbestos joined.

Gradient layers were mentioned under preparation of the layers, and may consist in a blend of two different adsorbents ranging from 100% A on one edge of the plate to 100% B on the opposite edge. The gradient on the layer may be one of activity of a single adsorbent by gradation of the amount of moisture present [183, 184], of impregnation, of pH, or of a complexing agent. This type of gradient is useful for finding the most effective adsorbent for separating a given mixture.

In 1965 Dallas [185] investigated the effect of the preadsorption of solvent vapors from the tank on the R_f values, and Prinzler and Tauchmann [186] called attention to the fact that the change in the stationary phase, because of the adsorption of solvent vapors, should be taken into account. De Zeeuw [187, 188] and Van Dijk et al. [189] investigated the effect of pretreating the adsorbent with various solvent vapors, and de Zeeuw developed equipment and techniques for vapor-programed thin-layer chromatography in which the adsorbent layer is exposed in sections to various solvent vapors so as to form a gradient layer [190, 191]. Equipment for vapor programming is available from Desaga and CAMAG.

Another form of gradient is that of thickness of the layer forming a wedge-shaped layer. Such a gradient may be used to clean up samples [164, 165] or to apply more sample at the origin and permits the detection of trace amounts of components in relatively large amounts of extract [166]. Abbott and Thomson [165] used a modified spreader for preparing wedge-shaped layers or they may be prepared by using the Chromflex TLC Gradient Plate (Kontes). The latter is a wedge-shaped glass plate with side guides for spreading the slurry.

A still further type of gradient is that of temperature, either an increasing or a decreasing temperature during development [192–197]. Kaufmann et al. [197] were able to separate critical pairs of lipids within individual classes by applying temperature programming and low-temperature thin-layer chromatography in reverse-phase systems. Blasius et al. [198] designed equipment for high-voltage ionophoresis in a temperature gradient. Peltier elements were used for cooling and setting the temperature gradients. Improved separations were obtained over those obtained at constant temperature.

Pore gradient polyacrylamide gels have been prepared for electrophoretic molecular sieving [199–202]. Margolis and Wrigley [203] prepared a pore gradient acrylamide gel in which the degree of cross-linking was increased at high gel concentrations. This helped to eliminate the effect of charge differences in determining molecular weights.

Partition Chromatography

Partition chromatography has become best known through its application in paper partition chromatography introduced by Martin and Synge [204]. In partition chromatography a liquid phase is absorbed on the support so that during the development the materials being separated are partitioned between the liquid phase on the support and the moving solvent. In order not to change the stationary phase, the developing solvent is usually saturated with that phase.

Aside from the thin layers of cellulose, to which the same solvents and procedures that have been used in paper chromatography can be applied, other thin-layer materials have been used as supports for thin-layer partition chromatography. Silica gel layers have been used [205–211], and kieselguhr, because of its low adsorbtive strength, has also been used as a support for partition chromatography [205, 212–215].

If the support is saturated with a nonpolar solvent, then the separation is called a reverse-phase partition. Details on the impregnation of the layers have been given in Chapter III, Section 7. Besides the usual cellulose, silica gel, and kieselguhr that have been used as a support for thin-layer reverse-phase chromatography, alumina [216], gypsum [217, 218], starch [219, 220], and zinc carbonate [43] have also been used as supports.

Thin-Layer Gel Chromatography

In gel filtration, molecules are separated according to their size. Most commonly used for this type of work are the dextran gels (Sephadex) with various degrees of cross-linking present. The size of the pores in the gel depends on the amount of cross-linking. (For preparation of layers see Chapter III, Section 6.) In more recent times Hjertén [221] developed a cross-linked polyacrylamide gel for gel filtration, and a graded series of polyacrylamide beads with different porosities for thin-layer gel filtration (TLG) is available from Bio-Rad Laboratories. Molecular sieve effects (gel filtration) can also occur with silica gel layers, especially when the adsorbent is prewetted with the developing solvent [222–225].

After application of the sample, which must be accomplished with the plate in a horizontal position, development is carried out in a descending fashion with the plate at an angle of 10 to 20° from the horizontal. The rate of solvent flow can be controlled by adjusting the angle. A sandwich layer or a special apparatus for supporting the plate (Pharmacia) may be used for this work. Jaworek [226] and James et al. [227] give details of apparatus for gel filtration. Equipment for measuring optical density of separated fractions is also described [227]. Special precautions were taken to produce uniform layers for the quantitative work. A horizontal plate may also be used, and in this case the solvent flow is regulated by adjusting the height of the solvent feed. Circular technique can be used and a time advantage has been claimed over normal thin-layer chromatography [228].

Water solutions are used for development; however, methylated Sephadex has been developed for use in polar organic solvents [229, 230]. Slagt and Sipman [231] applied the method to thin-layer gel filtration. Methylation of the gel was carried out with dimethyl sulfate in aqueous sodium hydroxide. The Pharmacia apparatus could not be used with organic solvents because it is constructed of plastic.

Because Sephadex gel layers tend to loosen during the staining procedure Pomeranz [232] obtained greater stability by carefully spraying the layer after development with a 1.5% aqueous suspension of agar gel which had been warmed and then cooled to 55°C.

Mention has already been made (Chapter III, Section 6) of the use of gel filtration for the determination of molecular weights. Thin-layer gel filtration has the advantage that very small samples, even smaller than 0.1 mg, can be used, and though not as precise as molecular weight determination by column filtration, the method has comparable accuracy [233]. Anomalous molecular weights occur in gel chromatography if the sample proteins differ significantly in shape from the proteins used for calibration. These anomalies disappear if the proteins are dissolved in 5 to 8 *M* guanidine hydrochloride [234], because they adopt a random coil configuration. Klaus et al. [235] and Heinz and Prosch [236] have applied this technique to the determination of molecular weights of polypeptide chains using thin-layer gel filtration. Urea and sodium dodecyl sulfate solutions have been used for the same purpose [237].

Some additional representative examples of gel filtration are enzyme

[238–240], dyes [241, 242], globulins [243, 244], and proteins [245–248]. Otocka [249] discussed the theoretical aspects of molecular weight distributions of polymer fractions on silica gel. General information on gel filtration can also be found in Determann [250], Wieland and Determann [251], and Williams [252].

Thin-Layer Electrophoresis

The concept of thin-layer electrophoresis was introduced by Consden et al. [253] in 1946. They used 1.4-mm-thick layers of agar or silica gel to separate amino acids and peptides.

Wieme and Rabaey [254] were the next to use this technique. They used a 2-mm-thick layer of agar gel on a glass plate. Wieme [255], Baur [256, 257], and Ramsey [258] have discussed the advantages of thin-layer electrophoresis over the normal electrophoresis techniques. These advantages include (1) the ability to provide a more efficient cooling system and thus decrease the temperature effects, (2) greater sensitivity of detection for some things than in the normal thin-layer electrophoresis, and (3) elimination of the need to slice the gel after the electrophoresis. Kowalczyk, in a series of papers [259–268], has discussed the various factors that can change during electrophoresis and can thus affect the course of the electrophoresis, such as changes of pH, temperature, or concentration, and has discussed means of minimizing these changes.

Besides the usual glass base, a number of other substances have been used to advantage as a support for the thin layer. Cellulose sheeting has been used and gives adequate support to the gel layer during handling and staining [269, 270]. Baur [256, 257] has found cellulose casing 80F (Union Carbide, Film Packaging Division) to be ideal for this purpose. Eriksen [271] has used plastic troughs for his thin-layer electrophoresis, and Weiner and Zak [272] have used Teflon-covered glass paper as a support. The latter also used quartz slides as supports for the layers when uv densitometry of the electropherograms was planned. Ground-glass plates have been used [273, 274] because they provide better adhesion of the gel so that there is a minimum destruction of the gel during staining and washing. Zak et al. [275] pointed out that the underlying base is an overlooked variable in gel electrophoresis. Although vegetable parchment paper and Teflon-coated paper were good bases for serum protein and nucleobase separations, respectively, they gave poor resolution when agarose gel was underlaid in the attempt to separate adenine nucleotides.

Agar layers have been used for the electrophoresis of proteins [254, 255, 276–286], mucopolysaccharides [287], organophosphorous pesticides [288], hemoglobin [289–291], nucleic acid constituents [292], porphyrin pigments [271], inorganic ions [293], and viruses [294]. In some cases sulfate-free agar (agarose) has been used to avoid interference by the sulfate ion [287, 294]. Methods for separating agar and agarose and preparing granulated beads have been published [295–297], and agarose is available commercially from Bio-Rad Laboratories and from Pharmacia. Porath et al. [298] and Lääs [299] have shown that both agarose and agar can be greatly improved by cross-linking using epichlorohydrin and still further improved by reduction of the cross-linked product. Their resulting products showed decreased sulfur content and adsorptive capacity over the commercial products. Khramov and Galaev [300] purified agar gel for microelectrophoresis by treating with EDTA. Russell et al. [294] used a series of thin layers of agarose gel of increasing concentrations to separate viruses by particle size classification; however, it was found that electrophoretic mobility also had to be considered.

Agarose has been used for the microelectrophoresis of RNA [301] and for microimmunoelectrophoresis [302, 303]. Kostner and Holasek [304] increased the sensitivity of electroimmunodiffusion and two-dimensional immunoelectrophoresis by incorporating 3 to 5% dextran T70 or polyethylene glycol in the agarose gel. Laurell [305] used agarose gel containing antibodies for the quantitative estimation of proteins. Hazama and Uchimura [306] used a mixture of agarose and acrylamide gel for the ultramicroelectrophoresis of 1 pg of protein in 1 pl of human serum.

The agar layers can be prepared from a 1% solution of agar in an appropriate buffer; uniform layers can be prepared by pouring a given quantity onto the supporting plate containing in a suitable vessel. After gelling, the excess agar can be cut away from the edges of the plate.

Starch gel thin layers have been used for the separation of enzymes [258, 307–309], hemoglobin [257, 270, 310–314], and proteins [270, 314–322].

For the preparation of starch gel layers, 10 g of hydrolyzed starch (Connaught Medical Research Laboratories) is heated with 100 ml of an appropriate buffer over a water bath with constant stirring, and as with the agar solution, the hot solution must be deaerated under reduced pressure. A measured quantity of a starch solution is then poured onto the supporting plate to give a uniform layer, or as in some cases, guides may be used to give the proper depth of gel. Baur [257] presses a glass plate firmly on top of the starch to provide a uniform layer; metal discs at the sides of the glass plate give the proper spacing. After the gel is thoroughly cooled the glass cover plate is removed and the excess gel on the edges can be trimmed off. Daams [316] uses a similar technique in the preparation of micro starch gel plates, except that he covers the upper plate with a piece of wet cellophane. After the gel has set the glass cover plate is carefully removed leaving the cellophane in place; this is subsequently withdrawn from the gel.

In the matter of application of the sample to either agar or starch gel layers, several techniques are available. One method is to simply introduce the solution by means of a capillary pipet into a slit in the gel. Ramsey [258] removes a narrow strip of gel 0.5 mm wide by means of a special cutter made from two razor blades. After the sample is introduced into the slot, the latter is sealed with melted petrolatum which is just warm enough to flow freely. Korngold [321] punches small holes into the gel with a pipet and removes the gel by suction. Samples can also be applied with filter paper strips which are placed on the gel surface [257], or they may be inserted in slits in the gel or in preformed slots prepared during the casting of the gel. Wieme [285] and Wieme and Rabaey [254] have also demonstrated in working with biological material that it is possible to use a fragment of tissue which is placed directly on the gel surface.

In general electrophoresis on thin layers can be carried out in most of the equipment that is available for paper electrophoresis. A general diagram for the setup is shown in Fig. 5.20. The setup for electrophoresis on thin layers of agar or starch gel has usually included a cell filled with buffered gel interspersed between the electrode chamber and the thin layer, but with other types of thin-layer materials this connecting cell is usually dispensed with. For connecting the thin layers with the electrode cell, filter paper or sponges saturated with buffer solution can be used. Burns and Turner [322a] used strips of Miracloth (Calbiochem) encased in strips of washed dialysis tubing folded so that the Miracloth did not come into direct contact with the thin layer. Wieme placed his agar-covered slides face down so that the agar layer contacted the agar cells which in turn were in contact with the electrolyte in the electrode cells.

Numerous designs have been published for constructing electrophoresis equipment and only representative examples can be cited. Either vertical [323–326] or horizontal [327–330] equipment may be used, and there is considerable variation in design. In general the horizontal apparatus is simpler to construct.

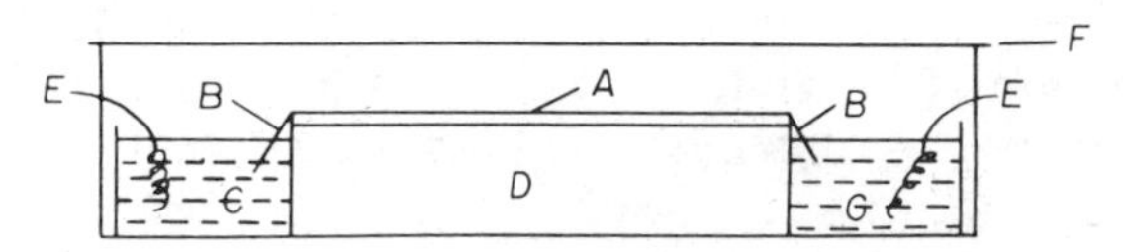

Fig. 5.20 General diagram of the setup for thin-layer electrophoresis. (*A*) thin-layer plate, (*B*) filter paper to connect layer to electrode cell, (*C*) electrode cell, (*D*) support, (*E*) electrodes, (*F*) cover plate.

In order to keep the temperature from rising during the electrophoresis run, the thin-layer plate may be submersed in petroleum ether. Honegger [129] cooled his thin layers for ionophoresis by placing them on a water-cooled surface, and this is the most commonly used method for cooling during an electrophoresis run. Commercial equipment using this procedure is available from CAMAG, Pharmacia, K. Marggraf, Research Specialties, Gelman, Desaga, Shandon, Turner, Analtech, E-C Apparatus, and Warner-Chilcott. The latter apparatus is shown in Fig. 5.21.

Raymond [322] has designed an electrophoresis apparatus that can be operated in either a vertical or a horizontal position. The vertical position can be used only with a gel that is impermeable to the mass flow of liquid in order to keep the upper electrode cell from draining into the lower cell. This apparatus provides for a 3-mm-thick layer of gel which is formed between two parallel water-cooled plates that may be taken apart after electrophoresis in order to remove the gel. With the double cooling setup the equipment can be operated above 500 V at 200 mA without exceeding the cooling capacity. The apparatus is commercially available from the E-C Apparatus Corp.

Because the resistance in an electrophoresis system changes during the run Schaffer and Johnson [331] designed a power supply that would supply constant power as this resistance changed.

Thin layers of silica gel for electrophoresis work can be made in the

Fig. 5.21 Warner-Chilcott water-cooled electrophoresis apparatus for thin-layer plates. Courtesy of Warner-Chilcott Laboratories.

usual manner, using either water or a satisfactory buffer solution for slurrying the silica gel. After drying, the layers may be saturated with water if they have been previously formed with a buffer solution or with a buffer solution if they have been prepared with a water slurry. Contact between the layers and the electrolyte cells is made with strips of filter paper saturated with the same buffer solution which is used for the layers. Pastuska and Trinks [332] have prepared their layers in a slightly different manner. Filter paper strips are glued to two edges of the grease-free glass plates, using the adhesive in only a few spots so as not to interfere with the uptake of the electrolyte solution. The paper should cover about 10 mm of the end of the plate, and numerous parallel cuts are made in the ends of the paper that dip into the electrolyte in order to improve capillary action. After this initial preparation the plates are then coated with the thin-layer material.

Silica gel has been used as a thin-layer material for the electrophoresis of amino acids [129, 253, 333], coal tar food colors [334, 335], amines [129, 333], inorganic ions [336, 337], naphthols [333], phenols [332], phenolcarboxylic acids [332], alkaloids [338], peptides [253, 339, 340], and carbohydrates [341].

Raymond and Weintraub [342] initiated the use of acrylamide gel as a supporting medium for thin-layer electrophoresis. This gel is formed by polymerizing a mixture of acrylamide and *N,N*-methylenebisacrylamide in buffer solutions to form a transparent, flexible, insoluble gel suitable for gel electrophoresis. The process does not require heat in that gels are formed by a polymerization cross-linking reaction. (During the polymerization the solution must be protected from atmospheric oxygen.) The gels can be formed in the presence of various buffer solutions so that they are ready to use as soon as the gel has set. The mixture of organic monomers for preparation of the gel is sold by American Cyanamid under the name Cyanogum 41. The product is also handled by the E-C Apparatus Corp. (See also under Isoelectric Focusing for purification of materials and preparation of the gel.)

Thin-layer acrylamide layers have been used for the separation of proteins [130, 131, 322, 343], enzymes [344], and globulins [345, 346].

The mobility of molecules of different weight can be varied in both starch and acrylamide gels by varying the concentration of the gels. This effect is so great that the order of migration of proteins may be reversed by increasing the concentration of a starch gel [347]. Ornstein [348] has shown with polyacrylamide that this difference is due to the difference in pore size and that the pore size can be varied by changing the concentration of the polymer. Raymond and Aurell [130] and Raymond and Nakamichi [131] have used this gel concentration effect for two-dimensional work. (See Chapter V, Section 5a.)

Sephadex has been used as a medium for the electrophoresis of proteins [125, 349–351]. Enzymes which cannot be separated by electrophoresis on paper or starch layers without loss of activity can be recovered after electrophoresis on Sephadex without any loss in activity [349]. Johansson and Rymo [125] and Fasella et al. [350] have applied Sephadex to two-dimensional work by using thin-layer gel filtration in the first direction followed by thin-layer electrophoresis. For the preparation of the Sephadex layers the dry powder may be mixed with an excess of the proper buffer solution (1:7.5) and allowed to stand for 24 hr. The excess fluid is then removed and the resulting gel is poured onto the glass plate. The layer is leveled off by means of a glass rod and suitable guide strips. The gel layer may be covered with a sheet of glass to prevent loss of moisture during electrophoresis.

Vendrely et al. [351] gave greater rigidity to their Sephadex layers by incorporating agarose.

Cellulose layers have been used for the separation of nucleosides [352, 353], acids [354, 355], alkaloids [356], peptides [357], and amino acids [358]. The separations were comparable to those obtained with paper electrophoresis. Mixed layers of cellulose and silica gel have also been used [359, 360].

Kieselguhr has been used in the thin-layer electrophoresis of coal tar food colors [333, 334], amines and amino acids [129], and inorganic ions [335].

Alumina has been used in this field for the separation of coal tar food colors [334], proteins [361], and amines and amino acids [129].

Plaster of Paris has been used as a support in thin-layer electrophoresis [362, 363].

Two-dimensional electrophoresis may be employed. One way to do this is to use different concentrations of gel [364] which may be cast at one time or, after electrophoresis in one direction, a strip of the gel containing the compounds may be cut from the layer and embedded in a new layer of different concentration [365, 366]. Different combinations may also be used, for example, agar in one dimension and starch gel in the second dimension [367]. For ribosomal proteins the electrophoresis was run in an alkaline buffer in the first dimension, and in an acid buffer in the second dimension [368, 369]. Two-dimensional immunoelectrophoresis may be applied to proteins [370, 371]. In this case the second dimension is run in gel containing antiserum.

Isoelectric Focusing

Although Ikeda and Suzuki [372] first described the isolation of glutamic acid by electrolysis in 1912, Williams and Waterman [373] first stated the principles on which isoelectric focusing is based. The technique became practical when Vesterberg and Svensson [374, 375] first produced suitable ampholytes for establishing a pH gradient.

It is a method for the analysis and characterization of enzymes, hormones, and other ampholytes of biological interest. Compounds are separated because of their difference in isoelectric points. The technique can be described as electrophoresis on a layer having a pH gradient. The ampholytes migrate until they reach a point where the pH is equal to their isoelectric point, and there they become concentrated in a sharp zone. Thus in addition to providing a means for separating these compounds it affords an opportunity to determine their isoelectric points.

This technique will separate compounds that cannot be separated by electrophoresis, although the reverse situation is also true; that is, electrophoresis will sometimes separate compounds that cannot be separated by electric focusing. This makes a powerful combination for two-dimensional work.

By using isoelectric focusing, proteins and other ampholytes which had been considered homogeneous have been shown to be heterogeneous. This is because the method permits a degree of resolution not attainable by other means. Molecules whose p*I* values differ by as little as 0.005 pH units may be resolved [376]. This is possible, because in contrast to electrophoresis where the bands become more diffuse, the isoelectric bands become sharper as the separation progresses. For this same reason it is possible to detect minor components and 0.2% of an impurity has been detected [377]. It is thus a valuable technique for final purification.

In general, thin-layer isoelectric focusing may be carried out in equipment used for thin-layer electrophoresis. Either vertical [323–326, 378, 379] or horizontal [327–329, 380–384] apparatus may be used. (See the preceding section K for commercial manufacturers.)

Awdeh et al. [385] devised an apparatus in which the thin-layer plate is placed gel side down on carbon electrodes, thus eliminating the need for electrode wells. Before the plate was positioned the cathode was moistened with 5% (vol/vol) ethylenediamine solution and the anode with 5% (vol/vol) phosphoric acid. With this type of equipment it is desirable to employ a cold chamber in order to keep the temperature from increasing too much.

Avital and Elson [384] also eliminated the need for filter paper bridges to connect the buffer wells with the horizontal gels by casting the latter with bent ends to dip into the electrode chambers. Another and simpler apparatus which eliminates the need for bridges has been described by Leaback and Rutter [382]. In this case the gel on each end of the layer forms one wall of an electrode well, thus placing the gel in direct contact with the electrode solution. Since this equipment has a gel-coated glass plate it can be cooled by placing it on a metal block cooled to 4°C. (When low temperatures are used it is advisable to cover the gel surface with a sheet of plastic to prevent moisture condensation on the surface.) Delincée and Radola [380] used this technique to cool their plates, but made contact with their electrodes by placing them on filter pads soaked in electrode solution and resting on the gel surface. Criddle et al. [386] designed a chamber cooled with refrigerated air.

In the preparation of gel plates there are a number of techniques that can be applied to improve the adherence of the gel to the supporting plate. The plates of course should be thoroughly cleaned with alcoholic potassium hydroxide. Finlaysen and Chrambach [387] found that the gel adhesion to glass surface could be increased by coating the glass with a high-molecular-weight, linear polyacrylamide (1% solution of Gelamide 250, American Cyanamid). They also recommend substituting ethylene diacrylate wholly or in part for *N,N'*-methylenebisacrylamide in the formulation in order to increase the adhesion to glass. Wada and Snell [388] used a ground-glass plate to achieve this purpose.

Polyacrylamide gel has been used a great deal in this work, because of its low electroendosmotic flow. Kohnert et al. [389] and Loerning [390] recommend purifying the reagents before preparing the gel to improve the separation profile. They accomplished this by twice recrystallizing the acrylamide and the *N,N'*-methylenebisacrylamide from acetone and chloroform, respectively.

For proteins with molecular weights up to 500,000, gels containing 4% acrylamide may be used [380]. These can be prepared from stock solutions which can be kept in the dark at 4°C for a month. The stock solutions are prepared as follows:

1. *Catalyst stock solution: N,N,N',N'*-tetramethylethylenediamine, 1.12 ml; riboflavin, 11.2 mg; and water to make 100 ml.
2. *Acrylamide stock solution: N,N'*-methylenebisacrylamide, 0.64 g; acrylamide, 24 g; and water to make 100 ml.

To prepare the gel for a 20 × 10 × 1 mm layer, mix 6 ml of the acrylamide solution, 1.6 ml of the catalyst solution, 0.7 ml of a 40% solution of carrier ampholytes (LKB Producter-AB), and dilute to 36 ml. Been and Rasch [378] recommend filtering of the acrylamide mixtures just before casting in order to avoid flecks of dust or lint within sample tracks to be used for photographing or densitometric determinations.

Righetti and Drysdale [391] added 5 to 10% glycerol to the gel to improve the consistency, and Bates and Deyoe [392] added 3.5% dimethyl sulfoxide to their gel to improve firmness and resilience.

The gels are polymerized by the action of light and two 40-W daylight fluorescent lamps at a distance of six in. for 30 min should provide sufficient energy.

For low molecular weights of less than 200,000, higher concentrations of gel (5 to 6%) may be used (391). On the other hand, for still higher molecular weights (above 500,000) more dilute gels should be used. Florini et al. [393] used a 2.6% gel for analyzing myosin. However, the use of riboflavin for initiating the polymerization in these dilute solutions resulted in gels that were too soft to handle. Therefore, they used 0.1% (wt/vol) ammonium persulfate as activator for the polymerization. Use of the latter brings up the possibility of oxidation of sensitive compounds and excess persulfate should be washed from the gel. Robinson [394] washed the gels in running distilled water for 1 hr.

With washed gels the carrier ampholytes have to be added after the gels are cleared. Robinson found that spreading the Ampholine across the surface of the gel gave uneven distribution and resulted in distortion of the pH gradient. This could be overcome by first completely drying the gels and then soaking them in 2% Ampholine.

One advantage of thin-layer isoelectric focusing is in keeping the amount of carrier ampholytes to a minimum, because the commercial product Ampholine is expensive. It is a synthetic mixture of low-molecular-weight polyamino–polycarboxylic acids of unknown composition.

Because of the expense of commercial Ampholine Vinogradov et al. [395] investigated the possibility of producing these in the laboratory, and obtained a product that gave good results in the pH range 4–8. Their method was as follows:

Reagents: acrylic acid (Aldrich) freshly distilled under vacuum just before use. Pentaethylenehexamine (PEHA) (Union Carbide) also vacuum distilled before use.

Procedure: under nitrogen, acrylic acid is added dropwise to a well stirred solution of 0.15 moles of the PEHA in 35 ml of water till a nitrogen: carboxyl ratio of 2: 1 is reached over a period of 45 to 60 min. The reaction is then brought to 70°C and stirred overnight (16 to 20 hr). After cooling to room temperature, the mixture is diluted with deionized water to 40% (wt/wt).

Lundblad et al. [396] synthesized more alkaline ampholytes extending the range up to a pH of 11.1, and more recently, Righetti et al. [397, 398] investigated the preparation and fractionation of carrier ampholytes.

The concentration of the carrier ampholytes is important. Kohnert et al. [389] examined some of the factors affecting isoelectric focusing. With a concentration of 1% ampholytes they were unable to obtain stable linear pH gradients and had to increase the concentration to 2%. They also found it necessary to change from an electrolyte system of sodium hydroxide and sulfuric acid to 1% phosphoric acid at the anode and 1% ethylenediamine solution at the cathode to give a smooth pH gradient. The use of higher concentrations of ampholytes did not improve the separation and decreased the protein migration rate.

When thin layers of polyacrylamide gels are prepared, they may be cast between two glass plates held apart by a suitable spacer to give the desired thickness and to contain the gel. Bours [399] siliconized one of the plates so that the gel would not stick. He gives the following two methods of doing this:

1. Expose the plates to the vapors of dichloromethylsilane in a sealed box placed in a fume hood for two weeks. The plates are then thoroughly rinsed with water.
2. Dip the well cleaned plates in a 2% solution of dichlorodimethylsilane in carbon tetrachloride and allow the carbon tetrachloride to evaporate.

Another way to accomplish the same result is to cover the glass plate with a sheet of thin plastic. The glass can then be lifted off, leaving the thin plastic to be peeled off without disturbing the gel.

Lewin [400] recommends pre-electrofocusing of the Ampholine prior to application of the sample in order to obtain improved repeatable patterns and interpretations. It is also advisable to use a single batch of Ampholine for a given investigation because different batches have been found to give slightly different patterns [401]. Different concentrations of the same batch of ampholytes also gave different pH gradient patterns [392].

Samples, which should be desalted to prevent distortion of the gradients [402], can be applied in a number of ways. They can be applied to slits in the gel as in electrophoresis or in pockets cast in the gels. Solutions may also be slowly applied to the surface of the gel. Strips of filter paper soaked in the sample solution may be laid on the gel surface or, better, strips of cellulose acetate may be used [403], because some proteins are not readily eluted from filter paper [404]. An advantage of thin layers is that the sample can be placed as near as possible to its isoelectric point to avoid denaturing a protein that is sensitive to pH [394, 405, 406].

One of the effects that must be guarded against is the oxidation of sensitive compounds, especially proteins containing an SH group. This can arise from oxygen produced at the anode and also can occur if the gel formation has been catalyzed by persulfate, for it may be difficult to remove all traces of the latter. Attempts to remove persulfate by pre-electrophoresis at pH 4.3 were not successful [407, 408]. The effects of persulfate can be counteracted by adding ascorbic acid as a reducing agent [409]. Pre-electrophoresis with dithionate [410], hydroquinone, cysteine, or thioglycolate [408] are other ways to remove oxidizing compounds. Cooling the gel can also decrease the possibility of oxidation; in addition it has been shown that cooling in the basic region improves resolution [394].

Jordan and Raymond [411] have proposed a catalyst composed of ascorbic acid, 0.0025% ferrous sulfate, and 0.03% hydrogen peroxide from 30% stock for catalyzing polyacrylamide gels in acid systems.

It is important not to carry the isoelectric focusing for too long a period of time. If this is done there is a progressive flattening of the pH versus distance curve [387]. This is known as the "plateau phenomenon," but it develops very slowly. Vesterberg [409] mentions two methods of avoiding this problem. One way is to add the carrier ampholytes after polymerization of the gel. The other is to add hemoglobin or some other easily visible protein at two places, one near the anode and the other near the cathode. When these two bands join, this protein is at its focus point. The time is then increased by 25% to take care of compounds that move slower. Even if the carrier ampholytes are added after polymerization, it is advisable not to run the experiment too long.

Artifacts sometimes form and thin layers provide a convenient way of discovering them [412]. The results may be checked for artifacts, by rerunning individual bands and by applying the sample at different places on the gel.

Thin layers are also convenient for running multiple samples or for comparing samples alongside one another under exactly the same conditions.

After equilibrium has been reached the pH must be measured at frequent points along the gel. This may be done in one of two ways. Gel samples (4-mm discs) may be eluted in 0.75 ml of distilled water for 2 hr before the pH is determined [394]. Care must be taken not to dilute the sample too much; otherwise the buffering capacity of the ampholyte will be exceeded. It has been recommended [380] that the volume of water should be less than seven times the gel volume. To minimize the absorption of carbon dioxide when measuring in an alkaline medium, boiled water should be used or the eluates should be flushed with nitrogen. In order to have adequate conductivity a little (10 m*M*) sodium chloride can be added [380]. The alternative method of measuring pH, and probably the simplest, is to use a flat membrane electrode directly on the surface of the gel. (Suitable electrodes are the combination microelectrode Cat.

#14153 of Instrumentation Laboratory, Inc. and type LOT 403-30-M8 Ingold, A.G.).

In general most enzymes do not lose their activity after electrofocusing; however, some metalloenzymes do lose activity. Usually the activity can be restored by furnishing the metal ion in the assay mixture [413]. For the detection of enzymes using the techniques employed in gel electrophoresis it may be necessary to add a suitable buffer to offset the effect of the ampholytes.

Isoelectric focusing has been used successfully on a micro scale using 8.2 × 8.2 cm or 9 × 9 cm layers [399] and on 7.5 × 2.5 mm microscope slides [403]. In this respect some of the techniques for microelectrophoresis used by Daams [316] should be of interest.

Because of the different factors concerned in the separation by isoelectric focusing and by electrophoresis it is only natural to combine these in a two-dimensional technique. Wrigley and Shepherd [414] subjected the protein from a single grain of wheat to gel electrofocusing and then embedded this in a 12% starch gel slab for electrophoresis in the second dimension. The number of gliadin components separated in this manner was twice that separated by either technique by itself. Although the isoelectric focusing was performed in a tube rather than in a thin layer the principle still applies. Latner [415] and Dale and Latner [416] applied this two-dimensional technique with acrylamide gel in both directions in clinical work to show the difference in the proteins of normal and pathological sera. As a further example of the usefulness of this technique Macko and Stegemann [417] were able to differentiate between several plant varieties by separating the soluble proteins from potato tubers.

The combination of gel electrofocusing with gradient gel electrophoresis carried this technique a step further [418], giving values for both the p*I* and the molecular weight. In this respect the techniques used for the preparation of gradient layers for electrophoresis may be of interest [199–203, 370, 378].

Another type of two-dimensional technique incorporating isoelectric focusing is that employing immunodiffusion. In this case isoelectric focusing in polyacrylamide gel is used to obtain a separation, and then a slice of the gel is embedded in agarose containing an antiserum. Electrophoresis at right angles to the first direction carries the proteins into the agarose where migrating zones of antigen–antibody appear [419].

Agarose gels have been tried as an electrofocusing medium [420–422], but are not entirely satisfactory because they do not give stable pH gradients, mainly because of too high an electroosmotic flow.

Radola [423, 424] has used granular gels such as Sephadex G-75 "Superfine" or G-200 "Superfine" (Pharmacia) and Bio-Gel P-60 minus 400 mesh (Bio-Rad Laboratories) for electrofocusing. In this way he avoided the possibility of molecular sieving effects from the use of polyacrylamide gels. Glass plates were coated with 7.5, 4, or 5.4 g/100 ml, respectively, of the three gels mentioned above. One percent (wt/vol) of Ampholine was used as a carrier ampholyte with 0.05 to 0.1% each of arginine and lysine to stabilize the pH in the alkaline region. The gels were then dried to about 80% water content yielding a 0.6 mm thick layer. (After proper drying the edges of the gel showed 1 to 3 mm fissures.) Cellulose acetate strips soaked in 0.2 *M* sulfuric acid and 0.4 *M* ethylenediamine connected the gel layer with the corresponding electrode solution. In the pH zone 4 to 5 it was sometimes necessary to increase the Ampholine concentration to 2% to obtain the desired stability. This method can be easily adapted to preparative quantities in the gram range by simply increasing the thickness of the layer and still retain the resolution.

There are a number of staining procedures that can be used to locate the protein bands separated by isoelectric focusing, but some of the stains require the removal of the carrier ampholytes prior to staining. Bours [399] washed the gels six or seven times with decreasing concentrations (10 to 3%) of trichloroacetic acid using 3 hr for each washing period. This was followed by repeated washing with methanol–acetic acid–water (45:9:46) until the pH of the gel reached 4.0 to prevent precipitation of the dye at the lower pH. Humphreys [403] found sulfosalicyclic acid to be less corrosive to handle than trichloroacetic acid, and therefore used it to fix the gel and remove carrier ampholytes. He washed for 48 hr in four changes of the acid. Hayes and Wellner [425] used a mixture of 5% trichloroacetic and 5% sulfosalicylic acids with a final washing in water.

Li and Li [426] used two different techniques to remove Ampholine. In one, the fractions were dialyzed against a suitable buffer. Following this the dialyzing sac was subjected to reverse dialysis in a saturated ammonium sulfate solution to precipitate the protein, which was then removed by centrifugation. In the other method the volume of the fraction was first reduced to 2 to 3 ml by ultrafiltration and then dialyzed *in situ* against the buffers until it had increased two to three times. The volume was again reduced to 2 to 3 ml under reduced pressure and ultimately to the volume desired depending on the protein concentration.

A number of staining techniques have been developed to eliminate the long prewashing period. Spencer and King [427] stained their gel layers overnight in a mixture of 100 ml of 10% trichloroacetic acid, 100 ml of 10% sulfosalicylic acid, 2 ml of 1% Coomassie Blue dye, and 40 ml of methanol. Righetti and Drysdale [376] modified this procedure to improve the sensitivity and decrease the background staining. This consisted in storing the gel for 4 hr or more at room temperature in a 0.05% solution of Coomassie Blue and 0.1% cupric sulfate in a mixture of acetic acid–ethanol–water (2:5:13), followed by 4 hr with the dye concentration reduced to 0.01%. In this case, after staining the excess dye is removed by an acetic acid–ethanol–water (1:1:8) wash.

Frater [412] used a mixture of 0.05% Fast Acid Blue B (C.I. 44035) and 0.05% Coomassie Violet R (C.I. 42650) in 5% acetic acid. Destaining of the background was accomplished in 4 to 5 hr with 5% acetic acid. Because various proteins stain differently it is advisable to experiment with various stains to obtain the best results.

Righetti and Drysdale [376] have published a review of isoelectric focusing in gels. A book on this subject has been published [427a] and contains a number of articles on thin-layer work.

Automated Thin-Layer Chromatography

The first instrument for automation of the entire thin-layer chromatographic process was introduced in 1972 [428] and was marketed by J. T. Baker. Samples were automatically applied to adsorbents coated on a reel of 35-mm Mylar film. The filmstrip was automatically carried through the entire procedure of development, drying, spraying of the visualizing agent, and quantitative measurement of the spots. Apparently because of technical difficulties, this instrument was withdrawn from the market.

Another instrument for automation of thin-layer chromatography has been placed on the market by the Lightner Instrument Co. Samples are spotted on individual thin-layer chromatographic sheets by means of a capillary from the 100-sample storage magazine. The capillary is then rinsed. Standard solutions may be spotted on the same sheet. After spotting the sheets are moved to one of the 10 developing tanks located on a turn table. The sheets are developed for a programmed interval, then removed and dried. They are then scanned by an optical detector in one of three modes and the results are recorded on a strip chart or digitized. The instrument can measure fluorescence, visible absorption, or fluorescent quench. In this case there is no provision for spraying a visualizing agent. The company claims standard deviations of 4 to 6%.

There are other portions of the TLC process that have been automated by various workers. Fosslien [429] developed an automatic lipid extraction and spot application system. Snyder and Smith [430] used an automated zonal scraper and collector with an automated liquid-scintillation spectrometer, a direct data-transfer system, a card punch, a computer, and an electroplotter to facilitate radioactive work. The scraper will automatically remove 1-, 2-, or 5-mm zones from the origin to the solvent front for analysis. Fosslien and co-workers [431] developed a computer-controlled scraper that would sequentially scrape selected spots from any area on a standard plate. Another zonal scraper (semiautomatic) has been designed by Kasang et al. [432].

Automated development has also been considered by a number of workers, and perhaps the simplest is the apparatus developed by Ismail and Harkness [433] where they modified an alarm clock to turn a stopcock that added solvent to the tank after 4 hr of exposure to solvent vapors for equilibration of the tank. Saunders and Snyder [434] coated adsorbent on a drum and revolved the drum slowly as development took place so as to keep a constant distance between the solvent front and the solvent level in the tank. Halpaap and Bausch [435] used a similar technique moving thin-layer flexible sheets by means of rollers placed above and in the tank. These latter two devices keep the rate of development constant; however, they are of little use if slow-moving component is present with a medium- or high-R_f component. An automated development tank controlled by timers, a photocell, relays, and solenoid valves has been constructed by Schneck and co-workers [436]. The run may be terminated by either the photocell or timers, and either one- or two-dimensional thin-layer chromatography may be carried out. It is so constructed that a motor will rotate the apparatus so that development can take place in the second dimension. Multiple or stepwise development may also be used. Contrary to expectations the setup is quite simple and ingenious. Philp [437] has designed another automatic developing unit. In this case solvent is added to the tank by gravity flow, and when the run is completed the solvent is drained off and pumped back to the solvent reservoir. The solvent is then removed from the plate using nitrogen or air, and the development cycle may be repeated any number of times with the same or different solvents to obtain multiple or stepwise development. Provision has also been made for rotating the plate so that two-dimensional work may be accomplished. The entire operation is controlled automatically by a 12-position punched tape and 20 digital preselectors.

Perry and co-workers [438, 439] have developed a unit for automatically carrying out multiple development. The chromatographic plate, after sample application, is combined with a glass plate and suitable spacers so as to form a sandwich layer, which is then placed in the solvent trough. The development time is variable and controlled by a programmer. After the selected time period, the solvent on the plate is evaporated by means of a heater, a flow of inert gas, or a combination of the two. The plate remains in the solvent trough during this evaporation, the solvent trough being protected by a shield. The heat is then shut off and the solvent allowed to advance for another timed period, this time a little longer, and then the evaporation cycle is repeated. The equipment can be programmed for as many as 99 cycles if desired. As an example of the automation possible with this equipment, a run was made for 72 hr, during which a total of 68 multiple developments was completed [440]. The equipment is manufactured by Regis Chemical Co. In this procedure as in regular multiple development, the spots are concentrated by the repeated development, because as the developing solvent re-enters the layer it reaches the bottom of the spot and starts to move it forward before the top of the spot is moistened by the solvent. In a modification of the procedure [439, 440] the layer is covered by a reflecting shield

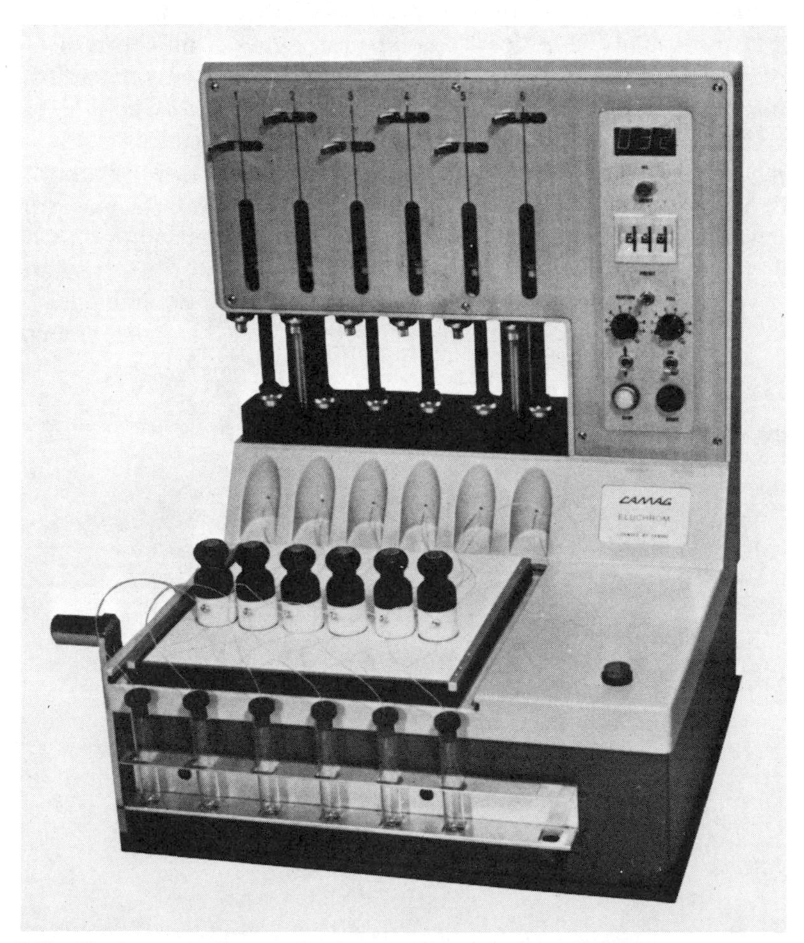

Fig. 5.22 Eluchrom equipment for automatic elution from thin-layer chromatographic plates. Courtesy of CAMAG Inc.

containing a narrow 3-mm-wide slit over the line of developing spots. When the layer is exposed to heat, the solvent is removed from beneath this slit first and consequently solvent in the sides of the layer move toward this center area carrying the spot material toward the center. This technique has been designated centered programmed multiple development.

A different bit of automation was introduced with the apparatus designed by Falk and Krummen [441]. With this equipment the compounds may be eluted directly from the chromatographic plate without first scraping off the spot. The equipment, known as the Eluchrom, is shown in Fig. 5.22 and is commercially available from CAMAG. In the operation of this equipment a series of as many as six rings are milled around the desired spots. This isolates the compound spots from the rest of the plate. The elution chambers are then clamped in place over the isolated spots and from 1 to 5 ml of an eluting solvent is gradually applied to one side of each chamber by means of a syringe. The eluate is collected at the opposite side in a suitable container. The rate of flow of the solvent can be adjusted to the rate of desorption of the compounds. Compressed air blows out the residual solvent and dries the layer. It takes only about 10 min to set up the system for automatic elution and the attendant can then give his attention to other matters while the compounds are being eluted. Compound recoveries of better than 99% have been reported [442].

Computerization is a form of automation, but because it is used mainly for quantitative analysis it is dealt with in that chapter.

References

1. W. Trappe, *Biochem. Z.,* **305,** 150 (1940).
2. J. J. Wren, *J. Chromatogr.,* **4,** 173 (1969).
3. H. H. Strain, *Chromatographic Adsorption Analysis,* Interscience, New York, 1945 p. 66.
4. H. P. Kaufmann and Z. Makus, *Fette, Seifen, Anstrichm.,* **62,** 1014 (1960).
5. T. I. Bulenkov, *Zh. Anal. Khim.,* **23,** 848 (1968).
6. J. A. Attaway, L. J. Barabas, and R. W. Wolford, *Anal. Chem.,* **37,** 1289 (1965).
7. J. A. Attaway, *J. Chromatogr.,* **31,** 231 (1967).
8. N. A. Izmailov and M. S. Schraiber, *Farmatsiya (Sofia),* **1938,** 1.
9. M. O'l. Crowe, *Anal. Chem.,* **13,** 845 (1941).
10. J. M. Miller and J. G. Kirchner, *Anal. Chem.,* **24,** 1480 (1952).

11. H. S. Knight and S. Groennings, *Anal. Chem.*, **26**, 1549 (1954).
12. D. L. Massart, *J. Chromatogr.*, **79**, 157 (1973).
13. D. L. Massart and H. De Clercq, *Anal. Chem.*, **46**, 1988 (1974).
13a. H. De Clercq, E. Blockeel, E. Defrise-Gussenhoven, and D. L. Massart, *Anal. Chem.*, **47**, 2275 (1975).
13b. H. De Clercq and D. L. Massart, *J. Chromatogr.*, **115**, 1 (1975).
13c. A. C. Moffat, K. W. Smalldon, and C. Brown, *J. Chromatogr.*, **90**, 1 (1974).
13d. A. C. Moffat and K. W. Smalldon, *J. Chromatogr.*, **90**, 9 (1974).
14. S. Turina, M. Trbojević, and M. Kaštelan-Macan, *Anal. Chem.*, **46**, 988 (1974).
15. A. G. Baker, P. J. Carr, and G. Nickless, *Chem. Ind. (London)*, **2**, 901 (1972).
16. D. J. Panova, M. F. Mincheva, and A. D. Minchev, *Compt. Rend. Acad. Bulg. Sci.*, **25**, 1245 (1972).
16a. G. Rouser, *J. Chromatogr. Sci.*, **11**, 60 (1973).
17. P. F. Lofts, S. J. Purdy, and E. V. Truter, *Lab. Pract.*, **18**, 1167 (1969).
18. J. Ościk and J. K. Różylo, *Chromatographia*, **4**, 516 (1971).
19. L. R. Snyder, *J. Chromatogr.*, **63**, 15 (1971).
20. *Ibid.*, **25**, 274 (1966).
21. J. A. Thoma, *Anal. Chem.*, **37**, 500 (1965).
22. J. A. Thoma and C. R. Perisho, *Anal. Chem.*, **39**, 745 (1967).
23. J. A. Thoma, "Polar Solvents, Supports, and Separation," in *Advances in Chromatography*, Vol. 6, J. C. Giddings and R. A. Keller, Eds., Marcel Dekker, New York, 1968, p. 61.
24. A. Waksmundzki and J. K. Różylo, *J. Chromatogr.*, **49**, 313 (1970).
25. J. K. Różylo, *J. Chromatogr.*, **85**, 136 (1973).
26. B. W. Nichols, *Biochim. Biophys. Acta*, **70**, 417 (1963).
27. F. Kunz and D. Kosin, *Clin. Chim. Acta*, **27**, 185 (1970).
28. E. Roeder, "Azeotropic Mixtures as Chromatographic Solvents in TLC," *Progress in Thin-Layer Chromatography and Related Methods*, Vol. II, A. Niederwieser and G. Pataki, Eds., Ann Arbor Science, Ann Arbor, Mich., 1971, p. 93.
29. J. A. Manthey and M. E. Amundson, *J. Chromatogr.*, **19**, 522 (1965).
30. R. B. Holtz, P. Swenson, M. Abel, and T. A. Walter, *Lipids*, **6**, 523 (1971).
31. P. W. Parodi and R. J. Dunstan, *Aust. J. Dairy Technol.*, **23**, 20 (1968).
32. Y. Asakawa, F. Genjida, and T. Matsuura, *Anal. Lett.*, **2**, 333 (1969).
33. C. Bucke, *J. Chromatogr.*, **31**, 247 (1967).
34. D. G. Crosby and N. Aharonson, *J. Chromatogr.*, **25**, 330 (1966).
35. D. C. Abbott, H. Egan, E. W. Hammond, and J. Thomson, *Analyst (London)*, **89**, 480 (1964).
36. D. Rosi and P. Hamilton, *J. Chromatogr.*, **9**, 388 (1962).
37. M. Brenner and A. Niederwieser, *Experientia*, **16**, 378 (1960).
38. E. Urion, M. Metche, and J. P. Haluk, *Brauwissenschaft*, **16**, 211 (1963).
39. N. Nybom, *J. Chromatogr.*, **14**, 118 (1964).
40. G. H. Sloane-Stanley and L. M. Bowler, *Lab. Pract.*, **11**, 769 (1962).

41. E. G. Wollish, M. Schmall, and M. Hawrylyshyn, *Anal. Chem.*, **33**, 1138 (1961).
42. M. A. Ford, *Lab. Pract.*, **16**, 322 (1967).
43. H. T. Badings and J. G. Wassink, *Neth. Milk Dairy J.*, **17**, 132 (1963).
44. I. Sankoff and T. L. Sourkes, *Can. J. Biochem. Physiol.*, **41**, 1381 (1963).
45. J. Jensen, *J. Chromatogr.*, **10**, 236 (1963).
46. N. Karpitschka, *Mikrochim. Acta*, **1963**, 157.
47. S. Hara, M. Takeuchi, and N. Matsumoto, *Bunseki Kagaku*, **13**, 359 (1964).
48. D. Jaenchen, *J. Chromatogr.*, **14**, 261 (1964).
49. E. Stahl, "Instruments Used in Thin-Layer Chromatography and their Operation," in *Thin-Layer Chromatography*, E. Stahl, Ed., Academic Press, New York, 1965, p. 18.
50. R. Wasicky, *Naturwissenschaften*, **50**, 569 (1963).
51. Z. Procházka, *Chem. Listy*, **55**, 974 (1961).
52. J. Davídek and E. Davídková, *Pharmazie*, **16**, 352 (1961).
53. C. Lapp and K. Erali, *Bull. Sci. Pharmacol.*, **47**, 49 (1940).
54. E. A. Mistryukov, *J. Chromatogr.*, **9**, 311 (1962).
55. B. G. Johansson and L. Rymo, *Acta Chem. Scand.*, **16**, 2067 (1962).
56. W. L. Stanley and S. H. Vannier, *J. Assoc. Off. Agric. Chem.*, **40**, 582 (1957).
57. J. G. Kirchner, J. M. Miller, and G. J. Keller, *Anal. Chem.*, **23**, 420 (1951).
58. W. L. Stanley, S. H. Vannier, and B. Gentili, *J. Assoc. Offic. Chem.*, **40**, 282 (1957).
59. M. K. Seikel, M. A. Millett, and J. F. Saeman, *J. Chromatogr.*, **15**, 115 (1964).
60. L. Birkofer, C. Kaiser, H. A. Meyer-Stoll, and F. Suppan, *Z. Naturforsch.*, **17B**, 352 (1962).
61. L. Goeldel, W. Zimmerman, and D. Lommer, *Z. Physiol. Chem.*, **333**, 35 (1963).
62. P. M. Reisert and D. Schumacher, *Experientia*, **19**, 84 (1963).
63. N. Zoellner and G. Wolfram, *Klin. Wochenschr.*, **40**, 1098 (1962).
64. E. A. Mistryukov, *Collect. Czech. Chem. Commun.*, **26**, 2071 (1961).
65. G. Hesse and M. Alexander, *Journees Intern. Etude Methodes Separation Immediate Chromatog., Paris 1961*, 1962, p. 229.
66. J. E. Meinhard and N. F. Hall, *Anal. Chem.*, **21**, 185 (1949).
67. L. H. Bryant, *Nature*, **175**, 556 (1955).
68. A. E. Sherwood, *Lab. Pract.*, **15**, 1391 (1966).
69. G. J. Litt and R. G. Johl, *J. Chromatogr.*, **20**, 605 (1965).
70. W. J. van Ooij, *J. Chromatogr.*, **42**, 432 (1969).
71. A. Zlatkis and R. E. Kaiser, Eds., *HPTLC High Performance Thin-Layer Chromatography*, Elsevier, Amsterdam, 1977.
72. H. Halpaap and J. Ripphahn, "High Performance Thin-Layer Chromatography: Development, Data and Results," in *HPTLC High Performance Thin-Layer Chromatography*, A. Zlatkis and R. Kaiser, Eds., Elsevier, Amsterdam, 1977.
73. M. Mottier, *Mitt. Geb. Lebensm. Hyg.*, **49**, 454 (1958).

74. M. Mottier and M. Potterat, *Anal. Chim. Acta,* **13,** 46 (1955).
75. E. V. Truter, *J. Chromatogr.,* **14,** 57 (1964).
76. L. M. Libbey and E. A. Day, *J. Chromatogr.,* **14,** 273 (1964).
77. M. H. Anwar, *J. Chem. Educ.,* **40,** 29 (1963).
78. G. Nedlkovitch, "Contribution au dosage de la vitamine D par chromatographie sur couche mince en continu," in *Int. Symp. Chromatogr., Electrophor., 4th, 1967,* Ann Arbor Science, Ann Arbor, Mich., 1968, p. 521.
79. G. Cavina and G. Moretti, *J. Chromatogr.,* **22,** 41 (1966).
80. G. H. Stewart and C. T. Wendel, *J. Chromatogr. Sci.,* **13,** 105 (1975).
81. G. H. Stewart and T. D. Gierke, *J. Chromatogr. Sci.,* **8,** 129 (1970).
82. M. Brenner and A. Niederwieser, Swiss Pat. 364,130 (August 31, 1962).
83. M. Brenner and A. Niederwieser, *Experientia,* **17,** 237 (1961).
84. T. M. Lees, M. J. Lynch, and F. R. Mosher, *J. Chromatogr.,* **18,** 595 (1965).
85. S. Hara and K. Mibe, *J. Chromatogr.,* **66,** 75 (1972).
86. S. Hara, S. Yamazaki, and H. Ichikawa, *Chem. Ind. (London),* **1969,** 1657.
87. R. D. Bennett and E. Heftmann, *J. Chromatogr.,* **12,** 245 (1963).
88. L. Lábler, "Thin-Layer Chromatography on Loose Layers of Alumina," in *Thin-Layer Chromatography,* G. B. Marini-Bettòlo, Ed., Elsevier, Amsterdam, 1964, p. 32.
89. L. Johansson and S. Kashemsanta, *J. Chromatogr.,* **45,** 471 (1969).
90. J. H. van Dijk, *Z. Anal. Chem.,* **236,** 326 (1968).
91. Ibid., "Horizontal Centripetal Thin-Layer Chromatography with Continuous Elution and Collection of Separated Components," in *Column Chromatog., Int. Symp. Sep. Methods, 5th 1969,* E. S. Kovats, Ed., Sauerlander AG., Aarau, Switzerland, 1970, p. 234.
92. V. J. R. De Deyne and A. F. Vetters, *J. Chromatogr.,* **103,** 177 (1975).
93. S. Turina, V. Marjanovic-Krajovan, and M. Obradovic, *Anal. Chem.,* **36,** 1905 (1964).
94. S. Turina and V. Marjanovć-Krajovan, "Continuous Separation on the TLC by the Principle of the Rotary Disc," in *Int. Symp. Chromatogr., Electrophor., 4th, 1967,* Ann Arbor Science, Ann Arbor, Mich., 1968, p. 149.
95. R. Visser, *Anal. Chim. Acta,* **38,** 157 (1967).
96. C. P. Dietrich, *Anal. Biochem.,* **51,** 345 (1973).
97. A. Jeanes, C. S. Wise, and R. J. Dimler, *Anal. Chem.,* **23,** 415 (1951).
98. J. A. Thoma, *Anal. Chem.,* **35,** 214 (1963).
99. J. A. Thoma, *J. Chromatogr.,* **12,** 441 (1963).
100. L. Stárka and R. Hampl, *J. Chromatogr.,* **12,** 347 (1963).
101. H. P. Lenk, *Z. Anal. Chem.,* **184,** 107 (1961).
102. H. Halpaap, *Chem.-Ing.-Tech.,* **35,** 488 (1963).
102a. H.-J. Petrowitz, *Chem.-Ztg.,* **93,** 329 (1969).
102b. G. Goldstein, *Anal. Chem.,* **42,** 140 (1970).
103. J. M. Miller and J. G. Kirchner, *Anal. Chem.,* **23,** 428 (1951).
104. H. Weicker, *Klin. Wochenschr.,* **37,** 763 (1959).
105. E. Stahl, *Arch. Pharm.,* **292/64,** 411 (1959).
106. H. Kaunitz, E. Gauglitz, Jr., and D. G. McKay, *Metabolism,* **12,** 371 (1963).

107. E. Stahl and U. Kaltenbach, *J. Chromatogr.*, **5**, 458 (1961).
108. T. A. Dyer, *J. Chromatogr.*, **11**, 414 (1963).
109. K. Teichert, E. Mutschler, and H. Rochelmeyer, *Dtsch. Apoth.-Ztg.*, **100**, 283 (1960).
110. N. Zoellner and G. Wolfram, *Klin. Wochenschr.*, **40**, 1100 (1962).
111. U. Baumann, *Z. Anal. Chem.*, **173**, 458 (1960).
112. H. Gaenshirt, F. W. Koss, and K. Morianz, *Arzneimittel-Forsch.*, **10**, 943 (1960).
113. A. Vacíková, V. Felt, and J. Malíková, *J. Chromatogr.*, **9**, 301 (1962).
114. F. Hernández Alvarado, *J. Chromatogr.*, **42**, 144 (1969).
115. G. Székely, *J. Chromatogr.*, **42**, 543 (1969).
116. S. Narasimhulu, I. Keswani, and G. L. Flickinger, *Steroids*, **12**, 1 (1968).
117. J. M. Iacono and T. T. Ishikawa, *J. Chromatogr.*, **40**, 175 (1969).
118. T. Furukawa, *J. Sci. Hiroshima Univ. Ser.*, **A21**, 285 (1958); *Chem. Abstr.*, **53**, 809 (1959).
119. A. Tiselius, *Ark. Kemi*, **14B**, No. 22 (1940).
120. A. Niederwieser and M. Brenner, *Experientia*, **21**, 50 (1965).
121. M. Viricel, C. Gonnet, and A. Lamotte, *Chromatographia*, **7**, 345 (1974).
122. R. Frache and A. Dadone, *Chromatographia*, **6**, 274 (1973).
123. A. Niederwieser and C. C. Honegger, "Gradient Techniques in Thin-Layer Chromatography, in *Advances in Chromatography*, Vol. 2, J. C. Giddings and R. A. Keller, Eds., Marcel Dekker, New York, 1966, p. 123.
124. L. D. Bergel'son, E. V. Dyatlovitskaya, and V. V. Voronkova, *Izv. Akad. Nauk SSSR, Otd. Khim. Nauk*, **1963**, 954.
125. B. G. Johansson and L. Rymo, *Biochem. J.*, **92**, 5P (1964).
126. C. Vergani, R. Stabilini, and A. Agostoni, *J. Chromatogr.*, **28**, 135 (1967).
127. M. Calderon and W. J. Baumann, *Biochim. Biophys. Acta*, **210**, 7 (1970).
128. H. P. Kaufmann, Z. Makus, and F. Deicke, *Fette, Seifen, Anstrichm.*, **63**, 235 (1961).
129. C. G. Honegger, *Helv. Chim. Acta*, **44**, 173 (1961).
130. S. Raymond and B. Aurell, *Science*, **138**, 152 (1962).
131. S. Raymond and M. Nakamichi, *Anal. Biochem.*, **7**, 225 (1964).
132. F. Jimeno de Osso, *J. Chromatogr.*, **60**, 272 (1971).
133. J. D. Pollack, D. S. Clark, and N. L. Somerson, *J. Lipid Res.*, **12**, 563 (1971).
134. E. Zappi, M. Schmidt, and F. Prange, *J. Chromatogr.*, **43**, 543 (1969).
135. E. F. L. J. Anet, *J. Chromatogr.*, **9**, 291 (1962).
136. J. G. Kirchner and V. P. Flanagan, Gordon Research Conference, Colby Junior College, New London, N.H., August 1962.
137. J. G. Kirchner and V. P. Flanagan, 147th Meeting of The American Chemical Society, Philadelphia, Pa., April 1964.
138. J. A. Berger, G. Meyniel, P. Blanquet, and J. Petit, *Compt. Rend.*, **257**, 1534 (1963).
139. J. A. Berger, G. Meyniel, and J. Petit, *Compt. Rend.*, **255**, 1116 (1962).
140. J. A. Berger, G. Meyniel, J. Petit, and P. Blanquet, *Bull. Soc. Chim. Fr.*, **1963**, 2662.

141. D. C. Abbott, J. A. Bunting, and J. Thomson, *Analyst (London),* **91,** 94 (1966).
142. V. Lakshminarayana and P. K. Menon, *Pestic. Sci.,* **2,** 103 (1971).
143. D. R. Gilmore and A. Cortes, *J. Chromatogr.,* **21,** 148 (1966).
144. A. Váradi, *J. Chromatogr.,* **110,** 166 (1975).
145. R. Sable-Amplis, R. Agid, and D. Abadie, *J. Chromatogr.,* **94,** 287 (1974).
146. A. M. Pumphrey, *Biochem. J.,* **102,** 30P (1967).
147. P. S. Bond, *J. Chromatogr.,* **34,** 554 (1968).
147a. F. Kreuzig, *J. Chromatogr.,* **142,** 441 (1977).
148. H. J. McDonald, E. W. Bermes, and H. G. Shepherd, *Naturwissenschaften,* **44,** 9 (1957).
149. H. J. McDonald, L. V. McKendell, and E. W. Bermes, *J. Chromatogr.,* **1,** 259 (1958).
150. J. F. Herndon, H. E. Appert, J. C. Touchstone, and C. N. Davis, *Anal. Chem.,* **34,** 1061 (1962).
151. B. P. Korzun and S. Brody, *J. Pharm. Sci.,* **53,** 454 (1964).
152. J. Rosmus, M. Pavlicek, and Z. Deyl, "XII. Centrifugal Thin-Layer Chromatography," in *Thin-Layer Chromatography,* G. B. Marini-Bettólo, Ed., Elsevier, Amsterdam, 1964, p. 119.
153. A. Affonso, *J. Chromatogr.,* **22,** 1 (1966).
154. A. Lepoivre, *Bull. Soc. Chim. Belg.,* **81,** 213 (1972).
155. Z. Deyl, J. Rosmus, and M. Pavlíček, *Chromatogr. Rev.,* **6,** 44 (1964).
156. J. G. Marchal and T. Mittwer, *Proc. K. Ned. Acad. Wet.,* **54C,** 391 (1951).
157. J. G. Marchal and T. Mittwer, *Compt. Rend. Soc. Biol.,* **145,** 417 (1951).
158. M. Mottier, *Mitt. Geb. Lebensm. Hyg.,* **47,** 372 (1956).
159. J. W. Copius-Peereboom and H. W. Beekes, *J. Chromatogr.,* **9,** 316 (1962).
160. T. Furukawa, *Nippon Kagaku Zasshi,* **80,** 45 (1959); *Chem. Abstr.,* **54,** 4107 (1960).
161. H. Hausser, *Arch. Kriminol.,* **125,** 72 (1960).
162. V. Prey, H. Berbalk, and M. Kausz, *Mikrochim. Acta,* **1961,** 968.
163. H. Bayzer, *Experientia,* **20,** 233 (1964).
164. D. C. Abbott and J. Thomson, *Chem. Ind. (London),* **481,** (1964).
165. D. C. Abbott and J. Thomson, *Analyst (London),* **89,** 613 (1964).
166. N. G. Bazán, Jr., and C. D. Joel, *J. Lipid Res.,* **11,** 42 (1970).
167. T. Wieland and H. Determann, *Experientia,* **18,** 431 (1962).
168. A. F. Hofmann, *Biochim. Biophys. Acta,* **60,** 458 (1962).
169. S. M. Rybicka, *Chem. Ind. (London),* **1962,** 308.
170. *Ibid.,* p. 1947.
171. L. Valentine, *Peint., Pigm. Vernis,* **39,** 295 (1963).
172. L. Luzzatto and C. N. Okoye, *Biochem. Biophys. Res. Commun.,* **29,** 705 (1967).
173. R. G. Stickland, *Anal. Biochem.,* **10,** 108 (1965).
174. J. C. Hegenauer, K. D. Tartof, and G. W. Nace, *Anal. Biochem.,* **13,** 6 (1965).
175. E. A. Peterson and J. Rowland, *J. Chromatogr.,* **5,** 330 (1961).
176. E. A. Peterson and H. A. Sober, *Anal. Chem.,* **31,** 857 (1959).

177. A. Niederwieser and C. G. Honegger, *Helv. Chim. Acta,* **48,** 893 (1965).
178. K. Randerath and E. Randerath, *J. Chromatogr.,* **16,** 111 (1964).
179. G. E. Tarr, *J. Chromatogr.,* **52,** 357 (1970).
180. A. Niederwieser, *Chromatographia,* **2,** 362 (1969).
181. S. Turina, Z. Šoljić, and V. Marjanović, *J. Chromatogr.,* **39,** 81 (1969).
182. S. Turina and V. Jamnicki, *Anal. Chem.,* **44,** 1892 (1972).
183. F. Giess, H. Schlitt, and A. Klose, *Z. Anal. Chem.,* **213,** 331 (1965).
184. F. Geiss and H. Schlitt, "'KS' une nouvelle chambre chromatographique pour la CCM, à applications multiples," in *Int. Symp. Chromatogr., Electrophor., 5th, 1968,* Ann Arbor-Humphrey Science, Ann Arbor, Mich., 1969, p. 109.
185. M. S. J. Dallas, *J. Chromatogr.,* **17,** 267 (1965).
186. H. W. Prinzler and H. Tauchmann, *J. Chromatogr.,* **29,** 142 (1967).
187. R. A. de Zeeuw, *J. Chromatogr.,* **32,** 43 (1968).
188. R. A. de Zeeuw, *Anal. Chem.,* **40,** 915 (1969).
189. J. H. Van Dijk and W. J. Mijs, *Z. Anal. Chem.,* **236,** 419 (1968).
190. R. A. de Zeeuw, *Anal. Chem.,* **40,** 2134 (1968).
191. R. A. de Zeeuw, "Vapor-Programmed Thin-Layer Chromatography, Development and Applications," in *Progress in Separation and Purification,* Vol. 3, E. S. Perry and C. J. Van Oss, Eds., Wiley-Interscience, New York, 1970, p. 1.
192. T. Hodisan, S. Gocan, and C. Liteanu, *Stud. Univ. Babes-Bolyai, Ser. Chem.,* **17,** 63 (1972); through *Chem. Abstr.,* **78,** 79440k (1973).
193. T. Hodisan, S. Bandeanu, and C. Liteanu, *Stud. Univ. Babes-Bolyai, Ser. Chem.,* **17,** 119 (1972); through *Chem. Abstr.,* **78,** 105610n (1973).
194. T. Hodisan and C. Liteanu, *Stud. Univ. Babes-Bolyai, Ser. Chem.,* **17,** 73 (1972); through *Anal. Abstr.,* **25,** 360 (1973).
195. C. Liteanu and T. Hodisan, *Rev. Roum. Chim.,* **17,** 1985 (1972).
196. S. Gocan and C. Liteanu, *Rev. Roum. Chim.,* **17,** 661 (1972).
197. H. P. Kaufmann, K. D. Mukherjee, and Q. Khalid, *Nahrung,* **11,** 631 (1967).
198. E. Blasius, H. Augustin, and G. Klemm, *J. Chromatogr.,* **108,** 53 (1975).
199. J. E. Caton and G. Goldstein, *Anal. Biochem.,* **42,** 14 (1971).
200. J. Margolis and K. G. Kenrick, *Anal. Biochem.,* **25,** 347 (1968).
201. H. Foissy, *J. Chromatogr.,* **106,** 51 (1975).
202. A. C. Arcus, *Anal. Biochem.,* **37,** 53 (1970).
203. J. Margolis and C. W. Wrigley, *J. Chromatogr.,* **106,** 204 (1975).
204. A. J. Martin and R. L. M. Synge, *Biochem. J.,* **35,** 1358 (1941).
205. D. I. Cargill, *Analyst (London),* **87,** 865 (1962).
206. L. Hoerhammer, H. Wagner, and H. Koenig, *Dtsch. Apoth.-Ztg.,* **103,** 502 (1963).
207. T. Kagawa, T. Fukinbara, and Y. Sumiki, *Agr. Biol. Chem. (Tokyo),* **27,** 598 (1963).
208. H. P. Kaufmann and A. K. Sen Gupta, *Fette, Seifen, Anstrichm.,* **65,** 529 (1963).
209. E. Knappe and D. Peteri, *Z. Anal. Chem.,* **188,** 184 (1962).
210. *Ibid.,* p. 352.

211. J. A. Petzold, W. Camp, Jr., and E. R. Kirch, *J. Pharm. Sci.,* **52,** 1106 (1963).
212. E. Knappe and D. Peteri, *Z. Anal. Chem.,* **190,** 380 (1962).
213. R. D. Bennett and E. Heftmann, *J. Chromatogr.,* **9,** 353 (1962).
214. S. Patton, P. G. Kenney, and E. N. Boyd, *Mfg. Confect.,* **44,** 35 (1964).
215. M. Yawata and E. M. Gold, *Steroids,* **3,** 435 (1964).
216. A. F. Hofmann, "Thin-Layer Chromatography of Bile Acids and their Derivatives," and "Hydroxyapatite as an Adsorbent for Thin-Layer Chromatography: Separations of Lipids and Proteins," in *New Biochemical Separations,* A. T. James and L. J. Morris, Eds., Van Nostrand, New York, 1964, pp. 261, 283.
217. H. P. Kaufmann and T. H. Khoe, *Fette, Seifen, Anstrichm.,* **64,** 81 (1962).
218. H. P. Kaufmann and B. Das, *Fette, Seifen, Anstrichm.,* **65,** 398 (1963).
219. J. Davídek, "Chromatography on Thin Layer of Starch with Reversed Phases," in *Thin-Layer Chromatography,* G. B. Marini-Bettòlo, Ed., Elsevier, Amsterdam, 1964, p. 117.
220. J. Davídek and G. Janíček, *J. Chromatogr.,* **15,** 542 (1964).
221. S. Hjertén, *Arch. Biochem. Biophys., Suppl.,* **1,** 147 (1962).
222. H. Halpaap and K. Klatyk, *J. Chromatogr.,* **33,** 80 (1968).
223. B. G. Belenkii and E. S. Gankina, *J. Chromatogr.,* **53,** 3 (1970).
224. N. Donkai and H. Inagaki, *J. Chromatogr.,* **71,** 473 (1972).
225. E. P. Otocka, M. Y. Hellman, and P. M. Muglia, *Macromolecules,* **5,** 227 (1972).
226. D. Jaworek, *Chromatographia,* **2,** 289 (1969).
227. A. N. James, E. Pickard, and P. G. Shotton, *J. Chromatogr.,* **32,** 64 (1968).
228. K. Konishi and S. Yamaguchi, *Anal. Chem.,* **38,** 1755 (1966).
229. E. Nystroem and J. Sjoevall, *Anal. Biochem.,* **12,** 235 (1965).
230. *Ibid.,* **17,** 574 (1965).
231. C. Slagt and W. A. Sipman, *J. Chromatogr.,* **74,** 352 (1972).
232. Y. Pomeranz, *Chemist-Analyst,* **54,** 57 (1965).
233. C. Calzolari, L. Favretto, and B. Stancher, *J. Chromatogr.,* **54,** 373 (1971).
234. C. Tanford, K. Kawahara, and S. Lapanje, *J. Am. Chem. Soc.,* **89,** 729 (1967).
235. G. G. B. Klaus, D. E. Nitecki, and J. W. Goodman, *Anal. Biochem.,* **45,** 286 (1972).
236. F. Heinz and W. Prosch, *Anal. Biochem.,* **40,** 327 (1971).
237. Z. Wasyl, E. Luchter-Wasyl, and W. Bielański, Jr., *Biochim. Biophys. Acta,* **285,** 279 (1972).
238. N. S. Dizik and F. W. Knapp, *J. Food Sci.,* **35,** 282 (1970).
239. J. W. C. Peereboom and H. W. Beekes, *J. Chromatogr.,* **39,** 339 (1969).
240. D. Jaworek, "Duennschicht-Chromatographie und Elektrophorese an Molekularsieben," in *Int. Symp. Chromatogr. Electrophor., 6th 1970,* Ann Arbor Science, Ann Arbor, Mich., 1971, p. 118.
241. J. R. Parrish, *J. Chromatogr.,* **33,** 542 (1968).
242. R. W. Horobin and J. Gardiner, *J. Chromatogr.,* **43,** 545 (1969).
243. Z. Stránský and M. Srch, *J. Chromatogr.,* **28,** 146 (1967).

244. M. H. Blessing and B. Gebele, *Res. Exp. Med.,* **162,** 143 (1974).
245. B. J. Radola, *J. Chromatogr.,* **38,** 61 (1968).
246. *Ibid.,* **38,** 78 (1968).
247. K. J. Hruska and M. Franek, *J. Chromatogr.,* **93,** 475 (1974).
248. J. N. Miller, O. Erinle, J. M. Roberts, and C. Thirkettle, *J. Chromatogr.,* **105,** 317 (1975).
249. E. P. Otocka, *Adv. Chem. Ser.,* No. 125, 55 (1973).
250. H. Determann, *Gel Chromatography,* Springer-Verlag, Berlin, 1968, p. 51.
251. T. Wieland and H. Determann, *J. Chromatogr.,* **28,** 2 (1967).
252. K. W. Williams, *Lab. Pract.,* **22,** 306 (1973).
253. R. Consden, A. H. Gordon, and A. J. P. Martin, *Biochem. J.,* **40,** 33 (1946).
254. R. J. Wieme and M. Rabaey, *Naturwissenschaften.,* **44,** 112 (1957).
255. R. J. Wieme, *Clin. Chim. Acta,* **4,** 317 (1959).
256. E. W. Baur, private communication.
257. E. W. Baur, *J. Lab. Clin. Med.,* **61,** 166 (1963).
258. H. A. Ramsey, *Anal. Biochem.,* **5,** 83 (1963).
259. J. Kowalczyk, *Chem. Anal. (Warsaw),* **9,** 21 (1964).
260. *Ibid.,* **10,** 29 (1965).
261. *Ibid.,* **9,** 29 (1964).
262. *Ibid.,* p. 213.
263. J. Kowalczyk, *J. Chromatogr.,* **14,** 411 (1964).
264. J. Kowalczyk, *Chem. Anal. (Warsaw),* **8,** 659 (1963).
265. *Ibid.,* p. 823.
266. *Ibid.,* p. 835.
267. *Ibid.,* **9,** 891 (1964).
268. *Ibid.,* p. 899.
269. E. Correni, *Naturwissenschaften,* **51,** 40 (1964).
270. W. G. Dangerfield, *Nature,* **202,** 520 (1964).
271. L. Eriksen, *Scand. J. Clin. Lab. Invest.,* **10,** 39 (1958).
272. L. M. Weiner and B. Zak, *Clin. Chim. Acta,* **9,** 407 (1964).
273. H. Wada and E. E. Snell, *Anal. Biochem.,* **46,** 548 (1972).
274. R. Holmes, *Biochim. Biophys. Acta,* **133,** 174 (1967).
275. B. Zak, L. M. Weiner, and E. Baginski, *J. Chromatogr.,* **20,** 157 (1965).
276. J. K. Herd and L. Motycka, *Anal. Biochem.,* **53,** 514 (1973).
277. U. S. V. Acharya, M. Swaminathan, A. Sreenivasan, and V. Subrahmanyan, *Indian J. Med. Res.,* **52,** 224 (1964).
278. Y. Davlyatov, *Uzb. Biol. Zh.,* **7,** 45 (1963); through *Chem. Abstr.,* **59,** 13140 (1963).
279. H. J. van der Helm and M. G. Holster, *Clin. Chim. Acta,* **10,** 483 (1964).
280. D. Pette, *Klin. Wochenschr.,* **36,** 1106 (1958).
281. L. Popadiuk, *Arch. Immunol. Terapii Dosw.,* **9,** 139 (1961).
282. A. N. Ramanathan, *Antiseptic (Madras, India),* **60,** 1017 (1963).
283. V. Vaiciuvenas, *Lab. Delo,* **9,** 7 (1963); through *Chem. Abstr.,* **59,** 13091 (1963).
284. V. Vaiciuvenas, *Mater. 1-go (Pervogo) Soveshch. Aktual'n. Vpor. Klin. Biokhim., Riga, Sb.,* **1962,** 151; through *Chem. Abstr.,* **59,** 11869 (1963).

285. R. J. Wieme, *Behringwerk-Mitt.*, **34,** 27 (1958).
286. R. J. Wieme, *J. Chromatogr.*, **1,** 166 (1958).
287. C. van Arkel, R. E. Ballieux, and F. L. J. Jordan, *J. Chromatogr.*, **11,** 421 (1963).
288. P. Bruaux, S. Dormal, and G. Thomas, *Ann. Biol. Clin. (Paris),* **22,** 375 (1964).
289. R. Dalgelite, L. Juhnjaviciute, and V. Vaiciuvenas, *Lab. Delo,* **9,** 5 (1963).
290. M. van Sande and G. van Ros, *Ann. Soc. Belge Med. Trop.*, **43,** 537 (1963).
291. V. J. Yakulis, P. Heller, A. M. Josephson, L. Singer, and L. Hall, *Am. J. Clin. Pathol.*, **34,** 28 (1960).
292. K. Dose and S. Risi, *Z. Anal. Chem.*, **205,** 394 (1964).
293. B. Pfrunder, R. Zurflueh, H. Seiler, and H. Erlenmeyer, *Helv. Chim. Acta,* **45,** 1153 (1962).
294. B. Russell, J. Levitt, and A. Polson, *Biochim. Biophys. Acta,* **79,** 622 (1964).
295. A. M. Egorov, A. Kh. Vakhabov, and V. Ya. Chernyak, *J. Chromatogr.*, **46,** 143 (1970).
296. S. Bengtsson and L. Philipson, *Biochim. Biophys. Acta,* **79,** 399 (1964).
297. M. Duckworth and W. Yaphe, *Anal. Biochem.*, **44,** 636 (1971).
298. J. Porath, J.-C. Janson, and T. Låås, *J. Chromatogr.*, **60,** 167 (1971).
299. T. Låås, *J. Chromatogr.*, **66,** 347 (1972).
300. V. A. Khramov and Y. V. Galaev, *Lab. Delo,* **1968,** 311.
301. B. Daneholt, U. Ringborg, E. Egyhazy, and B. Lambert, *Nature,* **218,** 292 (1968).
302. D. R. Davies, E. D. Spurr, and J. B. Versey, *Clin. Sci.*, **40,** 411 (1971).
303. J. Dony, B. Beys, A. Rappe, and H. Muldermans, "Quelques applications de l'analyse immunoelectrophoretique a des medicaments d'origine animale," in *Int. Symp. Chromatog. Electrophor., 5th, 1968,* Ann Arbor-Humphrey Science, Ann Arbor, Mich., 1969, p. 552.
304. G. Kostner and A. Holasek, *Anal. Biochem.*, **46,** 680 (1972).
305. C.-B. Laurell, *Anal. Biochem.*, **15,** 45 (1966).
306. H. Hazama and H. Uchimura, *Michrochem. J.*, **17,** 318 (1972).
307. M. Baudler and F. Stuhlmann, *Naturwissenschaften,* **51,** 57 (1964).
308. E. W. Baur, private communication.
309. E. W. Baur, *Science,* **140,** 816 (1963).
310. E. W. Baur, *Clin. Chim. Acta,* **9,** 252 (1964).
311. E. W. Baur, N. M. Rowley, and A. G. Motulsky, Annual Meeting of the American Society of Human Genetics, Boulder, Col., August 26–28, 1964.
312. P. Berkeš-Tomašević, J. Rosić, and I. Berkeš, *Acta Pharm. Jugosl.*, **13,** 69 (1963).
313. G. Efremov, B. Vaskov, H. Duma, and M. Andrejeva, *Acta Med. Iugosl.*, **17,** 252 (1963); through *Chem. Abstr.*, **61,** 12305 (1964).
314. C. L. Marsh, C. R. Jolliff, and L. C. Payne, *Tech. Bull. Regist. Med. Technol.*, **34,** 1 (1964).
315. P. Berkeš-Tomašević, J. Rosić, and M. Ignjatović, *Arh. Farm. (Belgr.),* **13,** 9 (1963).
316. J. H. Daams, *J. Chromatogr.*, **10,** 450 (1963).

317. E. Espinosa, *Anal. Biochem.*, **9,** 146 (1964).
318. H. v. Euler, *Acta Biochim. Pol.*, **11,** 311 (1964).
319. J. Groulade, J. N. Fine, and C. Ollivier, *Nature,* **191,** 72 (1961).
320. J. Groulade and C. Ollivier, *Ann. Biol. Clin. (Paris),* **18,** 595 (1960).
321. L. Korngold, *Anal. Biochem.*, **6,** 47 (1963).
322. S. Raymond, *Ann. N.Y. Acad. Sci.*, **121,** 350 (1964).
322a. D. J. W. Burns and N. A. Turner, *J. Chromatogr.*, **30,** 469 (1967).
323. W. V. Styvesant, *Nature,* **214,** 405 (1967).
324. M. S. Reid and R. L. Bieleski, *Anal. Biochem.*, **22,** 374 (1968).
325. R. M. Roberts and J. S. Jones, *Anal. Biochem.*, **49,** 592 (1972).
326. R. De Wachter and W. Fiers, *Anal. Biochem.*, **49,** 184 (1972).
327. W. J. Ritschard, *J. Chromatogr.*, **16,** 327 (1964).
328. L. Cunningham, E. M. Rasch, A. L. Lewis, and R. Heitsch, *J. Histochem. Cytochem.*, **18,** 853 (1970).
329. J. H. Daams, *J. Chromatogr.*, **10,** 450 (1963).
330. H. A. Ramsey, *Anal. Biochem.*, **5,** 83 (1963).
331. H. E. Schaffer and F. M. Johnson, *Anal. Biochem.*, **51,** 577 (1973).
332. G. Pastuska and H. Trinks, *Chem.-Ztg.*, **85,** 535 (1961).
333. *Ibid.*, **86,** 135 (1962).
334. W. J. Criddle, G. J. Moody, and J. D. R. Thomas, *J. Chem. Educ.*, **41,** 609 (1964).
335. W. J. Criddle, G. J. Moody, and J. D. R. Thomas, *Nature,* **203,** 1327 (1964).
336. A. Moghissi, *Anal. Chim. Acta,* **30,** 91 (1964).
337. R. Frache and A. Dadone, *Chromatographia,* **6,** 430 (1973).
338. S. Agurell, *Acta Pharm. Suecica,* **2,** 357 (1965).
339. J. R. Sargent and B. P. Vadlamudi, *Anal. Biochem.*, **25,** 583 (1968).
340. H.-P. Nast and H. Fasold, *J. Chromatogr.*, **27,** 499 (1967).
341. V. Stefanovich, *J. Chromatogr.*, **31,** 466 (1967).
342. S. Raymond and L. Weintraub, *Science,* **130,** 711 (1959).
343. S. Raymond, *Clin. Chem.*, **8,** 455 (1962).
344. A. Baumgarten, *Blood,* **22,** 466 (1963).
345. A. Baumgarten, *Nature,* **199,** 490 (1963).
346. S. Raymond and Y.-J. Wang, *Anal. Biochem.*, **1,** 391 (1960).
347. O. Smithies, *Arch. Biochem. Biophys. Suppl.* **1,** 125 (1962).
348. L. Ornstein, *Ann. N.Y. Acad. Sci.*, **121,** 321 (1964).
349. K. Dose and G. Krause, *Naturwissenschaften,* **49,** 349 (1962).
350. P. Fasella, A. Giartosio, and C. Turano, "Applications of Thin-Layer Chromatography on Sephadex to the Study of Proteins," in *Thin-Layer Chromatography,* G. B. Marini-Bettòlo, Ed., Elsevier, Amsterdam, 1964, p. 205.
351. R. Vendrely, Y. Coirault, and A. Vanderplancke, *Compt. Rend.*, **258,** 6399 (1964).
352. K. Keck and U. Hagen, *Biochim. Biophys. Acta,* **87,** 685 (1964).
353. G. Augusti-Tocco, C. Carestia, P. Grippo, E. Parisi, and E. Scarano, *Biochim. Biophys. Acta,* **155,** 8 (1968).
354. H. Bayzer, *J. Chromatogr.*, **27,** 104 (1967).
355. P. Nygaard, *J. Chromatogr.*, **30,** 240 (1967).

356. A. S. C. Wan, *J. Chromatogr.*, **60**, 371 (1971).
357. D. J. W. Burns and N. A. Turner, *J. Chromatogr.*, **30**, 469 (1967).
358. Ch. Montant and J. M. Rouze-Soulet, *Bull. Soc. Chim. Biol.*, **42**, 161 (1960).
359. R. L. Bieleski and N. A. Turner, *Anal. Biochem.*, **17**, 278 (1966).
360. A. R. Cook and R. L. Bieleski, *Anal. Biochem.*, **28**, 428 (1969).
361. D. V. Lopiekes, F. R. Dastoli, and S. Price, *J. Chromatogr.*, **23**, 182 (1966).
362. F. Dobici and G. Grassini, *J. Chromatogr.*, **10**, 98 (1963).
363. A. Affonso, *J. Chromatogr.*, **31**, 646 (1967).
364. M. Szylit, *Bull. Soc. Chim. Biol.*, **49**, 1884 (1967).
365. S. Raymond and B. Aurell, *Science*, **138**, 152 (1962).
366. S. Raymond and M. Nakamichi, *Anal. Biochem.*, **7**, 225 (1964).
367. E. Espinosa, *Anal. Biochem.*, **9**, 146 (1964).
368. E. Kaltschmidt and H. G. Wittmann, *Anal. Biochem.*, **36**, 401 (1970).
369. S. Avital and D. Elson, *Anal. Biochem.*, **57**, 287 (1974).
370. H. G. M. Clarke and T. Freeman, *Clin. Sci.*, **35**, 403 (1963).
371. C. A. Converse and D. S. Papermaster, *Science*, **189**, 469 (1975).
372. K. Ikeda and S. Suzuki, U.S. Pat. 1,015,891 (January 30, 1912).
373. R. R. Williams and R. E. Waterman, *Proc. Soc. Exp. Biol. Med.*, **27**, 56 (1929/30).
374. O. Vesterberg and H. Svensson, *Acta Chem. Scand.*, **20**, 820 (1966).
375. O. Vesterberg, *Acta Chem. Scand.*, **23**, 2653 (1969).
376. P. G. Righetti and J. W. Drysdale, *J. Chromatogr.*, **98**, 271 (1974).
377. J. W. Drysdale, P. G. Righetti, and H. F. Bunn, *Biochem. Biophys. Acta*, **229**, 42 (1971).
378. A. C. Been and E. M. Rausch, *J. Histochem. Cytochem.*, **20**, 368 (1972).
379. *Ibid.*, **18**, 675 (1970).
380. H. Delinceé and B. J. Radola, *Biochem. Biophys. Acta*, **200**, 404 (1973).
381. Z, L, Awdeh, A. R. Williamson, and B. A. Askonas, *Nature*, **219**, 447 (1968).
382. D. H. Leaback and A. C. Rutter, *Biochem. Biophys. Res. Commun.*, **32**, 447 (1968).
383. O. Vecter, *Biochem. Biophys. Acta*, **257**, 11 (1972).
384. S. Avital and D. Elson, *Anal. Biochem.*, **57**, 274 (1974).
385. Z. L. Awdeh, A. R. Williamson, and B. A. Askonas, *Nature*, **219**, 66 (1968).
386. W. J. Criddle, G. J. Moody, and J. D. R. Thomas, *Analyst (London)*, **94**, 461 (1969).
387. G. R. Finlayson and A. Chrambach, *Anal. Biochem.*, **40**, 292 (1971).
388. H. Wada and E. E. Snell, *Anal. Biochem.*, **46**, 548 (1972).
389. K.-D. Kohnert, E. Schmid, H. Zuehlke, and H. Fiedler, *J. Chromatogr.*, **76**, 263 (1973).
390. U. E. Loerning, *Biochem. J.*, **102**, 251 (1967).
391. P. G. Righetti and J. W. Drysdale, *Ann. N.Y. Acad. Sci.*, **209**, 163 (1973).
392. L. S. Bates and C. W. Deyoe, *J. Chromatogr.*, **73**, 296 (1972).
393. J. R. Florini, R. P. Brivio, and B.-A. Battelle, *Ann. N.Y. Acad. Sci.*, **209**, 299 (1973).
394. H. K. Robinson, *Anal. Biochem.*, **49**, 353 (1972).

395. S. N. Vinogradov, S. Lowenkron, M. R. Andonian, J. Bagshaw, K. Felgenhauer, and S. J. Pak, *Biochem. Biophys. Res. Commun.*, **54**, 501 (1973).
396. G. Lundblad, O. Vesterberg, R. Zimmerman, and J. Ling, *Acta Chem. Scand.*, **26**, 1711 (1972).
397. P. G. Righetti, M. Pagani, and E. Gianazza, *J. Chromatogr.*, **109**, 341 (1975).
398. E. Gianazza, M. Pagani, M. Luzzana, and P. G. Righetti, *J. Chromatogr.*, **109**, 357 (1975).
399. J. Bours, *J. Chromatogr.*, **60**, 225 (1971).
400. S. Lewin, *Biochem. J.*, **118**, 37p (1970).
401. G. R. Finlayson and A. Chrambach, *Anal. Biochem.*, **40**, 292 (1971).
402. R. Blaich, *Naturwissenschaften*, **58**, 55 (1971).
403. K. C. Humphreys, *J. Chromatogr.*, **49**, 503 (1970).
404. C. J. Smyth and T. Wadstroem; see Ref. 376.
405. S. Lewin, *Biochem. J.*, **117**, 41p (1970).
406. D. Graesslin, A. Trautwein, and G. Bettendorf, *J. Chromatogr.*, **63**, 475 (1971).
407. K. H. Fantes and I. G. S. Furminger, *Nature*, **215**, 750 (1967).
408. R. F. Peterson, *J. Agric. Food Chem.*, **19**, 595 (1971).
409. O. Vesterberg, *Ann. N.Y. Acad. Sci.*, **209**, 23 (1973).
410. H. F. Bunn, *Ann. N.Y. Acad. Sci.*, **209**, 345 (1973).
411. E. M. Jordan and S. Raymond, *Anal. Biochem.*, **27**, 205 (1969).
412. R. Frater, *J. Chromatogr.*, **50**, 469 (1970).
413. A. L. Latner, M. E. Parsons, and A. W. Skillen, *Biochem. J.*, **118**, 298 (1970).
414. C. W. Wrigley and K. W. Shepherd, *Ann. N.Y. Acad. Sci.*, **209**, 154 (1973).
415. A. L. Latner, *Ann. N.Y. Acad. Sci.*, **209**, 281 (1973).
416. G. Dale and L. L. Latner, *Clin. Chim. Acta*, **24**, 61 (1969).
417. V. Macko and H. Stegemann, *Z. Physiol. Chem.*, **350**, 917 (1969).
418. K. G. Kenrick and J. Margolis, *Anal. Biochem.*, **33**, 204 (1970).
419. N. Catsimpoolas, *Ann. N.Y. Acad. Sci.*, **209**, 144 (1973).
420. N. Catsimpoolas, *Sci. Tools*, **16**, 1 (1969).
421. R. F. Riley and M. K. Coleman, *J. Lab. Clin. Med.*, **72**, 714 (1968).
422. R. Quast, *J. Chromatogr.*, **54**, 405 (1971).
423. B. J. Radola, *Biochim. Biophys. Acta*, **295**, 412 (1973).
424. B. J. Radola, *Ann. N.Y. Acad. Sci.*, **209**, 127 (1973).
425. M. B. Hayes and D. Wellner, *J. Biol. Chem.*, **244**, 6636 (1969).
426. Y.-T. Li and S.-C. Li, *Ann. N.Y. Acad. Sci.*, **209**, 187 (1973).
427. E. M. Spencer and T. P. King, *J. Biol. Chem.*, **246**, 201 (1971).
427a. J. P. Arbuthnott and J. A. Beeley, Eds., *Isoelectric Focusing*, Butterworths, Woburn, Mass., 1975.
428. K. Brandt, *Amer. Lab.*, **4**, 69 (1972).
429. E. Fosslien, *J. Chromatogr.*, **63**, 59 (1971).
430. F. Snyder and D. Smith, "An Automated System for Sample Collection and Computer Analysis of Thin-Layer Radiochromatograms," in *Separation Techniques in Chemistry and Biochemistry*, R. A. Keller, Ed., Marcel Dekker, New York, 1967, p. 331.

431. E. Fosslien, F. Musil, D. Domizi, L. Blickenstaff, and J. Lumeng, *J. Chromatogr.,* **63,** 131 (1971).
432. G. Kasang, G. Goeldner, and N. Weiss, *J. Chromatogr.,* **59,** 393 (1971).
433. A. A. Ismail and R. A. Harkness, *Biochem. J.,* **99,** 717 (1966).
434. D. L. Sanders and L. R. Snyder, *J. Chromatogr. Sci.,* **8,** 706 (1970).
435. H. Halpaap and H. Bausch, Ger. Pat. 2,026,304 (December 15, 1971).
436. L. Schneck, M. Pourfar, and A. Benjamin, *J. Lipid Res.,* **11,** 66 (1970).
437. J. M. Philp, *Proc. Soc. Anal. Chem.,* **9,** 293 (1972).
438. J. A. Perry, T. H. Jupille, and L. J. Glunz, *Anal. Chem.,* **47,** 65A (1975).
439. J. A. Perry, *J. Chromatogr.,* **110,** 27 (1975).
440. *Ibid.,* **113,** 267 (1975).
441. H. Faulk and K. Krummen, *J. Chromatogr.,* **103,** 279 (1975).
442. R. K. Vitek, C. J. Seul, M. Baier, and E. Lau, *Am. Lab.,* **6,** 109 (1974).

Chapter **VI**

REACTIONS ON PLATES

The inert character of the thin-layer material makes it ideally suited for use with stronger corrosive reagents. Miller and Kirchner [1] in 1953 originated and developed the idea of carrying out chemical reactions directly on thin-layer plates. With this technique the sample can be spotted onto the plate and then covered with a reagent. After completion of the reaction, development in a suitable solvent separates the products of the reaction. In cases where this technique is not suitable the reagent and compound can be mixed on a microscale in a small test tube, or in capillaries as proposed by Mathis and Ourisson [2]. The crude mixture can then be applied directly to the chromatoplate. The R_f value of the original compound coupled with the chromatographic results of the reaction often are enough to positively identify a compound, and in other cases the results can offer valuable clues to the identity of the compound. As an example, a sample of citral spotted on a silica gel plate and covered by a drop of 30% hydrogen peroxide was then exposed to ultraviolet light for 10 min in order to catalyze the oxidation to geranic acid. A second spot of citral was covered by a drop of a 10% solution of lithium aluminum hydride in ether in order to reduce it to geraniol. After chromatographing,

Table 6.1 Results of Reactions for Chromatostrip Identification of Terpenes and Other Essential Oil Constituents on Silicic Acid Chromatostrips[a,b]

		Reduction				Derivatives			
Compound	*Oxodation,[c] CrO_3*	*Aluminum Isopropoxide[d]*	*Lithium Aluminum Hydride*	*Dehydration,[c] H_2SO_4*	*Hydrolysis,[d] KOH*	*Semicarbazone[c]*	*3,5-Dinitrobenzoate[d]*	*Phenylhydrazone[c]*	*Phenyl Isocyanate[d]*
Carveol	Carvone			Hydrocarbon					Carbamate
Linalool	Citral			Hydrocarbon			No reaction		Carbamate
Geraniol	Citral			Hydrocarbon			Benzoate		Carbamate
α-Terpineol	No reaction	No reaction		Hydrocarbon			No reaction		No reaction
Nopol	No reaction								Carbamate
Methyl heptenol	Methyl heptenone			No reaction			Benzoate		Carbamate
Octyl alcohol				No reaction					
Nerol	Citral			Hydrocarbon			Benzoate		Carbamate
Pulegone	No reaction	No reaction	Pulegol[c]	No reaction		No reaction		Hydrazone	
Methyl heptenone	No reaction	Methyl heptenol		No reaction		Semicarbazone		Hydrazone	
Carvone		Carveol	Carveol[c]	No reaction		No reaction		No reaction	
Citral	No reaction	Geraniol	Geraniol[c]	No reaction		2 Semicarbazones		Hydrazone	
Lauric aldehyde	No reaction			No reaction		Semicarbazone		Hydrazone	
Cinnamaldehyde		Cinnamyl alcohol	Cinnamyl alcohol[c]	No reaction		Semicarbazone		Hydrazone	
Citronellol				No reaction				Hydrazone	
Furfural						Semicarbazone			
Linalool monoxide			Reduction[c]						
Terpinyl acetate		No reaction	Terpineol[c]	Trace of hydrocarbon	Terpineol				
Linalyl acetate				Trace of hydrocarbon	Linalool				
Carvyl acetate				No reaction	Carveol				
Geranyl acetate					Geraniol				
Neryl acetate			Nerol[d]		Nerol				

[a] From Miller and Kirchner [1]; reproduced with permission of the Americal Chemical Society.

[b] "No reaction" indicates that no reaction products were observed. In oxidation of citral to geranic acid, results of oxidation are not visible because of interference from acetic acid; therefore it is marked no reaction. In dehydration reaction, it means no hydrocarbons are formed.

[c] Reaction directly on chromatostrip.

[d] Reaction on microscale in test tube, then mixture chromatographed.

the R_f values of the two reaction products along with the R_f value of the original compound served fairly well to establish the identity of the latter. The numerous examples given in this work (Table 6.1) illustrate not only the versatility of this special technique but also the wide applicability of thin-layer chromatography to many different types of compounds, if indeed it needs to be demonstrated with all the examples of separation on various adsorbents available in the literature.

In many cases the reactions on plates do not go to completion so that there is present a mixture of the original compound and the resulting reaction products.

Many times specific reactions may be used to give a great deal of information about an unknown compound with the expenditure of a very small amount of material. The procedures for chemical reactions on thin layers may be summarized, and no doubt these will serve to remind the reader of many alternatives that can be used.

1 OXIDATION

The spot of sample that is to be oxidized at the origin is covered with a drop of a saturated solution of chromic anhydride in glacial acetic acid. Some compounds can be oxidized by the application of 30% hydrogen peroxide and exposure to UV light for 10 min [1].

Kofoed et al. [3, 4] oxidized phenothiazines to the corresponding sulfoxides after development on silica gel by applying 10 to 20% hydrogen peroxide to the spot. There was indication that 30% hydrogen peroxide followed by drying with hot air produced a second product which had a lower R_f value than the sulfoxide and probably corresponded to the sulfone.

Ognyanov [5, 6] treated the thin-layer plate with ozone after the chromatogram had been developed in order to assist in making the compounds visible. Naturally, this same reaction could be applied to the sample prior to the development. After the sample is spotted, the plate is placed in a desiccator which contains 10 to 15% ozone. Fifteen to 20 min later, it is exposed to air to remove excess ozone.

Fuhrmann and Jeralomon [7] examined this reaction for the detection of a wide variety of compounds using various detecting agents for detecting the oxidation products. For paraffinic hydrocarbons and aliphatic amides the exposure to ozone was carried out at 55 to 90°C instead of at room temperature. Before spraying on the detecting reagent it is essential that the excess ozone be removed from the plate.

Weidemann [8] oxidized 2-desoxypolyols on silica gel layers with 0.1% periodic acid in 20% *o*-phosphoric acid and heating at 50°C for 15 min.

Excess periodic acid was then destroyed with sulfur dioxide before spraying with the detecting agent.

Frijns [9] oxidized reserpine and rescinamine by exposure of the developed chromatogram to direct daylight for spontaneous air oxidation of the compounds to the 3-hydro compounds. Wilk et al. [10] oxidized naphthylamine on silica gel layers by air oxidation at 100°C for 10 min after spraying with trifluoroacetic acid.

Kaess and Mathis [11] have oxidized alkaloids on the layer with 10% chromic acid in glacial acetic acid. Oxidations of alkaloids with *p*-nitrobenzoic were carried out in capillaries before applying to the thin-layer plate.

In order to separate strychnine from brucine, Rusiecki and Henneberg [12] applied potassium dichromate to the sample at the starting point. The dichromate oxidized the brucine to its *o*-quinone which remained fixed at the starting point in the butanol–36% hydrochloric acid solvent that was used to move the strychnine. Malins and Mangold [13] and Mangold [14] separated saturated fatty acids and esters in the presence of unsaturated components by developing the chromatogram in a solvent composed of peracetic acid–acetic acid–water (2:15:3) by incorporating the oxidizing agent in the solvent. In this manner the unsaturated components were oxidized and moved with the solvent front.

Hamman and Martin [15] oxidized 17-hydroxycorticosteroids to 17-ketosteroids by spraying the plate twice with 10% aqueous sodium periodate solution and then heating the plate to approximately 50°C.

Gardner [6] in a two-dimensional technique for organophosphorus pesticides oxidized the compounds with bromine vapor after the first development and then developed in the second direction before detecting with the enzyme inhibition test (T-113).

Mathis and Ourisson [2] used *p*-nitroperbenzoic acid, sodium hypobromite, and osmium and ruthenium tetraoxides for oxidations in capillaries.

2 REDUCTION

Reduction with lithium aluminum hydride can be carried out by adding a drop of a 10% solution in ether to the sample spot. Care must be taken to avoid too great an excess of reagent because it reacts vigorously with the moisture in the solvent and in some cases with the solvent itself. Any excess reducing agent can be withdrawn from the layer by means of a medicine dropper. In the reduction of esters by this reagent it is necessary to carry out the reaction in a small test tube or capillary, after which a sample of the reaction mixture may be spotted on the layer. This same procedure for carrying out the reaction prior to spotting the sample is used with the reduction with aluminum isopropoxide. A drop of a solution of 5 g of aluminum isopropoxide in 50 ml of benzene is added to a drop of the sample in a small test tube. The mixture is heated until a distillate appears on the cool part of the test tube. This is centrifuged back and a drop of the mixture is then applied to the thin layer.

Hamman and Martin [15] used a freshly prepared solution of equal volumes of 10% ethanolic sodium borohydride and 0.1 *N* sodium hydroxide to reduce numerous steroids. After reacting for 30 min, the excess reagent was neutralized with 25% acid. Smith and Price [17] found that a 5-min exposure to a 1% solution of sodium borohydride in methanol was sufficient to reduce 7-ketocholesterol. The same method was used for the reduction of sterol hydroperoxides [18]. Polesuk and Ma [19] and Kaess and Mathis [11] have reduced a number of alkaloids by spraying with sodium borohydride solution. Glaser et al. [20] used a spray of freshly prepared 0.4% sodium borohydride solution in 95% ethanol to reduce disulfides. Reduction was allowed to proceed for 15 to 20 min, and then the excess reagent was destroyed with an acid solution of glacial acetic acid–6 *N* hydrochloric acid–acetone (8:2:90).

Kaufman and Khoe [21], Kaufmann et al. [22], and Knappe and Peteri [23] used a catalytic hydrogenation directly on the thin-layer plate. In order to carry this out a drop of 2% colloidal palladium solution is placed on the thin layer, then dried at 80 to 90°C for 1 hr. After the sample is spotted directly on the palladium layer the hydrogenation is conducted for 1 hr in a desiccator filled with hydrogen. Kaufmann and co-workers [21, 22] applied this technique in the two-dimensional separation of critical pairs of fatty acids (Fig. 6.1). They ran a partition chromatographic separation in one direction and after applying the catalyst and hydrogenating they partitioned in the second direction. Wieland and Ottenheym [24] in determining the configuration of amino acids coupled a few milligrams of the amino acid with a carbobenzyloxy-L-amino acid azide and then removed the protecting group by hydrogenation at the sample application spot after moistening with palladium chloride solution.

Graf and Hoppe [25], in a procedure for making nitro compounds visible on the plates, first reduced the nitro group to an amine group by spraying with a freshly prepared solution of 3 ml of 15% zinc chloride and 15 ml of hydrochloric acid in 180 ml of water. Here again, this reaction could be applied to sample spots prior to development. Yasuda [26] carried out a reduction of nitro groups by incorporating a zinc reductor directly in the thin layer. Thawley [27] incorporated finely powdered tin (5% wt/wt) in the silica gel layer for the reduction of 3,5-dinitrobenzoates and 2,4-dinitrophenylhydrazones. After development,

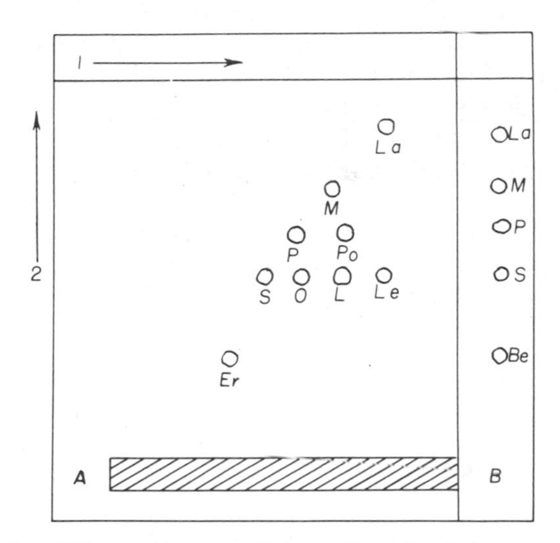

Fig. 6.1 Separation of fatty acid mixture by two-dimensional chromatography employing hydrogenation on the plate. Adsorbent, kieselguhr G impregnated with undecane; solvent, acetic acid–acetonitrile (3:2) 80% saturated with undecane; detecting agent, Rhodamine B. Development in the second direction after hydrogenation using a colloidal palladium catalyst (hatched area). Sample size, 5 μg each. Starting point *A*: lauric, myristic, palmitic, palmitolic, stearic, oleic, linolenic, linoleic, and erucic acids. Starting point *B*: lauric, myristic, palmitic, stearic, and behenic acids. From Kaufmann et al. [9]; reproduced with permission of the authors and Industriverlag von Hernhaussen K. G.

the plate was placed overnight in a tank containing hydrogen chloride vapors. Incorporation of powdered zinc or tin in cellulose layers did not give consistent results, as was also shown by Barton [28]. Barton found that for cellulose layers the zinc powder should be sprinkled or dusted evenly on the layer. He used the method for the reduction of 3-hydroxyflavanones. Even with silica gel layers the zinc cannot be incorporated in the layer if acid developing agents are to be used. After development the zinc powder may be dusted on or sprayed on as a suspension in acetone. Schuetz and Schindler [29] reduced nitro groups of pesticides by spraying with an acidic titanium (III) chloride solution.

Tyrer et al. [30] reduced tetrazolium salts to colored formazans by adding a drop of ammonium sulfide solution to the sample spot. The plate was then warmed to evaporate the excess reagent. Ammonium sulfide vapors were also used as was an alkaline ascorbate solution.

Nakamura and Tamura [31] reduced *S*-alkylthiosulfates and *S*-arylthiosulfates, the so-called "Bunte salts," by spraying with a freshly prepared solution of 154.2 mg of dithiothreitol in 50 ml of 0.05 *M* tris–hydrochloric acid buffer (pH 9.20) containing 93.1 mg ethylenediaminetetraacetic acid disodium salt. The Bunte salts were reduced to hydrogen sulfite and the corresponding thiols on standing for at least 5 min. They also used a freshly prepared 2% solution of potassium cyanide in 95% methanol for the reduction of disulfides. After spraying, the plate is allowed to stand for 5 min.

Scotney and Truter [32] reduced 11β-hydroperoxylanostenyl acetate to 11-oxolanostenyl acetate by spraying the sample spot on a silica gel layer with 5% aqueous ferrous ammonium sulfate–methanol–ether (2:1:1).

3 DEHYDRATION

Terpene alcohols have been converted to hydrocarbons by adding a drop of concentrated sulfuric acid to the sample spot. The thin-layer plate is then developed with hexane, and since oxygenated compounds do not move in hexane, hydrocarbons that are formed during this reaction are readily moved out away from the reaction zone [1]. Phosphorus oxychloride in pyridine has been used in capillaries[15]. The pyridine–chromic acid complex may be used in a similar manner [33]. To the suspension obtained with 40 μl of anhydrous pyridine and 4 mg of finely powered chromium(VI) oxide is added 5 mg of the test substance in 50 μl of anhydrous pyridine. The mixture is allowed to stand for 12 hr at room temperature before 10 μl is applied to a thin-layer plate.

It has been shown [9, 34] that certain alkaloids can be dehydrated by heating (105 to 120°C) on a thin-layer plate.

4 HYDROLYSIS

Baggiolini and Dewald [35] carried out the hydrolysis of *p*-aminobenzoic acid esters and sulfonamides by placing the thin-layer plate in an atmosphere of concentrated hydrochloric acid vapor maintained at 100°C. Kartnig and Wegschaider [36] used the same procedure for the hydrolysis of some saponins and glycosides [37]. Glycopeptides have been hydrolyzed on thin layers heated at 100 to 105°C in a tight glass chamber containing 4 *N* hydrochloric acid [38]. Nitriles can be hydrolyzed to their corresponding ammonium salts [39] by spraying the plate with 2 *N* sulfuric acid–hydrogen peroxide (9:1), drying 40 to 60 min in air, covering with a glass plate, and heating (120°C) for 7 to 8 min, or they may be hydrolyzed by exposing to an atmosphere of hydrogen chloride for 12 hr at 50°C [40]. Schmid and Mangold [41] cleaved aldehydrogenic lipids by placing the layer face down 15 cm above a dish of concentrated hydrochloric acid heated to 40 to 50°C for 5 min. Visanathan et al. [42] applied a slight modification of this method for the quantitative assessment of aldehydes liberated from a mixture of alkenyl acyl and diacyl ethanolamine phosphatides. Anderson et al. [43] examined this reaction and found that it was not complete so that for quantitative work the reaction should be carried out by shaking in a tube with concentrated hydrochloric acid; however the lipids do not need to be eluted from the silica gel first. Owens [44] made use of a mercuric chloride spray for the specific hydrolysis of plasmalogens to (2-acyl) lysophospholipids.

Steroid sulfate cleavage was accomplished by Payne and Mason [45] by exposing the thin-layer plate to an atmosphere of hydrochloric acid–dioxane (90:10) for 3 hr. Hurwitz [46] used a solution of 0.1 ml concentrated sulfuric acid in 9.9 ml of either dioxane or acetone for the same type of steroid reaction. Sodium testosterone sulfate was not affected by the acetone solution and only slightly by the dioxane solution. On the other hand, sodium estrone sulfate was hydrolyzed completely by the dioxane solution and quite extensively by the acetone reagent. Schuetz [47] carried out a combined hydrolysis and reduction of nitrazepam and its major metabolities on thin layers by spraying with an acidic titanium(III) chloride solution. The layer was covered with a glass plate and heated for 10 min at 100°C. Curtius and Mueller [48] hydrolyzed the trimethylsilyl ethers of steroids by spraying the layer with methanol–concentrated sulfuric acid (1:1).

Carbamates have been hydrolyzed to the corresponding phenols by spraying the layers with 5% potassium hydroxide solution [49]. Marcus and Ma [50] hydrolyzed flavone esters by spraying with a 2 *N* sodium hydroxide solution, leaving the plate stand at room temperature for 45 min, heating at 110°C for 5 min, and then spraying with acetic acid–water (1:1). Hydrolysis of esters can be carried out by mixing a drop of the compound with a drop of potassium hydroxide in ethylene glycol (6 g in 100 ml) [51] in a small test tube and heating until a ring of liquid condenses on the cool part of the tube. The liquid condensate is then centrifuged back into the mixture for spotting on the thin-layer plate. Kaess and Mathis [9] saponified a number of alkaloids by heating in a capillary with 0.1 *N* ethanolic potassium hydroxide at 100°C for 1 hr.

Barney [52] used a solution of 11.2 ml of 57% hydriodic acid (free of inhibitors) and 50 ml of glacial acetic acid diluted to 100 ml with water as a hydrolysis reagent for numerous organophosphorus compounds. The hydrolysis was accomplished by heating the layers to 250°C for 15 min after they had been sprayed. Askew et al. [53, 54] modified this procedure by clipping a second glass plate over the sprayed plate before heating. In this way it was possible to hydrolyze at the lower temperature of 180°C.

Trimethyl, triethyl, and tributyl phosphates could be hydrolyzed in this fashion but not triphenyl or tricresyl phosphates.

5 HALOGENATION

Cargill [55] brominated a sample of sterols in order to effect a separation of cholestanol from cholesterol by means of thin-layer chromatography. To accomplish this the sample was spotted on the plate followed by a 0.1% (wt/vol) solution of bromine in chloroform. The bromine added was equivalent to two to three times the weight of the sample. After development in benzene–ethyl acetate (2:1) the cholestanol could be clearly differentiated from the reaction products of cholesterol with bromine. Schuetz and Schuetz [56] separated phenylbutazone and prenazone by using the same technique. De Zeeuw and Wijsbeek [57] applied bromination at the origin to the identification and separation of closely related barbiturates and thiobarbiturates. They pointed out that many times the reactions do not go to completion or are chain reactions. This can give rise to a pattern of products which are specific in nature and can aid in identification. Schuetz and Schuetz [58, 59] have divided barbiturates into groups according to the type of reaction with bromine. Kaufmann and co-workers [21, 22, 60] employed a slightly different technique for bromination on thin-layer plates. They used acetic acid-acetonitrile (1:1) containing 0.5% bromine as a developing agent for separating critical pairs of fatty acids; thus the bromination takes place during development.

Copius Peereboom and Beekes [61], using the same technique of bromine addition to the developing solvent, could differentiate between cholesterol and its critical pair, brassicasterol. Koehler et al. [62] used the technique to separate saturated from unsaturated fatty acids prior to gas chromatographic analysis.

Polesuk and Ma [63] chlorinated acetanilide, *p*-chloroacetanilide, and 2,5-dichloroacetanilide on silica gel. The plate was placed in a tank filled with chlorine for 20 sec and then in a circulating air oven at 60°C for 5 min. Acetanilide reacted completely yielding five products, whereas *p*-chloroacetanilide reacted incompletely and gave only three products. About 50% of the 2,5-dichloroacetanilide reacted forming two polychlorinated products.

Wilk and Brill [64] treated alkaloids on thin layers with iodine vapor to form many derivatives whose sequence after development could be used for identification. Wilk et al. [10] exposed thin-layer plates containing polynuclear hydrocarbons to iodine vapor in a dark tank for several minutes to several hours until the reaction was complete as shown by the intense color of the spots. After removal of the excess iodine, the layers were developed in an appropriate solvent. The same procedure was applied to a number of aromatic amines. Schmidt [65] found that 23 out of 43 pharmaceutical compounds which he tested reacted with iodine on a thin-layer plate. Wilk and Taupp [66] examined the dehydration of cholesterol adsorbed on thin layers and treated with iodine vapors. They isolated five compounds: a derivative of a dimerization of cholesterol, a derivative of a trimerization, a trimer, 1-isopropyl-4-methylpicene, and 10-isopropyl-7-11*H*-indeno-2,1-phenanthrene. Brown and Turner [67] investigated the action of iodine on phenolic steroids adsorbed on silica gel layers. Estrone yielded two compounds, 2-iodoestrone the major component and 2,4-diiodoestrone the minor product. In preparative runs the iodination was found to be practically complete after 6 hr at room temperature. Even after only 20 min the yield of the major product was 10%, showing that some reaction would occur within the time required to locate the zones. These results indicate that care must be taken in using this reagent for locating compounds in quantitative work. In contrast acetylation of the phenolic hydroxyl group prevented the iodination of both estrone acetate and its 2-iodo derivative.

6 ENZYMATIC ACTION

Randerath and Randerath [68] carried out an enzymatic reaction directly on an anion-exchange layer of cellulose impregnated with polyethylenimine. A buffered solution of phosphodiesterase was applied to the sample spot of cytidine diphosphate glucose. This was then covered with Parafilm and allowed to stand for 45 to 60 min at 23°C. Chromatography of the degradation products yielded cytidine 5′-monophosphate and glucose 1-phosphate. Further degradation of the monophosphates could be accomplished by following the first enzyme reaction with a solution of prostate phosphomonoesterase with the production of cytidine, orthophosphate, and glucose.

Tschiersch and Schwabe [69] separated sugars and aglycones from glycosides and glycoside mixtures by using a two-dimensional separation with an enzymatic treatment of the layer after the first development

7 ESTERIFICATION

In order to increase the mobility of steroidal sapogenines, Bennett and Heftmann [70] found that it was possible to esterify the three hydroxy steroids directly on thin-layer plates by means of trifluoroacetic anhydride. After treating the compounds with the anhydride, it was necessary to dry the plate in the hood for several minutes in order to remove the trifluoroacetic acid which was formed as a by-product. As an alternative, the sapogenins could be esterified prior to spotting on the layer by adding 2 ml of trifluoroacetic anhydride to 22 ml of a 0.01-0.1% solution of the sapogenins. The solution was shaken for 1 min, then the acid was neutralized with 1 ml of 2 *N* aqueous sodium carbonate. Riess [71] methylated organophosphorus acids directly on chromatographic plates with an ether solution of diazomethane. Carboxylic acids containing a piperidine ring have been esterified in the same manner [72]. Elgamal and Fayez [73] applied a 5% solution of potassium carbonate in acetone (containing a little water). The plate was dried and then sprayed with 50% methyl iodide in acetone. It was then placed in a tank saturated with vapor from a mixture of methyl iodide–acetone (1:4) for 3 hr at 50°C. Kaufmann et al. [74] applied a 12% methanolic potassium hydroxide solution for the methanolysis of lipids. Viswanathan et al [42, 75, 76] used a modification of this in their two-dimensional techniques for the quantitative analysis of lipids using both an acid hydrolysis and a methanolysis. Holloway and Challen [77] esterified by treating the sample spot with boron trifluoride–methanol and then heating with a hot air drier. The reagent may be prepared [78] by bubbling boron trifluoride (Matheson, Coleman and Bell) into 1 liter of methanol cooled in an ice bath (fume hood) without allowing white fumes to emerge from the flask until 125 g has been absorbed. This reagent keeps for 2 years. Oette and Doss [79] applied sodium methylate to the transesterification of lipids. The samples were applied to plates that had been predeveloped with chloroform–methanol (2:1) to remove impurities and activated at 110°C. Transesterification was carried out by spraying with 2 *N* sodium methylate. The reaction was complete in 5 min. Other workers [80, 81] have used this same technique.

It has also been shown that the adsorbent may be scraped off the layer and the adsorbed compounds methylated directly without eluting them first [82–84]. Probably the simplest method here is to place the adsorbed material in ethyl acetate–methanol (9:1) or diethyl ether–methanol (9:1) and pass in diazomethane until a persistent yellow color develops [85]. Mancha Perelló [86] esterified fatty acids adsorbed on silica gel by heating with methanol–benzene–concentrated sulfuric acid (345:115:4) for 30 min on a water bath.

Marcus and Ma [50] acetylated flavones by heating in a sealed capillary with anhydrous pyridine and acetic anhydride. Kaess and Mathis [22] acetylated with acetic anhydride and also with acetyl chloride by direct application to the layer. Dallas [87] acetylated polyglycerols directly on the layers by spraying with 60% (vol/vol) acetic anhydride in anhydrous pyridine. The sprayed area was covered with a glass plate and placed upside down on a hot plate at 95 to 100°C for 15 min. The cover plate was then removed and the excess reagent evaporated by letting it stand for at least several hours. Holloway and Challen [77] acetylated by applying acetyl chloride to the sample spot with a capillary pipet. Excess reagent was then removed with a hot air drier.

8 NITRATION

Klesment [88] nitrated phenolic compounds on aluminum oxide layers by exposing them to nitric acid vapors or after previous saturation of the layer with 5% sodium nitrite, nitrosated in hydrogen chloride vapors. Wilk et al. [10] nitrated polycyclic hydrocarbons by placing the silica gel layer in a tank containing a layer of phosphorus pentoxide and allowing 2–3 ml of concentrated nitric acid to drip into the tank. The nitration was complete within 10 min. Polesuk and Ma [34, 89] carried out the nitration of compounds by spraying with 90% nitric acid (excess acid should be avoided). The layer above the spotted area was protected from the spray, and after spraying the plate was heated in an oven at 105°C for 30 min.

9 PREPARATION OF DERIVATIVES

There are a number of derivatives that have been prepared directly on the thin layer prior to developing the chromatogram [1]. Phenylhydrazones can be prepared by adding a drop of phenylhydrazine to the compound spot. The same is true for semicarbazide hydrochloride in water neutralized with sodium hydroxide. Tumlinson et al. [90] applied carbonyl reagents directly to the layer at the exit of a GLC unit so that the carbonyl compounds eluting from the column would be derivatized as they reached the layer. They applied solutions of 2,4-dinitrophenylhydrazine in orthophosphoric acid–ethanol, *p*-nitrophenylhydrazine (0.5 g in 30 ml of ethanol, 6 ml of methylal, 5 drops of concentrated H_3PO_4), and 2,4-dinitrophenylsemicarbazide (0.15 g in 20 ml of boiling ethanol and 6 drops of concentrated hydrochloric acid). The spot was kept moist while the peak was being adsorbed. If more than 2 μl of compound was being derivatized, more than one spot was used to adsorb the peak in order to avoid reagent caking and overloading of the adsorbent. The direct derivatization on the plate with 2,4-dinitrophenylhydrazine has been carried out with phenolic aldehydes [91], with 14 benzaldehyde derivatives [92], with *p*-benzoquinone and its chloro derivatives [93]; and also to the preparation of the *p*-bromophenacyl esters of chlorobenzoic acids [94].

Lisboa [95] applied a 0.1% solution of Girard's reagent (trimethylacetohydrazide ammonium chloride) in 10% (vol/vol) glacial acetic acid in methanol to the starting point on a thin-layer plate. Steroid ketones were then applied and allowed to stand for 15 hr in a tank saturated with acetic acid vapor. The acetic acid was then removed by heating the plate for 10 min at 80°C before developing the chromatogram.

Lisboa and Diczfalusy [96] applied the Boute reaction [97] for the preparation of nitroso derivatives from estrogens. This is accomplished by exposing the spotted compound to ammonia vapor and then to nitrogen dioxide which is prepared from metallic copper and concentrated nitric acid.

Pataki [98] prepared the dinitrophenyl derivatives of amino acids directly on the plates prior to a second dimensional run. This is accomplished by spraying the dried plate with a buffered solution containing 8.4 g of sodium bicarbonate and 2.5 ml of 1 *N* sodium hydroxide per 100 ml of solution, and then with a 10% (wt/vol) solution of dinitrofluorobenzene in methanol. After being covered with a glass plate supported by two strips of polyethylene along the edges of the chromatogram, this "sandwich" is heated in a dark oven at 40°C for 1 hr. It is cooled and then placed in an ether bath for 10 min. A final drying completes the operation. Pataki et al. [99–100] reported further work on this technique.

Parihar et al. [101] prepared the *p*-toluenesulfonates of some amines directly on a number of different adsorbent layers. After spotting the compounds a solution of the reagent in pyridine was added and then the plates were heated at 60°C for 4 hr.

For preparing derivatives of alcohols directly on thin layers placed under the exit of a GC unit, Minyard et al. [102] used solutions of *o*-nitrophenylisocyanate (10 g in 100 ml of benzene) and 3, 5-dinitrobenzoyl chloride (1.0 g in 7.5 ml of *p*-xylene and 1.0 ml of tetrahydrofuran). It was necessary to control the exhaust temperature of the GC column for optimum results. For the DNBs an exhaust temperature of 185°C was best for primary alcohols and terpene alcohols, whereas better yields of secondary alcohols were obtained with a temperature of 155°C. An outlet temperature of 185°C was better for the other reagent. After preparation of the DNBs, interference of the excess reagent in the subsequent development was eliminated by applying 10% sodium hydroxide to the spot.

Lin and Narasimhachari [103] prepared isothiocyanate derivatives of amines by spraying the plate with a 10% solution of carbon disulfide in ethyl acetate, and letting the plate stand for 30 min in a tank saturated with carbon disulfide vapors. This was followed by a spray of methanol–sulfuric acid (1 : 1) after which the plates were heated at 100°C for 10 min.

For the ultra-microanalysis of serum protein in a few nanoliters of sample, Nakajima et al. [104] spotted 10 μl of a 0.05% solution of 5-

dimethylaminonaphthalene-1-sulfonyl chloride in hexane at the origin. After evaporation of the solvent, 5 to 20 nl of the sample was applied to the same spot. The plate was then exposed for 10 hr to the vapor of 0.1 *M* triethylamine bicarbonate buffer (pH 8.5). Excess reagent, its bi-products, and derivatives of low-molecular-weight compounds were then removed from the origin by successive development with butanol–acetic acid–water (4:1:5) and isopropanol–methyl acetate–28% ammonia (9:7:4), leaving the protein derivative at the origin.

Cohen et al. [105] prepared the dinitrophenyl ether derivatives of phenols by spraying the layer with a saturated solution of sodium methoxide in methanol followed by a 4% (wt/vol) solution of 1-fluoro-2,4-dinitrobenzene in acetone. The plate was covered with another glass plate and heated in an oven for 40 min at 190°C.

Przybylski [106] formed derivatives of aflatoxins B_1 and G_1 directly on the thin-layer plate by spotting the samples with trifluoroacetic acid–benzene (1:1) (prepared fresh daily). After 5 min the plate was dried with a hot air blower for 10 min keeping the temperature under 40°C.

Inglis et al. [107] converted the thiazolinones obtained from a protein sequenator to the phenylthiohydantoins by spotting the compounds on a thin-layer plate followed by heptafluorobutyric acid. It was then heated for 10 min at 140°C followed by rapid cooling to room temperature. The method could not be used for threonine, serine, *S*-carboxymethylcysteine, tryptophan, N^{ϵ}-phenylthiocarbamyllysine, or glutamic acid.

10 MISCELLANEOUS REACTIONS

Mottier [108] added a drop of 1 *N* alkali to the origin on an aluminum oxide layer before adding the acidic dye extract. The plate was then heated to 104°C for 3 min in order to convert the dyes to the salts so that they could be chromatographed. Randerath and Weimann [109] have applied 0.005 ml of a solution of polyuridylic acid (6 mg/ml) to the starting sample spots containing deoxyriboligonucleotides and polyribonucleotides on layers of polyethylenimine cellulose. The polyuridylic acid forms a complex with complementary deoxyadenosine oligonucleotides and thus prevents their migration during the chromatographing. In this procedure there is a temperature effect that has to be taken into account. In the inorganic field Seiler and Kaffenberger [110] have applied the halides as sodium or potassium salts to the silica gel layer; then by using ammonium hydroxide in the developing agent, these are converted to the ammonium salts so that the chloride, bromide, and iodine are separated as ammonium salts. Only the fluoride remains at the origin. Seiler and Rothweiler [111] have spotted the alkali salts of strong acids along with barium acetate at the origin. The alkali salts are converted to acetates and can be separated with the barium remaining at the origin. Weicker and Brossmer [112] found that in chromatographing hexoses, pentoses, and disaccharides on silica gel layers with solvents containing ammonia, amino sugars were formed apparently because of a catalytic action of the silica gel. Stahl [113] has also used the reaction of samples on thin layers in his so-called SRS (separation–reaction–separation) technique in which the reaction on the thin layer occurs after the first development. If the spots are not on an approximate diagonal of the plate after developing in the same solvent in the second direction, it indicates that those spots which are not on this diagonal were affected by the reaction. Stahl demonstrated this by the separation of pyrethrins. In this case the pyrethrins were exposed to ultraviolet light which catalyzed the oxidation of the compounds.

Prandi [114] has dealkylated *N*-alkylated amines by heating on thin layers of silica gel or alumina to a temperature of 110 to 160°C for periods varying from 1 to 6 hr. Normally the main reaction products were the dealkylation products, but sometimes other side-reaction products were formed. The author recommends a two-dimensional treatment for impure products or mixtures. An initial separation is first carried out followed by dealkylation and then separation in a second direction. Diazotization of amine and coupling with a phenol has been used for some time as a means of locating amines on thin layers. Marcus et al. [115] have presented this as a method of synthesizing dyes on a micro scale. Using a protecting cover plate they sprayed the sample area with the diazotizing solution (500 mg sodium nitrite in 50 ml of 1 *N* hydrochloric acid) until it was well saturated. Excess nitrous acid was then removed by heating at 105°C for 5 min. The phenol was added as a 5% spray in methanol. The same authors [116] prepared the 2,4-dinitro-5-aminophenyl derivatives of amino acids in capillary tubes. The reaction mixture was applied to one corner of a thin-layer plate and developed in a suitable solvent. The developed, dried plate was then covered to protect the unused portion of the layer and sprayed with 0.1 *M* methanolic sodium hydroxide so that it was just wet. It was allowed to stand for 10 min and then was sprayed with a freshly prepared 0.2% solution of *p*-nitrobenzenediazonium fluoborate in methanol–water (1:1) until the area was wet. The layer was uncovered and allowed to stand at room temperature for 15 min. The dyes that were formed were then separated in the second dimension.

Bierl et al. [117] used a thin-layer reaction with phosphoric acid to determine if epoxides were present in an unknown mixture and also to help determine the configuration of the epoxide if any were present. On silica gel plates prewashed with ether 5 μl of 10% aqueous phosphoric acid was spotted. After drying, another 5 μl of acid was added and allowed to dry for 20 min. Sample (20 to 100 μg) was applied on this spot and also on a section of the layer that had not been treated with acid. After the plate was dried for 1 hr, it was developed with a suitable mixture of ether–hexane. The reaction products of the epoxides are more polar and remain near the origin in contrast to the epoxides in the original sample. For determination of the configuration, the procedure was the same except that 25 to 35 μg of compound was used and the reaction was allowed to proceed for only 5 min. Cis epoxides, 1,2-epoxides, and trisubstituted epoxides were completely or almost completely cleaved in this time. Trans epoxides or hindered epoxides (alkyl substituted alpha to epoxide) reacted only partially or in trace amounts during this time period. Visualization was accomplished with iodine vapor and/or by spraying with 2% cupric acetate containing 8% by vol of 85% orthophosphoric acid.

Moore and Babb [118] applied cellulose layers to investigate cellulose-reactive materials used in the textile industry. After the sample application, the plates were heated (cured) in a preheated oven at the specified temperature for 5 min. The plates were then developed, air-dried, and sprayed with suitable visualizing agents. All the compounds tested appeared to be cured between 125 and 175°C.

References

1. J. M. Miller and J. G. Kirchner, *Anal. Chem.*, **25,** 1107 (1953).
2. C. Mathis and G. Ourisson, *J. Chromatogr.*, **12,** 94 (1963).
3. J. Kofoed, C. Fabrierkiewicz, and G. H. W. Lucas, *J. Chromatogr.*, **23,** 410 (1966).
4. J. Kofoed, C. Fabrierkiewiez, and G. H. W. Lucas, *Nature,* **211,** 147 (1966).
5. I. Ognyanov, *Compt. Rend. Acad. Bulg. Sci.*, **16,** 161 (1963).
6. *Ibid.*, p. 265.
7. R. Fuhrmann and D. Jeralomon, *J. Chromatogr,* **22,** 468 (1966).
8. G. Weidemann, *J. Chromatogr,* **54,** 141 (1971).
9. J. M. G. J. Frijns, *Pharm. Weekbl. Ned.*, **106,** 605 (1971); through *Chem. Abstr.*, **75,** 121441b (1971).
10. M. Wilk, U. Hoppe, W. Taupp, and J. Rochlitz, *J. Chromatogr,* **27,** 311 (1967).
11. A. Kaess and C. Mathis, *Chromatogr. Electrophor., Symp. Int., 4th 1966,* Ann Arbor Science, Ann Arbor, Mich., 1968, p. 525.
12. W. Rusiecki and M. Henneberg, *Ann. Pharm. Fr.*, **21,** 843 (1963).
13. D. C. Malins and H. K. Mangold, *J. Am. Oil Chem. Soc.*, **37,** 576 (1960).

14. H. K. Mangold, *Fette, Seifen, Anstrichm.*, **61,** 877 (1959).
15. B. L. Hamman and M. M. Martin, *Steroids*, **10,** 169 (1967).
16. A. M. Gardner, *J. Assoc. Off. Anal. Chem.*, **54,** 517 (1971).
17. L. L. Smith and J. C. Price, *J. Chromatogr.*, **26,** 509 (1967).
18. J. E. Van Lier and L. L. Smith, *J. Chromatogr.*, **41,** 37 (1969).
19. J. Polesuk and T. S. Ma, *J. Chromatogr.*, **57,** 315 (1971).
20. C. B. Glaser, H. Maeda, and J. Meienhofer, *J. Chromatogr.*, **50,** 151 (1970).
21. H. P. Kaufmann and T. H. Khoe, *Fette, Seifen, Anstrichm.*, **64,** 81 (1962).
22. H. P. Kaufmann, Z. Makus, and T. H. Khoe, *Fette, Seifen, Anstrichm.*, **64,** 1, (1962).
23. E. Knappe and D. Peteri, *Z. Anal. Chem.*, **190,** 380 (1962).
24. T. Wieland and H. Ottenheym, *Peptides, Proc. Eur. Peptide Symp., 8th, 1966*, North-Holland, Amsterdam, 1967, p. 195.
25. E. Graf and W. Hoppe, Dtsch. Apoth.-Ztg., **102,** 393 (1962).
26. S. K. Yasuda, *J. Chromatogr.*, **13,** 78 (1964).
27. A. R. Thawley, *J. Chromatogr.*, **38,** 399 (1968).
28. G. M. Barton, *J. Chromatogr.*, **34,** 562 (1968).
29. H. Schuetz and A. Schindler, *Z. Anal. Chem.*, **270,** 356 (1974).
30. J. H. Tyrer, M. J. Eadie, and W. D. Hooper, *J. Chromatogr.*, **39,** 312 (1969).
31. H. Nakamura and Z. Tamura, *J. Chromatogr.*, **104,** 389 (1975).
32. J. Scotney and E. V. Truter, *J. Chem. Soc. C*, **1968,** 1911.
33. K. Schreiber, O. Aurich, and G. Osske, *J. Chromatogr.*, **12,** 63 (1963).
34. J. Polesuk and T. S. Ma, *Mikrochim. Acta*, **1970,** 677.
35. M. Baggiolini and B. Dewald, *J. Chromatogr.*, **30,** 259 (1967).
36. T. Kartnig and O. Wegschaider, *Planta Med.*, **21,** 144 (1972).
37. T. Kartnig and O. Wegschaider, *J. Chromatogr.*, **61,** 375 (1971).
38. E. Moczar, *J. Chromatogr.*, **76,** 417 (1973).
39. H. G. Eulenhoefer, *J. Chromatogr.*, **36,** 198 (1968).
40. G. Valk, M. Peters, and L. Husung, *Z. Anal. Chem.*, **249,** 245 (1970).
41. H. H. O. Schmid and H. K. Mangold, *Biochim. Biophys. Acta*, **125,** 182 (1966).
42. C. V. Viswanathan, F. Phillips, and W. O. Lundberg, *J. Chromatogr.*, **35,** 66 (1968).
43. R. O. Anderson, R. D. Garrett, M. L. Blank, and F. Snyder, *Lipids*, **4,** 327 (1969).
44. K. Owens, *Biochem. J.*, **100,** 354 (1966).
45. A. H. Payne and M. Mason, *Anal. Biochem.*, **26,** 463 (1968).
46. A. R. Hurwitz, *Anal. Biochem.*, **46,** 338 (1972).
47. H. Schuetz, *J. Chromatogr.*, **94,** 159 (1974).
48. H.-C. Curtius and M. Mueller, *J. Chromatogr.*, **32,** 222 (1968).
49. R. J. Ashworth and T. J. Sheets, *J. Agric. Food Chem.*, **20,** 407 (1972).
50. B. J. Marcus and T. S. Ma, *Mikrochim. Acta*, **1969,** 815.
51. C. E. Redemann and H. J. Lucas, *Ind. Eng. Chem. Anal Ed.*, **9,** 521 (1937).
52. J. E. Barney II, *J. Chromatogr.*, **20,** 334 (1965).

53. J. Askew, J. H. Ruzicka, and B. B. Wheals, *Analyst (London),* **94,** 275 (1969).
54. J. Askew, J. H. Ruzicka, and B. B. Wheals, *J. Chromatogr.,* **37,** 369 (1968).
55. D. I. Cargill, *Analyst (London)* **87,** 865 (1962).
56. C. Schuetz and H. Schuetz, *Arzneimittel-Forsch.,* **23,** 428 (1973); through *Chem. Abstr.,* **78,** 164161q (1973).
57. R. A. De Zeeuw and J. Wijsbeek, *J. Chromatogr.,* **48,** 222 (1970).
58. C. Schuetz and H. Schuetz, *Arch. Toxikol.,* **28,** 286 (1972).
59. C. Schuetz and H. Schuetz, *Dtsch. Apoth.-Ztg.,* **113,** 1559 (1973).
60. H. P. Kaufmann, H. Wessels, and B. Das, *Fette, Seifen, Anstrichm.,* **64,** 723 (1962).
61. J. W. Copius Peereboom and H. W. Beekes, *J. Chromatogr.,* **9,** 316 (1962).
62. W. R. Koehler, J. L. Solan, and H. T. Hammond, *Anal. Biochem.,* **8,** 353 (1964).
63. J. Polesuk and T. S. Ma, *Mikrochim. Acta,* **1971,** 662.
64. M. Wilk and U. Brill, *Arch. Pharm. (Weinheim),* **301,** 282 (1968).
65. F. Schmidt, *Krankenhaus-Apoth.,* **23,** 10 (1973); through *Chem. Abstr.,* **79,** 45866j (1973).
66. M. Wilk and W. Taupp, *Z. Naturforsch. B.* **24,** 16 (1969); through *Chem. Abstr.,* **70,** 78238y (1969).
67. W. Brown and A. B. Turner, *J. Chromatogr.,* **26,** 518 (1967).
68. K. Randerath and E. Randerath, *Angew. Chem. Int. Ed. Engl.,* **3,** 442 (1964).
69. B. Tschiersch and K. Schwabe, *Pharmazie,* **29,** 484 (1974).
70. R. D. Bennett and E. Heftmann, *J. Chromatogr.,* **9,** 353 (1962).
71. J. Riess, *J. Chromatogr.,* **19,** 527 (1965).
72. Y. Maruyama, *Igaku To Seibutsugaku,* **73,** 20 (1966); through *Chem. Abstr.,* **69,** 92757c (1968).
73. M. H. A. Elgamal and M. B. E. Fayez, *Z. Anal. Chem.,* **226,** 408 (1967).
74. H. P. Kaufmann, S. S. Radwan, and A. K. S. Ahmad, *Fette, Seifen, Anstrichm.,* **68,** 261 (1966).
75. C. V. Viswanathan, M. Basilo, S. P. Hoevet, and W. O. Lundberg, *J. Chromatogr.,* **34,** 241 (1968).
76. S. P. Hoevet, C. V. Viswanathan, and W. O. Lundberg, *J. Chromatogr.,* **34,** 195 (1968).
77. P. J. Holloway and S. B. Challen, *J. Chromatogr.,* **25,** 336 (1966).
78. L. D. Metcalfe, A. A. Schmitz, and J. R. Pelka, *Anal. Chem.,* **38,** 514 (1966).
79. K. Oette and M. Doss, *J. Chromatogr.,* **32,** 439 (1968).
80. C. V. Viswanathan, F. Phillips, and W. O. Lundberg, *J. Chromatogr.,* **38,** 267 (1968).
81. S. Saha and J. Dutta, *Lipids,* **8,** 653 (1973).
82. H. G. Sammons and S. M. Wiggs, *Analyst (London),* **85,** 417 (1960).
83. P. Husek, *Z. Klin. Chem. Klin. Biochem.,* **7,** 627 (1969).
84. E. P. Tsyganov, *Lab Delo,* **1971,** 490.

85. J. D. Mann, N. G. Porter, and J. E. Lancaster, *J. Chromatogr.,* **92,** 177 (1974).
86. M. Mancha Perelló, *Grasas Aceites,* **18,** 231 (1967).
87. M. S. J. Dallas, *J. Chromatogr.,* **48,** 225 (1970).
88. I. R. Klesment, *Gazov. Khromatogr.,* No. 4, 102 (1966); through *Chem. Abstr.,* **66,** 26752a (1967).
89. J. Polusk and T. S. Ma, *Mikrochim. Acta,* **1969,** 352.
90. J. H. Tumlinson, J. P. Minyard, P. A. Hedin, and A. C. Thompson, *J. Chromatogr.,* **29,** 80 (1967).
91. P. Froment and A. Robert, *Chromatographia,* **4,** 173 (1971).
92. I. Wilczynska, *Chem. Anal. (Warsaw),* **17,** 21 (1972); through *Chem. Abstr.,* **77,** 42872f (1972).
93. *Ibid.,* **16,** 69 (1971); through *Chem. Abstr.,* **75,** 5401q (1971).
94. E. D. Stedman, *Analyst (London),* **94,** 594 (1969).
95. B. P. Lisboa, *J. Chromatogr.,* **24,** 475 (1966).
96. B. P. Lisboa and E. Diczfalusy, *Acta Endocrinol.,* **40,** 60 (1962).
97. J. Boute, *Ann. Endocrinol.,* **14,** 518 (1953).
98. G. Pataki, *J. Chromatogr.,* **16,** 541 (1964).
99. G. Pataki, J. Borko, and A. Kunz, *Z. Klin. Chem. Klin. Biochem.,* **6,** 458 (1968).
100. G. Pataki, J. Borko, H. C. Curtius, and F. Tancredi, *Chromatographia,* **1,** 406 (1968).
101. D. B. Parihar, S. P. Sharma, and K. C. Tewari, *J. Chromatogr.,* **24,** 443 (1966).
102. J. P. Minyard, J. H. Tumlinson, A. C. Thompson, and P. A. Hedin, *J. Chromatogr.,* **29,** 88 (1967).
103. R.-L. Lin and N. Narasimhachari, *Anal. Biochem.,* **57,** 46 (1974).
104. T. Nakajima, H. Endou, F. Sakai, and Z. Tamura, *Chem. Pharm. Bull. (Tokyo),* **18,** 1935 (1970).
105. I. C. Cohen, J. Norcup, J. H. A. Ruzicka, and B. B. Wheals, *J. Chromatogr.,* **44,** 251 (1969).
106. W. Przybylski, *J. Assoc. Off. Anal. Chem.,* **58,** 163 (1975).
107. A. S. Inglis, P. W. Nicholls, and P. McK. Strike, *J. Chromatogr.,* **107,** 73 (1975).
108. M. Mottier, *Mitt. Geb. Lebensm. Hyg.,* **47,** 372 (1956).
109. K. Randerath and G. Weimann. *Biochim. Biophys. Acta,* **76,** 129 (1963).
110. H. Seiler and T. Kaffenberger, *Helv. Chim. Acta,* **44,** 1282 (1961).
111. H. Seiler and W. Rothweiler, *Helv. Chim. Acta,* **44,** 941 (1961).
112. H. Weicker and R. Brossmer, *Klin. Wochenschr.,* **39,** 1265 (1961).
113. E. Stahl, *Arch. Pharm.,* **293/65,** 531 (1960).
114. C. Prandi, *J. Chromatogr.,* **48,** 214 (1970).
115. B. J. Marcus, A. Fono, and T. S. Ma, *Mikrochim. Acta,* **1967,** 960.
116. B. J. Marcus and T. S. Ma, *Mikrochim. Acta,* **1971,** 267.
117. B. A. Bierl, M. Beroza, and M. H. Aldridge, *Anal. Chem.,* **43,** 636 (1971).
118. D. R. Moore and R. M. Babb, *Text. Res. J.,* **42,** 500 (1972).

Chapter **VII**

DETECTION OF COLORLESS COMPOUNDS

In general the detection of colored compounds causes no problem. The same holds true of compounds which exhibit fluorescence or phosphorescence under ultraviolet light. In some cases where the color is not very intense, a spray reagent may be used to increase the sensitivity of the method of detection, such as in the case of the colored carotenoid aldehydes. Winterstein and Hegedues [1, 2] found that the sensitivity of detection of these compounds could be increased by spraying the chromatogram first with an alcoholic solution of rhodanine followed by a concentrated aqueous ammonia or sodium hydroxide solution.

There are numerous spray reagents which can be used to make the various colorless compounds visible on the chromatogram. These can be divided into two classes: (1) those which are general reagents and will detect a large number of different types of compounds, and (2) those which are more specific in nature, indicating the type of compound or functional group that is present, such as the use of a glacial acetic acid solution of *o*-dianisidine for the detection of aldehydes [3].

In applying spray reagents, a finely atomized spray must be used so as to cover the chromatogram with a uniform coating of the reagent. For the very corrosive reagents such as concentrated acids and oxidizing solutions, one of the all-glass sprayers should be used. The sprayer design described by Morris [4] is available from most supply houses. For less corrosive reagents, an extremely useful sprayer is the airbrush used by artists. The spray tips on this are adjustable to any degree of fineness desired, and the spray gun itself can be rapidly interchanged to different reagents kept in the screw-capped glass containers which are available. Especially useful also is the small colored cup which fits onto the artist's spray brush and holds a few milliliters of solution. The reagent may be poured into the cup and after spraying the cup may be rinsed quickly for use with another reagent. I have used this type of spray gun extensively, even for such reagents as hydrochloric acid solutions of 2,4-dinitrophenylhydrazine. If a small quantity of water or other solvent is run through the sprayer immediately after use, the gun will last for a long time. Also available on the market are the spray guns using containers of fluorinated hydrocarbons for providing the driving force for the reagents which are kept in screw-capped jars. Aerosol-type containers are available with numerous ready-mixed reagents; however, these have a tendency to give a coarse spray [5].

The use of loose layers of adsorbent presents a particularly difficult problem as far as spray reagents are concerned. The spray gun must be kept at some distance from the layers so as not to blow away the adsorbent particles, and the air pressure for the spray gun should be kept at a minimum. A development technique has been used in which the visualizing agent has been allowed to soak across the layer by capillary attraction in a direction perpendicular to that used for the compound separation [6, 7]. Another technique that has been used has been to spray the loose-layer plate by means of a reagent-wetted toothbrush rubbed on a metal sieve [8].

1 UNIVERSAL OR GENERAL REAGENTS

The original and one of the more generally used spray reagents of this nature is concentrated sulfuric acid, first used by Kirchner et al. [3]. Some compounds appear in the cold whereas others appear only after heating the sprayed plate. It was for use with these corrosive-type reagents that I developed the gypsum-bound layers. For extremely unreactive compounds such as camphor, 5% nitric acid was added to the concentrated sulfuric acid to increase the oxidizing power of the reagent, so that the compounds on the plate would char when heated. Various modifications of this basic technique have been used, such as 50% sulfuric [9–15]. Lisboa and Diczfalusy [16] used 2% sulfuric acid in a 50:50 ethanol–water mixture as a spray, followed by heating the chromatoplate at 100°C. Anthony and Beher [17] used a freshly prepared solution of

15% concentrated sulfuric acid in anhydrous *n*-butanol (vol/vol) for the detection of bile acids. After the dried plates were sprayed, they were heated at 110°C for 25 to 30 min for conjugated acids or 45 to 50 min for free acids. Sulfuric acid–acetic anhydride mixtures have been employed in ratios of 5:95 [17], and 1:4 (Liebermann–Burchard reagent) [18]. Matsumoto [19] has used an acetic anhydride spray followed, after drying, by one of concentrated sulfuric acid. Besides nitric acid, other oxidizing agents have been added to sulfuric acid. These include a saturated solution of potassium dichromate in concentrated sulfuric acid [20] or a saturated solution of potassium dichromate in 80% sulfuric acid [21], a solution of 3 g of sodium dichromate in 20 ml of water and 10 ml concentrated sulfuric acid [22], a saturated solution of chromic acid in concentrated sulfuric acid [23], and 0.5 g of potassium permanganate in 15 ml concentrated sulfuric acid (**Caution:** mix only small quantities, for manganous heptoxide is explosive) [22]. For all these reagents, the plates are dried to remove the solvent, sprayed with the reagent, and finally heated to develop the color or the charred spot, as the case may be. A spray of 50% phosphoric acid solution followed by heating has also been used for charring compounds [24], as well as concentrated nitric acid [15]. Both 25% [25] and 70% [6] perchloric acid have been used as detecting agents. In the latter case, characteristic colors were formed in the cold with vitamins on aluminum oxide.

A number of methods have been developed to avoid the problems associated with spraying corrosive sulfuric acid solutions. Zimiński and Borowski [26] used a 20% aqueous solution of ammonium sulfate or of a 1:1 mixture of ammonium sulfate and ammonium hydrogen sulfate. When the layer is heated sulfuric acid is liberated; however higher temperatures are required than for a sulfuric acid-sprayed layer. Walker [27] incorporated the ammonium sulfate in the layer and Korolczuk and Kwaśniewska [28] incorporated 1 to 5% cupric sulfate in the layer as a charring agent. Jones et al. [29] exposed the thin layers to sulfuryl chloride or sulfur trioxide vapor and then to steam to generate the sulfuric acid directly on the layer. Martin and Allen [30] designed equipment for exposing the layer to sulfur trioxide while the layer was being heated. Another method of avoiding the spray problem is to impregnate the silica gel layer with 4% sulfuric acid in methanol [31] prior to development. In this case solvents must be used that will not wash the acid from the layer. The same holds true for silica gel layers impregnated with phosphomolybdic acid [32].

Antimony trichloride and pentachloride are also useful as general reagents, in many cases giving characteristic colors with various compounds. The pentachloride is used as a 10 to 20% [33] solution in carbon tetrachloride. Antimony trichloride is used as a saturated solution in alcohol-free chloroform. After either of these reagents is sprayed, the plate is heated to 100 to 120°C for 5 to 10 min. Reichelt and Pitra [34] used a saturated solution of antimony trichloride in alcohol-free chloroform containing 20% acetic anhydride.

Another useful general reagent is the fluorescein–bromine test which will disclose unsaturated or other compounds which will react with bromine readily. This test, originated by Kirchner et al. [3], consists in spraying the solvent-free chromatoplate with a solution of 0.05% fluorescein in water. The plate is then exposed to bromine vapor by blowing gently across the top of a bottle of bromine. The fluorescein is converted to the red dye eosin, except where compounds are located which take up the bromine, thus leaving the fluorescein with its normal yellow color.

Iodine has found considerable use as a general reagent. It is only necessary to place the dried chromatogram in a closed chamber containing a few crystals of iodine [35]. Most organic compounds appear as brown spots. The position of the spots should be marked because they tend to fade rapidly. In lieu of iodine vapor, the chromatograms may be sprayed with a 1% solution of iodine in methanol or ethanol. Boyle and Nelson [36] used a strip technique by depositing a thin film of iodine on the inside of a glass trough which was then inverted over the thin layer.

A relatively new technique that has been introduced by Cotgreave and Lynes [37] for the detection and quantitation of compounds separated on a chromatostrip is that of pyrolysis of the compounds directly on the chromatogram and detection of the pyrolysis products by a flame ionization detector. Various modifications, such as the use of adsorbent-coated wires [38], silicic acid–copper oxide layers on the inside of glass tubes [39], and silica gel sintered glass ceramic sticks [40] have been developed. More details of this technique are discussed under quantitative analysis in Chapter XI, Section 7.

A technique that is convenient, especially for locating compounds for quantitative analysis, is one in which water is used as a spray agent [41–43]. The chromatograms are simply sprayed with water until the chromatoplate is translucent, and then the water-insoluble compounds appear as white opaque spots against a dark background.

A method introduced by Heidbrink [44] appears to be very promising as a general reagent. This consists in exposing the complete and solvent-free chromatogram to various gases and vapors, followed by pyrolysis. The reagents that have been used are iodine, bromine, chlorine, formaldehyde, and nitrogen dioxide. After the plate is exposed to the atmosphere of one of these gases for 1 to 2 min, it is then placed on an aluminum or stainless steel block and heated from above with an electric heater whose heating coils are covered by a quartz plate to protect them from corrosion. With this arrangement, the chromatoplates can be heated to 800 to 900°C without breaking the supporting glass plate. The heating is continued for at least 15 min to see if a reaction will take place. Variously colored spots appear with a wide variety of substances. Substances which would normally vaporize readily from the plate are retained by this treatment and give colored spots. If it is not known whether a compound will give a reaction with one of the five reagents, the reaction should be checked independently with all five of the vapors.

Conway et al. [45] have devised a means of detecting volatile hydrolysis products on a thin-layer plate. After the developed layer is sprayed with the hydrolyzing agent it is covered with a thin sheet of polyethylene (food wraps "Handi-Wrap" and "Cut-Rite Wrap" are suitable) followed by a sheet of filter paper saturated with a solution of the detecting agent. A glass cover plate is then secured in place with rubber bands and the assembly is heated for 20 min in an oven. In the examples cited, chlormezanone was hydrolyzed with 2 *N* sodium hydroxide at 50°C. The detection reagent was 0.5% dinitrophenylhydrazine and 0.1 *N* *p*-toluenesulfonic acid in methyl Cellosolve. The *p*-chlorobenzaldehyde that was formed diffused through the membrane to form a yellow spot of the hydrazone. A follow-up spray of 2.5% tetramethylammonium hydroxide intensified the spot. In order to detect ammonia from the hydrolysis of β-hydroxypropionamide the membrane was replaced with a 0.045-in. sheet of polyethylene perforated with $\frac{1}{16}$ in. holes to permit the ammonia to pass through the barrier. In this case the compound was hydrolyzed at 100°C and Nessler's reagent was used to detect the ammonia. Sensitivity in both cases was 3 μg.

Smith [45a] introduced a technique for inducing fluorescence for the quantitative thin-layer chromatography of organic compounds which can be used for the detection of compounds, and Segura and Gotto [45b] modified this technique (see Chapter XI, Section 1). Shanfield et al. [45c, 45d] have investigated an extension of this work by the use of electrically activated gases combined with subsequent treatment with ammonium carbonate vapors at 130°C to induce fluorescence in compounds adsorbed on thin-layer plates. The latter process could be applied only to layers bonded with sintered glass and no investigation was made of the effect of developing solvents on the procedure, that is, the tests were conducted only on spotted samples and not on normally developed chromatograms. The technique is interesting and invites further experimentation.

2 FLUORESCENT LAYERS

Another useful technique introduced for thin-layer work by Kirchner et al. [3, 46] is the use of fluorescent layers. When examined under ultraviolet light, these layers have a bright fluorescence and whenever compounds are present which absorb the ultraviolet light being used, they appear as dark spots on a fluorescent background. In order to produce these fluorescent layers, 1% of each of the inorganic fluorescent materials used by Sease [47, 48] for column chromatography was mixed with the thin-layer adsorbent material. The fluorescent compounds used were Du Pont zinc sulfide #62 and zinc silicate #609. Reitsema [49] incorporated Rhodamine 6G as a fluorescing agent, and Stahl [50] has prepared the adsorbent slurry with 0.04% aqueous sodium fluorescein instead of water. However, the inorganic fluorescing materials provide by far the better fluorescent medium. After development of the chromatogram, fluorescent reagents have also been sprayed on to provide a fluorescent layer. These include an aqueous sodium fluorescein solution [51], a 0.005 to 0.01% solution of morin in methanol [52], 0.2% 2′,7′-dichlorofluorescein in ethanol [23], and a 0.2% water solution of Rhodamine B. For spraying Rhodamine B solution, a minimum thickness of adsorbent layer is required and should be between 0.35 and 0.5 mm [53].

3 MORE SPECIFIC SPRAY REAGENTS

Throughout the section on applications of thin-layer chromatography, some of the reagents used with each type of compound have been given. For convenience these reagents and others are listed below. No attempt has been made to list all the references in which these spray reagents have been applied, but representative references have been given so as to cover the various types of compounds for which a given spray reagent has been used.

Tests for Individual Compounds or Types

T-1. Acid Violet (Violuric acid)
Reagent: 1.5% solution of Acid Violet (keep below 60°C in preparing solution).
Procedure: Heat sprayed chromatogram for 20 min at 100°C.
Use: Alkali and alkaline earth metals [54].

T-2. Alizarin
Reagent: Saturated solution of alizarin in alcohol.
Procedure: Spray with alizarin solution, then with 25% ammo-

nium hydroxide. A follow-up spray of glacial acetic acid will lighten the background.

Use: Inorganic ions, Ba, Ca, Mg, Al, Ti, Fe, Zn, Li, Th, NH_4, Se, Ag, Hg, Pb, Cu, Cd, Bi, Cr, Mn, Co, Ni, Ga, In, Be, Zr, Ce, Sc, Pd, Pt, U, and the rare earths. Violet to red spots [55].

T-3. Amido Black stain

Reagent: Saturated solution of Amido Black 10B in methanol–water–acetic acid (5:5:1).

Procedure: (1) Starch gel. Immerse in stain for 30 sec, then in four consecutive rinses in the solvent for 1 min each [56].

(2) *Agar*. Dry carefully, then immerse for $\frac{1}{2}$ hr in stain, wash in three changes of solvent for periods of 10 min, 2 hr, and $\frac{1}{2}$ hr, respectively [57].

Use: For staining electropherograms of serum.

T-4. Amido Black stain

Reagent: 0.5% Amido Black 10B in 5% acetic acid.

Use: Proteins on agar gel [58].

T-5. *p*-Aminobenzoic acid

Reagent: 5% *p*-aminobenzoic acid in methanol. Cyanogen chloride, freshly prepared: 20 ml of 28% chloramine suspension in water, 20 ml of 1 *N* hydrochloric acid and 10 ml of 10% potassium cyanide solution (**Caution:** poison).

Procedure: Spray with aminobenzoic acid, then place in atmosphere of cyanogen chloride.

Use: Nicotinic acid, red spot [59]. Nicotinic acid amide, orange-red.

Sensitivity: 0.1 γ nicotinic acid.

T-6. *p*-Aminodiethylaniline–sulfur dioxide

Reagent: 0.5% *p*-aminodiethylaniline–sulfur dioxide in 5% sodium bicarbonate [60].

Procedure: Spray plate and leave overnight.

Use: Δ^4-3-keto-C_{21}-steroids.

T-7. *o*-Aminodiphenyl-orthophosphoric acid

Reagent: 0.3 g of *o*-aminodiphenyl + 5 ml of orthophosphoric acid (sp gr 1.88) in 95 ml of EtOH.

Procedure: After spraying heat at 110°C for 15 to 20 min.

Use: Carbohydrates, brown spots [61, 62].

Sensitivity: 0.1 γ.

T-8. Aminoethyl diphenylboric acid

Reagent: 1 g of 2-(aminoethyl)diphenylboric acid in 50 ml of each of methanol and propanol.

Procedure: Spray and observe in visible and in ultraviolet light.
Use: Anthocyanidins, flavonols, flavones, glycoflavones, and hydroxy and methoxycinnamic acids [63].
Sensitivity: 1.0 μg.

T-9. Aminoguanidine–dichromate
Reagent: (*a*) 2.5 g aminoguanidine sulfate monohydrate in 100 ml of water. Stable at room temperature.
(*b*) 1 ml of 1% potassium dichromate in water added to 100 ml of concentrated sulfuric acid. Stable for 1 month.
Procedure: Dry plate and spray with (*a*) followed by (*b*). Heat for 10 min at 110°C.
Use: Sugars, blue or blue-gray spots. Fucose, lilac. β-D-Galacturonic acid is also violet, but R_f value differs [64].
Sensitivity: 0.1 μg.

T-10. *p*-Aminohippuric acid
Reagent: (*a*) 0.3% *p*-aminohippuric acid in ethanol.
(*b*) Same as (*a*) but with 3% phthalic acid added.
Procedure: Air-dry plate and spray with (*a*) for hexoses and pentoses or with (*b*) if reducing sugars are present. Heat at 140°C for 8 min. Orange spots.
Use: Sugars [65, 66].
Sensitivity: 1 μg under ultraviolet light.

T-11. *o*-Aminophenol
Reagent: 1% *o*-aminophenol in methanol + 10 ml of phosphoric acid + 5 ml of water.
Procedure: After spraying heat to 110 to 120°C for 10 min.
Use: Amino sugars [67].

T-12. Ammonium ceric nitrate
Reagent: (*a*) 1% ammonium ceric nitrate in 0.2 *N* nitric acid.
(*b*) 1.5 g of *N,N*-dimethyl-*p*-phenylenediamine dihydrochloride dissolved in a mixture of 128 ml of methanol, 25 ml of water, and 1.5 ml of acetic acid.
(*c*) 25% solution of *N,N,N′,N′*-tetramethyl-4,4′-diaminodiphenylmethane in acetone [68].
Procedure: (1) Spray freshly mixed (*a*) and (*b*) (1:1). Heat plate at 105°C for 5 min.
Or (2) spray freshly mixed (*a*) and (*c*) (1:1). Heat plate at 105°C for 5 min.
Use: Polyalcohols, (1) yellow-green on red background or (2) white to pale blue on blue.

T-13. Ammonium ceric nitrate
Reagent: 6% ammonium ceric nitrate in 2 *N* nitric acid.

Procedure: Dry chromatogram for 5 min at 105°C. Cool before spraying.
Use: Polyalcohols, brown spots on yellow [68].

T-14. Ammonium molybdate–copper
Reagent: Place 0.08 g of metallic copper in 1 ml of water containing 0.25 g of ammonium molybdate. Chill and add 1 ml of concentrated sulfuric acid and mix thoroughly. Keep at room temperature for 2 hr with occasional shaking. Remove copper and add 3.2 ml of concentrated sulfuric acid.
Procedure: Spray with reagent and place plate in oven at 65 to 70°C for 5 min, spray again, and return to oven for 5 to 6 min.
Use: Phospholipids, blue against light blue background [69].
Sensitivity: Lecithin 1 μg.

T-15. Ammonium molybdate–perchloric acid
Reagent: 3 g of ammonium molybdate in 25 ml of water, 30 ml of 1 *N* hydrochloric acid, 15 ml of 60% perchloric acid.
Procedure: After spraying, dry at 105°C for 20 min.
Use: General reagent for lipids [70], blue-black spots. Hexose- and triosephosphates [71]: also observe under ultraviolet light.

T-16. Ammonium molybdate (Hanes–Isherwood reagent [72])
Reagent: 0.5 g ammonium molybdate tetrahydrate in 5 ml of water + 1.5 ml of hydrochloric acid and 2.5 ml of perchloric acid. After cooling dilute to 50 ml with acetone. Prepare 1 day ahead of use. Stable for approximately 3 weeks.
Procedure: After spraying, dry plate under infrared lamp held 30 cm away for 2 min. Expose dried plate to long-wave ultraviolet (360 mμ) for 7 min.
Use: Organophosphorus pesticides [73]. Mono- and diphosphoric acids [74].

T-17. Ammonium molybdate–benzidine
Reagent: (*a*) Dissolve 2.0 g of ammonium molybdate in the minimum amount of 1 *M* hydrochloric acid and dilute to 100 ml. Stable for several months.

(*b*) Dissolve 0.050 g of benzidine in 10 ml of acetic acid with stirring and addition of minimum amount of water. Add 22.5 g of sodium acetate trihydrate and dilute to 100 ml. Stable for minimum of 2 weeks.

(*c*) Add 11.2 ml of 57% periodic acid (free of inhibitors) to 50 ml of acetic acid and dilute to 100 ml. Stable for several months.

(*d*) Dissolve 5.7 g ammonium persulfate in smallest amount of water and dilute to 100 ml with water. Stable for 1 day.

Procedure: (1) Spray with (*c*) and heat so that the surface of the plate reaches 250° for 15 min. Cool and spray with (*a*), then after 3 min spray with (*b*). Phosphate and thiophosphate pesticides, and related phosphorus acids give blue spots on a white background.

(2) Spray with (*c*) and heat as in (1), then cool and spray with (*d*); heat again for 15 min and then cool. Spray with (*a*) followed by (*b*) after 3 min. Phosphonate and phosphate pesticides, related thio compounds, and related phosphorus acids give blue spots on a white background. Note: orthophosphate is positive in both (1) and (2) methods, but will also give a blue spot without using reagent (*c*). Organophosphonates in general are negative with (1).

Use: Organophosphorus compounds [75].

Sensitivity: Less than 1 μg.

T-18. Ammonium molybdate–oxalic acid–stannous chloride.

Reagent: (*a*) 1% ammonium molybdate in 0.125 *M* sulfuric acid.

(*b*) Saturated oxalic acid.

(*c*) 1% stannous chloride in 10% hydrochloric acid.

Procedure: Spray in succession with (*a*), (*b*), and (*c*).

Use: Specific for silicates [76].

Sensitivity: 0.006 μg silica.

T-19. Ammonium molybdate–stannous chloride

Reagent: (*a*) Dissolve 2 g of ammonium molybdate ($NH_4Mo_7O_{24}\cdot 4H_2O$) in 20 ml of water–hydrochloric acid (1:1) with gentle heat, then dilute to 100 ml. Solution is stable for several weeks.

(*b*) Dissolve 1 g stannous chloride ($SnCl_2\cdot 2H_2O$) in 10 ml of hydrochloric acid with heating. Then add 40 ml of water and 50 ml of acetone.

(*c*) Mix 25 ml of periodic acid (sp gr 1.7) in 25 ml of acetic acid and 50 ml of water.

Procedure: For phosphate and phosphite ions spray with (*a*), dry and spray with (*b*). For organophosphorus pesticides and other compounds that need to be hydrolyzed spray the dry plate with (*c*), cover with a glass plate and heat in an oven for 30 min at 180°C. Spray with (*a*) and heat again for 5 min. Cool and spray with (*b*), then place in a tank filled with ammonia vapor to bleach the background. Blue spots appear.

Use: Phosphate and phosphite ions [77]. Organophosphorus pesticides [78].

Sensitivity: Pesticides 1 μg.

T-20. Ammonium thiocyanate–ferrous sulfate

Reagent: (*a*) 0.2 g of ammonium thiocyanate in 15 ml of acetone.
(*b*) 4% ferrous sulfate solution.
Procedure: Add 10 ml of (*b*) to (*a*) just prior to spraying.
Use: Peroxides [50]. Immediate appearance of brownish red spots.

T-21. Ammonium vanadate (Mandelin's reagent)
Reagent: Add 1 g ammonium vanadate to 100 ml of concentrated sulfuric acid. Prior to use shake thoroughly.
Procedure: After spraying, record colors and then heat to 85°C.
Use: Antihistamines [79].

T-22. Ammonium vanadate–sulfuric acid
Reagent: (*a*) Dissolve 1.62 g of anhydrous ammonium vanadate (recrystallize twice from 5% ammonia, wash with cold absolute ethanol, then anhydrous ether, and vacuum dry at 135 to 140°C for 2 hr) in 125 ml of concentrated sulfuric acid, cool, and add **to** 125 ml of ice-cold water.
(*b*) Dilute (*a*) 1:10 as needed for spraying.
Procedure: Spray with (*b*) and heat at 110°C for 3 min.
Use: Steroids, alkaloids, phenolic compounds [80].
Sensitivity: Varies with compounds, 0.25 to 2.5 μg.

T-23. Aniline phosphate
Reagent: Add 20 ml of aniline to 200 ml of water, then add 180 ml of acetic acid followed by 10 ml of phosphoric acid. Store at 4°C.
Procedure: Dilute 2 parts of reagent with 3 parts of acetone for spraying. Heat plate at 100°C for 2 to 5 min.
Use: Carbohydrates. Pentoses red-brown, aldoses and sorbose yellow or yellow-brown.

T-24. Aniline phthalate
Reagent: 0.93 g of aniline + 1.66 g of phthalic acid in 100 ml of moist butanol [81].
Procedure: Spray, then heat for 10 min at 105°C.
Use: Reducing sugars [82, 83].

T-25. *p*-Anisaldehyde
Reagent: Prepare fresh: 0.5 ml *p*-anisaldehyde in 5 ml of 70% perchloric acid, 10 ml of acetone, and 40 ml of water.
Procedure: Spray, heat plate at 75 to 80°C for 4 to 5 min. Observe colors in visible and ultraviolet light over period of 1 hr.
Use: Digitalis glycosides [84].
Sensitivity: 0.1 μg visible. 0.02 μg ultraviolet.

T-26. Anisaldehyde (modified Kagi–Mischer reagent)

Reagent: 0.5 ml of anisaldehyde in 1 ml of sulfuric acid and 50 ml of glacial acetic acid.
Procedure: Spray and heat 10 min at 125°C.
Use: Bile acids [85]. Other steroids [86].
Sensitivity: 1 μg.

T-27. Anisaldehyde
Reagent: Mix 1 ml of anisaldehyde with 9 ml of 95% ethanol, cool in ice bath, and add 1 ml of concentrated sulfuric acid.
Procedure: Spray and then heat plate at 90°C for up to 10 min.
Use: Prostaglandins, sugars.
Sensitivity: 0.1 nM, 0.05 μg, respectively.

T-28. Anisidine phthalate
Reagent: 0.1 *M* solution of *p*-anisidine and phthalic acid in 96% ethanol.
Use: Sugars [87]. Hexoses, green. Pentoses, red-violet. Methyl pentoses, yellow-green. Uronic acids, brown.
Sensitivity: Methyl pentoses and hexoses: 0.5 μg. Pentoses and uronic acid: 0.1 to 0.2 μg.

T-29. Anthrone
Reagent: (*a*) 10% sulfuric acid.
(*b*) 1% (wt/vol) anthrone in benzene.
Procedure: Spray with (*a*) followed by (*b*). Heat at 100°C for 10 min. Colors are not stable and disappear on cooling. Reappear on reheating.
Use: Cerebrosides. Other lipids do not interfere [88].

T-30. Antimony pentachloride
Reagent: 10 to 20% solution of antimony pentachloride in carbon tetrachloride [33].
Procedure: Spray with reagent and heat plate to 120°C. Observe in both daylight and ultraviolet light.
Use: A general reagent.

T-31. Antimony trichloride (Carr–Price reagent)
Reagent: Saturated solution of antimony trichloride in alcohol-free chloroform.
Procedure: Spray and heat plate to 100°C for 10 min.
Use: General reagent giving varied colors with many compounds.

T-32. Antimony trichloride–acetic anhydride
Reagent: Saturated solution of antimony trichloride in alcohol-free chloroform containing 20% acetic anhydride [34].
Procedure: Spray and heat plate to 130°C for 5 to 10 min.
Use: Steroids.

T-33. Antimony trichloride–thionyl chloride

Reagent: Saturated solution of antimony trichloride in alcohol-free chloroform containing 10% thionyl chloride [89].
Procedure: Spray and heat to 110 to 120°C.
Use: Steroids containing a Δ^4 double bond.

T-34. Beam reagent [90]
Reagent: 5% potassium hydroxide in 99% ethanol.
Procedure: Heat for 5 min at 105°C.
Use: Cannabidolic acid and cannabidiol [91]. Blue-violet spots.

T-35. Benzidine
Reagent: 0.2% benzidine in acetic acid.
Procedure: Spray and heat plates for 10 min at 100°C.
Use: Aldoses [92]. Monoaldoses give brown spots. Chlorinated pesticides (use 0.1% solution and after spraying expose to short-wave ultraviolet for 5 to 10 sec). Yellow-green to blue-green spots on cream-colored background [93].
Sensitivity: Pesticides 0.5 to 1 μg.

T-36. Benzidine, tetra-azotized [94]
Reagent: (*a*) 5 g benzidine in 14 ml of hydrochloric acid, dilute to 1 liter (stable for 1 week).
(*b*) 10% sodium nitrite.
Procedure: Mix equal volumes of (*a*) and (*b*) shortly before use; solution is stable for 2 to 3 hr.
Use: Phenols. Phloroglucinol–resorcinol type gives red azo dyes.
Sensitivity: 2 to 3 μg.

T-37. Benzidine–potassium iodide
Reagent: (*a*) 5 ml of commercial household bleach in 50 ml benzene (freshly prepared).
(*b*) 0.5 g of benzidine and 1 crystal of potassium iodide in 50 ml of 50% ethanol; filter and keep out of direct light (stable for 2 hr).
Procedure: Spray dried chromatogram with (*a*), then dry to remove excess chlorine and spray with (*b*).
Use: Sphingolipids [95]. Blue spots.
Sensitivity: 5 to 10 μg.

T-38. Benzidine: Stain for thin-layer starch gels [96]
Reagent: All solutions prepared with distilled water.
(*a*) Acetate buffer 0.1 *M*, 4.7.
(*b*) Dissolve 1 g of benzidine dihydrochloride and 1 g of sodium nitroprusside in 500 ml of an aqueous solution of 1% acetic acid at room temperature, using a magnetic stirrer (pH $\approx$2.8). Prepare fresh daily. The solution is light sensitive.

(*c*) 3% aqueous solution of sodium pyrophosphate, freshly prepared (pH ≈10.3).

(*d*) Aqueous solution of 15 vol % glycerol and 2 vol % acetic acid in which 10 g of sodium nitrate per 1000 ml is dissolved shortly before use.

Procedure: (1) Prerinse the gel in solution (*a*) for 20 min renewing the fluid after 10 min.

(2) Immediately before use add 0.2 ml of 30% hydrogen peroxide per 100 ml of staining solution (*b*), and immerse the gel in this solution for 7 min.

(3) Discard staining solution thoroughly and immerse gel for 5 min in an aqueous solution of 3% sodium pyrophosphate. Gently rub the surface of the gel clean of the grayish precipitate, removing also all filter paper strips if such were used for sample application.

(4) Immerse the gel for 4 min in absolute methanol.

(5) Rinse the gel in a large amount of distilled water for 30 min to remove the methanol, renewing the water after 15 min.

(6) Immerse the gel for 30 min in solution (*d*).

(7) Dry the gel on a glass plate, in a warm air stream, at not more than 50°C for 48 to 72 hr. The gel should face the glass surface and be covered by the supporting sheet. Weight the borders for the first few hours of the drying process.

T-39. Bials reagent

Reagent: 40.7 ml concentrated hydrochloric acid, 0.1 g of orcinol, 1 ml of 1% ferric chloride solution diluted to 50 ml with water [97].

Procedure: Place chromatogram in hydrogen chloride atmosphere for 1.5 hr at 80°C, then spray with the reagent and heat again at 80°C until the color develops.

Use: Glycolipids [70], violet spots on white background.

T-40. Boute reaction [98]

Reagent: Nitrogen dioxide (prepare from concentrated nitric acid and copper).

Procedure: Expose plate to ammonia vapor and then to nitrogen dioxide.

Use: Phenolic OH groups [16, 99]. Yellow color stable for several days.

T-41. Brilliant Green

Reagent: 0.5% Brilliant green in acetone.

Procedure: Spray dry plate with reagent and immediately expose to bromine vapor.

Use: Triazine herbicides [100], deep green spots on off-white background. Organophosphorus compounds.

T-42. Bromocresol green

Reagent: 0.3% bromocresol green in 80% (by volume) methanol + 8 drops of 30% sodium hydroxide per 100 ml.

Use: Acids [3], yellow spots on green background.

T-43. Bromocresol purple

Reagent: 0.04 g bromocresol purple in 100 ml of 50% ethanol. Adjust to pH 10 with alkali.

Use: Halogen anions (except fluoride) [101]. Dicarboxylic acids [102], yellow spots on blue background. Herbicides (halogenated phenoxy acids) [103].

Sensitivity: Herbicides 1.0 to 2.0 μg.

T-44. Bromine–ferric chloride–sulfosalicylic acid [90, 104]

Reagent: (*a*) 0.1% ferric chloride in 80% ethanol.

(*b*) 1% sulfosalicylic acid in 80% ethanol.

Procedure: Expose plates to bromine vapor for 10 min. Spray with (*a*) and let dry 15 min. Spray with (*b*).

Use: Organophosphorus and thiophosphates [105]. Pesticides, white spots on mauve background.

Sensitivity: 5 μg.

T-45. Bromine–potassium iodide

Reagent: 2% potassium iodide in water.

Procedure: Compounds are separated on silica gel G layers containing 1.5 g of amylopectin/28.5 g of silica gel G. After development, dry layers at 100°C for 10 min, then cool and expose to bromine vapors for 3 sec. Remove excess bromine with cold air stream (hair dryer) for 2 min. Spray with potassium iodide.

Use: nitrophenols and halogenated phenols, dark blue spots [106].

Sensitivity: 0.025 to 0.1 μg.

T-46. Bromosuccinimide–fluorescein

Reagent: (*a*) 0.4 g of *N*-bromosuccinimide in 100 ml of acetic acid.

(*b*) 0.01 g of fluorescein in 100 ml of 96% ethanol.

Procedure: Spray plate with (*a*), then heat for 10 min at 120°C. Cool and spray with (*b*). Examine in daylight and under long-wavelength ultraviolet light.

Use: Barbiturates with a double bond in side chain at C-5. Some other pharmaceuticals also react with this reagent [107].

T-47. Bromothymol blue

Reagent: 40 mg in 100 ml of 0.01 *N* sodium hydroxide.

Procedure: Spray and then expose to ammonia vapor.

Use: Lipids [108], blue-green spots.
Sensitivity: 0.1 to 1 γ.

T-48. Butyl hypochlorite–iodine
Reagent: (*a*) The silica gel G layers are prepared with 0.5% solution of soluble starch.
(*b*) 1% *tert*-butyl hypochlorite in cyclohexane.
(*c*) Freshly prepared 1% solution of analytical-grade potassium iodide in acetone–water (3:1).
Procedure: The chromatograms must be run on the starch-containing layers. After drying spray evenly with (*b*) then aerate 15 to 20 min with cool dry air to remove excess (*b*). Spray with (*c*). Spots may be intensified by spraying with water acidified with hydrochloric acid.
Use: Ninhydrin negative nitrogen compounds such as imides, amides, and cyclic peptides [109].

T-49. Calcium nitrate
Reagent: 5% calcium nitrate in 95% ethanol.
Procedure: Spray with reagent then expose to ultraviolet light.
Use: Diphenylamine [110], yellow-green spot.

T-50. Ceric ammonium nitrate–hydroxylamine
Reagent: (*a*) 5% solution of ceric ammonium nitrate $[(NH_4)_2Ce(NO_3)_6]$ in acetone. Freshly prepared and filtered.
(*b*) 5% hydroxylamine hydrochloride in 80% acetone.
Procedure: Spray with (*a*) followed by (*b*). Dry with hot air. Heating the plates at 110°C for 5 min can increase the sensitivity in some cases.
Use: Free phenolic groups and indole derivatives. (Aliphatic alcohols or amines, sugars, saturated cycloalcohols, carboxylic acids, aldehydes, and ketones do not react [111].
Sensitivity: 1 to 10 μg.

T-51. Ceric ammonium sulfate
Reagent: Heat 1 g of ceric ammonium sulfate in 99 g of syrupy phosphoric acid until solution takes place [112].
Use: Alkaloids [113].

T-52. Ceric ammonium sulfate–molybdate reagent
Reagent: (*a*) Powder and grind 10 g of ceric ammonium sulfate into a plaste with 17.5 ml of concentrated sulfuric acid. Continue grinding while slowly adding 20 ml of water on the sides. Dilute to 100 ml and filter.
(*b*) Molybdenum blue reagent (T-167).
Procedure: Spray with a 1:1 mixture of (*a*) and (*b*).

Use: Bile acids [114].
Sensitivity: 1 μg.

T-53. Ceric sulfate
Reagent: Saturated solution of ceric sulfate in 60% sulfuric acid.
Procedure: Spray and heat at 120°C for 15 min.
Use: Alkaloids [115]. Gibberellins (saturated ceric sulfate–concentrated sulfuric acid 1:1) [116].

T-54. Ceric sulfate
Reagent: 0.3% ceric sulfate in concentrated nitric acid.
Procedure: Spray and observe under 366-mμ ultraviolet.
Use: Polyphenyls [117].

T-55. Ceric sulfate–trichloroacetic acid
Reagent: Boil 0.1 g ceric sulfate in 4 ml of water containing 1 g of trichloroacetic acid adding concentrated sulfuric acid (*d* 1.84) dropwise until solution clarifies [118].
Procedure: Spray and heat to 110°C.
Use: Tocopherols [119]. Triterpenoids, organic iodides, colchicine, brucine, papaverine, apomorphine, and physostigmine.

T-56. Chloramine-T
Reagent: 10% chloramine-T in water.
Procedure: Spray plate with chloramine-T reagent and then with 1 *N* hydrochloric acid. Heat at 96 to 98°C to remove chlorine, then expose to ammonia.
Use: Detection of caffeine [33]. Rose-red color.

T-57. Chloramine-T–trichloroacetic acid
Reagent: (*a*) Freshly prepared solution of 3% chloramine-T in water.
(*b*) 25% trichloroacetic acid in ethanol. Stable for several days.
Procedure: Spray plate with a mixture of 10 ml of (*a*) and 40 ml of (*b*). Heat at 110°C for 5 to 10 min and examine under long-wavelength ultraviolet light.
Use: Cardenolides and bufadienolides.
Sensitivity: 0.01 μg.

T-58. Chloranil
Reagent: 0.2% chloranil (2,3,5,6-tetrachloro-1,4-benzoquinone) in monochlorobenzene.
Procedure: Spray with freshly prepared reagent. Effective on silica gel layers, but does not work on cellulose.
Use: Primary and secondary aromatic amines [120].
Sensitivity: 0.01 μg.

T-59. Chlorine–benzidine

Reagent: Add 2 ml of 10% potassium iodide solution to 100 ml of a 0.5% benzidine solution in ethanol.
Procedure: Expose plate to chlorine gas (may be generated from potassium permanganate and hydrochloric acid) for 5 min. Remove excess chlorine by heating to 105°C. Spray with benzidine reagent.
Use: Sedatives, caffeine [121, 122].

T-60. Chlorine–potassium iodide–starch
Procedure: Expose to chlorine for ½ hr. Allow excess chlorine to vaporize, then spray with starch potassium iodide solution [123].
Use: Carbobenzoxyamino acid esters [124].

T-61. Chlorine–toluidine
Reagent: 80 mg of *o*-toluidine and 0.5 g of potassium iodide in 15 ml glacial acetic acid and diluted to 250 ml [125].
Procedure: Moisten plate over boiling water and then expose to chlorine for 15 to 20 min. Dry in air 2 to 3 min. Spray corner of plate with toluidine reagent; if blue, wait longer before spraying.
Use: Carbobenzoxyamino acids [126]. Phenylthiohydantoin amino acids [127]. Thiamine, riboflavin, pyridoxine, cyanocobalamin, and nicotinamide [128].
Biotin [129]. Alkylated purine and pyrimidine bases [130].
Sensitivity: Biotin 0.3 μg. Purine and pyrimidine bases 200 ng.

T-62. 9-Chloroacridine
Reagent: Dissolve 21.3 mg of 9-chloroacridine in 100 ml of 95% ethanol. Prepare fresh daily and keep in refrigerator. [Recrystallize reagent from light petroleum (30 to 60°C) before use.]
Procedure: Spray in well-ventilated hood, dry for 5 min at 110°C, and observe colors. Finally observe under long-wavelength ultraviolet to detect blue fluorescence (arylhydroxylamines) or fluorescent quenching (arylamines).
Use: Primary arylamines and arylhydroxylamines, yellow to orange spots. Specific detection of arylhydroxylamines in presence of arylamines [131].
Sensitivity: 0.5 ng by fluorescence or quenching.

T-63. Chlorosulfonic acid–acetic acid
Reagent: Chlorosulfonic acid–acetic acid (1:2) [132].
Procedure: Spray and heat to 130°C for 1 min.
Use: Olefins [133]. Sapogenins [132]. Digitalis glycosides, green spots which fluoresce brown-violet in ultraviolet light [134].

T-64. Chromotropic–sulfuric acid
Reagent: Stock solution: 10% (wt/vol) solution of sodium 1,8-dihydroxy naphthalene-3,6-disulfonate. Spray reagent: 1 vol of

stock solution in 5 vol of sulfuric acid-water (5:3) (prepare fresh daily).

Procedure: Spray solvent-free plate, record colors, heat ½ at 105 to 110°C. Again record colors.

Use: Insecticide synergists (3,4-methylenedioxyphenyl compounds) [135].

Sensitivity: 0.1 to 5 μg [136].

T-65. Cinnamaldehyde

Reagent: 5 ml of cinnamaldehyde in ethanol–concentrated hydrochloric acid (95:5).

Use: Hydroxyskatoles [137]. Psychotropic drugs.

T-66. Clorox–benzidine [138]

Reagent: (*a*) To 5 ml of Clorox brand bleach in 50 ml of benzene, add 5 ml of glacial acetic acid (use reagent immediately).

(*b*) 0.5 g of benzidine + 1 small crystal potassium iodide in 50 ml of 50% ethanol. Filter solution. (Keep out of direct light during preparation and storage). Stable for 2 hr.

Procedure: Spray with (*a*), remove excess chlorine by drying in hood. Then spray with (*b*).

Use: Sphingolipids. Blue on white background. Acid polysaccharides (replace benzene with water in the above reagent.)

T-67. Cobalt acetate–lithium hydroxide (Zwikker's reagent)

Reagent: (*a*) 0.5% methanolic cobalt acetate

(*b*) 0.5% methanolic lithium hydroxide

Procedure: Spray (*a*) followed by (*b*).

Use: Barbiturates [51].

T-68. Cobalt chloride

Reagent: 1% solution of cobalt chloride (water-free) in acetone.

Procedure: Spray reagent. (Small quantities of organophosphorus esters require heating to 40 to 50°C.)

Use: Organic phosphorus compounds including trialkylphosphates [139]. Aromatic amines and imidazoles [140].

Sensitivity: Imidazoles 50 $\mu g/cm^2$.

T-69. Cobalt nitrate

Reagent: 2.5 g of cobalt nitrate and 1.25 g of ammonium thiocyanate in 10 ml of ethanol.

Use: Organic phosphorus compounds [141]. Imidazoles [140].

T-70. Cobaltinitrite

Reagent: Mix 11.4 g of cobalt acetate [$Co(CH_3COO)_2 \cdot 4H_2O$], 16.2 g of lead acetate [$Pb(CH_3COO)_2 \cdot 3H_2O$], 20 g of sodium nitrite; and 2 ml of glacial acetic acid. Dilute to 150 ml, centrifuge and filter.

Procedure: Spray with reagent–methanol (3:1) [142]. (Merkus [143] recommends spraying with saturated barium nitrate solution first to precipitate sulfate ions before using this spray.)
Use: Potassium.
Sensitivity: 2 μg.

T-70a. Coomassie Brilliant Blue
Reagent: (*a*) 5% trichloroacetic acid.
(*b*) 1% stock solution of Coomassie Brilliant Blue (C.I. #42660, CoLab) For use dilute 10 ml with 7 ml of glacial acetic acid and 25 ml of methanol, then dilute to 100 ml with water.
(*c*) 7% acetic acid.
Procedure: Immerse gel sheets in (*a*) for 15 min at 4°C, then immerse in freshly prepared and filtered (*b*) for 60 to 72 hr at 4°C. Destain for 2 hr in multiple rinses of (*c*), then rinse in water.
Use: Staining total proteins in polyacrylamide layers [186a].

T-71. Cupric acetate–dithiooxamide [144]
Reagent: (*a*) 20 ml of saturated cupric acetate in 1000 ml of solution.
(*b*) 0.1% alcoholic dithiooxamide solution.
Procedure: (1) Stabilization of the chromatogram for washing [445]: place developed and dried chromatogram in vacuum desiccator at 300 mm with 3 to 5 ml of dichlorodimethylsilane for 15 min. Remove and expose to air for 30 min.
(2) Dip plate in water.
(3) Immerse plate 10 min in (*a*).
(4) Wash 30 min in running water, then rinse in distilled water.
(5) Immerse 10 min in (*b*).
(6) Wash with distilled water.
Use: Fatty acids. Green spots on white background.

T-72. Cupric acetate–silver nitrate
Reagent: Dissolve 1.7 g of silver nitrate and 1.8 g of copper acetate in 20 ml of concentrated ammonium hydroxide. Dilute to 100 ml with absolute ethanol [146].
Procedure: Spray and heat 20 to 30 min at 100 to 120°C.
Use: trithiofluorobenzaldehydes.

T-73. Cupric chloride
Reagent: (*a*) 0.5% cupric chloride.
(*b*) Saturated alcoholic cupric acetate.
Use: Oximes [147] α-Benzaldoxime required spraying with (*b*) followed by heating for 10 min at 100°C.

T-74. Cupric chloride

Reagent: 2 g of cupric chloride in 11 ml of ethanol + 2.5 ml of concentrated hydrochloric acid.
Use: Systox and Meta-Systox [148].

T-75. Cupric sulfate
Reagent: 10% cupric sulfate–2% ammonia (5:1).
Procedure: Spray, then heat plate at 110°C for 10 min. Blue or brown spots with substances containing amido groups.
Use: Diuretics [149].

T-76. 3,5-Diaminobenzoic acid–phosphoric acid
Reagent: 1 g of 3,5-diaminobenzoic acid dihydrochloride in a mixture of 25 ml of 80% phosphoric acid and 60 ml of water.
Procedure: Spray and heat plate to develop colors and observe in ultraviolet light.
Use: 2-Deoxysugars [150], green-yellow fluorescence.

T-77. *o*-Dianisidine
Reagent: Saturated solution of *o*-dianisidine in glacial acetic acid.
Use: Aldehydes [3].

T-78. Diazotization reagent
Reagent: (*a*) 3% (vol/vol) pentyl nitrite in diethyl ether + 3% of 98% formic acid. Stable for 2 months.
(*b*) 1% 2-naphthol dissolved in 5% sodium hydroxide. Prepare monthly.
Procedure: Spray with (*a*) (1 ml/10 cm^2). Dry plate at room temperature in a current of air for 10 min. Spray cautiously with (*b*).
Use: Primary arylamines. In general, monoamines give red dyes, diamines purple colors [151].

T-79. 2,6-dibromobenzoquinone-4-chloroimide
Reagent: 1% solution of 2,6-dibromo-*N*-chloro-*p*-quinoneimine.
Use: Detection of mono and diesters of phosphoric and phosphorothioic acids. Thiolo-S and sulfhydryl groups give yellow spots, thiono-S red spots, and thioureas brown spots [152]. Aromatic amines and carbazoles [153].
Sensitivity: 0.5 μg.

T-80. 2,3-Dichloro-5,6-dicyano-1,4-benzoquinone
Reagent: (*a*) 0.1% solution of the reagent in benzene, prepared immediately before use.
(*b*) 2% solution of the reagent in benzene [154].
Procedure: For sulfur-containing pesticides, dry plate at 85°C for 30 min, cool, and expose to bromine vapor. Pass cool air over the surface for 1 min to remove excess bromine and then spray with (*a*). Allow to stand 2 min before spraying lightly with 90%

ethanol. Expose to ultraviolet light for 30 min. For sulfoxides, sulfones, and sulfides, spray with (*b*), note colors, then expose to ammonia vapor and note colors again. Sulfoxides are orange to crimson, sulfones are lilac to violet, and sulfides are pink to blue, changing to pink to orange after exposure to ammonia. For carbazoles, spray with (*a*).
Use: Sulfur compounds. Carbazoles.
Sensitivity: Pesticides 0.05 μg [155]. Carbazoles 0.1 μg [156].

T-81. 2′,7′-Dichlorofluorescein
Reagent: 0.2% ethanolic solution of 2′,7′-dichlorofluorescein [157].
Procedure: Examine under ultraviolet light after spraying.
Use: Lipids. Bright yellow-green fluorescence.

T-82. 2,6-Dichlorophenol–indophenol
Reagent: (*a*) Cold-saturated aqueous solution of 2,6-dichlorophenol–indophenol [158].
(*b*) 0.1% solution of 2,6-dichlorophenol–indophenol in 95% alcohol [159].
Procedure: Spray with either (*a*) or (*b*). A short heating period may help to bring out the color.
Use: Organic acids, pink spots on sky-blue background.

T-83. Dichloroquinone chloroimide
Reagent: (*a*) 0.4% solution of dichloroquinone chloroimide.
(*b*) 10% sodium carbonate in 30% methanol.
Procedure: Spray successively with (*a*) and then (*b*).
Use: Aloe compounds [160].

T-84. Dichloroquinone chloroimide
Reagent: 0.1% dichloroquinone chloroimide in ethanol.
Procedure: After spraying plate, expose to ammonia vapors.
Use: Vitamin B_6, blue color [161]. Phenolic terpenes, varied colors [162].
Sensitivity: Biotin 0.1 μg [129].

T-85. Dichloroquinone chloroimide
Reagent: 1% 2,6-dichloroquinone chloroimide in ethanol.
Procedure: 15 min after spraying the reagent, spray with 2% borax solution (in some cases the borax spray causes a color change).
Use: Antioxidants [163, 164].

T-86. Dicobaltoctacarbonyl
Reagent: Reagent: (*a*) 0.5% solution of dicobaltoctacarbonyl in petroleum ether (120 to 135°C).

(*b*) 0.5% α-nitroso-β-naphthol in acetic acid–water (1:1).
Procedure: Spray dry plate with reagent (*a*). After 10 min spray with 1 *N* hydrochloric acid and dry again. Spray with Neatan and after thoroughly hardening soak off the chromatogram and wash thoroughly for approximately 2 hr. Then press between filter paper to remove excess moisture and expose to bromine vapor for 1 min. Dip in reagent (*b*) and wash off excess reagent with 0.5% ammonia.
Use: Polyacetylene compounds [165], brown-red spots on yellow background.

T-87. 9-Dicyanomethylene-2,4,7-trinitrofluorene
Reagent: 2 g of 9-dicyanomethylene-2,4,7-trinitrofluorene in 100 ml of acetone. Freshly prepared prior to use.
Use: Aromatic ethers [166].

T-88. *p*-Diethylaminobenzaldehyde
Reagent: 0.25% *p*-diethylaminobenzaldehyde in 0.25 *N* hydrochloric acid (prepared from concentrated hydrochloric acid) in absolute ethanol.
Procedure: Zinc reducer incorporated in layers (3 g of zinc dust for 30 g of silica gel).
Use: Nitro derivatives [167]. Nitrose diphenylamines [168].
Sensitivity: 1 to 4 μg.

T-89. *p*-Dimethylaminobenzaldehyde–ferric chloride (van Urk's reagent)
Reagent: 0.125 g of *p*-dimethylaminobenzaldehyde and 0.1 ml of 5% ferric chloride in 100 ml of 65% sulfuric acid.
Use: Phenothiazine compounds [169]. Hydroxyskatoles [136]. *N,N*-Diethyl-D-lysergamide (LSD) [170].
Sensitivity: LSD 0.05 μg.

T-90. *p*-Dimethylaminobenzaldehyde (Ehrlich's reagent)
Reagent: 1 g of *p*-dimethylaminobenzaldehyde in 25 ml of concentrated hydrochloric acid and 75 ml of methanol.
Use: Hydroxyskatoles [136]. Alkaloids [171]. Sulfonamides, lemon yellow spots [172]. Fast bases [173].

T-91. *p*-Dimethylaminobenzaldehyde (modified)
Reagent: 1 g of *p*-dimethylaminobenzaldehyde in 30 ml of ethanol, 3 ml of concentrated hydrochloric acid, 180 ml of 1-butanol [174].
Procedure: Spray and heat at 120°C for 30 min.
Use: Phenylurea derivatives. Carbamate and urea herbicides [175].
Sensitivity: 0.2 to 0.4 μg.

T-92. *p*-Dimethylaminobenzaldehyde (modified)
Reagent: 1% *p*-dimethylaminobenzaldehyde in 5% hydrochloric acid [176].
Use: Sulfonamides.

T-93. Dimethylglyoxime
Reagent: (*a*) 10% dimethylglyoxime in ammoniacal ethanol.
(*b*) 1% dimethylglyoxime in 95% ethanol.
Procedure: For thin-layer chromatogram, spray (*a*). For ionophoresis agar layers, expose layer 3 min to ammonia vapor then soak in (*b*) for 1 to 2 min.
Use: Nickel, red spot [177].

T-94. Dimethyl-*p*-phenylenediamine dihydrochloride
Reagent: (*a*) 1.5 g of *N,N*-dimethyl-*p*-phenylenediamine dihydrochloride dissolved in 128 ml of methanol, 25 ml of water, and 1 ml of acetic acid [178].
(*b*) 0.5 g of the diamine in 100 ml of ethanol containing 1 g of sodium (freshly prepared) [179].
Use: Organic peroxides [reagent (*a*)], purple-red spots. Chlorinated pesticides [reagent (*b*)]. After spraying with the reagent, moisten by spraying with water and then expose for 1 min to ultraviolet light. Dirty violet to green spots. Carbromal [180].
Sensitivity: Pesticides, <0.5 μg; carbromal 5 μg; peroxides 0.1 μg.

T-95. *m*-Dinitrobenzene–potassium hydroxide (Zimmerman reagent)
Reagent: 1% *m*-dinitrobenzene in ethanol–5 *N* potassium hydroxide (2:1).
Procedure: Spray, then dry in current of hot air.
Use: 17- and 3-Keto steroids [181]. Methylene groups activated by a keto group in the *ortho* position [182].

T-96. 2,4-Dinitrophenylhydrazine
Reagent: 0.4 g of 2,4-dinitrophenylhydrazine in 100 ml of 2 *N* hydrochloric acid.
Use: Carbonyl compounds [49], yellow to red spots. Testosterone derivatives.

T-97. Diphenylamine
Reagent: 2.3 g diphenylamine in 100 ml of water-saturated *n*-butanol.
Procedure: Dry in air and then dry for 20 min at 130°C.
Use: Aldoses and ketoses [183], blue spots.

T-98. Diphenylamine
Reagent: Dilute 20 ml of a 10% solution of diphenylamine in

alcohol with 100 ml concentrated hydrochloric acid and 80 ml glacial acetic acid.
Procedure: Spray lightly. Heat at 110°C after covering with another glass plate until spots appear (30 to 40 min).
Use: Glycolipids [108, 184], blue spots.

T-99. Diphenylamine
Reagent: 1% solution of diphenylamine in 95% ethanol.
Procedure: Spray and expose plate to short-wave ultraviolet light.
Use: Nitrate esters [110], yellow-green spots on colorless background. Explosives (5% reagent solution used) [185], varied colors. Chlorinated pesticides (dilute reagent 1:10 with ethanol) [93].
Sensitivity: Pesticides 0.5 μg.

T-100. Diphenylamine–palladium chloride
Reagent: (*a*) 1.5% diphenylamine in ethanol
(*b*) 0.1% palladium chloride in 0.2% saline solution.
Procedure: Spray lightly with a mixture of 5 parts of (*a*) and 1 part of (*b*). Expose moist plate to 240-mμ ultraviolet light.
Use: Nitrosamines [186], blue to violet spots on colorless background. Quinoid systems and nitro compounds interfere.
Sensitivity: 0.5 μg, 1 to 2 μg for volatile compounds.

T-101. Diphenylamine–zinc chloride
Reagent: 0.5 g each of diphenylamine and zinc chloride in 100 ml of acetone.
Procedure: Spray the dried plate and then heat for 5 min at 200°C.
Use: Chlorinated pesticides [187, 188]. Varied colors.
Sensitivity: 1 to 5 μg.

T-102. Diphenylboric acid-β-aminoethyl ester
Reagent: 1% diphenylboric acid-β-aminoethyl ester in methanol.
Procedure: Observe fluorescence under 366-mμ ultraviolet light.
Use: Flavonols, coumarins, and derivatives [189].

T-103. *s*-Diphenylcarbazide
Reagent: 1% *s*-diphenylcarbazide in 95% ethanol.
Procedure: For ionophoresis of metal ions: after washing with water, expose to ammonia vapor and dip in reagent for 10 min. For chromatograms, spray with reagent followed by 25% ammonia hydroxide.
Use: Heavy metal ions [190].
Sensitivity: <0.5 μg.

T-104. *s*-Diphenylcarbazone
Reagent: 0.1% *s*-diphenylcarbazone in 95% ethanol.

Use: Acetoxymercuric-methoxy derivatives of unsaturated esters and barbiturates [191, 192]. Purple spots on light rose background. Dialkyltin salts (trialkyl do not react) (use 0.01% solution of reagent in chloroform) [193], red-violet spots.

T-105. Diphenylpicrylhydrazyl
Reagent: 15 mg of diphenylpicrylhydrazyl in 25 ml of chloroform.
Procedure: Spray reagent, then heat 5 to 10 min at 110°C.
Use: Terpene hydrocarbons, alcohols, carbonyls, oxides, esters, and ethers [194]. Yellow spots on purple background.

T-106. 2,5-Diphenyl-3-(4-styrylphenyl)tetrazolium chloride
Reagent: (*a*) 1% methanolic solution of 2,5-diphenyl-3-(4-styrylphenyl)tetrazolium chloride.
(*b*) 3% sodium hydroxide (aqueous).
Procedure: Mix 1 vol (*a*) with 10 vol (*b*) and spray immediately for alumina plates. For silica gel layers, spray first with 2 *N* sodium hydroxide to provide an alkaline condition.
Use: Steroids [195]. Strong purple spots on yellow background.
Sensitivity: 0.1 μg or less.

T-107. Dipicrylamine
Reagent: 0.2 g of dipicrylamine in 50 ml of acetone + 50 ml of water.
Procedure: Spray heavily until color produced.
Use: Detection of choline tartrate, deep red on yellow background; color intensifies on heating at 80°C for 5 min [129].
Sensitivity: 3 μg.

T-108. Dipyridyl–ferric chloride (Emmerie–Engel reagent)
Reagent: 0.2% (wt/vol) ferric chloride in 95% ethanol–0.5% (wt/vol) α,α'-dipyridyl in 95% ethanol (1:1).
Use: Hydroquinones [196]. Tocopherols [197].

T-109. Dithiooxamide (rubeanic acid)
Reagent: 0.1% dithiooxamide in ethanol–*n*-butanol (1:1).
Procedure: Spray and dry 20 min at 100°C
Use: Metal ions, Cu, Co, Ni [198].
Sensitivity: 0.03 to 0.05 μg.

T-110. Dithizone
Reagent: 0.01% dithizone in carbon tetrachloride or chloroform.
Procedure: Spray reagent and note colors, then spray with 25% ammonia and again note colors.
Use: Heavy metal ions [199]. Organic tin salts [193]. Organomercury fungicides (use 0.5% solution of reagent) [200].
Sensitivity: Fungicides 0.5 to 20 μg.

T-111. Dragendorff reagent according to Thies and Reuther [201] (as modified by Vágújfalvi) [202]

Reagent: Dissolve 2.6 g of bismuth carbonate and 7.0 g of sodium iodide in 25 ml of boiling glacial acetic acid. Allow to stand 12 hr before filtering off the sodium acetate. The stock solution (kept in a brown bottle) is prepared by adding 8 ml of ethyl acetate to the filtered solution. Spraying solution: stock solution–acetic acid–ethyl acetate (1:2.5:6).

Use: Alkaloids: Orange spots (these can be intensified by spraying with *N*/20 sulfuric acid [202]). Adenine: citron yellow changing to blood-red color when sprayed with acid (specific) [203].

T-112. Dragendorff reagent (Munier modification) [204]

Reagent: (*a*) 17 g of basic bismuth nitrate and 200 g of tartaric acid in 800 ml of water.

(*b*) 160 g of potassium iodide in 400 ml of water. For use take 25 ml of the mixed solutions (*a*) + (*b*), 50 g of tartaric acid and 250 ml of water. The spray solution is stable for a week, but the stock solution will keep for a month or more.

Use: Alkaloids [205]. Cyclohexylamines [206]. Polyethylene glycols and derivatives [207]. Polyethene oxide compounds [208]. Lactams [209]. Lipids [210]. Steroids (α,β-unsaturated) [182]. Systemic fungicides [211].

Sensitivity: 0.25 to 1.0 μg.

T-113. Esterase inhibition test

Reagent: (*a*) 0.01 *M* tris(hydroxymethyl)aminomethane–maleate–trismaleate) buffer with 0.01 *M* nicotinamide, pH 7.2.

(*b*) 0.05 *M* tris(hydroxymethyl)aminomethane buffer, pH 8.3.

(*c*) Oxidizing solution containing 0.05 *M* each of potassium ferrocyanide and ferricyanide.

(*d*) Enzyme spray solution. Homogenize (2 min in a VirTis homogenizer with rheostat set at 50) 50 g of fresh beef liver + 180 ml of (*a*), then centrifuge at 4°C for 5 min at 2000*g*. To 100 ml of the supernatant add (*a*) to make 500 ml adding magnesium chloride to a final concentration of 0.1 *M*. Let stand for 24 hr at 1 to 9°C and then centrifuge for 15 min at 300*g*. The enzyme supernatant may be frozen and stored for several months.

(*e*) 13 ml of (*b*) mixed with 2 ml of (*c*) and then mixed with 15 mg of 5-bromoindoxyl acetate dissolved in 5 ml of ethanol.

Procedure: Air-dry the plates and then expose to bromine (if the plate becomes yellow it indicates overexposure). Allow to stand in the air until all odor of bromine has disappeared and then spray with (*d*) gradually and evenly until the gel is thoroughly wet. Dry the plates at room temperature for 20 min. Spray with (*e*). White spots appear on a blue background in 1 to 30 min [212].

Use: Organophosphorus pesticides.

Sensitivity: 0.1 to 1000 ng. (Note: overloading of chromatograms with plant extracts should be avoided because excessively large aliquots of pesticide-free plant extracts can produce zones of inhibition [213].)

T-114. Ethylenediamine

Reagent: Mix equal volumes of ethylenediamine and water or dilute sodium hydroxide.

Procedure: Spray, then heat layer for 20 min at 50 to 60°C. View under ultraviolet light.

Use: Catecholamines [214].

Sensitivity: 0.003 to 0.005 μg.

T-115. Fast Blue Salt B (tetraazotized di-*o*-anisidine)

Reagent: 0.5% solution of Fast Blue Salt B.

Procedure: Spray reagent followed by 0.1 *N* sodium hydroxide.

Use: Phenols [215]. Cannabidiol [91].

T-116. Fast Blue Salt BB (diazotized 1-amino-4-benzoylamido-2,5-diethoxybenzene)

Reagent: 0.5% solution of Fast Blue Salt BB.

Procedure: Spray reagent followed by 0.1 *N* sodium hydroxide.

Use: Phenols [215]. Anthraquinones (spray with alcoholic potassium hydroxide, then with reagent) [216].

T-117. Fast Red Salt B (diazotized 5-nitro-2-aminoanisole)

Reagent: 0.5% solution of Fast Red Salt B

Procedure: Spray reagent followed by 0.1 *N* sodium hydroxide.

Use: Phenols [215]. Hydroxybenzophenones [217].

T-118. Ferric chloride

Reagent: (*a*) 5% ferric chloride–2 *N* acetic acid (1:1).

(*b*) 2% ferric chloride (aqueous).

(*c*) Saturated solution of water-free ferric chloride in methanol.

(*d*) Dissolve 16.7 g $FeCl_3 \cdot 6H_2O$ in 10 ml concentrated hydrochloric acid and dilute with methanol to 1 liter.

Use: (*a*) Pyrazolones [44].

(*b*) Differentiate phenothiazines (red to violet) from phenothiazine sulfoxides (no reaction) [218]. Ferrocyanide, ferricyanide, and thiocyanate ions [219].

(*c*) Terpene phenols [162].

(*d*) Hydroxamic acids [220].

T-119. Ferric chloride–perchloric–nitric reagent
Reagent: 5% ferric chloride–20% perchloric acid–50% nitric acid (1:9:10) [221].
Use: Phenothiazines.

T-120. Ferric chloride–potassium ferricyanide
Reagent: 0.1 *M* ferric chloride–0.1 *M* potassium ferricyanide (1:1) freshly prepared.
Use: Aromatic amines [222], blue spots. Tryptamine [223]. Phenols and phenolic steroids [18].

T-121. Ferric chloride–potassium ferricyanide
Reagent: (*a*) Dissolve 1.3 g of ferric chloride in 100 ml of 2 *N* hydrochloric acid.
(*b*) Dissolve 0.7 g of potassium ferricyanide in 100 ml of water.
Procedure: Mix equal volumes of freshly prepared (*a*) and (*b*) prior to use. Avoid excessive heat and light as this darkens the background.
Use: Fat-soluble vitamins and antioxidants. To detect α-tocopherol acetate mix 2 vol of the reagent with 1 vol concentrated hydrochloric acid.

T-122. Ferric chloride–potassium ferricyanide–arsenite
Reagent: (*a*) 2.7% ferric chloride ($FeCl_3 \cdot 6H_2O$) in 2 *N* hydrochloric acid.
(*b*) 3.5% potassium ferricyanide.
(*c*) Dissolve 3.8 g of arsenic trioxide in 25 ml of 2 *N* sodium hydroxide, then cool to 5°C and add 50 ml of 2 *N* sulfuric acid (precooled to 5°C). Dilute to 100 ml.
Procedure: Mix (*a*), (*b*), and (*c*) in a ratio of 5:5:1 immediately before use. Spray and after color develops wash carefully with water. (Use on starch-bound layers.)
Use: Iodoamino acids [224].

T-123. Ferric chloride–sodium molybdate
Reagent: (*a*) Saturated solution of ferric chloride.
(*b*) 0.1 *M* sodium molybdate.
Procedure: Spray with (*a*) and follow immediately with (*b*). Heat at 140° 3 to 5 min.
Use: differentiate saturated and unsaturated methyl esters. Saturated esters give orange spots and unsaturated blue-purple spots on a brown background [225].

T-124. Ferric chloride–sulfosalicylic acid
Reagent: (*a*) Dissolve 0.1 g of ferric chloride ($FeCl_3 \cdot 6H_2O$) in 1 ml of 1 *N* hydrochloric acid and then dilute to 100 ml with 80% ethanol.

(*b*) 1% sulfosalicylic acid in 80% ethanol [73].
Procedure: Spray (*a*) and (*b*) successively.
Use: Organic phosphorus pesticides.

T-125. Ferric chloride–sulfosalicylic acid
Reagent: Dissolve 0.1 g of ferric chloride ($FeCl_3 \cdot 6H_2O$) and 7.0 g of sulfosalicylic acid in 25 ml of water and dilute to 100 ml with 95% ethanol.
Use: To detect phosphate groups in lipids [226]. White fluorescent spots on purple background.

T-126. Ferric sulfate–potassium ferricyanide
Reagent: (*a*) 0.5% ferric sulfate in 1 *N* sulfuric acid.
(*b*) 0.2% potassium ferricyanide [162].
Procedure: Mix equal volumes of (*a*) and (*b*) for spraying. Observe colors before (10 min) and after heating to 110°C.
Use: Phenolic compounds.

T-127. Ferrous ammonium sulfate–potassium thiocyanate
Reagent: Mixture of the two reagents.
Use: Stable peroxides not reacting with potassium iodide [227].

T-128. Fluorescamine
Reagent: (*a*) 10 mg of fluorescamine (Fluram, Hoffmann-La-Roche) in 100 ml of reagent-grade acetone. (Use clean volumetric flask and anhydrous conditions for longevity of the reagent solution). Keep refrigerated.
(*b*) 10% triethylamine in pesticide-grade methylene chloride. Keep refrigerated.
Procedure: Dry plate 30 to 60 min. Spray with (*a*) and dry with blower 2 min and then spray with (*b*) which stabilizes the fluorescence for quantitative work. Allow to stand 40 min before making quantitative measurements.
Use: Primary amino acids, peptides, proteins [228]. For detection of aldehydes, spot plate with aldehyde, superimpose spots of 5 μg of aniline and then developed with an appropriate solvent. Visualize as above [229]. Primary amines, aquamarine or yellow fluorescence; secondary amines appear dark purple, quenching fluorescence, and become fluorescent yellow if followed by 5-dimethylaminonaphthalene-1-sulfonyl chloride (DNS-Cl) [230].
Sensitivity: 4 to 80 ng of amino acids, peptides, and proteins. Aldehydes 20 to 200 ng.

T-129. Fluorescamine–glacial acetic
Reagent: 1 mg/ml of fluorescamine (Fluram, Hoffmann-LaRoche) in glacial acetic acid. Stable for weeks at room temperature.
Procedure: Spray the dried plate and again allow to dry.

Use: Aromatic amines only. Forms yellow stable products except for dichlorobenzidine product which fades in 5 min [231].
Sensitivity: Nanomole quantities.

T-130. Fluorescein–bromine
Reagent: 0.05% of sodium fluorescein solution.
Procedure: After spraying, expose to bromine vapor. (Avoid excess bromine.)
Use: Ethylenic unsaturated or other types of compounds that react with bromine. Yellow spots on pink background [3].

T-131. Folin–Ciocalteu's reagent
Reagent: (*a*) Dissolve 10 g of sodium tungstate and 2.5 g sodium molybdate in 70 ml of water; then add 5 ml of phosphoric acid and 10 ml of concentrated hydrochloric acid. Reflux 10 hr and add 15 g of lithium sulfate, 5 ml of water, and 1 drop of bromine. Reflux again for 15 min, cool, and dilute to 100 ml. Reagent should not be green.
(*b*) 20% sodium carbonate [232].
Procedure: Spray with (*b*), dry briefly, then spray (*a*)–water (1:3).
Use: Methylatedhydrazines [233]. Phenols, phenol carboxylic acids, estrogens.
Sensitivity: Hydrazines 0.12 $\mu g/cm^2$ [234].

T-132. Formaldehyde–hydrochloric acid (Procházka reagent) [170]
Reagent: 35% formaldehyde–25% hydrochloric acid–95% ethanol (1:1:2) (freshly prepared).
Procedure: Spray and heat plates for 5 min at 100°C.
Use: Indole derivatives, varied colors in daylight, fluorescent in ultraviolet light [171]. May be intensified by spraying with aqua regia (3 vol concentrated hydrochloric acid + 1 vol concentrated nitric acid).
Sensitivity: 0.01 μg.

T-133. Formaldehyde–sulfuric acid
Reagent: 40% formaldehyde–water–sulfuric acid (1:45:55).
Procedure: After spraying, heat chromatogram at 120°C for 10 min. Sensitivity can be increased by following this with the Dragendorff reagent of Thies and Reuther as modified by Vágújfalvi (which see) [237].
Use: Phenothiazines.

T-134. Formaldehyde–sulfuric acid
Reagent: 1.0 ml of formaldehyde (37%) in 50 ml of concentrated sulfuric acid [238].
Use: Hydrocarbons and heterocyclics (more sensitive than tetracyanoethylene).

T-135. Formic acid vapor
Procedure: Exposure to formic acid vapors for 1 min.
Use: Detection of quinine and quinidine [239]. Intense blue fluorescence under ultraviolet light.

T-136. Furfural–sulfuric acid
Reagent: Furfural (freshly distilled)–concentrated sulfuric acid (1:50).
Procedure: Heat the sprayed plates 30 min at 105 to 110°C.
Use: 3,4-Methylenedioxylphenyl synergists [135].

T-137. Fuchsin–sulfurous acid (Schiff's reagent)
Reagent: (*a*) Pass sulfur dioxide through 0.1% fuchsin solution until colorless.
(*b*) Add 1 ml of (*a*), 1 ml of 0.05 *M* mercuric chloride, and 10 ml of 0.05 *M* sulfuric acid to water to make 100 ml of solution [226].
Use: Aldehyde groups, violet spots on pale violet background.

T-138. Gentian violet–bromine
Reagent: 0.1% gentian violet in methanol.
Procedure: Spray with reagent and then expose to bromine vapor.
Use: Lipids, blue spots on yellow background [240].

T-139. Glucose–phosphoric acid reagent
Reagent: 2 g of glucose + 10 ml of 85% phosphoric acid + 40 ml of water + 30 ml of ethanol + 30 ml of *n*-butanol.
Procedure: Spray and heat at 115°C for 10 min.
Use: Aromatic amines, erythromycins [241].

T-140. Hydrazine sulfate
Reagent: (*a*) Saturated hydrazine sulfate–4 *N* hydrochloric acid (9:1) [242].
(*b*) 1% hydrazine sulfate in 1 *N* hydrochloric acid [243].
Procedure: (1) Spray and observe in daylight and ultraviolet light.
(2) Heat to 100°C and again observe in ultraviolet light.
Use: Aldehydes.

T-141. Hydrochloric acid vapors
Procedure: Exposure in hydrogen chloride vapor chamber.
Use: Chalcones [244]. Red spots. 4-Dimethylaminoazobenzene and metabolites [245].

T-142. Hydrogen sulfide
Procedure: Expose to well-washed hydrogen sulfide.
Use: Inorganic ions [246].

T-143. *p*-Hydroxybenzaldehyde–sulfuric acid (Komarowsky's reagent).
Reagent: 2% methanolic *p*-hydroxybenzaldehyde–50% (vol/vol) sulfuric acid (10:1) freshly prepared [195].

Procedure: Spray and heat 10 min at 60°C.
Use: Steroids.

T-144. Hydroxylamine–ferric reagent
Reagent: (*a*) 1 g of hydroxylamine hydrochloride in 9 ml of water.
(*b*) 2 g of sodium hydroxide in 8 ml of water.
(*c*) 4 g of ferric nitrate [$Fe(NO_3)_3 \cdot 9H_2O$] in 60 ml of water and 40 ml of acetic acid.
Procedure: (1) Spray with mixture of 1 part (*a*) and 1 part (*b*).
(2) Dry plates at 110°C for 10 min.
(3) Spray mixture of 45 ml of (*c*) with 6 ml of concentrated hydrochloric acid.
Use: Esters, fatty acid esters [226]. Sugar acetates [43]. Lactones [209].

T-145. 8-Hydroxyquinoline
Reagent: 10% 8-hydroxyquinoline in ammoniacal ethanol.
Use: Mo, Zn, Mn, Co, Fe, Ga [177, 247]. Cr, Ni, Al [248, 249]. Alkaline earths.

T-146. Iodine
Reagent: (*a*) Place iodine crystals in closed chamber. (Chamber may be warmed to increase vaporization.)
(*b*) Saturated solution of iodine in hexane [250].
Use: General reagent.

T-147. Iodine–azide
Reagent: (*a*) 3.5 g of sodium azide in 100 ml of 0.1 *N* iodine solution.
(*b*) 0.5% starch solution.
Procedure: Spray with (*a*) followed by (*b*) (better results can be obtained by spraying a mixture of the reagents [251]).
Use: Phenylthiohydantoins [252]. Penicillins [253]. Thiophosphoric esters [254].

T-148. Iodine–potassium iodide
Reagent: 5% iodine in 10% potassium iodide solution–water–2 *N* acetic acid (2:3:5).
Use: Alkaloids [51].
Note: Silver nitrate-impregnated layers must first be sprayed with saturated potassium bromide [255]. Steroids [60]. Use 0.3% iodine in 0.5% potassium iodide solution. A final spray of ether [256] modifies the reaction of some of the steroids.

T-149. Iodoplatinate
Reagent: 5 ml of 10% platinum chloride in 250 ml of 2% potassium iodide solution.

Use: Alkaloids and other nitrogen compounds. Water-soluble vitamins.

Sensitivity: (On alumina) vitamin B_1 0.01 μg, vitamin C 0.20 μg, nicotinic acid 0.40 μg, biotin 2.00 μg, choline chloride 2.00 μg, D-pantothenate 0.40 μg [257].

T-150. Iodoplatinate (modified)

Reagent: (*a*) Combine 1 ml of 0.2% platinum chloride solution, 0.1 ml of 3 to 4% hydrochloric acid, and 20 ml of acetone (reagent grade).

(*b*) 20% potassium iodide solution.

Procedure: Combine (*a*) with 0.1 ml of (*b*) and spray. Solutions (*a*) and (*b*) stable for several weeks. Mixture stable for 1 day.

Use: Penicillins [258].

Sensitivity: 0.05 to 0.3 μg for some of the common penicillins, less sensitive but still good for penicillin esters, sulfoxides, and propenyl penicillins.

T-151. Isatin

Reagent: 0.4% isatin in concentrated sulfuric acid [259].

Procedure: Observe sprayed plate for colored spots, then heat at 120°C several minutes.

Use: Thiophene derivatives.

T-152. Isatin–cadmium acetate

Reagent: Dissolve 0.5 g of cadmium acetate in 50 ml of water and 20 ml of acetic acid, then dilute to 500 ml with propanone.

Procedure: To the portion of stock solution required for use add isatin (weigh to ±1 mg) to provide a 0.2% (wt/vol) solution.

Use: Amino acids; also imino acids. Blue color with proline and hydroxyproline [260].

T-153. Isatin–zinc acetate [135]

Reagent: Warm on a water bath at 70 to 80°C until solution takes place and then cool quickly:

(*a*) 1 g of isatin and 1.5 g of zinc acetate in 100 ml of isopropanol and 1 ml of pyridine.

(*b*) As in (*a*), except use 1 ml of acetic acid + 95 ml of isopropanol + 5 ml of water as the solvent.

Procedure: Spray heavily with either (*a*) or (*b*) and dry 30 min at 90°C. To obtain better color differentiation let development take place at room temperature (20 hr).

Use: Amino acids [262].

T-154. Isonicotinic acid hydrazide

Reagent: 2 g of isonicotinic acid hydrazide in 500 ml of water containing 2.5 ml of concentrated hydrochloric acid.

Procedure: Allow to stand after spraying. Reactions with some compounds may take as long as 16 hr.
Use: Steroids [263].

T-155. Kedde reagent
Reagent: Freshly prepared mixture of equal volumes of 2% methanolic 3,5-dinitrobenzoic acid and 2 *N* aqueous potassium hydroxide.
Use: Steroid glycosides [264]. *Strophanthus* glycosides, purple spots in visible light [265].

T-156. Lead acetate
Reagent: 25% basic lead acetate solution.
Use: Flavonoids [266].

T-157. Lead tetraacetate
Reagent: Dissolve 3 g of red lead oxide in 100 g of glacial acetic acid. Let stand 2 hr, filter the solution.
Use: Sugars and sugar alcohols [267].

T-158. Liebermann–Buchard reagent
Reagent: 4 vol acetic anhydride + 1 vol concentrated sulfuric acid (reagent can be used on starch-bound layers) [18].
Use: Unsaturated sterols.

T-159. Malachite green.
Reagent: (*a*) 1 g of potassium hydroxide in 10 ml of water diluted to 100 ml with 95% ethanol.
(*b*) 1 ml of saturated acetone solution of malachite green oxalate in 51 ml of water, 45 ml of acetone, and 4 ml of pH-7 buffer (Beckman #3581).
Procedure: Spray chromatogram with (*a*) and heat 5 min at 150°C. Wash cooled plate with acetone to remove organic residues. Spray with (*b*) to detect potassium sulfite.
Use: Organic sulfite pesticides [268]. White spots on blue background.

T-160. Manganous chloride
Reagent: Dissolve 50 mg of manganous chloride ($MnCl_2 \cdot 4H_2O$) in 15 ml of water and 0.5 ml of concentrated sulfuric acid.
Procedure: Spray and place in an oven at 100 to 110°C for 10 to 15 min. Note colors formed.
Use: Bile acids. Cholesterol gives a pink color that fades in 5 min. Bile acid colors deepen in 5 min. All colors fade on longer exposure to air [269].
Sensitivity: Bile acids 2 μg, cholesterol 1 μg.

T-161. Mercurous nitrate

Reagent: Dissolve 1 g of mercurous nitrate in concentrated nitric acid and dilute to 100 ml with distilled water.
Use: Detection of methyprylon, ethinamate, carbromal [270].

T-162. *p*-Methoxybenzaldehyde
Reagent: 1 ml of *p*-methoxybenzaldehyde and 1 ml of sulfuric acid in 18 ml of ethanol [271].
Procedure: Spray reagent and heat plate to 110°C.
Use: Sugar phenylhydrazones: yellow-green spots in 2 to 3 min. Sugars: green, blue, or violet spots in 10 min.

T-163. Methylene blue
Reagent: 25 mg methylene blue dissolved in 100 ml of 0.05 *N* sulfuric acid by repeated grinding in a mortar with small quantities of the acid (keep in dark).
Procedure: Spray mixture of equal volume of reagent and acetone.
Use: Steroid sulfates [272].

T-164. Methylene blue (reduced)
Reagent: Filter through glass wool a mixture of 20 ml of 0.001 *M* methylene blue, 2 ml of concentrated sulfuric acid, and 1 g of zinc dust.
Use: Quinones (but not naphthoquinones) [196], blue spots.

T-165. Methylumbelliferone
Reagent: (*a*) 0.5% iodine in ethanol.
(*b*) A solution of 0.075 g of 4-methylumbelliferone in 100 ml of 50:50 ethanol–water. The solution is made alkaline with 10 ml of 0.1 *N* ammonium hydroxide.
Procedure: Spray first with (*a*) and record spots, then spray with (*b*) and observe under ultraviolet light.
Use: Organic phosphorus pesticides [73].

T-166. Millon's reagent
Reagent: Dissolve 5 g of mercury in 10 g of fuming nitric acid and dilute with 10 ml of water.
Use: Arbutin, hydroquinone, caffeine. Barbiturates [180].

T-167. Molybdenum blue reagent [273]
Reagent: (*a*) Boil gently with occasional shaking 40.11 g of molybdenum trioxide in 1010 ml of 25 *N* sulfuric acid (without evolution of white fumes) until dissolved. Cool and bring volume up to 1 liter with water.
(*b*) Boil (gently as above with the same precautions) 1.78 g molybdenum powder in 500 ml of (*a*) for 15 min. Cool to room temperature and decant from any residue and make up to 500 ml again.

Procedure: Spray (*a*)–(*b*)–water (1:1:2) mixture. This solution is greenish yellow when correctly diluted. Too much water gives a yellow solution and too little, a blue color. Stable for months.
Use: Phospholipids [274]. Blue spots on white or light blue-gray background.
Sensitivity: 0.005 μM.

T-168. Morin (2′,3,4′,5,7-pentahydroxyflavone)
Reagent: 0.005 to 0.05% morin in methanol.
Procedure: Dry at 100°C for 2 min and examine under ultraviolet light immediately.
Use: General reagent. Yellow-green fluorescent or dark spots on fluorescent background [52, 275].

T-169. Naphthalene black (stain for Sephadex layers)
Reagent: 1 g of naphthalene black in 100 ml of a mixture of 50 ml of methanol, 40 ml of water, and 10 ml of glacial acetic acid.
Procedure: Cover completed Sephadex plate with filter paper (Whatman No. 3 MM) taking care to exclude air bubbles. Dry for 30 min at 80 to 90°C. Immerse in dye bath for 30 min, then wash in same solvent mixture to remove excess dye [276].
Use: Proteins.

T-170. Naphthalene black (stain for electrophoresis)
Reagent: Dissolve 0.25 g of naphthalene black in 25 ml of glacial acetic acid and 500 ml of distilled water.
Procedure: Plasticized starch-gel layer is immersed for 2 hr in the reagent. Wash out excess dye in 5% acetic acid until washings are colorless. Add 50% glycine to the final wash [277].
Use: Proteins.

T-171. Naphthoresorcinol
Reagent: 200 mg of naphthoresorcinol in 100 ml of ethanol plus 10 ml of phosphoric acid.
Procedure: After spraying, heat plates to 110°C for 5 to 10 min.
Use: Carbohydrates [278].

T-172. Naphthoquinone–perchloric acid reagent [279]
Reagent: 0.1% solution of 1,2-naphthoquinone-2-sulfonic acid in a mixture of ethanol–60% perchloric acid–40% formaldehyde–water (2:1:0.1:0.9).
Procedure: Spray uniformly and then dry at 70 to 80°C observing the color formation. Too long a heating period converts all the spots to a brown-black color.
Use: Sterols
Sensitivity: Cholesterol 0.03 μg.

T-173. α-Naphthylamine

Reagent: 1 g α-naphthylamine in 100 ml of ethanol.
Use: 3,5-Dinitrobenzoates; yellow to orange spots [280]. 3,5-Dinitrobenzamides [281]. Chlorinated pesticides, dilute reagent 1:10 with ethanol and after spraying, expose to short-wave ultraviolet light for 30 sec. Reddish-brown to reddish-yellow spots, unstable in daylight [93].

T-174. *N*-(1-Naphthyl)ethylenediamine (Bratton–Marshall reagent)
Reagent: (*a*) 1 *N* hydrochloric acid
(*b*) 5% sodium nitrite
(c) 100 mg *N*-(1-naphthyl)ethylenediamine dihydrochloride in 100 ml of water.
Procedure: Spray (*a*), then (*b*), and mark any yellow spots. Dry at 100°C to remove excess nitrous acid. Spray with (*c*).
Use: Sulfonamides [282]. Reddish purple spots.
Procedure: For folic acid expose layer to ultraviolet light for 30 min, then spray.
Sensitivity: 0.2 μg [129].

T-175. Nile Blue A (reduced form)
Reagent: Mix 1 g of zinc dust with 20 ml of 0.001 *M* Nile Blue A and 2 ml of concentrated sulfuric acid. Filter through glass wool.
Use: Quinones and especially naphthoquinones [196].

T-176. Ninhydrin–cadmium
Reagent: Dissolve 0.5 g of cadmium acetate in 50 ml of water and 20 ml of acetic acid, then dilute to 500 ml with propane. To the portion of reagent required for use, add ninhydrin (weigh to ±1 mg) to give a concentration of 0.2% (wt/vol).
Procedure: Spray until layer is transparent. Heat at 60°C for 30 min and cool. Note color and allow to stand overnight in ammonia-free atmosphere. Note color again.
Use: Amino acids and dipeptides [283].

T-177. Ninhydrin–collidine
Reagent: 0.3 g of ninhydrin in 95 ml of isopropanol, 5 ml of 2,4,6-collidine, and 5 ml of acetic acid.
Procedure: Spray and then dry at 90°C.
Use: Amino acids [169, 284]. Amino sugars [83]. Blue spots.

T-178. Ninhydrin–cupric nitrate [285]
Reagent: (*a*) 0.2% ninhydrin in a mixture of 50 ml of absolute ethanol, 10 ml of glacial acetic, and 2 ml of 2,4,6-collidine.
(*b*) 1% cupric nitrate ($Cu(NO_3)_2 \cdot 3H_2O$) in absolute alcohol.
Procedure: Spray dried plates with a fresh mixture of 25 ml of (*a*) and 1.5 ml of (*b*). Heat plates 1.5 to 2 min at 105°C.
Use: Amino acids [286].

T-179. β-Ocimene–2,4-dinitrophenylhydrazine

Reagent: (*a*) 2% β-ocimene in hexane purified immediately before use (pass through a Florisil column, 10 mm × 100 mm for 100 ml of solution).

(*b*) 2% 2,4-dinitrophenylhydrazine in 30% perchloric acid.

Procedure: Spray with (*a*) and allow to stand at room temperature 1 hr or heat at 50°C for 10 to 15 min. Spray with (*b*).

Use: Antioxidants, white spots on yellow-brown background [287].

Sensitivity: BHA 1 ng.

T-180. *p*-Nitroaniline (diazotized)

Reagent: (*a*) 1 g of *p*-nitroaniline in 200 ml of 2 *N* hydrochloric acid.

(*b*) 5% sodium nitrite solution.

Procedure: Add solution (*b*) with stirring to 10 ml of (*a*) until mixture is colorless, then spray. Most spots can be intensified by spraying with sodium carbonate solution [162]. (For detecting plasticizers, first spray chromatogram with 0.5 *N* ethanolic potassium hydroxide and heat 15 min at 60°C. Follow with diazonium spray [164].)

Use: Phenols [162]. Plasticizers [164]. Chloroquin [288]. Catecholamines [289].

Sensitivity: Catecholamines 0.5 μg [289]. Adrenaline 0.003 μg [214].

T-181. *p*-Nitrobenzenediazonium fluoroborate

Reagent: (*a*) 1.5 *N* sodium hydroxide in methanol.

(*b*) 0.01% *p*-nitrobenzenediazonium fluoroborate in 50:50 diethyl ether–methanol or in acetone.

Procedure: Spray chromatogram with (*a*) and then (*b*).

Use: Phenolic compounds [290, 291]. Catecholamines [292]. Aryl-*N*-methyl carbamates [293]. Indoles and imidazoles [292].

Sensitivity: Catecholamines 0.01 to 0.05 μg. Carbamates 0.1 μg. Phenolic compounds 0.1 to 0.5 μg [293]. Indoles and imidazoles 0.1 to 2 μg.

T-182. 4-(*p*-Nitrobenzyl)pyridine

Reagent: (*a*) 5% solution of *p*-toluenesulfonyl chloride in anhydrous pyridine–toluene (1:1).

(*b*) 2% solution of 4-(*p*-nitrobenzyl)pyridine in acetone.

(*c*) 1 *M* sodium carbonate.

Procedure: Spray with (*a*) and allow to dry, then spray with (*b*); after 1 min apply heat gun for 1 min and then spray with fine mist of (*c*). Deep blue or purple spot indicates positive test.

Alternative procedure: Mix 10 to 50 μg of the alcohol in 50 μl of acetone with 50 μl of a 15% solution of *p*-toluenesulfonyl chloride in pyridine and allow to stand at 0°C for 1.5 hr. Spot on cellulose layers and develop with hexane–ethyl acetate (4:1) for 10 cm. Air-dry and proceed as above, beginning with spraying (*b*).
Use: Test for hydroxyl groups [294].
Sensitivity: Microgram range. Amino, ester, and ether groups do not interfere, but carboxylate and phenolic groups interfere.

T-183. 4-(*p*-Nitrobenzyl)pyridine–tetraethylenepentamine
Reagent: (*a*) 2% 4-(*p*-nitrobenzyl)pyridine (Aldrich) in purified acetone (refluxed 1 hr with 1 g of potassium permanganate/liter, then distill).
(*b*) 10% tetraethylenepentamine (Eastman Kodak #T5902) in purified acetone.
Procedure: Spray the dried chromatogram heavily with (*a*), evaporate acetone, then heat 5 min at 110°C in forced draft. Spray lightly with (*b*).
Use: Organophosphate pesticides, blue spots on white background (diazinon gives red spot).
Sensitivity: 0.5 to 2 μg [295].

T-184. Nitrogen trioxide (for diazotizing) [296]
Reagent: (*a*) Fresh solution of nitrogen trioxide in toluene: pour 10 ml of toluene on top of 10 ml of 6 *N* hydrochloric acid. Then slowly add 2 g of sodium nitrite in 10 ml of water and swirl the solution to facilitate the extraction.
(*b*) Toluene solution containing 0.1 mole of a phenol and 0.2 mole of an amine per 100 ml of solution for coupling with diazonium salts.
Procedure: Spray with solution (*a*) and then (*b*).
Use: (2,4-dinitrophenyl)amino acids or aromatic amines.
Sensitivity: 10^{-9} moles.

T-185. 4-(4′-Nitrobenzyl)pyridine
Reagent: (*a*) 5% solution of 4-(4′-nitrobenzyl)pyridine in acetone.
(*b*) Dilute sodium hydroxide.
Procedure: Spray with (*a*); violet color intensified after spraying with (*b*).
Use: Acetylene derivatives. (since it is not specific this should be used in combination with ultraviolet spectroscopy) [297].

T-186. Osmium tetroxide
Procedure: Expose plate to the vapors of osmium tetroxide in a sealed chamber [298]; 5 to 10 min for isolated double bond; 1 hr or more for conjugated double bond [182].

Use: Compounds with double bonds. Has been used for thin-layer chromatography of lipids [298] steroids [182]. Brown to black spots.

T-187. Ozone–indigo reaction [299]

Reagent: Dissolve 130 mg of indigo in 1 ml of concentrated sulfuric acid by heating on a water bath for 1 hr. Dilute mixture to 500 ml.

Procedure: Expose chromatogram in chamber containing 10 to 15% ozone for 15 to 20 min. Air plate to remove excess ozone, then spray with reagent.

Use: Unsaturated compounds. White, yellow, or brown spots on blue background.

T-188. Palladium–calcein

Reagent: Prepare a 0.02 *M* phosphate buffer [NaH_2PO_4–Na_2HPO_4 (1:1)] containing 1.0×10^{-4} *M* of Pd^{2+} and 7.0×10^{-5} *M* of calcein (G. F. Smith). Allow to stand 12 hr before use.

Procedure: Spray layer and observe under ultraviolet light.

Use: Detection of metal ions separated as diethyldithiocarbamates [300].

Sensitivity: 0.1 to 1 ng.

T-189. Palladium chloride

Reagent: 0.5% palladium chloride in dilute hydrochloric acid.

Use: Thiophosphoric acid esters [179]. Yellow on pale brown background.

Sensitivity: $<5\ \gamma$. Phenothiazines [301]. Varied colors.

T-190. Paraformaldehyde–phosphoric acid

Reagent: 0.03 g of paraformaldehyde in 100 ml of concentrated phosphoric acid ($d = 1.7$). Stable for 1 week [255].

Use: Alkaloids.

T-191. Periodic–perchloric acid

Reagent: 10 g of periodic acid and a few milligrams of vanadium pentoxide in 100 ml of 70% perchloric acid [302].

Procedure: Shake reagent before spraying well dried plate.

Use: Thiophosphoric acid esters.

T-192. Periodic acid–Schiff reagent [108]

Reagent: (*a*) 0.5 g periodic acid in 100 ml of 90% acetic acid.

(*b*) Mix equal volumes of cold (0°C) 30% sodium metabisulfite and 3 *N* hydrochloric acid.

(*c*) 200 mg fuchsin and 5 ml of 10% sodium metabisulfite in 85 ml of water. (Keep for 12 hr, then treat with carbon and filter.)

Procedure: Spray lightly with (*a*) followed by (*b*) and (*c*), consecutively. Heat for 15 min at 90°C.

Use: Unsaturated monoglycerides, violet. Polyene acids, gray-green.

T-193. Phenol–sulfuric acid

Reagent: 3 g of phenol + 5 ml of concentrated sulfuric acid in 95 ml of ethanol.

Procedure: Spray and heat 10 to 15 min at 110°C. Spots may be intensified by additional heating.

Use: Carbohydrates [62], brown spots.

T-194. *m*-Phenylenediamine hydrochloride [220]

Reagent: 0.2 *M* *m*-phenylenediamine hydrochloride in 76% ethanol.

Procedure: Spray and heat to 110°C for 5 min. Examine under ultraviolet light.

Use: Hexose and triosphosphates [71].

T-195. Phosphomolybdic acid

Reagent: (*a*) From 2 to 20% solution of phosphomolybdic acid in ethanol or methyl Cellosolve.

(*b*) For neutral plates: 4 ml of concentrated hydrochloric acid + 100 ml of 10% phosphomolybdic acid in 95% ethanol [18].

Procedure: (1) Gypsum-bound layers: spray (*a*) and heat to 100°C for 20 min. (For saturated triglycerides expose to iodine vapor 5 min prior to spraying). Ammonia vapors may be used to intensify spots and lighten the background [304].

(2) Starch-bound layers: spray (*a*) (10%), evaporate solvent with hot air blower, then heat at 100°C until solvent front appears (10 min or less) [18].

Use: Lipids, steroids, antioxidants. A general reagent.

Sensitivity: Terpenes 0.05 to 1 μg. Phenothiazine 1 μg.

T-196. Phosphomolybdic acid–stannous chloride [144]

Reagent: (*a*) 1% solution of phosphomolybdic acid in a 50:50 chloroform–ethanol mixture.

(*b*) 1% stannous chloride in 2 *N* hydrochloric acid.

Procedure: Stabilize layers with dichlorodimethylsilane (see under reagent T-71, cupric acetate–dithiooxamide) then spray with (*a*). Wash with running water and then dip in (*b*).

Use: Phospholipids.

T-197. Phosphotungstic acid

Reagent: 10% phosphotungstic acid in 90% ethanol.

Procedure: After spraying, heat at 90 to 100°C for 15 min.

Use: Cholesterol and cholesterol esters [108]. Red spots on white background.

Sensitivity: 2 to 4 γ.

T-198. *o*-Phthalaldehyde–mercaptoethanol
Reagent: (*a*) 0.1% *o*-phthalaldehyde and 0.1% 2-mercaptoethanol in acetone. Stable for several days at room temperature in the dark.
(*b*) 1% triethylamine in acetone.
Procedure: Spray generously with (*a*) and 5 min later with (*b*). After 10 min view under 350-nm ultraviolet light.
Use: Amino acids and peptides. Spots disappear within a few minutes on silica gel at room temperature. On cellulose they are a little more stable. Proline is detected by heating at 100°C for 1 hr although the other spots disappear. *N*-protected groups (*tert*-butyloxycarbonyl, 2-phenylisopropyloxycarbonyl, and *p*-methoxybenzoyloxycarbonyl) are detected at 500-p*M* level by drying plate at 100°C for 2 hr before spraying with reagent.
Sensitivity: 50 to 100 p*M* amino acid. Two to five times more sensitive on silica gel, but aromatic amino acids (especially cystine) are more easily detected on cellulose [305].

T-199. Ponceau S (stain for Sephadex layers)
Reagent: 0.2% Ponceau S in 10% acetic acid.
Procedure: Stain for 30 min, then wash with water. For complete details, see under T-128, naphthalene black.
Use: Proteins [276].

T-200. Potassium ferricyanide–potassium hydroxide
Reagent: (*a*) 0.1% potassium ferricyanide.
(*b*) 15% potassium hydroxide.
Procedure: Mix 1.5 ml of (*a*), 10 ml of (*b*), and 20 ml of water prior to spraying. Observe chromatogram in long-wave ultraviolet light before and after spraying and heating.
Use: Catecholamines [306].

T-201. Potassium ferricyanide–ferric chloride
Reagent: (*a*) 0.2% potassium ferricyanide + 0.01% ferric chloride hexahydrate in 2 *N* hydrochloric acid [307].
(*b*) Equal parts of 0.1 *M* ferric chloride and 0.1 *M* potassium ferricyanide.
Use: (*a*) 2,4-Dinitrophenylhydrazones. Rate of color formation and color changes depends on compounds (see original work).
(*b*) aromatic amines [222].

T-202. Potassium ferricyanide–ferric ammonium sulfate [303]
Reagent: (*a*) 0.1% potassium ferricyanide in 0.25% sodium carbonate.
(*b*) 200 mg of ferric ammonium sulfate in 100 ml of water and 5 ml of 85% phosphoric acid.

Procedure: Spray with (*a*) and heat for 30 min at 80°C. Cool and spray with (*b*).

Use: Steroids [60].

Sensitivity: 2 to 5 μg.

T-203. Potassium ferrocyanide

Reagent: 1% solution of potassium ferrocyanide.

Use: Uranium and iron [247]. Copper and mercury [246].

T-204. Potassium hydroxide

Reagent: (*a*) 2 vol 5% potassium hydroxide and 1 vol acetone [309].

(*b*) 2 *N* alcoholic potassium hydroxide [308].

Procedure: (1) Spray with (*a*) or (*b*) and heat to 80 to 100°C.

(2) For nonacetylated citrate esters, first spray with acetic anhydride–concentrated phosphoric acid–dioxane (5:0.5:5) and heat for 30 min at 100°C. Cool and proceed with (1).

Use: Aromatic nitro compounds and amines [309]. Acetylated and nonacetylated citrate esters [308], yellow fluorescent spots.

T-205. Potassium hypochlorite–*o*-toluidine (Reindel–Hoppe reagent modified according to Greig and Leaback) [310].

Reagent: (*a*) 2 g of potassium hypochlorite per 100 ml of solution.

(*b*) Saturated solution of *o*-toluidine in 2% acetic acid–85% potassium iodide (1:1).

Procedure: Spray (*a*) lightly and after standing 1 to 1.5 hr spray with (*b*).

Use: Amino acids [262]. Blue-black spots on white background.

T-206. Potassium iodide–ammonia–hydrogen sulfide [249]

Reagent: 2% potassium iodide.

Procedure: Spray with reagent, dry, and expose to ammonia vapors and then hydrogen sulfide.

Use: Inorganic ions of the hydrogen sulfide group.

T-207. Potassium iodide–starch

Reagent: (*a*) 4% potassium iodide–acetic acid (1:4) (1% sodium sulfite added dropwise to decolorize reagent).

(*b*) 1% starch solution.

Procedure: Spray with (*a*) then after 5 min with (*b*).

Use: Peroxides [311], blue spots.

T-208. Potassium iodobismuthate

Reagent: (*a*) 1.7 g of bismuth nitrate and 20 g of tartaric acid in 80 ml of water.

(*b*) 16 g of potassium iodide in 40 ml of water.

(*c*) Mix equal parts of (*a*) and (*b*).

(*d*) Prepare fresh daily. 10 g of tartaric acid in 50 ml of water + 10 ml of (*c*).
Procedure: Spray with (*d*) and then expose to bromine vapor.
Use: Systemic fungicides [312].
Sensitivity: 0.25 to 0.6 μg.

T-209. Potassium iodoplatinate
Reagent: 2 ml of 10% platinum chloride and 25 ml of 8% potassium iodide diluted to 100 ml.
Use: Alkaloids [169]. Phenothiazines [218]. Biotin [161]. Lipids [313].
Sensitivity: Alkaloids 10 μg. Lipids 4 to 10 μg.

T-210. Potassium permanganate (acidic)
Reagent: Mix equal volumes of 0.1 *N* potassium permanganate and 2 *N* acetic acid [51].
Use: General reagent.

T-211. Potassium permanganate (alkaline)
Reagent: Freshly prepared mixture of equal parts of 2% permanganate and 4% sodium bicarbonate.
Use: General reagent.
Sensitivity: Steroids 2 μg [60].

T-212. Potassium permanganate
Reagent: 0.25 to 2% in water.
Use: General reagent.

T-213. Potassium permanganate–sulfuric acid
Reagent: 500 mg of potassium permanganate in 15 ml of concentrated sulfuric acid (**Caution:** mix only small quantities—manganous heptoxide is explosive) [22].
Use: General reagent. White spots on pink background. Phosphinoxides (sensitivity = 1 γ).

T-214. Potassium permanganate–bromophenol blue [314]
Reagent: (*a*) 0.5% potassium permanganate.
(*b*) 0.2% aqueous bromophenol.
Procedure: Spray with (*a*) and after 10 to 15 min spray with (*b*).
Use: General reagent.
Sensitivity: Amaromycin 0.1 μg. Ten times more sensitive than potassium permanganate alone.

T-215. Potassium thiocarbonate
Reagent: 0.1 *M* potassium thiocarbonate.
Procedure: Spray. Various colors develop.
Use: Detection of metal ions [315].
Sensitivity: Varies from 0.7 for Ru^{3+} to 12.1 μg for Sb^{3+}.

T-216. Pyrene

Reagent: 0.05% solution of pyrene (phenanthrene) (purified) in petroleum ether (60 to 80°C).

Procedure: Spray silica gel layer and observe under ultraviolet light. Silica gel inhibits green fluorescence of pyrene, but bile acids reverse this and exhibit a green fluorescent spot. Nondestructive; after marking spots, develop within 0.5 hr in diethyl ether–petroleum ether (60 to 80°C) to move pyrene to the solvent front.

Use: Bile acids [316].

Sensitivity: Bile acids 0.2 g, bile salts 0.1 μg.

T-217. Procatechol violet (pyrocatecholsulfophthalein)

Reagent: 100 mg pyrocatechol violet in 100 ml of alcohol.

Procedure: Expose plate to ultraviolet for 20 min, then spray with reagent [317].

Use: Organotin compounds. Deep blue to violet spots on bright gray-brown background.

T-218. 1-(2-Pyridylazo)-2-naphthol [278]

Reagent: (*a*) 0.4% 1-(2-pyridylazo)-2-naphthol in ethanol.

(*b*) 0.8 g cobalt nitrate in 100 ml of water.

(*c*) 2 *M* Sodium acetate buffer pH 4.6 (iron-free).

(*d*) A mixture of 4 ml of (*b*) and 2 ml of (*c*) diluted to 50 ml.

Procedure: Spray with (*a*) followed by (*d*) with drying between sprays.

Use: Glucosiduronates. Violet on yellow background.

T-219. 1-(2-Pyridylazo)-2-naphthol

Reagent: 0.25% 1-(2-pyridylazo)-2-naphthol in ethanol.

Use: UO_2^{2+} [248].

Sensitivity: 1 γ.

T-220. Quercetin

Reagent: 0.01 *M* quercetin.

Use: Ti, Zr, Th, Mo, W, U, Fe.

Sensitivity: 0.02 to 4.20 μg [318].

T-221. Resorcinol–sulfuric acid

Reagent: (*a*) 20% resorcinol in ethanol (containing a little zinc chloride).

(*b*) 4 *N* sulfuric acid.

Procedure: (1) Spray with (*a*), heat 10 min at 150°C, then spray with (*b*) and heat 20 min at 120°C. Finally, spray with 40% potassium hydroxide [164].

(2) Spray with 50:50 mixture of (*a*) and (*b*) and heat to 120°C for 10 min. Cool and expose to ammonia vapors [308].

Use: Plasticizers (phthalate esters).
Sensitivity: 20 γ.

T-222. Rhodamine B
Reagent: (*a*) 0.05% Rhodamine B in ethanol [35].
(*b*) 0.2% Rhodamine B in water [319].
Procedure: (1) Spray and observe in daylight and ultraviolet light (3% hydrogen peroxide enhances color) [320]. Herbicides (halogenated phenoxy acids), spray 0.02% reagent [103]; for carbamate and urea herbicides 0.005% reagent used [175].
(2) Glycerides: Spray (*a*), then 10 *N* potassium hydroxide (sensitivity sometimes increased by repeating potassium hydroxide spray after a few minutes [35]).
Use: Food preservatives. Pink to purple spots intensified by spraying with 3% hydrogen peroxide [320]. Lipids: purple spots on pink [319]. Glycerides: bright spots on pink-red to blood-red background [35]. Herbicides [103, 175].
Sensitivity: Herbicides 0.3 to 1.5 μg.

T-223. Rhodanine
Reagent: Alcoholic rhodanine solution.
Procedure: Spray with reagent followed by concentrated ammonium hydroxide or sodium hydroxide [1].
Use: Polyene aldehydes.
Sensitivity: 0.03 γ.

T-224. Silver acetate
Reagent: 1% silver acetate.
Procedure: Air-dry plates 10 to 30 min, then spray and dry again.
Use: Barbiturates, chalky white spots [321].
Sensitivity: 0.5 μg.

T-225. Silver nitrate–ammonium hydroxide–sodium methoxide [322]
Reagent: (*a*) 300 mg of silver nitrate in 100 ml of methanol.
(*b*) Saturated solution of ammonia in methanol.
(*c*) 7 g of sodium in 100 ml of methanol.
Procedure: Spray a 5 : 1 : 2 mixture of (*a*) : (*b*) : (*c*).
Use: Acyl sugar derivatives [323].

T-226. Silver nitrate (ammoniacal)
Reagent: 0.1 to 0.5 *N* silver nitrate with added ammonium hydroxide until precipitate dissolves.
Procedure: Spray and heat plate 110 to 120°C for 10 min. Exposure to ultraviolet light for 10 min may be necessary.
Use: Chlorinated herbicides [324]. α-Glycol groupings [325]. Terpenic phenols [162]. Halogen ions (except fluorides) [101]. Sulfur-containing anions. [326]. Arsenate, phosphite, phosphate, arsen-

ite [219]. Reducing sugars [327]. **Caution:** the solution should be made up fresh and should not be stored owing to the formation of sensitive, explosive compounds. Do not expose solution to direct sunlight [328].

T-227. Silver nitrate (nitric acid)
Reagent: 0.1% silver nitrate in 3 *N* nitric acid [329].
Procedure: Spray and dry 5 min at 80°C. Expose to daylight 10 to 15 hr.
Use: Herbicides.

T-228. Silver nitrate–diphenylcarbazone
Reagent: (*a*) 1% silver nitrate.
(*b*) 0.1% diphenylcarbazone in 95% ethanol.
Procedure: Spray with (*a*), then (*b*).
Use: Barbiturates [330], purple-blue spots.

T-229. Silver nitrate–bromophenol blue
Reagent: (*a*) 0.5% silver nitrate in ethanol (wt/vol).
(*b*) 0.2% bromophenol blue + 0.15% silver nitrate in ethanol–ethyl acetate (1 : 1).
Procedure: Spray (*a*) and dry at 100°C for 5 min, then spray (*b*) and dry at 100°C for 10 min.
Use: Chlorinated pesticides [331], yellow spots on blue background. Carboxin and other fungicides [332].

T-230. Silver nitrate–formaldehyde
Reagent: (*a*) 0.05 *N* silver nitrate.
(*b*) 37% formaldehyde.
(*c*) 2 *N* potassium hydroxide.
(*d*) Concentrated nitric acid–30% hydrogen peroxide (1 : 1).
Procedure: Spray with (*a*) and dry, then spray with (*b*), and while still moist follow with (*c*). Dry 30 min at 130 to 135°C. Cool and spray with (*d*), then expose to daylight or ultraviolet light.
Use: Pesticides [333].

T-231. Silver nitrate–phenoxyethanol [334]
Reagent: Dissolve 0.1 g of silver nitrate (0.425 to 1.7 g has also been used) in 1 ml of water, add 10 ml of 2-phenoxyethanol, and dilute the mixture to 200 ml with reagent acetone, then add 3 drops of 30% hydrogen peroxide. Store in dark until used.
Procedure: Spray, dry for 5 min in hood and then at 75°C for 15 min. Expose to ultraviolet light for a minimum period of time using standards as controls (not over 15 min for silica gel and up to 50 min for alumina) (layers should be prewashed to remove interfering chlorides) [335].

Use: Chlorinated pesticides.
Sensitivity: 0.01 to 0.1 μg.

T-232. Sodium bichromate–sulfuric acid
Reagent: 3 g of sodium bichromate in a mixture of 20 ml of water and 10 ml of concentrated sulfuric acid [22].
Use: General reagent.

T-233. Sodium hydroxide
Reagent: 2% sodium hydroxide in 90% ethanol.
Use: 2,4-dinitrophenylhydrazones [336].
Sensitivity: $<0.1\ \gamma$.

T-234. Sodium molybdate
Reagent: Dissolve 10 g of sodium molybdate in 100 ml of 3 to 4 *N* hydrochloric acid and 1.0 g of hydrazine hydrochloride in 100 ml of water. Mix the two solutions and heat on a boiling water bath for 5 min. Cool and dilute to 1 liter. Stable for several months.
Use: Phospholipids [337].

T-235. Sodium 1,2-naphthoquinone-4-sulfonate (NQS reagent)
Reagent: (*a*) 1 *N* sodium hydroxide.
(*b*) saturated solution of sodium 1,2-naphthoquinone-4-sulfonate in ethanol–water (1:1).
Procedure: Spray with (*a*), then with (*b*).
Use: Thiazide diuretics. Stable orange spots in 15 min. Basic compounds with primary amine groups or reactive methylene groups. Barbiturates do not react. [338].

T-236. Sodium nitrite–β-naphthol
Reagent: (*a*) Freshly prepared 1% sodium nitrite in 0.1 *N* hydrochloric acid.
(*b*) 0.2% β-naphthol in 0.1 *N* sodium hydroxide.
Procedure: Spray with (*a*) and dry 5 min at 100°C, then spray with (*b*).
Use: Sulfonamides [339]. Aromatic amines [222, 340].

T-237. Sodium nitroprusside
Reagent: (*a*) 0.5% solution of sodium nitroprusside.
(*b*) 10% solution of sodium nitrite.
Procedure: Spray with (*a*) followed by (*b*). Dry and spray with glacial acetic acid. Further colors develop on exposure to ammonia vapors.
Use: Pyrazoles [341].

T-238. Sodium nitroprusside–acetaldehyde
Reagent: (*a*) 5% sodium nitroprusside in 10% acetaldehyde.
(*b*) 1% sodium carbonate.

Procedure: Spray with a 50:50 mixture of (*a*) and (*b*).
Use: Secondary aliphatic amines. Morpholine, diethanolamine [342].

T-239. Sodium nitroprusside–potassium ferricyanide
Reagent: (*a*) 10% sodium hydroxide.
(*b*) 10% sodium nitroprusside.
(*c*) 10% potassium ferricyanide.
Procedure: Mix equal parts of (*a*), (*b*), and (*c*) with 3 parts of water and allow to stand 30 min before spraying (stable several weeks under refrigeration).
Use: Cyanamide derivatives. Arginine, creatine, creatinine, etc [262]. Streptomycins and neomycins, vermilion spots on yellowish background [343].

T-240. Sodium pentacyanoferrate (Ferron's reagent)
Reagent: (*a*) 1% sodium pentacyanoferrate in water.
(*b*) 20% sodium hydroxide.
Procedure: Mix 15 ml of (*a*) with 5 ml of (*b*) and add 1 drop of 30% hydrogen peroxide. Stable for 24 hr.
Use: Diuretics [344].

T-241. Sodium periodate–benzidine
Reagent: (*a*) 0.1% sodium metaperiodate.
(*b*) 0.5% benzidine in butanol–acetic acid (4:1).
Procedure: Spray with (*a*) and then after 4 min with (*b*).
Use: Bivalent sulfur compounds [345], white spots on dark blue background.
Sensitivity: DL-methionine 5 to 10 γ. Aromatic sulfur compounds 20 to 30 γ.

T-242. Sodium periodate–benzidine–silver nitrate [346]
Reagent: (*a*) 0.1% sodium metaperiodate.
(*b*) Add a mixture of 70 ml of water, 30 ml of acetone, and 1.5 ml of 1 *N* hydrochloric acid to 2.8 g of benzidine in 80 ml of 95% ethanol.
(*c*) 1 ml of saturated silver nitrate added to 20 ml of acetone with stirring. Add water dropwise until the precipitate just dissolves.
Procedure: Spray with (*a*) and air-dry before spraying with (*b*). Then place in ammonia atmosphere for 5 min before spraying with (*c*).
Use: Sugars and sugar alcohols [346].

T-243. Sodium periodate–Schiff reagent [347]
Reagent: (*a*) 0.5% sodium periodate.

(*b*) 0.5% *p*-rosaniline decolorized with sulfur dioxide.
(*c*) 1% perchloric acid.

Procedure: Spray with (*a*) and after 5 min (while still damp) expose to sulfur dioxide and spray with (*b*). After 1 hr lighten background by spraying with (*c*).

Use: Phospho- and glycolipids [210]. Blue and purple spots on yellowish background.

T-244. Sodium pyrophosphate–dimethylaminobenzaldehyde

Reagent: (*a*) 2 to 4 mg/ml of sodium pyrophosphate in 30% hydrogen peroxide.
(*b*) Acetic anhydride–petroleum ether (80 to 100°C)–benzene (1:4:5).
(*c*) Modified Ehrlich's reagent. 1 g of dimethylaminobenzaldehyde in 70 ml of ethanol + 30 ml of carbitol (diethyleneglycol monoethyl ether) + 1.5 ml of concentrated hydrochloric acid.

Procedure: Spray (*a*) on air-dried plates and heat 15 min at 90 to 100°C. Cool, spray with (*b*), and repeat heating. Follow with (*c*). [Omit spray (*a*) for detecting *N*-oxides.]

Use: Test for pyrrolizidine alkaloids [348]. Only alkaloids with an unsaturated (3-pyrroline) ring will respond.

Sensitivity: Approximately 0.2 μg. Ester alkaloids containing a 3-pyrroline ring, 1 to 20 μg.

T-245. Sodium tungstate–trichloroacetic acid

Reagent: (*a*) 6 ml of 10% sodium tungstate solution + 6 ml of 5% trichloroacetic acid + 3 ml of 0.5 *N* hydrochloric acid.
(*b*) Freshly prepared 5% sodium nitrite solution.
(*c*) 0.05 *N* sodium hydroxide.

Procedure: Add (*b*) to (*a*) and spray plate; after 3 min spray with (*c*). (Polyamide layers are dried at 30°C for a few minutes to remove water before applying the second spray.)

Use: Phenolic compounds, *o*-dihydroxy compounds except quercetin turn red after the second spray. Phenol, naringin, and hesperidin do not react [349].

T-246. Stannous chloride.

Reagent: (*a*) Dissolve 5 g of stannous chloride in 10 ml of concentrated hydrochloric acid and dilute to 35 ml.
(*b*) 1% β-naphthol in 2 *N* sodium hydroxide.

Procedure: Dry plate and spray with (*a*), then expose to nitrous oxide fumes. Spray with (*b*).

Use: Detect methanol in ethanol (separate 2,4-dinitrobenzoates) [350]. Can be used to detect other alcohols as the 2,4-dinitrobenzoates [351].

Sensitivity: 0.04 μg methanol.

T-247. Stannous chloride–hydrochloric acid

Reagent: (*a*) 10% stannous chloride in concentrated hydrochloric acid.

(*b*) Dilute (*a*) 200-fold with 0.5 *M* sulfuric acid.

Procedure: (1) For aromatic nitro compounds, spray with (*a*).

(2) For triose- and hexosephosphates, spray (*b*) on still warm plates after first using reagent T-15.

T-248. Stannous chloride–monochloroacetic acid

Reagent: 5 g of stannous chloride and 5 g of monochloroacetic acid in 90 ml of chloroform.

Procedure: Spray and heat plate for 3 min at 100°C.

Use: Terpenes [353].

Sensitivity: 0.5 to 2 μg.

T-248a. Sudan black B

Reagent: 0.7% Sudan black B in 85% ethylene glycol.

Procedure: Immerse gel layers in reagent for 5 hr. Rinse for 1 hr in at least five changes of 70% glycol. (Note: if also staining for total protein with test T-181a, do test T-248 first and then T-181a.)

Use: Staining lipoproteins on polyacrylamide gel layers [186a].

T-249. Sulfanilic acid (diazotized) (Pauly's reagent)

Reagent: (*a*) 0.5% each of sulfanilic acid and sodium nitrite in 1 *N* hydrochloric acid.

(*b*) 1 *N* potassium hydroxide.

Procedure: Spray with (*a*) followed by (*b*).

Use: Phenols [354]. Aromatic amines [288]. Sugar phenylosazones [355]. Catecholamines [214].

T-250. Tannic acid

Reagent: Mix 3 ml of 85% phosphoric acid, 1 ml of 0.33% tannic acid (USP) in acetic acid (wt/vol), and 26 ml of acetone just prior to use.

Procedure: Spray and develop color by heating 15 min at 80 to 90°C.

Use: Differentiate pyrethrins (pink) from piperonyl butoxide (blue and blue-black) [356].

Sensitivity: 0.5 to 1 μg pyrethrin.

T-251. Tetrabromophenolphthalein, ethyl ester

Reagent: (*a*) Stock solution of 1 g tetrabromophenolphthalein, ethyl ester in 100 ml of acetone.

(*b*) Dilute 10 ml of (*a*) to 50 ml with acetone.

(*c*) Dissolve 0.5 g of silver nitrate in 25 ml of distilled water and dilute to 100 ml with acetone.

(*d*) Dissolve 5 g of citric acid in 50 ml of water and dilute to 100 ml with acetone.

Procedure: Spray with (*b*) and then with (*c*). After 2 min spray with (*d*). Blue or purple spots on a yellow background. After 10 min respray with (*d*). Evaluate within 10 min of respraying because spots will fade.

Use: Organothiophosphate residues (note: plates must be prewashed prior to chromatographic use to remove chlorides) [357].

Sensitivity: 0.05 μg.

T-252. Tetracyanoethylene

Reagent: (*a*) Saturated solution of tetracyanoethylene in benzene.

(*b*) 2% solution of tetracyanoethylene in benzene.

(*c*) 0.5% solution of tetracyanoethylene in ethyl acetate.

Procedure: Spray with (*c*) and heat at 80°C.

Use: For hydrocarbons, heterocyclics, aromatic amines, and phenols (*a*) has been used [8, 238, 358]. For chlorophenols and their esters as pesticides [359] and for sulfides, sulfones, and sulfoxides [360] (*b*) has been used.

Sensitivity: Indoles 0.5 to 1.0 μg.

T-253. Tetracyanoquiondimethanide [361]

Reagent: 0.5% lithium tetracyanoquinodimethanide in ethanol–water (1:1).

Procedure: Spray and observe after 15 to 20 min. Dry at 80°C or expose to ammonia vapor (if acidic solvents have been used) before spraying reagent.

Use: Mono- and divalent post-transition metal ions.

T-254. *N,N,N′,N′*-tetramethyl-4,4′-diaminodiphenylmethane–diammonium ceric nitrate

Reagent: (*a*) 0.25% *N,N,N′,N′*-tetramethyl-4,4′-diaminodiphenylmethane in acetone.

(*b*) 1% diammonium ceric nitrate in 0.2 *N* nitric acid.

Procedure: Spray freshly prepared mixture of (*a*) and (*b*) (1:1). Heat for 5 min at 105°C.

Use: Polyalcohols [68]. White to light blue on blue background.

T-255. Tetrazolium blue

Reagent: 0.5% tetrazolium blue in 2.5 *N* sodium hydroxide.

Use: Reducing steroids [362].

Sensitivity: 0.5 to 1 μg [363].

T-256. 4-(2-thiazolylazo)resorcinol

Reagent: 0.1% 4-(2-thiazolylazo)resorcinol in 95% ethanol.
Procedure: Spray with reagent, then expose to ammonia vapor for 30 min. Dry in air stream at 40°C for 30 to 45 min.
Use: Ni, Cu, Co.
Sensitivity: 0.01 to 0.02 μg [364].

T-257. Thiobarbituric acid–ferric chloride
Reagent: (*a*) 3 g sodium dichromate in 100 ml of 30% (vol/vol) sulfuric acid.
(*b*) 2.8 g of 2-thiobarbituric acid + 2.12 g of sodium carbonate in 100 ml of water.
(*c*) 1 g of ferric chloride in 100 ml of 1% (vol/vol) hydrochloric acid.
Procedure: Spray moderately heavy with (*a*) (if too light, insufficient oxidation takes place, and if too heavy, the aldehydes are destroyed). After about 30 sec, spray with (*b*) and heat at 45°C for 20 min. Then spray with (*c*) and heat for 15 min at 45°C. (Overspraying with (*b*) or (*c*) causes a brown mottled appearance.)
Use: Identification of polyhydric alcohols [365].
Sensitivity: 100 to 300 μg.

T-258. Thymol–sulfuric acid
Reagent: (*a*) 20% thymol in ethanol.
(*b*) 4 *N* sulfuric acid.
Procedure: Spray with (*a*) and heat 10 min at 90°C, then spray with (*b*) and heat 10 to 15 min at 120°C.
Use: Plasticizers (varied colors) [164].

T-259. Thymol–sulfuric acid
Reagent: 1 g of thymol in a mixture of 10 ml of concentrated sulfuric and 190 ml of ethanol.
Procedure: Spray and heat at 120°C for 15 to 20 min.
Use: Carbohydrates [62]. Dark pink changing to faint violet with further heating.

T-260. *o*-Toluidine
Reagent: 0.5% *o*-toluidine in ethanol (*o*-dianisidine may be substituted).
Procedure: Spray and dry, then expose to short-wave ultraviolet light.
Use: Chlorinated pesticides [366]. Green spots on white background.
Sensitivity: 0.5 to 1 μg.

T-261. Toluidine blue–Lissamine green stain [367]

Reagent: (*a*) Mixture of formol–methanol (1:4).
(*b*) 0.1% Cetavlon in physiological saline.
(*c*) 0.04% Toluidine blue in water–dry acetone (1:4).
(*d*) 0.3 g Lissamine green in 100 ml of 1% acetic acid.

Procedure for staining agar gel layers: Immerse for 15 min in (*a*) and then for 1 hr in (*b*). Dry at 37°C under filter paper. Immerse dried layer in (*c*) for 15 min, then rinse in 1% acetic acid until background is colorless. Immerse in (*b*) for 5 min and again rise in acetic acid.

Use: Combined staining of mucopolysaccharides (red-purple) and proteins (green).

T-262. Trichloroacetic acid

Reagent: (*a*) 4% (wt/vol) trichloroacetic acid in chloroform.
(*b*) 25% trichloroacetic in chloroform.

Procedure: For (*a*) spray and let stand 10 to 30 min. For (*b*), spray and heat 2 min at 100°C and observe in ultraviolet light.

Use: (*a*) Menthofuran [368] pink.
(*b*) *Strophanthus* glycosides [369], sensitivity 0.4 μg. Yellow fluorescence.

T-263. Trichloroacetic acid–chloramine-t [370]

Reagent: Freshly prepared mixture of 25% trichloroacetic acid in 95% ethanol and 3% chloramine-T (4:1).

Procedure: Spray and then heat for 5 to 10 min at 110°C. Observe in daylight and ultraviolet light.

Use: Cardenolides [34, 369, 371].

Sensitivity: 0.2 μg.

T-264. *N*,2,6-trichloro-*p*-quinone imine (Gibb's reagent)

Reagent: (*a*) 2 g of *N*,2,6-trichloro-*p*-benzoquinone imine in 100 ml of ethanol.
(*b*) 10% sodium carbonate.
(*c*) Modified reagent: 0.1% *N*,2,6-trichloro-*p*-benzoquinone imine in chloroform–dimethyl sulfoxide (9:1) (previously saturated with sodium bicarbonate). Stable for 4 months; unstable to basic vapors. If solution darkens or discolors, discard.

Procedure: Spray with (*a*), air-dry, then spray with (*b*).

Use: Hydroxyskatoles [137], phenols (in general, *p*-substituted phenols do not react).

For drugs of many classes spray with (*c*). Amphetamines, anesthetics, antihistamines, barbiturates, decongestants, diuretics, narcotics, psychoactive drugs, and sulfa drugs [372].

Sensitivity: Barbiturates, subnanogram quantities [270].

T-266. 1,3,5-Trinitrobenzene

Reagent: (*a*) 1 g of potassium hydroxide in 10 ml of water and diluted to 100 ml with 95% alcohol.

(*b*) Saturated solution of 1,3,5-dinitrobenzene in acetone.

Procedure: Spray chromatogram with (*a*) and heat 5 min at 150°C. Then wash cooled plate with acetone to remove organic residues. Spray with (*b*) to detect potassium sulfite.

Use: Organic sulfite pesticides [268]. Pink to red spots.

T-267. Trinitrobenzenesulfonic acid

Reagent: 0.1% 2,4,6-trinitrobenzenesulfonic acid in acetone–ethanol–water (2:2:1).

Use: Amino acids [373].

T-268. 2,3,5-Triphenyltetrazolium chloride

Reagent: Freshly prepared by mixing equal volumes of methanolic solutions of 4% 2,3,5-triphenyltetrazolium chloride and 1 *N* sodium hydroxide.

Procedure: Spray with reagent and then heat for 5 to 10 min at 110°C.

Use: Steroids [263]. Red spots.

T-269. Uranyl nitrate

Reagent: 5 g uranyl nitrate in 95 ml 10% sulfuric acid (vol/vol) [374].

Procedure: Spray and heat plate 6 to 7 min at 110°C.

Use: Steroids.

T-270. Urea

Reagent: 1 g of urea in 48 ml of water saturated with butanol and 4.5 ml of 85% orthophosphoric acid.

Procedure: Spray, then heat plate at 105 to 110°C for 30 min. Intense blue-gray spots on almost colorless background.

Use: Sucrose esters [375].

Sensitivity: 0.3 μg.

T-271. Urea–1-naphthol–bromine (Sakaguchi reagent)

Reagent: (*a*) 16% urea–0.2% α-naphthol in ethanol (5:1).

(*b*) 3.3 ml bromine in 500 ml 5% sodium hydroxide.

Procedure: Spray with (*a*), dry at 40°C, then spray with (*b*).

Use: Specific for guanido group. Arginine and homoarginine turn pink to red.

T-272. Vanillin–phosphoric acid

Reagent: 1% vanillin in 50% phosphoric acid.

Procedure: Spray and heat 10 to 23 min at 120°C.

Use: Steroids [25, 376].

T-273. Vanillin–silver nitrate

Reagent: (*a*) 3.0 g vanillin in 100 ml of ethanol + 0.5 ml of sulfuric acid (*d* 1.84).

(*b*) 2.5% aqueous silver nitrate–0.5% bromophenol blue in acetone (1 : 1).

Procedure: Spray with (*a*), then dry at 90°C for 5 min and spray with (*b*). Air-dry for 10 min and then at 90°C for 1½ min.

Use: Adenyl compounds. Pink spot on gold background [377].

Sensitivity: Adenine 0.0001 to 0.05 μM tested directly at the spotting location.

T-274. Vanillin–sulfuric acid

Reagent: 0.5% vanillin in sulfuric acid–ethanol (4 : 1) [378]. (Prepare fresh daily).

Procedure: Spray, then heat for 5 min at 100°C (extended heating turns all spots brown).

Use: Steroids [378]. Terpenes [379]. Plasticizers [164]. Carbamate pesticides ;380].

Sensitivity: Steroids 5 $\mu g/cm^2$.

T-275. Zinc chloride

Reagent: 30% methanolic solution of zinc chloride (the solution should be filtered through a sterimat) [195].

Procedure: Spray and heat for 60 min at 105°C. Use glass cover plate on removing from oven to prevent moisture pickup. Examine under 366-mμ ultraviolet light.

Use: Steroids.

Sensitivity: 0.1 μg or less.

T-276. Zinc uranyl acetate

Reagent: Saturated solution of zinc uranyl acetate in 2 *N* acetic acid.

Procedure: Spray and observe in ultraviolet light. Bright fluorescent spots.

Use: Sodium, lithium (potassium visible only in higher concentration) [381].

It should be remembered that in some cases the reaction of a given spray reagent depends, to some extent, on the adsorbent in the chromatographic layer. For example, in detecting caffeine by exposing the layer to chlorine and then spraying with potassium iodide–benzidine solution, the test is negative on silica gel G layers but positive on aluminum oxide G [122]. It is also well to remember that the sensitivity of a reagent is a relative value, because the sensitivity of a given reagent will be greater for compounds with low R_f values than for those with high R_f values. As the compound travels over the layer, it spreads out so that the concentration per unit area is less for the higher R_f values.

More than one spray reagent can be used on the same chromatogram by spraying lightly and then removing the sprayed layer by means of Scotch tape, and then spraying the freshly exposed surface with a different reagent [382]. Another variation that is useful, especially where the sample must not be contaminated with the spray reagent, is to press a piece of filter paper [383] or another chromatoplate [384] on the developed chromatogram in order to obtain a "print." This print is then sprayed with an appropriate reagent.

In some cases, successive reagents can be applied to reveal additional spots. Iodine vapor can be used and after marking the spots the iodine is removed in a vacuum so that additional reagents may be used [210]. As another example of this technique with organic pesticides [73] iodine in ethanol was sprayed first, followed in succession by fluorescein, 4-methylumbelliferone, and finally silver nitrate. The chromatogram was observed after each spray.

4 RADIOACTIVE METHODS

Radioaudiographs may be made by placing X-ray film in direct contact with the adsorbent layer. Kodak "No-Screen Medical X-ray Film" may be used for this purpose. The activity level of the material will, of course, control the exposure which may vary from 1 hr to 9 days [385]. Sheppard and Tsien [386] give directions for coating photographic plates with Kodak nuclear track emulsion type NPB for use in making radioaudiographs of thin-layer plates. A thin plastic film is sometimes inserted between the X-ray film and the chromatogram, but in this case the exposure most be increased [387]. Polaroid type 57 film may be used in place of X-ray film for autoradiography [388, 389] (See Chapter XI, Section 6 for more information on radioactive detection methods).

Radioactive compounds may be located directly on the plate by using a thin-end-window Geiger counter [390–392] or a gas-flow counter [393–395].

5 BIOLOGICAL DETECTION METHODS

These methods can be used for the detection of zones where antibiotics are separated. In principle, the method is to contact the chromatographic plate with an agar layer which has been inoculated with a suitable microorganism. In practice, there are a number of ways in which this has been accomplished. Nicolaus et al. [396, 397] incorporated triphenyltetrazolium chloride in their inoculated agar medium. The microorganisms reduce the triphenyltetrazolium chloride to triphenylformazan, which has a deep reddish brown color, but wherever the antibiotic is present this reduction does not take place and the light yellow spots appear against the dark background. The medium, which contains 0.7 ml of a 5% solution of triphenyltetrazolium chloride in 50% methanol per 50 ml of medium, is poured gently onto the surface of the chromatoplate contained in a shallow plastic tray. After cooling, the layer is protected from atmospheric oxygen by pouring a thin coating of sterile agar solution on top of the coated plate. After this protective layer of agar has set, the plate is kept in a closed container in the refrigerator for 1 hr at 0°C in order to allow the antibiotic to diffuse into the agar. The plate is then incubated at 37°C for 16 hr or longer to allow the microorganisms to multiply. For inoculating the agar medium, various organisms were used depending on the antibiotic being tested; *Sarcina lutea* for rifomycin, *Staphylococcus aureus* for penicillins, and *Bacillus subtilis* for penicillins and tetracyclines. Begue and Kline [398] tested a number of tetrazolium salts and found that INT (*p*-iodonitrotetrazolium violet) (2-[*p*-iodophenyl]-3-*p*-nitrophenyl-5-phenyltetrazolium chloride) was the most serviceable. Not all tetrazolium salts give a satisfactory result in a given situation, so that the best salt may have to be determined experimentally.

In screening for antibiotics with antitumor activity by means of paper chromatography, Schuurmans et al. [399] incorporated reazurin dye, Sarcoma 180, and Detroit 6 cells into the agar medium. After the incubation period of 2 days under special condition, the antitumor activity was detected by the zone of inhibition of cellular dehydrogenase activity. The technique should work just as well with thin-layer chromatography. Perlman et al. [400], in assaying for antitumor compounds, incorporated Eagle's KB cells, Earle's L cells, and L-1210 cells in an agar medium. The agar plates were used to prepare bioautographs of thin-layer plates.

Tóth and Piorr [401] applied a bioassay to the detection and assay of chlorinated insecticides. After chromatography and evaporation of the developing solvent, a framework with 22 transverse compartments to cover progressive R_f zones was placed on the plate. *Drosophila* were put into these compartments and covered with a glass plate to test for insecticidal activity.

Brodasky [402] prepared trays of agar medium seeded with *Bacillus pumilus* for the detection of neomycins. To allow the antibiotics to diffuse into the agar gel, the chromatograms were pressed firmly on the agar surface for a period of 5 to 10 min. They were then removed and the agar layers were incubated for 16 hr at 28°C. Bickel et al. [403] placed a filter paper on the inoculated agar layer before pressing the chromatogram against this with a 2-kg weight (for a 20 × 20 cm plate) for 20 min in order to allow the antibiotic to diffuse into the agar. Meyers and Smith [404] used a slightly different method for loose-layer chromatograms. In this case the solvent-free chromatogram was covered with a moist filter paper, followed by a clean glass plate. This "sandwich" was inverted so that the chromatoplate was upside down and the edges of the filter paper were then folded back over the layer support. The extra glass plate could then be removed and the paper covered layer was pressed onto the seeded agar layer for diffusion of the antibiotic.

The use of ChromAR Sheets (Mallinckrodt), a silicic acid-impregnated glass fiber sheet, was found to prevent the adherence of silica particles to the agar surface and to require less sample application than the use of coated plates for bioautography [405].

Narisimhachari and Ramachandran [406] used the "Scotch tape" technique for transferring the surface adsorbent from the thin-layer plate to the nutrient agar plate. A control strip was also taken from a blank area of the TLC plate. Reusser [407] dipped the developed and dried plate into a collodion solution and then dried it in a stream of air. The layer was then stripped from the plate and applied to the agar test plate.

Kline and Golab [408] developed a simple but elegant technique for detecting antibiotics. This technique involves spraying the completed chromatogram with an agar medium at 100°C. This was accomplished with a Devilbiss paint spray gun using 27 lb/in.^{2} pressure. This thin layer of agar was then allowed to cool before pouring the inoculated agar medium (cooled to 48°C) directly on the prepared plate. This procedure eliminated the need for using a filter paper cover over the chromatographic plate and also prevented the spreading of the antibiotic by pouring the inoculated medium directly onto the chromatographic plate.

Homan and Fuchs [409] sprayed the plates directly with a conidial suspension of *Cladosporium cucmerinum* for the detection of fungitoxic compounds after proper incubation. Care must be taken not to allow the plates to become too wet. For the detection of antibacterial substances in *Xanthomonas pruni,* which does not reduce tetrazolium dyes, Hamilton and Cook [410] incorporated 0.4% gelatin into the nutrient agar. After incubation the ability of *X. pruni* to hydrolyze gelatin (also starch) was utilized by flooding the layer with 10% mercuric sulfate in 2.5 *N* hydrochloric acid, causing the unhydrolyzed gelatin to form a white precipitate.

Meyers and Erickson [411] found that the addition of 0.1% potassium nitrate as an oxidant improved the growth of *Staphylococcus aureus* 209 P used as a test organism when the microorganism was covered by a glass plate.

Ono [412] used a bioautographic method for the detection of vitamin B_{12} by incorporating *Lactobacillus leichmannii* into the vitamin B_{12} assay agar medium.

Kreuzig [413] used *Pediococcus cerevisiae* to detect folic acid in extracts of mold mycelium on dual layer plates. Sensitivity was 50 pg.

In all these methods, if the development of the chromatographic plate has been carried out with an acidic solvent the completed chromatogram should be exposed to ammonia vapor in order to neutralize the acid which would inhibit the growth of the microorganisms [396, 397].

Aszalos et al. [414] list the test organisms for detecting numerous antibiotics and Betina [415] has reviewed the use of bioautography in thin-layer and paper chromatography.

References

1. A. Winterstein and B. Hegedues, *Chimia (Aarau)*, **14**, 18 (1960).
2. A. Winterstein and B. Hegedues, *Z. Physiol. Chem.*, **321**, 97 (1960).
3. J. G. Kirchner, J. M. Miller, and G. J. Keller, *Anal. Chem.*, **23**, 420 (1951).
4. R. T. Morris, *Anal. Chem.*, **24**, 1528 (1952).
5. M. B. Naff and A. S. Naff, *J. Chem. Educ.*, **40**, 534 (1963).
6. J. Blattná and J. Davídek, *Experientia*, **17**, 474 (1961).
7. J. Davídek and J. Blattná, *J. Chromatogr.*, **7**, 204 (1962).
8. J. Janák, *J. Chromatogr.*, **15**, 28 (1964).
9. L. J. Morris, R. T. Holman, and K. Fontell, *J. Am. Oil Chem. Soc.*, **37**, 323 (1960).
10. A. A. Akhrem, and A. I. Kuznetsova, *Proc. Acad. Sci. USSR, Chem. Technol. Sect. (Engl. Transl.)*, **138**, 507 (1961).
11. O. S. Privett and M. L. Blank, *J. Lipid Res.*, **2**, 37 (1961).
12. R. D. Bennett and E. Heftmann, *J. Chromatogr.*, **9**, 348 (1962).
13. R. H. Anderson, T. E. Huntley, W. M. Schwecke, and J. H. Nelson, *J. Am. Oil Chem. Soc.*, **40**, 349 (1963).
14. D. Abramson and M. Blecher, *J. Lipid Res.*, **5**, 628 (1964).
15. A. Berg and J. Lam, *J. Chromatogr.*, **16**, 157 (1964).
16. B. P. Lisboa and E. Diczfalusy, *Acta Endocrinol.*, **40**, 60 (1962).
17. W. L. Anthony and W. T. Beher, *J. Chromatogr.*, **13**, 567 (1964).
18. L. L. Smith and T. Foell, *J. Chromatogr.*, **9**, 339 (1962).
19. N. Matsumoto, *Chem. Pharm. Bull. (Tokyo)*, **11**, 1189 (1963); *Chem. Abstr.*, **59**, 15559 (1963).
20. E. Ehrhardt and F. Cramer, *J. Chromatogr.*, **7**, 405 (1962).
21. O. S. Privett and M. L. Blank, *J. Am. Oil Chem. Soc.*, **39**, 520 (1962).
22. H. Ertel and L. Horner, *J. Chromatogr.*, **7**, 268 (1962).
23. H. K. Mangold, *J. Am. Oil Chem. Soc.*, **38**, 708 (1961).

24. C. B. Barrett, M. S. J. Dallas, and F. B. Padley, *J. Am. Oil Chem. Soc.*, **40,** 580 (1963).
25. H. Metz, *Naturwissenschaften,* **48,** 569 (1961).
26. T. Ziminski and E. Borowski, *J. Chromatogr.*, **23,** 480 (1966).
27. B. L. Walker, *J. Chromatogr.*, **56,** 320 (1971).
28. J. Korolczuk and I. Kwaśniewska, *J. Chromatogr.*, **88,** 428 (1974).
29. D. Jones, D. E. Bowyer, G. A. Gresham, and A. N. Howard, *J. Chromatogr.*, **24,** 228 (1966).
30. T. T. Martin and M. C. Allen, *J. Am. Oil Chem. Soc.*, **48,** 752 (1971).
31. J. C. Touchstone, T. Murawec, M. Kasparow, and W. Wortmann, *J. Chromatogr. Sci.*, **10,** 490 (1972).
32. J. C. Touchstone, A. K. Balin, T. Murawec, and M. Kasparow, *J. Chromatogr. Sci.*, **8,** 443 (1970).
33. H. Gaenshirt and A. Malzacher, *Arch. Pharm.*, **293/65,** 925 (1960).
34. J. Reichelt and J. Pitra, *Collect. Czech. Chem. Commun.*, **27,** 1709 (1962).
35. L. Anker and D. Sonanini, *Pharm. Acta Helv.*, **37,** 360 (1962).
36. P. H. Boyle and P. H. Nelson, *Chem. Ind. (London),* **1967,** 1220.
37. T. Cotgreave and A. Lynes, *J. Chromatogr.*, **30,** 117 (1967).
38. J. J. Szakasits, P. V. Peurifoy, and L. A. Woods, *Anal. Chem.*, **42,** 351 (1970).
39. E. Haati and J. Jaakonmaki, *Ann. Med. Exp. Biol. Fenn.*, **47,** 175 (1969).
40. T. Okumura, T. Kadono, and A. Iso'o, *J. Chromatogr.*, **108,** 329 (1975).
41. H. Gaenshirt, *Arch. Pharm.*, **296,** 73 (1963).
42. R. J. Gritter and R. J. Albers, *J. Chromatogr.*, **9,** 392 (1962).
43. M. E. Tate and C. T. Bishop, *Can. J. Chem.*, **40,** 1043 (1962).
44. W. Heidbrink, *Fette, Seifen, Anstrichm.*, **66,** 569 (1964).
45. W. D. Conway, R. W. Piontek, and J. M. Shekosky, *J. Chromatogr.*, **27,** 317 (1967).
45a. B. R. Smith, *J. Chromatogr.*, **82,** 95 (1973).
45b. R. Segura and A. M. Gotto, Jr., *J. Chromatogr.*, **99,** 643 (1974).
45c. H. Shanfield, F. Hsu, and A. J. P. Martin, *J. Chromatogr.*, **126,** 457 (1976).
45d. H. Shanfield, K. Y. Lee, and A. J. P. Martin, *J. Chromatogr.*, **142,** 387 (1977).
46. J. M. Miller and J. G. Kirchner, *Anal. Chem.*, **26,** 2002 (1954).
47. J. W. Sease, *J. Am. Chem. Soc.*, **69,** 2242 (1947).
48. *Ibid.*, **70,** 3630 (1948).
49. R. H. Reitsema, *Anal. Chem.*, **26,** 960 (1954).
50. E. Stahl, *Chem.-Ztg.*, **82,** 323 (1958).
51. G. Machata, *Mikrochim. Acta,* **1960,** 79.
52. V. Cěrný, J. Joska, and L. Lábler, *Collect. Czech. Chem. Commun.*, **26,** 1658 (1961).
53. W. D. Loomis, private communication.
54. H. Seiler and W. Rothweiler, *Helv. Chim. Acta,* **44,** 941 (1961).
55. H. Hammerschmidt and M. Mueller, *Papier,* **17,** 448 (1963).
56. J. H. Daams, *J. Chromatogr.*, **10,** 450 (1963).
57. R. J. Wieme and M. Rabaey, *Naturwissenschaften,* **44,** 112 (1957).

58. L. Popadiuk, *Arch. Immunol. Ter. Dows.*, **9,** 139 (1961).
59. E. Nuernberg, *Dtsch. Apoth.-Ztg.*, **101,** 142 (1961).
60. B. P. Lisboa, *J. Chromatogr.*, **16,** 136 (1964).
61. V. Prey, H. Scherz, and E. Bancher, *Mikrochim. Acta,* **1963,** 567.
62. S. Adachi, *J. Chromatogr.*, **17,** 295 (1965).
63. B. H. Somaroo, M. L. Thakur, and W. F. Grant, *J. Chromatogr.*, **87,** 290 (1973).
64. P. M. Martins and Y. P. Dick, *J. Chromatogr.*, **32,** 188 (1968).
65. W. M. Connors and W. K. Boak, *J. Chromatogr.*, **16,** 243 (1964).
66. L. Sattler and F. W. Zerban, *Anal. Chem.*, **24,** 1862 (1953).
67. H. Brockmann, E. Spohler, and T. Waehneldt, *Chem. Ber.*, **96,** 2925 (1963).
68. E. Knappe, D. Peteri, and I. Rohdewald, *Z. Anal. Chem.*, **199,** 270 (1963).
69. S. K. Goswami and C. H. Frey, *J. Lipid Res.*, **12,** 509 (1971).
70. H. Wagner, L. Hoerhammer, and P. Wolff, *Biochem. Z.* **334,** 175 (1961).
71. P. P. Waring and Z. Z. Ziporin, *J. Chromatogr.*, **15,** 168 (1964).
72. C. S. Hanes and F. A. Isherwood, *Nature,* **164,** 1107 (1949).
73. C. W. Stanley, *J. Chromatogr.*, **16,** 467 (1964).
74. M. Baudler and F. Stuhlmann, *Naturwissenschaften,* **51,** 57 (1964).
75. J. E. Barney II, *J. Chromatogr.*, **20,** 334 (1965).
76. M. E. Q. Pilson and R. J. Fragala, *Anal. Chim. Acta,* **52,** 553 (1970).
77. H. Seiler, *Helv. Chim. Acta,* **44,** 1753 (1961).
78. J. Askew, J. H. Ruzicka, and B. B. Wheals, *Analyst (London),* **94,** 275 (1969).
79. W. W. Fike and I. Sunshine, *Anal. Chem.*, **37,** 127 (1965).
80. M. Malaiyandi, J. P. Barrette, and M. Lanquette, *J. Chromatogr.*, **101,** 155 (1974).
81. J. Kowalczyk, *J. Chromatogr.*, **14,** 411 (1964).
82. L. D. Bergel'son, E. V. Diatlovitskaya, and V. V. Voronkova, *Dokl. Akad. Nauk SSSR,* **149,** 1319 (1963).
83. H. Weicker and R. Brossmer, *Klin. Wochenschr.*, **39,** 1265 (1961).
84. I. Sjoeholm, *Svensk Farm. Tidskr.*, **66,** 321 (1962).
85. D. Kritchevsky, D. S. Martak, and G. H. Rothblat, *Anal. Biochem.*, **5,** 388 (1963).
86. B. Johannesen and A. Sandal, *Medd. Norsk Farm. Selsk.*, **23,** 105 (1961).
87. A. Schweiger, *J. Chromatogr.*, **9,** 374 (1962).
88. C. M. Van Gent, O. J. Roseleur, and P. Van der Bijl, *J. Chromatogr.*, **85,** 174 (1973).
89. S. Heřmánek, V. Schwarz, and Z. Čekan, *Collect. Czech. Chem. Commun.*, **26,** 1669 (1961).
90. H. F. MacRae and W. P. McKinley, *J. Assoc. Off. Agric. Chem.*, **44,** 207 (1961).
91. F. Korte and H. Sieper, *J. Chromatogr.*, **13,** 90 (1964).
92. E. Bancher, H. Scherz, and K. Kaindl, *Mikrochim. Acta,* **1964,** 652.
93. V. M. Adamović, *J. Chromatogr.*, **23,** 274 (1966).
94. C. M. Kottemann, *J. Assoc. Off. Anal. Chem.*, **49,** 954 (1966).
95. M. D. Bischel and J. H. Austin, *Biochim. Biophys. Acta,* **70,** 598 (1963).

96. E. W. Baur, private communication.
97. E. Klenk and H. Langerbeins, *Z. Physiol. Chem.*, **270,** 185 (1941).
98. J. Boute, *Ann. Endocrinol. (Paris),* **14,** 518 (1953).
99. E. Klenk and W. Gielen, *Z. Physiol. Chem.*, **333,** 162 (1963).
100. D. C. Abbott, J. A. Bunting, and J. Thomson, *Analyst (London),* **90,** 356 (1965).
101. H. Seiler and T. Kaffenberger, *Helv. Chim. Acta,* **44,** 1282 (1961).
102. E. Knappe and D. Peteri, *Z. Anal. Chem.*, **188,** 184 (1962).
103. H. Thielemann, *Z. Anal. Chem.*, **272,** 286 (1974).
104. H. E. Wade and D. M. Morgan, *Nature,* **171,** 529 (1953).
105. M. Salamé, *J. Chromatogr.*, **16,** 476 (1964).
106. G. Tadema and P. H. Batelaan, *J. Chromatogr.*, **33,** 460 (1968).
107. S. Goenechea, *J. Chromatogr.*, **40,** 182 (1969).
108. H. Jatzkewitz and E. Mehl, *Z. Physiol. Chem.*, **320,** 251 (1960).
109. S. D. Killilea and P. O'Carra, *J. Chromatogr.*, **54,** 284 (1971).
110. L. D. Hayward, R. A. Kitchen, and D. J. Livingstone, *Can. J. Chem.*, **40,** 434 (1962).
111. J. Chrastil, *J. Chromatogr.*, **115,** 273 (1975).
112. M. Gorman, M. Neuss, and K. Biemann, *J. Am. Chem. Soc.*, **84,** 1058 (1962).
113. N. J. Cone, R. Miller, and N. Neuss, *J. Pharm. Sci.*, **52,** 688 (1963).
114. G. S. Sundaram and H. S. Sodhi, *J. Chromatogr.*, **61,** 370 (1971).
115. C. Kump and H. Schmid, *Helv. Chim. Acta,* **45,** 1090 (1962).
116. G. Sembdner, R. Gross, and K. Schreiber, *Experientia,* **18,** 584 (1962).
117. F. Geiss and H. Schlitt, "Analyse von polyphenylgemischen mit der Dünnschchtchromatographie," Joint Research Center, Chemistry Department, Ispra Brussels, November 1961, 17 pages.
118. O. E. Schultz and D. Strauss, *Arzneimittel-Forsch.*, **5,** 342 (1955).
119. A. Seher, *Mikrochim. Acta,* **1961,** 308.
120. L. M. Pires and A. N. Roseira, *J. Chromatogr.*, **56,** 59 (1971).
121. R. Lindfors, *Ann. Med. Exp. Biol. Fenn. (Helsinki),* **41,** 355 (1963).
122. H. Gaenshirt, *Arch. Pharm.*, **296,** 73 (1963).
123. N. H. Rydon and P. W. G. Smith, *Nature,* **169,** 922 (1952).
124. H. G. Zachau and W. Karau, *Chem. Ber.*, **93,** 1830 (1960).
125. F. Reindel and W. Hoppe, *Chem. Ber.*, **87,** 1103 (1954).
126. G. Pataki, *J. Chromatogr.*, **12,** 541 (1963).
127. M. Brenner, A. Niederwieser, and G. Pataki, *Experientia,* **17,** 145 (1961).
128. H. Thielemann, *Z. Chem.*, **13,** 15 (1973).
129. R. T. Nuttall and I. E. Bush, *Analyst (London),* **96,** 875 (1971).
130. H. J. Issaq and E. W. Barr, *J. Chromatogr.*, **132,** 121 (1977).
131. J. T. Stewart and L. A. Sternson, *J. Chromatogr.*, **92,** 182 (1974).
132. R. Tschesche, W. Freytag, and G. Snatzke, *Chem. Ber.*, **92,** 3053 (1959).
133. A. S. Gupta and S. Dev, *J. Chromatogr.*, **12,** 189 (1963).
134. R. Tschesche, W. Freytag, and G. Snatzke, *Chem. Ber.*, **92,** 3053 (1959).
135. M. Beroza, *J. Agric. Food Chem.*, **11,** 51 (1963).
136. S. W. Gunner and T. B. Hand, *J. Chromatogr.*, **37,** 356 (1968).

137. R. A. Heacock and M. E. Mahon, *Can. J. Biochem. Physiol.*, **41**, 487 (1963).
138. M. D. Bischel and J. H. Austin, *Biochim. Biophys. Acta*, **70**, 598 (1963).
139. R. Donner and Kh. Lohs, *J. Chromatogr.*, **17**, 349 (1965).
140. S. M. Aharoni and M. H. Litt, *Anal. Chem.*, **42**, 1467 (1970).
141. M. Geldmacher-Mallinckrodt and U. Weigel, *Arch. Toxikol.*, **20**, 114 (1963).
142. M. H. Hashmi, M. A. Shahid, A. A. Ayaz, F. R. Chughtai, and N. Hassan, *Anal. Chem.*, **38**, 1554 (1966).
143. F. W. H. M. Merkus, "Progress in Inorganic Thin-Layer Chromatography," in *Progress in Separation and Purification*, Vol. 3, E. S. Perry and C. J. Van Oss, Eds., Wiley-Interscience, 1970, p. 234.
144. H. P. Kaufmann and T. H. Khoe, *Fette, Seifen, Anstrichm.*, **64**, 81 (1962).
145. H. P. Kaufmann, Z. Makus, and T. H. Khoe, *Fette, Seifen, Anstrichm.*, **63**, 689 (1961).
146. E. Campaigne and M. Georgiadis, *J. Org. Chem.*, **28**, 1044 (1963).
147. M. Hranisavljević-Jakovljević, I. Pejković-Tadić, and A. Stojiljković, *J. Chromatogr.*, **12**, 70 (1963).
148. M. Geldmacher-Mallinckrodt, *Dtsch. Z. Ges. Gerichtl. Med.*, **54**, 90 (1963).
149. K. C. Guven and S. Cobanlar, *Eczacilik. Bull.*, **9**, 98 (1967).
150. G. Weidemann and W. Fischer, *Z. Physiol. Chem.*, **336**, 189 (1964).
151. G. R. N. Jones, *J. Chromatogr.*, **77**, 357 (1973).
152. J. Stenerson, *J. Chromatogr.*, **54**, 77 (1971).
153. J. H. Ross, *Anal. Chem.*, **40**, 2138 (1968).
154. L. Fishbein and J. Fawkes, *J. Chromatogr.*, **22**, 323 (1966).
155. R. W. Frei and P. E. Belliveau, *Chromatographia*, **5**, 296 (1972).
156. S. Roy and D. P. Chakraborty, *J. Chromatogr.*, **96**, 266 (1974).
157. H. K. Mangold and D. C. Malins, *J. Am. Oil Chem. Soc.*, **37**, 383 (1960).
158. J. Franc, M. Hájková, and M. Jehlicka, *Chem. Zvesti.*, **17**, 542 (1963).
159. C. Passera, A. Pedrotti, and G. Ferrari, *J. Chromatogr.*, **14**, 289 (1964).
160. H. Boehme and L. Kreutzig, *Apoth.-Ztg.*, **103**, 505 (1963).
161. H. Gaenshirt and A. Malzacher, *Naturwissenschaften*, **47**, 279 (1960).
162. M. H. Klouwen and R. ter Heide, *Parfuem. Kosmet.*, **43**, 195 (1962).
163. A. Seher, *Fette, Seifen, Anstrichm.*, **61**, 345 (1959).
164. J. W. Copius-Peereboom, *J. Chromatogr.*, **4**, 323 (1960).
165. K. E. Schulte, F. Ahrens, and E. Sprenger, *Pharm. Ztg., Ver. Apoth.-Ztg.*, **108**, 1165 (1963).
166. J. E. Forrest, R. Richard, and R. A. Heacock, *J. Chromatogr.*, **65**, 439 (1972).
167. S. K. Yasuda, *J. Chromatogr.*, **13**, 78 (1964).
168. *Ibid.*, **14**, 65 (1964).
169. E. Nuernberg, *Arch. Pharm.*, **292/64**, 610 (1959).
170. L. A. Dal Cortiva, J. R. Broich, A. Dihrberg, and B. Newman, *Anal. Chem.*, **38**, 1959 (1966).
171. S. Agurell and E. Ramstad, *Lloydia*, **25**, 67 (1962).
172. B. G. Osborne, *J. Chromatogr.*, **70**, 190 (1972).
173. J. Perkavec and M. Perpar, *Mikrochim. Acta*, **1964**, 1029.
174. H. G. Henkel, *Chimia (Aarau)*, **18**, 252 (1964).

175. D. Spengler and A. Jumar, *J. Chromatogr.,* **49,** 329 (1970).
176. N. Karpitschka, *Mikrochim. Acta,* **1963,** 157.
177. J. E. Meinhard and N. F. Hall, *Anal. Chem.,* **21,** 185 (1949).
178. E. Knappe and D. Peteri, *Z. Anal. Chem.,* **190,** 386 (1962).
179. J. Baeumler and S. Rippstein, *Helv. Chim. Acta,* **44,** 1162 (1961).
180. *Ibid., Arch. Pharm.,* **296,** 301 (1963).
181. L. Stárka and R. Hampl, *J. Chromatogr.,* **12,** 347 (1963).
182. B. P. Lisboa, *J. Chromatogr.,* **13,** 391 (1964).
183. H. Grasshof, *J. Chromatogr.,* **14,** 513 (1964).
184. C. G. Honegger, *Helv. Chim. Acta,* **45,** 281 (1962).
185. J. Hansson, *Explosivstoffe,* **10,** 73 (1963).
186. R. Preussmann, D. Daiber, and H. Hengy, *Nature,* **201,** 502 (1964).
187. D. Katz, *J. Chromatogr.,* **15,** 269 (1964).
188. L. J. Faucheux, Jr., *J. Assoc. Off. Agric. Chem.,* **48,** 955 (1965).
189. R. New, *Naturwissenschaften,* **44,** 181 (1957).
190. B. Pfunder, R. Zurflueh, H. Seiler, and H. Erlenmeyer, *Helv. Chim. Acta,* **45,** 1153 (1962).
191. H. K. Mangold and R. Kammereck, *Chem. Ind. (London),* **1961,** 1032.
192. J. Lehmann and V. Karamustafaoglu, *Scand. J. Clin. Lab. Invest.,* **14,** 554 (1962).
193. M. Tuerler and D. Hoegl, *Mitt. Geb. Lebensm. Hyg.,* **52,** 132 (1961).
194. G. Bergstroem and C. Lagercrantz, *Acta Chem. Scand.,* **18,** 560 (1964).
195. P. J. Stevens, *J. Chromatogr.,* **14,** 269 (1964).
196. R. A. Dilley, *Anal. Biochem.,* **7,** 240 (1964).
197. O. R. Braekkan, G. Lambertsen, and H. Myklestad, *Fiskeridir. Skr. Ser. Teknol. Unders.,* **4,** 3 (1963).
198. H. Seiler, *Helv. Chim. Acta,* **46,** 2629 (1963).
199. P. Kuenzi, dissertation, Basel University, 1962.
200. F. Geike and I. Schuphan, *J. Chromatogr.,* **72,** 153 (1972).
201. H. Thies and F. W. Reuther, *Naturwissenschaften,* **41,** 230 (1954).
202. D. Vágújfalvi, *Planta Med.,* **8,** 34 (1960).
203. Tyihák, *J. Chromatogr.,* **14,** 125 (1964).
204. R. Munier, *Bull. Soc. Chim. Biol.,* **35,** 1225 (1953).
205. K. Teichert, E. Mutschler, and H. Rochelmeyer, *Dtsch. Apoth.-Ztg.,* **100,** 477 (1960).
206. H. Feltkamp and F. Koch, *J. Chromatogr.,* **15,** 314 (1964).
207. K. Thoma, R. Rombach, and E. Ullmann, *Arch. Pharm.,* **298,** 19 (1965).
208. K. Buerger, *Z. Anal. Chem.,* **196,** 259 (1963).
209. F. Korte and J. Vogel, *J. Chromatogr.,* **9,** 381 (1962).
210. M. Lepage, *J. Chromatogr.,* **13,** 99 (1964).
211. P. B. Baker, J. E. Farrow, and R. A. H. Hoodless, *J. Chromatogr.,* **81,** 174 (1973).
212. C. E. Mendoza, D. L. Grant, B. Braceland, and K. A. McCully, *Analyst (London),* **1969,** 805.
213. J. E. Cassidy, D. P. Ryskiewich, and R. T. Murphy, *J. Agric. Food Chem.,* **17,** 558 (1969).

214. R. Segura-Cardona and K. Soehring, *Med. Exp.*, **10,** 251 (1964).
215. G. Pastuska, *Z. Anal. Chem.*, **179,** 355 (1961).
216. L. Hoerhammer, H. Wagner, and G. Bittner, *Pharm. Ztg., Ver. Apoth.-Ztg.*, **108,** 259 (1963).
217. E. Knappe, D. Peteri, and I. Rohdewald, *Z. Anal. Chem.*, **197,** 364 (1963).
218. J. Cochin and J. W. Daly, *J. Pharmacol. Exp. Therap.*, **139,** 160 (1963).
219. B. N. Sen, *Anal. Chim. Acta,* **23,** 152 (1960).
220. E. Knappe and K. G. Yekundi, *Z. Anal. Chem.*, **203,** 87 (1964).
221. A. Noirfalise and M. H. Grosjean, *J. Chromatogr.*, **16,** 236 (1964).
222. M. Gillio-Tox, S. A. Previtera, and A. Vimercati, *J. Chromatogr.*, **13,** 571 (1964).
223. J. N. Ebie and R. M. Brooker, *Experientia,* **18,** 524 (1962).
224. S. J. Patterson and R. L. Clements, *Analyst (London),* **89,** 328 (1964).
225. T. W. Hammonds and G. Shone, *J. Chromatogr.*, **15,** 200 (1964).
226. W. D. Skidmore and C. Entenman, *J. Lipid Res.*, **3,** 471 (1962).
227. K. Maruyama, K. Onoe, and R. Goto, *Nippon Kagaku Zasshi,* **77,** 1496 (1956); through *Chem. Abstr.*, **52,** 2665 (1958).
228. J. Sherma and G. Marzoni, *Am. Lab.*, **6,** 21 (1974).
229. J. C. Young, *J. Chromatogr.*, **130,** 392 (1977).
230. R. L. Ranieri and J. L. McLaughlin, *J. Chromatogr.*, **111,** 234 (1975).
231. E. Rinde and W. Troll, *Anal. Chem.*, **48,** 542 (1976).
232. O. Folin and V. Ciocalteau, *J. Biol. Chem.*, **73,** 627 (1929).
233. E. S. Fiala and J. H. Weisburger, *J. Chromatogr.*, **105,** 189 (1975).
234. R. W. Keith, D. Le Tourneau, and D. Mahlum, *J. Chromatogr.*, **1,** 534 (1958).
235. Ž. Procházka, *Chem. Listy,* **47,** 1637 (1953).
236. E. Stahl and H. Kaldewey, *Z. Physiol. Chem.*, **323,** 182 (1961).
237. W. Awe and W. Schultz, *Pharm. Ztg., Ver. Apoth.-Ztg.*, **107,** 1333 (1962).
238. N. Kucharczyk, J. Fohl, and J. Vymetal, *J. Chromatogr.*, **11,** 55 (1963).
239. K. H. Mueller and H. Honerlagen, *Arch. Pharm.*, **293/65,** 202 (1960).
240. C. B. Rollins and R. D. Wood, *J. Chromatogr.*, **16,** 555 (1964).
241. F. Micheel and H. Schweppe, *Mikrochim. Acta,* **1954,** 53.
242. M. H. Klouwen, R. ter Heide, and J. G. J. Kok, *Fette, Seifen, Anstrichm.*, **65,** 414 (1963).
243. E. Sundt and A. Saccardi, *Food Tech.*, **16,** 89 (1962).
244. W. L. Stanley, *J. Assoc. Off. Agric. Chem.*, **44,** 546 (1961).
245. J. C. Topham and J. W. Westrop, *J. Chromatogr.*, **16,** 233 (1964).
246. B. N. Sen, *Anal. Chim. Acta,* **12,** 154 (1955).
247. P. Markl and F. Hecht, *Mikrochim. Acta,* **1963,** 889.
248. H. Seiler and M. Seiler, *Helv. Chim. Acta,* **44,** 939 (1961).
249. *Ibid.*, **43,** 1939 (1960).
250. G. Adam and K. Schreiber, *Z. Chem.*, **3,** 100 (1963).
251. E. Cherbuliez, B. Baehler, and J. Rabinowitz, *Helv. Chim. Acta,* **47,** 1350 (1964).
252. *Ibid.*, **43,** 1871 (1960).
253. R. Fischer and H. Lautner, *Arch. Pharm.*, **294/66,** 1 (1961).

254. R. Fischer and W. Klingelhoeller, *Pflanzenschutz Ber.*, **27,** 165 (1961).
255. K. Schreiber, O. Aurich, and G. Osske, *J. Chromatogr.*, **12,** 63 (1963).
256. W. J. McAleer and M. A. Kozlowski, *Arch. Biochem. Biophys.*, **66,** 125 (1957).
257. M. H. Hashmi, F. R. Chugtai, A. S. Adil, and T. Qureshi, *Mikrochim. Acta*, **1967,** 1111.
258. M. Pokorny, N. Vitezić, and M. Japelj, *J. Chromatogr.*, **77,** 458 (1973).
259. R. F. Curtis and G. T. Phillips, *J. Chromatogr.*, **9,** 366 (1962).
260. J. G. Heathcote and C. Haworth, *J. Chromatogr.*, **43,** 84 (1969).
261. J. Barrollier, J. Heilman, and E. Watzke, *Z. Physiol. Chem.*, **304,** 21 (1956).
262. E. von Arx and R. Neher, *J. Chromatogr.*, **12,** 329 (1963).
263. J. Vaedtke and A. Gajewska, *J. Chromatogr.*, **9,** 345 (1962).
264. B. Goerlich, *Planta Medica*, **9,** 442 (1961).
265. L. Hoerhammer, H. Wagner, and H. Koenig, *Dtsch. Apoth.-Ztg.*, **103,** 502 (1963).
266. L. Hoerhammer, H. Wagner, and K. Hein, *J. Chromatogr.*, **13,** 235 (1964).
267. L. Wasserman and H. Hanus, *Naturwissenschaften*, **50,** 351 (1963).
268. R. C. Blinn and F. A. Gunther, *J. Assoc. Off. Agric. Chemists*, **46,** 204 (1963).
269. S. K. Goswami and C. F. Frey, *J. Chromatogr.*, **53,** 389 (1970).
270. C. J. Umberger, *J. Chromatogr.*, **60,** 95 (1971).
271. H. H. Stroh and W. Schueler, *Z. Chem.*, **4,** 188 (1964).
272. O. Crépy, O. Judas, and B. Lachese, *J. Chromatogr.*, **16,** 340 (1964).
273. C. Zinzadze, *Ind. Eng. Chem.*, **7,** 227 (1935).
274. J. C. Dittmer and R. L. Lester, *J. Lipid Res.*, **5,** 126 (1964).
275. P. Schellenberg, *Angew. Chem. Int. Ed. Engl.*, **1,** 114 (1962).
276. C. J. O. R. Morris, *J. Chromatogr.*, **16,** 167 (1964).
277. J. Groulade, J. N. Fine, and C. Ollivier, *Nature*, **191,** 72 (1961).
278. V. Prey, H. Berbalk, and M. Kausz, *Mikrochim. Acta*, **1961,** 968.
279. E. Richter, *J. Chromatogr.*, **18,** 164 (1965).
280. J. H. Dhont and C. de Rooy, *Analyst (London)*, **86,** 527 (1961).
281. K. Teichert, E. Mutschler, and H. Rochelmeyer, *Dtsch. Apoth.-Ztg.*, **100,** 283 (1960).
282. S. Klein and B. T. Kho, *J. Pharm. Sci.*, **51,** 966 (1962).
283. J. G. Heathcote, R. J. Washington, and B. J. Keogh, *J. Chromatogr.*, **104,** 141 (1975).
284. L. J. Morris, R. T. Holman, and K. Fontell, *J. Lipid Res.*, **1,** 412 (1960).
285. E. D. Moffat and R. I. Lytle, *Anal. Chem.*, **31,** 926 (1959).
286. M. Brenner and A. Niederwieser, *Experientia*, **16,** 378 (1960).
287. E. F. L. J. Anet, *J. Chromatogr.*, **63,** 465 (1971).
288. J. Baeumler and M. Luedin, *Arch. Toxikol.*, **20,** 96 (1963).
289. J. D. Sapira, *J. Chromatogr.*, **42,** 136 (1969).
290. M. Schulz, H. Seeboth, and W. Wieker, *Z. Chem.*, **2,** 279 (1962).
291. M. Chiba and H. V. Morley, *J. Assoc. Off. Agric. Chem.*, **47,** 667 (1964).
292. R. Roser and P. M. Tocci, *J. Chromatogr.*, **72,** 207 (1972).

293. J. M. Finocchiaro and W. R. Benson, *J. Assoc. Off. Anal. Chem.*, **50,** 888 (1967).
294. J. G. Pomonis, R. F. Severson, and P. J. Freeman, *J. Chromatogr.*, **40,** 78 (1969).
295. R. R. Watts, *J. Assoc. Off. Agric. Chem.*, **48,** 1161 (1965).
296. R. S. Ratney, *J. Chromatogr.*, **11,** 111 (1963).
297. K. E. Schulte and G. Ruecker, *J. Chromatogr.*, **49,** 317 (1970).
298. N. Zoellner and G. Wolfram, *Klin. Wochenschr.*, **40,** 1100 (1962).
299. I. Ognyanov, *Compt. Rend. Acad. Bulg. Sci.*, **16,** 161 (1963).
300. R. M. Cassidy, V. Miketukova, and R. W. Frei, *Anal. Lett.*, **5,** 115 (1972).
301. J. Baeumler and S. Rippstein, *Pharm. Acta Helv.*, **36,** 382 (1961).
302. H. Petschik and E. Steger, *J. Chromatogr.*, **9,** 307 (1962).
303. N. R. Stephenson, *Can. J. Biochem. Physiol.*, **37,** 391 (1959).
304. H. Meyer, *Dtsch. Lebensm-Rundsch.*, **57,** 170 (1961).
305. E. G. G. Lindeberg, *J. Chromatogr.*, **117,** 439 (1976).
306. N. H. Choulis, *J. Pharm. Sci.*, **56,** 904 (1967).
307. A. Mehlitz, K. Gierschner, and T. Minas, *Chem.-Ztg.*, **87,** 573 (1963).
308. F. Jaminet, *Farmaco Ed. Prat.*, **18,** 633 (1963).
309. T. Furukawa, *Nippon Kagaku Zasshi,* **78,** 1185 (1957); *Chem. Abstr.*, **52,** 13364 (1958).
310. C. G. Greig and D. H. Leaback, *Nature,* **188,** 310 (1960).
311. E. Stahl, *Arch. Pharm.*, **293/65,** 531 (1960).
312. P. B. Baker, J. E. Farrow, and R. A. Hoodless, *J. Chromatogr.*, **81,** 174 (1973).
313. J.-T. Huang, H.-C. Hsiu, and K.-T. Wang, *J. Chromatogr.*, **29,** 391 (1967).
314. E. Akita and T. Ikekawa, *J. Chromatogr.*, **2,** 250 (1963).
315. K. N. Johri and N. K. Kaushik, *Ind. J. App. Chem.*, **33,** 173 (1970).
316. M. A. Eastwood and D. Hamilton, *Biochem. J.*, **105,** 37c (1967).
317. K. Buerger, *Z. Anal. Chem.*, **192,** 280 (1962).
318. K. N. Johri and H. C. Mehra, *Mikrochim. Acta,* **1971,** 317.
319. S. David and H. Hirshfeld, *Bull. Soc. Chim. Fr.*, **1963,** 1011.
320. J. W. Copius-Peereboom and H. W. Beekes, *J. Chromatogr.*, **14,** 417 (1964).
321. S. J. Mulé, *J. Chromatogr.*, **39,** 302 (1969).
322. R. Cadenas and J. O. Deferrari, *Analyst (London),* **86,** 132 (1961).
323. J. O. Deferrari, R. M. de Lederkremer, B. Matsuhiro, and J. F. Sproviero, *J. Chromatogr.*, **9,** 283 (1962).
324. D. C. Abbott, H. Egan, E. W. Hammond, and J. Thomson, *Analyst (London),* **89,** 480 (1964).
325. L. D. Bergel'son, E. V. Dyatlovitskaya, and V. V. Voronkova, *Proc. Acad. Sci. USSR, Chem. Sect. (Engl. Transl.),* **141,** 1076 (1961).
326. H. Seiler and H. Erlenmeyer, *Helv. Chim. Acta,* **47,** 264 (1964).
327. E. C. Bate-Smith and R. G. Westall, *Biochim. Biophys. Acta,* **4,** 427 (1950).
328. E. A. Wallington, *J. Med. Lab. Tech.*, **22,** 220 (1965).
329. H. G. Henkel and W. Ebing, *J. Chromatogr.*, **14,** 283 (1964).
330. J. A. Petzold, W. Camp, Jr., and E. R. Kirch, *J. Pharm. Sci.*, **52,** 1106 (1963).

331. D. C. Abbott, H. Egan, and J. Thomson, *J. Chromatogr.*, **16,** 481 (1964).
332. R. K. Tripathi and G. Bhaktavatsalam, *J. Chromatogr.*, **87,** 283 (1973).
333. T. Salo, K. Salminen, and K. Fiskari, *Z. Lebensm. Untersuch.-Forsch.*, **117,** 369 (1962).
334. M. F. Kovacs, Jr., *J. Assoc. Off. Agric. Chem.*, **46,** 884 (1963).
335. *Ibid.*, **48,** 1018 (1965).
336. E. F. L. J. Anet, *J. Chromatogr.*, **9,** 291 (1962).
337. V. E. Vaskovsky and V. I. Svetashev, *J. Chromatogr.*, **65,** 451 (1972).
338. R. Maes, M. Gijbels, and L. Lauruelle, *J. Chromatogr.*, **53,** 408 (1970).
339. T. Bicán-Fišter and V. Kajganović, *J. Chromatogr.*, **11,** 492 (1963).
340. J. Baeumler and S. Rippstein, *Helv. Chim. Acta,* **44,** 2208 (1961).
341. J. P. Peyre and M. Reynier, *Ann. Pharm. Fr.*, **27,** 749 (1969).
342. J. Kloubek and A. Marhoul, *Collect. Czech. Chem. Commun.*, **28,** 1016 (1963).
343. T. Katayama and H. Ikeda, *Sci. Pap. Inst. Phys. Chem. Res., Tokyo,* **60,** 85 (1960).
344. R. Neidlein, H. Kruell, and M. Meyl, *Dtsch. Apoth.-Ztg.*, **105,** 481 (1965).
345. R. Stephan and J. G. Erdman, *Nature,* **203,** 749 (1964).
346. D. Waldi, *J. Chromatogr.*, **18,** 417 (1965).
347. J. Baddiley, J. Buchanan, R. E. Handschumacher, and J. F. Prescott, *J. Chem. Soc.*, **1956,** 2818.
348. A. R. Mattocks, *J. Chromatogr.*, **27,** 505 (1967).
349. I. S. Bharia, J. Singh, and K. L. Bajaj, *J. Chromatogr.*, **79,** 350 (1973).
350. J. Vamos, A. Bratner, G. Szaz, and A. Vegh, *Acta Pharm. Hung.*, **38,** 378 (1968).
351. *Ibid.*, **40,** 135 (1970).
352. T. Furukawa, *Nippon Kagaku Zasshi,* **78,** 1185 (1957); through *Chem. Abstr.*, **52,** 13364 (1958).
353. J. C. Kohli, *J. Chromatogr.*, **121,** 116 (1976).
354. G. Wagner, *Pharmazie,* **10,** 302 (1955).
355. H. J. Haas and A. Seeliger, *J. Chromatogr.*, **13,** 573 (1964).
356. B. M. Olive, *J. Assoc. Off. Anal. Chem.*, **56,** 915 (1973).
357. M. F. Kovacs, Jr., *J. Assoc. Off. Agric. Chem.*, **47,** 1097 (1964).
358. F. Geiss, H. Schlitt, F. J. Ritter, and W. M. Weimar, *J. Chromatogr.*, **12,** 469 (1963).
359. L. Fishbein, *J. Chromatogr.*, **24,** 245 (1966).
360. L. Fishbein and J. Fawkes, *J. Chromatogr.*, **22,** 323 (1966).
361. L. F Druding, *Anal. Chem.*, **35,** 1582 (1963).
362. O. Adamec, J. Matis, and M. Galvanek, *Steroids,* **1,** 495 (1963).
363. J. L. McCarthy, A. L. Brodsky, J. A. Mitchell, and R. F. Herrscher, *Anal. Biochem.*, **8,** 164 (1964).
364. R. W. Frei and V. Miketukova, *Mikrochim. Acta,* **1971,** 290.
365. M. A. Nisbet, *Analyst (London),* **1969,** 811.
366. I. Kawashiro and Y. Hosogai, *Shokuhin Eiseigaku Zasshi,* **5,** 54 (1964); *Chem. Abstr.*, **61,** 6262 (1964).

367. C. van Arkel, R. E. Ballieux, and F. L. J. Jordan, *J. Chromatogr.,* **11,** 421 (1963).
368. J. Battaile, R. L. Dunning, and W. D. Loomis, *Biochim. Biophys. Acta,* **51,** 538 (1961).
369. G. Lukas, *Sci. Pharm.,* **30,** 47 (1962).
370. K. B. Jensen, *Acta Pharmacol. Toxicol.,* **9,** 99 (1953).
371. G. R. Duncan, *J. Chromatogr.,* **8,** 37 (1962).
372. J. A. Vinson and J. E. Hooyman, *J. Chromatogr.,* **105,** 415 (1975).
373. R. L. Munier, A. Peigner, and C. Thommegay, *Chromatographia,* **3,** 205 (1970).
374. D. B. Gower, *J. Chromatogr.,* **14,** 424 (1964).
375. M. Ranny, *Veda Vyzk. Prum. Potravin,* **18,** 191 (1968); through *Chem. Abstr.,* **72,** 62550z (1970).
376. H. Sander, *Naturwissenschaften,* **48,** 303 (1961).
377. M. E. Wright and D. G. Satchell, *J. Chromatogr.,* **55,** 413 (1971).
378. J. S. Matthews, *Biochim. Biophys. Acta,* **69,** 163 (1963).
379. M. Ito, S. Wakamatsu, and H. Kawahara, *J. Chem. Soc. Japan, Pure Chem. Sect.,* **75,** 413 (1954); *Chem. Abstr.,* **48,** 13172 (1954).
380. J. M. Finocchiaro and W. R. Benson, *J. Assoc. Off. Anal. Chem.,* **50,** 888 (1967).
381. F. W. H. M. Merkus, "Progress in Inorganic Thin-Layer Chromatography," in *Progress in Separation and Purification,* Vol. 3, E. S. Perry and C. J. Van Oss, Eds., Wiley-Interscience, 1970, 234.
382. R. Neher, "Thin-Layer Chromatography of Steroids," in *Thin-Layer Chromatography,* G. B. Marini-Bettòlo, Ed., Elsevier, Amsterdam, 1964, p. 75.
383. K. Dose and G. Krause, *Naturwissenschaften,* **49,** 349 (1962).
384. M. Dobiasova, *J. Lipid Res.,* **4,** 481 (1963).
385. R. H. Reitsema, F. J. Cramer, N. J. Scully, and W. Chorney, *J. Pharm. Sci.,* **50,** 18 (1961).
386. H. Sheppard and W. H. Tsien, *Anal. Chem.,* **35,** 1992 (1963).
387. R. A. Schwane and R. S. Nakon, *Anal. Chem.,* **37,** 315 (1965).
388. C. O. Tio and S. F. Sisenwine, *J. Chromatogr.,* **48,** 555 (1970).
389. K. M. Prescott and G. S. David, *Anal. Biochem.,* **57,** 232 (1974).
390. A. Breccia and F. Spalletti, *Nature,* **198,** 756 (1963).
391. A. Massaglia and U. Rosa, *J. Chromatogr.,* **14,** 516 (1964).
392. J. Rosenberg and M. Bolgar, *Anal. Chem.,* **35,** 1559 (1963).
393. P. Karlson, R. Maurer and M. Wenzel, *Z. Naturforsch.,* **18,** 219 (1963).
394. P.-E. Schulze and M. Wenzel, *Angew. Chem. Int. Ed. Engl.,* **1,** 580 (1962).
395. M. L. Borke and E. R. Kirch, *J. Am. Pharm. Assoc. Sci. Ed.,* **42,** 627 (1953).
396. B. J. R. Nicolaus, C. Coronelli, and A. Binaghi, *Farmaco Ed. Prat.,* **16,** 349 (1961); *Chem. Abstr.,* **56,** 7428 (1962).
397. *Ibid., Experientia,* **17,** 473 (1961).
398. W. J. Begue and R. M. Kline, *J. Chromatogr.,* **64,** 182 (1972).
399. D. M. Schuurmans, D. T. Duncan, and B. H. Olson, *Cancer Res.,* **24,** 83 (1964).

400. D. Perlman, W. L. Lummis, and H. J. Geiersbach, *J. Pharm. Sci.*, **58,** 633 (1969).
401. L. Tóth and W. Piorr, *Z.-Lebensm.-Forsch.*, **133,** 322 (1967).
402. T. F. Brodasky, *Anal. Chem.*, **35,** 343 (1963).
403. H. Bickel, E. Gaeumann, R. Huetter, W. Sackmann, E. Vischer, W. Voser, A. Wettstein, and H. Zaehner, *Helv. Chim. Acta,* **45,** 1396 (1962).
404. E. Meyers and D. A. Smith, *J. Chromatogr.*, **14,** 129 (1964).
405. G. H. Wagman and J. V. Bailey, *J. Chromatogr.*, **41,** 263 (1969).
406. N. Narasimhachari and S. Ramachandran, *J. Chromatogr.*, **27,** 494 (1967).
407. P. Reusser, *Z. Anal. Chem.*, **231,** 345 (1967).
408. R. M. Kline and T. Golab, *J. Chromatogr.*, **18,** 409 (1965).
409. A. L. Homans and A. Fuchs, *J. Chromatogr.*, **51,** 327 (1970).
410. P. B. Hamilton and C. E. Cook, *J. Chromatogr.*, **35,** 295 (1968).
411. E. Meyers and R. C. Erickson, *J. Chromatogr.*, **26,** 531 (1967).
412. T. Ono and M. Kawasaki, *Bitamin,* **30,** 280 (1964); through *Chem. Abstr.*, **62,** 1957 (1965).
413. F. Kreuzig, *Z. Anal. Chem.*, **255,** 126 (1971).
414. A. Aszalos, S. Davis, and D. Frost, *J. Chromatogr.*, **37,** 487 (1968).
415. V. Betina, *J. Chromatogr.*, **78,** 41 (1973).

Chapter **VIII**

DOCUMENTATION

1 TRACING OR PRESERVING THE LAYER

There are several ways in which the records of the thin-layer results may be kept. Transparent paper may be laid over the plate and tracings made of the spot pattern [1], or in some cases acetate sheets have been used [2, 3]. In the latter case individual sheets may be used for specific groups of spots and several acetate sheets may be put together to give the overall pattern. Transparent polyethylene sheet (0.0015-in. thick; Crystal X Corporation) which is transparent to ultraviolet light has also been used [4]. It is preferable to use a sheet of glass between the paper or plastic sheet, and a light box then assists in seeing the spots [5]. Nybom [6] has developed a code for indicating colors when making tracings.

One of the earlier and simpler techniques [7] is to press a piece of Scotch tape onto the surface (the ends of the tape should project beyond the ends of the plate). The back of the tape is then firmly rubbed with a smooth object so as to bring the adhesive film into close contact with the layer. When the tape is peeled from the plate the upper surface of the layer remains firmly attached to the tape which may be fastened on a card by means of the projecting ends of the tape.

Mottier and Potterat [8], in working with loose layers of aluminum oxide, allowed melted paraffin to flow onto the plate after the chromatogram had been developed. Barrollier [9] poured a solution of 4% collodion containing 7.5% glycerol over the chromatogram. After the solvent had evaporated the resulting film could be peeled from the plate. As a further improvement plastic sprays were developed [10–12]. Lichtenberger [12] found that a 15% water dispersion of vinylidene chloride provided a very stable layer but was slightly yellow in color. On the other hand, a vinyl propionate gave a mechanically weaker layer but was pure white in color. There are now a number of plastic aerosol preparations on the market for spraying thin-layer chromatograms. In using these sprays, a very light coating should be sprayed and allowed to dry in order to keep the spots from running. The layer is then sprayed again to saturate the adsorbent. After drying, a piece of colorless transparent tape is applied to the top surface. The plastic layer along the edge of the plate is immersed in water in order to loosen the layer. After this treatment the layer can be peeled from the supporting surface. The plates cannot be soaked in water if water-soluble compounds are present. Since the bottom surface of the layer still contains loose particles, the layer may be turned over and given a quick light spray with the plastic in order to fix these.

Although it is rather bulky the entire thin-layer plate may be saved by placing another sheet of glass on top and binding the edges of the plates together with tape [13, 14]. In this respect, layers on the plastic sheets have an advantage because they may be covered with another sheet of plastic and pasted directly in a notebook.

2 PHOTOGRAPHIC AND OTHER REPRODUCTION TECHNIQUES

Of course the thin-layer plates can be photographed with either black and white or color film. Brown and Benjamin [15] have used both reflected and transmitted light in photographing their results because the trace contaminants were more readily seen by this procedure. Ulshoeffer and Doss [16] give the details for photographing porphyrin spots in black and white or color. Color photographs of ultraviolet fluorescing compounds have also been used by others [17–20, 23]. The Polaroid camera is especially convenient for this work [21–23] because the results are immediately visible while the chromatogram is still available for further photographing. Heinz and Vitek [23] discuss and illustrate the use of filters with the Polaroid camera for enhancing the contrast in photographing and obtaining proper reproduction of various types of spots on chromatograms. Jackson [24] has given details for recording short-wave ultraviolet absorption by spots on thin-layer chromatograms using a pinhole camera.

The Reprostar camera stand produced by CAMAG contains proper lighting for photographing in either ultraviolet or visible light.

For radioactive compounds, audioradiographs may be obtained by direct contact of the film with the thin layer (Fig. 8.1). Kodak "No-Screen Medical X-ray Film" may be used for this purpose. Exposures of course will vary and may be as short as 1 hr for high-level activity or as

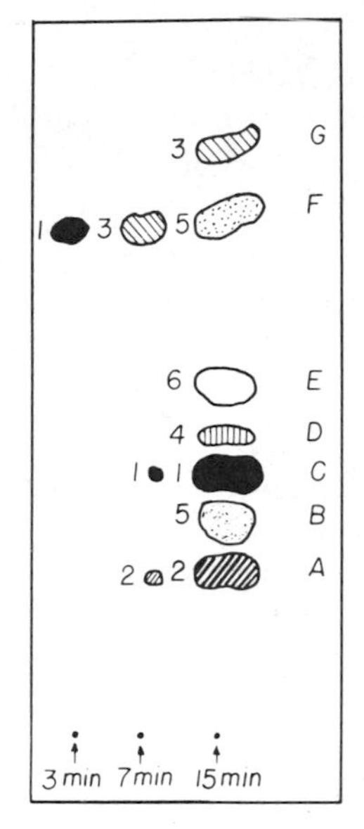

Fig. 8.1 A drawing representing an audioradiograph of chromatogram of oils produced by mint plants exposed to $^{14}CO_2$ atmosphere for 3, 7, and 15 min. Although the spots were quite distinct, the contrast in the spots was inadequate for reproduction, *A*, unidentified spot at R_f 0.23; *B*, menthol; *C*, piperitone; *D*, pulegone; *E*, menthone; *F*, unidentified spot at R_f 0.77; *G*, hydrocarbons. The numbers 1 to 6 indicate relative intensity of the spots, which decreases with increasing number. From Reitsema et al. [25]; reproduced with permission of the authors and the American Pharmaceutical Association.

long as 9 days for weaker activity [25]. Richardson et al. [26] succeeded in differentiating ^{14}C and tritium by placing a shield of cellophane (3.35 mg/cm^2) between the film and the thin layer. This shield absorbs practically all of the tritium activity and passes about 50% of the ^{14}C-activity. Sheppard and Tsien [27] used Kodak nuclear track emulsion type NTB for their photographic plates, and they give directions for coating this emulsion onto glass plates.

Direct prints of various kinds can be made from the thin-layer chromatograms. Copying machines such as the Xerox 914 Office Copier [28] and others may be used. Felici et al. [29] found that the Xerox Model 422 gave the best results. Diazo paper may also be used for copying [30–32]. This is best accomplished by placing a cellophane sheet over the thin layer followed by a sheet of diazo paper and covering the whole with a glass plate. After exposure to light, the diazo print is developed by exposure to ammonia vapors. The cellophane sheet is useful for any copying process where the reagents, which have been sprayed on the plate, may affect the copying medium [33]. McSweeney [34] used photographic bromide paper for making a contact print of ultraviolet-absorbing and fluorescent compounds through the glass supporting plate. This method was used with cellulose, DEAE, or PEI. Palmork [35] has used photographic paper (Kodak Bromide WSG.3S) which was sensitized to ultraviolet light by a bath of less than 2 min in 1% citric acid in ethanol [36]. Hefendehl [37] has made the layers transparent by spraying with a mixture of equal parts of paraffin and ether before obtaining a photoprint. Hassur and Whitlock [38] have used Kodak SWR (short-wave radiation) film or Fine-Grain Positive Film 5302 or 7302 for direct contact photography of unstained nucleic acid bands in polyacrylamide gels. The resulting prints may be scanned for quantitative work. Because of their lower absorption of ultraviolet light, proteins are best stained with Coomassie Brilliant Blue before photographing by this direct contact method. Sprenger [39] makes use of zinc oxide-coated papers which can be given an electrostatic charge. After the papers are charged in the dark room, the finished chromatogram is placed on top and given a brief exposure to a suitable light source. The positive-tone-powder (Philip A. Hunt Co.) is then applied to the paper and clings only where the charge still remains and in proportion to the amount of light received at any given point. The image is fixed by warming at 120°C for 30 sec. Blueprint paper has also been used for making copies of thin-layer chromatograms [40].

3 OTHER METHODS

Jakovljevic and Bishara [41] reported a computerized storage system for keeping thin-layer and paper chromatographic data. This method gave quick access to the data and eliminated the photographing or tracing of chromatograms, thus decreasing the bulk storage.

In thin-layer electrophoresis using starch gels, a satisfactory record of the thin layers may be obtained by plastifying the thin-layer itself [42–47]. In this procedure it is best to use a cellophane support for the preparation of the layer. After staining and rinsing the thin layer the following plasticizing procedure may be used [43]: the gel layer is immersed in absolute methanol for 5 min, but this step may be omitted if the layer has been treated with methanol during the staining process. The gel layer and a cover sheet of cellophane is then immersed in an aqueous solution of 15% glycerol and 2% acetic acid (vol/vol) for 30 min (for acetate buffers pH 5, it is necessary to increase the glycerol content to 20%). The gel and cellophane covering sheet are withdrawn from the plasticizing solution and placed upon a sheet of glass which has been wet with the plasticizing solution. The sheet is smoothed out to remove all air bubbles and the edges of the covering sheet are held down by a weight to prevent curling. Drying is carried out in a warm air stream not exceeding 50°C. This procedure produces a dry transparent record which may be trimmed and conveniently filed. Although there is a slight contraction during drying this is uniform and does not interfere with interpretation of the pherogram. Acrylamide gels may also be dried for preservation [48, 49]. Here again shrinkage is uniform in both directions, and the film can be rehydrated to its original dimensions.

Criddle et al. [50] propose the use of freeze drying for electropherograms in order to remove the moisture. Freezing is accomplished by simply placing the electropherogram for 10 to 20 sec on dry ice, after which the moisture can be removed under vacuum in a desiccator fitted with a cold condenser. This procedure prevents the migration of zones during the drying period and also helps to preserve sensitive compounds.

References

1. C. G. Honegger, *Helv. Chim. Acta,* **45,** 281 (1962).
2. J. S. Amenta, *J. Chromatogr.,* **11,** 263 (1963).
3. R. W. Scora, *J. Chromatogr.,* **13,** 251 (1964).
4. J. E. Ciardi and E. P. Anderson, *Anal. Biochem.,* **22,** 398 (1968).
5. H. H. Berlet, *J. Chromatogr.,* **21,** 485 (1966).
6. N. Nybom, *J. Chromatogr.,* **26,** 520 (1967).
7. J. E. Meinhard and N. F. Hall, *Anal. Chem.,* **21,** 185 (1949).
8. M. Mottier and M. Potterat, *Mitt. Gebiete Lebensm. Hyg.,* **43,** 123 (1952).
9. J. Barrollier, *Naturwissenschaften,* **48,** 404 (1961).
10. P. R. Bhandari, B. Lerch, and G. Wohlleben, *Pharm. Ztg., Ver. Apoth.-Ztg.,* **107,** 1618 (1962).
11. A. S. Csallany and H. H. Draper, *Anal. Biochem.,* **4,** 418 (1962).
12. W. Lichtenberger, *Z. Anal. Chem.,* **185,** 111 (1962).
13. H. Hausser, *Arch. Kriminol.,* **125,** 72 (1960).
14. G. Pastuska and H. J. Petrowitz, *Chem.-Ztg.,* **86,** 311 (1962).
15. T. L. Brown and J. Benjamin, *Anal. Chem.,* **36,** 446 (1964).
16. B. Ulshoeffer and M. Doss, *J. Chromatogr.,* **44,** 407 (1969).
17. W. Specht and A. Stier, *Phot. Korr.,* **100,** 187 (1964).
18. R. Jackson, *J. Chromatogr.,* **20,** 410 (1965).
19. J. F. Gonnet, *J. Chromatogr.,* **86,** 192 (1973).
20. H. L. Kay, "Cinematographie de separations chromatographiques sur couches minces," in *Chromatogr. Electrophor. Symp. Int., 4th, 1966,* Ann Arbor Science, Ann Arbor, Mich., 1968, p. 407.
21. I. D. Jones, L. S. Bennett, and R. C. White, *J. Chromatogr.,* **30,** 622 (1967).
22. E. Hansbury, J. Langham, and D. G. Ott, *J. Chromatogr.,* **9,** 393 (1962).

23. D. E. Heinz and R. K. Vitek, *J. Chromatogr., Sci.,* **13,** 570 (1975).
24. R. Jackson, *J. Chromatogr.,* **29,** 252 (1967).
25. R. H. Reitsema, F. J. Cramer, N. J. Scully, and W. Chorney, *J. Pharm. Sci.,* **50,** 18 (1961).
26. G. S. Richardson, I. Weliky, W. Batchelder, M. Griffith, and L. L. Engel, *J. Chromatogr.,* **12,** 115 (1963).
27. H. Sheppard and W. H. Tsien, *Anal. Chem.,* **35,** 1992 (1963).
28. J. Hilton and W. B. Hall, *J. Chromatogr.,* **7,** 266 (1962).
29. R. Felici, E. Franco, and M. Cristalli, *J. Chromatogr.,* **90,** 208 (1974).
30. F. Eisenberg, Jr., *J. Chromatogr.,* **9,** 390 (1962).
31. H. E. Sprenger, *Z. Anal. Chem.,* **199,** 241 (1963).
32. B. B. Zeitman, *J. Lipid Res.,* **5,** 628 (1964).
33. H. R. Getz and D. D. Lawson, *J. Chromatogr.,* **7,** 266 (1962).
34. G. P. Sweeney, *J. Chromatogr.,* **33,** 548 (1968).
35. H. Palmork, *Acta Chem. Scand.,* **17,** 1456 (1963).
36. *Handbook of Chemistry and Physics,* Chemical Rubber Publishing Co., Cleveland, Ohio, 1951–52, p. 2774.
37. F. W. Hefendehl, *Planta Med.,* **8,** 65 (1960).
38. S. M. Hassur and H. W. Whitlock, Jr., *Anal. Biochem.,* **62,** 609 (1974).
39. H. E. Sprenger, *Z. Anal. Chem.,* **199,** 338 (1963).
40. N. S. Radin, *J. Lipid Res.,* **6,** 442 (1965).
41. I. M. Jakovljevic and R. H. Bishara, *J. Chromatogr.,* **110,** 398 (1975).
42. E. W. Baur, *J. Lab. Clin. Med.,* **61,** 166 (1963).
43. E. W. Baur, *Nature,* **202,** 520 (1964).
44. P. Berkěs-Tomašević, J. Rosić, and M. Ignjatović, *Arhiv Farm. (Belgr.),* **13,** 9 (1963).
45. W. G. Dangerfield, *Nature,* **202,** 520 (1964).
46. J. Groulade, J. N. Fine, and C. Ollivier, *Nature,* **191,** 72 (1961).
47. J. Groulade and C. Ollivier, *Ann. Biol. Clin. (Paris),* **18,** 595 (1960).
48. S. Raymond and L. Weintraub, *Science,* **130,** 711 (1959).
49. R. Quast, *Z. Klin. Chem. Klin. Biochem.,* **9,** 175 (1971).
50. W. J. Criddle, G. J. Moody, and J. D. R. Thomas, *J. Chromatogr.,* **18,** 530 (1965).

Chapter **IX**

REPRODUCIBILITY OF R_f VALUES

Kirchner et al. [1, 2] in 151 first showed that by careful control of conditions (standardization), the reproducibility of R_f values could be held to ±0.05. Two of the more important factors for control of the R_f values are the careful preparation of the layer so as to have uniform thickness and control of the activity of the layer by standardized drying conditions and handling of the plate. They pointed out the sensitivity of silicic acid layers to the moisture of the atmosphere. Because of these various factors that can affect the R_f values, the need for running standard compounds along with the unknown material was *clearly* pointed out at that time. Other workers have confirmed the variability of R_f values with a variation in the thickness of the layer [3–14]. Yamamoto and Furukawa [12] and Furukawa [5] very nicely demonstrated the effect of thickness of the layer by making wedge-shaped layers which were thicker on one end than on the other. By keeping the solvent travel distance constant so as to eliminate this factor, they demonstrated the effect of layer thickness on R_f value. Table 9.1 (values from Pataki and Keller [8]) shows the nature of the R_f variation with layer thickness. Dallas [15] pointed out that the heat of adsorption would be greater and less readily dissipated in thicker layers than in thin layers, and this would have an effect on the R_f value.

The amount of moisture in the adsorbent, which of course is directly related to its activity, has a very pronounced effect on R_f values. Tables 9.2 and 9.3 illustrate the effect of the activity of the adsorbent on the R_f values [6, 16]. The effect of moisture is so great that care must be taken in handling the plates not to expose them too long to the atmosphere, especially if they have been dried at other than atmospheric conditions. Dallas [4] found that "more than half the total amount of moisture adsorbed by silica gel at equilibrium (in an atmosphere of about 50% relative humidity) was taken up within about 3 min, and that even breathing on a plate during the spotting process could markedly affect R_f values." Geiss and Schlitt [17] have shown that aluminum oxide layers exposed

Table 9.1 Variation in R_f Value with Different Layer Thickness[a,b]

	0.25 *mm*		0.5 *mm*		0.75 *mm*		1 *mm*	
Dye	R_f	SR_f	R_f	SR_f	R_f	SR_f	R_f	SR_f
Indophenol	0.056	0.007	0.059	0.004	0.074	0.009	0.078	0.007
Sudan red	0.142	0.008	0.152	0.008	0.171	0.009	0.185	0.011
Butter yellow	0.395	0.011	0.402	0.014	0.419	0.016	0.432	0.014

[a] From G. Pataki and M. Keller [8]; reproduced with permission of the authors and Verlag Helvetica Chimica Acta.
[b] Average R_f from 80 individual runs.

to the atmosphere with a relative humidity of 65% gained 2% moisture within 6 min and 3% moisture within a period of 20 min. Although the moisture pickup depends on the atmospheric conditions, it is best not to expose the layers for too long a period prior to running the chromatogram. (For ways to minimize this exposure when spotting the sample see Chapter IV).

Geiss and Schlitt [17] have also shown that the R_f values of a substance can be varied by as much as 300%, between development in an atmosphere at 1% humidity and an atmosphere of 80% relative humidity. Because of this result Geiss et al. [18] carried out an extensive investigation of the effect of moisture on the R_f values of polyphenyl compounds, and of the resulting effect on the resolution of mixtures of these compounds. A special chamber was designed in order to be able to carefully control the temperature and humidity during development. In contrast to these results Badings [19] found that under humid conditions a series of 2,4-dinitrophenylhydrazones could not be separated.

Table 9.2 Effect of the Activity of an Adsorbent (Aluminum Oxide) on the $R_f \times 100$ Value of Dyes[a]

	Activity According to Brockmann and Schodder			
Azo Dye	II	III	IV	V
Azobenzene	59	74	85	95
p-Methoxyazobenzene	16	49	69	89
Sudan yellow	1	25	57	78
Sudan red	0	10	33	56
p-Aminoazobenzene	0	3	3	19

[a] From S. Heřmánek et al. [6]; reproduced with permission of the authors and VEB Verlag Volk und Gesundheit.

Table 9.3 Comparison of $R_f \times 100$ Values of compounds on Silica Gel Layers Dried under Different Conditions[a]

		R_f Values[b] ($\times$ 100)		
Compound	*Solvent*	*A*	*B*	*C*
Indophenol	Benzene	5.0	3.6	2.2
Sudan Red G	Benzene	13.0	11.6	10.4
Dimethylazobenzene	Benzene	37.0	34.2	32.7
PTH–glycine[c]	Chloroform	12.5	11.0	9.7
Methylamine·HCl	*n*–BuOH–AcOH–H_2O (4:1:1)	16.2	12.5	10.2
Carbobenzoxyglycine	*n*–PrOH–H_2O (7:3)	59.9	53.5	51.1
Carbobenzoxyalanine	*n*–PrOH–H_2O (7:3)	61.4	55.2	53.1
Glycine	Phenol–H_2O 75:25 wt/wt)	22.4	20.0	18.6

[a] From J. Kelemen and G. Pataki [16]; reproduced with permission of the authors and Springer-Verlag.

[b] On silica gel G, (*A*) dried overnight in the air; (*B*) warm-air-dried, then 30 min at 110°C; (*C*) warm-air-dried, then 30 min at 120°C.

[c] PTH = Phenylthiohydantoin.

Numerous investigators have examined the effect of humidity on R_f values [4, 20–30] and found that in order to have reproducible values the humidity must be kept constant.

Another factor that affects the R_f value is the degree of saturation or lack of saturation of the developing chamber atmosphere with the solvent vapors [4, 7, 31–40]. This arises from the fact that if the chamber is not saturated, then evaporation of the solvent from the layer takes place and in effect increases the amount of solvent that travels over the layer. Consequently, R_f values in an unsaturated atmosphere will be higher than those in a saturated atmosphere; however, if conditions are kept standardized reproducible R_f values can be obtained in unsaturated chambers [1, 41]. Honegger [7], in checking the effect of the saturation of the chamber, the separation temperature, and the activity of the adsorbent (silica gel G) on the R_f values of increasingly thicker layers, found that the activity of the adsorbent influenced the separation the most.

The effect of temperature on the R_f value in thin-layer chromatography has been investigated [4, 7, 12, 42–45] and has not been found to have a very great effect except in some cases. When it does occur, the effect is an increase in R_f value with increasing temperature, and this is probably because of the increased evaporation from the thin layer, again causing an increase in the amount of solvent that travels over the layer. This effect of course would be greater for unsaturated or only partly saturated chambers and should be scarcely noticeable in a fully saturated chamber which is carefully sealed against loss of solvent vapors.

The effect of the distance from the solvent level to the point of spot application on the R_f value depends on the adsorbent used, the compounds being separated, and the solvent system in use [11, 35, 44, 50]. Furukawa [44] has shown that with a two-component solvent system in separating a mixture composed of higher-R_f-value components and low-R_f-value components, this distance does have an effect on the R_f value. This is because there is actually a separation of the components of the solvent on the layer, and because the higher-R_f components which travel with the less polar solvent are not affected by the distance of the spot above the solvent level, but the lower-R_f components which are moved by the solvent mixture are affected by the spotting distance from the solvent level. The higher the sample is spotted, the longer it will take the solvent mixture to reach the spot and start moving it, because of the separation effect of the layer on the solvent mixture.

Naturally the total distance that the solvent travels will affect the R_f value [10, 35, 50], because the greater the distance the solvent travels, the greater will be the distance through which the spots are moved.

The size of the sample applied will have an effect on the R_f value and this again will vary depending on the type of compounds present [1, 3, 42, 43, 51–53]. With steroids, Čĕrný et al. [53] found that R_f values were independent of concentration between the values of 50 and 200 μg, but below 50 μg the R_f values were affected by the concentration. Since the effect varies in different situations, each case must be considered by itself [42]. Another closely related factor that can have an effect is that of applying the sample either as one application or as a series of separate applications allowing the solvent to evaporate between applications [51]. A series of applications in one spot causes some radial chromatography to take place and can affect both the shape and the R_f value of a spot. In applying the sample it is best to use as nonpolar a solvent as possible under the circumstances. The use of alcohol as a sample solvent in the separation of terpenes [1] has been found not only to affect the R_f value but also to cause streaking of the spots. Impurities and multicomponents in the sample can affect the R_f value so that this value may not be exactly the same as the R_f value of the pure compound run as a standard.

The particle size of the adsorbent has a considerable influence upon the R_f value, a finer particle having a tendency to increase the R_f value

[11, 55–58]. Ebing [57] recommends using sieved adsorbents covering a range of only 10 μm somewhere in the 20- to 60-μm size.

The R_f value is affected not only by the particle size, but also by the specific surface area and pore diameter [58–61]. For reproducible R_f values it is important to use a narrow pore distribution and a narrow particle size distribution [61]. Adsorbents from different sources vary in particle size, pore volume and diameter, surface area, particle size distribution, and the number of active sites. Even different batches from the same source have variations. All these factors affect the R_f values, and it is little wonder that the R_f values vary from laboratory to laboratory. French et al. [62] and Rábek [63] found inversions of R_f values when using precoated layers from different sources.

Landmark et al. [64] investigated the effect of adding scintillators to the adsorbent. Anthracene displaced the R_f values a little, whereas zinc silicate altered the R_f values as well as the resolving properties of the layer.

Depending on the solvents used, their reuse can affect the R_f value. This can be caused by a number of factors; in the case of volatile solvents it can be caused by the loss of the more volatile components [65], or it can be due to dilution of the solvent with volatile components of the mixture being separated [1]. Even the less volatile developing mixtures can have their composition changed by selective adsorption of one of the components on the thin layers.

One factor that must be taken into consideration in reporting R_f values is the method of development; that is, ascending technique will give a different R_f value than a descending one or a horizontal development.

Davídek and Janíček [66] investigated the effect of paraffin-oil concentration in impregnated starch layers on the R_f values and arrived at a figure of 10% as optimum for the separation of oil-soluble dyes. This of course could vary with different compounds and different solvents. These authors also investigated the effect of drying time in removing the impregnating solvent (petroleum ether) on the R_f values and concluded that 20 min at room temperature was adequate.

As far as the actual reproducibility of R_f values is concerned in thin-layer chromatography, Brenner et al. [42] have shown by statistical analysis that the R_f values obtained in thin-layer chromatography are as reproducible as those obtained in paper chromatography. This of course holds only if careful attention is given to all the details involved so that the conditions under which the thin-layer chromatograms are run are as identical as possible in order to avoid those factors which have been mentioned above.

Dhont et al. [67–69] have shown that by using two reference standards and the equation of Galanos and Kapoulas [70] the R_f values can be corrected where conditions are not completely standarized.

Smith et al. [71] have listed the details which they feel should be given whenever R_f values are published in order for other workers to be able to duplicate the work. These include the following:

1. Make and dimensions of the developing tank.
2. How the tank was sealed.
3. Whether the tank was lined for saturation.
4. If saturated with solvent vapor, the conditions for saturating.
5. Temperature of the run.
6. Humidity and relative humidity.
7. Supplier of thin-layer materials.
8. Precoated or home-made layers.
9. Thickness of layer, grams of adsorbent per milliliter of water or other coating solvent, backing for the layer.
10. Binders, fluorescent agents, buffers, impregnating agents.
11. How layer was activated.
12. Technique used, for example, descending, ascending.
13. Sample applicator used to apply sample.
14. Spotting distance from edge of plate.
15. Number of plates per tank.
16. Running time and distance of solvent travel.
17. Solvent composition and purity of solvents. Any purification methods used.
18. Size of spot or streak.
19. Sample solvent.
20. Amount of sample applied.
21. How applied, such as in open air, under nitrogen, covered by glass plate.

It must be remembered that even though a spot agrees in R_f value with that of a known compound, it is only an indication that it could be identical and confirmation must be made by other means.

References

1. J. G. Kirchner, J. M. Miller, and G. J. Keller, *Anal. Chem.*, **23,** 420 (1951).
2. J. M. Miller and J. G. Kirchner, *Anal. Chem.*, **25,** 1107 (1953).
3. L. Birkofer, C. Kaiser, H. A. Meyer-Stoll, and F. Suppan, *Z. Naturforsch.*, **17B,** 352 (1962).

4. M. S. J. Dallas, *J. Chromatogr.*, **17**, 267 (1965).
5. T. Furukawa, *Nippon Kagaku Zasshi*, **80**, 45 (1959); *Chem. Abstr.*, **54**, 4107 (1960).
6. S. Hěrmánek, V. Schwarz, and Z. Čekan, *Pharmazie*, **16**, 566 (1961).
7. C. G. Honegger, *Helv. Chim. Acta*, **46**, 1772 (1963).
8. G. Pataki and M. Keller, *Helv. Chim. Acta*, **46**, 1054 (1963).
9. E. J. Shellard, *Res. Develop. Ind.*, **1963**, No. 21, 30.
10. E. Stahl, G. Schroeter, G. Kraft, and R. Renz, *Pharmazie*, **11**, 633 (1956).
11. L. Stárka and R. Hampl, *J. Chromatogr.*, **12**, 347 (1963).
12. K. Yamamoto and T. Furukawa, *J. Fac. Educ. Hiroshima Univ.*, **4**, 37 (1956).
13. M. N. Shcherbakova, *Tr. 1-go (Pervogo) Mosk. Med. Inst.*, **61**, 197 (1968); through *Chem. Abstr.*, **71**, 129123n (1969).
14. F. Geiss, *J. Chromatogr.*, **33**, 9 (1968).
15. M. S. J. Dallas, *J. Chromatogr.*, **33**, 193 (1968).
16. J. Kelemen and G. Pataki, *Z. Anal. Chem.*, **195**, 81 (1963).
17. F. Geiss and H. Schlitt, *Naturwissenschaften*, **50**, 350 (1963).
18. F. Geiss, H. Schlitt, F. J. Ritter, and W. M. Weimar, *J. Chromatogr.*, **12**, 469 (1963).
19. H. T. Badings, *J. Chromatogr.*, **14**, 265 (1964).
20. S. Sandroni and F. Geiss, *Chromatographia*, **2**, 165 (1969).
21. K. Chmel, *J. Chromatogr.*, **97**, 131 (1974).
22. R. A. De Zeeuw, *J. Chromatogr.*, **33**, 227 (1968).
23. F. Geiss and M. T. Van der Venne, *Int. Symp. Chromatogr., Electrophor., Lect. Pap., 4th 1966,* Ann Arbor Science, Ann Arbor, Mich., 1968, p. 153.
24. W. L. Reichel, *J. Chromatogr.*, **26**, 304 (1967).
25. F. Geiss, H. Schlitt, and A. Klose, *Z. Anal. Chem.*, **213**, 21 (1965).
26. H. W. Prinzler and H. Tauchmann, *J. Chromatogr.*, **29**, 142 (1967).
27. J. Pitra, *J. Chromatogr.*, **33**, 220 (1968).
28. C. Versino, L. Fogliano, and F. Giaretti, *Riv. Combust.*, **20**, 527 (1966); through *Chem. Abstr.*, **66**, 72254w (1967).
29. H. W. Peter and H. U. Wolf, *J. Chromatogr.*, **82**, 15 (1973).
30. F. Geiss, H. Schlitt, and A. Klose, *Z. Anal. Chem.*, **213**, 331 (1965).
31. C. G. Honegger, *Helv. Chim. Acta*, **46**, 1730 (1963).
32. B. P. Lisboa and E. Diczfalusy, *Acta Endocrinol.*, **40**, 60 (1962).
33. G. Rouser, G. Kritchevsky, D. Heller, and E. Lieber, *J. Am. Oil Chem. Soc.*, **40**, 425 (1963).
34. I. Sankoff and T. L. Sourkes, *Can. J. Biochem. Physiol.*, **41**, 1381 (1963).
35. E. J. Shellard, *Lab. Pract.*, **13**, 290 (1964).
36. D. Jaenchen, *J. Chromatogr.*, **33**, 195 (1968).
37. A. Lamotte, *Bull. Soc. Chim. Fr.*, **1971**, 1509.
38. H. Naghizadeh-Nouniaz and A. Lamotte, *Bull. Soc. Chim. Fr.*, **1971**, 1515.
39. F. Geiss and H. Schlitt, *Chromatographia*, **1**, 387 (1968).
40. R. A. De Zeeuw, H. Compaan, F. J. Ritter, J. H. Dhont, C. Vinkenborg, and R. P. Labadie, *J. Chromatogr.*, **47**, 382 (1970).
41. R. De Zeeuw, *J. Chromatogr.*, **33**, 222 (1968).

42. M. Brenner, A. Niederwieser, G. Pataki, and A. R. Fahmy, *Experientia,* **18,** 101 (1962).
43. N. J. Cone, R. Miller, and N. Neuss, *J. Pharm. Sci.,* **52,** 688 (1963).
44. T. Furukawa, *J. Sci. Hiroshima Univ. Ser.,* **A21,** 285 (1958); *Chem. Abstr.,* **53,** 809 (1959).
45. J. G. Kirchner and V. P. Flanagan, 147th Meeting of the American Chemical Society, April, 1964, Philadelphia, Pa.
46. H. E. Berlet and A. Voelkl, *J. Chromatogr.,* **71,** 376 (1972).
47. G. H. Pastuska, R. Krueger, and V. Lehmann, *Chem.-Ztg.,* **95,** 414 (1971).
48. F. Geiss and H. Schlitt, *J. Chromatogr.,* **33,** 208 (1968).
49. E. J. Singh and L. L. Gershbein, *J. Chromatogr., Sci.,* **8,** 162 (1970).
50. T. Furukawa, *J. Fac. Educ. Hiroshima Univ.,* **3,** 53 (1955).
51. D. C. Abbott, H. Egan, E. W. Hammond, and J. Thompson, *Analyst (London),* **89,** 480 (1964).
52. M. Brenner and A. Niederwieser, *Experientia,* **16,** 378 (1960).
53. V. Cěrný, J. Joska, and L. Lábler, *Collect. Czech. Chem. Commun.,* **26,** 1658 (1961).
54. K. R. Brain, *J. Chromatogr.,* **75,** 124 (1973).
55. J. Vaedtke, A. Gajewska, and A. Czarnocka, *J. Chromatogr.,* **12,** 208 (1963).
56. K. Yamamoto, T. Furukawa, and M. Matsukura, *J. Fac. Educ. Hiroshima Univ.,* **5,** 77 (1957).
57. W. Ebing, *J. Chromatogr.,* **44,** 81 (1969).
58. H. Halpapp, *J. Chromatogr.,* **78,** 77 (1973).
59. A. Waksmundzki and Różylo, *Ann. Univ. Mariae Curie-Sklodowska Lublin-Polonia Sect. AA., 1965,* **20,** 93 (1967); through *Chem. Abstr.,* **69,** 49053h (1968).
60. A. Waksmundzki and J. Różylo, *J. Chromatogr.,* **33,** 90 (1968).
61. M. Vanhaelen, *Ann. Pharm. Fr.,* **26,** 565 (1968).
62. W. N. French, F. Matsui, and S. J. Smith, *J. Chromatogr.,* **86,** 211 (1973).
63. V. Rábek, *J. Chromatogr.,* **33,** 186 (1968).
64. L. H. Landmark, A. K. Hognestad, and S. Prydz, *J. Chromatogr.,* **46,** 267 (1970).
65. J. Battaile, R. L. Dunning, and W. D. Loomis, *Biochim. Biophys. Acta,* **51,** 538 (1961).
66. J. Davídek and G. Janíček, *J. Chromatogr.,* **15,** 542 (1964).
67. J. H. Dhont, C. Vinkenborg, H. Compann, F. J. Ritter, R. P. Labadie, A. Verweij, and R. A. De Zeeuw, *J. Chromatogr.,* **71,** 283 (1972).
68. *Ibid.,* **47,** 376 (1970).
69. J. H. Dohnt and G. J. C. Mulders-Dijkman, *Chem. Tech. (Amsterdam),* **24,** 761 (1969); through *Chem. Abstr.,* **72,** 93635k (1970).
70. D. S. Galanos and V. M. Kapoulas, *J. Chromatogr.,* **13,** 128 (1964).
71. I. Smith, A. D. Baitsholts, A. A. Boulton, and K. Randerath, *J. Chromatogr.,* **82,** 159 (1973).

Chapter X

PREPARATIVE CHROMATOGRAPHY

1 USE OF COLUMNS

The first example of the application of thin-layer techniques to preparative work was by Miller and Kirchner in 1951 [1]. In this case round or square bars of adsorbent held together with plaster of Paris were formed around glass rods. The glass rods thus added mechanical stability to the envelope-free columns. After spotting the sample on one end of the column, the column was pressed into a loose bed of calcium sulfate which in turn was supported by a special distributor.

The column was developed by ascending capillary attraction. Figure 10.1 illustrates the setup for developing this "chromatobar." Columns as large as 4 in.^{2} have been made without difficulty. Since the columns contain no glass envelope the bar made without difficulty. Since the columns contain no glass envelope the bar may easily be removed from the development and sprayed with a locating agent in order to check the development. The surface that has been sprayed may then be scraped to remove the spraying reagent, and the bar once more returned to the solvent for further development. At the conclusion of the run after the column is sectioned with a coping saw, the surface may be scraped free of locating reagent to prevent contamination of the sample. Figure 10.2 shows the chromatobar separation of some terpenes. This technique has been used by a number of other workers [2–5].

Miller and Kirchner in 1952 [6] also introduced another application of thin-layer chromatography to preparative work which is especially useful where very large quantities of material must be handled in order to isolate

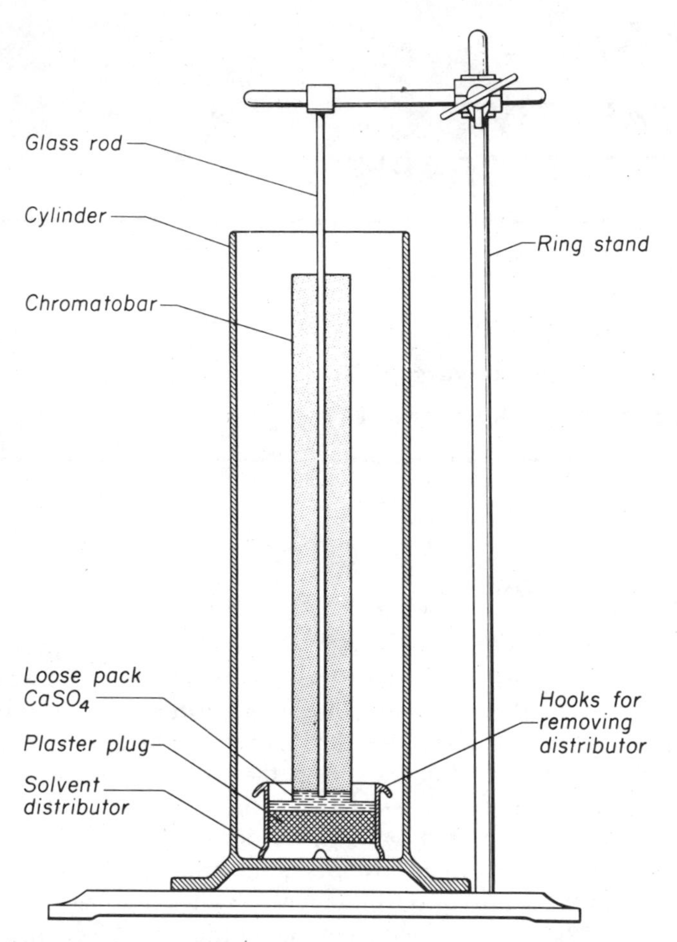

Fig. 10.1 Apparatus for developing chromatobar (see text for description). From J. M. Miller and J. G. Kirchner [1]; reproduced with permission of the American Chemical Society.

several smaller fractions. This procedure consists in using thin layers following the separation of components on chromatographic columns. Prior to running the column, the best solvent and the best adsorbent are determined by preliminary checks on chromatostrips (thin layers on narrow glass strips). Effluent from the column itself is collected in suitable fractions by means of a fraction cutter. After every fifth fraction or so, the efficiency of the separation is checked by spotting a small amount of the eluate onto a chromatostrip and developing in the same solvent as used for the column. These samplings then determine the location of

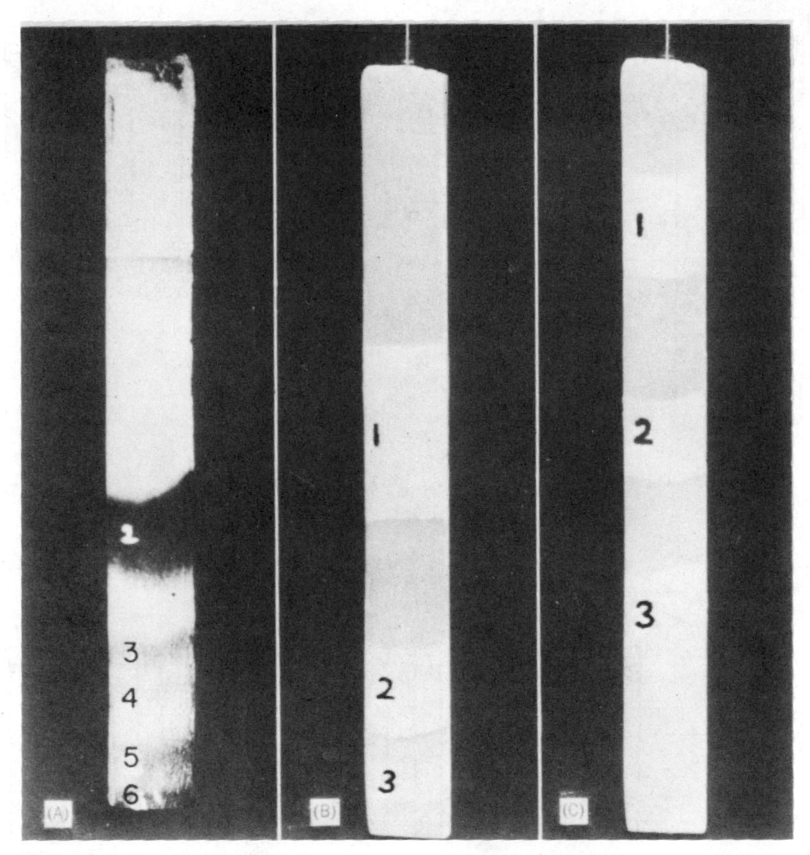

Fig. 10.2 Some examples of separations on chromatobars. (*A*) Chromatograph of crude isoeugenol on fluorescent silicic acid chromatobar using hexane–ethyl acetate (85:15). (1) solvent front, (2) isoeugenol, (3–6) unknown impurities. (*B*) Separation of limonene (1), terpinyl acetate (2), and α-terpineol (3) by step development. First development in hexane for 4 in. Second development in hexane–ethyl acetate (85:15). Total solvent travel 10 in. Detection by fluorescein–bromine test [2]. (*C*) Separation of 17 mg α-pinene (1), 10 mg terpinyl acetate (2), 10 mg α-terpineol (3). Detection by fluorescein–bromine test [2]. From J. M. Miller and J. G. Kirchner [1]; reproduced with permission of the American Chemical Society.

materials in the eluate portions. When a fraction which contains more than one compound is found, additional chromatostrips are run on both sides in order to find the starting points of the pure fractions. (Details of this procedure are illustrated in Fig. 10.3.) In order to avoid the typical coning of the chromatographic zones that occurs with the ordinary dry-packed columns, slurry-packed columns should be used for this work [6].

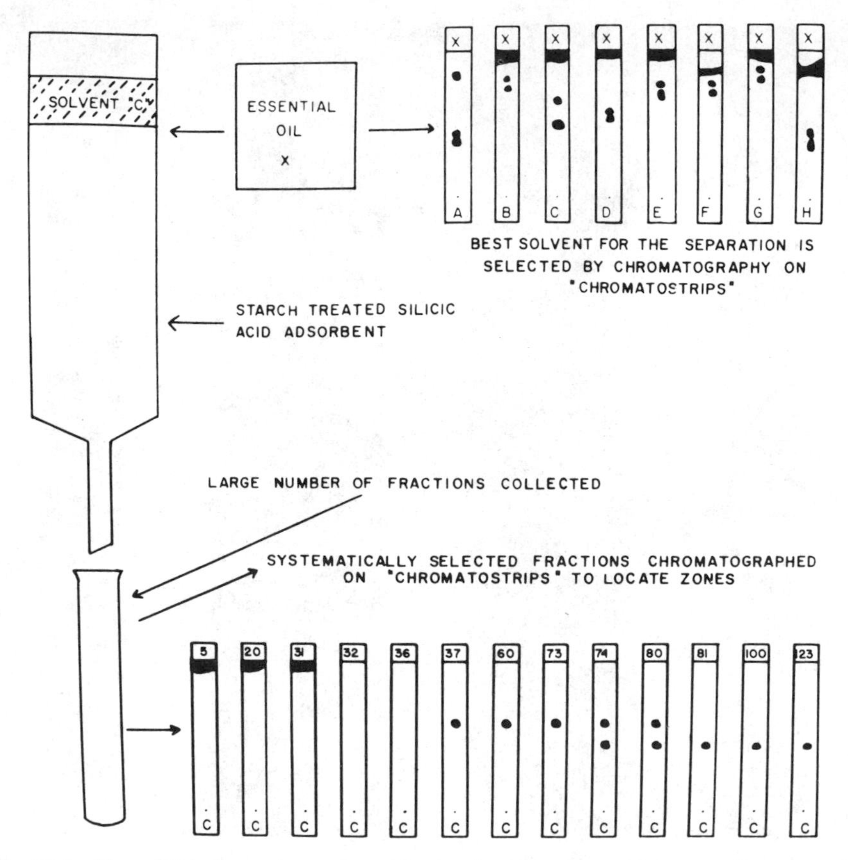

Fig. 10.3 Use of thin-layer chromatography in checking preparative separations on columns. From J. M. Miller and J. G. Kirchner [5]; reproduced with permission of the American Chemical Society.

A somewhat similar procedure has been used by Dahn and Fuchs [7]. In this case the authors used a cellophane tubing to contain the adsorbent material. The column was dry packed and then the sample was added on a small amount of the adsorbent in the same fashion, topped by another short section of dry adsorbent. The column was developed in a horizontal position with transport of the solvent by capillary action. It was sectioned after first locating the zones by viewing in visible or ultraviolet light or by the addition of fluorescent compounds to the adsorbent. In the case of amino acids, the column was first slit open and then a piece of filter paper was pressed against the adsorbent to effect transfer of some of the amino acids to the paper. The latter was then developed with ninhydrin reagent. Sectioned zones could be further checked by thin-layer chromatography.

2 ON REGULAR THIN LAYERS

Thin-layer plates have been applied directly to preparative chromatography. For this purpose thicker layers are usually used. Ritter and Meyer [8] prefer layers 1 mm thick, since thicker layers do not always give as good a resolution. The usual equipment can be used for the preparation of the layers although modifications can be made to the slurry for better results. Honegger [9] recommends that the adsorbent-water ratio be gradually lowered from 1:2 to 1:1.57 for layers 1 to 5 mm thick in the case of silica gel G. For aluminum oxide G, the ratio should be 1:0.9. He also added 2% more plaster of Paris to the silica gel in order to diminish cracking in the thicker layers. The drying time was increased to 60 min and was assisted by infrared radiation from above. The activation time at 110°C was increased from $\frac{1}{2}$ to 24 hr. Even with these precautions approximately 50% of the 5-mm-thick layers developed cracks on drying. Korzun et al. [10], in order to avoid the cracking of thick layers, allowed their plates to set overnight before drying at 80°C in an oven. Mistryukov [11] and Cěrný et al. [12] have used loose layers of alumina for preparative thin-layer work. These authors also describe the use of loose layers of silica gel for preparative work.

Horobin [13] used thin-layer sheets for preparative work, developing in a descending manner. Johansson and Kashemsanta [14] also used a descending technique, but used thin-layer plates. Because of the need in these techniques to run numerous plates to obtain preparative amounts of matierial Von Arx [15] designed tanks and holders for developing as many as 40 1-m plates at one time. Jordan [16, 17] used cylindrical layers which were coated on the outside of large test tubes. Kalis et al. [18] used loose layers 0.7 cm thick contained in a metal box and inclined at an angle in the developing tank, and Clark [19] cast 5-mm layers also in metal boxes. In both the latter cases the layer is accessible from only one side, and the zones progress at an angle rather than perpendicular to the plate. Brendel et al. [20] used a conical layer coated in a glass cone frustum. Stutz et al. [21] after developing two plates simultaneously by clamping them together face to face, eluted the compounds by running the eluting solvent perpendicular to the direction of development. Visser [22] developed an apparatus for continuous preparative thin-layer chromatography in which the thin layer is continuously prepared and dried on a moving belt. As the layer moves along the sample is deposited, the chromatogram developed, the separated compounds eluted, and finally the used adsorbent scraped from the belt.

Bockemueller and Kaiser [23] have used blocks of calcium sulfate as layers for preparative electrophoresis, separating as much as 1 g of material at one time, and Affonso [24–26] has used 1 to 5-mm plates of calcium sulfate for chromatography, but this limits the separation possibilities, because of the adsorption capabilities of the calcium sulfate.

Sample application to the thin layer in preparative work is somewhat more difficult than with normal thin-layer work. A number of methods have been devised to minimize this difficulty. One method is to simply draw the pipet along a straight edge as the sample is allowed to flow onto the plate. However, unless the flow and the rate of movement of the pipet are carefully controlled the sample application is apt to be irregular. Honegger [9] applied his samples to a V-shaped groove, 1 to 2 mm wide and half the depth of the layer. Care must be taken not to remove the adsorbent all the way to the glass plate for this would interfere with the solvent movement. A somewhat different technique has been used by Connolly et al. [27]. Using very light strokes, two cuts are made in the adsorbent clear to the glass plate leaving a ridge of adsorbent approximately 3 mm wide between the knife cuts. After the loose material in the cuts is blown out the solution is applied to this ridge as uniformly as possible with a medicine dropper or by other means. The solvent is allowed to evaporate and then the cuts are filled with dry adsorbent. This is accomplished by placing an aluminum foil mask with a 1-mm slit over the cuts; a spatula is then used to press the dry adsorbent into the cuts. After the foil is removed the chromatogram can then be run in the normal manner without this loose adsorbent falling out.

Halpaap [28] has applied the thin-line technique, previously described under sample application techniques, to preparative plates. Briefly, this is the technique of moving the starting line to a new location about 1 cm above the original application line by using a solvent in which all components readily travel. Then after the solvent is removed, the plate may be developed in the usual manner.

Attention is called to Chapter IV "Sample Application," in which equipment for spotting line samples is described and illustrated.

The quantity of sample that can be applied to a given preparative plate depends on the size of the plate, the thickness of the layer, the adsorbent used, and the specific sample which is to be separated. Honegger [9] applied from 5 to 25 mg of sample per millimeter of layer thickness for a 20 × 20 cm plate of silica gel G. Korzun et al. [10] have applied 100 to 500 mg of compounds to a 10 × 15 in. plate covered with a 1-mm layer.

Halpaap [28] uses plates 20 cm high and 1 m long, and has applied as much as 10 g of a sample to a plate coated with a 1.5-mm-thick layer of silica gel.

De Zeeuw and Wijsbeek [29] have applied vapor programming to preparative layers 1 mm thick.

One problem that does arise with preparative work is the detection of the zones. All the numerous reagents which are so readily applicable to normal thin-layer chromatography cannot be used because of contamination of the sample. Of course if the compounds are colored or if they fluoresce in ultraviolet light there is no problem; fluorescent layers [30] can be used to advantage where the compounds absorb in the ultraviolet region. Compounds which contain a radioactive atom can of course be located by taking a radioaudiograph or by means of a counter [31].

One method which has been used to advantage is to spray silica gel plated with water [32, 33]. The water-moistened plate becomes translucent and the bands show up as white areas. Sharper zones can sometimes be obtained by allowing the saturated plate to dry until the zones become distinct. Iodine has been used as a detecting agent for lipids [34]; however, Nichaman et al. [35] proved that there is a decrease in the unsaturated lipids when iodine is used as a detecting agent. This loss is presumably due to the iodination of the double bonds.

For loose layers, Mistryukov [11] has applied narrow strips of moistened paper against the layer. The strips are then removed and the adhering particles of adsorbent are sprayed with a suitable spray reagent. A somewhat similar technique has been used by Dobiasova [36]. A narrow strip of a plate coated with silicic acid is pressed against the side of the preparative chromatogram in order to obtain a "print"; this is then developed in the usual manner. As an alternative, the major portion of the plate may be covered with a sheet of glass or with a layer of foil in order to isolate one or both edges of the chromatogram that can be sprayed with a suitable reagent. This then gives the general location of the zones, but unless the zones are well separated or are unusually straight, this can lead to some mixing of the bands in the removal process.

Once the zones have been located they may simply be scraped from the plate into a sintered glass funnel and eluted with an appropriate solvent. A number of devices have been made for removing the zones all based on the vacuum suction principle. The first of these was by Mottier and Potterat [37] in 1955 for picking out the spots from loose-layer plates by the application of vacuum to one end of a 6 to 8-mm glass tube with a constriction in it to hold a cotton plug. The adsorbent was thus sucked into the tube and retained by the cotton plug. Glass wool [38, 39] can be substituted for the cotton plug, or an asbestos mat supported by 200-

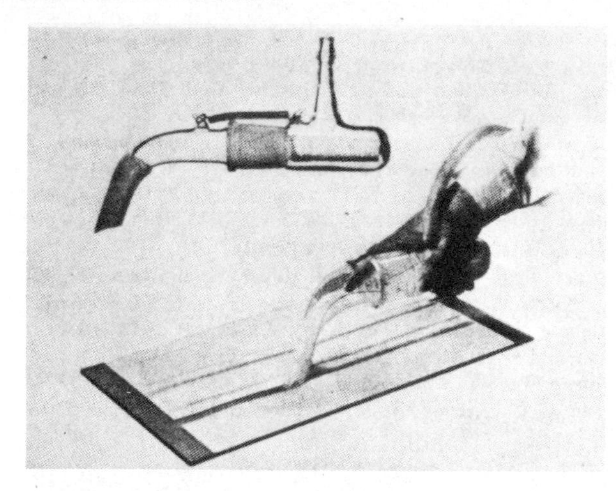

Fig. 10.4 Vacuum pickup for removing thin-layer bands. Courtesy of Brinkmann Instruments.

mesh stainless steel screen may be used [40]. Goldrick and Hirsch [41], Hansson [42], and Matthews et al. [43] have used sintered glass discs to retain the adsorbent. Another modification of this simple apparatus consists in adding a glass tube with a drawn out tip on it so as to increase the suction velocity at the inlet. This may be attached by a short length of plastic or rubber tubing [44].

Hay et al. [45] and Spikner and Towne [46] aspirate the adsorbent spots directly into small test tubes or volumetric flasks respectively which contain an eluting solvent. The solution can then be made up to volume and aliquots taken for analysis.

Ritter and Meyer [8] designed a collector containing a Soxhlet extractor thimble as the receiver for the adsorbent material. The inlet tube was fitted with a piece of plastic tubing so that close contact could be made with the glass plate without scratching the latter. The accumulated material from a number of plates could then be extracted in a Soxhlet apparatus. Equipment of this design is now offered for sale by Desaga in both a micro and macro version (Fig. 10.4).

3 USE OF VERY THICK LAYERS

Kirchner [47] developed a method for the production and development of thick layers ranging from $\frac{1}{8}$ to $\frac{1}{2}$ in. thick for preparative work. These layers do not have a supporting back plate, but are contained in a stainless steel framework with thin stainless steel wires stretched across the frame

Fig. 10.5 Stainless steel frame with parallel supporting wires. From Kirchner [47]; reproduced with permission of the Elsevier Scientific Publishing Co.

Fig. 10.6 Stainless steel frame with diagonal supporting wires. From Kirchner [47]; reproduced with permission of the Elsevier Scientific Publishing Co.

to help support the adsorbent layer (Figs. 10.5, 10.6). These thin wires do not interfere with the separation. Both sides of the layer are available for application of the sample and for observation of the separated zones. Cross sections of developed layers show that application of the sample from both sides coupled with diffusion distribute the zones across even the $\frac{1}{2}$-in.-thick layers. The layers are composed of silica gel with 20% (based on the amount of silica gel present) gypsum as a binder although other adsorbents may be used. A $\frac{1}{4}$-in. plate contains 147 g of adsorbent–binder mixture (when prepared with Mallinckrodt SilicAR) which contrasts with only 4.4 g on an average 250-μ plate. Thus the $\frac{1}{4}$-in. layer is equivalent to 33 normal plates and a $\frac{1}{2}$-in. layer would be equivalent to 66 normal plates. Since the layers are cast in a stainless steel mold machined

Fig. 10.7 Separation of 100 mg of each of six dyes on a $\frac{1}{4} \times 8 \times 8$ in. silica gel self-supporting layer. Top to bottom: Yellow OB, Sudan I, Sudan III, Sudan II, Methyl Red, and Crystal Violet. (All dyes previously purified by thin-layer chromatography). Development details: (1) in unsaturated tank (u.t.) to the top with benzene–chloroform (10:1); (2) in u.t. for 120 mm with same solvent; (3) in u.t. for 120 mm with benzene; (4) in s.t. to the top with benzene–chloroform (10:1); (5) twice in u.t. with benzene to the top; and (6) twice in u.t. for 63 mm with benzene–chloroform mixture with ethanol added in the ratio of 10:3. Adsorbent: Mallinckrodt SilicAR® TLC-7F with 20% (by wt of the adsorbent) gypsum as binder. From Kirchner [47]; reproduced with permission of the Elsevier Scientific Publishing Co.

Fig. 10.8 Purification of 500 mg of Sudan II on $\frac{1}{4} \times 8 \times 8$ in. silica gel layer (Mallinckrodt SilicAR® TLC-7F with 20% gypsum binder) contained in a stainless steel framework. The layer has no backing plate and is accessible from both sides. Developed in an unsaturated tank twice with benzene–chloroform (20 : 1). From Kirchner [47]; reproduced with permission of the Elsevier Scientific Publishing Co.

to exact dimensions the layers are perfectly flat and uniform. These layers are especially useful for preparative work because after both sides are sprayed with a detecting agent and the location of the bands is marked, the thin surface film containing the detection agent may be scraped away so as not to contaminate the eluates. Development is carried out in a special tank that holds the layer in a vertical position to ensure uniform development across the thickness of the layer rather than at an angle, and also to provide a smaller vapor chamber than the normal tanks. Figures 10.7 and 10.8 show the separation of some dyes by this technique.

References

1. J. M. Miller and J. G. Kirchner, *Anal. Chem.*, **23** 428 (1951).
2. B. Balogh, *Anal. Chem.*, **36,** 2498 (1964).
3. G. Gogroef, *Pharmazie*, **12,** 38 (1957).

4. L. D. Hayward, R. A. Kitchen, and D. J. Livingstone, *Can. J. Chem.*, **40,** 434 (1962).
5. B. Balogh, *Anal. Chem.*, **36,** 2498 (1964).
6. J. M. Miller and J. G. Kirchner, *Anal. Chem.*, **24,** 1480 (1952).
7. H. Dahn and H. Fuchs, *Helv. Chim. Acta,* **45,** 261 (1962).
8. F. J. Ritter and G. M. Meyer, *Nature,* **193,** 941 (1962).
9. C. G. Honegger, *Helv. Chim. Acta,* **45,** 1409 (1962).
10. B. P. Korzun, L. Dorfman, and S. M. Brody, *Anal. Chem.*, **35,** 950 (1963).
11. E. A. Mistryukov, *J. Chromatogr.*, **9,** 311 (1962).
12. Cěrný, J. Joska, and L. Lábler, *Collect. Czech. Chem. Commun.*, **26,** 1658 (1961).
13. R. W. Horobin, *J. Chromatogr.*, **37,** 354 (1968).
14. L. Johansson and S. Kashemsanta, *J. Chromatogr.*, **45,** 473 (1969).
15. E. Von Arx, *J. Chromatogr.*, **64,** 297 (1972).
16. D. M. Jordan, *J. Chromatogr.*, **57,** 427 (1971).
17. *Ibid.*, **63,** 442 (1971).
18. V. Kalis, A. S. Karsakevich, and A. Kirsteins, *Latv. PSR Zinat. Akad. Vestis Kim. Ser.*, **1967,** 572.
19. G. W. Clark, *J. Chromatogr.*, **34,** 262 (1968).
20. K. Brendel, R. S. Steele, and E. A. Davidson, *J. Chromatogr.*, **30,** 232 (1967).
21. M. Stutz, W. D. Ludemann, and S. Sass, *Anal. Chem.*, **40,** 258 (1968).
22. R. Visser, *Anal. Chim. Acta,* **38,** 157 (1967).
23. W. Bockemueller and P. Kaiser, *J. Chromatogr.*, **18,** 86 (1965).
24. A. Affonso, *J. Chromatogr.*, **21,** 332 (1966).
25. *Ibid.*, **22,** 452 (1966).
26. *Ibid.*, **27,** 324 (1967).
27. J. P. Connolly, P. J. Flanagan, R. O. Dorchaí, and J. B. Thomson, *J. Chromatogr.*, **15,** 105 (1964).
28. H. Halpaap, *Chem.-Ing.-Tech.*, **35,** 488 (1963).
29. R. A. de Zeeuw and J. Wijsbeek, *Anal. Chem.*, **42,** 90 (1970).
30. J. G. Kirchner, J. M. Miller, and G. J. Keller, *Anal. Chem.*, **23,** 420 (1951).
31. P. E. Schulze and M. Wenzel, *Angew. Chem., Int. Ed. Engl.*, **1,** 580 (1962).
32. E. Campaigne and M. Georgiadis, *J. Org. Chem.*, **28,** 1044 (1963).
33. R. J. Gritter and R. J. Albers, *J. Chromatogr.*, **9,** 392 (1962).
34. D. C. Malins and H. K. Mangold, *J. Am. Oil Chem. Soc.*, **37,** 576 (1960).
35. Z. Nichaman, C. C. Sweeley, N. M. Oldham, and R. E. Olson, *J. Lipid Res.*, **4,** 484 (1963).
36. M. Dobiasova, *J. Lipid Res.*, **4,** 481 (1963).
37. M. Mottier and M. Potterat, *Anal. Chim. Acta,* **13,** 46 (1955).
38. M. Beroza and T. P. McGovern, *Chemist-Analyst,* **52,** 82 (1963).
39. J. Janák, *Nature,* **195,** 696 (1962).
40. M. A. Millett, W. E. Moore, and J. F. Saeman, *Anal. Chem.*, **36,** 491 (1964).
41. B. Goldrick and J. Hirsch, *J. Lipid Res.*, **4,** 482 (1963).
42. J. Hansson, *Explosivstoffe,* **10,** 73 (1963).

43. J. S. Matthews, A. L. Pereda-V., and A. Aguilera-P., *J. Chromatogr,* **9,** 331 (1962).
44. M. Mottier, *Mitt. Geb. Lebensm. Hyg.,* **49,** 454 (1958).
45. G. W. Hay, B. A. Lewis, and F. Smith, *J. Chromatogr.,* **11,** 479 (1963).
46. J. E. Spikner and J. C. Towne, *Chemist-Analyst,* **52,** 50 (1963).
47. J. G. Kirchner, *J. Chromatogr.,* **63,** 45 (1971).

Chapter **XI**

QUANTITATION

There are many factors that can affect the quantitative results obtained by means of thin layers [1], and there are certain parameters that affect all methods whether by direct densitometry, fluorimetry, an elution method, or otherwise. An important factor and a possible source of large errors is in the application of the sample to the thin-layer plate. One source of error is that due to creep back on the tip of the syringe [2–4]; part of the drop curls back around the tip of the syringe and a variable amount remains behind after the drop is discharged. Evaporation takes place concentrating this solution that remains behind, and subsequent drops wash this material off, significantly increasing the concentration in a particular drop. Errors from this source can be minimized by using as fine a tip as possible and by coating it with silicone [4]. Samuels [5] recommends that the shape of the microsyringe needle be modified by grinding a reverse bevel so that the needle is pointed in the center instead of at the side.

This creep-back error cannot be avoided by touching the tip of the needle to the layer, because the capillary attraction withdraws additional fluid from the syringe. Equipment such as the Chromatocharger and Linomat of CAMAG and the Chromatoplot of Burkard Scientific, which eject the drops, avoids both the curl back and the capillary problem.

Using a group of students, Brain [6] investigated the "operator effect" in the application of samples for quantitative work. Under various conditions of use he tested the Microcap (Drummond) (the Microcap is a capillary cut to hold a precise amount) and a repeating dispenser using a Hamilton syringe.

There was a great deal of variation among individual operators. Best results were obtained with the repeating dispenser. With unskilled operators, the mean coefficient of variation was 5.6%, but with skilled operators it was only 3.5%. In contrast to this, the single application of the contents of the Microcap resulted in a mean coefficient of variation of 9.5% with individual coefficients of variation ranging from 1.2% to as high as 53.4%. Attempts to decrease the error with the Microcap by (*a*) rinsing the micropipet, (*b*) applying the charge as a series of five applications instead of one continuous application, or (*c*) expelling the contents by means of a rubber bulb all served to increase the error. Fairbairns and Relph [4] examined the "operator effect" with 10 experienced operators in applying samples with various instruments and found errors as large as ±25%.

The error also varied with the layer material ranging from a mean coefficient of variation of 6.8% for silica gel to 13.7% for polyethylenimine cellulose [6].

Brain summed up his work with the comment, "It is essential that any person needing quantitative sample application should check carefully the errors in the application of samples, by themselves under their experimental conditions, and not simply accept the published figures of other workers."

The sample spot should be as small as possible and should not be overloaded; otherwise streaking will occur. It is also important when making multiple applications that the amount of volume delivered each time be constant so as to obtain spots of equal area [6a]. Great care must be exercised so as not to disturb the surface of the layer because this causes distortion of the shape of the spot [7].

Klaus [8] has used an internal standard to help avoid the problem of accurate sample application.

Another factor to be considered is the variation from plate to plate. Good correlation can be obtained when using a single plate for repeat samples, but variations can occur between different plates. Table 11.1 shows that the deviation between spots measured on the same plate was less than the deviation between spots measured on different plates. Errors of this type can be minimized by running two standards on each plate, one a multiple of the other. In this way, the regression line relating the amount measured in the final spot to the amount applied can be checked.

Table 11.1 Reproducibility of Scanning of Different Spots in the Same and on Different Chromatograms.[a,b]

	On the Same Chromatogram		*On Different Chromatograms*	
Substance	*Peak Area* (mm^2), *Mean Values* ($n = 6$)	s (%)	*Peak Area* (mm^2), *Mean Values* ($n = 6$)	s (%)
PTH–proline[c]	2300	5.8	1820	11.9
DNP–proline[d]	2080	7.4	2400	8.6
DNP–proline[e]	640	4.4	630	5.4
DANS–proline[f]	1550	3	1710	9.1
DANS–proline[g]	2045	6.2	2640	14.4

[a] From Pataki [9]; reproduced with the permission of the author and Friedrich Vieweg & Sohn GmbH.

[b] Time delay between drying and scanning was standardized in each case.

[c] 2 μg; quenching (silica gel F).

[d] 2 μg; quenching (silica gel).

[e] 2 μg; reflectance (silica gel).

[f] 2 μg; fluorescence (silica gel).

[g] 2 μg; fluorescence (silica gel), after spraying with triethanolamine–isopropanol (1:4).

Of course development must be in the same solvent and under the same conditions, in order to avoid any slight variation which could affect the results. The most desirable solvent for quantitative work is a single component solvent that will give an R_f value between 0.25 and 0.75 [10]. Interference may be encountered from impurities carried by the solvent front if the R_f values are too high. Another point to remember is that the higher the R_f value the greater the amount of diffusion of the sample. With mixed solvents, demixing may occur and if the compound travels with the β-front, then lateral diffusion of the spot will occur with a decrease in precision [11]. Also in this case, adsorbent impurities could be carried by the β-front.

The theoretical aspects of quantitative thin-layer chromatography have been discussed by Grimm [12], Boulton and Pollak [13–17], Klaus [18], Seiler and Moeller [19, 20], and Goldman and Goodall [21–23]. Kaiser [24–27] gives a good, easily followed discussion of the calculation of the errors in quantitative analysis, and Pollak and Boulton [28] give a mathematical treatment of the nonlinearity effects in optical scanning. Two books have been published on quantitative thin-layer chromatography [28a, 28b].

1 ELUTION METHODS

Because assay methods were already available for compounds in solution and the techniques and instrumentation had not been developed for the direct thin-layer plate analysis the first quantitative thin-layer work evolved around elution techniques.

In order to obtain satisfactory solutions for analysis, a number of conditions must be observed; these are as follows:

1. The desired component must be sharply separated on the chromatogram from interfering substances.
2. The material must be completely recoverable from the adsorbent or at least consistently recoverable so that a correction can be applied for the loss.
3. The adsorbent must be checked for interfering compounds, and if present, they must be removed prior to the analysis.
4. The chromatographic zones must be detectable by some noninterfering method.
5. Standard samples must be run to check the validity of the analysis.

Several techniques have been introduced to decrease the amount of impurities from the adsorbent in the final eluate. Mulder and Veenstra [29] applied their samples as streaks and used continuous horizontal chromatography to separate the compounds and to move the bands to the top of the plate for concentration by evaporation. (A number of methods for doing this are mentioned under continuous development in Chapter V.) The plates were then turned sideways, and by the use of a narrow strip of paper to transfer solvent to the required areas the bands were concentrated into spots. De Deyne and Vetters [30] used a different technique; after obtaining the desired separation of their spots, they turned the plates through 90° and removed the adsorbent from above the spots. Small squares of ash-free filter paper shaped to a point were then placed in contact with the spots and held in place by a glass plate. The tips of the paper were bent out, away from the glass support. Then by means of horizontal chromatography in the second direction and a paper bridge to apply solvent to the remaining adsorbent, the spots were carried out to the tips of the papers. It was then a simple matter to extract the compounds from the paper. An alternative method was to shape the adsorbent above the spot into a point where the products could be concentrated. Additional methods for removing fine particles of silica gel from eluates are mentioned under infrared techniques in Chapter XII.

Iron in silica gel can interfere especially in colorimetric methods, and this can be removed prior to the preparation of the layers as mentioned in Chapter III, or the iron may be removed from precoated plates by predevelopment in methanol–concentrated hydrochloric acid (9:1) followed by drying and reactivation of the layers.

The recovery of material from the adsorbent is seldom 100%, but Eberle et al. [31] were able to increase their recovery of a pesticide from silica gel layers from 45 to 55% by ordinary extraction to 85 to 95% by exposing the silica gel–solvent slurry to ultrasound.

Loss during chromatography and elution can be compensated for by running different concentrations of standard solutions through the entire process. A graph can be constructed relating the quantity in the applied sample to the results obtained in the assay.

Because there is a possibility of losses when removing spots by scraping, Hastings and Wong [32] scraped the adsorbent away from around the spot, which was then moistened with water. After the plate was placed in a freezer, the spot could be lifted off in one piece. Turchinsky and Shershneva [33] used a similar procedure, but used nitrocellulose solution to fix the spot. An alternative to scraping is to use the Eluchrom apparatus (Fig. 5.12), designed by Falk and Krummen [33a] and marketed by CAMAG. With this equipment elution is carried out directly on the plate as described under automation in Chapter V.

Court [34] has reviewed quantitative thin-layer elution techniques.

Spectrophotometric Methods

The first application of thin-layer chromatography to quantitative analysis was by Kirchner et al. in 1954 [35]. This was an application in the determination of biphenyl in citrus fruits and fruit products and served not only as an analytical tool but also to show the reliability of the thin-layer chromatographic method when conditions and procedures are *carefully standardized.*

Since biphenyl-treated packaging material is used by the citrus industry to prevent molding of citrus fruits during shipping and storage, it is essential that a method be available for determining the amount of biphenyl vapors that have been absorbed by the fruit. Because of the low concentrations present in citrus juice it was necessary to have a method of concentrating the biphenyl material. This was accomplished by distilling the slurry of fruit and water in a modified [35] Clevenger distillation apparatus [36]. During the distillation procedure the biphenyl and the citrus oils were collected in a small volume of heptane, which was then carefully transferred to a volumetric flask and made up to a standard volume. For the thin-layer supports, glass strips 13 × 136 mm were selected with a micrometer so as to be uniform in thickness. The layers of silicic acid containing a starch binder and zinc cadmium sulfide and zinc silicate as fluorescing agents were spread by hand with a spatula using guide strips to regulate the thickness of the layer. Commercial spreading equipment does not give satisfactory layers for this type of work since slight variations in the thickness of the layer cause variations in the background adsorption obtained from the silicic acid layer. Ultraviolet-absorbing background impurities were diminished by predeveloping the activated strips in 95% ethanol which carried ultraviolet-absorbing impurities to the top of the strip. The strips were then dried at 85°C for 4 min in a mechanical convection oven and cooled in a desiccator prior to use. The sample was applied to the chromatostrip by means of a syringe microburet [37] capable of delivering 0.1 μl of solution. (This equipment is now available from general laboratory supply houses.) A water-soluble grease [38] was applied to both the glass joint of the delivery tip and the plunger of the syringe so as to prevent evaporation losses. The strips were developed for a distance of 10 cm with purified petroleum ether (30 to 60°C) giving the biphenyl an R_f value of 0.45 which was well separated from that of the citrus oil hydrocarbons that move close to the solvent front. The biphenyl appearing as a dark spot on the fluorescent strip under ultraviolet light was marked, and a standard area, 22 mm long and the width of the strip, with the biphenyl spot in the center of the area, was transferred to a sintered glass funnel by scraping with a spatula. Ninety-five percent alcohol was used to elute the biphenyl directly into a 3-ml volumetric flask, and after diluting to volume the ultraviolet absorption was measured in a spectrophotometer at 248 mμ. The reading was corrected by taking the average of several strips run through the procedure as blank runs, and the amount of biphenyl was then determined from the corrected density value and the standard curve shown in Fig. 11.1. This standard curve was prepared by adding known amounts of biphenyl to various citrus samples and thus includes the recovery of actual amounts throughout the entire process. The analysis was good for amounts as low as 0.1 ppm in the juice and up to 600 ppm in the peel samples. The average error was found to be ±2.8% with a maximum error of 9.3% in 57 analyses. Only 7 of the 57 analyses were more than 5% in error. Stanley et al. [39] have adopted the technique with slight modifications and Baxter [40] has compared this method with a chemical method for the analysis of biphenyl.

Even though prewashed adsorbents are used, there is still spectrophotometric absorbing material in the blank solutions. The accuracy of an assay can be improved by running a blank [35], and Harris et al. [41] found that better blanks could be obtained by using equal weights of adsorbent from a properly developed and dried plate rather than an equal

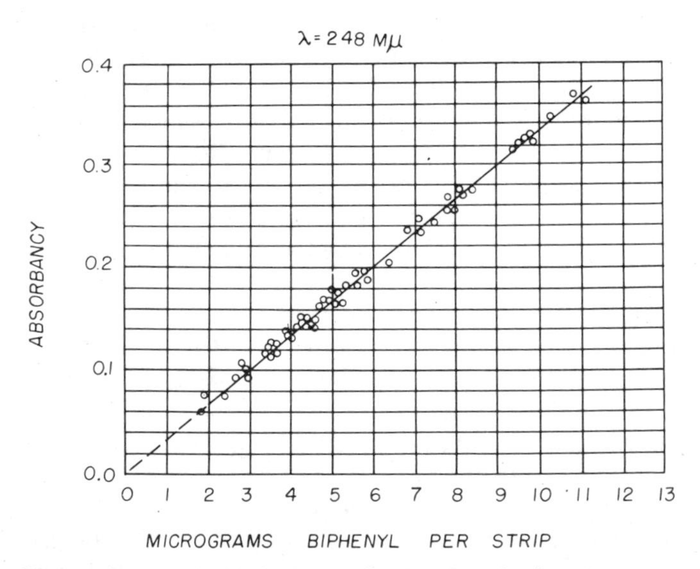

Fig. 11.1 Spectrophotometric standard curve for the determination of biphenyl using thin-layer chromatography. λ = 248 mμ, 10-mm cell. From Kirchner et al. [35]; reproduced with permission of the American Chemical Society.

area from the plate. Another way to minimize the adsorption due to impurities from the adsorbent is to convert the desired compound to a colored complex that can be measured in the visible range, for the adsorbent impurities interfere in the 200 to 300-nm wavelengths range.

Boerning and Reinicke [42] and Coffey and Newburgh [43] have applied a direct spectrophotometric method to the analysis of nucleotides. DEAE cellulose layers are used for the separation. It is best to wash this material in order to remove ultraviolet-absorbing material; this can be accomplished by stirring 10 g of DEAE cellulose for 30 min at a time in four separate washes of 250 ml of 1 *N* hydrochloric acid. After the final hydrochloric acid wash is filtered off, the cellulose is washed with water until free of any traces of acid. The cellulose may then be applied to the plates in the usual manner. Coffey and Newburgh found that 10% calcium sulfate as a binder altered the adsorption so as to effect a better separation of deoxyribomononucleotides with ammonium isobutyrate as a solvent. With calcium sulfate-free layers, hydrochloric acid gave the best separation; however, as Boernig and Reinicke showed, the concentration of hydrochloric acid is critical since in one case 0.022 *N* hydrochloric acid gave a poor separation whereas 0.025 *N* hydrochloric acid gave a good separation. Each separate batch of plates must be checked to determine the optimum concentration of hydrochloric acid. After drying the chromatograms to a point where the adsorbent is still slightly damp, the zones are removed and eluted with 1 *M* sodium chloride, and at the same time, a cellulose blank is eluted from a separate strip. The eluate is filtered through a G-4 fritted filter and measured at 260 mμ.

Weiner and Zak [44] have applied a direct spectrographic method to the eluates after the electrophoresis of nucleotides on an agar gel.

With these as examples, the direct spectrophotometric measurement of eluates from thin-layer chromatographic separations has been used for hydrocarbons [45–48], food colors [49–50], carbonyl compounds [51–53], quinones [54], pesticides [39, 55–58], acids [59–61], glycosides [61–64], antibiotics [65–67], drugs [68–72], pigments [73–75], vitamins [76–80], alkaloids [81–85], explosives [86], terpenes [87–89], esters [90], antiferments in food [91], 2,4-dinitrophenylhydrazones [51–53, 92–94], steroids [95–97], lipids [98–100], pentachlorophenol [101], amino acids [102, 103], indole compounds [104], amines [105], nitro compounds [106], nitrosamines [107], nucleotides and oligonucleotides [108–110], bile acids [111], and phenols [112].

Colorimetric Methods

These are determinations in which the compounds, after chromatographing, are treated in some manner so as to obtain a colored solution that can be measured in a colorimeter (in some cases in a spectrophotometer so that a narrow wavelength band may be used). The same general rules apply here as for the spectrophotometric determinations. As an example of this method can be cited the determination of *O,O*-dimethyl dithiophosphorylphenyl acetate [113] and Cygon [*O,O*-dimethyl-*S*-(*N*-methylcarbamoylmethyl) phosphorodithioate] [114] and the corresponding oxygen analog residues by oxidation with perchloric and nitric acids and subsequent determination of the phosphorus by the molybdenum blue method. Dimethoate (generic name) is an organic phosphate insecticide and the analytical method must be capable of detecting not only the parent compound but also the plant metabolic oxygen analog [*O,O*-dimethyl-*S*-(*N*-methylcarbamoylmethyl) phosphorothioate].

For this determination 2 × 8 in. glass plates are covered with a 0.5-mm layer of a 1:1 mixture of silica gel G and silica gel HF by the usual procedures. They are dried at room temperature overnight and then prewashed by developing in acetone.

The sample is obtained by macerating 100 g of the plant tissue and thoroughly extracting with acetone, after which the acetone is removed from the extract at 25°C. The residue is then extracted thoroughly with hexane in order to remove plant waxes. Care must be taken in extracting with hexane not to shake too vigorously at first because of the formation of emulsions. The water phase is then saturated with sodium sulfate and extracted with chloroform. This final concentrate is evaporated down to 100 to 200 μl for application to the thin-layer plate. Development is carried out with acetone–chloroform (75:25) in a saturated atmosphere. The completed chromatogram is dried at room temperature and sprayed with alcoholic palladium chloride reagent. After standing overnight the bands are scraped into sintered glass funnels where the elution is accomplished with 1 *N* nitric acid. The samples are then oxidized with 70% perchloric acid (this step must be handled with care because of hazards associated with perchloric acid). After cooling and diluting, the color development is carried out by treating with ammonium molybdate and Fiske–Subvarow reagent in an acidic medium. The absorbance is measured at 820 mμ and corrections are applied by carrying an untreated material through the entire procedure. Calibration curves are prepared from standard solutions of dimethoate and its oxygen analog. The authors developed this procedure with a view to its use, with minor modifications, in the determination of other organophosphorus compounds.

Heusser [115] determined the standard deviation for a number of colorimetric methods and found them ranging from ±1.2 to ±4.2%.

Colorimetric methods in thin-layer chromatography have been applied to the analysis of amino acids and proteins [116], pesticides [57, 114], glycosides [61, 117], vitamins [76, 118–120], drugs [121–123], terpenes [124], phenols [125–127], alkaloids [128–130], alcohols [131, 132], carbohydrates [133–135], lipids [136–143], acids [144], antioxidants [145, 146], steroids [96, 147–153], and cardenolides [154].

Fluorimetric Methods

Bruinvels [155] has applied this type of determination to the analysis of aldosterone, hydrocortisone, and corticosterone. Silica gel containing 3% of a fluorescent drug (S.5.Gruen/1 5319157, Leuchtstoffwerk) was used for the preparation of the plates. After the compounds were spotted, the plates were developed in an 8% ethanol (96%) in chloroform solution in a closed chamber at 38.5°C for 1 hr by the ascending technique. The spots that were visible in ultraviolet light were transferred to centrifuge tubes. The aldosterone spot was mixed with 1 ml of concentrated sulfuric acid and allowed to stand for 1 hr. After centrifuging the silica gel, the solution was then heated for 1 hr at 100°C after which it was cooled and the fluorescence measured. An ethanol–sulfuric acid mixture was used to elute the hydrocortisone and the corticosterone. Recovery was 95 ± 3.4% and quantities from 0.5 to 4 μg could be measured in this manner.

Mueller and Honerlagen [156] and Braeckman et al. [157] have used a fluorimetric method for the determination of quinine and quinidine in cinchona bark.

Vannier and Stanley [158] have combined a descending chromatostrip method using silica gel layers, with the fluorescence of 7-geranoxycoumarin in alkaline solution for the determination of as little as 0.5% grapefruit oil in lemon oil. This compound occurs in grapefruit oil but is absent in normal lemon oil.

Applications of a fluorimetric method have been made by Sawicki et al. [46] to the analysis of benzo(α)pyrene in atmospheric pollutants, by Genest and Farmilo [159] to the determination of lysergic acid diethylamide in narcotic products, and by Kutáček and Prochazka [160] to the determination of indole compounds in Cruciferae.

Fluorimetric determinations after elution have been used for bile acids [161], catecholamines [162], carbohydrates [163], and porphyrins [164, 165].

Gravimetric Analysis

This method of direct analysis has been applied to the determination of lipids [166–169], pesticides [170], alkaloids [171, 172], plasticizers [173], detergents [173a], and steroids [173b]. It is not surprising that the method has not been used in more cases, because of the nature of the problems involved in its application. Chief among these perhaps are the difficulty of determining blank value corrections and the very small quantities of material in an individual spot.

Volumetric Analysis

Several analyses of thin-layer eluates have been made by this method. Vacíková et al. [174] eluted free fatty acids from the zones and determined these by titration. Similarly, the phospholipids and the triglycerides were hydrolyzed and then the liberated fatty acids were estimated by titration. Ikram and Bakhsh [175] determined the alkaloids from *Datura alba* leaves and roots by first chromatographing on loose layers of alumina and then eluting the spot corresponding to tropane with chloroform. After removal of the solvent the residue was taken up in 0.01 *N* sulfuric acid and back-titrated with standard sodium hydroxide solution. Within the range of 0.20 and 1.30 mg of alkaloids the error ranged from 0 to 2%. Pastuska [176] separated sugars on silica gel layers and determined them by a titrimetric method. The spots were scraped into a flask and heated with a twofold excess of potassium dichromate solution for 60 min on a water bath at 95°C. A blank run was taken from a corresponding section of adsorbent at the same height as the sugar spot. After cooling, 20 ml of water was added to the dichromate extract along with 5 ml of a 5% potassium iodate solution. The liberated iodine was titrated with 0.01 *N* thiosulfate solution using a starch indicator. Similarly, Dobici and Grassini [177] applied the thiosulfate titration method to the determination of periodate and iodate after separation on layers of plaster of Paris. Methods and titrants have been determined for automatic conductometric titration, and a small conductance cell has been developed in order to titrate spots in situ [177a].

Potentiometric titrations have been used for steroids [178], tin stabilizers [179], pharmaceuticals [180], and inorganics [181].

Gas Chromatography

Gas chromatography can be applied to the quantitative determination of lipids which have been separated by thin-layer chromatography. Vioque and Holman [138] applied the method to fatty esters which were separated on silica gel G by means of various mixtures of diethyl ether in hexane. The zones were then eluted with diethyl ether and made up to a known volume from which an aliquot was injected into the gas chromatographic equipment. Bowyer et al. [182, 183] determined the serum lipids in this manner by first extracting approximately 5 mg of lipids from the serum and applying this to the silicic acid layer. After the chromatogram was developed the spots were eluted with 10% sulfuric acid (wt/vol) and then esterified by adding dry methanol and heating for

1 hr at 80°C (for sphingomyelin, 16 hr). A crystal of hydroquinone acts as an antioxidant during this procedure. After methylation, water is added to the sample and the esters are extracted with petroleum ether (40 to 60°C). The petroleum ether extract is dried over anhydrous sodium sulfate–sodium bicarbonate (4:1) and after being concentrated is ready to be applied to the gas chromatographic column. A succinic acid–ethylene glycol–polyester column may be used for the separation. With this method, care must be taken not to use iodine as a locating agent because this causes a partial loss of unsaturated acids [184]. A similar determination has been applied to the analysis of steroids [185, 186]. In this case the trimethylsilyl ethers can be prepared by a reaction with hexamethyldisilazane. Using this method the 15 μg of testosterone from a 24-hr sample of urine could be determined with an error of $\pm 7\%$ [185].

Hoffsommer [187] used gas chromatography with a ^{63}Ni electron capture detector for determining nitro compounds eluted in the micro- to picogram range from a thin-layer chromatographic plate.

From the examples given it can be seen that compounds that are volatile enough may be eluted and injected directly into a gas chromatographic unit for quantitation. Less volatile material may be treated to prepare a suitable volatile derivative.

Polarographic Methods

Polarographic determinations have been applied to a number of eluates from thin-layer chromatograms, for example, for pharmaceuticals [188], pesticides [189, 189a], and nitro compounds [190]. Errors ranged from ± 1 to 5%. Kutáček and Procházka [160] have applied an oscillopolarographic method to the determination of glucobrassicin, an indole compound in Cruciferae. Huber [191] found that cathode-ray polarography increased the sensitivity of a determination by a factor of 10 to 100 over that of the spectrophotometric method, and in addition there was no need for a blank correction. He cites three examples: a hydrophobic compound, a hydrophilic compound, and a hydrophilic compound that had to be converted to a polarographic active derivative. A polarographic electrode has been patented that permits direct determinations on the layer without elution of the spot [191a].

2 DIRECT QUANTITATIVE METHODS

Direct measurement on the chromatographic plate has the advantage that there is less chance of loss of material, it is faster, and smaller quantities can be determined. It is advisable in all types of direct measurement to scan the TLC plates prior to use and to select the most uniform plates for quantitative work.

Direct Fluorimetry

If a compound fluoresces when exposed to light of a given wavelength then a measure of the amount of light emitted can be used as a quantitative evaluation of the amount of compound present. Since the beam does not pass through the adsorbent and the glass support, there is no great loss due to scattering as in the case of direct densitometry, although fluorescent measurements may also be made from the back side. Naturally, measurement from the illuminated side yields more light. Fluorescent measurements are less affected by optical background noise than are absorption measurements [192].

Direct fluorescent measurements are more sensitive than density reflectance measurements so that nanogram amounts can be measured instead of micrograms, and Varga and Richards [193] have used fluorescent scanning to determine picomole levels of amino acids. Ripphahn and Halpaap [194] have used silica gel of pore size 60 Å in high-performance thin-layer chromatography (HPTLC) to determine 10 pg to 100 ng of fluorescent compounds.

There are a number of factors that can affect the fluorescent analysis; for example, there can be a change of linearity of the spot versus the amount when different exciting wavelengths are used [195]. Therefore it is important to make sure that an exciting wavelength is used that will give linear results. On the other hand, fluorimetry has the advantage that there is a much wider range of concentration over which the response is nearly linear [192].

The adsorbent used affects the amount of fluorescence; polyamide layers were found to give a higher fluorescence than silica gel layers [196]. The moisture content of the layer also has an influence on the intensity of the fluorescence. This effect can be minimized by operating on a time schedule and also by spraying the plate with triethanolamine–isopropanol (1 : 4) [197, 198].

Smith [199] has introduced a method for inducing fluorescence in a great variety of compounds. This consists in adding 1.5 g of ammonium sulfate to 60 ml of water for slurrying 30 g of silica gel. After the plates are developed and dried they are exposed to a saturated atmosphere of *tert*-butyl hypochlorite vapors for 15 min. They are then heated for 15 min at 150 to 180°C depending on the compound. A similar technique [200] has been proposed for normally prepared layers. After development and drying of the layers they are heated at 110 to 150°C in a closed chamber with ammonium hydrogen carbonate or ammonium hydroxide.

X-ray fluorescent spectrometry has been applied to the quantitative determination on thin layers. A scanner was designed [201] to fit into an X-ray spectrometer and tested on various compounds of phosphorus, sulfur, and halogens, providing an element-specific instrument with limits of detection around 2 μg. Libby [202] examined phosphates, phospholipids, bromosalicylamides, and organosulfur compounds using this technique and describes three calibration methods. X-ray fluorescence has been used for the the determination of phospholipids [203] and for pesticides [204].

Ebel and Herold [205] describe the on-line coupling with a programable desk calculator for the fluorimetric evaluation of TLC plates.

Connors and Boak [206] adopted the use of a flexible layer support so that the chromatograms could be bent to fit on the rotating drum of a Turner Model III Fluorimeter (G. K. Turner Associates). The supports were cut (2 × $8\frac{1}{2}$ in.) from 0.005 in. stainless steel shim stock which was roughened on one side with coarse emergy cloth to provide a holding surface for the thin layer. Two types of layers were used: (1) a mixture of 18 g of silica gel G with 27 g of Celite (Johns Manville) and (2) cellulose powder with a calcium sulfate binder. The silica gel was preferred because of the low fluorescent background.

The method was applied to the determination of sugars separated with ethyl acetate–pyridine–water (2:1:2) or *n*-butanol–acetic acid–water (4:1:2). After development of the chromatograms, they were air dried and then sprayed with an alcohol solution of *p*-aminohippuric acid [207]. The fluorescent compounds were formed on heating for 8 min at 140°C.

For measuring, the chromatoplates were fastened to the drum by means of masking tape. A rectangular slit (3 × 25 mm) was used for scanning every 0.5 cm after adjusting the instrument to zero on a blank section of the chromatogram. Standard calibration curves were prepared for the operating range between 1 and 20 μg.

Fluorescent Quenching

In the case of fluorescent quenching, the layer is made fluorescent by incorporating a suitable agent in the layer [208]. Compounds that absorb ultraviolet light without emitting light can then be measured. Since only a fraction of the depth of a spot is used in quenching and also in fluorescent measurements, it is important that the size of the spots be consistent and that the application of the sample be made in a consistent manner [209].

Spectral Reflectance

Since the first edition of this book was written, the use of reflectance spectroscopy has increased to a considerable extent. This is because of improvement in both instrumentation and techniques.

In reflectance techniques, the incident beam penetrates the layer to a certain depth, and then that which is not absorbed or scattered outside the range of the measuring system is measured as reflected light.

Reflectance measurements can be handled in one of two ways. In one case the spot together with sufficient material to give a constant weight are intimately ground in a mortar and packed uniformly in a cup or cell [210]. The corresponding blank is made of adsorbent treated in the same manner. The other method is to measure the reflectance directly on the layer. The first method is more precise, but is more time consuming.

In using direct plate reflectance measurements, the thickness of the layer does not have as great an effect on the measurement as in direct densitometry; however, it is well to use as uniform and reproducible plates as possible. Uniformity of the plates can be improved by keeping the particle size of the adsorbent as uniform as possible, and paying strict attention to slurry concentrations so as to obtain uniform packing of the adsorbent on the layers. To assist in producing uniformity it is advisable to use a white background beneath the layer during measurements [211, 212].

Jork [213] found that the sensitivity could be increased by a factor of 10 by using finely granulated adsorbents having a small pore diameter. Treiber et al. [214] found that increased sensitivity and reproducibility could be obtained by making simultaneous transmission and reflectance measurements on the same track resulting in a constant base line. This was confirmed by Jork [215] and Hezel [216]. This increased the detection limit by a factor of 10 to 100. In this case nanogram quantities of material could be measured with a reproducibility of $s = 5$ to 7%. Frei [212] recommends the use of a double-beam instrument whenever chromogenic reagents have been used for visualizing the spots. When carefully standardized procedures are used, relative deviations have been quoted of 1.5 to 5.3% [217, 218] when the determinations have been carried out on a single plate. On different chromatograms the coefficient of variation is higher, being about 4 to 6% [217].

Yamamato et al. [219] used a dual-wavelength point zigzag scanning method to improve quantitative results. This technique is especially effective for spots of irregular shape or different concentrations. In this procedure the spot is scanned in a zigzag manner with dual-wavelength light pulses having a minute cross-section. One of the wavelengths is of a nature such that it is not absorbed by the compound under investigation, but experiences the same scattering fluctuations as the other wavelength of light. In this way the second beam compensates for irregularities in the light scattering ability across the layer.

Bethke et al. [220] have developed a "data pair" technique for reducing errors in reflectance measurements. This consists in spotting reference and standard solutions so as to give a pair for each concentration, with the two spots being separated by a distance of one-half a plate width. The average of two forward and two backward scans is then taken with a relative standard deviation of about ±1.5%.

Lieu and Frodyma [221] have published a method for selecting the optimum concentration for reflectance measurements.

Ebel and Herold [222] have reported on the on-line coupling with a programable electronic desk calculator for reflectance measurements, and Kynast [223] has published a computer program for the direct evaluation of reflectance values using an Olivetti Programma 101.

Frei [224] has published a review on thin-layer reflectance spectroscopy.

Densitometry

In direct densitometry the light that falls on the plate is divided in a number of ways; part of it is reflected, some is scattered, and the balance passes through to be absorbed to a greater or less extent by the sample and the adsorbent. For this reason the layers must be homogeneous and very uniform in thickness. This homogeneity can be attained by thorough mixing of the adsorbent mixture without incorporation of air bubbles.

In order to obtain uniform layers, Hara et al. [2, 225] found it was necessary to add water carefully to the adsorbent–binder mixture so as to obtain a suspension of a definite density which had to be applied in a carefully dried applicator to dry plates.

If it is necessary to treat the sample spot with a reagent to produce a colored product, certain conditions must be observed: the colored product must not diffuse out into the surrounding area, and the color must be proportional to the amount of sample and must be reproducible and stable. In applying the reagent it is important to see that it is distributed uniformly across the plate. One way to diminish this source of error is to apply alternate spots of sample and standard solutions; Graham et al. [11] have shown that in spraying a chromogenic reagent the average difference between any two adjacent spots was 2.5% compared to 5.2% maximum error for a whole plate.

In making the actual measurements, a number of factors have to be considered. Dallas [226] found that scanning with a 1-mm spot gave a more nearly constant peak area than scanning with a slit; however, Shellard and Alam [227] preferred the use of a slit. They found that when the slit width was smaller than the width of the spot, the coefficient of variation was smaller than when the slit width was greater than the width of the spot. In addition, better results were obtained with a slit width of

0.5 mm than with one of 0.3 mm. The position of this slit with relation to the spot is also important, and Pataki [9] has demonstrated the error produced by shifting the slit 1 mm to each side of the optimum point. He recommends shifting the slit slightly to each side of the initial scan in order to make sure that the initial scan is at the real peak maximum.

The amount of moisture in the layer can affect the results; Dallas [226] found a change of the order of 1% in the peak area with a 3% change in the relative humidity when he measured the plates before and after storing the same plate under different relative humidity conditions.

When a series of spots are measured, the position of the scanner must be adjusted for each spot because slight variations in the R_f value can affect the results.

The plate may be made more transparent by spraying with paraffin–ether mixture (1:1) [228]; however, with a sensitive instrument this technique was found to give large errors [2]. Thomas et al. [229] found that spraying with mineral oil in ether to reduce light scattering resulted in a considerable reduction in sensitivity; they preferred to reduce the effect of light refraction by the method of Blank et al. [230], wherein a narrow slit is placed below as well as above the chromatoplate.

Scanning of the spot may be made parallel to the direction of development or 90° to this direction. In the parallel direction the base line may not return to the same level, because (*a*) the solvent leaves a narrow trail in the center, (*b*) the spots are not completely separated, or (*c*) impurities are deposited between the separated spots. Therefore, the direction of scan should be chosen to suit the occasion. Turano and Turner [231] found it advantageous to average the results obtained by scanning in both directions. This technique has also been found useful for irregularly shaped zones [219]. Averaging of repeated scans also decreases the error in the determination.

A recorder fitted with an integrator may be used in scanning the spot; this gives a count that is proportional to the area of the curve. However, if the recorder does not return to the original base line, because of previously mentioned factors, the count and consequently the area will be in error. It is then best to measure the area with a planimeter even though it takes more time. For planimetry measurements the chart speed should be high [226]. Errors in the use of the planimeter may also be minimized by averaging repetitive measurements. Other methods of measuring the area such as with millimeter-squared paper are less accurate. Waksmundzki and Różylo [232] found that the spot areas as well as the peak areas of the densitometric curves depend on the specific surface area of the adsorbent, the mobile phase, and the viscosity coefficient of the solvent, as well as the rate of development.

Bethke and Frei [233] have applied the data pair technique they used for reflectance measurements [220] in which reference and standard solutions are spotted to give a pair of spots for each concentration of solution with the two spots placed about half a plate distance apart. Each spot was measured twice in a forward and twice in a reverse scan parallel to the direction of chromatography. The four readings were averaged for that particular solution. Relative standard deviations of ±1.4% were reported.

Treiber et al. [234–236] developed an equation incorporating both the Lambert–Beer and the Kubelka–Munk theories. In applying this, and using a 1-mm circular aperture to scan the spot in parallel lines 2 mm apart with integration of the results, Treiber [236] obtained a standard deviation of ±0.46% compared with ±3.2% for the Lambert–Beer equation and ±2.4% for the Kubelka–Munk equation.

For densitometry, a dual-wavelength spectrometer is desirable. In this case two beams of light of different wavelengths, one at the peak of absorption and the other slightly to one side of the peak, are alternately flashed through the spot. With properly designed electronics the background optical density is suppressed without decreasing the sensitivity of the instrument. Salganicoff et al. [237] describe a dual-wavelength spectrophotometer for paper chromatograms that is based on the work of Chance [238, 239]. This is in contrast to the dual-beam instruments that split a beam of light so that one section passes through the spot and the other through a blank area of the plate where the absorption and scattering characteristics of the layer may be different.

Small computers can be applied to advantage in thin-layer quantitative work. Pollak [240] discusses the various aspects which include analog versus digital, determination of the area under a curve, and matching of actual distribution of concentration with an ideal distribution. This latter technique leads to calculations of low concentrations with increased accuracy and to calculations where the chromatographic zones overlap. A number of papers illustrating the usefulness of these techniques for densitometry have appeared [241–251]. Ebel and Hocke [252] describe a system for automatically scanning TLC plates wherein the peak areas measured by an electronic integrator are fed into an on-line coupled desk calculator for immediate results.

A soft laser scanning densitometer has appeared [253] which is claimed to give superior resolution over densitometers using conventional light sources. This increased resolution is obtained by being able to narrow the light beam to 3 to 10 μm without the use of a slit.

Other methods of spot densitometry may be used. For example, the spot may be photographed or photostated and the resulting print can be used in the densitometer [254–261]. Blank et al. [257] made an audioradiograph of a chromatoplate containing radioactive lipids and then measured the X-ray film by densitometry. Hefendehl [254] and Vioque and Vioque [258] made their TLC plates transparent before photographing them, the negative being measured in a densitometer. Polaroid film has been used [259, 260] and Jacobsohn [261] describes a one-step photographic method.

Goodall [262] has discussed the significant factors in preparing and scanning thin-layer chromatograms especially for ultraviolet work.

Pollak [262a] and Treiber [263] have published on the development of spectrodensitometers for thin-layer work.

3 SPOT AREA METHODS

Semiquantitative Determination by Visual Comparison of Spots

For a quick semiquantitative determination of the amount of components present in thin-layer spots, a visual comparison of the size of the spot, the intensity of the spot, or a combination of the two in comparison to known standard spots can be used. In this case a series of standard spots are run so as to blanket the expected quantity in the unknown sample. The error in this type of analysis may run all the way from ±5 to 30%, depending on each specific instance and on the care with which each analysis is made. The analysis has been used for carbohydrates [264], alcohols [265], acids [59], alkaloids [266–268], explosives [269], terpenes [270, 271], drugs [272, 273], pigments [274], vitamins [275], antibiotics [276], steroids [277–279], lipids [280–282], pesticides [283], mineral oil [284], and inorganic ions [285–287].

Spot Area and Weight Relationships

The quantitative method of relating spot size to the quantity of compound as applied to paper chromatography was naturally extended to thin-layer work. A number of variations of the general idea have been applied at different times, and from the results it would appear that no one method is applicable in every case.

There are a number of factors which affect the size of the spot and consequently the accuracy of the resulting analysis [287, 288]. These include the following:

1. Particle size and activity of the adsorbent, thus calling for a homogeneous adsorbent.
2. Thickness of the layer, thus requiring a very uniformly prepared plate.
3. Application of uniformly sized starting spots from solutions whose concentrations are not too different.
4. Use of a precision microburet or capillary pipet.

5. Constant vapor phase in the separating chamber.
6. Solvents which give a sharp separation of the substances to be determined.
7. Choice of reagents that give sharp contrasting spots.
8. Application of the reagent in a uniform spray.
9. Running the standard at the same time the unknown is run.

Petrowitz [289] found a straight-line relationship between the area of the spot and the quantity of two insecticides, gammexan and DDT. This relationship also held true for six tar oil components. Seiler [287] applied the same direct relationship for the determination of inorganic ions, and Seher [290, 291] used an area versus weight graph for the analysis of antioxidants, although in the latter case it was not a straight-line relationship. Brenner and Niederwieser [292], analyzing amino acids, plotted the area of the spot against the logarithm of the weight of the sample and applied the section of the curve that was linear to their determination. Aurenge et al. [293], analyzing various phenols, obtained a straight-line relationship by plotting the square of the surface area against the weight of the material.

Purdy and Truter [294, 295] investigated 16 different compounds and found a linear relationship between the square root of the area and the log of the weight of the compound. The areas of the spots are measured by placing a sheet of transparent paper over the chromatogram and tracing the outline of the spots. This tracing is then transferred to millimeter graph paper where the squares within the spot may be counted. In order to keep the method accurate the standard samples must be processed at the same time as the unknown samples.

To avoid the need for setting up a calibration graph for each experiment, Purdy and Truter used an algebraic method which requires running only three samples. These consist of a solution of the mixture to be analyzed, a dilute solution of the same mixture, and a solution of the standard. If desired, a dilute solution of the standard may also be run. The samples are all spotted on the same plate and after development the areas of the various spots are determined as described above. For substitution in the equation, W and A are the weight and area of the unknown and W_s and A_s are the weight and area of the standard. A_d and d are the area of the diluted unknown spot and the dilution factor, respectively. The authors have designated this analysis the W_s method.

$$\log W = \log W_s + \frac{\sqrt{A} - \sqrt{A_s}}{\sqrt{A_d} - \sqrt{A}} \log d$$

Of the two Purdy and Truter methods, the first or graphical method is the least accurate with an error of $\pm 4.3\%$. The W_s is the most accurate method with an error of ±2.7% for a chromatographic run and 3.6% for a partition separation.

Oswald and Flueck [288] examined all the various methods of relating area of spot to the quantity of material on thin-layer chromatograms. They were able to confirm the findings of Purdy and Truter on numerous chromatograms. However, in many cases none of the various area–weight relationships were linear. They therefore set up a procedure based on the fact that short sections of any curve may be found where the response is linear. They examined the variation of spot size with amounts of different substances and found that by staying within ±20% of a given quantity of substance they could find a region which was satisfactorily linear and above which the spot size decreased greatly with increasing amount of substance. On the other hand, below this point the smallness of the spot size greatly increased the error. They determined, empirically, the optimum region of sample size for the alkaloids they were investigating. In measuring the area of spots they adapted a planimeter, they determined that for the millimeter-cross-section method of determining the area, the spot had to be 100 mm or greater in order to reduce the error to a minimum acceptable value.

Nybom [296] examined the area–weight relationship and found that different relationships existed between the area and the weight of the material depending on the thickness of the layer. A thin layer showed a linear relationship between log weight and area, whereas a thick layer showed a linearity between weight and the square root of the area. He also found that the visualizing agent had an effect on the area–weight relationship. If β-alanine spots were visualized with isatin a straight-line relationship existed between the log of the area and the log of the amount. On the other hand, the same quantities when visualized with ninhydrin showed a straight-line relationship between the area of the spot and the log of the amount.

4 LIMITING SENSITIVITY

This method of quantitative analysis was one of the first applied to thin-layer chromatography by Kirchner et al. [35] in 1954. Like the spectrophotometric method it was designed for the determination of biphenyl in citrus fruits and fruit products. Preliminary preparation of the sample is the same as described under spectrophotometric methods. The heptane–oil–biphenyl solution obtained from this preparation is then diluted with *n*-heptane to make dilutions of 1 : 10 and 1 : 100. The thin layers are prepared from silicic acid with a 2½% starch binder and contain zinc–

cadmium sulfide and zinc silicate as fluorescent agents. A slurry of this mixture with water is coated on 13 × 136 mm glass strips for individual spotting, or if desired a larger plate may be prepared for multiple spots. The sample spots may be applied with a 0.01-ml pipet, but a syringe microburet is more convenient because it decreases the number of dilutions needed. A series of spots are applied containing multiples of 0.01 ml of the various dilutions. The chromatograms are then run in purified petroleum ether (30 to 60°C) until the solvent reaches a line 5 cm from the origin (approximately 7 min), whereupon the strips are removed and the excess solvent is evaporated in the air. Examination of the chromatograms under ultraviolet light discloses dark spots of biphenyl on a fluorescent background. Additional dilutions are then run until the chromatogram is found which just shows a faint but definite dark biphenyl spot, and this is the minimum detectable amount. Because this amount may vary with individuals, each analyst should determine the value for himself; it will probably be around 0.55 γ of biphenyl. With the dilution factor and the amount of the solution spotted on the chromatogram known, the quantity present in the sample can be easily calculated. Because of the concentrating effect of the steam distillation in the preparation of the sample, this method allows 0.05 ppm to be detected in a 1500-ml sample of citrus juice or 1.5 ppm with a 50-ml sample of juice from a single fruit. With 19 samples ranging from 0.5 to 575 ppm the average error was found to be 11.3%. This method has been applied to other thin-layer determinations.

Broadbent et al. [297] have used this method to detect the amount of aflatoxin B in ground nuts and ground nut products. The sample in this case is prepared by an extraction of ground nut meal with methanol and the extracted material is then transferred into chloroform solution for drying and concentration. The extract is applied to aluminum oxide layers containing a plaster of Paris binder, and after development with 1.5% of methanol in chloroform the plates are examined under fluorescent light for the blue-purple fluorescent spot of aflatoxin B. By running additional concentrations the minimum detectable spot is determined.

Hasselquist and Jaarma [298] applied this method to the determination of ascorbic acid in potato tubers. They used a phosphomolybdic acid spray for the detection of the ascorbic acid spots. In this case the minimum detectable amount was 0.2 γ of ascorbic acid. Lagoni and Wortmann [299] applied the method to the determination of β-carotene and vitamin A in fats and oils.

Pilný et al. [299a] developed a photographic method for detecting the minimum amount of exposure needed for a given spot not to appear in the photograph.

5 BIOASSAYS

Although the very nature of a biological assay prevents the attainment of the precision and accuracy that is possible with a chemical method, there are times when such an assay is essential. Such methods have been applied to the determination of various substances contained in thin-layer spots. Using *Bacillus pumilus* seeded in agar plates, Brodasky [300] has developed an assay for the determination of neomycin sulfates after separation on thin-layer carbon plates. Separation of the neomycins is accomplished on thin-layer plates of carbon prepared from slurries containing 30 g of Nuchar (C-190-N) vegetable carbon black and 1.5 g of plaster of Paris in 220 ml of distilled water, preferably adjusted to pH 2 with sulfuric acid. The acidified carbon slurry is allowed to stand for 16 hr prior to preparing the plates. After spotting and developing, the plates are exposed to an ammonia atmosphere for a minimum of 5 min in order to neutralize any excess acidity which would inhibit the growth of the microorganism. The antibiotics are detected and analyzed by placing the chromatogram face down on a tray of streptomycin agar seeded with 2×10^5 cells per milliliter of *B. pumilus*. Contact is maintained for 5 to 10 min, and after the chromatographic plate is removed the tray is incubated for 16 hr at 28°C. At this time no attempt is made to remove carbon particles adhering to the agar, but after incubation the surface of the agar may be moistened and cleared with a microscope slide gently drawn across the area. The quantity is determined by comparing the zone diameter with a standard curve constructed by plotting the log of the quantity versus zone diameters. When the spots are not circular, greater accuracy can probably be obtained by using the area instead of the zone diameter. Peterson and Edgington [301] estimated the amount of the fungicide benomyl by plotting the diameter of the zones of inhibition against the log of the amount.

Horton and Thompson [302] have used a bioassay of prostaglandins by eluting the spots from the thin-layer plate and applying the solutions to the atropinized rabbit duodenum and the hamster colon. Pure prostaglandin E_1 was run simultaneously for comparison.

Kaldewey and Stahl [303] have applied the well-known Avena test to the assay of auxins isolated by thin-layer chromatography. The silica zone material is scraped from the plate and spread uniformly over the agar blocks that are used in the test. These are kept in darkness for at least $\frac{1}{2}$ hr so as to allow the auxins to diffuse into the agar blocks before the test is run. Recovery of added indoleacetic acid was on the order of 85%. A synergistic effect of silica gel and calcium on the growth of coleoptiles of *Triticum* has been observed [304, 305], so that in an assay of this type such activity must be taken into account.

Cima and Mantovan [306] have compared a microbiological assay of vitamin B_{12} using *Lactobacillus leichmanii* ATCC 7830 with a spectrophotometric assay after isolation on thin layers of silica gel.

Salo et al. [307] have used *Drosophila melanongaster* for evaluating the toxicity of insecticide residues.

Billow and Speaker [308] report on a method whereby the seeded agar is poured over chromatographic strips, and after incubation the agar layer is lifted from the chromatogram with a blunt forceps to be placed on a glass plate for densitometry. No supporting data were given.

6 RADIOACTIVE METHODS

There are a number of ways in which radioactivity may be utilized for quantitative determinations in thin-layer work. Samples may be tagged by direct reaction with the components such as the esterification of acids with radioactive diazomethane [309], or by the isotope dilution technique of the addition of a known quantity of radioactive component to the mixture to be separated [186, 310]. Schulze and Wenzel [311] have used the Wilzbach technique [311, 312] for marking compounds with tritium. Once marked, these compounds separated on layers may be assayed by various methods: (1) autoradiography, (2) autoscintillography, (3) elution with subsequent radioassay, (4), strip scanning with Geiger–Müller detectors, (5) strip scanning with phototube detectors, (6) two-dimensional scanning using a new type of electron multiplier detector, (7) two-dimensional scanning with a β-camera, (8) combustion methods, (9) zonal profile scanning using liquid scintillation detection, and finally (10) the so-called sublimation autography. A number of reviews have been published on radioactive techniques and materials as related to thin layers: Snyder [313], Catch [314], Figge et al. [315], and Koss and Jerchel [316].

Autoradiography

The actual quantitative determination for the amount of activity present depends on the equipment available to the individual investigator. A radioaudiograph may be prepared and the density of the radiogram may be determined as explained under spot densitometry. In making the radioaudiogram, the X-ray film may be placed in direct contact with the adsorbent layer or a thin plastic film may be inserted between the two. In the latter case, the exposure time must be increased because of the absorption of the rays by the thin sheet of plastic. Schwane and Nakon

[317] give the percentage of beta radiation absorbed by several plastic films as follows: 0.00025 Mylar (22%), 0.0005 Mylar (39%), Saran Wrap (45%), and Handiwrap (24%). Where the film is in direct contact with the thin layer, it must be ascertained that the photographic activity is not increased by sublimation of the compound into the photographic film [318], thus giving the procedure designated sublimation autography. In the latter case the darkening of the photographic plate may be due to chemial activity as well as radioactivity.

The layer may be impregnated directly with a photographic emulsion after the chromatogram is run [319]; however, this has the disadvantage of contaminating the spots with photographic emulsion if the compounds are to be recovered and also places the emulsion in direct contact with the compounds so that chemical reactions can ensue.

Polaroid film type 57 film may be used in place of X-ray film for autoradiography [320, 321]. This film is comparable in exposure time to the X-ray film and in some cases, as with radioiodine [321], it is more sensitive. A cassette for using Polaroid film has been described [320].

Autoscintillography

One problem with autoradiography is that because of the low beta energies, and especially for tritium, a large part of the energy is absorbed by the chromatographic layer so that only a few electrons are able to escape to be registered on the film. One way to overcome this is to incorporate a scintillator in the layer to give off light as it absorbs the beta particles. This light, which is absorbed to a much less extent than the beta rays, can then be registered by the film. This technique is known as autoscintillography, scintography, or scintillation fluorography. The scintillator can be applied in a number of ways: by dipping, by spraying, or in a slurry during the preparation of the layer.

Randerath [322] has determined that the sensitivity of this method mainly depends on five factors; (1) the concentration of the scintillator, (2) the solvent used to apply the scintillator to the chromatogram, (3) the method of applying the scintillator, (4) the temperature during exposure of the film, and (5) the film used, although additional factors do have a bearing on the sensitivity [322, 323]. The optimum concentration of the scintillator varies with different scintillators and is greater for radiocarbon than for tritium [324]. Low temperature at the time of exposure increases the sensitivity to radiocarbon about twice and about 50 to 100 times for tritium at a temperature of −78°C [325]. The latter reference gives a discussion of the reasons for this enhancement. Benzene has been found to be an efficient solid scintillator at low temperatures and offers the advantage that it can be evaporated from the layer after returning to room temperature, leaving the chromatogram free of scintillator contamination [326]. Of course sensitivity can be increased by making the compound spots as compact as possible, and since fast commercial film has an emulsion on both sides of the supporting sheet only one side of which is used for scintillography, some of the background density can be reduced by preventing the development solution from contacting the emulsion on the back side of the film. This is done by taping an additional supporting film with waterproof tape to the back of the photographic film during development [327].

Strip Scanning

The radioactivity may be measured directly on the thin-layer plate either by means of a thin-end-window Geiger counter [328, 329] or by means of a gas flow counter [309, 311, 330, 331]. The latter is of course the more sensitive and is very useful for measuring low-energy beta-emitting radioisotopes. Commercial instruments are available from a number of sources, and Snyder [332, 333] in two reviews illustrates a number of these. Prydz [334] in a summary of the state of the art in radiochromatography tabulates a buyer's guide to scanners. Ravenhill and James [331] describe a simple, sensitive scanner employing an open window counter and automatic count integration.

A variation of the direct measurement of the radioactivity on a chromatoplate is to treat the layer with a plastic binder [317, 335, 337]. The chromatogram is then removed from the supporting plate and can be cut into strips and measured in the instruments used for scanning paper chromatograms. If the plastic solution is sprayed on, a piece of transparent tape is applied to the surface to assist in peeling off the adsorbent. The layer should then be turned over and given a light spray on the reverse side to prevent loss of particles of adsorbent [335]. Squibb [338] has used a plastic support for the layer. After spotting, developing, and drying, the chromatogram is given a spray of plastic to bind it to the supporting plastic strip. The entire plate can then be cut into strips for scanning in standard equipment. For coating chromatograms of radioactive water-soluble substances, Schwane and Nakon [317] use a solution of 11.2 g of polystyrene in 100 ml of benzene with an added plasticizer. The amount of plasticizer varies with the adsorbent that is being used; for silica gel G and kieselguhr G the amounts are 0.023 and 0.014 ml, respectively, for each milliliter of polystyrene solution. The radioactivity is measured from the smooth side of the film that was in contact with the glass surface so as to prevent absorption of the energy from weak emitters by the adhesive tape [335]. The commercially prepared Chromagram sheets are also convenient for use in scanning instruments.

Two-Dimensional Scanning

There are a number of instruments that have been designed for two-dimensional scanning. The β-camera (Baird Atomic) uses 1622 Geiger–Müller tubes to provide an instantaneous pattern of the radioactivity of an entire TLC plate (20 × 20 cm) on a cathode-ray tube. A permanent record can be made by using a Polaroid camera. Although expensive the instrument is good for short-lived isotopes that need to be handled quickly and for the recovery of compounds without contamination from a scintillator.

Arnoff [339] developed a two-dimensional scanner using a single Geiger–Müller tube in connection with an electrosensitive paper recorder, and Gilbert and Keene [340] used a series of counters to detect the radioactivity of a chromatographic strip. Wenzel and Hoffmann [341] used a normal chromatogram scanner in two methods for scanning two-dimensional chromatograms. Hariharan et al. [342] designed a scanner using 12 Geiger–Müller end-window halogen counters for the rapid detection of radioactivity in wet-paper electropherograms at 0°C. Two-dimensional capability is attained by use of a supporting slide that is moved mechanically, and recording is accomplished by using electrosensitive paper. Although designed for paper there is no reason why it could not be used for thin layers. For measuring tritium the counter would have to be replaced with the windowless type.

Prydz et al. [343], employing a new type of electron multiplier detector, designed an instrument for one- and two-dimensional scanning, and added facilities for computerized contour mapping of the results [344]. Lowe [345] described a unit that transmits the data to a punched tape which in turn can be fed to a computer for processing.

Wedzicha [346] discusses a method for the enhancement of radioactive data by accumulation and Fourier transformation.

Tykva [347] designed an instrument for nondestructive measurement based on semiconductography using a silicon barrier detector [348]. Because of the low background count (0.02 ± 0.01 cpm) a high sensitivity can be obtained. The instrument can be used for the simultaneous determination of β-nuclides in multilabeled substances [349].

The two-dimensional detection with a spark chamber [350] has sometimes been referred to as a quantitative method; however, the chamber itself is strictly a qualitative method.

Liquid Scintillation Methods

Another instrument that lends itself to thin-layer work is a scintillation counter [116, 186, 351–358]. There are two general methods of using this.

In one, the compounds to be determined are eluted from the adsorbent, the solvent removed preferably under nitrogen if it is sensitive to oxidation, and the residue mixed with the scintillator solution. Normal procedures are then used for measuring in an appropriate instrument. On the other hand, the scintillation counter can be used to measure material which is adsorbed on the silica gel without first eluting it if proper precautions are taken. In order to prevent self-quenching the silica gel is kept from settling by adding a gelling agent to the liquid scintillator solution. Snyder and Stephens [356] suspended their zone adsorbent material in 15 ml of a 4% (wt/vol) of Cab-O-Sil (thixotropic gel powder, Packard) in a toluene scintillation solution which contained 5 g of 2,5-diphenyloxazole (PPO) and 0.3 g of *p*-bis[2-(4-methyl-5-phenyloxazolyl)]benzene (POPOP) per liter of toluene. In order to avoid the loss of some of the radioactivity by having it trapped underneath the cap of the counting vial, they capped the vials with Scotch tape trimmed to the circumference of the opening before shaking the assembly. Polyethylene vials should be used whenever phosphatides are analyzed to avoid adsorption of some of the polar lipids on the surface of a glass counting vial. With this method they were able to obtain a recovery of the radioactivity of 99 ± 2.2%. If fatty acids are separated on silica gel by this procedure with a nonacidic solvent, then 50 μl of acetic acid must be added directly to the counting vial to prevent a decrease in the counting efficiency. Riondel et al. [352] used the same type of thixotropic gel powder in their scintillation measurements where necessary. This was necessary when measuring nonacetylated thiosemicarbazones of steroids which normally adsorb on the glass or plastic of the counting vials. This adsorption was prevented by addition of ethanol and the thixotropic gel to the scintillation fluid. Roucayrol and Taillandier [354] measured scintillation directly on the gel plate after appropriate preparation. The glass support of the chromatogram was cemented to an aluminum sheet with the edges turned up so as to act as a container for the scintillator fluid which was poured on the thin layer until it became translucent. Then 10 ml of a gellable scintillator made by adding a few grams of aluminum 2-ethylhexanoate to 200 g of a toluene base scintillator was added.

Rivlin and Wilson [353] reported a simple method for separating steroids from liquid scintillation phosphor by chromatographing on thin layers of silica gel with ethyl acetate. The method can probably be used for other nonsteroid compounds so as to salvage isotopic material.

Snyder [359] has compared strip scanning, in which the radioactive material is assayed directly on the thin layer, with zonal scanning, in which the adsorbent material is scraped off in successive zones and analyzed by liquid scintillation methods. Both sensitivity and degree of resolution for detecting the separations were greater in the zonal method. Manual [360] and automatic [361] (Analabs) scrapers have been developed, and the entire system has been automated including coupling to an on-line computer [362]. Fosslien et al. [363] developed a computer-controlled scraper to remove individual spots in sequence from a thin-layer plate with automatic transfer to scintillation vials or other containers.

Quenching, caused by the adsorption of beta energies of polar compounds that remain on the adsorbent, must be corrected. Krichevsky et al. [364] discuss the indirect and direct methods of doing this and present a computer-aided system for quench correction in liquid scintillation counting. Boeckx et al. [365] published a Fortran IV program for the analysis of doubly labeled and variably quenched liquid scintillation samples, and Gruenstein and Smith [366] describe a method of double label autoradiography whereby two compounds labeled with ^{3}H and ^{14}C, respectively, may be identified on the chromatogram. An autoradiographic film exposed for 24 hr visualizes the ^{14}C, and then application of a scintillator fluor to the chromatogram makes the ^{3}H visible on the autoradiograph in 4 hr, whereas the ^{14}C would still require 24 hr. In order to eliminate quenching caused by the adsorbent, Shaw et al. [367] dissolved the silica gel in the scintillation vial with hydrofluoric acid. Bollinger et al. [368] used a urea type inclusion complex of a toluene cocktail for suspending insoluble material for gel-scintillation counting. The method is simple and very rapid.

J. T. Baker Chemical Company has published [369] a guide to liquid scintillation counting that contains a number of useful tables including one for helping to solve problems that arise in the use of this technique.

Combustion Methods

Radioactive samples may also be ignited in an oxygen atmosphere with the combustion products passing over copper oxide to convert the carbon and hydrogen to carbon dioxide and water, respectively. The resulting products can then be either measured continuously or collected and measured by liquid scintillation. Combustion furnaces and continuous measuring detectors are available commercially. Griffiths and Mallinson [370] give details on construction of a furnace. In this case the water is condensed at low temperature and the carbon dioxide is absorbed in phenylethylamine. Haahti et al. [371] carried out chromatographic separations on silicic acid–copper oxide (2:3) layers coated on the inside of glass tubes. After development and complete removal of the developing solvent, the tube was placed in a combustion furnace which gradually moved over the length of the tube through which helium passed. Water from the combustion was absorbed by calcium chloride or phosphorus pentoxide, and in the case of radioactive carbon this was detected in a proportional flow detector. For inactive materials a thermal conductivity detector was used. Modifications of this technique (reported in the following section) have been made and applied to inactive compounds; with proper detectors these could be applied to radioactive work.

7 MISCELLANEOUS

Enzymatic analysis has been applied to the determination of thin-layer zones. Scheig et al. [372] have used this method for analyzing oxidized and reduced diphosphopyridine and triphosphopyridine neucleotides.

The flask combustion method has been used for the quantitative analysis of halogen-containing compounds [373]. In order to achieve complete combustion it was necessary to use a minimum combustion flask capacity of 500 ml and not more than 2 to 3 cm^2 of spot area. Final determination was by the usual titration methods. Similarly, sulfur compounds have been determined [374], and in this case, the sulfate formed was treated with barium chromate solution and then measured at 370 nm.

In 1966 Konaka and Terabe [375] introduced a quantitative method wherein the spots were scraped into a platinum boat and placed in a combustion furnace. After displacing the air with helium containing 2% oxygen, the samples were burned with the combustion products being swept through cupric oxide, reduced copper, anhydrone, manganese dioxide, and finally through a carbon column activated at 100°C. Gas chromatographic separation of the nitrogen and carbon dioxide took place and the nitrogen area was determined by an electronic integrator. With a slight change in the combustion furnace this technique was applied to the determination of carbon [376]. To keep the blank value low, the method of De Deyne and Vetters [30] was used to concentrate the sample in a small area on the thin layer.

Cotgreave and Lynes [377] developed a technique where the chromatogram produced on a 25-mm-wide plate was introduced directly into a combustion tube which was heated by a traveling furnace. With the furnace operating at 700°C the volatile products were swept by nitrogen through a flame ionization detector. In order to keep background impurities at a minimum, the thin-layer plate was heated in the furnace at 600°C prior to the chromatographic run. Padley [378, 379] used a 0.5-mm glass rod as a support for the thin layer. After development and drying the chromatogram was drawn directly through a flame ionization detector. Here again to keep the background low, the layer was drawn through the flame ionization detector flame prior to spotting and running the chromatogram. Szakasits et al. [380] also designed equipment for this technique, but preferred to use a wire as a support for the thin layer.

A still different combustion technique was introduced by Haahti and Jaakonmaki [381]. In this case the chromatograms were separated on silicic acid–copper oxide (2:3) layers coated on the inside of 5-mm-i.d. glass tubes. The completed chromatograms were heated by a traveling combustion furnace, and after the water was trapped the carbon dioxide was determined by a catharometer. Further developments were made [382] so that the carbon dioxide could be determined with a thermal conductivity detector or by proper use of trapping absorbents, nitrogen could be determined. Use of a proportional flow detector permitted determination of radioactive carbon. Still further modifications [383–386] permitted the use of all-silica-gel layers as well as layers of silica gel–copper oxide mixture, and the products from the pyrolysis could be detected directly by the flame ionization detector or they could be diverted through a combustion furnace for conversion to carbon dioxide. The carbon dioxide could also be passed through a nickel catalyst using hydrogen as a carrier gas to convert the carbon dioxide to methane for detection in the flame ionization detector. By using a nitrogen–oxygen gas mixture as a carrier through the pyrolysis tube some lipids, steroids, and terpenoids not pyrolyzed in nitrogen or hydrogen were completely vaporized.

References

1. J. G. Kirchner, *J. Chromatogr.*, **123,** 5 (1976).
2. S. Harra, H. Tanaka, and M. Takeuchi, *Chem. Pharm. Bull.*, **12,** 626 (1964).
3. J. W. Fairbairn and S. J. Relph, *J. Chromatogr.*, **33,** 494 (1968).
4. J. W. Fairbairn, "Factors Involved in Producing Uniform Spots," in *Quantitative Paper and Thin-Layer Chromatography*, E. J. Shellard, Ed., Academic Press, New York, 1968, p. 1.
5. S. Samuels, *J. Chromatogr.*, **32,** 751 (1968).
6. K. R. Brain, " 'Operator Effect' in Quantitative Sample Application," in *Quantitative Thin-Layer Chromatography*, Fisons Scientific Apparatus, Loughborough, 1971, p. 39.

6a. E. J. Shellard and M. Z. Alam, *J. Chromatogr.*, **33,** 347 (1968).

7. E. V. Truter, *Thin-Film Chromatography*, Interscience, New York, 1963.
8. R. Klaus, *J. Chromatogr.*, **40,** 235 (1969).
9. G. Pataki, *Chromatographia*, **1,** 492 (1968).
10. J. C. Touchstone, S. S. Levine, and T. Murawec, *Anal. Chem.*, **43,** 858 (1971).

11. R. J. T. Graham, L. S. Bark, and D. A. Tinsley, *J. Chromatogr.*, **39,** 211 (1969).
12. W. Grimm, *J. Chromatogr.*, **89,** 39 (1974).
13. V. Pollak, *J. Chromatogr.*, **105,** 279 (1975).
14. A. A. Boulton and V. Pollak, *J. Chromatogr.*, **45,** 189 (1969).
15. V. Pollak and A. A. Boulton, *J. Chromatogr.*, **45,** 200 (1969).
16. *Ibid.*, **50,** 19 (190).
17. *Ibid.*, p. 39.
18. R. Klaus, *J. Chromatogr.*, **16,** 311 (1964).
19. N. Seiler and H. Moeller, *Chromatographia*, **2,** 470 (1969).
20. *Ibid.*, p. 319 (1969).
21. J. Goldman and R. R. Goodall, *J. Chromatogr.*, **32,** 24 (1968).
22. *Ibid.*, **40,** 345 (1969).
23. J. Goldman, *J. Chromatogr.*, **78,** 7 (1973).
24. R. Kaiser, *Chromatographia*, **4,** 123 (1971).
25. R. Kaiser, *Chromatographia*, **4,** 215 (1971).
26. *Ibid.*, p. 361.
27. *Ibid.*, p. 479.
28. V. Pollak and A. A. Boulton, *J. Chromatogr.*, **50,** 30 (1970).
28a. J. C. Touchstone, Ed., *Quantitative Thin-Layer Chromatography*, Wiley, New York, 1973.
28b. E. J. Shellard, Ed., *Quantitative Paper and Thin-Layer Chromatography*, Academic Press, New York, 1968.
29. J. L. Mulder and G. J. Veenstra, *J. Chromatogr.*, **24,** 250 (1966).
30. V. J. R. De Deyne and A. F. Vetters, *J. Chromatogr.*, **31,** 261 (1967).
31. D. O. Eberle, R. G. Delley, G. G. Székely, and K. H. Stammbach, *J. Agric. Food Chem.*, **15,** 213 (1967).
32. P. Hastings and J. T. Wong, *Anal. Biochem.*, **22,** 169 (1968).
33. M. F. Turchinsky and L. P. Shershneva, *Anal. Biochem.*, **54,** 315 (1973).
33a. H. Falk and K. Krummen, *J. Chromatogr.*, **103,** 279 (1975).
34. W. E. Court, "Quantitative Thin-Layer Chromatography Using Elution Techniques," in *Quantitative Paper and Thin-Layer Chromatography*, E. J. Shellard, Ed., Academic Press, New York, 1968, p. 29.
35. J. G. Kirchner, J. M. Miller, and R. G. Rice, *J. Agric. Food Chem.*, **2,** 1031 (1954).
36. J. F. Clevenger, *J. Assoc. Off. Agric. Chem.*, **17,** 371 (1934).
37. A. Lazarow, *J. Lab. Clin. Med.*, **35,** 810 (1950).
38. C. C. Meloche and W. G. Frederick, *J. Am. Chem. Soc.*, **54,** 3265 (1932).
39. W. L. Stanley, S. H. Vannier, and B. Gentili, *J. Assoc. Off. Agric. Chem.*, **40,** 282 (1957).
40. R. A. Baxter, *J. Assoc. Off. Agric. Chem.*, **40,** 249 (1957).
41. M. J. Harris, A. F. Stewart, and W. E. Court, *Planta Med.*, **16,** 217 (1968).
42. H. Boernig and C. Reinicke, *Acta Biol. Med. Ger.*, **11,** 600 (1963).
43. R. G. Coffey and R. W. Newburgh, *J. Chromatogr.*, **11,** 376 (1963).
44. L. M. Weiner and B. Zak, *Clin. Chim. Acta*, **9,** 407 (1964).
45. F. J. Ritter, P. Canonne, and F. Geiss, *Z. Anal. Chem.*, **205** 313 (1964).

46. E. Sawicki, J. W. Stanley, W. C. Elbert, and J. D. Pfaff, *Anal. Chem.*, **36,** 497 (1964).
47. G. Chatot, R. Dangy-Caye, and R. Fontanges, *Chromatographia,* **5,** 460 (1972).
48. G. Biernoth, *Fette, Seifen, Anstrichm.*, **70,** 217 (1968).
49. J. Davídek, J. Pokorný, and G. Janíček, *Z. Lebensm. Unters. Forsch.*, **116,** 13 (1961).
50. G. Janíček, J. Pokorný, and J. Davídek, *Sb. Vys. Sk. Chem.-Technol. Praze, Technol. Paliv,* **6,** 75 (1962).
51. J. Lacharme, *Bull. Trav. Soc. Pharm. Lyon,* **7,** 55 (1963).
52. G. M. Nano and P. Sancin, *Ann. Chim.* (*Rome*), **53,** 677 (1963); *Chem. Astr.*, **59,** 12189 (1963).
53. M. Pailer, H. Kuhn, and I. Gruenberger, *Fachlich Mitt. Oesterr. Tabakregie, Pt 3,* **1962,** 33.
54. G. Pettersson, *J. Chromatogr.*, **12,** 352 (1963).
55. D. C. Abbott and J. Thomson, *Chem. Ind.* (*London*), **481** (1964).
56. D. C. Abbott and J. Thomson, *Analyst,* **89,** 613 (1964).
57. R. C. Blinn, *J. Assoc. Off. Agric. Chem.*, **47,** 641 (1964).
58. Y. Doi, *Kagaku Keisatsu Kenkyusho Hokoku,* **16,** 51 (1963).
59. R. W. Bailey, *Anal. Che.*, **36,** 2021 (1964).
60. M. A. Millett, W. E. Moore, and J. F. Saeman, *Anal. Chem.*, **36,** 491 (1964).
61. M. Schantz, L. Ivars, I. Kukkoven, and A. Ruuskanen, *Planta Med.*, **10,** 98 (1962).
62. H. Sieper, R. Longo, and F. Korte, *Arch. Pharm.*, **296,** 403 (1963).
63. W. Steidle, *Ann. Chem.*, **662,** 126 (1963).
64. W. Steidle, *Planta Med.*, **9,** 435 (1961).
65. P. A. Nussbaumer, *Pharm. Acta Helv.*, **38,** 245 (1963).
66. P. A. Nussbaumer, *Pharm. Acta Helv.*, **38,** 758 (1963).
67. M. Nekola, *Z. Anal. Chem.*, **268,** 272 (1974).
68. J. Baeumler and S. Rippstein, *Helv. Chim. Acta,* **44,** 2208 (1961).
69. H. Gaenshirt, *Arch. Pharm.*, **296,** 129 (1963).
70. K. W. Gerritsma and M. C. B. van Rheede, *Pharm. Weekblad,* **97,** 765 (1962); *Chem. Abstr.*, **58,** 6646 (1963).
71. E. Ragazzi, *Boll. Chim. Farm.*, **100,** 402 (1961); *Chem. Abstr.*, **56,** 10283 (1962).
72. F. Schlemmer and E. Link, *Pharm. Ztg. Ver. Apoth.-Zig.*, **104,** 1349 (1959).
73. K. Egger, *Planta,* **58,** 664 (1962).
74. E. C. Grob, W. Eichenberger, and R. P. Pflugshaupt, *Chimia* (*Aarau*), **15,** 565 (1961).
75. R. W. Balek and A. Szutka, *J. Chromatogr.*, **17,** 127 (1965).
76. E. Castren, *Farm. Aikak.*, **71,** 351 (1962).
77. M. Covello and O. Schettino, *Farmaco Ed. Prat.*, **19,** 38 (1964).
78. G. Lambertsen, H. Myklestad, and O. R. Braekkan, *J. Sci. Food Agric.*, **13,** 617 (1962).
79. H. Wagner, L. Hoerhammer, and B. Dengler, *J. Chromatogr.*, **7,** 211 (1962).
80. Z. Zloch, *Mikrochim. Acta,* **1975,** 213.

81. F. Korte and H. Sieper, *J. Chromatogr.*, **14,** 178 (1964).
82. A. Liukkonen, *Farm. Aikak.*, **71,** 329 (1962); *Chem. Abstr.*, **58,** 8850 (1963).
83. R. Paquin and M. Lepage, *J. Chromatogr.*, **12,** 57 (1963).
84. K. Teichert, E. Mutschler, and H. Rochelmeyer, *Z. Anal. Chem.*, **181,** 325 (1961).
85. E. J. Shellard and M. Z. Alam, *J. Chromatogr.*, **32,** 472 (1968).
86. J. Hansson, *Explosivstoffe,* **10,** 73 (1963).
87. Y. Kuroiwa and H. Hashimoto, *Rep. Res. Lab. Kirin Brewery Co., Ltd.*, **3,** 5 (1960).
88. Y. Kuroiwa and H. Hashimoto, *J. Inst. Brewing,* **67,** 347 (1961).
89. W. L. Stanley and S. H. Vannier, *J. Assoc. Off. Agric. Chem.*, **40,** 582 (1957).
90. H. Gaenshirt and K. Morianz, *Arch. Pharm.*, **293/65,** 1065 (1960).
91. M. Covello and O. Schettino, "The Application of Thin-Layer Chromatography to Investigations of Antifermentatives in Foodstuffs," in *Thin-Layer Chromatography,* G. B. Marini-Bettòlo, Ed., Elsevier, Amsterdam, 1964, p. 215.
92. J. Dancis, J. Hutzler, and M. Levitz, *Biochim. Biophys. Acta,* **78,** 85 (1963).
93. G. M. Nano, "Thin-Layer Chromatography of 2,4-Dinitrophenylhydrazones of Aliphatic Carbonyl Compounds and Their Quantitative Determination," in *Thin-Layer Chromatography,* G. B. Marini-Bettòlo, Ed., Elsevier, Amsterdam, 1964, p. 138.
94. A. Roudier, *Assoc. Tech. Ind. Papet. Bull.*, **17,** 314 (1963).
95. H. L. Bird, H. F. Brickley, J. P. Comer, P. E. Hartsaw, and M. L. Johnson, *Anal. Chem.,* **35,** 346 (1963).
96. G. Cavina and C. Vicari, "Qualitative and Quantitative Analysis of Natural and Synthetic Corticosteroids by Thin-Layer Chromatography," in *Thin-Layer Chromatography,* G. B. Marini-Bettòlo, Ed., Elsevier, Amsterdam, 1964, p. 180.
97. J. S. Matthews, A. L. Pereda-V., and A. Aguilera-P., *J. Chromatogr.*, **9,** 331 (1962).
98. W. A. Harland, J. D. Gilbert, and C. J. W. Brooks, *Biochim. Biophys. Acta,* **316,** 378 (1973).
99. R. C. Hunt and D. J. Ellar, *Biochim. Biophys. Acta,* **339,** 173 (1974).
100. E. Bende, A. Szabo, and V. Somogyi, *Sutoipar,* **21,** 226 (1971); through *CAMAG Bibliogr. Serv.*, **36,** 047 (1975).
101. R. Deters, *Chem.-Zig.*, **86,** 388 (1962).
102. K. Esser, *J. Chromatogr.*, **18,** 414 (1965).
103. J. M. Davis, *J. Chromatogr.*, **69,** 333 (1972).
104. J. Opieńska-Blauth, H. Kraczkowski, H. Brzuskiewicz, and Z. Zagórski, *J. Chromatogr.*, **17,** 288 (1965).
105. F. Matsui, J. R. Watson, and W. N. French, *J. Chromatogr.*, **44,** 109 (1969).
106. J. A. Hurlbut, *J. Chem. Educ.*, **48,** 411 (1971).
107. G. Eisenbrand, K. Spaczynski, and R. Preussmann, *J. Chromatogr.*, **51,** 503 (1970).
108. G. H. Buteau, Jr., and J. E. Simmons, *Anal. Biochem.*, **37,** 461 (1970).

109. D. Sofrova, R. Miksanova, and S. Leblova, *Vnitr. Lek.*, **15**, 395 (1969); through *Chem. Abstr.*, **71**, 27810s (1969).
110. J. Gangloff, G. Keith, and G. Dirheimer, *Bull. Soc. Chim. Biol.*, **52**, 125 (1970).
111. A. Gerstmeyer, *Zentr. Pharm., Pharmakother. Laboratoriumsdiagn.*, **111**, 821 (1972).
112. R. Franiau and R. Mussche, *J. Inst. Brewing*, **78**, 450 (1972).
113. B. Bazzi, R. Santi, G. Canale, and M. Radice, Montecatini, *Ist. Ric. Agrar. Contrib.*, **1963**, 12 pp.
114. W. A. Steller and A. N. Curry, *J. Assoc. Off. Agric. Chem.*, **47**, 645 (1964).
115. D. Heusser, *J. Chromatogr.*, **33**, 400 (1968).
116. D. Myhill and D. S. Jackson, *Anal. Biochem.*, **6**, 193 (1963).
117. T. Rahaṇdraha, M. Chanez, and P. Boiteau, *Ann. Pharm. F.*, **21**, 561 (1963).
118. R. A. Dilley, *Anal. Biochem.*, **7**, 240 (1964).
119. G. Katsui, Y. Ichimura, and Y. Nishimoto, *Bitamin*, **23**, 35 (1961).
120. L. M. Kuznetsova and V. M. Koval'ova, *Ukr. Biokhim. Zh.*, **36**, 302 (1964).
121. T. I. Bulenkov, *Med. Prom. SSSR*, **17**, 26 (1963); through *Chem. Abstr.*, **60**, 5280 (1964).
122. A. Fioro and M. Marigo, *Nature*, **182**, 943 (1958).
123. M. Marigo, *Arch. Kriminol.*, **128**, 99 (1961); through *Chem. Abstr.*, **56**, 5068 (1962).
124. T. Akazawa and K. Wada, *Agr. Biol. Chem. (Tokyo)*, **25**, 30 (1961).
125. T. G. Lipina, *Tr. Khim. Khim. Tekhnol.*, **1962**, 424.
126. H. Seeboth, *Monatsber. Dtsch. Akad. Wiss. Berlin*, **5**, 693 (1963).
127. H. Seeboth and H. Goersch, *Chem. Tech. (Berlin)*, **15**, 294 (1963).
128. J. L. McLaughlin, J. E. Goyan, and A. G. Paul, *J. Pharm. Sci.*, **53**, 306 (1964).
129. W. Poethke and W. Kinze, *Arch. Pharm.*, **297**, 593 (1964).
130. P. Haefelfinger, *J. Chromatogr.*, **33**, 370 (1968).
131. T. G. Lipina, *Metody Opred. Vredn. Veschestv Vozukhe, Moscow*, **1961**, 41; through *Chem. Abstr.*, **59**, 4471 (1963).
132. T. P. Lipina, *Zavodsk. Lab.*, **26**, 55 (1960).
133. G. W. Hay, B. A. Lewis, and F. Smith, *J. Chromatogr.*, **11**, 479 (1963).
134. S. Singh and B. E. Stacey, *Analyst (London)*, **97**, 977 (1972).
135. C. W. Raadsveld and H. Klomp, *J. Chromatogr.*, **57**, 99 (1971).
136. H. P. Kaufmann and C. V. Viswanathan, *Fette, Seifen, Anstrichm.*, **65**, 607 (1963).
137. B. M. Phillips and N. Robinson, *Clin. Chim. Acta*, **8**, 832 (1963).
138. E. Vioque and R. T. Holman, *J. Am. Oil Chem. Soc.*, **39**, 63 (1962).
139. F. Parker, V. Rauda, and W. H. Morrison, *J. Chromatogr.*, **34**, 35 (1968).
140. G. J. Nelson, "Quantitative Analysis of Blood Lipids," in *Blood Lipids and Lipoproteins: Quantitation, Composition, and Metabolism*, G. J. Nelson, Ed., Wiley-Interscience, New York, 1972, p. 51.
141. A. Yamamoto and G. Rouser, *Lipids*, **5**, 442 (1970).
142. N. Neskovič, L. Sarlieve, J. L. Nussbaum, D. Kostič, and P. Mandel, *Clin. Chim. Acta*, **38**, 147 (1972).
143. J. Harris and J. D. Klingman, *J. Neurochem.*, **19**, 1267 (1972).

144. I. Sankoff and T. L. Sourkes, *Can. J. Biochem. Physiol.*, **41**, 1381 (1963).
145. M. R. Sahasrabudhe, *J. Assoc. Off. Agric. Chem.*, **47**, 888 (1964).
146. A. Rutkowski, H. Kozlowska, and J. Szerszynski, *Rocz. Panstw. Zakl. Hig.*, **14**, 361 (1963).
147. H. O. Bang, *J. Chromatogr.*, **14**, 520 (1964).
148. W. Bernauer and L. Schmidt, *Arch. Ex. Pathol. Pharmakol.*, **246**, 68 (1963).
149. B. Frosch and H. Wagener, *Klin. Wochenschr.*, **42**, 901 (1964).
150. H. Struck, *Mikrochim. Acta*, **1961**, 634.
151. T. Bićan-Fišter, *J. Chromatogr.*, **22**, 465 (1966).
152. M. Luisi, C. Fassorra, C. Levanti, and F. Franchi, *J. Chromatogr.*, **58**, 213 (1971).
153. B. W. Grunbaum and N. Pace, *Microchem. J.*, **15**, 103 (1970).
154. T. Bićan-Fišter and J. Merkaš, *J. Chromatogr.*, **41**, 91 (1969).
155. J. Bruinvels, *Experientia*, **19**, 551 (1963).
156. K. H. Mueller and H. Honerlagen, *Arch. Pharm.*, **193/65**, 202 (1960).
157. P. Braeckman, R. van Severen, and L. De Jaeger- Van Moeseke, *Pharm. Tijdschr. Belg.*, **40**, 113 (1963); through *Chem. Abstr.*, **60**, 1541 (1964).
158. S. H. Vannier and W. L. Stanley, *J. Assoc. Off. Agric. Chem.*, **41**, 432 (1958).
159. K. Genest and C. G. Farmilo, *J. Pharm. Pharmacol.*, **16**, 250 (1964).
160. M. Kutáček and Z. Procházka, *Colloq. Int. Centre Natl. Rech. Sci. (Paris)*, **123**, 445 (1963, (Pub. 1964); through *Chem. Abstr.*, **61**, 9760 (1964).
161. A. Gerstmeyer, *Zentr. Pharm., Pharmakother. Laboratoriumsdiagn.*, **111**, 821 (1972); through *Anal. Abstr.*, **24**, 1745 (1973).
162. S. Takahashi and L. R. Gjessing, *Clin. Chim. Acta*, **36**, 369 (1972).
163. G. Saglietto and G. Mantovani, *Z. Zuckerind.*, **20**, 17 (1970).
164. M. Doss, *Z. Klin. Chem. Biochem.*, **8**, 208 (1970).
165. M. Doss, D. Look, H. Henning, G. J. Lueders, W. Doelle, and G. Strohmeyer, *Z. Klin. Chem. Biochem.*, **9**, 471 (1971).
166. E. Vioque, L. J. Morris, and R. T. Holman, *J. Am. Oil Chem. Soc.*, **38**, 489 (1961).
167. L. Angelelli, G. Cavina, G. Moretti, and P. Siniscalchi, *Farmaco Ed. Prat.*, **21**, 493 (1966).
168. J. S. Huang, G. L. Downes, F. O. Belzer, *J. Lipid Res.*, **12**, 622 (1971).
169. B. I. Mitsner, Z. S. Syrtsova, V. M. Kopylov, E. N. Zvonkova, and K. A. Andrianov, *Zh. Obshch. Khim.*, **40**, 942 (1970).
170. B. Bazzi, R. Fabbrini, and M. Radice, *J. Assoc. Off. Anal. Chem.*, **54**, 1313 (1971).
171. V. Preininger, A. D. Cross, J. W. Murphy, F. Šantavý, and T. Toube, *Collec. Czech. Chem. Commun.*, **34**, 875 (1969).
172. H. Potešilová, H. Šantavý, A. El-Hamidi, and F. Šantavý, *Collect. Czech. Chem. Commun.*, **34**, 3540 (1969).
173. H. Steuerle and W. Pfab, *Dtsch. Lebensm.-Rundsch.*, **65**, 113 (1969).
173a. H. Koenig, *Z. Anal. Chem.*, **251**, 359 (1970).
173b. Z. Procházka and Nguyen Gia Chan, *Collect. Czech. Chem. Commun.*, **35**, 2209 (1970).
174. A. Vacíková, V. Felt, and J. Malíková, *J. Chromatogr.*, **9**, 301 (1962).

175. M. Ikram and M. K. Bakhsh, *Anal. Chem.*, **36,** 111 (1964).
176. G. Pastuska, *Z. Anal. Chem.*, **179,** 427 (1961).
177. F. Dobici and G. Grassini, *J. Chromatogr.*, **10,** 98 (1963).
177a. W. Boardmen and B. Warren, *Proc. Soc. Anal. Chem. Conf., Nottingham, England,* **1965,** 151.
178. E. Csizér and S. Goeroeg, *J. Chromatogr.*, **76,** 502 (1973).
179. J. Udris, *Analyst (London)*, **96,** 130 (1971).
180. J. J. A. Wijnne, E. Bletz, and J. M. Frijns, *Pharm. Weekbl.*, **102,** 959 (1967).
181. V. Vukcevic-Kovacevic and S. Seremet, *Bull. Sci. Conseil Acad. RPF Yougosl.*, **14,** 377 (1969).
182. D. E. Bowyer, W. M. F. Leat, A. N. Howard, and G. A. Gresham, *Biochem. J.*, **89,** 24P (1963).
183. D. E. Bowyer, W. M. F. Leat, A. N. Howard, and G. A. Gresham, *Biochim. Biophys. Acta,* **70,** 423 (1963).
184. M. Z. Nichaman, C. C. Sweeley, N. M. Oldham, and R. E. Olson, *J. Lipid Res.*, **4,** 484 (1963).
185. W. Futterweit, N. L. McNiven, L. Narcus, C. Lantos, M. Drosdowsky, and R. I. Dorfman, *Steroids,* **1,** 628 (1963).
186. R. Guerra-Garcia, S. C. Chattoraj, L. J. Gabrilove, and H. H. Wotiz, *Steroids,* **2,** 605 (1963).
187. J. C. Hoffsommer, *J. Chromatogr.*, **51,** 243 (1970).
188. J. Volke and H. Oelschlaeger, *Sci. Pharm. Proc., 25th 1965,* **2,** 105 (1966).
189. A. Kotarski and K. Mosinska, *Chem. Anal. (Warsaw),* **12,** 329 (1967).
189a. J. Kováč, *J. Chromatogr.*, **11,** 412 (1963).
190. R. J. T. Graham, A. E. Nya, D. A. Tinsley, *Int. Symp. Chromatogr. Electrophor., Lect. Pap., 6th 1970,* Ann Arbor Science, Ann Arbor, Mich., 1971, p. 105.
191. W. Huber, *Chromatographia,* **1,** 212 (1968).
191a. R. R. Fike, U.S. Pat. 3,752,744 (August 14, 1973).
192. V. Pollak and A. A. Boulton, *J. Chromatogr.*, **72,** 231 (1972).
193. J. M. Varga and F. F. Richards, *Anal. Biochem.*, **53,** 397 (1973).
194. J. Ripphahn and H. Halpaap, *J. Chromatogr.*, **112,** 81 (1975).
195. J. C. Touchstone, S. S. Levine, and T. Muravec, *Anal. Chem.*, **43,** 858 (1971).
196. G. Pataki and K.-T. Wang, *J. Chromatogr.*, **37,** 499 (1968).
197. M. Seiler and M. Weichmann, *Z. Anal. Chem.*, **220,** 109 (1966).
198. D. Janchen and G. Pataki, *J. Chromatogr.*, **33,** 391 (1968).
199. B. R. Smith, *J. Chromatogr.*, **82,** 95 (1973).
200. R. Segura and A. M. Gotto, Jr., *J. Chromatogr.*, **99,** 643 (1974).
201. P. M. Houpt, *X-Ray Spectrom.*, **1,** 37 (1972).
202. R. A. Libby, *Anal. Chem.*, **40,** 1507 (1968).
203. L. R. Chapman, *J. Chromatogr.*, **66,** 303 (1972).
204. J. Alter, D. Diehlmann, R. Beydatsch, P. Kohler, D. Quass, and W. Spichale, *Z. Anal. Chem.*, **233,** 188 (1968).
205. S. Ebel and G. Herold, *Z. Anal. Chem.*, **266,** 281 (1973).
206. W. M. Connors and W. K. Boak, *J. Chromatogr.*, **16,** 243 (1964).

207. L. Sattler and F. W. Zerban, *Anal. Chem.*, **24,** 1862 (1952).
208. J. G. Kirchner, J. M. Miller, and G. J. Keller, *Anal. Chem.*, **23,** 420 (1951).
209. G. Pataki, *Chromatographia,* **1,** 492 (1968).
210. M. M. Frodyma, R. W. Frei, and D. J. Williams, *J. Chromatogr.*, **13,** 61 (1964).
211. M. S. J. Dallas, *J. Chromatogr.*, **33,** 337 (1968).
212. R. W. Frei, *J. Chromatogr.*, **64,** 285 (1972).
213. H. Jork, *J. Chromatogr.*, **48,** 372 (1970).
214. L. R. Treiber, R. Nordberg, S. Lindstedt, and P. Stoellnberger, *J. Chromatogr.*, **63,** 211 (1971).
215. H. Jork, *J. Chromatogr.*, **82,** 85 (1973).
216. U. Hezel, *Angew. Chem. Int. Ed. Engl.*, **12,** 298 (1973).
217. H. Zuercher, G. Pataki, J. Boroko, and R. W. Frei, *J. Chromatogr.*, **43,** 457 (1969).
218. S. Ebel and G. Herold, *Chromatographia,* **8,** 35 (1975).
219. H. Yamamoto, T. Kurita, J. Suzuki, R. Hira, K. Nakano, H. Makabe, and K. Shibata, *J. Chromatogr.*, **116,** 29 (1976).
220. H. Bethke, W. Santi, and R. W. Frei, *J. Chromatogr. Sci.*, **12,** 392 (1974).
221. V. T. Lieu and M. M. Frodyma, *Talanta,* **13,** 1319 (1966).
222. S. Ebel and G. Herold, *Z. Anal. Chem.*, **270,** 19 (1974).
223. G. Kynast, *Chromatographia,* **3,** 425 (1970).
224. R. W. Frei, "Reflectance Spectroscopy in Thin-Layer Chromatography," in *Progress in Thin-Layer Chromatography & Related Methods,* Vol. II, A. Niederwieser and G. Pataki, Eds., Ann Arbor Science, Ann Arbor, Mich., 1971, p. 1.
225. S. Hara, M. Takeuchi, M. Tachibana, and G. Chichrara, *Chem. Pharm. Bull. (Tokyo),* **12,** 483 (1964).
226. M. S. J. Dallas, *J. Chromatogr.*, **33,** 337 (1968).
227. E. J. Shellard and M. Z. Alam, *J. Chromatogr.*, **33,** 347 (1968).
228. N. M. Neskovic, *Arch. Inst. Pasteur Tunis,* **43,** 513 (1966).
229. A. E. Thomas, III, J. E. Scharoun, and H. Ralston, *J. Am. Oil Chem. Soc.*, **42,** 789 (1965).
230. M. L. Blank, J. A. Schmit, and O. S. Privett, *J. Am. Oil Chem. Soc.*, **41,** 371 (1964).
231. P. Turano and W. J. Turner, *J. Chromatogr.*, **64,** 347 (1972).
232. A. Waksmundzki and J. K. Różylo, *J. Chromatogr.*, **78,** 55 (1973).
233. H. Bethke and R. W. Frei, *J. Chromatogr.*, **91,** 433 (1974).
234. L. R. Treiber, B. Oertengren, R. Lindsten, and T. Oertegren, *J. Chromatogr.*, **73,** 151 (1972).
235. L. R. Treiber, *J. Chromatogr.*, **69,** 399 (1972).
236. *Ibid.*, **100,** 123 (1974).
237. L. Salganicoff, M. Kraybill, D. Mayer, and V. Legallais, *J. Chromatogr.*, **26,** 434 (1967).
238. B. Chance, *Rev. Sci. Instr.*, **22,** 619 (1951).
239. B. Chance, in *Methods in Enzymology,* Vol. 12, S. P. Colowick and N. O. Kaplan, Eds., Academic Press, New York, 1957, p. 273.

240. V. Pollak, *J. Chromatogr.*, **63,** 145 (1971).
241. A. A. Boulton and V. Pollak, *J. Chromatogr.*, **63,** 75 (1971).
242. S. Turina and M. Kaštelan-Macan, *J. Chromatogr.*, **48,** 35 (1970).
243. J. Goldman and R. R. Goodall, *J. Chromatogr.*, **40,** 345 (1969).
244. *Ibid.*, **47,** 386 (1970).
245. R. R. Goodall, *J. Chromatogr.*, **78,** 153 (1973).
246. *Ibid.*, **103,** 265 (1975).
247. S. Ebel and H. Kussmaul, *Z. Anal. Chem.*, **268,** 268 (1974).
248. *Ibid.*, **269,** 10 (1974).
249. S. Ebel and G. Herold, *Z. Anal. Chem.*, **270,** 19 (1974).
250. J. M. Owen, *J. Chromatogr.*, **79,** 165 (1973).
251. A. Franck, *Chem.-Ztg.*, **92,** 115 (1969).
252. S. Ebel and J. Hocke, *Chromatographia*, **9,** 78 (1976).
253. R. A. Zeineh, W. P. Nijm, and F. H. Al-Azzawi, *Am. Lab.*, **7,** 51 (1975).
254. F. W. Hefendehl, *Planta Med.*, **8,** 65 (1960).
255. S. M. Rybicka, *Chem. Ind.* (*London*), **1962,** 1947.
256. H. Rasmussen, *J. Chromatogr.*, **27,** 142 (1967).
257. M. L. Blank, J. A. Schmit, and O. S. Privett, *J. Am. Oil Chem. Soc.*, **41,** 371 (1964).
258. E. Vioque and A. Vioque, *Grasas Aceites (Seville, Spain)*, **15,** 125 (1964).
259. P. C. Kelleher, *J. Chromatogr.*, **52,** 437 (1970).
260. A. N. Siakotos and G. Rouser, *Anal. Biochem.*, **14,** 162 (1966).
261. G. M. Jacobsohn, *Anal. Chem.*, **36,** 2030 (1964).
262. R. R. Goodall, *J. Chromatogr.*, **123,** 5 (1976).
262a. V. Pollak, *J. Chromatogr.*, **123,** 11 (1976).
263. L. R. Treiber, *J. Chromatogr.*, **123,** 23 (1976).
264. J. Wright, *Chem. Ind.* (*London*), **1963,** 1125.
265. V. Castagnola, *Boll. Chim. Farm.*, **102,** 784 (1963).
266. D. Groeger, V. E. Tyler, Jr., and J. E. Dusenberry, *Lloydia*, **24,** 97 (1961).
267. T. Hohmann and H. Rochelmeyer, *Arch. Pharm.*, **297,** 186 (1964).
268. E. Nuernberg, *Arch. Pharm.*, **292/64,** 610 (1959).
269. J. G. L. Harthon, *Acta Chem. Scand.*, **15,** 1401 (1961).
270. N. J. Gurvich, *Vses. Nauchin.-Issled., Inst. Maslichn. Efiromasl. Kult. Vses. Akad. S/Kh. Nauk, Kratkii Otchet*, **1956,** 154; through *Chem. Abstr.*, **54,** 25595 (1960).
271. E. Stahl, *Chem.-Zig.*, **82,** 323 (1958).
272. E. Tyihák, I. Sárkańy-Kiss, and J. Máthe, *Pharm. Zentralhalle*, **102,** 128 (1963).
273. D. Waldi, *Arch. Pharm.*, **295,** 125 (1962).
274. A. Winterstein and B. Hegedues, *Z. Physiol. Chem.*, **312,** 97 (1960).
275. J. P. Vuilleumier, G. Brubacher, and M. Kalivoda, *Helv. Chim. Acta*, **46,** 2983 (1963).
276. P. A. Nussbaumer, *Pharm. Acta Helv.*, **37,** 161 (1962).
277. W. Bernauer, L. Schmidt, and G. Ullman, *Med. Exp.*, **9,** 191 (1963).
278. M. J. D. Van Dam, G. J. De Kleuver, and J. G. de Heus, *J. Chromatogr.*, **4,** 26 (1960).

279. D. Waldi, *Lab. Sci. (Milan)*, **11,** 81 (1963).
280. C. G. Honegger, *Helv. Chim. Acta,* **45,** 2020 (1962).
281. H. Meyer, *Rev. Int. Chocolat.,* **17,** 270 (1962).
282. M. J. D. Van Dam, *Bull. Soc. Chim. Belg.,* **70,** 122 (1961).
283. M. F. Kovacs, Jr., *J. Assoc. Off. Agric. Chem.,* **46,** 884 (1963).
284. F. Radler and M. V. Grncarevic, *J. Agric. Food Chem.,* **12,** 266 (1964).
285. P. Kuenzi, inaugural dissertation, Basel University, 1962.
286. M. K. Seikel, M. A. Millett, and J. F. Saeman, *J. Chromatogr.,* **15,** 115 (1964).
287. H. Seiler, *Helv. Chim. Acta,* **46,** 2629 (1963).
288. N. Oswald and H. Flueck, *Pharm. Acta Helv.,* **39,** 293 (1964).
289. H.-J. Petrowitz, *Mitt. Dtsch. Ges. Holzforsch.,* **48,** 57 (1961).
290. A. Seher, *Mikrochim. Acta,* **1961,** 308.
291. A. Seher, *Nahrung,* **4,** 466 (1960).
292. M. Brenner and A. Niederwieser, *Experientia,* **16,** 378 (1960).
293. J. Aurenge, M. DeGeorges, and J. Normand, *Bull. Soc. Chim. Fr.,* **1963,** 1732.
294. S. J. Purdy and E. V. Truter, *Chem. Ind. (London)*, **1962,** 506.
295. S. J. Purdy and E. V. Truter, *Analyst,* **87,** 802 (1962).
296. N. Nybom, *J. Chromatogr.,* **28,** 447 (1967).
297. J. H. Broadbent, J. A. Cornelius, and G. Shone, *Analyst (London)*, **88,** 214 (1963).
298. H. Hasselquist and M. Jaarma, *Acta Chem. Scand.,* **17,** 529 (1963).
299. H. Lagoni and A. Wortmann, *Milchwissenschaft,* **11,** 206 (1956).
299a. J. Pilný, E. Svojtková, M. Juřicová, and Z. Deyl, *J. Chromatogr.,* **78,** 161 (1973).
300. T. F. Brodasky, *Anal. Chem.,* **35,** 343 (1963).
301. C. A. Peterson and L. V. Edgington, *J. Agric. Food Chem.,* **17,** 898 (1969).
302. E. W. Horton and C. J. Thompson, *Brit. J. Pharmacol.,* **22,** 183 (1964).
303. H. Kaldewey and E. Stahl, *Planta,* **62,** 22 (1964).
304. G. Collet, *Compt. Rend.,* **259,** 871 (1964).
305. J. Dubouchet and P.-E. Pilet, *Ann. Physiol. Veg.,* **5,** 175 (1963).
306. L. Cima and R. Mantovan, *Farm. (Pavia), Ed. Prat.,* **17,** 473 (1962).
307. T. Salo, K. Salminen, and K. Fiskari, *Z. Lebensm. Unters.-Forsch.,* **117,** 369 (1962).
308. J. Billow and T. J. Speaker, *J. Chromatogr.,* **67,** 191 (1972).
309. H. K. Mangold, R. Kammereck, and D. C. Malins, *Microchem. J., Symp. Ser. 1961,* **2,** 697 (1962).
310. A. Vermeulen and J. C. M. Verplancke, *Steroids,* **2,** 453 (1963).
311. P.-E. Schulze and M. Wenzel, *Angew. Chem., Int. Ed. Engl.,* **1,** 580 (1962).
312. M. Wenzel and P.-E. Schulze, *Tritium-Markierung,* W. de Gruyter, Berlin, 1962, 176 pp.
313. F. Snyder, "Thin-Layer Radiochromatography and Related Procedures," in *Progress in Thin-Layer Chromatography & Related Methods,* Vol. 1, A. Niederwieser and G. Pataki, Eds., Ann Arbor Science, Ann Arbor, 1970, p. 53.

314. J. R. Catch, *Purity and Analysis of Labeled Compounds,* U.K. Atomic Energy Authority, Radiochemistry Centre, 1968, 24 pp.
315. K. Figge, H. Picter, and H. Ossenbrueggen, *Glas-Instrum.-Tech.,* **14,** 900, 907, 913, 1013, 1019, 1025 (1970).
316. F. W. Koss and D. Jerchel, *Radiochim. Acta,* **3,** 220 (1964).
317. R. A. Schwane and R. S. Nakon, *Anal. Chem.,* **37,** 315 (1965).
318. A. T. Wilson and D. J. Spedding, *J. Chromatogr.,* **18,** 76 (1965).
319. J. Chamberlain, A. Hughes, A. W. Rogers, and G. H. Thomas, *Nature,* **201,** 774 (1964).
320. C. O. Tio and S. F. Sisenwine, *J. Chromatogr.,* **48,** 555 (1970).
321. K. M. Prescott and G. S. David, *Anal. Biochem.,* **57,** 232 (1974).
322. K. Randerath, *Anal. Chem.,* **41,** 991 (1969).
323. K. Randerath, *Anal. Biochem.,* **34,** 188 (1970).
324. S. Prydz, T. B. Melo, E. L. Eriksen, and J. F. Koren, *J. Chromatogr.,* **47,** 157 (1970).
325. S. Prydz, T. B. Melo, and J. F. Koren, *Anal. Chem.,* **45,** 2106 (1973).
326. S. Prydz, T. B. Meloe, J. F. Koren, and E. L. Ericksen, *Anal. Chem.,* **42,** 156 (1970).
327. A. T. Wilson and D. J. Spedding, *J. Chromatogr.,* **18,** 76 (1965).
328. A. Breccia and F. Spalletti, *Nature,* **198,** 756 (1963).
329. J. Rosenberg and M. Bolgar, *Anal. Chem.,* **35,** 1559 (1963).
330. J. Rosenberg and M. Bolgar, *Anal. Chem.,* **35,** 1559 (1963).
331. J. R. Ravenhill and A. T. James, *J. Chromatogr.,* **26,** 89 (1967).
332. F. Snyder, *Advances in Tracer Methodology,* Vol. 4, Plenum, New York, 1968, p. 81.
333. F. Snyder, *Isotope Radiat. Technol.,* **6,** 381 (1969).
334. S. Prydz, *Anal. Chem.,* **45,** 2317 (1973).
335. A. S. Csallany and H. H. Draper, *Anal. Biochem.,* **4,** 418 (1962).
336. A. Moghissi, *J. Chromatogr.,* **13,** 542 (1964).
337. R. J. Redgwell, N. A. Turner, and R. L. Bieleski, *J. Chromatogr.,* **88,** 25 (1974).
338. R. L. Squibb, *Nature,* **198,** 317 (1963).
339. S. Arnoff, *Nucleonics,* **14,** 92 (1956).
340. G. W. Gilbert and J. P. Keene, *Radioisotopes Sci. Res., Proc. Int. Conf. Paris,* **1,** 698 (1957).
341. M. Wenzel and K. Hoffmann, *Anal. Biochem.,* **44,** 97 (1971).
342. P. V. Hariharan, G. Poole, and H. E. Johns, *J. Chromatogr.,* **32,** 356 (1968).
343. S. Prydz, T. B. Melo, and J. F. Koren, *J. Chromatogr.,* **59,** 99 (1971).
344. T. O. Seim and S. Prydz, *J. Chromatogr.,* **73,** 183 (1972).
345. A. E. Lowe, "Bidimensional Radiochromatogram Scanning," in *Quantitative Paper and Thin-Layer Chromatography,* E. J. Shellard, Ed., Academic Press, London, 1968, p. 119.
346. B. L. Wedzicha, *J. Chromatogr.,* **101,** 79 (1974).
347. R. Tykva, *Advances in Physical and Biological Radiation Detectors,* International Atomic Energy Agency, Vienna, 1971, p. 211.
348. R. Tykva, *Excerpta Med. Int. Congr. Ser.,* No. 301, 455 (1973).

349. R. Tykva and I. Votruba, *J. Chromatogr.*, **93,** 399 (1974).
350. B. R. Pullan, "The Evaluation of Radiochromatograms Using a Spark Chamber," in *Quantitative Paper and Thin-Layer Chromatography,* E. J. Shellard, Ed., Academic Press, London, 1968, p. 128.
351. J. L. Brown and J. M. Johnston, *J. Lipid Res.*, **3** 480 (1962).
352. A. Riondel, J. F. Tait, M. Gut, S. A. S. Tait, E. Joachim, and B. Little, *J. Clin. Endocrinol. Metab.*, **23,** 620 (1963).
353. R. S. Rivlin and H. Wilson, *Anal. Biochem.*, **5,** 267 (1963).
354. J.-C. Roucayrol and P. Taillandier, *Compt. Rend.*, **256,** 4653 (1963).
355. S. Seno, W. V. Kessler, and J. E. Christian, *J. Pharm. Sci.*, **53,** 1101 (1964).
356. F. Snyder and N. Stephens, *Anal. Biochem.*, **4,** 128 (1962).
357. R. Manning and D. N. Brindley, *Biochem. J.*, **130,** 1003 (1972).
358. S. Mookerjea, D. E. C. Cole, A. Chow, and P. Letts, *Can. J. Biochem.*, **50,** 1094 (1972).
359. F. Snyder, *Sep. Sci.*, **1,** 655 (1966).
360. F. Snyder, T. J. Alford, and H. Kimble, *U.S. At. Energy Comm. ORINS Rep.*, **44.**
361. F. Snyder and H. Kimble, *U.S. At. Energy Comm. ORINS Rep.*, **47.**
362. F. Snyder, "Liquid Scintillation Radioassay of Thin-Layer Chromatograms," in *Current Status of Liquid Scintillation Counting,* E. D. Bransome, Jr., Ed., Grune and Stratton, New York, 1970, p. 248.
363. E. Fosslien, F. Musil, D. Domizi, L. Blickenstaff, and J. Luming, *J. Chromatogr.*, **63,** 131 (1971).
364. M. I. Krichevsky, S. A. Zaveler, and J. Bulkeley, *Anal. Biochem.*, **22,** 442 (1968).
365. R. L. Boeckx, D. J. Protti, and K. Dakshinamurti, *Anal. Biochem.*, **53,** 491 (1973).
366. E. Gruenstein and T. W. Smith, *Anal. Biochem.*, **61,** 429 (1974).
367. W. A. Shaw, W. R. Harlan, and A. Bennett, *Anal. Biochem.*, **43,** 119 (1971).
368. J. N. Bollinger, W. A. Mallow, J. W. Register, Jr., and D. E. Johnson, *Anal. Chem.*, **39,** 1508 (1967).
369. J. T. Baker Chemical Company, *Chemist-Analyst,* **65,** 2 (1976).
370. M. H. Griffiths and A. Mallinson, *Anal. Biochem.*, **22,** 465 (1968).
371. E. Haahti, R. Vihko, and I. Jaakonmaeki, *J. Chromatogr. Sci.*, **8,** 370 (1970).
372. R. L. Scheig, R. Annunziata, and L. A. Pesch, *Anal. Biochem.*, **5,** 291 (1963).
373. Y. Imai, K. Yamauchi, S. Terabe, R. Konaka, and J. Sugita, *J. Chromatogr.*, **36,** 545 (1968).
374. N. H. Paik and B. K. Kim, *Yakhak Hoeji,* **13,** 84 (1969); through *Chem. Abstr.*, **73,** 1812k (1970).
375. R. Konaka and S. Terabe, *J. Chromatogr.*, **24,** 236 (1966).
376. Y. Tsujino, Y. Imai, K. Yamauchi, and J. Sugita, *J. Chromatogr.*, **42,** 419 (1969).
377. T. Cotgreave and A. Lynes, *J. Chromatogr.*, **30,** 117 (1967).
378. F. B. Padley, *Chem. Ind.* (*London*), **1967,** 874.

379. F. B. Padley, *J. Chromatogr.*, **39**, 37 (1969).
380. J. J. Szakasits, P. V. Peurifoy, and L. A. Woods, *Anal. Chem.*, **42**, 351 (1970).
381. E. Haahti and I. Jaakonmaki, *Ann. Med. Exp. Biol. Fenn.*, **47**, 175 (1969).
382. K. D. Mukherjee, H. Spaans, and E. Haahti, *J. Chromatogr.*, **61**, 317 (1971).
383. K. O. Mukherjee, H. Spaans, and E. Haahti, *J. Chromatogr. Sci.*, **10**, 193 (1972).
384. K. D. Mukherjee, *J. Chromatogr.*, **96**, 242 (1974).
385. H. K. Mangold and K. D. Mukherjee, *J. Chromatogr. Sci.*, **13**, 398 (1975).
386. E. H. P. Kaufmann and K. D. Mukherjee, *Fette, Seifen, Anstrichm.*, **71**, 11 (1969).

Chapter **XII**

COMBINATION OF THIN-LAYER CHROMATOGRAPHY WITH OTHER TECHNIQUES

1 COLUMN CHROMATOGRAPHY

Thin-layer chromatography can be used to select the best combination of solvent and adsorbent for a given column separation. Later, as the column is being run, the eluent can be checked rapidly by thin-layer chromatography so that by the time the separation has been completed the arbitrary fractions with similar components can be combined. This technique, which was thoroughly developed by Miller and Kirchner [1], has been described in Chapter X and illustrated in Fig. 10.3. This combination technique with column chromatography has been used for many different types of compounds, and there are many individual papers in which this procedure has been employed.

Van Dijk [2] has coupled column chromatography directly with thin-layer chromatography by using a variable splitter with a variable drive to control the application of a portion of the column eluate to the TLC plate.

With the advent of HPLC, thin-layer chromatography can be used to select the proper solvent and to optimize the separation conditions for HPLC [3–6]. This may be easily and quickly accomplished by using the HPRC equipment of CAMAG (see Chapter V, Section 4).

2 GAS CHROMATOGRAPHY

There are a number of ways in which thin-layer chromatography can be combined to advantage with gas chromatography. The spots obtained from thin-layer chromatography may be eluted, concentrated, and then subjected to gas chromatographic analysis. This method has been applied by Ikeda et al. [7, 8] in the analysis of citrus oils and other essential oils [9]. Morris et al. [10] have separated the esters of unsaturated hydroxy acids by thin-layer chromatography and then subjected the isolated compounds to gas chromatography. Litchfield et al. [11] separated triglycerides according to the number of double bonds by chromatographing on thin layers of silver nitrate impregnated with silicic acid. These fractions were then recovered and separated according to molecular weight by using gas chromatography. The method has been used for vitamins [12, 13], azaheterocyclics [14], pesticides [15,16], and polynuclear hydrocarbons [17].

Janák [18] combined centripetal thin-layer chromatography with gas chromatography for the analysis of ultratrace amounts of lindane in cabbage extracts. Eluates from the centripetal fractionation were injected into the gas chromatograph.

Anderson et al. [19] have used thin-layer chromatography to isolate the oxidation products of BHA(3-*tert*-butyl-4-hydroxyanisole)and BHT(3,5-di-*tert*-butyl-4-hydroxytoluene) in breakfast cereals from interfering fats so that the oxidation products could be injected into a gas column.

Janák [20] has interspersed thin-layer chromatography between two gas chromatographic runs. In this way, the fractions from the first gas chromatogram were subjected to thin-layer chromatography and the fractions resulting from the latter separation were then rechromatographed on a gas column. The method proved to be especially useful for high-boiling substances.

The fractions from the chromatoplates can also be altered prior to running a gas chromatographic separation. In this way, the mono-and diglycerides from lipolyzed milk fat, after being recovered from thin-layer plates, were converted to methyl and butyl esters prior to gas chromatographic analysis. Similar techniques have been applied to tissue lipids [21, 22]. Malins [23] separated alkoxydiglycerides from other lipids by thin-layer silicic acid chromatography. This was then followed by saponification and methylation to yield fatty acid methyl esters and glyceryl ethers. The methyl esters were analyzed by gas chromatography. In order to analyze the glyceryl ethers, they were first oxidized and the resulting aldehydes were separated by gas chromatography. Privett and co-workers [24–27] have analyzed unsaturated methyl esters of fatty acids by first separating the esters according to degree of unsaturation on silver nitrate-impregnated silica gel. The separated fractions were eluted and treated to reductive ozonolysis, and the resulting fragments were then analyzed by gas chromatography using a flame ionization detector. Details of the method are given under the methyl esters section under lipids.

The lipids can be converted to their methyl esters by treating with methanolic boron trifluoride [28] or they may be converted to the trimethylsilyl derivatives [29, 30]. Gedam et al. [31] separated the methyl esters of the fatty acids from sardine oil and then hydrogenated the fractions before submitting them to gas chromatographic analysis.

Trimethylsilyl derivatives have also been used for the gas chromatographic analysis of TLC fractionation of steroids [32, 33] as have acetylated derivatives [34]. There are many mixtures of steroids that contain isomers which are difficult to separate, and thin-layer and gas chromatography complement one another so that complete separations can be made [35]. Lisboa [36] gives some good examples in which the combination of thin-layer chromatography with gas chromatography–mass spectrometry is indispensable for the complete identification of steroids.

For the determination of subnanogram levels of cardenolides Watson et al. [37] first separated them by thin-layer chromatography and then converted the fractions to the fluorobutyrates by reacting with heptafluorobutyric anhydride. The derivatives were then determined by gas chromatography using an electron capture detector.

Morris [38] separated cholesterol esters on silver nitrate-impregnated layers. The individual fractions were converted to methyl esters by transmethylation and then analyzed by gas chromatography.

Mangold and Kammereck [39] used a combination method in fractionating the methyl esters of fatty acids. The hydroxylated acid esters which would interfere in the gas chromatographic separation were first removed by thin-layer chromatography. Then the purified esters were treated with mercuric acetate in order to prepare the acetoxymercuric methoxy derivatives of the unsaturated esters. These were separated into classes by thin-layer chromatography on silicic acid according to the degree of unsaturation. The regenerated esters were then subjected to gas chromatography. Pairs of esters which could not be resolved by gas chromatography could be separated by the thin-layer chromatography of their mercuric derivatives and vice versa. Jensen and Sampugna [40] applied the method to the identification of milk fatty acids.

The reverse procedure—that is, the examination of gas chromatographic fractions by thin-layer chromatography—has also been employed [41–44].

De Klein [45] has used gas chromatography in a different manner to quantitate fractions obtained by thin-layer chromatography. He scraped the spots from the TLC plate into ampuls, which were then introduced into the ampul-crusher injection port of a gas chromatograph. Yields of 96 to 100% were obtained by using temperature programming after crushing the ampul. By using this technique there is less chance for loss of sample, there are no interfering solvent peaks, and the sealed ampuls may be stored. Franc and Šenkýřová [46] scraped the adsorbent containing the compound directly into a reaction gas unit, and the gaseous products of the reaction were then determined by gas chromatography.

Another application of the combination of thin-layer chromatography with gas chromatography has been the direct application of the compounds to the thin-layer plate as they emerge from the exit tube of the gas chromatographic apparatus. The first application of this was by Casu and Cavallotti [47]. By means of a motor and a system of gears, a chromatostrip was carried along underneath the exit tube at a distance of 1 mm at the same rate as the recorder chart. These workers, however, did not apply thin-layer chromatography as such to the strips. They used the strips as a means of locating specific types of compounds. For example, aliphatic and aromatic hydrocarbons were located by using a strip wet with sulfuric acid-formaldehyde.

At approximately the same time, three independent groups have investigated the direct coupling of thin-layer chromatography and gas chromatography. Nigam et al. [48] applied the method to the essential oils of *Mentha* and *Eucalyptus*, and this was accomplished by placing the thin-layer plate directly under a 1-mm (i.d.) platinum needle inserted in the exit of the vapor-phase unit, whenever a peak appeared on the strip-chart recorder. The chromatoplate was then developed in hexane–diethyl ether (10:1). A more elegant procedure is that described by Janák [49, 50] and Kaiser [51]. In this case a thin-layer plate is moved along under the outlet from the gas chromatographic unit. This movement is variable so that the effluent from the column can be laid down in a strip as the plate moves along at the same rate as the chart recorder for the gas chromatograph; or it may be activated by the recorder, so that the individual peaks are deposited as individual spots during which time the plate motion is stopped. This latter procedure gives a more concentrated application of the sample. After the plate has been spotted, it is developed in a normal fashion. As pointed out by Kaiser, a number of situations can arise in which the two systems supplement one another, especially since the separations are based on different characteristics. Gas chromatography separates according to the relative volatility and thin-layer chromatography according to the functional groups present. Janák et al.

[52], who have used loose layers in this technique, have applied a coupling device in which the thin-layer plate is moved logarithmically with time, while the gas chromatogram operates under isothermal conditions. By using this procedure, they were able to apply the Kováts retention indices [53] for identifying the zones and also to correlate the zones with the boiling points. The applicability of coupled gas–thin-layer chromatography can be seen from Fig. 12.1 taken from Kaiser [51].

Tumlinson et al. [54] have used this technique for the identification of carbonyl compounds by directing the GC exit stream onto a TLC starting line which had been treated with 2,4-dinitrophenylhydrazine or other carbonyl reagent. The derivatives were then chromatographed. Conversely, Curtius and Mueller [55], after adsorbing trimethylsilyl ethers of steroids on silica gel plates, sprayed the layers with 1% hydrochloric acid in methanol to free the steroids before chromatographing; the trimethylsilyl ethers were also chromatographed. Parker et al. [56] have adapted coupled GC-TLC to forensic investigations, thus providing a better separation technique as well as confirmatory evidence for the presence or absence of a compound. Kaiser [57] and Janák [58] give a general dis-

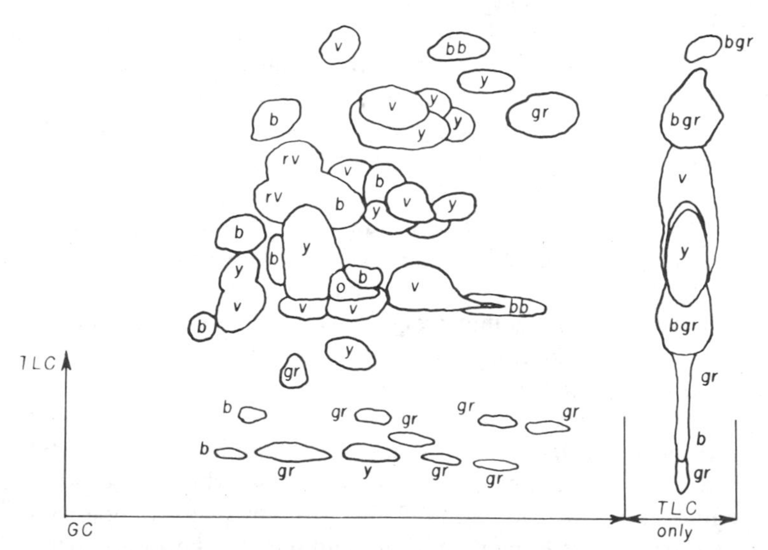

Fig. 12.1 Separation of a very close distillation fraction of xylenols in which sulfur compounds are present as well as nonphenolic. GC-TLC-coupled and TLC-control chromatograms. Visualized with dibromoquinonechloroimide and exposure to ammonia vapors, then hydrochloric acid. More than 42 components. *v*, violet; *rv*, red-violet; *bb*, bright blue; *b*, blue; *y*, yellow; *gr*, gray; *o*, orange. Silica gel GF with chloroform. From Kaiser [51]; reproduced with permission of the author and Springer-Verlag.

cussion of the subject of coupled GC-TLC techniques. Humphrey [59] describes an ingenious way of moving the TLC plate by taping it to the GC recorder paper hanging over the edge of a flat-plate recorder. Commercially CAMAG supplies the "Dichrome" unit for moving the TLC plate along the GC exit at varying speeds and also by stepwise motion if desired.

Another technique employing GC detectors has been developed and is discussed in Chapter XI, Section 7.

3 SPECTROPHOTOMETRY

Mass spectrometry and thin-layer chromatography can be combined directly by placing the silica gel adsorbed spots in the ion source as proposed by Heyns and Gruetzmacher [61]. Fétizon [62] has discussed the applications of mass spectrometry with thin-layer separations. Deverse et al. [63] describe an inlet system for the introduction of samples where the sample-flow rate is regulated by the temperature control. Nilsson et al. [64] also describe a simple device for introducing either sample-containing adsorbent or liquids (eluted samples). Down and Gwyn [65] examined the parameters affecting the mass spectrometry of adsorbed drugs. Diffuse sample spots could not always be accommodated in the sample tip, and they confirmed the findings of Heyns and Gruetzmacher [61] that the threshold below which poor spectra were obtained lay near the point where the compound to silica gel ratio is 1 to 2%. The number and intensity of the peaks depended on the solvent used for the development of the chromatogram. Some solvents gave more intense background peaks than others and indicated the need to use highly purified solvents (See Chapter V, Section 1). Samples developed with methanol gave peaks characteristic of phthalate plasticizers; light petroleum ethers, intense high-molecular-weight hydrocarbon peaks; chloroform, some background peaks up to about *m/e* 100 below 100 to 150°C, which disappeared almost completely at higher temperatures; ammonia–methanol mixtures, very low background interference; solvents containing acetic acid, large background peaks at *m/e* 60 and below, at temperatures up to 120°C, which dropped off in intensity above this temperature. With probe temperatures of 250 to 300°C all of the drugs examined gave high-quality spectra except codeine and methadone. Background spectra may be run to subtract from the sample spectrum to improve the lower working limits. Volatilization from silica gel usually caused one or two main peaks to disappear and sometimes one or two new major peaks to appear.

The compounds may also be eluted from the adsorbent placed in a capillary tube and the eluate evaporated on the solid sample probe [66,

67]. In order to get rid of some background peaks, Székely [68] passed the eluate by capillary attraction through a capillary filled with potassium bromide; this helps to remove fine particles of silica gel.

Infrared techniques using ultramicro potassium bromide pellets with a beam condensing system can be combined nicely with thin layers. As an example of this, Nash et al. [69] used this procedure to identify the rotenone spot on a chromatogram in a sample of technical-grade rotenone. The spots were located under ultraviolet light on fluorescent silica gel layers and were then scraped into centrifuge tubes where they were eluted with spectro-grade ethyl acetate. The centrifuged eluate was then carefully transferred to a 10 × 30 mm stainless steel vial where the solvent was removed by an air stream. The last traces of solvent were removed in an Abderhalden apparatus prior to mixing the residue with spectro grade potassium bromide for pelletizing. Simple procedures have been described [70, 71] for preparing micro pellets of potassium bromide, and Anderson and Wilson [72] described a method for making pellets 0.15 to 0.35 mmm in diameter by using watch jewels as dies. These pellets are for samples less than 0.5 μ and are used in micro infrared analysis using Fourier transform infrared spectrometry. In this case very special care is needed to prevent contamination.

In general centrifugation is insufficient to remove extremely fine particles of silica gel. Spencer and Beggs [73] found that centrifuging the eluate at 3000 rpm for 5 min would not eliminate the very fine particles, but high-speed centrifugation at 10,000 rpm for 30 min would do a fairly effective job.

A number of techniques have been tried in order to eliminate the fine silica gel because some gel preparations contain an impurity giving bands at 3.4 and 7.2 μ [74]. Use of an ultrafine fritted funnel has been recommended [74], if a high vacuum is used so that filtration does not take *too* long. Spencer and Beggs [73] found that this was fairly effective, but was slow and the filter soon became clogged and could not be cleaned. They finally turned to membrane filtration which was quite effective. Masaracchia and Gawienowski [75] also found the membrane filter very effective and used an Alpha-8 Metrical ® filter (Gelman) with pore size of 0.20 μ in combination with the Gelman Swinny type of hypodermic adapter as a convenient method of eluting samples. Another technique is to filter the eluate through potassium bromide packed in a capillary [76]. Amos [77] packed the silica gel containing the spot on top of a layer of potassium bromide in the cone of an 18-gauge hypodermic needle. A 1-ml glass syringe filled with acetone served as a reservoir as the compound was eluted and applied dropwise to a 10-mg pile of ground potassium bromide. Each drop was evaporated before applying the next drop.

The potassium bromide was then thoroughly mixed and pressed into a 1.5-mm disc.

Rice [78] scraped away the adsorbent around the desired spot on the thin-layer plate leaving the spot within a "teardrop" shaped area. The glass surrounding this area was carefully wiped with a clean cloth moistened with a volatile pure solvent, and then 15 to 20 mg of spectrograde potassium bromide was placed in a line 0.2 × 0.6 cm on the cleaned area. The end of the powder touched the point of the teardrop. Solvent was added dropwise to wash the compound from the spot into the potassium bromide. De Klein [79] modified this technique by placing the potassium bromide in a semicircle around the tip of the teardrop, but separated from it by 1 to 2 mm. Garner and Packer [80] placed the eluate from a TLC spot at the base of a triangular potassium bromide pellet (Harshaw's Wick-Stick) and allowed the solvent to ascend the tip carrying the solute with it. As the solvent evaporated from the tip fresh solvent was added to concentrate all of the solute at the tip, which was then broken off, ground, and pressed into a pellet. Goodman [81] used a somewhat similar procedure by clamping the TLC plate with the spot in contact with the base of a triangular layer of potassium bromide formed on another glass plate.

Thin-layer plates used for infrared work should be predeveloped or washed with a suitable solvent to remove impurities adsorbed on the layer [82]. Amos [77] investigated other sources of contamination and found that the plastic tip on the vacuum pickup devices such as marketed by Desaga was a large source of contamination. If precoated plates are used a check should be made to be sure the binder will not contribute infrared-absorbing impurities to the eluates. Amos examined various solvents and found high-purity solvents to be fairly satisfactory for infrared work although they are relatively expensive. Simple distillation of analytical-grade solvents proved fairly effective in providing satisfactory reagents. Laboratory glassware proved to be another source of contamination and could not be cleaned with chromic acid, nitric acid, or hot caustic soda, or by soaking in 2% aqueous solutions of five detergents for 24 hrs. Proper cleaning was achieved by treating the glassware for 30 min an an ultrasonic bath containing one of the three detergents "Extran," "R.B.S. 25," or "Decon 75."

Percival and Griffiths [83] used a silver chloride plate as a support for their adsorbent so that they could take the infrared spectra directly on the plate. A Fourier spectrometer was used and ratio-recording applied in order to compensate for the spectral bands owing to the adsorbent layer.

Multiple internal reflection spectroscopy has also been used in thin-layer work. The eluate from the spot is evaporated directly on the plate [84] or mirror [85] used for the reflectance measurements. Beyermann and Dietz [86] volatilized the silica gel layer by treating it with hydrogen fluoride. The aluminum plate which formed the support for the layer then served directly as the mirror for infrared reflectance spectroscopy. Of course this procedure could not be used for volatile substances or compounds that would be altered by hydrogen fluoride. Hermann et al. [87] have applied multiple extractions using solvent of increasing polarity combined with thin-layer chromatography, multiple internal reflection spectrometry, mass spectrometry, and gas chromatography to identify components of mixtures. For multiple internal reflection spectrometry more than 10 μg of compound was sufficient to obtain an interpretable spectrum without the need for scale expansion.

Adams and Gardner [88] have investigated the application of in situ Raman spectroscopy to the analysis of thin-layer chromatograms. Silica gel layers without binder were used, although those with calcium sulfate may be used. Developing solvents were used which gave the most concentrated spots after development. Spectrum accumulation was used to improve the signal to noise ratio when necessary.

4 MISCELLANEOUS COMBINATIONS

The versatility of thin-layer chromatography lends itself to the checking of other separation procedures and purification processes. It has been applied to the checking of distillation fractions and to checking the progress of purification by molecular distillation [89]. It has been used to check the progress of purification by recrystallizing [90] as well as for the elucidation of the composition of the mother liquors from recrystallization [91]. Wolfrom and Groebke in one instance obtained the first seed for recrystallization through thin-layer chromatography [90]. Paper chromatography has been combined with thin-layer chromatography in certain purifications. Peifer et al. [92] have used micro-thin-layer chromatography to determine the completeness of extraction of lipids from fish, and Rigby and Bethune [93] have used chromatoplates to check countercurrent extraction fractions. Ott and Gunther [94] adopted the Technicon Auto-Analyzer to the analysis of TLC spot scrapings for cholesterase-inhibiting pesticides.

A very useful technique for the examination of natural and other products consists in the controlled heating of the sample in a tube so that the vapors are deposited on a thin-layer plate as it is programmed across the exit tube as a function of the temperature. An inert gas may be swept through the heated tube to assist in removing less volatile materials.

Reactive gases, such as ammonia or water vapor, may be used as a carrier gas. The method has been dubbed the "TAS" method by Stahl [95], who claims to have originated the method; however, the first paper on this unique technique was first published by Rogers [96] in 1967 and is really based on the sublimation technique of Wasicky and Akisue [97].

References

1. J. M. Miller and J. G. Kirchner, *Anal. Chem.,* **24,** 1480 (1952).
2. J. H. van Dijk, *Z. Anal. Chem.,* **247,** 262 (1969).
3. A. G. Netting, *J. Chromatogr.,* **53,** 507 (1970).
4. H. Schlitt and F. Geiss, *J. Chromatogr.,* **67,** 261 (1972).
5. R. Amos and S. G. Perry, *J. Chromatogr.,* **83,** 245 (1973).
6. B. Coq, J. P. Nicolas, A. Lamotte, and M. Porthault, *J. Chromatogr.,* **97,** 137 (1974).
7. R. M. Ikeda, W. L. Stanley, L. A. Rolle, and S. H. Vannier, *J. Food Sci.,* **27,** 593 (1962).
8. R. M. Ikeda, W. L. Stanley, S. H. Vannier, and L. A. Rolle, *Food Technol.,* **15,** 379 (1961).
9. R. M. Ikeda, W. L. Stanley, S. H. Vannier, and E. M. Spitler, *J. Food Sci.,* **27,** 455 (1962).
10. L. J. Morris, R. T. Holman, and K. Fontell, *J. Am. Oil Chem. Soc.,* **37,** 323 (1960).
11. C. Litchfield, M. Farquhar, and R. Reiser, *J. Am. Oil Chem. Soc.,* **41,** 588 (1964).
12. H. D. M. Touw, B. M. C. Kroese, and H. M. Molenaar, *J. Assoc. Off. Anal. Chem.,* **55,** 622 (1972).
13. W. Wachs and H.-U. Melchert, *Dtach. Lebensm.-Rundsch.,* **67,** 221 (1971).
14. D. Brocco, A. Cimmino, and M. Possanzini, *J. Chromatogr.,* **84,** 371 (1973).
15. M. Kotakemori and A. Kawagisi, *Bunseki Kagaku,* **20,** 709 (1971); through *Chem. Abstr.,* **75,** 709 (1971).
16. T. Stijve and E. Cardinale, *Mitt. Geb. Lebensm. Hyg.,* **63,** 142 (1972).
17. D. Brocco, V. Cantuti and G. P. Cartoni, *J. Chromatogr.,* **49,** 66 (1970).
18. J. Janák, V. Martinū, and J. Rǒžičková, *J. Chromatogr.,* **78,** 127 (1973).
19. R. H. Anderson, T. E. Huntley, W. M. Schwecke, and J. H. Nelson, *J. Am. Oil Chem. Soc.,* **40,** 349 (1963).
20. J. Janák, *Nature,* **195,** 696 (1962).
21. D. E. Bowyer, W. M. F. Leat, A. N. Howard, and G. A. Gresham, *Biochem. J.,* **89,** 24P (1963).
22. M. Dobiasova, *J. Lipid Res.,* **4,** 481 (1963).
23. D. C. Malins, *Chem. Ind. (London),* **1960,** 1359.
24. O. S. Privett, M. L. Blank, and O. Romanus, *J. Lipid Res.,* **4,** 260 (1963).
25. O. S. Privett and E. C. Nickell, *J. Am. Oil Chem. Soc.,* **41,** 72 (1964).

26. O. S. Privett and E. C. Nickell, *J. Lipid Res.*, **4**, 208 (1963).
27. O. S. Privett and C. Nickell, *J. Am. Oil Chem. Soc.*, **39**, 414 (1962).
28. E. A. Moscatelli, *Lipids,* **7**, 268 (1972).
29. K. Samuelsson, *Scand. J. Clin. Lab. Invest.*, **27**, 381 (1971).
30. O. Hirayama and H. Matsuda, *Agric. Biol. Chem. (Tokyo),* **36**, 2593 (1972).
31. P. H. Gedam, M. R. Subbaram, and J. S. Aggarwal, *Fette, Seifen, Anstrichm.*, **73**, 748 (1971).
32. W. Futterweit, N. L. McNiven, L. Narcus, C. Lantos, M. Drosdowsky, and R. I. Dorfman, *Steroids,* **1**, 628 (1963).
33. R. Guerra-Garcia, S. C. Chattoraj, L. J. Gabrilove, and H. H. Wotis, *Steroids,* **2**, 605 (1963).
34. H. H. Wotiz and S. C. Chattoraj, *Anal. Chem.*, **36**, 1466 (1964).
35. F. Berthou, L. Bardou, and H. H. Floch, *J. Chromatogr.*, **93**, 149 (1974).
36. B. P. Lisboa, *J. Chromatogr.*, **48**, 364 (1970).
37. E. Watson, P. Tramell, and S. M. Kalman, *J. Chromatogr.*, **69**, 157 (1972).
38. L. J. Morris, *J. Lipid Res.*, **4**, 357 (1963).
39. H. K. Mangold and R. Kammereck, *Chem. Ind. (London),* **1961**, 1032.
40. R. G. Jensen and J. Sampugna, *J. Dairy Sci.*, **45**, 435 (1962).
41. J. A. Attaway and R. W. Wolford, 5th International Symposium on Gas Chromatography, Brighton, England, September, 1964.
42. E. Demole, *Compt. Rend.*, **243**, 1883 (1956).
43. A. S. Gupta and S. Dev, *J. Chromatogr.*, **12**, 189 (1963).
44. L. J. Morris, R. T. Holman, and K. Fontell, *J. Lipid Res.*, **1**, 412 (1960).
45. W. J. de Klein, *Z. Anal. Chem.*, **249**, 81 (1970).
46. J. Franc and J. Šenkýřová, *J. Chromatogr.*, **78**, 123 (1973).
47. B. Casu and L. Cavallotti, *Anal. Chem.*, **34**, 1514 (1962).
48. I. C. Nigam, M. Sahasrubudhe, and L. Levi, *Can. J. Chem.*, **41**, 1535 (1963).
49. J. Janák, *J. Gas Chromatogr.*, **1**, 20 (1963).
50. J. Janák, *J. Chromatogr.*, **15**, 15 (1964).
51. R. Kaiser, *Z. Anal. Chem.*, **205**, 284 (1964).
52. J. Janák, I. Klimeš, and K. Hána, *J. Chromatogr.*, **18**, 270 (1965).
53. E. Kováts, *Z. Anal. Chem.*, **181**, 351 (1961).
54. J. H. Tumlinson, J. B. Minyard, P. A. Hedin, and A. C. Thompson, *J. Chromatogr.*, **29**, 80 (1967).
55. H.-C. Curtius and M. Mueller, *J. Chromatogr.*, **32**, 222 (1968).
56. K. D. Parker, J. A. Wright, and C. H. Hine, *Forensic Sci. Soc. J.*, **7**, 162 (1967).
57. R. Kaiser, "Gas Chromatography and Thin-Layer Chromatography," in *Ancillary Techniques of Gas Chromatography,* L. S. Ettre and W. H. McFadden, Eds., Wiley-Interscience, New York, 1969, p. 299.
58. J. Janák, "Two-Dimensional Chromatography Using Gas Chromatography as One Dimension," in *Progress in Thin-Layer Chromatography and Related Methods, Vol. II,* A. Niederwieser and G. Pataki, Eds., Ann Arbor Science, Ann Arbor, Mich., 1971, p. 63.
59. A. M. Humphrey, *J. Chromatogr.*, **53**, 375 (1970).
60. J. Janák, V. Martinů, and J. Růžičková, *J. Chromatogr.*, **78**, 127 (1973).

61. K. Heyns and H. F. Gruetzmacher, *Angew. Chem., Int. Ed. Engl.,* **1,** 400 (1962).
62. M. Fétizon, "Spectrometrie de Masse et Chromatographie en Couche Mince," in *Thin-Layer Chromatography,* G. B. Marini-Bettolo, Ed., Elsevier, Amsterdam, 1964, p. 69.
63. F. T. Deverse, E. Gripstein, and L. G. Lesoine, *Inst. News,* **18,** 16 (1967).
64. C.-A. Nilsson, A. Norstroem, and K. Andersson, *J. Chromatogr.,* **73,** 270 (1972).
65. G. J. Down and S. A. Gwyn, *J. Chromatogr.,* **103,** 208 (1975).
66. R. L. Clark, *Chem. Ind. (London),* **1971,** 1434.
67. M. J. Rix, B. R. Webster, and I. C. Wright, *Chem. Ind. (London),* **1969,** 452.
68. G. Székely, *J. Chromatogr.,* **48,** 313 (1970).
69. N. Nash, P. Allen, A. Bevenue, and H. Beckman, *J. Chromatogr.,* **12,** 421 (1963).
70. W. J. de Klein and K. Ulbert, *Anal. Chem.,* **41,** 682 (1969).
71. F. Bissett, A. L. Bluhm, and L. Long, Jr., *Anal. Chem.,* **31,** 1927 (1959).
72. D. H. Anderson and T. E. Wilson, *Anal. Chem.,* **47,** 2482 (1975).
73. R. D. Spencer and B. H. Beggs, *J. Chromatogr.,* **21,** 52 (1966).
74. J. G. Graselli and M. K. Snavely, *Progr. Infrared Spectrosc.,* **3,** 55 (1967).
75. R. A. Masaracchia and A. M. Gawienowski, *Steroids,* **11,** 718 (1968).
76. W. E. Court and M. S. Habib, *J. Chromatogr.,* **73,** 274 (1972).
77. R. Amos, *J. Chromatogr.,* **48,** 343 (1970).
78. D. D. Rice, *Anal. Chem.,* **39,** 1906 (1967).
79. W. J. de Klein, *Anal. Chem.,* **41,** 667 (1969).
80. H. R. Garner and H. Packer, *Appl. Spectrosc.,* **22,** 122 (1967).
81. G. W. Goodman, "The Application of Spectroscopy to Thin-Layer Chromatography," in *Quantitative Paper and Thin-Layer Chromatography,* E. J. Shellard, Ed., Academic Press, New York, 1968, 91.
82. J. G. Kirchner, J. M. Miller, and R. G. Rice, *J. Agric. Food Chem.,* **2,** 1031 (1954).
83. C. J. Percival and P. R. Griffiths, *Anal. Chem.,* **47,** 154 (1975).
84. E. I. Saier, H. F. Acevedo, and B. M. Dick, *Anal. Biochem.,* **37,** 345 (1970).
85. K. Beyermann and E. Roeder, *Z. Anal. Chem.,* **230,** 347 (1967).
86. K. Beyermann and J. Dietz, *Z. Anal. Chem.,* **256,** 349 (1971).
87. T. S. Hermann, R. L. Levy, L. J. Leng, and A. A. Post, *Dev. Appl. Spectrosc.,* **1968** (7B), 128 (1970).
88. D. M. Adams and J. A. Gardner, *J. Chem. Soc., Perkin Trans.,* **2,** 2278 (1972).
89. D. D. Lawson and H. R. Getz, *Chem. Ind. (London),* **1961,** 1404.
90. M. L. Wolfrom and W. Groebke, *J. Org. Chem.,* **28,** 2986 (1963).
91. A. A. Akhrem and A. I. Kuznetsova, *Proc. Acad. Sci. USSR, Chem. Sect. (Engl. Transl.),* **138,** 507 (1961).
92. J. J. Peifer, F. Janssen, R. Muesing, and W. O. Lundberg, *J. Am. Oil Chem. Soc.,* **39,** 292 (1962).
93. F. L. Rigby and J. L. Bethune, *Am. Soc. Brewing Chem. Proc.,* **1955,** 174.

94. D. E. Ott and F. A. Gunther, *J. Assoc. Off. Anal. Chem.,* **49,** 669 (1966).
95. E. Stahl, *Analyst (London),* **94,** 723 (1969).
96. R. N. Rogers, *Anal. Chem.,* **39,** 730 (1967).
97. R. Wasicky and G. Akisue, *Rev. Fac. Farm. Bioquim. Univ. Sao Paulo,* **4,** 85 (1966).

Part **II**

APPLICATIONS OF THIN-LAYER CHROMATOGRAPHY

Chapter **XIII**

ACIDS

1 ALIPHATIC MONOCARBOXYLIC ACIDS

Separation of the normal acids with even-numbered carbon atoms from decanol to docosanoic can be effected on layers of kieselguhr G using cyclohexane as the developing solvent [1]. Separation of the acids through hexacosanol can be achieved through 10 cm of development, but this is insufficient for separating the higher homologs. These can be separated by using a kieselguhr layer dried at 110°C for 30 min in combination with a continuous development using the slotted lid technique described by Truter [2]. For separation of the lower members the kieselguhr layer is dried at room temperature for 3 hr.

Prey et al. [3] have separated formic, acetic, and lactic acids on silica gel G layers using a solvent mixture of pyridine–petroleum ether (1:2) with R_f values of 0.52, 0.58, and 0.63, respectively. Separation with ethanol–ammonia–water (80:4:16) gave R_f values of 0.64, 0.66, and 0.51. The best detecting agent was the disulfuric acid ester of dihydroindanthroazine.

Lynes [4] has separated the lower carboxylic acids on neutral layers of silica gel, with a mixture of methyl acetate–2.5% ammonium hydroxide (95:5). If fresh solvent is used a double development is required to give sufficient separation. Use of a solvent that has stood for 24 hr gives a different R_f value so that standards must be run at the same time as the unknowns (Table 13.1). To detect the acids the plates are sprayed with

Table 13.1 $R_f \times 100$ Values of Low-Molecular-Weight Straight-Chain Carboxylic Acids and Some Branched-Chain Acids on Silica Gel[a,b]

Acid	*Double Run, Fresh Solvent*	*Double Run, Solvent Aged 24 hr*	*Single Run, Solvent Aged 24 hr*
Formic	05	07	03
Acetic	10	13	06
Propionic	15	30	15
n-Butyric	24	40	22
n-Valeric	39	50	30
n-Hexanoic	52	57	34
n-Heptanoic	55	60	39
n-Octanoic	58	66	43
n-Nonanoic	61	69	45
Trimethylacetic	57	71	47
α-Methylbutyric (DL)	39	65	39
β-Methylbutyric	34	53	31
Isobutyric	27	57	32

[a] From A. Lynes [4]; reproduced with permission of the author and the Elsevier Publishing Co.
[b] Solvent: methyl acetate–2.5% ammonia (95:5).

an alcoholic solution of methyl red, after which they are heated in an oven at 105°C to remove the ammonia, at which time the acids appear as dark red spots on an orange background. Braun and Vorendohre [5] used methyl ethyl ketone–ethanol–25% ammonia (65:15:28). Lupton [6] used silica gel impregnated with 10% ethylene glycol for the partition chromatography of acids using petroleum ether (40–60°C)–acetone (2:1) saturated with ethylene glycol as the solvent in a two-stage development. In this case some of the sample always remained at the origin. Bayzer [7] separated the first five aliphatic acids on silica gel using propanol–ammonia solvent for thin-layer chromatography followed by electrophoresis in the second dimension. In this way it was possible to separate formic and acetic acids, which were not separable by one-dimensional thin-layer chromatography.

Guetlbauer [8] used *n*-butanol–diethylamine–water (85:1:14) to chromatograph 19 straight- and branched-chain acids on cellulose. Ninhydrin was used to detect the acids as their diethylamine salts in limits of 0.5 μg of acid. Dittmann [9] used *sec*-butanol–2 *N* ammonia (80:20) with cellulose layers. Bachur [10] was able to detect acids with a p*K* of less than 7 on cellulose at the 0.0025-μmole level by exposing the dried cellulose layers to pyridine vapor for 15 min and then placing under a short-wavelength ultraviolet light for 15 min. The spots were brown on a white background. In applying this test it is necessary to remove any traces of volatile acids used in the developing solvent.

Reverse-phase chromatography on silanized silica gel has been used with mixtures of methanol–water (10:90 to 60:40) [11], dioxane–water–formic acid (60:35:5) [12], and with methyl cyanide–acetic acid–water (7:1:2.5) [13]; the latter solvent in a ratio of 6:3:1 has also been used for the methyl esters [13]. Methyl esters may also be separated with acetone–methanol–water–acetic acid (60:35:5) [12]. Some critical pairs cannot be separated by this method, but these can be resolved by using the silanized plates in combination with solvents saturated with silver nitrate [14].

Formic through valeric acids have been separated on a maize starch layer using butanol–1.5 *N* ammonia (30:20) [15].

Urea–Celite (3:1) layers have been used to separate the methyl esters of branched-chain fatty acids from those of straight-chain unsaturated acids [16]. After sample application, the layers were left in an atmosphere of methanol overnight to promote clathrate formation. Separation was achieved using petroleum ether as the solvent. Another way to accomplish this separation is to develop the sample on silica gel plates for 15 min with 7% methanolic urea, dry in the air for 15 min, and then complete the development with hexane–ethyl ether (1:1) [17]. Urea layers bound with calcium sulfate may also be used for this type of separation [18]. *Cis–trans* fatty acids may be separated in this manner.

Silica gel layers containing silver nitrate can be used for the separation of saturated and unsaturated acids using benzene–petroleum ether (80:30) [19]. The methyl esters of these two classes of compounds can be separated using benzene–petroleum ether–ethyl ether (80:20:5) [19] or *n*-hexane–benzene (1:1) [20].

Hromatka and Aue [21, 22] have shown that the log R_f values of monocarboxylic acids plotted against the number of carbon atoms form a straight line for the even-numbered acids, with acetic acid deviating somewhat. A second straight line holds for the odd-numbered acids with formic and propionic deviating from the line. Similar relationships also hold true for the dicarboxylic acids but the deviations are greater.

Yamamoto and Furukawa [23] separated the anilides and phenylhydrazides of the lower fatty acids by means of chromatostrips of silicic acid bound with plaster of Paris. Separations were carried out using *n*-hexane–ethyl acetate (1:1), *n*-hexane-butyl acetate (1:1), benzene–butyl acetate (1:1), and ethyl acetate. Detection of the derivatives was carried out by spraying with sulfuric–nitric acid mixture (1:1). Phenylhydrazides were detected in the cold, but the chromatostrips were heated for detection of the anilides. Thompson and Hedin [24] separated the 2,4-dinitrophenylhydrazides on aluminum oxide with chloroform–methanol–diethylamine (99:1:0.5) and on silica gel using a solvent ratio of 98:2:0.5. For polyamide layers they used a solvent mixture of methanol–water (9:1). The *p*-bromoanilides and *p*-toluidides have also been used for the separation of the C_2 to C_7 acids on silica gel layers and up to C_{11} by reverse-phase separation on silica gel impregnated with paraffin oil [25]. Detection limits using ultraviolet light were 0.5 and 2 μg for the bromoanilides and the toluidides, respectively.

Some polyene fatty acids have been separated by conversion to their mercuric adducts before chromatography on a silicic acid–silica gel mixture (3:7) with isobutanol–formic acid–water (100:0.5:15.7) [26]. The R_f values were as follows: C_{22}-hexenoic, 0.06; C_{20}-pentenoic, 0.15; C_{16}-tetrenoic and C_{22}-pentenoic, 0.23; C_{18}-tetrenoic, 0.30; C_{20}-tetrenoic, 0.38; C_{16}-trienoic and C_{22}-tetrenoic, 0.46; C_{18}-trienoic, 0.54; C_{20}-trienoic, 0.61; C_{16}-dienoic, 0.66; C_{18}-dienoic, 0.73; C_{16}-monoenoic, 0.74; higher monoenoic acids, 0.77 to 0.82; and saturated acids, 0.91 to 0.96. The separated compounds were reconverted to methyl esters and identified by a number of methods including chromatography on paraffin-impregnated silica layers with formic acid–acetonitrile–acetone (2:2:1). The mercury adducts of the methyl esters have been separated on silica gel after which they were oxidatively cleaved into mono- and dicarboxylic acids with potassium permanganate in anhydrous acetic acid [27]. Identification was by gas chromatography.

Knappe and Yekundi [28] separated the carboxylic acids by conversion to the corresponding hydroxamic acids before chromatographing. For the preparation of the derivatives, 100 mg of the acid is mixed with 3 ml of hydroxylamine reagent (if necessary 3 ml of tetrahydrofuran may be added) and heated for 15 to 30 min under reflux. The product is filtered from insoluble material and diluted before application to the plates. For the reagent two standard solutions are prepared: (1) 69.0 g hydroxylamine hydrochloride in 1000 ml of methanol and (2) 56.0 g potassium hydroxide in 1000 ml of methanol. For use, one part of the first solution is freshly mixed with two parts of the second solution. Separation is carried out on impregnated kieselguhr layers which are prepared by mixing 30 g of kieselguhr G with 12 g of diethylene glycol adipate polyester (or triethylene glycol adipate polyester), 60 ml acetone, and 0.05 g sodium diethyl dithiocarbamate (the latter to prevent formation of peroxides). After the plates are coated they are dried in the air for 10 min and then for 30 min at 105°C. Separation is achieved with a solvent mixture composed of diisopropyl ether–petroleum ether–carbon tetrachloride–formic acid–

water (50:20:20:8:1). The compounds appear as violet spots on a yellow background if they are sprayed with a solution composed of 16.7 g of ferric chloride and 10 ml concentrated hydrochloric acid in 1 liter of methanol. The R_f values are given in Table 13.2 Heusser [13] separated hydroxamic acids on silanized silica gel using methanol–dioxane–chloroform–glycocol buffer (pH 3.0) (4:3:1:4).

Knappe and Rohdewald [29] have applied the use of polyester-impregnated layers to the separation of substituted acetyl acetic acid amides.

Vioque and Maza [30] have separated the *p*-phenylazophenacyl esters of fatty acids by reverse-phase thin-layer chromatography. The *N, N*-dimethyl-*p*-aminobenzeneazophenacyl esters have also been used for the separation of the acids [31–33]. On silica gel benzene–ethyl acetate (20:1), carbon tetrachloride–acetonitrile (9:1), or hexane–acetone (3:1) may be used. For dimethylformamide-impregnated layers [33] benzene–toluene (5:1) may be used as a solvent. Sensitivity of these colored derivatives is approximately 0.05 μg. Churáček and Pechová [33] list the R_f values of the first nine aliphatic acid 2,4-dinitrobenzyl esters in eight solvent systems. Detection of these latter esters was with an acetone solution of sodium ethanolate. The *p*-nitrobenzyl and the *p*-bromophenacyl esters

Table 13.2 Average R_f of the Hydroxamic Acid Derivatives of the C_2 to C_{12} Fatty Acids on Polyester-Impregnated Kieselguhr (1:3)[a,b]

	With Diethylene Glycol Adipate		*With Triethylene Glycol Adipate*	
Hydroxamic Acid of	*Average R_f Value*	*Average Difference*	*Average R_f Value*	*Average Difference*
Acetic acid	0.15		0.09	
Propionic acid	0.25	0.10	0.13	0.04
Butyric acid	0.34	0.09	0.19	0.06
n-Valeric acid	0.43	0.09	0.26	0.07
Caproic acid	0.52	0.09	0.33	0.07
Heptoic acid	0.63	0.11	0.43	0.10
Caprylic acid	0.73	0.10	0.53	0.10
Nonylic acid	0.84	0.11	0.63	0.10
Capric acid	0.92	0.08	0.76	0.13
Undecylic acid	0.94	0.02	0.87	0.11
Lauric acid	0.96	0.02	0.96	0.09

[a] From E. Knappe and K. G. Yekundi [28]; reproduced with permission of the authors and Springer-Verlag.
[b] Solvent mixture: diisopropyl ether–petroleum ether–carbon tetrachloride–formic acid–water (50:20:20:8:1).

have been used for the separation of the first 12 acids [34]. Kosuge et al. [34a] separated the 4-hydroxyphenacyl esters of some aliphatic acids with acetic acid–cyclohexane on silica gel. Andreev et al. [34b] used neutral alumina with benzene–isoamyl acetate (4:1) for the separation of normal C_1 to C_7 acids as their aromatic amides. The spots were photographed in ultraviolet light and evaluated with a microphotometer. Sensitivity was 3×10^{-9} mole.

2 KETO ACIDS

For the separation of keto and carboxylic acids of biological interest, Passera et al. [35] purified silica gel G by washing with hydrochloric acid–water (1:1) and a 0.1% solution of EDTA in order to remove interfering ions. After purification, 13% plaster of Paris was added to the silica gel to replace what had been removed by the hydrochloric acid. The compounds were separated using a travel distance of 13 cm, with mixtures of propanol–28° Bé ammonium hydroxide (7:3) and ethanol–chloroform–28° Bé ammonium hydroxide–water (7:4:2:0.2) as the best solvents. A 0.1% solution of 2,6-dichloroindophenol in 95% alcohol was used as an indicating reagent, yielding pink spots on a sky-blue background. The definition of the spots could be increased by exposing to ammonia vapors after spraying with the reagent. On the other hand, Rink and Herrmann [36] first converted ketocarboxylic acids by reacting with rhodanine and then separating the resulting derivatives on acetyl cellulose with various concentrations of *n*-propanol–*n*-butanol–ammonium carbonate solutions.

The 2,4-dinitrophenylhydrazone derivatives have been used for the separation of keto acids [37–43] and the R_f values for some separations are given in Table 13.3. Ronkainen [40] pointed out that some of the derivatives yielded two spots due to isomers in some of the acidic solvents. Haekkinen and Kulonen [44] call attention to an artifact in the chromatographic separation of 2,4-dinitrophenylhydrazones of keto acids. It appears to arise from the 2,4-dinitrophenylhydrazine and can be mistaken for a keto acid derivative. Berlet [42] used a two-dimensional separation in the dark at 3°C on silica gel; first with butanol–ethanol–0.5 *N* ammonia (7:1:2) followed by benzene–tetrahydrofuran–acetic acid (57:35:8) in the second direction. Quantitative analysis of these compounds can be achieved by eluting the compounds and measuring at 370 nm [40, 41]. Separation on cellulose layers [39] was accomplished with butanol–propanol–benzene–6% ammonia (3:10:3:4, 3:11:3:3, or 3:3:4:3).

Table 13.3 R_{st} Values of 2,4-Dinitrophenylhydrazones of Keto Acids in Different Systems[a]

	Silica Gel G		*Acidic Silica Gel G Prepared from 30 g Silica Gel G + 5 ml Propionic Acid + 60 ml Water*[b]
	0.3-mm Layer Dried at 110°C Overnight[b]	*0.1-mm Layer Dried at 110–120°C for 30 min*[b]	
Keto Acid	*Isoamyl Alcohol*–0.25 *N Ammonium Hydroxide* (20:1) [41]	*Petroleum Ether* (60–80°C)-*Ethyl Formate* (13:7) + 0.104 *M Propionic Acid per* 100 *ml* [40]	
Oxaloacetic	0.0		
α-Ketoglutaric	0.0	0.05	0.02
α-Keto-β-methylvaleric		0.58	0.48
α-Ketocaproic		0.56	0.48
Acetoacetic	0.12		
Pyruvic (isomer 1)	0.15	0.28	0.20
Pyruvic (isomer 2)	0.38		
α-Ketoisovaleric		0.59	0.39
α-Ketobutyric		0.48	0.29
Levulinic		0.56	0.49
Phenylpyruvic (isomer 1)	0.30		
Phenylpyruvic (isomer 2)	0.63		
2,4-Dinitrophenylhydrazine	1.00	1.00	1.00

[a] $R_{st} = R_f$ compound/R_f 2,4-dinitrophenylhydrazine.
[b] Solvent travel distance = 11.5 cm (saturated atmosphere).

3 HYDROXY ACIDS

Dittmann [9] separated a group of hydroxy acids by using cellulose layers and a group of three solvents; amyl alcohol—formic acid—water (4:40:2), *n*-butanol—formic acid—water (60:10:20), and *sec*-butanol–2 *N* ammonia (80:20). Guetlbauer [8] chromatographed 10 hydroxy acids as their diethylamine salts on cellulose using *n*-butanol–diethylamine–water (85:1:14) as the solvent. Detection was with ninhydrin, permitting 0.5 μg of acid to be detected. The 4-(4-dimethylaminophenylazo)phenacyl esters of six hydroxy acids have been separated on silica gel [32].

There have been quite a number of papers published on the separation of hydroxy acids such as citric, malic, glyceric, glycolic, lactic, and tartaric acids, which are found in fruits and other foods [8, 32, 45–53], just to mention a few. Munier et al. [45] give the R_f values obtained with 15 solvents on cellulose. Baraldi [46] used a two-dimensional technique on cellulose listing four solvents that could be used. Buchbauer [47] used a two-dimensional technique on silica gel G with isobutanol–formic acid–water (8:1:1) in the first direction and ethanol–ammonia–water (25:4:3) in the second direction. Trop and Levinger [48] used a silica gel–kieselguhr (1:1) layer with three solvents; benzene–95% ethanol–ammonia (10:20:5), butyl acetate–methanol–ammonia (15:20:5), and butyl acetate–acetic acid–water (30:20:10). The acids were detected as brown spots on a white background by spraying with 10% ceric ammonium nitrate in absolute ethanol and then with 0.25% indole in the same solvent. Quantitative results were obtained by measuring the carbon dioxide evolved when treated with nitric acid and ceric ammonium citrate reagent. Stoll [49] using the same adsorbent with benzene–ethanol–25% ammonia (2:4:1) determined the acids by gas chromatography after converting them to the methyl esters after the separation. Bourzeix et al. [50] separated these acids on cellulose using butanol–formic acid–water (4:2:5) and determined them quantitatively by densitometry.

4 DICARBOXYLIC ACIDS

Dicarboxylic acids have been separated by chromatography on silicic acid [54, 55], by chromatography on a silica gel G–kieselguhr G mixture [56], and by partition chromatography with kieselguhr G impregnated with polyethylene glycol M 1000[57, 58]. Bancher and Scherz [59] examined a number of adsorbents and found the best results with cellulose layers using an acidic mobile phase. Petrowitz and Pastuska [55] used solvent mixtures of benzene–methyl alcohol–acetic acid (45:8:4) and benzene–dioxane–acetic acid (90:25:4) for separation on silicic acid layers. Knappe and Peteri [57, 58] used a polyethylene glycol-impregnated kieselguhr layer (1.0:0.5) combined with a solvent mixture of diisopropyl ether–formic acid–water (90:7:3) that had been saturated with the polyethylene glycol (Fig. 13.1). Braun and Geenen [54] applied an alkaline solvent consisting of 96% alcohol–water–25% ammonium hydroxide (100:12:16) to separations of the ammonium salts on silica gel layers. As detecting agents a number of indicator dyes may be used, for example, bromophenol blue and bromocresol blue. Table 13.4 gives the R_f values of some of these separations.

Pastuska and Petrowitz [60] have shown that the cis–trans acids can be separated on layers of silica gel with benzene–methanol–acetic acid

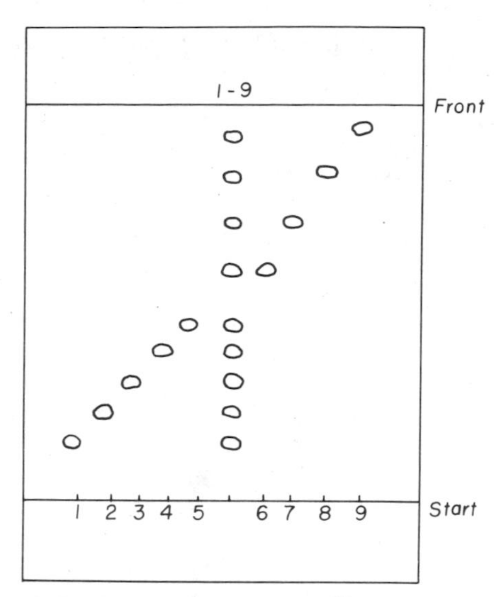

Fig. 13.1 Separation of dicarboxylic acids on kieselguhr G impregnated with polyethylene glycol M 1000 (1:0.5). Solvent, diisopropyl ether–formic acid–water (90:7:3). Development distance, 12 cm. (1) oxalic, (2) malonic, (3) succinic, (4) glutaric, (5) adipic, (6) pimelic, (7) suberic, (8) azelaic, (9) sebacic. From Knappe and Petri [57]; reproduced with permission of the authors and Springer-Verlag.

(45:8:4) or benzene–dioxane–acetic acid (90:25:4) (Table 13.5). In this case the spots were detected with an alkaline potassium permanganate spray. Knappe and Peteri [61] have also separated unsaturated aliphatic dicarboxylic acids on polyethylene glycol M 4000 impregnated on kieselguhr, using a solvent composed of isopropyl ether–formic acid–water (90:7:3). To assist in the identification of mixed acids they have applied a two-dimensional method by running the unsaturated acids in one direction, then hydrogenating the spots and developing the saturated acids in the second direction. The R_f values for the unsaturated acids are as follows: maleic 0.20, citraconic 0.32, glutaconic 0.36, itaconic 0.38, fumaric 0.61, and mesaconic 0.75.

Suryaraman and Cave [62] have applied ethanol–ammonium hydroxide–tetrahydrofuran (7:3:3) to the separation of mono- or dicarboxylic acids on silica gel G plates. Goebell and Klingenberg [63] have used a two-dimensional technique to obtain a good separation of the substrates of the tricarboxylic acid cycle. A separation which is made on cellulose layers is run in the first direction using a solvent mixture of 95% ethanol–25% ammonium hydroxide–water (8:2:1) in which the acids move as the

Table 13.4 $R_f \times 100$ Values of Some Dicarboxylic Acids

Acid	*Silica Gel G Impregnated with Polyethylene Glycol M* 1000 (1.0:0.5)[a]	*Silica Gel G*[b]			*Silica Gel G–Kieselguhr G*[c] (1:1)		
		A	*B*	*C*	*D*	*E*	*F*
Oxalic	14	5	0	0	6	10	16
Malonic	21	14	13	5	13	20	55
Succinic	28	30	28	23	20	30	74
Glutaric	36	39	35	28	27	38	77
Adipic	43	43	42	34	32	47	80
Pimelic	55	53	47	36	—	—	—
Suberic	67	54	50	40	—	—	—
Azelaic	82	56	53	43	—	—	—
Sebacic	92	67	55	47	—	—	—
Malic	20	—	13	6	8	20	54
Citric	—	5	2	2	4	12	49
Tartaric	—	8	—	—	4	14	42

[a] Solvent: diisopropyl ether–formic acid–water (90:7:3) saturated with polyethylene glycol M 1000. Saturated atmosphere. Developing distance 12 cm [57].

[b] Developing distance 10 cm. (*A*) Solvent: 96% ethanol–water–25% ammonium hydroxide (100:12:16). Acids as the ammonium salts. Unsaturated atmosphere [54]. (*B*) Solvent: benzene–methanol–acetic acid (45:8:4) [55]. (*C*) Solvent: benzene–dioxane–acetic acid (90:25:4) [55].

[c] Developing distance 6 cm [56]. (*D*) Solvent: benzene–ethanol–ammonium hydroxide (10:20:5). (*E*) Solvent: butyl acetate–methanol–ammonium hydroxide (15:20:5). (*F*) Solvent: butyl acetate–acetic acid–water (30:20:10).

ammonium salts. The plate is then turned at right angles and developed in the second direction with isobutanol–5 *M* formic acid (2:3) in which the substrate travels as the undissociated acids. Quantitative determinations may be made with appropriate enzyme reactions or by means of autoradiography.

Nygaard [64] employed a two-dimensional method with electrophoresis on a cellulose in the first direction followed by chromatography with isoamyl alcohol–5 *M* formic acid (2:1) in the second direction.

Table 13.5 $R_f \times 100$ Values for the Separation of Some cis–trans Acids on Silica Gel Layers 0.3 mm Thick, Using a Development Distance of 10 cm[a]

Acid	*Benzoic–Methanol–Acetic Acid* (45:8:4)	*Benzene–Dioxane–Acetic Acid* (90:25:4)
Isocrotonic	70	71
Crotonic	73	73
Tiglic	71	79
Maleic	13	6
Fumaric	43	22
Citraconic	18	7
Mesaconic	55	53
Itaconic	46	34
cis-Aconitic	3	4
trans-Aconitic	12	4

[a] From G. Pastuska and H.-J. Petrowitz [60]; reproduced with permission of the authors and Elsevier Publishing Co.

5 AROMATIC CARBOXYLIC ACIDS

Frankenfeld [65] chromatographed a group of 12 aromatic acids in benzene–pyridine (17:3) on silica gel layers.

The separation and determination of salicylsalicylic acid in salicylic acid has been accomplished on thin layers of aluminum oxide G using 0.1 *N* hydrochloric acid in absolute ethanol as the developing solvent [66]. The quantitative analysis is accomplished by eluting the spots and then measuring by ultraviolet spectrophotometry.

3-Methoxy-4-hydroxymandelic acid, an important metabolic product of noradrenaline and adrenaline, has been isolated and determined in urine [67]. The acid is extracted from urine with ethyl acetate and, after concentrating and drying, the ethyl acetate solution can be applied directly to the silica gel G plates. Development takes place in about 90 min using isopropanol–ethyl acetate–ammonia–water (45:30:17:8). A semiquantitative determination can be made by comparison of spots with a standard after spraying with a 0.1% ethanol solution of 2,6-dichloroquinone-4-chloroimide. Quantitative determinations can be made by eluting the spot and measuring at 510 mμ after diazotizing with *p*-nitroaniline.

Homovanillic acid is a normal constituent of urine. Its determination can be of value in assessing diseased conditions. Sankoff and Sourkes

[68] have worked out a procedure for the analysis of homovanillic acid (4-hydroxy-3-methoxyphenylacetic acid) in urine. Five milliliters of urine is acidified with hydrochloric acid to pH 1 to 2 and then extracted with 6 ml of ethyl acetate after saturating with sodium chloride. A second extraction of 5 ml of ethyl acetate is then made, the combined extracts are evaporated under reduced pressure, and the residue is transferred to 0.5 ml of methanol. The methanol solution is spotted directly on silica gel G plates and developed with the upper layer of a mixture of benzene–acetic acid–water (2:3:1). The acid is detected by spraying with a 1 *N* solution of Folin's phenol reagent followed by 10% sodium carbonate. A quantitative determination may be made by eluting the spot and measuring the color at 750 mμ when treated with Folin's phenol reagent. Tautz et al. [69] have also published, recently, on the quantitative determination of this acid and related compounds.

Maugras et al. [70] have separated stereoisomers of 2-methyl-3-(6-methoxy-2-naphthyl)pentanoic and 2-methyl-2-propyl-3-(6-methoxy-2-naphthyl)pentanoic acids as well as the corresponding nitriles and methyl esters using aluminum oxide and silica gel layers.

Millett et al. [71] determined furoic acid in the presence of hydroxymethylfuroic acid.

Kolesinska et al. [72, 73] have chromatographed a group of benzenecarboxylic acids and Knappe and Stuck [74] have used partition methods for the separation of substituted benzoic acids. Tabak et al. [75] employed silica gel plates containing silver oxide for the separation of a group of substituted benzoic acids.

Bories [76] separated some phenylalkanoic acids by chromatographing their anilides on alumina with hexane–ethyl acetate (10:3) as developing solvent.

6 PHENOLCARBOXYLIC ACIDS

These acids can be separated on silica gel or on silica gels that have been treated with various complexing agents. Furukawa [77] has used chromatostrips of silica gel bound with plaster of Paris for separating a group of these compounds. Hexane–ethyl acetate (4:1 or 3:2), benzene–ethyl ether (4:1), or benzene were used as solvents. Halmekoski [78] examined the effect of chelate-forming salts on the separations of phenocarboxylic acids, using five different solvents (see Table 3.4). The best separation of catechol homologs was obtained with a solvent composed of ethyl acetate–isopropanol–water (65:24:11). On the other hand, the best separation of the guaiacol homologs was obtained with a solvent of *n*-butyl ether (saturated with water)–acetic acid (10:1).

A group of phenolic acids found in plants have been separated on silica gel layers [79]. Dichloromethane–acetic acid–water (2:1:1) was used for acids containing only hydroxyl groups and benzene–acetic acid (45:4) for those containing both OH and MeO groups and also for salicylic acid. Quantitative results were obtained by eluting and measuring in a spectrophotometer. Jangaard [80] examined a large number of these components as well as other phenolics found in plants using cellulose layers with 2% formic acid, 20% potassium chloride, isopropanol–ammonia–water (8:1:1), and acetic acid as solvents.

Lyman et al. [81] used silicic acid chromatostrips to separate a large group of phenolcarboxylic acids and their derivatives. Using fluorescent layers as described by Kirchner et al. [82], they were able to locate the compounds either by their own fluorescence or by their absorption of ultraviolet light, thus forming a dark spot on the fluorescent background. Pastuska [83] has separated a group of 22 phenols and phenolcarboxylic acids on silica gel G layers. He investigated five diazotized reagents as coupling salts to produce colored compounds and lists the R_f values and the colors produced with these reagents. Kratzl and Puschmann [84] likewise used silica gel G plates for the separation of phenolcarboxylic acids and other compounds which occurred as degradation products of lignin. They applied the method to preparative separations. Table 13.6 lists some R_f values for various phenolcarboxylic acids.

Wang and Lin [85] have used polyamide layers for the separation of phenolic acids and related compounds occurring naturally. The layers are prepared by dissolving 20 g of polyamide resin in 100 ml of 80% formic acid. The solvent is removed in a filter paper covered chamber at a temperature of 25°C in the presence of water vapor. After standing overnight the plates are dried at 130°C for 15 min. Solvents used for the development were benzene, chloroform, ethyl acetate, and acetone. The R_f values are listed but the compounds are not identified.

Wang [86] has obtained the R_f values of some phenolic acids on polyamide with ethyl acetate–acetic acid (95:5) as follows: *o*-, *m*-, and *p*-hydroxybenzoic acids 0.57, 0.38, 0.42, respectively, tannic acid 0.06, and gallic acid 0.10.

Urion et al. [87] have also used thin layers of polyamide for the separation of the phenolic acids in barley. In this case the layers are prepared from granular material with a starch binder following the procedure of Kirchner et al. [82]. For the separation the solvent of Wang and Lin [85] was used and also acetic–water (30:7) (see Table 13.6).

Ramaut [88, 89] studied the separation of depsides and depsidones of orcinol and applied the knowledge to the separation of lichenic acids. Benzene–dioxane–acetic acid (90:25:4) was used with silica gel HF for

Table 13.6 $R_f \times 100$ Values of Some Phenolic Carboxylic Acids (see also Table 3.6)

	Ref. 81[a]			*Ref.* 83[b]		*Ref.* 87[c]	
Acid	*A*	*B*	*C*	*D*	*E*	*F*	*G*
Salicylic	72	88	48			84	37
Protocatechuic	32	55	10	32	39	35	42
Gentisic	35	65	17	30	40	32	35
2,4,6-Trihydroxybenzoic	15	50	10				
p-Hydroxyphenylacetic	35	72	14				
m-Hydroxyphenylacetic	32	65	19				
2,5-Dihydroxyphenylacetic	67	85	43				
α-Resorcylic	21	61	73				
β-Resorcylic	57	85	19	54	52		
γ-Resorcylic	10	15	10				
Vanillic	46	76	29	54	61	75	48
Ferulic	42	76	16	50	58	79	30
Isoferulic						75	26[d] 44[e]
Syringic	33	57	16	48	60	82	59
m·Hydroxybenzoic				49	51		
p-Hydroxybenzoic						60	46
Gallic				18	23	11	37
o-Coumaric						60	21
p-Coumaric				49	52	64	26[d] 40[e]
Caffeic						35	25[d] 41[e]
Dihydrocaffeic						46	51

[a] On chromatostrips of silicic acid with starch binder, 0.5-mm layer, 10 to 11 cm development distance. (*A*) Ether–Skellysolve B (7:3). (*B*) Ethyl acetate–Skellysolve B (3:1). (*C*) Acetone–Skellysolve B (1:3).

[b] Silica gel G, 0.5-mm layer, 10-cm development distance. (*D*) Benzene–dioxane–acetic acid (90:25:4). (*E*) Benzene–methanol–acetic acid (90:16:8).

[c] Polyamide with starch binder, 0.15-mm layer, 15-cm development. (*F*) Ethyl acetate–acetic acid (95:5). (*G*) Acetic acid–water (30:70).

[d] Main spot.

[e] Secondary spot.

the separation. At the same time Bachmann [90] investigated the lichen acids of the β-orcinol group in extracts of *Parmelia robusta* on silica gel with the same solvent system. Culberson and Kristinsson [91] and Culberson [92] have worked out a systematic method for identifying more than 200 natural products and derivatives from lichens. This was based on the use of Merck Silica Gel F_{254} plates with three solvent systems, benzene–dioxane–acetic acid (90:25:4), hexane–ethyl ether–formic acid (5:4:1), and toluene–acetic acid (85:15). Included is a punch card system to help in identifying components. However, since this work was published, the formulation of these plates has been altered so that relative R_f values have been changed considerably. Some changes to compensate for these differences have been published [93] and this original work should be consulted before using this method.

Phenolcarboxylic acids have also been separated by electrophoresis [94]. This was conducted on layers of silica gel G or kieselguhr G. The silica gel plates were prepared by mixing 4.5 g of silica gel with 8 ml of 3% boric acid solution. For the electrolyte a mixture of 80 ml of ethanol, 30 ml of water, 4 g of boric acid, and 2 g of crystalline sodium acetate was used after adjusting to pH 4.5 with acetic acid. For the preparation of the kieselguhr G layers a slurry of 6.5 g of the diatomaceous earth was mixed with 9 ml of the 3% boric acid solution. The electrolyte for this

Table 13.7 Electrophoresis of Phenolic Carboxylic Acids on Silica Gel and Kieselguhr Layers[a,b]

Kieselguhr				*Silica Gel*		
Color[d]	*Direction*	M_g[c]	*Compound*	M_g[c]	*Direction*	*Color*[d]
(1)	Anode	0.99	Salicylic acid	0.49	Anode	(2)
Yellow-brown	Anode	1.12	Protocatechuic acid	0.83	Anode	Bright brown
Yellow-brown	Anode	0.92	Gentisic acid	0.56	Anode	Bright red
Yellow	Anode	1.00	*m*-Hydroxybenzoic acid	1.00	Anode	Yellow
Yellow	Anode	0.86	*p*-Hydroxybenzoic acid	0.77	Anode	Yellow
Red-yellow	Anode	0.90	β-Resorcylic acid	0.64	Anode	Red-brown
Orange	Anode	1.05	Gallic acid	0.64	Anode	Orange
Yellow	Anode	0.86	*p*-Coumaric acid	0.72	Anode	Yellow-brown
Yellow-brown	Anode	1.07	Caffeic acid	0.66	Anode	Brown
Brown	Anode	0.87	Vanillic acid	0.86	Anode	Yellow-brown
Yellow-red	Anode	0.83	Syringic acid	0.74	Anode	Yellow-red
Brown	Anode	0.77	Ferulic acid	0.73	Anode	Yellow-brown
Red	Anode	0.78	Isoferulic acid	0.76	Anode	Red

[a] From G. Pastuska and H. Trinks [94]; reproduced with permission of the authors and Alfred Huethig Verlag.
[b] See text for preparation of layers.
[c] M_g = Distance traveled relative to distance traveled by *m*-hydroxybenzoic acid.
[d] Color developed with diazotized benzidine unless otherwise indicated. (1) Detected with alkaline permanganate solution. (2) Detected with antimony pentachloride.

situation was the same except that the pH was adjusted to 5.5 with acetic acid. Both separations were carried out at a field strength of 20 V/cm. The M_g values referred to *m*-hydroxybenzoic acid are listed for the six phenols and phenolcarboxylic acids. Table 13.7 lists the separations for the phenolcarboxylic acids.

Haluk et al. [95] have used a two-dimensional technique for separating phenolcarboxylic acids with thin-layer chromatography in one direction followed by electrophoresis in the other. With silica gel, thin-layer chromatography was performed with benzene–dioxane–acetic acid (90:25:4) for the benzoic acid series and with benzene–methanol–acetic acid (45:8:4) for the cinnamic acid series followed by thin-layer electrophoresis in the second direction at a pH of 5.3 with a buffer of pyridine–acetic acid–water (25:10:24) [96]. On polyamide, thin-layer chromatography was carried out with ethyl acetate–acetic acid (95:5) then thin-layer electrophoresis using a pH of 8.9 (Aronsson buffer) for benzoic acids and at a pH of 3.55 using pyridine–acetic acid–water (1:6:90) for the cinnamic series. On cellulose the thin-layer electrophoresis was carried out first at a pH of 5.3 followed by thin-layer chromatography with 30% acetic acid.

Phenolic urinary acids have been separated [96] two-dimensionally on cellulose with isopropanol–25% ammonia–water (8:1:1) and benzene–acetic acid–water (2:3:1).

References

1. S. J. Purdy and E. V. Truter, *J. Chromatogr.*, **14,** 62 (1964).
2. E. V. Truter, *J. Chromatogr.*, **14,** 57 (1964).
3. V. Prey, H. Berbalk, and M. Kausz, *Mikrochim. Acta,* **1962,** 449.
4. A. Lynes, *J. Chromatogr.*, **15,** 108 (1964).
5. D. Braun and G. Vorendohre, *Chromatographia,* **1,** 405 (1968).
6. C. J. Lupton, *J. Chromatogr.*, **104,** 223 (1975).
7. H. Bayzer, *J. Chromatogr.*, **27,** 104 (1967).
8. F. Guetlbauer, *J. Chromatogr.*, **45,** 104 (1969).
9. J. Dittmann, *J. Chromatogr.*, **34,** 407 (1968).
10. N. R. Bachur, *Anal. Biochem.*, **13,** 463 (1968).
11. J. F. Rodrigues De Miranda and T. D. Eikelboom, *J. Chromatogr.*, **114,** 274 (1975).
12. W. O. Ord and P. C. Bamford, *Chem. Ind. (London),* **1966,** 1681.
13. D. Heusser, *J. Chromatogr.*, **33,** 62 (1968).
14. W. O. Ord and P. C. Bamford, *Chem. Ind. (London),* **1967,** 277.
15. V. D. Canić and N. Perisić-Janjić, *Teh. Belgr.*, **25,** 330 (1970); through *Chem. Abstr.*, **73,** 31374m (1970).

16. J. Hradec and P. Menšík, *J. Chromatogr.*, **32,** 502 (1968).
17. E. Myannik and O. Ikonopisteva, *Izv. Akad. Nauk Est. SSR, Khim. Geol.*, **21,** 53 (1972); through *Anal. Abstr.*, **24,** 1631 (1973).
18. V. M. Bhatnagar and A. Liberii, *J. Chromatogr.*, **18,** 177 (1965).
19. A. Strocchi and R. T. Holman, *Riv. Ital. Sostanze Grasse*, **48,** 617 (1971); through *Anal. Abstr.*, **23,** 1968 (1972).
20. R. G. Ackman, S. N. Hooper, and J. Hingley, *J. Chromatogr., Sci.*, **10,** 430 (1972).
21. O. Hromatka and W. A. Aue, *Monatsh. Chem.*, **93,** 497 (1962).
22. O. Hromatka and W. A. Aue, *Monatsh. Chem.*, **93,** 503 (1962).
23. K. Yamamoto and T. Furukawa, *Hiroshima Univ. J. Fac. Educ.*, **5,** 85 (1957).
24. A. C. Thompson and P. A. Hedin, *J. Chromatogr.*, **21,** 13 (1966).
25. J. P. Lebacq, M. Severin, J. Casimir, and M. Renard, *Bull. Rech. Agron. Gembloux*, **4,** 130 (1969).
26. H. Wagner and P. Pohl, *Biochem. Z.*, **340,** 337 (1964).
27. P. Pohl, H. Glasl, and H. Wagner, *J. Chromatogr.*, **42,** 75 (1969).
28. E. Knappe and K. G. Yekundi, *Z. Anal. Chem.*, **203,** 87 (1964).
29. E. Knappe and I. Rohdewald, *Z. Anal. Chem.*, **208,** 195 (1965).
30. E. Vioque and M. P. Maza, *Grasas Aceites (Seville, Spain)*, **15,** 63 (1964).
31. H. Esterbauer, W. Just, and H. Sterk, *Fette, Seifen, Anstrichm.*, **74,** 13 (1972).
32. I. M. Seligman and F. A. Doy, *Anal. Biochem.*, **46,** 62 (1972).
33. J. Churáček and H. Pechová, *J. Chromatogr.*, **48,** 250 (1970).
34. J. H. Dhont, G. G. Kuijpers, and J. C. de Beauveser, *Chem. Tech. (Amsterdam)*, **26,** 473 (1971).
34a. S. Kosuge, M. Furuta, and Y. Takikawa, *Jap. Anal.*, **18,** 235 (1969).
34b. L. V. Andreev, Z. I. Finkel'shtein, and S. S. Belyaev, *Prikl. Biokhim. Mikrobiol.*, **10,** 308 (1974); through *Chem. Abstr.*, **81,** 9487u (1974).
35. C. Passera, A. Pedrotti, and G. Ferrari, *J. Chromatogr.*, **14,** 289 (1964).
36. M. Rink and S. Herrmann, *J. Chromatogr.*, **14,** 523 (1964).
37. J. Cotte, C. Collombel, M. Culivre, and L. Padis, *Rev. Fr. Etud. Clin. Biol.*, **12,** 496 (1967).
38. P. Ronkainen, *J. Chromatogr.*, **28,** 263 (1967).
39. D. Chiari and M. Roehr, *Mikrochim. Acta*, **1967,** 140.
40. P. Ronkainen, *J. Chromatogr.*, **11,** 228 (1963).
41. J. Dancis, J. Hutzler, and M. Levitz, *Biochim. Biophys. Acta*, **78,** 85 (1963).
42. H. H. Berlet, *Anal. Biochem.*, **22,** 525 (1968).
43. N. Ariga, *Anal. Biochem.*, **49,** 436 (1972).
44. H. M. Haekkinen and E. Kulonen, *J. Chromatogr.*, **18,** 174 (1965).
45. R. L. Munier, A. M. Drapier, and B. Faivre, *Chromatographia*, **6,** 466 (1973).
46. D. Baraldi, *J. Chromatogr.*, **42,** 125 (1969).
47. G. Buchbauer, *Sci. Pharm.*, **40,** 259 (1972); through *Anal. Abstr.*, **25,** 448 (1973).
48. M. Trop and I. M. Levinger, *J. Assoc. Off. Anal. Chem.*, **53,** 621 (1970).
49. U. Stoll, *J. Chromatogr.*, **52,** 145 (1970).

50. M. Bourzeix, J. Guitraud, and F. Champagnol, *J. Chromatogr.*, **50,** 83 (1970).
51. G. A. Ravdel, G. F. Zhukova, and L. A. Shchukina, *J. Anal. Chem. USSR (Engl. Transl.),* **26,** 2023 (1971).
52. G. Lehmann and P. Martinod, *Z. Lebensm.-Unters.-Forsch.*, **130,** 269 (1966).
53. A. Schweiger, *Z. Lebensm.-Unters.-Forsch.*, **124,** 20 (1963).
54. D. Braun and H. Geenen, *J. Chromatogr.*, **7,** 56 (1962).
55. H.-J. Petrowitz and G. Pastuska, *J. Chromatogr.*, **7,** 128 (1962).
56. E. Bancher, H. Scherz, and V. Prey, *Mikrochim. Acta,* **1963,** 712.
57. E. Knappe and D. Peteri, *Z. Anal. Chem.*, **188,** 184 (1962).
58. E. Knappe and D. Peteri, *Z. Anal. Chem.*, **188,** 352 (1962).
59. E. Bancher and H. Scherz, *Mikrochim. Acta,* **1964,** 1159.
60. G. Pastuska and H. J. Petrowitz, *J. Chromatogr.*, **10,** 517 (1963).
61. E. Knappe and D. Peteri, *Z. Anal. Chem.*, **190,** 380 (1962).
62. M. G. Suryaraman and W. T. Cave, *Anal. Chim. Acta,* **30,** 96 (1964).
63. H. Goebell and M. Klingenberg, *Chromatogr., Symp., 2nd, Brussels,* **1962,** 153.
64. P. Nygaard, *J. Chromatogr.*, **30,** 240 (1967).
65. J. W. Frankenfeld, *J. Chromatogr.*, **18,** 179 (1965).
66. R. W. Bailey, *Anal. Chem.*, **36,** 2021 (1964).
67. E. Schmid and N. Henning, *Klin. Wochenschr.*, **41,** 566 (1963).
68. I. Sankoff and T. L. Sourkes, *Can. J. Biochem. Physiol.*, **41,** 1381 (1963).
69. N. A. Tautz, G. Voltmer, and E. Schmid, *Klin. Wochenschr.*, **43,** 233 (1965).
70. M. Maugras, Ch. Robin, and R. Gay, *Bull. Soc. Chim. Biol.*, **44,** 887 (1962).
71. M. A. Millett, W. E. Moore, and J. F. Saeman, *Anal. Chem.*, **36,** 491 (1964).
72. J. Kolesinska, T. Urbanski, and A. Wielopolski, *Chem. Anal. (Warsaw),* **10,** 1107 (1965).
73. *Ibid.*, **11,** 473 (1966).
74. E. Knappe and J.-I. Stuck, *Z. Anal. Chem.*, **227,** 353 (1967).
75. S. Tabak, A. E. Mauro, and A. Del'Acqua, *J. Chromatogr.*, **52,** 500 (1970).
76. G. F. Bories, *J. Chromatogr.*, **36,** 377 (1968).
77. T. Furukawa, *Nippon Kagaku Zasshi,* **80,** 387 (1959); *Chem. Abstr.*, **54,** 13938 (1960).
78. J. Halmekoski, *Suomen Kemistil.*, **35B,** 39 (1962).
79. H. Schmidtlein and K. Herrmann, *J. Chromatogr.*, **115,** 123 (1975).
80. N. O. Jangaard, *J. Chromatogr.*, **50,** 146 (1970).
81. R. L. Lyman, A. L. Livingston, E. M. Bickoff, and A. N. Booth, *J. Org. Chem.*, **23,** 756 (1958).
82. J. G. Kirchner, J. M. Miller, and G. J. Keller, *Anal. Chem.*, **23,** 420 (1951).
83. G. Pastuska, *Z. Anal. Chem.*, **179,** 355 (1961).
84. K. Kratzl and G. Puschmann, *Holzforschung,* **14,** 1 (1960).
85. K.-T. Wang and Y.-T. Lin, *J. Chin. Chem. Soc. (Taiwan),* **10,** 146 (1963).
86. K.-T. Wang, *J. Chin. Chem. Soc. (Taiwan),* **8,** 241 (1961).
87. E. Urion, M. Metche, and J. P. Haluk, *Brauwissenschaft,* **16,** 211 (1963).
88. J. L. Ramaut, *Bull. Soc. Chim. Belg.*, **72,** 97 (1963).
89. J. L. Ramaut, *Bull. Soc. Chim. Belg.*, **72,** 316 (1963).
90. O. Bachmann, *Oesterr. Bot. Z.*, **110,** 103 (1963).

91. C. F. Culberson and H.-D. Kristinsson, *J. Chromatogr.,* **46,** 85 (1970).
92. C. F. Culberson, *J. Chromatogr,* **72,** 113 (1972).
93. *Ibid.,* **97,** 107 (1974).
94. G. Pastuska and H. Trinks, *Chem.-Ztg.,* **85,** 535 (1961).
95. J. P. Haluk, C. Duval, and M. Metche, *Int. Symp. Chromatog. Electrophoresis, 6th,* **1970,** Ann Arbor Science, Ann Arbor, Mich., 1971, p. 431.
96. R. Humbel, *Rev. Roum. Biochim.,* **7,** 45 (1970); through *Chem. Abstr.,* **73,** 5383lk (1970).

Chapter **XIV**

ALCOHOLS AND GLYCOLS

1 DIRECT SEPARATION OF ALCOHOLS

Even-numbered alcohols from decanol through hexacosanol can be separated by chromatographing on kieselguhr G with cyclohexane as a developing solvent [1]. The development is carried out in a saturated atmosphere by lining the developing tank with filter paper soaked in the solvent. Attaway and Wolford [2] have reported the R_f values of the lower alcohols on silica gel (Table 14.1). Kučera [3] has separated these compounds (Table 14.1) on alumina using several different solvent systems, but alcohols lower than butanol cannot be chromatographed on alumina because of their volatility. As detecting agents, iodine, ammoniacal silver nitrate, or concentrated sulfuric acid with subsequent heating may be used. In using the ammoniacal spray, the adsorbent must be carefully purified in order to obtain a white background. After the plate is sprayed with 5% silver nitrate solution in 10% aqueous ammonia, it is dried for 3 to 5 min in a drying oven at 140°C in the dark. Singh and Gershbein [4] separated 13 higher alcohols on silica gel with 1-butanol saturated with water; a few of these, along with two other alcohols, were run in other solvents. Nobuhara [5] gives the R_f values for 14 saturated and unsaturated C_6 to C_{12} alcohols on silica gel G with acetone–water (6:4) and dioxane–water (6:4). Vinyl acetylenic carbinols and acetylenic keto alcohols as well as the esters [6], and monoenoic, dienoic, and trienoic alcohols [7] have been chromatographed on thin layers. Morris et al. [8] have chromatographed positional isomers of long-chain aliphatic alcohols. Dunphy et al. [8a] separated *cis–trans* isomers of isoprenoid alcohols by reverse-phase chromatography on kieselguhr.

Kučera [3] has proposed the chromatography of glycols as a means of

Table 14.1 $R_f \times 100$ Values of Alcohols[a]

	Aluminum Oxide (Loose Layer) [3]			
Alcohol	*Hexane-Acetone* (3:1)	*Ether-Ethanol* (99:1)[b]	*Hexane-Acetone* (4:1)	*Silica Gel G* [2], *in Methylene Chloride*
n-Butanol	25	92	25	18
Isobutanol				19
n-Pentanol	30	93	26	20
Isopentanol				21
n-Hexanol	35	95	27	20
3-Hexene-1-ol				25
Methylheptenol				26
n-Octanol	43	96	30	23
n-Nonanol				24
n-Decanol				26
Benzyl alcohol	20	95	17	
Diacetone alcohol		86	36	

[a] Development distance: aluminum oxide 20 cm, silica gel F 15 cm.
[b] 100 ml shaken with 10 ml of water and upper layer used.

determining the alcohol content in ether, because the R_f values are greatly affected by small changes in ethanol concentration in an alcohol–ether solvent system. Akhrem et al. [9] have separated acetylenic alcohols on loose layers of aluminum oxide using an ether–benzene mixture as a developing solvent.

Wassermann and Hanus [10] separated some sugar alcohols on a kieselguhr–silica gel G (3:2) plate with isopropanol–ethyl acetate–water (27.0:3.5:2.0). Although good separations were achieved the R_f values were not very reproducible. Prey et al. [11] used butanol–water (9:1) for a separation on silica gel G, and Grasshof [12, 13] has found magnesium silicate to be a good adsorbent using *n*-propanol–water (5:5) and *n*-propanol–water–*n*-propylamine (5:3:2) as solvents. The R_f values for some sugar alcohols are given in Table 14.2. Hay et al. [14] give the following R_f values for sugar alcohols on silica gel in *n*-butanol–acetic acid–ethyl ether–water (9:6:3:1): erythritol 0.52, D-mannitol 0.38, D-glucitol 0.39, galactitol 0.36, glycerol 0.58, and malitol 0.10. Waldi [15] has separated several sugar alcohols and Castagnola [16] has used isopropanol-0.1 *N* boric acid (17:3), in order to separate and determine mannitol in commercial sorbitol.

Table 14.2 $R_f \times 100$ Values of Some Sugar Alcohols

	Silica Gel G	Magnesium Silicate (Woelm)			
	Ref. 11	Ref. 12		Ref. 13	
Alcohol	*n-Butanol–Water* (9:1)	*n-Propanol–Water* (1:1)	*n-Propanol–Water–n-Propylamine* (5:3:2)	*n-Propanol–Water–Chloroform* (6:2:1)[a]	*n*-Propanol–Water–Ethyl Methyl Ketone (2:1:1)[a]
L-Arabitol	—	70	55	36	52
D-Sorbitol	50	65	48	27	43
D-Mannitol	50	70	51	30	46
Dulcitol	—	67	50	28	43
Glycol	45	—	—	65	—
Glycerine	38	74	61	53	72

[a] Unsaturated chamber. Developing distance 10 cm.

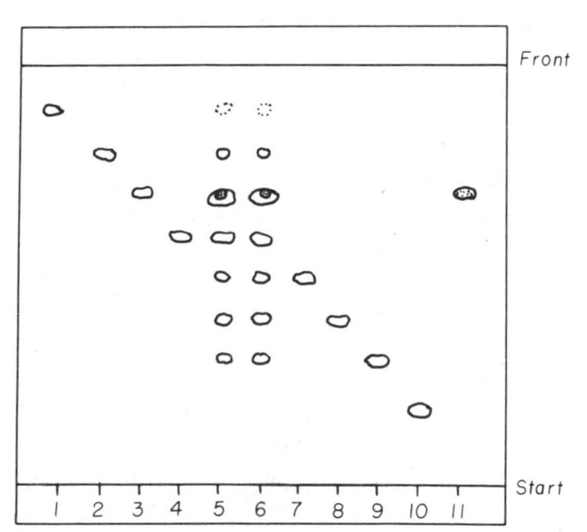

Fig. 14.1 Separation of wax alcohols from wax esters of skin lipids. Adsorbent: kieselguhr G. Impregnating agent: paraffin oil (10% in 50–60°C petroleum ether). Developing solvent: 85% acetic acid. Running time: $2\frac{1}{2}$ hr. Visualizing agent: phosphomolybdic acid. Application: synthetic materials always 2γ, unknown material always 10γ. (1) lauryl alcohol, (2) myristyl alcohol, (3) palmityl alcohol, (4) stearyl alcohol, (5) wax alcohol from healthy individual, (6) wax alcohol from sick individual, (7) arachic alcohol, (8) behenyl alcohol, (9) lignoceric alcohol, (10) cerotyl alcohol, (11) oleyl alcohol. From Kaufmann and Viswanathan [21]; reproduced with permission of the authors and Industrieverlag von Hernhaussen K. G.

Valentinis and Renzo [17] determined sorbitol and mannitol in wine by two-dimensional thin-layer chromatography on boric acid-impregnated silica gel. Solvents were dioxane–methanol–water (5:3:2) in the first direction and methanol–water (6:4) in the second.

Hara and Takeuchi [18] have separated several bile alcohols on silica gel layers dried at 130°C using a chloroform–methanol (9:1) solvent. Solvent travel distance was 15 cm and the most sensitive detection method was charring by heating with concentrated sulfuric acid. The compounds and R_f values were as follows: 3α,24-dihydroxy-5β-cholane 0.82, 3α,12α,24-trihydroxy-5β-cholane 0.34, and 3α,7α,12α,24-tetrahydroxy-5β-cholane 0.10. Kazuno and Hoshita [19] chromatographed 40 bile alcohols.

In the field of lipids, Kaufmann and Das [20] and Kaufmann and Viswanathan [21] separated a series of wax alcohols (Figs. 14.1 and 14.2). This was accomplished on kieselguhr G which had been impregnated with a 10% solution of paraffin in petroleum ether (50 o 70°C). Solvents for the separation were either 85% acetic acid or 90% acetone. Kieselguhr

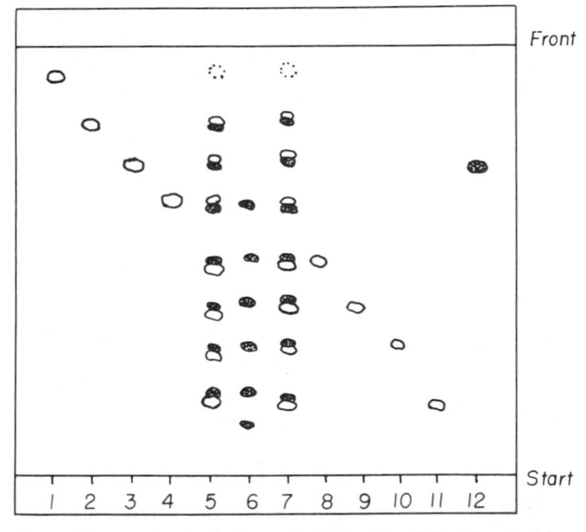

Fig. 14.2 Separation of wax alcohols from hair lipids. Adsorbent: kieselguhr G. Impregnating agent: paraffin oil (10% in 50–70°C petroleum ether). Developing solvent: 90% acetone. Running time: 2 hr. Visualizing agent: phosphomolybdic acid. Application: synthetic material always 2γ, unknown material 10γ. (1) lauryl alcohol, (2) myristyl alcohol, (3) palmityl alcohol, (4) stearyl alcohol, (5) wax alcohols from wax esters of hair, (6) free wax alcohols from hair lipids, (7) wax alcohols from wax esters of hair, (8) arachic alcohol, (9) behenyl alcohol, (10) lignoceric alcohol, (11) cerotyl alcohol, (12) oleyl alcohol. From Kaufmann and Viswanathan [21]; reproduced with permission of the authors and Industrieverlag von Hernhaussen K. G.

impregnated with a 240 to 250°C petroleum fraction was also used with a solvent mixture of isopropanol–ethanol–acetic acid–water (8:3:4:2) saturated with the layer-impregnating material. For observation of the spots the plates were sprayed with phosphomolybdic acid reagent or with Rhodamine B. Hashimoto and Mukai [22] in working with the higher alcohols used silica gel layers and found the best solvents for the C_{10-18} alcohols to be petroleum ether–ether (4:1), hexane–ether (7:3), hexane–ether–acetic acid (70:3:0.1), and xylene–ether (4:1). The method was applied to whale wax, sperm head oil, wool wax, and beeswax. Hashimoto et al. [23] have also used layers of activated bleaching earth. Reutner [24] has analyzed waxes and wax mixtures with silica and benzene, and Kolattukudy [25] has examined the surface lipids of pea leaves. Scholz [26] in separating wax alcohols and acids used silica gel impregnated with 0.5% silicone oil and 10% tetradecane in petroleum ether. The C_{14-24} compounds were separated at 42°C with 90% acetic acid saturated with tetradecane. The C_{22-40} compounds were separated at 60°C with acetic acid–tetradecane (96:4). Bandyopadhyay and Chakrabarty [27] separated 26 C_7 to C_{24} fatty alcohols on kieselguhr G impregnated with 10% liquid paraffin using acetone–water (3:1) as solvent, of which 80% was previously saturated with liquid paraffin. Detection was accomplished by heating at 110°C after spraying with 1% alcoholic phosphomolybdic acid. The acetates could also be separated using a (9:1) mixture of the same solvent. Subbarao et al. [28], working with oxygenated fatty compounds, included a number of alcohols whose R_f values are listed in Table 14.3. Subbarao and Achaya [29] have also separated the 6-,7-, (8-?), 9-, 10-, 12-, and 18-hydroxy isomers of stearic acids, alcohols, and esters on silica gel G. For the acids and alcohols the solvent was diethyl ether-petroleum ether (2:3) with added 2% acetic or formic acid. For the methyl esters the solvent ratio was 1:3, again with added acid. Hara et al. [30] have determined the lactic, malic, and tartaric acid content of

Table 14.3 $R_f \times 100$ Values of Some Fatty Alcohols Separated on Silica Gel G. Development distance 15 cm [28]

	Ether–Petroleum Ether		
Alcohol	3:7	8:2	9:1
Stearyl alcohol	72	96	93
Oleyl alcohol	55		93
Undecenyl alcohol		89	
Ricinoleyl alcohol	10		83
cis-9,10-Epoxystearyl alcohol	32		

wine using thin-layer chromatography. Schweiger [31] has chromatographed lactic acid as well as other carboxylic acids on cellulose layers. Morris et al. [32] have reported on some unsaturated hydroxy acids in seed oils, and Sgoutas and Kummerow [33] have separated some hydroxy stearic acids on silica gel using chloroform–methanol–acetic acid (90:10:2).

2 SEPARATION OF ALCOHOL DERIVATIVES

Alcohols may also be separated as their derivatives. Malins et al. [34, 35] have used the nitrate derivatives for this purpose. The nitrates were prepared as follows: "two drops of 70% nitric acid (about 0.05 ml) were added to absolute acetic anhydride (0.3 ml) in a test tube. The solution was cooled in an ice bath during the addition of the nitric acid. About 50 mg of hydroxy compound, or a mixture of hydroxy compounds, was added after the reagent had warmed up to room temperature. The resulting solution was then maintained at this temperature for ten minutes. After being cooled it was made up to 5.0 ml with diethyl ether." This ether solution was applied directly to the thin-layer silica gel G plates for separation using *n*-hexane as a developing solvent. These spots were located by spraying with 2′, 7′-dichlorofluorescein and observing under ultraviolet light or by staining the chromatograms with iodine vapor.

Katz and Keeney [36] have used the *p*-phenylazobenzoates for analysis of fatty alcohols obtained by the reduction of methyl esters and naturally occurring glycerides. The alcohols are esterified with a freshly prepared solution of 0.4% (wt/vol) of *p*-phenylazobenzoyl chloride in anhydrous ether. This esterification may be carried out as follows [36]: "Two milliliters of the esterification solution are added per milligram of long chain solvent free alcohol in a 150 × 18 mm test tube. For quantities of alcohol less than 1 mg, 2 ml of the esterification solution are constantly used. Twenty-five lambdas of pyridine are added for each 8 mg of *p*-phenylazobenzoyl chloride present. The test tubes are briefly mixed, tightly stoppered with a rubber stopper, and placed in a ½-inch deep 34°C water bath for 30 min. At the end of the reaction time the ether and pyridine are removed on a steam bath under a stream of nitrogen gas." The residues from the preparation of the phenylazobenzoates are taken up in hexane and for preliminary purification they may be passed through an aluminum oxide column using *n*-hexane–chloroform (2.5:1) as a solvent. For the thin-layer separation the plates are coated with silica gel G and activated by heating for 2 hr at 115°C. The plates are then prepared for reverse-phase chromatography by immersing for 5 min in a 5% solution of *n*-dodecane in *n*-hexane. After evaporation of the hexane the samples are spotted in *n*-hexane and developed with a solution of acetonitrile–2-butanone (8:2) saturated with *n*-dodecane. The spots may be visualized by spraying with a 0.2% solution of 2′,7′-dichlorofluorescein in methanol. Observation under long-wave ultraviolet light discloses the esters as purple spots on an orange background. If it is desired to run a quantitative analysis on the esters after the separation, the eluted material may be passed through an alumina column with hexane–chloroform (2.5:1) as a solvent in order to free the material of 2′,7′-dichlorofluorescein. Analysis is made by a spectrophotometric method.

Churáček et al. [37] have used the *N,N*-dimethyl-*p*-aminophenylazobenzoates because they are easily prepared, the reagent is stable, and less than 0.5 μg can be detected. Because of their color no visualization agent needs to be used although the intensity of the spots can be increased by spraying with 0.01 *N* sulfuric acid. They list R_f values for 15 derivatives using eight solvent systems; however, not all compounds can be separated in a single system.

A number of workers have used the 3,5-dinitrobenzoates for differentiating alcohols by thin-layer chromatography [38–45]. Mehlitz et al. [46] used these derivatives for the differentiation of saturated and unsaturated alcohols, and Schantz et al. [43] separated a group of terpene alcohol dinitrophenylbenzoates.

Lipina [47, 48] has used the 3,5-dinitrobenzoates for the determination of the vapors of butyl and isooctyl alcohols in the air. The alcohol vapors are adsorbed from the air by silica gel, and 100 mg of silica gel are eluted with 6 ml of benzene in order to recover the adsorbed alcohols. After converting to the esters of 3,5-dinitrobenzoic acid, the esters are separated on silica gel layers using a 7 to 10% solution of ether in gasoline as the developing solvent. For the preparation of the 3,5-dinitrobenzoate [39] a weighed amount of alcohol is refluxed for 30 to 60 min with a slight excess of 3,5-dinitrobenzoyl chloride in 10 ml of benzene to which 0.1 ml of dry pyridine has been added. After cooling, the reaction mixture is successively extracted with 25 ml of 0.1 *N* sulfuric acid, 25 ml of 0.5% solution of sodium carbonate, and water. The benzene solution is then dried with a little anhydrous sodium sulfate after which the solution is concentrated to a few milliliters. The R_B values (the ratio of the R_f value of the dinitrobenzoate to the R_f value of butter yellow) are given for a series of alcohols in Table 14.4.

Vámos et al. have used the 2,4-dinitrobenzoates for the detection of higher alcohols [49] and of methanol [50] in ethanol. Using test T-246 0.04 μg of methanol could be detected.

Other derivatives have been used for separating alcohols by thin-layer chromatography; the 2,4-dinitrobenzenesulfenates [51], the xanthogen-

Table 14.4. R_B Values (R_f/R_f of butter yellow) of Some Alcohol 3,5-Dinitrobenzoates

Alcohol DNB	$R_B{}^a$					
	A [41]	*B* [41]	*C* [41]	*D* [41]	*E* [40]	*F* [40]
Methanol	0.44	0.42	0.44	0.70		
Ethanol	0.63	0.61	0.59	0.77		
1-Propanol	0.78	0.75	0.74	0.88	0.68	2.01
2-Propanol	0.84	0.78	0.76	0.88	0.79	2.08
1-Butanol	0.94	0.87	0.84	0.96	0.82	1.90
2-Butanol					0.92	2.02
Isobutanol	0.96	0.86	0.81	0.98		
1-Pentanol	1.00	0.98	—	1.03	0.96	1.71
2-Pentanol	1.17	1.02	1.02	1.11	1.05	1.89
3-Pentanol					1.05	1.91
2-Methyl-1-butanol					1.01	1.76
3-Methyl-1-butanol	1.06	0.94	0.92	1.02	0.96	1.84
1-Hexanol	1.11	0.95	0.99	1.08	1.04	1.57
3-Hexen-1-ol					0.94	1.73
2-Methyl-1-pentanol	1.13	1.01	1.01	1.10		
2,4-Dimethyl-1-pentanol	1.13	1.04	1.01	1.10		
1-Heptanol	1.13	1.02	1.02	1.09	1.08	1.42
2-Heptanol	1.17	1.12	1.08	1.13	1.15	1.77
3-Heptanol	1.26	1.18	1.14	1.17		
4-Heptanol					1.42	1.72
1-Octanol	1.22	1.03	1.07	1.15	1.15	1.28
2-Octanol					1.23	1.54
1-Nonanol					1.19	1.17
1-Decanol	1.27	1.10	1.10	1.21	1.24	1.03
1-Undecanol					1.28	0.91
1-Dodecanol	1.27	1.10	1.10	1.21	1.35	0.78
Phenylethyl alcohol					0.61	1.54
Cinnamyl alcohol					0.66	1.31
α-Terpineol					1.15	1.17
Geraniol					1.04	1.44
Citronellol					1.08	1.40
Linalool					1.07	1.46
Menthol					1.46	1.44

[a] Adsorbent and solvent:
A = silica gel G with cyclohexane–ethyl acetate (9:1); *B* = silica gel G with cyclohexane–methyl acetate (9:1); *C* = silica gel G with carbon tetrachloride–cyclohexane–ethyl acetate (80:15:5); *D* = silica gel with benzene–cyclohexene (9:1); *E* = silica gel G with benzene–petroleum ether (38–50°C) (1:1); *F* = polyamide with methanol–water (9:1)

ates [52], the acetates [53], the *p*-nitrobenzoates [54], the β-alkoxypropionitriles [55], the dinitrophenyl ethers [56], the *N*-(1-phenylethyl)urethanes [57], the α-naphthylurethanes [54], and the *p*-toluenesulfonates (tosylates) [58].

Pomonis et al. [58] have worked out a spot test for alcohols using the reaction of 4-(*p*-nitrobenzyl)pyridine with the tosylates as the basis for the test (T-182). On cellulose layers the sensitivity was in the microgram range.

3 GLYCOLS

Wright [59] has published a method for the detection of humectants in tobacco by means of thin-layer chromatography on silica gel that was activated for 1 hr at 110°C. For this analysis a single cigarette was immersed in 5 ml of water for 30 min and then centrifuged; 4 μl of the extract was applied to the silica gel layer. Development was carried out either in acetone or in butanol–acetone–water (4:5:1). In the first solvent the R_f values for some glycols were as follows: ethanediol 0.49, 1,2-propanediol 0.61, 2,3-butanediol 0.68, glycerol 0.30. The corresponding values for the second solvent were 0.54, 0.60, 0.64, and 0.49. As little as 2 μg of the glycols and 1 μg of glycerol could be detected by spraying with 1% lead tetraacetate in dry benzene. The spots appeared white against a brown background and were intensified by heating for 5 min at 110°C. As an alternative locating method, which is as sensitive as the tetraacetate reagent, three successive sprays were used: potassium periodate (0.5%), potassium iodide (5%), and finally, starch solution.

Kučera [3] has separated some glycols on loose layers of alumina with an activity grade of III–IV (Brockmann) using a hexane–acetone (4:1) solvent or a mixture of ether–ethanol (99:1) saturated with water. Several detecting agents were used, the most sensitive being the exposure to iodine vapor for 30 to 60 min at 20°C followed by exposure of the plate to the air until the brown background disappeared. This test, however, was not as sensitive as the lead tetraacetate reaction mentioned above, the sensitivity of the iodine test for 1,2-propanediol was 20 μg. Table 14.5 gives the R_f values for these compounds. According to Kučera, "the separation of 1,2-glycols from other glycols of similar R_f values can be carried out easily on alumina impregnated with 3% ammonium borate; thus, the R_f values of 1,2-glycols are appreciably reduced, probably in consequence of complex formation with boric acid."

Knappe et al. [60] have separated 17 commercially important polyhydric alcohols on three different layer systems (Table 14.5). For the preparation of impregnated kieselguhr G, 5 g of polyamide (Ultramid 1C)

Table 14.5 $R_f \times 100$ Values of Some Polyalcohols

Alcohol	*Aluminum Oxide G* $CHCl_3$–*PhMe*–*HCOOH* (80:17:3) [60]	*Silica Gel G* *n-BuOH Satd. with 1.5 N* NH_4OH [60]	*Acetone*–H_2O (49:1) [61]	*Polyamide-Impregnated kieselguhr* $CHCl_3$ [60]	*Loose-Layer Alumina* $(C_2H_5)_2O$–C_2H_5OH (99:1) (*Satd. with* H_2O) [3]
Pentaerythritol	0	27		0	
Glycerin	2	31	52	16	
Tetramethylcyclohexanol	4	47		9	
2,2-Bis(hydroxymethyl)-1-propanol	12	57		19	
Ethylene glycol	13	56	74	52	25
2,2-Bis(hydroxymethyl)-1-butanol	19	66		29	
1,3-Propanediol	26	64	80	55	
Diethylene glycol	33	52	65	92	
1,4-Butanediol	35	77	85	45	47
1,2-Propanediol	36	67	82	72	38
Triethylene glycol	40	42	60	94	
1,6-Hexanediol	42	77		68	
1,3-Butanediol	50	77	88	84	55
2,2-Dimethyl-1,3-propanediol	60	87		80	
2,3-Butanediol	—	—		—	62
1,5-Pentanediol	—	—		—	63
Dipropylene glycol-1,2	66	72		91	
1,2,6-Hexanetriol	75	83		62	
2-Methyl-2,4-pentanediol	80	90		93	84

soaked overnight in 60 ml of a mixture of benzene in methanol (1:1). It was then warmed slightly on a water bath to form a homogeneous solution followed by thorough mixing with 30 g of kieselguhr G in a porcelain mortar prior to spreading on the plates. Drying was carried out at 105°C for 30 min. For locating the spots a series of spray reagents were used based on a combination of strong oxidants and aromatic diamines.

Nisbet [61] separated polyhydric alcohols with acetone–water (49:1) on silica gel and differentiated compounds with close R_f values by the colors produced with test T-257.

Seher [62, 63] used a solvent consisting of ethyl acetate–isopropanol–water (65:22.7:12.3) to separate some polyglycerines. The layers were silica gel G prepared with 0.02 *M* sodium acetate solution and dried at 105°C for 1 to 1½ hr.

Bergel'son et al. [64] used a series of four solvents to separate nine compounds containing α-glycol groupings on silica gel layers. The R_f values for these separations have been published separately [65]. The same authors [66] have also described a descending thin-layer method for the separations of polyhydroxy compounds. In this case a thin layer of cellulose was used with the three different solvents: butanol–pyridine–water (10:3:3), butanol 25% ammonium hydroxide–water (16:1:2), and phenol–butanol–acetic acid–water (5:5:2:10). For dihydroxy acids a solvent consisting of butanol–8% ammonium hydroxide–aqueous borax (8:1:2) was used.

Grasshof [13] has separated glycols on magnesium silicate layers using *n*-propanol–water–chloroform (6:2:1) as a solvent. As a detecting agent a 1% potassium permenganate solution in water was used.

The separation of the methyl esters of polyhydroxy acids has been investigated by Morris [67]. Impregnation of silica gel layers with glycol complexing agents aided in the separation. Boric acid, sodium borate, or sodium arsenite impregnation permitted the differentiation of *threo*- and *erythro*-dihydroxy isomers, and the tri- and tetrahydroxystearates could be separated on sodium arsenite-impregnated layers.

Akhrem et al. [9] have used an aluminum oxide layer with an ether–benzene solvent for the separation of some glycols. Iodine vapor was used as a detecting agent.

Subbarao and Achaya [29] have applied thin-layer chromatography to the separation of a series of hydroxy derivatives from castor oil. These consisted of undecylenic alcohol, dihydroxyundecane, trihydroxyundecane, and the monoglyceride of undecylenic acid. Clear separations were obtained according to the number of hydroxyl groups present.

Hromatka and Aue [68] have chromatographed ethylene glycol, 1,3-propylene glycol, 1,6-hexanediol, 1,7-heptanediol, 1,9-nonanediol, 1,10-

decanediol, and 1,13-tridecanediol on silica gel G with absolute ethanol. The spots were located with potassium permanganate solution. They observed that a linearity existed between the log of the R_f value and the number of carbon atoms. However, if the R_f values were plotted against the number of carbon atoms, two straight lines were obtained, one for the even-numbered carbon atoms and the other for the odd-numbered carbon atoms.

De Simone and Vicedomini [69] compared the R_f values of 10 polyhydric alcohols including some sugar alcohols in seven solvent systems on silica gel and silica gel impregnated with lead nitrate. On the treated layers the mobility of the polyols decreased with an increase in the number of adjacent hydroxyl groups. Sahasrabudhe [70] used silica gel impregnated with 4% boric acid with benzene–methanol (8:3) for free polyglycerols and their esters.

Polyoxyethylene glycols have been separated on silica gel with ethanol–methanol–ammonia (12:4:2), (12:3:2), ethanol–methanol–water (12:4:2) [71], and toluene–ethyl acetate (4:6), methyl ethyl ketone–water–acetic acid (95:4:1), butanol–methanol–water–acetic acid (77:13:8:2) [72]; on silica impregnated with sodium acetate with ethyl acetate–isopropanol–water (65:23:12) [73]; on aluminum oxide with chloroform–ethanol (98:2) [74]; on kieselguhr–silica gel (1:1) prepared with 0.5% sodium metabisulfite using ethyl acetate–isopropanol–acetone–methanol–water (50:15:15:4:16) for separating the higher linear polymers [75]; and on kieselguhr–silica (1:1) prepared with 0.045 *M* calcium chloride with ethyl acetate–isopropanol–water (110:61:29) for separating nonlinear compounds from linear isomers of the lower compounds [75]. These compounds may be separated from polysorbates on silica gel G with methanol–chloroform–acetic acid (5:4:2) [76].

Kremer [77] has a separate group of isomeric pentitols and hexitols on cellulose MN 300 with methyl ethyl ketone–acetic acid–0.75 *M* boric acid (40:10:9), *n*-butanol–0.75 *M* boric acid (85:15), and isopropanol–acetic acid–0.75 *M* boric acid (7:1:2).

Quantitatively glycerol has been determined polarographically after separation from oligoglycerols by thin-layer chromatography [78]. It can also be determined spectrophotometrically with chromotropic acid after eltion with water followed by oxidation with potassium iodate [79].

Glycols may also be separated in the form of their derivatives. Polyethylene glycols and polyethylene glycol monoethers have been separated by chromatography on thin layers of kieselguhr impregnated with 20% formamide in acetone [80]. Senegal gum was used as a binder for the layers.

Polyols may be separated as their acetates on silica gel with benzene–

Table 14.6 $R_f \times 100$ of Esters of 2-Butyne-1,4-diol on Silica Gel[a,b]

Compound	*Benzene*	*Chloroform*	*Isopropyl ether–Isooctane* (1:1)	*Isopropyl Ether*	*Ethyl Acetate–Isooctane* (1:1)
2-Butyne-1,4-diol	0	0	0	5	8
Monoformate	0	4	7	24	28
Diformate	12	27	24	48	51
Monoacetate	0	4	7	20	27
Diacetate	8	22	23	41	48
Monopropionate	0	4	10	28	34
Dipropionate	13	27	38	59	60
Monobutyrate	3	4	13	32	38
Dibutyrate	18	36	45	64	65
Monovalerate	3	5	14	35	40
Divalerate	23	44	51	69	69

[a] From M. Naff et al. [89]; reproduced with permission of the authors and Elsevier Publishing Co.

[b] Developing distance 15 cm. Temperature 27 ± 1°C. All R_f's the average of three determinations. Paper-lined tank.

ethyl acetate–ethanol (89:10:1) [81] or on kieselguhr G–silica gel G (1:1) with cyclohexane–ether–ethanol–acetic acid (5:51:3:1) [82]. Trimethylsilyl derivatives have been used for separations [83, 84]. Thoma et al. [85] used a two-dimensional separation for a group of polyethylene glycol stearates on silica gel. The solvent for the first direction was *n*-butanol–ethanol–25% ammonia (14:3:5) and that in the second chloroform–methanol–water (3:25:5). Favretto et al. [86–88] examined a number of derivatives for the determination of molecular-weight distribution of polyethylene glycols.

Naff et al. [89] have separated 2-butyne-1,4-diol and its mono- and diesters of formic, acetic, propionic, butyric, and valeric acids by chromatographing on silica gel plates. The solvent systems used were benzene, chloroform, isopropyl ether, isopropyl ether-isooctane (1:1), and ethyl acetate–isooctane (1:1). The R_f values are given in Table 14.6.

References

1. S. J. Purdy and E. V. Truter, *J. Chromatogr.*, **14**, 62 (1964).
2. J. A. Attaway and R. W. Wolford, 5th International Symposium on Gas Chromatography, Brighton, England, September, 1964.

3. J. Kučera, *Collect. Czech. Chem. Commun.*, **28**, 1341 (1963).
4. E. J. Singh and L. L. Gershbein, *J. Chromatogr.*, **23**, 180 (1966).
5. A. Nobuhara, *J. Chromatogr.*, **30**, 235 (1967).
6. S. A. Melkonyan, L. G. Grigoryan, V. N. Zhamagortsyan, and S. A. Vartanyan, *Arm. Khim. Zh.*, **19**, 199 (1966).
7. A. Hashimoto, A. Hirotani, and K. Mukai, *Yukagaku,* **15,** 206 (1966); through *Chem. Abstr.*, **65,** 3726b (1966).
8. L. J. Morris, D. M. Wharry, and E. W. Hammond, *J. Chromatogr.*, **33,** 471 (1968).
8a. P. J. Dunphy, J. D. Kerr, J. F. Pennock, and K. J. Whittle, *Chem. Ind. (London),* **1966,** 1549.
9. A. A. Akhrem, A. I, Kuznetsova, Y. A. Titov, and I. S. Levina, *Izv. Akad. Nauk SSSR, Otd. Khim. Nauk,* **1962,** 657; through *Chem. Abstr.*, **57,** 4003 (1962).
10. L. Wassermann, and H. Hanus, *Naturwissenschaften,* **50,** 351 (1963).
11. V. Prey, H. Berbalk, and M. Kausz, *Mikrochim. Acta,* **1962,** 449.
12. H. Grasshof, *J. Chromatogr.*, **14,** 513 (1964).
13. H. Grasshof, *Deut. Apoth.-Ztg.*, **103,** 1396 (1963).
14. G. W. Hay, B. A. Lewis, and F. Smith, *J. Chromatogr.*, **11,** 479 (1963).
15. D. Waldi, *J. Chromatogr.*, **18,** 417 (1965).
16. V. Castagnola, *Boll. Chim. Farm.*, **102,** 784 (1963).
17. G. Valentinis and M. Renzo, *Ind. Aliment. (Pinerolo, Italy),* **7,** 73 (1968).
18. S. Hara and M. Takeuchi, *J. Chromatogr.*, **11,** 565 (1963).
19. T. Kazuno and T. Hoshita, *Steroids,* **3,** 55 (1964).
20. H. P. Kaufmann and B. Das, *Fette, Seifen, Anstrichm.*, **65,** 398 (1963).
21. H. P. Kaufmann and C. V. Viswanathan, *Fette, Seifen, Anstrichm.*, **65,** 607 (1963).
22. A. Hashimoto and K. Mukai, *Yukagaku,* **12,** 613 (1963); through *Chem. Abstr.*, **60,** 9883 (1964).
23. A. Hashimoto, A. Hirotani, and K. Mukai, *Yakugaku,* **14,** 343 (1965).
24. F. Reutner, *Fette, Seifen, Anstrichm.*, **70,** 162 (1968).
25. P. E. Kolattukudy, *Lipids,* **5,** 398 (1970).
26. G. H. Scholz, *Fette, Seifen, Anstrichm.*, **69,** 333 (1967).
27. C. Bandyopadhyay and M. M. Chakrabarty, *J. Chromatogr.*, **32,** 297 (1968).
28. R. Subbarao, M. W. Roomi, M. R. Subbaram, and K. T. Achaya, *J. Chromatogr.*, **9,** 295 (1962).
29. R. Subbarao and K. T. Achaya, *J. Chromatogr.*, **16,** 235 (1964).
30. S. Hara, K. Morinaga, and K. Otsuka, *Hakko Kogaku Zasshi,* **42,** 426 (1964).
31. A. Schweiger, *Z. Lebensm. Unters.-Forsch.*, **124,** 20 (1963).
32. L. J. Morris, R. T. Holman, and K. Fontell, *J. Am. Oil Chem., Soc.*, **37,** 323 (1960).
33. D. Sgoutas and F. A. Kummerow, *J. Am. Oil Chem., Soc.*, **40,** 138 (1963).
34. D. C. Malins, J. C. Wekell, and C. R. Houle, *Anal. Chem.*, **36,** 658 (1964).
35. J. C. Wekell, C. R. Houle, and D. C. Malins, *J. Chromatogr.*, **14,** 529 (1964).
36. K. Katz and M. Keeney, *Anal. Chem.*, **36,** 231 (1964).
37. J. Churáček, M. Huškova, H. Pechová and J. Říha, *J. Chromatogr.*, **49,** 511 (1970).

38. L. Labat and A. L. Montes, *An. Asoc. Quim. Arg.*, **41,** 166 (1953); *Chem. Abstr.*, **48,** 3637 (1954).
39. J. H. Dhont and C. de Rooy, *Analyst, (London)* **86,** 527 (1961).
40. J. P. Minyard, J. H. Tumlinson, A. C. Thompson, and P. A. Hedin, *J. Chromatogr.*, **29,** 88 (1967).
41. M. Severin, *J. Chromatogr.*, **26,** 101 (1967).
42. W. Diemair, K. Pfeilsticker, and I. Hoelscher, *Z. Anal. Chem.*, **234,** 418 (1968); also *J. Chromatogr.*, **39,** D11 (1968).
43. M. v. Schantz, S. Juvonen, A. Oksanen, and I. Hakamaa, *J. Chromatogr.*, **38,** 364 (1968).
44. V. D. Canic, N. V. Perisić-Janjić, and M. J. Babin, *Z. Anal. Chem.*, **264,** 415 (1973).
45. J. Vámos and A. Brantner, *Int. Symp. Chromatog. Electrophor., Lect. Pap., 6th,* **1970,** Ann Arbor Science, Ann Arbor, Mich., 1971, p. 228.
46. A. Mehlitz, K. Gierschner, and T. Minas, *Chem.-Ztg.*, **89,** 175 (1965).
47. T. G. Lipina, *Metody Opred. Vredn. Veschestv. Vozukhe, Moscow,* **1961,** 41; through *Chem. Abstr.*, **59,** 4471 (1963).
48. T. P. Lipina, *Zavodsk. Lab.*, **26,** 55 (1960).
49. J. Vámos, A. Bratner, G. Szasz, and A. Vegh, *Acta Pharm. Hung.*, **40,** 135 (1970).
50. *Ibid.*, **38,** 378 (1968).
51. G. Lefebvre, J. Berthelin, M. Maugras, R. Gay, and E. Urion, *Bull. Soc. Chim. Fr.*, **1966,** 266.
52. M. Roehr and D. Chiari, *Microchim. Acta,* **1967,** 137.
53. T. Takahashi and H. H. O. Schmid, *Chem. Phys. Lipids,* **4,** 243 (1970).
54. D. W. Connell and C. R. Strauss, *J. Chromatogr.*, **72,** 391 (1972).
55. M. M. Buzlanova, V. N. Ul'yanova, and S. I. Obtemperanskaya, *Zh. Anal. Khim.*, **23,** 1425 (1968).
56. D. B. Parihar, S. P. Sharma, and K. C. Tewari, *J. Chromatogr.*, **21,** 261 (1966).
57. W. Freytag and K. H. Ney, *J. Chromatogr.*, **41,** 473 (1969).
58. J. G. Pomonis, R. F. Stevenson, and P. J. Freeman, *J. Chromatogr.*, **40,** 78 (1969).
59. J. Wright, *Chem. Ind. (London),* **1963,** 1125.
60. E. Knappe, D. Peteri, and I. Rohdewald, *Z. Anal. Chem.*, **199,** 270 (1963).
61. M. A. Nisbet, *Analyst, (London)* **1969,** 811.
62. A. Seher, *Fette, Seifen, Anstrichm.*, **67,** 24 (1965).
63. A. Seher, *Fette, Seifen, Anstrichm.*, **66,** 371 (1964).
64. L. D. Bergel'son, E. V. Dyatlovitskaya, and V. V. Voronkova, *Proc. Acad. Sci. USSR, Chem. Sect. (Engl. Transl.),* **141,** 1076 (1961).
65. L. D. Bergel'son, E. V. Dyatlovitskaya, and V. V. Voronkova, *Dokl. Akad. Nauk SSSR,* **141,** 84 (1961); also *J. Chromatogr.*, **10,** D17 (1963).
66. L. D. Bergel'son, E. V. Diatlovitskaya, and V. V. Voronkova, *Dokl. Akad. Nauk SSSR,* **149,** 1319 (1963).
67. L. J. Morris, *J. Chromatogr.*, **12,** 321 (1963).
68. O. Hromatka and W. A. Aue, *Monatsh. Chem.*, **93,** 503 (1962).

69. V. De Simone and M. Vicedomini, *J. Chromatogr.*, **37,** 538 (1968).
70. M. R. Sahasrabudhe, *J. Am. Oil Chem. Soc.*, **44,** 376 (1967).
71. K. Obruba, *Collect. Czech. Chem. Commun.*, **27,** 2968 (1962); through *Chem. Abstr.*, **58,** 9337 (1963).
72. D. Falgoux, P. Mangin, J. Engel, and C. Granger, *Z. Anal. Chem.*, **236,** 228 (1968).
73. W. Gerhardt and R. Holzbauer, *Chromatographia*, **2,** 468 (1969).
74. I. A. Vakhtina, P. A. Okunev, and O. G. Tarakanov, *Zh. Anal. Khim.*, **21,** 630 (1966).
75. M. J. S. Dallas and M. F. Stewart, *Analyst (London)*, **92,** 634 (1967).
76. A. L. Thakkar, P. B. Kuehn, and N. A. Hall, *Am. J. Pharm.*, **139,** 122 (1967).
77. B. P. Kremer, *J. Chromatogr.*, **110,** 171 (1975).
78. M. Jaworski, J. Bogaczek, and K. Walczyk, *Chem. Anal. (Warsaw)*, **14,** 313 (1969).
79. J. Čoupek, S. Pokorný, E. Mareš, L. Žežulková, N-T. Luan, and J. Pikorný, *J. Chromatogr.*, **120,** 411 (1976).
80. H. Gauthier and G. Mangency, *J. Chromatogr.*, **14,** 209 (1964).
81. C. Dumazert, C. Ghiglione, and T. Pugnet, *Bull. Soc. Pharm. Marseille*, **12,** 337 (1963).
82. M. S. J. Dallas, *J. Chromatogr.*, **48,** 225 (1970).
83. N. L. Gregory, *J. Chromatogr.*, **36,** 342 (1968).
84. K. C. Leibman and E. Ortiz, *J. Chromatogr.*, **32,** 757 (1968).
85. K. Thoma, R. Rombach, and E. Ullmann, *Arch. Pharm. (Weinheim, Ger.)*, **298,** 19 (1965).
86. L. Favretto, G. Pertoldi Marletta, and L. Favretto Gabrielli, *J. Chromatogr.*, **46,** 255 (1970).
87. *Ibid.*, **50,** 304 (1970).
88. L. Favretto, L. Favretto Gabrielli, and G. Pertoldi Marletta, *J. Chromatogr.*, **66,** 167 (1972).
89. M. Naff, S. Naff, and J. A. Strite, *J. Chromatogr.*, **11,** 496 (1963).

Chapter **XV**

ALKALOIDS

1 SYSTEMATIC ANALYSIS

The speed with which thin-layer chromatography can be run is a distinct advantage in many cases [1]. It can be applied in toxicology to the isolation and determination of alkaloids where the 30 to 60-min runs give a great advantage [2–4] in comparison to the 12 to 24 hr required for paper.

Farnsworth and Euler [5] have worked out a method for simply and quickly detecting alkaloids in 2-g samples of plant material. A preliminary extraction and purification was carried out to remove interfering pigments and then the extracts were applied to thin layers of silica gel G. After development with *n*-butanol–acetic acid–water (4:1:1) the alkaloids were made visible by spraying the layers with the Munier and Macheboeuf [6] modification of Dragendorff's reagent. The method was tested on 28 alkaloid plants and eight alkaloid-free plants as controls. Because of the Dragendorff reagent's ability to react with certain nonalkaloid compounds, two of the eight control plants gave false positive results. However, this feature is not a serious disadvantage. Neto and Mancini [7] give a slightly different method of extraction for the rapid detection of alkaloids. Although it is not rapid, Bush and Jeffreys [8] use a methanol percolation method to extract alkaloids and then remove the alkaloids from the methanol with ion-exchange resins which can then be washed to remove interfering pigments before eluting the alkaloids from the resins.

Waldi et al. [9] have set up a systematic procedure for the analysis of alkaloids by means of thin-layer chromatography. Using this procedure the alkaloids are first separated into two groups. This is accomplished by applying increasing concentrations of the alkaloids (the solution to be analyzed should contain 0.05 to 5% alkaloids) to a silica gel plate along with Rhodamine B or a solution of reserpine as a standard. Development is accomplished with a mixture of cyclohexane–chloroform–diethylamine (5:4:1). The plate is dried in the air and then sprayed with iodoplatinate solution after first examining in ultraviolet light at 365 mμ. Arbitrarily, compounds which have an R_f value lower than 0.30 are assigned to group I and compounds whose R_f values are greater than 0.30 are assigned to group II (Tables 15.1 and 15.2). Group I alkaloids are then chromatographed on silica gel G using the solvent systems chloroform–acetone–diethylamine (5:4:1) and chloroform–diethylamine (9:1). The R_f values and the colors, both in ultraviolet light and after spraying with the iodoplatinate reagent, are compared to those in Tables 15.1 and 15.2 in order to identify the compounds. If necessary the material may be chromatographed in additional systems for aid in identification, especially on

Table 15.1 R_f Values of Some Alkaloids[a,b]

	Silica Gel G					*Aluminum Oxide G*		*Silica Gel G Prepared with 0.1N NaOH*
Alkaloid	*Chloroform–Acetone–Diethylamine (5:4:1)*	*Chloroform–Diethylamine (9:1)*	*Cyclohexane–Chloroform–Diethylamine (5:4:1)*	*Cyclohexane–Diethylamine (9:1)*	*Benzene–Ethyl Acetate–Diethylamine (7:2:1)*	*Chloroform*	*Cyclohexane–Chloroform (3:7) + 0.5% Diethylamine*	*Methanol*
Group I								
Narceine	0.03	0.00	0.00	0.00	0.00	0.00	0.00	0.00
Cupreine	0.03	0.00	0.00	0.00	0.00	0.00	0.00	0.46
Sarpagine	0.12	0.04	0.00	0.00	0.00	0.00	0.00	0.00
Ergometrine	0.14	0.06	0.00	0.00	0.02	0.03	0.00	0.64
Morphine	0.10	0.08	0.00	0.00	0.03	0.03	0.00	0.34
Dihydroergotamine	0.21	0.12	0.00	0.00	0.03	0.07	0.00	0.61
Serpentine	0.24	0.15	0.00	0.00	0.04	0.00	0.00	0.00
Ergotamine	0.24	0.16	0.00	0.00	0.03	0.10	0.05	0.59
Boldine	0.16	0.16	0.03	0.00	0.05	0.24	0.06	0.58
Dihydromorphinone	0.24	0.23	0.08	0.01	0.11	0.05	0.08	0.16
Ergometrinine	0.42	0.25	0.03	0.00	0.08	0.12	0.10	0.62
Ephedrine[c]	—	—	—	—	—	—	—	—
Quinine	0.19	0.26	0.07	0.00	0.17	0.09	0.18	0.43
Dihydroergocristine	0.42	0.30	0.03	0.00	0.07	0.15	0.07	0.69
Hordenine	0.33	0.36	0.14	0.05	0.28	0.00	0.15	0.35

Ergocristine	0.51	0.38	0.14	0.05	0.13	0.46	0.15	0.70
Quinidine	0.33	0.40	0.15	0.00	0.25	0.12	0.18	0.50
Atropine	0.38	0.40	0.16	0.05	0.12	0.00	0.10	0.17
Colchicine	0.47	0.41	0.04	0.00	0.04	0.11	0.00	0.57
Ajmaline	0.47	0.42	0.12	0.03	0.30	0.06	0.13	0.56
Cinchonine	0.38	0.44	0.17	0.07	0.27	0.00	0.22	0.40
Homatropine	0.37	0.45	0.15	0.05	0.23	0.04	0.24	0.15
Ergotaminine	0.24	0.51	0.00	0.00	0.14	0.42	0.15	0.68
Pilocarpine	0.41	0.52	0.09	0.00	0.13	0.32	0.25	0.55
Codeine	0.38	0.53	0.16	0.04	0.26	0.12	0.27	0.35
Dihydrocodeine	0.38	0.54	0.18	0.06	0.28	0.10	0.30	0.25
Serpentinine[c]	0.53	0.56	0.08	0.00	0.10	0.00	0.03	0.12
Ergocristinine	0.61	0.57	0.13	0.00	0.20	0.00	0.27	0.70
Scopolamine	0.56	0.60	0.19	0.03	0.34	0.30	0.00	0.52
Yohimbine	0.63	0.62	0.18	0.03	0.37	0.33	0.15	0.60
Brucine	0.42	0.63	0.18	0.00	0.19	0.50	0.54	0.12
Cephaeline	0.56	0.63	0.19	0.02	0.23	0.25	0.17	0.37
Rauwolscine	0.55	0.63	0.18	0.04	0.36	0.36	0.15	0.68
Dihydrocodeinone	0.51	0.65	0.21	0.04	0.30	0.48	0.43	0.18
Apoatropine	0.54	0.67	0.40	0.20	0.26	0.15	0.40	0.16
Strychnine[c]	0.53	0.76	0.28	0.05	0.38	0.57	0.60	0.22
Reserpine	0.72	0.80	0.20	0.00	0.46	0.63	0.35	0.69
Group II								
Physostigmine	0.65	0.9	0.32	0.04	0.44	0.59	0.50	0.46
Aconitine	0.68	0.9	0.35	0.03	0.49	0.36	0.60	0.65
Bulbocapnine	0.65	0.9	0.35	0.07	0.54	0.78	0.70	0.48
Emetine	0.67	0.9	0.40	0.06	0.45	0.38	0.58	0.50
Papaverine	0.67	0.9	0.42	0.03	0.47	0.85	0.84	0.70
Cotarnine[c]	0.60	0.9	0.43	0.31	0.45	0.00	0.25	0.00
Scopoline	0.60	0.9	0.44	0.20	0.44	0.46	0.50	0.37

Table 15.1 (Continued)

	Silica Gel G					*Aluminum Oxide G*		*Silica Gel G Prepared with 0.1N NaOH*
Alkaloid	*Chloroform–Acetone–Diethylamine* (5:4:1)	*Chloroform–Diethylamine* (9:1)	*Cyclohexane–Chloroform–Diethylamine* (5:4:1)	*Cyclohexane–Diethylamine* (9:1)	*Benzene–Ethyl Acetate–Diethylamine* (7:2:1)	*Chloroform*	*Cyclohexane–Chloroform* (3:7) + 0.5% *Diethylamine*	*Methanol*
Lobeline	0.68	0.9	0.48	0.14	0.48	0.55	0.60	0.55
Narcotine	0.72	0.9	0.51	0.10	0.57	0.81	0.79	0.72
Thebaine	0.65	0.9	0.51	0.16	0.50	0.71	0.76	0.40
Aspidospermine	0.65	0.9	0.54	0.20	0.49	0.50	0.60	0.65
Tropacocaine	0.65	0.9	0.56	0.34	0.45	0.58	0.78	0.35
Arecoline	0.66	0.9	0.56	0.34	0.48	0.00	0.00	0.00
Hydrastinine[c]	0.66	0.9	0.58	0.41	0.50	0.00	0.25	0.00
Neopsicaine	0.66	0.9	0.60	0.35	0.53	0.83	0.82	0.59
Cocaine	0.73	0.9	0.65	0.36	0.58	0.84	0.77	0.62
Sparteine	0.70	0.9	0.68	0.68	0.55	0.00	0.55	0.05

[a] From D. Waldi, K. Schnackerz, and F. Munter [9]; reproduced with permisison of the authors and the Elsevier Publishing Co.

[b] All development distances 10 cm.

[c] Gives elongated spots.

Table 15.2 Color Reactions of Some Alkaloids[a]

Alkaloid	*Color of Fluorescence in Ultraviolet* (365 mμ)	*Color after Spraying with Iodoplatinate Reagent*	*Number of Secondary Spots*
Group I			
Narceine	—	Deep blue	—
Cupreine	Brownish yellow	Red-brown	—
Sarpagine	—	Beige	—
Ergometrine	Violet-blue	White[b]	—
Morphine	—	Deep blue	—
Dihydroergotamine	Violet-blue	Brownish	—
Seprentine	Dark brown	Red-brown	—
Ergotamine	Violet-blue	Pink	—
Boldine	Violet	Beige	—
Dihydromorphinone	—	Brownish yellow	—
Ergometrinine	Violet-blue	Violet-blue	—
Ephedrine	—	Light brown	—
Quinine	Blue	Yellow-white	—
Dihydroergocristine	Violet-blue	Brownish	—
Hordenine	—	White[b]	—
Ergocristine	Violet-blue	Beige-light brown	—
Quinidine	Blue	Light yellow	1
Atropine	—	Violet-blue	—
Colchicine	—	Light gray	—
Ajmaline	Bluish	Beige	2
Cinchonine	—	Beige-brown	—
Homatropine	—	Violet-blue	—
Ergotaminine	Violet-blue	Pink	—
Pilocarpine	—	Light brown	—
Codeine	—	Heather	—
Dihydrocodeine	Blue	Violet-blue	—
Sepretinine	Yellow-green	Yellow-brown	—
Ergocristinine	Violet-blue	Light brown	—
Scopolamine	—	Violet	—
Yohimbine	Blue-green	Light yellow	3
Brucine	—	Violet-brown	1
Cephaeline	Violet-blue	White[b]	3
Rauwolscine	Yellow-green	Pale beige	1
Dihydrocodeinone	—	Violet	2
Apoatropine	—	Violet-blue	—
Strychnine	—	Yellow	—
Reserpine	Green-yellow	White[b]	3
Group II			
Physostigmine	—	Pink	1
Aconitine	—	Red-brown	3

Table 15.2 (Continued)

Alkaloid	Color of Fluorescence in Ultraviolet (365 mμ)	Color after Spraying with Iodoplatinate Reagent	Number of Secondary Spots
Bulbocapnine	Blue	White[b]	—
Emetine	Blue	Red-brown	3
Papaverine	Yellowish	Yellow	—
Cotarnine	Green-yellow	Violet	1
Scopoline	—	White[b]	—
Lobeline	—	Red-brown	—
Narcotine	Blue	Light yellow	—
Thebaine	—	Red-brown	—
Aspidospermine	—	White[b]	—
Tropacocaine	—	Violet	—
Arecoline	—	White[b]	—
Hydrastinine	Steel blue	Violet-blue	1
Neopsicaine	—	Yellow	—
Cocaine	—	Violet	—
Sparteine	—	Violet	—

[a] From D. Waldi, K. Schnackerz, and F. Munter [9]; reproduced with permission of the author's and the Elsevier Publishing Co.

[b] Pink background.

alumina G with chloroform and on alkaline silica gel G (prepared with 0.1 *N* sodium hydroxide solution instead of water) and methanol.

For identification, the group II alkaloids are run on silica gel G layers with the solvents cyclohexane–diethylamine (9:1) and /or benzene–ethyl acetate–diethylamine (7:2:1). Again the colors and the R_f values are compared with those in Tables 15.1 and 15.2.

The values for the 54 alkaloids (Table 15.1) are given with all the separation systems so that any of the additional R_f values may be used in doubtful cases. It is also advisable to use more than one sample spot so that other well-known alkaloid reagents can be applied. Baiulescu and Constantinescu [10] list some in situ microreactions for the identification of alkaloids in mixtures along with the results for 17 alkaloids.

Noirfalise and Mees [11] have also established a systematic procedure for the analysis of 34 alkaloids and some amine bases.

Teichert et al. [12] give a general discussion of the use of thin layers for the separation of various alkaloid groups.

Mulé [13] has worked out a procedure for the quantitative extraction and analysis of narcotic analgesics in human biological materials using ultraviolet spectrophotometry, thin-layer chromatography, and gas chromatography. The method was applied to 31 compounds including opium alkaloids.

Tyihák and Held [14] in their review on pharmacognosy list 103 references using thin-layer chromatography for the analysis of alkaloids in pharmaceuticals covering the period up to 1969. Tyihák and Vágújfalvi [14a] include a bibliography of the 1050 publications up to 1969 in their review on the thin-layer chromatography of alkaloids. Bibliographies on the analysis of drug abuse, including some alkaloids, cover the period up to 1972 [15] and from there to 1974 [16]. Additional references may also be found in the general bibliographies mentioned in the introduction to this edition.

The most preferred adsorbent for alkaloids has been silica gel, although alumina [17], cellulose [18], polyamide [19], kieselguhr [20], talc [21], magnesium oxide [22], and calcium carbonate [23] have been used, as well as some mixed adsorbents such as silica gel–magnesia (1:1) [24] or calcium carbonate–magnesium oxide–calcium hydroxide (29.5:6:5) [25]. Sometimes the adsorbent is mixed with an alkalie to form an alkaline adsorbent.

Electrophoresis [26–28] has been employed as well as a two-dimensional combination of thin-layer chromatography and electrophoresis [26]. Two-dimensional thin-layer chromatography can also be used to good advantage [29, 30].

Solvents for the separation of alkaloids have been many and varied and examples of these are given under the various alkaloid types.

The most commonly used detecting agent is the well-known Dragendorff reagent or modifications of it. Vágújfalvi [31] used a 1:4 dilution of this reagent, but found that the sensitivity could be increased by following with 10% sulfuric acid. Other detecting agents that have been used include dansyl chloride, Wagner's reagent, and tetrazotized benzidine spray [32], ceric ammonium sulfate [33], Munier reagent [34], potassium ferrocyanide [35], Van Urk's reagent [36], a mixture of perchloric acid and ferric chloride [37], and iodoplatinate [38], as well as numerous others. Peuch et al. [39] used Dragendorff reagent followed by a solution of sodium nitrite to obtain a sensitivity of 0.1 μg. Tyihák and Vágújfalvi [14a] found a sensitivity of 0.01 to 0.05 μg for the dilute sulfuric acid spray for some opium alkaloids.

Quantitatively a number of methods are available both in situ and after elution, for example, reflectance scanning [40, 41], spot area [42], densitometry [42, 43], fluorescence [44], radioactive scanning [45], titration after elution [46], catalytic polarography [47], spectrophotometry after elution [48], and gas chromatography after elution [49].

Table 15.3 $R_f \times 100$ of Some Purine Alkaloids

Alkaloid	*Silica Gel Prepared with 0.2 M Sörenson buffer of pH 6.8*	*Silicic Acid*[a] [50]		*Loose-Layer Aluminum Oxide*[a] [55]	
	Chloroform–96% Ethanol (9:1) [54]	*Ethyl Acetate–Methanol–Acetic Acid* (8:1:1)	*Ethyl Acetate–Methanol–12 N HCl* (18:2:0.05)	*Chloroform–n-butanol* (98:2)	*Chloroform–Acetone* (1:1)
Theobromine	22	36	25	15	9
Theophylline	37	50	41	30	15
Caffeine	57	41	36	55	60
Aminophylline				30	15
Methyl caffeine				55	60

[a] R_f values with additional solvents are given in the original papers.

2 THE PURINE ALKALOIDS

These compounds can be separated on silicic acid, silica gel [4, 50–53], silica gel prepared with 0.2 *M* buffer solution [54], and aluminum oxide [55]. The R_f values for some of these compounds are given in Table 15.3. Baehler [50] used a microsublimation method for locating the spots and also for recovering the compounds. the spots may also be located by spraying first with an alcoholic iodine–potassium iodide solution followed by 25% hydrochloric acid–96% ethanol (1:1). Fincke [56] applied thin-layer chromatography to the detection of both caffeine and theobromine in cocoa butter.

Quantitation of caffeine, theobromine, and theophylline can be made by elution and measurement at 276 nm [53] or by a spot area method [51].

3 THE PHENYLALKYLAMINE ALKALOIDS

This is the group to which ephedrine belongs. The alkaloid constituents of *Catha edulis* have been investigated by Ristić and Thomas [57]. Silica gel plates were used for the separation with a solvent consisting of isopropanol with 5% ammonium hydroxide. Cathine (*d*-pseudonorephedrine) was the main alkaloid with some *l*-ephedrine and a third minor alkaloid being present. The R_f values for the three alkaloids were found to be 0.62, 0.30, and 0.92, respectively. Ephedrine would not move [9] in the basic solvent systems containing diethylamine but had an R_f value of 0.32 in acetone–methanol–acetic acid (5:4:1). More recently [58] propanol–5% ammonia (9:1) and butanol–7.5% ammonia (9:1) have been used.

Hordenine has been found in *Lophophora williamsii* [59], and Todd [60] examined the alkaloids in the cactaceae of this family in several locations in Mexico. Lundstroem and Agurell [61, 62] have also examined this interesting group of compounds. Sato et al. [63] examined *Pelecyphora aselliformis,* another of the "peyote" cacti. *Ariocarpus kotschoubeyanus* [64] and *A. fissuratus* [65] have also been investigated. Cochin and Daly [66] list the R_f values of mescaline in six different separation systems. Tyihák et al. [67–69] have studied the capsaicin isomers in the *Capsicum* genus.

4 THE PYRIDINE ALKALOIDS

Tschesche et al. [70] examined the alkaloids in *Lobelia syphilitica,* and of the numerous alkaloids present, seven were isolated in a pure form by partition and adsorption chromatography. Partition chromatography was carried out on paper, but the adsorption chromatography was carried out on thin layers of silica gel with chloroform–methanol (3:1) and on acidic alumina with chloroform–ethanol (19:1). Five of the seven alkaloids were identified and the R_f values are given.

Rother et al. [71] applied thin-layer chromatography on silica gel and alumina in the synthesis and structural determination of anaferine.

Similarly, Pailer and Libiseller [72, 73] used thin-layer chromatography for the isolation, purification, and structural determination of the alkaloid evonine from *Euonymus europaeus.*

Moll [74] has applied a solvent consisting of chloroform–absolute ethanol–25% ammonium hydroxide (9:1:1) to the separation of hemlock alkaloids and other piperidine bases on silica gel G using a saturated developing chamber. Detection was accomplished with iodine vapor or by spraying first with a solution of 0.5% 1-chloro-2,4-dinitrobenzene in ethanol, followed by 0.05% bromothymol blue solution in ethanol. The bases appeared as blue spots on a yellow background.

5 THE PYRROLIDINE ALKALOIDS

The alkaloids of *Achillea arten* have been investigated [75] with the assistance of silica gel layers using a chloroform–methanol (1:2) solvent. Potassium iodoplatinate or concentrated sulfuric acid with the application of heat were used as locating agents. The main alkaloid achillein which was difficult to separate from glycocolbetaine by paper chromatography was readily separated on silica gel layers. *l*-Stachydrine was also identified among the components.

Vidic [76] has separated the synthetic pyrrolidine compound Jetrium from morphine and other bases on silica gel layers by means of 0.1 methanolic ammonia solution (0.17% ammonia, water-free). Jetrium had an R_f value of 0.85.

6 THE PYRIDINE–PYRROLIDINE AND THE DIPYRIDINE ALKALOIDS

Nicotine has been separated from other alkaloids and drugs by a number of investigators [66, 77, 78]. Table 15.4 gives the R_f values in 11 different systems. Several means of detection may be used; it may be observed as a dark spot on a fluorescent plate or as a red-violet spot using Dragendorff reagent. Fejér-Kossey [79] published on the separation of 10 tobacco alkaloids and a method is available for their spectrophotometric determination after separation by thin-layer chromatography [80]. Martin [81] has a method for the determination of nicotine residues in foods.

Anabasine, which belongs to the same group, has been run in three different solvents systems [78] on alkaline silica gel plates. The R_f values in 11:1, 9:1, and 8:2 mixtures of chloroform–96% ethanol are 0.14, 0.16, and 0.33, respectively.

The alkaloids of *Nicotiana glauca* and *N. paniculata* have been examined by paper and thin-layer chromatography [82]. *dl*-Anabasine is the main component of *N. glauca* with lesser amounts of *l*-nornicotine, ni-

Table 15.4 $R_f \times 100$ Values of Nicotine in Various Systems

Adsorbent	*Solvent System*	$R_f \times 100$	*Ref.*
Alkaline silica gel	Chloroform–96% ethanol (11:1)	38	78
Alkaline silica gel	Chloroform–96% ethanol (9:1)	44	78
Alkaline silica gel	Chloroform–96% ethanol (8:2)	62	78
Silica gel G	Methanol	50	4
Silica gel G	Methanol–acetone–triethanolamine (1:1:0.03)	56–58	76
Silica gel G	Ethanol–pyridine–dioxane–water (10:4:5:1)	58	77
Silica gel G	Ethanol–acetic acid–water (6:3:1)	27	77
Silica gel G	Ethanol–dioxane–benzene–ammonium hydroxide (1:8:10:1)	90	77
Silica gel G	Methanol–*n*-butanol–benzene–water (12:3:2:3)	44	77
Alumina G	*n*-Butanol–*n*-butyl ether–acetic acid (4:5:1)	72	77
Alumina G	*n*-Butanol–*n*-butyl ether–ammonium hydroxide (5:14:1)	90	77

cotine, and piperidine, whereas the main component of *N. paniculata* is *l*-nicotine together with smaller amounts of *l*-nornicotine, pyridine, and *β*-nicotine.

7 THE CONDENSED PIPERIDINE–PYRROLIDINE ALKALOIDS

In this group in the belladonna alkaloids, quantitative methods for the determination of atropine have been set up [83]. After separation of the atropine on alumina using methanol as the developing solvent the atropine spot may be eluted with chloroform. The solvent is removed under reduced pressure and the residue is transferred to 2.0 ml of alcohol and 5.0 ml of 0.01 *N* sulfuric acid. The excess acid is then back-titrated with 0.01 *N* sodium hydroxide using methyl red as an indicator. As an alternative method, the area of the spot obtained on silica gel G plates may be determined by measurement of the spot size with a planimeter [84]. Using the latter method the following alkaloids have been separated with methyl ethyl ketone–methanol–7.5% ammonium hydroxide (3:1.5:0.5): aposcopolamine, scopolamine, osine, apoatropine, belladonnine, atropine, and tropine. The isomers *l*-hyoscyamine and atropine (*dl*-hyoscyamine) could not be separated. Khanna et al. [85] and Khafagy et al. [86] have applied thin-layer chromatography to the separation of *Withania* alkaloids using silica gel layers with ethanol–ammonium hydroxide (8:2) and chloroform–cyclohexane–diethylamine (7:3:1). Vegh et al. [87] have separated atropine from papaverine and aminopyrine. Teijgeler [88] mentions the separation of atropine from apoatropine using chloroform–absolute alcohol–25% ammonium hydroxide (9:1:1). Neumann and Schroeter [89] have used silica gel G with a solvent mixture of ethanol–25% ammonia (8:2) to obtain the following separations (R_f): tropine 0.26, pseudotropine 0.44, and tropinone 0.75. Steinegger and Gebistorf [90] have worked out a simple extraction process combined with thin-layer and paper chromatography for the rapid determination of 5 to 10% adulterants in the leaf drugs of stramonium, belladonna, and hyoscyamus.

Verzár-Petri and Haggag [91] examined the alkaloids in *Datura innoxia* and separated cuskohygrine, hyoscyamine, 3-tropoilteloidine, meteloidine, and scopolamine (R_f 0.25, 0.30, 0.34, 0.50, 0.56, and 0.70, respectively) using methyl Cellosolve–*tert*-amyl alcohol–water–ammonia (6:12:6:0.2). DeMaggio and Lott [92] found that the use of ultra sound during the extraction of *Datura stramonium* alkaloids increased the yields over conventional procedures.

A fluorescent method has been worked out for the quantitative determination of scopolamine [93], and a direct densitometric method has been used for both scopolamine and hyoscyamine [94]. Cocaine has been separated from atropine, homoatropine, and hyoscine [95] and from codeine, heroin, 6-monoacetylmorphine, morphine, and quinine [96]. Methods have been worked out for the detection of cocaine metabolites in urine [97–100].

Polesuk and Ma [101] investigated 23 solvent systems for separating and identifying tropane alkaloids and related compounds.

8 THE QUINOLINE ALKALOIDS

The Cinchona Alkaloids

Thin-layer chromatography has been applied to the separation of eight cinchona alkaloids on alkaline silica gel G plates [102]. A solvent composed of chloroform–methanol–diethylamine (80:20:1) was used and the R_f values tabulated. The method was applied to the examination of 14 commercial preparations. [103]. They were also chromatographed on straight silica gel G layers [104] using a solvent mixture of kerosene–diethylamine–acetone (23:9:9). On silica gel G plates Oswald and Flueck [84] have separated small quantities (less than 1 μg) of quinine, cinchonine, and quinidine with isopropanol–benzene–diethylamine (2:4:1). In order to get a clear separation of these three alkaloids with larger amounts, it was necessary to use a solvent composed of benzene–ether–diethylamine (20:12:5) and to use a multiple development of two times, developing a distance of 15 cm each time. Although the R_f values of quinidine and cinchonine are very close together, the two may be differentiated by the typical fluorescence of quinidine in long-wave ultraviolet light and the dark color reaction of the cinchonine with the iodoplatinate reagent. If the cinchonine is sprayed with 98% formic acid prior to the iodoplatinate reagent, it takes on a dark blue instead of the beige-brown color found with the straight iodoplatinate reagent. Braeckman et al. [105] used a combination of thin-layer chromatography and fluorimetry to determine the quinine content of cinchona bark. Some R_f values for these compounds are given in Table 15.1 for the general separation of alkaloids. Šaršúnová and Hrivňák [106] have separated the cinchona alkaloids on silica using chloroform–acetone–dimethylamine (5:4:1) and obtained quantitative evaluation by gas chromatography after the compounds were eluted.

Forty-nine quinoline alkaloids were examined on silica and alumina with benzene–ethyl acetate (6:4), toluene–ethyl acetate–formic acid (5:4:1), and ethyl acetate–methanol (10:1) [107].

9 THE ISOQUINOLINE ALKALOIDS

The Opium Alkaloids

Because of their importance the opium alkaloids have been rather thoroughly investigated by thin-layer chromatographic methods. As early as 1953, Borke and Kirsch [108] showed the applicability of thin-layer chromatography to the separation of opium alkaloid mixtures by applying the method as originated by Kirchner et al. [1]. They used a mixture of equal parts of silicic acid and magnesium oxide held on the plates with a plaster of Paris binder. At this time they also demonstrated the usefulness of a buffer with which the adsorbent is mixed instead of pure water. In this case they used a phosphate buffer with a pH of 6.6. Development was accomplished with dioxane as the solvent. Mariani and Mariani-Marelli [109] on the other hand used a buffered layer (pH 5 acetate buffer) of aluminum oxide for the separation of opium alkaloids. The aluminum oxide was activated at 500°C and then allowed to adsorb 5% of the buffer solution. Extracts of various opiums could be separated in 15 min. The spots were observed under ultraviolet light. Neubauer and Mothes [110] applied a benzene–methanol (8:2) solvent to the separation of 10 of the more important opium alkaloids on silica gel G layers. The Munier modification of the Dragendorff reagent was used for locating the spots. The method was also used as a quantitative method by photographing the evenly sprayed plates and then measuring the spots photometrically. Bayer [111] has used a solvent mixture of xylene–methyl ethyl ketone–methanol–diethylamine (20:20:3:1) for separation of a group of five opium alkaloids on silica gel layers. Good separation was obtained between the various alkaloids (Table 15.5). Heusser [112] has used a thin-layer method for thc determination of various alkaloids in medicinal plants including morphine in powdered opium. Poethke and Kinze [113–115] have separated opium alkaloids by two-dimensional chromatography on aluminum oxide. Quantitative determinations were made photometrically on eluates and by the use of an integrator with an extinction registering device to evaluate the spots [116]. Moiseev [117] separated opium alkaloids on silica gel using benzene–ethanol (9:1) for the solvent. Using a 15-cm development the following R_f values were obtained: morphine 0.05, codeine 0.11, thebaine 0.26, and papaverine 0.41. Kupferberg et al. [118] have applied a fluorometric method for the identification of submicrogram amounts of morphine and related compounds and Szasz et al. [119] have included some morphine alkaloids in separating various basic substances by thin-layer chromatography. Penna-Herreros [120] has separated morphine, normorphine, and nalorphine on silica gel layers.

Table 15.5 $R_f \times 100$ Values of Some Opium Alkaloids in Various Systems[a,b]

Alkaloid	*A* *u* [108]	*B* *v* [109]	*C* *w* [110]	*D* *x* [129]	*E* *x* [129]	*F* *y* [129]	*G* *w* [111]
Morphine	39	50	11	3	14	28	12
Codeine	62	45	21	16	64	85	26
Laudanine			26				
Laudanosine			42				
Papaverine	88	100	63	60	85	96	59
Thebaine			40	38	85	100	45
Narcotoline			58				
Narcotine	92		68	100	92	100	74
Narceine	0	20		0	0	0	
Cryptopine			34				
Protopine			38				

[a] Solvent: *A* = dioxane; *B* = ethanol–*n*-butanol–water (1:9:1); *C* = benzene–methanol (8:2); *D* = *n*-hexane–acetone (3:1); *E* = chloroform–ethyl acetate (3:1); *F* = chloroform–ethyl acetate (1:1); *G* = xylene–methyl ethyl ketone–methanol–diethylamine (20:20:3:1). Development distance 13.5 cm.

[b] Adsorbent: *u* = silicic acid–heavy magnesium oxide (1:1) with $CaSO_4$ binder; *v* = aluminum oxide buffered to pH 5; *w* = silica gel G; *x* = heavy hydrated magnesium oxide prepared with 2.5% $CaCl_2 \cdot 6H_2O$ (5:9) (wt/vol); *y* = heavy hydrated magnesium oxide prepared with 2% $MgSO_4 \cdot 7H_2O$ (5:9) (wt/vol).

Circular chromatography on silica gel G has been applied to the separation of morphine quinine, and papaverine by Schantz [121]. Ikram et al. [122] gives the R_f values in three different solvents for morphine, codeine, papaverine, and narcotine on loose layers of alumina. R_f values for some of the more important morphine alkaloids are given in Table 15.5 and additional values may be found in Table 15.1

Teichert et al. [12, 78] have separated opium alkaloids and synthetic morphine derivatives on neutral and on alkaline silica gel plates as well as on formamide-impregnated cellulose layers (Table 15.6).

Mulé [13] has used cellulose plates prepared by mixing 15 g of cellulose powder (MN 300 G) with 90 ml of 0.1 *M* phosphate buffer (pH 8.0) for the separation of some morphine derivatives in two solvents as well as silica gel G in seven solvents. The R_f values are listed for these and also for other narcotic analgesics (see Table 26.6). Sherma et al. [123] have developed an in situ spectrodensitometric determination of morphine and

Table 15.6 $R_f \times 100$ Values of Some Opium Alkaloids and Synthetic Morphine Derivatives [78][a,b]

Alkaloid	*A* *w*	*B* *x*	*B* *y*	*C* *z*
Morphine	10	2	27	0
Dilaudid	13	5	27	6
Dicodid	28	10	34	63
Codeine	33	12	41	37
Dionin	37	14	44	57
Acedicon	59	24	—	90
Eucodal	70	47	79	75
Papaverine	78	74	86	89
Narcotine	81	78	92	94
Thebaine	—	—	—	85

[a] Adsorbent: *A* = silica gel G prepared with 0.2 *N* potassium hydroxide; *B* = silica gel G; *C* = cellulose layers impregnated with 20% formamide in acetone.

[b] Solvent: *w* = chloroform–ethanol (8:2); *x* = chloroform–ethanol (9:1); *y* = dimethylformamide–diethylamine–ethanol–ethyl acetate (5:2:20:75); *z* = benzene–heptane–chlorofrom–diethylamine (6:5:1:0.02).

amphetamine on thin layers. Gorodetzký [124] gives the sensitivity of detection for 16 opioids, cocaine, and quinine relative to their detection in urine. Steele [125] lists eight solvent systems for the separation and identification of 26 compounds found in narcotic seizures, and found that ethyl acetate–benzene–acetonitrile–ammonia (50:30:15:5) and (25:30:40:5) gave a good separation of morphine, codeine, thebaine, papaverine, and narcotine. Sensitive techniques have been developed for the determination of morphine [126] and heroin [127] in the urine, and Christopoulos and Kirch [128] have improved the techniques for the isolation and identification of morphine in postmortem tissues.

Ragazzi et al. [129] used magnesium oxide layers impregnated with either calcium fluoride or magnesium sulfate for separation of a group of alkaloids including some of the opium group (Table 15.5). The original paper gives the R_f values for other alkaloids as well as additional R_f values in different solvents for the opium alkaloids.

Huang et al. [130] separated seven opium alkaloids on polyamide using cyclohexane–ethyl acetate–*n*-propanol–dimethylamine (30:2.5:0.9:0.1) and water–absolute ethanol–dimethylamine (88:12:0.1).

Other Papaveraceae Alkaloids

Eighty-two alkaloids of the genus *Papaver* have been chromatographed on silica gel G in benzene–acetone–methanol (7:2:1) and on alumina with heptane–chloroform–ether (4:5:1) by Pfeifer, who tabulates these values as well as references to original work on the alkaloids [131, 132]. For alkaloids with low R_f values he recommends benzene–acetone–ether–10% ammonia (4:6:1:0.3) or benzene–acetone–ether–isopropanol–3% ammonia (20:15:10:7.5:2.5) for additional resolution. Detection was with a modified Dragendorff reagent [133].

Miscellaneous Isoquinoline Alkaloids

In the synthesis of isoquinoline alkaloids Bobbitt et al. [134] have used preparative thin-layer chromatography for separating the acetates of the alkaloids.

For the separation of bisbenzylisoquinoline alkaloids, alkaline silica gel G plates were used [135] with a solvent system of chloroform–ethyl acetate–methanol (2:2:1). The R_f values are given for the separation of six compounds. Bhatnagar and Bhattacharji [136] have also used thin-layer chromatography for this group of alkaloids.

10 THE INDOLE ALKALOIDS

The Ergot Alkaloids

The important drug ergot is composed of the mycelia of the fungus *Claviceps' purpurea*. From this a large group of alkaloids has been isolated. Klavehn and Rochelmeyer [137] and Klavehn et al. [138] used neutral silica gel with chloroform–ethanol mixtures for the separation and determination of ergot alkaloids. For detecting the compounds, Van Urk's reagent for detecting indole compounds was used. This reagent was also the basis for the quantitative method and can be used either as a semiquantitative determination by comparing the intensity of the test spots with standard spots or in a quantitative fashion by eluting the compound, and after reacting with Van Urk's reagent determining the color in a colorimeter. Teichert et al. [54] have used a formamide-impregnated cellulose plate combined with a stepwise development to separate the same alkaloids. The cellulose plates were impregnated by dipping in a 20% solution of formamide in acetone after which the acetone was allowed to evaporate. The development was carried out for a distance of 15 cm first in benzene–heptane–chloroform (6:5:3), and then a second development was carried out in the same direction with benzene–heptane (6:5) (Table 15.7). Hohmann and Rochelmeyer [139] have applied the same impregnated layer for the separation of hydrogenated ergot alkaloids, in this case using a solvent composed of ethyl acetate–heptane–diethylamine (5:6:0.02). Reichelt and Kudrnáč [140] separated 16 ergot alkaloids on silica gel impregnated with formamide using either diisopropyl ether–tetrahydrofuran–toluene–diethylamine (75:15:15:0.1) or diisopropyl ether–toluene–ethanol–diethylamine (75:20:5:0.1).

Table 15.7 *$R_f \times 100$* Values of Ergot Alkaloids and Some Hydrogenated Derivatives[a]

	SPO	*Silica Gel G*				*Cellulose Impregnated with Dimethylformamide*	
Alkaloid	*t*	*u*	*v*	*w*	*x*	*y*	*z*
Ergotamine	1.0	31.3	30.8	13	51	6	11
Ergosine	1.6	35.0	31.4	13	51	11	17
Ergocristine	9.3	54.4	55.8	28	69	67	41
Ergocornine	9.5	57.9	58.8	28	69	73	50
Ergocryptine	14.8	59.8	55.4	28	69	83	61
Ergotaminine	6.5	68.2	63.8	34	75	50	
Ergosinine	11.5	74.7	67.7	34	75	65	
Ergocristinine	37.9	79.6	74.3	45	81	90	
Ergocorninine	31.0	83.3	72.6	45	81	93	
Ergocryptinine	45.7	85.4	74.6	45	81	97	
Ergonovine	0.0	17.3	12.0				
Ergometrinine	0.8	44.1	38.4		30		
Ergometrine					17		
Dihydroergotamine					11		9
Dihydroergocrystine					14		30
Dihydroergocornine							38
Dihydroergocryptine							50

[a] Solvent: t = chloroform–ethanol–water (3:1:1), development distance 17 cm (142); u = ethyl acetate-*N,N*-dimethylformamide–ethanol (13:1.9:0.1), development distance 17 cm (142); v = benzene-*N,N*-dimethylformamide (13:2), development distance 17 cm (142); w = chloroform–ethanol (19:1), development distance 10 cm (137,138); x = chloroform–ethanol (9:1), development distance 10 cm (137,138); y = first development 15 cm with benzene–heptane–chloroform (6:5:3), second development 15 cm with benzene–heptane (6:5) (54); z = ethyl acetate–heptane–diethylamine (5:6:0.02) (139).

Groeger and Erge [141] have discussed a series of eight different separating systems for use with ergot alkaloids. McLaughlin et al. [142] examined the separation of 12 ergot alkaloids in three different systems consisting of silica gel G with benzene–dimethylformamide (13:2) or ethyl acetate–dimethylformamide–ethanol (13:1.9:0.1) and aluminum oxide G with chloroform–ethanol–water (3:1:1) (Table 15.7). In addition to running the individual compounds they also ran mixtures which demonstrated the effect of the mixture on the R_f values of the individual components. The alkaloids were detected by spraying with Ehrlich's reagent (5% *p*-dimethylaminobenzaldehyde in concentrated hydrochloric acid) and by examination in ultraviolet light. For the quantitative determination, the work was carried out under subdued light because of the effect of light on the ergot alkaloids. Quantitative determination was carried out by elution of the spot and reacting the solution with *p*-dimethylaminobenzaldehyde solution for 30 min. The absorbance was then measured at 590 mμ. Eich and Schunack [143] used fluorescent measurements for determining these alkaloids; sensitivity ranged from 15 to 100 ng. Groeger et al. [144] used starch-bound layers of silicic acid to investigate the alkaloids from *Claviceps paspali,* comparing material obtained in Australia and in Arkansas. Zinser and Baumgaertel [145] used silica gel plates with benzene–chloroform–ethanol (2:4:1) and heptane–carbon tetrachloride–pyridine (1:3:2) for testing the purity of natural and hydrogenated ergot alkaloids. The procedure was used for testing the stability of ergot alkaloids in aqueous preparations.

The related clavine alkaloids can be separated on silica gel as shown by Klavehn and Rochelmeyer [137] (Table 15.8). Mixtures of chloroform and ethanol were used as solvents as well as ethyl acetate–ethanol–dimethylformamide (85:10:5). Best resolution was obtained by developing first in chloroform–ethanol (95:5) followed by a second development in the dimethylformamide solvent. Using silica gel G with ethyl acetate–ethanol–dimethylformamide (13:1:1), Agurell and Ramstad [146] examined the clavine alkaloids in *Pennisetum* ergot and obtained R_f values for 18 compounds (Table 15.8).

The seeds of the ancient Aztec drug Ololiuquie *(Rivea corymbosa)* and of *Ipomoea tricolor Cav.* were investigated by Hofmann [147] and Hofmann and Tscherter [148]. Separations were accomplished on alumina layers using chloroform–methanol (95:5). For positive identification work a combination of column chromatography and preparative thin-layer chromatography yielded crystalline compounds which were identified as lysergic acid amide, isolysergic acid amide, chanoclavine, elymoclavine, and lysergol. Beyerman et al. [149] examined the seeds and sometimes the leaves of 25 Convolvulaceae species for lysergic acid

Table 15.8 $R_f \times 100$ of Clavine Alkaloids on Silica Gel

	Solvent			
Alkaloid	*Ethyl Acetate–Ethanol–Dimethylformamide* (13:1:1) [146]	*Chloroform–Ethanol* (95:5) [137][a]	*Chloroform–Ethanol* (90:10) [137][a]	*Chloroform–Ethanol* (80:20) [137][a]
Agroclavine	54	15	38	45
Elymoclavine	23	2	13	15
Penniclavine	38	4	17	20
Isopenniclavine	71	8	30	35
Setoclavine	68	18	42	50
Isosetoclavine	81	20	46	60
Secaclavine	5			
Lycergene	75			
Lysergol	27			
Isolysergol	57			
Festuclavine	36			
Pyroclavine	51		20	
Lysergine	64			
Isolysergine	78			
Costaclavine	16			
Dihydroelymoclavine	12			
Fumigaclavine A	74			
Fumigaclavine B	40			

[a] Development distance 10 cm.

amide and ergot alkaloids. Of these only *Ipomoea rubrocoerulea* Hook var. *Praecox* (morning glory) and the variety *Ipomoea* "Pearly-Gates" showed the presence of indole alkaloids. The thin-layer chromatograms were practically identical with those found by Hofmann [147].

Genest and Farmilo [150] have applied a thin-layer chromatographic method for the detection and determination of lysergic acid diethylamide in the presence of heroin or other legally controlled drugs. Spectrophotofluorometric analysis was carried out on the eluted spots. In checking lysergic acid against 14 other related ergot alkaloids apt to be encountered in illicit drug traffic, Fowler et al. [151] found that no single solvent in the 18 investigated could separate the lysergide from all the other alkaloids.

Rauwolfia Alkaloids

These alkaloids have been separated on loose layers of alumina [122] and on formamide-impregnated cellulose layers [78] (Table 15.9). Additional R_f values may be found in Table 15.2.

Schlemmer and Link [152, 153] applied a separation with methanol–methyl ethyl ketone–heptane (8.4:33.6:58) on silica gel for the separation of reserpine. A quantitative evaluation was obtained by eluting the spot with 96% ethanol–dioxane (1:1) and measuring in an ultraviolet spectrophotometer. Liukkonen [154] applied the method to the separation of reserpine, rescinnamine, ajmaline, yohimbine, and raubasine.

Ullmann and Kassalitzky [155] separated reserpine and rescinnamine from pharmaceutical solutions containing polyethylene oxide solubilizers. A preliminary separation is achieved by mixing 4 ml of the fluid extract with 3.0 g of 10% sodium carbonate solution, then triturating with 4 g of kieselguhr. After drying the mixture thoroughly in a desiccator it is extracted with chloroform. The chloroform solution concentrated to 5 ml at low temperatures can then be applied directly to the formamide-impregnated silica gel plates. For the separation, *n*-heptane–methyl ethyl ketone (2:1) saturated with water is used. For locating the spots the Dragendorff reagent may be used and has a sensitivity of 0.5 γ. The sensitivity can be increased to 0.05 γ by following the Dragendorff reagent

Table 15.9 $R_f \times 100$ Values of Some Rauwolfia Alkaloids

Alkaloid	*Formamide Impregnated Cellulose Layers* [78] *Heptane–Methyl Ethyl Ketone* (1:1)[a]	*Loose-Layer Alumina* [122] *Chloroform–Acetone* (85:15)[b]	*Loose-Layer Alumina* [122] *Absolute Ethanol*[b]	*Loose-Layer Alumina* [122] *Chloroform–Ethanol–Acetone* (90:5:5)[b]
Sarpagine	3.0			
Serpentine	6.0	2.4	75	34
Ajmaline	28	2.4	87	51
Yohimbine	33			
Rescinnamine	51			
Reserpine	59	60		89
Reserpinine	89			
Ajmalicine		77		
Serpentinine			86	73

[a] Development carried out in an ammonia atmosphere.
[b] Development distance 33 cm.

spray with a mixture of 100 ml of 5% perchloric acid and 3 ml of 0.05 *M* ferric chloride solution. The quantitative determination is made according to Schlemmer and Link [152, 153]. Court [156] separated Rauwolfia alkaloids into two groups by extracting a 1 *M* hydrochloric acid solution of the alkaloids with chloroform at two different concentrations of ammonia. Using silica gel the weaker bases were separated with methanol–methyl ethyl ketone–heptane (1:3:6) and the stronger bases with *n*-butanol–acetic acid–water (4:1:1). For quantitative work 12 common Rauwolfia alkaloids were separated with five solvent systems.

The Harmala Alkaloids

Bernauer [158] has isolated harmin and 1,2,3,4-tetrahydroharmin from Indian cold drugs by means of extraction and thin-layer chromatography on aluminum oxide with acetone–ethanol (85:15) as the developing solvent. By using preparative thin layer, harmin could be isolated in 1.3% and the tetrahydro derivative in 0.2% yield. Groeger [159] separated harman, harmine, harmaline, harmol, and harmalol on silica gel with chloroform–methanol–10% ammonia (80:20:1.5) and chloroform–methanol (3:1) with R_f values of 0.65, 0.5; 0.66, 0.44; 0.35, 0.02; 0.46, 0.30; and 0.13, 0.02, respectively. Lutomski et al. [160] determined harman and harmine spectroscopically after treating the eluates with methanol–sulfuric acid (100:1).

The Strychnos Alkaloids

Strychnine has been separated from brucine [161] by chromatographing on silica gel G. Potassium dichromate solution is applied to the sample spot thereby oxidizing the brucine to its *o*-quinone, which remains fixed at the starting point. The strychnine can then be developed away from the origin and can be detected by means of Bouchardat reagent which gives a brown spot. The method was used for detecting strychnine and brucine in the viscera of poisoned rats.

Grandolini et al. [162] have separated strychnos alkaloids on silica gel with butanol–hydrochloric acid (95:5) equilibrated with water, and with chloroform–methanol (4:1). A two-dimensional technique was employed using the chloroform mixture in one direction followed by the acidic solvent for the second direction. The same workers were also able to separate some stereoisomeric alkaloids from *Skyganthus acutus* belonging to the cyclopentanolpiperidine group. The R_f values were given for the various alkaloids investigated. Weissmann et al. [163] used thin-layer chromatography on large plates in separating some reaction products of strychanone.

Thirty alkaloids found in the genus *Strychnos* were chromatographed in seven different solvent systems on silica gel and on alumina, and the correlations between R_f value and structure were examined [164].

Vinca Alkaloids (*Cantharanthus* Alkaloids)

Jakovljevič et al. [165] separated eight *Vinca rosa* alkaloids on alumina layers prepared with 0.5 *N* lithium hydroxide in order to eliminate tailing in the straight alumina layers. Solvents used were ethanol–acetonitrile (5:95) and benzene–acetonitrile (7:3). Ceric ammonium sulfate 1% in 85% phosphoric acid was used for detection and gives many varied colors with the different alkaloids. A 50% dilution of the reagent was sprayed on plates warmed at 100°C for 5 min. Color development was poor if the chromatograms were kept more than 30 min after development. Benzene, benzene–ethyl ether (1:1), and benzene–ethyl acetate (4:1) have also been used with alumina [166]. Cone et al. [167] give the R_f values of 26 *Vinca* alkaloids on silica and alumina with nine solvents. Two additional solvents are given for several of the more difficult separations. With complex mixtures two-dimensional chromatograms were used to advantage. Farnsworth et al. [168] examined 63 *Cantharanthus* alkaloids on silica gel with three solvents.

Oxindole Alkaloids

Shellard et al. [169] chromatographed 18 oxindole alkaloids on silica and alumina and have published a series of papers on this group of alkaloids including quantitative techniques for some *Mitragyna* oxindoles by densitometry [170], ultraviolet spectrophotometry [171], and colorimetry [172]. On silica gel some of the solvents that can be used are ether, chloroform–acetone (5:4), ethyl acetate–chloroform (95:5), ethyl acetate–17% ammonia–isopropanol (9:4:7), and ether–diethylamine (19:1). On alumina chloroform or ethyl acetate–chloroform (95:5) may be used. More recently Shellard and Houghton [173] have been examining the *Mitragyna* species of Asia.

Miscellaneous Indole Alkaloids

Kump and co-workers [174–176] have worked on the isolation of the *Pleiocarpa* alkaloids using 7% methanol in chloroform as a solvent on silica gel G. The R_{st} values (using pleiocarpine as a standard) were as follows: pleiocarpinine 0.69, kopsinine 0.62, eburnamenine 0.90, pleiomutine 0.37, pleiomutinine 0.17, pleiocarpamine 0.48, and pleiocarpinidine 0.46. With 4% methanol in chloroform, pleiocarpinilam had a corresponding value of 1.55 and kopsinilam had a value of 1.35. As a visualizing agent they used a saturated solution of ceric sulfate in 60% sulfuric acid.

Walser and Djerassi [177] investigated the alkaloids of *Vallesia dichotoma* and give the R_f values of 22 indole and dihydroindole alkaloids on silica gel in methanol–methylene chloride (5:95) and methanol–ethyl acetate (5:95) along with the colors produced with ceric sulfate reagent (T-53) [178].

Augrell et al. [179] investigated the indole alkaloids in *Virola* species and other South America plants. Two new compounds were isolated from *Riccardia sinuata* [180] and a method has been worked out for the isolation and identification of the indole alkaloid ibogaine from biological material [181]. Phillipson and Shellard [182] have used thin-layer chromatography on alumina with 15% calcium sulfate to differentiate between indole alkaloids with cis and trans C/D and D/E ring junctions.

11 PYRROLIZIDINE ALKALOIDS

Chalmers et al. [183] reported the R_f values of a group of 58 compounds including pyrrolizidine alkaloids and basic derivatives on silica gel G plates prepared from 30 g of adsorbent–binder mixture with 60 ml of 0.1 *N* sodium hydroxide. Methanol was used as a solvent. In addition R_f values on paper are given as well as retention times on gas chromatography. Sharma et al. [184] examined eight species of *Crotalaria*. The pyrrolizidine bases were separated on silica gel with chloroform–methanol–ammonia (85:14:1). These alkaloids have been examined by thin-layer chromatography in *Liguaria* and *Senecia* [185, 186]. Mattocks [187] has devised a more sensitive test for pyrrolizidine alkaloids which is positive only to alkaloids having an unsaturated (3-pyrroline) ring in the basic moiety (T-244).

12 MISCELLANEOUS ALKALOIDS

Schlaremann [188] separated the acridine alkaloids from callus cultures of *Ruta graveolens* on a mixed adsorbent layer of calcium carbonate–magnesium oxide–calcium hydroxide (29.5:6:5) with petroleum ether–benzene–methanol–methyl ethyl ketone (70:20:2:8) and on polyamide with the same solvent in a 50:40:5:5 mixture.

Hěrmánek et al. [189] give a list of R_f values for 37 alkaloids separated on layers of alumina using eight different solvents including chloroform, ethanol, benzene, and mixtures of ethanol in chloroform. Paris and Paris [190] give the R_f values for 20 alkaloids on silica gel and aluminum oxide. Doepke [122] gives the R_f values for the separation of 42 *Amaryllidaceae* alkaloids using silica gel G layers and a solvent composed of ethyl acetate–chloroform–methanol (2:2:1).

Máthé and Tyihák [191] have examined the alkaloids of *Colchicum hungaricum*. The solvents used were chloroform, methanol, and ethyl acetate–benzene (5:95) with silica gel as the thin layer material. The alkaloid content was compared to that of *Colchicum autumnale* and found to be quite similar. The lumiderivatives of some colchicine alkaloids from the subfamily *wurmbaeoideae* were separated on silica gel G with benzene–ethyl acetate–diethylamine ((5:4:1 and 7:2:1), 8% methanol, and cyclohexane–diethylamine (4:1) [192]. Two-dimensional thin-layer chromatography was applied to the separation of colchicine alkaloids from galenic cholchicum preparations [193].

The alkaloids of *Senecio cineraria* have been separated [194] on silica gel with chloroform–methanol (9:1). Two-dimensional thin-layer chromatography on silica was applied to the alkaloids of *S. vulgaris* [195] using *n*-butanol–acetic acid–water (1:1:1) in the first direction and methanol in the second direction.

Fish and Waterman [196] found six alkaloids in *Fagara macrophylla* by chromatographing the chloroform soluble fraction on silica gel layers. The alkaloids from *Fagara rubescens* and *F. leprieurii* were also examined [197].

The alkaloids in ipecacuanha roots were examined on TLC layers of silica gel G with toluene–benzene–ethyl acetate–diethylamine–methanol (35:35:20:10:2) [198], and a photodensitometric method for the principal alkaloids has been published [199]. Habib and Harkiss [198] have also published on the quantitative determination of emetine and cephaeline.

Among the quinolizadine alkaloids, Cho and Martin [200] separated 22 lupine alkaloids on silica with mixtures of chloroform–methanol–ammonia in ratios varying from 95:4:1 to 85:14:1; R_f values are also given for compounds on alumina with benzene–acetone–methanol (34:3:3) and for cellulose layers with butanol–concentrated hydrochloric acid–water (70:7.5:13.5).

Lepri et al. [201] reported on the use of thin layers of anion and cation exchangers with 48 alkaloids. They give the R_f values for 41 of these on Bio-Rad AG 1-X4 and Cellex D with four buffers of different pH values and also those for microcrystalline cellulose with one buffer solution. A modified Dragendorff reagent was used for detection.

References

1. J. G. Kirchner, J. M. Miller, and G. J. Keller, *Anal. Chem.*, **23,** 420 (1951).
2. H. Kozuka, *Kagaku Keisatsu Kenkyusho Hokoku,* **16,** 39 (1963); through *Chem. Abstr.*, **59,** 15121 (1963).
3. V. I. Lobanov, *Sudebno-Med. Ekspertiza, Min. Zdravookhr. SSSR,* **6,** 42 (1963); through *Chem. Abstr.*, **61,** 7330 (1964).
4. G. Machata, *Mikrochim. Acta,* **47,** 79 (1960).
5. N. R. Farnsworth and K. L. Euler, *Lloydia,* **25,** 186 (1962).
6. R. Munier and M. Macheboeuf, *Bull. Soc. Chim. Biol.*, **33,** 846 (1951).
7. J. J. Neto and B. Mancini, *Rev. Fac. Farm. Odontol. Araraquara,* **2,** 29 (1968); through *Chem. Abstr.*, **71,** 16016m (1969).
8. L. P. Bush and J. A. D. Jeffreys, *J. Chromatogr.*, **111,** 165 (1975).
9. D. Waldi, K. Schnackerz, and F. Munter, *J. Chromatogr.*, **6,** 61 (1961).
10. G. E. Baiulescu and T. Constantinescu, *Anal. Chem.*, **47,** 2156 (1975).
11. A. Noirfalise and G. Mees, *J. Chromatogr.*, **31,** 594 (1967).
12. K. Teichert, E. Mutschler, and H. Rochelmeyer, *Z. Anal. Chem.*, **181,** 325 (1961).
13. S. J. Mulé, *Anal. Chem.*, **36,** 1907 (1964).
14. E. Tyihák and G. Held, "Thin-Layer Chromatography in Pharmacognosy," in *Progress in Thin-Layer Chromatography and Related Methods,* Vol II, A. Niederwieser and G. Pataki, Eds., Ann Arbor Science, Ann Arbor, Mich., 1971, p. 183.

14a. E. Tyihák and D. Vágújfalvi, "Thin-Layer Chromatography of Alkaloids," in *Progress in Thin-Layer Chromatography and Related Methods,* Vol. III, A. Niederwieser and G. Pataki, Eds., Ann Arbor Science, Ann Arbor, Mich., 1972, p. 71.

15. Anonymous, *J. Chromatogr., Sci.*, **12,** 328 (1974).
16. *Ibid.*, **10,** 352 (1972).
17. R. G. Achari and W. E. Court, *Planta Med.*, **24,** 176 (1973).
18. D. Giacopello, *J. Chromatogr.*, **19,** 172 (1965).
19. H.-C. Hsiu, J.-T. Huang, T.-B. Shih, K.-L. Yang, K.-T. Wang, and A. L. Lin, *J. Chin. Chem. Soc. (Taiwan),* **14,** 161 (1967).
20. N. Ivanov and A. Boneva, *Bulg. Tytutyun,* **11,** 30 (1966); through *Chem. Abstr.*, **65,** 8668f (1966).
21. I. Zarebska and A. Ozarowski, *Farm. Pol.*, **22,** 518 (1966); through *Chem. Abstr.*, **66,** 40741m (1967).
22. E. Ragazzi and G. Veronese, *Mikrochim. Acta,* **1965,** 966.
23. S. Ebel, E. Bahr, and E. Plate, *J. Chromatogr.*, **59,** 212 (1971).
24. M. L. Borke and E. R. Kirch, *J. Am. Pharm. Assoc. Sci. Ed.*, **42,** 627 (1953).
25. W. Scharlemann and *Z. Naturforsch.*, **27b,** 806 (1972).
26. S. Agurell, *Acta Pharm. Succica,* **2,** 357 (1965); through *Chem. Abstr.*, **64,** 6404g (1966).

27. J. M. Calderwood and F. Fish, *J. Pharm. Pharmacol.*, **21** (Suppl.), 126 (1969).
28. A. S. C. Wan, *J. Chromatogr.*, **60**, 371 (1971).
29. N. Seiler and L. Demisch, *Biochem. Pharmacol.*, **23**, 259 (1974).
30. M. Haag-Berrurier and M. C. Mathis, *Ann. Pharm. Fr.*, **31**, 457 (1973).
31. D. Vágújfalvi, *Planta Med.*, **13**, 79 (1965).
32. K. M. K. Hornemann, J. M. Neal, and J. L. McLaughlin, *J. Pharm. Sci.* **61**, 41 (1972).
33. G. Aynilian, N. R. Farnsworth, R. L. Lyon, and H. H. S. Fong, *J. Pharm. Sci.*, **61**, 298 (1972).
34. M. R. Castagnou and S. Larcebau, *Bull. Soc. Pharm. Bord.*, **103**, 201 (1964); through *Chem. Abstr.*, **62**, 10798f (1965).
35. K. C. Guven and N. Guven, *Eczacilik Bull.*, **14**, 75 (1972); through *Anal. Abstr.*, **24**, 3040 (1973).
36. T. Niwaguchi and T. Inoue, *J. Chromatogr.*, **43**, 510 (1969).
37. K. Teichert, E. Mutschler, and H. Rochelmeyer, *Dtsch. Apoth.-Ztg.*, **100**, 477 (1960).
38. K. Haisová, J. Slávik, and L. Dolejš, *Collect. Czech. Chem. Commun.*, **38**, 3312 (1973).
39. A. Puech, M. Jacob, and D. Gaudy, *J. Chromatogr.*, **68**, 161 (1972).
40. E. J. Shellard and K. Sarpong, *J. Pharm. Pharmacol.*, **22**, (Suppl), 34S (1970).
41. G. Smith, *Proc. Soc. Anal. Chem.*, **8**, 66 (1971).
42. Z. P. Kostennikova and V. E. Chichiro, *Farmatsiya (Moscow)*, **18**, 39 (1969); through *Chem. Abstr.*, **71**, 128782q (1969).
43. E. J. Shellard and P. J. Houghton, *Planta Med.*, **20**, 82 (1971).
44. W. Messerschmidt, *J. Chromatogr.*, **33**, 551 (1968).
45. P. J. Wakelyn, R. D. Stipanovic, and A. A. Bell, *J. Agric. Food Chem.*, **22**, 567 (1974).
46. K. Karatodorov, *Farmatsiya (Sofia)*, **16**, 41 (1966); through *Chem. Abstr.*, **66**, 68970j (1967).
47. C.-C. Chen, S.-L. Chang, and J.-L. Tai, *Yao Hsueh Hsueh Pao*, **13**, 131 (1966); through *Chem. Abstr.*, **65**, 8674c (1966).
48. A. R. Saint-Firmin and R. R. Paris, *J. Chromatogr.*, **31**, 252 (1967).
49. M. Šaršúnová and J. Hrivnák, *Pharmazia*, **29**, 608 (1974).
50. B. Baehler, *Helv. Chim. Acta*, **45**, 309 (1962).
51. U. M. Senanayake and R. O. B. Wijesekera, *J. Chromatogr.*, **32**, 75 (1968).
52. F. Luedy-Tenger, *Pharm. Acta Helv.*, **45**, 254 (1970).
53. C. Franzke, K. S. Grunert, and H. Griehl, *Z. Lebensm.-Unters.-Forsch.*, **139**, 85 (1969).
54. K. Teichert, E. Mutschler, and H. Rochelmeyer, *Dtsch. Apoth.-Ztg.*, **100**, 283 (1960).
55. M. Săršúnova and V. Schwarz, *Pharmazie*, **18**, 207 (1963).
56. A. Fincke, *Fette, Seifen, Anstrichm.*, **65**, 647 (1963).
57. S. Ristić and A. Thomas, *Arch. Pharm.*, **295**, 524 (1962).

58. G. Ruecher, H. Kroger, M. Schikarski, and S. Quedan, *Planta Med.*, **24**, 61 (1973).
59. J. L. McLaughlin and A. G. Paul, *J. Pharm. Sci.*, **54**, 661 (1965).
60. J. S. Todd, *Lloydia,* **32,** 395 (1969).
61. J. Lundstroem and S. Agurell, *J. Chromatogr.*, **30,** 271 (1967).
62. S. Agurell, *Lloydia,* **32,** 206 (1969).
63. J. M. Neal, P. T. Sato, W. N. Howald, and J. L. McLaughlin, *Science,* **176,** 1131 (1972).
64. J. M. Neal, P. T. Sato, C. L. Johnson, and J. L. McLaughlin, *J. Pharm. Sci.,* **60,** 477 (1971).
65. D. G. Norquist and J. L. McLaughlin, *J. Pharm. Sci.,* **59,** 1840 (1970).
66. J. Cochin and J. W. Daly, *Experientia,* **18,** 294 (1962).
67. E. Tyihák, A. Gulyás, and K. Juhász, *Herba Hung.*, **5,** 225 (1966); through *Chem. Abstr.,* **68,** 4312r (1968).
68. E. Tyihák, K. Juhász, A. Gulyás, and J. Szoeke, *Abh. Dtsch. Akad. Wiss. Berlin Kl. Chem. Geol. Biol.,* **1966,** 623.
69. K. Juhász and E. Tyihák, *Acta Agron. Acad. Sci. Hung.,* **18,** 113 (1969).
70. R. Tschesche, K. Kometani, F. Kowitz, and G. Snatzke, *Chem. Ber.,* **94,** 3327 (1961).
71. A. Rother, J. M. Bobbitt, and A. E. Schwarting, *Chem. Ind. (London),* **1962,** 654.
72. M. Pailer and R. Libiseller, *Monatsh. Chem.,* **93,** 403 (1962).
73. Ibid., 511.
74. F. Moll, *Arch. Pharm.,* **296/68,** 205 (1963).
75. M. Pailer and W. G. Kump, *Arch. Pharm.,* **293/65,** 646 (1960).
76. E. Vidic, *Arch. Toxikol.,* **19,** 254 (1961).
77. J. Baeumler and S. Rippstein, *Pharm. Acta Helv.,* **36,** 382 (1961).
78. K. Teichert, E. Mutschler, and H. Rochelmeyer, *Dtsch. Apoth.-Ztg.,* **100,** **477** (1960).
79. O. Fejér-Kossey, *J. Chromatogr.,* **31,** 592 (1967).
80. M. Y. Lovkova and N. S. Minozhed, *Prikl. Biochim. Mikrobiol.,* **5,** 487 (1969); through *Chem. Abstr.,* **71,** 105239d (1969).
81. R. J. Martin, *J. Assoc. Off. Anal. Chem.,* **49,** 766 (1966).
82. A. G. González and F. Diaz Rodríguez, *An. Real Soc. Espan. Fis. Quim. (Madrid) Ser. B.,* **58,** 431 (1962); through *Chem. Abstr.,* **58,** 3686 (1963).
83. M. Ikram and M. K. Bakhsh, *Anal. Chem.,* **36,** 111 (1964).
84. N. Oswald and H. Flueck, *Pharm. Acta Helv.,* **39,** 293 (1964).
85. K. L. Khanna, A. E. Schwarting, A. Rother, and J. M. Bobbitt, *Lloydia,* **24,** 179 (1961).
86. S. Khafagy, A. M. El-Moghazy, and F. Sandberg, *Svensk Farm. Tidskr.,* **66,** 481 (1962).
87. A. Vegh, R. Budvari, G. Szasz, A. Brantner, and P. Gracza, *Acta Pharm. Hung.,* **33,** 67 (1963).
88. C. A. Teijgeler, *Pharm. Weekblad,* **97,** 507 (1962).
89. D. Neumann and H.-B. Schroeter, *J. Chromatogr.,* **16,** 414 (1964).

90. E. Steinegger and J. Gebistorf, *Pharm. Acta Helv.*, **37**, 343 (1962).
91. G. Verzár-Petri and M. Y. Haggag, *Herba Hung.*, **15**, 87 (1976).
92. A. E. DeMaggio and J. A. Lott, *J. Pharm. Sci.*, **53**, 945 (1964); through *Chem. Abstr.*, **61**, 11847e.
93. W. Messerschmidt, *Dtsch. Apoth.-Ztg.*, 109, 199 (1969); through *Anal. Abstr.*, **18**, 4295 (1970).
94. B. L. W. Chu, E. S. Mika, M. J. Solomon, and F. A. Crane, *J. Pharm. Sci.*, **58**, 1073 (1969).
95. A. Fiebig, J. Felczak, and S. Janicki, *Pharm. Pol.*, **25**, 971 (1969).
96. J. Paul and F. Conine, *Microchem. J.*, **18**, 142 (1973).
97. S. Koontz, D. Besemer, N. Mackey, and R. Phillips, *J. Chromatogr.*, **85**, 75 (1973).
98. M. L. Bastos and D. B. Hoffman, *J. Chromatogr., Sci.*, **12**, 269 (1974).
99. M. L. Bastos, D. Jukofsky, and S. J. Mulé, *J. Chromatogr.*, **89**, 335 (1974).
100. J. E. Wallace, H. E. Hamilton, H. Schwertner, D. E. King, J. L. McNay, and K. Blum, *J. Chromatogr.*, **114**, 433 (1975).
101. J. Polesuk and T. S. Ma, *Mikrochim. Acta*, **1970**, 670.
102. A. Suszko-Purzycka and W. Trzebny, *J. Chromatogr.*, **16**, 239 (1964).
103. *Ibid.*, **17**, 114 (1965).
104. K. H. Mueller and H. Honerlagen, *Arch. Pharm.*, **293/65**, 202 (1960).
105. P. Braeckman, R. van Severen, and L. De Jaeger-Van Moeske, *Pharm. Tijdschr. Belg.*, **40**, 113 (1963); through *Chem. Abstr.*, **60**, 1541 (1964).
106. M. Saršúnová and J. Hrivňák, *Pharmazie*, **29**, 608 (1974).
107. Z. Rózsa, K. Szendrei, I. Novák, E. Minker, M. Koltai, and J. Reisch, *J. Chromatogr.*, **100**, 218 (1974).
108. M. L. Borke and E. R. Kirch, *J. Am. Pharm. Assoc. Sci. Ed.*, **42**, 627 (1953).
109. A. Mariani and O. Mariani-Marelli, *Rend. 1st Super. Sanita*, **22**, 759 (1959); *Chem. Abstr.*, **54**, 11374 (1960).
110. D. Neubauer and K. Mothes, *Planta Med.*, **9**, 466 (1961).
111. I. Bayer, *J. Chromatogr.*, **16**, 237 (1964).
112. D. Heusser, *Planta Med.*, **12**, 237 (1964).
113. W. Poethke and W. Kinze, *Arch. Pharm.*, **297/69**, 593 (1964).
114. W. Poethke and W. Kinze, *Pharm. Zentralhalle*, **101**, 685 (1962).
115. *Ibid.*, **102**, 692 (1963).
116. *Ibid.*, **103**, 577 (1964).
117. R. K. Moiseev, *Aptechn. Delo*, **13**, 29 (1964); through *Chem. Abstr.*, **62**, 6341 (1965).
118. H. J. Kupferberg, A. Burkhalter, and E. L. Way, *J. Chromatogr.*, **16**, 558 (1964).
119. G. Szasz, L. Khin, and R. Budvari, *Acta Pharm. Hung.*, **33**, 245 (1963).
120. A. Penna-Herreros, *J. Chromatogr.*, **14**, 536 (1964).
121. M. Von Schantz, "Uber die Anwendung der Zirkulartechnik Beim Chromatographieren auf Kieselgel-Dünnschichten, Trennung und Reindarstellung von Morphin, Papaverin und Chinin aus Deren Gemischen," in *Thin-*

Layer Chromatography, G. B. Marini-Bettòlo, Ed., Elsevier, Amsterdam, 1964, p. 122.
122. M. Ikram, G. A. Miana, and M. Islam, *J. Chromatogr.,* **11,** 260 (1963).
123. J. Sherma, M. F. Dobbins, and J. C. Touchstone, *J. Chromatogr., Sci.,* **12,** 300 (1974).
124. C. W. Gorodetzký, *Toxicol. Appl. Pharmacol.,* **23,** 511 (1972).
125. J. A. Steele, *J. Chromatogr.,* **19,** 300 (1965).
126. J. E. Wallace, J. D. Biggs, J. H. Merritt, H. E. Hamilton, and K. Blum, *J. Chromatogr.,* **71,** 135 (1972).
127. A. Viala and M. Estadieu, *J. Chromatogr.,* **72,** 127 (1972).
128. G. N. Christopoulos and E. R. Kirch, *J. Chromatogr.,* **65,** 507 (1972).
129. E. Ragazzi, G. Veronese, and C. Giacobazzi, "Thin-Layer Chromatography of Alkaloids on Magnesia Chromatoplates," in *Thin-Layer Chromatography,* G. B. Marini-Bettòlo, Ed., Elsevier, Amsterdam, 1964, p. 149.
130. J.-T. Huang, H.-C. Hsiu, and K.-T. Wang, *J. Chromatogr.,* **29,** 391 (1967).
131. S. Pfeifer, *J. Chromatogr.,* **24,** 364 (1966).
132. *Ibid.,* **41,** 127 (1969).
133. J. Zarnack and S. Pfeifer, *Pharmazie,* **17,** 431 (1962).
134. J. M. Bobbitt, R. Ebermann, and M. Schubert, *Tetrahedron Lett.,* **1963,** 575.
135. W. Doepke, *Arch. Pharm.,* **295/67,** 605 (1962).
136. A. K. Bhatnagar and S. Bhattacharji, *Indian J. Chem.,* **2,** 43 (1965).
137. M. Klavehn and H. Rochelmeyer, *Deut. Apoth.-Ztg.,* **101,** 477 (1961).
138. M. Klavehn, H. Rochelmeyer, and J. Seyfreid, *Deut. Apoth.-Ztg.,* **101,** 75 (1961).
139. T. Hohmann and H. Rochelmeyer, *Arch. Pharm.,* **297/69,** 186 (1964).
140. J. Reichelt and S. Kudrnáč, *J. Chromatogr.,* **87,** 433 (1973).
141. D. Groeger and D. Erge, *Pharmazie,* **18,** 346 (1963).
142. J. L. McLaughlin, J. E. Goyan, and A. G. Paul, *J. Pharm. Sci.,* **53,** 306 (1964).
143. E. Eich and W. Schunack, *Planta Med.,* **27,** 58 (1975).
144. D. Groeger, V. E. Tyler, Jr., and J. E. Dusenberry, *Lloydia,* **24,** 97 (1961).
145. M. Zinser and C. Baumgaertel, *Arch. Pharm.,* **297/69,** 158 (1964).
146. S. Agurell and E. Ramstad, *Lloydia,* **25,** 67 (1962).
147. A. Hofmann, *Planta Med.,* **9,** 354 (1961).
148. A. Hofmann and H. Tscherter, *Experientia,* **16,** 414 (1960).
149. H. C. Beyerman, A. van de Linde, and G. J. Henning, *Chem. Weekblad,* **59,** 508 (1963).
150. K. Genest and C. G. Farmilo, *J. Pharm. Pharmacol.,* **16,** 250 (1964).
151. R. Fowler, P. J. Gomm, and D. A. Patterson, *J. Chromatogr.,* **72,** 351 (1972).
152. E. Link, *Pharm. Ztg. Ver. Apoth-Ztg.,* **104,** 646 (1959).
153. F. Schlemmer and E. Link, *Pharm. Ztg. Ver. Apoth-Ztg.,* **104,** 1349 (1959).
154. A. Liukkonen, *Farm. Aikakauslehti,* **71,** 329 (1962); *Chem. Abstr.,* **58,** 8850 (1963).

155. E. Ullmann and H. Kassalitzky, *Arch. Pharm.*, **295/67,** 37 (1962).
156. W. E. Court, *Can. J. Pharm. Sci.*, **1,** 76 (1966).
157. W. E. Court and M. S. Habib, *J. Chromatogr.*, **80,** 101 (1973).
158. K. Bernauer, *Helv. Chim. Acta.*, **47,** 1075 (1964).
159. D. Groeger, *Pharmazie,* **23,** 210 (1968).
160. J. Lutomski, Z. Kowalewski, K. Drost, and M. Kucharska, *Herba Pol.*, **14,** 23 (1968); through *Anal. Abstr.*, **19,** 3206 (1969).
161. W. Rusiecki and M. Henneberg, *Ann. Pharm. Fr.*, **21,** 843 (1963).
162. G. Grandolini, C. Galeffi, E. Montalvo, C. C. Casinovi, and G. B. Marini-Bettòlo, "Some Applications of Thin-Layer Chromatography for the Separation of Alkaloids," in *Thin-Layer Chromatography,* G. B. Marini-Bettòlo, Ed., Elsevier, Amsterdam, 1964, p. 155.
163. C. Weissman, H. Schmid, and P. Karrer, *Helv. Chim. Acta,* **45,** 62 (1962).
164. J. D. Phillipson and N. G. Bisset, *J. Chromatogr.*, **48,** 493 (1970).
165. I. M. Jakovljevič, L. D. Seay, and R. W. Shaffer, *J. Pharm. Sci.*, **53,** 553 (1964).
166. J. Mokrý, I. Kompiš, and G. Spiteller, *Collect. Czech. Chem. Commun.*, **32,** 2523 (1967).
167. N. J. Cone, R. Miller, and N. Neuss, *J. Pharm. Sci.*, **52,** 688 (1963).
168. N. R. Farnsworth, R. N. Blomster, D. Damratoski, W. A. Meer, and L. V. Cammarato, *Lloydia,* **27,** 302 (1964).
169. E. J. Shellard, J. D. Phillipson, and D. Gupta, *J. Chromatogr.*, **32,** 704 (1968).
170. E. J. Shellard and M. Z. Alam, *J. Chromatogr.*, **33,** 347 (1968).
171. *Ibid.*, **32,** 472 (1968).
172. *Ibid.*, p. 489.
173. E. J. Shellard and P. J. Houghton, *Planta Med.*, **24,** 341 (1973).
174. W. G. Kump, M. B. Patel, J. M. Rowson, and H. Schmid, *Helv. Chim. Acta,* **47,** 1497 (1964).
175. C. Kump and H. Schmid, *Helv. Chim. Acta,* **45,** 1090 (1962).
176. W. G. Kump and H. Schmid, *Helv. Chim. Acta,* **44,** 1503 (1961).
177. A. Walser and C. Djerassi, *Helv. Chim. Acta,* **48,** 391 (1965).
178. A. Walser and C. Djerassi, *J. Chromatogr.*, **21,** D5 (1966).
179. S. Agurell, B. Holmstedt, J. Lindgren, and R. E. Schultes, *Acta Chem. Scand.*, **23,** 903 (1969).
180. V. Benesova, Z. Šamek, V. Herout, and F. Šorm, *Collect. Czech. Chem. Commun.*, **34,** 1807 (1969).
181. H. I. Dhahir, N. C. Jain, and J. I. Thornton, *Forensic Sci. Soc. J.*, **12,** 309 (1972).
182. J. D. Phillipson and E. J. Shellard, *J. Chromatogr.*, **24,** 84 (1966).
183. A. H. Chalmers, C. C. J. Culvenor, and L. W. Smith, *J. Chromatogr.*, **20,** 270 (1965).
184. R. K. Sharma, G. S. Khajuria, and C. K. Atal, *J. Chromatogr.*, **19,** 433 (1965).
185. A. Klasek, T. Reichstein, and F. Šantavý, *Helv. Chim. Acta,* **51,** 1088 (1968).

186. A. Klasek, P. Sedmera, and F. Šantavý, *Collect. Czech. Chem. Commun.*, **36**, 2205 (1971).
187. A. R. Mattocks, *J. Chromatogr.*, **27**, 505 (1967).
188. W. Scharlemann and *Z. Naturforsch.*, **27b**, 806 (1972).
189. S. Hěrmánek, V. Schwarz, and Z. Čekan, *Pharmazie*, **16**, 566 (1961).
190. R. R. Paris and M. Paris, *Bull. Soc. Chim. Fr.*, **1963**, 1597.
191. I. Máthé and E. Tyihák, *Herba Hungarica*, **2**, 35 (1963).
192. H. Potešilova, J. Wiedermannová, and F. Šantavý, *Collect. Czech. Chem. Commun.*, **34**, 3642 (1969).
193. M. Haag-Berrurier and M. C. Mathis, *Ann. Pharm. Fr.*, **31**, 457 (1973); through *CAMAG Bibliogr., Serv.*, **33**, 33290 (1974).
194. A. A. M. Habib, *Planta Med.*, **26**, 279 (1974); through *CAMAG Bibliogr., Serv.*, **35**, 35154.
195. A. Kerry, *Acta Agron. Acad. Sci. Hung.*, **24**, 3 (1975); through *CAMAG Bibliogr., Serv.*, **36**, 36148.
196. F. Fish and P. G. Waterman, *J. Pharm. Pharmacol.*, **23**, 67 (1971).
197. *Ibid.*, **23**, *Suppl.*, 132S (1971).
198. M. S. Habib and K. J. Harkiss, *Planta Med.*, **18**, 270 (1970).
199. V. Massa, F. Gal, P. Susplugas, and G. Maestre, *Trav. Soc. Pharm. Montp.* **30**, 301 (1970); through *Chem. Abstr.*, **75**, 25448p (1971).
200. Y. D. Cho and R. O. Martin, *Anal. Biochem.*, **44**, 49 (1971).
201. L. Lepri, P. G. Desideri, and M. Lepori, *J. Chromatogr.*, **116**, 131 (1976).

Chapter **XVI**

AMINES

1 ALIPHATIC AMINES

Teichert et al. [1] ran a series of alkylamines including several aromatic amines on several adsorbents using various solvents (Tables 16.1 and 16.2). These alkylamines may be chromatographed on silica gel, buffered silica gel, or cellulose layers. Solvents for their separation are given in the tables and it should be noted that the more volatile amines cannot be chromatographed with the ammoniacal solvent. These compounds may be detected with ninhydrin and with iodine.

Primary, secondary, and tertiary amines have been separated on silica gel with chloroform–ammonia (39:1) and determined titrimetrically after locating with iodine vapor [2]. Chloroform–methanol–17% ammonia (2:2:1) may also be used for separation of their hydrochlorides [3]. Lauckner et al. [4] separated C_8 to C_{16} amines on silica or silica impregnated with sodium acetate using cyclohexane saturated with formic acid. Aures et al. [5] used a two-dimensional method on cellulose to separate many amines occuring in biological systems. Solvents were butanol–0.1 *N* hydrochloric acid in the first direction followed by isopropanol–5 *N* ammonia–water (8:1:1). Lane [6] used aluminum oxide for differentiating primary, secondary, and tertiary long-chain amines.

Feltkamp and Koch [7] chromatographed a large number of stereoisomers including 24 cyclohexylamines, eight amino decalins, *cis* and *trans*-2-isopropylcyclopentylamines, bornylamine, and isobornylamine. The separations were accomplished on silica gel G layers using a solvent composed of 8 parts of petroleum ether and 17 parts of a mixture of 2 parts of concentrated ammonia and 98 parts of acetone. The bases were

Table 16.1 $R_f \times 100$ Values of Amines on Neutral and Buffered Silica Gel Plates in Various Solvents[a]

(*a*) Neutral plates

Amine	*95% Ethanol–25% Ammonia* (4:1)	*Phenol–Water* (8:3)	*Butanol–Acetic Acid–Water* (4:1:5)
Methylamine	—	11	10
Ethylamine	—	15	13
n-Propylamine	—	30	19
Isoamylamine	—	49	39
Cadaverine	3	6	2
Putrescine	3	4	2
Ethanolamine	30	18	11
Histamine	41	26	2
Tyramine	56	44	38
Phenylethylamine	66	56	37
Benzylamine	70	49	36
Tryptamine	60	54	43

(*b*) Buffered plates

Amine	*Phosphate-Buffered (pH* 6.8) *Silica Gel with 70% Alcohol*	*Sodium Acetate-Buffered Silica Gel with BuOH–HOAc–H_2O* (4:1:5)
Methylamine	10	7
Ethylamine	20	14
n-Propylamine	30	23
Isoamylamine	45	44
Cadaverine	1	1
Putrescine	1	1
Ethanolamine	10	10
Histamine	3	3
Tyramine	55	40
Phenylethylamine	55	42
Benzylamine	50	42
Tryptamine	90	45

[a] From K. Teichert, E. Mutschler, and H. Rochelmeyer [1]; reproduced with permission of the authors and Deutsche Apotheker-Verlag.

Table 16.2 $R_f \times 100$ Values of Amines on Cellulose Layers with Amyl Alcohol–Acetic Acid–Water (4:1:5)[a]

Amine	$R_f \times 100$
Methylamine	12
Ethylamine	19
n-Propylamine	31
n-Butylamine	43
Isoamylamine	58
Ethanolamine	7
Histamine	2
Tyramine	28

[a] From K. Teichert, E. Mutschler, and H. Rochelmeyer [1]; reproduced with permission of the authors and Deutsche Apotheker-Verlag.

located with a mixture of equal parts of 0.1 *N* iodine solution and 10% sulfuric acid or by means of the Munier modification of the Dragendorff reagent.

Polyamines have been separated on silica gel and on cellulose with diethylene glycol monoethyl ether (Carbitol)–propanol–water (14:3:3) saturated with sodium chloride or ethylene glycol monoethyl ether (methyl Cellosolve)–propanol–water (14:3:3) saturated with sodium chloride. For quantitative work, cellulose layers were used and after reaction with ninhydrin the spots were eluted and measured at 575 nm [8]. Direct densitometry has also been used for these compounds [9]. In this case they were separated on silica gel with *tert*-butanol–pyridine–26% ammonia (1:1:1) saturated with ammonia gas and made visible with 0.08 to 1% phenol red in 60% ethanol. Abe and Samejima [10] preferred to use sintered glass silica gel plates with *n*-butanol–acetic acid–pyridine–water (12:3:6:4). Spearmine and spermadine were determined by direct fluorescence after developing (rather than spraying) with fluorescamine in acetone (20 mg/100 ml).

Beckett and Choulis [11, 12] examined the phenomenon of multiple spot formation in paper and cellulose thin-layer chromatography of amines when using neutral or weakly acidic solvent systems. They concluded that this effect was caused by the presence of carboxy groups in the cellulose. The multiple spot formation could be eliminated by pretreatment of the cellulose with diazomethane. Wesley-Hadzija [13] has examined the production of double-spot formation in the case of fenfluramine hydrochloride and other amines on silica gel and alumina. Double spots were not formed with the sulfate. After careful investigation it was concluded that double spots were the result of "(*a*) partial hydrolysis of the amine salts, depending on the ionic strength of the acid; (*b*) volatilization of the liberated hydrochloric acid; and (*c*) the presence of charged groups in silica gel and alumina thin layers."

Udenfriend et al. [14] proposed the use of fluorescamine (4-phenylspiro-[furan-2(3*H*),1′-phthalan]-3,3′dione) as a reagent (T-128) for the determination in the picomole range of compounds containing a primary amine group.

Lepri et al. [15] examined the chromatographic behavior of 30 aliphatic amines on layers of Dowex 50-X4 (Na^+ and H^+), sodium carboxymethyl cellulose, and Rexyn 102 (Na^+) ion exchangers using hydrochloric acid and various buffer and salt solutions as developing agents.

A group of amines have been separated by thin-layer electrophoresis [16] on layers of silica gel G prepared with 3% boric acid solution. As electrolyte, a mixture of 80 ml of ethanol, 30 ml of distilled water, and 2 g of crystalline sodium acetate was used. The pH value was adjusted to 12 with 40% sodium hydroxide. A field strength of 10 V/cm was used for the separation. The tertiary amine, triethanolamine, was detected with alkaline permanganate solution and the remaining amines with ninhydrin. Honegger [17] has used a two-dimensional method combining electrophoresis in one direction and chromatography in the second direction to separate a group of amines.

A large number of derivatives have been proposed for assisting in separating and identifying primary and secondary amines on thin layers. Among these is the derivative formed by reaction with 5-(dimethylamino)naphthalene-1-sulfonyl chloride (dansyl or DANS-Cl). This derivative permits the detection and determination of very small quantities in the range of a fraction of a nanomole; however, there are a number of disadvantages to its use: (*a*) it is not a specific reagent, for it reacts with hydroxyl groups of phenolic compounds, with certain amino acids, and with some alcohols; (*b*) there are fluorescent side products of the dansylation procedure; and (*c*) if basic amino acids are present, as for example in natural products, dansylated fragments are produced. In spite of its faults it is a valuable reagent for the determination of primary and secondary amines, and Seiler [18, 19] gives a good account of its advantages, disadvantages, and techniques for its use. Derivatives may be prepared by reacting the amine in 1 vol of water with 3 vol of an acetone–water (3:1) solution of the reagent. The reaction mixture is saturated with sodium bicarbonate. The reaction is essentially complete after a few minutes, but Seiler [19] recommends storing the reaction mixture at room temperature in the dark 3 to 4 hr or overnight to ensure completion for quantitative work. If very small amounts of amines are present with large amounts of polyamines the dansylated products may be run through a small column of Sephadex LH 20 with chloroform in order to separate the DANS-amines from the DANS-polyamines [18] before thin-layer chromatography. DANS-amines have been separated on silica gel with chloroform–triethylamine (5:1) [20], and chloroform–ethyl acetate (3:1) or cyclohexane–ethyl acetate (7:3) [21]. Quantitation was accomplished by in situ fluorimetry directly of the DANS derivative or after spraying the DANS derivative spot with 20% 2,2′,2″-nitrioltriethanol in isopropanol [21]. Two-dimensional separation is also very useful in separating these derivatives using chloroform–methanol (3:2) followed by chloroform–triethylamine (5:1) [18], ethyl acetate–cyclohexane (3:4) with benzene–triethylamine (8:1) in the second direction [22], or ethyl acetate–cyclohexane (3:2) followed by benzene–methanol–cyclohexane (17:1:2) [23]. In the latter case the DANS derivatives of ammonia and diamine were eluted and rechromatographed two-dimensionally in chloroform–methanol (90:6) and chloroform–triethylamine (5:1).

Another reagent (7-chloro-4-nitrobenzoxa-1,3-diazole), which itself is not fluorescent, has been proposed as being better than the dansyl reagent [24, 25]. It does not produce fluorescent hydrolysis products, and it is more specific than DANS-Cl for primary and secondary amines because it does not produce fluorescent products with phenols, thiols, and anilines [26]. Amine derivatives can be prepared by heating 1 to 20 μg of amine in 25 to 550 μl of methanol with four to five equivalents of a 0.05% methanol solution of the reagent and 50 to 100 μl of 0.1 *M* aqueous sodium bicarbonate solution at 55°C for 1.5 hr [25]. The reaction mixture may be chromatographed on silica gel by developing twice with cyclohexane–ethyl acetate (1:1) and then at right angles twice with heptane–ethyl acetate–*n*-butanol (8:1:1) or on polyamide-11 foils (Macherey, Nagel) three times with the last-named solvent. Quantitative analysis was accomplished by in situ fluorimetry on the polyamide layers.

Among the other derivatives that have been used in TLC work have been the 3,5-dinitrobenzamides [1]. These may be prepared in the normal manner by reacting with 3,5-dinitrobenzoyl chloride. They have been separated on silica gel with chloroform–96% ethanol (99:1) (Table 16.3) and visualized under ultraviolet light or by spraying with either a chloroform solution of iodine or with an α-naphthylamine solution.

The 2.4-dinitrophenyl derivatives are prepared by reacting 5 mmole of amine hydrochloride and 5 mmole 2,4-dinitrofluorobenzene in 50 ml of

Table 16.3 $R_f \times 100$ Values of Some Amine Derivatives on Silica Gel[a]

3,5-Dinitrobenzamide of	*$CHCL_3$–96% EtOH* (99:1)
Methylamine	14
Dimethylamine	47
Ethylamine	22
Diethylamine	68
n-Propylamine	35
Isobutylamine	42
Isoamylamine	50

[a] From K. Teichert, E. Mutschler, and H. Rochelmeyer [1]; reproduced with permission of the authors and Deutsche Apotheker-Verlag.

acetone with 10 mmole of sodium bicarbonate in 50 ml of water at 60°C for 40 min. Separations can be achieved on silica gel using pentane–benzene–triethylamine (45:45:10) [27]. Michalec [28] has used a two-dimensional method in which the first development was with chloroform–methanol (95:5) along a silica gel layer only part of which was impregnated with silver nitrate, followed by development with the upper phase of methanol–tetralin–water (9:1:1) in the second dimension across the balance of the layer which had been impregnated with a 5% solution of tetralin in ether. This technique separated saturated, monoenoic, and dienoic long-chain bases and their degradation products.

Other derivatives that have been used include the 4-(dimethylamino)-3,5-dinitrobenzamides [29, 30] separated on kieselguhr–Carbowax (4:1) with hexane–ethyl acetate (7:3); the 4-(4-nitrophenylazo)benzamides [31, 32] separated on silica gel with chloroform–acetone (20:1); the 9-isothiocyanatoacridine derivatives [33] chromatographed on cellulose with dimethylformamide–chloroform–28% ammonia (5:13:3); the 4-(phenylazo)benzenesulfonamides [34] on alumina with ethyl acetate–petroleum ether (62–82°C) (1:4); the *p*-(*N*,*N*-dimethylamino)benzene-*p*′-azobenzamides [35] on silica gel in a number of solvents including cyclohexane–methyl ethyl ketone (7:3); the enamine derivatives (β-aminovinyl *o*-hydroxyphenyl ketones) [36] on silica gel with ethyl acetate–benzene (1:5), chloroform–xylene (4:1), or xylene–acetone (9:1); and the *p*-toluenesulfonates [37] on silica gel and on silica gel buffered with 0.5 *N* oxalic acid using (*a*) chloroform, (*b*) chloroform–xylene (4:1), (*c*) chloroform–xylene (19:1), on neutral alumina with (*a*), (*b*), or (*d*) petroleum ether–ether (1:1), on basic or acidic alumina with (*b*), (*c*), or (*d*).

2 AROMATIC AMINES

Yamamoto [38] separated *o*- and *p*-phenylenediamines by radial chromatography on Japanese acid clay.

A silica gel precipitated in the presence of methylphenylamine [39] showed a preferential adsorption for this compound and for methyl orange over ethylphenylamine and ethyl orange. The reverse effect could be obtained by preparing the gel in the presence of ethylphenylamine. The separation of a group of diphenylamines [40, 41] is described under explosives.

Ortho, meta-, and *para*-nitroanilines have been separated on loose layers of aluminum oxide [42] and also on silica gel layers [43]. Schwartz et al. [44] have separated an homologous series of dinitrophenylamines by thin-layer partition chromatography and Keswani and Weber [45] separated nitroanilines and related compounds on silica gel G with hexane–acetone (3:1). Visualization reagents included (*a*) 5% methanolic 1-naphthol, (*b*) 0.1 *M* ferric chloride + 0.1 *M* potassium ferricyanide, and (*c*) successive applications of 5% sodium nitrite in 1 *N* hydrochloric acid, 5% phenol, and 7% sodium carbonate. Gillio-Tos et al. [46] undertook a survey of isomeric ring-substituted anilines on silica gel G in five different solvent systems. Table 16.4 gives the R_f values for these compounds. Sherma and Marzoni [47] separated a group of eight anilines and three aminophenols on commercially prepared adsorbent sheets of two silica gels, one aluminum oxide, and two polyamides. Solvents for the silica gel systems were benzene–ethanol (95:5), benzene–methanol (4:2), and benzene–acetone (7:2); for the alumina benzene–methanol (85:15); and for the polyamide benzene–methanol (9:1). Detection was accomplished with fluorescamine in acetone (0.1 mg/ml) and a follow up spray of 10% triethylamine in methylene chloride was used to stabilize the fluorescence for quantitative determination by direct fluorimetry on the silica gel layers. Sensitivity ranged from 4 to 80 ng. Rinde and Troll [48], in working out a colorimetric assay for aromatic amines, used a 1 mg/ml solution of fluorescamine in glacial acetic acid, because this solution reacts only with aromatic amines and forms stable yellow products. The reagent solution is stable at room temperature for weeks. Separations were accomplished on silica gel with chloroform–acetic acid–methanol (18:1:1). Tabak et al. [49] chromatographed 10 substituted anilines on silica gel and silica gel impregnated with silver oxide as a complexing agent using petroleum ether–ethyl acetate (1:1) and petroleum ether–ethyl acetate–benzene (6:3:1). Yasuda has chromatographed aromatic amines on silica gel layers impregnated with manganese salts [50] and cadmium acetate [51] as well as halogenated anilines on zinc salt-impregnated layers [52].

Table 16.4 $R_f \times 100$ Values of Some Aromatic Amines on Silica Gel[a]

Amine	Solvent[b]				
	A	*B*	*C*	*D*	*E*
o-Toluidine	62	42	64	17	84
m-Toluidine	54	29	63	10	83
p-Toluidine	40	20	59	5	80
o-Aminophenol	34	24	58	0	80
m-Aminophenol	29	13	53	0	75
p-Aminophenol	6	1	12	0	62
o-Aminobenzoic acid	62	47	74	44	98
m-Aminobenzoic acid	50	28	61	12	95
p-Aminobenzoic acid	59	37	68	29	97
o-Anisidine	60	42	70	15	81
m-Anisidine	51	30	62	9	80
p-Anisidine	11	2	17	2	58
o-Nitroaniline	69	55	77	52	93
m-Nitroaniline	64	44	71	36	92
p-Nitroaniline	58	37	67	29	91
o-Phenylenediamine	0	0	0	0	63
m-Phenylenediamine	0	0	0	0	53
p-Phenylenediamine	0	0	0	0	40
o-Bromoaniline	81	78	85	69	95
m-Bromoaniline	70	58	75	44	93
p-Bromoaniline	61	47	67	27	89
o-Chloroaniline	78	75	82	66	96
m-Chloroaniline	68	51	75	40	94
p-Chloroaniline	60	41	64	22	89

[a] From M. Gillio-Tos, S. A. Previtera, and A. Vimercati [46]; reproduced with permission of the authors and the Elsevier Publishing Co.

[b] A = dibutyl ether–ethyl acetate–acetic acid (10:10:1); *B* = dibutyl ether–ethyl acetate–acetic acid (15:5:1); *C* = dibutyl ether–ethyl acetate–acetic acid (5:15:1); *D* = dibutyl ether–acetic acid–*n*-hexane (20:1:4); *E* = acetic acid–*n*-butanol–water (1:4:5).

Pastuska and Petrowitz [53] chromatographed nitrogen derivatives of benzene, phenol, toluene, chlorobenzene, azobenzene, benzotriazole, phenazine, and benzidine. R_f values are given in benzene, benzene-methanol (4:1), and benzene–dioxane–acetic acid (90:25:4). Lepri et al. [54] chromatographed 45 primary aromatic amines on a number of anion-exchange thin layers; AG 1-X4 (CH_3COO^-), AG 1-X4 (ClO^-), Dowex 50-X4 (H^+), Cellex D(CH_3COO^-), sodium carboxymethyl cellulose, and on microcrystalline cellulose. Twenty of the amines were examined by electrophoresis on AG 1-X4 (CH_3 COO^-) thin layers. Duncan et al. [55] chromatographed a group of nine aromatic amines on cellulose layers impregnated with a liquid ion exchanger, bis(2-ethylhexyl) phosphate.

Derivatives of aromatic amines have also been used for separating and identifying these compounds. For example Parihar et al. [56] prepared the 2,4-dinitrophenyl derivatives for chromatographing a group of 11 aromatic amines. The R_f values were determined on silica gel; buffered silica; neutral, basic, and acidic alumina; and on kieselgel G with nine solvent systems, although not every adsorbent was used with each solvent. The dansyl derivatives may be used and Chapman et al. [57] chromatographed 128 phenylalkyl derivatives on silica gel with ethyl acetate–cyclohexane (9:12), carbon tetrachloride–triethylamine (5:1), and triethylamine–*n*-butyl acetate (1:5). The mass spectra of these derivatives were also determined. Coupling with 1- and 2-naphthols or other compounds after diazotization of the amine has also been used to prepare colored derivatives for chromatographing [58, 59].

There are a number of detecting reagents that can be used when the amines are chromatographed directly. Barney et al. [60] examined a group of reagents and found 2,3-dichloro-1,4-naphthoquinone to be the most generally useful reagent although it reacts with other classes of compounds. They found *p*-(dimethylamino)benzaldehyde would detect most aromatic amines and a few aliphatic amines with the best sensitivity of any of the reagents they tested. Diazotization and coupling can be carried out directly on the layer (test T-78) [61], and this reaction has been compared with 12 other reagents. Ratney [62] exposed the plates to nitrogen dioxide to diazotize the amines and then sprayed with a solution of 0.1 M β-naphthol and 0.1 M triethylamine in benzene. Sensitivity ranged from 0.1 to 0.3 μg although some required larger amounts. Stewart and Sternson [63] used 9-chloroacridine as a sensitive test for the detection and differentiation of arylamines and arylhydroxylamines (test T-62).

3 NITROGEN HETEROCYCLICS

Mistryukov [64] has separated 34 primary and secondary amines, mainly decahydroquinoline and perhydropyrindine derivatives, in six different solvent systems on loose layers of aluminum oxide with a Brockmann activity of III [65]. In order to get the more tightly adsorbed compounds to move it was necessary to use solvents containing ammo-

Table 16.5 $R_f \times 100$ Values of Some Quinoline and Other Nitrogen Heterocyclics in Various Systems

	Silica Gel G, Development Distance 10 cm				*Loose-Layer Silica Gel PHH (Spolana N.E.)*	
Nitrogen Heterocycle	*Chloroform* [68]	*Benzene* [67, 69]	*Benzene–Methanol* (95:5) [67, 69]	*Methanol* [67, 69]	*Chloroform* [70]	*Benzene* [70]
Quinoline	11	9	30			2
Isoquinoline	12	7	22			3
2-Methylquinoline	16	4	27	89	40[a]	4
4-Methylquinoline	13	2	22	82	40[a]	4
6-Methylquinoline	18	5	26	86		
7-Methylquinoline	16					3
8-Methylquinoline	31	17	53	94	38[a]	3
1-Methylisoquinoline					29[a]	2
3-Methylisoquinoline	11					2
2,4-Dimethylquinoline	8					
2,6-Dimethylquinoline	11					4
2,8-Dimethylquinoline	47					
Acridine	19					0
3,4-Benzoquinoline	19					
5,6-Benzoquinoline	13					
7,8-Benzoquinoline	47					1
3,4-Benzacridine	75					
Indole					63	65
2-Methylindole					60	65
3-Methylindole					70	65
5-Methylindole					65	65
7-Methylindole					65	64
Carbazole					75	74
2-Methylcarbazole					70	77

[a] Tailing from the origin.

nia. Petrowitz [66–68] has separated some quinoline derivatives as well as various other compounds found in tars [69] on silica gel G and Janák et al. [70, 71] have used a more polar silica gel for quinoline and other nitrogen heterocyclics. Chloroform, benzene, and other more polar solvents can be used for the separations (Table 16.5). For the location of the compounds, Dragendorff's reagent [68] or tetracyanoethylene [70, 72] may be used.

Bender et al. [73, 74] and Sawicki et al. [75] have applied thin-layer chromatography and spectrophotofluorometry to the characterization of carbazoles and polynuclear carbazoles obtained from air polluted by coal tar pitch fumes. Table 16.6 gives the R_f values of these compounds on aluminum oxide using pentane–chloroform [3:2] and ammonium hydroxide as solvents. The pentane–chloroform solvent is especially useful in separating the carbazoles as a group from the polynuclear aromatic hydrocarbons, which travel with the solvent front, and the phenolic compounds which stay near the origin. The ammonium hydroxide as a solvent is useful for separating the 4*H*-benzo(*def*)carbazole type of compound from other carbazoles. In addition, 25% aqueous *N,N*-dimethylformamide can be used with cellulose plates to separate carbazole with an R_f value of 0.4 and 4*H*-benzo(*def*)carbazole with an R_f value of 0.24 from the remaining benzocarbazoles which have R_f values of less than 0.14. The compounds may be detected by their fluorescence in ultraviolet light; however, the sensitivity can be increased in most cases by applying a 29% solution of tetraethylammonium hydroxide in methanol, which is

Table 16.6 Approximate $R_f \times 100$ Values for Some Polynuclear Carbazoles [74]

	Aluminum Oxide G			
Compound	*Pentane–Chloroform* (3:2)	*Reagent-Grade Ammonium Hydroxide*	*Fluorescent Color*[a]	*Sensitivity*[a] (μg)
Carbazole	50	<2	Blue	0.02
11*H*-Benzo(*a*)carbazole	46	<2	Blue	0.02
5*H*-Benzo(*b*)carbazole	39	<2	Yellow	0.006
7*H*-Benzo(*c*)carbazole	37	<2	Blue	0.2
4*H*-Benzo(*def*)carbazole	49	60	Blue	0.01
7*H*-Dibenzo(*c,g*)carbazole	30	<2	Blue	0.025

[a] With 29% tetramethylammonium hydroxide in methanol.

Table 16.7 Some Solvent–Adsorbent Combinations for Various Types of Nitrogen Heterocyclics

Compound Type	*Adsorbent*	*Solvents*	*Ref.*
Pyrroles	Silica	C_6H_{14}–$CHCl_3$ (9:1) 3×; ether–2% HAc in hexane (1:1)	79, 80
Pyrrole acids	Silica	$CHCL_3$–96% HAc (1:1), Bz–MeOH–HAc (45:8:4)	81
Pyridines	Silica	Bz–MeOH (25:1); EtAc–MeOH–HAc (15:4:1); ether–Me_2 formamide (99:1)	82, 83, 84[a]
Pyridines	Silica–Ag_2O	Me_2CO–C_6H_6 (2:3), MeCOEt–*i*-PrOH (4:1), $CHCl_3$–MeOH (3:2)	84a[a]
Piperidines	Alumina	$CHCl_3$ satd. with NH_4OH	85
Quinolines	Silica	EtAc–*i*-PrOH–NH_4OH (9:6:4); Bz–EtAc (1:1)	86,[a] 87
Pyrazoles	Silica	EtAc (satd. with H_2O); $CHCl_3$–CH_3COCH_3 (7:3), methyl ethyl ketone	88, 89
Pyrazolones	Silica	Cyclohexane–$CHCl_3$–EtOH (4:10:1), cyclohexane–$CH_3COC_2H_5$ (4:5)	90
Pyrazolidines	Alumina	Benzene, chloroform	91
Piperazines	Silica	EtAc–MeOH (4:1), $CHCl_3$–MeOH–HAc (14:2:1)	92, 93[a]
Diaminopyrimidines	Silica	$CHCl_3$–MeOH (3:1)	94
Imidazoles	Silica	$CHCl_3$–CH_3COCH_3–HAc (34:4:3); EtAc satd. with NH_4OH	95, 96[a]
Imidazolines	Silica	Bz–CH_3COCH_3–25% NH_4OH (4:17:1)	97
Carbazoles	Silica	Bz; EtAc–MeOH–HCOOH–pyridine (75:7.5:7.5:10); EtOH	98, 99,[a] 100
Carbzaoles	Alumina	Pet. ether (40–60°C)–$CHCl_3$ (10:1), pet. ether–HAc (10:1)	101
Azines	Alumina	Bz–$CHCl_3$ (1:1)	102
Triazoles	Alumina	Hexane–Bz (1:1), 1% adipate in xylene–HCOOH (49:1)	103[a]
Indoles	Silica	BuOH–HAc–H_2O (2:1:1); ether–HAc (100:1); C_6H_5OH–H_2O (4:1)	104, 105,[a] 106[a]
Indoles	Silica	*i*-PrOH–25% NH_4OH–H_2O (20:1:2); BuOH–EtOH–cyclohexyl NH_2 (76:3:6)	107,[a] 108
Indoles	Silica	CH_3COCH_3–$CHCL_3$–HAc–H_2O (8:8:4:1); MeCOEt–MeOH–conc. NH_4OH (40:10:1)	105,[a] 106[a]
Indoles	Cellulose	BuOH–HAc–H_2O (12:3:5); Bz–dioxane–H_2O (1:1:1), Bz–pyridine–H_2O (1:1:1)	109, 110[a]
Indoles	Polyamide	$CHCl_3$–EtAc–HAc (7:2:1); $CHCL_3$–cyclohexane–BuOH–HAc–H_2O (1:1:1:1:0.2)	111,[a] 112
Indole esters	Silica	Octanol–pet. ether (110–115°C) (1:5)	113

[a] Additional solvent combinations are listed in the reference.

applied as a very small microdrop. Sawicki et al. [76] have also separated aza heterocyclic compounds from polynuclear aromatic hydrocarbons on cellulose layers with formic acid–water (1:1). In this system the hydrocarbons remain at the origin and the heterocyclic compounds have R_f values of 0.6 to 0.9. Other adsorbent–solvent combinations may also be used for the same purpose. Twenty-two aza heterocyclics were separated from one another using the following three systems: (1) cellulose with dimethylformamide–water (35:15), (2) cellulose with acetic acid–water (3:7), and (3) aluminum oxide with hexane–diethyl ether (19:1). Trifluoroacetic acid was used as a detecting agent. The same group of workers [77, 78] have carried out more extensive investigations on these compounds and others found in various combustion products.

There are numerous heterocyclic compounds and many combinations of solvents and adsorbents that have been used in the separation of these by thin-layer chromatography. Some of these are listed in Table 16.7.

In working with indole compounds, Sági [114] pointed out that if silica gel plates containing the samples are dried for more than 2 hr there is a considerable loss for example of indoleacetic acid. Where long delays cannot be avoided he recommends using cellulose layers which do not show this effect. Two-dimensional thin-layer chromatography is also useful in differentiating the indoles. Pairs of solvents used on silica gel include *(a)* isopropanol–25% ammonia–water (20:1:2), *(b)* *n*-butanol–acetic acid–water (15:3:5) [107]; *(a)* methyl acetate–isopropanol–25% ammonia (9:7:4), *(b)* chloroform–methanol–96% acetic acid (9:4:1) [115]; and *(a)* benzene–dioxane (13:7), *(b)* diisopropyl ether–*N,N*-dimethylformamide (4:1) [116]. The indole acids may be methylated with diazomethane before chromatographing [113, 116]. On cellulose, propanol–water–ammonia (75:25:2) and propanol–water–acetic acid (75:25:2) may be used for two-dimensional work [117]. Johri [117a] has used thin-layer electrophoresis for the separation of some indole derivatives.

With this group of reactive compounds there are a number of reagents that can be used for detection. Pyrroles may be visualized with Ehrlich's reagent (T-90). Pyridines have been detected with the iodide–azide test (T-147) and with Dragendorff's reagent. A group of pyridines and their R_f values are given in Table 16.8. Pyrazoles have been visualized with sodium nitroprusside (T-237) [88]; pyrazolones with 1% mercuric nitrate, iodine–potassium iodide, or 10% ferrocyanide–12.5% hydrochloric acid (1:1) [90]; and imidazoles with iodoplatinate reagent. Indoles can be detected with Ehrlich's (T-90), Van Urk's (T-89), or Prochazka's (T-132) reagents. Table 16.9 gives some R_f values for a number of indoles along with the colors obtained with Van Urk's reagent.

Table 16.8 $R_f \times 100$ Values of Pyridine Compounds on Silica Gel G in Several Solvents[a]

Compound	*Solvent*[b]		
	Chloroform	*Ethyl Acetate*	*Acetone*
Pyridine	4	29	54
2-Methylpyridine	3	30	54
3-Methylpyridine	6	35	55
4-Methylpyridine	4	27	48
2,4-Dimethylpyridine	4	28	49
2,6-Dimethylpyridine	6	36	59
2,4,6-Trimethylpyridine	2	26	51
2-Ethylpyridine	3	42	62
2-*n*-Propylpyridine	12	47	64
2-Hydroxypyridine	0	6	20
3-Hydroxypyridine	0	23	53
4-Hydroxypyridine	0	0	2
2-Aminopyridine	>0	27	50
3-Aminopyridine	>0	18	45
4-Aminopyridine	>0	5	14
Pyridine-2-carbinol	>0	18	45
Pyridine-3-carbinol	>0	13	39
Pyridine-4-carbinol	>0	4	39
Pyridine-2-aldehyde	11	51	67
Pyridine-3-aldehyde	4	33	58
Pyridine-4-aldehyde	5	36	56
Pyridine-2-carboxylic acid	0	2	4
Pyridine-3-carboxylic acid	0	6	6
Pyridine-4-carboxylic acid	0	5	5
Pyridine-2,6-dicarboxylic acid	4	3	5
2-Acetylpyridine	20	57	69
2-Benzoylpyridine	9	62	71
2-Fluoropyridine	24	62	69
2-Chloropyridine	28	61	70
2-Bromopyridine	26	63	72
3-Chloropyridine	16	56	65
3-Bromopyridine	13	57	67
3-Iodopyridine	13	58	70

[a] From H.-J Petrowitz, G. Pastuska, and S. Wagner [68]; reproduced with permission of the authors and Alfred Huethig Verlag.
[b] Development distance 10 cm in all solvents.

Table 16.9 $R_f \times 100$ Values of Some Indole Derivatives on Silica Gel G in Various Solvents[a]

Indole Derivative	*Solvent*[b]							*Color with van Urk's Reagent* [119]
	A	*B*	*C*	*D*	*E*	*F*	*G*	
Indole	84	73						Dark red to violet
Skatole	87	78						Blue
3-Hydroxymethylindole	84	45						Pink
Indole-3-aldehyde	81	20	29			68		Yellow[c]
Indole-3-acetaldehyde	86	46						Reddish brown
Indole-3-carboxylic acid			4	10	57	17	34	Red
Indole-3-acetic acid	31	28	0	1	29	6	11	Blue-violet
β-Indole-(3)-propionic acid	38	34	1	3	37	10	19	Blue
γ-Indole-(3)-butyric acid	40	38	2	7	44	14	27	Blue
β-Indole-(3)-acrylic acid	33	29						Pink
Indole-(3)-acetonitrile	85	46	49	70	75	73	78	Gray
Indole-(3)-acetamide			16	29	59	54	57	Blue-violet
5-Hydroxyindole-(3)-acetic acid	19	4						Blue to violet
Ethyl indole-(3)-acetate			61	77	75	80	80	Violet
Indole-(3)-ethanol			30	55	70	67	70	Blue-gray (yellow rim)
Tryptamine	77	0						Blue-green
Serotonin	65	0						Gray
Gramine	77	0						Yellow to beige
DL-Tryptophan	23	0						Blue-green
DL-5-Methyltryptophan	28	0						Blue
DL-5-Hydroxytryptophan	14	0						Blue-gray
Isatin	75	27						Stays orange

[a] Layer thickness 0.5 μg, solvent travel 10 cm, saturated chamber.

[b] *A* = Methyl acetate–isopropanol–25% ammonium hydroxide (9:7:4) [118]; *B* = chloroform–96% acetate acid (95:5) [118]; *C* = chloroform–carbon tetrachloride–methanol (5:4:1) [119]; *D* = chloroform–96% ethanol (9:1) [119]; *E* = ethyl acetate–isopropanol–water (65:24:11) [119]; *F* = chloroform–carbon tetrachloride–methanol (2:1:1) [119]; *G* = chloroform–96% ethanol (13:7) [119].

[c] Stahl and Kaldewey [118] obtained a pink color.

4 MISCELLANEOUS NITROGEN COMPOUNDS

Bird and Stevens [120] used thin-layer chromatography in checking the purity of nitrofurazone. This compound sometimes has 5-nitro-2-furfuraldazine in it as an impurity which is readily differentiated because it travels with the solvent front, in contrast to an R_f of 0.23 for the nitrofurazone. A system of benzene–acetone (3:2) was used on silica gel layers.

Because of the possible presence of nitroso carcinogens in foods the detection and determination of *N*-nitrosamines has taken on a new importance. Preussmann et al. [121] were able to separate alkyl and arylnitrosamines on silica gel with hexane–ether–dichloromethane (4:3:2) and cyclic nitrosamines with the same solvent mixture in a ratio of 5:7:10. A new spray reagent was formulated which detected 0.5 μg of nitrosamine (1 to 2 μg of volatile nitrosamines) (T-100). In investigating the trace analysis of these compounds in foods, Eisenbrand et al. [122] established four principles to be followed: *(a)* use dry dichloromethane for spotting the samples, *(b)* concentrate dilute extracts in a Kuderna–Danish evaporator to avoid losses of volatile nitrosamines, *(c)* develop the chromatogram in the dark at 4°C, and *(d)* see that the layer thickness is not less than 0.6 mm. As an additional factor they determined that the nitrosamines could be quantitatively recovered from the adsorbent by steam distillation. Vasundhara et al. [123] found the resolution of nitrosamines to be better on magnesium silicate using light petroleum–cyclohexane (19:1) or toluene–dichloromethane (9:1). Sen and Dalpe [124] were able to determine these compounds added to alcoholic beverages in amounts of 0.25 to 0.1 ppm with recoveries ranging from 20 to 100%.

Quaternary ammonium compounds have been separated by Bayzer [125] on cellulose layers using the wedge-strip technique. Excellent separation was achieved with chloroform–methanol–water (75:22:3). Dragendorff reagent was used for detection. Waldi [126] obtained an R_f value of 0.62 for choline with cyclohexane–chloroform–ethanol–acetic acid (4:3:2:1) on aluminum oxide layers. With cyclohexane–chloroform–acetic acid (9:9:2), the quaternary ammonium bases, neostigmine, berberine, and 1 [1-methyl-2-(10-phenothiazinyl)ethyl]trimethylammonium salt had R_f values of 0.47, 0.55, and 0.60, respectively. Newhall and Pieringer [127] found two quaternary ammonium compounds that decomposed partially to their amine precursors when chromatographed on alumina. Two-dimensional separations using electrophoresis in the first direction followed by thin-layer chromatography at 90°C has proved useful [128].

Vernin and Metzger [129] chromatographed a group of thiazole derivatives having Cl, SMe, OH, NH_2, or SH groups in the 2-position and phenyl, *p*-bromophenyl, *p*-chlorophenyl, *p*-methoxyphenyl, or *p*-nitrophenyl in the 4-position.

Schmid et al. [130] have examined the separation of substances involved in catecholamine and serotonin metabolism. Potter et al. [131] have also chromatographed the catecholamines and their metabolites and Schneider and Gillis [132] have studied the catecholamine biosynthesis. Meythaler et al. [133] have investigated the accumulation of amines by platelets. Partition chromatography on thin layers of cellulose was used by Vahidi and Sankar [134] for separating the catecholamines and their metabolites. Three solvent systems were used: *n*-butanol–acetic acid–water (5:1:3), ethyl acetate–acetic acid–water (5:1.5:5.3), and *n*-butanol–ethyl acetate–acetic acid–water (2:3:1:3). In each case the lower phase was used as the stationary phase. The compounds were detected with a 20% solution of Folin phenol reagent in water. Acetylated derivatives of these compounds have also been separated on silica gel layers [135, 136]. Dworzak and Huck [137] ran quantitative determinations using the fluorescent quenching principle and also by reflectance [138] after exposing the plates to diffuse daylight for 2 days for a photochemical reaction to take place. Johnson and Boukma [139] separated dopa, dopamine, and noradrenaline on cellulose MN 33 GF with butanol–ethanol–1 *M* acetic acid (7:2:2). Baumann et al. [140] used methyl ethyl ketone–acetone–2.5 *N* acetic acid (2:1:1) for a group of these compounds on cellulose. Sapira [141] used Woelm polyamide layers with isobutanol–acetic acid–cyclohexane (80:7:10) and detected the compounds with tests T-249, T-180, and T-114. Seiler and Wiechmann [142] separated the DANS derivatives of these compounds on silica gel layers using a two-dimensional separation. Two solvent systems were used: *(a)* twice in the first direction with chloroform followed by twice at 90° with butyl acetate–cyclohexane–ethyl acetate–triethylamine (11:10:4:4), or *(b)* first with diisopropyl ether to remove any traces of the dansyl chloride, then in the same direction with butyl acetate–triethylamine (5:1), followed by two developments in triethylamine–diisopropyl ether (5:1) at right angles.

Alkanolamines have been separated on magnesium silicate and on neutral alumina using *n*-propanol–water–chloroform (6:2:1) [143]. On neutral silica gel they may be separated with methylene chloride–ethanol–95% ammonia (43:43:15) [144]. They can be detected with ninhydrin, and an overlay spray of 0.2% by wt of alizarin in acetone acids further in identification. Phosphatidyl ethanolamines can be fractionated according to the double bonds on silver nitrate-impregnated silica with chloroform–methanol–0.5% acetic acid (65:25:4) [145].

References

1. K. Teichert, E. Mutschler, and H. Rochelmeyer, *Deut. Apoth.-Ztg.*, **100,** 283 (1960).
2. P. Voogt, *Fette, Seifen, Anstrichm.*, **68,** 825 (1966).
3. R. Gnehm, H. U. Reich, and P. Guyer, *Chimia (Aarau),* **19,** 585 (1965).
4. J. Lauckner, E. Helm, and H. Fuerst, *Chem. Tech. (Berlin),* **18,** 372 (1966).
5. D. Aures, R. Fleming, and R. Hakanson, *J. Chromatogr.,* **33,** 480 (1968).
6. E. S. Lane, *J. Chromatogr.,* **18,** 426 (1965).
7. H. Feltkamp and F. Koch, *J. Chromatogr.,* **15,** 314 (1964).
8. J. E. Hammond and E. H. Herbst, *Anal. Biochem.,* **22,** 474 (1968).
9. I. Wiesner and L. Wiesnerová, *J. Chromatogr.,* **114,** 411 (1975).
10. F. Abe and K. Samejima, *Anal. Biochem.,* **67,** 298 (1975).
11. A. H. Beckett and N. H. Choulis, 23rd Intern. Kongr. der Pharmaz. Wissenschaften, Muenster, September 9–14, 1963.
12. A. H. Beckett and N. H. Choulis, *J. Pharm. Pharmacol.,* **15,** 236T (1963).
13. B. Wesley-Hadzija, *J. Chromatogr.,* **79,** 243 (1973).
14. S. Udenfriend, S. Stein, P. Boehlen, W. Dairman, W. Leimgruber, and M. Weigele, *Science,* **178,** 871 (1972).
15. L. Lepri, P. G. Desideri, and V. Coas, *J. Chromatogr.,* **79,** 129 (1973).
16. G. Pastuska and H. Trinks, *Chem.-Ztg.,* **86,** 135 (1962).
17. C. G. Honegger, *Helv. Chim. Acta,* **44,** 173 (1961).
18. N. Seiler, *J. Chromatogr.,* **63,** 97 (1971).
19. N. Seiler and M. Wiechmann, "TLC Analysis of Amines as Their DANS-Derivatives," in *Progress in Thin-Layer Chromatography and Related Methods,* Vol. 1, A. Niederwieser and G. Pataki, Eds., Ann Arbor Science, Ann Arbor, Mich., 1970, p. 95.
20. J. H. Fleisher and D. H. Russell, *J. Chromatogr.,* **110,** 335 (1975).
21. E. H. Gruger, Jr., *J. Agric. Food Chem.,* **20,** 781 (1972).
22. C. R. Creveling, K. Kondo, and J. W. Daly, Clin. Chem., **14,** 302 (1968).
23. A. Askar, K. Rubach, and J. Schormueller, *Chem. Mikrobiol. Technol. Lebensm.,* **1,** 187 (1972).
24. P. B. Gosh and M. W. Whitehouse, *Biochem. J.,* **108,** 155 (1968).
25. H.-J. Klimisch and L. Stadler, *J. Chromatogr.,* **90,** 141 (1974).
26. J. F. Lawrence and R. W. Frei, *Anal. Chem.,* **44,** 2046 (1972).
27. H.-I. Ilert and T. Hartmann, *J. Chromatogr.,* **71,** 119 (1972).
28. Č. Michalec, *J. Chromatogr.,* **41,** 267 (1969).
29. A. Zeman and I. P. G. Wirotama, *Z. Anal. Chem.,* **259,** 351 (1972).
30. I. P. G. Wirotama and K. H. Ney, *J. Chromatogr.,* **61,** 166 (1971).
31. G. Neurath and E. Doerk, *Chem. Ber.,* **97,** 172 (1964).
32. K. Heyns, H. P. Harke, H. Scharmann, and H. F. Gruetzmacher, *Z. Anal. Chem.,* **230,** 118 (1967).
33. J. E. Sinsheimer, D. D. Hong, J. T. Stewart, M. L. Fink, and J. H. Burckhalter, *J. Pharm. Sci.,* **60,** 141 (1971).

34. A. Jart and A. J. Bigler, *J. Chromatogr.*, **29,** 255 (1967).
35. J. Churáček, H. Pechová, D. Tocksteinová, and Z. Zíkova, *J. Chromatogr.*, **72,** 145 (1972).
36. K. Kostka, *J. Chromatogr.*, **49,** 249 (1970).
37. D. B. Parihar, S. P. Sharma, and K. C. Tewari, *J. Chromatogr.*, **24,** 443 (1966).
38. D. Yamamoto, *Nippon Kagaku Zasshi,* **79,** 1030 (1958); through *Chem. Abstr.*, **53,** 8847 (1959).
39. H. Erlenmeyer and H. Bartels, *Helv. Chim. Acta,* **47,** 46 (1964).
40. J. Hansson, *Explosivstoffe,* **10,** 73 (1963).
41. J. Hansson and A. Alm, *J. Chromatogr.*, **9,** 385 (1962).
42. S. Heřmánek, V. Schwarz, and Z. Čekan, *Pharmazie,* **16,** 566 (1961).
43. A. Waksmundzki, J. Rozylo, and J. Oscik, *Chem. Anal. (Warsaw),* **8,** 965 (1963).
44. D. P. Schwartz, R. Brewington, and O. W. Parks, *Microchem. J.*, **8,** 402 (1964).
45. C. L. Keswani and D. J. Weber, *J. Chromatogr.*, **30,** 130 (1967).
46. M. Gillio-Tos, S. A. Previtera, and A. Vimercati, *J. Chromatogr.*, **13,** 571 (1964).
47. J. Sherma and G. Marzoni, *Am. Lab.*, **6,** 21 (1974).
48. E. Rinde and W. Troll, *Anal. Chem.*, **48,** 542 (1976).
49. S. Tabak, A. E. Mauro, and A. Del'Acqua, *J. Chromatogr.*, **52,** 500 (1970).
50. K. Yasuda, *J. Chromatogr.*, **87,** 565 (1973).
51. *Ibid.*, **72,** 413 (1972).
52. *Ibid.*, **74,** 142 (1972).
53. G. Pastuska and H.-J. Petrowitz, *Chem.-Ztg.*, **88,** 311 (1964).
54. L. Lepri, P. G. Desideri, and V. Coas, *J. Chromatogr.*, **90,** 331 (1974).
55. G. Duncan, L. Kitching, and R. J. T. Graham, *J. Chromatogr.*, **47,** 232 (1970).
56. D. B. Parihar, S. P. Sharma, and K. K. Verma, *J. Chromatogr.*, **26,** 292 (1967).
57. D. I. Chapman, J. R. Chapman, and J. Clark, *Int. J. Biochem.*, **3,** 66 (1972).
58. H. Thielemann, *Pharmazie,* **24,** 703 (1969).
59. H. Geissbuehler and D. Gross, *J. Chromatogr.*, **27,** 296 (1967).
60. J. E. Barney II, S. R. Harvey, and T. S. Hermann, *J. Chromatogr.*, **45,** 82 (1969).
61. G. R. N. Jones, *J. Chromatogr.*, **77,** 357 (1973).
62. R. S. Ratney, *J. Chromatogr.*, **26,** 299 (1967).
63. J. T. Stewart and L. A. Sternson, *J. Chromatogr.*, **92,** 182 (1974).
64. E. A. Mistryukov, *J. Chromatogr.*, **9,** 314 (1962).
65. H. Brockmann and H. Schodder, *Chem. Ber.*, **74,** 73 (1941).
66. H.-J. Petrowitz, *Mitt. Deut. Ges. Holzforsch.*, **48,** 57 (1961).
67. H.-J. Petrowitz, *Chem.-Ztg.*, **85,** 143 (1961).
68. H.-J. Petrowitz, G. Pastuska and S. Wagner, *Chem.-Ztg.*, **89,** 7 (1965).
69. H.-J. Petrowitz, *Materialpruefung,* **2,** 309 (1960).
70. J. Janák, *J. Chromatogr.*, **15,** 15 (1964).

71. J. Janák, *Nature*, **195**, 696 (1962).
72. P. V. Peuryfoy, S. G. Slaymaker, and M. Nager, *Anal. Chem.*, **31**, 1740 (1959).
73. D. F. Bender, E. Sawacki, and R. M. Wilson, Jr., *Air Water Pollut.*, **8**, 633 (1964).
74. D. F. Bender, E. Sawacki, and R. M. Wilson, Jr., *Anal. Chem.*, **36**, 1011 (1964).
75. E. Sawicki, T. W. Stanley, and H. Johnson, *Mikrochem. J.*, **8**, 257 (1964).
76. E. Sawicki, T. W. Stanley, J. D. Pfaff, and W. C. Elbert, *Anal. Chim. Acta*, **31**, 359 (1964).
77. E. Sawicki, W. C. Elbert, and T. W. Stanley, *J. Chromatogr.*, **17**, 120 (1965).
78. E. Sawicki, T. W. Stanley, and W. C. Elbert, *J. Chromatogr.*, **18**, 512 (1965).
79. M. Prost, M. Urbain, and R. Charlier, *Helv. Chim. Acta*, **52**, 1134 (1969).
80. M. W. Roomi, *J. Chromatogr.*, **65**, 580 (1972).
81. F. Binns, R. F. Chapman, N. C. Robson, G. A. Swan, and A. Waggott, *J. Chem. Soc. C*, **1970**, 1128.
82. M. Kiessling, J. Lauckner, S. Braeuer, and H. Fuerst, *Chem. Tech. (Berlin)*, **19**, 364 (1967); through *Anal. Abstr.*, **15**, 5408 (1968).
83. D. Stefanescu, D. Moraru, I. Predescu, P. Barza, T. Stanciu, and F. Coman, *Farmacia (Bucharest)*, **17**, 115 (1969); through *Chem. Abstr.*, **71**, 30331s (1969).
84. C. Di Bello and G. Galiazzo, *J. Chromatogr.*, **26**, 309 (1967).
84a. S. Tabak and M. R. Machado Verzola, *J. Chromatogr.*, **51**, 334 (1970).
85. G. T. Katvalyan and E. A. Mistryukov, *Izv. Akad. Nauk SSSR, Ser. Khim.* **1969**, 1809; through *Chem. Abstr.*, **72**, 12497s (1970).
86. R. Kido, T. Noguchi, T. Tsuji, H. Kaseda, and Y. Matsumura, *Wakayama Med. Rep.*, *11*, 115 (1966); through *Anal. Abstr.*, **15**, 1548 (1968).
87. C. Troszkiewicz, J. Suwiński, and W. Zieliński, *Chem. Anal. (Warsaw)*, **13**, 3 (1968).
88. J. P. Peyre and M. Reynier, *Ann. Pharm. Fr.*, **27**, 749 (1969).
89. K. C. Guven and B. Tekinalp, *Eczacilik Bul.*, **10**, 26 (1968); through *Anal. Abstr.*, **16**, 3218 (1969).
90. K. C. Guven, *Eczacilik Bul.*, **6**, 117 (1964); through *Chem. Abstr.*, **62**, 12981h (1965).
91. V. G. Yakutovich, B. L. Moldauer, Y. P. Kitaev, and Z. S. Titova, *Izv. Akad. Mauk SSSR, Ser. Khim.*, **1970**, 2086; through *CAMAG Bibliogr.*, **25**, 22 (1971).
92. D. H. Giao, A. Verdier, and A. Lattes, *J. Chromatogr.*, **41**, 107 (1969).
93. J. W. Westley, V. A. Close, D. N. Nitecki, and B. Halpern, *Anal. Chem.*, **40**, 1888 (1968).
94. W. S. Simmons and R. L. DeAngelis, *Anal. Chem.*, **45**, 1538 (1973).
95. J. A. F. De Silva, N. Munno, and N. Strojny, *J. Pharm. Sci.*, **59**, 201 (1970).
96. E. R. Cole, G. Crank, and A-S. Sheikh, *J. Chromatogr.*, **78**, 323 (1973).
97. S. Goenechea, *J. Chromatogr.*, **36**, 375 (1968).

98. E. Zielinska, *Chem. Anal (Warsaw),* **14,** 397 (1969).
99. A. Waksmundzki, J. K. Rozyło, and A. Scisłowicz, *J. Chromatogr.,* **46,** 204 (1970).
100. J. A. F. De Silva, N. Strojny, and K. Stika, *Anal. Chem.,* **48,** 144 (1976).
101. J. Dutta, M. Hoque, and D. P. Chakraborty, *J. Chromatogr.,* **42,** 555 (1969).
102. L. H. Klemm, C. E. Klopfenstein, and H. P. Kelly, *J. Chromatogr.,* **23,** 428 (1966).
103. V. Kapisinšká and M. Karvaš, *Chem. Prum.,* **21,** 129 (1971).
104. F. Y.-H. Wu and D. McCormick, *Biochim. Biophys. Acta,* **236,** 479 (1971).
105. I. L. Hansen and M. A. Crawford, *J. Chromatogr.,* **22,** 330 (1966).
106. A. S. Agathopulos, *Chim. Chron.,* **30,** 213 (1965).
107. J. Opieńska-Blauth, H. Kraczkowski, H. Brzuszkiewicz, and Z. Zagórski, *J. Chromatogr.,* **17,** 288 (1965).
108. M. E. Mahon and G. L. Mattok, *Anal. Biochem.,* **19,** 180 (1967).
109. H. G. Schlossberger, H. Kuch, and I. Buhrow, *Z. Physiol. Chem.,* **333,** 152 (1968).
110. R. K. Raj, and O. Hutzinger, *Anal. Biochem.,* **33,** 471 (1970).
111. K. T. Wang, Y. C. Tung, and H. H. Lai, *Nature,* **213,** 213 (1967).
112. U. Freimuth, M. Buechner, B. Zawta, W. Hubl, and D. Keibel, *Dtsch. Gesundheitswes.,* **21,** 2039 (1966).
113. U. Hornemann and H. G. Floss, *Anal. Biochem.,* **26,** 469 (1968).
114. F. Sági, *J. Chromatogr.,* **39,** 334 (1969).
115. D. S. Young, *Clin. Chem.,* **16,** 681 (1970).
116. K.-W. Glombitza, *J. Chromatogr.,* **25,** 87 (1966).
117. A. Niederwieser and P. Giliberti, *J. Chromatogr.,* **61,** 95 (1971).
117a. B. N. Johri, *J. Chromatogr.,* **50,** 340 (1970).
118. E. Stahl and H. Kaldewey, *Z. Physiol. Chem.,* **323,** 182 (1961).
119. G. Ballin, *J. Chromatogr.,* **16,** 152 (1964).
120. R. F. Bird and S. G. E. Stevens, *Analyst, (London)* **87,** 362 (1962).
121. R. Preussmann, D. Daiber, and H. Hengy, *Nature,* **201,** 502 (1964).
122. G. Eisenbrand, K. Spaczynski, and R. Preussmann, *J. Chromatogr.,* **51,** 503 (1970).
123. T. S. Vasundhara, S. Jayaraman, and D. B. Parihar, *J. Chromatogr.,* **115,** 535 (1975).
124. N. P. Sen and C. Dalpe, *Analyst (London),* **97,** 216 (1972).
125. H. Bayzer, *Experientia,* **20,** 233 (1964).
126. D. Waldi, *Naturwissenschaften,* **50,** 614 (1963).
127. W. F. Newhall and A. P. Pieringer, *J. Agric. Food Chem.,* **15,** 488 (1967).
128. H. Bayzer, *J. Chromatogr.,* **24,** 372 (1966).
129. G. Vernin and J. Metzger, *Chim. Anal. (Paris),* **46,** 487 (1964).
130. E. Schmid, L. Zicha, J. Krautheim, and J. Blumberg, *Med. Exp.,* **7,** 8 (1962).
131. W. P. Potter, R. F. Vochten, and A. F. Schaepdryver, *Experientia,* **21,** 482 (1965).
132. F. H. Schneider and C. N. Gillis, *Biochem. Pharmacol.,* **14,** 623 (1965).

133. C. Meythaler, E. Schmid, J. Blumberg, L. Zicha, and S. Witte, *Med. Exp.*, **7,** 232 (1962).
134. A. Vahidi and D. V. Sankar, *J. Chromatogr.,* **43,** 135 (1969).
135. D. Waldi, *Arch. Pharm.* **295/32,** 125 (1962).
136. J. E. Forrest and R. A. Heacock, *J. Chromatogr.,* **44,** 638 (1969).
137. E. Dworzak and H. Huck, *J. Chromatogr.,* **61,** 162 (1971).
138. H. Huck and E. Dworzak, *J. Chromatogr.,* **74,** 303 (1972).
139. G. A. Johnson and S. J. Boukma, *Anal. Biochem.,* **18,** 143 (1967).
140. P. Baumann, B. Scherer, W. Kraemer, and N. Matussek, *J. Chromatogr.,* **59,** 463 (1971).
141. J. D. Sapira, *J. Chromatogr.,* **42,** 134 (1969).
142. N. Seiler and M. Wiechmann, *J. Chromatogr.,* **28,** 351 (1967).
143. H. Grasshof, *J. Chromatogr.,* **20,** 165 (1965).
144. A. Lynes, *J. Chromatogr.,* **23,** 316 (1966).
145. S. M. Hopkins, G. Sheehan, and R. L. Lyman, *Biochim. Biophys. Acta,* **164,** 272 (1968).

Chapter **XVII**

AMINO ACIDS, PROTEINS, AND PEPTIDES

1 DIRECT SEPARATION OF AMINO ACIDS

The speed of thin-layer chromatography has a very distinct advantage over that of paper chromatography in the analysis of amino acids. A two-dimensional analysis that requires several days to run on paper chromatography can be accomplished within 4 to 5 hr by thin-layer chromatography. Another factor in favor of the thin-layer work is the increased sensitivity which varies from twice as sensitive for histamine to as much as fifty times for cysteic acid [24]. In two-dimensional work the difference in sensitivity is even greater, and for some individual amino acids may range as high as 250 to 500 times as sensitive.

Niederwieser [1] has reviewed the thin-layer chromatography of amino acids and their derivatives, and Pataki [2] has published a book on this technique as applied to amino acids and peptides.

Amino acids have been separated on thin layers of starch, cellulose, acetylated cellulose, aluminum oxide, calcium oxide, iron oxide hydrate, charcoal, calcium hydroxide, calcium phosphate, calcium sulfate, magnesium oxide, magnesium trisilicate, alginic acid, polyacrylonitrile, polyamide, Sephadex, kieselguhr, anion and cation exchangers, and silica gel. The two most commonly used are silica gel and cellulose.

Desalting of Samples

In separating amino acids from natural sources it is sometimes necessary to remove interfering compounds prior to chromatographic separation in order to prevent tailing and deformation of the spots. This occurs especially with urine samples and hydrolysates of proteins or peptides where high salt concentrations occur. A number of methods have been proposed and used for desalting: *(a)* electrophoresis [3–5], *(b)* ion-exchange resins [6–11], *(c)* Sephadex [12–14], *(d)* neutral polystyrene resin (Poropak Q) [15, 16], *(d)* ion-retardation resins [17, 18], and *(e)* solvent extraction [19, 20]. Heathcote et al. [18] compared amino acid recoveries from ion-retardation resins (Bio-Rad AG 11A8) with those from electrophoresis, solvent extraction, and ion-exchange methods, and found the ion retardation method gave the best recoveries. However, this resin is not suitable for removing citrates and acetates [17], and it is also inadequate for desalting soil samples and rock hydrolysates [21]. For soil samples Pollock and Miyamoto [21] used successive treatments with HF, NaOH, and HCl to remove Al, Ca, Mg, Na, Fe, and K ions and then finished off with a Dowex AG 50W-X8 H^+ column.

Thompson et al. [22] obtained good recoveries from ion-exchange resins by operating at 0 to 6°C. Basic amino acids were retained on the ammonium form of Dowex 50-X4 and the others on the hydrogen form. For successful use of this procedure the resins must be properly prepared.

Jacobson [22a] recommends sandwich dialysis for procedures that use polyacrylamide, starch, or agar gels. A small piece of cellulose or glass filter paper impregnated with the sample is sandwiched between two pieces of 5% polyacrylamide gel (3 mm thick). Proteins of molecular weight 50,000 or more will not diffuse appreciably into the gel. The salt and smaller molecules diffuse into the gel which may be replaced after a few minutes if high concentrations of salt are present. After 15 min the sample is ready to place at the origin. For small proteins a 10 to 20% gel is recommended.

For concentrating dilute protein solutions see the introduction to Chapter IV.

Separation on Silica Gel

Brenner and Niederwieser [23] and Fahmy et al. [24] have made an extensive study of the separation of amino acids on silica gel G and report the R_f values on 26 compounds in various solvents (Table 17.1). Separations were carried out on air-dried silica gel plates in a saturated atmosphere. Two-dimensional chromatography with different solvents was applied effectively for pairs which were difficult to separate. After drying the chromatograms at 110°C for 10 min they detected the amino acid spots by test T-178. The colors that form are characteristic for the individual amino acids. The development distance in this case was 10 cm; Bancher et al. [25] have published on a method using smaller plates and a development distance of 6 cm. They report the R_f values for 16 amino acids in five solvents using mixtures of silica gel and kieselguhr (4:11 and 1:1).

There are numerous solvent systems that have been used for the separation of amino acids on silica gel and some of these are listed in Table 17.2.

Pataki [33] examined the separations of some amino acids on five different silica gels and found that under identical conditions some gels gave better separation of certain groups of acids than did others, which shows that for optimum results not only the solvent but also the brand of silica gel must be selected.

In investigating aminoaciduria, Opienska-Blauth et al. [34], applying the methods of Brenner et al. [23], reported the R_f values of 33 amino acids in three solvents (Table 17.1). In addition to those solvents listed in the table, a number of special solvents were applied. A solvent composed of phenol–*m*-cresol–borate buffer of pH 9.3 (1:1:1) was used to separate the leucine and valine groups. For basic amino acids they used

Table 17.1 $R_f \times 100$ Values of Amino Acids in Various Solvents on Silica Gel G

	Solvent[a] [23]					Solvent[a] [34]		
Amino Acid	*A*	*B*	*C*	*D*	*E*	*F*	*G*	*C*
α-Alanine	47	37	27	39	40	49	25	32
β-Alanine	33	26	27	30	29	49	20	33
α-Aminobutyric acid						54	25	32
β-Aminobutyric acid						48	32	36
γ-Aminobutyric acid						38	30	39
α-Aminoisobutyric acid						57	26	35
β-Aminoisobutyric acid						46	29	38
α-Aminocaprylic acid	66	65	60	58	60			
Arginine	4	2	8	10	6	6	14	13
Asparagine						46	19	22
Aspartic acid	55	33	21	9	7	56	5	26
Citrulline						46	29	26
Cysteic acid	69	50	14	17	21	61	5	20
Cystine	39	32	16	27	22	8	9	7
Dihydroxyphenylalanine			45					
Glutamic acid	63	35	27	14	15	55	7	32
Glutamine						55	28	24
Glycine	43	32	22	29	34	50	18	28
Histidine	33	20	6	38	42	33	24	10
Hydroxyproline	44	34	20	28	31	63	33	26
Isoleucine	60	53	46	52	58	60	36	47
Leucine	61	55	47	53	58	63	37	53
Lysine	3	2	5	18	11	5	8	10
Methionine	59	51	40	51	60	62	36	43
Norleucine	61	57	49	53	59	66	54	55
Norvaline	56	50	38	49	57	67	56	54
Ornithine						6	5	8
Phenylalanine	63	58	49	54	60	63	41	54
Proline	35	26	19	37	30	48	45	24
Sarcosine	31	22	17	34	31			
Serine	48	35	22	27	31	52	19	29
Taurine						59	22	29
Threonine	50	37	25	37	40	66	18	28
Tryptophan	65	62	56	55	58	69	45	56
Tyrosine	65	62	56	55	58	65	36	50
Valine	55	45	35	48	56	56	29	38

[a] *A* = 96% ethanol–water (7:3); *B* = *n*-propanol–water (7:3); *C* = *n*-butanol–acetic acid–water (4:1:1); *D* = *n*-propanol–34% ammonium hydroxide (7:3); *E* = 96% ethanol–34% ammonium hydroxide (7:3); *F* = *n*-propanol–water (1:1); *G* = phenol–water (3:1). Development distance, 10 cm in all solvents.

Table 17.2 Some Additional Solvents for Chromatographing Amino Acids on Silica Gel

Solvent System	*Ref.*
Ethyl acetate–isopropanol–water–ammonia (20:20:25:1:5)	26
Acetone–water–acetic acid–formic acid (50:15:12:3)	26
Phenol–water–acetic acid (25:25:25:2.5)	26
Chloroform–methanol–ammonia (30:30:7.5)	26
Ethyl acetate–pyridine–acetic acid–water (30:20:6:11)	27
Toluene–acetic acid–water (30:20:1) (develop twice)	28
96% Ethanol–water–diethylamine (70:29:1)	29
Chloroform–formic acid (20:1)	30
Chloroform–methanol (9:1)	30
Isopropanol–water (7:3)	31
Isopropanol–5% ammonia (7:3)	31
Phenol–0.06 *M* borate buffer pH 9.30 (9:1)	32

acetone–pyridine–*n*-butanol–water–diethylamine (15:9:15:8:10), as well as the Brenner–Niederwieser solvent of propanol–water (7:3).

Carisano [35] has used a thin-layer separation of 3-methylhistidine, 1-methylhistidine, and histidine in order to detect whale extract in soup products. This is possible because whale extract contains a considerable quantity of β-alanyl-3-methylhistidine along with carnosine (β-alanyl-histidine) in contrast to beef extract which contains the latter plus a small amount of anserine (β-alanyl-l-methylhistidine) [36]. After hydrolysis of the extract these three compounds are separated by chromatographing on silica gel G in methanol–pyridine–water–acetic acid (6:6:4:1). The R_{st} values referred to histidine are 3-methylhistidine 0.84, and 1-methylhistidine 0.86. To obtain a better separation, especially when larger quantities of histidine have a tendency to tail and interfere with the 3-methylhistidine spot, a two-dimensional method may be employed by developing in the first direction with phenol–ethanol–water–ammonia (3:1:1:0.1) and then in the second direction with methanol–pyridine–water–acetic acid (6:6:4:1). The 3-methylhistidine spot can be readily differentiated from the 1-methylhistidine using the polychromatic developer (T-178). After the plates are sprayed with this reagent, they are heated in an oven for 1.5 to 2 min at 105°C at which time the 3-methylhistidine spot becomes intensely blue-violet in color with a yellowish halo, while the 1-methylhistidine appears as a light yellow color sometimes tinged with green.

Rokkones [37] has given the R_f values of 34 ninhydrin-positive substances isolated from urine and separated on silica gel plates. The chromatograms were developed first with chloroform–methanol–17% ammonium hydroxide (2:2:1) and then with phenol–water (3:1).

Euler et al. [38] compared the amino acids in normal and tumor serums from rats and found four times as much glycine in the tumor serum as in the normal serum.

Squibb [39] has separated and determined, on silica gel layers, the amino acids in avian liver samples. For the separation, a step development was applied using butanol–acetic acid–water (3:1:1) for the first development and 75% phenol for the second development in the same direction. Direct densitometry was applied at 525 mμ for the quantitative determinations of the spots which were visualized with 0.25% ninhydrin in acetone. Squibb [40] has also used silica gel layers on plastic plates. After development and staining of the chromatograms, they were sprayed with a plastic spray, Tuffilm Spray #543 (M. Grumbacher), in order to bind the layer to the support. The chromatograms were then cut into strips for scanning in standard equipment.

Marcucci and Mussini [41] chromatographed proline, dehydroproline, some hydroxyprolines and the corresponding nitroso derivatives on silica gel layers. Myhill and Jackson [42] separated proline and hydroxyproline as well as the corresponding nitroso derivatives on cellulose layers with a butanol–acetic acid–water (63:27:10) solvent.

Voigt et al. [43] chromatographed γ-aminobutyric acid and the other important free amino acids occurring in the brain on silica gel layers, and reported the R_f values in six solvents.

There are many combinations of solvents that can be used for two-dimensional work on silica gel. but some of them that have been used successfully with silica gel are: ethyl acetate–isopropanol–water–ammonia (40:40:50:3) in the first direction and acetone–water–acetic acid–formic acid (50:15:12:3) or chloroform–methanol–ammonia (4:4:1) in the second [26]; *tert*-amyl alcohol–methyl ethyl ketone–water (6:2:2), followed by isopropanol–formic acid–water (20:1:5) at 90° to the first direction [44]; butanol–acetic acid–water (4:1:1) followed by phenol–water (3:1) (wt/wt) to which 0.5 ml of 5% sodium cyanide/100 g has been added [45]; and propanol–water (16:9), with butanol–acetic acid–water (3:1:1) in the second direction [46].

Shellard and Jolliffe [29] have examined the effect on R_f value of chromatographing amino acids in grass pollen extracts preserved in 50% glycerol. The pattern of the R_f values in four solvents was used to help identify the individual amino acids even though the values had changed.

Separation on Cellulose Layers

Thin layers of cellulose have also been applied to the separation of amino acids [47–53]. In general the solvents and spray reagents used in paper chromatography may be applied here. Von Arx and Neher [54] have worked out a technique using a combination of cellulose layers in four solvent systems together with five color reactions for the separation and identification of 52 amino acids (Table 17.3). In applying this procedure, six 20 × 20 cm cellulose coated plates are prepared (the authors use a total of four color reactions on the two-dimensional combination I + II) and dried overnight at room temperature in a horizontal position. The amino acids are spotted in a water solution in one corner of the plates, 25 mm from the edges, in 1-μl amounts or less; the total quantity of a given amino acid should be kept below 10 μg.

All six plates are developed in solvent system I for the first-dimensional separation (note: it is important for the separation in all systems, that the chamber atmosphere is not saturated with solvent vapors). The solvent is allowed to travel to the top of the plate and then the layer is dried at 90°C for 5 to 10 min or at room temperature for 12 hr. For development in the second direction a control mixture is used consisting of 0.5 μg glycine, 1 μg tyrosine, and 1 μg norleucine, all contained in 1 μl. This is applied in line with the first-dimensional run. The plates are then developed in the second dimension at right angles to the first run. Four of the plates are developed in solvent system II and one each in solvent systems III and IV. After the second run, the plates are dried at 90°C for 20 min before spraying with the color reagents. The five color reactions that are used are tests T-177, T-153, T-249, T-205, and T-239.

In addition to the solvents listed in Table 17.3, 24 other solvent mixtures, which can be used for the separation of amino acids on cellulose or silica gel with an additional three solvent mixtures for use only on silica gel, are given in the original work.

Haworth and Heathcote [55, 56] have developed a new system for separating and identifying 76 amino acids and other nitrogen containing metabolites found in biological fluids. Separations were accomplished by two-dimensional development for 13 cm each way on cellulose MN 300 layers. The cellulose was purified beforehand as described in Chapter III, Section 2. Solvents for the separation consisted of 2-propanol–butanone–1 *N* hydrochloric acid (12:3:5) in the first direction and 2-methylpropanol–2-butanone–propanone–methanol–water–(0.88) ammonia (40:20:20:1:14:5) in the second direction. Later [56a] the second solvent was modified to 2-methylbutan-2-ol-butanone–propanone–methanol–water–ammonia (sp gr 0.88) (10:4:2:1:3:1). After the first development

Table 17.3 Two-Dimensional Separation of Amino Acids on Cellulose Layers, $R_f \times 100$ Values and Color Reactions.[a,b]

	Solvents[b,c]					*Detecting Reagents*[e]				
Amino Acid	I	II	III	IV	μg[d]	*NC*	*IS*	*PD*	*RH*	*NPK*
Glycocyamine	0–7	34	8	71	—	—	Pink		(+)	+
Arginine	3–8	13	2	90	0.2	Violet	Pink		+	+
Creatine	2–9	34	8	86	—	—	Yellow	(+)	+	+
α,α-Diaminopimelic acid	2–10	5	6	36	2.0	Violet	Red		Gray	
Aspartic acid	8–14	21	20	20	0.05	Green	Violet		(+)	
Lanthionine	9–17	8	22	35	1.0	Violet	Orange		Pink	
Canavanine sulfate	12–17	7	11	66	0.2	Violet	Brownish		+	+
Dihydroxyphenylalanine	10–22	16	10	13	10.0	Gray	Lilac	+	(Yellow)	
Glutamic acid	15–20	30	21	25	0.05	Violet	Lilac		—	
Hydroxyglutamic acid	16–21	30	15	25	0.05	Violet	Pink		(+)	
Cystine	17–22	4	26	35	1.0	Brown	Pink		(Yellow)	
Citrulline	19–24	16	15	68	0.1	Violet	Pink		+	
Asparagine	19–25	12	21	47	2.0	Yellow	Pink		+	
Cysteic acid	23–27	6	41	9	0.05	Violet	Yellow		(+)	
Methionine sulfoxide	24–28	19	26	82	0.2	Violet	Pink		(Brown)	
Glutamine	25–29	14	18	59	0.5	Violet	Pink		(+)	
p-Aminohippuric acid	44–49	65	74	89	10.0	Violet	Yellow		+	
Tyrosine	45–51	38	56	71	0.5	Brown	Red	+	(Pink)	
Taurine	46–52	21	44	51	0.2	Violet	Yellow		+	
Valine	49–55	55	44	87	0.05	Violet	Pink		(+)	
ϵ-Amino-*n*-caproic acid	52–57	55	20	87	1.0	Violet	Pink		+	
Norvaline	52–58	55	45	87	0.05	Violet	Red		(+)	
Kynurenine	54–60	39	47	80	0.5	Brown	Pink		(Yellow)	
Allothreonine	54–60	31	55	59	0.5	Violet	Pink		(+)	
Tryptophan	56–62	41	60	88	0.5	Violet	Lilac		(Brown)	

Table 17.3 (Continued)

Amino Acid	Solvent[b,c] I	II	III	IV	μg[d]	Detecting Reagents[e] NC	IS	PD	RH	NPK
Methionine	56–62	50	54	88	0.5	Violet	Pink		—	
Isoleucine	58–64	65	62	88	0.1	Violet	Pink		(+)	
Alloisoleucine	58–64	65	62	88	0.1	Violet	Pink		(+)	
Leucine	59–66	65	60	88	0.5	Violet	Pink		—	
Norleucine	59–66	65	60	88	0.5	Violet	Lilac		—	
Threonine	60–67	31	67	56	0.5	Violet	Pink		(+)	
α-Phenylalanine	61–67	54	73	88	0.2	Violet	Lilac		(Yellow)	
Diiodotyrosine	62–69	50	73	71	5.0	Violet	Lilac	(+)	+	
Thyronine	63–69	62	82	88	2.0	Brown	Brown		—	
α-Phenylglycine	63–69	76	81	77	5.0	Yellow	Yellow	(+)	(Brown)	
β-Hydroxyvaline	62–69	41	81	65	0.5	Violet	Pink		(+)	
Thyroxine	69–76	73	84	86	2.0	Brown	Yellow	—	(+)	
Hydroxyproline	26–30	22	19	66	1.0	Yellow	Blue		(+)	
Glycine	28–33	22	19	46	0.05	Brown	Pink		+	
Ornithine	28–33	8	5	82	0.2	Violet	Red		+	
Hydroxylysine	29–34	8	11	71	0.5	Violet	Lilac		+	
Lysine	32–37	10	6	82	0.5	Violet	Red		+	
α-Alanine	33–38	34	22	61	0.05	Violet	Violet		—	
β-Alanine	33–39	37	14	66	0.05	Green	Lilac		+	
Creatinine	33–39	36	25	97	—	—	Yellow		+	+
Sarcosine	36–41	29	17	80	0.1	Gray	Yellow		+	
α,γ-Diaminobutyric acid	36–42	8	20	73	2.0	Violet	Pink		+	
Histidine	36–41	11	34	87	0.3	Gray	Lilac	+	(Gray)	
Methionine sulfone	37–42	20	36	73	0.2	Violet	Pink		(+)	
Dimethylcysteine	40–45	10	53	73	0.2	Violet	Pink		—	

Proline	41–46	35	19	87	0.5	Yellow	Blue	(+)
β-Aminobutyric acid	41–46	43	21	87	0.05	Lilac	Yellow	+
α-Aminoisobutyric acid	41–47	50	25	78	0.05	Violet	Lilac	+
α-Amino-*n*-butyric acid	42–47	42	31	78	0.05	Violet	Lilac	(+)
γ-Aminobutyric acid	42–48	46	13	78	0.2	Violet	Lilac	+
β-Aminoisobutyric acid	42–48	47	22	78	0.05	Violet	Lilac	+
Serine	42–49	21	38	41	0.2	Violet	Orange	Yellow

[a] From E. von Arx and R. Neher [54]; reproduced with permission of the authors and Elsevier Publishing Co.

[b] In all systems the chamber atmosphere is *not* saturated with solvent vapor. Development distance 16 cm.

[c] Column I: approximate range of $R_f \times 100$ values in one-dimensional development in first solvent; columns II, III, IV: $R_f \times 100$ values after development in second dimension in respective solvent. Solvents: (I) *n*-butanol–acetone–diethylamine–water (10:10:2:5); (II) isopropanol–formic acid–water (20:1:5); (III) *sec*-butanol–methyl ethyl ketone–dicyclohexylamine–water (10:10:2:5); (IV) phenol–water (75:25) + 7.5 mg sodium cyanide. The chamber atmosphere in this case is in equilibrium with the vapors of 3% ammonium hydroxide.

[d] μg in the practical region of detection for the ninhydrin reaction on two-dimensional chromatogram using combination of I and II. The actual limit of detection is lower but varies somewhat with the drying conditions after applying the ninhydrin reagent.

[e] Detecting reagents (see text for preparation): except for *NC* a color reaction indicates presence of 0.5 to 5γ of amino acid. (+) indicates a weak color reaction. *NC* = Ninhydrin-collidine reagent; *IS* = isatin; *PD* = Pauly's diazo reagent, + signifies a yellow to reddish color; *RH* = Reindel–Hoppe reagent modified, + indicates a bluish black color; *NPK* = sodium nitroprusside–potassium ferricyanide reagent, + indicates a red color.

the plates were dried in a stream of cold air for 15 min and then in a convection oven at 60°C for 15 min to remove all traces of hydrochloric acid. Separation was achieved in a saturated atmosphere obtained by allowing the solvent from the previous day's run to remain in the tank [57]; prior to the run this was replaced with fresh solvent. Identification was completed by the use of one or more of 12 color reagents. The R_f values and the numerous color reactions are given in the original papers.

Some of the numerous solvents that have been used for the separation of amino acids on cellulose are listed in Table 17.4. Combinations that have been used for two-dimensional work include: *n*-butanol–pyridine–water (1:1:1) followed by 88% phenol–25% ammonia–water (10:0.8:1) containing 1 mg of O-oxyquinoline per 200 ml (58); isopropanol–formic acid–water (20:1:5) in the first direction and *tert*-butanol–2-butanone–0.9 N ammonia–water (25:15:7:3) at right angles (59), both developments in an unsaturated tank; similarly in unsaturated tanks *n*-propanol–water (7:3) for 12 cm, than at right angles for 10 cm with isopropanol–water (4:1), followed by another development in the first direction with the first solvent again (60, 61); and methanol–chloroform–ammonia (2:2:1) in the first direction followed by methanol–water–pyridine (20:5:1) in the second direction for general purposes, but for the separation of leucine and isoleucine, *tert*-amyl alcohol–methyl ethyl ketone–water (6:2:2) is used in the first direction (62).

Amino acids chromatographed in the presence of trichloracetic acid (used in deproteinizing serum samples) show anomalous behavior. This

Table 17.4 Some Additional Solvents for Chromatographing Amino Acids on Cellulose

Solvent System	*Ref.*
Ethyl acetate–pyridine–acetic acid–water (5:5:1:3)	51
n-Butanol–acetic acid–water (4:1:1, 10:3:9)	
n-Butanol–acetic acid–water–ethanol (10:1:3:0.3, 4:1:10:1)	52
sec-Butanol–dimethyl ketone–acetic acid–water (10:10:2:5)	53
n-Butanol–dimethyl ketone–ammonia–water (20:20:4:1)	
Phenol–methanol–water (7:10:3)	
Collidine–*n*-butanol–acetone–water (2:10:10:5)	
Collidine–methanol–acetone–water (2:10:10:5)	
n-Propanol–water (4:1)	
Ethanol–acetic acid–water (2:1:2)	
Cyclohexanol–acetone–diethylamine–water (10:5:2:5)	53a
tert-Butanol–acetic acid–water (5:1:1)	

interference can be almost completely removed by predevelopment (2×) in ether saturated with formic acid (63).

Wieland and Buku (63a) have developed a micro method for determining the configuration of an amino acid. This consists in coupling it with an *N*-protected L-amino acid by nonracemizing method, removing the *N*-protection, and chromatographing the LL or LD dipeptide that was formed.

Separation of Miscellaneous Adsorbents

Llosa et al. [64] have studied the separation of 19 amino acids on diethylaminoethyl cellulose using nine different solvent systems. They have tabulated the R_f values and the color reactions with ninhydrin. Lepri et al. (65) chromatographed 17 aromatic amino acids on DEA cellulose (Cellex D, ClO_4^- form, Bio-Rad Labs.) using 0.1 *M* sodium perchlorate as solvent. Vercaemst et al. [66] used DEAE cellulose layers with a triple development with *n*-butanol–acetic acid–water (4:1:5) for one-dimensional chromatography of 18 amino acids. Two-dimensional separations were made with two developments with the same solvent followed by a pyridine–water (4:1) development at right angles. The pairs glutamic acid and glycine, and aspartic acid and serine are well separated by this technique.

Other ion-exchange resins have been investigated. Tyihák et al. [67] and Hazai et al. [68] used Fixion 50-X8 commercial plates which contain Dowex 50-X8 type resin. Using the Na^+ form with three different buffer solvents the R_f values of 30 amino acids have been reported [67], and in the H^+ form with acidic solvents 15 R_f values [68]. The ion-exchange resin AG 1-X4 (ClO_4^-) and Dowex 50-X4 (H^+ and Na^+) have been used to separate aromatic amino acids [65]. Alginic acid as an ion-exchange agent was tested with 31 amino acids [69].

Dicarboxylic amino acids can be separated on aluminum oxide with 2 *N* acetic acid as solvent [70]. Monocarboxylic amino acids do not separate and are concentrated at the solvent front.

Affonso [71] has separated 12 amino acids on calcium sulfate (plaster of Paris) layers in acetone–chloroform–acetic acid–acetate buffer pH 4 (4:1:0.5:1.5). Ninhydrin was added directly to the developing solvent without affecting the R_f values.

Petrović and Petrović [72] separated 22 amino acids on rice starch layers bound with gypsum. First-dimensional separation was with *n*-butanol–acetone–diethylamine–water (10:10:2:5); second-dimensional solvents were: phenol–water (3:1) in the presence of ammonia vapor, isopropanol–water–formic acid (20:5:1), isopropanol–water–acetic acid–pyridine (25:20:5:2), and *n*-butanol–acetic acid–water–pyridine

(20:5:25:1). The layers were slower than cellulose or silica gel but still faster than paper. Sensitivity with ninhydrin–lutidine is good at 0.1 μg (0.5 μg for asparagine) although for two dimensional work it is recommended that 0.5 μg of each acid be applied. Maize starch was unsatisfactory.

A mixture of cellulose–silica gel (5:2) has been used with excellent results [73–75]. Solvent combinations for two-dimensional development were: phenol–water (4:1 wt/vol) with overnight drying at 40°C, followed by two runs in butanol–acetic acid–water (5:1:4 top phase) in the second direction [73], *sec*-butanol–80% formic acid–water (15:3:2) first and then *n*-propanol–34% ammonia (67:33) [74], or isopropanol–formic acid–water (20:1:5) with *tert*-butanol–ethyl acetate–ammonia (0.88)–water (5:3:1:1) [75]. The mixed layers gave better results than the individual adsorbents. Tested on 27 acids [73] the spots were compact and well separated and the sensitivity to ninhydrin reagent was twice that on cellulose and five times that on silica gel. The layers are firmer than silica gel G alone. Other compounds present in crude extracts and urine samples did not interfere with the separation.

Separation of Iodoamino Acids

Zappi [76] has reviewed the separation of iodoamino acids by thin-layer chromatography.

Schorn and Winkler [77] have chromatographed the iodoamino acids, 3-monoiodotyrosine, 3,5-diiodotyrosine, 3,5-diiodothyronine, 3,3′,5-triiodothyronine, and thyroxine in 50 different solvent systems on silica gel layers that were air-dried overnight. A number of useful solvent systems were found. For separating mono- and diiodotyrosine, *n*-butanol–acetone–1 *N* ammonium hydroxide (1:4:1) gave the best separation with fairly good separation also of 3,3′,5-triiodothyronine, thyroxine, and iodide ion. Better separation of the latter three could be obtained with acetone–methyl *p*-hydroxybenzoate–1 *N* ammonium hydroxide (16:1:3); a two-dimensional separation could be used to give excellent resolution of all the compounds. Quantitatively, the compounds were determined in a scintillation counter. Schneider and Schneider [78] investigated acetic acid–benzene-xylene mixtures on silica gel layers dried at 105°C and found that a 6:2:2 ratio gave the best separation. Alkaline solvents were also tried and in this case a phenol-acetone–1 *N* sodium hydroxide (2:7:1) mixture in an ammonia atmosphere gave the best separation in an alkaline solvent. The acidic solvent gave the best separation of diiodotyrosine and diiodothyronine, whereas the basic solvent gave the best separation of diiodothyronine and triiodothyronine. For a detecting agent they used the ceric sulfate–arsenite reagent of Mandl and Block [79]. Stahl and

Pfeifle [80] also chromatographed these compounds on starch-bound silica gel layers. Detection was made possible by spraying the dried chromatogram with 50% acetic acid and a 10-min exposure to short-wavelength ultraviolet light, followed by an additional spray of 10% acetic acid. The iodine liberated by the photoreaction appeared as a typical blue spot. Berger et al. [81] have used multiple layers to separate a mixture of iodide ion, thyroxine, diiodotyrosine, and monoiodotyrosine. The multiple layer consisted of a narrow strip of silver chloride with the balance of the plate covered with the anion-exchange resin Dowex 1 × 2 (OH^-). In operation the sample was spotted on the silver chloride layer. On development with 3 *N* potassium hydroxide in methanol, the iodide ion remained at the origin, the thyroxine was carried to the boundary of the two layers, and the mono- and diiodo derivatives were separated on the ion-exchange layer. Massaglia and Rosa [82] have achieved the same separation by using silica gel layers with butanol–acetic acid–water (4:1:5) as the developing solvent. The tyrosine was visualized with ninhydrin while the other radioactive compounds were detected with a GM counter.

Lipids in the methanol extracts of biological materials interfere with the separation of thyroid hormones. Zappi et al. [83] solved this problem on cellulose layers by applying the sample equidistant from the sides of the plate and 4 cm from one edge. From the opposite edge a development was made with chloroform. When this solvent reached the sample near the upper edge it dissolved the lipids and moved them to the top leaving the amino acids at the spotting location. The lipid impurities were scraped from the layer and the plate was then developed from this edge with acetone–0.1 *N* acetic acid (1:4).

Hollingsworth et al. [84] have used thin layers of cellulose to separate the iodoamino acids: 3-monoiodo-L-tyrosine, 3,5-diiodo-L-tyrosine, 3,5-diiodo-L-thyronine, 3′,3,5-triiodo-L-thyronine, and L-thyroxine as well as L-tyrosine. Excellent separations were obtained (except for 3,5-diiodo-L-thyronine and 3′,3,5-triodo-L-thyronine) in a mixture of *tert*-butanol, 2 *N* ammonium hydroxide, and chloroform. Patterson and Clements [85] have used chromatography on paper sheets as well as starch-bound cellulose plates for the identification of thyroxine in a feed additive. Detection was made with test T-122.

Ouellette and Balcius [86] used a mixture of cellulose MN 300 and silica gel G (4:1) to separate 5-monoiodotyrosine, 3,5-diiodotyrosine, 3,3′,5-triiodothyronine, and 3,3′,5,5′-tetraiodothyronine. The best solvents were *tert*-amyl alcohol–dioxane–1 *N* ammonia (2:2:1) and ethanol–methyl ethyl ketone–2 *N* ammonia (1:4:1).

Hamada and Ingbar [87] used electrophoresis on a layer of Pevikon C-

870 (Mercer) to separate thyroxine and 3,3′,5-triiodo-L-thyronine either free or in normal serum.

Separation by Electrophoresis

Honegger [88] demonstrated the separation of amino acids by electrophoresis on thin layers. The thin-layer electrophoresis was carried out on layers of silica gel G, kieselguhr G, or aluminum oxide prepared with sodium citrate buffer (0.1 *M*, pH 3.8). A current of 460 V at 12.6 mA was used for the separation, with an electrolyte composed of 2 *N* acetic acid–0.6 *N* formic acid (1:1). (For details of the electrophoretic setup see Chapter V, Section 5.) A combination electrophoresis–chromatographic method was also applied by running the electrophoresis first in one direction and then drying the plate and running the chromatographic separation in the second direction. Nybom [42] has also published on a combined electrophoresis–chromatographic method for amino acid analysis.

Bieleski and Turner [89] used a mixed layer of cellulose–silica gel (12.5:5 or 5:2) for a two-dimensional electrophoresis-TLC method. The extract (of plant tissue) was applied near one corner of the plate as a 2.5-cm streak. A pH-2.0 buffer (17 ml 90% formic acid + 57 ml acetic acid per liter) was then sprayed over the layer, wetting the origin last. After excess buffer was blotted away and the layer connected with the electrode chambers by means of wicks, electrophoresis was carried out at 12 to 18°C for 20 min at 1000 V (55 V/cm) and 20 to 30 mA. The plate was then dried at 40°C for 15 min in a forced air draught in an ammonia-free atmosphere. The plate was dipped into water at right angles to the electrophoretic run so that the entire plate was covered up to within 1 cm of the ends of the electrophoretic bands. After 5 min the ascending water moved the bands into compact spots near the margin of the plate. (If arginine or other basic amino acids are present 1% acetic acid is used instead of water). After drying again the plate was developed in the second dimension with methyl ethyl ketone–pyridine–water–acetic acid (70:15:15:2) in order to separate threonine and glutamic acid as well as to remove interfering materials. A second development was made with *n*-propanol–water–*n*-propylacetate–aceticacid–pyridine(120:60:20:4:1).

Munier et al. [89a] used a two-dimensional separation on cellulose layers (20 × 40 cm). Chromatography was carried out in the shorter direction with pyridine–ethanol–water (2:2:1) until a dye marker (Brilliant Black BN) reached the edge of the plate. After the plate was dry it was sprayed with 0.025 *M* sodium tetraborate. Electrophoesis was at right angles for 3 hr at 410 V/cm. Detection was with test T-267. When peptides and their constituent amino acids were run the thin-layer chromatography was stopped when the solvent reached the edge of the plate. References are given to a number of earlier papers on this subject.

Montant and Touze-Soulet [90] describe the application of electrophoresis using a thin layer of cellulose with a pyridine–acetic acid–water buffer of pH 3.9 for the separation of amino acids. Pastuska and Trinks [91] have separated a small group of amino acids by electrophoresis on silica gel G impregnated with a borax buffer. A mixture of 80 ml of ethanol and 30 ml of distilled water containing 2 g of crystalline sodium acetate was used as the electrolyte solution. Phosphorylated cellulose [92] and Sephadex G25 [93] have been used as supports for the electrophoresis of amino acids.

2 SEPARATION OF AMINO ACID DERIVATIVES

The separation of amino acid derivatives is very important in the structural determination of peptides. This follows from the reaction of proteins and peptides with 2,4-dinitrofluorobenzene, 1-dimethylaminoaphthalene-5-sulfonyl chloride, or phenylisothiocyanate with the subsequent degradation to the dinitrophenyl amino acids, 1-dimethylaminonaphthalenesulfonyl (DANS or DNS) amino acids, and phenylthiohydantoins, respectively. Preparation of these derivatives has been adequately described in the literature (DNP [94–97], DANS [98–101], PTH [102–106] and is not discussed at this point. Rosmus and Deyl [106a–106c] have reviewed the methods for identification of *N*-terminal amino acids in peptides and proteins.

Dinitrophenyl (DNP) Amino Acids

Brenner et al. [107] have investigated the separation of these derivatives on silica gel. Those derivatives which are acid- and water-soluble and are not extracted from solution by ether can be separated by chromatographing in *n*-propanol-34% ammonia (7:3) (Table 17.5). Of the seven derivatives listed in the table only two, DNP–cysteic acid and mono-DNP–cystine, cannot be separated; however, this is not serious since they seldom occur together and also they may be differentiated by their difference in color on reaction with ninhydrin. Since the sample is applied in an acid solution, the excess acid must be evaporated thoroughly by heating to 60°C for 10 min prior to running the chromatogram.

For separating the more numerous acid-insoluble DNP amino acids which are extractable with ether, a group of five solvents were chosen (Table 17.6). Solvent 1, which is a two-phase system, is composed of toluene–pyridine–ethylene chlorohydrin–0.8 *N* ammonium hydroxide (10:3:6:6) [108]. The upper phase is used for development and the lower

Table 17.5 Identification of Acid- and Water-Soluble DNP Amino Acids Separated on Silica Gel with *n*-Propanol–34% Ammonium Hydroxide (7:3)[a]

DNP Amino Acid	$R_f \times 100$	*Color*	*Absorption in UV*[b]	*Color with Ninhydrin*
Mono-DNP-$(Cys)_2$	29	Yellow	+	Brown
DNP-$CySO_3H$	29	Yellow	+	Yellow
α-DNP-Arg	43	Yellow	+	Yellow
ε-DNP-Lys	44	Yellow	+	Brown
O-DNP-Tyr	49	Colorless	+	Violet
α-DNP-his	57	Yellow	+	Yellow
Di-DNP-His	65	Yellow	+	Yellow

[a] From M. Brenner, A. Niederwieser, and G. Pataki [107]; reproduced with permission of the authors and Birkhaeuser Verlag.
[b] Visible on fluorescent layers as dark spots.

phase for pretreatment of the layer as follows: The plates are placed in a developing tank with the lower phase of the solvent in the bottom. To prevent the plates from touching the solvent they are allowed to rest on glass rods. The tank is lined with filter paper in order to obtain a saturated atmosphere, and the layers can be prevented from touching this liner either by scoring the liner at the point where the plate touches it or by using glass rod inserts. The plates are allowed to equilibrate in this atmosphere overnight. They are then removed from the conditioning atmosphere and covered immediately with a sheet of glass leaving only the lower spotting area exposed. This prevents loss of the solvent vapors which is very rapid and noticeable in the effect on the R_f values, even after 2 min. After spotting the compounds the plate is developed immediately in the upper phase of the first solvent system. Despite the fact that this system has a tendency to cause "tailing," it does provide good separations.

Since complete separation cannot be accomplished for all the derivatives by one-dimensional chromatography, the solvents in Table 17.6 may be used to carry out two-dimensional work. As an example, a combination of solvents 1 and 4 separates the isomeric leucine derivatives and the isomeric valine derivatives. In running the two-dimensional work, care must be taken to guard against the oxidation of the derivatives, and therefore removal of the solvent after the first development is kept to a minimum of time. The authors recommend 10 min exposure in air followed by 10 min drying at 60°C, and finally cooling for 10 to 15 min.

Figure 17.1 shows a two-dimensional separation using the "toluene" system in the first direction and a chloroform–methanol–acetic acid (95:5:1) mixture for the second dimension.

Drawert et al. [109] have applied the method to radioactive-labeled amino acids using a two-dimensional method with solvent 1 in the first dimension and solvent 2 for the second dimension. Quantitative determinations were made by impregnating the layers with collodion, after which the layers were removed from the supporting plate for counting.

Blass [110], by using continuous development (7 hr) in the first direction on silica gel with toluene–2-chloroethanol–pyridine–25% ammonia (20:12:6:1) and also in the second direction (8 hr) with chloroform–*tert*-amyl alcohol–acetic acid (95:5:1), was able to separate DNP-leucine and DNP-isoleucine. This can not be accomplished in the system of Brenner et al. [107].

Walz et al. [111] have applied the separation of dinitrophenyl derivatives to the detection of amino acids in urine. The general methods of

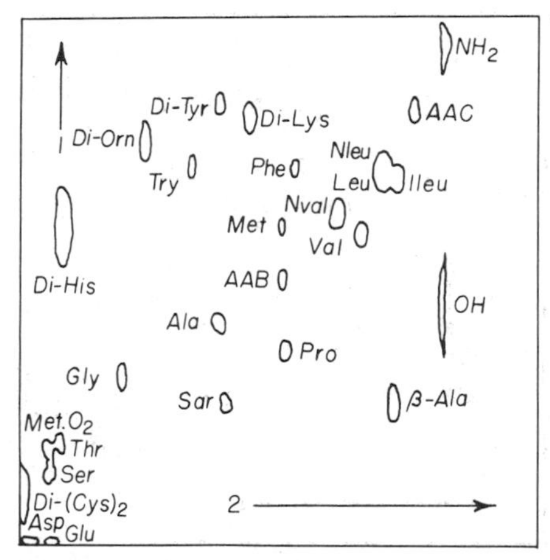

Fig. 17.1 Two-dimensional chromatogram of 1-μg amounts of DNP amino acids. First direction in toluene mixture (see text), second direction chloroform–methanol–acetic acid (95:5:1). OH = 2,4-dinitrophenol, Di = di-DNP derivative, NH_2 = 2,4-dinitroaniline, AAB = α-aminobutyric acid, AAC = α-aminocaprylic acid, Ala = alanine, Asp = aspartic acid, $(Cys)_2$ = cystine, Glu = glutamic acid, Gly = glycine, His = histidine, Ileu = isoleucine, Leu = leucine, Nleu = norleucine, Lys = lysine, Met = methionine, $MetO_2$ = methionine sulfone, Orn = ornithine, Phe = phenylalanine, Pro = proline, Sar = sarcosine, Ser = serine, Thr = threonine, Try = tryptophan, Tyr = tyrosine, Val = valine, Nval = norvaline. Original 13 × 13 cm. From Brenner, Niederwieser, and Pataki [107]; reproduced with permission of the authors and Birkhaeuser Verlag.

Table 17.6 $R_f \times 100$ Values of Ether-Soluble, Acid-Insoluble DNP Amino Acids on Silica Gel[a]

	Solvent[b]										
	1[e]	2 Ascending		3 Ascending		4[d]			5[d]		
							Horizontal			Horizontal	
DNP Amino Acid	*Ascending*		*Indirect*[c]		*Indirect*[e]	*Ascending*		*Indirect*[e]	*Ascending*		*Indirect*[e]
DNP-α-AB[f]	46	72	44	73	42	52	52	55	79	85	75
DNP-α-AC[g]	79	92	66	83	57	105	108	109	108	101	106
DNP-Ala	34	54	35	60	34	32	33	38	59	66	58
DNP-β-Ala	27	71	57	73	50	89	98	100	99	95	102
DNP-Asp	2	13	8	9	13	6	5	11	7	6	6
DNP-Glu	1	26	17	31	21	12	12	23	12	12	14
DNP-Gly	27	32	22	40	23	17	18	22	31	38	31
DNP-Ileu	64	83	63	81	57	107	107	107	100	101	104
DNP-Leu	66	82	62	80	54	100	100	100	100	100	100
DNP-Nleu	69	82	60	80	52	86	90	88	101	100	98
DNP-Met	55	70	39	69	38	43	43	47	72	81	74
DNP-Met·O_2[h]	17	—	—	—	4	3	3	2	10	10	7
DNP-Phe	67	75	46	74	41	44	46	52	81	86	76
DNP-Pro	29	65	41	67	38	58	59	62	78	84	75
DNP-Sar	23	56	35	57	32	34	35	41	59	65	60
DNP-Ser	15	11	10	11	10	9	10	14	7	8	7

DNP-Thr	20	17	13	15	12	12	14	20	9	11	11
DNP-Try	65	69	38	69	31	23	25	33	54	61	49
DNP-Val	53	79	56	77	51	76	81	85	91	98	86
DNP-Nval	56	77	52	76	48	65	70	75	86	95	89
Di-DNP-$(Cys)_2$	—	3	2	1	1	0	0	2	0	2	2
Di-DNP-His	53	11	9	8	4	5	4	8	12	16	14
Di-DNP-Lys	74	56	35	60	30	12	13	19	66	73	65
Di-DNP-Orn	70	34	23	40	20	6	6	10	39	46	39
Di-DNP-Tyr	76	58	35	60	30	17	16	19	57	65	57
2,4-DNP-OH[i]	41	100	76	83	55	22	21	23	148	102	111
2,4-DNP-NH_2[j]	90	90	84	72	63	115	128	129	131	101	115

[a] From M. Brenner, A. Niederwieser, and G. Pataki [107]; reproduced with permission of the authors and Birkhaeuser Verlag.

[b] *1* = Toluene–pyridine–ethylene chlorohydrin–0.8 *N* ammonium hydroxide (10:3:6:6) [107]; see text. *2* = Chloroform–benzyl alcohol–acetic acid (70:30:3). *3* = Chloroform–*tert*-amyl alcohol (70:30:3). *4* = Benzene–pyridine–acetic acid (40:10:1). *5* = Chloroform–methanol–acetic acid (95:5:1).

[c] See text for use of solvent *1*.

[d] R_{st} value based on R_f of DNP-Leu.

[e] Value after developing first in solvent 1, drying and then redeveloping in solvent listed.

[f] αAB = α-aminobutyric acid.

[g] αAC = α-aminocaprylic acid.

[h] Met·O_2 = methionine sulfone.

[i] 2,4-DNP-OH = 2,4-dinitrophenol.

[j] 2,4-DNP-NH_2 = 2,4-dinitroaniline.

Brenner et al. [107] were followed. In addition to the solvents used in this previous work, four additional solvents were applied: toluene–pyridine–ethylene chlorohydrin–25% ammonium hydroxide (50:35:15:7), chloroform–methanol–acetic acid (70:30:5), pyridine, and *n*-butanol saturated with 25% ammonium hydroxide at room temperature. In addition to two-dimensional work, it was found advantageous to use multiple development.

Pataki [112, 113] has applied a novel method for the detection of amino acids in the sequential analysis of peptides. The amino acids are chromatographed on silica gel G using *n*-propanol–water (7:3) as the solvent. After the plates are dried, they are sprayed with a buffered solution containing 8.4 g of sodium bicarbonate and 2.5 ml of 1 *N* sodium hydroxide per 100 ml of solution and finally with 10% (wt/vol) solution of dinitrofluorobenzene in methanol. The chromatogram is then covered with a second blank glass plate which is supported by two strips of polyethylene placed along the two edges of the chromatogram which have been cleared of adsorbent material. The protected chromatogram is then heated in a dark oven at 40°C for 1 hr. The chromatogram is then cooled and placed in an ether bath for 10 min after which it is dried so that the spots can be marked. After separating the amino acids in one direction and converting them to the dinitrophenyl derivatives, the latter can be chromatographed in a second dimension with one of the solvents already mentioned.

Munier and Sarrazin [114] have used a two-dimensional separation on cellulose for the separation of the ether-soluble DNP amino acids and a combination electrophoresis–TLC separation of the acid- and water-soluble derivatives. Grant and Wicken [115] used a mixed layer of silica gel–cellulose (2:5) with development for 15 min in the first direction with isopropanol–acetic acid–water (15:2:3), then after drying, full development in the same direction with the upper phase of butanol–0.15 *N* ammonia and finally at right angles with 1.5 *M* sodium phosphate buffer (pH 6.0). Wang et al. [116, 117] has applied two-dimensional chromatography on polyamide layers with benzene–acetic acid (4:1) first followed by carbon tetrachloride–acetic acid (4:1). Results with additional solvents were also listed.

Because of the photosensitivity of these derivatives it is advisable to carry out their preparation and chromatography in the absence of direct illumination.

Ratney [118] has diazotized the DNP derivatives on the thin layer by spraying with a toluene solution of nitrogen trioxide followed by toluene solution of phenol and diethylamine.

Phenylthiohydantoin (PTH) Amino Acids

These derivatives are formed when the reaction products of proteins or peptides with phenylisothiocyanate are degraded. They are very useful in the analysis of peptide structures and are also useful for separating amino acids from interfering material. In addition, the improved automated sequencing of amino acids has increased the need for a rapid method of identifying these compounds, for 10 to 25 cycles can be run with one derivative completed every 72 to 90 min. The technique of gas chromatography is used, but in the case of some derivatives ambiguities remain, thus requiring confirmation by an additional method. Brenner et al. [107] have investigated the separation of these derivatives in four different solvents on silica gel layers. Pataki [119], using this method, has reported the R_f values of 34 amino acid derivatives (Table 17.7). Pataki [120] has modified the method so that the PTH amino acids can be separated for the most part by running two two-dimensional chromatograms and a third single-dimensional chromatogram. One chromatogram is run in chloroform–methanol (9:1) followed by chloroform–formic acid (20:1) in the second direction. The second two-dimensional chromatogram is run first in chloroform followed by *n*-heptane–1,2-dichloroethane–formic acid–propionic acid (30:10:7:6) in the second direction. The one-dimensional chromatogram is run in chloroform–methanol–formic acid (35:15:1).

Cherbuliez et al. [121] have applied the separation of these derivatives to the investigation of the structure of peptides. They recommend the use of a silica deactivated by incorporation of 15% water.

Silica gel layers on plastic sheets have been used [122–124]. Solvents used on these systems include: heptane–propionic acid–dichlorocthane (58:17:25) and heptane–butanol–75% formic acid (50:30:9) [122]; the upper layer of butyl acetate–formamide–propionic acid–water (160:6:3:3) is useful for acidic amino acid PTH's, xylene–methanol (8:1) is useful for the identification of proline, and xylene–isopropanol (7:2) is the most useful general-purpose solvent of these three, all of which are used in unsaturated tanks on Eastman Chromogram sheet 6060 [123]. Walz and Reuterby [124], using heptane–dichlorethane–propionic acid (9:5:6) and xylene–methanol (8:1) combined with ultraviolet examination, iodine azide reagent, and 1.7% ninhydrin–collidine–acetic acid (15:2:5), were able to identify seven steps from an automated Edman degradation in 1 hr and have the residue identified.

Inglis and Nicholls [125, 126] used precoated (kieselgel 60F 254) aluminum plates which gave smaller and more discrete spots than silica gel

Table 17.7 $R_f \times 100$ of PTH Amino Acids on Silica Gel G. Development distance 18 cm. Values are averages of six determinations.[a]

	Solvent		
PTH Derivatives of	*Chloroform*	*Chloroform–Methanol* (9:1)	*Chloroform–Formic Acid* (20:1)
α-Aminobutyric acid	26	79	54
α-Aminocaprylic acid	44	84	67
α-Aminoisobutyric acid	27	80	56
Alanine	18	77	44
Arginine	0	1	0
Aspartic acid	0	2	16
Asparagine	0	34	9
Citrulline	0	34	8
Cysteic acid	0	0	0
Glutamic acid	1	5	18
Glutamine	0	40	11
Glycine	11	68	35
Histidine	1	40	1
Hydroxyproline	5	64	28
Isoleucine	39	83	62
Tyrosine	3	59	22
Leucine	39	84	63
Lysine	12	78	34
Methionine	34	81	54
Methylglutamic acid	23	82	50
Methionine sulfoxide	0	54	15
Methionine sulfone	2	59	17
Methylserine	1	51	18
Norleucine	40	83	62
Norvaline	34	81	57
Ornithine	7	72	30
Phenylalanine	30	81	54
Proline	60	89	70
Serine	1	43	10
Threonine	1	58	17
Tryptophan	14	71	41
Valine	33	81	58
Phenyl thiourea	12	65	32
Diphenyl thiourea	42	82	71

[a] From G. Pataki [119]; reproduced with permission of the author and H. R. Sauerlaender and Co.

layers. Separation between a number of the derivatives was improved and the color reactions for the nonpolar PTH amino acids were more sensitive. The solvent system was a modified Edman H system: ethylene chloride–acetic acid (60:7).

Inglis et al. [126a] prepared the PTH derivatives directly on the TLC layer by adding 1 μl of heptafluorobutyric acid to the amino acid spot and heating in an oven at 140°C for 10 min.

Polyamide layers have also been used for the separation of PTH derivatives [127–132]. Summers et al. [128] developed a technique for the subnanomolar separation and detection of these derivatives by incorporating butyl-PBD [2-(4-*tert*-butylphenyl)-5-(4-biphenyl)-1,3,4-oxadiazole] as a fluorescent agent in the first solvent of a two-dimensional system. By using double-faced polyamide sheets (5 × 5 cm) two samples could be run in two dimensions in less than 30 min. Kulbe [130] modified the solvent system for this technique using toluene–*n*-pentane–acetic acid (30:15:8) containing 250 mg of butyl-PBD per liter for the first-dimensional solvent and 25% acetic acid or 40% aqueous pyridine–acetic acid (9:1) for the second. The lower limits of detection at 254 nm is 0.05 to 0.2 nmole. By this procedure 10 to 20 samples per hour can be run compared to two or three by gas chromatography, including derivatives which must be modified before they can be run on gas chromatography. Kulbe [132] developed two solvent systems either of which may be used for separating the PTH or the MTH (methylthiohydantoin) derivatives of arginine, histidine, and cysteic acid. These are ethyl acetate–*n*-butanol–acetic acid (35:10:1) and ethyl acetate–*tert*-butanol–acetic acid (35:10:1). Each solvent contains the fluorescent indicator mentioned above.

1-Dimethylamino-5-naphthalenesulfonyl (DANS or DNS) Amino Acids

Seiler and Weichmann [133] have investigated the separation of the 1-dimethylamino-5-naphthalenesulfonyl (DANS) amino acids [134] on thin layers of silica gel. Using two-dimensional techniques with methyl acetate–isopropanol–concentrated ammonia (9:7:4) in the first direction and chloroform–methanol–acetic acid (15:5:1) or chloroform–ethyl acetate–methanol–acetic acid (30:50:20:1) in the second direction, they obtained almost complete separation with a mixture of 30 amino acid derivatives. Prior to the second development the plates are dried for 10 min at 100°C. Because of the size of the group added to the amino acid the method is very sensitive, and as little as 10^{-10} mole of acid can be detected on a two-dimensional chromatogram by means of the yellow fluorescence which is visible on the still moist chromatograms under ultraviolet light.

Cole et al. [135] used benzene–pyridine–acetic acid (40:10:1) in the first dimension on silica gel for 1.5 hr followed by *n*-butanol saturated with 0.2 *N* sodium hydroxide at right angles. To separate leucine and isoleucine derivatives the second development was with *n*-butanol–chloroform (3:97).

Stehelin and Duranton [136] applied stepwise development on silica gel layers divided into narrow strips. The solvents in order of use in an unsaturated tank were: toluene–pyridine–acetic acid (70:30:0.8), chloroform–*tert*-butanol–acetic acid (12:8:3), and toluene—2-chloroethanol–25% ammonia (30:50:2.2). If it was necessary to separate valine, leucine, and isoleucine completely, the plate was turned through 180°C and developed with chloroform–*tert*-butanol–acetic acid (60:40:0.2) for 1 hr.

Polyamide layers may also be used for the separation of these derivatives. Casola and Di Matteo [137] used 3% formic acid in the first direction for 5 min and with benzene–acetic acid (45:5) for the same time. In order to resolve DANS-aspartic, DANS-glutamic, and DANS-asparagine, as well as separate DANS-glutamine from DANS-OH and DANS-alanine from DANS-NH_2, a third run was made in the second dimension with ethyl acetate–methanol–acetic acid (20:1:1). Lee and Safille [138] used the same first two solvents by changing the proportions to (1.5%) and (4.5:1), respectively. A third solvent, benzene–acetic acid–ethyl acetate (22.5:5:1) was used after the first development to separate DANS-alanine and DANS-NH_2. The big advantage of this adsorbent is its speed compared with silica gel.

Munier et al. [139–141] have used thin-layer chromatography on cellulose as well as two-dimensional procedure with chromatography in the first direction, and electrophoresis at right angles. Arnott and Ward [142] have also used the same technique on cellulose. Electrophoresis on cellulose can be performed with a pH-4.4 buffer (0.4% pyridine and 0.8% acetic acid) at 1000 V keeping the temperature at 12 to 15°C. [143].

Deyl and Rosmus [144] used a multiple-layer system (silica gel, alumina, polyamide) to obtain three R_f values at one time. Solvents were chloroform–benzyl alcohol–acetic acid (70:30:3) and *n*-butanol–pyridine–acetic acid–water (15:10:3:12).

The R_f values of some DANS-amino acids are given in Table 17.8.

Mention can be made at this point of the micro technique of Schmer and Kreil [147] for detecting formyl and acetyl groups in proteins and peptides. The lyophylized sample of 1 to 2 mg is dissolved in 0.5 ml of 0.1 *M* hydrochloric acid and dried in vacuo. The residue is heated for 17 hr at 100°C in a sealed tube containing anhydrous hydrazine. Excess hydrazine is removed under reduced pressure, and the residue treated with 0.3 ml of 0.2 *M* sodium citrate buffer (pH 3.0) and a five- to ten-fold excess of dansyl chloride (wt/wt) in 0.3 ml of ethanol. This mixture is

Table 17.8 $R_{DANS-NH_2}$ Values × 100 of DANS Amino Acids in Various Solvents on Silica Gel

	Solvent[a]					
	Ref. 145		*Ref. 146*			
DANS Amino Acid	*A*	*B*	*C*	*D*	*E*	*F*
DANS-NH_2	100.0	100.0	100.0	100.0	100.0	100.0
DANS-ileu	90.5	83.5	92	100.0	100.0	89
DANS-leu	87.5	67.5	89	100.0	99	63
DANS-val	82.5	75.0	85	100	97	75
DANS-pro	78.5	51.5	79	93	96	50
DANS-phe	73.0	71.0	69	94	99	42
DANS-met	66.0	69.0	65	92	99	33
DANS-ala	62	56	61	89	96	29
Di-DANS-lys	55.0	82.0	48	93	96	33
Di-DANS-orn	49.0	79.0				
Di-DANS-tyr	47.0	49.0				
DANS-gly	44.0	46.0	38	65	93	12
DANS-try	41.0	64.5	34	87	95	25
DANS-thr	25.0	45.0	18	25	57	2
DANS-hypro	30.0	36.0				
DANS-glu	24.0	0.5	18	42	85	8
DANS-ser	19.0	36.0	15	20	52	2
DANS-met-SO	15.0	58.0				
DANS-metSO_2	14.0	57.0				
DANS-asp	8.0	0	6	2	60	0
DANS-asn	8.0	20.0	2	2	10	3
DANS-gln	8.0	29.0	3	7	14	0
DANS-orn	0	14.0				
ε-DANS-lys	0	20.0	2	8	5	4
DANS-cySO_3H	0	4.0				
DANS-OH	0	67.0				
Di-DANS-$(cys)_2$	5.0	9.0				
Di-DANS-his	4.0	42.0				
DANS-arg	0	29.0	0	0	0	0

[a] Solvents: *A* = toluene–pyridine–acetic acid (150:50:3.5); *B* = toluene–2-chloroethanol–25% ammonia (100:80:6.7); *C* = benzene–pyridine–acetic acid (40:10:1); *D* = chloroform–*tert*-amyl alcohol–acetic acid (70:30:3); *E* = chloroform–*tert*-amyl alcohol–formic acid (70:30:1); *F* = chloroform–*tert*-amyl alcohol–acetic acid (70:30:0.5).

incubated for 24 hr at 37°C (under these conditions the α-amino group reaction is negligible). After drying, the residue is dissolved in 0.3 ml of water and extracted three times with chloroform. The extract is concentrated and aliquots are chromatographed on silica gel G to detect the formyl- and acetyl-dansylhydrazines. Solvents are: *(a)* benzene–pyridine–acetic acid (8:2:0.5), *(b)* methyl acetate–isopropanol–ammonia (9:7:4), and *(c)* chloroform–*n*-butanol–acetic acid (6:3:1). The R_f values in the order formyl, acetyl are *(a)* 0.76, 0.70; *(b)* 0.82, 0.92; and *(c)* 0.75, 0.65. Two-dimensional separation may be applied with *(c)* in the first direction and *(b)* in the second or with *(a)* followed by *(c)*.

Carbobenzoxy (Cbo) and *tert*-Butyloxycarbonyl (BOC) Amino Acids

Ehrhardt and Cramer [148] have demonstrated the use of thin-layer chromatography in separating carbobenzoxy (Cbo) amino acids as well as the corresponding derivatives of peptides and peptide esters. These derivatives are very useful in the synthesis of peptides, and consequently their separation from each other and from unreacted components used in their preparation is very useful. For their separation various mixtures of *n*-butanol–acetic acid–5% ammonium hydroxide and *n*-butanol–acetic acid–water–pyridine or ammonium hydroxide can be used. With two-phase mixtures, both phases were placed in the developing tank; however, the silica gel layers were allowed to contact only the upper phase. Ehrhardt and Cramer [148] give a list of compounds with their R_f values (Table 17.9). For detecting the compounds, the plates are dried for 10 to 15 min at 120 to 150°C, followed by spraying the still hot plates with a ninhydrin solution consisting of 0.2% ninhydrin in 95% *n*-butanol with added 5% 2 *N* acetic acid. This will disclose the amino acids, peptides, and amino acid ester hydrochlorides. The plates are then heated again to 120 to 150°C and sprayed with a saturated solution of potassium dichromate in concentrated sulfuric acid. Dark green spots appear at the locations of the Cbo compounds and peptides or amino acids which contain a phenyl radical. (It is sometimes necessary to heat the plates on a hot plate after spraying with the oxidizing mixture.)

Schellenberg [149] has proposed the use of morin (T-168) as a reliable detection medium for *N*-protected amino acid an peptide derivatives on silica gel layers. Pataki [150] has used a modified chlorine-toluidine reaction (T-61) for the detection of carbobenzoxy amino acids because it is more sensitive than potassium dichromate in concentrated sulfuric acid.

Wang et al. [151] have used Amilan 1001 polyamide (Toyo Rayon Co.) to form a solvent-resistant layer on poly(ethylene terepthalate) film for separating these derivatives. The layers (available commercially from

Table 17.9 $R_f \times 100$ Values of Carbobenzoxy (Cbo) Amino Acids, Cbo Peptides, Cbo Peptide Esters, and Nonderivatized Materials on Silica Gel G[a]

	Solvent[b]			
Cbo Derivative	*A*	*B*	*C*	*D*
Cbo-Gly	81	76	74	64
Cbo-Digly	66	65	56	56
Cbo-Trigly	57	56	41	50
Cbo-Tetragly	51	43	26	45
Cbo-Digly ethyl ester	82	73	75	74
Cbo-Trigly ethyl ester	76	67	68	69
Cbo-DL-Ala-Gly ethyl ester	81	75	79	77
Cbo-DL-Ala	77	77	74	61
Cbo-DL-Diala	72	72	68	59
Cbo-Gly-DL-Ala	68	68	63	56
Cbo-DL-Ala-Gly	68	68	61	57
Cbo-Gly-DL-Ala-Gly	61	59	50	54
Cbo-DL-Phe-Gly ethyl ester	86	83	80	76
Cbo-Gly-DL-Phe ethyl ester	84	79	77	76
Cbo-Gly-Gly-DL-Phe ethyl ester	81	75	69	76
Cbo-Gly-DL-Phe-Gly ethyl ester	83	78	76	74
Cbo-DL-Ala-DL-Phe	78	78	72	61
Cbo-Gly-DL-Phe	72	76	69	65
Cbo-Gly-L-Ileu	75	73	74	67
Cbo-Gly-L-Leu	74	74	69	65
Cbo-Gly-Gly-L-Leu	71	69	63	65
Cbo-Gly-L-Glu	72	71	65	62
Cbo-DL-Phe	74	76	77	75
Gly-Gly	17	25	4	15
DL-Ala-Gly	21	30	9	22
Gly-L-Leu	43	44	24	41
Gly	22	29	10	21
DL-Ala	27	34	14	27
DL-Phe	45	42	31	46
HCl-DL-Phe ethyl ester	59	52	44	65
HCl-Gly-ethyl ester	43	42	24	53

[a] From E. Ehrhardt and F. Cramer [148]; reproduced with permission of the authors and the Elsevier Publishing Co.

[b] *A* = *n*-Butanol–acetone–acetic acid–5% ammonium hydroxide–water (4.5:1.5:1:1:2); *B* = *n*-butanol–acetic acid–5% ammonium hydroxide (5.5:3:1:5); *C* = *n*-butanol–acetic acid–5% ammonium hydroxide–water (6:1:1:2); *D* = *n*-butanol–acetic acid–pyridine–water (15:3:10:12); running time 2 hr for 8 to 10 cm.

Chen Chin Trading Co. and Gallard-Schlesinger) are faster than silica gel layers.

Schwyzer et al. [152] have adopted the use of *tert*-butyloxycarbony derivatives of amino acids and peptides for certain synthetic work because of the ease of removing the protective group without alteration of the parent compound (which sometimes happens in the removal of the carbobenzoxy and other protective groups). Silica gel and aluminum oxide layers were applied in chromatographing these derivatives with a number of solvents including *n*-propanol–ethyl acetate–water (7:1:2), *tert*-amyl alcohol–isopropanol–water (100:40:55), *n*-butanol–methyl ethyl ketone–dicyclohexylamine–water (10:10:2:5), *sec*-butanol–isopropanol–chloroacetic acid–water (70 ml:10 ml:3 g:40 ml), *sec*-butanol–isopropanol–5% sodium diethylbarbiturate–water (20:3:2:12), and chloroform–methanol–17% ammonium hydroxide (20:20:9).

The *tert*-butyloxycarbonyl amino acids give a negative ninhydrin test; however, if the derivatives on the thin-layer plate are heated in an oven at 125 to 130°C for 25 min and then sprayed while still hot with a 0.25% solution of ninhydrin in butanol, a positive test is obtained [153].

Dinitropyridyl (DNPyr) and Nitropyrimidyl (NPm) Amino Acids

Because of the noticeable destruction of the *N*-terminal dinitrophenyl (DNP) amino acid during the hydrolysis step in analyzing peptides and proteins, a number of other reagents have been proposed [154]. Bello and Signor [155] have separated the amino acid derivatives of two of these, namely, the dintropyridyl (DNPyr) and the nitropyrimidyl (NPm) derivatives on thin layers of silica gel. Quantitative recovery of these derivatives is readily accomplished by a 15 to 20 min hydrolysis of the terminal peptide bond with 6 *N* hydrochloric acid containing 30% formic acid. Six different solvent systems were investigated for separating those derivatives which could be extracted from the acidic hydrolysate. For two-dimensional work chloroform–methanol–acetic acid (95:5:1) was used in the first direction and *n*-propanol–30% ammonium hydroxide (7:3) in the second direction. The DNPyr-amino acids are readily visible in either daylight or ultraviolet light, and the NPm-amino acids can be detected by spraying with 1% potassium permanganate followed by 1 *N* hydrochloric acid.

Esters of Amino Acids

Mussini and Marcucci [156] have separated the *n*-butyl esters of 18 amino acids on silica gel G layers. Development was carried out with benzene–*n*-butanol (3:1), although the basic amino acids, lysine, histidine, and ornithine, did not move in this solvent. For these esters, *n*-

butanol–acetic acid–water (12:3:5) was used. Further work with the same system has separated the butyl derivatives of proline, dehydroproline, and hydroxyprolines [41]. At the same time the nitroso derivatives of these imino acids were separated using a solvent system composed of *n*-butanol–acetic acid–water (12:3:5). Myhill and Jackson [42] also chromatographed the nitroso derivatives of L-proline and L-hydroxyproline.

3 SEPARATION OF PROTEINS AND PEPTIDES

Chromatographic Separations

Mention has been made of the separation of some of these compounds by chromatographing derivatives in which active groups have been blocked; in addition, it should be noted that the direct separation of the compounds themselves has also been accomplished.

Feltkamp and Pfrommer [157] chromatographed a group of diastereomeric dipeptides on both cellulose and silica gel layers. Tripeptides have been chromatographed on silica gel layers with *n*-butanol–acetic acid–water (4:1:1 or 4:1:5) and *n*-butanol–formic acid–water (5:1:1) [158]. Numerous other solvent combinations including those used for amino acids may be used for peptides.

Schwyzer et al. [159, 160] have applied separations on thin layers of silica gel and of aluminum oxide for chromatographing synthetic corticotropic active polypeptides.

Heathcote et al. [161] used cellulose layers in two-dimensional separations with isopropanol–butanol–1 *N* hydrochloric acid (12:3:5) or *n*-butanol–butanone–water–0.88 sp gr ammonia (80:5:17:3) for the first dimension and 2-methyl-2-butanol–propanone–methanol–water–0.88 sp gr ammonia (10:4:2:1:3:1) in the second direction. The peptides were detected with test T-176. Burns and Turner [162] applied the technique developed for amino acids [89, see discussion under amino acids]. Electrophoresis was carried out on purified cellulose MN 300 at 15°C for 25 min at 50 V/cm. The buffer consisted of glacial acetic acid–98% formic acid–water (17:5:280, pH 2). The wicks were Miracloth (Calbiochem) strips washed in buffer to remove ninhydrin-positive substances and enclosed in a strip of washed dialysis tubing, slit and folded over so that the Miracloth did not come into direct contact with the layer. The plate was dried at 30°C (temperatures above this increased adsorption of the higher peptides) and then dipped in 1% acetic acid (see electrophoresis under amino acids) to consolidate the bands into spots. After again drying chromatography was carried out at right angles using several solvents: *n*-butanol–acetic acid–pyridine–water (5:1:4:4), isobutanol–pyridine–

water (7:7:6), isoamyl alcohol–pyridine–water (7:7:6), and *n*-butanol–acetic acid–water (5:1:4 upper phase). Multiple or stepwise development sometimes aided in the separation. Munier et al. [163, 164] also used a two-dimensional separation with low-voltage (10 V/cm) electrophoresis along a 40 × 20 cm cellulose layer and chromatography across the shorter dimension. Numerous solvents are listed.

Furlan and Beck [164a] compared the use of fluorescamine with ninhydrin as a detection agent in the peptide mapping on thin-layer cellulose plates. In the example used, a fragment "d" of fibrinogen, at least 35 spots could be detected with fluorescamine and only 20 with ninhydrin.

Ménard et al. [165] applied thin-layer chromatography in the separation and identification of two new basic peptides from *Zizyphus oenoplia*. Wieland and Georgopoulos [166] have applied two-dimensional development in the separation of peptides on thin layers of silica gel–gypsum and silica gel–starch. They also employed a two-dimensional system using electrophoresis in one direction followed by chromatography in the other direction. Simonianova and Rybak [167] have applied thin layers of phosphocellulose, and Hofmann [168] has applied hydroxyapatite layers to the separation of proteins. In the latter case, the layers were bound with an alcohol-soluble polyamide.

Gel filtration methods using Sephadex have been applied in thin-layer work to the separation of amino acids, peptides, and proteins. This is a very logical application since the Sephadex separates the compounds according to their molecular weights. Determann [169] has separated tyrosine, tyrosyl-leucyl-glycylglutamyl-phenylalanine, the condensation product Plastein, and bovin serum albumin on thin layers of Sephadex G 25 using 0.05 *M* ammonium hydroxide. In the preparation of these layers the Sephadex must not be overdried in order to avoid irregular development. The layers are dried only until the gel grains are visible. Wieland and Determann [170] applied a gradient elution method to the separation of two lactic dehydrogenases of bovine heart on layers of DEAE Sephadex.

Johansson and Rymo [171] separated an artificial mixture of bovine serum albumin, β-lactoglobulin, and α-lactalbumin on Sephadex G-75. The spots were visualized by staining in a saturated solution of Amido Black in methanol–acetic acid–water (8:1:1). After immersion in the dye bath for 1 hr, the excess dye was removed by repeated washing with methanol–acetic acid–water (70:15:15). The same authors [172] separated high-molecular-weight proteins on Sephadex G-100 and G-200. Morris [173] has also applied Sephadex G-100 and G-200 to the separation of proteins having a molecular weight of up to 180,000. In preparing the plates, the Sephadex gel was allowed to stand in the solvent for 48 hr in order to allow for the swelling of the gel; after the plates were coated they were placed in a horizontal position in a closed vessel containing the solvent for another 18 hr. After spotting the protein sample, the plates were then developed with 0.5 *M* sodium chloride in a continuous, descending fashion at an angle of 10 to 20°C with a pad of filter paper placed at the end to soak up the excess solvent as it left the plate. On completion of the run the layer was carefully covered by a sheet of Whatman #3 MN Filter paper, without trapping air bubbles beneath the paper, and the covered plate was transferred to an oven for drying at 80 to 90°C for 30 min. The proteins were located by staining with 0.2% Ponceau S in 10% acetic acid for 30 min with subsequent water washing to remove the excess dye; or for greater sensitivity either 1% Naphthalene Black 12B or 0.01% Nigrosine in methanol–water–acetic acid (5:4:1) followed by washing with the mixed solvent. This separation gave compact zones with little trailing. Cytochrome C, ovalbumin, and thyroglobulin were separated successfully on Sephadex G-100 and lysozyme; hemoglobin and gamma globulin were chromatographed on Sephadex G-200 (Table 17.10). Fasella et al. [174] have also investigated the separation of pep-

Table 17.10 R_{Hb}[b] × 100 Values of Protein on Sephadex G-100 and G-200 Using 0.5 *M* Sodium Chloride[a]

		R_{Hb}	
Protein	*Molecular Weight* × 10^{-3}	*Sephadex G-100*	*Sephadex G-200*
Cytochrome C	13.0	68	74
Ribonuclease	13.6	68	74
Lysozyme	14.5	65	70
Myoglobin	16.9	79	80
α-Chymotrypsin	22.5	87	87
Trypsin	23.8	83	86
Ovomucoid	27.0	94	103
Pepsin	35.0	99	104
Ovalbumin	45.0	103	104
Hemoglobin	68.0	100	100
Bovine serum albumin	65.0	114	122
Bovine γ-globulin	180.0	128	154
Thyroglobulin	650.0	133	183
Macroglobulins	1000	—	186

[a] From C. J. O. Morris [173]; reproduced with permission of the author and the Elsevier Publishing Co.

[b] $R_{Hb} = R_f$ protein/R_f hemoglobin.

tides and proteins on buffer-treated Sephadex. Two-dimensional work was carried out by chromatographing in one direction and then running an electrophoretic separation in the second direction. By running standards of known molecular weight a rough estimate of the molecular weight of the unknown components may be obtained.

For the detection of separated substances on Sephadex layers the Sephadex may be treated directly, as previously described, or a sheet of damp filter paper may be pressed against the layer and allowed to remain there for 40 to 50 min. The paper is then removed and dried and sprayed with detecting agents. Using this method, approximately 20% of the material is transferred from the Sephadex onto the paper [174].

Roberts [175] used a modification of the chlorination technique [176] for detecting proteins on Sephadex. The layers were dried at 37°C for 5 to 10 min and then exposed to chlorine for 10 to 15 min. After standing in a stream of air they were sprayed with a solution containing 20% ammonium sulfate and 5% sodium carbonate (wt/vol) to destroy excess chlorine. After 15 min they were sprayed with starch–iodide (1% each wt/vol) solution to obtain blue spots on a white background. Sensitivity was less than 1 μg of protein.

Miller et al. [177] dyed the proteins with Procion Brilliant Orange 2RS (Dylon) prior to gel filtration on Sephadex. The proteins (5 to 10 mg/ml) in pH-7.6 phosphate buffer was mixed with an equal volume of the dye (10 mg/ml) in the same buffer. After standing overnight the dye–protein mixture could be applied directly to the thin layer. Orange dextran (molecular weight 2×10^6) was used as a void volume marker [178]. The dye had no significant effect on either the molecular weight or the R_f value.

Radola [178a, 178b] has a good discussion of the various aspects of the gel filtration of proteins with some examples.

Separation by Electrophoresis and Isoelectric Focusing

The introduction of electrophoresis of these materials goes back to 1946 when Consden et al. [179] first demonstrated the use of thin-layer electrophoresis with 1.4-mm-thick layers of agar and of silica gel to separate amino acids and peptides. For the preparation of various layers and the techniques of operation, the reader is referred to Chapter V, Section 5.

In 1957, Wieme and Rabaey [180] introduced an ultramicro electrophoretic technique for the separation of proteins. This was accomplished on microscope slides covered to a depth of 1.5 to 2 mm by agar gel. By this technique amounts of protein of the order of 0.1 μg could be measured. Spotting of the sample is accomplished by introducing the protein solution into a small hole or linear groove in the gel, and if only a small fragment of tissue is available, it is possible to apply this directly to the gel without an intermediate extraction step. After the electrophoresis, which takes from 10 to 30 min, the plate is fixed for 30 min in a 5% acetic acid solution in 70% ethanol (vol/vol). The excess fixing solution is removed with a blotting paper after which the plate is carefully dried at 37°C. Staining and washing can then be carried out in the usual fashion to obtain the completed pherogram. As an example of the procedure, the authors cite the resolution of the proteins in the aqueous humor of the eye using 0.1 ml of 0.02% protein solution. Applying a modification of this technique, Wieme [181] has resolved normal human serum into at least nine fractions: one (or two) prealbumin, albumin, α_1-globulin, two α_2-globulins, three β-globulins, and a continuous spectrum of γ-globulins. With pathological human serum a greater number of fractions could be obtained. The modifications consisted of using a high voltage gradient (15 V/cm) so as to decrease the running time and diminish the effect of diffusion. The high voltage gradient in turn required a cooling system which was accomplished by immersing the plate in petroleum ether. In order to prevent a discontinuity at the junction with the electrode vessels, contact with the buffer solution in the electrode vessels was made by means of large agar blocks.

Wieme [182] has separated soluble proteins from human liver punctures without homogenization by electrophoresis in agar gel. Acharya et al. [183], Ramanathan [184], and Popadiuk [185] have separated serum proteins by agar-gel electrophoresis on microscope slides. Pette [186] obtained a separation of cerebral spinal fluid to a total of nine components in 60 min, using (microscope slide) agar gel electrophoresis.

Been and Rasch [186a] developed a vertical microsystem for discontinuous electrophoresis of proteins in polyacrylamide gels. The gels were cast in multiple layers of varying pore size. After electrophoresis at 150 V, the gels were stained for total protein with test T-181a. Glycoproteins were detected with the periodic acid–Schiff reaction (T-192) and lipoproteins with Sudan B (T-248a). Nonspecific esterases were detected by using Fast Blue RR as a coupler and α-naphthol acetate as a substrate. Excellent results were obtained using two different concentrations of gel (8 and 12%) to form steps of small-pore gel within a single sheet.

Ritschard [187] has developed a two-dimensional procedure for making peptide maps, which consists of a chromatographic separation in one direction followed by electrophoresis in the second direction. Silica gel G layers are used for the procedure; prior to the run a preliminary check is made to determine the best solvent for the separation of the enzyme digests of proteins. Eight solvent systems are listed for use in the technique. For neutral systems *n*-propanol or 96% ethanol with water (7 : 3) may be used; for basic solvents *n*-propanol, or 96% ethanol with 34% ammonium hydroxide in a ratio of 7:3, or chloroform–methanol–34% ammonium hydroxide (2:2:1) may be used. The acid solvents consist of *n*-propanol or 96% ethanol with water and acetic acid in the ratio of 7:2:1, or *n*-butanol–water–acetic acid in a 4:1:1 ratio. After the best solvent for the separation is chosen, the sample is chromatographed on a 200 × 200 mm thin-layer plate. Preferably, this is done in an ascending technique, but if the separation is not adequate the plate may be developed by the continuous technique of Brenner and Niederwieser [188] or in the modified apparatus described by Ritschard. The plate is then removed and after drying for 10 min at 100°C it is cooled and sprayed with the appropriate buffer for the electrophoretic separation which is run at 950 to 1000 V and 30 mA. With this high current density the plate must be adequately cooled (see Chapter V, Section 5). Prior to the electrophoretic run a check is made to determine the concentration of the sample and the time needed for the run. This is accomplished by spotting a series of varying amounts of the peptide mixture across the middle of a 200 × 200 mm thin-layer plate. After spraying with buffer solution the plate is placed in the electrophoresis apparatus and run for 30 min. This sample plate is then dried at 100°C and sprayed with ninhydrin reagent so that the optimum concentration and the optimum separation time can be determined. The buffer recommended for the electrophoresis consists of a mixture of 1 ml of pyridine in 10 ml of glacial acetic acid diluted to 500 ml, pH 3.5. As an example of this technique, a tryptic digest of myosin gave a peptide map of more than 60 spots.

After the separation of proteins by agar gel electrophoresis, van der Helm and Holster [189] determined the fractions by an elution method. An agar layer was placed on a plastic sheet, dried with filter papers, and then stained with a solution consisting of 0.5 g of Amido Black, 5 g of mercuric chloride, and 5 ml of acetic acid in 100 ml of aqueous solution. After treatment for 30 min, the film was washed with 5% acetic acid and dried. Following this, the protein bands were cut and eluted with 1 ml of Complexone (500 mg/liter) in 0.5 *N* sodium hydroxide.

Because of molecular sieve effects, electrophoresis in starch gels usually yields more fractions than the corresponding electrophoresis in agar gel. Thin-layer starch gel electrophoresis also has an advantage over the conventional starch gels [190]. In certain cases fractions can be detected on the thin-layer starch gel which cannot be detected in conventional starch gels. An additional advantage is the elimination of the need for slicing the gel after electrophoresis.

Thin-layer starch gel electrophoresis has been applied to the separation of proteins in serum [191–193]. For visualizing the protein fractions the starch gel layers are stained with Amido Black 10 B. Immunoelectrophoresis is a technique for detecting and identifying the zones separated by electrophoresis. Korngold [194] has applied a method whereby the zones obtained by starch-gel electrophoresis are allowed to diffuse into agar layers where they react with the antiserum that has been applied. To accomplish this, after the electrophoretic separation has been completed, the starch layer is loosened with a razor blade and carefully transferred to a 1.2% agar gel layer. The surface of the gel that was in contact with the glass is placed next to the agar taking care to exclude any air bubbles. One hour is allowed for the diffusion process to take place, after which narrow slits are cut into the two gels. These slits or trenches can be cut by using a cutter made from two razor blades. The antiserum (0.05 to 0.2 ml) is applied to the trenches and allowed to stand for 24 hr at room temperature. At the end of this time the starch gel may be removed so that shadow graphs can be made of the precipitin lines that have appeared in the agar gel.

Starch gel electrophoresis has also been used for the separation and identification of hemoglobin proteins [195–197]. Raymond and co-workers [198–200] have fostered the use of acrylamide gel for electrophoretic separations. As with starch gel, increased resolving power is possible with this medium as compared to agar gel. Concentrations of gels ranging from 3 to 25% have been investigated in the separation of serum samples. These same authors have described a two-dimensional technique for electrophoresis in which the electrophoresis is run in one direction at a given gel concentration, and then a strip of the gel containing the separated proteins is imbedded in a second layer of gel of a different concentration, and the electrophoresis is repeated in the second dimension. Espinosa [201] applied the samc principle using agar gel in one direction and starch gel in the second direction.

In the discussion of amino acids in this chapter, some mention has been made of the use of Sephadex for the separation of proteins. Additional work along these lines has been reported by Johansson aud Rymo [202] and Dose and Krause [203]. One of the advantages in using Sephadex is the fact that biological preparations may be separated without the loss of activity which occurs when separated on paper or starch layers.

Isoelectric focusing has been an important addition to the techniques for separating proteins, and many references relating to its use with proteins have been cited in Chapter V, Section 5. Righetti and Drysdale [204] in their general review on isoelectric focusing in gels cite further examples in thin-layer work especially under two-dimensional procedures.

Radola [205] examined various granulated gels for thin-layer isoelectric focusing with a series of proteins and found Sephadex G-75 Superfine gave the best results. Zones were detected by making a paper print and staining the print with one or more of the dyes used in electrophoresis work.

Polyacrylamide gel is the most widely used media for isoelectric focusing of proteins, and the techniques for preparation of the layers and carrying out the separations have been described in Chapter V, Section 5. Only a few of the numerous examples of its use can be cited here: bacterial proteins and Australia antigens [206], soluble proteins in eye lens extracts [207], L-amino oxidase [208], insulin [209], and glycoprotein hormones [210].

Isoelectric focusing in thin-layers has a number of advantages over other forms of this technique [211]: it provides ease of sample application, uses simple apparatus, conserves expensive carrier ampholytes, permits simultaneous separation of several mixtures under identical conditions, and still retains high resolving power. The acrylamide gels may also be stained and subjected to densitometric analysis and finally dried for storage.

Various combinations of techniques can be used in two-dimensional techniques for proteins and peptides: gradient gel electrophoresis with isoelectric focusing [212], gel filtration with isoelectric focusing [213], gel filtration with electrophoresis [214], two-dimensional electrophoresis with different buffer systems [215–218], chromatography in two dimensions with different solvents [219–221], and chromatography combined with either electrophoresis or isoelectric focusing. Wright et al. [222] evaluated one- and two-dimensional procedures for the separation of complex protein mixtures by counting the number of bands separated and obtained more bands by using a two-dimensional acrylamide gel electrophoresis than by discontinuous or continuous-gradient acrylamide gel electrophoresis, gel isoelectric focusing, or gel isoelectric focusing followed by continuous-gradient gel electrophoresis.

4 QUANTITATIVE ANALYSIS

In Situ Densitometry

Direct measurements of amino acids have been made on thin layers of cellulose [223] by scanning at 523 nm after spraying with ninhydrin. The relative standard deviation was 10%, with 0.05 μg the detection limit. Reflectance measurements have been made at 490 nm [224–226] on cellulose layers using test T-176 as a detection agent and at 620 nm for proline and hydroxyproline using test T-152. Sensitivity was 0.5 nmole for most amino acids. Urea was determined in an analogous manner by using Ehrlich's reagent (T-90) [227]. Frodyma and Frei [228, 229] have determined amino acids by reflectance measurements by removing the silica gel spots from the layer and packing the material in cells. In order to eliminate one drying step the ninhydrin reagent was dissolved in the developing solvent.

PTH-amino acids have been quantitated on silica gel layers by reflectance measurements at 269 nm [230]. As little as 0.5 nmole could be determined.

Proteins on polyacrylamide gels have been determined densitometrically after staining with Coomassie Blue [231], Fast Green [232], or Amido Black [233].

In Situ Fluorescent and Quenching Measurements

The fluorescence of DANS amino acids under ultraviolet light has been utilized for their determination. With separations on silica gel this method has a sensitivity of 10^{-11} moles [234–237]. Measurement was made at 490 to 530 nm with excitation at 313 nm with a reproducibility of 3 to 5%. Since the yields of these derivatives vary with the conditions of preparation, they must be carefully controlled. These compounds are also sensitive to light and Pouchan and Passeron [238] made a detailed study of this effect. To obtain reproducible results the exposure to the scanning light should be carefully standardized, and the readings of the unknown and the standard spots should follow in close succession. Other factors that must be kept constant are: position of the scanner, time between drying and scanning, volume of the sample, length of solvent travel, and layer thickness [239, 240]. The moisture content of the layer also has an influence on the fluorescence [241], but this effect can be minimized by spraying with triethanolamine–isopropanol (1:4) [241, 242]. Measurements were more sensitive on polyamide layers, but were subject to the quality of the polyamide used.

The DNP amino acids may also be determined by in situ fluorimetry [239, 243].

The introduction of fluorescamine as a spray reagent for thin-layer chromatography (T-128) has provided another method for in situ fluorescent determination of amino acids [244]. With this reagent determinations as low as 10 ng may be obtained on silica gel layers [245, 246]. Ragland et al. [246a] labeled proteins with fluorescamine for separation by disc gel electrophoresis with subsequent fluorescent scanning. As little as 6 ng of myoglobin could be detected.

Fluorescent quenching has been used for the determination of the DNP and PTH amino acids [239]. Relative standard deviations ranged from 5 to 7% for measurements on a single plate. If spots from different plates were compared the deviations were greater.

Spectrophotometric Measurements After Elution

For spectrophotometric measurement of amino acids eluted from the thin layers, the layers are first sprayed with ninhydrin reagent and the eluted colored reaction compound is measured at 570 nm [89, 247]. Ruestow and Hock [248] developed their chromatograms with solvents containing 1,3-dimethylalloxan. The red spots made visible by heating the layers at 70°C for 90 min are more stable than the colors produced with ninhydrin. The colors can be extracted with 30% water in acetone and measured at 530 nm.

Various amino acid derivatives have been eluted and measured spectrophotometrically: the PTH amino acids [249], the trinitrophenyl amino acids at 340 nm [250], the 3-(nitro-2-pyridyl) amino acids at 420 nm, the 5-(nitro-2-pyridyl) amino acids at 370 nm [251], and the DNP amino acids at 360 nm [252, 253]. The fluorescence of the extract of the DNP amino acids has also been measured quantitatively at 510 nm with excitation at 340 nm [254].

Radioactive Determinations

Naturally radioactive tracers can be used for the quantitative evaluation of amino acids and their derivatives, and a few examples can be cited here. By using ^{14}C-labeled dansyl chloride and separations on polyamide layers [255] autoradiography or scintillation counting may be used to determine these derivatives in the 10^{-14} *M* range, and by using ^{3}H-labeled dansyl chloride the sensitivity may be extended by one order of magnitude [256]. Drawert et al. [257] measured ^{14}C-labeled dinitrophenyl amino acids by impregnating the chromatograms with a thin film of collodion, removing the film and measuring the spots. Iodoamino acids labeled with ^{131}I have been counted on layers [258].

Miscellaneous Quantitative Methods

Spot area measurements have been used with an accuracy of ±4 to ±6% [259–261] and Sen et al. [262] have compared the relative color intensities of the unknown with varying amounts of standard.

References

1. A. Niederwieser, "Thin-Layer Chromatography of Amino Acids and Derivatives," in *Methods of Enzymology,* C. H. W. Hirs, Ed., **25B,** 1972, p. 60.
2. G. Pataki, *Techniques of Thin-Layer Chromatography in Amino Acid and Peptide Chemistry,* Walter de Gruyter, Berlin, 1968.
3. R. L. Bieleski and N. A. Turner, *Anal. Biochem.,* **17,** 278 (1966).

4. M. R. Stevens, *Anal. Chem.*, **45,** 1543 (1973).
5. J. Dayman, W. F. Coulson, I. Smith, M. J. Smith, and J. B. Jepson, *Clin. Chim. Acta,* **40,** 335 (1972).
6. R. Frentz, *Ann. Biol. Clin.*, **23,** 1145 (1965).
7. C. Nicot, R.-I. Cheftel, and J. Moretti, *J. Chromatogr.*, **31,** 565 (1967).
8. W. Lazarus, *J. Chromatogr.*, **87,** 169 (1973).
9. M. N. Copley and E. V. Truter, *J. Chromatogr.*, **45,** 480 (1969).
10. A. Dréze, S. Moore, and E. Bigwood, *Anal. Chim. Acta,* **11,** 554 (1954).
11. F. J. Sowden, *Soil Sci.*, **107,** 364 (1969).
12. S. F. Contractor and P. Jomain, *Clin. Chim. Acta,* **14,** 535 (1966).
13. R. L. Munier and A.-M. Drapier, *Chromatographia,* **2,** 340 (1969).
14. M. W. Neal and J. R. Florini, *Anal. Biochem.*, **55,** 328 (1973).
15. A. Niederwieser, *J. Chromatogr.*, **61,** 81 (1971).
16. *Ibid.*, **54,** 215 (1971).
17. C. Rollins, L. Jensen, and A. N. Schwartz, *Anal. Chem.*, **34,** 711 (1962).
18. J. G. Heathcote, D. M. Davies, and C. Haworth, *Clin. Chim. Acta,* **32,** 457 (1971).
19. P. Boulanger and G. Biserte, *Bull. Soc. Chim. Biol.*, **31,** 696 (1949).
20. T. Z. Nowakowski and M. Byers, *J. Sci. Food Agric.*, **23,** 1313 (1972).
21. G. E. Pollock and A. K. Miyamoto, *J. Agric. Food Chem.*, **19,** 104 (1971).
22. J. F. Thompson, C. J. Morris, and R. K. Gering, *Anal. Chem.*, **31,** 1028 (1959).
22a. K. B. Jacobson, *Anal. Biochem.*, **38,** 555 (1970).
23. M. Brenner and A. Niederwieser, *Experientia,* **16,** 378 (1960).
24. A. R. Fahmy, A. Niederwieser, G. Pataki, and M. Brenner, *Helv. Chim. Acta,* **44,** 2022 (1961).
25. E. Bancher, H. Scherz, and V. Prey, *Mikrochim. Acta,* **1963,** 712.
26. K. Chmel, *Collect. Czech. Chem. Commun.*, **37,** 3034 (1972), also *J. Chromatogr.*, **84,** D105 (1973).
27. E. Kraas, E. Stark, F. S. Tjoeng, E. Breitmaier, and G. Jung, *Chem. Ber.*, **108,** 1111 (1975).
28. G. van Kerckhoven, V. Blaton, D. Vandamme, and H. Peeters, *J. Chromatogr.*, **100,** 215 (1974).
29. E. J. Shellard and G. H. Jolliffe, *J. Chromatogr.*, **31,** 82 (1967).
30. S. R. Mardashev, L. A. Semina, V. N. Prozorovskii, and A. M. Sokhina, *Biokhimiya,* **32,** 761 (1967).
31. S. Voigt, M. Solle, and K. Konitzer, *J. Chromatogr.*, **17,** 180 (1965).
32. J. S. Cohen and I. Putter, *Biochim. Biophys. Acta,* **222,** 515 (1970).
33. G. Pataki, *J. Chromatogr.*, **17,** 580 (1965).
34. J. Opienska-Blauth, H. Kraczkowski, and H. Brzuszkiewicz, "The Adaptation of the Technique of Thin-Layer Chromatography to Aminoaciduria Investigation," in *Thin-Layer Chromatography,* G. B. Marini-Bettòlo, Ed., Elsevier, Amsterdam, 1964, p. 165.
35. A. Carisano, *J. Chromatogr.*, **13,** 83 (1964).
36. F. Pocchiari, L. Tentori, and G. Vivaldi, *Sci. Rept. Ist. Super. Sanita,* **2,** 188 (1962).
37. T. Rokkones, *Scand. J. Clin. Lab. Invest.*, **16,** 149 (1964).

38. H. v. Euler, H. Hasselquist, and I. Limnell, *Ark. Kemi,* **21,** 259 (1963); through *Chem. Abstr.,* **59,** 15706 (1963).
39. R. L. Squibb, *Nature,* **199,** 1216 (1963).
40. *Ibid.,* **198,** 317 (1963).
41. F. Marcucci and E. Mussini, *J. Chromatogr.,* **18,** 431 (1965).
42. D. Myhill and D. S. Jackson, *Anal. Biochem.,* **6,** 193 (1963).
43. S. Voigt, M. Solle, and K. Konitzer, *J. Chromatogr.,* **17,** 180 (1965).
44. H. Seydel and R. Voigt, *Pharmazie,* **24,** 531 (1969).
45. E. Ploechl, *Clin. Chim. Acta,* **21,** 271 (1968).
46. R. W. Frei, I. T. Fukui, V. T. Lieu, and M. M. Frodyma, *Chimia (Aarau),* **20,** 23 (1966).
47. K. Teichert, E. Mutschler, and H. Rochelmeyer, *Dtsch. Apoth.-Ztg.,* **100,** 283 (1960).
48. P. Wollenweber, *J. Chromatogr.,* **9,** 369 (1962).
49. T. Dittmann, *Z. Klin. Chem.,* **1,** 190 (1963).
50. L. Hoerhammer, H. Wagner, and F. Kilger, *Dtsch. Apoth.,* **15,** 1 (1963).
51. E. Moczar, L. Robert, and M. Moczar, *Eur. J. Biochem.,* **6,** 213 (1968).
52. V. R. Villaneuva and M. Barbier, *Bull. Soc. Chim. Fr.,* **10,** 3992 (1967).
53. D. Chiari, M. Roehr, and G. Widtmann, *Mikrochim. Acta,* **1965,** 669, also *J. Chromatogr.,* **23,** D3 (1966).
53a. P. F. Lofts, S. J. Purdy, and E. V. Truter, *Lab. Pract.,* **18,** 1167 (1969).
54. E. von Arx and R. Neher, *J. Chromatogr.,* **12,** 329 (1963).
55. C. Haworth and J. G. Heathcote, *J. Chromatogr.,* **41,** 380 (1969).
56. J. G. Heathcote, R. J. Washington, C. Haworth, and S. Bell, *J. Chromatogr.,* **51,** 267 (1970).
56a. J. G. Heathcote and S. J. Al-Alawi, *J. Chromatogr.,* **129,** 211 (1976).
57. I. Sankoff and T. L. Gourkes, *Can. J. Biochem. Physiol.,* **41,** 1381 (1963).
58. S. K. Wadman, H. F. de Jong, and P. K. de Bree, *Clin. Chim. Acta,* **25,** 87 (1969).
59. G. Ogner, *Acta Chem. Scand.,* **23,** 2185 (1969).
60. J. Dittmann, *Z. Klin. Chem.,* **1,** 190 (1963).
61. Ibid., **3,** 59 (1965).
62. E. Bujard and J. Mauron, *J. Chromatogr.,* **21,** 19 (1966).
63. H. Stuebchen-Kirchner, *J. Chromatogr.,* **64,** 103 (1972).
63a. T. Wieland and A. Buku, *Anal. Biochem.,* **26,** 378 (1968).
64. P. de la Llosa, C. Tertrin, and M. Jutisz, *J. Chromatogr.,* **14,** 136 (1964).
65. L. Lepri, P. G. Desideri, and V. Coas, *J. Chromatogr.,* **88,** 331 (1974).
66. R. Vercaemst, V. Blaton, and H. Peeters, *J. Chromatogr.,* **43,** 132 (1969).
67. E. Tyihák, S. Ferenczi, I. Hazai, S. Zoltán, and A. Patthy, *J. Chromatogr.,* **102,** 257 (1974).
68. I. Hazai, S. Zoltán, J. Salát, S. Ferenczi, and T. Dévényi, *J. Chromatogr.,* **102,** 245 (1974).
69. D. Cozzi, P. G. Desideri, L. Lepri, and V. Coas, *J. Chromatogr.,* **40,** 138 (1969).
70. L. P. Rasteikiene and T. A. Pranskiene, *Liet. TSR Mokslu Akad., Darbai, Ser. B,* **1963,** 5; through *Chem. Abstr.,* **60,** 6217 (1964).
71. A. Affonso, *J. Chromatogr.,* **22,** 452 (1966).

72. S. M. Petrović and S. E. Petrović, *J. Chromatogr.*, **21,** 313 (1966).
73. N. A. Turner and R. J. Redgwell, *J. Chromatogr.*, **21,** 129 (1966).
74. G. Molnár and F. Sztarickai, *Kiserl. Orvostud.*, **19,** 534 (1967); through *Chem. Abstr.*, **68,** 3657p (1968).
75. S. Pal and O. Takács, *Kiserl. Orvostud.*, **20,** 360 (1968); through *Anal. Abstr.*, **17,** 1604 (1969).
76. E. Zappi, "Identification of Circulating Iodoamino Acids by Thin-Layer Chromatography," in *Progress in Thin-Layer Chromatography and Related Methods,* Vol. I, A Niederwieser and G. Pataki, Eds., Ann Arbor Science, Ann Arbor, Mich., 1970, p. 147.
77. H. Schorn and C. Winkler, *J. Chromatogr.*, **18,** 69 (1965).
78. G. Schneider and C. Schneider, *Z. Physiol. Chem.*, **332,** 316 (1963).
79. R. H. Mandl and R. J. Block, *Arch. Biochem. Biophys.*, **81,** 25 (1959).
80. E. Stahl and J. Pfeifle, *Z. Anal. Chem.*, **200,** 377 (1964).
81. J. A. Berger, G. Meyniel, P. Blanquet, and J. Petit, *Compt. Rend.*, **257,** 1534 (1963).
82. A. Massaglia and U. Rosa, *J. Chromatogr.*, **14,** 516 (1964).
83. E. Zappi, M. Schmidt, and F. Prange, *J. Chromatogr.*, **43,** 543 (1969).
84. D. Hollingsworth, M. Dillard, and P. K. Bondy, *J. Lab. Clin. Med.*, **62,** 346 (1963).
85. S. J. Patterson and R. L. Clements, *Analyst (London),* **89,** 328 (1964).
86. R. P. Ouellette and J. F. Balcius, *J. Chromatogr.*, **24,** 465 (1966).
87. S. Hamada and S. H. Ingbar, *J. Chromatogr.*, **61,** 352 (1971).
88. C. G. Honegger, *Helv. Chim. Acta,* **44,** 173 (1961).
89. R. L. Bieleski and N. A. Turner, *Anal. Biochem.*, **17,** 278 (1966).
89a. R. L. Munier, A. Peigner, and C. Thommegay, *Chromatographia,* **3,** 205 (1970).
90. Ch. Montant and J. M. Touze-Soulet, *Bull. Soc. Chim. Biol.*, **42,** 161 (1960).
91. G. Pastuska and H. Trinks, *Chem.-Ztg.*, **86,** 135 (1962).
92. J. Chudzik and A. Klein, *J. Chromatogr.*, **36,** 262 (1968).
93. J. Jentsch, *Z. Naturforsch.*, **24b,** 264 (1969).
94. F. Sanger, *Biochem. J.*, **39,** 507 (1945).
95. A. L. Levy and D. Chung, *J. Am. Chem. Soc.*, **77,** 2899 (1955).
96. S. Walz, A. R. Fahmy, G. Pataki, A. Niederwieser, and M. Brenner, *Experientia,* **19,** 213 (1963).
97. K. R. Rao and H. R. Sober, *J. Am. Chem. Soc.*, **76,** 1328 (1954).
98. Z. Deyl and J. Rosmus, *J. Chromatogr.*, **20,** 514 (1965).
99. N. Seiler and J. Wiechmann, *Experientia,* **20,** 559 (1964).
100. W. R. Gray and B. S. Hartley, *Biochem. J.*, **89,** 60P (1963).
101. *Ibid.*, p. 379.
102. P. Edman, *Acta Chem. Scand.*, **4,** 277, 283 (1950).
103. *Ibid.*, **7,** 700 (1953).
104. J. Sjoequist, *Ark. Kemi,* **11,** 129 (1959).
105. J. Sjoequist, *Biochim. Biophys. Acta,* **41,** 20 (1960).
106. E. Cherbuliez, J. Marszalek, and J. Rabinowitz, *Helv. Chim. Acta,* **46,** 1445 (1963).
106a. J. Rosmus and Z. Deyl, *Chromatogr. Rev.*, **13,** 163 (1971).

106b. J. Rosmus and Z. Deyl, *J. Chromatogr.*, **70**, 221 (1972).
106c. Z. Deyl, *J. Chromatogr.*, **127**, 91 (1976).
107. M. Brenner, A. Niederwieser, and G. Pataki, *Experientia*, **17**, 145 (1961).
108. G. Biserte and R. Osteux, *Bull. Soc. Chim. Biol.*, **33**, 50 (1951).
109. F. Drawert, O. Bachmann, and K.-H. Reuther, *J. Chromatogr.*, **9**, 376 (1962).
110. J. Blass, *Chromatographia*, **2**, 178 (1969).
111. D. Walz, A. R. Fahmy, G. Pataki, A. Niederwieser, and M. Brenner, *Experientia*, **19**, 213 (1963).
112. G. Pataki, *J. Chromatogr.*, **16**, 541 (1964).
113. G. Pataki, J. Borko, H. C. Curtius, and F. Tancredi, *Chromatographia*, **1**, 406 (1968).
114. R. L. Munier and G. Sarrazin, *J. Chromatogr.*, **22**, 347 (1966).
115. W. D. Grant and A. J. Wicken, *J. Chromatogr.*, **47**, 124 (1970).
116. K.-T. Wang, J. M. K. Huang, and I. S. Y. Wang, *J. Chromatogr.*, **22**, 362 (1966).
117. K.-T. Wang and I. S. Y. Wang, *J. Chromatogr.*, **24**, 460 (1966).
118. R. S. Ratney, *J. Chromatogr.*, **11**, 111 (1963).
119. G. Pataki, *Chimia (Aarau)*, **18**, 23 (1964).
120. G. P. Pataki, dissertation, Basil University, 1962.
121. E. Cherbuliez, B. Baehler, and J. Rabinowtiz, *Helv. Chim. Acta*, **47**, 1350 (1964).
122. J. S. Jeppsson and J. Sjoequist, *Anal. Biochem.*, **18**, 264 (1967).
123. T. Inagami and K. Murakami, *Anal. Biochem.*, **47**, 501 (1972).
124. D. A. Walz and J. Reuterby, *J. Chromatogr.*, **104**, 180 (1975).
125. A. S. Inglis and P. W. Nicholls, *J. Chromatogr.*, **79**, 344 (1973).
126. *Ibid.*, **97**, 289 (1974).
126a. A. S. Inglis, P. W. Nicholls, and P. McK. Strike, *J. Chromatogr.*, **107**, 73 (1975).
127. K.-T. Wang, I. S. Y. Wang, A. L. Lin, and C.-S. Wang, *J. Chromatogr.*, **26**, 323 (1967).
128. M. R. Summers, G. W. Smythers, and S. Oroszlan, *Anal. Biochem.*, **53**, 624 (1973).
129. K. D. Kulbe, *Anal. Biochem.*, **44**, 548 (1971).
130. *Ibid.*, **59**, 564 (1974).
131. K. D. Kulbe and Y. M. Nogueira-Hattesohl, *Anal. Biochem.*, **63**, 624 (1975).
132. K. D. Kulbe, *J. Chromatogr.*, **115**, 629 (1975).
133. N. Seiler and J. Weichmann, *Experientia*, **20**, 559 (1964).
134. W. R. Gray and B. S. Hartley, *Biochem. J.*, **89**, 59P (1963).
135. M. Cole, J. C. Fletcher, and A. Robson, *J. Chromatogr.*, **20**, 616 (1965).
136. D. Stehelin and H. Duranton, *J. Chromatogr.*, **43**, 93 (1969).
137. L. Casola and G. Di Matteo, *Anal. Biochem.*, **49**, 416 (1972).
138. M.-L. Lee and A. Safille, *J. Chromatogr.*, **116**, 462 (1976).
139. R. L. Munier, C. Thommegay, and A. M. Drapier, *Chromatographia*, **1**, 95 (1968).
140. R. L. Munier and A. M. Drapier, *Chromatographia*, **5**, 306 (1972).

141. R. L. Munier, *Chromatographia,* **1,** 95 (1968).
142. M. S. Arnott and D. N. Ward, *Anal. Biochem.,* **21,** 50 (1967).
143. B. P. Sloan, *J. Chromatogr.,* **42,** 426 (1969).
144. Z. Deyl and J. Rosmus, *J. Chromatogr.,* **67,** 368 (1972).
145. J. O. Zanetta, G. Vincendon, P. Mandel, and G. Gombos, *J. Chromatogr.,* **51,** 441 (1970).
146. D. Morse and B. L. Horecker, *Anal. Biochem.,* **14,** 429 (1966).
147. G. Schmer and G. Kreil, *Anal. Biochem.,* **29,** 186 (1969).
148. E. Ehrhardt and F. Cramer, *J. Chromatogr.,* **7,** 405 (1962).
149. P. Schellenberg, *Angew. Chem. Int. Ed. Engl.,* **1,** 114 (1962).
150. G. Pataki, *J. Chromatogr.,* **16,** 553 (1964).
151. K.-T. Wang, K.-Y. Chen, and B. Weinstein, *J. Chromatogr.,* **32,** 591 (1968).
152. R. Schwyzer, B. Iselin, H. Kappeler, B. Riniker, W. Rittel, and H. Zuber, *Helv. Chim. Acta,* **46,** 1975 (1963).
153. Y. Wolman and Y. S. Klausner, *J. Chromatogr.,* **24,** 277 (1966).
154. A. Signor, E. Scoffane, L. Biondi, and S. Bezzi, *Gazz. Chim. Ital.,* **93,** 65 (1963).
155. C. di Bello and A. Signor, *J. Chromatogr.,* **17,** 506 (1965).
156. E. Mussini and F. Marcucci, *J. Chromatogr.,* **17,** 576 (1965).
157. H. Feltkamp and H. Pfrommer, *J. Chromatogr.,* **18,** 403 (1965).
158. B. Brtnick, A. Trka, and M. Zaoral, *Collect. Czech. Chem. Commun.,* **40,** 179 (1975); through *Chem. Abstr.,* **82,** 171425q (1975).
159. R. Schwyzer, B. Riniker, and H. Kappeler, *Helv. Chim. Acta,* **46,** 1541 (1963).
160. R. Schwyzer and P. Sieber, *Nature,* **199,** 172 (1963).
161. J. G. Heathcote, R. J. Washington, and B. J. Keogh, *J. Chromatogr.,* **104,** 141 (1975).
162. D. J. W. Burns and N. A. Turner, *J. Chromatogr.,* **30,** 469 (1967).
163. R. L. Munier, B. Faivre, and A. Lebreau, *Chromatographia,* **6,** 14 (1973).
164. *Ibid.,* p. 71.
164a. M. Furlan and E. A. Beck, *J. Chromatogr.,* **101,** 244 (1974).
165. E. L. Ménard, J. M. Mueller, A. F. Thomas, S. S. Bhadnagar, and N. J. Dastoor, *Helv. Chim. Acta,* **46,** 1801 (1963).
166. T. Wieland and D. Georgopoulos, *Biochem. Z.,* **340,** 476 (1964).
167. E. Simonianova and M. Rybak, *Biochem. Biophys. Acta,* **92,** 194 (1964).
168. A. F. Hofmann, *Biochim. Biophys. Acta,* **60,** 458 (1962).
169. H. Determann, *Experientia,* **18,** 430 (1962).
170. T. Wieland and H. Determann, *Experientia,* **18,** 431 (1962).
171. B. G. Johansson and L. Rymo, *Acta Chem. Scand.,* **16,** 2067 (1962).
172. *Ibid.,* **18,** 217 (1964).
173. C. J. O. R. Morris, *J. Chromatogr.,* **16,** 167 (1964).
174. P. Fasella, A. Giartosio, and C. Turano, "Applications of Thin-Layer Chromatography on Sephadex to the Study of Proteins," in *Thin-Layer Chromatography,* G. B. Marini-Bettòlo, Ed., Elsevier, Amsterdam, 1964, p. 205.
175. G. P. Roberts, *J. Chromatogr.,* **22,** 90 (1966).
176. H. M. Rydon and P. W. G. Smith, *Nature,* **169,** 922 (1952).

177. J. N. Miller, O. Erinle, J. M. Roberts, and C. Thirkettle, *J. Chromatogr.*, **105,** 317 (1975).
178. J. N. Miller, *J. Chromatogr.*, **65,** 355 (1972).
178a. B. J. Radola, *J. Chromatogr.*, **38,** 61 (1968).
178b. *Ibid.*, p. 78.
179. R. Consden, A. H. Gordon, and A. J. P. Martin, *Biochem. J.*, **40,** 33 (1946).
180. R. J. Wieme and M. Rabaey, *Naturwissenschaften,* **44,** 112 (1957).
181. R. J. Wieme, *Clin. Chim. Acta,* **4,** 317 (1959).
182. R. J. Wieme, *Behringwerk-Mitt.,* **34,** 27 (1958).
183. U. S. V. Acharya, M. Swaminathan, A. Sreenivasan, and V. Subrahmanyan, *Indian J. Med. Res.*, **52,** 224 (1964).
184. A. N. Ramanathan, *Antiseptic (Madras, India),* **60,** 1017 (1963).
185. L. Popadiuk, *Arch. Immunol. Ter. Dosw.*, **9,** 139 (1961).
186. D. Pette, *Klin. Wochenschr.*, **36,** 1106 (1958).
186a. A. C. Been and E. M. Rasch, *J. Histochem. Cytochem.*, **20,** 368 (1972).
187. W. J. Ritschard, *J. Chromatogr.*, **16,** 327 (1964).
188. M. Brenner and A. Niederwieser, *Experientia,* **17,** 237 (1961).
189. H. J. van der Helm and M. G. Holster, *Clin. Chim. Acta,* **10,** 483 (1964).
190. E. W. Baur, private communication.
191. J. Groulade and C. Ollivier, *Ann. Biol. Clin. (Paris),* **18,** 595 (1960).
192. J. H. Daams, *J. Chromatogr.*, **10,** 450 (1963).
193. H. Mouray, J. Moretti, and J. M. Fine, *Bull. Soc. Chim. Biol.*, **43,** 993 (1961).
194. L. Korngold, *Anal. Biochem.*, **6,** 47 (1963).
195. E. W. Baur, *J. Lab. Clin. Med.*, **61,** 166 (1963).
196. P. Berkeš-Tomašević, J. Rosić, and I. Berkeš, *Acta Pharm. Jugosl.*, **13,** 69 (1963).
197. C. L. Marsh, C. R. Jolliff, and L. C. Payne, *Tech. Bull. Regist. Med. Technol.*, **34,** 1 (1964).
198. S. Raymond, *Ann. N.Y. Acad Sci.*, **121,** 350 (1964).
199. S. Raymond and B. Aurell, *Science,* **138,** 152 (1962).
200. S. Raymond and M. Nakamichi, *Anal. Biochem.*, **7,** 225 (1964).
201. E. Espinosa, *Anal. Biochem.*, **9,** 146 (1964).
202. B. G. Johansson and L. Rymo, *Biochem. J.*, **92,** 5P (1964).
203. K. Dose and G. Krause, *Naturwissenschaften,* **49,** 349 (1962).
204. P. G. Righetti and J. W. Drysdale, *J. Chromatogr.*, **98,** 271 (1974).
205. B. J. Radola, *Biochim. Biophys. Acta,* **295,** 412 (1973).
206. T. Wadstroem, *Ann. N.Y. Acad. Sci.*, **209,** 405 (1973).
207. J. Bours, *J. Chromatogr.*, **60,** 225 (1971).
208. M. B. Hayes and D. Wellner, *J. Biol. Chem.*, **244,** 6636 (1969).
209. K.-D. Kohnert, E. Schmid, H. Zuehlke, and H. Fiedler, *J. Chromatogr.*, **76,** 263 (1973).
210. D. Graesslin, A. Trautwein, and G. Bettendorf, *J. Chromatogr.*, **63,** 475 (1971).
211. D. H. Leaback and A. C. Rutter, *Biochem. Biophys. Res. Commun.*, **32,** 447 (1968).

212. K. G. Kenrick and J. Margolis, *Anal. Biochem.*, **33,** 204 (1970).
213. F. Drawert and W. Mueller, *Chromatographia,* **4,** 23 (1971).
214. C. P. Dietrich, *Anal. Biochem.*, **51,** 345 (1973).
215. J. H. Cronenberger and V. A. Erdmann, *Anal. Biochem.*, **70,** 499 (1976).
216. H. Wada and E. E. Snell, *Anal. Biochem.*, **46,** 548 (1972).
217. L. J. Mets and L. Bogorad, *Anal. Biochem.*, **57,** 200 (1974).
218. T. Hultin and A. Sjoeqvist, *Anal. Biochem.*, **46,** 342 (1972).
219. H. Tichy, *Anal. Biochem.*, **69,** 552 (1975).
220. J. G. Heathcote, R. J. Washington, and B. J. Keogh, *J. Chromatogr.*, **92,** 355 (1974).
221. J. G. Heathcote, B. J. Keogh, and R. J. Washington, *J. Chromatogr.*, **78,** 181 (1973).
222. G. L. Wright, Jr., K. B. Farrell, and D. B. Roberts, *Biochim. Biophys. Acta,* **295,** 396 (1973).
223. G. Kynast and J. W. Dudenhausen, *Z. Klin. Chem. Klin. Biochem.*, **10,** 573 (1972).
224. J. G. Heathcote, D. M. Davies, C. Haworth, and R. W. A. Oliver, *J. Chromatogr.*, **55,** 377 (1971).
225. J. G. Heathcote and C. Haworth, *J. Chromatogr.*, **43,** 84 (1969).
226. J. G. Heathcote and C. Haworth, *Biochem. J.*, **114,** 667 (1969).
227. J. G. Heathcote, D. M. Davies, and C. Haworth, *J. Chromatogr.*, **67,** 325 (1972).
228. M. M. Frodyma and R. W. Frei, *J. Chromatogr.*, **15,** 501 (1964).
229. *Ibid.*, **17,** 131 (1965).
230. G. K. Zwolinski and L. R. Treiber, *J. Chromatogr.*, **107,** 311 (1975).
231. W. N. Fishbein, *Anal. Biochem.*, **46,** 388 (1972).
232. M. A. Gorovsky, K. Carlson, and J. L. Rosenbaum, *Anal. Biochem.*, **35,** 359 (1970).
233. Y. Y. Gofman, *Biokhimiya,* **30,** 1160 (1965).
234. V. A. Spivak, V. V. Shcherbukhin, V. M. Orlov, and Y. M. Varshavskii, *Anal. Biochem.*, **39,** 271 (1971).
235. V. A. Spivak, V. M. Orlov, V. V. Shcherbukhin, and Y. M. Varshavskii, *Anal. Biochem.*, **35,** 227 (1970).
236. V. A. Spivak, M. I. Levjant, S. P. Katrukha, and J. M. Varshavsky, *Anal. Biochem.*, **44,** 503 (1971).
237. V. A. Spivak, V. A. Fedoseev, V. M. Orlov, and J. M. Varshavsky, *Anal. Biochem.*, **44,** 12 (1971).
238. M. I. Pouchan and E. J. Passeron, *Anal. Biochem.*, **63,** 585 (1975).
239. G. Pataki and K.-T. Wang, *J. Chromatogr.*, **37,** 499 (1968).
240. G. Pataki, *Chromatographia,* **1,** 492 (1968).
241. M. Seiler and M. Wiechmann, *Z. Anal. Chem.*, **220,** 109 (1966).
242. D. Jaenchen and G. Pataki, *J. Chromatogr.*, **33,** 391 (1968).
243. J. Borko and R. W. Frei, *Mikrochim. Acta,* **1969,** 1144.
244. S. Undenfriend, S. Stein, P. Bochlen, W. Dairman, W. Leimgruber, and M. Weigele, *Science,* **178,** 871 (1972).
245. A. M. Felix and M. H. Jimenez, *J. Chromatogr.*, **89,** 361 (1974).

246. J. Sherma and J. C. Touchstone, *Anal. Lett.*, **7**, 279 (1974).
246a. W. L. Ragland, J. L. Pace, and D. L. Kemper, *Anal. Biochem.*, **59**, 24 (1974).
247. M. E. Clark, *Analyst (London)*, **93**, 810 (1968).
248. B. Ruestow and A. Hock, *Pharmazie*, **24**, 453 (1969).
249. G. F. Smith and M. Murray, *Anal. Biochem.*, **23**, 183 (1968).
250. D. E. Nitecki, I. M. Stoltenberg, and J. W. Goodman, *Anal. Biochem.*, **19**, 344 (1967).
251. E. Celon, L. Biondi, and E. Bordignon, *J. Chromatogr.*, **35**, 47 (1968).
252. J. P. Thornber and J. M. Olson, *Biochemistry*, **7**, 2242 (1968).
253. D. Bogdanovsky, E. Interschick-Niebler, K. H. Schleifer, F. Fiedler, and O. Kandler, *Eur. J. Biochem.*, **22**, 173 (1971).
254. J. P. Zanetta, G. Vincendon, P. Mandel, and G. Gombos, *J. Chromatogr.*, **51**, 441 (1970).
255. V. Neuhoff and M. Weise, *Arzneim.-Forsch.*, **20**, 368 (1970).
256. S. R. Burzynski, *Anal. Biochem.*, **65**, 93 (1975).
257. F. Drawert, O. Bachmann, and K. H. Reuther, *J. Chromatogr.*, **9**, 376 (1962).
258. C. Wu and R. C. Ling, *Anal. Biochem.*, **37**, 313 (1970).
259. V. V. Nesterov, B. G. Belen'kii, and D. P. Erastov, *Biokhimiya*, **33**, 537 (1968).
260. G. Pataki and M. Keller, *Klin. Wochenschr.*, **43**, 227 (1965).
261. E. Bancher, J. Washuettl, and M. D. Olfat, *Mikrochim. Acta*, **1968**, 773.
262. N. P. Sen, E. Somers, and R. C. O'Brien, *Anal. Biochem.*, **28**, 345 (1969).

Chapter **XVIII**

ANTIBIOTICS

1 CLASSIFICATION SYSTEMS

Because of its high resolving power and speed, thin-layer chromatography lends itself well to the separation and identification of antibiotics. Ikekawa et al. [1] have used silica gel layers to investigate a large number of these compounds (Table 18.1). In addition to the values given in the table the following $R_f \times 100$ values were given for butanol–acetic acid–water (3:1:1): olendomycin 29, ammaromycin 38, erythromycin 39, picromycin 41, carbomycin 55, leucomycin 58, tylosin 59, tertiomycin B 63, tertiomycin A 86, unamycin A 36, amphotericin A 33, pentamycin 67, actinomycin C 68, actinomycin J 73, telomycin 44, amphomycin 53, acidomycin 74, and enteromycin 73. Blastmycin had an R_f value of 0.78 in chloroform–methanol–17% ammonium hydroxide (2:2:1). In general, for macrolide, peptide, or antifungal antibiotics, the butanol-acetic acid–water (3:1:1) solvent was useful. Basic antibiotics could be separated by the upper layer of chloroform–methanol–17% ammonium hydroxide (2:1:1) or by propanol–pyridine–acetic acid–water (15:10:3:12). Po-

Table 18.1 $R_f \times 100$ Values of Some Antibiotics on Silica Gel G [1]

Antibiotic	Solvent[a]					
	A	*B*	*C*	*D*	*E*	*F*
Streptothricin		26	52			
Neomycin B		51	46			
Catenulin		60	54			
Zygomycin A		62	56			
Kanamycin		65	55			
Paromomycin		68	40			
Amminosidin		68	40			
Viomycin		11				
Blasticidin S		70				
Nystatin	18			18		
Pimaricin	34			34		
Amphotericin				33		
Pentamycin	67			67		
Trichomycin	17			45		
Bacitracin				58	13	
Antimycin A				72	81	
Blastmycin				82		
Homomycin				42		
Novobiocin				82		
Etamycin	66				80	
Pyridomycin	38				18	
Thiolutin	65				64	70
Aureothricin	58				57	63
Mitomycin C	54				45	38
Porfiromycin	50				50	48
Althiomycin					78	
Chromomycin A_3						
Spiramycin	8					65

[a] *A* = Butanol–acetic acid–water (3:1:1) (see text for R_f values for additional antibiotics in this solvent); *B* = chloroform–methanol–17% ammonium hydroxide (2:1:1) upper phase; *C* = *n*-propanol–pyridine–acetic acid–water (15:10:3:10); *D* = ethanol–ammonium hydroxide–water (8:1:1); *E* = ethanol–water (4:1); *F* = ethyl acetate–methanol (20:3).

lyene antibiotics can be developed in ethanol–ammonium hydroxide–water (8:1:1), ethanol–water (4:1), or ethyl acetate–methanol (20:3). For nucleotide antibiotics a solvent consisting of ethyl acetate–methanol (2:1) was used. For detection of the spots test T-214 was used.

Ito et al. [2] preferred cellulose layers (cellulose MN 300) for the separation of a similar group of antibiotics (Table 18.2). A solvent mixture

Table 18.2 R_f and R_G[b] × 100 Values of Water-Soluble Antibiotics on Cellulose with Propanol–Pyridine–Acetic Acid–Water (15:10:3:12) and their Color Reactions[a]

Antibiotic	$R_f \times 100$	$R_G \times 100$	*Detecting Agent*[c]	*Color*
Glebomycin[d]	41	84	ON	Red
Streptomycin[e]	44	90	ON	Red
Dihydrostreptomycin[e]	44	90	ON	Red
Hydroxystreptomycin[d]	32	66	ON	Red
Netropsin[d]	51	105	ON	Red
Amidinomycin[e]	53	108	N	Reddish violet
Gentamicin[e]	35	72		
	28	57	N	Violet
	20	41		
Streptothricin[e]	26	53	N	Violet
Viomycin[e]	21	43	N	Brownish violet
Kanamycin A[e]	17	35	N	Violet
B[f]	15	31	N	Violet
C[f]	23	47	N	Violet
Paromomycin[e]	15	31	N	Violet
Zygomycin[e]	15	31	N	Violet
Catenulin[d]	15	31	N	Violet
Neomycin[e]	10	20	N	Violet
Fradiomycin[e]	10	20	N	Violet
Glucosamine[d]	49	100	N	Reddish violet

[a] From Y. Ito, M. Namba, N. Naghama, T. Yamaguchi, and T. Okuda [2]; reproduced with permission of the authors and The Japan Antibiotic Research Association.

[b] $R_G = R_f$ value of antibiotic/R_f value of glucosamine hydrochloride.

[c] ON: oxidized nitroprusside reagent; N: ninhydrin reagent.

[d] Hydrochloride.

[e] Sulfate.

[f] Free base.

of *n*-propanol–pyridine–acetic acid–water (15:10:3:12) was used. Another solvent that was useful was water-saturated butanol containing 2% of *p*-toluenesulfonic acid.

Aszalos et al. [3] set up a system for classifying crude antibiotics based on whether a given antibiotic moves or does not move in a specified solvent system. Initially a set of three solvents divides the antibiotics into four main groups. These are further divided into 15 subgroups by 11 additional solvent systems. A total of 84 antibiotics were included in the study. With only a small number of antibiotics in each subgroup these can be differentiated by chromatographing on layers buffered at pH 2 or 11 and by microbiological and chemical tests. The only use made of R_f values is to indicate that the compound has moved from the origin. For this work Eastman Chromagram sheets, silica gel type 13181 (formerly 6060) were used. Group I antibiotics are those that do not move in any of the three initial solvent systems: (*a*) methanol, (*b*) 10% methanol in chloroform, and (*d*) chloroform. Group II antibiotics move only in solvent (*a*), group III move in both (*a*) and (*b*), and group IV move in all three solvents. Each group is then subdivided by its own group of solvents. For group I these are (Ia) pyridine–water (1:1), (Ib) pyridine–water–absolute ethanol (1:1:1) and (Ic) pyridine–water–absolute ethanol (1:1:3). Group II solvents are (IIa) butanol–methanol (1:1), (IIb) chloroform–methanol (1:1), and IIc) absolute ethanol. Group III solvents are (IIIa) methanol–benzene (3:22), (IIIb) methanol–benzene (3:47), and IIIc) methanol–benzene (1:24). Group IV solvents are (IVa) methanol–benzene (1:99) and (IVb) methanol–benzene–chloroform (1:49:50).

The method was designed to be applied to samples of crude antibiotics in order to rapidly assess the probability that the antibiotic in question is a known one. It is especially helpful in crude extracts where R_f values are affected by other components in the mixture. This system does not necessarily place antibiotics with close chemical relationship in the same group, and as pointed out by the authors this can be advantageous because those that are not closely related chemically can be differentiated by chemical and microbiological tests more easily than those that are. Issaq et al. [3a] used the same principle for classifying 151 antibiotics having antitumor properties. The chromatograms of these compounds were developed on silica gel with the three initial solvent systems mentioned above and in addition ethyl acetate was included in this first group of solvent systems. The compounds were divided into five groups as to whether or not they moved in these four solvents, then 14 additional solvent systems divided them into 19 subgroups. Bioautographic tests were also applied.

Most of the chromatographic classification systems for antibiotics have been based on paper chromatography; however, Paris and Theallet [4] classified 22 antibiotics into seven groups using a combination of paper chromatography and electrophoresis. Schmitt and Mathis [5] separated 42 antibiotics into four groups using silica gel G with (I) chloroform–methanol–acetic acid (45:4:1), (II) chloroform–methanol–water (80:20:2.5), (III) butanol–acetic acid–water (2:1:1), and (IV) water–sodium citrate–citric acid (100:20:5); solvent (III) was used with silica gel G buffered with pH 3 phosphate buffer. Antibiotics of closely related chemical composition usually fell in the same chromatographic group; for example, macrolides in group II did not move in solvent I, tetracyclines in group III did not move in solvents I and II, and streptomycin type antibiotics did not move in I, II, or III, but moved readily in IV. Group I antibiotics moved in solvents I, II, and III, and in some cases in IV. Wayland and Weiss [6] have set up a systematic procedure for identifying antibiotics in sensitivity discs. The tests include chemical, spectrophotometric, thin-layer chromatography, and paper chromatography on 28 antibiotics.

Huettenrauch and Schulze [7] have separated a series of antibiotics with glycosidic structure on layers of a 50:50 mixture of silica gel G and aluminum oxide. Using a solvent mixture of *n*-propanol–ethyl acetate–water–25% ammonium hydroxide (5:1:3:1), they found the following R_f values referred to paromomycin: streptomycin 0.07, neomycin B+C 0.52, kanamycin 0.70, and paromomycin 1.00. Separations were carried out in a continuous fashion using a modification of the apparatus of Brenner and Niederwieser [8].

Zuidweg et al. [9] separated 17 antibiotics on Sephadex G-15. Their technique eliminated the use of organic solvents which sometimes cause false inhibition zones in bioautographic tests by their incomplete removal. The plates were run by descending chromatography in a sandwich layer at an angle of 30°C. After development the plate was pressed onto a seeded agar plate covered with a sheet of lens tissue for 30 min. Incubation was carried out at the optimum temperature for the test organism.

Wagman and Weinstein [10] published a book, *Chromatography of Antibiotics,* which is essentially a data book containing information on separations by paper chromatography, thin-layer chromatography, electrophoresis, gas chromatography, and countercurrent distribution). For thin-layer chromatography the R_f values are listed. The compounds are listed in alphabetical order.

Aszalos and Frost [10a] have reviewed the thin-layer chromatography of antibiotics.

2 ACTINOMYCETES METABOLITES

Thin-layer chromatography has been applied in the isolation, purification, and identification of various metabolic products of the actinomycetes [11, 12]. Beck et al. [13] report that the R_f values for nonactin and some of its homologs on silica gel G, using chloroform-ethyl acetate (1:2), are as follows: nonactin 0.62, monactin 0.48, dinactin 0.32, and trinactin 0.15. Bickel et al. [14] give the R_f values of acumycin and a number of reference antibiotics in three different solvents on silica gel G (Table 18.3). The spots were located by development either with sulfuric acid or bioautographically. Bioautographic development was accomplished by placing a thin layer of filter paper on the surface of a bacteria-inoculated agar layer. The 20 × 20 cm plate was then pressed against the paper with a weight of approximately 2 kg for a period of 20 min. The agar layer was incubated at 37°C for 16 to 18 hr.

Cassani et al. [15] have chromatographed the actinomycins on both silica gel and alumina. To separate the C group from the F group, the actinomycins may be chromatographed on silica gel with a number of different solvents, one of the better ones being butanol–acetic acid–water (10:1:3) in which the F group has an R_f value of 0.50 and the C group 0.70, for a development distance of 15 cm. Individual compounds may be separated on alumina with the lower layer of a solvent mixture of

Table 18.3 $R_f \times 100$ Values of Acumycin and Reference Antibiotics of the Marcrolide Group on Silica Gel G[a]

Antibiotic	*Methanol*	*Chloroform–Methanol* (95:5)	*Chloroform–Methanol* (1:1)
Acumycin	66	35	82
Angolamycin	65	18	82
Tylosin	68	7	81
Carbomycin	75	40	88
Foromacidin A	32	2	59
Foromacidin B	34	5	61
Foromacidin C	37	6	64
Erythromycin	16	3	29
Narbomycin	22	12	41
Picromycin	22	7	36
Lancamycin	74	37	87

[a] From Bickel et al. [14]; reproduced with permission of the authors and Verlag Helvetica Chimica Acta.

ethyl acetate–(*sym*-tetrachloroethane)–water (3:1:3) in which (using a travel distance of 12.5 cm) the following R_f values are obtained: $C_1$0.44, $C_2$0.51, $C_3$0.58, $F_1$0.21, and $F_2$0.35. These compounds can be readily located by their bright orange color or by viewing under ultraviolet light.

Kondo et al [16] have used active carbon for the separation of water-soluble basic antibiotics produced by streptomyces. Four types of plates were used consisting of neutral and acidic activated carbon with and without a gypsum binder. Best results were accomplished with the acidified carbon. For the unbound layers, the slurry was prepared by mixing 10 g of active carbon, 30 ml of 0.5 *N* hydrochloric acid, and 30 ml of methanol. For the gypsum-bound layers, 0.5 g of gypsum was added and sulfuric acid was substituted for the hydrochloric acid. Six solvents were investigated with the best results being obtained with a mixture of methanol–0.5 *N* acid (1:4), and using hydrochloric or sulfuric acid depending on which acid was used for preparing the thin-layer slurry. The antibiotics were separated into four groups: streptomycin, streptothricin, fradiomycin, and kanamycin. Nussbaumer and Schorderet [17] have used thin layers of silica with *n*-butanol–water–methanol (40:20:10) + *p*-toluenesulfonic acid for the identification of streptomycin and dihydrostreptomycin.

Katayama and Ikeda [18] applied two-dimensional separations to the streptomycins by using thin-layer chromatography on silica gel followed by electrophoreis. Complete separation of all the components could be achieved by thin-layer chromatography in four stages: (*1*) with water-saturated butanol containing 2% each of *p*-toluenesulfonic acid and piperidine (solvent S_1) followed by electrophoresis in 1% sodium tetraborate, (*2*) with 3% sodium acetate (solvent S_2) followed by electrophoresis in 1% sodium tetraborate, (*3*) with S_2 followed by electrophoresis in Michaelis–Veronal buffer pH 8.0, and (*4*) with S_2 followed by electrophoresis in 0.04 *M* ammonium formate buffer pH 3. The compounds were detected with test T-239.

Streptomycin, dihydrodesoxystreptomycin, and dihydrostreptomycin can be separated on silica gel using 3 or 4% sodium acetate [19]. Streptomycin and other tuberculostatic antibiotics (kanamycin, viomycin, cycloserine, rifamycin SV, and capreomycin) can be separated by developing first with acetone–2% sodium acetate (9:1) and then with *n*-butanol–pyridine–methanol–acetic acid–water (30:20:20:1:30) using silica gel G [20].

Borowiecka [21] separated streptomycin, neomycin, kanamycin, paromomycin, gentamycin, mycerin (a form of neomycin), framycetin, and lincomycin on silica gel–kieselguhr (1:2) with methanol–ethyl acetate–water–25%ammonia–pyridine–3.85%ammoniumacetate(10:2:6:2:1:20).

Keller-Schierlein and Roncari [21] have investigated the hydrolysis products of lancamycin.

3 ERYTHROMYCINS

Anderson [22] separated these antibiotics on thin layers of silica gel using a solvent composed of methylene chloride–methanol–benzene–formamide (80:20:20:2–5). The humidity condition in the laboratory was the controlling factor for the percentage of formamide in the solvent. The higher the humidity, the less formamide was required for separation; thus at 20 RH 5 volumes were required, and at 30 to 40% RH, 3 to 2 volumes. The spots could be located either by spraying with 10% phosphomolybdic acid in ethanol or with 50% aqueous sulfuric acid. In both cases the plates were heated on a hot plate after spraying the reagent. The phosphomolybdic acid spray was the most sensitive reagent, but the spots faded after 1 or 2 hr; therefore, for permanent records, charring with sulfuric acid was desired.

Malczewska-Konecka et al. [24] separated erythromycins A and C and anhydroerythromycin on silica gel with the upper layer of ethyl acetate–isopropanol–15% ammonium acetate adjusted to pH 10.1 (9:7:8). The compounds were detected with a sensitivity of 0.05 μg with a 0.15% solution of xanthydrol in hydrochloric acid–acetic acid (12:1). Richard et al. [25] used methanol–0.02 *N* sodium acetate (4:1) on silica gel G prepared with 0.02 *N* sodium acetate to separate erythromycin, erythromycin estolate, and erythromycin ethylsuccinate from some degradation products and pharmaceutical excipients. The R_f values of 23 other antibiotics were determined. Radecka et al. [26] applied a direct densitometric method to determine erythromycin or erythromycin estolate in capsules using the same separation media. Visualization was accomplished with test T-139. The coefficient of variation ranged from 2.1 to 3.4% with a sensitivity of 5 μg.

Meyers and Smith [27] have described the use of a bioautographic technique for loose layers with erythromycin and several other antibiotics. After the chromatograms are completed and dried they are covered with a moistened filter paper supported by a clean glass plate. After inversion of the chromatoplate so that it is upside down, the edges of the filter paper are folded back over the layer support. On removing the extra glass plate, the paper covered layer is pressed onto a seeded agar layer (the authors give details for preparing the latter using *Streptococcus lactis*) and incubated overnight at 37°C.

Easterbrook and Hersey [28] have worked out a thin-layer autobiographic assay for erythromycin and its esters using *Sarcina lutea* ATCC

9341 as the test organism. The area of the zones of inhibition were measured by projecting the images on a screen at 12× magnification. There was a linear relationship between the logarithm of concentration and the square root of the area. Sensitivity on aluminum-backed layers ranged from 0.03 μg for the base and the stearate to 0.05 μg for the estolate. The separation system was that of Richard et al. [25] using prewashed plates.

4 NEOMYCINS

Brodasky [29] has applied layers of activated carbon to the separation of neomycin sulfates. Both neutral and acidified layers were used. For the preparation of the layers, 30 g of Nuchar (C–190–N) and 1.5 g of plaster of Paris were slurried with either 220 ml of distilled water or water adjusted to pH 2 with sulfuric acid. The acidified carbon slurry was allowed to stand for 16 hr before preparing the plates. The prepared plates were air-dried. For detecting the zones the agar diffusion method was employed. Using an agar seeded with *B. pumilus,* this technique was also used as a quantitative method, although the accuracy was not as great as the radioactive tracer techniques. (See Chapter XI, Section 6 for a description of the technique.) The R_f values of the neomycin sulfates are given in Table 18.4.

The neomycins A, B, and C have been separated on thin layers of cellulose [30] using methyl ethyl ketone–isopropanol–6.5 *N* ammonia (8:2:3). A can be separated from B and C by using silica gel H with 3.85% ammonium acetate [31]; C can be separated from A and B by using 3.4% ammonia as the solvent. Stretton et al. [32] separated neomycin sulfate, polymyxin B sulfate, and zinc bacitracin on silica gel G

Table 18.4 $R_f \times 100$ Values of Neomycin Sulfates with Various Chromatographic Systems[a]

	Acidified Carbon Developed with		*Untreated Carbon Developed with*	
Neomycin	H_2O	0.5 *N* H_2SO_4	H_2O	0.5 *N* H_2SO_4
A	60	61	0	54
B	10	21	0	24
C	10	43	0	45

[a] From T. F. Brodasky [29]; reproduced with permission of the author and the American Chemical Society.

with *n*-butanol–acetic acid–pyridine–water (15:11:3:19) (R_f = 0.14, 0.50, 0.61) and with *n*-butanol–water–pyridine–acetic acid–ethanol (60:10:6:15:5) (R_f = 0.05, 0.34, 0.66).

5 PENICILLINS

Nussbaumer [34,35] has separated the penicillins after an acid hydrolysis. In analyzing penicillin products, the influence of 40 different ingredients on the identification of penicillins by means of thin-layer chromatography was studied. The same author [36, 37] investigated the spectrophotometric determination of penicillin in various pharmaceutical preparations. Polyethylene glycols and sodium stearate interfere in the direct determination, but a preliminary separation on thin-layer chromatography separates the penicillins from the interfering materials. For the separation, 20% rice starch was mixed with silica gel G in a phosphate buffer of pH 5.8 for the preparation of the layers which were used with butyl acetate–*n*-butanol–acetic acid–phosphate buffer pH 5.8–methanol (80:15:40:24:5).

Fischer and Lautner [38] examined a series of penicillin preparations using silica gel G layers in a saturated atmosphere with the two solvents acetone–methanol (1:1) and isopropanol–methanol (3:7). The results are shown in Table 18.5. As the detecting agent, the iodine–azide reaction (T-147) was employed. The reagent colors the plate brown, but the catalytic effect of the sulfur groups in the penicillins causes white spots to appear against the dark background. In some cases the preparations give a second spot which is often yellow and can thus aid in the identification. Nicolaus et al. [39, 40] have investigated the separation of penicillins V and G, 2,6-dimethoxybenzylpenicillin, and 6-aminopenicillanic acid. For detection of the spots a bioautographic method is used. For the detection of 6-aminopenicillanic acid, it is converted to penicillin G on the layer before the microbiological test is run. The R_f values of 6-aminopenicillanic acid and penicillin V are given for butanol buffered to pH 4.6 as 0.09 and 0.23, respectively, and for butanol–water–acetic acid (40:40:1) as 0.26 and 0.56, respectively. In the biological test the completed chromatographic plate is placed in a plastic tray, cut to fit the plate so that inoculated (*Bacillus subtilis*) agar may be poured directly onto the thin layer. Since triphenyltetrazolium chloride is used in the agar medium it must be protected from the oxygen by coating the inoculated medium with a layer of sterile agar. To permit the antibiotics to diffuse into the agar layer, the coated plate is placed in the refrigerator at 0°C for 1 hr, after which it is incubated at 37°C for 16 hr. Antibiotic zones are light yellow in contrast to the red-brown zones which permit growth. The

Table 18.5 R_f Values of Various Penicillin Products on Silica Gel G. Solvent Travel Distance 12 cm in Saturated Atmosphere[a]

Penicillin	*Acetone–Methanol* (1:1) White Spot	*Acetone–Methanol* (1:1) Yellow Spot	*Isopropanol–Methanol* (3:7) White Spot	*Isopropanol–Methanol* (3:7) Yellow Spot
Penicillin G potassium salt	0.16		0.22 0.48	
Penicillin V potassium salt	0.68		0.64	
Penicillin V acid	0.68	0.54	0.74	0.24
α-Phenoxyethylpenicillin potassium salt	0.72		0.68	
Penicllin P potassium salt	0.62		0.58	
Phenylmercaptomethylpenicillin	0.04 0.58		0.08 0.54	
Procaine–penicillin	0.62	0.52	0.60	0.42
N,N'-Dibenzylethylene-diamine–di–penicillin G	0.58	0.52	0.56	0.24
N,N'-Dibenzylethylene-diamine–di–penicillin V	0.64	0.52	0.64	0.24
Diethylaminoethanol ester hydroiodide of penicillin V	0.72		0.04 0.60	
Cl-Benzylpyrrolidyl–methyl-ben-zimidazole penicillin G	0.64	0.91	0.60 1.00	0.84

[a] From R. Fischer and H. Lautner [38]; reproduced with permission of the authors and Verlag Chemie.

method is sensitive to 0.01 to 0.1 γ. Kline and Golab [41] have found it desirable to spray the dried chromatogram with agar at 100°C using a DeVilbiss paint spray gun with an air pressure of 27 lb/in.^{2}. This coating prevents the antibiotic spots from spreading when the plates are placed in the trays and covered with the inoculated medium.

McGilveray and Strickland [42] separated a group of penicillins on cellulose MN 300 layers and on silica gel G layers. Some solvents and R_f values are given in Table 18.6. Acetone–acetic acid (19:10) was also useful for separating ampicillin and hetacillin on silica gel. The systems did not separate dicloxacillin from nafcillin or phenethicillin from phenoxymethylpenicillin. Nafcillin could be differentiated by the intense yellow it gave with 50% sulfuric acid; the only other penicillin to give a yellow color (methicillin) appeared at a different R_f value. Biagi et al. [43,44] investigated the influence of pH in the reverse-phase thin-layer

Table 18.6 Some R_M and $R_f \times 100$ Values for Penicillins

	R_M[a] [43, 44]			$R_f \times 100$[b] [42]		
Compound	*pH* 2.6	*pH* 7.4	*pH* 9.4	*A*	*B*	*C*
Dicloxacillin	1.76	1.62	1.43	47	22	65
Cloxacillin	1.67	1.34	1.21	65	38	64
Nafcillin	1.43	1.39	1.20	47(37)[c]	22	64
Oxacillin	1.39	1.05	0.96	74	49	65(37)
Phenethicillin	1.35	1.03	0.91	84	73(55)	66
Phenoxymethylpenicillin	1.17	0.89	0.89	82	76(56)	66
Benzylpenicillin	0.84	0.55	0.45	90	90	61(28)
Methicillin	0.78	0.47	0.41	93	93	52(18)
Ampicillin	0.11	0.07	0.05	97	98(91)	12
Carbenecillin	−0.37	−0.46	−0.50			
Methyleneampicillin		0.28				
Hetacillin				96	98	30(12)

[a] Silica gel impregnated with 5% Silicon DC 200 in ether, mobile phase sodium acetate–Veronal buffers at the indicated pH values with or without varying amounts of acetone. R_M values calculated for 0% acetone [43, 44].

[b] Adsorbent: A = cellulose MN 300 with 0.1 M sodium chloride solution as developing solvent [42]; B = cellulose MN 300 with 0.3 M citric acid solution saturated with n-butanol as developing solvent [42]; C = silica gel with isoamyl acetate–methanol–formic acid–water (13:4:1:2) [42].

[c] Numbers in parentheses indicate secondary spots.

chromatography of penicillins and cephalosporins. The extrapolated R_M values for the penicillins are listed in Table 18.6. Auterhoff and Kienzler [45] separated penicillins on silica gel F with benzene–ethyl formate–formic acid (80:15:6). The degradation products of oxacillin [46] and of phenoxymethylpenicillin [47] have been chromatographed by Korchagin et al. Vandamme and Voets [48] used four solvent systems to separate the spontaneous, chemical, and enzymatic degradation products of penicillins and two cephalosporins on silica gel G. These were: n-butanol–water–ethanol–acetic acid (5:2:1.5:1.5), n-butanol–water–acetic acid (4:1:1), acetone–acetic acid (19:1), and 85% acetone.

Manni et al. [49] used an in situ reflectance method for determining potassium penicillin G in pharmaceutical preparations. After chromatography on silica gel with acetone–chloroform–acetic acid (10:9:1) the spots were measured at 230 nm. The standard deviation was 5% for zones containing 1 μg of sample and 1.5% for zones of 10-μg content. Sinsheimer et al. [50] used an in situ fluorescent method for determining traces of penicillin. Benzylpenicillin was hydrolyzed with penicillinase and then chromatographed on silica gel G with dimethylformamide–chloroform–28% ammonia (10:5:4). The spot at R_f 0.50 was measured at 510 nm with excitation at 410 nm. For phenoxymethylpenicillin the spot at R_f 0.88 was measured at 480 nm with excitation at 260 nm. Methicillin was hydrolyzed chemically and the spot at R_f 0.73 was measured at 480 nm with excitation at 350 nm. The sensitivity limits in the order given above were 3 μg, 0.76 μg, and 0.82 μg. Penicillanic acid has been determined by fluorescent densitometry at 440 nm with excitation at 350 nm of the spot with R_f value at 0.45 [51]. In this case after development on silica gel with chloroform–ethyl acetate–formic acid (60:40:1) the plate was exposed to ammonia vapor for 3 min to form a fluorescent product. Semiquantitative determinations of penicillins and their degradation products have been applied by visual comparison of the size and color of the spots [52, 53].

Cephalosporin C and some of its derivatives have been chromatographed on silica gel G with *n*-butanol–acetic acid (10:1) saturated with water, and *n*-butanol–pyridine–acetic acid–water (17:12:4:15) [54]. Various combinations of acetone with benzene (1:4, 2:3) and with toluene (2:3, 1:4, 3:7, 1:9) as well as toluene–ethyl acetate (1:1) were used for specific compounds. Biagi et al. [55] separated 13 cephalosporins by reverse-phase chromatography on silicone-impregnated silica gel using sodium acetate–Veronal buffer (pH 7.4) containing 0 to 24% acetone. The R_M values were calculated. For separating cephalosporin C, cephalothin, and cephaloridine, Buri [56] used silica gel layers buffered with phosphate buffer (pH 5.8) with isopropanol–methanol–phosphate buffer pH 5.8 (2:7:1).

6 RIFAMYCINS

These antibiotics have been run in a number of alcohols and in acetone on silica gel G layers [39, 57]. Acetone was the best solvent; with it rifamycin B could be separated from rifamycin O, rifamycin SV from rifamycin S, and rifamycin B from rifamycin SV. These compounds are strongly colored so that they can be seen on the plates without use of a reagent if the concentration is high enough. They can be detected, however, in much lower concentrations with the microbiological method. In 2 to 20 μg amounts Kolos and Eidus [58] gives the R_f values for rifamycin and desacetylrifamycin as 0.55 and 0.37, respectively, in chloroform–ethanol—0.1 *N* hydrochloric acid (84:15.9:0.1) on Eastman silica gel Chromagram sheets No. 6060. Maggi et al. [59] chromatographed a group of semisynthetic rifamycins on silica gel with acetone.

7 TETRACYCLINES

Nicolaus et al. [39, 40] have also investigated the separation of tetracyclines on thin layers of silica gel G. They examined a large number of solvents; four of the better ones are given in Table 18.7 with the corresponding R_f values. In addition to the separations shown in the table, hydroxytetracycline and dimethyltetracycline could be separated from tetracycline, chlortetracycline, and desoxytetracycline by using 10% citric acid solution saturated with butanol. For detection, the plates may be sprayed with 1 *N* hydrochloric acid followed by heating at 50°C. The tetracyclines appear as yellow spots. Desoxytetracycline can be visualized by coupling with a diazonium salt. If the microbiological test is to be applied, the agar is inoculated with *Bacterium subtilis*.

Ascione et al. [60] separated tetracycline, chlortetracycline, demethylchlortetracycline, and the anhydro- and epi-derivatives on acid-washed kieselguhr G impregnated with 20% polyoxyethylene glycol 400 solution in glycerol—0.1 *M* EDTA at pH 7 (1:19). The developing solvent

Table 18.7 $R_f \times 100$ of Some Tetracyclines in Various Thin-Layer Systems

	Silica Hel[a]				*Kieselguhr*[a,b]		
Compound	*A*	*B*	*C*	*D*	*E*	*F*	*G*
Tetracycline	36	38	61	50	53	36	Pk
Chlortetracycline	30	43	72	60	76	60	Pk
Anhydrotetracycline	0–20	45	68	50	93	83	Pk
Oxytetracycline	58	41	70	55	60	20	Y
Anhydrochlortetracycline	0–20	46	75	52	83	57	Pk
Demethylchlortetracycline					73	44	Pk
Methacycline					44	29	Pk-Br
Doxycycline					53	57	Pk-Br
4-Epitetracycline					20	12	Y
Epianhydrotetracycline					47	50	Y-Pk
4-Epichlortetracycline					33	21	Y-Pk

[a] Solvent: *A* = 10% citric acid [39]; *B* = *n*-butanol–methanol–10% citric acid (4:1:2) [39]; *C* = *n*-butanol–methanol–10% citric acid (4:2:2) [39]; *D* = *n*-butanol–methanol–10% citric acid (4:3:2) [39]; *E* = methyl ethyl ketone saturated with McIlvaine's pH 4.7 buffer [61]; *F* = dichlormethane–ethyl formate–ethanol (9:9:2) saturated with same buffer [61]; *G* = color test (T-115). Pk = pink, Y = yellow, Br = brown. Sensitivity 1 μg [61].

[b] Kieselguhr G slurried in 0.1 *M* EDTA–20% polyethylene glycol 400 in glycerine (vol/vol) (19:1). Plates dried over 4 hr at room temperature, then predeveloped in solvent *E* and again dried [61].

was the organic layer of ethyl acetate—0.1 *M* EDTA pH 7 (6:1). By examination under long-wavelength ultraviolet 0.05 μg of antibiotic could be detected. Gyanchandani et al. [61] have separated tetracyclines and their degradation products on kieselguhr-impregnated with EDTA; the R_f values and solvents are given in Table 18.7.

The quantitative determination of anhydrotetracyclines in degraded tetracycline tablets has been made by eluting the TLC zones and measuring the extinction at 428 nm [62]. For the determination of epitetracycline and chlortetracycline the epitetracycline zone was scraped off and eluted with 0.1 *N* hydrochloric acid [63]. Extinction measurement was at 356 nm. The chlortetracycline, after elution with 0.2 *N* sodium hydroxide, was measured by fluorescence at 414 with excitation at 364 nm. Radecka and Wilson [64] used an in situ method for determining tetracycline in pharmaceutical preparations. Although less precise than the spectrophotometric method of Alvarez Fernandez et al. [65] it was more sensitive and rapid and was more accurate than the official microbiological method. Van Hoeck et al. [66] developed an in situ fluorometric method for this group of compounds.

8 PEPTIDE ANTIBIOTICS

The polymyxin antibiotics are peptides with a very similar structure. In addition to the amino acids they contain a fatty acid component. Iglóy and Miszei [67, 68] separated polymyxin B, D, and M on silica gel G with acetone–water–acetic acid–2 *N* ammonium hydroxide (15:5:1:2). Polymyxin E (colistin) could not be separated from polymyxin B. Howlett and Selzer [69] developed a method for differentiating these two by hydrolyzing the sample with 5 *N* hydrochloric acid in a sealed tube for 6 hr at 120°C. The amino acid composition was then differentiated by chromatographing in a saturated chamber on silica Gel MN G with a solvent composed of 68 mg potassium iodide in 84 ml of 90% phenol and 16 ml of water. Visualization with ninhydrin shows L-2,4-diaminobutyric acid, L-threonine, phenylalanine, and leucine for polymyxin B, whereas colistin did not show the phenylalanine spot. Haemers and Moerloose [70] preferred hydrolysis at 110°C for 22 hr to avoid the appearance of ghost spots.

The quinomycin antibiotics are peptide lactones with a quinoxaline ring differing only in their *N*-methyl amino acid moiety. Shōji [71] chromatographed the A, B, B_0, C, D, and E compounds by circular chromatography on aluminum oxide with the lower layer of ethyl acetate–1,1,2,2-tetrachloroethane–water (3:1:3). All the components were separated except B and B_0. The compounds were detected under ultraviolet light on the fluorescent layer.

Studer et al. [72–76] in determining the structure of enniatin B synthesized a number of cyclic peptides with antibiotic activity. Both enniatin A and B had an R_f value of 0.39 in benzene–ether–methanol (17:2:1) on silica gel.

9 POLYENE MACROLIDE ANTIBIOTICS

Ikekawa et al. [1] separated six polyenes on silica gel G with *n*-butanol–acetic acid–water (3:1:1) with the following R_f values: nystatin 0.18, pimaricin 0.34, unamycin A 0.36, amphotericin A 0.33, pentamycin 0.67, and trichomycin 0.17. Bergy and Eble [77] used silica gel HF buffered with a mixture of 0.2 *M* potassium dihydrogen phosphate–0.2 *M* disodium hydrogen phosphate (1:1) and solvent of methylene chloride–methanol (17:3) to separate the filipin complex into four components with the following R_f values: I 0.8, II 0.7, III 0.6, and IV 0.5. Ochab [78] separated trichomycin, fumagillin, natamycin, nystatin, amphotericin A, and amphotericin B on silica gel G. Best results were obtained with ethanol–ammonia–water–dioxane (8:1:1:1) and *n*-butanol–pyridine–water (3:2:1). Martin and McDaniel [79] separated the candihexin complex on silica gel G into five components with the following R_f values: A 0.26, B 0.29, D 0.36, E 0.39, and F 0.43. The candidin complex was separated into two fractions: A 0.27 and B 0.32. A third component at R_f 0.36 as reported by Borowski et al. [80] in a crude candidin preparation did not appear in their purified candidin. Separation was achieved in the lower phase of chloroform–methanol–20% ammonium hydroxide (2:2:1). Quantitative determinations were made by in situ densitometry at 340 nm for candihexin and 360 nm for candidin.

10 MISCELLANEOUS

Bird and Stevens [81] have used thin-layer chromatography for the detection of impurities in nitrofurazone, a synthetic antibacterial compound. A mixture of benzene–acetone (3:2) on silica gel G gave an R_f value of 0.23 for nitrofurazone. 5-Nitro-2-furaldazine, one of the impurities, traveled with the solvent front and a second impurity had an R_f value of 0.65.

Maehr and Schaffner [82], by thin-layer chromatography on silica gel with chloroform–methanol–28% ammonium hydroxide–water (1:4:2:1) supplemented with paper chromatography for several acetyl derivatives, has shown that there are 16 antibiotics in the gentamicin complex. The

Table 18.8 Some $R_f \times 100$ Values for Antibiotics in the Gentamicin Complex

	Solvent[a]		
Antiobiotic	*X*	*Y*	*Z*
A_1	10		
A	16	60	18
B	22		
X	28		
B_1	40		
C_1[b]		71	54
C_2[b]		76	58
D[b]		69	42

[a] *X* = Chloroform–methanol–28% ammonium hydroxide (1:1:1) [83]; *Y* = chloroform–methanol–28% ammonium hydroxide–water (1:4:2:1) [82]; *Z* = chloroform–methanol–28% ammonium hydroxide (2:1:1) [82].

[b] Major component. Also D = C_{1a}.

R_f values of the major components are given in Table 18.8. Wagman et al. [83] has separated four of the minor components (Table 18.8). Wilson et al. [84] has chromatographed some of the other minor components in the same system.

Patulin, a furopyranone antibiotic isolated from a number of fungi, has been chromatographed in a number of solvents (Table 18.9). Scott and Kennedy [85] have developed a quantitative method for this antibiotic in apple juice based on the minimum detectable fluorescence when the silica gel layer is sprayed with a 0.5% solution of 3-methyl-2-benzothiazolinone hydrazone and heated for 15 min at 130°C.

Table 18.9 Some $R_f \times 100$ Values of Patulin in Various Solvents on Silica Gel

	Ref.				
Solvent	86	87	88	89	85
Ethanol–water (4:1)	71	76			
Toluene–ethyl acetate–90% formic acid (6:3:1)	37		41		
Benzene–methanol–acetic acid (24:2:1)	13		21		
Benzene–propionic acid–water (2:2:1)	64			66	
Chloroform	4				
Toluene–ethyl acetate–90% formic acid (5:4:1)					5

Fischer and Riegelman [90] developed an in situ fluorescent method for the determination of griseofulvin and griseofulvin-4′-alcohol in plasma. Separation was carried out on silica gel containing 6.7% of its weight of colloidal boehmite alumina (Du Pont) using anhydrous ether–acetone (3:2) as a solvent. The antibiotics were extracted from the plasma with ether. Kadner et al. [91] used silica gel G–neutral alumina activity I (30:2), and after separation of the compounds they were eluted for fluorescent measurement at 385 nm with 365 nm excitation.

Bergy and Eble [92] separated the filipin complex into four fractions, filipin II, filipin III, filipin IV, and filipin complex I, the latter being a mixture of at least five components. The separation was achieved on silica gel HF buffered with a mixture of 0.2 *M* potassium dihydrogen phosphate and 0.2 *M* disodium hydrogen phosphate and developed in methylene chloride–methanol (17:3). The R_f values were 0.7, 0.6, 0.5, and 0.8, respectively.

Niphimycin has been separated into four components, A_1, A_2, B_1, and B_2 by a double development in ethyl acetate–ethanol–water (150:45:28) on silica gel GF layers. Variamycin has been separated on silicic acid with methanol–chloroform (9:1) [94]. Quantitative results were obtained by measuring the extinction of the eluate at 280 nm. Clindamycin and its metabolites, clindamycin sulfoxide, *N*-demethylclindamycin, and *N*-demethylclindamycin sulfoxide were separated on Eastman silica gel 6061 sheets with acetone–methyl ethyl ketone–water (20.1:72.1:7.8) [95]. The spots were detected and analyzed by bioautography.

The technique for the use of bioautographic detection has been described in Chapter XI, Section 5. Aszalos et al. [3] list the microorganisms used for the detection of 84 antibiotics.

References

1. T. Ikekawa, F. Iwami, E Akita, and H. Umezawa, *J. Antibiot. (Tokyo), Ser. A.*, **16**, 56 (1963).
2. Y. Ito, M. Namba, N. Naghama, T. Yamaguchi, and T. Okuda, *J. Antibiot. (Tokyo), Ser. A*, **17**, 218 (1964).
3. A. Aszalos, S. Davis, and D. Frost, *J. Chromatogr.*, **37**, 487 (1968).
3a. H. J. Issaq, E. W. Barr, T. Wei, C. Meyers, and A. Aszalos, *J. Chromatogr.*, **133**, 291 (1977).
4. R. Paris and J. P. Theallet, *Ann. Pharm. Fr.*, **20**, 436 (1962).
5. J. P. Schmitt and C. Mathis, *Ann. Pharm. Fr.*, **23**, 205 (1970).
6. L. G. Wayland and P. J. Weiss, *J. Pharm. Sci.*, **57**, 806 (1968).
7. R. Huettenrauch and J. Schulze, *Die Pharmazie*, **19**, 334 (1964).
8. M. Brenner and A. Niederwieser, *Experientia*, **17**, 237 (1961).

9. M. H. J. Zuidweg, J. G. Oostendorp, and C. J. K. Bos, *J. Chromatogr.*, **42,** 552 (1969).
10. G. H. Wagman and M. J. Weinstein, *Chromatography of Antibiotics,* Elsevier, Amsterdam, 1973.
10a. A. Aszalos and D. Frost, "Thin-Layer Chromatography of Antibiotics," in *Methods in Enzymology,* Vol. 43, *Antibiotics*, J. H. Hash, Ed., Academic Press, New York, 1975, p. 172.
11. V. Prelog, A. M. Gold, G. Talbot, and A. Zamojski, *Helv. Chim. Acta,* **45,** 4 (1962).
12. V. Prelog and A. Walser, *Helv. Chim. Acta,* **45,** 631 (1962).
13. J. Beck, H. Gerlach, V. Prelog, and W. Voser, *Helv. Chim. Acta,* **45,** 620 (1962).
14. H. Bickel, E. Gaeumann, R. Huetter, W. Sackmann, E. Viseher, W. Voser, A. Wettstein, and H. Zaehner,*Helv. Chim. Acta,* **45,** 1396 (1962).
15. G. Cassani, A. Albertini, and O. Ciferri, *J. Chromatogr.*, **13,** 238 (1964).
16. S. Kondo, M. Sezaki, and M. Shimura, *Penishirin Sono Ta Koseibusshitsu,* **17,**1 (1964); through *Chem. Abstr.*, **61,** 4681 (1964).
17. P. A. Nussbaumer and M. Schorderet, *Pharm. Acta Helv.*, **40,** 205 (1965).
18. T. Katayama and H. Ikeda, *Sci. Pap. Inst. Phys. Chem. Res. Tokyo,* **63,** 49 (1969).
19. T. Sato and H. Ikeda, *Sci. Pap. Inst. Phys. Chem. Res. Tokyo,* **59,** 159 (1965).
20. R. Voigt and A. G. Maa Bared, *J. Chromatogr.*, **36,** 120 (1968).
21. B. Borowiecka, *Diss. Pharm. Pharmacol.*, **22,** 345 (1970).
22. W. Keller-Schierlein and G. Roncari, *Helv. Chim. Acta,* **45,** 138 (1962).
23. T. T. Anderson, *J. Chromatogr.*, **14,** 127 (1964).
24. W. Malczewska-Konecka, Z. Piekarska, and Z. Kowszyk-Gindife, *Chem. Anal. (Warsaw),* 14, 1093 (1969); through *Anal. Abstr.*, **19,** 4345 (1970).
25. G. Richard, C. Radecka, D. W. Hughes, and W. L. Wilson, *J. Chromatogr.*, **67,** 69 (1972).
26. C. Radecka, W. L. Wilson, and D. W. Hughes, *J. Pharm. Sci.*, **61,** 430 (1972).
27. E. Meyers and D. A. Smith, *J. Chromatogr.*, **14,** 129 (1964).
28. S. M. Easterbrook and J. A. Hersey, *J. Chromatogr.*, **121,** 390 (1976).
29. T. F. Brodasky, and *Anal. Chem.*, **343** (1963).
30. P. V. Deshmukh, S. I. Mehta, and M. G. Vaidya, *Hindustan Antibiot. Bull.* **12,** 68 (1969); through *Chem. Abstr.*, **73,** 69890q (1970).
31. C. Vickers, *J. Pharm. Pharmacol., Suppl.*, **18,** 17 (1966).
32. R. J. Stretton, J. P. Carr, and J. Watson-Walker, *J. Chromatogr.*, **45,** 155 (1969).
33. J. S. Pitton, *Antibiot. Advan. Res., Prod. Clin Use Proc. Congr., Prague* **1964,** 49 (Pub. 1965).
34. P. A. Nussbaumer, *Pharm. Acta Helv.*, **37,** 65 (1962).
35. *Ibid.*, p. 161.
36. *Ibid.*, **38,** 245 (1963).

37. *Ibid*., p. 758.
38. R. Fischer and H. Lautner, *Arch. Pharm.*, **294/66,** 1 (1961).
39. B. J. R. Nicolaus, C. Coronelli, and A. Binaghi, *Farmaco Ed. Prat.*, **16,** 349 (1961); *Chem. Abstr.* **56,** 7428 (1962).
40. B. J. R. Nicolaus, C. Coronelli, and A. Binaghi, *Experientia,* **17,** 473 (1961).
41. R. M. Kline and T. Golab. *J. Chromatogr.*, **18,** 409 (1965).
42. I. J. McGilveray and R. D. Strickland, *J. Pharm. Sci.*, **56,** 77 (1967).
43. G. L. Biagi, A. M. Barbaro, M. F. Gamba, and M. C. Guerra, *J. Chromatogr.*, **41,** 371 (1969).
44. G. L. Biagi, A. M. Barbaro, and M. C. Guerra, *J. Chromatogr.*, **51,** 548 (1970).
45. H. Auterhoff and H. Kienzler, *Dtsch. Apoth. Ztg.*, **112,** 297 (1972).
46. V. B. Korchagin, L. I. Serova, A. D. Tomashchik, M. A. Panina, and A. M. Zhidkova, *Antibiotiki (Moscow),* **17,** 316 (1972); through *Anal. Abstr.*, **23,** 4134 (1972).
47. V. B. Korchagin, L. I. Serova, S. P. Dement'eva, I. N. Navol'neva, I. I. Inozemtseva, D. M. Trakhtenberg, and N. I. Kotova, *Antibiotiki (Moscow),* **16,** 8 (1971); through *Anal. Abstr.*, **21,** 3696 (1971).
48. E. J. Vandamme and J. P. Voets, *J. Chromatogr.*, **71,** 141 (1972).
49. P. E. Manni, M. F. Bourgeois, R. A. Lipper, J. M. Blaha, and S. L. Hem, *J. Chromatogr.*, **85,** 177 (1973).
50. J. E. Sinsheimer, D. D. Hong, and J. H. Burckhalter, *J. Pharm. Sci.*, **58,** 1041 (1969).
51. A. Ciegler and C. P. Kurtzman, *J. Chromatogr.*, **51,** 511 (1970).
52. V. B. Korchagin, L. I. Serova, S. P. Dement'eva, I. N. Navol'neva, I. I. Inozemtseva, D. M. Trakhtenberg, and N. I. Kotova, *Antibiotiki (Moscow),* **16,** 410 (1971); through *Anal. Abstr.*, **22,** 1801 (1972).
53. V. B. Korchagin, L. I. Serova, S. P. Dement'eva, T. N. Sergeeva, S. P. Kurakina, and I. T. Strukov, *Antibiotiki (Moscow),* **16,** 516 (1971); through *Anal. Abstr.*, **22,** 2626 (1972).
54. B. Fechtig, H. Peter, H. Bickel, and E. Vischer, *Helv. Chim. Acta,* **51,** 1108 (1968).
55. G. L. Biagi, A. M. Barbaro, M. C. Guerra, and M. F. Gamba, *J. Chromatogr.*, **44,** 195 (1969).
56. P. Buri, *Pharm. Acta Helv.*, **42,** 344 (1967).
57. P. Sensi, C. Coronelli, and B. J. R. Nicolaus, *J. Chromatogr.*, **5,** 519 (1961).
58. O. T. Kolos and L. L. Eidus, *J. Chromatogr.*, **68,** 294 (1972).
59. N. Maggi, V. Arioli, and P. Sensi, *J. Med. Chem.*, **8,** 790 (1965).
60. P. P. Ascione, J. B. Zagar, and G. P. Chrekian, *J. Pharm. Sci.*, **56,** 1393 (1967).
61. N. D. Gyanchandani, I. J. McGilveray, and D. W. Hughes, *J. Pharm. Sci.*, **59,** 224 (1970).
62. D. L. Simmons, H. S. L. Woo, C. M. Koorengevel, and P. Seers, *J. Pharm. Sci.*, **55,** 1313 (1966).
63. I. C. Dijkhuis and M. R. Brommet, *J. Pharm. Sci.*, **59,** 558 (1970).
64. C. Radecka and W. L. Wilson, *J. Chromatogr.*, **57,** 297 (1971).

65. A. Alvarz Fernandez, V. Torre Noceda, and E. Sanchez Carrera, *J. Pharm. Sci.*, **58,** 443 (1969).
66. G. Van Hoeck, L. Kapetanidis, and A. Mirimanoff, *Pharm. Acta Helv.*, **47,** 316 (1972).
67. M. Iglóy and A. Mizsei, *J. Chromatogr.*, **28,** 456 (1967).
68. *Ibid.*, **34,** 546 (1968).
69. M. R. Howlett and G. B. Selzer, *J. Chromatogr.*, **30,** 630 (1967).
70. A. Haemers and P. de Moerloose, *J. Chromatogr.*, **52,** 154 (1970).
71. J. Shōji, *J. Chromatogr.*, **26,** 306 (1967).
72. R. O. Studer, W. Lergier, and K. Vogler, *Helv. Chim. Acta,* **46,** 612 (1963).
73. R. O. Studer, K. Vogler, and W. Lergier, *Helv. Chim. Acta,* **44,** 131 (1961).
74. K. Vogler, R. O. Studer, W. Lergier, and P. Lanz, *Helv. Chim. Acta,* **43,** 1751 (1960).
75. P. Quitt, R. O. Studer, and K. Vogler, *Helv. Chim. Acta,* **46,** 1715 (1963).
76. Pl. A. Plattner, K. Vogler, R. O. Studer, P. Quitt, and W. Keller-Schierlein, *Helv. Chim. Acta,* **46,** 927 (1963).
77. M. E. Bergy and T. E. Eble, *Biochemistry,* **7,** 653 (1968).
78. S. Ochab, *Diss. Pharm. Pharmacol.*, **22,** 351 (1970); through *Chem. Abstr.*, **74,** 79732z (1971).
79. J. F. Martin and L. E. McDaniel, *J. Chromatogr.*, **104,** 151 (1975).
80. E. Borowski, L. Falkowski, J. Golik, J. Zielinski, T. Ziminski, W. Mechlinski, E. Jereczek, P. Kolodziejczyk, H. Adlercreutz, C. P. Schaffner, and S. Neelakantan, *Tetrahedron Lett.*, **1971,** 1987.
81. R. F. Bird and S. G. E. Stevens, *Analyst (London),* **87,** 362 (1962).
82. H. Maehr and C. P. Schaffner, *J. Chromatogr.*, **30,** 572 (1967).
83. G. H. Wagman, J. A. Marquez, J. V. Bailey, D. Cooper, J. Weinstein, R. Tkach, and P. Daniels, *J. Chromatogr.*, **70,** 171 (1972).
84. W. L. Wilson, G. Richard, and D. W. Hughes, *J. Chromatogr.*, **78,** 442 (1973).
85. P. M. Scott and B. P. C. Kennedy, *J. Assoc. Off. Anal. Chem.*, **56,** 813 (1973).
86. J. Reiss, *Chromatographia,* **4,** 576 (1971).
87. F. A. Norstadt and T. M. McCalla, *Science,* **140,** 410 (1963).
88. P. M. Scott, J. W. Lawrence, and W. van Walbeck, *Appl. Microbiol.*, **20,** 839 (1970).
89. S. W. Tanenbaum and E. W. Bassett, *Biochim. Biophys. Acta,* **28,** 21 (1958).
90. L. J. Fischer and S. Riegelman, *J. Chromatogr.*, **21,** 268 (1966).
91. H. Kadner, H. Jacobi, and S. Engel, *Pharmazie,* **26,** 94 (1971).
92. M. E. Bergy and T. E. Eble, *Biochemistry,* **7,** 653 (1968).
93. S. Againa, M. Kalushkova, and I. Georgieva, *J. Chromatogr.*, **109,** 177 (1975).
94. Y. V. Zhdanovich, G. B. Lokshin, and Z. V. Petrakushina, *Antibiotiki (Moscow),* **19,** 687 (1974); through *Anal. Abstr.*, **28,** 4E23 (1975).
95. T. F. Brodasky, C. Lewis, and T. E. Eble, *J. Chromatogr.*, **123,** 33 (1976).

Chapter **XIX**

CARBOHYDRATES

1 SUGARS

Since publication of the first two papers mentioning the separation of sugars by thin-layer chromatography in 1960 [1, 2], numerous papers have been published on sugars and sugar derivatives. Mansfield [2a], Scherz et al. [2b] and Ghebregzabher et al. [2c] have reviewed the thin-layer chromatography of carbohydrates, and Moczar and Moczar [2d] have published on the thin-layer chromatography of carbohydrate side chains of glycoproteins. Sugars have been separated on a number of different layer materials including magnesium silicate, calcium silicate, calcium sulfate, polyamide, Sephadex, silica layers impregnated with complexing agents, magnesium oxide, silica gel, aluminum oxide, cellulose, buffered layers of silica gel or kieselguhr and even on mixtures of chromic oxide with Celite. As can be seen from Table 19.1, no one single system is satisfactory for all separations; a change in solvent and in some cases a change in both solvent and adsorbent is required. Pastuska [3]

Table 19.1 $R_f \times 100$ Values for Sugars in Various Systems[a]

	Magnesium Silicate (Woelm) Air-Dried. Unsaturated Chamber[b]				*Silica Gel G*[c]	*Silica Gel G Prepared with 0.1 N Boric Acid*[b]		*Kieselguhr G*[b,d]	*Acidic Aluminum Oxide*[b]		*Silicic Acid Prepared with 1/15 M Phosphate Buffer pH 8*		*Cellulose*
Sugar	*A*	*B*	*C*	*D*	*E*	*F*	*G*	*H*	*I*	*J*	*K*	*L*	*M*
D-Glucose	66	67	28	42	49	63	42	17	31	24	22	41	19
D-Fructose	65	52	31	42	51	52	31	25	33	26	26	46	24
D-Xylose	66	65	36	53	57	59	39	39	40	37	40	58	
L-Xylose													30
D-Ribose	58	60	34	45	59			49			37	58	39
D-Galactose	62	67	23	34		55	32	18	24	16	16	30	17
Lactose	61	61	12	23	9	56	25	4			6	22	4
L-Arabinose	61	65	28	40	57	62	42	28	35	28			29
D-Arabinose	61	65	28	39							28	47	27
L-Rhamnose	67	65	50	64		67	52	62	58	49	55	72	
L-Sorbose	66	55	31	45		51	24	26					23
D-Mannose	66	66	32	46	55	58	32	23	36	26	28	49	23
Maltose	59	63	18	30	29			6			12	35	6
Sucrose	70	56	22	40	25	63	29	8			14	36	10
Raffinose	64	43	9	20	13						4	12	
Trehalose	62	50	20	32									
D-Lyxose													33
Cellobiose													5

[a] *A* = *n*-Propanol–water (5:5) [59]; *B* = *n*-propanol–water–*n*-propylamine (5:3:2) [59]; *C* = *n*-propanol–water–chloroform (6:2:1) [60]; *D* = *n*-propanol–water–methyl ethyl ketone (2:1:1) [60]; *E* = *n*-butanol–acetic acid–ethyl ether–water (9:6:3:1) [24]; *F* = benzene–acetic acid–methanol (1:1:3) [3]; *G* = methyl ether ketone–acetic acid–methanol (3:1:1) [3]; *H* = ethyl acetate–isopropanol–water (130:57:23) [27]; *I* = *n*-butanol–acetone–water (4:5:1) [57]; *J* = *n*-butanol–acetone–water (7:2:1) [57]; *K* = *n*-butanol–acetone–water (4:5:1) [13]; *L* = *n*-butanol–dioxane–water (4:5:1) [13]; *M* = formic acid–methyl ethyl ketone–*tert*-butanol–water (3:6:8:3) [39].

[b] Development distance 10 cm.

[c] Dried at 135°C.

[d] Prepared with 0.02 *M* sodium acetate.

has used layers of silica gel impregnated with 0.1 *N* boric acid solution with benzene–acetic acid–methanol (1:1:3) and methyl ethyl ketone–acetic acid–methanol (3:1:1) (Table 19.1). The sugars were detected by spraying either with aniline phthalate or with a 1:1 mixture of 20% sulfuric acid and 0.2% naphthoresorcinol solution in ethanol. After spraying, the plates were dried at 105°C. The sensitivity of detection was under 4 μg. Porges and Porgesová [4], using Czechoslovakian silica gel which had been treated with hydrochloric acid to remove iron, obtained somewhat higher R_f values with the same solvent, benzene–acetic acid–methanol (1:1:3), as used by Pastuska. Ethyl acetate–isopropanol–acetic acid–water (4:2:1:1) [5] and chloroform–methanol (6:4) [6] have also been used with this adsorbent. Additional solvents have been given in Refs. 6 to 9. Silica gel layers impregnated with 0.02 *M* borate buffer (pH 8.0) have also been used with *n*-butanol–acetic acid–water (5:4:1) (10), *n*-butanol–water–acetic acid–ether (3:2:2:1) [11], and chloroform–ethanol (5:1) or benzene–ethanol (10:1) [12]. Ragazzi and Veronese [13] obtained excellent chromatograms even in the presence of appreciable amounts of salt. These workers used a silica gel impregnated with a 1/15 *M* phosphate buffer of pH 8 with two solvents, *n*-butanol–acetone–water (4:5:1) and *n*-butanol–dioxane–water (4:5:1). With these systems, samples of sugars between 1 and 200 μg could be used. Using multiple development Lombard [14] separated five sugars on the same layers even in the presence of acid with methyl ethyl ketone–acetic acid–water saturated with boric acid (9:1:1). The R_f values for seven monosaccharides have been given with 10 solvents on silica impregnated with various concentrations of phosphate buffer [15]. Lato et al. [16] list the R_f values of 16 monosaccharides, seven oligosaccharides, and two methyl derivatives on phosphate-impregnated layers as well as acetate-impregnated layers with 42 solvents, of which 23 proved to be useful. Using a 33-cm run on monosodium phosphate-impregnated silica with acetone–water–chloroform–methanol (8:0.5:1:1) the following sugars could be separated: raffinose, lactose, maltose, galactose, sucrose, levulose, arabinose, ribose, and rhamnose. Weidemann and Fischer [17] have used silica gel layers buffered with sodium acetate for the separation and detection of 2-deoxy sugars using ethyl acetate–2-propanol–water (65:23:12) as a solvent. Washuettl [18] used ethyl acetate–dimethylformamide–water (12:6:1) with acetate-impregnated layers.

Pifferi [19] was able to separate sugars from anthocyanins and anthocyanidins on silica gel layers impregnated with basic lead acetate when the sample solutions were kept slightly acid.

Cotte et al. [20] and Masera and Kaeser [21] have used buffered silica gel layers for the identification of urine and blood sugars. For the separation of maltose and glucose, especially with larger quantities of maltose, a 50:50 mixture of silica gel and aluminum oxide was used [21].

Waldi [22] has used layers prepared by mixing 20 g of kieselguhr G with 40 ml of phosphate buffer (equal parts 0.1 *M* phosphoric acid solution and 0.1 *M* disodium hydrogen phosphate) with a solvent mixture of *n*-butanol–acetone–phosphate buffer solution (4:5:1). Talukder [23] was able to separate fucose from seven other neutral monosaccharides on kieselguhr G buffered with 0.15 *M* monosodium phosphate with ethyl acetate–methanol–*n*-butanol–water (16:3:3:2); xylose and ribose were not separated with this combination.

Hay et al. [24] have investigated an extensive series of sugars and sugar derivatives on layers of unimpregnated silica gel G. As a solvent for the sugars they used a mixture of *n*-butanol–acetic acid–ethyl ether–water (9:6:3:1) and as detecting agents they used concentrated sulfuric acid or a 0.5% solution of potassium permanganate in 1 *N* sodium hydroxide. After spraying with the detecting agent the plates were heated at 100°C. In addition to the R_f values listed in Table 19.1 the following values were given: cellobiose 0.32, isomaltose 0.16, laminaribiose 0.26, and melibiose 0.15. Bergel'son et al. [25] included a number of sugars in examining the detection of α-glycol groupings on thin-layer chromatoplates by various detecting agents. Silica gel layers were used with four different solvents. The R_f values are given with the limits of detection for four different reagents. The most sensitive agent was a 5% solution of silver nitrate with 25% ammonium hydroxide.

Garbutt [26] compared the separation of some carbohydrates on kieselguhr G, Filter-Cel, and Hyflo Super-Cel (Johns Manville). The filter aids were used with calcium sulfate as a binding material. Filter-Cel and combinations of Filter-Cel with Hyflo Super-Cel gave better resolution of monosaccharides than either kieselguhr G or Hyflo Super-Cel. *n*-Butanol–pyridine–water (15:3:2) was employed for the separation. The R_f values are given for glucose, maltose, xylose, and two tentatively identified sugars, isomaltose and panose. Stahl and Kaltenbach [27] used kieselguhr G layers prepared with 0.02 *M* sodium acetate solution for the separation of a group of sugars with ethyl acetate–isopropanol–water (130:57:23) as the developing solvent. The R_f values they obtained are given in Table 19.1.

Mixtures of kieselguhr and silica gel have also been used for separating sugars; Wassermann and Hanus [28] have used a 3:2 mixture, respectively, with isopropanol–ethyl acetate–water (27:3.5:1) in a saturated atmosphere. Prey et al. [29, 30] investigated the variation of R_f values of sugars on different mixtures of kieselguhr and silica gel, ranging from pure kieselguhr to pure silica gel, checking numerous solvent combinations for separating efficiency. The best separations were achieved with a silica gel–kieselguhr mixture (1:2) impregnated with 2% polyvinyl alcohol solution or with silica gel–kieselguhr (1:4) impregnated with 0.02 *M* sodium acetate solution. The best solvents with these mixtures were ethyl acetate–acetone–water (20:20:3) and ethyl acetate–dimethylformamide–water (15:3:1), respectively. With the sodium acetate–buffered layers, a spray reagent was used of 2% ethanolic naphthoresorcinol solution containing 10% phosphoric acid. This detecting agent yields a deep red color with ketoses, a blue-violet with aldopentoses and blue with aldohexoses. It could not be used with the layers impregnated with polyvinyl alcohol solution so an aniline–diphenylamine–phosphoric acid spray was used. In this case the aldopentoses give gray-green spots, the methylaldopentoses yellow-green spots, the aldohexoses bluish gray spots, and the ketohexoses reddish brown colors. These authors used a wedge-strip technique to increase the sharpness of the separations. A two-dimensional method was used to separate glucose, fructose, and sucrose [29]. In this case silica gel G with a boric acid buffer was used with methyl ethyl ketone–acetic acid–methanol (3:1:1) in the first direction, and butanol–acetone–water (4:5:1) in the second direction. Grundschober and Prey [31] used 0.1 *N* boric acid-buffered silica gel. These workers also used thin-layer chromatography for investigating the reaction products of carbohydrates with olefins [32], and have published [33] a faster method for running sugars using mixtures of silica gel and kieselguhr on 75 × 75 mm plates. Since the developing time increases with the size of the plate, using the small plates has reduced the running time to 6 to 10 min. With this technique, the amount of material applied to the layer must be kept small (0.5 to 1 μg) in order to obtain good separations.

Weicker and Brossmer [34] have observed that in chromatographing sugars on silica gel layers using propanol–ammonium hydroxide–water (6:2:1), hexosamines and other Elson–Morgan reactive substances are formed, probably by a silica gel-catalyzed amination. No amino sugars were formed with pyridine solvents.

A number of papers have appeared on the use of cellulose layers for the separation of sugars. Bergel'son et al. [35] used a descending technique with three different solvents: butanol–pyridine–water (10:3:3), butanol–25% ammonium hydroxide–water (16:1:2), and phenol–butanol–acetic acid–water (5:5:2:10). The development time was relatively long, 7 to 18 hr. Aniline phthalate was used as a detecting agent giving a sensitivity of 0.5 to 1.0 γ. Schweiger [36] has applied thin layers of cellulose to the separation of seven sugars. Layers of cellulose powder MN 300 (0.25 mm thick) that had been dried at 100°C were applied to galactose, glucose, mannose, xylose, ribose, and rhamnose using both phases of a mixture of ethyl acetate–pyridine–water (2:1:2). Multiple development (2×) was used in order to accomplish the separation. The mannose and arabinose which were very close together could be separated by rechromatographing in 1% ammonia in phenol saturated with water. Metche et al. [37] have applied the Schweiger method for the separation of glucose, galactose, mannose, and arabinose. R_f values are given showing the effect of layer thickness on the values. Grau and Schweiger [38] have used this method for the detection of swelling agents (polysaccharides) in meat products. In order to prepare the sample for the chromatogram, 5 g of the meat was homogenized and then extracted two or three times with 20 ml each of petroleum ether in order to remove the fat. After removing the simple sugars with three extractions of 20 ml each of 50% ethanol, the residue was refluxed for 3 hr with 20 ml of 5% sulfuric acid. After neutralizing the hydrolysate with 2 M barium hydroxide it was filtered. Ten milliliters of the filtrate was evaporated at 30 to 50°C (in vacuum) to dryness and the residue dissolved in 1 ml of 40% methanol. One-half of this solution was then purified by running through a Merck-I acid cation-exchange (H^+) column using water as the eluting solvent. The first 15 ml of eluate was evaporated to dryness and taken up in 0.5 ml of aqueous methanol. This solution was then ready for application to the cellulose thin-layer plates. The intensities of the spots were tabulated for the various polysaccharides that were investigated.

Vomhof and Tucker [39] examined the separation of simple sugars on layers of cellulose 300 MN in nine different solvents and found that a mixture of formic acid–methyl ethyl ketone–*tert*-butanol–water (3:6:8:3) gave the best results yielding sharper spots without streaking. Damonte et al. [40] modified this solvent to a ratio of (3:5:7:5) for separating oligosaccharides and which would still separate sucrose, glucose, and fructose. Wolfrom et al. [41] have used a microcrystalline cellulose, Avirin (American Viscose), with pyridine–ethyl acetate–acetic acid–water (5:5:1:3) and butanol–acetic acid–water (3:1:1) for a group of 11 sugars as well as various sugar derivatives. For the separation of pentoses on microcrystalline cellulose, Linek et al. [42] used ethyl acetate–pyridine–water (8:1:1); other mixtures of this solvent have been used by other workers: (6:3:2) [43] and (2:1:2) [44]. An acidic component was added to this solvent by Kamp and Van Oort [45], who separated 11 sugars with pyridine–ethyl acetate–acetic acid–water (8:14:4:5). Petre et al. [46] used a triple development with water–acetic acid–*tert*-butanol–ethyl methyl ketone (3:3:8:6) to separate lactose, maltose, sucrose, galactose, mannose, fructose, arabinose, xylose, fucose, ribose, rhamnose, galacturonic acid, glucuronic acid, and mannuronic acid. Petrović and Canić [47] found that rice starch layers gave poor separations, but obtained excellent results with cellulose layers both one- and two-dimensionally. Best separation was achieved with two-dimensional work using *n*-butanol–acetone–diethylamine–water (10:10:2:5) in the first direction and phenol–water (3:1) in an atmosphere of 25% ammonia in the second direction, and in this way separated 12 sugars. By using *n*-butanol saturated with water–trichloroethylene–95% ethanol (3:1:1) with cellulose layers in continuous horizontal development in a B-N chamber, Berger and Agate [48] were able to separate monosaccharides even in the presence of large quantities of sucrose (1:100); however, complete separation required about 19 hr.

As miscellaneous adsorbents for the separation of carbohydrates, Zhdanov et al. [49] have used layers of gypsum with various mixtures of chloroform–methanol. Affonso [50] used acetone–butanol–chloroform–acetate buffer pH 3.6 (5:4:1:1) with strips of plaster of Paris to separate glucose, fructose, arabinose, lactose, xylose, and mannose in 30 min. Hesse and Alexander [51] used layers of chromic oxide combined with Celite (6:1) for the separation of 10 sugars. After the development and because of the dark color of the adsorbent, the layers were sprayed with a thin layer of silica followed by aniline phthalate as a detecting reagent. The colors were developed by heating at 110°C for 15 min. Graphite layers were tried but the spots showed tail formation. Birkofer et al. [52] separated a mixture of hexoses and pentoses on a layer composed of kieselguhr G (6 g), aluminum oxide G (1 g), and polyacrylonitrile (5 g) made up with 0.02 *M* secondary sodium phosphate solution. As a developing agent a mixture of *n*-amyl alcohol–isopropanol–ethyl acetate–water (11:3:22:6) was used. Mixtures of silica gel–kieselguhr (3:1) have been used for the separation of malto- and meglosaccharides [53, 54]. Solvent systems of *n*-butanol–ethanol–water (5:3:2) and various proportions of *n*-propanol–ethanol–water depending on the size of the polymers were used.

Marais [55] chromatographed nine sugars on polyamide layers using ethyl formate–methanol (8:1). Careful control of temperature was necessary. Grafe and Engelhardt [56] examined the separation of sugars on six commercial samples of polyamide with butanol–acetone–water–acetic acid (6:2:1:1). For two-dimensional work, methyl ethyl ketone–acetone–methanol–acetic acid–trimethyl borate (16:1:2:2:2) was employed as the second solvent.

Stroh and Schueler [57] have used acidic aluminum oxide with two solvents for the separation of a number of sugars (Table 19.1). Kochetkov et al. [58] have used aluminum oxide for monosaccharides.

Grasshof [59, 60] carried out separations of sugars on magnesium silicate with four different solvents (Table 19.1). The addition of a primary amine in the solvent decreases the R_f values for ketoses and leaves unchanged or increases the R_f values for aldoses. If the solvent containing *n*-propylamine is used, great care must be taken to remove all the amine in order not to interfere with the color reactions. This requires a drying period of 2 hr in the air followed by 1 hr at 130°C. Tore [61] has applied calcium silicate to the separation, getting best results with *n*-butanol–water and *n*-butanol butylacetate–water mixtures. Both buffered and unbuffered layers were used; in the former case 11 g of calcium silicate (Silene EF) was mixed with 3 g of Celite 535 and 700 mg of sodium acetate. These buffered plates were dried at 110°C for 20 hr. Good separations were obtained by multiple developing (8×) over a developing distance of 16.9 cm.

DeSouza and Panek [62] have used thin-layer chromatography on kieselguhr G plates in order to follow the hydrolysis of starch by α- and β-amylase. Samples were taken every 5 min and the increase in reducing sugars could be readily observed. The α-amylase produced glucose, dextrins, and maltose, whereas β-amylase produced only glucose and maltose.

To improve the separation of carbohydrates two-dimensional thin-layer chromatography may be employed. On silica gel chloroform–methanol–water (16:9:2) has been used in the first direction and ethyl acetate–methanol–acetate acid–water (6:3:3:2) in the second [63] or two developments with butanol–ethanol–water (5:1:4) followed by butanol–acetic acid–ether–water (9:6:3:1) at 90°C [64]. Impregnated layers have also been used to advantage in two-dimensional work. Silica gel and kieselguhr layers impregnated with borate have been used with isobutanol–acetic acid–water (5:4:1) followed by *n*-propanol–water (8.5:1.5) or *n*-propanol–dimethyl sulfoxide–water (8.6:0.5:0.9) in the second dimension [65]. With boric acid as an impregnant for silica gel layers, isopropanol–water (4:1) and *n*-butanol–ethyl acetate–isopropanol–acetic acid–water (7:20:12:7:6) as well as others have been used as combination solvents for two-dimensional work [9]. Mezzetti et al. [66] investigated a number of impregnants for separating carbohydrates on dual layers of silica gel. Excellent results were obtained and some of the better combinations were (*a*) silica gel impregnated with a mixture of equal volumes of 0.132 *M* sodium tetraborate, 0.204 *M* boric acid, and 0.06 *M* sodium tungstate (layer A) and the second layer impregnated with 0.036 boric acid (layer B). Solvents were ethyl acetate–isopropanol–water or isopropanol–*n*-propanol–water (14:14:5) in the first direction along layer A followed by ethyl acetate–acetic acid–methanol–water (12:3:3:2) or *n*-butanol–ethyl acetate–isopropanol–acetic acid–water (7:20:12:7:6) at right angles across layer B; (*b*) silica gel impregnated with 0.24 *M* sodium acetate (layer C) and layer A as the dual layer with solvent acetone–water (9:1) for layer C followed by methyl acetate–isopropanol–water (2:2:1) for layer A. The authors list additional combinations that give good separations.

In the field of oligosaccharides some separations have been included in the work cited for sugars up to this point, but there have been investigations specifically aimed at separating these compounds. Spitschan [67] used cellulose MN 300 with freshly prepared acetic acid–ethyl acetate–pyridine–water (1:7:5:3) to separate the sugars obtained by hydrolysis of starch. In order to remove salts which caused tailing he extracted the freeze-dried hydrolysate mixture with pyridine which dissolved the sugars. Damonte et al. [40] also used cellulose for oligosaccharides and found the best solvent to be acetic acid–methyl ethyl ketone–*tert*-butanol–water (3:5:7:5); it was used in a triple development.

As with simple sugars, silica gel is a useful adsorbent for the separation of oligosaccharides. Powing and Irzykiewicz [68] used various combinations of *n*-propanol, isopropanol, water, ammonia, *tert*-butanol, *n*-butanol, isoamyl alcohol, and ethanol in unsaturated tanks, but the most satisfactory for multiple development was *n*-propanol–water–ammonia (70:30:1) with which they were able to separate hexasaccharides and indicated that octasaccharides could also be separated. Cello-oligosaccharides including cellohexose were separated in a saturated atmosphere using isopropanol–water–ethyl acetate (1:2:1) [69]. The spots were located with test T-23. Hansen [70] found that nearly all monosaccharides moved ahead of lactic acid when it was included in the solvent system thus providing for a more efficient separation of the oligosaccharides. His solvent consisted of isopropanol–acetone–1 *M* lactic acid (4:4:2).

Buffered silica gel layers have also been applied to the separation of oligosaccharides. Silica gel layers prepared with a 1:1 mixture of 0.2 *M* orthophosphoric acid and 0.05 *M* sodium tungstate were developed twice with isopropanol–ethyl acetate–water (2:2:1) and then at right angles with butanol–pyridine–water (8:4:3) separating a group of nine oligosaccharides [71]. Ovodov et al. [15] examined the separation of five oligosaccharides using impregnations with various concentrations of sodium mono- and dihydrogen phosphate with four different solvent systems. Mezzetti et al. [72] found that impregnation with boric acid was not very selective with oligosaccharides and examined 107 solvent systems with silica gels impregnated with various concentrations of sodium tungstate and molybdic acid at various pH values. Best results were obtained with low-pH plates impregnated with phosphoric acid–sodium tungstate or with saturated molybdic acid. Molybdic acid impregnation gave more highly resolved multiple development chromatograms than the phosphotungstic acid impregnation, but the latter gave the best two-dimensional separations. Best separations (five mono + nine oligosaccharides) were obtained with two developments with ethyl acetate–isopropanol–water (10:6:3) on molybdic acid-impregnated layers.

Collins and Chandorkar [73], in examining the oligofructosans in chicory, used various mixtures of *n*-propanol–ethyl acetate–water depending on the degree of polymerization (DP) for separations on kieselguhr G. A ratio of 4:5:1 was used for DP 7 to 8 and 6:2:2 for homologs of DP 2 to 20.

Because of the use of glycerol as a preservative of saccharides in biological fluids Shellard and Jolliffe [73a] examined the effect of this compound on the rate of movement of simple sugars on silica gel and cellulose layers. In the presence of glycerol, buffered silica gel was preferred to cellulose, and it was recommended that samples be diluted to contain 10 to 20% glycerol before chromatographing.

2 SUGAR DERIVATIVES

Phenylhydrazones, Osazones, and Related Compounds

Stroh and Schueler [57], in addition to separating sugars, have also separated sugar phenylhydrazones on acidic layers of aluminum oxide with the same solvents (Table 19.2). For detecting these compounds the layers were sprayed with a solution of *p*-methoxybenzaldehyde–sulfuric acid–ethanol (1:1:18). Yellow-green spots appeared after heating for 2 to 3 min at 110°C. Rink and Herrmann [74] have applied the separation of phenylosazones to the separation and identification of sugars in urine. In order to form the derivatives, 10 ml of urine is heated on a boiling water bath for 30 min with 0.4 g of phenylhydrazine hydrochloride and 0.6 g of sodium acetate. After the crystalline product was cooled and washed with water, it was dissolved in a mixture of dioxane–methanol (1:1). For the preparation of the thin layers, 30 g of kieselguhr G was mixed with 60 ml of 0.05 *M* sodium tetraborate solution and dried at 80°C for 30 min after being spread on the plates. The separations were accomplished with chloroform–dioxane–tetrahydrofuran-0.1 *M* sodium tetraborate (40:20:20:1.5). The R_f values that were obtained are given in Table 19.2. In order to preserve the location of the spots they were marked immediately on drying because the yellow color of the osazones disappears very quickly. Tore [75] has used calcium silicate (Silene EF) in separating phenylosazones with two solvents (Table 19.2). Haas and Seeliger [76] have used polyamide layers combined with dimethylformamide–benzene (3:97) to separate a group of these derivatives using a solvent develop-

Table 19.2 $R_f \times 100$ of Some Sugar Derivatives in Various Systems

	Phenylhydrazone		*Phenylosazones*		
	Acidic Aluminum Oside, Developing Distance 10 *cm* [57]		*Kieselguhr G Prepared with* 0.05 *M Sodium Tetraborate* [74]	*Silene EF (Hydrated Calcium Silicate), Dried* 24 *hr at* 110°C [75]	
Sugar	*Butanol–Acetone–Water* (4:5:1)	*Butanol–Acetone–Water* (7:2:1)	*Chloroform–Dioxane–Tetra-hydrofuran–0.1 M Sodium Tetraborate* (40:20:20:1.5)	*Chloroform–Acetone–95% Ethanol* (5:3:3)	*Chloroform–Acetone–Ethanol–Water* (10:10:6:1)
Arabinose	75	63	91	65	
Xylose	77	68	72		
Galactose	69	61	52	29	
Mannose	71	62			
Glucose	68	64	39	22	
Rhamnose	83	76		78	
Fructose			39		
Lactose			2		50
Sorbose			21		
Ribose			91		
Maltose			12		57
Cellobiose					57

[a] Dried at 80°C for 30 min, developing distance 10 cm.

ment distance of 14 cm. The oligosaccharide derivatives could be separated from those of the monosaccharides by using pyridine–water (3:17). Brown or reddish brown spots were obtained by spraying the chromatogram with a freshly prepared solution of diazotized sulfanilic acid in 2 *N* sodium carbonate.

Applegarth et al. [77] chromatographed a group of *p*-bromophenylosazones on silica gel with benzene–methanol (9:1). The osazones had the following approximate R_f values: pentoses 0.10, hexoses 0.04, monomethylpentoses 0.24, monomethylhexoses 0.18, and dimethylpentoses 0.29.

Sugar mercaptals have been separated on silica gel G using benzene–ethanol (20:3). Detection was accomplished with *p*-anisidine·HCl in 5 ml of sulfuric acid and 100 ml of butanol [78].

Esters

Various esters of sugars have been chromatographed by thin-layer chromatography. Among these are the acetates, the benzoates, and the fatty acid esters; the majority of which have been on silica gel layers. Sucrose fatty acid esters have been separated by Linow et al. [79], Ranny [80], Weiss et al. [81]. Sahasrabudhe and Chadha [82], and by Kinoshita [83, 84]. With benzene–ethanol (3:1) [44] a sucrose palmitate sample has been separated into 11 components. Kinoshita examined many solvents for thc separation of these materials and some of the better solvents proved to be mixtures of methanol–chloroform–acetic acid (3:16:1 and 1:8:1) as well as methanol–chloroform–acetic acid–water (5:40:4:1). Gee [85] has applied the narrow chromatostrips of Kirchner et al. [86]. Silica gel containing 5% starch binder was used as the adsorbent. Of the solvents tried a mixture of toluene–ethyl acetate–95% ethanol (2:1:1) gave the best separations. As an indicating agent the layers were sprayed with a 0.2% solution of dichlorofluorescein in 95% ethanol. The last-named solvent was also used [81] for the separation of sucrose palmitates in quantitative estimations. The lower esters (mono-, di-, tri-, and tetra-) can be separated better by two-dimensional thin-layer chromatography with chloroform–methanol (4:1); however, this can not be used for the quantitative work. The higher esters may be separated with petroleum ether–ethyl ether–acetic acid (75:25:1). For quantitative determination by densitometry the esters were detected with test T-270, with a sensitivity of approximately 0.3 μg. Sorbitan fatty acid esters have been separated on silica gel impregnated with 4% boric acid using benzene–ethyl ether–methanol (15:4:1) [82].

Acetylated sugars have been investigated by a number of workers using silica gel layers [24, 87–91]. Dumazert et al. [89], using benzene–

ethanol (95:5) as a solvent with silica gel G plates that had been dried for 30 min at 140°C, obtained the following R_f values referred to glucose pentaacetate: maltose octaacetate 0.75, sucrose octaacetate 0.60, lactose octaacetate 0.56, and cellobiose octaacetate 0.40. They also included the acetates of a number of sugar alcohols. Micheel and Berendes [90] using cyclohexane–isopropyl ether–pyridine (4:4:2) obtained the following R_f values on silica gel: α-octaacetyllactose 0.30, 2,3,4,6-tetraacetyl-D-glucose 0.39, β-octaacetylmaltose 0.43, β-tetraacetyl-D-xylose 0.69, hexaacetyl-D-xylose 0.54, α-pentaacetyl-D-altrose 0.55, and α-pentaacetyl-D-glucose 0.68. These spots were located by charring with concentrated sulfuric acid at 110°C. Tate and Bishop [91] applied solutions of methanol in benzene to the separation of some sugar acetates. The concentration of methanol varied from 2 to 10% depending on the compounds under investigation. Nonpolar low-molecular-weight acetates required only 2% methanol whereas fully acetylated amino sugars required as high a concentration as 10%. For the detection of the acetates the ferric hydroxamate reaction (test T-144) was employed. With this treatment the acetates showed up as dark purple spots on a yellow background. Hay et al. [24] applied silica gel G layers to the separation of a group of derivatives including acetates; Deferrari et al. [88] used Mallinckrodt silicic acid with a starch binder for similar derivatives. The R_f values for a group of acetates obtained by these two groups of workers are listed in Table 19.3. Deferrari et al. applied the silver nitrate–ammonia–sodium methylate reagent of Cadenas and Deferrari [92] for locating the acetylated derivatives. Dumazert et al. [87] separated acetates with benzene–ethanol (95:5) on layers of silica gel G that had been dried for 30 min at 140°C. Referred to pentaacetylglucose as 1, the R_{st} values of the octaacetyl derivatives of maltose, sucrose, lactose, and cellobiose are reported as 0.75, 0.60, 0.56, and 0.40, respectively. As little as 5 γ of the acetates could be detected by spraying with a hydroxylamine reagent.

Deferrari et al. [88] also separated a group of benzoyl derivatives. Solvents for the separation on starch-bound silicic acid included chloroform–benzene (3:7), 0.5% methanol in benzene, and ethyl acetate–benzene (4:6 and 3:97). It is interesting to note that although the R_f values of anomeric acetates differed by only a small amount, the R_f values of anomeric benzoates differed by quite a bit.

Hay et al. [24] give a group of R_f values for acetal and mercaptal derivatives of sugars and polyols separated in the same systems as for the acetates.

Wolfrom et al. [93] applied thin layers of Magnesol (magnesium silicate), that had been treated (see Chapter II, Section 2) to produce a neutral product, to the separation of some acetates. As a solvent they

Table 19.3 $R_f \times 100$ Values for Some Sugar Acetates

	Mallinckrodt Silicic Acid + 10% Starch Binder [88][a]		*Silica Gel G, Dried Overnight at 135°C* [24]	
Sugar Acetate	*Ethyl Acetate–Benzene* (3:7)	*Methanol–Benzene* (2:98)	*Upper Phase of Benzene–Ethanol–Water–Ammonium Hydroxide* (200:47:15:1)	*n-Butanol–Acetic Acid–Ethyl Ether–Water* (9:6:3:1)
Penta-*O*-acetyl-α-D-galactopyranose	65			
Penta-*O*-acetyl-α-D-galactofuranose	52			
Penta-*O*-acetyl-β-D-galactopyranose	56		81	77
Penta-*O*-acetyl-β-D-galactofuranose	49			
Penta-*O*-acetyl-β-D-glucopyranose	68	63	81	79
Penta-*O*-acetyl-β-D-mannopyranose	57	52		
Penta-*O*-acetyl-α-D-glucopyranose	66		82	79
Tetra-*O*-acetyl-α-D-lyxopyranose	78	74		
Tetra-*O*-acetyl-β-D-xylopyranose	71	69	84	84
Hexa-*O*-acetylxylobiose			72	84
Octa-*O*-acetyl-α-maltose			79	85
Octa-*O*-acetyl-β-cellobiose			64	84
Octa-*O*-acetyl-β-laminaribiose			62	81
Octa-*O*-acetyl-sucrose			63	77
Octa-*O*-acetyl-gentiobiose	27	22		

[a] Dried at 110°C for 2 hr, development distance 13 cm.

used ethyl acetate–benzene (1:1) which they allowed to develop to a distance of 15 to 17 cm.

Hexose phosphates and triose phosphates can be separated by thin-layer chromatography on cellulose powder using two-dimensional development [94]. The best resolution was obtained with cellulose MN 300 powder. The sample, spotted on plates dried at 105°C for 2 hr, was first developed in the direction of the slurry application in the organic phase of a mixture of 60 ml of *tert*-amyl alcohol and 30 ml of water containing 2 g of *p*-toluenesulfonic acid. This development took 6 to 8 hr for 16 to 18 cm, after which the plates were air-dried at room temperature overnight. The second development, at right angles to the first, was with isobutyric acid–concentrated ammonium hydroxide–water (66:1:33). For detecting this group of compounds three consecutive spray reagents were employed. The first of these was test T-194. After drying, this was followed by T-15 and heating at 110 to 120°C for 5 to 8 min. Heating was stopped when the cellulose edges began to char and while the plates were still warm, T-247 was applied. Variously colored spots appeared, but the contrast diminished as the plates dried and the background became pale violet. By following this with a spray of concentrated ammonium hydroxide the spots became dark blue on a light background. Although the phosphates were well separated, 3-phosphoglyceric acid and 2-phosphoglyceric acid could not be separated. Grassetti et al. [95] also used cellulose layers with acetone–acetonitrile–1 *N* hydrochloric acid (32:13:5) for the separation of some sugar phosphates.

Dietrich et al. [96] have separated a group of sugar phosphates on ECTEOLA layers that had been prepared by slurrying 2 g of sieved ECTEOLA cellulose powder in 18 ml of 0.004 *M* ethylenediaminetetraacetic acid, pH 7.0. After the plates were dried at room temperature overnight they were sprayed with 0.1 *M* ammonium tetraborate and allowed to dry. The sugar phosphates were separated with 95% ethanol–0.1 *M* ammonium tetraborate pH 9.0 (3:2) and with the same solvent prepared with a pH-10 buffer (in the latter case the plates were sprayed with buffer of the same pH). Development was for a distance of 17 cm. Location of the spots was determined by spraying successively with benzidine–trichloroacetic acid and molybdate reagent.

Ethers

Prey et al. [97], in examining a series of solvents and conditions for the separation of methyl ethers, found that the 1-, 3-, and 6-*O*-methylfructoses could not be separated on silica gel G in butanol–acetone–water (4:5:1) or in ethanol–acetone–water (4:5:1), but that they separated nicely in the first solvent by using a silica gel G layer buffered with 0.1

Table 19.4 $R_f \times 100$ of Some *O*-Methyl Ethers of Sugars

	Silica Gel G			
Sugar Ether	*Upper phase of Benzene–Ethanol–Water–Ammonium Hydroxide* (sp gr 0.8) (200:47:15:1) [24]	*Isopropyl Ether–Methanol* (5:1) [98]	*n-Butanol–Acetone–Water* (4:5:1) [97]	*Silica Gel G Prepared with* 0.1 *N Boric Acid*
2-*O*-Methylglucose			72	
3-*O*-Methylglucose			72	
6-*O*-Methylglucose			64	
1-*O*-Methylfructose				33
3-*O*-Methylfructose				54
6-*O*-Methylfructose				70
2,3-Di-*O*-methyl-D-xylose	15			
2,3,4-Tri-*O*-methyl-D-xylose	28	50		
2,4-Di-*O*-methyl-D-glucose	5			
2,6-Di-*O*-methyl-D-glucose	5			
4,6-Di-*O*-methyl-D-glucose		11		
2,3,4-Tri-*O*-methyl-D-glucose		34		
3,4,6-Tri-*O*-methyl-D-glucose		27		
2,3,6-Tri-*O*-methyl-D-glucose	18	25		
2,4,6-Tri-*O*-methyl-D-glucose	13	23		
2,3,6-Tri-*O*-methyl-D-galactose		21		
2,3,4-Tri-*O*-methyl-D-galactose		16		
2,4,6-Tri-*O*-methyl-D-galactose	17	20		
2,3,4,6-Tetra-*O*-methyl-D-glucose	38	45		
2,3,4,6-Tetra-*O*-methyl-D-galactose		33		

N boric acid (Table 19.4). Tschesche and Wulff [98] and Hay et al. [24] resolved a number of methyl ethers on silica gel G layers using different solvents (Table 19.4). The separations can of course be improved by multiple development.

Wolfrom et al. [93] report the R_f values of a group of 9-*O*-methyl derivatives of sugars on Magnesol in a methanol–benzene (7:93) solvent system, as well as of another group of methyl derivatives [41] on layers of microcrystalline cellulose with the butanone–water azeotrope. Gee [99] chromatographed a group of methylated sugars on silica gel using ether–toluene (2:1) or methyl ethyl ketone–toluene (1:1). Methylated hexoses have been chromatographed on silica gel with chloroform–water–ammonia (500:6:3) [100]. Mied and Lee [101] separated the methyl ethers of D-arabinose with 2-butanone saturated with 3% ammonium hydroxide on silica gel.

A number of reagents can be used for locating these compounds, such as aniline hydrogen phthalate or chlorosulfonic acid in acetic acid (1:2) [98], 0.5% potassium permanganate in 1 *N* sodium hydroxide with subsequent heating, charring with sulfuric acid [24], or with the naphthoresorcinol–phosphoric acid reagent used by Prey et al. [97].

Prey et al. [97], Hay et al. [24], and Modi et al. [102] have all reported the R_f values of some isopropylidene derivatives obtained by chromatography on silica gel.

Tate and Bishop [103] used petroleum ether (65 to 110°C)–methanol (95:5) for separating benzyl derivatives of sugars on silica gel. Preparative thin-layer chromatography was used to purify the compounds for analysis and multiple development was applied to give higher degrees of resolution.

Lehrfeld [104] and Kaerkaeinen et al. [105] chromatographed trimethylsilyl ethers of some sugars on silica gel. Ethyl acetate will separate the derivatives of amino sugars from other TMS sugars and ethyl acetate–acetic acid–hexane (12:1:87) the neutral sugar derivative from the corresponding amino sugar and uronic acid derivatives.

Amino Sugars

Cellulose layers have been used for the separation of amino sugars. Guenther and Schweiger [106] separated glucosamine, galactosamine, and the corresponding *N*-acetyl derivatives. In order to obtain a separation of all four compounds, a two-dimensional technique was employed using a multiple development (2×) in the first direction with pyridine–ethyl acetate–acetic acid–water (5:6:1:3) or ethanol–pentanol–ammonia–water (8:2:2:1) as solvent. The plate was then sprayed with a borate buffer of pH 8.0 before developing in the second direction (again 2×) in ethyl–acetate–pyridine–tetrahydrofuran–water (7:3:2:2) or ethyl acetate–isopropanol–pyridine–water in the same proportions. Wolfrom et al. [41] used thin layers of microcrystalline cellulose, Avirin (American Viscose), with a solvent mixture composed of pyridine–ethyl acetate–acetic acid–water (5:5:1:3) for chromatographing a group of amino sugars.

The amino sugars can be detected with ninhydrin but the acetylated derivatives are made visible by spraying with silver nitrate–sodium hydroxide followed by dilute sodium thiosulfate solution [41]. An alternative method [106] is to spray the dried chromatogram with 0.1 *M* periodic acid in acetone and then after 10 min, with 3.5% sodium metaarsenite in 1 *N* hydrochloric acid. After an additional 2 min, the moist plate is given a spray of 0.6% alcoholic thiobarbituric acid solution and dried at 90°C for 5 min.

A few glycosamines have been separated on copper sulfate-impregnated silica gel with *n*-propanol–ammonia (4:1) [107].

A number of *N*-aryl glycosamines have been chromatographed on (*a*) silica gel, (*b*) kieselguhr, (*c*) calcium sulfate, and (*d*) synthetic zeolite X and the R_f values tabulated [108]. Distinct spots and high R_f values were obtained on silica gel with ether–toluene (2:1). The *N*-2,4-dinitrophenyl (DNP) derivatives of hexosamines were separated on silica gel using chloroform–methanol–acetic acid (90:7:3) [109]. The diphenylindenone sulfonylchloride (DIS) derivatives of hexosamines have been proposed as a sensitive means of detecting these compounds. The derivatives were separated on borate-buffered (pH 8.6) silica gel with chloroform–ethyl acetate–methanol–propioionic acid (70:40:22.5:0.5) or toluene–ethylene chlorohydrin–25% ammonia (3:5:2) [110]. After removal of the developing solvent the plates are treated with sodium ethylate and observed under ultraviolet light (365 nm). The amino sugars can be detected in quantities as low as 2×10^{-11} mole.

Sugar Acids

Uronic acids have been separated on sodium dihydrogen phosphate-impregnated silica gel [15]. Best results were obtained with layers impregnated with 0.2 *M* solutions. Solvent mixtures of *n*-butanol–ethanol–0.1 *M* phosphoric acid (1:10:5) and *n*-butanol–ethanol–0.1 *M* hydrochloric acid (1:10:5) were used for development. A group of eight sugar acids and their lactones were separated on silica gel with *n*-butanol–acetic acid–water (2:1:1) [24]. Similarly a group has been separated with acetone–water–benzyl alcohol–acetic acid (65:26:22:5) [111]. Kieselguhr impregnated with 0.1 *M* sodium dihydrogen phosphate buffer has been used to separate galacturonic, guluronic, and mannuronic acids using acetone–butanol–0.1 *M* sodium dihydrogen phosphate (8:5:7) as a solvent [112]. Detection with test T-171 gave a sensitivity of 0.5 μg. Glucuronic, galacturonic, guluronic, and mannuronic acids, mannuronic acid lactone, and guluronic acid lactone were separated by a double development with ethyl acetate–pyridine–water (2:1:2) on cellulose MN 300 HR [113].

3 DEXTRINS

Wiedenhof [114] has studied the separation of α- and β-cyclodextrins on microchromatoplates. The best separation was obtained on silica gel G by using a step technique. To accomplish this the plate was first developed in *n*-butanol–acetic acid–water–pyridine–dimethylformamide (6:3:1:2:4). The plates were then removed, dried, and developed one-third of the developing distance of the first solvent in a solvent composed of *n*-butanol–acetic acid–water (6:3:1). Separations on alusil layers (equal mixtures of silica gel G and aluminum oxide G) gave less satisfactory separations as the spots were not quite as sharp as when silica gel was used by itself. The spots were visualized by charring with a potassium dichromate–sulfuric acid mixture. Takeo et al. [115] obtained the best separations of the cyclodextrins with butanol–ethanol–water (4:3:3) on microcrystalline layers. Diemair and Koelbel [116] have separated dextrins from sugars by chromatographing on silica gel layers with ethanol–acetone–water (50:40:9) or on kieselguhr layers with ethyl acetate–isopropanol–water (65:23.5:11.5). The low-molecular-weight dextrins can be detected by spraying with triphenyltetrazolium chloride, aniline phosphate, or *m*-phenylenediamine hydrochloride reagents. The medium-molecular-weight dextrins can be detected with aniline–diphenylamine–phosphoric acid reagent, and the high molecular weights are detected with an iodine spray. Weill and Hanke [117] have separated maltodextrins up to 10 glucose units on kieselguhr G layers, with solvent mixtures of butanol–pyridine–water. Another promising solvent was a butanol–ethanol–water (5:3:2) mixture. Shannon and Creech [118] tested a large range of mixtures of *n*-butanol–pyridine–water for kieselguhr G separations. To obtain the optimum separations on a single plate, large (20 × 45 cm) plates were used with stepwise developments of different concentrations of *n*-butanol–pyridine–water; 18 cm with 39:39:22, 25 cm with the same mixture, 30 cm with 25:14:11, and finally 40 cm with 7:2:1. In each case the plate was dried before placing in the next solvent mixture. Dextrins with a degree of polymerization (DP) up to 27 units were separated. Huber et al. [54] and Covacevich and Richards [54a] used continuous development with various mixtures of *n*-propanol–ethanol–water or with *n*-propanol–nitromethane–water (5:2:3) in order to separate maltodextrins up to DP 35. Brown and Anderson [119] separated xylo- and cellodextrins on kieselguhr G with 65% isopropanol–ethyl acetate (1:1) and on kieselguhr F with isopropanol–ethyl acetate–water (42:35:23) [120]. The latter study also included mannodextrins. Test T-26 may be used for their detection.

Aspinall and Miller [121] applied gel filtration with Sephadex G-200 in thin layers to the separation of dextrins. Prior to the sample application the dextrins were dyed with Procion Brilliant Orange 2RS (Dylon) according to the technique of Dudman and Bishop [122]. Dextran 2000 dyed in the same manner was used as an internal standard for determining molecular weights.

4 QUANTITATIVE DETERMINATIONS

Colorimetric and Spectrophotometric Methods

Bancher et al. [123] determined sugars, separated by thin-layer chromatography, by use of a modification of the benzidine method of Jones and Pridham [124]. For this determination, the individual spots are scraped into a cuvet and mixed with 0.2 ml of 60% ethanol; 2 ml of 0.2% benzidine in acetic acid is added and the mixture is heated on a 100°C water bath for varying lengths of time, depending on the sugar being determined. For pentoses it is 15 min, hexoses 30 min, and disaccharides 60 min. (In determining disaccharides the chromatograms are first sprayed with concentrated hydrochloric acid before removing the individual spots.) After the mixtures are heated for the proper length of time, they are allowed to cool and then diluted to 2.5 ml with the benzidine reagent. Analysis is completed by centrifuging and measuring the absorbance at 350 mμ. A blank silica gel extract is used as a control, and standard curves are prepared by running known sugar solutions on the same plate with the test solutions.

Hay et al. [24] have applied the phenol–sulfuric acid method [125]. The absorbance of the solution is measured at 485 mμ. Kinoshita and Oyama [84] have used both the anthrone and the iron hydroxamic acid method for determining sucrose fatty acid esters after separation on thin layers, and Gee [85] has used the resorcinol–hydrochloric acid reagents of Roe [126] for the quantitative determination of sucrose fatty acid esters as well as for the free sugars. In order to avoid interference from the starch-bound layers of silica gel, Gee eluted the spots with dimethylformamide. This solvent not only failed to extract an interfering material from the starch binder but also gave a better elution of the sugars from the layers. Silica gel G layers could not be used because the recovery of sucrose materials was very poor. Scott [127] found that recovery of xylose varied with different silica gels; Silica gel H 93.1% recovery, Silica gel GF 94.6%, SilicAR-7 96%, and kieselguhr 97.7%.

Wolfrom et al. [128] used a borohydride-reduced microcrystalline cellulose (see Chapter III, Section 2) to eliminate interference from cellulose fibers in the aniline hydrogen phthalate method for reducing sugars.

Guinn [129] used the Park–Johnson [130] modification of the ferricyanide test for quantitation of reducing sugars. This test is very sensitive so that 0.01 μm of most reducing sugars can be determined. Cellulose fibers do not interfere in this test.

Reducing sugars may be eluted from layers and treated with tetrazolium blue (T-255). The formazans can be measured at 615 nm [130a].

Esser [131] has worked out a quantitative method for amino sugars by placing the ninhydrin-treated spots in methanolic cadmium acetate solution and measuring the color of the complex at 494 mμ.

In Situ Methods

A number of methods have been used to make the sugars visible on the thin layers so that direct densitometry may be employed in quantitative evaluations. The layers may be sprayed with a freshly prepared mixture of 1 g of aniline, 1 g of diphenylamine, 10 ml of phosphoric acid, and 100 ml of ethanol. After being air-dried they are heated at 110°C for 60 min and cooled for 20 min before scanning in the densitometer [53, 132].

Silica gel chromatograms have been sprayed with aniline–oxalic acid reagent consisting of equal volumes of 2.5% aqueous oxalic acid and 1.86% aniline in acetone. After drying the plates were then scanned in a densitometer [133] or measured by reflectance densitometry [134]. Of course this method is limited to reducing sugars, but has been used for hydrolysate samples. For aldoses, Champagnol and Bourzeix [135] found a mixture of 1.5 g of tartaric acid, 0.93 ml of aniline, and 100 ml of water-saturated butanol to be a superior spray reagent for densitometry.

Bukharov and Karneeva [136] examined a number of reagents for visualizing silica gel chromatograms of carbohydrates for densitometry. The best reagent which could be used for sugar alcohols, deoxy sugars, and aldopentoses consisted in spraying with 1% potassium iodate followed by heating at 100 to 120°C for 15 to 20 min and then spraying with a mixture of 1 part 0.1 *N* methanolic silver nitrate, 1 part ammonium hydroxide, and 2 parts 2 *N* sodium hydroxide. The plates were then heated under an infrared lamp and measured in reflected light using a 540-nm filter.

Charring of the spots has been one method used for various types of compounds. Pruden et al. [137] sprayed silica gel layers containing sugars with a solution of 1 g of ceric sulfate in 100 ml of 10% sulfuric acid. The plates were then heated at 110°C for 15 min to char the sugars. As in all determinations uniform spraying of the reagent is essential. Lehrfeld and Goodwin [138] examined four reagents for charring and recommended treatment with sulfuryl chloride as being the most consistent in results. This method consists in exposing the layers to sulfuryl chloride vapors [139] for 15 min, then for 15 min in water vapor. Charring was accomplished by heating at 150°C for 30 min.

Guebitz et al. [140] developed a fluorescent densitometric method for sugar acids in pharmaceutical preparations. The technique consists in dipping the completed chromatogram in a solution of 25 ml each of a 2% solution of lead tetraacetate in glacial acetic acid (wt/vol) and a 1% solution of 2,7-dichlorofluorescein in absolute ethanol (wt/vol) diluted to 1 liter with dry benzene. The plate is kept in the dark for 30 min and then dried for an equal period of time at 50°C in a vacuum drying oven. The fluorescence is measured at 530 nm. Another example of direct fluorescent determination of sugars is given in Chapter XI, Section 2.

Miscellaneous Methods

Pastuska [3] developed a volumetric method for the determination of sugars which has been described in Chapter XI, Section 1.

The direct comparison of unknown spots with standard, known quantities has been used for the estimation of raffinose in molasses [142] and for the determination of mannitol in sorbitol [142]. Lato et al. [143] have used this method for the determination of sugars in urine.

Nybom [144] found a rectilinear relationship between the areas of sugar spots and the log weight on thin layers (0.2 to 0.35 mm) but a linear relationship between the weight and the square root of the area on thick layers.

Radioactive methods may be used, and Shannon and Creech [118] and Gal [145] have applied scintillation methods to the determination of maltosaccharides and monosaccharides, respectively.

5 ELECTROPHORESIS OF CARBOHYDRATES

Arkel et al. [146] applied the microelectrophoresis method of Wieme [147, 148] to the separation of mucopolysaccharides. Since commercial agar exhibited staining with the dyes used in the method, a sulfate-free agar (agarose) was prepared according to a modified method [149] of Araki [150]. The electrophoresis was carried out in 0.9% agarose using a barbiturate buffer of pH 8.6 and applying a voltage of 20 V/cm. After the electrophoresis, which required about 7 min, the thin-layer slides were immersed in a solution of 0.1% Cetavlon for 1 hr to precipitate the mucopolysaccharides. Arkel et al. advised using Cetavlon in physiological saline solution in order to obtain the optimum precipitation. To locate the compounds the slides were stained in a Toluidine Blue solution made by dissolving 40 mg of Toluidine Blue in 20 ml of distilled water and 80 ml of dry acetone. Staining was carried out for 15 min, after which the slides were rinsed in 1% acetic acid solution until the background became colorless. These authors described an alternative stain and also a method for combined staining of both the proteins and the mucopolysaccharides.

Stefanovich [151] investigated the electrophoresis of a group of carbohydrates (mono-, di-, tri-, and polysaccharides) on silica gel G buffered with a pH-10.2 system made up of a boric acid–sodium carbonate mixture. Electrophoresis was conducted at 400 V and 80 mV, generally for 60 min. All carbohydrates moved toward the cathode, and this and other factors led to the conclusion that the movement was due to electroosmosis; however, the results indicated that this technique could be used for the separation of carbohydrates.

References

1. Z. Kowalewski, O. Schindler, H. Jaeger, and R. Reichstein, *Helv. Chim. Acta,* **43,** 1280 (1960).
2. M. Wyss-Huber, H. Jaeger, and E. Weiss, *Helv. Chim. Acta,* **43,** 1010 (1960).

2a. C. T. Mansfield, "Use of Thin-Layer Chromatography in Determination of Carbohydrates," in *Quantitative Thin-Layer Chromatography,* J. C. Touchstone, Ed., Wiley, New York, 1973, p. 79.

2b. H. Scherz, G. Stehlik, E. Bancher, and K. Kaindl, *Chromatogr. Rev.,* **10,** 1 (1968).

2c. M. Ghebregzabher, S. Rufini, B. Monaldi, and M. Lato, *J. Chromatogr.,* **127,** 133 (1976).

2d. E. Moczar and M. Moczar, "Thin-Layer Chromatography in Studies of Carbohydrate Side Chains of Glycoproteins," in *Progress in Thin-Layer Chromatography & Related Methods,* Vol. I, A. Niederwieser and G. Pataki, Eds., Ann Arbor-Humphrey Science, Ann Arbor, Mich., 1970, p. 168.

3. G. Pastuska, *Z. Anal. Chem.,* **179,** 427 (1961).
4. E. Porges and L. Porgesová, *Bratislav. Lekarske Listy,* **43–I,** 513 (1963).
5. T. Nakai, H. Demura, and M. Koyama, *J. Chromatogr.,* **66,** 87 (1972).
6. P. G. Pifferi, *Anal. Chem.,* **37,** 925 (1965).
7. C. B. Howard and P. C. Kelleher, *Clin. Chim. Acta,* **31,** 75 (1971).
8. D. H. Hettinga, A. H. Miah, E. G. Hammond, and G. W. Reinbold, *J. Dairy Sci.,* **53,** 1377 (1970).

9. M. Lato, B. Brunelli, G. Ciuffini, and T. Mezzetti, *J. Chromatogr.*, **34,** 26 (1968).
10. H. Jacin and A. R. Mishkin, *J. Chromatogr.*, **18,** 170 (1965).
11. J. M. M. Franken-Luykx and W. J. Klopper, *Brauwissenschaft*, **20,** 173 (1967).
12. K. Čapek, I. Tikal, J. Jarý, and M. Masojidková, *Collect. Czech. Chem. Commun.*, **36,** 1973 (1971).
13. E. Ragazzi and G. Veronese, *Farm. Pavia Ed. Prat.*, **18,** 152 (1963).
14. A. Lombard, *J. Chromatogr.*, **26,** 283 (1967).
15. Y. S. Ovodov, E. V. Evtushenko, V. E. Vaskovsky, R. G. Ovodova, and T. F. Solov'eva, *J. Chromatogr.*, **26,** 111 (1967).
16. M. Lato, B. Brunelli, G. Ciuffini, and T. Mezzetti, *J. Chromatogr.*, **39,** 407 (1969).
17. G. Weidemann and W. Fischer, *Z. Physiol. Chem.*, **336,** 189 (1964).
18. J. Washuettl, *Mikrochim. Acta*, **1969,** 1003.
19. P. G. Pifferi, *J. Chromatogr.*, **43,** 530 (1969).
20. J. Cotte, M. Mathieu, and C. Collombel, *Pathol. Biol. Sem. Hop.*, **12,** 747 (1964).
21. G. Masera and H. Kaeser, *Minerva Pediatr.*, **16,** 14 (1964).
22. D. Waldi, *J. Chromatogr.*, **18,** 417 (1965).
23. M. Q.-K. Talukder, *J. Chromatogr.*, **57,** 391 (1971).
24. G. W. Hay, B. A. Lewis, and F. Smith, *J. Chromatogr.*, **11,** 479 (1963).
25. L. D. Bergel'son, E. V. Dyatlovitskaya, and V. V. Voronkova, *Proc. Acad. Sci. USSR, Chem. Sect. Engl. Transl.*, **141,** 1076 (1961).
26. J. L. Garbutt, *J. Chromatogr.*, **15,** 90 (1964).
27. E. Stahl and U. Kaltenbach, *J. Chromatogr.*, **5,** 351 (1961).
28. L. Wassermann and H. Hanus, *Naturwissenschaften*, **50,** 351 (1963).
29. V. Prey, H. Berbalk, and M. Kausz, *Mikrochim. Acta*, **1961,** 968.
30. V. Prey, H. Scherz, and E. Bancher, *Mikrochim. Acta*, **1963,** 567.
31. F. Grundschober and V. Prey, *Monatsh. Chem.*, **92,** 1290 (1961).
32. V. Prey and F. Grundschober, *Chem. Ber.*, **95,** 1845 (1962).
33. E. Bancher, H. Scherz, and V. Prey, *Mikrochim. Acta*, **1963,** 712.
34. H. Weicker and R. Brossmer, *Klin. Wochenschr.*, **39,** 1265 (1961).
35. L. D. Bergel'son, E. V. Diatlovitskaya, and V. V. Voronkova, *Dokl. Akad. Nauk SSSR*, **149,** 1319 (1963).
36. A. Schweiger, *J. Chromatogr.*, **9,** 374 (1962).
37. M. Metche, J.-P. Haluk, Q.-H. Nguyen, and E. Urion, *Bull. Soc. Chim. Fr.*, **1963,** 1080.
38. R. Grau and A. Schweiger, *Z. Lebensm-Unter.-Forsch.*, **119,** 210 (1963).
39. D. W. Vomhof and T. C. Tucker, *J. Chromatogr.*, **17,** 300 (1965).
40. A. Damonte, A. Lombard, M. L. Tourn, and M. C. Cassone, *J. Chromatogr.*, **60,** 203 (1971).
41. M. L. Wolfrom, D. L. Patin, and R. M. de Lederkremer, *J. Chromatogr.*, **17,** 488 (1965).
42. K. Linek, L. Kuniak, and B. Alinče, *Chem. Zvesti*, **21,** 99 (1967).
43. R. S. Ersser and B. C. Andrew, *Med. Lab. Technol.*, **28,** 355 (1971).

44. G. Drews, Z. Naturforsch., **23b,** 671 (1968).
45. W. Kamp and A. Van Oort, *Pharm. Weekbl.,* **102,** 1295 (1967).
46. R. Petre, R. Dennis, B. P. Jackson, and K. R. Jethwa, *Planta Med.,* **21,** 81 (1972).
47. S. M. Petrović and V. D. Canić, *Mikrochim. Acta,* **1969,** 599.
48. P. D. Berger and A. S. Agate, *J. Chromatogr.,* **39,** 232 (1969).
49. Y. A. Zhdanov, G. N. Dorofeenko and S. V. Zelenskaya, *Dokl. Akad. Nauk SSSR,* **149,** 1332 (1963); through *Chem. Abstr.,* **59,** 3317 (1963).
50. A. Affonso, *J. Chromatogr.,* **27,** 324 (1967).
51. G. Hesse and M. Alexander, *Jour. Int. Etude Methodes Sep. Immed. Chromatogr., Paris,* **1961,** (pub. 1962), p. 229.
52. L. Birkofer, C. Kaiser, H. A. Meyer-Stoll, and F. Suppan, *Z. Naturforsch.,* **17B,** 352 (1962).
53. C. T. Mansfield and H. G. McElroy, Jr., *Anal. Chem.,* **43,** 586 (1971).
54. C. N. Huber, H. D. Scobell, H. Tai, and E. E. Fisher, *Anal. Chem.,* **40,** 207 (1968).
54a. M. T. Covacevich and G. N. Richards, *J. Chromatogr.,* **129,** 420 (1976).
55. J. P. Marais, *J. Chromatogr.,* **27,** 321 (1967).
56. I. Grafe and H. Engelhardt, *Chromatographia,* **5,** 307 (1972).
57. H. H. Stroh and W. Schueler, *Z. Chem.,* **4,** 188 (1964).
58. N. K. Kochetkov, B. A. Dmitriev, and A. I. Usov, *Dokl. Akad. Nauk SSSR,* **143,** 863 (1962); through *Chem. Abstr.,* **57,** 3995 (1962).
59. H. Grasshof, *J. Chromatogr.,* **14,** 513 (1964).
60. H. Grasshof, Deut. *Apoth.-Ztg.,* **103,** 1396 (1963).
61. J. P. Tore, *J. Chromatogr.,* **12,** 413 (1963).
62. N. O. de Souza and A. Panek, *J. Chromatogr.,* **15,** 103 (1964).
63. T. Kartnig and O. Wegschaider, *J. Chromatogr.,* **61,** 375 (1971).
64. A. Pastuszyn and H. Michl, *Mitt. Versuchsstn. Gaerungsgewerbe Inst. Angew. Mikrobiol.,* **20,** 1 (1966); through *Chem. Abstr.,* **66,** 65779t (1967).
65. K. Frigge, *Experientia,* **22,** 767 (1966).
66. T. Mezzetti, M. Ghebregziabhier, S. Rufini, G. Ciuffini, and M. Lato, *J. Chromatogr.,* **74,** 273 (1972).
67. R. Spitschan, *J. Chromatogr.,* **61,** 169 (1971).
68. R. F. Powning and H. Irzykiewicz, *J. Chromatogr.,* **29,** 115 (1967).
69. S. Saif-ur-Rahman, C. R. Krishnamurti, and W. D. Kitts, *J. Chromatogr.,* **38,** 400 (1968).
70. S. A. Hansen, *J. Chromatogr.,* **105,** 388 (1975).
71. T. Mezzetti, S. Rufini, and G. Ciuffini, *Minerva Pediatr.,* **23,** 1509 (1971); through *Anal. Abstr.,* **23,** 1898 (1972).
72. T. Mezzetti, M. Lato, S. Rufini, and G. Ciuffini, *J. Chromatogr.,* **63,** 329, (1971).
73. F. W. Collins and K. R. Chandorkar, *J. Chromatogr.,* **56,** 163 (1971).
73a. E. J. Shellard and G. H. Jolliffe, *J. Chromatogr.,* **24,** 76 (1966).
74. M. Rink and S. Herrmann, *J. Chromatogr.,* **12,** 415 (1963).
75. J. P. Tore, *Anal. Biochem.,* **7,** 123 (1964).
76. H. J. Haas and A. Seeliger, *J. Chromatogr.,* **13,** 573 (1964).

77. D. A. Applegarth, G. G. S. Dutton, and Y. Tanaka, *Can. J. Chem.*, **40,** 2177 (1962).
78. D. M. Bowker and J. R. Turvey, *J. Chromatogr.*, **22,** 486 (1966).
79. F. Linow, H. Ruttloff, and K. Taeufel, *Naturwissenschaften,* **21,** 689 (1963).
80. M. Ranny, *Veda Vyzk. Prum. Potravin.*, **18,** 191 (1968); through *Chem. Abstr.*, **72,** 62550z (1970).
81. T. J. Weiss, M. Brown, H. J. Zeringue, Jr., and R. O. Feuge, *J. Am. Oil Chem. Soc.*, **48,** 145 (1971).
82. M. R. Sahasrabudhe and R. K. Chadha, *J. Am. Oil Chem. Soc.*, **46,** 8 (1969).
83. S. Kinoshita, *Kogyo Kagaku Zasshi,* **66,** 450 (1963); *Chem. Abstr.*, **60,** 3207 (1964).
84. S. Kinoshita and M. Oyama, *Kogyo Kagaku Zasshi,* **66,** 455 (1963); *Chem. Abstr.*, **60,** 3208 (1964).
85. M. Gee, *J. Chromatogr.*, **9,** 278 (1962).
86. J. G. Kirchner, J. M. Miller, and G. J. Keller, *Anal. Chem.*, **23,** 420 (1951).
87. C. Dumazert, C. Ghiglione, and T. Pugnet, *Bull. Soc. Chim. Fr.*, **1963,** 475.
88. J. O. Deferrari, R. M. de Lederkremer, B. Matsuhiro, and J. F. Sproviero, *J. Chromatogr.*, **9,** 283 (1962).
89. C. Dumazert, C. Ghiglione, and T. Pugnet, *Bull. Soc. Pharm. Mars.*, **12,** 337 (1963).
90. F. Micheel and O. Berendes, *Mikrochim. Acta,* **1963,** 519.
91. M. E. Tate and C. T. Bishop, *Can. J. Chem.*, **40,** 1043 (1962).
92. R. Cadenas and J. O. Deferrari, *Analyst (London),* **86,** 132 (1961).
93. M. L. Wolfrom, R. M. de Lederkremer, and L. E. Anderson, *Anal. Chem.*, **35,** 1357 (1963).
94. P. P. Waring and Z. Z. Ziporin, *J. Chromatogr.*, **15,** 168 (1964).
95. D. R. Grassetti, J. F. Murray, Jr., and J. L. Wellings, *J. Chromatogr.*, **18,** 612 (1965).
96. C. P. Dietrich, S. M. C. Dietrich, and H. G. Pontis, *J. Chromatogr.*, **15,** 277 (1964).
97. V. Prey, H. Berbalk, and M. Kausz, *Mikrochim. Acta,* **1962,** 449.
98. R. Tschesche and G. Wulff, *Tetrahedron,* **19,** 621 (1963).
99. M. Gee, *Anal. Chem.*, **35,** 350 (1963).
100. A. Stoffyn, P. Stoffyn, and E. Martensson, *Biochim. Biophys. Acta,* **152,** 353 (1968).
101. P. A. Mied and Y. C. Lee, *Anal. Biochem.*, **49,** 534 (1972).
102. B. D. Modi, J. R. Patil, and J. L. Bose, *Indian J. Chem.*, **2,** 32 (1964).
103. M. E. Tate and C. T. Bishop, *Can. J. Chem.*, **41,** 1801 (1963).
104. J. Lehrfeld, *J. Chromatogr.*, **32,** 685 (1968).
105. J. E. Kaerkkaeinen, E. O. Haahti, and A. A. Lehtonen, *Anal. Chem.*, **38,** 1316 (1966).
106. H. Guenther and A. Schweiger, *J. Chromatogr.*, **17,** 602 (1965).
107. M. D. Martz and A. F. Krivis, *Anal. Chem.*, **43,** 790 (1971).
108. N. I. Suzdaleva, B. N. Stepanenko, and V. V. Zelenkova, *Prikl. Biokhim. Mikrobiol.*, **4,** 320 (1968); through *Chem. Abstr.*, **69,** 92754g (1968).

109. H. J. Haas and A. Weigerding, *Carbohydrate Res.*, **12,** 211 (1970).
110. Y. Vladovska-Yukhnovska, C. P. Ivanov, and M. Malgrand, *J. Chromatogr.*, **90,** 181 (1974).
111. H. J. Haas and G. Schwiersch, *J. Chromatogr.*, **42,** 124 (1969).
112. W. Ernst, *Anal. Chim. Acta,* **40,** 161 (1968).
113. H. Guenther and A. Schweiger, *J. Chromatogr.*, **34,** 498 (1968).
114. N. Wiedenhof, *J. Chromatogr.*, **15,** 100 (1964).
115. K. Takeo, Y. Kondo, and T. Kuge, *Agric. Biol. Chem.*, **34,** 954 (1970).
116. W. Diemair and R. Koelbel, *Z. Lebensm. Unters.-Forsch.*, **124,** 157 (1964).
117. C. E. Weill and P. Hanke, *Anal. Chem.*, **34,** 1736 (1962).
118. J. C. Shannon and R. G. Creech, *J. Chromatogr.*, **44,** 307 (1969).
119. W. Brown and O. Anderson, *J. Chromatogr.*, **57,** 255 (1971).
120. W. Brown and K. Chitumbo, *J. Chromatogr.*, **66,** 370 (1972).
121. P. T. Aspinall and J. N. Miller, *Anal. Biochem.*, **53,** 509 (1973).
122. W. F. Dudman and C. T. Bishop, *Can. J. Chem.*, **46,** 3079 (1968).
123. E. Bancher, H. Scherz, and K. Kaindl, *Mikrochim. Acta,* **1964,** 652.
124. J. K. N. Jones and J. B. Pridham, *Biochem. J.*, **58,** 288 (1954).
125. M. Dubois, K. A. Gilles, J. K. Hamilton, P. A. Rebers, and F. Smith, *Anal. Chem.*, **28,** 350 (1956).
126. J. H. Roe, *J. Biol. Chem.*, **107,** 15 (1934).
127. R. W. Scott, *J. Chromatogr.*, **49,** 473 (1970).
128. M. L. Wolfrom, R. M. de Lederkremer, and G. Schwab, *J. Chromatogr.*, **22,** 474 (1966).
129. G. Guinn, *J. Chromatogr.*, **30,** 178 (1967).
130. J. T. Park and M. J. Johnson, *J. Biol. Chem.*, **181,** 149 (1949).
130a. C. W. Raadsveld and H. Klomp, *J. Chromatogr.*, **57,** 99 (1971).
131. K. Esser, *J. Chromatogr.*, **18,** 414 (1965).
132. B. L. Welch and N. E. Martin, *J. Chromatogr.*, **72,** 359 (1972).
133. J. W. Mizelle, W. J. Dunlap, and Simon H. Wender, *J. Chromatogr.*, **28,** 427 (1967).
134. K. Kringstad, *Norsk Skogind.*, **21,** 210 (1967); through *Anal. Abstr.*, **15,** 4826 (1968).
135. F. Champagnol and M. Bourzeix, *J. Chromatogr.*, **59,** 472 (1971).
136. V. G. Bukharov and L. N. Karneeva, *Izv. Akad. Nauk SSSR, Ser. Khim.*, **1970,** 1473; through *Anal. Abstr.*, **20,** 3903 (1971).
137. B. B. Pruden, G. Pineault, and H. Loutfi, *J. Chromatogr.*, **115,** 477 (1975).
138. J. Lehrfeld and J. C. Goodwin, *J. Chromatogr.*, **45,** 150 (1969).
139. D. Jones, D. E. Bowyer, G. A. Gresham, and A. N. Howard, *J. Chromatogr.*, **24,** 226 (1966).
140. G. Guebitz, R. W. Frei, and H. Bethke, *J. Chromatogr.*, **117,** 337 (1976).
141. V. Prey, W. Braunsteiner, R. Goller, and F. Stressler-Buchwein, *Z. Zuckerind.*, **14,** 135 (1964).
142. V. Castagnola, *Boll. Chim. Farm.*, **102,** 784 (1963).
143. M. Lato, B. Brunelli and G. Ciuffini, *J. Chromatogr.*, **36,** 191 (1968).
144. N. Nybom, *J. Chromatogr.*, **28,** 447 (1967).

145. A. E. Gal, *J. Chromatogr.*, **34,** 266 (1968).
146. C. van Arkel, R. E. Ballieux, and F. L. J. Jordan, *J. Chromatogr.*, **11,** 421 (1963).
147. R. J. Wieme, *Clin. Chim. Acta*, **4,** 317 (1959).
148. R. J. Wieme and M. Rabaey, *Naturwissenschaften*, **44,** 112 (1957).
149. S. Hjerten, *Biochim. Biophys. Acta*, **53,** 514 (1961).
150. C. Araki, *Int. Congr. Biochem., 4th, Vienna, 1958*, **1,** 15 (1959).
151. V. Stefanovich, *J. Chromatogr.*, **31,** 466 (1967).

Chapter **XX**

CARBONYL COMPOUNDS

1 DIRECT SEPARATION

Marcuse and co-workers [1, 2] have carried out an extensive study of the separation of alkyl aldehydes and ketones on silica gel G activated at 120°C for 30 min. Table 20.1 gives the R_f values of a group of these compounds. For a series of aldehydes the following R_f values are given in benzene as a developing solvent: pentanal 0.53, heptanal 0.55, octanal 0.56, nonanal 0.59, decanal 0.62, undecanal 0.63, dodecanal 0.64, tridecanal 0.65, and tetradecanal 0.66. In the same solvent for a series of alkan-3-ones, the following R_f values are given: nonanone 0.47, decanone 0.50, undecanone 0.52, dodecanone 0.54, tridecanone 0.56, hexadecanone 0.62, octadecanone 0.66, and eicosanone 0.68. Better resolution can be obtained by using an undecane-impregnated layer. Various solvent mixtures were tried and it was found that a mixture of methanol–water (7:3) gave the best separation for C_{10} to C_{14} aldehydes. The water content was decreased for the separation of higher aldehydes. For separating ketones on the impregnated layers the best solvent was methanol–water (9:1). Other useful solvents were mixtures of acetonitrile and water or acetic acid. To get a separation between aldehydes and ketones and to further separate homologs in mixtures of the two, a two-dimensional method was employed. The development in the first direction took place on unimpregnated silica gel layers; after the solvent was removed from the plate the unused portion of the plate above the line of the compounds was impregnated with undecane. The second development was then made in a different solvent across the impregnated layer. For the two-dimen-

Table 20.1 $R_f \times 100$ Values of Some Aliphatic Carbonyl Compounds on Silica Gel G[a]

Compound	*Petroleum Ether–Ethyl Ether* (98:2)[b]	*Benzene*[b]
Tridecanal	32	65
Pentadecan-2-one	21	37
Tridecan-3-one	32	56
Tridecan-4-one	35	61
Dodecan-5-one	38	64
Tridecan-6-one	42	67
Tridecan-7-one	45	68
Pentadecan-8-one	47	73
Heptadecan-9-one	50	77
Heneicosan-10-one		80[c]
Heneicosan-11-one	57	84
Tricosan-12-one	60	87
Heptacosan-13-one		81[c]
Heptacosan-14-one	64	87

[a] From R. Marcuse, U. Mobech-Hanssen, and P.-O. Goethe [2]; reproduced with permission of the authors and Industrieverlag von Hernhaussen K. G.
[b] Development distance 10 cm.
[c] From Marcuse [1].

sional separation of C_{10} to C_{14} aldehydes, development in the first direction was with petroleum ether–ethyl ether–acetic acid (97:2:1), and in the second direction with methanol–water–acetic acid (7:3:2). The C_{10} to C_{20} alkan-3-ones could be separated using benzene in the first direction and methanol in the second direction. For the two-dimensional separation of aldehydes and ketones simultaneously with a separation of individual homologs, a number of solvent combinations were used. For the first direction, benzene–acetic acid (99:1) or petroleum ether–diethyl ether (98:2) was used, and for the second direction, a mixture of acetonitrile–acetic acid (3:1) or methanol–water (9:1). The authors reviewed the application of many reagents for detecting carbonyl groups and selected a 10% phosphomolybdic acid in ethanol spray, followed by heating at 120°C.

Mahadevan et al. [3] used thin-layer chromatography on silica gel G with toluene to separate aldehyde dimethyl acetals from methyl esters and higher fatty aldehydes. In order to separate the methyl esters and the aldehydes the plates were developed in a second dimension with petroleum ether (30 to 60°C)–ethyl ether–acetic acid (90:10:1). The saturated aldehydes were then separated from the unsaturated by using silver nitrate-impregnated silica gel layers with petroleum ether–ethyl ether (19:6) with the saturated aldehydes moving with the solvent front. The unsaturated aldehydes were separated according to the degree of unsaturation with the lower unsaturated compounds having the higher R_f values. The saturated aldehydes were removed with ether and separated by reverse-phase partition chromatography on silica gel impregnated with silicone. Development was carried out three times with 85% aqueous acetone saturated to 90% with silicone.

Barbier et al. [4] have used a starch-bound silica gel layer with benzene–ethyl acetate as a solvent to separate a group of β-dicarbonyl compounds. For separating aliphatic components a benzene–ethyl acetate ratio of 7:3 proved to be satisfactory and for the aromatic compounds a ratio of 1:1 was suitable. Kučera [5], using loose layers of alumina with hexane–acetone (4:1), reports the following R_f values for some diketones: diacetyl 0.76, acetylacetone 0.03, acetonylacetone 0.40, and phorone 0.86. Saitoh and Suzuki [6] chromatographed 10 β-diketones in carbon tetrachloride, toluene, benzene, dichloromethane, and diethyl ether.

Dhar and Misra [7] examined a group of 93 α,β-unsaturated carbonyl compounds on silica gel G with petroleum ether (40 to 60°C)–ethyl acetate (5:1). Chromatography was carried out in a saturated tank. Another group of 20 compounds were separated using benzene–petroleum ether (40 to 60°C) (2:3) [8].

Petrowitz [9] separated a group of hydroxyaldehydes on silica gel plates using a benzene–methanol (95:5) solvent for a development of 12 cm and obtained the R_f values shown in Table 20.2. Heřmánek et al. [10] separated hydroxyaldehydes on loose layers of alumina (Table 20.2). Prey et al. [11] have used a silica gel layer buffered with 0.1 *N* boric acid for separating some hydroxyaldehydes and ketones. The solvents for the buffered layers consisted of butanol–water (9:1) and butanol–acetone–water (4:5:1). An ammoniacal silver nitrate solution was used to detect these compounds.

Chawla et al. [12] separated six hydroxyxanthones on silica gel G impregnated with 2% nitrobenzene. Development was in a saturated tank with benzene–xylene–ethyl formate–formic acid (3:7:8:2). The compounds could be detected with iodine, alcoholic ferric chloride, or 15% sulfuric acid. Saleh [13] chromatographed 25 naturally occurring hydroxyxanthones on silica gel DS-5 (CAMAG) in 35 solvent systems for which the R_f values are reported. Arends [14] preferred to use a highly purified silica gel with the addition of 2% of the disodium salt of EDTA for xanthones so as to prevent tailing. A 2:1 mixture of cornstarch and tapioca flour was used as a binder. Methylene chloride–ethyl acetate (19:1) was used as a developing solvent.

Knappe et al. [15] have separated a mixture of substituted 2-hydroxybenzophenones on layers of kieselguhr, aluminum oxide, cellulose powder, or silica gel which were impregnated with adipic acid triethylene glycol polyester. The impregnating agent was mixed in with the slurry prior to the preparation of the layers. For the inorganic materials, 30 g of the adsorbent was mixed with 12 g of polyester solution (80 to 82% adipic acid triethylene glycol polyester in methyl glycol), 25 ml water, 25 ml ethanol, and 0.05 g of sodium diethyldithiocarbamate. After spreading the layers they were dried at 105°C for 30 min. In the case of the cellulose support, 15 g of cellulose powder MN 300 G was blended with 6.2 g of polyester solution, 0.05 g of sodium diethyldithiocarbamate, 35 ml of water, and 35 ml of alcohol. The drying of the layers was carried out at the same temperature. As a solvent for the separation, a mixture of *m*-xylene and formic acid (98:2) that had been saturated with polyester was used. Ďurišinova and Belluš [16] give the R_f values for 38 derivatives of 2-hydroxybenzophenone on silica gel D with carbon tetrachloride–ethanol (120:1); polyamide with chloroform–acetic acid (4:1) and silica gel D (impregnated with boric acid) with heptane–ethanol (3:1), carbon tetrachloride–ethanol (120:1), and chloroform–benzene (4:1). Diazotized sulfanilic acid was used as a detection reagent. Libosvar et al. [17] used thin layers of aluminum oxide to control the classical chloramphenicol synthesis. The purity of *p*-nitrophenone was checked as well as the course of its bromination, the hydroxymethylation of α-acetamido-*p*-nitroacetophenone, the reduction of α-acetamido-β-hydroxy-*p*-nitroacetophenone, and the dichloroacetylation of D-*threo*-1-*p*-nitrophenyl-1,3-propanediol.

Kheifitis et al. [18] separated a group of 15 alkylcyclohexanones on a layer of neutral alumina using benzene-petroleum ether (1:3) as a solvent. Visualization was accomplished using iodine vapors.

Kore et al. [19] separated some acetals and the corresponding aldehydes on thin layers of silica gel, using a solvent of 15% ethyl acetate in hexane. The compounds were located by spraying with a saturated solution of 2,4-dinitrophenylhydrazine in 2 *N* hydrochloric acid. Mahadevan et al. [3] separated the dimethyl acetals of higher fatty aldehydes on silica gel G plates impregnated with silicone and using 85% aqueous acetone saturated to 90% with silicone. Separation according to degree of unsaturation was achieved on silver nitrate-impregnated silica gel layers with petroleum ether (30 to 60°C)–ethyl ether (22:3). Cyclic acetals of long-chain aldehydes were separated from free aldehydes and impurities on

Table 20.2. R_f Values of Substituted Benzaldehydes in Various Systems

	Silica Gel G[a] [22]			*Silica Gel G, Development Distance 12 cm* [9]	*Loose-Layer Aluminum Oxide (Brockmann activity III)* [10]			
Benzaldehyde	*Benzene–Ethyl Acetate–Acetic Acid* (90:5:5)	*Chloroform*	*Benzene–Pyridine* (9:1)	*Benzene–Methanol* (95:5)	*Benzene–Ethanol* (98:2)	*Benzene–Ethanol* (95:5)	*Benzene–Ethanol* (90:10)	*Chloroform*
Benzaldehyde					0.80	0.84		
o-Hydroxybenzaldehyde	1.50	1.52	1.42	0.19	0.32	0.44	0.52	0.26
m-Hydroxybenzaldehyde	0.66	0.24	0.90		0.03	0.30	0.47	0.11
p-Hydroxybenzaldehyde	0.51	0.16	0.78	0.17	0.02	0.15	0.24	0.05
2,4-Dihydroxybenzaldehyde	0.66	0.20	0.79					
2,5-Dihydroxybenzaldehyde	0.59	0.20	0.76					

3,4-Dihydroxybenzaldehyde	0.17	0.02	0.32	0.06				
2-Hydroxy-3-methoxybenzaldehyde	1.33	1.25	0.87	0.53	0.03	0.04	0.04	0.06
2-Hydroxy-3-ethoxybenzaldehyde	1.54	1.48	1.19					
3-Hydroxy-4-methoxybenzaldehyde	0.62	0.55	0.71		0.04	0.21	0.32	0.26
3-Hydroxy-4-ethoxybenzaldehyde	0.80	0.77	0.89					
4-Hydroxy-3-methoxybenzaldehyde	0.81	0.71	0.80	0.27	0.04	0.13	0.16	0.13
4-Hydroxy-3-ethoxybenzaldehyde	1.00	1.00	1.00					
4-Hydroxy-3,5-dimethoxybenzaldehyde	0.56	0.50	0.43	0.14				
3,4-Dihydroxy-5-methoxybenzaldehyde	0.21	0.10	0.18					
o-Methoxybenzaldehyde	1.58	1.49	1.38					
m-Methoxybenzaldehyde	1.72	1.58	1.52					
p-Methoxybenzaldehyde	1.47	1.47	1.41	0.65	0.19	0.39		
2,3-Dimethoxybenzaldehyde	1.42	1.55	1.36					
2,4-Dimethoxybenzaldehyde	1.16	1.30	1.09					
2,5-Dimethoxybenzaldehyde	1.60	1.54	1.42					
3,4-Dimethoxybenzaldehyde	0.98	1.43	1.13	0.45				
3,4-Methylenedioxybenzaldehyde	1.51	1.56	1.40					
3,5-Dimethoxybenzaldehyde	1.73	1.68	1.53					
3,4,5-Trimethoxybenzaldehyde	1.10	1.36	1.23					

[a] R_f values referred to 4-hydroxy-3-ethoxybenzaldehyde.

silica gel with xylene [20]. Kul'nevich et al. [21] separated the cyclic acetals of 2-furaldehyde and related aldehydes on loose layers of alumina with acetone–benzene (1:10). Several cis and trans isomers were separated. Viswanathan et al. [21a] chromatographed the dimethyl acetals of dodecanal, hexadecanal, and *cis,cis,cis*-9,12,15-octadecatrienal; and the diethyl, dipropyl, diallyl, dibutyl, diisobutyl, dipentyl, and diisopentyl acetals of hexadecanal. Silica gel G was used with toluene as a solvent. Four critical pairs were found which were separated by other TLC techniques. Kore et al. [19] chromatographed 22 acetals and aldehydes using the chromatostrip technique of Kirchner [24]. Separations were on silica gel bound with plaster of Paris (the so-called silica gel G) using 15% ethyl acetate in hexane as the solvent.

2 SEPARATION OF CARBONYL DERIVATIVES

2,4-Dinitrophenylhydrazones

The earliest separation of aliphatic 2,4-dinitrophenylhydrazones by thin-layer chromatography was that of Onoe [23] in 1952. Using the chromatostrip technique introduced by Kirchner et al. [24], Onoe chromatographed a group of 2,4-dinitrophenylhydrazones including those of the *n*-aliphatic aldehydes up to C_{10}. The separations were carried out on silica gel layers bound with 2.5% polyvinyl alcohol. A number of different systems have since been used for the separation of these compounds. Labat and Montes in 1953 [25] used silica gel–bentonite layers (4:1) and applied a step technique with three different solvent systems. Rosmus and Deyl [26] applied separations on loose layers of alumina to a group of carbonyl 2,4-dinitrophenylhydrazones using ether and benzene–hexane (1:1) as solvents. Bordet and Michel [27] used silica gel with five different solvents to separate 0.02 μmole of acetone, methyl ethyl ketone, and methyl propyl ketone; Pailer et al. [28] also applied silica gel to the separation and identification of the derivatives of low-molecular-weight carbonyl compounds in cigarette smoke. Rink and Herrmann [29] separated the 2,4-dinitrophenylhydrazones of acetone and acetoacetic acid obtained from urine. This was accomplished on acetylated cellulose with methanol–water–25% ammonium hydroxide (90:10:3) and with *n*-propanol-ammonium carbonate solution (2.5:1). The ammonium carbonate solution was prepared by mixing two volumes of 10% ammonium carbonate with one volume of 5 *N* ammonium hydroxide. Dhont and de Rooy [30] separated a group of 2,4-dinitrophenylhydrazones of nine aldehydes occurring in foods (Table 20.3). The R_B values (R_f value referred to butter yellow) are reported on silica gel in two different solvents. The two solvents can be used advantageously in two-dimensional work.

Mehlitz et al. [31] investigated a group of solvents for the separation of 2,4-dinitrophenylhydrazones on silica gel plates. They found that the derivatives of the aliphatic aldehydes and ketones up to about C_{10} could be separated with carbon tetrachloride–hexane–ethyl acetate (10:2:1); those of the higher carbonyl compounds could be separated in petroleum ether (40 to 60°C)–diisopropyl ether (22:3), and those of the aromatic carbonyl compounds in benzene–hexane (3:2). In naturally occurring food aromas there are often carbonyl combinations which cannot be separated chromatographically; hence these authors looked for a color reaction that would help to differentiate these compounds. As an example of such a group they cite caproic aldehyde, 2-hexenal, and methyl propyl ketone whose 2,4-dinitrophenylhydrazone R_B values (referred to butter yellow) are 0.88 to 0.89 in a carbon tetrachloride–hexane–ethyl acetate mixture. They found that by spraying the chromatogram with a mixture of 0.2% potassium ferricyanide and 0.01% ferric chloride hexahydrate in 2 N hydrochloric acid (T-201), these compounds could be differentiated. With this reagent the methyl propyl ketone spot became greenish blue, quickly turning blue; the 2-hexenal spot did not change color but only became more intensified and the caproic aldehyde spot became olive green after a longer period of time and finally turned blue after standing

Table 20.3 R_B[a] Values on Silica Gel G of a Group of 2,4-Dinitrophenylhydrazones of Nine Aldehydes Found in Foods[b]

Aldehyde	*Benzene–Petroleum Ether* (3:1)	*Benzene–Ethyl Acetate* (95:5)
Vanillin	0.06	0.17
Vertraldehyde	0.0	0.45
Ethylvanillin	0.0	Streaks
Salicylaldehyde	0.50	0.85
Cinnamaldehyde	0.83	1.04
Benzaldehyde	1.06	1.03
α-Ionone	1.40	1.14
β-Ionone	1.41	1.14
Anisaldehyde	—	0.88

[a] $R_B = R_f$ of compound/R_f of butter yellow.
[b] From J. H. Dhont and C. de Rooy [30]; reproduced by permission of the authors and W. Heffer and Sons.

about 15 min. An examination of the data obtained from the entire series of compounds resulted in the following findings [31]: "1. The DNPH derivatives of saturated ketones give an immediate blue color reaction. 2. The DNPH derivatives of saturated aldehydes react slower and for the most part become olive green (after standing fifteen minutes they finally become blue). 3. The DNPH of unsaturated carbonyl compounds up to a carbon content of C_{10} do not react, but deepen in color intensity. Higher molecular weight compounds react after a long time giving a weak color reaction." Table 20.4 gives the R_f values of the compounds investigated along with the results of the color reaction.

Auvinen and Favorskaya [32] have reported on a micro method for the preparation of 2,4-dinitrophenylhydrazones coupled with thin-layer chromatography for the identification of aldehydes and ketones. For the preparation of the derivatives, 13 to 17 mg of the carbonyl is refluxed on a water bath for 25 to 30 min with 25 ml of a solution of purified 2,4-dinitrophenylhydrazine. The reagent is prepared by dissolving 100 mg of the hydrazine and 550 mg of oxalic acid in methanol to make 50 ml of solution. R_f values were given for a series of compounds. In the preparation of DNPH derivatives of phenolic carbonyl compounds in low concentrations, Wildenhain and Henseke [33] overcame the difficulties resulting from the hindering action of the OH group by etherifying the OH groups [34] before preparing the derivatives. Ronkainen and Brummer [35] investigated the preparation of DNPH derivatives in very dilute solutions, especially those containing alcohol. Bishydrazones of dicarbonyls could be precipitated, but the monohydrazones of aldehydes and keto acids were isolated by adsorption on activated carbon with subsequent elution in a Soxhlet apparatus, first with methyl formate and dichloromethane in succession for the aldehyde derivatives and then under nitrogen at reduced pressure (17 mm Hg) with an azeotropic mixture of pyridine–water for the extraction of the keto acid derivatives. In this latter step the oxalacetic acid derivative is decarboxylated mainly to pyruvic acid hydrazone. Craske and Edwards [36] examined the technique of Schwartz et al. [37] of preparing quantitative yields of the DNPH derivatives by passing a solution containing the carbonyl compounds through a dinitrophenylhydrazine–phosphoric acid–Celite column. The need for *stable* carbonyl-free solvents was emphasized, because some samples of carbonyl-free petroleum ether still produced some DNPH derivative on passing through the Schwartz column. The TLC method of Craske and Edwards [38] (as explained later in the text) was used to separate the derivatives after their recovery from unchanged reagent and extraneous material by the ion-exchange method of Szonyi [39]. Schwartz et al. [40] also freed 2,4-DNPH's of excess reagent by passing them through freshly treated cation-exchange resin AG 50W-X4 using purified methanol and carbonyl-free benzene. Recovery of model 2,4-DNPH's was 98 to 101% and 1 g of moisture-free resin retained 100 mg of reagent.

The method of Auvinen and Favorskaya [32] has been extended by Shevchenko and Favorskaya [41] to the separation of DNPH's of isomeric ketones. Aluminum oxide layers were used with cyclohexanediethyl ether (4:1). The following R_f values are given: isobutyl methyl ketone 0.31, diethyl ketone 0.32, methyl butyl ketone 0.30, methyl propyl ketone 0.27, isopropyl ethyl ketone 0.41, ethyl propyl ketone 0.39, cyclopentanone 0.20, pinacolone 0.41, cyclohexanone 0.24, and isopropyl methyl ketone 0.31. Two-dimensional chromatography was used where the compounds could not be separated in a single dimension.

Activated magnesia has also been used for the separation of these derivatives. Schwartz and Parks [42] used layers composed of 15 g of Seasorb 43 (activated magnesia, Fisher) and 6 g of Celite. The mixture was applied to the glass plates by slurrying with 45 ml of 95% ethanol. The finished plates were dried in air first, and then at 100°C for 16 to 20 hr. Using a mixture of chloroform–hexane (17:3), the DNPH compounds could be separated into classes and were readily distinguishable by their colors. The methyl ketone derivatives are gray and have the highest R_f values, those of the saturated aldehydes are tan and have next highest R_f values, next come the 2-enals of a rust-red color, and finally with the lowest R_f values are the 2,4-dienals with a lavender color. There is some overlapping in the groups, but they can be differentiated by means of the color difference. Schwartz et al. [43] have extended this method to the separation of dicarbonyl bis(2,4-dinitrophenylhydrazones). In this case the magnesia was conditioned by heating in a muffle furnace at 525 ± 25°C for 16 hr, and the Celite 545 with which it was blended was dried at 100°C for 24 hr. The compounds were applied to the plate in an ethyl acetate solution and the plate was developed in a direction parallel to that of the application of the slurry. The solvent was an acetone–benzene–methanol mixture whose composition was picked to give the best separation with each lot of magnesia; these mixtures ranged from 75:23:2 to 75:15:10. Here again a difference in color in the different classes was an aid to the identification; the 2,3-diketone derivatives are violet and the α-keto aldehydes and glyoxal are blue. The method is very sensitive; 0.01 to 0.02 μg of diacetyl bis(2,4-dinitrophenylhydrazone) can be detected. Schwartz et al. [44] modified their procedure using magnesia–Celite (3:7) as the adsorbent. For monocarbonyl compounds the plates were air-dried for 48 hr and development was with hexane–chloroform (19:5). For the dicarbonyl derivatives the plates were air-dried 30 min and then heated at 100°C for 1 hr with development in chloroform–

Table 20.4. R_B[a] Values and Color Reactions of 2,4-Dinitrophenylhydrazones of Carbonyl Compounds Separated on Silica Gel G[b]

Compound	*Carbon Tetra-chloride–Hexane–Ethyl Acetate* (10:2:1)	*Petroleum Ether* (40 to 60°C) *Diisopropyl Ether* (22:3)	*Time (sec) for Color Reaction*[c]	*Color*	*Remarks*
Methanal	0.40	0.23	20	Green	120 sec pine green, 180 sec blue
Ethanal	0.46	0.26	30	Green	120 sec pine green, 180 sec blue
Propanal	0.64	0.31	30	Green	120 sec pine green, 180 sec blue
n-Butanal	0.74	0.56	30	Soft green	120 sec pine green, 180 sec blue
Iso butanal	0.82	0.65	50	Soft green	120 sec green, 180 sec pine green, 230 sec blue
n-Pentanal	0.81	0.68	30	Soft green	100 sec green, 180 sec blue
Isopentanal	0.82	0.69	40	Soft green	100 sec green, 180 sec blue
Hexanal	0.88	0.75	45	Soft green	120 sec green, 170 sec pine green, 230 sec blue
Heptanal	0.93	0.79	45	Soft green	120 sec green, 170 sec pine green, 230 sec blue
Octanal	0.98	0.84	45	Soft green	120 sec green, 170 sec pine green, 230 sec blue
Nonanal	0.99	0.86	50	Soft green	120 sec green, 170 sec pine green, 230 sec blue
Decanal	1.03	0.89	50	Soft green	120 sec green, 170 sec pine green, 230 sec blue
Dodecanal	1.07	0.91	50	Soft green	120 sec green, 170 sec pine green, 230 sec blue
Acetone	0.61	0.25	15	Greenish blue	25 sec blue
Methyl ethyl ketone	0.78	0.69	15	Greenish blue	25 sec blue

Methyl propyl ketone	0.89	0.81	25	Greenish blue	25 sec blue
Cyclohexanone	0.75	0.61	10	Green	30 sec blue
β-Ionone	1.11	1.23	70	Dirty green	
l-Carvone	1.13	1.25	10	Color deepening	
d-Carvone	1.15	1.22			
α-Ionone	1.20	1.32	45	Dirty green	
Acrolein	0.62	0.42	270	Green	
Crotonaldehyde	0.67	0.45	10	Color deepening	
trans-2-Hexenal	0.88	0.75	10	Color deepening	
2-Hexenal	0.89	0.76	10	Color deepening; after 5 min very weak coloring	
2-Decen-1-al	1.06	0.97	10	Color deepening	
2-Dodecen-1-al	1.10	1.04	300	Very weak green	
Citral	0.98	0.79	10	Color deepening	270 sec dirty yellow-green
	1.05	0.96			
Diaectyl	0.26			No color change	
Furfural	0.43	0.21	10	Color deepening	120 sec gray, 240 sec blue
Anisaldehyde	0.50	0.21	10	Color deepening	
Benzaldehyde	0.76		10	Color deepening	
Acetophenone	0.90		10	Color deepening	

[a] $R_B = R_f$ compound/R_f butter yellow.

[b] From A. Mehlitz, K. Gierschner, and T. Minas [31]; reproduced with permission of the authors and Alfred Heuthig Verlag.

[c] Spray reagent: 0.2% $K_3[Fe(CN)_6]$ + 0.01% $FeCl_3 \cdot 6H_2O$ in 2 *N* HCl.

methanol (19:1). Microcel T-38 (Johns Manville) layers were prepared from slurries of 15 g of adsorbent with a solution of 12.5 ml of polyoxyethylene in 70 ml of ethanol. The plates were air-dried and then heated at 100°C for 5 min. The 1,2-diketone derivatives were separated with benzene–hexane (3:2); the other α-ketone derivatives with a 1:3 mixture; and the homologous series of methy ketones, saturated aldehydes, 2-enals, and 4-enals with hexane (all these solvents were saturated with the impregnating agent). Hydrazones of the C_5 to C_{11} carbonyls were separated on air-dried layers of Microcel T-38 prepared with the polyoxyethylene dissolved in 65 ml of 1% methanolic potassium hydroxide. The solvent was hexane saturated with polyoxyethylene.

Cobb [45] also worked with the dicarbonyl bis-derivatives. In this case mixtures of Seasorb 43–silica gel G (1:1) and Seasorb 43–Celite–calcium sulfate (10:8.5:1.5) were used as adsorbents. As solvents he used chloroform–tetrahydrofuran–methanol (15:4:1) and benzene saturated with ethanolamine-8% methanol. Colors of the spots were intensified by spraying with 10% ethanolic potassium hydroxide. Cobb et al. [46] also applied the reverse-phase technique of Libbey and Day [47] on mineral oil-impregnated silica gel layers using dioxane–water (6:4) as the solvent. Continuous development by exposure of the end of the plate provided good separation of homologous series of the derivatives of 2,3-diketones and α-ketoalkanals.

Two procedures for the systematic separation of 2,4-dinitrophenylhydrazones appeared at approximately the same time (48, 49). Although the systems are different they have some aspects in common, that is, the use of a combination of partition and adsorption chromatography and the use of silver nitrate-impregnated layers. For the separation into classes, Urbach (49) used layers of aluminum oxide G dried at 115°C for 15 min and then stored, open to the atmosphere, until used. Using a mixture of petroleum ether (30 to 40°C)–diethyl ether (96:4), the *n*-alkanals, *n*-alkan-2-ones, and the *n*-1-alken-3-ones could be separated into their respective groups. However, the first member of each of the ketone series had R_f values that were appreciably lower then the remaining members of the series. Thus the methyl vinyl ketone had the same R_f value as the methyl ethyl ketone, and the acetone derivative had the same values as the alkanal derivatives. In order to separate the aldehydes into classes, advantage was taken of the complexing action of silver nitrate with unsaturated compounds.

Using Urbach's procedure isomeric compounds having the same number of carbon atoms but different degrees of unsaturation were separated from one another. Urbach prepared the thin layers by using 30 g of aluminum oxide G and 7.5 g of silver nitrate in 50 ml of water. After allowing the plaster of Paris binder to set, the plates were oven dried at 115 to 135°C. Here again, the plates were equilibrated to the atmosphere by allowing to stand overnight in the open, but protected from bright light. The silver nitrate-impregnated layers were developed with petroleum ether (30 to 40°C)–diethyl ether (21:4). Using a multiple development (2×) in this system, the 2,4-DNPHs of alkanals, 2-alkenals, and 2,4-alkadienals could be separated according to their group. For the separation it is also best to remove ketone derivatives by first chromatographing on an aluminum oxide plate. For the separation of members of a given homologous series partition chromatography was applied. Kieselguhr layers were impregnated by dipping in a solution of 10% 2-phenoxyethanol in acetone. After removal of the acetone and spotting of the compounds in ether solution, the plates were developed in petroleum ether (100 to 120°C). Multiple development was employed to increase the separation and to this extent three developments were carried out for a distance of 9 cm and a fourth development a distance of 11 cm. This procedure separates the individual members of any given homologous series. However, where mixtures contain members of more than one homologous series there may be overlapping of spots in a one-dimensional system. In order to carry out the two-dimensional analysis of a mixture, a solution of the mixed 2,4-dinitrophenyl-hydrazones was applied in the usual manner in the corner of an aluminum oxide plate for two-dimensional work. Development in petroleum ether (30 to 40°C)–diethyl ether (96:4) was allowed to run to the edge of the plate, thus giving a development distance of approximately 16 cm. The plate was removed and allowed to stand till the solvent had evaporated, after which it was again placed in the same solvent and developed a second time in the same direction for the same distance. It was then removed and after the solvent was evaporated the plate was impregnated by dipping in a 10% phenoxyethanol solution in acetone. Care was taken during the impregnation not to wash off the spots of the first development. The acetone was evaporated and then the plate was developed in the second direction with petroleum ether (100 to 120°C). This first plate permitted the identification of the ketones except for acetone which, being the first member of the series, was out of line and was found with the lower aldehydes. In order to separate and identify the aldehydes a second aluminum oxide G plate was prepared, and the original mixture was developed on this with petroleum ether (30 to 40°C)–ether (96:4). Since the purpose of this plate is to separate the ketones from the aldehydes, the development may be repeated two or more times in order to get a sharp separation between these two classes. After this has been accomplished, the aldehydes are removed from the adsorbent by extracting with diethyl ether. The ether extract is concentrated and applied to an aluminum oxide plate impregnated with silver nitrate. Development in the first direction is carried out with petroleum ether (30 to 40°C)–diethyl ether (21:4). For the second direction the plate is first impregnated with 10% phenoxyethanol in acetone and then developed with petroleum ether (100 to 120°C).

Badings and Wassink [48] have applied layers of basic zinc carbonate containing amylopectin as a binding agent and layers of kieselguhr impregnated with silver nitrate for separating 2,4-dinitrophenylhydrazones into classes. In using the zinc carbonate plates, it was found essential to have a perfectly dry atmosphere because small amounts of moisture inactivated the plates. In order to eliminate the moisture the plates were reactivated at 110°C for $\frac{1}{2}$ hr just prior to use; in addition, after spotting with the sample, the plate was placed in the developing chamber and dry air was passed through the chamber for approximately $\frac{1}{2}$ to 1 hr. For the zinc carbonate layers the best separating solvent proved to be a mixture of petroleum ether–benzene–pyridine (7:1:2) containing 0.1% absolute ethanol. With the kieselguhr plates impregnated with silver nitrate, petroleum ether (60 to 70°C) was used as the mobile phase. In order to separate the classes into individual components according to chain length, partition chromatography was employed using zinc carbonate impregnated with Carbowax 400 and petroleum ether (100 to 120°C) as the separating solvent. They also found that kieselguhr impregnated with Carbowax with the same solvent could be used for separations according to chain length. The R_f values for the different classes examined according to chain length are given in Table 20.5.

In order to separate a complex mixture a preliminary separation is first carried out by thin-layer partition chromatography as described above. The compounds are then recovered from the plate in three groups, those of short chain length (low R_f values), medium chain length (medium R_f values), and long chain length (high R_f values). Each of these three groups is separated into classes by adsorption chromatography on zinc carbonate layers or on kieselguhr layers impregnated with silver nitrate. A third separation is now carried out by combining eluted fractions of similar R_f values from the second separation and applying these to another zinc carbonate plate impregnated with Carbowax 400 for separation into the individual components according to chain length, using partition chromatography. This separation of complex mixtures is illustrated in Fig. 20.1. Badings has investigated the oxidation products of ammonium linoleate [50] via the 2,4-DNPHs on zinc carbonate layers. He has published on a controlled-atmosphere chamber [51] for separations such as those just described for carbonyl derivatives.

Table 20.5 R_f Values for a Number of 2,4-DNPHs. Data Refers to Plates of Zinc Carbonate (250 μ) Impregnated with Carbowax 400 (Degree of Impregnation 25%). Mobile Phase Petroleum Ether b.p. 100 to 120°C[a]

Carbon Chain Length	*Class of 2,4-DNPH*				
	n-Alkanals	*n-2-Alkenals*	*n-2,4-Alkadienals*	*n-Alkan-2-ones*	*n-2-Alken-4-ones*
2	0.05	—	—	—	—
3	0.09	—	—	0.18	—
4	0.13	0.06	—	0.29	—
5	0.17	0.10	—	0.32	0.18
6	0.23	0.15	0.06	0.46	0.43
7	0.28	0.20	0.09	0.55	0.50
8	0.35	0.25	—	0.63	—
9	0.42	0.31	0.12	0.69	—
10	0.49	0.37	—	—	0.77
11	—	0.41	0.22	—	—

[a] From H. T. Badings and J. G. Wassink [48]; reproduced with permission of the authors and publishers.

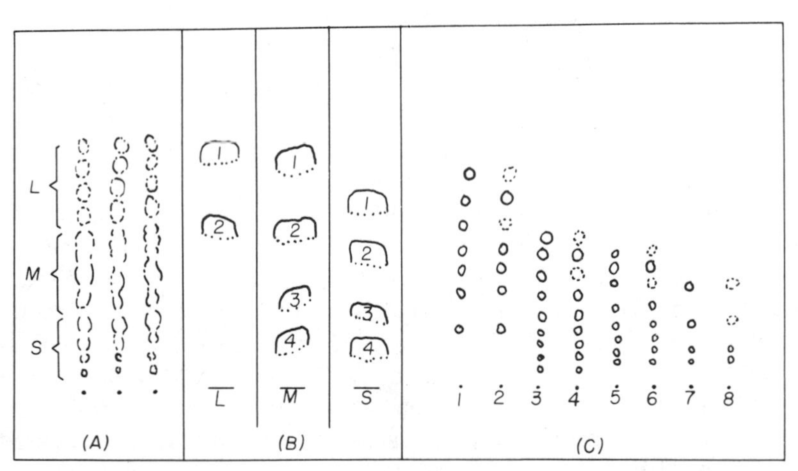

Fig. 20.1 Separation of complex misture of 2,4-DNPHs. *(A)* Preseparation by thin-layer chromatography (partition) into long*(L)*-, medium*(M)*-, and short*(S)*-chain 2,4-DNPHs. *(B)* Class separation by thin-layer chromatography (adsorption) of the fractions *L*, *M*, and *S*. *(C)* Determination of chain length by thin-layer chromatography (partition) of the members of each of the separated classes of 2,4-DNPHs: *1*. test mixture of alkanones (C_3 to C_9 inclusive); *2*. combined alkanone fractions ($L_1 + M_1 + S_1$); *3*. test mixture of alkanals (C_2 to C_{10} inclusive); *4*. combined alkanal fractions ($L_2 + M_2 + S_2$); *5*. test mixture of 2-alkenals (C_4 to C_{11} inclusive); *6*. combined 2-alkenal fractions ($M_3 + S_3$); *7*. test mixture of 2,4-alkadienals (C_6, C_7, C_9, C_{11}); *8*. combined 2,4-alkadienal fractions ($M_4 + S_4$). From Badings and Wassink [48]; reproduced with permission of the authors and publishers.

Beyer and Kargl [52] also used zinc carbonate layers but with straight pyridine for separating a group of 20 derivatives into three classes: (I) those giving a yellow-brown color on the layer (R_f from 0.63 to 0.95), (II) those giving a pink-purple color on the layer (R_f from 0.07 to 0.68 showing extensive streaking), and (III) those exhibiting red to blue to violet colors on zinc carbonate (R_f from 0 to 0.13). The third group were all bis-derivatives. Class I was then separated on silica gel with ethyl acetate–hexane (1:4) (*a*) in the first direction followed by *p*-xylene in the second dimension. Class II was further separated by using (*a*) in the first direction followed by di-*n*-butyl ether at right angles. For class III the second-dimensional solvent was lutidine–hexane (1:4).

Meijboom [53] adopting the method of Badings and Wassink [48], changed the Carbowax 400 concentration to 33.3% on kieselguhr G. This improved the separation of the DNPH derivatives of C_{10} to C_{12}-saturated aldehydes and C_8, C_9, and C_{11} methyl ketones. The R_H (relative migration rate with respect to hexanal–DNPH) was determined for 12 alkanals, 62 unsaturated aldehydes, and 8 ketones. Meijboom [53a] separated unsaturated aldehydes and methyl ketone 2,4-dinitrophenylhydrazones on the same adsorbent with carbon tetrachloride–hexane–ethyl acetate (10:2:1) and then extracted the bands and separated again on silver nitrate-impregnated silica gel with benzene.

Craske and Edwards [54] used magnesium oxide–Microcell T 38 (2:3) in a two-dimensional technique to separate 2,4-DNPHs first by class and then by chain length. After the layer was activated at 60°C for 1 hr the sample was applied as a streak in the direction of spreading. Development was at right angles to the spreading direction with petroleum ether (40 to 70°C)–chloroform (17:7) in a saturated tank. The bands were then concentrated by using the right-hand edge as the bottom of the chromatogram and developing with chloroform–SVR (4:1) until the solvent reached the edge of the original streak. This was repeated a second time thus compacting the DNPs into bands at the top of the plate. A third development was then given in the same direction with 20% Carbowax 400 in chloroform until the solvent moved just past the line of DNPs. Impregnated in this manner, the plates were dried and then turned 180° and developed with petroleum ether to separate according to chain length. Prior to this latter development known compounds may be spotted along an unused edge of the layer. If the magnesium oxide was dried overnight at 110°C prior to the preparation of the slurry the tendency for the spots to tail was decreased during the chain-length separation. The order of the R_f values was as follows: ketones>alkanals>2-alkenals>2,4-alkadienals. The classes can be further differentiated by spraying with 10% potassium hydroxide in 80% ethanol; alkanones are gray-brown to brown, alkanals brown to reddish brown, 2-alkenals pink-brown, and 2,4-alkadienals mauve. As usual, class separation is not absolute and the shortest chain-length members have R_f values approximately equal to that of the next lower class.

Meijboom and Jurriens [55] applied 30% silver nitrate-impregnated silica gel layers to the separation of the positional and geometrical isomers of 2,4-dinitrophenylhydrazones of hexenals and heptenals. Benzene was employed as the solvent. Denti and Luboz [56] have used silver nitrate-impregnated silica gel and alumina as well as the untreated adsorbents for separating a group of carbonyl derivatives with mixtures of benzene–petroleum ether (40 to 70°C) (3:2), benzene–*n*-hexane (1:1), chloroform–petroleum ether (3:1), and cyclohexane–nitrobenzene–petroleum ether (6:3:2).

Nano and Sancin [57] have separated a group of low-molecular-weight carbonyl derivatives on neutral alumina (Woelm) using cyclohexane–nitrobenzene (2:1) or hexane–chloroform–nitrobenzene (8:2:1), and Nano [58] has extended the method to other compounds. A quantitative determination was also employed. This involved eluting the separated spots with chloroform and measuring the absorption. Table 20.6 gives the R_f values of the various compounds investigated in six different solvents.

Jart and Bigler [59] give the R_s values referred to butanal for 70 DNPHs on alumina using ethyl acetate–petroleum ether (62 to 82°C) saturated with water (3:20). Kore et al. [60] used *n*-hexane–ethanol (17:3) or *n*-heptane–ethanol (4:1) with aluminum oxide layers of activity grade II or IV to separate a group of derivatives.

Byrne [61] emphasized that it is important to have small starting spots. Also, when highly insoluble compounds like mesoxaldehyde tris-DNPH were applied in volatile solvents the compounds were often "precipitated" on the adsorbent and on subsequent development remained at the origin or streaked badly. This was overcome by using nitrobenzene as a sample solvent. The nitrobenzene was then removed by repeated development in petroleum ether (80 to 100°C), which moved the nitro compound to the top of the plate from where it was removed before developing the chromatogram. Byrne chromatographed a group of 41 carbonyl derivatives on silica gel G and on alumina G with petroleum ether (80 to 100°C)–ethyl ether (7:3), benzene–tetrahydrofuran (98:2 to 7:3), and benzene–tetrahydrofuran–acetic acid (15:9:1). Continuous horizontal thin-layer chromatography was used for compounds which were not completely separated by normal development. The plates were sprayed

Table 20.6 R_{st}[a] Values of 2,4-Dinitrophenylhydrazones of Carbonyl Compounds in Various Solvents[b]

	R_{st} *Value*					
2,4-DNPH of	*Nitrobenzene–Chloroform–Hexane* (1:2:8)	*Ether–Petroleum Ether* (b.p. 40–70°C) (2.5:7)	*Ethyl Acetate–Hexane* (1:9)	*Nitrobenzene–Carbon Tetrachloride* (1:9)	*Nitrobenzene–Cyclohexane* (1:2)	*Nitrobenzene–Carbon Tetrachloride* (1:2)
Formaldehyde	0.70	0.52	0.52	0.74	0.80	0.84
Acetaldehyde	0.86	0.69	0.69	0.84	0.94	0.92
Propionaldehyde	0.97	0.95	0.96	0.96	0.98	1.00
Butyraldehyde	1.10	1.11	1.09	1.05	—	—
n-Valeraldehyde	1.16	1.19	1.21	1.08	—	—
Caproicaldehyde	1.22	1.26	1.26	1.12	—	—
Oenanthal	1.27	1.30	1.31	1.14	—	—
Caprylaldehyde	1.30	1.34	1.34	1.17	—	—
Glycolaldehyde	0.04	0	0	0.02	0.06	0.05
Glyoxal[c]	0.10	0	0	0.17	0.67	0.42
Glycollic acid	0	0	0	0	0	1
Acrylaldehyde	0.86	0.92	0.91	0.93	—	1.01
Glyceraldehyde	—	—	—	0.05	—	0.07
Dihydroxyacetone[c]	—	—	—	—	0.72	0.65
Acetone	1	1	1	1	1	1
Pyruvic acid	0	0	0	0	0	0
Mesoxalic acid (diethyl ester)	—	—	—	0.72	—	0.87

[a] $R_{st} = R_f$ of compound/R_f of acetone derivative.
[b] From G. M. Nano [58]; reproduced with permission of the author and Elsevier Publishing Co.
[c] Values are for osazones.

with ethanolamine to give distinctive colors for different classes which are more lasting than those colors produced with sodium or potassium hydroxide.

Libbey and Day [47] applied reverse-phase thin-layer chromatography to the separation of some *n*-alkanals and *n*-alkan-2-ones on silica gel layers impregnated with mineral oil. The coating of the layers was carried out by immersing them in a 10% solution of mineral oil in petroleum ether. Dioxane–water (13:7) was used as a developing solvent. When additional resolution was needed the plates were developed in a continuous manner by allowing them to project through a slit in the Saran cover.

Anet [62] has separated the derivatives of some hydroxycarbonyl compounds in connection with the study of the degradation of carbohydrates. Aluminum oxide and silica gel plates were used in the separations with toluene and toluene–ethyl acetate mixtures; R_f values are listed for 16 different compounds.

The chromatographic separation of substituted benzaldehyde DNPHs have been examined on Eastman Chromagram silica gel sheets and on silica gel G on glass plates using benzene as a solvent [63] and on zinc carbonate with carbon disulfide–chloroform (9:1) [64]. Yamazaki et al. [65] used a two-dimensional technique on kieselguhr for separating these compounds. For the first dimension the layers were impregnated with 5 to 20% polyethylene glycol (PEG 200) and developed with cyclohexane–ethyl ether (7:3); the plates were then impregnated with 10% tetralin in petroleum ether for development in the second direction with 90% methanol–benzene–acetic acid–dimethyl formamide (10:1:1:1). Impregnation with dimethyl formamide for the first direction and tetralin for the second were also used with the same solvent combinations. Takeuchi et al. [66] examined the effect of preadsorbed solvents on the R_f values of these compounds.

Stereoisomers of substituted benzophenone DNPHs have been separated on silica gel [67, 68] and on silica gel impregnated with dimethylformamide [69]. Alpha-substituted carbonyl DNPHs were chromatographed on silver nitrate-impregnated alumina [70]. Some branched-chain derivatives could be separated from other derivatives having the same chain length. Denti and Luboz [71] give the R_f values of 17 derivatives on alumina and silica gel with and without silver nitrate impregnation in four solvents. Bruemmer and Mueller-Penning [72] give the R_S values referred to butter yellow for 247 derivatives on silica gel G with petroleum ether (40 to 60°C)–chloroform–ethyl acetate (30:3:1). Other separations of 2,4-DNPH derivatives include: bishydrazones of dicarbonyls [73], 32 aromatic ketones and aldehydes (Table 20.7) [74], cycloketones [75, 76],

Table 20.7 R_f and R_{st} (referred to Veratraldehyde) × 100 Values of 2,4-Dinitrophenylhydrazones of Some Aromatic Carbonyl Compounds on Starch-Bound Silica Gel Layers. Solvent: Ethyl Acetate–Petroleum Ether (75 to 120°C) (1:2). Distance 14 cm[a]

2,4-DNPH of	R_f	R_{st}
Benzaldehyde	60	177
Salicylaldehyde	48	141
m-Hydroxybenzaldehyde	32	94
p-Hydroxybenzaldehyde	30	88
Protocatechuic aldehyde	2	6
2,4-Dihydroxybenzaldehyde	23	68
2,5-Dihydroxybenzaldehyde	21	62
Anisaldehyde	48	141
o-Methoxybenzaldehyde	49	144
m-Methoxybenzaldehyde	50	147
Vanillin	23	68
Syringaldehyde	5	15
3-Ethoxy-4-hydroxybenzaldehyde	32	94
Isovanillin	23	68
o-Vanillin	32	94
Veratraldehyde	34	100
2,4-Dimethoxybenzaldehyde	41	121
3,5-Dimethoxybenzaldehyde	47	138
2,5-Dimethoxybenzaldehyde	45	133
2,3-Dimethoxybenzaldehyde	47	138
p-Ethoxybenzaldehyde	53	156
Acetylvanillin	33	97
4-Ethoxy-3-methoxybenzaldehyde	42	124
Cinnamaldehyde	60	177
p-Coumaraldehyde	33	97
Coniferaldehyde	27	80
Sinapaldehyde	15	44
p-Hydroxybenzylacetone	32	94
p-Methoxybenzylacetone	49	144
Acetovanillone	23	68
Acetosyringone	0	0
2,4-Dihydroxyacetophenone	26	76

[a] From G. Ruffini [74]; reproduced with permission of the author and The Elsevier Publishing Co.

aliphatic unsaturated aldehydes [77], carbonyl compounds in alcoholic beverages [78–82], and phenolic carbonyl compounds [33, 83] (See also Chapter XXX, Section 3).

Edwards [83a] separated the stereoisomers of the DNPHs of the normal C_3 to C_9 aldehydes on kieselguhr G impregnated with 2-phenoxyethanol using heptane as the solvent.

Oximes

Pejković-Tadić et al. [84–87] examined the separation of oximes of isomeric pairs (Table 20.8). This proved to be a convenient method for separating pure isomers, which are often difficult to obtain by other methods such as crystallization. The isomeric compounds were separated on layers of silica gel G using benzene–ethyl acetate (5:1). Application of the samples to the plate was made in tetrahydrofuran. In order to detect the oximes, the plates were sprayed with a 0.5% cupric chloride solution. α-Benzaldoxime did not give a color reaction with this reagent

Table 20.8 $R_f \times 100$ Values of Isomeric Oximes on Silica Gel G in Benzene–Ethyl Acetate (5:1). Development Distance 14 cm (2X for aliphatics) [84–86]

Compound	*syn Form*	*anti Form*
p-Tolualdoximes	54	33
p-Anisaldoximes	42	27
p-Cuminaldoximes	54	37
o-Nitrobenzaldoximes	47	40
m-Nitrobenzaldoximes	52	35
p-Nitrobenzaldoximes	53	34
Benzaldoximes	50	32
Benzoin oximes	14	37
Anisoin oximes	5	23
Formaldoxime	45.0	
Acetaldoxime	16.4	9.3
Propionaldoxime	48.6	42.8
Butyraldoxime	50.7	44.3
Valeraldoxime	54.3	46.4
Hexanaloxime	57.1	47.8
Heptanaloxime	60.0	50.0
Octanaloxime	62.9	51.4
Nonanaloxime	65.7	53.6
Decanaloxime	68.6	55.7
Isobutyraldoxime	57.1	47.1
Isovaleraldoxime	52.8	45.7

but the green color of the complex was obtained by spraying with a saturated alcoholic solution of copper acetate monohydrate and then heating the plate at 100°C for 10 min. The aliphatic oximes were detected with iodine except for the *syn* form of acetaldoxime which was visualized with alcoholic silver nitrate. Toul et al. [88] list 13 solvents for separating various groups of monoximes and dioximes of diketones on silica gel. Unterhalt has published a number of papers on the TLC of unsaturated oximes including Refs. 89 to 92. Silica gel G with benzene–ethyl acetate (3:1) was used.

Miscellaneous Derivatives

Kore and Ivanova [93] chromatographed the bisulfite derivatives on silica gel with *n*-hexane–ethanol (17:3); detection was with 2,4-dinitrophenylhydrazine. Camp and O'Brien [94] separated semicarbazones of common aldehydes on silica gel with nine different solvent systems. Isonicotinohydrazones have been separated on silica gel with six solvents [95]. One of the better solvents was benzene–ethanol–ethyl acetate (2:1:1). The compounds were located with 0.5% iodine in chloroform. A 1% ferric chloride solution was used to differentiate between the isomeric hydroxybenzaldehydes. A group of isonicotinohydrazones have also been separated on alumina with hexane–acetone (5:3) [96]. The 4-nitrophenylhydrazones of 28 aromatic carbonyls have been chromatographed on silica gel with dichloromethane, benzene–methanol (19:1), benzene, and ethyl acetate–hexane (1:1) [97]. Although visible in ordinary light, spraying with potassium hydroxide yielded characteristic colors. Another group of these derivatives, mostly aliphatic, have been separated on silica gel with benzene–petroleum ether (38 to 50°C) (4:1) and on polyamide with methanol–water (19:1) [98]. The 2-diphenylacetyl-1,3-indandione 1-hydrazones of 49 carbonyl compounds of all types were chromatographed on silica gel [99]. The R_f values are given for aliphatic aldehydes, substituted benzaldehydes and acetophenones, 9-acridone, and acetone with chloroform–hexane (1:1) as the solvent. The remaining compounds were chromatographed with a 2:1 mixture. Simple azines have been chromatographed on silica gel with benzene, hexane–ethyl acetate (1:1), and benzene–paraffin–acetic acid–water (11:4:9:6) [100]. The hydrazones of 2-hydrazinobenzothizole were used to differentiate aldehydes from ketones [101]. Separations of the derivatives were made on silica gel with petroleum ether (30 to 40°C)–ethyl acetate–acetic acid (44:5:1) in a saturated tank. The layers were sprayed with an ethanol solution of 0.1% *p*-nitrobenzenediazonium fluoroborate followed by 10% alcoholic potassium hydroxide. The colored reaction products are not given by ketones. The 2,4-dinitrophenylsemicarbazones of 25 carbonyl compounds were chromatographed on polyamide with methanol–water (19:1) and on silica gel G with heptane–benzene (4:1) [98].

References

1. R. Marcuse, *J. Chromatogr.*, **7,** 407 (1962).
2. R. Marcuse, U. Mobech-Hanssen, and P.-O. Goethe, *Fette, Seifen, Anstrichm.*, **66,** 192 (1964).
3. V. Mahadevan, C. V. Viswanathan, and W. O. Lundberg, *J. Chromatogr.*, **24,** 357 (1966).
4. M. Barbier, L. P. Vinogradova, and S. I. Zav'yalov, *Izv. Akad. Nauk SSSR., Otd. Khim. Nauk,* **1961,** 162; through *Chem. Abstr.*, **55,** 16077 (1961).
5. J. Kučera, *Collec. Czech. Chem. Commun.*, **28,** 1341 (1963).
6. K. Saitoh and N. Suzuki, *J. Chromatogr.*, **92,** 371 (1974).
7. D. N. Dhar and S. S. Misra, *J. Chromatogr.*, **69,** 416 (1972).
8. D. N. Dhar, *J. Chromatogr.*, **67,** 186 (1972).
9. H.-J. Petrowitz, *Z. Anal. Chem.*, **183,** 432 (1961).
10. S. Heřmánek, V. Schwarz, and Z. Čekan, *Pharmazie,* **16,** 566 (1961).
11. V. Prey, H. Berbalk, and M. Kausz, *Mikrochim. Acta,* **1962,** 449.
12. H. M. Chawla, S. S. Chibber, and U. Khera, *J. Chromatogr.*, **111,** 246 (1975).
13. N. A. M. Saleh, *J. Chromatogr.*, **92,** 467 (1974).
14. P. Arends, *J. Chromatogr.*, **47,** 550 (1970).
15. E. Knappe, D. Peteri, and I. Rohdewald, *Z. Anal. Chem.*, **197,** 364 (1963).
16. Ľ. Ďurišinova and D. Belluš, *J. Chromatogr.*, **32,** 584 (1968).
17. J. Libosvar, J.Nedbal, and V. Hach, *Cesk. Farm.*, **11,** 73 (1962).
18. L. A. Kheifitis, G. I. Moldovanskaya, and L. M. Shulov, *Zh. Analit. Khim.*, **18,** 267 (1963).
19. S. A. Kore, E. I. Shepelenkova, and E. M. Chernova, *Maslob.-Zhir. Prom.*, **28,** 32 (1962); through *Chem. Abstr.*, **57,** 4037 (1962).
20. P. V. Rao, S. Ramachandran, and D. G. Cornwell, *J. Lipid Res.*, **8,** 380 (1967).
21. V. G. Kul'nevich, Z. I. Zelikman, and N. F. Efendieva, *Izv. Vyssh. Ucheb. Zaved., Pishch. Tekhnol.*, **1971,** 168; through *Anal. Abstr.*, **23,** 1574 (1972).
21a. C. V. Viswanathan, F. Phillips, and W. O. Lundberg, *J. Chromatogr.*, **30,** 237 (1967).
22. M. H. Klouwen, R. ter Heide, and J. G. J. Kok, *Fette, Seifen, Anstrichm.*, **65,** 414 (1963).
23. K. Onoe, *J. Chem. Soc. Japan, Pure Chem. Sect.*, **73,** 337 (1952).
24. J. G. Kirchner, J. M. Miller, and G. J. Keller, *Anal. Chem.*, **23,** 420 (1951).
25. L. Labat and A. L. Montes, *Anales Asoc. Quim. Arg.*, **41,** 166 (1953); *Chem. Abstr.*, **48,** 3637 (1954).

26. J. Rosmus and Z. Deyl, *J. Chromatogr.*, **6**, 187 (1961).
27. C. Bordet and G. Michel, *Compt. Rend.*, **256**, 3482 (1963).
28. M. Pailer, H. Kuhn, and I. Gruenberger, *Fachliche Mitt. Oesterr Tobakregie*, **1962**, 33.
29. M. Rink and S. Herrmann, *J. Chromatogr.*, **2**, 249 (1963).
30. J. H. Dhont and C. de Rooy, *Analyst (London)* **86, 74** (1961).
31. A. Mehlitz, K. Gierschner, and T. Minas, *Chem.-Ztg.*, **87**, 573 (1963).
32. E. M. Auvinen and I. A. Favorskaya, *Vestn. Leningr. Univ.*, **18**, *Ser. Fiz. i Khim.*, 122 (1963); through *Chem. Abstr.*, **59**, 1187 (1963).
33. W. Wildenhain and G. Henseke, *Chimia*, **20**, 357 (1966).
34. H. Zahn and A. Wuerz, *Z. Anal. Chem.*, **134**, 183 (1951).
35. P. Ronkainen and S. Brummer, *J. Chromatogr.*, **27**, 374 (1967).
36. J. D. Craske and R. A. Edwards, *J. Chromatogr.*, **57**, 265 (1971).
37. D. P. Schwartz, H. S. Haller, and M. Keeney, *Anal. Chem.*, **35**, 2191 (1963).
38. J. D. Craske and R. A. Edwards, *J. Chromatogr.*, **51**, 237 (1970).
39. C. Szonyi, referred to in Ref. 36.
40. D. P. Schwartz, A. R. Johnson, and O. W. Parks, *Microchem. J.*, **6**, 37 (1962).
41. Z. A. Shevchenko and I. A. Favorskaya, *Vestn. Leningr. Univ.*, **19**, *Ser. Fiz. Khim.*, 107 (1964); through *Chem. Abstr.*, **61**, 8874 (1964).
42. D. P. Schwartz and O. W. Parks, *Microchem. J.*, **7**, 403 (1963).
43. D. P. Schwartz, M. Keeney, and O. W. Parks, *Microchem. J.*, **8**, 176 (1964).
44. D. P. Schwartz, J. Shamey, C. R. Brewington, and O. W. Parks, *Microchem. J.*, **13**, 407 (1968).
45. W. Y. Cobb, *J. Chromatogr.*, **14**, 512 (1964).
46. W. Y. Cobb, L. M. Libbey, and E. A. Day, *J. Chromatogr.*, **17**, 606 (1965).
47. L. M. Libbey and E. A. Day, *J. Chromatogr.*, **14**, 273 (1964).
48. H. T. Badings and J. G. Wassink, *Neth. Milk Dairy J.*, **17**, 132 (1963).
49. G. Urbach, *J. Chromatogr.*, **12**, 196 (1963).
50. H. T. Badings, *J. Am. Oil Chem. Soc.*, **36**, 648 (1959).
51. H. T. Badings, *J. Chromatogr.*, **14**, 265 (1964).
52. C. F. Beyer and T. E. Kargl, *J. Chromatogr.*, **65**, 435 (1972).
53. P. W. Meijboom, *J. Chromatogr.*, **24**, 427 (1966).
53a. P. W. Meijboom, *Fette, Seifen, Anstrichm.*, **70**, 477 (1968).
54. J. D. Craske and R. A. Edwards, *J. Chromatogr.*, **51**, 237 (1970).
55. P. W. Meijboom and G. Jurriens, *J. Chromatogr.*, **18**, 424 (1965).
56. E. Denti and M. P. Luboz, *J. Chromatogr.*, **18**, 325 (1965).
57. G. M. Nano and P. Sancin, *Experientia*, **19**, 323 (1963).
58. G. M. Nano. "Thin-Layer Chromatograph of 2,4-Dinitrophenylhydrazones of Aliphatic Carbonyl Compounds and their Quantitative Determination," in *Thin-Layer Chromatography*, G. B. Marini-Bettòlo, Ed., Elsevier, Amsterdam, 1964, p. 138.
59. A. Jart and A. J. Bigler, *J. Chromatogr.*, **23**, 261 (1966).
60. S. A. Kore, B. Ya. Reingach, E. I. Shepelenkova, E. M. Chernova, and R. P. Ivanova, *Tr. Vses. Nauch. Issled. Inst. Sin. Natur. Dushinstnykh Veshchestv.* **7**, 141 (1965); through *Chem. Abstr.*, **66**, 43507a (1967).
61. G. A. Byrne, *J. Chromatogr.*, **20**, 528 (1965).

62. E. F. L. J. Anet, *J. Chromatogr.*, **9,** 291 (1962).
63. K. T. Finley and R. E. Gilman, *J. Chromatogr.*, **22,** 36 (1966).
64. W. C. Eisenberg, R. E. Gilman, and K. T. Finley, *J. Chromatogr.*, **44,** 569 (1969).
65. Y. Yamazaki, Y. Suzuki, and T. Takeuchi, *Bunseki Kagaku,* **21,** 1223 (1972); through *Chem. Abstr.*, **78,** 11275e (1973).
66. T. Takeuchi, Y. Suzuki, and H. Okazaki, *Bunseki Kagaku,* **21,** 1149 (1972); through *Chem. Abstr.*, **78,** 11270z (1973).
67. L. Tschetter, *Proc. S. D. Acad. Sci.*, **43,** 165 (1964); through *Chem. Abstr.*, **63,** 8165b (1965).
68. E. Reimann, *Proc. S. D. Acad. Sci.*, **43,** 170 (1964); through *Chem. Abstr.*, **63,** 8165b (1965).
69. H.-J. Petrowitz, *J. Chromatogr.*, **40,** 462 (1969).
70. D. Sloot, *J. Chromatogr.*, **24,** 451 (1966).
71. E. Denti and M. P. Luboz, *J. Chromatogr.*, **18,** 325 (1965).
72. J.-M. Bruemmer and T. J. Mueller-Penning, *J. Chromatogr.*, **27,** 290 (1967).
73. P. Ronkainen, *J. Chromatogr.*, **27,** 380 (1967).
74. G. Ruffini, *J. Chromatogr.*, **17,** 483 (1965).
75. R. D. Slentz, R. E. Gilman, and K. T. Finley, *J. Chromatogr.*, **44,** 563 (1969).
76. H.-J. Petrowitz, *Chem.-Ztg.*, **94,** 883 (1970).
77. P. W. Meijboom and G. A. Jongenotter, *Fette, Seifen, Anstrichm.*, **77,** 135 (1975).
78. P. Ronkainen and H. Suomalainen, *Suom. Kemistil. B,* **39,** 280, (1966).
79. P. Ronkainen, V. Arkima, and H. Soumalainen, *J. Inst. Brewing,* **73,** 567 (1967).
80. P. P. Ronkainen, S. Brummer, and H. Suomalainen, *J. Chromatogr.*, **28,** 270 (1967).
81. S. R. Palamand, G. D. Nelson, and W. A. Hardwick, *Proc. Am. Soc. Brew. Chem.*, **1970,** 186; through *Anal. Abstr.*, **22,** 1178 (1972).
82. S. R. Palamand, K. S. Markl, and W. A. Hardwick, *Proc. Meet. Am. Soc. Brew. Chem.*, **1971,** 211; through *Anal. Abstr.*, **74,** 1196 (1973).
83. P. Froment and A. Robert, *Chromatographia,* **4,** 173 (1971).
83a. H. M. Edwards, Jr., *J. Chromatogr.*, **22,** 29 (1966).
84. M. Hranisavljević,-Jakovljević, I. Pejković-Tadić, and A. Stojilkjović, *J. Chromatogr.*, **12,** 70 (1963).
85. I. Pejković-Tadić, M. Hranisavljević-Jakovljević, and S. Nešić, "Thin-Layer Chromatography of Isomeric Oximes, II," in *Thin-Layer Chromatography,* G. B. Marini-Bettòlo, Ed., Elsevier, Amsterdam, 1964, p. 160.
86. I. Pejković-Tadić, M. Hranisavljević-Jakovljević, and S. Nešić, *J. Chromatogr.*, **21,** 239 (1966).
87. S. Nešić, Z. Nikić, I. Pejković-Tadić, and M. Hranisavljević-Jakovljević, *J. Chromatogr.*, **76,** 185 (1973).
88. J. Toul, J. Padrta, and Okáč, *J. Chromatogr.*, **57,** 107 (1971).
89. B. Unterhalt and U. Pindur, *Arch. Pharm. (Weinheim, Ger.),* **306,** 813 (1973).
90. B. Unterhalt, *Arch. Pharm. (Weinheim, Ger.),* **299,** 274 (1966).

91. *Ibid.*, p. 626.
92. *Ibid.*, **303,** 661 (1970).
93. S. A. Kore and R. P. Ivanova, *Tr. Vses. Nauch. Issled, Inst. Sin. Natur. Issled. Inst. Sin. Natur. Dushistnykh Veshchestv.*, **7,** 145 (1965); through *Chem. Abstr.*, **66,** 43508b (1967).
94. B. J. Camp and F. O'Brien, *J. Chromatogr.*, **20,** 178 (1965).
95. G. S. Vasilikiotis and A. Canavis, *Anal. Lett.*, **963** (1968).
96. M. H. Hashmi and M. A. Shahid, *Mikrochim. Acta.*, **1968,** 1045.
97. E. D. Barber and E. Sawicki, *Anal. Chem.*, **40,** 984 (1968).
98. J. H. Tumlinson, J. P. Minyard, P. A. Hedin, and A. C. Thompson, *J. Chromatogr.*, **29,** 80 (1967).
99. D. J. Pietrzyk and E. P. Chan, *Anal. Chem.*, **42,** 37 (1970).
100. E. D. Barber, *J. Chromatogr.*, **27,** 398 (1967).
101. F. C. Hunt, *J. Chromatogr.*, **40,** 465 (1969).

Chapter **XXI**

DYES

1 OIL-SOLUBLE FOOD DYES

Adsorption Chromatography

Oil-soluble food dyes were among the first compounds that were run on loose-layer chromatoplates. Mottier and Potterat used both radial [1] and ascending [2] chromatography with loose layers of alumina to separate some food dyes, both synthetic and natural. Lagoni and Wortmann [3–5] also used radial chromatography on loose layers of alumina to separate the following fat-soluble food colors: dimethylaminoazobenzene, bixin, β-carotene, vitamin A, Martius Yellow, Ceres Orange GN, Ceres Orange R, Ceres Red BB, Ceres Red G, and Ceres Yellow 3G.

In order to detect these dyes in the food material it is necessary to first isolate them; Janíček et al. [6] give a method for their isolation from butter. Ten grams of butter is heated with 100 ml of 10% ethanolic potassium hydroxide solution on a water bath for 30 min. The solution is then allowed to come to room temperature after the addition of 100 ml of distilled water. The dye material can then be separated by extracting four times with 50 ml of hexane. The combined hexane extracts are washed twice with 100 ml of water and dried over sodium sulfate. The hexane is then distilled off and the residue taken up in a small quantity of benzene for application to the thin-layer plates. For the separation, a loose layer of aluminum oxide was used with a petroleum ether–carbon tetrachloride (1:1) mixture as a solvent. The same authors have also published the R_f values for the same dyes in a series of 12 different solvents [7].

Reiners [8] mixed 5 g of the dyed fat or oil with 1 g of activated ferric hydroxide, made into a slurry with light petroleum ether. This was filtered through a sintered-glass filter and washed with petroleum ether to remove the fat. The dye was liberated by using warm 18% hydrochloric acid to dissolve the ferric hydroxide and then extracting with ether. After washing and concentrating the dyes were separated on kieselguhr G–0.5 *N* oxalic acid (2:5) with hexane or petroleum ether (50–70°C). R_f values are given for 22 fat-soluble dyes.

Montag [9] has used silica gel G plates for the separation of Martius Yellow, dimethylaminoazobenzene, Ceres yellow, Ceres Orange, Sudan Red G, Ceres Red BB, and indophenol. The initial fat was dissolved in petroleum ether and chromatographed on an aluminum oxide column to remove the fat and concentrate the dyes, which were then eluted with alcohol and applied to the thin-layer plates.

Fujii and Kamikura [10] studied the separation of 15 oil-soluble dyes on silica gel using 12 different solvents. They found that azobenzene, *p*-aminoazobenzene, butter yellow, *p*-methoxyazobenzene, *p*-hydroxyazobenzene, and Sudan G could be separated with 1,2-dichloroethane. Xylene or pentachloroethane would separate Oil Yellow AB and Oil Yellow OB. For the separation of Sudan III, Sudan IV, and Oil Red OS, the solvents dichloroethane, chloroform, or 1,1,2-trichloroethane could be used. Chloroform separated Quinoline Yellow SS from the remaining dyes.

Copius-Peereboom [11] examined a series of 12 dyes along with several natural pigments on silica gel G, aluminum oxide G, and kieselguhr G using six different solvent systems.

Electrophoresis has been applied to the separation of 26 coal tar food colors permitted in the United Kingdom [12]. The results in six different electrolyte solutions on paper, kieselguhr, silica gel, and alumina are tabulated.

Quite a number of research workers have published on oil-soluble dyes: Ciasca and Casinovi [13] give R_f values for 15 dyes on silica gel in two different solvents; Barrett and Ryan [14] used silica gel with five different solvents for 16 water-insoluble dyes; Canuti and Magrassi [15] analyzed a mixture containing 14 dyes permitted in Italy; Synodinos et al. [16] separated dyes permitted in Greece on calcium carbonate layers; and Naff and Naff [17] demonstrated the separation of dyes on microscope slides. The R_f values of a number of oil soluble dyes are given in Table 21.1.

Wollenweber [18] has applied the separation of dyes on thin layers of cellulose, both with and without a plaster of Paris binder. Layers prepared from either cellulose powder MN 300 or MN 300 G were dried at 105°C for 10 min. The three solvents used in the separation were 2.5% sodium citrate solution–25% ammonia hydroxide (4:1), *n*-propanol–ethyl acetate–water (6:1:3), and *tert*-butanol–propionic acid–water (50:12:38), the latter with the addition of 0.4% potassium chloride. Salo and Salminen [19] have applied this method to the separation of a series of food colors permitted in Finland. The results are shown in Table 21.2 along with some values on the separation by means of silica gel [13]. (It should be remembered that the food laws in various countries differ and that these are continually changing so that what may be permitted in one country today may not be permitted tomorrow.)

Rai [20] chromatographed 21 fat-soluble dyes on silica gel G plates using benzene, xylene, petroleum ether (40–60°C)–acetone (9:1), and petroleum ether (40–60°C)–chloroform (3:1) as solvents. The R_f values are listed.

Copius-Peereboom and Beekes [21] chromatographed 19 oil-soluble dyes in a number of systems; on polyamide with chloroform–methanol–water (5:15:1), acetone–ethanol–water (6:3:1), methanol–acetone–

Table 21.1 $R_f \times 100$ of Some Oil-Soluble Dyes

		Silica Gel[b] [10]					*Silica Gel G* [11]			*Aluminum OxideG* [11]	*Kieselguhr G* [11]
Dye	*C.I.*[a]	*1,2-Dichloroethane*	*Xylene*	*Pentachloroethane*	*Chloroform*	*1,1,2-Trichloroethane*	*Hexane–Ethyl Acetate* (9:1)	*Petroleum Ether–Ether–Acetic Acid* (70:30:1)	*Petroleum Ether–Ether–Ammonia* (70:30:1)	*Hexane–Ethyl Acetate* (98:2)	*Cyclohexane*
Martius Yellow	10315	—	—	—	—	—	0	28	0	0	0
Azobenzene		64	63	45	88	65	—	—	—	—	—
Solvent Yellow	11000	28	6	3	37	18	—	—	—	—	—
Butter Yellow	11020	49	27	10	70	47	68	68	57	59	85
p-Methoxyazobenzene		54	40	22	70	55	—	—	—	—	—
Yellow AB	11380	62	50	34	82	64	25	46	41	22	88
Yellow OB	11390	62	55	40	82	64	27	50	49	27	87
Solvent Yellow 7	11800	12	4	1	11	8	—	—	—	—	—
Sudan I	12055	42	23	16	64	41	68	77	70	56	63
Ceres Yellow	12700	—	—	—	—	—	54	74	75	56	54
Yellow XP	12740	—	—	—	—	—	60	81	80	68	40
Quinoline Yellow SS	47007	3	0	0	18	5	—	—	—	—	—
Sudan G	11920	5	2	0	5	4	14	36	37	0	0
Oil Orange SS	12100	42	20	16	64	41	—	—	—	—	—
Sudan II	12140	42	20	16	64	41	72	78	67	62	44
Ceres Red G	12150	—	—	—	—	—	18	30	36	19	16
Sudan III	2610	41	140	7	59	34	56	68	61	41	15
Sudan IV	2610	42	145	10	63	38	56	68	61	38	15
Oil Red OS	26125	35	10	7	58	34	—	—	—	—	—

[a] Color Index for the Society of Dyers and Colourists, Bradford, 1956. [b] Development distance 10 cm.

Table 21.2 $R_f \times 100$ of Some Oil-Soluble Dyes on Cellulose Powder and Silica Gel

		Cellulose MN 300 [19]		*Silica Gel* [13]		*Cellulose MN* 300 *G* [18]	
Dye	*C.I.*[a]	*2% Sodium Citrate in 5% Ammonia*	*tert-Butanol–Propionic Acid–Water (50:12:38) with 0.4% KCl*	*n-Butanol–Ethanol–Water (20:20:5)*	*n-Butanol–Ethanol–Water–HCl (0.3 N) (20:20:5:1)*	*Propanol–Ethyl Acetate–Water (50:12:38)*	*tert-Butanol–Propionic Acid–Water (50:12:38) with 0.4% KCl*
New Blue	42045	68	76	—	—	—	—
Indantrene Blue RS	69800	0	0	0	0	0	0
Indigo Carmine	73015	18	11	0	0–3	22	11
Guinea Green	42085	17	73	—	—	—	—
Patent Green	42053	86	62	—	—	—	—
Fast Yellow	13105	58	42	51	46	51	45
Tartrazine	19140	70	9	(13–33)[b]	4	20	10
Chrysoidine	14270	—	—	65	64	60	78
Quinoline Yellow	47005	—	—	68	56	33	18
Orange GGN	15980	47	43	51	28	54	51
Orange S	15985	45	41	(51)	(0–28)	52	48
Azorubine Red #2	14720	19	49	—	24–36	56	63
Naphthol Red	16045	25	44	—	—	55	56
Amaranth	16185	42	5	17	0	15	7
Ponceau 4 R	16255	65	20	37	12	28	23
Erythrosin	45430	5	86	—	—	(74)	100
Ponceau SX	14700	37	43	—	—	—	—
Scarlet GN	14815	81	59	55	0	—	—
Brilliant Black	28440	13	1	(0)	(0)	10	2
Acid Violet	42640	10	73	—	—	—	—
Violet	42650	17	71	—	—	—	—

[a] Color Index of the Society of Dyers and Colourists, Bradford, 1956.

[b] Parentheses indicate streaking. (Note: some dyes contain impurities registering as minor spots. Only the main spots are included here.)

acetic acid (6:2:2), and petroleum ether (80–100°C)–acetic acid (75:25:0.5); on silica gel with hexane–ethyl acetate (9:1), chloroform, petroleum ether–acetic acid–ethyl ether (7:1:3), petroleum ether–ammonia–ethyl ether (7:1:3), and trichlorethylene (4× development); on kieselguhr G with cyclohexane; on aluminum oxide G with hexane–ethyl acetate (49:1); and on silver nitrate-impregnated silica gel G with chloroform–petroleum ether–acetic acid (75:25::0.5), chloroform–carbon tetrachloride–acetic acid (80:20:0.2), and benzene–ethyl ether–acetic acid (90:10:0.1). R_f values are listed for Sudan I, II, and III, bixin, Martius Yellow, chlorophyll, α-carotene, Butter Yellow, Ceres Red G, Ceres Orange GN, Ceres Yellow 3G, Oil Yellow XP, Yellow OB, Yellow Ab, Orange SS, and 8′-apo-β-carotenal.

Quantitative Determination of Oil-Soluble Dyes

Davídek et al. [6, 7] have used a spectrophotometric method for measuring the dye after it is eluted from the thin-layer plate. For this determination the spot can be scraped into a sintered-glass funnel and eluted with 8 ml of benzene, after which the solution is made up to 10 ml and measured in the spectrophotometer. In contrast to paper chromatography, the entire analysis including the extraction of the dye from the food material to the final spectrophotometric determination can be accomplished in 2 hr; the development time for paper chromatography runs 6 to 10 hr. On a series of runs the error ranged from ±1.3 to ±4.0%. As mentioned previously (Chapter X1, Section 2), Frodyma et al. [22] have used a spectroreflectance method for determining dyes. Brain et al. [23, 24] have used a direct densitometric method for the evaluation of some food dyes. They pointed out that the impurities in samples from different sources vary greatly and therefore the test samples must be compared with standards of the subsidiary dye impurities of known concentration.

Reverse-Phase Chromatography

Davídek and Janíček [25] have used a reverse-phase chromatographic method supporting the paraffin-oil phase on a layer of starch. These layers are prepared by adding 10 g of starch to a 10% solution of paraffin oil in petroleum ether to form a suspension that can be spread easily. Solvents used for the separation are 50, 70, and 100% methanol, and methanol–water–glacial acetic acid mixtures (16:3:1) and (8:1:1). The best results are obtained with the solvent mixtures; multicomponent mixtures of dyes are readily separated (Table 21.3).

Ramamurthy and Bhalerao [26] used methanol–water–ammonia (20:5:1) with calcium carbonate bound with starch and impregnated with liquid paraffin to separate a number of oil- and water-soluble food dyes.

Table 21.3 $R_f \times 100$ Values of Oil-Soluble Dyes Separated by Reverse-Phase Chromatography on Starch Layers Impregnated with Paraffin Oil[a]

	Solvent System				
Dye	50% *Methanol*	70% *Methanol*	100% *Methanol*	*Methanol–Water–Glacial Acetic Acid* (16:3:1)	*Methanol–Water–Glacial Acetic Acid* (8:1:1)
Yellow OB	4	29	59	57	61
Yellow AB	6	43	61	71	71
Orange SS	1	8	41	29	38
Oil Red OS	0	0	10	3	4
Sudan I	2	11	40	36	41
Sudan II	1	4	32	19	28
Sudan III	0	3	27	10	24
Sudan IV	0	2	18	4	10
Sudan Red G	4	26	54	53	58
Sudan Yellow 3G	2	14	48	33	48
Butter Yellow	4	23	57	59	66
Sudan GN		48	63	64	74
		75	95	94	98

[a] From J. Davídek and G. Janíček [25]; reproduced with permission of the authors and the Elsevier Publishing Co.

Hoodless et al. [27] impregnated microcrystalline cellulose layers with a 10% solution of liquid paraffin in petroleum ether (80–100°C). After evaporation of the petroleum ether, 10 oil-soluble dyes were separated with 2-methoxyethanol–methanol–water (11:3:6). Impregnated silica gel G did not give as good a separation, and impregnated alumina or kieselguhr G tended to fade the colors.

2 WATER-SOLUBLE FOOD DYES

In order to separate these dyes from the numerous water-soluble materials present in foods, Mottier and Potterat [28–30] have devised a means of extraction with quinoline. The water solution of the material is mixed with an equal volume of buffer of pH 3 and is then extracted with 10 to 20 ml of quinoline. After the mixture is shaken to extract the dye, the quinoline is separated by centrifuging, and if necessary the quinoline extraction may be repeated. The dyes are then transferred from the quinoline by extracting with ether after the addition of a little water. The ether solution can then be spotted on thin loose-layer alumina plates. Because the buffer solution during the extraction procedure converts the dyes into free acids, the sample spot is covered with a drop of 1 *N* alkali and then heated to 104°C in order to convert the dyes to the salts [30]. As solvents for the separation, mixtures of water–ethanol–*n*-butanol in 1:1:9–1:1:1 ratios are used. In general, a single development is sufficient to separate the dyes, but in the more difficult case multiple development may be employed. These authors have examined 50 water-soluble dyes.

Other methods have been developed for isolating the dyes from food and pharmacetucial components. Wool fiber can be boiled with the material in acid solution which fixes the dye in the wool [31]. After the wool is separated and washed, the dye is eluted with ammonia solution. In using this technique changes are apt to occur in the dye and the recovery may not be quantitative. Davídek [32, 33] adsorbed the dye from aqueous solution onto polyamide powder. Sugars, glycols, organic acids, and ethanol do not interfere with the adsorption of the dyes and most of the natural anthocyanines are not adsorbed. Adsorbed anthocyanines can be removed by washing with 10% methanolic acetic acid before the dye is eluted with methanol–25% ammonia (19:1). Lehmann et al. [34] describe their modification of this technique used on various food products. Gilhooley et al. [35] modified the technique still further for application to quantitative work. Food products were divided into three groups, each group being treated in a slightly different manner. For samples, such as jams, jellies, and sweets, which are completely soluble in water, 5 g is dissolved in 50 ml of water with heating. This is poured through a 15 × 20 mm column of polyamide, followed by six 10-ml portions of water and three 5-ml portions of acetone. The dyes are eluted with the minimum volume of acetone–water–ammonia (sp gr 0.88) (40:9:1). Ammonia is removed by an air current, the volume is reduced by one-half, then brought up to the original volume, and acidified to pH 5 to 6 with hydrochloric acid. The solution is again adsorbed on a 10 × 20 mm column of polyamide, and washed five time with 5 ml of hot water. Elution with the acetone–ammonia solution is followed by evaporation to near dryness. The residue is dissolved in a few drops of 0.1 *N* hydrochloric acid and subjected to thin-layer chromatography. If erythrosine is suspected the residue is dissolved in water. For bakery products, 5 g of chopped sample is dried at 100°C for 30 min and then extracted three times with 30-ml portions of petroleum ether (40 to 60°C) by stirring and decanting. The sample is then ground to a coarse powder, mixed with 4 g of Celite, and poured into a chromatographic tube. Thirty milliliters of acetone is run through the column using slight pressure after the entire column is wet. The acetone eluate is discarded. The column is eluted with 50 ml of methanol–water–tetramethylammonium hydroxide (40:9:1) and the pH of the eluate is adjusted to 6 with dilute hydrochloric acid. Then 5 ml of a 1% solution of polyoxyethylene sorbitan monooleate is added and the volume reduced by one-half on a steam bath. The solution is diluted to the original volume and cooled, then poured onto a 10 × 20 mm column of polyamide (6 mm of sand on top of the column keeps the polyamide from being disturbed). The column is washed three times with 5 ml of acetone, five times with 5 ml of chloroform–absolute ethanol–water–formic acid (100:90:10:1), again three times with 5 ml of acetone, and finally three times with 10 ml of water. The dyes are eluted with the minimum amount of the acetone–ammonia solution and the technique proceeds as above. For canned meat and sausage products, 25 g of chopped product is ground with 5 g of acid-washed sand and then mixed with 10 g of Celite. After extraction with chloroform for 2 hr in a Soxhlet thimble the solvent is evaporated from the residual sample. The powdered sample is then placed in a 22 × 300 mm column and extracted with methanol–water–ammonia (sp gr 0.88) until all the dye has come through. Five milliliters of 1% polyoxyethylene sorbitan monooleate is added, and the solution is evaporated on a steam bath in an air current until the ammonia and methanol are removed. After diluting with an equal volume of water it is acidified to pH 6 with hydrochloric acid. This is then poured through a chromatographic column containing a 15 × 20 mm column of polyamide, followed by three washings of 10 ml each of water, two of 5 ml of acetone, two of 5 ml of chloroform–water–formic acid (100:90:10:1), and two more with 5 ml of acetone. Elution and concentration is the same as for the jam and jelly procedure.

Takeshita et al. [36] has isolated water-soluble acid dyes from food products by mixing the properly prepared sample with DEAE Sephadex (acetate form) in water. After the Sephadex containing the adsorbed dye was isolated and washed, the dye was eluted with 2 *N* hydrochloric acid (or sulfuric acid)–isopropanol (1:1). This acidic eluate was neutralized with 4 *N* ammonia–isopropanol (1:1) and applied directly to a polyamide layer for separation of the dyes. The best solvent for the separation was pyridine–methanol–28% aqueous ammonia (3:3:1:1) using 8-cm development. Isolation of the dyes by this technique did away with the concentration steps in previous methods.

Alary et al [37] isolated the dyes in excipients by dissolving the capsule in water and precipitating the sucrose and gelatin with methanol. The solution was then evaporated to dryness and extracted with methanol to

Table 21.4 $R_f \times 100$ of Some Food Colors[a] in Various Systems

Food Color	*C.I.*[b]	*System*[c]				
		A	*B*	*C*	*D*	*E*
Chocolate Brown FB		0	0	Str		
Chocolate Brown HT	20285	0	0	Str		
Brown FK		57;73	19;42;58	Str		
Black PN	28440	8	15	40		
Violet BNP		20	39	70		
Indigo Carmine	73015	0	0	20		0 to 40
Blue VRS		15	40			
Green S	44090	0	17	90		41
Yellow RY		0	4;1			
Yellow 2G	18965	0;56	19	90		
Tartrazine	19140	30	15	80		17
Eosin	45380				55	
Sunset Yellow	15985	55	38	60		35
Orange G	16230	55	44	80		
Phloxine	45410			57		
Yellow RFS		62	44;66			
Naphthol Yellow	10316	62	52	60		
Orange RN	15970	71	70	40;50		
Ponceau 4R	16255	13	17	70		32
Amaranth	16185	20	17	60	69	18
Ponceau MX	16150	23;73	25;53	20	76	
Ponceau 3R	16155	25	39	20	86[d]	54
Rose Bengal	45440			37		
Red 6B	18055	33	32	40		
Red 2G	18050	37	33	60		47
Acid Red	45100				64	
Red 10B	17200	48	39	20		52
Oil Red XO	12140				67	
Ponceau SX	14700	49;73	37;44;54	40	74	51
Fast Red E	16045	57	44	40		
Coralline	43800				80	
Carmoisin	14720	61	34	30		51
Red FB	14780	73	53	0		
Erythrosine BS	45430	76	67	10	44	92
Rhodamine	45170				62	

[a] Food laws differ in various countries and are subject to change.

[b] Color Index for the Society of Dyers and Colourists, Bradford, 1956.

[c] *A* = *n*-BuOH–NH_4OH (sp gr 0.088) (90:1) on polycarbonate foil for 5 cm [45]. *B* = Upper phase *n*-BuOH–EtMeCO–NH_4OH (0.88)–H_2O (5:3:1:1) 12.5 cm on silica [46]. *C* = TriNa citrate–H_2O–ammonia (0.88) (wt/vol/vol) 15 cm on microcrystalline cellulose [47]. *D* = 5% NH_4Cl–*i*-PrOH (1.5:4) for 10 cm on polyamide –silica (7:52) [48]. *E* = HAc–*i*-BuOH–H_2O (2:5:2) on silica gel G [20].

[d] Tailing.

eliminate the remainder of the sugar. Logar and Perpar [38] removed the coatings on tablets mechanically prior to dissolving in 50% ethanol and adsorption wool.

Hayes et al. [39] give the R_f values of 14 dyes on 24 solvent-adsorbent combinations. Davídek and Davídková [40] list the R_f values of 10 dyes in 11 combinations of ammonia–methanol–water on polyamide layers showing the effect of ammonia and methanol concentration. Massart and De Clercq [41] illustrate the use of numerical taxonomy for selecting the optimal sets of solvents for a TLC separation using R_f values of dyes. Tewari et al [42] give the R_f values of the components found in 27 dyes used in alcoholic drinks as well as those separated from 10 alcoholic drinks. Separations were on silica gel G with development in *n*-butanol–acetic acid–water (4:1:5) for 15 cm. Other alcoholic beverage dye separations have been reported [43, 44].

The R_f values of a group of food colors in various systems are given in Table 21.4. Additional R_f values are given in the original references [20, 46–48].

Hoodless et al. [47] have developed a system for the rapid identification of any of 49 dyes. Preliminary identification is obtained without determining the R_f value of the dye, but instead it is given a four-letter code depending on the dye's location on the chromatograms relative to the location of Orange G and Amaranth in four solvents. Having thus limited the number of possibilities further identification is obtained by determining R_f or R_x values. The R_f values in 9 solvents are given for each dye.

Turner and Jones [47a] chromatographed the triphenylmethane food dyes, Fast Green, FCF, Brilliant Blue FCF, Green S, Blue VRS, and Patent Blue V on DEAE cellulose and obtained the best separations with 1 *M* ammonium iodide and with 0.2 *M* ammonium benzoate.

3 INK PIGMENTS

Druding [49] has used thin-layer chromatography on silica gel plates to separate the pigments contained in various colored inks (Table 21.5). The inks were always applied in 95% alcohol solution and development was with the same solvent. Development distance was 10 cm on silica gel plates dried at 110°C for 1 hr. The inks used were Sheaffer's "Skrip" "washable" writing fluids. These inks also contained a fluorescent ingredient designated as RC-35. No identification of the various dyes was made. Perkavac and Perpar [50] also analyzed red, green, and blue writing inks by both paper and thin-layer chromatography, the latter giving the most reliable results. Sen and Ghosh [51] determined the R_f values for the ink components in 10 commercial blue-black inks. Purzycki et al. [52]

Table 21.5 $R_f \times 100$ of Dyes in "Skrip" "Washable" Writing Fluids on Silica Gel in 95% Ethanol. Development distance 10 cm.[a]

Ink	*Constituent*	R_f	*Ink*	*Constituent*	R_f
Black	RC-35	96	Green	RC-35	95
	Yellow	94		Yellow	93
	Red	92		Blue	78
	Blue	80	Purple[b]	Blue	85
Blue	RC-35	93		Red	84
	Blue	92		Purple	74
	Dark blue	85		Dark blue	0
	Blue	0			

[a] From L. F. Druding [49]; reproduced with permission of the author and the American Chemical Society.

[b] Although the ink itself fluoresced under an ultraviolet light, no trace of an ultraviolet constituent could be found for this particular sample on the chromatographic plates. RC-35 = fluorescent component.

separated the colored components of Polish blue and blue-black inks on silica gel G with chloroform–methanol–formic acid (15:10:1). Verma and Rai [53] found that *n*-butanol–3 *N* acetic acid–ethanol (5:5:2) gave good results and could be used for all ink colors except red inks, which showed better results with methyl ethyl ketone–pyridine–water (90:5:5). Separations were on silica gel.

An excellent paper by Nakamura and Shimoda [54] has appeared on the use of thin-layer chromatography for the identification of ball-point inks taken from documents. A hypodermic needle was used to punch out small discs of writing and, after elution with pyridine, separation was carried out on silica gel microplates using a solvent mixture of *n*-butanol–ethanol–water (10:2:3). The R_f values of a large number of dyes used in inks are listed as well as a number of spot tests to aid in their identification.

Reiners [55] examined ball-point inks on kieselguhr G layers with acetone–toluene (9:1) and Van Dessel and Van Regenmortel [56] analyzed blue ball-point inks with propanol–toluene (7:3) on silica gel layers.

Smalldon [57] placed small fragments of paper containing a straight portion of ink writing 2 mm long face down on a 4 × 9 cm silica gel G layers. This was covered by another plain glass plate to hold the paper in place and subjected to thin-layer chromatography with isoamyl alcohol–acetone–water–ammonia (50:50:30:0.04). A second solvent composed of n-butanol–3 N acetic acid–ethanol (5:5:2) was also useful. Many blue inks gave excellent results using the first solvent combined with an alkaline silica gel layer. Rettie and Haynes [58] obtained good results especially with blue inks on silica gel with n-butanol–ethanol–water–acetic acid (12:4:4:0.1).

Kelly and Cantu [59] proposed standard methods for ink identification which includes TLC identification. Samples cut from the paper with a punch made from a No. 16 hypodermic needle are eluted with ethanol, pyridine, or ethanol–water (1:1), depending on the solubility. Samples taken where one or more lines cross yield more ink than single lines. The samples are chromatographed on cellulose with n-butanol–isopropanol–distilled water (2:1:1) and on silica gel with n-butanol–ethanol–10% oxalic acid solution (10:2:3) for a distance of 7.5 cm.

4 HISTOLOGICAL STAINS

Waldi [60] separated a series of dyes on silica gel G layers using a solvent consisting of chloroform–acetone–isopropanol–sulfurous acid (5 to 6% SO_2) (3:4:2:1). The R_f values for these dyes are as follows: Acridine Orange 0.41, Alkali Blue 0.16 and 0.34, Brilliant Green 0.59, Brilliant Cresyl Blue 0.21 and 0.52, Erichromazurol S 0.39, Gentian Violet 0.43 and 0.48, Crystal Violet 0.43, Light Green 0.11, Malachite Green 0.35, Metanil Yellow 0.39, Methylene Blue B 0.9, Methylene Green 0.18, and Victoria Blue 0.51.

Stier and Specht [61] have examined a series of xanthene stains for histological work. The stains were examined on silica gel G layers using the same solvent which Waldi used [60] as well as n-propanol–formic acid (8:2) and 75% sodium acetate solution–1% hydrochloric acid–methanol (4:1:4). They also used an activated phosphorylated cellulose layer with acetate buffer (Veronal)–ethanol (5:3) solvents having a pH of 3 and 8. The stains could be divided into four groups: (1) mainly Pyronine G, (2) a mixture of pyronine, Pyronine Y, Rhodamine S, and Acridine Red, (3) mainly Rhodamine B, and (4) mainly Rhodamine G. Similarly named stains of different origin sometimes differed greatly in their composition.

Horobin and Goldstein [62] have investigated the impurities in 11 Alcian Blue, Astra Blue, Alcian Yellow, and Alcian Green samples by chromatographing on cellulose with n-butanol–water–acetic acid (3:3:1) in closed tanks or (4:4:1) in open tanks. Dobres and Moats [63] chromatographed Azure A, B, and C. Methylene Blue, Toluidine Blue O, thionine, Pyronine B and Y, Methyl Green, Crystal Violet, Amido Black 10B, and Buffalo Black (NBR) on silica gel G with n-propanol–n–butanol–ammonia–water (4:4:1:1) and *n*-propanol–ammonia–water (8:1:1). On Adsorbosil plates the latter solvent was used in a ratio of (7:2:1). Loach [64] examined Methylene Blue, Methylene Violet, Azure A, B, and C, and thionine on silica gel. Triple development in 95% ethanol–chloroform–acetic acid (17:2:1) with Brinkmann MN Sil-N-Hr/UV (254) sheets gave the best separation of the dyes and their oxidation and photodecomposition products. Separation was carried out in a saturated atmosphere in the dark. Horobin and Murgatroyd [65] compared paper chromatography, thin-layer chromatography on silica gel HF 254, and thin-layer electrophoresis on agar for the resolution of histological dye components. Thin-layer chromatography was superior to paper and in some cases resolved more components than electrophoresis, whereas with other dyes the latter was superior. The solvent systems used for thin-layer chromatography were *n*-butanol–acetic acid–water (4:1:5), 1% hydrochloric acid (wt/vol)–0.75% sodium acetate (wt/vol)–methanol (4:1:4), benzene–propionic acid (4:1), chloroform–methanol (4:1), acetic acid–ethyl Cellosolve acetate–water (1:2:2), concentrated ammonia, dimethylformamide, and water. Unfortunately no R_f values were listed nor is there any indication as to which solvent gave the best results with each dye. Duplicate samples from different suppliers were also run on some of the dyes showing the differences in the composition of some of these. The electrophoresis was carried out on an 0.8-mm agar layer at pH 4 and 9. Horobin [66] has reviewed the impurities present in biological dyes. Horobin and Murgatroyd [67] chromatographed 27 commercial samples of pyronine and Rhodamine on Eastman silica gel sheets with chloroform–methanol (4:1). Seven different red dyes were detected and separated. Lansink [68] separated Sudan Black B into its components on silica gel with chloroform–benzene (1:1). Cramer et al. [68a] used silica gel with the solvents of Dobres and Moats [63] for analysis of thiazine dyes used in Giesma stains.

The tetrazolium salts, which in themselves are relatively colorless compounds, are used in histochemistry to pinpoint the sites of tissue oxidations, because of their being able to be reduced to highly colored formazans. Tyrer et al. [69] chromatographed a group of 10 of these and also the formazans produced by their reduction. Silica gel G was used with *n*-butanol–water–acetic acid (78:17:5) for the salts and with hexane–dichloromethane (2:3) for the formazans. Grossman and Wagner [70] examined nitroblue tetrazolium from four different sources by thin-layer gel filtration on Sephadex G-25 Superfine with 0.9% sodium chloride. Jones [71] chromatographed a group of nine bistetrazolium salts on silica gel F with 3-methyl-1-butanol–formic acid–water (8:1:1). The spots were located by spraying with an alkaline sodium ascorbate solution.

Balogh et al. [71a] used silica gel H–cellulose MN 300 (2: 1) with *n*-butyl acetate–*n*-butanol–acetic acid–water (4:1:2:1) for monotetrazolium salts. For ditertrazolium salts the solvent combination was 7:3:5:3. Visualization was accomplished with an alkaline solution of sodium dithionate, glyceraldehyde, or glycolic aldehyde.

5 INDICATORS

Waldi [60] also separated a series of indicators on a 1:1 mixture of silica gel G and alumina G using a solvent composed of ethyl acetate–methanol–5 *N* ammonium hydroxide (6:3:1) and obtained the following R_f values (in some cases minor spots were also found but only the main spots are listed here): Chlorophenol Red 0.27, Bromocresol Purple 0.40, Cresol Red 0.42, *m*-Cresol Purple 0.43, Bromophenol Blue 0.48, Bromochlorophenol Blue 0.48, Bromothymol Blue 0.65, Benzyl Orange 0.67, Methyl Orange 0.67, Thymol Blue 0.74, phenolphthalein 0.83, and *p*-ethoxychrysoidine 0.84. Aliotta and Roso [72] chromatographed Phenol Red, Chlorophenol Red, Bromophenol Red, Bromochlorophenol Blue, Bromophenol Blue, Cresol, Iodophenol Blue, *m*-Cresol Purple, Bromocresol Purple, Bromocresol Green, Thymol Blue, and Bromothymol Blue on silica gel or silica gel impregnated with pH-6.0 phthalate buffer using water-saturated butanol as a solvent. Rettie and Haynes [58] separated Thymol Blue, Bromophenol Blue, and Phenol Red on silica gel G with amyl alcohol–ethanol–ammonia (10:9:1), ethyl acetate–pyridine–water (6:3:1), and benzene–isopropanol–acetic acid (60:40:1). Marowski and Fabricius [73] examined Phenol Red from different origins on silica gel with *n*-butanol–ethanol–5% ammonia (4:1:1). Chiang and Yeh [74] chromatographed 12 common indicator dyes on polyamide–silica gel (1:5.2) using both acidic and basic solvent systems containing 0.5 to 7% sodium chloride to reduce hydrogen bonding. Fluorescein dyes which are used as adsorption indicators have been separated on silica gel with chloroform–acetic acid (9:1), benzene–acetic acid (9:1) [75], and with toleuen–acetic acid (13:7) [76].

6 COSMETIC DYES

Pinter and Kramer [77] examined blue and green dyes and those to be used on the skin, but not on the eyes or mouth, by heating 0.2 g of the sample with two successive 5-ml portions of ligroin. The residue was then extracted twice with 2 ml of 50% ethanol. Chromatography was on silica gel with chloroform–acetone–*n*-propanol–sulfurous acid (9% SO_2) (2:5:2:1). Dyes with an R_f less than 0.20 were then chromatographed with n-butanol–acetic acid–water (25:6:19) containing 0.4% potassium chloride. The color reaction when sprayed with 4% hydrochloric acid or sodium hydroxide could be used to help distinguish between dyes incompletely separated in the first solvent. Dyes C.I. 60730 and 62085 could not be completely separated but could be distinguished by the lilac and blue color, respectively, in the second solvent system. The $R_f \times 100$ values for the two solvents were: 42095, 0 and 14, 7; 10020, 15, 10; 63000, 0 and 17, 0 and 41; 19140, 18, 35; 44090, 20, 19; 4535, 24, 23; 52015, 28, 19; 42100, 33 and 39, 31 and 28; 42080, 49, 40; 61570, 53, 36; 62085, 76, 60; 60730, 77, 67; 42040, 78, 60; 43820, 82, 67; 42140, 83, 67.

Perdih [78] separated the cosmetic azo pigments Hansa Yellow G, Hansa Yellow 10G, Hansa Yellow GR, Hansa Yellow 3R, Permanent Red GG, Permanent Red F4RH, Permanent Carmine FB, Permanent Brown FG, Flaming Red, and Toluidine Red on air-dried silica gel G with benzene–nitrobenzene (5:1).

Lipstick Dyes

Deshusses and Desbaumes [79] used n-propanol–ammonia (9:1) with silica gel G to separate 11 dyes used in lipsticks. Ruedt [80] used the same system to separate 12 xanthene dyes used in lipsticks, quantitative results were obtained by eluting and measuring the adsorption spectrum in the region 450 to 550 nm. Davídek et al. [81] have likewise applied a loose-layer technique with aluminum oxide to the separation of lipstick dyes. They give the R_f values for some dyes and the results of the analysis of some commercial lipsticks.

Cotsis and Garey [82, 83] have used calcium-bound silica gel layers (Adsorbosil-1) in the separation and determination of lipstick dyes. The dyes were obtained by refluxing the lipsticks with a 3:1 mixture of benzene and acetone. Three solvent mixtures were employed for the separations: benzene–methanol–ammonium hydroxide (65:30:4), benzene–n-amyl alcohol–concentrated hydrochloric acid (65:30:5), and benzene–n-propanol–ammonium hydroxide (6:3:1). Quantitative evaluations were made by eluting the spots and measuring in a spectrophotometer. Silk [84] has also used silica gel layers for determining lipstick dyes. In this case the samples were added directly by rubbing the softened lipstick on a silica gel plate of 375-μ thickness (this thick layer helped to adsorb the oils present). Most of the dyes could be separated with two solvents, *(a)* dichloromethane and *(b)* ethyl acetate–methanol–ammonium hydroxide (diluted 3 to 7) (15:3:3); however, when D & C Red No. 7 was present a phosphate-buffered plate was used. With the buffered plates, the development was first made with a mixture of n-butanol–95% ethanol–ammonium hydroxide (20:4:3) followed by development with solvent *(b)*. Spectrophotometric determinations were used for quantitative work. Perdih [85] also applied the lipsticks directly to the layer. Using 13 stationary phases with 41 solvent systems they separated more than 150 dyes, of which 37 were found in lipsticks. By using the direct application method with a series of selective mobile phases they could resolve most commercial formulations in three- or five-run procedures. For more complicated mixtures and for the concentration of minute amounts of subsidiary dyes a dimethylformamide–85% phosphoric acid (20:1) extraction procedure was used. By combination of this with several phase separations and selective silica gel and polyamide separations, the dyes were separated into three groups which were then subjected to thin-layer chromatography. Lehmann and Recktenwald [86] dissolved the sample in warm dimethylformamide from which the fats and waxes were removed by extraction with petroleum ether. The dimethylformamide solution was diluted (1:1) with water and decolorized with polyamide. After washing the polyamide with water and methanol the dyes were eluted with methanol–ammonia (19:1). Jork et al. [87] determined eosin by fluorescent scanning after separating on silica with ethyl acetate–methanol–25% ammonia (5:2:1).

Hair Dyes

Brown [88] reviewed the chemistry of hair dyes and gives petroleum ether–acetone–ammonia (sp gr 0.88) (30:10:1) with silica gel as suitable for the separation of the simpler bases used in oxidation hair dyes. For larger molecules of the *N*-substituted nitro-*p*-phenylenediamine type the acetone content can be increased to 20. The oxidation hair dyes are mixtures of dye intermediates, surfactants, hair conditioners, sulfites, ammonia, and perfumes. Their analysis is difficult because of the easily oxidized dye intermediates. Hordynska and Legatowa [89] mixed 4 g of the sample with 1 ml of 5% sodium bisulfite and 6 ml of ethanol. If necessary the mixture was filtered and 20 μl of the solution added to the silica gel G layer. The chromatogram was developed in the first direction with the lower phase of ammonia–water–light petroleum ether–chloroform–*n*-butanol (4:13:10:50:5). Development in the second direction was with the same solvent used by Kottemann [90], that is, chloroform–ethyl acetate–ethanol–acetic acid (35:10:6:2). The amines and aminophenols were detected with 1% ethanolic 4-dimethylaminobenzaldehyde. Polyphenols were detected with diazotized benzidine solution (T-36). Kottemann had used the same solvent for the first direction except that 2 ml of 50% hypophosphorus acid had been added for each 20 ml of ammonia. She chromatographed 30 dye intermediates and examined five unknown dye intermediate solutions. The procedure could detect about

0.02% of an intermediate in the original dye solution. Hydroquinone could not be detected below 0.1% because of its ease of oxidation. Other solvents used for these compounds include *n*-butyl acetate–benzene (2:1) [91], and isopropyl ether–acetone–isopropanol (10:1:1) [92]. Takemura [91] gives the R_f values and colors (with diaminobenzaldehyde) for 20 compounds. Zelazna and Legatowa [93] extracted the dyes from emulsified hair dyes with acetic acid after breaking down the emulsion by freezing. Solvent for the separation was *n*-butanol–acetic acid–ethanol–water (50:1:10:40) with visualization with 1% ferric chloride. Goldstein et al. [94] examined 28 dyes with 37 different chromatographic systems.

7 FLUORESCENT DYES

For the separation of optical bleaches derived from stilbene, Saenz Lascano Ruiz and Laroche [95] have used layers prepared from equal weights of aluminum oxide and 2.5% sodium carbonate solution. Solvents used were butanol–ethanol–water (2:1:1) and diethylene glycol monoethyl ether–2% ammonium hydroxide (4:1). Latinák [96] used two methods of chromatographing these compounds on silica gel layers: direct and indirect. For the direct separation the following solvents were used: *n*-propanol–5% sodium bicarbonate (2:1), *n*-butanol–pyridine–water (1:1:1), *n*-butanol–pyridine–25% ammonia (1:1:1), and *n*-pentanol–pyridine–25% ammonia (1:1:1). The R_f values were affected by the substituents on the triazine ring, the values increasing in the following order: $OH < NH_2 < CH_3O < C_6H_5NH$. The cis isomers had lower R_f values than the corresponding trans isomers. In the indirect method the compounds were first oxidized with alkaline permanganate, after which the resulting aldehydes were chromatographed.

Schlegelmilch et al. [97] separated derivatives of flavonic acid used as fluorescent whiteners by thin-layer chromatography on cellulose acetate with chloroform–methanol–buffer solution (1:3:1), on cellulose with *n*-butanol–acetic acid–10% sodium chloride (4:1:1), and on silica gel or alumina with methyl ethyl ketone–diethylamine–25% ammonia (3:1:1). Theidel and Schmitz [98] separated 16 optical brighteners using polyamide layers with methanol–water–ammonia (10:1:4) and methanol–6 *N* hydrochloric acid (10:2), and on silica gel G using *n*-hexanol–pyridine–ethyl acetate–ammonia–methanol (5:5:5:5:3) and benzene–chloroform (2:3). For extraction of the dyes the following solvents were used: ethylene glycol–ammonia (7:3) and pyridine–water (1:1). The R_f values (including *cis* and *trans* isomers) are listed. Figge [99] classified 18 brighteners according to their solubilities in various solvents. Nonionic brighteners were separated from ionic brighteners by chromatographing on silica gel with alkaline polar solvents. The individuals of each group were then separated by one or two-dimensional thin-layer chromatography. Nineteen solvent systems are listed.

8 TEXTILE DYES

Aside from the food dyes, cosmetic dyes, inks, fluorescent dyes, histological stains, and indicators that have already been mentioned, the remaining dyes, mostly for textiles, can be divided into 10 classes [100], namely: acid, azoic dyes and components, basic, direct, disperse, reactive mordant, solvent, sulfur, and vat dyes. Needless to say, some of the dyes already mentioned also fall into these same classes. Rettie and Haynes [58] and Menon [101] have reviewed the application of thin-layer chromatography to the separation of dyes. Rayburn et al . [102] point out the value of thin-layer chromatography for the dye industry and give some examples of its use: lightfastness and fastness to gas fading, surface-active agents, and textile resins can all be determined by thin-layer chromatography.

Sweeny [103] gives a list of solvents for extracting dyes from various fibers, dimethyl sulfoxide being one of the best; however, it will dissolve acetate fibers. Monochlorobenzene is good for acetate fibers. After extraction, the dye solution should be concentrated at low temperature under a vacuum to minimize decomposition. For spotting, 2 to 4 μl of a 1% soltuion is satisfactory. There is no really good solvent for extracting vat dyes from cotton; dimethyl sulfoxide, pyridine, nitrobenzene, and dimethylformamide may be tried, giving fair to bad results depending on the dyes encountered [103].

Direct Dyes

Sweeny [103] recommends a number of solvents for various dye classes, and for direct dyes he lists: propanol–water–acetic acid (5:5:3:1), butyl acetate–pyridine–water (5:5:2), butanol–acetone–water (5:5:3), and the organic layer from butanol–water–ammonia (2:1:1) for use with silica gel layers. Brown [104] gives *n*-butanol–ethanol–ammonia–pyridine–water (8:3:4:4:3) as a suitable solvent. Meckel et al. [105] chromatographed nine direct azo dyes into a number of constituents on silica gel and alumina layers prepared with 2.5% sodium carbonate; although longer running, silica gel gave the sharpest separations. Solvent for the separations was butyl acetate–pyridine–water (6:9:5). Raban [106] separated the azo dyes Congo Red (C.I. 22120), Trypan Blue (C.I. 23850), Direct Blue (C.I. 22610), Direct Green B (C.I. 30295), benzopurpurin (C.I. 23500), and Direct Red F (C.I. 22310) into their components on alumina with a solvent mixture of ethanol and water. The exact solvent proportions varied with the type of alumina used. Logar et al. [106a] give the R_f values for 61 direct dyes on silica gel with *n*-butyl acetate–pyridine–quinoline–water (3:3:1:3). Wada et al. [106b] found 95% ethanol-dilute ammonia (1:9)–*n*-butanol (9:2:3) to be the best developer, of those tested, for 190 acid and direct leather dyes on silica gel.

Acid Dyes

For the separation of acid dyes on silica gel the following solvents may be used: propanol–acetone–water–acetic acid (5:5:3:1), pyridine–butyl acetate–water (9:6:5), *n*-butanol–ethanol–water (2:1:1), and propanol–28% ammonia (1:1) [103]. Rettie and Haynes [58] have found benzene–propionic acid (4:1) to be the best solvent for fluorescein dyes on silica gel; however, for mono- and dihalogenated derivatives they preferred benzene–chloroform–propionic acid (2:2:1). Meckel et al. [105] separated Benzyl Fast Red GRG, Palatine Fast Green GN, and Palatine Fast Red on silica gel prepared with 2.5% sodium carbonate with butyl acetate–pyridine–water (6:9:5). Cellulose or silica gel can be used for anthraquinone dyes using *n*-butyl acetate–pyridine–water (2:2:1) [58]. Loga et al. [106] used kieselguhr layers with *n*-butanol–acetic acid–water (2:1:5).

Basic Dyes

For basic dyes butanol–ethanol–water–acetic acid (10:1:1:1) or benzene–methanol (9:1) may be used with silica gel [103]. Rettie and Haynes [58] prefer chloroform–methanol (4:1) for the basic azo dyes. Logar et al. [107, 108] chromatographed 19 basic dyes for polyacrylonitrile fibers on alumina with ethanol–water (5:2) and another 11 dyes on silica gel G with pyridine–water (1:2). Arsov et al. [109] chromatographed 23 basic dyes on silica gel with chloroform–methyl ethyl ketone–acetic acid–formic acid (8:6:1); chloroform–*n*-propanol–pyridine–acetic acid–water (8:6:1:1:2); chloroform–methyl ethyl ketone–formic acid (6:8:1); and chloroform–isopropanol–pyridine–acetic acid–water in 6:12:3:1:1 and 6:8:3:1:2 ratios. Three additional solvent ratios of the last solvent were also tested. Takeshita et al. [110] chromatographed 15 basic dyes, namely, fuchsine base, Crystal Violet, Ethyl Violet, Malachite Green, Night Blue, Bismark Brown, Neutral Red, Safranine O, Acridine Orange, Methylene Blue, New Methylene Blue, Nile Blue, Pyronine G, Rhodamine 6G, and Rhodamine B. Polyamide layers were used with 5 solvents including methanol, ethanol, 28% ammonia–methanol (1:8), benzene–methanol (5:1), and carbon tetrachloride–methanol (4:1).

Disperse Dyes

Sweeny [103] recommends benzene–acetone (9:1) and benzene–chloroform–acetone (5:2:1) for use with silica gel for chromatographing disperse dyes. Wollenweber [18] separated anthraquinone dyes on acetylated cellulose powder with a 10% acetyl content. With cellulose MN 300 Ac or MN 300 G/Ac and ethyl acetate–tetrahydrofuran–water (6:35:47) as a solvent the R_f values were, respectively, 1,4-dihydroxyanthraquinone (Qinizarin; C.I. 58050) 0.15, 0.06; 4-amino-1-hydroxyanthraquinone (Disperse Red 15; C.I. 60710) 0.22, 0.11; and 1,4-diaminoanthraquinone (Disperse Violet 1; C.I. 61100) 0.32, 021. Although obtaining separations on this adsorbent with tetrahydrofuran–water–4 *N* acetic acid mixtures, Rettie and Haynes [58] preferred silica gel with chloroform, or acetone–chloroform mixtures for anthraquinone disperse dyes and intermediates. The 1-amino-, 2-amino-, 1,2-diamino-, and 1,4-diaminoanthraquinones could be easily separated with chloroform–acetone (9:1). Gemzova and Gasparič [111] chromatographed 30 disperse azo dyes on alumina with benzene and with chloroform. Dousheva et al. [111a] chromatographed 27 disperse dyes on silica gel (type 60). Best solvent for the separation was *n*-hexane–ethyl acetate–acetone (5:4:1). Franc and Hájková [112, 113] have chromatographed some amino anthraquinones on loose layers of aluminum oxide (activity III). The separations were carried out in cyclohexane–ether (1:1) and quantitative measurements were made by direct measurement of the intensity of the spots with a Zeiss densitometer having a percision of ±6%. The R_f values for the various anthraquinones were as follows: 1-amino, 0.62, 2-amino 0.23, 1,2-diamino 0.65, 1,4-diamino 0.10, 1,5-diamino 0.46, 1,8-diamino 0.35 2,6-diamino 0.00, 1,6-diamino 0.08, and 1,7-diamino 0.14. Bansho et al. [114] have chromatographed a group of 17 amino anthraquinones on aluminum oxide in four different solvents. The results are tabulated. (See also Chapter XXVI, Sections 8 and 19 for some naturally occurring anthraquinones.)

Reactive Dyes

Brown [104] gives the following procedure for extracting reactive dyes from cellulose:

"1. Shake 100 mg of the dyeing with 2 ml of 90% (vol/vol) sulfuric acid for 20 min.
2. Pour into 50 ml of cold distilled water.
3. Boil gently for 2 min.
4. Cool to room temperature.
5. Add 0.4 g of aminoethyl cellulose flock.
6. Stir for 5 min or until all dye has been adsorbed.

7. Filter (vacuum) on to a sintered glass crucible (porosity 3).
8. Wash with 200 ml of cold distilled water.
9. Disconnect vacuum and replace filter flask by clean flask.
10. Add 25 ml of 10% (vol/vol) ammonia (0.880), stir, and allow to stand for 2 min.
11. Filter and evaporate filtrate.
12. Dissolve dye in dimethylformamide."

Cee [115] has reviewed the paper and thin-layer chromatography of reactive dyes. Perkavac and Perpar [116] used isobutanol–*n*-propanol–ethyl acetate–water (2:4:1:3), dioxane–acetone (1:1), and *n*-propanol–ethyl acetate–water (6:1:3) for the separation of a number of reactive dyes having triazine rings, pyrimidine rings, vinyl sulfone groups, or sulfonamide groups. Szuchy et al. [117] used isopropanol–ammonia (2:1) for some Procion reactive dyes on silica gel. These included P. Brilliant Red MX-8B, P. Turquoise H-A, and P. Yellow MX-2R. Cee and Gasparič [118] give the R_f values for 41 Remazol Reactive dyes in three forms, the vinyl sulfone, the 2-hydroxyethyl sulfone, and the 2-hydroxyethyl sulfone sulfoester on silica gel G in *n*-butyl acetate–pyridine–water (2:2:1) and *n*-butyl acetate–acetic acid–water (2:2:1). The same combination was used for dichloroquinoxaline dyes [119]. Duschewa et al. [120] chromatographed 13 monochlorotriazo reactive dyes on silica gel in 12 solvent systems.

Mordant Dyes

Schetty [121–125] and Schetty and Kuster [126] have carried out the separation of some chromium and cobalt complexes of azo and azomethine dyes. The separations were accomplished on aluminum oxide layers dried at 120°C using methanol as the developing solvent.

In studying the metal complexes of 29 azo dyestuffs, Pollard et al. [127] checked the purity of the azo dyes (prior to complexing) by chromatographing on silica gel G (containing starch to improve the binding of the layer). The most successful solvents were petroleum ether (40–60°C)–diethyl ether–ethanol (10:10:1), *n*-butanol–ethanol–2 *N* ammonia (3:1:1) and isopropanol–methyl ethyl ketone–ammonia (sp gr 0.88) (4:3:3). Toul and Mouková [128] have chromatographed another group of these compounds, the *o*-hydroxy-substituted monoazo dyes, many of which are used in analytical and *N*-heterocyclic coordination chemistry. Deactivated silica gel G layers were used with benzene–acetic acid (4:1), benzene–chloroform–acetic acid (25:5:6.2), carbon tetrachloride–acetic acid (15:4), carbon tetrachloride–ethyl acetate–acetic acid (30:3:2), and carbon tetrachloride–acetic acid (15:2). In this case the use of acidic solvents suppressed the formation of metal complexes with trace metal impurities in the adsorbent. The R_f values are listed for both saturated and unsaturated tanks.

Azoic Dyes and Components

These require a coupling compound with which the dye is formed directly in or on the fiber. Perkavac and Perpar [129] chromatographed a group of seven of these Fast bases on silica gel with *n*-butanol–acetic acid–water (16:4:5) and *n*-butanol–pyridine–water (1:2:1). The spots were visualized with test T-90. A group of nine Fast salts (stabilized diazonium salts) were also separated on silica gel with *n*-butanol–acetic acid–water (1:1:5). Gasparič [130] chromatographed 32 Fast bases on aluminum oxide layers using benzene as the solvent. A total of 86 samples were run, as dyes from different manufacturers were checked. Again, the spots were detected with test T-90. Thielemann [131] separated α- and β-naphthol by reacting them with Fast Black Salt B, Fast Black Salt G, Fast Violet Salt B, Fast Red Salt ITR, Fast Blue Salt BB, Fast Blue Salt B, Fast Garnet Salt GBC, and Fast Corinth Salt V. The dyes thus formed were chromatographed on silica gel with benzene, benzene–ethyl acetate (3:2), dichloromethane–ethyl acetate (3:2), and dichloromethane–ethyl acetate–diethylamine (92:5:2) as solvents.

Solvent Dyes

In general solvent dyes can be chromatographed on silica gel with benzene, chloroform, benzene–carbon tetrachloride (1:4), benzene–acetone (1:1), chloroform–methanol (9:1), hexane–ethyl acetate (9:1), and petroleum ether–diethyl ether–acetic acid (70:30:1); and on alumina with hexane–ethyl acetate (49:1) or carbon tetrachloride. A number of oil-soluble dyes have been mentioned under the food dyes, the cosmetic dyes, and the inks. Solvent dyes are also used in petroleum products, wood stains and plastics.

In order to be able to differentiate between various gasolines, dyes are added to give distinctive colors. Hausser [132] has investigated the gasolines used in Germany. In order to concentrate the dyes, 1 to 10 ml of gasoline is evaporated to approximately one-third and then added with petroleum ether to a column of aluminum oxide. After the column is thoroughly washed with petroleum ether it is dried and the colored zone eluted with a little acetone; the acetone solution is then concentrated to a few drops for addition to the thin-layer plate. Benzene is used for the developing solvent on silica gel layers; wedge-shaped layers can be used to advantage to improve the separation. Hausser lists the R_f values and colors of the various spots from the different gasolines examined.

Table 21.6 $R_f \times 100$ Values of Some Azo Dyes on Silica Gel [133]

Dye	*C.I.*[a]	1,1-*Dichloro-ethane*	1,1,2-*Trichloro-ethane*
Pigment Yellow 1	11680	39	38
Pigment Orange 1	11725	25	30
Permanent Orange	12075	31	30
Pigment Red	12085	47	45
Toluidine Red	12120	23	26
Pigment Red 22	12315	16	18
Pigment Red 18	12350	12	16

[a] Color Index for the Society of Dyers and Colourists, Bradford, 1956.

Fujii and Kamikura [133] studied the separation of some organic azo pigments in 17 different solvents using silica gel as the adsorbent. The R_f values obtained with the two most promising solvents, 1,1-dichloroethane and 1,1,2-trichloroethane, are given in Table 21.6.

9 MISCELLANEOUS

Gasparič [134] examined a group of azo pigments on silica gel layers. Dimethylformamide was used as a sample solvent and in the case of the lakes it was necessary to warm the solution. It was important to evaporate the spotting solvent completely to avoid affecting the separation. The pigments consisted of five types: *(1)* those containing acetoacetarylamides, *(2)* those containing 2-naphthol, *(3)* pyrazolone azo dyes, *(4)* those containing arylamides of 2-hydroxy-3-naphthoic acid, and *(5)* lakes of azo dyes containing carboxy and/or sulfo groups. The first four groups were developed with benzene, sometimes a multiple development, and in some cases with toluene. The fifth group was chromatographed with *n*-butyl acetate–acetic acid–water (4:2.5:1), 1-propanol–ammonia (2:1), 1-propanol–*n*-butyl acetate–ammonia (2:1:1), and 1-propanol–ammonia (2:1). The latter solvent was especially useful for differentiating lakes with one and two sulfo and/or carboxy groups.

Masschlein-Kleiner [135] applied thin-layer chromatography to the differentiation of old and modern dyes by their quinone content. Samples were chromatographed on acetylated cellulose with ethyl acetate–tetrahydrofuran–water (6:35:47). Chromatograms are shown for cochineal, madder, kermes, lac, bedstraw, sandal, henna, walnut, alkanet, safflower, and madder, purpurin, cochineal, and alizarin lakes.

Table 21.7 Separation of Dyes by Electrophoresis on Thin Layers of Silica Gel[a,b]

	Acidic Electrophoresis			*Basic Electrophoresis*		
Color	*Distance Traveled (mm)*	*Direction of Movement*	*Dye*	*Direction of Movement*	*Distance Traveled (mm)*	*Color*
Yellow-red	25	Anode	Methyl Orange		0	Orange
				Anode	7	
Red	0		Methyl Red	Anode	34	Yellow
Yellow	0		Dimethyl Yellow		0	Yellow
Blue	0		Congo Red	Anode	30	Red
Black	0		Eriochrome Black T		0	Red
Red	65	Anode	Crystal Ponceau	Anode	35	Red
Red	80	Anode				
Red with NaOH	14	Cathode	Phenolphthalein	Cathode	16	Red
Colorless	0		Thymolphthalein		0	Blue
Yellow	51	Anode	Bromophenol Blue	Anode	15	Blue
Yellow-green	28	Anode	Fluorescein	Anode	22	Yellow
	0		Rhodamine B		0	Red
Red	25	Anode				
Visible in UV	37	Anode				
Visible in UV	56	Anode				
Visible in UV	75	Anode				
	0		Alizarine S		0	Visible in UV
Visible in UV	52	Anode				
Blue	0		Neocarmine W		0	Blue
Yellow	78	Anode		Anode	28	Yellow
Red	85	Anode		Anode	38	Red
				Anode	54	Red

[a] From G. Pastuska and H. Trinks [137]; reproduced with permission of the authors and Alfred Huethig Verlag.
[b] See text for details on preparation of layers and electrolytes.

10 ELECTROPHORESIS OF DYES

Pastuska and Trinks [13] have applied thin-layer electrophoresis to the separation of some dyes in both an acid and a basic medium (Table 21.7). For the acidic separation, the silica gel layers were prepared with 3% boric acid, and the dye solutions were applied to the middle of the plate after 20 min. The electrolyte for the separation was prepared by mixing 80 ml of ethanol, 30 ml of distilled water, 4 g of boric acid, and 2 g of crystalline sodium acetate and acidifying to a pH of 4.5 with acetic acid. Electrophoresis was carried out for 120 min at 10 V/cm. For the basic separation the silica gel layers were mixed with 3% boric solution. In this case the electrolyte was adjusted to a pH of 12 with sodium hydroxide.

Criddle [137] reported the migration distances for the electrophoresis of 26 coal tar food colors on kieselguhr, silica gel, and alumina as support layers. Electrophoresis was carried out in 1 *N* acetic acid, 0.1 *N* ammonia, and 4.0, 6.0, 8.0, and 9.2 pH buffers.

De Zeeuw [138] illustrated the use of vapor programmed thin-layer chromatography for improving the separation of some dyes.

References

1. M. Mottier and M. Potterat, *Mitt. Geb. Lebensm. Hyg.*, **43**, 123 (1952).
2. M. Mottier and M. Potterat, *Mitt. Geb. Lebensm. Hyg.*, **43**, 118 (1952).
3. H. Lagoni and A. Wortmann, *Int. Dairy Congr., 14th Rome,* **1956.**
4. H. Lagoni and A. Wortmann, *Milchwissenschaft,* **11**, 206 (1956).
5. *Ibid.*, **10**, 360 (1955).
6. G. Janíček, J. Pokorný, and J. Davídek, *Sb. Vys. Skoly Chem.-Technol. Praze, Technol. Oddil Fak. Potravin. Technol.*, **6**, 75 (1962).
7. J. Davídek, J. Pokorný, and G. Janíček, *Z. Lebensm. Unters.-Forsch.*, **116**, 13 (1961).
8. W. Reiners, *Z. Anal. Chem.*, **229**, 409 (1967).
9. A. Montag, *Z. Lebensm. Unters.-Forsch.*, **116**, 413 (1962).
10. S. Fujii and M. Kamikura, *Shokuhin Eiseigaku Zasshi,* **4**, 96 (1963); *Chem. Abstr.*, **59**, 11691 (1963).
11. J. W. Copius-Peereboom, *Chem. Weekb.*, **57**, 625 (1961).
12. W. J. Criddle, G. J. Moody, and J. D. R. Thomas, *J. Chromatogr.*, **16**, 350 (1964).
13. M. A. Ciasca and C. G. Casinovi, "Thin-Layer Chromatography on Silica Gel of Food Colours", in *Thin-Layer Chromatography,* G. B. Marini-Bettólo, Ed., Elsevier, Amsterdam, 1964, p. 212.
14. J. F. Barrett and A. J. Rayn, *Nature,* **199**, 372 (1963).
15. A. Canuti and B. L. Magrassi, *Chim. Ind. (Milan),* **46**, 284 (1964); through *Chem. Abstr.*, **61**, 1171 (1964).

16. E. Synodinos, G. Lotakis, and E. Kokkoti-Kotaki, *Chim. Chron. (Athens)*, **28,** 77 (1963); through *Chem. Abstr.*, **60,** 1089 (1964).
17. M. B. Naff and A. S. Naff, *J. Chem. Ed.*, **40,** 534 (1963).
18. P. Wollenweber, *J. Chromatogr.*, **7,** 557 (1962).
19. T. Salo and K. Salminen, *Suom. Kemistil. B,* **35,** 146 (1962).
20. J. Rai, *Chromatographia,* **4,** 211 (1971).
21. J. W. Copius-Peerboom and H. W. Beekes, *J. Chromatogr.*, **20,** 43 (1965).
22. M. M. Frodyma, R. W. Frei, and D. J. Williams, *J. Chromatogr.*, **13,** 61 (1964).
23. K. R. Brain, T. D. Turner, and B. E. Jones, *J. Pharm. Pharmacol.*, **23,** 250S (1971).
24. K. R. Brain, B. E. Jones, and T. D. Turner, *J. Chromatogr.*, **109,** 383 (1975).
25. J. Davídek and G. Janíček, *J. Chromatogr.*, **15,** 542 (1964).
26. M. K. Ramamurthy and V. R. Bhalerao, *Analyst (London),* **89,** 740 (1964).
27. R. A. Hoodless, J. Thomson, and J. E. Arnold, *J. Chromatogr.*, **56,** 332 (1971).
28. M. Mottier and M. Potterat, *Mitt. Geb. Lebensm. Hyg.*, **44,** 293 (1953).
29. M. Mottier and M. Potterat, *Anal. Chim. Acta,* **13,** 46 (1955).
30. M. Mottier, *Mitt. Geb. Lebensm. Hyg.*, **47,** 372 (1956).
31. *Official Method of Analysis,* Association of Official Agricultural Chemists, Washington, D. C., 7th ed. 1950.
32. J. Davídek, *Z. Lebensm. Unters. Forsch.*, **142,** 410 (1970).
33. J. Davídek and E. Davídková, *Z. Lebensm. Unters. Forsch.*, **131,** 99 (1966).
34. G. Lehmann, P. Collet, H.-G. Hahn, and M. R. F. Ashworth, *J. Assoc. Off. Anal. Chem.*, **53,** 1182 (1970).
35. R. A. Gilhooley, R. A. Hoodless, K. G. Pitman, and J. Thomson, *J. Chromatogr.*, **72,** 325 (1972).
36. R. Takeshita, T. Yamashita, and N. Itoh, *J. Chromatogr.*, **73,** 173 (1972).
37. J. Alary, C. Luu Duc, and A. Coeur, Bull. *Trav. Soc. Pharm. Lyon,* **10,** 78 (1966).
38. S. Logar and M. Perpar, *Farm. Vestn.*, **18,** 116 (1967); through *Chem. Abstr.*, **68,** 62723a (1968).
39. W. P. Hayes, N. Y. Nyaku, and D. T. Burns, *J. Chromatogr.*, **71,** 585 (1972).
40. J. Davídek and E. Davídková, *J. Chromatogr.*, **26,** 529 (1967).
41. D. L. Massart and H. De Clercq, *Anal. Chem.*, **46,** 1988 (1974).
42. S. N. Tewari, S. C. Sharma, and V. K. Sharma, *Chromatographia,* **7,** 308 (1974).
43. G. E. Martin and D. M. Figert, *J. Assoc. Off. Anal. Chem.*, **57,** 217 (1974).
44. G. Lehmann, P. Collet, and M. Moran, *Z. Lebensm. Unters. Forsch.*, **143,** 191 (1970).
45. R. J. T. Graham and A. E. Nya, *J. Chromatogr.*, **43,** 547 (1969).
46. R. J. T. Graham and A. E. Nye, *Int. Symp. Chromatogr., Electrophoresis, 5th, 1968,* Ann Arbor–Humphrey Science, Ann Arbor, Mich., 1969, p. 486.
47. R. A. Hoodless, K. G. Pitman, T. E. Stewart, J. Thomson, and J. E. Arnold, *J. Chromatogr.*, **54,** 393 (1971).
48. H.-C. Chiang, *J. Chromatogr.*, **40,** 189 (1969).
49. L. F. Druding, *J. Chem. Educ.*, **40,** 536 (1963).

50. J. Perkavec and M. Perpar, *Kem. Ind. (Zagreb).*, **12,** 829 (1963).
51. N. K. Sen and P. C. Ghosh, *Indian J. Appl. Chem.*, **33,** 357 (1973).
52. J. Purzycki, A. Szwarc, and M. Owoc, *Chem. Anal. (Warsaw)*, **10,** 485 (1965); through *Chem. Abstr.*, **64,** 5298e (1966).
53. M. R. Verma and J. Rai, *Int. Symp. Chromatogr., Electrophoresis, 4th, 1966*, Ann Arbor-Humphrey Science, Ann Arbor, Mich., 1968, p. 549.
54. G. R. Nakamura and S. C. Shimoda, *J. Criminal Law, Criminol. Police Sci.*, **56,** 113 (1965).
55. W. Reiners, *Z. Anal. Chem.*, **219,** 272 (1966).
56. L. Van Dessel and P. Van Regenmortel, *Z. Anal. Chem.*, **247,** 280 (1969).
57. K. W. Smalldon, *J. Forensic Sci. Soc.*, **9,** 151 (1969).
58. G. H. Rettie and C. G. Haynes, *J. Soc. Dyers Colour.*, **80,** 629 (1964).
59. J. D. Kelly and A. A. Cantu, *J. Assoc. Off. Anal. Chem.*, **58,** 122 (1975).
60. H. Gaenshirt, D. Waldi, and E. Stahl, "Synthetic Organic Materials," in *Thin-Layer Chromatography*, English ed., E. Stahl, Ed., Academic Press, 1965, p. 347.
61. A. Stier and W. Specht, *Naturwissenschaften*, **50,** 549 (1963).
62. R. W. Horobin and D. J. Goldstein, *Histochem. J.*, **4,** 391 (1972).
63. H. L. Dobres and W. A. Moats, *Stain Technol.*, **43,** 27 (1968).
64. K. W. Loach, *J. Chromatogr.*, **60,** 119 (1971).
65. R. W. Horobin and L. B. Murgatroyd, *Histochemie*, **11,** 141 (1967).
66. R. W. Horobin, *Histochem. J.*, **1,** 231 (1969).
67. R. W. Horobin and L. B. Murgatroyd, *Stain Technol.*, **44,** 297 (1969).
68. A. G. Lansink, *Histochemie*, **16,** 68 (1968).
68a. A. D. Cramer, E. R. Rogers, J. W. Parker, and R. J. Lukes, *Am. J. Clin. Pathol.*, **60,** 148 (1973).
69. J. H. Tyrer, M. J. Eadie, and W. D. Hooper, *J. Chromatogr.*, **39,** 312 (1969).
70. H. Grossmann and H. Wagner, *J. Chromatogr.*, **35,** 301 (1968).
71. G. R. N. Jones, *J. Chromatogr.*, **39,** 336 (1969).
71a. S. Balogh, J. Tamas, and *J. Hegedus-Vajda, Magy. Kem. Foly.*, **81,** 227 (1975); through *Chem. Abstr.*, **83,** 125828w (1975).
72. G. Aliotta and E. Roso, *An. Asoc. Quim. Argent.*, **59,** 437 (1971).
73. B. Marowski and W. Fabricius, *Z. Klin. Chem. Klin. Biochem.*, **9,** 419 (1971); through *Chem. Abstr.*, **76,** 43543r (1972).
74. H.-C. Chiang and C.-C. Yeh, *T'ai-Wan Yao Hsueh Tsa Chih*, **21,** 6 (1969); through *Chem. Abstr.*, **75,** 147554r (1971).
75. R. C. Walker and M. Beroza, *J. Assoc. Off. Agr. Chem.*, **46,** 250 (1963).
76. L. F. Druding, *J. Chem. Educ.*, **40,** 536 (1963).
77. I. Pinter and M. Kramer, *Parfuem. Kosmet.*, **50,** 129 (1969).
78. A. Perdih, *Kem. Ind.*, **16,** 344 (1967); through *Chem. Abstr.*, **68,** 16069u (1968).
79. J. Deshusses and P. Desbaumes, *Am. Perfum. Cosmet.*, **83,** 37 (1968).
80. U. Ruedt, *Fette, Seifen, Anstrichm.*, **71,** 982 (1969).
81. J. Davídek, J. Pokorný, and V. Pokorná, *Cesk. Hyg.*, **7,** 548 (1962).
82. T. P. Cotsis and J. C. Garey, *Drug Cosmetic Ind.*, **95,** 172 (1964).
83. T. P. Cotsis and J. C. Garey, *Proc. Sci. Sect. Toilet Goods Assoc.*, **41,** 3 (1964).

84. R. S. Silk, *J. Assoc. Off. Agr. Chem.*, **48,** 838 (1965).
85. A. Perdih, *Z. Anal. Chem.*, **260,** 278 (1972).
86. G. Lehmann and U. Recktenwald, *Z. Lebensm. Unters. Forsch.*, **146,** 147 (1971).
87. H. Jork, G. Lehmann, and U. Recktenwald, J. Chromatogr., **107,** 173 (1975).
88. J. C. Brown, *J. Soc. Cosmet. Chem.*, **18,** 225 (1967).
89. S. Hordynska and B. Legatowa, *Rocz. Panstw. Zakl. Hig.*, **20,** 673 (1969); through *Anal. Abstr.*, **20,** 1771 (1971).
90. C. M. Kottemann, *J. Assoc. Off. Anal. Chem.*, **49,** 954 (1966).
91. I. Takemura, *Bunseki Kagaku,* **19,** 899 (1970); through *Anal. Abstr.*, **22,** 229 (1972).
92. M. Nanjo, E. Isohata, S. Kano, and N. Kobayashi, *Eisei Shikenjo Hokoku,* **85,** 90 (1967); through *Chem. Abstr.*, **69,** 61494k (1968).
93. K. Zelazan and B. Legatowa, *Rocz. Panstw. Zak. Hig.*, **22,** 427 (1971); through *Chem. Abstr.*, **76,** 6623w (1972).
94. S. Goldstein, A. A. Kopf, and R. Feinland, *Proc. Joint Conf. Cosmet. Sci.*, **19,** 38 (1968).
95. I. Saenz Lascano Ruiz and C. Laroche, *Bull. Soc. Chim. Fr.*, **1963,** 1594.
96. J. Latinák, *J. Chromatogr.*, **14,** 482 (1964).
97. F. Schlegelmilch, H. Adbelkader, and M. Eckelt, *Text. Ind. (Muenchen-Gladbach, Ger.),* **73,** 274 (1971).
98. H. Theidel and G. Schmitz, *J. Chromatogr.*, **27,** 413 (1967).
99. K. Figge, *Fette, Seifen, Anstrichm.*, **70,** 680 (1968).
100. A. J. Paradiso, *Chem Tech,* **1971,** 292.
101. V. K. Menon, *Colourage,* **17,** 38 (1970).
102. J. A. Rayburn, L. B. Arnold, Jr., J. Hoffman, W. H. Lawson, J. E. Nettles, A. L. Smith, and E. White, *Text. Chem. Color,* **1,** 58 (1969).
103. C. D. Sweeny, *Am. Dyst. Rep.*, **61,** 70 (1972).
104. J. C. Brown, *J. Soc. Dyers Colour.*, **80,** 185 (1964).
105. L. Meckel, H. Milster, and U. Krause, *Textil-Praxis,* **16,** 1052 (1961).
106. P. Raban, *Nature,* **199,** 596 (1963).
106a. S. Logar, N. Mesiček, J. Perkavec, and M. Perpar, *Kem. Ind. (Zagreb),* **17,** 473 (1968); through *Chem. Abstr.*, **70,** 79120r (1969).
106b. K. Wada, L. Shirai, and H. Okamura, *Hikaku Kagaku,* **13,** 149 (1958); through *Chem. Abstr.*, **68,** 88227s (191968).
107. S. Logar, J. Perkavec, and M. Perpar, *Mikrochim. Acta,* **1964,** 712.
108. *Ibid.*, **1967,** 496; also *J. Chromatogr.*, **30,** D14 (1967).
109. A. M. Arsov, B. K. Mesrob, and A. G. Gateva, *J. Chromatogr.*, **81,** 181 (1973).
110. R. Takeshita, N. Itoh, and Y. Sakagami, *J. Chromatogr.*, **57,** 437 (1971).
111. I. Gemzova and J. Gasparič, *Collect. Czech. Chem. Commun.*, **32,** 2740 (1967).
111a. M. Dousheva, A. Arsov, V. Kostova, and B. Mesrob, *J. Chromatogr.*, **121,** 131 (1976).
112. J. Franc and M. Hájková, *J. Chromatogr.*, **16,** 345 (1964).
113. J. Franc, M. Hájková, and M. Jehlicka, *Chem. Zvesti.*, **17,** 542 (1963).

114. Y. Bansho, I. Saito, and S. Suzuki, *Kogyo Kagaku Zasshi,* **64,** 1061 (1961); through *Chem. Abstr.,* **57,** 4041 (1962).
115. A. Cee, *Sb. Ved. Pr., Vys. Sk, Chemickotechnol., Pardubice* **1967,** 205 through *Chem. Abstr.,* **71,** 40189r (1969).
116. J. Perkavec and M. Perpar, *Z. Anal. Chem.,* **206,** 356 (1964).
117. L. Szuchy, J. Mityko, and M. Gajdacs, *Kolor. Ert.,* **16,** 31 (1974).
118. A. Cee and J. Gasparič, *Collect. Czech. Chem. Commun.,* **33,** 1091 (1968); also *J. Chromatogr.,* **39,** D14 (1969).
119. A. Cee and J. Gasperič, *Mikrochim. Acta,* **1968,** 452.
120. M. Duschewa, L. Jankow, and K. Dímov, *Melliand Textilber.,* **56,** 147 (1975).
121. G. Schetty, *Helv. Chim. Acta,* **45,** 809 (1962).
122. *Ibid.,* p. 1026.
123. *Ibid.,* p. 1095.
124. *Ibid., Helv. Chim. Acta,* **46,** 1132 (1963).
125. G. Schetty, *Helv. Chim. Acta,* **52,** 1016 (1969).
126. G. Schetty and W. Kuster, *Helv. Chim. Acta,* **44,** 2193 (1961).
127. F. H. Pollard, G. Nickless, T. J. Samuelson, and R. G. Anderson, *J. Chromatogr.,* **16,** 231 (1964).
128. J. Toul and N. Mouková, *J. Chromatogr.,* **67,** 335 (1972).
129. J. Perkavec and M. Perpar, *Mikrochim. Acta,* **1964,** 1029.
130. J. Gasparič, *Z. Anal. Chem.,* **218,** 113 (1966); also *J. Chromatogr.,* **25,** D2 (1966).
131. H. Thielemann, *Mikrochim. Acta,* **1972,** 718.
132. H. Hausser, *Arch. Kriminol.,* **125,** 72 (1960).
133. S. Fujii and M. Kamikura, *Shokuhin Eisegaku Zasshi,* **4,** 135 (1963); through *Chem. Abstr.,* **60,** 9883 (1964).
134. J. Gasparič, *J. Chromatogr.,* **66,** 179 (1972).
135. L. Masschlein-Kleiner, *Mikrochim. Acta,* **1967,** 1080; also *J. Chromatogr.,* **34,** D45 (1968).
136. G. Pastuska and H. Trinks, *Chem.-Ztg.,* **86,** 135 (1962).
137. W. J. Criddle, G. J. Moody, and J. D. R. Thomas, *J. Chromatogr.,* **16,** 350 (1964).
138. R. A. de Zeeuw, *Anal. Chem.,* **40,** 2134 (1968).

Chapter **XXII**

HYDROCARBONS

In 1952 Kirchner and Miller [1] published a chromatographic method for the preparation of terpeneless essential oils using chromatostrips to follow the separation. This method was based on the results obtained by chromatographing large numbers of hydrocarbons and oxygenated constituents on thin layers of silicic acid using hexane as the developing solvent. Using this solvent, oxygenated compounds remain at the origin in contrast to the hydrocarbons which move quite readily (Table 22.1).

Kucharczyk et al. [2] separated a group of compounds including a large number of hydrocarbons (Table 22.2) on layers of both silica gel and aluminum oxide using hexane and carbon tetrachloride as solvents. The spots were located either by spraying with 0.2 ml of a 37% formaldehyde solution in 10 ml of concentrated sulfuric acid or 10% solution of tetracyanoethylene in benzene, or by examining the fluorescent layers under ultraviolet light. Janák [3, 4], in applying a combination gas chromatographic and thin-layer technique, ran a number of hydrocarbons in three solvents (Table 22.2). The thin-layer chromatograms were run on loose layers of silica gel in a layer thickness of 0.6 to 0.9 mm. As a detecting agent he used a saturated solution of tetracyanoethylene in benzene. Because of the loose layers, the detecting agent was dropped from a capillary pipet or sprayed on by rubbing a wet tooth brush on a metal sieve. The colored complexes appeared after evaporation of the benzene in a drying oven at 100°C. Ognyanov [5] also separated some of the same compounds on loose-layer aluminum oxide of Brockmann activity between I and II with hexane as the mobile phase. He reported, however, that constant R_f values could not be obtained because the aluminum oxide was partially inactivated through moisture adsorption from the air during the plate preparation and the sample application. This is borne out by the fact that some of the compounds have R_f values identical with those of Kucharczyk et al. [2], whereas other R_f values differ. Some

Table 22.1 $R_f \times 100$ Values of Some Hydrocarbons on Silica Acid "Chromatostrips" Using 5% Starch Binder[a,b]

Compound	R_f	*Compound*	R_f
Anthracene	34	β-Methylnaphthalene	49
sec-Butylbenzene	61	Myrcene	56
tert-Butylbenzene	70	Naphthalene	35
Camphene	74	4-Nonene	94
Cedrene	82	Octadecene	88
Cumene	62	α-Pinene	83
p-Cymene	38	β-Pinene	80
Bicyclopentadiene	72	Pyrene	27
Biphenyl	30	Stilbene	20
Heptene	87	Styrene	55
Hexyne	43	Terpinolene	64
Limonene	41	Xylene	65
Mesitylene	64		

[a] From J. G. Kirchner and J. M. Miller [1]; reproduced with permission of the American Chemical Society.

[b] Layers dried at 105°C for 15 min followed by $\frac{1}{2}$ hr at 3 mm Hg over phosphorus pentoxide. Solvent is hexane (benzene-free), development distance 10 cm.

additional compounds that were run are given with their R_f values as follows: *p*-diisopropylbenzene 0.89, 3,4,5,11-tetrahydroacenaphthene 0.74, acenaphthylene 0.49, triphenylmethane 0.43, and diphenylmethane 0.18. For detecting unsaturated compounds, Ognyanov [6] placed the chromatographic plate in a chamber containing 10 to 15% ozone for 20 min. Then after exposure to air to remove the excess ozone, the plate was sprayed with indigosulfonic acid solution (0.13 g of indigo + 1 ml of concentrated sulfuric acid heated for 1 hr and then diluted to 500 ml). After heating, the white or yellow to brown spots appeared on a blue background.

Schulte et al. [7, 8] have investigated naturally occurring polyacetylene compounds on silica gel layers. For detection, the plates were first sprayed with a 0.5% solution of dicobaltoctacarbonyl in petroleum ether (120 to 135°C) and allowed to stand for 10 min. They were then sprayed with 1 *N* hydrochloric acid and allowed to dry. The chromatograms, after treatment with Neatan, were soaked loose from the support and washed for 2 hr to remove excess reagent. The excess moisture was blotted off with filter paper, and they were then exposed to bromine vapor for 1 min. Finally, they were dipped into a 0.5% solution of α-nitroso-β-naphthol in acetic acid–water (1 : 1), and then washed thoroughly with 0.5%

ammonia. The compounds appeared as red colored chelate complexes on an almost white background.

Prey et al. [9] have separated a series of low-molecular-weight olefins by chromatographing their mercuric acetate addition products on silica gel G (Table 22.3), using a solvent mixture of propanol–triethylamine–water (50:25:25). The mercuric acetate addition products appear as blue-violet spots when sprayed with a 2% alcoholic diphenylcarbazone solution, followed by heating at 80°C for a short period. Braun and Vorendohre [10] have also used the mercuric acetate derivatives for separating unsaturated polymerizable compounds. After preparing the derivatives they removed excess mercuric acetate by adding saturated hydrazine sulfate solution; without this treatment the mercuric acetate streaked and gave an additional spot on the chromatogram. Separation on silica gel was achieved with methyl ethyl ketone–*n*-propanol–ethanol–ammonium hydroxide (10:1:4:7) in a saturated atmosphere. Detection was carried out by drying and then exposing to concentrated hydrochloric acid vapors for 5 to 10 minutes after which the plates were sprayed with a 0.1% dithizone solution in carbon tetrachloride.

Suryaraman and Cave [11] chromatographed a group of long-chain (C_{10} to C_{16}) hydrocarbons as well as the corresponding alcohol, aldehyde, and diol derivatives on silica gel.

Hyyrylaeinen [12] and Mani and Lakshminarayana [13] have applied thin layers to the detection and determination of the adulteration of vegetable oil with mineral oil. Radler and Grncarevic [14] have likewise determined the addition of mineral oil to dried fruit which is sometimes used to prevent the fruit from sticking together. Krieger [15] applied radial thin-layer chromatography to the detection and characterization of mineral oil in soil samples.

Biernoth [16, 16a] and Howard et al. [17] have used spectrophotometry to determine the polycyclic aromatic hydrocarbons in edible oils after separation of the hydrocarbons by thin-layer chromatography. Alumina [16] or silica gel [18] with cyclohexane or dimethylformamide-impregnated cellulose with 2,2,4-trimethylpentane [17] may be used for the separation. As little as 0.1 μg per kilogram of benzo [*a*] pyrene could be detected. Chand et al. [19] used silica gel impregnated with silver nitrate with benzene as a solvent for the detection of adulteration of vegetable oils with mineral oils. Visualization was achieved by heating after spraying with 50% ethanolic phosphoric acid. Howard et al. [20] determined the polycyclic hydrocarbons in solvents used for the extraction of edible oils. Potthast and Eigner [21] developed a method for the extraction of polycyclic aromatic hydrocarbons from smoked meat products with propylene carbonate prior to thin-layer chromatography, and White et al.

Table 22.2 R_f Values × 100 of Some Aromatic Hydrocarbons

	Silica Gel G	Silica Gel PHH (Loose-Layer)				Neutral Al_2O_3 (Loose-Layer) (Activity I–II)	
Hydrocarbon	*Diisobutylene* [3a]	*n-Hexane* [3]	*Cyclohexane* [3]	*n-Hexane* [2]	CCL_4 [2]	*n-Hexane* [2]	CCl_4 [2]
Naphthalene	50	50	40	59	76		85
1-Methylnaphthalene	45	40	46	54	74	55	78
2-Methylnaphthálene	44	46	47	50	87	52	85
Hexadecylbenzene	74						
Dodecylbenzene	70						
Octylbenzene	69						
Hexylbenzene	65						
1,3-Dimethylnaphthalene	47	42	53				
1,4-Dimethylnaphthalene	48	55	54				
1,5-Dimethylnaphthalene	48	44	47	50	72	46	85
1,6-Dimethylnaphthalene	45	47	47	54	75	46	83
1,7-Dimethylnaphthalene	47	50	52				
2,3-Dimethylnaphthalene	40	36	40	48	75	39	81
2,6-Dimethylnaphthalene	42	46	44	50	72	52	85
2,7-Dimethylnaphthalene		45	44	53		42	80
1,3,5-Triethylbenzene	65						
Phenylcyclohexane	61						
1,2,4-Trimethylbenzene	60						
1-Propyl-2,4,6-trimethylbenzene	57						
Pentamethylbenzene	53						
2-Ethylnaphthalene		55	52				
2,3,5-Trimethylnaphthalene		50	47				
2,4,6-Trimethylnaphthalene	43						

1-Ethylnaphthalene	45						
2-Octadecylnaphthalene	65						
1-Phenylnaphthalene	39						
2-Phenylnaphthalene	33	34	38				
Biphenyl	40	45	45	45	74	58	79
4,4′-Dimethylbiphenyl	42	39	44				
4-Methylbiphenyl	43						
3-Methylbiphenyl	42						
Acenaphthene	48	52	53	44	83	44	70
Fluorene	38	40	40	32	66	35	73
Diphenylmethane	35						
1-Methylphenanthrene		29	32				
Anthracene	34	40	42				
2-Methylanthracene	38	40	40				
9-Methylanthracene	38						
9,10-Dimethylanthracene	38						
Pyrene	33	35	38	32	64	10	65
2,3-Benzofluorene	29						
1,2-Benzofluorene	29						
1,4-Diphenylbenzene	29						
3-Methylpyrene	33						
Fluoranthene	31	30	25	29	66	10	55
Chrysene	32	18	18	0	11	0	0
3,4-Benzo[*a*]pyrene				17	65	0	10
Tetrahydronaphthalene	49			—	—	—	75
2,2′-Binaphthyl	28						
1,2-Dihydronaphthalene	46						
9,10-Dihydrophenanthrene	35						
Perylene	29						
1,2,4,5-Tetramethylbenzene	51			62	76	—	85

Table 22.3 $R_f \times 100$ Values of the Mercuric Acetate Addition Compounds of the Lower Olefins on Silica Gel. Solvent: Propanol–Triethylamine–Water (50:25:25)[a]

Compound	$R_f \times 100$	*Compound*	$R_f \times 100$
Ethene	7	1-Pentene	29
2-Pentene	22	1-Butene	17
Propene	13	1-Hexene	31

[a] From V. Prey, A. Berger, and H. Berbalk [9]; reproduced with permission of the authors and Springer-Verlag.

[22] examined the polycyclic hydrocarbons in liquid smoke flavors. Rohrlich [23] determined the 3,4-benzopyrene in cereals and Howard et al. [24] estimated the polycyclic aromatic hydrocarbon content in total diet composite. In the latter case an additional chromatographic system was cellulose acetate layers with ethanol–toluene–water (17:4:4). Grimmer and Duevel [25] investigated the polynuclear hydrocarbon content of vegetables, cereals, and vegetable oils using alumina with cyclohexane–isooctanol–xylene (1:1:1). Genest and Smith [26] applied a simple thin-layer chromatographic method to the detection of benzo[*a*]pyrene in smoked foods. Howard et al. [27] developed a method for the extraction and estimation of polycyclic aromatic hydrocarbons in smoked foods and modified [28] the Genest and Smith method for the determination of benzo[*a*]pyrene to obtain a lower limit of detection of 0.5 ppb. Rhee and Bratzler [29] investigated the polycyclic hydrocarbon composition of wood smoke.

The aromatic hydrocarbon content of cigarette smoke has also been investigated by thin-layer chromatography [30, 31].

Because of the importance of polycyclic aromatic hydrocarbons in air polution, Sawicki et al. [32–34] have applied thin-layer chromatography to their separation and identification. The adsorbents included aluminum oxide, cellulose, cellulose acetate, and silica gel; the best solvents for this separation consisted of mixtures of water with dimethylformamide, acetic acid, or formic acid. Decreasing the percentage of water favored the separation of the smaller hydrocarbons, and conversely, increasing the percentage of water favored the separation of the larger hydrocarbons. The best all-around separation of polynuclear aromatic hydrocarbons was obtained on cellulose layers with a dimethylformamide–water (1:1) solvent. The spots were detected by their fluorescence under ultraviolet light, and a technique was worked out for quantitative analysis using spectrophotofluorometry. Sawicki and Johnson [35, 36] worked out characterization tests based on fluorescence, and Bender and Sawicki [37] have given some sensitivity limits using concentrated sulfuric acid as the detecting agent. Elbert and Stanley [38] examined the R_f values of benzo[*a*]pyrene and five other carcinogens in 18 different solvents. Ikan et al. [39] chromatographed the monomethyl fluorenes as well as a group of variously substituted 2-fluorenes. Wieland et al. [40] and Badger et al. [41] have also used acetylated cellulose as an adsorbent for polycyclic aromatic hydrocarbons. Here again, water-base solvents were used, for example, methanol–ether–water (4:4:1) and toluene–ethanol–water (4:17:1). Schaad [42] also preferred acetylated cellulose (20%) MN 300 AC as an adsorbent. On development with ethanol–dichloromethane–water (20:10:1), the following R_f values were obtained: benzo[*a*]pyrene 0.32, benzo[*b*]fluoranthene 0.49, benzo[*k*] fluoranthene 0.51, benzo[*a*]anthracene 0.62, and benzo[*e*]pyrene 0.70. Chatot et al. [43] used a mixed layer, acetylated cellulose–alumina (1:2) for a two-dimensional separation. The first-dimension solvent was pentane–ether (98.5:1.5) and the second-dimension, ethanol–toluene–water (17:4:4).

Air pollution and the attention being focused on carcinogens has increased the interest in the thin-layer chromatography of polycyclic aromatic hydrocarbons, and numerous publications have appeared. White and Howard [44] reported the R_f values of 20 of these compounds on cellulose impregnated with dimethylformamide using isooctane as the mobile phase, and on acetylated cellulose with ethanol–toluene–water (17:4:4). Raaen [45] separated a group of nine compounds on a mixed layer of Aviamide-6 (FMC) and Fluoroglide 200 TWO218 (Chemplast) in a ratio of 4:1. The solvent was *n*-propanol. The compounds were visualized by the varicolored fluorescence produced at liquid nitrogen temperature under 254-nm ultraviolet light [35, 36]. Keefer [46] chromatographed 11 compounds on unactivated magnesium hydroxide layers with benzene. Janák and co-workers [47, 48] examined the use of Poropak Q and Poropak T (Waters) in a number of solvents as an adsorbent for these compounds. Ritter et al. [49] give the R_f values for 24 compounds on Benton 38–Celite (1:1) using *n*-heptane as a solvent. Matsushita and Suzuki [50] used dual-band plates with 26% acetylated cellulose as one band and alumina as the other. The polynuclear hydrocarbons were chromatographed with hexane–ethyl ether (19:1) along the alumina. The plate was then turned through 90° for development with methanol–ethyl ether–water (4:4:1) over the cellulose acetate band. Peurifoy et al. [51] used a three-stage plate composed of adjacent layers of silica gel, alumina, and Florisil in that order. The layers were used for the separation of monoaromatics, polyaromatics, and resins from petroleum fractions. By development in cyclohexane–ethyl acetate–benzene (42:1:1), starting with the Florisil at the bottom, the three fractions were separated with the resins on the Florisil, the polyaromatics on the alumina, and the monoaromatics on the silica layer.

Petrowitz [52] has separated polynuclear hydrocarbons on silica gel G using heptane as the solvent. Detection was accomplished with antimony pentachloride in carbon tetrachloride. Matsushita et al. [53] also used silica gel layers using a mixed solvent, *n*-hexane–*o*-dichlorobenzene–pyridine (10:1:0.5) for development in an atmosphere of 25% relative humidity. Using a 12-cm development distance, the following R_f values were obtained: anthracene 0.51, pyrene 0.41, 1,2-benzanthracene 0.32, 3.4-benzopyrene 0.23, perylene 0.19, and 1,12-benzoperylene 0.15. Arro [54] found that aluminum oxide would separate 1,2-benzanthracene from 3,4-benzopyrene, but would not give a complete separation of the latter from 1,2-benzopyrene. Koehler et al. [55] list the R_f values for 15 dibenzanthracene, benzopyrene, and related compounds on aluminum oxide and on acetyl cellulose. The results were applied to a two-dimensional separation using mixed layers composed of 7.5 g of acetyl cellulose and 15 g of aluminum oxide. Development in the first direction was with *n*-hexane–toluene–*n*-pentane (90:5:5) and in the second direction with methanol–ether–water (4:4:1). Pavelko and D'Ambrosio [56] used cyclohexane with aluminum oxide plates to separate coronene, benzo[*ghi*]perylene, benzo[*a*]-pyrene, benzo[*e*]pyrene, fluoranthene, and pyrene.

Berg and Lam [57] have investigated aluminum oxide and silica gel layers impregnated with complexing agents for the separation of 21 polycyclic, aromatic hydrocarbons. (For preparation of the layers see Chapter III, Section 4.) Caffeine on silica gel and 2,4,7-trinitrofluorene on alumina gave good separations with petroleum ether containing small amounts of ether, pyridine, aniline, or Tetralin as a solvent. In many cases multiple development was required. The silica gel plates were activated at 120°C for 2 hr and the alumina at 150°C for 3 hr.

Franck-Neumann and Joessang [58] used silica gel impregnated with trinitrobenzene to separate pyrene, anthracene, acenaphthene, octahydroanthracene, and napthalene with cyclohexane as a solvent. With 5% ethyl acetate in cyclohexane, chrysene, retinene, and pimanthrene were separated. Harvey and Halonen [59] chromatographed 53 aromatic hydrocarbons on silica gel layers impregnated with 2,4,7-trinitrofluorenone and with 1,3,5-trinitrobenzene with benzene–heptane (1:4). Short and Young [60] obtained good results with pyromellitic dianhydride as an impregnating agent for silica gel. Brightly colored and highly fluorescent complexes were formed. The plates were developed with cyclohexane. Kessler and Mueller [61] used picric acid as an impregnating agent for separating 21 hydrocarbons. Benzene was used as the developing solvent.

A number of methods have been used for the quantitative analysis of aromatic hydrocarbons. The most used has been the use of fluorescent spectrophotometry after eluting from the thin-layer plate [62–66] or by ultraviolet spectrophotometry [65–70]. The compounds may be eluted from the thin layers with benzene or diethyl ether. Matsushita and Arashidani [71] preferred dimethyl sulfoxide as the solvent in which to measure fluorescence because the fluorescence was stronger than when using benzene. Hood and Winefordner [72] used fluorescent measurement at -196°C of 16 polycyclic hydrocarbons in the range of 0.1 μg.

Hunter [73] compared gravimetry with direct densitometry (after charring) and obtained comparable accuracy except when standards were not available; in this case the gravimetric method was more accurate. Direct fluorometric scanning of thin-layer chromatograms has also been used for quantitative work [74–76]. The limit of detection in this method has been estimated to be 0.1 to 0.001 μg on the average, depending on the intensity of the fluorescence [76]. De Wiest et al. [77] have combined liquid scintillation spectrometry with fluorimetry as a means of assessing the accuracy of the latter. Majer et al. [78] and Gilchrist et al. [79] have applied mass spectrometry to the determination of polycyclic aromatic hydrocarbons separated by thin-layer chromatography.

Inscoe [80] has investigated the photochemical changes taking place in thin-layer chromatograms of polycyclic aromatic hydrocarbons. It was determined that most of the changes were the result of photochemical reactions accelerated by exposure to long-wavelength ultraviolet or to 253.7-mμ light; therefore, in working with these compounds care should be taken to minimize these changes. Lam and Berg [81] found that plates impregnated with caffeine protected these compounds from the oxidation under the influence of light. Geiss and co-workers [82–85] have carried out a series of investigations on polyphenyls, including an extensive study of the effect of humidity in the developing chamber on the separations achieved. Preconditioning the plates at low relative humidities favored the separation of terphenyls, whereas preconditioning at higher humidities favored the separations of the higher-boiling polyphenyls (see also Chapter IX). The most sensitive visualizing agent was 0.3% ceric sulfate in concentrated nitric acid. Schlitt [86] has chromatographed 12 polyphenyls on silica gel with hexane–carbon tetrachloride (1:1) as a solvent.

Lambertsen and Holman [87] have investigated the hydrocarbons in herring oil using thin-layer and vapor-phase chromatography.

For the separation and determination of biphenyl in citrus products by Kirchner et al. [88], see Chapter XI, Sections 1 and 12.

Sawicki [89] reviews some TLC work of aromatic hydrocarbons in reviewing the general subject of analyzing aerotoxicants, and Schaad [90] includes a section on thin-layer chromatography in his general review of the chromatography of polycyclic aromatic hydrocarbons.

References

1. J. G. Kirchner and J. M. Miller, *Ind. Eng. Chem.*, **44,** 318 (1952).
2. N. Kucharczyk, J. Fohl, and J. Vymětal, *J. Chromatogr.*, **11,** 55 (1963).
3. J. Janák, *J. Chromatogr.*, **15,** 15 (1964).

3a. A. Pawlowska-Marzec, *Chem. Anal. (Warsaw),* **13,** 471 (1968); also *J. Chromatogr.*, **39,** D1 (1969).

4. J. Janák, I. Klimeš, and K. Hána, *J. Chromatogr.*, **18,** 270 (1965).
5. I. Ognyanov, *Compt. Rend. Acad. Bulgare Sci.*, **16,** 265 (1963).
6. I. Ognyanov, *Compt. Rend. Acad. Bulgare Sci.*, **16,** 161 (1963).
7. K. E. Schulte, *Congr. Sci. Farm., Conf., Comun. Pisa,* **21,** 798 (1961; pub. 1962).
8. K. E. Schulte, F. Ahrens, and E. Sprenger, *Pharm. Ztg., Ver. Apoth.-Ztg.*, **108,** 1165 (1963).
9. V. Prey, A. Berger, and H. Berbalk, *Z. Anal. Chem.*, **185,** 113 (1962).
10. D. Braun and G. Vorendohre, *Z. Anal. Chem.*, **199,** 37 (1963).
11. M. G. Suryaraman and W. T. Cave, *Anal. Chim. Acta,* **30,** 96 (1964).
12. M. Hyyrylaeinen, *Farm. Aikak.*, **72,** 161 (1963); through *Chem. Abstr.*, **59,** 6623 (1963).
13. V. V. S. Mani and G. Lakshminarayana, *Indian J. Technol.*, **3,** 416 (1965).
14. F. Radler and M. V. Grncarevic, *J. Agric. Food Chem.*, **12** 266 (1964).
15. H. Krieger, *Gas-Wasserfach,* **104,** 695 (1963).
16. G. Biernoth, *Fette Seifen, Anstrichm.*, **70,** 217 (1968).

16a. G. Biernoth, *J. Chromatogr.*, **36,** 325 (1968).

17. J. W. Howard, E. W. Turicchi, R. H. White, and T. Fazio, *J. Assoc. Off. Anal. Chem.*, **49,** 1236 (1966).
18. C. Franzle and W. Fritz, *Fette, Seifen, Anstrichm.*, **71,** 23 (1969).
19. S. Chand C. Srinivasulu, and S. N. Mahapatra, *J. Chromatogr.*, **106,** 475 (1975).
20. J. W. Howard, T. Fazio, and R. H. White, *J. Agric. Food Chem.*, **16,** 72 (1968).
21. K. Potthast and G. Eigner, *J. Chromatogr.*, **103,** 173 (1975).
22. R. H. White, J. W. Howard, and C. J. Barnes, *J. Agric. Food Chem.*, **19,** 143 (1971).
23. M. Rohrlich, *Qual. Plant. Mater. Veg.*, **16,** 334 (1968); through *Chem. Abstr.*, **70,** 95509e (1969).
24. J. W. Howard, T. Fazio, R. H. White, and B. A. Klimeck, *J. Assoc. Off. Anal. Chem.*, **51,** 122 (1968).

25. G. Grimmer and D. Duevel, *Z. Naturforsch.*, **25b,** 1171 (1970).
26. C. Genest and D. M. Smith, *J. Assoc. Off. Agric. Chem.*, **47,** 894 (1964).
27. J. W. Howard, R. T. Teague, Jr., R. H. White, and B. E. Fry, Jr., *J. Assoc. Off. Anal. Chem.*, **49,** 595 (1966).
28. J. W. Howard, R. H. White, B. E. Frey, Jr., and E. W. Turicchi, *J. Assoc. Off. Anal. Chem.*, **49,** 611 (1966).
29. K. S. Rhee and L. J. Bratzler, *J. Food Sci.*, **33,** 626 (1968).
30. I. Schmeltz, R. L. Stedman, and W. J. Chamberlain, *Anal. Chem.*, **36,** 2499 (1964).
31. H.-J. Klimisch and E. Kircheim, *Chromatographia,* **9,** 119 (1976).
32. E. Sawicki, J. W. Stanley, W. C. Elbert, and J. D. Pfaff, *Anal. Chem.*, **36,** 497 (1964).
33. E. Sawicki, T. R. Stanley, J. D. Pfaff, and W. C. Elbert, *Chemist-Analyst,* **53,** 6 (1964).
34. E. T. Sawicki, T. W. Stanley, and W. C. Elbert, *J. Chromatogr.*, **18,** 512 (1965).
35. E. Sawicki and H. Johnson, *Microchem. J.*, **8,** 85 (1964).
36. E. Sawicki and H. Johnson, *Mikrochim. Acta,* **1964,** 435.
37. D. F. Bender and E. Sawicki, *Chemist-Analyst,* **54,** 73 (1965).
38. W. C. Elbert and T. W. Stanley, *Chemist-Analyst,* **54,** 68 (1965).
39. R. Ikan, I. Kirson, and E. D. Bergmann, *J. Chromatogr.*, **18,** 526 (1965).
40. T. Wieland, G. Lueben, and H. Determann, *Experientia,* **18,** 432 (1962).
41. G. M. Badger, J. K. Donnelly, and T. M. Spotswood, *J. Chromatogr.*, **10,** 397 (1963).
42. R. E. Schaad, *Microchem. J.*, **15,** 208 (1970).
43. G. Chatot, M. Castegnaro, J. L. Roche, and R. Fontanges, *Chromatographia,* **3,** 507 (1970).
44. R. H. White and J. W. Howard, *J. Chromatogr.*, **29,** 108 (1967).
45. H. P. Raaen, *J. Chromatogr.*, **53,** 600 (1970).
46. L. K. Keefer, *J. Chromatogr.*, **31,** 390 (1967).
47. V. Martinu and J. Janák, *J. Chromatogr.*, **65,** 477 (1972).
48. J. Janák and V. Kubecová, *J. Chromatogr.*, **33,** 132 (1968).
49. F. J. Ritter, G. M. Meyer, and F. Geiss, *Pharm. Tijdschr. Belg.*, **42,** 125 (1965); through *Chem. Abstr.*, **63,** 17184h (1965).
50. H. Matsushita and Y. Suzuki, *Bull. Chem. Soc. Jap.*, **42,** 460 (1969).
51. P. V. Peurifoy, M. J. O'Neal, and L. A. Woods, *J. Chromatogr.*, **51,** 227 (1970).
52. H.-J. Petrowitz, "Zur Dünnschichtchromatographie Mehrkerniger Aromatischer Kohlenwasserstoffe," in *Thin-Layer Chromatography,* G. B. Marini-Bettòlo, Ed., Elsevier, Amsterdam, 1964, p. 132.
53. H. Matsushita, Y. Suzuki, and H. Sakabe, *Bull. Chem. Soc. Japan,* **36,** 1371 (1963); through *Chem. Abstr.*, **60,** 26g (1964).
54. I. Arro, *Eesti NSV Teaduste Akad. Toimetised, Tehniliste, Fuusikalis-Mat. Teaduste Seeria.*, **13,** 47 (1963); through *Chem. Abstr.*, **61,** 11946 (1964).
55. M. Koehler, H. Golder, and R. Schiesser, *Z. Anal. Chem.*, **206,** 430 (1964).
56. F. Pavelko and A. D'Ambrosio, *Centro Provencial Per La Studio Sugli*

Inquinamentic Atmosferici, Administrazione Provencial Di Milano, **1959,** 111.
57. A. Berg and J. Lam, *J. Chromatogr.,* **16,** 157 (1964).
58. M. Franck-Neumann and P. Joessang, *J. Chromatogr.,* **14,** 280 (1964).
59. R. G. Harvey and M. Halonen, *J. Chromatogr.,* **25,** 294 (1966).
60. G. D. Short and R. Young, *Analyst (London),* **94,** 259 (1969).
61. H. Kessler and E. Mueller, *J. Chromatogr.,* **24,** 469 (1966).
62. M. Koehler and H.-J. Eichoff, *Z. Anal. Chem.,* **232,** 401 (1967).
63. J. Jaeger, *Chem. Zvesti,* **21,** 321 (1967).
64. R. C. Pierce and M. Katz, *Anal. Chem.,* **47,** 1743 (1975).
65. E. Sawicki, T. W. Stanley, W. C. Elbert, and J. D. Pfaff, *Anal. Chem.,* **36,** 497 (1964).
66. J. Borneff and H. Kunte, *Arch. Hyg. Bakteriol,* **153,** 220 (1969); through *Anal. Abstr.,* **19,** 819 (1970).
67. L. E. Stromeberg and G. Widmark, *J. Chromatogr.,* **49,** 334 (1970).
68. G. Chatot, R. Dangy-Caye, and R. Fontanges, *Chromatographia,* **5,** 460 (1972).
69. G. Chatot, W. Jequier, M. Jay, and R. Fontanges, *J. Chromatogr.,* **45,** 415 (1969).
70. T. W. Stanley, M. J. Morgan, and J. E. Meeker, *Anal. Chem.,* **39,** 1327 (1967).
71. H. Matsushita and K. Arashidani, *Bunseki Kagaku,* **24,** 198 (1975); through *Anal. Abstr.,* **29,** 5H31 (1975).
72. L. V. S. Hood and J. D. Winefordner, *Anal. Chim. Acta,* **42,** 199 (1968).
73. L. Hunter, *Environ. Sci. Technol.,* **9,** 241 (1975).
74. E. Sawicki, T. W. Stanley, and H. Johnson, *Microchem. J.,* **8,** 257 (1964).
75. E. Sawicki, T. W. Stanley, and W. C. Elbert, *J. Chromatogr.,* **20,** 348 (1965).
76. L. Tóth, *J. Chromatogr.,* **50,** 72 (1970).
77. F. De Wiest, D. Rondia, and H. D. Fiorentina, *J. Chromatogr.,* **104,** 399 (1975).
78. J. R. Majer, R. Perry, and M. J. Reade, *J. Chromatogr.,* **48,** 328 (1970).
79. C. A. Gilchrist, A. Lynes, G. Steel, and B. T. Whitham, *Analyst (London),* **97** 880 (1972).
80. M. N. Inscoe, *Anal. Chem.,* **36,** 2505 (1964).
81. J. Lam and A. Berg, *J. Chromatogr.,* **20,** 168 (1965).
82. F. Geiss and M. J. Normand, *NASA, Doc., N*63-11764, 19 pp. (1962).
83. F. Geiss and H. Schlitt, "Analyse von Polyphenylgemischen mit der Dünnschichtchromatographie," EUR-I-1, Euratom, Brussels, November 1961, 17 pp.
84. F. Geiss, H. Schlitt, F. J. Ritter, and W. M. Weimar, *J. Chromatogr.,* **12,** 469 (1963).
85. F. J. Ritter, P. Canonne, and F. Geiss, *Z. Anal. Chem.,* **205,** 313 (1964).
86. H. Schlitt, Belgian Pat. 623,751 (April 18, 1963); through *Chem. Abstr.,* **61,** 2492 (1964).
87. G. Lambertsen and R. T. Holman, *Acta Chem. Scand.,* **17,** 281 (1963).

88. J. G. Kirchner, J. M. Miller, and R. G. Rice, *J. Agric. Food Chem.*, **2,** 1031 (1954).
89. E. Sawicki, "Analysis for Aerotoxicants," in *Critical Reviews in Analytical Chemistry,* Vol. 1, Chemical Rubber Company, Cleveland, Ohio, 1970, p. 275.
90. R. E. Schaad, *Chromatogr. Rev.,* **13,** 61 (1970).

Chapter **XXIII**

LIPIDS

1 FRACTIONATION ACCORDING TO CLASSES OF COMPOUNDS

The term lipid covers a great many different types of compounds such as esters, free acids, ethers, and mono-, di-, and triglycerides; the first thought that comes to mind is to carry out a separation to break a complex natural mixture of this kind into individual classes. This can be accomplished by thin-layer chromatography. This work has been the subject of a number of reviews [1–10]. Most of the thin-layer work on lipids, which has been quite extensive, has been on silica gel layers. To separate the classes of nonpolar lipids on silica gel, the nonpolar solvents such as petroleum ether, benzene, and carbon tetrachloride have been used as well as mixtures of these with very small amounts of more polar solvents such as diethyl ether and acetic acid. The solvent to be used will of course depend on the nature of the mixture to be separated. For the separation of neutral lipids several examples will be cited. Kaufmann and Viswanathan [11] used a petroleum ether (35 to 45°C)–benzene (4 : 7) solvent with silica gel G which resulted in the following separation in increasing order of R_f values: phosphatides remained at the origin followed in increasing order by free acids, cholesterol, triglycerides, and cholesterol esters. Mangold [12], with the same adsorbent, used petroleum ether (60 to 70°C)–diethyl ether (92:8) to chromatograph a lipid mixture with good resolution. Nichols [13], in examining the lipid extracts from lettuce and from cabbage, made a preliminary separation on a column of silicic acid with diethyl ether. The neutral lipids consisting of hydrocarbons, sterol esters, triglycerides, free fatty acids, diglycerides, and sterols were eluted with this solvent; whereas the polar lipids (glycolipids, phospholipids, sterol glycosides) remained on the column. These were eluted with ether–methanol (1:1) and with methanol. The neutral lipids were then separated on silicic acid using hexane–diethyl ether–acetic acid (70:30:1) as the developing solvent. Vogel et al. [14], in separating serum lipids, have used petroleum ether–diethyl ether–acetic acid (60:40:1) for fractionating the neutral lipids. Kaufmann and Makus [15] clearly separated into classes a mixture of fatty acids, keto acids, monoglycerides, and fatty aldehydes by using ethyl ether with silica gel plates. The same authors have used isopropyl ether on the same adsorbent to separate a mixture of triglycerides, diglycerides, monoglycerides, and fatty acids. A two-dimensional separation of 11 different classes is shown in Fig. 23.1.

Freeman and West [16] used a 34-cm layer of silica gel G in order to obtain a better separation for quantitative work. A stepwise development was used with ether–benzene–ethanol–acetic acid (40:50:2:0.3) for 60 min to separate cholesterol, diglycerides, monoglycerides, and phospholipids; the triglycerides and cholesterol esters moved near the solvent front. To separate the latter the second development in the same direction was carried out with ether–benzene (3:47) for 100 to 120 min. By this technique the 1,2- and 1,3-diglycerides were also resolved. Storry and Tuckley [17] using the same size plate used the single solvent mixture benzene–ether–ethyl acetate–acetic acid (40:5:5:0.1) in an S chamber arrangement to separate the same classes of lipids. Impurities in the silica gel were removed by predevelopment with methanol. Kelly [18] using 20-cm plates of silica gel G developed first with the least polar solvent for a distance of 16 cm to separate the cholesterol esters and the triglycerides. This solvent consisted of petroleum ether–ether–acetic acid (70:30:1). The second development in the same direction to a height of 11 cm was made with ether–petroleum ether–acetic acid (70:30:1) to separate the more polar neutral lipids. Better results were obtained by Manners et al. [19] by developing first to within 6 cm of the top of the plate with ether–petroleum ether (40 to 60°C)–acetic acid (65:35:0.5) and then to the top with ether–petroleum ether (6:94). Palmer et al. [20] modified the first

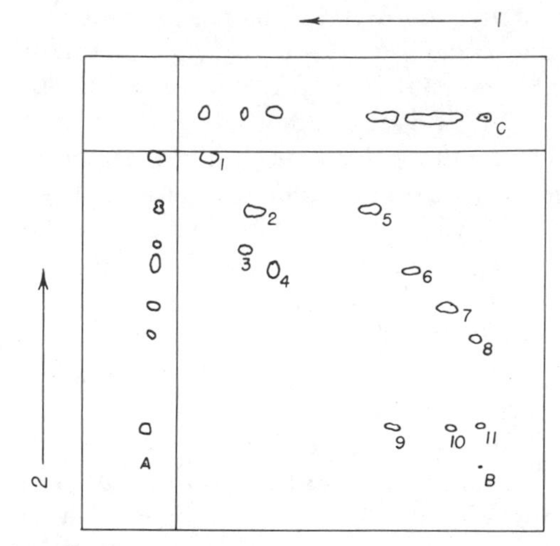

Fig. 23.1 Separation of compounds from 11 different classes. Application at the starting points *A*, *B*, and *C* (the same mixture, 3 μg of each): 1, tristearin; 2, myristic aldehyde; 3, distearin; 4, stearyl alcohol; 5, stearic acid; 6, 9,10-epoxystearic acid; 7, 12-hydroxystearic acid; 8, 9,10-12,13-diepoxystearic acid; 9, monostearin; 10, stearic acid amide; 11, 9,10-dihydroxystearic acid. Solvents: first development, ethyl ether; second development, isopropyl ether + 1.5% acetic acid. Running time: first development, 30 min; second development, 45 min. Detecting agent: phosphomolybdic acid. From Kaufmann and Makus [15]; reproduced with permission of the authors and Industrieverlag von Hernhaussen K. G.

solvent of Kelley to petroleum ether–ether–acetic acid (75:25:2.5) for a two-dimensional separation using the same solvent as Kelley for the second-dimensional solvent. Hubmann [21] developed a microscale method using 45 × 26 mm plates coated by dipping in a chloroform slurry of CAMAG D-O silica gel. With it as little as 1 ng of lipid extract could be separated. The sample was applied as a 1-mm spot near one corner of the plate and developed with chloroform–methanol–twice-distilled water (65:25:4) to within 1.5 cm of the top of the plate, using the long dimension of the plate first. After the plate was dried, it was developed in *n*-hexane–diethyl ether–acetic acid (85:20:2) in the same direcion, but this time to the top of the plate. The third development was made at 90° with the same solvent as for the second development. A fourth development was then carried out in the same direction on only the half of the plate containing the origin, isolating this section from the rest of the plate by scraping away some of the adsorbent. At the same time the lower edge of the other half of the layer was removed so that the solvent could not touch the adsorbent on that side. The solvent for the fourth development was *n*-butanol–acetic acid–water (3:1:1). Svetashev and Vaskovsky [22] for their micro plates ground the silica gel to a particle size of 2 to 7 μ as recommended by Belenky et al. [23]. Pollack et al. [24] developed a multidevelopment technique which has a number of advantages over other techniques. By adding trimethyl borate to their solvent mixtures they practically eliminated the interconversion of 1,2- and 1,3-dipalmitins and presumably that of 1- and 2-monoglycerides that normally occurs during resolution on silica gel G. In addition the method can resolve complex mixtures of lipids containing neutral lipids, glycolipids, and phospholipids. Using silica gel G plates precleaned by developing first with chloroform and then benzene–ethyl acetate (5:1), the first development was with chloroform–ethanol–trimethyl borate (100:1:6) (CEB) for 18 cm. The plates were rotated 90° for development in the second direction with benzene–ethyl acetate–trimethyl borate (100:20:7.2) (BEAB) for the same distance. The plates were again rotated 90° (180° from the first development position) and developed with heptane–benzene (3:2) (HB) for 18 cm. The fourth development in heptane (H) was for 10 cm after turning the plate another 90° (180° to the second development direction). Between each development the plates were air-dried in a hood at room temperature for 45 min. By this procedure 13 lipid standards were resolved: cholesteryl glucoside, monogalactosyl diglyceride, 1- and 2-monopalmitin, palmitic acid, cholesterol, 1,2- and 1,3-dipalmitin, tripalmitin, methyl palmitate, cholesteryl palmitate, squalene, and tetracosane (Fig. 23.2 *a,b*). Digalactosyl diglyceride, lecithin, and lysolecithin remained at the origin, and to resolve these or other phospholipids on the same plate, the fourth direction development was replaced by a development with chloroform–methanol–water (70:20:2.5) (CMW) in the second direction after isolation of the phospholipid lane by scraping away silica gel (Fig. 23.2 *c*). Gradient techniques have also been used for the separation of lipids [25–27]; these methods have been described in Chapter V, Section 5. It should be remembered that, as pointed out by Duthie and Atherton [28], the amount of lipid class from a given biological extraction varies with the source of that extraction and requires a particular solvent system for an efficient separation by thin-layer chromatography.

Weicker [29, 30] has applied a very good step technique in the separation of serum lipids. With silica gel as the adsorbent, the sample was developed in propanol–ammonia (2:1) for a distance of 3 cm. For this separation the solvent was not allowed to be more than 3 days old, and the solvent chamber was allowed to stand for 2 hr with the solvent so as to saturate the atmosphere before running the chromatogram. The developed chromatogram was then freed of solvent vapors and placed in chloroform–benzene (3:2) and developed for a distance of 10 cm. This

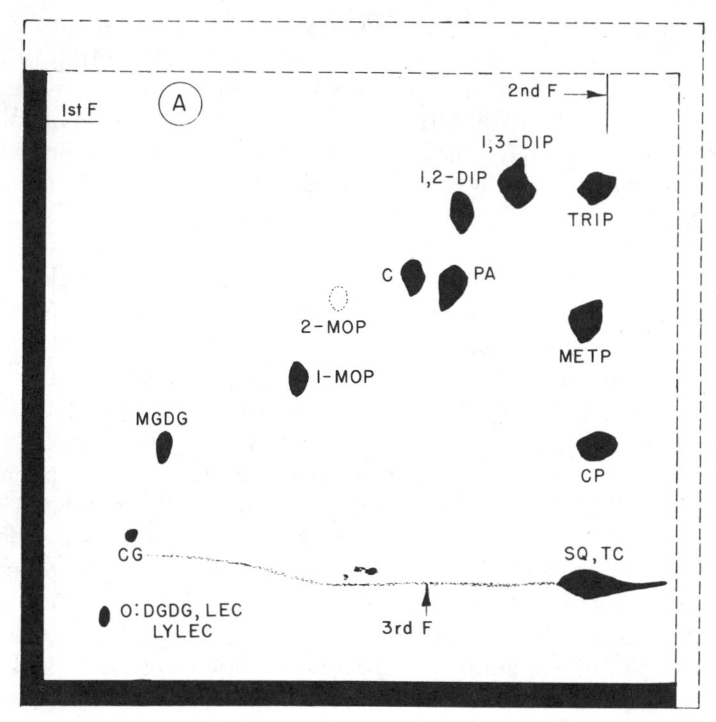

Fig. 23.2A Separation of neutral lipids, glycolipids, and phospholipids by four-directional development; plate after first and second development (see text for details). Abbreviations: LYLEC, lysolecithin; LEC, lecithin; DGDG, digalactosyl diglyceride; CG, cholesteryl glucoside; MGDG, monogalactosyl diglyceride; 1-MOP, 1-monopalmitin; 2-MOP, 2-monopalmitin; C, cholesterol; PA, palmitic acid; 1,2-DIP, 1,2-dipalmitin; 1,3-DIP, 1,3-dipalmitin; TRIP, tripalmitin; METP, methyl palmitate; CP, cholesteryl palmitate; SQ, squalene; TC, tetracosane; O, origin; F, front. From J. D. Pollack et al. [24]; reproduced with permission of the authors and the American Institute of Biological Sciences.

was followed by turning the chromatoplate through 180° and developing in the reverse direction with carbon tetrachloride for a distance of 10 cm. Cholesterol esters, free cholesterol, carotene, a group of fatty acids, lecithin, and some unknown compounds were separated. Laur [31] has also used a step technique in separating and identifying the lipid constituents of *Rhodymenia palmata, Gelidium sesquipedale,* and *Lemanea nodosa*. The samples spotted on silica gel plates were developed three or four times in the same direction with mixtures of hexane–ethyl ether–acetic acid ranging from 90:10:1 to 60:40:2, using increasing concentrations of ether with each step. Zoellner and Wolfram [32] separated plasma lipids with a petroleum ether (50 to 70°C)–methyl ethyl ketone–acetic

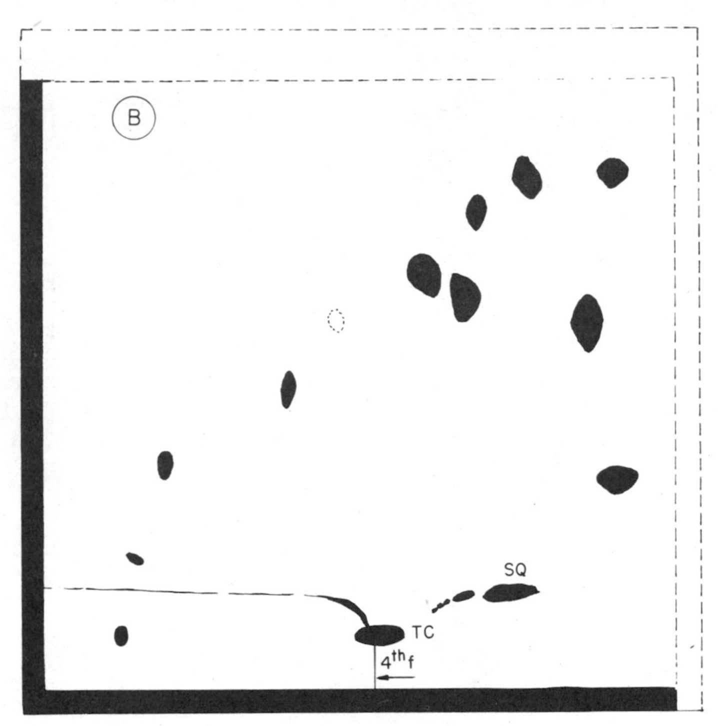

Fig. 23.2B Separation of neutral lipids, glycolipids, and phospholipids by four-directional development, after standard four directions (see abbreviations in Fig. 23.2A; see text for details). From J. D. Pollack et al. [24]; reproduced with permission of the authors and the American Institute of Biological Sciences.

acid (95:4:1) mixture; increasing the concentration of methyl ethyl ketone to obtain a mixture of 84:15:1 gave an even better separation. Araki [33] also used a two-step development to separate serum lipids on silica gel. The first solvent was a chloroform–methanol–acetic acid–water (25:10:3:2) mixture which separated the phospholipids, and the second solvent was hexane–ethyl ether–acetic acid (165:15:1) for the separation of the neutral lipids. Mangold and Malins [34] have separated the lipids of numerous oils including jojoba, castor, oiticica, shark liver, catfish liver, and fur seal-blubber oils with petroleum ether (60 to 70°C)–diethyl ether–acetic acid (90:10:1) on silicic acid. Castor and olive oils were resolved in the same solvents in a ratio of 70:30:2. Figure 23.3 from Malins and Mangold [35] illustrates a separation of lipid classes on silicic acid. For the analysis of bovine semen lipids Komarek et al. [36] used a stepwise development with ether and then with hexane–ether (9:1). Vacíková et al. [37] have separated serum lipids by using loose layers of

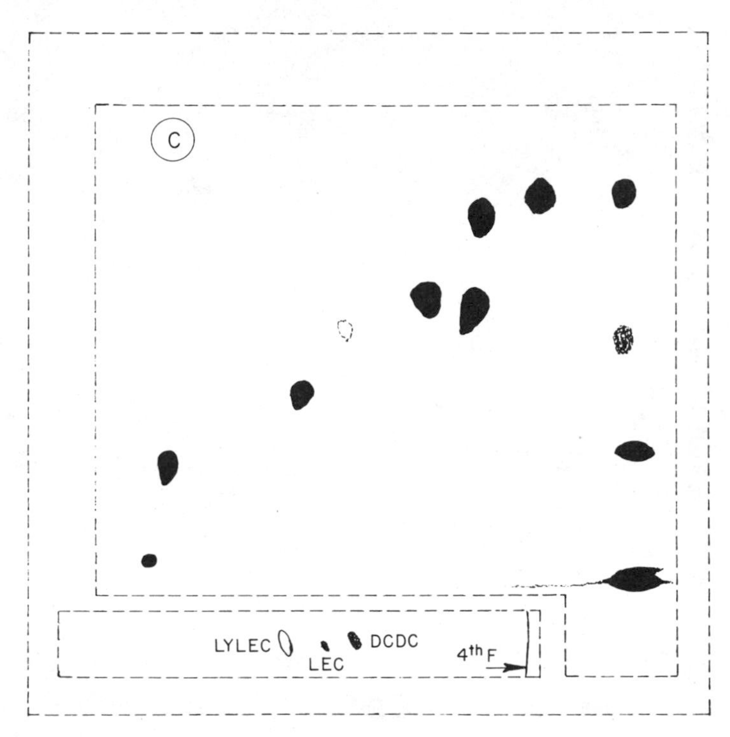

Fig. 23.2C Separation of neutral lipids, glycolipids, and phospholipids by four-directional development. First three directions, unlabeled spots, and fronts as in Fig. 23.2A. After three directions the polar lipid lane was isolated by scraping away silica gel before rechromatographing in the second direction with chloroform–methanol–water (70:20:2.5) (see text for details). From J. D. Pollack et al. [24]; reproduced with permission of the authors and the American Institute of Biological Sciences.

alumina. Development of the plate with petroleum ether–diethyl ether (95:5) separated the cholesterol esters and the triglycerides leaving the phospholipids, fatty acids, and cholesterol at the origin. After removing the two separated fractions, the plate above the starting line was cleaned of aluminum oxide and a fresh layer was applied. Development was then made with petroleum ether–diethyl ether–acetic acid (94.5:5:0.5) in order to separate the free fatty acids, phospholipids, and cholesterol. Table 23.1 gives the R_f values of these compounds in various solvents and the percent recovery.

Lie and Nyc [38] have used thin layers coated on the inside of test tubes for separating lipids with chloroform–methanol and hexane–ether mixtures.

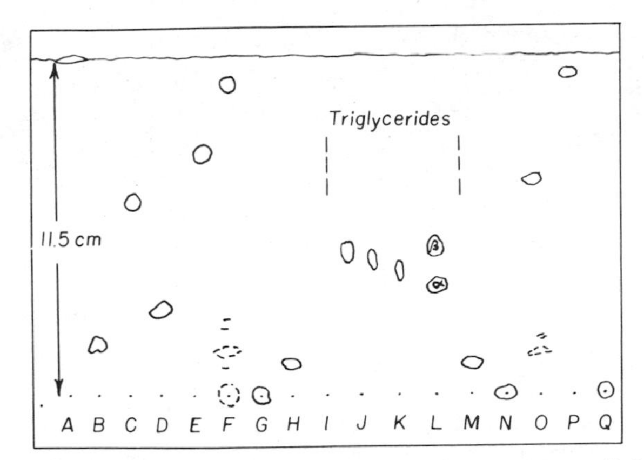

Fig. 23.3 Thin-layer adsorption chromatography of lipid classes on silicic acid. Solvent system: petroleum ether (60 to 70°C)–diethyl ether–acetic acid (90:10:1). Development time: 40 min. Indicator: dichlorofluorescein, 0.2% in ethanol. Amounts: 20 γ each. *A*, Octadecene-9; *B*, oleyl alcohol; *C*, oleylaldehyde; *D*, oleic acid; *E*, methyl oleate; *F*, cholesteryl oleate; *G*, monoolein; *H*, diolein, *I*, triolein; *J*, trilinolenin; *L*, tributyrin (α) plus tristearin (β); *M*, cholesterol; *N*, selachyl alcohol; *O*, chimyldiolein; *P*, oleyl oleate; *Q*, dioleyl lecithin. From Malins and Mangold [35]; reproduced with permission of the authors and the Americal Oil Chemists' Society.

Table 23.1 $R_f \times 100$ Values of Lipid Fractions on Loose Layers of Acidic Aluminum Oxide of Brockmann Activity IV in Various Solutions [37]

	Solvent				
Lipid	*Petroleum Ether–Ether* (95:5)	*Petroleum Ether–Ethanol* (98:2)	*Petroleum Ether–Ether–Acetic Acid* (94.5:5:0.5)	*Heptane–Acetic Acid* (98:2)	*Percent Recovery* [a]
Cholesterol	0	10	35	29	95
Cholesterol esters	91	87	—	—	96
Triglycerides	30	38	—	—	102
Fatty acids	0	6	47	37	98
Phospholipids	0	0	0	0	102

[a] Average recovery from aluminum oxide based on five runs except fatty acids, which is average of nine experiments.

As a rapid screening procedure for cholesterol and triglyceride in serum, Whitner et al. [39] applied 20 μl of serum directly to the layer of silica gel. The plates were developed in hexane–ether–acetic acid (80:20:1.5). If the cholesterol was not separated from other slower-moving lipids, a thicker layer (0.25 mm) was used and a triple development for 5 to 6 cm was made with chloroform–methanol (2:1) followed by the first mentioned solvent.

2 SEPARATION OF TRIGLYCERIDES

After an initial separation of the triglycerides as a class, by means of adsorption chromatography, this group can be broken down into individual components by using reverse-phase partition chromatography. Kaufmann and co-workers [40–43] separated 20 synthetic triglycerides and many natural glycerides using kieselguhr as a supporting layer for tetradecane which was applied as a 5% solution in petroleum ether. The separation was accomplished with a solvent composed of acetone–acetonitrile (4:1). For the separation of some natural triglycerides, the kieselguhr was impregnated with paraffin oil and the solvent was acetone–acetonitrile (7:4) saturated with paraffin oil; to get the proper separation the plates were given a multiple development (3×) for a distance of 14 cm. Soybean oil, corn oil, peanut oil, and linseed oil were examined with this system. In separating corn oil, linseed oil, and beef tallow on siliconized layers, the silica gel was impregnated with a silicone oil of viscosity 1.5 cP and the developing solvent, methanol–acetonitrile–propionitrile (5:4:1.5), was saturated with a silicone oil of viscosity 1000 cP. Litchfield [43] used silanized silica gel impregnated with 8% hexadecane for the separation of triglycerides. The nitroethane solvent was saturated with hexadecane. The fractions were hydrogenated before analyzing by gas chromatography. Thirty different triglyceride groups were obtained. Using a combination of adsorption chromatography and reverse-phase chromatography, Kaufmann and Wessels [44] examined the triglycerides of sunflower seed oil on 10 × 40 cm plates developed lengthwise. For the adsorption separation silica gel impregnated with silver nitrate was used with benzene–ether (4:1). In the preliminary separation, in order to get enough material to work with, 0.7-mm-thick layers were used, and the samples were applied as bands rather than spots so that up to 80 mg of lipids could be separated. The bands were then removed from the plates so that the compounds could be further separated by partition chromatography. For the separation of the saturated triglycerides, silica gel was impregnated with tetradecane to give an 80% saturation. As a solvent, acetone–acetonitrile (4:1) saturated with tetradecane was used. Visualization was accomplished by spraying with dichlorofluorescein solution and then examining under ultraviolet light. For separating the unsaturated triglycerides, the silica gel was impregnated with paraffin oil and used with the same solvent system with the exception that the solvent was 80% saturated with paraffin oil instead of tetradecane, and multiple development (2 to 3×) was used to increase the separation. The unsaturated compounds were located by an iodine spray followed by a spray of α-cyclodextrin. The silver-impregnated plates are very useful since a separation is achieved not only with respect to the number of double bonds, but also on the basis of different fatty acid composition; for example, using the numbers 0 to 3 to represent the number of cis double bonds in the fatty acid chains in triglycerides, Gunstone and Padley [45] found the order of separation of the triglycerides to be 333, 332, 331, 330, 322, 321, 320, 311, 222, 310, 221, 300, 220, 211, 210, 111, 200, 110, 100, and 000. De Vries and Jurriens [46] using benzene–ether (9:1) showed that the trans isomers had a higher R_f value than the corresponding cis compounds. The position of the double bond also affects the R_f value [46, 47]. Renkonen and Rikkinen [48] demonstrated the effect of the location of the unsaturated acid in the triglyceride. Wessels and Rajagopal [49] illustrated the use of reverse-phase separations and argentation chromatography to complement one another in separating critical pairs which occur in each method. In general a 5% level of impregnation with silver nitrate is sufficient; however, Morris et al. [50] found instances where 10% gave improved separations. They also found that triple development with toluene at -25°C gave improved resolution, and for the separation of positional isomers of monoenes they used a 30% impregnated layer. Wood and Snyder [51] in separating methyl esters of fatty acids on silver nitrate-impregnated layers obtained better resolution by impregnating with an ammoniacal silver nitrate solution. Burns et al. [52] have demonstrated a synergistic complexation effect in separating some triglycerides on silver nitrate-impregnated silica gel; certain triglycerides showing little or no movement in cyclooctene, cyclohexene, or benzene showed considerable movement in mixtures of benzene with the cyclic olefins.

Barrett et al. [47, 53] have used silica gel impregnated with 20% silver nitrate for the separation of some triglycerides. After spotting the glyceride mixtures in chloroform solution, the plates were developed with a mixture of carbon tetrachloride–chloroform–acetic acid (60:40:0.5) to which small amounts of ethanol were added, the amounts varying according to the type of glycerides to be separated. By the addition of 0.4% ethanol, unsaturated glycerides having up to three double bonds could be clearly separated; however, addition of 0.4% ethanol failed to give clear separations of the more unsaturated types. By increasing the ethanol to 1 to 1.5%, the unsaturated types could be more readily separated.

De Vries and Jurriens [54] have used horizontal chromatography to separate triglycerides on silver nitrate-impregnated silica gel. With benzene as the solvent, a mixture of 10 triglycerides was separated into seven fractions. Two dyes were added to the mixture in order to be able to follow the progress of the separation. Decanal dinitrophenylhydrazone had an R_f value similar to that of glyceryl tristearate, and Sudan III contained two dyes that corresponded to triglycerides having one and two cis double bonds, respectively. The three pairs SOO + EEO, EEE + SEO, and SSO + SEE (S = stearic, O = oleic, and E = elaidic) were not separated in this system.

There have been many reviews on the use of argentation thin-layer chromatography Boer [55], Jurriens [56], and more recently, Wessels and Rajagopal [49], Morris [57] covering the field up to 1966, and Morris and Nichols [58] covering the remaining literature to 1969.

Kaufmann and Wessels [43] applied the useful technique of enzymatic splitting. In this procedure, the hydrolysis was carried out by using pancreatic enzyme which splits the one and three positions leaving the two position intact. Thus the triglyceride fractions could be analyzed to see if they were a single isomer or a mixture. After the enzymatic splitting, the resulting mixture was chromatographed on silica gel with isopropyl ether containing 1% acetic acid to separate it into the three fractions: (1) unchanged triglycerides, (2) diglycerides and fatty acids, and (3) monoglycerides. The monoglycerides can be subjected to further analysis by hydrolyzing and determining the nature of the fatty acid component. Luddy et al. [59] have also investigated this enzyme on 25 to 50 mg samples, separating the reaction products with thin-layer chromatography.

Wolf and Dugan [60] have applied the pancreatic enzyme to the high-melting glycerides of the milk fat-globule membrane. They found that the individual fatty acids were not randomly distributed in the triglycerides. Jensen et al. [61, 62] examined the action of a concentrated milk lipase, β-esterase, on triglycerides. The enzyme did not show a preference for short- or long-chain acids esterified to the primary alcohol positions when short periods of digestion were used, but the primary ester positions were preferentially hydrolyzed.

Slakey and Lands [63], combining thin-layer chromatography on silica gel and silver nitrate-impregnated silica, gas chromatographic determination of methyl esters, and the use of pancreatic lipase and diglyceride kinase to rat liver triglycerides, determined the fatty acid composition at the 1, 2, and 3 positions. They found that the distribution of acids between the 1 and 3 positions were not random. In applying stereospecific analysis to triglycerides, Brockerhoff [64] partially degraded the triglyceride with methyl magnesium bromide to the 1,3-diglyceride which was isolated by chromatography on 3% boric acid-impregnated silica gel layers. Subsequent conversion to a phospholipid and treatment with phospholipase A permitted determination of the acids in the 1 and 3 positions.

Kaufmann and Khoe [65] have also found it advantageous to use calcium sulfate as a support for reverse-phase chromatography, because in using silica gel with reactions that need washing, the adsorbent had a tendency to slip off the plate. This can be offset to some extent by siliconizing; however, in some cases this prevents the color reactions from taking place, and removal of the dichlorodimethylsilane is very difficult. For the triglycerides the calcium sulfate layer was impregnated with a solution containing 5% tetradecane in petroleum ether to give an 80% saturation on the plate. The separation itself was carried out with acetone–acetonitrile (4:1) and the spots were visualized by spraying with 0.1% Sudan Black B in 50% ethanol.

Michalec et al. [66] chromatographed mixed triglycerides of oleic, linoleic, and linolenic acids as well as saturated triglycerides from various natural oils. The separations were carried out on paraffin oil-impregnated silica gel layers using acetic acid as the solvent.

Kaufmann and co-workers [65, 67] have separated critical pairs of unsaturated triglycerides by using a bromination technique. This was accomplished by applying a brominating solvent during the development. Both kieselguhr and calcium sulfate were used as supporting layers for the impregnating phase of a 240 to 250°C petroleum fraction. The solvent for this brominating separation was a mixture of propionic acid–acetonitrile (3:2), which was 80% saturated with the impregnating petroleum fraction and contained 0.5% of bromine. Previously brominated triglycerides could be separated on the same layers using acetone–acetonitrile (4:1), but the use of the brominating solvent with the triglycerides was much more convenient.

Krell and Hashim [68] have determined the triglyceride content of serum by a combination of thin-layer chromatography and infrared.

3 SEPARATION OF MONO- AND DIGLYCERIDES

Malins [69] isolated the alkoxy diglycerides of dogfish liver oil by using silicic acid plates and a solvent composed of petroleum ether (30 to 60°C)–diethyl ether–acetic acid (90:10:1). The alkoxy diglycerides thus obtained were saponified and converted to the methyl esters by using diazomethane. The methyl esters and glyceryl esters were then separated by using 5% diethyl ether in petroleum ether. Brown and Johnston [70] separated mono-, di-, and triglycerides on silica gel using *n*-hexane–diethyl ether–acetic acid–methanol (90:20:2:3) as the developing solvent. With this solvent system diglycerides were separated into two components, the 1,2 and the 1,3 isomers, Privett et al. [71] were able to separate mono-, di-, and tripalmitin with Skellysolve F–ether (7:3) on silica gel. However, to give proper space for the development of diglycerides and monoglycerides, these two were developed in the same solvent in proportions of 9:1 and 7:3, respectively. Privett and Blank [72] separated 1,2-distearin from 1,3-distearin using Skellysolve F–ether (3:2), but were unable to separate 1- and 2-monoglycerides. However, Rybicka [73, 74] was able to separate the 1- and 2-monoglycerides by using a gradient elution technique. In using this procedure to follow the glycerolysis of linseed oil, the separation on silica gel plates was started in petroleum ether (60 to 80°C)–ether at a ratio of 9:1, and gradually changed by a dropwise addition of ether to the final concentration of 2:3. This concentration was reached 5 min before the end of the development. The 2-monoglycerides remained at the starting point. In addition to the 1- and 2-monoglycerides, 1,2- and 1,3-diglycerides were also separated.

Jensen et al. [75] used preparative thin-layer chromatography in examining the composition of the diglycerides from lipolyzed milk fat.

Hofmann [76] has separated 1- and 2-monoglycerides on thin layers of hydroxyapatite with methyl isobutyl ketone at + 10°C. Typical R_f values for this solvent are given as 1-monoglycerides 0.42, 2-monoglycerides 0.59, and fatty acids 0.78. In contrast to silica gel layers hydroxyapatite does not give a good separation of 1,2- and 1,3-diglycerides.

Schewe and Coutelle [77] used silica gel G impregnated with boric acid and a solvent composed of chloroform–acetone–acetic acid (188:12:1) for the separation of mono- and diglycerides.

4 SEPARATION OF FATTY ACIDS

Kaufmann and Makus [15] used silica gel layers impregnated with undecane for the separation of fatty acids, using mixtures of acetic acid–water (24:1) and acetic acid–acetonitrile (1:1). The development time for a distance of 12 cm for the two solvents was 4 hr and 85 min, respectively. Kaufmann and Khoe [65] also used calcium sulfate (gypsum) as a support for impregnating with undecane in the separation of a mixture of saturated and unsaturated acids, using the acetic acid–acetonitrile (1:1) solvent. For the separation of critical pairs, such as lauric and linolenic, myristic and linoleic, palmitic and oleic, and erucic and arachidic, they applied the bromination and hydrogenation techniques. With either a calcium sulfate plate impregnated with undecane or a kieselguhr layer with the same impregnation [67] for the bromination, a two-dimensional chromatogram is run using acetic acid–acetonitrile (3:2) in the first direction and then, with the same solvent mixture in a 3:7 ratio and containing 0.5% of bromine, development is carried out in the second direction. During development, bromination of the unsaturated acids takes place, thus permitting a separation of critical pairs. As an alternative after the first development, the strip of the plate which was used for the initial separation of the compounds is sprayed with a 2% colloidal palladium solution, after which the plate is placed in a hydrogen atmosphere for 1 hr (see Chapter VI, Sections 2 and 5). Of course prior to the application of the colloidal palladium, the plate is dried in a desiccator to remove the excess acetic acid, and again after the reduction with hydrogen the plate is dried at 120°C for 15 min. The balance of the plate which has not been treated with the palladium is then reimpregnated with undecane prior to development in the second direction with the same solvent. This separation is illustrated in Fig. 6.1.

Malins and Mangold [35] have found a solvent system of acetic acid–water (17:3) satisfactory for separating C_{18} acids on siliconized silicic acid. Different proportions of acetic acid and water were used for acids of different chain lengths. A very useful technique for the separation of saturated and unsaturated acids can be applied; it consists in using a solvent mixture of peracetic acid–acetic acid–water (2:13:3) in combination with the siliconized plate. By this procedure the unsaturated compounds are converted to oxygenated derivatives and travel with the solvent front. These authors also applied a low-temperature technique to the separation of palmitic and oleic acid, which cannot be resolved on reverse-phase partition chromatograms at ordinary temperatures. At 4 to 6°C with siliconized silicic acid plates, oleic acid with an R_f of 0.1 could be separated from palmitic acid, which remained at the origin, using a mixture of formic acid–acetic acid–water (2:2:1).

Heusser [78] used dioxane–water–formic acid (12:7:1) for the separation of a group of fatty acids on silanized silica gel. Ord and Bamford [79] silanized their layers by exposing to the vapors of trimethylchlorosilane or hexamethyldisilazane. Different times of silanization (1 to 16 hr) produced different separations in some solvents. Increased time of exposure to silanization reduced the streaking which occurred with long-chain acids. Methyl cyanide–acetic acid–water (7:1:2.5) was used as a developing solvent.

For the separation of fatty acids from fossil lipids, Douglas and Powell

[80] used silica gel impregnated with 2 to 3% potassium hydroxide with ether–methanol (95:5) as the solvent. This system did not cause hydrolysis of esters present in the extracts.

Kaufmann et al. [81] were able to separate critical pairs of acids, esters, and alcohols using low-temperature thin-layer chromatography in reverse-phase systems by applying temperature programing.

Mizany [82] found it necessary to isolate very small quantities of neodecanoic acid (a mixture of di-α-branched decanoic acids) from other lipids after being used as a cotton harvest aid. This was accomplished by extracting an ether solution of the lipids with aqueous sodium hydroxide. The isolate acids were then esterified with boron trifluoride–methanol, except for the sterically hindered neodecanoic acid. The latter was then isolated by thin-layer chromatography on silica gel with isooctane–acetone–acetic acid (83:15:2). Another method of isolating small amounts of free fatty acids from larger amounts of neutral fats has been used by Winkler et al. [83] and Addison and Ackman [84]. The lipids were passed through a column of Sephadex LH-20 (swelled overnight in the eluting solvent) with 0.2% acetic acid in chloroform. The neutral fats flow through first, followed by the free acids.

Prostaglandins

The prostaglandins are a group of C_{20} fatty acids containing a cyclopentane ring with two side chains, although homologs have been synthesized *in vitro* from fatty acids whose chain lengths varied from 18 to 22 carbon atoms. The full significance of these compounds has not been established yet, but Bergstroem et al. [85] have published on their pharmacological properties. Care must be exercised in their extraction and estimation, because of the ease of modification of the compounds by these processes.

Silica gel layers have been used for the separation of these compounds. As an example, prostaglandin A_2, B_2, and E_2 have been separated on silica gel 60 F_{254} plates using ethyl ether–methanol–chloroform (13:3:4), and F_2 was isolated using chloroform–1,4-dioxane–water–acetic acid (35:60:2:3) [86]. The compounds were determined quantitatively by direct remission spectrodensitometry. Other useful solvent systems for the separation of prostaglandins include: chloroform–methanol–acetic acid (90:5:5) and (80:10:10), benzene–dioxane (5:4), and the upper phase of ethyl acetate–2,2,4-trimethylpentane–acetic acid–water–methanol (22:2:6:20:7). Silica gel impregnated with silver nitrate has also been used for separating prostaglandins E_1, E_2, F_1, F_2, A_1, and A_2 using benzene–dioxane–acetic acid (20:20:1) and ethyl acetate–methanol–2,2,4-trimethylpentane–water (11:3:3.5:1:10) [87]. Wickramasinghe and

Shaw [88] separated prostaglandins A, B, and C, which differ by the position of the double bond, on layers of silica gel impregnated by dipping twice in a 10% solution of ferric chloride in acetone with air drying after each impregnation. The layers were activated at 100°C for 90 min. The best separation was achieved with ethyl acetate–acetic acid–hexane (30:1:19).

A number of detection methods may be used: tests T-195, T-272, T-222, T-146, and T-31. Kiefer et al. [89] detected these compounds in subnanomol amounts by using test T-27.

Daniels [90] has reviewed the chromatography of these compounds and Shaw et al. [91] have reviewed their separation, identification, and estimation. Horton [92] has complied an extensive table of the prostaglandins.

5 SEPARATION OF DERIVATIVES OF FATTY ACIDS

Kaufmann and Ko [93] separated long-chain keto acids, hydroxy acids, and lactones on kieselguhr impregnated with a 240 to 250°C mineral oil fraction. As a solvent they used acetic acid–water (4:1) that was 80% saturated with the impregnating liquid. 2-Hydroxy acids were separated using the same general conditions except that the solvent ratio was changed to 3:1 and also 9:1. Kaufmann and Makus [15] separated epoxy and episulfido fatty acids on undecane-impregnated silica gel with 80% acetic acid. Subbaram et al. [94] used silica gel impregnated with silicone oil to chromatograph episulfido and epoxy fatty acids with acetonitrile–acetic acid–water (7:1:2). Malins and Mangold [35] separated hydroxylated acids from nonhydroxylated acids using a solvent system of petroleum ether–diethyl ether–acetic acid (70:30:1). Subbarao and Achaya [95] separated some close positional isomers of hydroxystearic acid and the corresponding alcohols and esters (Table 23.2) on layers of silica gel G. For the acids and the alcohols, a solvent consisting of diethyl ether–light petroleum ether (2:3) with 2% added acetic or formic acid was used. For the methyl esters the solvent ratio was 1:3, again with 2% added acid. Subbarao et al. [96] applied this method to a larger group of oxygenated derivatives of fatty acids and esters.

Roomi et al. [97] examined the separation of a group of purified fatty acids, their esters, and the corresponding alcohols by chromatographic and reverse-phase partition chromatography on thin layers. Silica gel G was used for the direct chromatographic work and silicone oil (Dow Corning silicone fluid 200) was used to impregnate the plate for the reverse-phase work. Impregnation was accomplished by developing the plate in 5% silicone oil in ether. The R_f values for these separations are

Table 23.2 $R_f \times 100$ Values of Hydroxystearic Acids, Alcohols, and Esters on Silica Gel G[a]

Hydroxystearic Isomer	*Acids*[b]	*Alcohols*[b]	*Methyl Esters*[c]
6-Hydroxy	36	22	32
7-Hydroxy	39	26	37
8-Hydroxy	43	30	40
9-Hydroxy	45	31	45
10-Hydroxy	50	35	50
12-Hydroxy	56	40	55
18-Hydroxy	40	28	37

[a] From R. Subbarao and K. T. Achaya [95]; reproduced with permission of the authors and The Elsevier Publishing Co.

[b] Solvent = light petroleum ether–ether (3:2) + 2% acetic acid.

[c] Solvent = light petroleum ether–ether (3:1) + 2% acetic acid.

given in Table 23.3. The same workers [98] chromatographed epoxy, hydroxy, halohydroxy, and keto fatty acids on boric acid-impregnated silica gel layers using various mixtures of ether–light petroleum ether as solvents. At the same time comparisons were made by direct chromatography on silica gel and by reversed-phase partition chromatography. *Erythro* and *threo* di- and tetrahydroxy fatty acids separate better on the boric acid-impregnated layers. Vioque [99] has developed a micro method for determining the geometrical configuration of epoxy groups. A few milligrams of the epoxy compound, in this case the methyl *cis*-or *trans*-9,10-epoxystearate, in 0.5 ml of benzene was stirred with 10 times the weight of Amberlyst XN-1005 (H^+) (Rohm & Haas) for 30 min. The *threo*-diol from the cis isomer and the *erythro*-diol from the trans isomer were then readily separated on boric acid-impregnated silic gel.

Morris et al. [100] used petroleum ether–ether (9:1) with added 1% acetic acid for the separation of some vicinally unsaturated hydroxy acids on silicic acid layers. The corresponding methyl esters were separated with the same solvent without acetic acid. Morris et al. [101, 102] have applied thin layers in the detection and evaluation of epoxy and hydroxy acids from natural products. Separations of the methyl esters were carried out on silica gel in mixtures of ethyl ether in petroleum ether (35 to 45°C); by the use of stepwise development with varying concentrations of these two solvents, the components could be separated by class, one at a time.

Table 23.3 $R_f \times 100$ Values for Some Fatty Acids and the Corresponding Esters and Alcohols on Silica Gel G Layers[a]

Compound			*Acid*		*Methyl Ester*		*Alcohols*				
								Reverse-Phase			
Chain Length	*Place of Unsaturation*	*Common Name of Acid*	*Direct, 5% Ether–Petroleum Ether*	*Reverse-Phase, Acetonitrile–Acetic Acid–Water (70:10:20)*	*Direct, 2% Ether–Petroleum Ether*	*Reverse-Phase, Acetonitrile–Acetic Acid–Water (70:10:20)*	*Direct, 2% Ether–Petroleum Ether*	*Acetonitrile–Acetic Acid–Water (70:10:20)*	*Acetic Acid–Water (70:30)*	*Acetic Acid–Water (80:20)*	*Acetic Acid–Water (90:10)*
					Acetylenic						
22	13,14	Behenolic	43	40	60	14	64	21	7	15	28
18	9,10	Stearolic	34	64	58	39	61	38	18	26	43
18	6,7	Tariric	34	62	58	40	61	38	18	26	43
11	10,11	Undecynoic	19	87	39	81	39	71	44	55	61
					Ethylenic						
22	13,14	Erucic	52	20	74	7	58	15	3	9	18
18	9,10	Oleic	41	50	73	29	54	28	12	18	34
18	6,7	Petroselinic	41	52	73	30	54	28	12	18	34
11	10,11	Undecenoic	26	75	57	70	34	55	34	44	51
18	9,10;12,13	Linoleic	—	62	—	40	54	38	18	26	43
18	9,10;12,13; 15,16	Linolenic	—	71	—	50	54	48	23	35	53
					Saturated						
22	nil	Behenic	52	3	74	0	58	0	0	0	4
18	nil	Stearic	45	40	71	14	54	21	7	15	28
11	nil	Undecanoic	28	66	57	60	34	38	26	32	38
20	nil	Arachidic	52	20	71	7	—	—	—	—	—
16	nil	Palmitic	44	50	73	29	—	28	12	18	34
14	nil	Myristic	44	64	73	39	—	38	18	26	43
9	nil	Pelargonic	28	75	59	70	—	—	—	—	—
12	nil	Lauric	—	71	—	50	—	48	23	35	53

[a] From M. W. Roomi, M. R. Subbarao, and K. T. Achaya [97]; reproduced with permission of the authors and The Elsevier Publishing Co.

Sgoutas and Kummerow [103] have separated some pairs of dihydroxy and tetrahydroxy stearic acids (Table 23.4) with chloroform–methanol–acetic acid (45:5:1) on silica gel.

Heusser [78] separated the fatty acid hydroxamates on silanized silica gel with methanol–dioxane–chloroform–glycine buffer (pH 3.0) (4:3:1:4) in an unsaturated tank. The derivatives were visualized with test T-118.

Bazán and Cellik [104] used a gradient thickness layer for the separation and quantitation of free fatty acids and other lipids. This permitted the separation of trace amounts of compounds from complex mixtures.

Malins and co-workers [105, 106] used the nitrate derivatives of fatty alcohols, hydroxy esters, glycerides, and glyceryl ethers as a means of separation and identification. (See Chapter XIV, Section 2 for the preparation of these derivatives.) The mono- and dinitrates of hydroxy compounds can be separated from other classes of compounds by chromatographing on thin layers of silicic acid using hexane as the developing solvent. The nitrate derivatives of hydroxy esters, glycerides, and glyceryl ethers require a more polar solvent such as *n*-hexane–diethyl ether (85:15). Glyceryl ethers which could not be separated from monoglycerides of similar chain length [34] could be separated by means of the nitrate derivatives. Malins and Houle [107] have used the same reagent, which is effective for nitrating alkenes [108], to nitrate methyl oleate. Three types of derivatives are formed: (**1**) isomeric nitro, (**2**) acetoxynitro, and (**3**) nitro-nitrate.

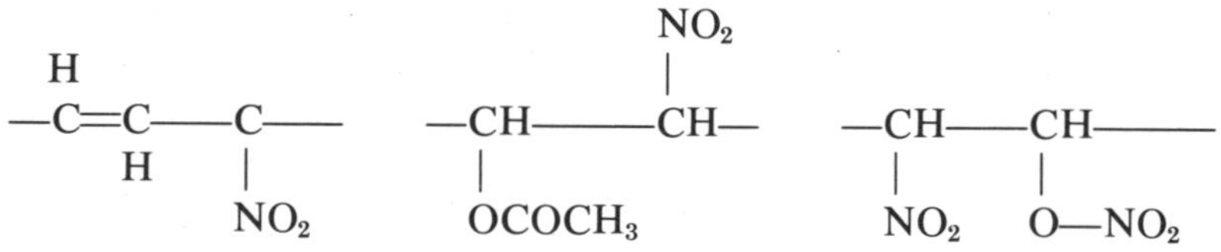

$$-\overset{H}{C}=\underset{H}{C}-\underset{NO_2}{\underset{|}{C}}- \qquad -\underset{OCOCH_3}{\underset{|}{CH}}-\overset{NO_2}{\overset{|}{CH}}- \qquad -\underset{NO_2}{\underset{|}{CH}}-\underset{O-NO_2}{\underset{|}{CH}}-$$

1 **2** **3**

These compounds can be separated on silicic acid with a petroleum ether (30 to 60°C)–diethyl ether (85:15) mixture.

Mangold and Kammereck [109] separated industrial synthetic derivatives of fatty acids containing nitrogen, sulfur, or phosphorus. This included a wide range of compounds, such as primary, secondary, and tertiary amines, mercaptans, thiocyanates, alkyl phenols, isocyanates, quaternary ammonium compounds, and surface-active agents. Four solvents that can be used for these compounds in stepwise manner with silica gel layers are petroleum ether (60 to 70°C)–benzene (95:5), benzene–ammonium hydroxide (benzene layer from equilibration of 100 ml of benzene with 10 ml of 1 *N* ammonium hydroxide at 20°C), ammoniacal

Table 23.4 $R_f \times 100$ Values of Di- and Tetrahydroxystearic Acids on Silica Gel G With Chloroform—Methanol—Acetic Acid (45:5:1). Development Distance 10 cm[a]

Substance	$R_f \times 100$ *Values*
threo-9,10-Dihydroxystearic acid	70
erythro-9,10-Dihydroxystearic acid	65
threo, threo, threo-9,10,12,13-Tetrahydroxystearic acid	48
threo, erythro, threo-9,10,12,13-Tetrahydroxystearic acid	44
erythro, threo, erythro-9,10,12,13-Tetrahydroxystearic acid	40
erythro, erythro, erythro-9,10,12,13-Tetrahydroxystearic acid	35

[a] From D. Sgoutas and F. A. Kummerow [103]; reproduced with permission of the authors and the American Oil Chemists' Society.

chloroform–methanol (97:3) (chloroform layer from equilibration of 10 ml of chloroform with 1 ml of 1 *N* ammonium hydroxide), and acetone–14 *N* ammonium hydroxide (9:1). The more strongly acidic compounds used as detergents (alkyl sulfates, sulfonates, phosphates, etc.) can be separated on silica gel layers (containing 10% ammonium sulfate) using mixtures of chloroform with methanol containing 5% 0.1 *N* sulfuric acid. The exact proportions depend on the polarity of the compounds, typical mixtures being in the ratio of 97:3 and 4:1, chloroform to methanol.

6 SEPARATION OF FATTY ACID METHYL ESTERS

Malins and Mangold [35] have used siliconized chromatoplates with acetonitrile–acetic acid–water (14:2:5) for separating some methyl esters derived from menhaden oil. Methyl esters of C_{18} esters were separated with 85% acetic acid again using siliconized silicic acid. The technique of separating saturated and unsaturated methyl esters by means of oxidizing the unsaturated acids was also employed, developing with peracetic acid–acetic acid–water (2:15:3) [110]. Applewhite et al. [111] have separated methyl esters on chromatostrips of silica gel G with various solvent mixtures (Table 23.5).

Hammonds and Shone [112] used a partition method for separating critical pairs of methyl esters. For this separation, kieselguhr G layers were impregnated with 10% liquid paraffin in petroleum ether (60 to 80°C). The impregnating solvent was removed at room temperature, and then the methyl esters were applied as petroleum ether solutions. Development was with a nitromethane–acetonitrile–acetic acid (15:2:2)

Table 23.5. $R_f \times 100$ Values for Some Fatty Acid Methyl Esters on Chromatostrips of Silica Gel G. Development Distance 10 cm[a]

Compound	Solvent Mixture					
	Skellysolve F–Diethyl Ether (9:1)	*Skellysolve F–Diethyl Ether* (7:3)	*Skellysolve F–Diethyl Ether* (1:1)	*Benzene–Diethyl Ether* (3:1)	*Benzene–Diethyl Ether* (1:1)	*Benzene–Methanol* (17:3)
Methyl palmitate[b]	62	89				
Methyl stearate[b]	60	89		95	97	81
Methyl oleate[b,d]	56	89				
Methyl linoleate[b,d]	58	88				
Methyl linolenate[b,d]	57	87				
Methyl eleostearate[b,c,d]	49[e]	82[e]				
Methyl 9-hydroxystearate[b]		43				
Methyl 12-hydroxystearate[b]		50			87	45
Methyl 9-hydroxy-10,12-octadecadienoate[b,c,d]		43	67		84	41
Methyl ricinoleate[b,d]		49	72			45
Methyl 12-ketostearate[b]		83			94	55
Methyl 9-keto-10,12-octadecadienoate[b,c,d]		61				

[a] From T. H. Applewhite, M. J. Diamond, and L. A. Goldblatt [111]; reproduced with permission of the authors and The American Oil Chemists' Society.
[b] 2,7-Dichlorofluorescein indicator.
[c] Quenching of fluorescent minerals.
[d] Fluorescein-bromine indicator.
[e] Major spot, minor impurity noted.

mixture for a distance of 10 cm. In this case the solvent was not equilibrated with the stationary phase. The spots were revealed by test T-123, differentiating saturated from unsaturated by the color difference. Using this method, methyl linolenate could be separated from the laurate, and linoleate from the myristate. Only a partial separation of methyl oleate could be achieved from the palmitate.

Ord and Bamford [80] also used a reverse-phase system for separating methyl esters. In this case they used a silica gel layer silanized with trimethylchlorosilane or hexamethyldisilazane with the bonded methyl groups forming the reverse phase. The solvent was acetonitrile–acetic acid–water (7:1:2.5). Using the same type of layer, Heusser [79] separated methyl esters with acetone–methanol–water (7:5:3.5). In both cases visualization was achieved by spraying with ethanolic phosphomolybdic acid and then heating.

Ruseva-Atanasova and Janák [113] separated methyl esters according to the number of carbon atoms by gas chromatography on a nonpolar stationary phase. The effluent, was deposited on a silver nitrate-impregnated silica gel layer (coupled gas–thin-layer chromatography). Development of the plate with petroleum ether–diethyl ether (7:3) then separated the components according to unsaturation. Methyl esters for gas chromatography can be quickly and conveniently prepared by transesterification directly on a thin-layer plate [114, 115]. After application of the sample to the silica gel layer or after an initial development to separate components, the layer is sprayed with 2 *N* sodium methylate. After 5 min the transesterification is complete and under these conditions only ester, and not acid–amide, linkages are split.

Morris [116] demonstrated the usefulness of silver nitrate-impregnated layers for the thin-layer separation of unsaturated methyl esters. Long-chain methyl esters were readily separated according to the degree of unsaturation, as well as the cis and trans monoethanoid esters. Boric acid impregnation was also used for the separation of dihydroxy esters in order to resolve *threo* and *erythro* isomers which could not be separated on untreated silica gel. Plates impregnated with both boric acid and silver nitrate permitted the simultaneous separation of isomers and vinylogs (Fig. 23.4). Morris [117] has extended these studies with the methyl esters of di-, tri-, and tetrahydroxy long-chain fatty acids. He tried a number of complexing agents, but found the best results with boric acid, sodium borate, or sodium arsenite. Silica gel impregnated with 10% of one of these agents gave clear separation of diastereoisomeric pairs of methyl dihydroxy esters, the lower melting *threo* isomer in each case having the higher R_f value. With the trihydroxy esters, separation on the boric acid and sodium borate layers was not very good, but the

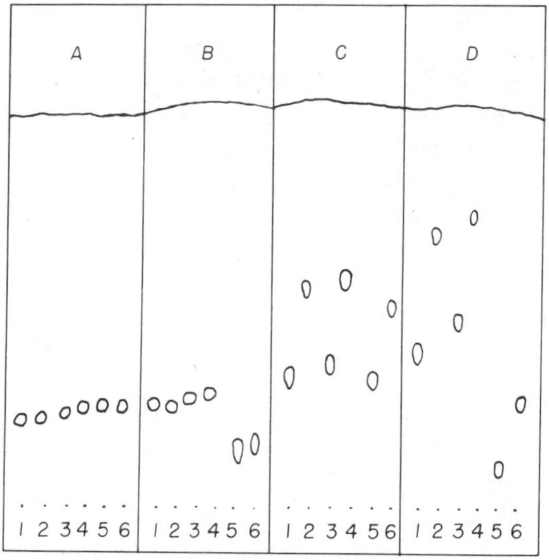

Fig. 23.4 Thin-layer chromatogram of methyl esters on silica gel. *A* = untreated, *B* = silver nitrate-impregnated, *C* = boric acid-impregnated, *D* = silver nitrate- and boric acid-impregnated. 1, *erythro*-9,10-dihydroxystearate; 2, *threo*-9,10,dihydroxystearate; 3, *erythro*-12,13-dihydroxystearate; 4, *threo*-12,13-dihydroxystearate; 5, *erythro*-12,13-dihydroxyoleate; 6, *threo*-12,13-dihydroxyoleate. Developing solvent: diethyl ether–hexane (60:40). Spots were located under ultraviolet light after spraying with 2′,7′-dichloroflourescein and reproduced by tracing. From Morris [116]; reproduced with permission of the author and The Society of Chemical Industry.

separations on the corresponding sodium arsenite layers showed some differentiation between *threo* and *erythro* isomers and also provided the differentiation of the four diastereoisomers of 9,10,12-trihydroxy stearate and of 9,12,13-trihydroxy stearate. Similarly with the tetrahydroxy compounds the use of sodium arsenite-impregnated gel affected the R_f values so that the higher melting isomers were observed to travel faster than the corresponding lower melting isomer. For the tetrahydroxy derivatives, the sodium borate impregnation gave some results which were better than the sodium arsenite-impregnated layers. For example, *erythro*-9,10-*erythro*-12,13-tetrahydroxystearic acid has the same R_f value on sodium arsenite as the *threo*-9,10-*erythro*-12,13-compound, but they can be separated on the sodium borate-impregnated layer. The solvents used in these separations were mixtures of methanol–chloroform. For the dihydroxy derivatives the ratio was 2:98, for the trihydroxy derivatives on sodium arsenite-impregnated layers the ratio was 1:99, and for the tetrahydroxy derivatives 1:24. For the separations of the tetrahydroxy derivatives on sodium borate layers a ratio of 1:9 was used. De Vries and Jurriens [46] have applied silver nitrate-impregnated silica gel to the separation of various geometric isomers. For the methyl esters of unsaturated fatty acids various mixtures of benzene and petroleum ether ranging in ratio from 7:3 to 9:1 were used.

Morris et al. [50] found improved separation of isomeric *cis*- and *trans*-octadecenoates on silver nitrate-impregnated (30%) layers by developing three times in toluene at -25°C. Wood and Snyder [51] obtained improved separation of methyl esters on layers impregnated with ammoniacal silver nitrate.

Bergel'son et al. [118] separated the isomeric methyl esters of monoolefinic acids by two-dimensional chromatography. The first development was on silica gel impregnated with dodecane using acetone–acetonitrile (7:10). The unused portion of the plate was then impregnated with silver nitrate and developed in the second direction with diethyl ether–petroleum ether (9:41).

This technique has been carried further [119] so as to provide a complete structural analysis of complex mixtures of fatty acids based on the two-dimensional separation of their methyl esters and subsequent identification of the unsaturated acids by oxidative cleavage directly on the adsorbent layer. Both positional isomers and stereoisomers of unsaturated fatty acids may be determined. For the two-dimensional separation gypsum-bound silica gel layers were impregnated by immersion in a 10% solution of dodecane in hexane. After application of the sample of methyl esters, the development in the first direction was with the solvent system of Kaufmann and Makus [15] using acetonitrile–acetone (1:1) 90% saturated with dodecane. The developed plate was then dried, successively, at room temperature for 30 min, at 90 to 95°C for 30 to 40 min, and overnight at room temperature. That portion of the plate which would contain the methyl esters of the unsaturated acids after the second development was then impregnated by spraying with 20% silver nitrate solution (the section where the saturated methyl esters would appear was not treated because of decreased sensitivity of the detecting agent on silver nitrate layers). The silver nitrate-treated plate was dried by gradually heating to 100°C over a period of 1 hr with an additional 1 hr at that temperature. After development in the second direction with dipropyl ether–hexane (2:3), the plates were dried at 70 to 80°C for 20 min. The saturated methyl esters were detected by spraying with a solution of 40 mg bromothymol blue in 100 ml of 0.01 *N* sodium hydroxide solution (note: the unsaturated ester section of the plate is not sprayed with this reagent). Heating at 70 to 80°C for 10 to 15 min disclosed the saturated esters as yellow spots on a blue background. The unsaturated methyl esters were detected by spraying with a solution of 40 ml of bromothymol blue in 100 ml of 20% ammonium hydroxide. They appeared as quickly fading light or dark blue spots on a grey background. For the further identification of the unsaturated acids by oxidative cleavage of the methyl esters, plates of cellulose powder with a gypsum binder were prepared. After the cellulose layer was impregnated with the 25% solution of dimethylformamide in benzene, it was dried for 20 min at room temperature and then at 60 to 70°C for a few minutes. The prepared plate was covered with glass to prevent evaporation of the impregnating material. The unsaturated methyl esters from the previously run silica gel plate were eluted with ether, concentrated, and then spotted on the cellulose layer. Oxidation was carried out directly at the sample spot by applying a mixture of 10 ml of 0.1 *M* sodium metaperiodate and 10 ml of a solution that was 0.1 *M* with respect to both potassium carbonate and potassium permanganate. The reaction was carried out at 55 to 60°C until the pink permanganate color disappeared. Repetition of the oxidation procedure insured the completion of the reaction. The alkali salts formed were converted to the free acid by adding a drop of 2 *N* hydrochloric acid solution, and then the separation of the oxidation products was carried out using a solvent composed of hexane–diethyl ether–dimethylformamide (40:20:1). The products of the oxidation (monocarboxylic acids and monomethyl esters of dicarboxylic acids) were detected by exposing to ammonia vapor and spraying with a mixture containing 200 mg of methyl red, 200 mg of bromothymol blue, 100 ml of formalin, 400 ml of ethanol, and 3 ml of 1 *N* sodium hydroxide. The acids appeared as yellow spots on a green background. The esters of acids, including arachidonate and higher, were oxidized directly on the silica gel plate, after which the reaction products were extracted with ether and applied to the cellulose plate. *n*-Alkanoic acids of more than 11 carbon atoms, which arise from the oxidation, travel with the solvent front on the cellulose layers. Their identification was made by carrying out the oxidation reaction on thin silica gel layers impregnated with dodecane and chromatographing in acetonitrile–acetic acid (4:1) saturated with dodecane. After drying at 110 to 120°C for 1 hr, the monocarboxylic acids were detected by spraying with a 10% solution of phosphomolybdic acid in ethanol and subsequent application of heat. In this case two oxidation reactions had to be carried out, one on the silica gel for the monocarboxylic acids, and the other on cellulose for the identification of the monomethyl esters of the dicarboxylic acids. The R_f values for the methyl esters of the saturated and unsaturated fatty acids are given in Table 23.6; R_f values for the monocarboxylic acids and monomethyl esters of dicarboxylic acids on dimethyl formamide-impregnated cellulose are given in Table 23.7. Figure 23.5

Table 23.6 $R_f \times 100$ Values of Methyl Esters of Saturated and Unsaturated Fatty Acids on Impregnated Silica Gel in Two-Dimensional Technique[a]

Compound	*First Direction*[b]	*Second Direction*[c]
Methyl esters of saturated acids		
Dodecanoic	78	80[d]
Tetradecanoic	67	80[d]
Hexadecanoic	56	80[d]
Octadecanoic	46	80[d]
Eicosanoic	36	80[d]
Docosanoic	28	80[d]
Methyl esters of unsaturated acids		
C_{16}		
cis-Hexadecen-9-oic	71	39
cis-Hexadecen-11-oic	71	46
C_{18}		
cis-Octadecen-7-oic	60	43
cis-Octadecen-9-oic	60	42
cis-Octadecen-11-oic	60	51
trans-Octadecen-9-oic	60	64
C_{20}		
cis-Eicosen-11-oic	50	52
C_{22}		
cis-Docosen-5-oic	40	39
cis-Docosen-11-oic	40	52
C_{26}		
cis-Hexacosen-9-oic	25	47
Octadeca-9,12-dienoic	71	8
Octadeca-9,12,15-trienoic	80	3
Eicosa-5,8,11,14-tetraenoic	89	0

[a] From Bergel'son et al. [120]; reproduced with permission of the authors and The Elsevier Publishing Co.

[b] Solvent: acetonitrile–acetone (1 : 1) 90% saturated with dodecane. Adsorbent: gypsum-bound silica gel plate impregnated with 10% dodecane in hexane.

[c] Solvent: diproproyl ether–hexane (2 : 3). Adsorbent: silica gel impregnated with 20% silver nitrate.

[d] Not impregnated with silver nitrate; see text for details.

shows separation of the oxidative-cleavage products of some unsaturated acids.

Vereschagin and Skvortsova [120] used silica gel impregnated with 2% dodecane to separate methyl esters with a solution of silver nitrate and dodecane in 70 to 90% aqueous methanol as the solvent. Ord and Bamford

Table 23.7 $R_f \times 100$ Values of Monocarboxylic Acids and Monomethyl Esters of Dicarboxylic Acids on Dimethylformamide-Impregnated Cellulose in Hexane–Diethyl Ether–Dimethylformamide (40:20:1)[a]

Compound	$R_f \times 100$	*Compound*	$R_f \times 100$
Monocarboxylic acid		Monomethyl esters of dicarboxylic acids	
Propionic	20	Glutaric	11
Butyric	31	Pimelic	17
Valeric	43	Azelaic	31
Caproic	54	Nonanedicarboxylic	54
Enantic	65		
Pelargonic	84		
Hendecanoic	94		

[a] From Bergel'son et al. [120]; reproduced with permission of the authors and The Elsevier Publishing Co.

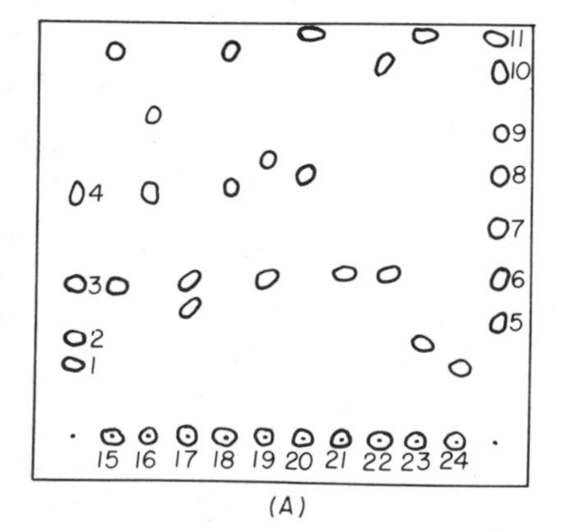

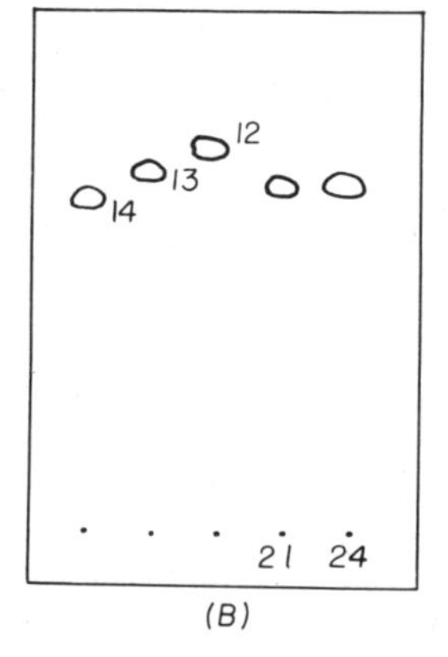

Fig. 23.5 Oxidative cleavage of unsaturated fatty acids. *(A)* Adsorbent: cellulose impregnated with a 25% solution of dimethylformamide in benzene Solvent: hexane–diethyl ether–dimethylformamide (40:20:1). Developing time: 20 min. *(B)* Adsorbent: silica gel impregnated with a 10% solution of dodecane in hexane. Solvent: acetic acid–acetonitrile (1:4) (saturated with dodecane). Developing time: 60 min. *Reference substances*: monomethyl esters of the acids: 1 = glutaric; 2 = pimelic; 3 = azelaic; 4 = nonane-1,9-dicarboxylic. Acids: 5 = propionic; 6 = butyric; 7 = valeric; 8 = caproic; 9 = enanthic; 10 = pelargonic; 11 = hendecanoic; 12 = palmitic; 13 = margaric; 14 = stearic. Methyl esters of unsaturated acids subjected to oxidation: 15 = oleic; 16 = *cis*-vaccenic; 17 = linolenic; 18 = *cis*-eicosen-11-oic; 19 = linoleic; 20 = *cis*-docosen-11-oic; 21 = *cis*-hexacosen-9-oic; 22 = elaidic; 23 = *cis*-octadecen-7-oic; 24 = *cis*-docosen-5-oic. From Bergel'son et al. [120]; reproduced with permission of the authors and The Elsevier Publishing Co.

[121] also used solvents containing silver nitrate for the separation of fatty acid methyl esters and glycerides on silanized silica gel layers (see Chapter III, Section 7 for preparation). For dichlorosilane-treated layers development was carried out in an unlined tank with 10% aqueous methanol saturated with silver nitrate or with water–methanol–acetonitrile (2.5:95:2.5) saturated with silver nitrate. For diethyldichlorosilane-treated layers, development was with water–methanol–nitromethane (2.5:95:2.5) or 10% aqueous acetone, both solutions saturated with silver nitrate. Paulose [122], after impregnating silica gel layers with a 5% solution of Dow Corning silicone fluid 200, sprayed the layers with a 10% solution of silver nitrate in 50% ethanol. The plates were dried at 105°C for 15 min before applying the sample. Development was with 95% aqueous methanol saturated with silicone oil and silver nitrate. Unsaturated esters of the same chain length but differing in the number of double bonds, such as oleate, linoleate, and linolenate, and saturated esters differing in chain length by two carbons were separated. The critical pairs linoleate and myristate and oleate and palmitate were separated. Positional esters, such as oleate and petroselinate were not resolved although the cis–trans isomers, oleate and elaidate, were separated. A number of pairs could not be separated: linolenate and ricinoleate, oleate and laurate, elaidate and myristate, and erucate and palmitate.

Mangold et al. [123] used radioactive diazomethane for the preparation of methyl esters to be used in the quantitative analysis of lipids. The preparation was as follows: "A solution of 10 mg (0.05 mmole) of *p*-tolylsulfonyl-methyl-^{14}C-nitrosamide (specific activity 0.6 mCi/mmole) in 1 ml of diethyl ether was allowed to react with 2 ml of an ice-cold solution of 0.1 g of sodium hydroxide in 10:1 ethanol–water in a micro gas generator. Diazomethane and ether were distilled from the reaction mixture by passing a slow stream of nitrogen into the reaction flask, which was immersed in a water bath held at 60 to 70°C. The diazomethane–ether solution was collected in two test tubes in series, each containing 1 to 2 ml of ether that was maintained at 0 to 5°C with ice water. The diazomethane solutions were combined at the end of the distillation. Aliquots of 0.5 to 1 ml of ether containing diazomethane were immediately added to solutions of 2 to 20 mg of fatty acids (0.01 to 0.1 mmole) in 90:10 diethyl ether–methanol [124, 125]. Lipids (10 to 20 mg) containing hydroxy or amino groups were labeled by reaction with 1:10 solution of acetic-1-^{14}C anhydride[$(CH_3{}^{14}CO)_2O$, specific activity 0.6 mCi/mmole] in pyridine. The reaction was conducted in a 5/150-mm sealed tube at 100°C for 30 to 60 min, using 20% excess reagent. After cooling, the tubes were opened and the reaction mixture was diluted with 10 ml of *N* sulfuric acid; the acetylated lipids were extracted with ether, washed with water, and dried. These radioactive compounds were also used in following the purification of various lipids.

Mangold and Kammereck [126] have reported on the use of the acetoxy–mercury–methoxy derivatives of methyl esters for separating the latter by thin-layer chromatography. Since the method was used in combination with gas chromatography, the methyl esters of hydroxylated acids were removed first in order to prevent the interference of some of these during the gas chromatography. For the preparation of these derivatives, the procedure of Jantzen and Andreas [127, 128] was used as follows: About 25 ml of the reagent, a solution of mercuric acetate (14 g) in methanol (250 ml), water (2.5 ml), and glacial acetic acid (1 ml), was added to the esters (1 g) and allowed to react in a stoppered flask in the dark, at room temperature. After 24 hr the methanol was evaporated at less than 30°C in vacuo, or by a stream of nitrogen, and the dry residue dissolved in chloroform (50 ml). The chloroform solution was washed with water (five times with 25 ml). The chloroform solution was washed with water (five times with 25 ml) to remove excess mercuric acetate and then dried with sodium sulfate. The derivatives were applied in chloroform solution to thin layers of silicic acid and chromatographed with a stepwise development. A mixture of petroleum ether (60 to 70°C)–diethyl ether (4:1) was used to separate the methyl esters from the acetoxy–mercury–methoxy derivatives of the unsaturated methyl esters. After the plates were dried, they were developed in *n*-propanol–acetic acid (100:1) for a distance of 12 to 14 cm. The mono-, di-, and trienoates were clearly separated by this procedure. The derivatives could be visualized as purple spots on a light rose background by spraying with a solution of 0.1% *s*-diphenylcarbazone in 95% ethanol.

White [129] used the methoxy, bromomercury adducts for the separation of unsaturated fatty acid esters, because better separation was obtained. These derivatives were formed by stirring the acetoxy–mercury–methoxy derivatives in chloroform solution with sodium bromide–methanol (5:95 wt/vol) in 10% excess for 30 sec. After treating with water the chloroform layer was separated and washed and was ready for application to the thin layer. The compounds were separated on silica gel G with heptane–dioxane (60:40). Minnikin and Smith [130] compared the corresponding chloro and iodo derivatives with the bromo derivatives. By using the latter two derivatives unsaturated esters can be fractionated according to number and stereochemistry of double bonds and also partially according to chain length. Hexane–diethyl ether (9:1) was suitable for separation on silica gel.

Privett and co-workers [131–133] have applied a reductive ozonolysis to methyl esters in the analysis and structural determination of unsaturated fatty acids. This procedure involves the quantitative and instantaneous formation of ozonides under carefully controlled conditions to reduce side reactions. The ozonides are then catalytically reduced to aldehydes. The ozonization is carried out in the following manner: "A solution of ozone is prepared by bubbling oxygen containing 2 to 3% ozone at about 100 ml per min through 10 ml of pentane in a 25-ml round-bottom flask immersed in a dry ice bath −60 to −70°C for 5 min. At the end of this time the solution has a dark blue color and contains about 0.3 mmole of ozone. The ozonization is performed by adding the sample (1 to 50 mg) dissolved in 2 to 3 ml of purified pentane, cooled as low as possible without causing crystallization of the sample, to the 10 ml of the ozone solution. As a guide for the ozonization of other esters, 10 ml of the ozone solution is enough to ozonize 50 mg of methyl oleate. A colorless solution at the end of the ozonization indicates that additional ozone must be supplied to give a faint bluish color to the solution. On the other hand, if the reaction is complete, a pale gray color will be evident. Thin-layer chromatography can be used to check the completeness of the ozonization by using petroleum ether (30 to 60°C)–diethyl ether (98 : 2) with silica gel layers. The reduction of the ozonides is carried out with hydrogen using Lindlar catalyst, and if the reduction products are at least five carbons long the reduction may be carried out in the pentane solution. Where shorter fragments occur, the pentane is removed and replaced by methyl caprylate or similar compounds having a long retention time on gas chromatographic analysis. The products of the reduction are analyzed by infrared, gas chromatography, and thin-layer chromatography. The same procedure has been reduced to an ultramicro scale [134]. In order to determine the position of double bonds in polyenoic fatty acid esters, Roehm and Privett [135] have applied a partial reduction with hydrazine followed by the reductive ozonolysis just described.

Kaufmann et al. [81a, 136] separated critical pairs of methyl esters, as well as the free acids and fatty alcohols, by applying isothermal procedures at low temperature in reverse-phase systems and by temperature programing.

Kannan et al. [137] have separated the carboxymethylthio derivatives formed by reacting mercaptoacetic acid with monounsaturated acids and esters. Separations were made on silica gel G with ether–light petroleum (2:3) and on siliconized gel with acetonitrile–acetic acid–water (7:1:2). Detection was by charring with sulfuric acid.

Hradec and Menšík [138] used thin layers made by suspending a urea–Celite 545 (120 mesh) (3:1) mixture un urea-saturated methanol. The layers were dried overnight at room temperature. Methyl esters of branched-chain acids were separated from those of straight-chain acids. After application of the sample, the plate was allowed to stand in an atmosphere of methanol overnight to allow time for the clathrate compounds to form. Development was carried out with petroleum ether. The branched-chain fatty acid esters had an R_f of 0.4 to 0.5 whereas the straight-chain compounds were near the solvent front.

Badings [139] has demonstrated the effect of a controlled atmosphere (oxygen-free) in separating some methyl esters. It is also advantageous to add an antioxidant to the solvent used in separations of methyl esters as well as other lipids. Wren and Szczepanowska [140] added 0.005% of 4-methyl-2,6-di-*tert*-butylphenol (BHT) to the developing solvent. Neudoerffer and Lea [141] preferred 1,4-dihydroxy-2-*tert*-butylbenzene (BHQ) (10 mg %) as an antioxidant.

Cornelius and Shone [142] separated the methyl esters of acids obtained by saponifying *Bombax oleaginum* seed oil. Methylation was carried out by the method of Metcalfe and Schmitz [143], that is, refluxing for 5 min with 12.5% boron trifluoride in methanol. Separation of the esters was accomplished on silver nitrate-impregnated silica gel.

Dhopeshwarkar and Mead [144, 145] showed that methyl esters do occur in the body. Kaufmann and Viswanathan [146] also found methyl and ethyl esters present in liver lipids of an alcoholic patient who had died of pneumonia. The amount of primary alcohol esters may be related to the alcohol content in the blood.

Firestone [147] separated monomer, dimer, and polymer methyl esters by both adsorption and reverse-phase partition chromatography. The samples were prepared by thermal polymerization. Rieche et al. [148] and Privett and Blank [149] have studied the autoxidation of methyl esters of unsaturated fatty acids which in some cases leads to the formation of polymers. Rieche et al. used low-molecular-weight compounds, separating and determining the products formed. Privett and Blank examined the initial stages of autoxidation and found that pro-oxygenic substances were formed prior to the formation of stable hydroperoxides. If care is taken to remove these, polyunsaturated acids of high stability can be prepared. Privett and Nickell [150] prepared methyl esters of high purity by reverse-phase column partition chromatography using thin-layer chromatography to follow the purification.

7 SEPARATIONS OF PHOSPHOLIPIDS, SPHINGOLIPIDS, AND GLYCOLIPIDS

Because of the polar nature of these compounds, a more polar solvent is required to move and develop them on silica gel layers. One of the widely used solvent mixtures is a mixture of chloroform–methanol–water. This has been used in a number of different proportions, such as 65:25:4 (151–154) and 60:35:8 [32, 152]. Vogel et al. [14] have used an 80:25:3 mixture for the separation of serum phospholipids and Nichols [13] has used the same mixture for separating phospholipids and glycolipids from lettuce and cabbage leaves. In addition the latter has used diisobutyl ketone–acetic acid–water (40:25:3.7), chloroform–methanol–7 *N* ammonium hydroxide (12:7:1), chloroform–methanol–acetic acid–water (65:25:8:4), and chloroform–methanol–acetic acid (65:25:8). Viswanathan [155] obtained a good separation of the lipids of *Tetrahymena pyriformis* into 10 polar components by using chloroform–methanol–concentrated ammonia (65:35:5) with silica gel G. Pohl et al. [156] developed a new solvent system for the separation of glycolipids, phospholipids, and neutral lipids. This consisted of acetone–benzene–water (91:30:8) for use with silica gel layers. Kunz and Kosin [157] developed a complicated but efficient solvent for the separation of phospholipids by adding small amounts of numerous inorganic salts to the system. The solvent consisted of: 32 ml of chloroform, 26 ml of *n*-butanol, 2.5 ml of *n*-propanol, 5.5 ml of ethanol, 9.4 ml of methanol, 10.5 ml of glacial acetic acid, 0.70 ml of 4% (wt/vol) of sodium bicarbonate, 0.25 ml of 7% (wt/vol) of sodium carbonate, 0.06 ml of 15% (wt/vol) of ammonium chloride, 0.05 ml of 3% (wt/vol) lithium hydroxide, 0.3 ml of 0.9% (wt/vol) lithium carbonate, 0.3 ml of ethyl acetate, 0.3 ml of concentrated ammonia, 0.05 of 20% (wt/vol) cesium chloride, 0.15 ml of 15% (wt/vol) potassium carbonate, 0.05 ml of 10% (wt/vol) potassium bromide, 0.13 ml of 2.5% (wt/vol) sodium fluoride, 0.12 ml of 12 ml of 12.5% (wt/vol) rubidium chloride, 0.08 ml of 22% (wt/vol) manganese acetate, 0.05 ml of 1.5% (wt/vol) mercuric chloride, 0.20 ml of "zinc acetate" (2 g of zinc acetate dissolved in 40 ml of distilled water and 1 ml of acetic acid), and 0.016 ml of "magnesium acetate" (1.5 g of mangesium oxide dissolved in 25 ml of water and 21.5 ml of acetic acid). This had to be prepared fresh just prior to use. The adsorbent was a mixture of 38.5 g of silica gel H and 4.5 g of silica gel G; the plates were activated at 115°C for 2-3 hr after air drying. Eighteen separate fractions were obtained in a one-dimensional development. Curri and Ninfo [158] have used a chloroform–methanol–water (14:6:1) mixture for the separation of phospholipids obtained from tissue extracts. Curri et al. [159–161] have also applied the same solvent to the separation of phospholipids directly from tissues, without the intermediate extraction step. In order to carry out this procedure the tissue sections were applied to microscope cover slides, which were in turn fastened to chromatographic plates by means of gum arabic or Apathy's serum. After the sample was applied in this manner, the plates were coated with silica gel G and dried in a desiccator. In a similar manner, rat liver mitochondria that had been properly washed in sucrose solution were applied as a suspension to glass paper. This application spot was then cut out and fastened to a previously prepared silica gel plate by means of a drop of the developing solvent. The separation of the phospholipids by this method was equal to that obtained from the separation of an extract.

A modification of the chloroform–methanol–water solvent has been made by the addition of acetic acid. Skipski et al. [162, 163] have used a chloroform–methanol–acetic acid–water (65:25:8:4) mixture with neutral silica gel plates and a 50:25:8:4 mixture with an alkaline silica gel plate. By using alkaline plates prepared with 0.01 *M* sodium carbonate or sodium acetate, these authors were able to separate phosphatidyl serine from other phospholipids. It was found that neutral silica gel plates prepared with a calcium sulfate binder exhibited a "load effect" with phosphatidyl serine, that is, the R_f value was affected by the quantity of material applied. This effect was eliminated by using silica gel (CAMAG) layers without a binder, or by using the alkaline plates. Nielsen [164, 165] investigated this load effect and found that acidic phospholipids, such as cardiolipin and phosphatidylinositol, exhibited different chromatographic behavior when applied as their monovalent salts from that exhibited by the divalent salts. Animal lipids contain both monovalent and divalent salts, and this must be taken into consideration when using acidic or neutral solvents. Nielsen recommends washing the lipid solution (in an organic solvent) with an aqueous magnesium chloride solution to convert all the salts to divalent. If monovalent salts are to be chromatographed in acidic or neutral solvents it is essential that the adsorbent be free of divalent metal ions. Gonzalez-Sastre and Folch-Pi [166] used silica gel H layers containing potassium oxalate as a sequestering agent for any calcium that might be present when chromatographing triphosphoinositides. When a solvent such as chloroform–methanol–7 *N* ammonium hydroxide (10:10:1) is used, both mono- and divalent salts are converted to the divalent ammonium salts [164]. Abramson and Blecher [167] used a two-dimensional separation for phospholipids in connection with a basic silica gel layer prepared by slurrying silica gel G with 0.01 *M* sodium carbonate solution. Using plates that were activated at 110°C, the first development was with chloroform–methanol–acetic acid–water (250:74:19:3) and the second development with a basic solvent, chloroform–methanol–7 *M* ammonium hydroxide (230:90:15). Using this system, the following compounds could be separated: lysolecithin, sphingomyelin, phosphatidyl choline, phosphatidyl inositol, phosphatidyl serine, phosphatidyl ethanolamine, phosphatidic acid, and cardiolipin. In the chromatography of bacterial lipids Minnikin and Abdolrahimzadeh [168] slurried 40 g of Merck silica gel PF_{254} in 100 ml of 0.2% sodium acetate for preparing their plates. With this impregnation phosphatidylglycerol appeared as a compact spot when chloroform–methanol–water (65:25:4) was used, in contrast to the elongated comet-shaped spot on unimpregnated layers. A two-dimensional system was used with chloroform–methanol–acetic acid–water (80:12:18:5) for the second-dimension solvent. For the separation of plant phospholipids and glycolipids, Lepage [169] has also used a two-dimensional separation. In this case, the first development was with chloroform–methanol–water (65:25:4) which gave a satisfactory separation of the phospholipids, and the second development at 90° to the first, was with diisobutyl ketone–acetic acid–water (8:5:1) which was more effective in separating glycolipids and the accompanying sterol glycosides. Among the detecting agents used was a perchloric acid–Schiff reagent [170]. The plates were first sprayed with a 0.5% sodium periodate solution and allowed to stand for 5 min, after which they were exposed to sulfur dioxide gas to remove the excess periodate. They were then sprayed with a 0.5% *p*-rosaniline solution (freshly decolorized with sulfur dioxide) and allowed to stand for the blue and purple spots to develop. To lighten the background they were given a spray of 1% perchloric acid solution. If this is followed by Zinzadze reagent (T-167), which detects phosphatidylethanolamine and cardiolipin, all polar lipids are detected and the spots are stable for several weeks [168]. Skidmore and Entenman [171] have also used a two-dimensional method for the separation of rat liver phosphatides. The first development on silica gel layers was carried in chloroform–methanol–7 *N* ammonium hydroxide (60:35:5). Development in the second direction was carried out with the same solvents in a ratio of 35:60:5. The R_f values in the first solvent were as follows: phosphatidic acid 0.73, phosphatidyl serine 0.19, phosphatidyl ethanolamine 0.58, phosphatidyl inositol 0.31, phosphatidyl choline 0.35, sphingomyelin 0.19, and lysophosphatidyl choline 0.11. The R_f values of typical hydrolysis products of the compounds were also obtained. Jatzkewitz [172] has used a two-step method for the separation of sphingolipids in the brain, and Payne [173] has used the same solvents in working with the lipids in nervous tissue. Using silicic acid layers the first development was with chloroform–methanol–water (14:6:1) for a distance of 15 cm, and then after drying, development was made to a distance of 10 cm with *n*-propanol–12.5% ammonium hydroxide (39:11). The R_f values for this group of compounds is given in Table 23.8. Adams and Sallee [174] used a two-step procedure on silica gel developing in a paper-lined tank with tetrahydrofuran–methanol (3:1) for 9 min. After solvent removal a second development in the same direction was made for 17 cm with chlo-

Table 23.8 Approximate $R_f \times 100$ Values for Lipids in Nervous Tissue. Stepwise Development on Silica Gel G[a,b]

Lipid	$R_f \times 100$
Ganglioside a	2.2
Ganglioside b	2.7
Ganglioside c	9
Ganglioside d	13
Lysocephalin	22
Phosphatidyl serine	25
Lysolecithin	27
Sphingomyelin a	31
Sphingomyelin b	33
Lecithin	46
Cerebroside sulfuric acid esters a	61
Cerebroside sulfuric acid esters b	64
Phosphatidyl ethanolamine	67–80[c]
Phrenosin	93
Kerasin	96
Cholesterol	99
Free fatty acids	100

[a] From S. N. Payne [173]; reproduced with permission of the author and The Elsevier Publishing Co.

[b] First development 15 cm in chloroform–methanol–water (14:6:1), second development 10 cm in *n*-propanol–12.5% ammonia (39:11).

[c] The first figure is the R_f value measured from the end of the spot and the second figure is the R_f value measured from the front of the spot.

roform–methanol–4 *M* ammonium hydroxide (75:37:7). Neskovic and Kostic [175] combined acidic and basic solvents in a two-step method. The adsorbent layers were prepared according to Rouser et al. [176]: 54 g of silica gel H and 6 g of finely ground Florisil slurried with 135 ml of water. The first development was with chloroform–methanol–30% ammonia–water (140:50:7:3) for 15 cm; the second development with chloroform–acetic acid–methanol–water (160:4:20:1.5) was run to the top of the plate. The use of magnesium silicate almost completely eliminates the spreading of acidic lipid spots [176]. Kaufmann et al. [177] used a chloroform–ether–acetic acid (97:2.3:0.5) solvent saturated with silver nitrate in the separation of lecithins on silver nitrate-impregnated silica gel layers. Further analysis of these compounds was accomplished by eluting and hydrolyzing the fractions from the silica gel layers. The component fatty acids were then separated on calcium sulfate layers [66]. Privett et al. [178, 179] have determined the structure of lecithins by their reductive ozonolysis method employing thin-layer chromatography to separate the final products.

Renkonen [180], in analyzing the glycerophosphatides, first carried out a dephosphorylation using a mixture of acetic acid and acetic anhydride. The diglyceride acetates thus obtained were fractionated according to the number of double bonds by using silver nitrate-impregnated silicic acid layers.

Because of the tendency of phospholipids to streak when chromatographed on silica gel, Mangold and Kammereck [109] prepared layers of silica gel containing 10% ammonium sulfate. This corrected the streaking tendency with the more commonly used solvents for this group. Kaulen [181] found that a small percentage (0.4%) of ammonium sulfate in the layer gave a clear-cut separation of phosphatidylserine from phosphatidylinositol, but with larger amounts up to 10% the R_f value of the former increased until it became identical with that of phosphatidylethanolamine. Horrocks [182], in examining brain phospholipids, tried a number of impregnating agents (Table 23.9). The developments were carried out in unlined but equilibrated tanks. With the exception of sodium borate, the impregnated layers were prepared by mixing 27 g of silica gel G with 3 g of the impregnating agent and 60 ml of water; for sodium borate, 20 ml of a saturated solution of sodium borate was diluted to 60 ml before adding the silica gel. Jatzkewitz and Mehl [183] have used anhydrous diethyl ether as a solvent, as well as mixtures of *n*-propanol–12.5% ammonium hydroxide (4:1) and *n*-propanol–17% ammonium hydroxide (7:3). Weicker et al. (184) have also used a propanol–1 *N* ammonium hydroxide–water solvent in a ratio of 6:2:1.

Cerebral lipids have been examined by Jatzkewitz [185, 186], Wagner et al. [154], Kochetkov et al. [187, 188], and Mueldner et al. [189]. Honegger [190, 191] has compared the brain lipids of patients suffering from multiple sclerosis with those of normal people, and Pliz and Jatzkewitz [192] have determined the C_{18} and C_{24} sphingomyelin content in normal and pathological brains. Jatzkewitz [172] and Kuhn et al. [193] have examined the gangliosides of beef brain, and four of them have been characterized by the following R_f values: 0.55, 0.35, 0.23, and 0.18, on silica gel G using *n*-propanol–water (7:3) as a developing solvent. Witting et al. [194], in examining the gangliosides in several species, modified the latter solvent to a 3:1 ratio in order to distribute the compounds over a 50-cm plate. Streaking and tailing were negligible. Dain et al. [195] found butanol–pyridine–water (3:2:1.5) and 78% phenol in water to be useful solvents for fractionating gangliosides. Wherrett and Cumings

Table 23.9 $R_f \times 100$ Values of Brain Lipids in Various Sysstems. Development Distance 10 cm[a]

	Silica Gel G			*Silica Gel Impregnated with*				
				Sodium Borate	*Sodium Acetate*	*Potassium Hydroxide*	*Oxalic Acid*	*Ammonium Sulfate*
Lipid	*Chloroform–Methanol–Water* (65:25:4)	*Chloroform–Methanol–Ammonium hydroxide* (62:25:4)	*Chloroform–Methanol–Ammonium Hydroxide* (75:25:4)	*Chloroform–Methanol–Ammonium Hydroxide* (62:25:4)		*Chloroform–Methanol–Water* (65:25:4)		
Cerebroside	89	96	58	44	90	61	82	77
	85	93	49	35	85	55	79	70
Phosphatidylethanolamine	78	90	47	59	92	45	55	65
Phosphatidylserine	—[b]	29	11	35	57	41[b]	78[b]	65
Lysophosphatidylethanolamine	55	46	21	—	70	—	29	37
Lysophosphatidylserine	—[b]	—	00	—	14	—	—	—
Phosphatidylcholine	55	63	43	45	77	35	35	45
Lysophosphatidylcholine	31	28	13	—	46	—	10	7
Sphingomyelin	47	47	24	30	58	20	19	30
Cerebroside sulfate	—	—	41	35	71	28	50	65
	—	—	36	30	67	23	35	55

[a] From L. A. Horrocks [182]; reproduced with permission of the authors and The American Oil Chemists' Society.
[b] Streaking.

[196] compared the chromatographic patterns of extracts of brain, spleen, and kidney tissue with those from purified ox cerebral cortex ganglioside preparations and tentatively identified eight of the tissue bands as gangliosides. Klenk and Gielen [197–199], Kuhn and Wiegandt [200], Johnson and McCluer [201], Wherrett et al. [202], Korey and Gonatas [203], and Sambasivarao and McCluer [204] have contributed to the structural analysis of human brain gangliosides.

Handa and Burton [205] applied an evaporative continuous technique to the separation of beef retina gangliosides. Separations were completed within 3 hr. Svennerholm [206] and Ledeen [207] have published reviews on the gangliosides.

Kochetkov et al. [208] found two-dimensional chromatography to be of value in separating sphingosine derivatives. A group of four solvents were used: butanol–ethyl acetate–11% ammonium hydroxide (15:8:2), butanol–ethyl acetate–5% formic acid (15:4:1), and chloroform–methanol in 3:2 and 2:3 ratios. R_f values in the acidic solvent were as follows: sphingosine 0.40, dihydrosphingosine 0.39, *O*-methyl ether of sphingosine 0.34, *O*-methyl ether of dihydrosphingosine 0.24, and psychosine and dihydropsychosine 0.15. The latter two could be separated slightly by chloroform–methanol (2:3) having R_f values of 0.20 and 0.14, respectively. Fujino and Zabin [209] and Weiss and Stiller [210] have separated sphingosine bases. Sambasivarao and McCluer [211] separated the free bases with chloroform–methanol–2 *N* ammonium hydroxide (40:10:1) on silica gel. Michalec [212, 213] has used a two-dimensional technique for separating the dinitrophenyl derivatives of sphingosine bases. Chloroform–methanol (9:1) was used in the first dimension on sodium tetraborate-impregnated silica gel G layers [212]. After the layer was impregnated with 5% Tetralin, methanol–Tetralin–water (9:1:1) (upper phase) was used in the second dimension. Aluminum oxide layers were also used [213] with chloroform–methanol (49:1) for the first dimension, followed by the same procedure as in the case of the silica gel layers. In order to help differentiate sphingolipid bases and complex sphingolipids, Kljaic et al. [214] used oxidation with osmium tetroxide to obtain products with different R_f values from those of the original compounds.

Eberlein and Gercken [215] examined the red cell glycosphingolipids in various mammalian species. Ceramide hexosides were separated from other lipids and fractionated into mono-, di-, and triglycosyl ceramides with tetrahydrofuran–water (5:1) on silica gel HR–magnesium silicate (4:1). Complex glycosphingolipids were separated from each other and from other lipids on kieselguhr G impregnated with 0.6 *M* boric acid–0.15 *M* disodium tetraborate buffer (pH 7.9) and air-dried for 24 hr. Solvent for this separation was chloroform–methanol–water (65:25:4).

Neskovic et al. [216] separated glycolipids from phospholipids with chloroform–acetone–pyridine–20% ammonia–water (10:15:30:1:1) as the first solvent in a two-step method on silica gel H–finely ground Florisil (27:3). Four systems were used for the second step, depending on the glycolipids to be separated. Cerebrosides, sulfatides, and monogalactosyl diglyceride were separated with chloroform–acetone–methanol–acetic acid–water (65:35:11:4:1.5). Neutral ceramide glycosides were separated with chloroform–methanol–water (65:25:4). Cerebrosides, sulfatides, ceramides, sphingosine, and psychosine were separated with chloroform–acetone–methanol–water (65:30:;2:2). Ceramides, monogalactosyl diglyceride, glucocerebrosides, galactocerebrosides, sulfatides, and psychosine were separated on layers partially impregnated with sodium borate. The solvent was the same as the first solvent but in a 68:26:12:5:3 ratio.

Hoevet et al. [217] used silver nitrate-impregnated silica gel layers with chloroform–methanol–water (70:25:3) to separate mixtures of alkenyl acyl- and diacylcholine phosphatides that varied in the unsaturation in the fatty acid moieties.

Viswanathan et al. [218] have developed a two-dimensional reaction thin-layer chromatographic method for the analysis of mixtures of alkenyl acyl-, alkyl acyl-, and diacylcholine phosphatides as well as for the analysis of phosphatide plasmalogens [219]. Owens [220] has also developed a two-dimensional procedure involving a reaction on the layer for the estimation of plasmalogens.

Hughes and Frais [221] have studied the phospholipids in normal and diseased muscle tissue, and Philippart and Menkes [222] have investigated the main spleenic glycolipids in Gaucher's diseases. Gray [223] and Curri et al. [224] have investigated the phospholipid composition of tumor cells.

Hausheer et al. [225] have published a method for the quantitative isolation of pure sphingomyelin from human and rat brains, and Wells and Dittmer [226] have reported on a preparative method for the isolation of brain cerebroside, sulfatide, and sphingomyelin.

Lipids have been investigated in the following animal organs and tissues: normal mouse liver [227], adrenal glands of rats [228], rabbit kidneys [229], beef brain and liver [230], pig brain [231], rat brain synaptic vessel [232], canine adrenal glands [233], tuna white muscle [234], chick embryo liver [235], and bovine erythrocytes [236]. Phosphatidylglycerol has been isolated from rat liver mitochondria [237].

Rouser et al. [238] obtained improved separations of phospholipids on layers of silica gel containing 10% magnesium silicate and mixed with 0.01 *M* potassium hydroxide. Two-dimensional separations were carried out with three different solvent systems: *(a)* first dimension, chloroform–methanol–25% ammonia (13:5:1), and second dimension, chloroform–methanol–acetic acid–water (3:4:1:1:0.5); *(b)* first dimension, as in *(a)* but in a 13:7:1 ratio, second dimension, as in *(a)* but in a ratio of 5:2:1:1:0.5; and *(c)* first dimension, chloroform–methanol–water (65:25:4), and second dimension, *n*-butanol–acetic acid–water (3:1:1).

Viswanathan et al. [239] isolated an artifact, an aldehydogenic phospholipid, formed by allowing a chloroform–methanol solution of phosphatide plasmalogens to stand for 6 weeks. Extracts that were worked up immediately did not contain the artifact.

Snyder [240] examined the behavior of glycerolipid analogs containing ether, ester, hydroxyl, or ketone groupings in 11 solvent systems, and has reviewed [241] the chemistry, physical properties, and chromatography of lipids containing ether bonds.

For the detection of phospholipids, Dittmer and Lester [242] have used a slight modification (T-167) of the Zinzadze [243] molybdate reagent. For sphingolipids the Bischel and Austin [244] modification of the benzidine method [245] may be used (T-37).

Renkonen and Varo [246] reviewed the thin-layer chromatography of phosphatides and glycolipids and Renkonen [247] reviewed the TLC analysis of subclasses and molecular species of polar lipids.

8 WAXES

Kaufmann and Das [248] chromatographed beeswax on silica gel G using trichloroethylene–chloroform (3:1) as the solvent at 22°C; shellac, carnauba wax, wool wax, and sunflower wax were chromatographed in the same system, but at 42°C. For the separation of the wax acids they used reverse-phase chromatography on layers of calcium sulfate (gypsum) impregnated with a 240 to 260°C petroleum fraction. The separation was run at 42°C using isopropanol–ethanol–acetic acid–water (8:3:4:1.3). (For separation of the wax alcohols see Chapter XIV.)

Haahti and co-workers [249, 250] have chromatographed the waxes and sterol esters of skin surface fat. Silica gel, aluminum oxide, and silver nitrate-impregnated silica gel were used in the separations. Nikkari and Haahti [251] found two types of diester waxes from the surface lipids of the rat. Thin-layer chromatography was on silica gel with benzene–hexane (1:1) and hexane–ether (9:1). Nicolaides et al. [252] also found diester waxes in surface lipids. A step method was used with development to half height of the plate with hexane–ether–acetic acid (80:20:1), followed by full development with hexane–ether (19:1) and a third development to the top with hexane.

Holloway and Challen [253] examined 60 natural waxes and 15 classes of wax constituents on several adsorbents; silica gel G was the most satisfactory adsorbent. Solvents included: carbon tetrachloride, dichloromethane, benzene–chloroform (7:3), and chloroform–ethyl acetate (1:1). At least two solvent systems were required to obtain satisfactory separations. Reutner [254] chromatographed 32 commercial waxes on silica gel at elevated temperatures. The following solvents and temperatures were used: benzene, benzene–acetic acid (99.5:0.5), and benzene–methyl acetate (98:2), all at 45°C,; tetrachloroethylene at 80°C; and tetrachloroethylene–acetic acid (99.5:0.5) at 70°C. After removal of the solvents by heating to 160°C, the plates were sprayed with 5% phosphomolybdic acid in methanol and heated at 160°C for 15 min.

Dietsche [255] chromatographed hydrocarbon waxes on silica gel–urea (3:2) using carbon tetrachloride–ethanol (7:3) as a solvent. The varying tendency of wax components to form inclusion compounds gave different R_f patterns from those obtained on silica gel alone. For visualization the plates were sprayed with 0.05% fluorescein and viewed under ultraviolet light.

Kaufmann et al. [256] chromatographed wax esters, diester waxes, and other unusual lipids on magnesium oxide and on magnesium oxide–silica gel layers with hexane–ether–ethyl acetate (60:40:1).

9 QUANTITATIVE ANALYSIS OF LIPIDS

Practically all of the methods have been applied to quantitative thin-layer analysis have been used for lipids. Vioque and Holman [257] have used a colorimetric method for the estimation of esters. After separation of the compounds, the spots are eluted with diethyl ether and then converted to hydroxamic acids by warming with 0.1 ml each of 2.5% sodium hydroxide and hydroxylamine hydrochloride solutions in 95% ethanol. The heating takes place on a 65 to 70°C water bath while the solvents are being evaporated to dryness. After cooling, the colored complexes are formed by adding 5 ml of ferric perchlorate reagent. After 30 min the color is measured in a spectrophotometer at 520 mμ. The ferric perchlorate reagent may be prepared as follows [257]: stock ferric perchlorate solution containing 5.0 g of ferric perchlorate (non-yellow) in 10 ml of 70% perchloric acid plus 10 ml of distilled water is diluted to 100 ml with cold absolute ethanol. Four milliliters of this solution plus 3 ml of 70% perchloric acid are diluted to 100 ml with chilled absolute ethanol daily prior to use as a reagent.

Walsh et al. [258] also used the hydroxamic acid reaction in determining lipid esters in barley and malt. In this case the esters were not eluted from the silica gel, but the reagents were applied directly to the adsorbed material. Prior to the colorimetric measurement, the silica gel was removed by centrifugation. The average recovery was 98.8% and the standard error was ±0.35%. For lipids without an ester group, these workers applied a modification of the dichromic acid colorimetric method of Johnson [259].

Zoellner and Kirsch [260] have used a modified sulfophosphovanillin reaction for the quantitative determination of plasma lipids.

Roughan and Batt [261] used a colorimetric method for the determination of sulfolipids by treating the samples, scraped from the layer, with phenol and sulfuric acid. This was more sensitive than the anthrone method [262]. The same reaction was used for galactolipids by lowering the concentration of phenol. Trisaturated glycerides were determined by Bandyopadhyay [263] by hydrolyzing the eluted material and assaying the glycerol content by the chromotropic acid method.

Another colorimetric determination that can be applied to phospholipids is the digestion of the material in perchloric acid and subsequent determination of the phosphorus content with ammonium molybdate solution. Curri et al. [161] digested the silica gel spots with 0.4 ml of 70% perchloric acid in a test tube by gently evaporating to dryness over a small flame. The residue was then mixed with 2.4 ml of 12% perchloric acid and heated in a boiling water bath for 10 min. This treatment not only digested the phospholipids but also insolubilized the silica gel. After cooling, the phosphorus was determined according to Wagner's method [153, 264, 265], which consists in adding 0.3 ml of 2.5% ammonium molybdate solution and 0.3 ml of a freshly prepared 10% ascorbic acid solution. The mixture is shaken thoroughly and kept at 38°C for 2 hr. The solution is centrifuged and its adsorption is measured at 820 mμ. Robinson and Phillips [266, 267] have used 1-amino-2-naphthol-4-sulfonic acid and Rastogi et al. [268] a solution of 2,4-dinitrophenylhydrazine in hydrochloric acid as a reducing agent in place of the ascorbic acid. Doizaki and Zieve [269] have used a somewhat different procedure for this determination, digesting the lipids with sulfuric acid with the addition of hydrogen peroxide.

Habermann et al. [151] removed the spots from the plate, dried and ashed them, and then treated them with molybdate.

It is sometimes necessary to elute phospholipids for analysis, but in some cases they are difficult to elute; French and Andersen [270] alternated two elutions of chloroform–methanol (2:1) with two of chloroform–methanol–water (3:5:2). Rouser et al. [271, 272] found it unnecessary to elute the phospholipids from the silica gel for the phosphorus determination, and Tichý [273] found that a very acidic medium with pH 0.5 to

0.7 was necessary for the development of the blue phosphomolybdate complex. Rouser et al. [274] reviewed the techniques available for the analysis of polar lipids. Kleinig and Lempert [275] described a microscale phospholipid analysis technique suitable for measuring 0.5 nmole of a phospholipid.

Measurement of the area of the spot and correlation of this with the sample weight or some function of it has been done [276–278].

Privett and Blank [279, 280] have adopted for quantitative work the general technique of Kirchner et al. [281] for the detection of all types of compounds. This consists in spraying the chromatogram with concentrated sulfuric acid containing an oxidizing agent with subsequent heating in order to char the spots. Since a variability in the thickness of the layer can affect the results of the analysis the prepared plate should be checked in the densitometer prior to applying the sample. After development of the chromatogram, the solvent is thoroughly evaporated from the plate before spraying lightly with a saturated solution of potassium dichromate in 80% (by weight) sulfuric acid. The chromatoplate is then heated at 180°C for 25 min in order to char the compound. Earlier work had been conducted with 50% aqueous sulfuric acid with subsequent heating at 360°C, but it was discovered that the amount of carbon in each spot was a result of a balance between evaporation and oxidation [280]. Lower-boiling compounds had a tendency to evaporate more than the higher-boiling ones before charring took place; on the other hand, the higher-boiling compounds were less susceptible to oxidation to free carbon. The charred spots are measured in a densitometer using a slit opening of 1 × 5 mm; a reading is taken over each 1 mm of travel over the length of the plate. The areas under the densitometer curves are directly proportional to the amount of sample. Since different amounts of carbon are obtained from equal amounts of structurally different compounds, it is important that a standard curve be prepared for the analysis of the type of compound that is being determined. As an example, a curve for triolein cannot be used for the analysis of a saturated tripalmitin compound. It should be stressed, however, that it is not necessary to have exactly the same compound for a standard as that which is being determined, but rather it should be of the same type. Other reagents have been used for charring (see Chapter VII, Section 1). Nutter and Privett [282] obtained more uniform charring by hydrogenating the unsaturated lipids using a platinum oxide catalyst (Adams catalyst, Matheson, Coleman, & Bell) prior to thin-layer chromatography. Although they are more difficult to separate, more common polar lipids between the desired R_f values of 0.3 and 0.8 can be separated by the judicious choice of solvents and adsorbents. Chobanov et al. [283] in separating triglycerides applied a similar principle *after* separating the compounds. In this case the developed plates were exposed to bromine vapor for 30 min to eliminate the double bonds. Downing [284] has published a method for analysis of neutral lipids using the charring technique wherein the necessity of reference mixtures is eliminated in determining the relative amounts of constituents in a mixture. Castellani et al. [285] have investigated the relationship between densitometric peak area and concentration of the lipid.

Araki [33] has applied spot densitometry to lipid spots sprayed with 5% ethanolic phosphomolybdate solution and made visible by heating at 180°C for 5 min. Neskovic [286] completely mineralized phospholipids by spraying with a mixture of 1 ml of 5% ammonium molybdate, 3 ml of concentrated nitric acid, and 16 ml of 65% perchloric acid, and then heating them, covered with a glass plate, for 1 hr in an oven (temperature rise, 120 to 160°C in 30 min). The glass cover was removed during the last 10 min. The layers were then sprayed with a mixture of 2 ml of 5% ammonium molybdate, 3 ml of 65% perchloric acid, and 5 ml of 1% ascorbic acid, and maintained at 50°C for 1 hr while covered with a glass plate. After they were made translucent with paraffin–ether (1:1) the color was measured in a photometer. Gangliosides have been determined [287] by densitometric measurement of the color produced by heating the layer sprayed with resorcinol–hydrochloric acid reagent.

All the various techniques for measuring radioactivity have been applied to the quantitative determination of lipid material separated by thin-layer chromatography. (For details of these techniques see Chapter XI, Section 6.) Kasang et al. [288] developed a semiautomatic scraper for removing zones from the layers, and automatic scrapers have been developed by Snyder and Kimble [289] and by Fosslien et al. [290] for the determination of radioactive lipids. These are discussed in Chapter V, Section 5. Fosslien [291] has also designed an automatic sample extractor and spot applicator for lipid work.

Nicolosi et al. [292] have used an *in situ* fluorescent method for determining components of five lipid classes. They found it necessary to hydrogenate the components prior to chromatographing in order to eliminate interference with the fluorescence of Rhodamine 6G by unsaturated compounds. However, Roch and Grossberg [293] found that incorporation of the Rhodamine 6G in the adsorbent layer instead of spraying the layer eliminated this difficulty. They were able to determine nanomole amounts of components in as little as 120 μg of total lipids. Heyneman et al. [294] used a 0.01% solution of magnesium 1-anilino-8-naphthalenesulfonate as a fluorescing agent for phospholipids.

Peter and Wolf [295] developed a sensitive *in situ* method for the determination of phospholipids down to 10^{-8} g using fluorescence. To obtain the necessary separation the silica gel plates were pretreated by exposure to an atmosphere of sulfuric acid–water (1:1) for 1 hr and then were exposed to the vapors of benzene–methanol varying from 2 to 16% methanol, depending on the compounds to be separated. The solvent for the development was light petroleum ether–chloroform–methanol–water (6:8:8:1). The samples were applied with the CAMAG Linomat, because the separations were better than when applied as spots. For developing the fluorescence the completed chromatogram was sprayed with concentrated sulfuric acid–ether (1:19) and heated at 100°C for 10 min in a drying chamber. The fluorescence was stable for several months. Measurement was at 400 nm on activation with 360-nm light.

Direct gravimetric analysis of lipid material has been applied by Komarek et al. [36], Vioque et al. [296], Williams et al. [297], and Dunn and Robson [298].

Vapor-phase chromatography can be conveniently used for the quantitative analysis of fractions separated by thin-layer chromatography [42, 299–301]. It is especially useful for the determination of the fatty acid composition. This is usually accomplished by converting to the methyl esters by one of several procedures. In using the gas chromatographic method for the determination, care must be taken not to use iodine vapor as a locating agent because the resulting gas chromatographic analyses are too low for the unsaturated compounds and too high for the saturated materials. This is due apparently to the reaction of the iodine with unsaturated compounds [302]. The technique of chromatographing on a tube coated on the inside with silica gel or silica gel–copper oxide and then vaporizing or pyrolyzing the separated components for determination by vapor-phase detectors has also been used and a description of the technique has been made in Chapter XI, Section 7.

Privett et al. [303] have reviewed the quantitative analysis of lipids and also the charring densitometry technique [304]. Kuksis [305] has reviewed the quantitative analysis of lipids by combined thin-layer–gas chromatography. Max and Quarles [306] reviewed the analysis of gangliosides. Rouser et al. [307] has reviewed the application of column and thin-layer chromatography to the quantitative analysis of polar lipids.

Vacíková et al. [37] estimated fatty acids by eluting the fractions and titrating.

10 MISCELLANEOUS

Fisher et al. [308] have studied the lipids of flours from seven varieties of wheat using two-dimensional chromatography. Twenty-three components could be detected. Among the samples examined there were varietal, seasonal, and environmental differences. Nelson et al. [309] have studied the lipids of whole wheat, and McKillican and Sims [310] have examined the endosperm lipids of three Canadian wheats.

Allen et al. [311] separated the lipids from spinach by column chromatography and monitored the fractions using silicic acid chromatoplates.

Kaufmann and Viswanathan [312] have analyzed the lipids of skin and hair, and Bey [313, 314] has analyzed the skin oils obtained from soiled clothing.

Wren and Szczepanowska [315] have investigated the use of 4-methyl-2,6-di-*tert*-butylphenol (BHT) as an antioxidant during the chromatography of lipids. For use, it is added in a 0.005% concentration to the chromatographic solvents. Although for many purposes it is not necessary to remove the antioxidant from the lipid fractions, this can be done in most cases by chromatographing in carbon tetrachloride; the authors also list other methods such as vacuum desiccation and steam distillation. Mitsev et al. [316] examined various antioxidants for the chromatography of lipids by spraying the layer with 1.5 ml of a 0.01% solution of the antioxidant and drying prior to use. The most effective antioxidants were the ethyl, propyl, octyl, and lauryl esters of gallic acid.

Thin-layer chromatography is a very useful tool in the detection of adulteration in lipids. Table 23.10 lists examples of the use of thin-layer chromatography for detecting adulteration of fats. Mani and Lakshminarayana [361] have reviewed the chromatographic detection of adulteration of oils and fats by various techniques.

The blue fluorescence occurring in some olive oil appears to originate with a fungus that grows in the oil. A fluorescent material was isolated from fungi growing in the olives [362].

Privett et al. [179] analyzed eight lipids by using thin-layer and gas chromatography. The fatty acids in lecithin were found to be primarily α-saturated, β-unsaturated. Acker and Greve [363] used thin-layer chromatography to determine the degree of oxidation in egg-dough products.

Purdy and Truter [364–367] have published a series of papers on the isolation and separation of the surface lipids from leaves, and Thirkell and Tristram [368] have examined lipids of alfalfa leaves.

The lipids (and steroids) in the peloids (mud) of the Euganean basin have been examined by Curri [369].

The method of Miller and Kirchner [370] for following the progress of a column chromatographic separation by means of thin-layer chromatography has been applied in the lipid field [61, 371–375].

Thin-layer chromatography has also been used to follow various types of reactions in the lipid field. Van Deenen et al. [376] followed the action of *Clostridium welchii* toxin on the phosphatides of red cell membranes.

Table 23.10 Thin-Layer Chromatography in the Detection of Adulteration of Lipids

Adulterant	*Lipid*	*Ref.*
Vegetable and other foreign fats	Butter	317–320
		321–324
		325–327
Cocoa butter substitutes	Cocoa butter	328
Extracted cocoa fat	Pressed cocoa fat	329
Mineral oils	Vegetable oils	330–333
Synthetic ester lubricants	Edible oils	334
Vegetable oils	Olive oil	335
Caster oil	Vegetable oils	336, 337
Solvent-extracted olive oil	Pressed olive oil	338
Husk oil	Olive oil	338, 339
Olive oil	Seed oil	338
Peanut, argemone, linseed, arachis oil	Mustard seed oil	340, 341
Coconut, palm kernel oils	Vegetable oils	342
Argemone oil	Vegatable oil	343
Animal fat	Vegetable oils	344–347
Vegetable oil	Coconut oil	348
Peanut oil	Caster oil	349
Nigerseed oil	Sesame seed oil	350
Peanut, safflower, linseed oils	Sesame seed oil	351
Mustard seed, soy bean, safflower oil	Peanut oil	352
Soybean oil	Sunflower seed oil	353
Sunflower oil	Olive oil	354
Apricot seed oil	Almond oil	355
Heavy fuel oil	Whale oil	356
Milk fat	Other fats	357
Lipids	Dried fruit coating	358
Liver	Meat preserves	359
Horse and kangaroo meat	Other meat	360

Gauglitz and Malins [377] used thin-layer chromatography to analyze the intermediates and the final ingredients in the preparation of polyunsaturated aldehydes, and Gauglitz and Lehman [378] monitored the preparation of alkyl esters from polyunsaturated triglycerides. Van Deenen and de Haas [379] examined the substrate specificity of phospholipase A from *Crotalus adamanteus*. Applewhite et al. [380] have used thin layers to check the purity in the preparation of some amides from castor-based acids.

Kaufmann and co-workers [381–383] have studied the lipids of the coffee bean, and Kaufmann and Viswanathan [384] analyzed mold lipids.

Additional application to the lipids in the medical field may be mentioned. Tuna et al. [385] have reevaluated the ^{131}I-triolein absorption test. Williams et al. [297] have used thin-layer chromatography to determine the types of lipids present in feces and fecaliths. Kaunitz et al. [386, 387] have studied the Shwartzman reaction as produced by a diet containing oxidized cod liver oil. Horning [388], in studying the lipids in patients with arteriosclerosis, used a combination of thin-layer chromatography and gas chromatography. Jaky et al. [389] have found a new lipid fraction in the blood serum of patients with arteriosclerosis and diabetes. Peiffer et al. [390] have investigated the lipid depressant activities of whole fish and their component oils.

Letters [391] separated the lipids of yeast on silicic acid and then chromatographed the neutral lipid fractions on silica gel G using redistilled chloroform as the solvent. The following R_f values were recorded: cholesterol palmitate 0.96, cholesterol 0.32, glycerol monopalmitate 0.15, glycerol dipalmitate 0.46, and glycerol tripalmitate 0.63.

Hughes [392] analyzed the lipid extract from the cultivated mushroom *Agaricus campestris* by thin-layer chromatography; the fatty acids were separated and analyzed by gas chromatography.

McKillican and Sims [393] surveyed the lipid material in Raja flax and Indian safflower at increasing stages of maturity. They found that the relative amounts of the different lipid classes varied as the seed matured, the greatest change occurring in the phospholipids. There were no free fatty acids or mono- and diglycerides present. Gunstone et al. [394] chromatographed *Jatropha curcas* seed oil on silica gel layers impregnated with silver nitrate. Thin layers were also used to monitor the crystallization from mixtures of acetone and methanolic silver nitrate.

Debuch et al. [395] cautions against allowing chromatograms to stand in tanks containing acetic acid prior to development, because the acetic acid vapor may hydrolyze the enol ether bonding of plasmalogens, which then appear as lysophosphatides. Rouser et al. [396] give procedures for preventing laboratory contamination in lipid chemistry. Sources of contamination include: laboratory soaps, cleansers, hand creams and lotions, hair tonics, laboratory greases, floor waxes, oil vapors, tobacco smoke, and hydrocarbon phases for gas chromatography.

References

1. K. Fontell, R. T. Holman, and G. Lambertsen, *J. Lipid Res.*, **1**, 391 (1960).
2. J. W. Copius-Peereboom, *Chem. Weekbl.*, **57**, 625 (1961).
3. H. K. Mangold, *J. Am. Oil Chem. Soc.*, **38**, 708 (1961).
4. E. Becker, *Ber. Getreidetag., Detmold,* **1963**, 77.

5. H. Kaneko and Y. Kawanishi, *Yukagaku,* **12,** 597 (1963); through *Chem. Abstr.,* **60,** 7119 (1964).
6. P. L. Hagony, *Olaj Szappan Kozmet.,* **13,** 46 (1964); through *Chem. Abstr.,* **62,** 807 (1965).
7. F. B. Padley, "Thin-Layer Chromatography of Lipids," in *Thin-Layer Chromatography,* G. B. Marini-Bettòlo, Ed., Elsevier, Amsterdam, 1964, p. 87.
8. C. V. Viswanathan, *Chromatogr., Rev.,* **11,** 153 (1969).
9. F. Shenstone, "Thin-Layer Chromatography of Lipids," in *Biochemistry and Methodology of Lipids,* A. R. Johnson, Ed., Wiley-Interscience, New York, 1971, p. 171.
10. F. B. Padley, *Chromatogr. Rev.,* **8,** 208 (1966).
11. H. P. Kaufmann and C. V. Viswanathan, *Fette, Seifen, Anstrichm.,* **65,** 538 (1963).
12. H. K. Mangold, *Fette, Seifen, Anstrichm.,* **61,** 877 (1959).
13. B. W. Nichols, *Biochim. Biophys. Acta,* **70,** 417 (1963).
14. W. C. Vogel, W. M. Doizaki, and L. Zieve, *J. Lipid Res.,* **3,** 138 (1962).
15. H. P. Kaufmann and Z. Makus, *Fette, Seifen, Anstrichm.,* **62,** 1014 (1960).
16. G. P. Freeman and D. West, *J. Lipid Res.,* **7,** 324 (1966).
17. J. E. Storry and B. Tuckley, *Lipids,* **2,** 501 (1967).
18. T. F. Kelley, *J. Chromatogr.,* **22,** 456 (1966).
19. M. J. Manners, D. E. Kidder, and P. M. Parsons, *J. Chromatogr.,* **43,** 276 (1969).
20. D. C. Palmer, J. A. Kintzios, and N. M. Papadopoulos, *J. Chromatogr. Sci.,* **10,** 107 (1972).
21. F.-H. Hubmann, *J. Chromatogr.,* **86,** 197 (1973).
22. V. I. Svetashev and V. E. Vaskovsky, *J. Chromatogr.,* **67,** 376 (1972).
23. B. G. Belenky, E. S. Gankina, S. A. Pryanishnikova, and D. P. Erastov, *Mol. Biol.,* **1,** 184 (1967).
24. J. D. Pollack, D. S. Clark, and N. L. Somerson, *J. Lipid Res.,* **12,** 563 (1971).
25. G. E. Tarr, *J. Chromatogr.,* **52,** 357 (1970).
26. A. Niederwieser, *J. Chromatogr.,* **21,** 326 (1966).
27. A. Niederwieser and C. C. Honegger, "Gradient Techniques in Thin-Layer Chromatography," in *Advances in Chromatography,* Vol. 2, J. C. Giddings and R. A, Keller, Eds., Marcel Dekker, New York, 1966, p. 123.
28. A. H. Duthie and H. V. Atherton, *J. Chromatogr.,* **51,** 319 (1970).
29. K. Huhnstock and H. Weicker, *Klin. Wochenschr.,* **38,** 1249 (1960).
30. H. Weicker, *Klin. Wochenschr.,* **37,** 763 (1959).
31. M. H. Laur, *Compt. Rend.,* **257,** 1501 (1963).
32. N. Zoellner and G. Wolfram, *Klin. Wochenschr.,* **40,** 1100 (1962).
33. E. Araki, *Nisshin Igaku,* **50,** 85 (1963); through *Chem. Abstr.,* **59,** 11866 (1963).
34. H. K. Mangold and D. C. Malins, *J. Am. Oil Chem. Soc.,* **37,** 383 (1960).
35. D. C. Malins and H. K. Mangold, *J. Am. Oil Chem. Soc.,* **37,** 576 (1960).
36. R. J. Komarek, R. G. Jensen, and B. W. Pickett, *J. Lipid Res.,* **5,** 268 (1964).

37. A. Vacíková, V. Felt, and J. Malíková, *J. Chromatogr.*, **9,** 301 (1962).
38. K. B. Lie and J. F. Nyc, *J. Chromatogr.*, **8,** 75 (1962).
39. V. S. Whitner, O. T. Grier, A. N. Mann, and R. F. Witter, *J. Am. Oil Chem. Soc.*, **42,** 1154 (1965).
40. H. P. Kaufmann and B. Das, *Fette, Seifen, Anstrichm.*, **64,** 214 (1962).
41. H. P. Kaufmann, Z. Makus, and T. H. Khoe, *Fette, Seifen, Anstrichm.*, **63,** 689 (1961).
42. H. P. Kaufmann and H. Wessels, *Fette, Seifen, Anstrichm.*, **66,** 13 (1964).
43. C. Litchfield, *Lipids,* **3,** 170 (1968).
44. H. P. Kaufmann and H. Wessels, *Fette, Seifen, Anstrichm.*, **66,** 81 (1964).
45. D. Gunstone and F. B. Padley, *J. Am. Oil Chem. Soc.*, **42,** 957 (1965).
46. B. de Vries and G. Jurriens, *Fette, Seifen, Anstrichm.*, **65,** 725 (1963).
47. C. B. Barrett, M. S. J. Dallas, and F. B. Padley, *Chem. Ind. (London),* **1962,** 1050.
48. O. Renkonen and L. Rikkinen, *Acta Chem. Scand.*, **21,** 2282 (1967).
49. H. Wessels and N. S. Rajagopal, *Fette, Seifen, Anstrichm.*, **71,** 543 (1969).
50. L. J. Morris, D. M. Wharry, and E. W. Hammond, *J. Chromatogr.*, **31,** 69 (1967).
51. R. Wood and F. Snyder, *J. Am. Oil Chem. Soc.*, **43,** 53 (1966).
52. D. T. Burns, R. J. Stretton, G. F. Shepherd, and M. S. J. Dallas, *J. Chromatogr.*, **44,** 399 (1969).
53. C. B. Barrett, M. S. J. Dallas, and F. B. Padley, *J. Am. Oil Chem. Soc.*, **40,** 580 (1963).
54. B. de Vries and G. Jurriens, *J. Chromatogr.*, **14,** 525 (1964).
55. F. C. den Boer, *Z. Anal. Chem.*, **205,** 308 (1964).
56. G. Jurriens, *Riv. Ital. Sostanze Grasse,* **42,** 116 (1965).
57. L. J. Morris, *J. Lipid Res.*, **7,** 717 (1966).
58. L. J. Morris and B. W. Nichols, "Argentation Thin-Layer Chromatography," in *Progress in Thin-Layer Chromatography and Related Methods,* Vol. 1, A. Niederwieser and G. Pataki, Eds., Ann Arbor-Humphrey Science, Ann Arbor, Mich., 1970, p. 75.
59. F. E. Luddy, R. A. Barford, S. F. Herb, and R. W. Riemenschneider, Am. Oil Chemists' Society Meeting, Minneapolis, September 30, 1963.
60. D. P. Wolf and L. R. Dugan, Jr., *J. Am. Oil Chem. Soc.*, **41,** 139 (1964).
61. R. G. Jensen, J. Sampugna, and R. M. Parry, Jr., *J. Dairy Sci.*, **45,** 842 (1962).
62. R. G. Jensen, J. Sampugna, R. M. Parry, Jr., K. M. Shahani, and R. C. Chandan, *J. Dairy Sci.*, **45,** 1527 (1962).
63. Sister P. M. Slakey and W. E. M. Lands, *Lipids,* **3,** 30 (1968).
64. H. Brockerhoff, *J. Lipid Res.*, **8,** 167 (1967).
65. H. P. Kaufmann and T. H. Khoe, *Fette, Seifen, Anstrichm.*, **64,** 81 (1962).
66. Č. Michalec, M. Šulc, and J. Měšťan, *Nature,* **193,** 63 (1962).
67. H. P. Kaufmann, Z. Makus, and T. H. Khoe, *Fette, Seifen, Anstrichm.*, **64,** 1 (1962).
68. K. Krell and S. A. Hashim, *J. Lipid Res.*, **4,** 407 (1963).
69. D. C. Malins, *Chem. Ind. (London),* **1960,** 1359.
70. J. L. Brown and J. M. Johnston, *J. Lipid Res.*, **3,** 480 (1962).

71. O. S. Privett, M. L. Blank, and W. O. Lundberg, *J. Am. Oil Chem. Soc.*, **38,** 312 (1961).
72. O. S. Privett and M. L. Blank, *J. Lipid Res.*, **2,** 37 (1961).
73. S. M. Rybicka, *Chem. Ind. (London),* **1962,** 308.
74. S. M. Rybicka, *Chem. Ind. (London),* **1962,** 1947.
75. R. G. Jensen, J. Sampugna, and G. W. Gander, *J. Dairy Sci.*, **44,** 1983 (1961).
76. A. F. Hofmann, *J. Lipid Res.*, **3,** 391 (1962).
77. T. Schewe and C. Coutelle, *Acta Biol. Med. Germ.*, **24,** 223 (1970).
78. D. Heusser, *J. Chromatogr.*, **33,** 62 (1968).
79. W. O. Ord and P. C. Bamford, *Chem. Ind.*, **1966,** 1681.
80. A. G. Douglas and T. G. Powell, *J. Chromatogr.*, **43,** 241 (1969).
81. H. P. Kaufmann, K. D. Mukherjee, and Q. Khalid, *Nahrung,* **11,** 631 (1967).
82. A. I. Mizany, *J. Chromatogr.*, **31,** 96 (1967).
83. L. Winkler, T. Heim, H. Schenk, B. Schlag, and E. Goetze, *J. Chromatogr.*, **70,** 164 (1972).
84. R. F. Addison and R. G. Ackman, *Anal. Biochem.*, **28,** 515 (1969).
85. S. Bergstroem, L. A. Carlson, and J. R. Weeks, *Pharmacol. Rev.*, **20,** 1 (1968).
86. M. Amin, *J. Chromatogr.*, **108,** 313 (1975).
87. H. A. Davis, E. W. Horton, K. B. Jones, and J. P. Quilliam, *Brit. J. Pharmacol.*, **42,** 569 (1971).
88. J. A. F. Wickramasinghe and S. R. Shaw, *Prostaglandins,* **4,** 903 (1973).
89. H. C. Kiefer, C. R. Johnson, K. L. Arora, and H. S. Kantor, *Anal. Biochem.*, **68,** 336 (1975).
90. E. G. Daniels, "Chromatography of Prostaglandins," in *Lipid Chromatographic Analysis,* 2nd ed., Vol. 2, Marcel Dekker, New York, 1968, p. 611.
91. J. E. Shaw and P. W. Ramwell, "Separation, Identification, and Estimation of Prostaglandins," in *Methods of Biochemical Analysis,* Vol. 17, D. Glick, Ed., Interscience, New York, 1969, p. 325.
92. E. W. Horton, *Prostaglandins,* Springer, Berlin, 1972, p. 21.
93. H. P. Kaufmann and Y. Su Ko, *Fette, Seifen, Anstrichm.*, **63,** 828 (1961).
94. M. R. Subbaram, M. W. Roomi, and K. T. Achaya, *J. Chromatogr.*, **21,** 324 (1966).
95. R. Subbarao and K. T. Achaya, *J. Chromatogr.*, **16,** 235 (1964).
96. R. Subbarao, M. W. Roomi, M. R. Subbaram, and K. T. Achaya, *J. Chromatogr.*, **9,** 295 (1962).
97. M. W. Roomi, M. R. Subbaram, and K. T. Achaya, *J. Chromatogr.*, **16,** 106 (1964).
98. *Ibid.*, **24,** 93 (1966).
99. E. Vioque, *J. Chromatogr.*, **39,** 235 (1969).
100. L. J. Morris, R. T. Holman, and K. Fontell, *J. Am. Oil Chem. Soc.*, **37,** 323 (1960).
101. L. J. Morris, H. Hayes, and R. T. Holman, *J. Am. Oil Chem. Soc.*, **38,** 316 (1961).
102. L. J. Morris, R. T. Holman, and K. Fontell, *J. Lipid Res.*, **2,** 68 (1961).

103. D. Sgoutas and F. A. Kummerow, *J. Am. Oil Chem. Soc.*, **40,** 138 (1963).
104. N. G. Bazán, Jr., and S. Cellik, *Anal. Biochem.*, **45,** 309 (1972).
105. D. C. Malins, J. C. Wekell, and C. R. Houle, *J. Am. Oil Chem. Soc.*, **41,** 44 (1964).
106. D. C. Malins, J. C. Wekell, and C. R. Houle, *Anal. Chem.*, **36,** 658 (1964).
107. D. C. Malins and C. R. Houle, *J. Am. Oil Chem. Soc.*, **40,** 43 (1963).
108. F. G. Bordwell and E. W. Garbisch, Jr., *J. Am. Chem. Soc.*, **82,** 3588 (1960).
109. H. K. Mangold and R. Kammereck, *J. Am. Oil Chem. Soc.*, **39,** 201 (1962).
110. H. K. Mangold, J. L. Gellerman, and H. Schlenk, *Federation Proc.*, **17,** 268 (1958).
111. T. H. Applewhite, M. J. Diamond, and L. A. Goldblatt, *J. Am. Oil Chem. Soc.*, **38,** 609 (1961).
112. T. W. Hammonds and G. Shone, *J. Chromatogr.*, **15,** 200 (1964).
113. N. Ruseva-Atanasova and J. Janák, *J. Chromatogr.*, **21,** 207 (1966).
114. K. Oette and M. Doss, *J. Chromatogr.*, **32,** 439 (1968).
115. S. Saha and J. Dutta, *Lipids*, **8,** 653 (1973).
116. L. J. Morris, *Chem. Ind. (London)*, **1962,** 1238.
117. L. J. Morris, *J. Chromatogr.*, **12,** 321 (1963).
118. L. D. Bergel'son, E. V. Dyatlovitskaya, and V. V. Voronkova, *Izv. Akad. Nauk SSSR, Otd. Khim. Nauk*, **1963,** 954.
119. L. D. Bergel'son, E. V. Dyatlovitskaya, and V. V. Voronkova, *J. Chromatogr.*, **15,** 191 (1964).
120. A. G. Vereshchagin and S. V. Skvortsova, *Dokl. Akad. Nauk SSSR*, **157,** 699 (1964); through *Chem. Abstr.*, **61,** 9757 (1964).
121. W. O. Ord and P. C. Bamford, *Chem. Ind. (London)*, **1967,** 277.
122. M. M. Paulose, *J. Chromatogr.*, **21,** 141 (1966).
123. H. K. Mangold, R. Kammereck, and D. C. Malins, *Microchem. J., Symp. Ser.*, **2,** 697 (1961, pub. 1962).
124. H. Schlenk and J. L. Gellerman, *Anal. Chem.*, **32,** 1412 (1960).
125. A. Stoll, J. Rutschmann, A. von Wartburg, and J. Renz, *Helv. Chim. Acta*, **39,** 993 (1956).
126. H. K. Mangold and R. Kammereck, *Chem. Ind. (London)*, **1961,** 1032.
127. E. Jantzen and H. Andreas, *Angew. Chem.*, **70,** 656 (1958).
128. E. Jantzen and H. Andreas, *Chem. Ber.*, **92,** 1427 (1959).
129. H. B. White, Jr., *J. Chromatogr.*, **21,** 213 (1966).
130. D. E. Minnikin and S. Smith, *J. Chromatogr.*, **103,** 205 (1975).
131. O. S. Privett and E. C. Nickell, *J. Am. Oil Chem. Soc.*, **39,** 414 (1962).
132. O. S. Privett and E. C. Nickell, *J. Lipid Res.*, **4,** 208 (1963).
133. O. S. Privett and E. C. Nickell, *J. Am. Oil Chem. Soc.*, **41,** 72 (1964).
134. O. S. Privett, M. L. Blank, and O. Romanus, *J. Lipid Res.*, **4,** 260 (1963).
135. J. N. Roehm and O. S. Privett, *J. Lipid Res.*, **10,** 245 (1969).
136. H. P. Kaufmann, K. D. Mukherjee, and Q. Khalid, *Fette, Seifen, Anstrichm.*, **69,** 820 (1967).
137. R. Kannan, M. R. Subbaram, and K. T. Achaya, *J. Chromatogr.*, **24,** 433 (1966).

138. J. Hradec and P. Menšík, *J. Chromatogr.*, **32**, 502 (1968).
139. H. T. Badings, *J. Chromatogr.*, **14**, 265 (1964).
140. J. J. Wren and A. D. Szczepanowska, *J. Chromatogr.*, **14**, 405 (1964).
141. T. S. Neudoerffer and C. H. Lea, *J. Chromatogr.*, **21**, 138 (1966).
142. J. A. Cornelius and G. Shone, *Chem. Ind. (London)*, **1963**, 1246.
143. L. D. Metcalfe and A. A. Schmitz, *Anal. Chem.*, **33**, 363 (1961).
144. G. A. Dhopeshwarkar and J. F. Mead, *J. Lipid Res.*, **3**, 238 (1962).
145. G. A. Dhopeshwarkar and J. F. Mead, *Proc. Soc. Exptl. Biol. Med.*, **109**, 425 (1962).
146. H. P. Kaufmann and C. V. Viswanathan, *Fette, Seifen, Anstrichm.*, **65**, 925 (1963).
147. D. Firestone, *J. Am. Oil Chem. Soc.*, **40**, 247 (1963).
148. A. Rieche, M. Schultz, H. E. Seyfarth, and G. Gottschalk, *Fette, Seifen, Anstrichm.*, **64**, 198 (1962).
149. O. S. Privett and M. L. Blank, *J. Am. Oil Chem. Soc.*, **39**, 465 (1962).
150. O. S. Privett and E. C. Nickell, *J. Am. Oil Chem. Soc.*, **40**, 189 (1963).
151. E. Habermann, G. Bandtlow, and B. Krusche, *Klin. Wochenschr.*, **39**, 816 (1961).
152. H. Wagner, *Fette, Seifen, Anstrichm.*, **62**, 1115 (1960).
153. *Ibid.*, **63**, 1119 (1961).
154. H. Wagner, L. Hoerhammer, and P. Wolff, *Biochem. Z.*, **334**, 175 (1961).
155. C. V. Viswanathan, *J. Chromatogr.*, **75**, 141 (1973).
156. P. Pohl, H. Glasl, and H. Wagner, *J. Chromatogr.*, **49**, 488 (1970).
157. F. Kunz and D. Kosin, *Clin. Chim. Acta*, **27**, 185 (1970).
158. S. B. Curri and V. Ninfo, *Riv. Anat. Patol. Oncol.*, **23**, 479 (1963); through *Chem. Abstr.*, **60**, 4445 (1964).
159. S. B. Curri and S. Mazzoni, *Acta Neurol. (Naples)*, **19**, 32 (1964); through *Chem. Abstr.*, **61**, 3408 (1964).
160. S. B. Curri, M. Raso, and C. R. Rossi, *Histochemie*, **4**, 113 (1964); through *Chem. Abstr.*, **61**, 13623 (1964).
161. S. B. Curri, C. R. Rossi, and L. Sartorelli, "Direct Analysis of Phospholipids of Mitochondria and Tissue Sections by Thin-Layer Chromatography," in *Thin-Layer Chromatography*, G. B. Marini-Bettòlo, Ed., Elsevier, Amsterdam, 1964, p. 174.
162. V. P. Skipski, R. F. Peterson, and M. Barclay, *J. Lipid Res.*, **3**, 467 (1962).
163. V. P. Skipski, R. F. Peterson, J. Sanders, and M. Barclay, *J. Lipid Res.*, **4**, 227 (1963).
164. H. Nielsen, *J. Chromatogr.*, **89**, 275 (1974).
165. *Ibid.*, **114**, 419 (1975).
166. F. Gonzalez-Sastre and J. Folch-Pi, *J. Lipid Res.*, **9**, 532 (1968).
167. D. Abramson and M. Blecher, *J. Lipid Res.*, **5**, 628 (1964).
168. D. E. Minnikin and H. Abdolrahimzadeh, *J. Chromatogr.*, **63**, 452 (1971).
169. M. Lepage, *J. Chromatogr.*, **13**, 99 (1964).
170. J. Baddiley, J. Buchanan, R. E. Handschumacher, and J. F. Prescott, *J. Chem. Soc.*, **1956**, 2818.
171. W. D. Skidmore and C. Entenman, *J. Lipid Res.*, **3**, 471 (1962).

172. H. Jatzkewitz, *Z. Physiol. Chem.*, **326,** 61 (1961).
173. S. N. Payne, *J. Chromatogr.*, **15,** 173 (1964).
174. G. M. Adams and T. L. Sallee, *J. Chromatogr.*, **49,** 552 (1970).
175. N. M. Neskovic and D. M. Kostic, *J. Chromatogr.*, **35,** 297 (1968).
176. G. Rouser, C. Galli, E. Lieber, M. L. Blank, and O. S. Privett, *J. Am. Oil Chem. Soc.*, **41,** 836 (1964).
177. H. P. Kaufman, H. Wessels, and C. Bondopadhyaya, *Fette, Seifen, Anstrichm.*, **65,** 543 (1963).
178. O. S. Privett and M. L. Blank, *J. Am. Oil Chem. Soc.*, **40,** 70 (1963).
179. O. S. Privett, M. L. Blank, and J. A. Schmit, *J. Food Sci.*, **27,** 463 (1962).
180. O. Renkonen, *Acta Chem. Scand.*, **18,** 271 (1964).
181. H. D, Kaulen, *Anal. Biochem.*, **45,** 664 (1972).
182. L. A. Horrocks, *J. Am. Oil Chem. Soc.*, **40,** 235 (1963).
183. H. Jatzkewitz and E. Mehl, *Z. Physiol. Chem.*, **320,** 251 (1960).
184. H. Weicker, J. A. Dain, G. Schmidt, and S. J. Thannhauser, *Federation Proc.*, **19,** 219 (1960).
185. H. Jatzkewitz, *Z. Physiol. Chem.*, **320,** 134 (1960).
186. H. Jatzkewitz, *Brain Lipids Lipoproteins Leucodystrophies, Proc. Neurochem. Symp., Rome,* **1961,** 147 (1963).
187. N. K. Kochetkov, I. G. Zhukova, and I. S. Glukhoded, *Biochim. Biophys. Acta,* **60,** 431 (1962).
188. N. K. Kochetkov, I. G. Zhukova, and I. S. Glukhoded, *Proc. Acad. Sci. USSR, Chem. Sect. (Engl. Transl.),* **139,** 716 (1961).
189. H. G. Mueldner, J. R. Wherrett, and J. N. Cumings, *J. Neurochem.*, **9,** 607 (1962).
190. C. G. Honegger, *Helv. Chim. Acta,* **45,** 281 (1962).
191. *Ibid.*, p. 2020.
192. H. Pliz and H. Jatzkewitz, *J. Neurochem,* **11,** 605 (1964).
193. R. Kuhn, H. Wiegandt, and H. Egge, *Angew. Chem.*, **73,** 580 (1961).
194. L. A. Whitting, R. S. Krishnan, A. H. Sakr, and M. K. Horwith, *Anal. Biochem.*, **22,** 295 (1968).
195. J. A. Dain, H. Weicker, G. Schmidt, and S. J. Thannhauser, *Cerebral Sphingolipidoses, Symp. Tay-Sachs' Disease Allied Disorders, New York, N.Y.*, **1961,** 289 (1962); through *Chem. Abstr.*, **58,** 739 (1963).
196. J. R. Wherrett and J. N. Cumings, *Biochem. J.*, **86,** 378 (1963).
197. E. Klenk and W. Gielen, *Z. Physiol. Chem.*, **323,** 126 (1961).
198. *Ibid.*, **326,** 144 (1961).
199. *Ibid.*, **333,** 162 (1963).
200. R. Kuhn and H. Wiegandt, *Chem. Ber.*, **96,** 866 (1963).
201. G. A. Johnson and R. H. McCluer, *Biochim. Biophys. Acta,* **70,** 487 (1963).
202. J. R. Wherrett, J. A. Lowden, and L. S. Wolfe, *Can. J. Biochem.*, **42,** 1057 (1964).
203. S. R. Korey and J. Gonatas, *Life Sci.*, **2,** 296 (1963).
204. K. Sambasivarao and R. H. McCluer, *J. Lipid Res.*, **5,** 103 (1964).
205. S. Handa and R. M. Burton, *Lipids,* **4,** 205 (1969).
206. L. Svennerholm, *J. Lipid Res.*, **5,** 145 (1964).

207. R. Ledeen, *J. Am. Oil Chem. Soc.,* **43,** 57 (1966).
208. N. K. Kochetkov, I. G. Zhukova, and I. S. Glukhoded, *Proc. Acad. Sci. USSR, Chem. Sect. (Engl. Transl.),* **147,** 987 (1962).
209. Y. Fujino and J. Zabin, *J. Biol. Chem.,* **237,** 2069 (1962).
210. B. Weiss and R. L. Stiller, *J. Lipid Res.,* **6,** 159 (1965).
211. K. Sambasivarao and R. H. McCluer, *J. Lipid Res.,* **4,** 106 (1963).
212. Č. Michalec, *J. Chromatogr.,* **24,** 228 (1966).
213. *Ibid.,* **20,** 594 (1965).
214. K. Kljaic, A. Gojmerac, and M. Prostenik, *Acta Pharm. Jugosl.,* **25,** 1 (1975); through *Anal. Abstr.,* **29,** 3D60 (1975).
215. K. Eberlein and G. Gercken, *J. Chromatogr.,* **61,** 285 (1971).
216. N. M. Neskovic, J. L. Nussbaum, and P. Mandel, *J. Chromatogr.,* **49,** 255 (1970).
217. S. P. Hoevet, C. V. Viswanathan, and W. O. Lundberg, *J. Chromatogr.,* **34,** 195 (1968).
218. C. V. Viswanathan, P. Phillips, and W. O. Lundberg, *J. Chromatogr.,* **38,** 267 (1968).
219. *Ibid.,* **35,** 66 (1968).
220. K. Owens, *Biochem. J.,* **100,** 354 (1966).
221. B. P. Hughes and F. F. Frais, *Biochem. J.,* **96,** 6P (1965).
222. M. Philippart and J. Menkes, *Biochem. Biophys. Res. Commun.,* **15,** 551 (1964).
223. G. M. Gray, *Biochem. J.,* **86,** 350 (1963).
224. S. B. Curri, F. E. Costantini, A. Carteri, and E. Manzin, *Riv. Anat. Patol. Oncol.,* **24,** 1132 (1963).
225. L. Hausheer, W. Pedersen, and K. Bernhard, *Helv. Chim. Acta,* **46,** 601 (1963).
226. M. A. Wells and J. C. Dittmer, *J. Chromatogr.,* **18,** 503 (1965).
227. G. J. Nelson, *J. Lipid Res.,* **3,** 256 (1962).
228. R. Angelico, G. Cavina, A. D'Antona, and G. Giocoli, *J. Chromatogr.,* **18,** 57 (1965).
229. T. E. Morgan, D. O. Tinker, and D. J. Hanahan, *Arch. Biochem. Biophys.,* **103,** 54 (1963).
230. G. Rouser, G. Kritchevsky, D. Heller, and E. Lieber, *J. Am. Oil Chem. Soc.,* **40,** 425 (1963).
231. Y. Kishimoto and N. S. Radin, *J. Lipid Res.,* **5,** 94 (1964).
232. R. M. Burton and J. M. Gibbons, *Biochim. Biophys. Acta,* **84,** 220 (1964).
233. L. C. Ta-Chuang and C. C. Sweeley, *Biochemistry,* **2,** 592 (1963).
234. C. Y. Shuster, J. R. Froines, and H. S. Olcott, *J. Am. Oil Chem. Soc.,* **41,** 36 (1964).
235. F. A. Manzoli and P. Carinci, *Boll. Soc. Ital. Biol. Sper.,* **40,** 1283 (1964).
236. D. J. Hanahan, J. Ekholm, and C. M. Jackson, *Biochemistry,* **2,** 630 (1963).
237. G. M. Gray, *Biochim. Biophys. Acta,* **84,** 35 (1964).
238. G. Rouser, G. Simon, and G. Kritchevsky, *Lipids,* **4,** 599 (1969).
239. C. V. Viswanathan, S. P. Hoevet, W. O. Lundberg, J. M. White, and G. A. Muccini, *J. Chromatogr.,* **40,** 225 (1969).

240. F. Snyder, *J. Chromatogr.*, **82,** 7 (1973).
241. F. Snyder, "The Chemistry, Physical Properties, and Chromatography of Lipids Containing Ether Bonds," in *Progress in Thin-Layer Chromatography and Related Methods,* Vol. II, A. Niederwieser and G. Pataki, Eds., Ann Arbor Science, Ann Arbor, Mich., 1971, p. 105.
242. J. C. Dittmer and R. L. Lester, *J. Lipid Res.,* **5,** 126 (1964).
243. C. Zinzadze, *Ind. Eng. Chem.,* **7,** 227 (1935).
244. M. D. Bischel and J. H. Austin, *Biochim. Biophys. Acta,* **70,** 598 (1963).
245. A. Bressler, *Biochim. Biophys. Acta,* **39,** 375 (1960).
246. O. Renkonen and P. Varo, *Lipid Chromatogr. Anal.,* **1,** 41 (1967).
247. O. Renkonen, "Thin-Layer Chromatographic Analysis of Subclasses and Molecular Species of Polar Lipids," in *Progress in Thin-Layer Chromatography and Related Methods,* Vol. II, A. Niederwieser and G. Pataki, Eds., Ann Arbor Science, Ann Arbor, Mich., 1971, p. 143.
248. H. P. Kaufmann and B. Das, *Fette, Seifen, Anstrichm.,* **65,** 398 (1963).
249. E. Haahti and T. Nikkari, *Acta Chem. Scand.,* **17,** 536 (1963).
250. E. Haahti, T. Nikkari, and K. Juva, *Acta Chem. Scand.,* **17,** 538 (1963).
251. T. Nikkari and E. Haahti, *Biochim. Biophys. Acta,* **164,** 294 (1968).
252. N. Nicolaides, H. C. Fu, and M. N. A. Ansari, *Lipids,* **5,** 299 (1970).
253. P. J. Holloway and S. B Challen, *J. Chromatogr.,* **25,** 336 (1966).
254. F. Reutner, *Fette, Seifen, Anstrichm.,* **70,** 162 (1968).
255. W. Dietsche, *Fette, Seifen, Anstrichm.,* **72,** 778 (1970).
256. H. P. Kaufmann, H. K. Mangold, and K. D. Mukherjee, *J. Lipid Res.,* **12,** 506 (1971).
257. E. Vioque and R. T. Holman, *J. Am. Oil Chem. Soc.,* **39,** 63 (1962).
258. D. E. Walsh, O. J. Banasik, and K. A. Gilles, *J. Chromatogr.,* **17,** 278 (1965).
259. J. Johnson, *J. Biol. Chem.,* **181,** 707 (1941).
260. N. Zoellner and K. Kirsch, *Z. Ges. Exp. Med.,* **135,** 545 (1962).
261. P. G. Roughan and R. D. Batt, *Anal. Biochem.,* **22,** 74 (1968).
262. G. B. Russell, *Anal. Biochem.,* **14,** 205 (1966).
263. C. Bandyopadhyay, *J. Chromatogr.,* **37,** 123 (1968).
264. H. Wagner, *Chromatogr., Symp., 2nd, Brussels,* **1962,** 243.
265. H. Wagner, *Congr. Sci. Farm., Conf. Commun., 21, Pisa,* **1961,** 911 (1962).
266. N. Robinson and B. M. Phillips, *Clin. Chim. Acta,* **8,** 385 (1963).
267. *Ibid.,* p. 832.
268. S. C. Rastogi, K. C. Srivastava, and R. D. Tiwari, *Z. Anal. Chem.,* **253,** 208 (1971).
269. W. M. Doizaki and L. Zieve, *Proc. Soc. Exptl. Biol. Med.,* **113,** 91 (1963).
270. J. A. French and D. W. Andersen, *J. Chromatogr.,* **80,** 133 (1973).
271. G. Rouser, A. N. Siakotos, and S. Fleischer, *Lipids,* **1,** 85 (1966).
272. G. Rouser, S. Fleischer, and A. Yamamoto, *Lipids,* **5,** 494 (1970).
273. J. Tichý, *J. Chromatogr.,* **78,** 89 (1973).
274. G. Rouser, G. Kritchevsky, C. Galli, and D. Heller, *J. Am. Oil Chem. Soc.,* **42,** 215 (1965).
275. H. Kleinig and U. Lempert, *J. Chromatogr.,* **53,** 595 (1970).

276. S. J. Purdy and E. V. Truter, *Analyst (London)*, **87,** 802 (1962).
277. R. J. M. Penick, M. H. Meisler, and R. H. McCluer, *Biochim. Biophys. Acta,* **116,** 279 (1966).
278. R. Duden and M. Czikajlo, *Z. Anal. Chem.,* **245,** 289 (1969).
279. O. S. Privett and M. L. Blank, *J. Am. Oil Chem. Soc.,* **39,** 520 (1962).
280. M. L. Blank, J. A. Schmit, and O. S. Privett, *J. Am. Oil Chem. Soc.,* **41,** 371 (1964).
281. J. G. Kirchner, J. M. Miller, and G. Keller, *Anal. Chem.,* **23,** 420 (1951).
282. L. J. Nutter and O. S. Privett, *J. Chromatogr.,* **35,** 519 (1968).
283. D. Chobanov, R. Tarandjiska, and R. Chobanova, *J. Am. Oil Chem. Soc.,* **53,** 48 (1976).
284. D. T. Downing, *J. Chromatogr.,* **38,** 91 (1968).
285. T. Castellani, P. Kjelgaard-Nielsen, and J. Wolff-Jensen, *J. Chromatogr.,* **104,** 123 (1975).
286. N. M. Neskovic, *J. Chromatogr.,* **27,** 488 (1967).
287. F. Šmíd and J. Reinišová, *J. Chromatogr.,* **86,** 200 (1973).
288. G. Kasang, G. Goeldner, and N. Weiss, *J. Chromatogr.,* **59,** 393 (1971).
289. F. Snyder and H. Kimble, *Anal. Biochem.,* **11,** 510 (1965).
290. E. Fosslien, F. Musil, D. Domizi, L. Blickenstaff, and J. Lumeng, *J. Chromatogr.,* **63,** 131 (1971).
291. E. Fosslien, *J. Chromatogr.,* **63,** 59 (1971).
292. R. J. Nicolosi, S. C. Smith, and R. F. Santerre, *J. Chromatogr.,* **60,** 111 (1971).
293. L. A. Roch and S. E. Grossberg, *Anal. Biochem.,* **41,** 105 (1971).
294. R. A. Heyneman, D. M. Bernard, and R. E. Vercauteren, *J. Chromatogr.,* **68,** 285 (1972).
295. H. W. Peter and H. U. Wolf, *J. Chromatogr.,* **82,** 15 (1973).
296. E. Vioque, L. J. Morris, and R. T. Holman, *J. Am. Oil Chem. Soc.,* **38,** 489 (1961).
297. J. A Williams, A. Sharma, L. J. Morris, and R. T. Holman, *Proc. Soc. Exp. Biol. Med.,* **105,** 192 (1960).
298. E. Dunn and P. Robson, *J. Chromatogr.,* **17,** 501 (1965).
299. D. E. Bowyer, W. M. F. Leat, A. N. Howard, and G. A. Gresham, *Biochem. Biophys. Acta,* **70,** 423 (1963).
300. J. S. Amenta, *J. Lipid Res.,* **5,** 270 (1964).
301. C. K. Huston and P. W. Albro, *J. Bacteriol.,* **88,** 425 (1964).
302. Z. Nichaman, C. C. Sweeley, N. M. Oldham, and R. E. Olson, *J. Lipid Res.,* **4,** 484 (1963).
303. O. S. Privett, M. L. Blank, D. W. Godding, and E. C. Nickell, *J. Am. Oil Chem. Soc.,* **42,** 381 (1965).
304. O. S. Privett, K. A. Dougherty, and W. L. Erdahl, "Quantitative Analysis of the Lipid Classes by Thin-Layer Chromatography via Charring and Densitometry," in *Quantitative Thin-Layer Chromatography,* J. C. Touchstone, Ed., Wiley, New York, 1973, p. 57.
305. A. Kuksis, *Chromatogr. Rev.,* **8,** 172 (1966).

306. S. R. Max and R. H. Quarles, "Quantitative Thin-Layer Chromatography of Gangliosides," in *Quantitative Thin-Layer Chromatography*, J. C. Touchstone, Ed., Wiley, New York, 1973, p. 235.
307. G. Rouser, G. Kritchevsky, C. Galli, and D. Heller, *J. Am. Oil Chem. Soc.*, **42,** 215 (1965).
308. N. Fisher, M. E. Broughton, D. J. Peel, and R. Bennett, *J. Sci. Food Agric.*, **15,** 325 (1964).
309. J. H. Nelson, R. L. Glass, and W. F. Geddes, *Cereal Chem.*, **40,** 337 (1963).
310. M. E. McKillican and R. P. Sims, *J. Am. Oil Chem. Soc.*, **41,** 341 (1964).
311. C. F. Allen, P. Good, H. F. Davis, and S. D. Fowler, *Biochem. Biophys. Res. Commun.*, **15,** 424 (1964).
312. H. P. Kaufmann and C. V. Viswanathan, *Fette, Seifen, Anstrichm.*, **65,** 607 (1963).
313. K.-H. Bey, *Fette, Seifen, Anstrichm.*, **65,** 611 (1963).
314. K.-H. Bey, *Am. Perfum. Cosmet.*, **79,** 35 (1964).
315. J. J. Wren and A. D. Szczepanowska, *J. Chromatogr.*, **14,** 405 (1964).
316. I. Mitsev, J. Slavčeva, and A. Popova, *Compt. Rend. Acad. Bulg. Sci.*, **20,** 693 (1967).
317. W. A. McGugan, *Int. Dairy Congr., 15th, London,* **1959,** 1534 (1960).
318. G. Lakshminarayana, *J. Proc. Oil Technol. Assoc., India, Kampur,* **21,** Part 2, 77 (1966).
319. M. K. Ramamurthy, K. M. Narayanan, V. R. Bhalerao, and N. N. Dastur, *Indian J. Dairy Sci.*, **20,** 11 (1967).
320. M. Bachmann, *Schweiz. Milchztg.*, **95,** 1018 (1965); through *Anal. Abstr.*, **19,** 2663 (1970).
321. M. M. Chakrabarty, C. Bandyopadhyay, D. Bhattacharyya, and A. K. Gayen, *J. Chromatogr.*, **36,** 84 (1968).
322. M. M. Chakrabarty, D. Bhattacharyya, and A. M. Gayen, *J. Chromatogr.*, **44,** 116 (1969).
323. R. Guillaumin, *Rev. Fr. Corp Gras,* **12,** 29 (1965).
324. H. Hendrikx and A. A. Huyghebaert, *Meded. Rijksfac. Landbouwet., Gent,* **33,** 331 (1968); through *Chem. Abstr.*, **71,** 2188g (1969).
325. *Ibid.*, p. 515 through *Chem. Abstr.*, **71,** 2190b (1969).
326. A. Huyghebaert and H. Hendrickx, *Meded. Rijksfac. Landbouwet., Gent,* **33,** 523 (1968); through *Chem. Abstr.*, **71,** 2191c (1969).
327. M. M. Chakrabarty and A. K. Gayen, *J. Oil Technol. Assoc. India,* **6,** 69 (1974).
328. H. P. Kaufmann, H. Wessels, and B. Das, *Fette, Seifen, Anstrichm.*, **64,** 723 (1962).
329. H. Meyer, *Rev. Int. Choc.*, **17,** 270 (1962).
330. M. Hyyrylaeinen, *Farm. Aikak.*, **72,** 161 (1963); through *Chem. Abstr.*, **59,** 5753 (1963).
331. F. G. Seitz, *Fette, Seifen, Anstrichm.*, **68,** 314 (1966).
332. S. Chand, C. Srinivasulu, and S. N. Mahapatra, *J. Chromatogr.*, **106,** 475 (1975).

333. V. V. S. Mani and G. Lakshminarayana, *Indian J. Technol.,* **2,** 416 (1964).
334. G. B. Crump, *Analyst (London),* **88,** 456 (1963).
335. J. Sliwiok, *Mikrochim. Acta,* **1965,** 294.
336. G. Lakshminarayana and V. V. S. Mani, *Indian J. Technol.,* **2,** 320 (1964).
337. G. Lakshminarayana, *Annual Reports, 1965–66, Reg. Res. Lab., Hyderabad, India,* p. 4.
338. G. Kotakis, *Rev. Fr. Corps Gras,* **14,** 143 (1967).
339. C. Dimoulas, *Rev. Fr. Corps Gras,* **16,** 721 (1969).
340. M. M. Chakrabarty, D. Bhattacharyya, and B. Mondal, *Indian J. Technol.,* **1,** 473 (1963).
341. M. M. Chakrabarty, K. Talapatra, and J. K. Roy, *Fette, Seifen, Anstrichm.,* **69,** 642 (1967).
342. P. Sengupta, A. Bose, and T. V. Mathews, *J. Inst. Chem. Calcutta,* **45,** 168 (1973).
343. M. R. Verma, J. Rai, and A. Ram, *Int. Symp. Chromatogr. Electrophor., Lect. Pap., 6th, 1970,* Ann Arbor Science, Ann Arbor, Mich., 1971, p. 396.
344. J. W. Copius-Peereboom and H. W. Beekes, *J. Chromatogr.,* **17,** 99 (1965).
345. *Ibid.,* **9,** 316 (1962).
346. U. Pallotta and L. Matarese, *Riv. Ital. Sostanze Grasse,* **40,** 579 (1963).
347. *Ibid.,* **41,** 210 (1964).
348. V. V. S. Mani and G. Lakshminarayana, *Indian J. Technol.,* **3,** 339 (1965).
349. C. Srinivasulu and S. N. Mahapatra, *J. Chromatogr.,* **86,** 261 (1973).
350. P. C. Pantulu, M. K. R. Murphy, and M. B. Rao, *J. Oil Technol. Assoc. India,* **6,** 23 (1974).
351. R. K. Shrivastava and P. G. Bhutey, *Indian Oil Soap J.,* **31,** 164 (1966).
352. T. N. B. Kaimal, V. V. S. Mani, K. T. Achaya, and G. Lakshminarayana, *J. Chromatogr.,* **100,** 243 (1974).
353. G. Biernoth, *Fette, Seifen, Anstrichm.,* **69,** 635 (1967).
354. J. Gracian Tous and J. Martel, *Grasas Aceites (Seville, Spain),* **26,** 207 (1974).
355. T. Gutfinger and A. Letan, *J. Agric. Food Chem.,* **21,** 1120 (1973).
356. L. V. Cocks and C. van Rede, *Laboratory Handbook for Oil and Fat Analysts,* Academic Press, London, 1966, p. 78.
357. J. Cerbulis and C. A. Zittle, *Fette, Seifen, Anstrichm.,* **67,** 273 (1965).
358. R. Ristrow, *Dtsch. Lebensm. Rundsch.,* **64,** 322 (1968).
359. J. Pokorný and G. Janiček, *Nahrung,* **11,** 381 (1967).
360. E. Payne, *J. Sci. Food Agric.,* **22,** 20 (1971).
361. V. V. S. Mani and G. Lakshminarayana, *Chromatogr. Rev.,* **10,** 159 (1968).
362. J. Gracián and J. Martel, *Grasas Aceites (Seville, Spain),* **13,** 128 (1962); through *Chem. Abstr.,* **58,** 7050 (1963).
363. L. Acker and H. Greve, *Fette, Seifen, Anstrichm.,* **65,** 1009 (1963).
364. S. J. Purdy and E. V. Truter, *Nature,* **190,** 554 (1961).
365. S. J. Purdy and E. V. Truter, *Proc. Roy. Soc. (London), Ser. B.,* **158,** 536 (1963).
366. *Ibid.,* 544 (1963).
367. *Ibid.,* 553 (1963).

368. D. Thirkell and G. R. Tristram, *J. Sci. Food Agric.*, **14,** 488 (1963).
369. S. B. Curri, *Anthol. Med. Santoriana,* **69,** 100 (1963); through *Chem. Abstr.*, **60,** 14335 (1964).
370. J. M. Miller and J. G. Kirchner, *Anal. Chem.*, **24,** 1480 (1952).
371. K. K. Carroll, *J. Am. Oil Chem. Soc.*, **40,** 413 (1963).
372. F. J. M. Daemen, G. H. de Haas, and L. L. M. van Deenen, *Rec. Trav. Chim.*, **82,** 487 (1963).
373. J. Hirsch, *Federation Proc.*, **20,** 269 (1961).
374. V. Mahadevan and W. O. Lundberg, *J. Lipid Res.*, **3,** 106 (1962).
375. N. Zoellner, *Z. Klin. Chem.*, **1,** 18 (1963).
376. L. L. M. van Deenen, J. de Gier, and G. H. de Haas, *Koninkl. Ned. Akad. Wetenschap. Proc. Ser. B,* **64,** 528 (1961).
377. E. J. Gauglitz, Jr. and D. C. Malins, *J. Am. Oil Chem. Soc.*, **37,** 425 (1960).
378. E. J. Gauglitz, Jr. and L. W. Lehman, *J. Am. Oil Chem. Soc.*, **40,** 197 (1963).
379. L. L. M. van Deenen and G. H. de Haas, *Biochem. Biophys. Acta,* **70,** 538 (1963).
380. T. H. Applewhite, J. S. Nelson, and L. A. Goldblatt, *J. Am. Oil Chem. Soc.*, **45,** 101 (1963).
381. H. P. Kaufmann and A. K. Sen Gupta, *Fette, Seifen, Anstrichm.*, **65,** 529 (1963).
382. H. P. Kaufmann and R. S. Hamsagar, *Fette, Seifen, Anstrichm.*, **64,** 206 (1962).
383. H. P. Kaufmann and R. Schickel, *Fette, Seifen, Anstrichm.*, **65,** 1012 (1963).
384. H. P. Kaufmann and C. V. Viswanathan, *Fette, Seifen, Anstrichm.*, **67, 7** (1965).
385. N. Tuna, H. K. Mangold, and D. G. Mosser, *J. Lab. Clin. Med.*, **61,** 620 (1963).
386. H. Kaunitz, E. Gauglitz, Jr., and D. G. McKay, *Metabolism.*, **12,** 371 (1963).
387. H. Kaunitz, D. C. Malins, and D. G. McKay, *J. Exp. Med.*, **115,** 1127 (1962).
388. E. C. Horning, *Mal. Cardiovasculari,* **4,** 7 (1963).
389. M. Jaky, A. Koranyi, and P. L. Hagony, *Nahrung,* **8,** 105 (1964).
390. J. J. Peifer, F. Janssen, R. Muesing, and W. O. Lundberg, *J. Am. Oil Chem. Soc.*, **39,** 292 (1962).
391. R. Letters, *J. Inst. Brewing,* **68,** 318 (1962).
392. D. H. Hughes, *Proc. Intern. Conf. Sci. Aspects Mushroom Growing, 5th, Philadelphia,* **1962,** 540; through *Chem. Abstr.*, **60,** 13576 (1964).
393. M. E. McKillican and R. P. A. Sims, *J. Am. Oil Chem. Soc.*, **40,** 108 (1963).
394. F. D. Gunstone, F. B. Padley, and M. I. Qureshi, *Chem. Ind. (London),* **1964,** 483.
395. H. Debuch, W. Mertens, and M. Winterfeld, *Z. Physiol. Chem.*, **349,** 896 (1968).
396. G. Rouser, Gene Kritchevsky, M. Whatley, and C. F. Baxter, *Lipids,* **1,** 107, (1966).

Chapter **XXIV**

NUCLEIC ACIDS AND NUCLEOTIDES

Throughout this section the following abbreviations are used: AMP adenosine monophosphate, ADP adenosine diphosphate, ATP adenosine triphosphate, DPN diphosphopyridine nucleotide, TPN triphosphopyridine nucleotide, DPNH diphosphopyridine nucleotide reduced, TPNH triphosphopyridine nucleotide reduced, UDPG uridine diphosphate glucose, GMP guanosine monophosphate, CMP cytidine monophosphate, UMP uridine monophosphate, DNA deoxyribonucleic acid, RNA ribonucleic acid, IMP inosine monophosphate, UMP uridine monophosphate, TMP thymidine monophosphate, ADPG adenosine diphosphate-glucose, GDPM guanosine diphosphate-mannose, CDPG cytidine diphosphate-glucose, UDPG uridine diphosphate-glucose, UDPGA uridine diphosphateglucuronic acid, and UDPAG uridine diphosphate-*N*-acetyl-glucosamine. The prefix d- indicates a deoxyribonucleotide.

Pataki [1] and Randerath and Randerath [1a] have reviewed the thin-layer chromatography of nucleosides, nucleotides, and nucleic acid bases, and Scheit [2] has reviewed the thin-layer chromatography of oligonucleotides. Saukkonen [3] has reviewed the general subject of chromatography of free nucleotides.

1 SEPARATION ON SILICA GEL LAYERS

Randerath [4] applied silica gel G layers to the separation of bases and nucleosides using distilled water as the developing solvent. He also used silica gel layers bound with collodion because of the possibility of forming stable calcium complexes on gypsum-bound layers [5]. Scheig et al. [6] separated oxidized and reduced diphosphopyridine (DPN) and triphosphopyridine (TPN) nucleotides on thin layers of silica gel G using a solvent composed of isobutyric acid–ammonium hydroxide–water (66:1:33). These compounds were also separated with a mixture of ethanol–1 *M* ammonium acetate (pH 7.5) (7:3). A 1:1 ratio of the same solvent mixture would not separate the oxidized and reduced triphosphopyridine.

Cerri and Maffi [7] separated a mixture of uric acid, xanthine, hypoxanthine, and 6-mercatopurine in an atmosphere of water-saturated butanol containing 2.5% ammonia. The solvent was butanol saturated with water. They also separated a mixture of guanosine, inosine, and adenosine, and resolved cytidine and uridine.

Using eight different solvent systems, Massaglia et al. [8] chromatographed a group of pyrimidine and related nucleosides labeled with radioactive iodine on silica gel and also on cellulose layers.

Tomashefski et al. [9] give the R_f values for 35 cyclic nucleotides, nucleosides, purines, pyrimidines, and nucleotides using acetonitrile-based solvents. These are various mixtures of acetonitrile with *n*-butanol, 0.1 *M* ammonium acetate, and 28% ammonium hydroxide. Development times ranged from 18 to 81 min. The results on microcrystalline cellulose with the same solvents are also given.

Numerous solvent combinations have been used for separations on silica gel and some of these include: chloroform–pyridine–methanol (18:1:1) and (8:1:1), isopropanol–ammonia–water (7:1:2), and chloroform–methanol (9:1) and (7:3) [10]; isopropanol–ammonia–water (6:3:1), *n*-butanol–acetic acid–water (5:2:3) [11]; ethanol–methanol 0.1 *M* sodium dodecylsulfate (5:2:3 to 1:6:3), ethanol–methanol–0.1 *M* ammonium acetate (5:2:3) [12]; *tert*-butanol–methyl ethyl ketone–formic acid–water (2:2:0.1:1), methanol–chloroform–triethylamine (4:6:0.5) [13]; and ethyl acetate and ethyl acetate–methanol (9:1) [14].

Two-dimensional techniques have also been employed on silica gel.

Using the lower layer of a mixture of chloroform–methanol–water (4:2:1) with an additional 5% methanol for the first direction and ethyl acetate–isopropanol–water (75:16:9) for the second dimension Cadet and Téoule [15] separated 54 nucleosides and their derivatives. Wimmer and Reichmann [16] used isopropanol–ammonia–water (7:1:2) or *n*-butanol–*n*-propanol–ammonia–water (65:5:10:20) for the first dimension and water-saturated butanol for the second. Deoxyribopolynucleotides have been separated with *n*-propanol–ammonium hydroxide–water (3:1:1) in the first dimension followed by isobutryic acid–1 *M* ammonium hydroxide–0.1 *M* EDTA (100:60:1.6) in the second dimension [17].

Munns and Sims [18] used a two-dimensional continuous development on silica gel–microcrystalline cellulose (2:3) to separate 17 methylated nucleoside constituents. First development was for 150 min with ethyl acetate–methanol–water–88% formic acid (100:25:20:1) and after drying, the second dimension was for the same time with acetonitrile–ethyl acetate–2-propanol–*n*-butanol–58% ammonium hydroxide–water (40:30:20:10:22:5). Filter paper was used to carry the solvent away to give the continuous effect.

Tetraborate-impregnated silica gel has been used to separate cyclic adenosine 3′5′-monophosphate from other adenine derivatives [19], and to separate cyclic 3′,5′-guanosine monophosphate from other guanine derivatives [20].

Hawrylshyn et al. [21] investigated a group of fluoropyrimidines and fluoronucleosides. R_f values for separation on silica gel GF are given in 12 different solvents. By using a two-dimensional technique with ethyl acetate–acetone–water (7:4:1) in the first direction and ethyl acetate–methanol–ammonium hydroxide (75:25:1) in the second direction the 10 compounds could be separated.

2 SEPARATION ON CELLULOSE

Randerath [4] and Randerath and Struck [22] have used thin layers of cellulose prepared by slurrying 10 g of MN 300 powder in 50 to 60 ml of acetone. No binder was used. The nucleo bases and nucleosides were separated by using distilled water as the developing solvent. Nucleotides were separated with *n*-butanol–acetone–acetic acid–5% ammonium hydroxide–water (4.5:1.5:1:1:2).

Di Jeso [23] avoided the use of plaster of Paris as a binder for cellulose layers in the chromatography of phosphagens and their hydrolysis products, the guanidine derivatives, and orthophosphates, because of the possible interference in the chromatography of the phosphate compounds in basic solvents.

Randerath [24] has made a direct comparison of the separation of nucleic acid derivatives on paper and on thin layers of cellulose containing a plaster of Paris binder under identical conditions. In the four solvents tested, that is, water, saturated ammonium sulfate–1 *M* sodium acetate–isopropanol (40:9:1), *tert*-amyl alcohol–formic acid–water (3:2:1), and *n*-butanol–acetone–acetic acid–5% ammonium hydroxide–water (3.5:2.5:1.5:1.5:1), the separations on the cellulose thin layers were superior to those on the paper either in resolution and sharpness of spots or developing time, or both. For example, using the *tert*-amyl alcohol solvent an excellent separation of nucleoside mono-, di-, and triphosphates was obtained on thin layers, but with paper chromatography the di- and triphosphates could not be resolved. Again with the *n*-butanol solvent, although comparable separations were obtained, those with the thin layer were obtained in 90 min compared to 6 to 8 hr for the paper chromatography.

Randerath [25] has separated nucleic acid bases on cellulose layers using a two-dimensional technique. He used methanol–hydrochloric acid–water (7:2:1) in the first dimension and *n*-butanol–methanol–water–ammonium hydroxide (60:20:20:1) for the second. More recently Randerath and Randerath [26] have used a two-dimensional separation for determining the base composition and sequence analysis of RNA by means of chemical post-labeling with tritium. Solvents for the resolution of complex mixtures of nucleoside trialcohols and labeled RNA digests were for the first development of 17 cm: acetonitrile–ethyl acetate–*n*-butanol–isopropanol–6 *N* ammonia (7:2:1:1:2.7), and *tert*-amyl alcohol–methyl ethyl ketone–acetonitrile–ethyl acetate–water–formic acid (4:2:1.5:2:1.5:0.18) for the second dimension for approximately 22.5 cm. The latter development was made possible by clipping a filter paper wick to the Chromogram sheet. The acetonitrile solvents gave excellent separations. Randerath et al. [27] also found that fibrous cellulose layers of the "MN-300" type (Baker-flex cellulose sheets #4468) gave superior results compared to the microcrystalline cellulose. In this case development in the first dimension was with acetonitrile–4 *N* ammonium hydroxide (3.4:1). For separating synthetic polydeoxyribonucleotides, Narang and Michniewicz [28] used isopropanol–5% ammonium hydroxide (2:1) for the first solvent and isobutyric acid–1 *M* ammonium hydroxide–0.1 *M* EDTA (100:60:1.6) or (75:60:1.6) for the second dimension on microcrystalline layers.

Other solvents for two-dimensional separations on cellulose layers include: *n*-butanol–water (43:7) in the first dimension and isopropanol–hydrochloric acid–water (170:41:39) in the second [29]; *tert*-butanol–methyl ethyl ketone–water–ammonium hydroxide (4:3:2:1) for the first solvent and isopropanol–water–hydrochloric acid (65:18.4:16.6) for the second solvent [30]; and isobutyric acid–0.5 *M* ammonia (5:3) as a first-dimensional solvent followed by *tert*-butanol–ammonium formate (pH 3.8) (1:1) at 90° to the first direction [31].

Ciardi and Anderson [32] give the R_f values for 26 purine and 22 pyrimidine derivatives in six different solvent systems on cellulose. For single-dimensional work isobutyric acid–0.5 *N* ammonia (10:6) and 0.1 *M* phosphate buffer, pH 6.8–saturated ammonium sulfate–*n*-propanol (50:30:1) gave the best results. Holguin-Hueso and Cardinaud [33] list the R_f values for 60 purine bases and deoxyribosides and 29 pyrimidine bases and deoxyribosides in six solvent systems.

Isomeric methylated deoxyguanosines obtained through the action of diazomethane or deoxyguanosine were separated on cellulose layers bound with plater of Paris [34]; 1-methyldeoxyguanosine and *O*-methyl-deoxyguanosine had R_f values of 0.70 and 0.78, respectively, in isopropanol–water (7:3), and 0.72 and 0.80 in isopropanol–water–ammonia (70:25:5).

Keck and Hagen [35] have used a two-dimensional combination of electrophoresis and chromatography to separate a mixture of deoxynucleotides, deoxynucleosides, and bases. The electrophoresis was carried out on a cellulose layer that had been sprayed with 0.05 *M* formate buffer (pH 3.4). Applying a current of 15 000 V and 25 mA at 0°C, the run was completed in 30 min. After the plate was dried, it was developed in saturated ammonium sulfate–1 *M* sodium citrate–isopropanol (40:9:1) for a distance of 10 cm. The location of the compounds was made by observation under ultraviolet light.

Bieleski [36], using a two-dimensional combination to separate phosphate esters, chromatographed twice, first with *n*-propanol–ammonium hydroxide (sp gr 0.90)–water (6:3:1) with 2.0 g of EDTA added per liter, and then applied electrophoresis in a pH-3.6 ammonium acetate buffer. A current of 1000 V and 35 mA was applied for 16 min. For separating RNA hydrolysates a 0.13 *M* formate buffer (pH 3.4) was used. Gassen [37] applied this technique to the fingerprinting of oligonucleotides and obtained good results up to the pentanucleotides. Mundry and Priess [38] applied the two-dimensional technique to the mapping of oligonucleotides. Electrophoresis was run first in 20% acetic acid with 8% formamide adjusted to pH 3.5 with ammonia. A current of 20 mA at 750 V was used for the separation. Chromatography was carried out with a 1:1 mixture of the electrophoresis buffer and tertiary butanol, adjusted to pH 4.5 with ammonia. The separated oligonucleotides were eluted from the layer, hydrolyzed, and resubmitted to electrophoresis to separate the hydrolysis products. Quantitative analysis was obtained from autoradiographs obtained by incorporation of ^{32}P-labeled tobacco mosaic virus RNA in the sample.

Table 24.1 gives the R_f values of nucleosides and bases in various systems.

Tyihák [39] has proposed a specific color reaction for the paper and thin-layer detection of adenine. Adenyl purines can be detected in the 0.01 to 0.05 μmole range by test T-273 [40]. Purines may also be detected by dipping the chromatogram in liquid nitrogen and viewing under shortwave ultraviolet light immediately on withdrawing [41]. Sensitivity is in the nanomole range but the phosphorescence only lasts for a few seconds.

Table 24.1 $R_f \times 100$ Values of Free Bases and Nucleosides in Various Systems[a,b]

	A			B	C			D	
Compound	*t*	*u*	*v*	*w*	*w*	*x*	*y*	*w*	*y*
Adenine	29	30	38	98	20	33	44	14	13
Guanine	33	37	38	73	10	31	40	23	0
Hypoxanthine	46	55	57						
Uracil	73	72	75	74	26	66	75	62	68
Cytosine				92	26	80	90	50	56
Thymine				83	41	74	85	52	54
Adenosine	56	53	75	91[c]	68	58	62		
Guanosine	50	58	80	59[c]	13	31[d]	39[d]		
Inosine	61	70	82						
2-Deoxyadenosine				97[c]	66	70	70		
2-Deoxyguanosine				73[c]	12	40	50		
2-Deoxycytidine				90[c]	93	82	90		
Cytidine	82	80	76	78[c]	96	79	94		
Thymidine				81[c]	49	81	90		
Uridine	84	81	85	63[c]	30	68	87		

[a] Solvent: *A* = water [1]; *B* = isobutyric acid–ammonium hydroxide (sp gr 0.90)–water (33:1:16), development distance 6 in. [45]; *C* = 0.005 *N* hydrochloric acid, development distance 6 in. [45]; *D* = saturated ammonium sulfate–1 *N* sodium acetate–isopropanol (40:9:1), development distance 6 in. [45].

[b] Adsorbent: *t* = ECTEOLA cellulose, *u* = cellulose, *v* = silica gel G, *w* = DEAE cellulose, *x* = DEAE cellulose with 5% calcium sulfate binder, *y* = DEAE cellulose with 10% calcium sulfate binder.

[c] With 10% calcium sulfate.

[d] Streaks.

3 SEPARATION ON MODIFIED CELLULOSE

Diethylaminoethyl Cellulose

Randerath [42, 43] has demonstrated the usefulness of this ion-exchange material for separating nucleotides. A 0.02 *N* hydrochloric acid solution gives a fairly good separation. Adenosine diphosphate (ADP) cannot be separated from 5′-uridine monophosphate (UMP) and cytidine diphosphate (CDP) overlaps these two somewhat. In order to separate those compounds which are near the origin, a 0.03 *N* solution may be used. This solution then gives a better resolution of CTP, UDP, ATP, GTP, and UTP, although the spots have a tendency to overlap. Good separation of the triphosphates is obtained by increasing the concentration of the solvent to 0.04 *N* and increasing the development distance from about 9 to 12.8 cm. Dyer [44] used a two-step development for the separation of guanylic, uridylic, adenylic, and cytidylic acids obtained from the alkaline hydrolysis of ribonucleic acid. The samples were obtained by hydrolyzing ribonucleic acid from yeast or from rye seedlings with 0.3 *N* potassium hydroxide at 37°C for 18 hr. The hydrolysate was freed of potassium by means of a cation-exchange resin or by precipitating as the perchlorate. The first development was made with 1-propanol–ammonium hydroxide (sp gr 0.880)–water (6:3:1) at 40°C for a distance of 7 cm. [Note: ordinary reagent-grade ammonium hydroxide was unsatisfactory unless the solvent was allowed to stand overnight. AnalaR ammonia (E. Merck, Ltd.) gave satisfactory results when freshly prepared.]. After the first development, the plate was dried and then placed in 0.24 *M* acetic acid at 20 to 25°C and allowed to develop to the same 7-cm distance. The following R_f values were obtained: guanylic acid 0.21, uridylic acid 0.34, cytidylic acid 0.82, and adenylic acid 0.53, 0.62 (the 2′- and 3′-phosphates of the latter were separated).

Coffey and Newburgh [45] examined the effect of varying amounts of calcium sulfate binder in DEAE cellulose layers with the results that the R_f values (Table 24.1) of nearly all compounds examined showed large differences with hydrochloric acid as a solvent. The best general separation of ribonucleotides was obtained with isobutyric acid–ammonium hydroxide (sp gr 0.90)–water (33:1:16); however, for the separation of 2′- from the corresponding 3′-phosphate isomer, a better separation was achieved using 0.005 *N* hydrochloric acid with a DEAE layer bound with 10% calcium sulfate. Since guanylic acid tends to streak very badly below pH 10, the monoribonucleotides were applied in a 0.1 *N* ammonium hydroxide solution. For the separation of the deoxyribonucleotides the most satisfactory condition was a combination of DEAE with 10% calcium sulfate binder and the same solvent as used for the ribonucleotides. The addition of the calcium sulfate binder permitted the separation of 2-deoxycytidine-5′-phosphate from 2-deoxyadenosine-5′-phosphate which overlapped when no binder was present. For the separation of both ribonucleosides and deoxyribonucleosides, the combination of dilute hydrochloric acid (0.005 *N*) with calcium sulfate-free DEAE cellulose layers was best. When calcium sulfate binder was used, the R_f values of uridine and thymidine were increased so as to interfere with the cytidine and the 2-deoxycytidine spots. The nucleosides were spotted from a neutral solution. The free bases applied in a 0.1 *N* hydrochloric acid solution were best separated by the solvent system of Randerath [42] using a calcium sulfate-free DEAE cellulose layer. For two-dimensional work Coffey and Newburgh [45] recommend hydrochloric acid and distilled water as solvents for the first dimension; if the isobutyrate solvent is used in the first direction then water can be used in the second direction but not hydrochloric acid. Because a high salt content causes streaking just as in the case of paper chromatography, the samples were desalted by using a charcoal column. Swanson [46] removed salts from a sample on a DEAE cellulose layer by a preliminary development in water. Stickland [47] separated the mono-, di-, and triphosphates of adenosine, cytidine, guanosine, and uridine by two-dimensional chromatography on DEAE cellulose, using ammonium bicarbonate and ammonium formate as solvents. A preliminary separation to remove interfering substances from the nucleotides was made on a dual-layer plate in which the first quarter of the plate was a mixture of Dowex-1 and Sephadex G-25 and the balance of the plate was only Sephadex. The R_f values relative to CMP have been given in tabular form [48].

Lapidot et al. [49] separated 16 ribonucleotidyl ($2' \rightarrow 5'$) ribonucleosides from their $3' \rightarrow 5'$ isomers on DEAE layers. All the adenyl and guanyl isomers were all separated with 0.1 *N* ammonium bicarbonate. The separation of dinucleoside monophosphates of the type C–N and U–N required different concentrations of ammonium bicarbonate.

Jaroszewicz et al. [50] used 0.08 *N* hydrochloric acid for the first dimension at 4°C and saturated ammonium sulfate–1 *M* sodium acetate–isopropanol (80:18:2) at room temperature for the second dimension for separating nucleotides. Prior to the second development it was necessary to remove acid by a predevelopment with 8% triethylamine in acetone followed by development with acetone (three times). Grippo et al. [51] used two-dimensional separations with various combinations of solvents to separate nucleotides, nucleosides, and nucleic acid bases, including: (1) 0.02 *N* hydrochloric acid and (2) 95% ethanol–1 *M* ammonium acetate saturated with tetraborate (1:1); (1) 0.04 *N* hydrochloric acid (2) same as above; (1) 0.005 *N* hydrochloric acid (2) 0.035 *N* hydrochloric aid; and (1) 0.005 *N* hydrochloric acid and (2) 0.02 *N* hydrochloric acid.

Kirnos et al. [51a] used a linear gradient elution with 0 to 0.35 *M* sodium chloride in 0.01 *M* sodium acetate buffer (pH 5.1 to 5.3) and 5 *M* urea to fractionate deoxyribonucleotides into individual isopliths (oligonucleotides with an equal number of nucleotide residues). The purines moved near the solvent front and the pyrimidine fragments were clearly separated into mono-, di-, tri-, tetra-, penta-, hexa-, and heptanucleotides. A stepwise gradient could also be used. In order to separate the pyrimidine nucleotides according to composition, the individual isopliths were eluted and spotted on another DEAE layer for resolution with the same gradient of sodium chloride but in a pH-3.2 buffer system.

ECTEOLA Cellulose

Randerath was the first to apply this ion-exchange material in the thin-layer separation of nucleic acid constituents [4, 5, 52]. Separations were achieved with distilled water, 0.15 *M* sodium chloride, and 0.01 *N* hydrochloric acid. Two-dimensional separations were carried out by developing first with the sodium chloride solvent and then in the second direction with the dilute hydrochloric acid. Not only could the mono-, di-, and triphosphates be fractionated into groups, but subfractionation within these groups could be obtained [42]. Nayar [53] separated 5′-AMP, 5′-ADP, and 5′-ATP on ECTEOLA cellulose MN-300 with a calcium sulfate binder, using a solvent system of *tert*-butanol–90% formic acid–water (5:4:5). Prior to spotting the samples, the plates were given a predevelopment in the same solvent system. Quantitative estimations were made by spectrophotometric measurement at 259 mμ after the spots were eluted with acidified water (three drops 1 *N* hydrochloric acid/5 ml).

Bauer and Martin [54] used a solvent consisting of 1.0 *M* ammonium hydroxide in 2 *M* sodium chloride and containing 0.01 *M* phosphate (pH 11.0) for chromatographing calf thymus deoxyribonucleic acid. Prior to chromatographing the sample, which was applied in 0.01 *M* phosphate buffer (pH 7), the plates were prewet by developing them to a distance of 1.5 cm in the developing solvent. Failure to do this prevented the DNA from moving away from the origin.

Panteleeva [55] has reported on the separation of 16 nucleotides using 0.01 to 0.07 *N* hydrochloric acid as the developing solvent. As in most cases, the spots were detected by observation under ultraviolet light.

Dietrich et al. [56] have used layers of ECTEOLA impregnated with ammonium tetraborate for the separation of sugar nucleotides and also some sugar phosphates. Two grams of sieved ECTEOLA cellulose powder (Serva-Entwicklungslabor) was slurried with 18 ml of 0.004 *M* ethylenediaminetetraacetic acid. After spreading and drying the layers at room temperature overnight, they were sprayed with 0.1 *M* ammonium tetraborate. (Note: the ammonium tetraborate could not be incorporated in the slurry.) Drying at 50°C for 30 min completed the preparation of the plates. Separation of the sugar nucleotides was accomplished with 95% ethanol–0.1 *M* ammonium tetraborate (pH 9.0) 3:2) or (1:1). The first mixture was also used for separating the sugar phosphates.

Table 24.2 gives the R_f values of nucleotides on DEAE and ECTEOLA cellulose.

Polyethylenimine Cellulose

Randerath [57] has also introduced the use of PEI cellulose. Good separations of mono- and diphosphates of uracil, cytosine, adenine, and guanine were obtained by a stepwise development using 0.6 *M* sodium chloride solution for 5 min as a first step, and then 0.8 *M* sodium chloride solution for the second step for a development distance of 6.6 cm. The plate was not dried between the two development steps. The corresponding triphosphates were separated in a single stage development with 1.25 *M* sodium chloride solution by using a development distance of 5.5 cm. The separation on PEI cellulose was sharper than the corresponding separations on ECTEOLA and DEAE cellulose layers. It was also somewhat more sensitive to smaller amounts, so that it was approximately 50 times more sensitive than paper chromatography for nucleotides.

The doexyribonucleoside 5′-phosphates and the alkaline hydrolysate of yeast ribonucleic acid were separated by two-dimensional chromatography (58). In order to remove impurities, the polyethylenimine cellulose layers were developed for a distance of 5 cm with 10% sodium chloride. Immediately afterward, they were placed in water as a solvent and developed to the top, twice. After drying at room temperature they were then ready for application of the sample. The stepwise development consisted in developing first with water up to the starting line, followed by 1 *N* formic acid (without intermediate drying) to a distance of 10 cm. The development in the second dimension was carried out with 60% saturated ammonium sulfate for a distance of 8 cm. Prior to eluting the compounds from the layer, the plate was soaked in methanol for 15 min to remove excess salts. The plate was then dried and the compounds were eluted with 60% saturated ammonium sulfate.

Foster et al. [59] have developed a technique for removing salt from nucleotide samples as the samples are applied to the PEI layer. Essentially, the sample dissolved in 10 ml of water is allowed to flow through

Table 24.2 $R_f \times 100$ Values of Nuelcotides on ECTEOLA Cellulose and on DEAE Cellulose in Various Solvents

	ECTEOLA			DEAE						
				DEAE (5% $CaSO_4$ Binder)	*DEAE (10% $CaSO_4$ Binder)*					
Nucleotide	0.15 *M Sodium Chloride* [4]	0.01 *N Hydrochloric Acid* [4]	*95% Ethanol–0.1 M Ammonium Tetraborate, pH 9.0 (3:2). Development distance 17 cm*[a] [56]	*0.005 N Hydrochloric Acid. Development distance 6 in.* [45]		*Isobutyric Acid–Ammonium Hydroxide (sp gr 0.90)–Water (33:1:16). Development Distance 6 in.* [45]	0.01 *N Hydrochloric Acid* [42]	0.02 *N Hydrochloric Acid* [42]	0.03 *N Hydrochloric Acid* [42]	0.04*N Hydrochloric Acid* [42]
5′-AMP	57	26					45	65		
AMP			42							
3′-AMP	48			4	24	63				
2′-AMP				4	32	63				
ADP	36	8	36				24	48	68	
ATP	21		32				6	11	20	56
5′-CMP	74	31					46	65		
3′-CMP	71			5	50	51				

2′-CMP				9	61	55				
CDP	51	11					31	53		
CTP	34						9	13	31	64
5′-GMP	55	14								
3′-GMP	44									
GDP	37	3								
GTP	17									
5′-IMP	74									
5′-TMP				5	48	47				
TMP			110							
5′-UMP	80	13								
UMP			53							
3′-UMP	75			2	42	36				
2′-UMP				2	48	36				
UDP	63	0	42				7	15	25	
UTP	44		37				4	4	8	18
2-d-5′-AMP				4	28	64				
2-d-5′-GMP				2	15	35				
2-d-5′-CMP				8	48	60				

[a] $R_{st} \times 100$ values referred to inorganic phosphate. Adsorbent sprayed with 0.1 *M* ammonium tetraborate (pH 9.0) and dried before development.

a capillary onto the thin layer, which is placed in contact near the origin with a pad of filter paper. The water, along with the salt, is removed by the filter paper while the sample concentrates at the origin. In order to avoid the salt problem and the loss of some sample by using a methanol wash [58], Santini and Ulrich [60] used a volatile solvent for the first development; maximum resolution of nucleotides was obtained with 0.1 *N* formic acid, pH 3.6. The second-dimensional solvent was 1 *M* lithium chloride, pH 7.0. They also found that optimum resolution was obtained if the sample was adjusted to pH 3.5. Koenigk [61] avoided the loss of sample in quantitative work by adding EDTA to the solvent to suppress the salt effect instead of washing with methanol. Nazar et al. [61a] avoided the salt problem by extracting nucleotides with acetic acid. The extract was then lyophilized and taken up in 0.6 ml of cold water. Application with applied heat to remove water concentrated the sample in a small area.

Randerath [62] has separated the deoxyribonucleotides from ribonucleotides by using lithium chloride solutions containing boric acid as a complexing agent. Without the boric acid the R_f values of the corresponding compounds are identical. The 5′-monophosphates were separated using a solvent mixture of 2% boric acid–2 *M* lithium chloride (2:1), and the nucleoside triphosphates were separated with 4% boric acid–4 *M* lithium chloride (4:3). R_f values [63] for the two groups were as follows: AMP 0.24, d-AMP 0.42, GMP 0.08, d-GMP 0.30, CMP 0.28, d-CMP 0.55, UMP 0.37, and TMP 0.63 in the 2% boric acid solvent; ATP 0.33, d-ATP 0.46, GTP 0.17, d-GTP 0.37, CTP 0.36, d-CTP 0.61, UTP 0.48, and TTP 0.70 in the 4% boric acid solvent. ADP had R_f values of 0.04 and 0.56, respectively, in the two solvents.

In examining the brain adenine nucleosides and nucleotides, Gonzales and Geel [64] also used a boric acid–lithium chloride solvent. Jacobsen and Kirchner [65] separated ribonucleosides from deoxyribonucleosides and arabinonucleosides on PEI layers that had been converted to the borate form by soaking the layers for 5 min in 750 ml of 0.4 *M* triethylammonium tetraborate (99 g boric acid + 112 ml triethylamine per liter). Either 0.1 *M* boric acid or 0.02 *M* ammonium formate, pH 4.7–ethanol (1:1) was used as a solvent. Buteau and Simmons [66] used essentially the same technique for a two-dimensional separation of the nucleosides of deoxyribonucleic acid. The first dimension was with 0.02 *M* ammonium acetate, pH 4.8–95% ethanol (1:1) and the second dimension was with 0.1 *M* boric acid.

Randerath and Randerath [63] have examined more fully the various factors affecting the separation of nucleotides on polyethylenimine cellulose thin layers. The layers were prepared from ion exchangers of their own preparation [67] (see Chapter III, Section 4), which gave different results from those made from commercial preparations. The 0.5-mm-thick layers have a capacity of approximately 1.5 meq. N/g of cellulose (if the layers are prepared from undialyzed polyethylenimine solution they have a lower capacity of 0.7 to 0.8 meq. N/g of cellulose). The quantity of polyethylenimine in the layers was also examined for effectiveness of resolution: 0.1% gave poor resolution and 0.5% gave good resolution. Concentrations above 1.5% (dialyzed) should not be used because it is difficult to remove low-molecular-weight impurities. After drying, the layers are scored along the sides, 4 to 5 mm from the edge, in order to prevent an edge effect during development. In addition parallel lines, 0.5 mm wide and 2 cm long, are scratched at intervals of 5 mm from the starting line to the lower edge of the plate. To remove impurities not removed during the dialysis of the polyethylenimine, the plates are given a preliminary development in distilled water and are then dried at room temperature for at least 12 hr. Applications of the sample are made from 0.002 *M* solutions of sodium or lithium salts of the nucleotides in distilled water, and the spots are dried for 3 min in a current of cold air.

The relationship was determined between the R_f values and the concentration of the lithium chloride used as a developing solvent. The R_f values of the nucleotides vary with the concentration of the developing solvent and the rate of change depends on the type of compounds, for example, diesters or monoesters. On the other hand, the R_f values of the nucleic acid bases and the nucleosides are independent of the salt concentration except at very high concentrations, where a decrease is apparently due to a salting-out effect. Stepwise development can be used to improve the separation between various groups of nucleotides; as an example of this, a mixture of adenine and uracil nucleotides was given a development in 0.1 *M* lithium chloride for 1 min followed by 5 min with 0.3 *M* lithium chloride, then 15 min with 0.7 *M* lithium chloride, and finally 25 min with 1.5 *M* lithium chloride. The development in each succeeding solvent was made without any intermediate drying so that a true gradient elution was obtained. Because large excesses of salts interfere with the separation, the sample should be desalted prior to running the chromatogram. Although these salts can be removed by adsorption of the nucleotides on activated charcoal their recovery from the charcoal is not quantitative, and for this reason a simpler and more quantitative method was used. This consisted in drying the plate after spotting the sample and then immersing it completely in anhydrous reagent-grade methanol for 10 min. The layers were then dried and chromatographed in the normal fashion. In this manner the excess salt was removed leaving more than 90% of the nucleotides on the plate. Table 24.3 gives the R_f

Table 24.3 $R_f \times 100$ Values of Nucleotides on Polyethylenimine Cellulose in Various Solvent Systems. Development Distance 10 cm[a]

Compound	Lithium Chloride in Water			Formic Acid–Sodium Formate Buffer pH 3.4				1.0 *N* Formic Acid	2.0 *N* Formic Acid–0.5 *M* Lithium Chloride (1:1)	2.0 *N* Formic Acid–2.0 *M* Lithium Chloride (1:1)
	0.25 *M*	1.0 *M*	1.6 *M*	0.5 *M*	1.0 *M*	2.0 *M*	4.0 *M*			
5′-AMP	11	52	65	68[b]	>80	>80	>80	>80	>80	>80
5′-IMP	13	59	74	40	60	73	>80	19	53	78
5′-GMP	6	40	51	28	45	57	65[c]	41	50[b]	72[b]
5′-CMP	15	64	75	70[b]	>80	—[d]	—	>80	>80	>80
5′-UMP	20	75	80	51	72	>80	—	20	64	>80
ADP	0	26	54	3	10	32	75[c]	3	29	70
IDP	0	30	63	0	4	14	49	0	8	55
GDP	0	17	45	0	2	9	34	0	13	61
CDP	0	33	64	8	20	45	>80	4	35	73
UDP	0	41	71	2	7	24	60	0	11	60
ATP	0	6	34	0	0	4	24	0	4	33
ITP	0	9	39	0	0	2	11	0	2	17
GTP	0	5	25	0	0	0	7	0	2	24
CTP	0	11	41	0	2	5	29	0	4	37
UTP	0	14	49	0	0	2	17	0	2	20

d-AMP	11	52	—							
d-GMP	6	41	—							
d-CMP	18	65	—							
TMP	24	74	—							
d-ATP	0	—	35							
d-GTP	0	—	26							
d-CTP	0	—	43							
TTP	0	—	52							
DPN	64	80	>80	68[b]	>80	>80		>80	>80	>80
DPNH	20	71	—				—			
TPN	3	60	—	14	43	>80	—	10	44[b]	80
TPNH	0	34	—							
ADPG	22	77	>80	30	57	>80	—	9	51[b]	>80
GDPM	12	72	>80	7	18	51	—	1	34	73
CDPG	27	>80	>80	45	67	>80	—	13	60	>80
UDPG	34	>80	>80	17	39[c]	77	—	1	27[c]	>80
UDPAG	44	>80	>80	32	55[c]	>80	—	2	38[c]	—
UDPGA	4	73	—	—	3	—	—	0	9	66

[a] From K. Randerath and E. Randerath [63]; reproduced with permission of the authors and The Elsevier Publishing Co.
[b] Spot in second front.
[c] Elongated spot.
[d] — = not investigated.

values for nucleotides on polyethylenimine cellulose layers in various solvent systems. By using the tables to select the proper solvents, depending on the mixtures involved, excellent separations can be obtained in most cases with two to five steps. For very complex mononucleotide mixtures, Randerath and Randerath [68] have applied two-dimensional anion-exchange chromatography on polyethylenimine cellulose thin layers. Figure 24.1 shows the effectiveness of this technique which makes use of stepwise development. The first development was made with 0.2 *M* lithium chloride for a period of 2 min; this was followed immediately (no intermediate drying) with 1.0 *M* lithium chloride for 6 min, and a final third development with 1.6 *M* lithium chloride to a total distance of 13.0 cm. These three developments were in the same direction without any intermediate drying between the steps. After drying in a current of warm air, below 50°C, the portions of the layer not needed for a development in the second dimension were scraped off the plate. The plate was then submerged in 1 liter of anhydrous methanol for 15 min in order to remove

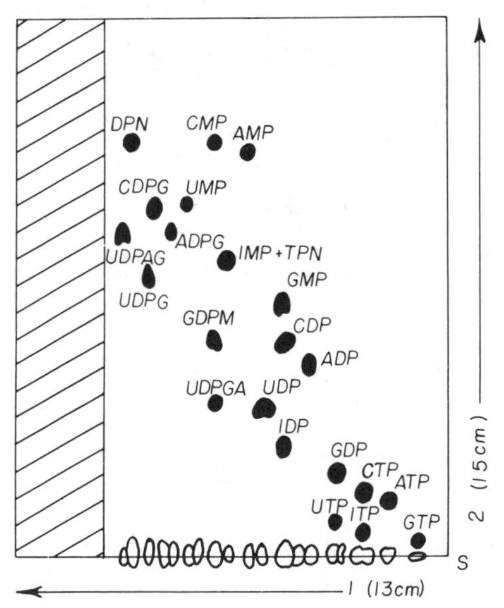

Fig. 24.1 Two-dimensional anion-exchange thin-layer chromatogram of mononucleotides. Layer thickness: 0.5 mm of PEI cellulose. A solution (0.1 ml) containing 23 ribonucleotides (10 to 15 μmoles each) was applied slowly in two 0.05-ml portions with intermediate drying from a micropipet to the layer at *S*. Development and detection were carried out as described in the text. The hatched area was removed after the development in the first dimension (see text). From Randerath and Randerath [68]; reproduced with permission of the authors and The Elsevier Publishing Co.

the lithium chloride. (Occasional agitation was used to help dissolve the salt.) After the methanol was removed by gentle drying, the plate was developed in three stages with formic acid–sodium formate buffers (pH 3.4). The first solvent of 0.5 *M* buffer was developed for 30 sec, the second solvent, 2.0 *M* buffer, was developed for 2 min, and the third solvent, 4.0 *M* buffer, was developed to a distance of 15 cm from the starting point. Here again, the three solvents were applied in succession without intermediate drying stages.

Raaen and Kraus [69] compared PEI cellulose MN-330 chromatofilms with PEI layers prepared in the laboratory from cellulose MN-300 and found the latter to be superior. They used these layers for the separation of complex mixtures of nucleosides and nucleotides. Two separate chromatograms were run. The first separation was with water in the first dimension and then with *n*-butanol–methanol–water–ammonium hydroxide (60:20:20:1). Nucleotides remained at the origin and thus simplified the separation and identification of the bases and nucleosides. Using the technique of Randerath and Randerath [68] the nucleotides were developed on the second plate in the first dimension for 2 min with 0.2 *M* lithium chloride, then for 6 min with 1.0 *M* lithium chloride, and finally to 13 cm from the origin with 1.6 *M* lithium chloride without intermediate drying of the layer (in other words, a gradient development). After drying they were washed for 15 min in 1 liter of absolute methanol. In a similar manner the second dimension was with formic acid–sodium formate buffer solutions of pH 3.4 as follows : 30 sec with 0.5 *M* buffer, 2 min with 2.0 *M* buffer, and finally to 15 cm from the origin with 4.0 *M* buffer. The methanol wash and the development in the second direction caused all the nucleobases and nucleosides except xanthosine to disappear.

Fullerton and Finch [70], in assaying pyrophosphate–ATP exchange reactions, incorporated urea into the developing solvent to attain a good separation of ATP from pyrophosphate. A two-step development on polyethyleminine first with 1 *M* lithium chloride for 6 cm to move any orthophosphate ahead of the ATP was followed without intermediate drying with 1.5 *M* lithium chloride containing 0.2 *M* sodium acetate and 4.0 *M* urea.

Cashel et al. [71] in working with nucleotide mixtures containing ^{32}P-labeled orthophosphate found that the severe "tailing" that occurred with sizable amounts of the latter could be eliminated by using orthophosphate developing solutions, and at the same time the size of the nucleotide spots were diminished.

Pataki and Niederwieser [72] have used gradient elution on polyethylenimine layers to separate nucleotides and their components. Using successively the following series of lithium chloride solutions: 0.6 ml of

0.1 *M*, 0.6 ml of 0.2 *M*, 0.6 ml of 0.5 *M*, 0.6 ml of 1 *M*, and 1.2 ml of 2 *M*, the individuals of a given group such as adenosine, uridine, and cytidine mono-, di-, and triphosphate were well separated; however, among the different groups the corresponding compounds had approximately the same R_f values. For example: adenine, guanine, uracil, cytosine, inosine, and thymine diphosphates all were approximately equal in R_f value. Separations were made in a sandwich type chamber. Schettino and La Rotonda [73] also used an elution gradient technique.

Randerath and Randerath [74, 75] have applied polyethylenimine thin-layer separations to the analysis of oligonucleotides, after digestion by chemical or enzymatic procedures. A novel mapping procedure was described which resolved all major mono-, di-, and trinucleotides from a pancreatic ribonuclease digest of RNA. Volckaert and Fiers [76] presented a very sensitive technique for the sequence determination of ^{32}P-labeled oligonucleotides using polyethylenimine layers. The oligonucleotide was treated with spleen exonuclease and after resolution of the intermediates on polyethylenimine layers it was possible to determine whether a Gp or an Ap residue had been removed. This was determined from a calibration grid and was based on the difference in increased mobility after removal of the residue. The sensitivity was 10 to 20 times greater than the electrophoretic technique on DEAE paper. Volckaert et al. [77] published a modified method for the fingerprinting of enzymatic digests of ^{32}P-labeled RNA on thin-layer plates. The first-dimensional separation was by conventional electrophoresis at pH 3.5 on cellulose acetate, and after transfer by the Southern technique [78] to a thin-layer sheet the development was carried out with a mixture of partially hydrolyzed RNA and urea. The T_1-ribonuclease digests were run on polyethylenimine layers and the pancreatic ribonuclease digests on DEAE layers. Excellent resolution was obtained and high sensitivity was achieved because of the small, compact spots. Litt and Hancock [79] have presented evidence that kethoxal (β-ethoxy-α-ketobutyraldehyde) can be a useful reagent for the determination of nucleotide sequences in single-stranded regions of transfer RNA. They used paper chromatography in their separations, but there is no reason why the technique could not be applied to thin-layer work.

Polyethylenimine cellulose has also been found to be superior to DEAE cellulose for the separation of deoxyribooligonucleotides [80].

Randerath and Weimann [81] have used thin layers of PEI cellulose to demonstrate the complex formation between deoxyribooligonucleotides and polyribonucleotides. By applying samples of deoxyadenosine oligonucleotides to spots of polyuridylic acid on thin layers of PEI cellulose, the oligonucleotides were retained at the origin under conditions which normally allow movement of the compounds on the chromatogram.

See Chapter VI, Section 6 for carrying out enzymatic reactions directly on polyethylenimine cellulose [82].

Polyphosphate Cellulose

Randerath and Randerath [83] have used this cation exchanger to separate the bases: uracil, cytosine, adenine, and guanine and the corresponding nucleosides.

4 SEPARATION ON SEPHADEX

Stickland [47] has used a combination plate of Dowex-1 and Sephadex G-25 for the first quarter and Sephadex alone for the remainder of the plate, in order to obtain a preliminary separation of nucleotides from interfering substances.

Wieland and Determann [84] separated adenosine mono-, di-, and triphosphates on the formate form of diethylaminoethyl Sephadex by means of gradient elution. Starting with a solution of 60 ml of 1 *N* formic acid, the composition of the solvent was gradually changed by the addition of a 2 *M* solution of ammonium formate in 50 ml of 10 *N* formic acid. The spots were visualized by spraying with a solution of fluorescein in methanol and observing under ultraviolet light. Hashizume and Sasaki [85] have also separated nucleotides on the same substrate by developing with a 1 *N* formic acid solution. The following R_f values were obtained: 5′-cytidine monophosphate 0.64, 3′-uridine monophosphate 0.05, 5′-uridine monophosphate, 0.03, uridine diphosphate 0.02, 3′-adenosine monophosphate 0.53, 5′-adenosine monophosphate 0.58, adenosine triphosphate 0.02, 3′-guanosine monophosphate 0.07, 5′-guanosine monophosphate 0.08, and 5′-inosine monophosphate 0.03.

Tortolani and Colosi [86] chromatographed adenine, thymine, uracil, xanthine, hypoxanthine, cytosine, and their nucleosides on mixtures (16:4) of fine Sephadex G-10 with silica gel GF or microcrystalline cellulose. The layers were developed in an ascending manner using 0.05 *M* phosphate buffer (pH 7.0).

5 SEPARATION ON ION-EXCHANGE LAYERS

Tomasz [87–91] has examined the application of Dowex 50-type resin layers for the separation of nucleotides and their components. They used Fixion 50-X8 plates (Chinoin) or Ionex 25 SA chromatosheets (Macherey,

Nagel) which consist of approximately one-third Dowex 50-type resin and two-thirds silica gel. The 5′-ribonucleotides could be separated on Fixion 50-X8 (H^+) with deionized water but not the isomeric ribonucleotides [87]. The eight principal ribo- and deoxyribonucleosides were resolved on resin layers (NH_4^+) by two-dimensional treatment with acetate buffers of pH 3.5 and 6.5 [88]. Continuous development with 2.8 *N* hydrochloric acid on Fixion 50-X8 (H^+) layers separated 5-methylcytosine, N^6-methyladenine, adenine, guanine, cytosine, 5-hydroxymethylcytosine, thymine, and uracil [89].

6 ELECTROPHORETIC SEPARATIONS

Dose and Risi [92] have used high-voltage electrophoresis on agar layers to separate nucleic acid constituents for their determination by spectroscopic measurement. Weiner and Zak [93] have applied electrophoresis to the separation of ribonucleic acid purines and pyrimidines on thin layers of agar gel coated on Teflon-glass paper (Fiberfilm T-20 A-60, Pallflex). For densitometric measurements, thin layers of agar were also used on quartz slides. The bases were applied to the agar gel by means of filter paper strips or #30 cotton thread soaked in buffer solutions of the bases. A voltage of 250 V was applied to obtain the separation.

Borkowski et al. [94] resolved the DNA bases, adenine, guanine, cytidine, and thymine by electrophoresis on agar gel layers using a sodium acetate–acetic acid 0.1 *M* buffer (pH 3.7). Electrophoresis was carried out at 5 to 7 V/cm for 45 min. The sample was obtained by hydrolyzing DNA with 72% perchloric acid for 60 min, and after hydrolysis the perchloric acid was removed by evaporation, leaving the bases salt free. Tsanev et al. [95] examined the influence of RNA concentration, pH, temperature, buffer, and gel concentration on the fractionation and mobility of RNA during gel electrophoresis. Agarose gel has also been used for the electrophoretic separation of AMP, ADP, ATP [96] and for the resolution of mixtures of adenine, adenosine, adenylic acid, and adenosine di- and triphosphates [97].

Nucleotides may also be separated by electrophoresis on polyacrylamide layers [98]. Peacock and Dingman [99] fractionated ribonucleic acid (RNA) from various rat tissues on a 3.5% gel prepared in a pH 8.3 buffer. Two-dimensional electrophoresis on polyacrylamide gel has been used to give improved separations. Ikemura and Dahlberg [100] have used a 10% gel in the first direction and a 20% gel in the second dimension for separating low-molecular-weight RNA. Varricchio and Ernst [101] used a 16% gel with 7 *M* urea at pH 8.3 in the first direction followed by a second-dimensional gel without the urea for separating *E. coli* and rat liver cytoplasmic tRNAs. De Wachter and Fiers [102] used a first-dimensional separation on a 12% gel at pH 3.5 and containing 6 *M* urea for separating some partial hydrolysates of purified RNA to obtain a separation according to both chain length and base composition. The second dimension was carried out at pH 8 giving a separation according to chain length. Righetti and Drysdale [103] have investigated the application of isoelectrofocusing on polyamide to the separation of nucleotides. Caton and Goldstein [104] used a 2.5 to 12% linear gradient polyacrylamide gel for the electrophoresis of ribonucleic acids. All RNA's from 4 to 28 *S* could be separated on a single gel with good resolution. Jeppesen [105] used linear gradients of 3.5 to 7.5% and 2.5 to 7.5% for separating DNA fragments. The bands were sharper than those obtained by conventional gel electrophoresis.

Thin-layer electrophoresis on cellulose layers has also been used for the separation of nucleotides [106–108]. (See also Section 2, "Separation on Cellulose.")

Brownlee and Sanger [109] separated labeled oligonucleotides of up to 50 residues long by first-dimensional ionophoresis on cellulose acetate followed by thin-layer chromatography on a mixed DEAE cellulose layer after transferring the bands to the layer.

7 QUANTITATIVE ANALYSIS

A number of techniques have been applied to the determination of nucleotides and their constituents.

In quantitative work with the thin-layer separation of adenine nucleotides on DEAE cellulose, Boernig and Reinicke [110] found that impurities present in DEAE cellulose showed an absorption in the same region as the nucleotides. This was removed by washing with hydrochloric acid (see Chapter III, Section 4). The developing solvent was 0.02 to 0.3 *N* hydrochloric acid and the exact concentration depended on the particular batch of plates. Each batch of plates was checked to determine the optimum concentration of acid. The quantitative determination of the nucleotides was made by eluting with 1 *M* sodium chloride and measuring the absorbance at 260 mμ against the eluant from a DEAE cellulose blank.

Change et al. [111] developed a spectrophotometric method for the determination of blood adenine nucleotide on a micro scale.

In situ reflectance spectroscopy can be used for the quantitative analysis of nucleic acid bases, nucleosides, and nucleotides after separation on thin layers [28, 112–115]. Sensitivity limits for reflectance measurements lay between 0.5 and 1.0 μg per spot [112] and the average accuracy ranged from 5.1 to 4.0%, although the inclusion of an internal standard improved the reproducibility to 3 to 4% relative standard deviation [114].

Pataki and coworkers [116, 117] have used in situ fluorescent measurements for the quantitation of nucleic acid bases, nucleosides, and nucleotides. Gissel and Stoll [118] used this method for the routine analysis of inosine and hypoxanthine in postmortem muscle tissue.

The most sensitive methods are of course those using radioactive isotopes. Randerath and Randerath [119, 120] described a tritium in vitro labeling technique for the quantitative analysis of ribose derivatives at the subnanomole level which is approximately a thousandfold more sensitive than conventional techniques. Measurement of the compounds was carried out by scintillation counting. Randerath et al. [121, 122] have applied the technique to the base analysis of ribonucleotides. Chang [123] has used a modification of the technique for the detection of one minor base in about 1800 to 2000 total bases. Radioactive phosphorus can also be used for the assay of nucleotides and their components [71, 124], and recently Johnson and Osuji [125] have proposed the use of ^{35}S-labeled *p*-hydrazinobenzenesulfonic acid as a reagent for preparing nucleoside derivatives. Of course ^{14}C labeling can be used [126], but is not as desirable because of the low level of activity.

8 MISCELLANEOUS

A number of desalting methods for nucleotides have been referred to, but mention should be made of others that may be useful at some time. Dougherty and Schepartz [127] recommend the use of a column (1 × 39.5 cm) of insoluble poly-(*N*-vinylpyrrolidine) (Polycar AT Powder, GAF) for the desalting of nucleotides, nucleosides, purines, and pyrimidines. The compounds were eluted quantitatively from the column with distilled water and in small volumes. Nucleotides elute just prior to sodium chloride, but cannot be separated from ammonium sulfate. Adenine, adenosine, uracil, guanine, and uridine are all eluted after the salts. In desalting oligonucleotide solutions Van den Bos et al. [128] stirred the oligonucleotides (20 to 200 μg) containing salts in 160 to 400 ml of water with 1 to 3 g of DEAE cellulose for 1 hr at room temperature. This was filtered and washed with water to remove residual salt and urea (if present). The oligonucleotides were eluted from the DEAE cellulose by stirring in 10 ml of 30% triethylamine carbonate (pH 10) for 30 min. The suspension was filtered and the triethylamine carbonate was removed under reduced pressure at 40°C. Recovery was better than 90%. Caron and Dugas [129] used prewashed Dowex 1-X2 (Cl^-) resin to remove salt. Essentially, the nucleotide solution to be desalted was shaken with 10 mg of the resin for

30 min, then filtered and the resin washed with water to remove adhering salt solution. The anion exchange was then converted to the formate form by shaking and rinsing the resin twice with 75 ml of 1.0 *M* ammonium formate for 20 min each time. The resin was then washed and the salt-free nucleotides recovered by shaking the resin with 25 ml of 10 *M* formic acid. The acid was removed by lyophilization. Recovery was in the range of 92 to 98%. Neal and Florini [130] developed a method for desalting small volumes of solution with Sephadex in a double centrifuge device. From 0.5 to 5.0 ml of solution could be desalted rapidly by centrifuging the solution through Sephadex G-25 medium beads (preswollen with 0.1 *M* ammonium sulfate) contained in a 40-ml polyallomer centrifuge tube pierced by three holes in the bottom and inserted in a 50-ml centrifuge tube. Good recoveries were obtained as long as the sample/bed volume ratio was between 0.17 and 0.27.

Derr et al. [131] have shown that there is an interaction between aminopurines or aminopyrimidines and fluorescent thin-layer plates. Apparently, the manganese-activated zinc silicate reacts with hydrochloric acid to give a nonfluorescent compound which retards the mobility of purine bases and may destroy them. Thus in applying samples they recommend the use of 0.05 *N* or lower concentration of hydrochloric acid, depending on the number of applications to be made at the spot.

Muni et al. [132] chromatographed the trimethylsilylated purines and pyrimidines on silic gel.

References

1. G. Pataki, "Thin-Layer Chromatography of Nucleic Acid Bases, Nucleosides, Nucleotides, and Related Compounds," in *Advances in Chromatography*, Vol. 7, J. C. Giddings and R. A. Keller, Ed., Marcel Dekker, New York, 1968, p. 47.

1a. K. Randerath and E. Randerath, "Thin-Layer Separation Methods for Nucleic Acid Derivatives," in *Methods in Enzymology*, Vol. XII A, L. Grossman and K. Moldave, Eds., Academic Press, New York, 1967, p. 323.

2. K. H. Scheit, "Layer-Chromatography of Oligonucleotides," in *Progress in Thin-Layer Chromatography and Related Methods*, Vol. 1, A. Niederwieser and G. Pataki, Eds., Ann Arbor-Humphrey Science, Ann Arbor, Mich., 1970, p. 197.

3. J. J. Saukkonen, *Chromatogr. Rev.*, **6**, 63 (1964).

4. K. Randerath, *Angew. Chem.*, **73**, 674 (1961).

5. *Ibid.*, p. 436.

6. R. L. Scheig, R. Annunziata, and L. A. Pesch, *Anal. Biochem.*, **5,** 291 (1963).
7. O. Cerri and G. Maffi, *Boll. Chim. Farm.*, **100,** 951 (1961).
8. A. Massaglia, U. Rosa, and S. Sosi, *J. Chromatogr.*, **17,** 316 (1965).
9. J. F. Tomashefski, Jr., R. J. Barrios, and O. Sudilovsky, *Anal. Biochem.*, **60,** 589 (1974).
10. J. Smrt, *Collect. Czech. Chem. Commun.*, **40,** 1043 (1975).
11. K.-H. Scheit, *Biochim. Biophys. Acta,* **134,** 218 (1967).
12. E. F. Nowoswiat and M. S. Poonian, *J. Chromatogr.*, **72,** 217 (1972).
13. E. D. Verdlov, G. S. Monastyrskaya, L. I. Guskova, T. L. Levitan, V. I. Sheichenko, and E. I. Budowsky, *Biochim. Biophys. Acta,* **340,** 153 (1974).
14. S. Neskow, T. Miyazaki, T. Khwaja, R. B. Meyer, and C. Heidelberger, *J. Med. Chem.*, **16,** 524 (1973); through *CAMAG Bibliog. Serv.*, **31,** 21 (1973).
15. J. Cadet and R. Téoule, *J. Chromatogr.*, **76,** 407 (1973).
16. E. Wimmer and M. E. Reichmann, *Nature,* **221,** 1122 (1969).
17. J. J. Michniewicz, C. P. Bahl, K. Itakura, N. Katagiri, and S. A. Narang, *J. Chromatogr.*, **85,** 159 (1973).
18. T. W. Munns and H. F. Sims, *J. Chromatogr.*, **111,** 403 (1975).
19. J. D. Upton, *J. Chromatogr.*, **52,** 171 (1970).
20. S. Hynie, *J. Chromatogr.*, **76,** 270 (1973).
21. M. Hawrylshyn, B. Z. Senkowski, and E. G. Wollish, *Microchem. J.*, **8,** 15 (1964).
22. K. Randerath and H. Struck, *J. Chromatogr.*, **6,** 365 (1961).
23. F. Di Jeso, *J. Chromatogr.*, **32,** 269 (1968).
24. K. Randerath, *Biochem. Biophys. Res. Commun.*, **6,** 452 (1962).
25. K. Randerath, *Nature,* **205,** 908 (1965).
26. K. Randerath and E. Randerath, *J. Chromatogr.*, **82,** 59 (1973).
27. K. Randerath, E. Randerath, L. S. Y. Chia, and B. J. Nowak, *Anal. Biochem.*, **59,** 263 (1974).
28. S. A. Narang and J. J. Michniewicz, *Anal. Biochem.*, **49,** 379 (1972).
29. G. R. Bjoerk and I. Stevensson, *Biochim. Biophys. Acta.*, **138,** 430 (1967).
30. M. Jacobson and C. Hedgcoth, *Anal. Biochem.*, **34,** 459 (1970).
31. F. Harada, F. Kimura, and S. Nishimura, *Biochim. Biophys. Acta,* **195,** 590 (1969).
32. J. E. Ciardi and E. P. Anderson, *Anal. Biochem.*, **22,** 398 (1968).
33. J. Holguin-Hueso and R. Cardinaud, *J. Chromatogr.*, **66,** 388 (1972).
34. G. N. Mahapatra and O. M. Friedman, *J. Chromatogr.*, **11,** 265 (1963).
35. K. Keck and U. Hagen, *Biochim. Biophys. Acta,* **87,** 685 (1964).
36. R. L. Bieleski, *Anal. Biochem.*, **12,** 230 (1965).
37. H. G. Gassen, *J. Chromatogr.*, **39,** 147 (1969).
38. K. W. Mundry and H. Priess, *Biochim. Biophys. Acta,* **269,** 225 (1972).
39. E. Tyihák, *J. Chromatogr.*, **14,** 125 (1964).
49. M. E. Wright and D. G. Satchell, *J. Chromatogr.*, **55,** 413 (1971).
41. R. T. Mayer, G. M. Holman, A. C. Bridges, *J. Chromatogr.*, **90,** 390 (1974).
42. K. Randerath, *Angew. Chem. Int. Ed. Engl.*, **1,** 435 (1962).
43. K. Randerath, *Nature,* **194,** 768 (1962).

44. T. A. Dyer, *J. Chromatogr.*, **11,** 414 (1963).
45. R. G. Coffey and R. W. Newburgh, *J. Chromatogr.*, **11,** 376 (1963).
46. P. D. Swanson, *J. Neurochem.*, **15,** 1159 (1968).
47. R. G. Stickland, *Anal. Biochem.*, **10,** 108 (1965).
48. R. G. Stickland, *J. Chromatogr.*, **23,** D8 (1966).
49. Y. Lapidot, I. Barzilay, and D. Salomon, *Anal. Biochem.*, **49,** 301 (1972).
50. K. Jaroszewicz, M.-J. Degueldre-Guillaume, and C. Liébecq, *J. Chromatogr.*, **24,** 279 (1966).
51. P. Grippo, M. Iaccarino, M. Rossi, and E. Scarano, *Biochim. Biophys. Acta*, **95,** 1 (1965).
51a. M. D. Kirnos, V. K. Vasilyev, and B. F. Vanyushin, *J. Chromatogr.*, **104,** 113 (1975).
52. K. Randerath, *Nature*, **194,** 768 (1962).
53. M. N. S. Nayar, *Life Sci.*, **3,** 1307 (1964).
54. R. D. Bauer and K. D. Martin, *J. Chromatogr.*, **16,** 519 (1964).
55. N. S. Panteleeva, *Vestn. Leningr. Univ. Ser. Biol.*, **19,** 73 (1964).
56. C. P. Dietrich, S. M. C. Dietrich, and H. G. Pontis, *J. Chromatogr.*, **15,** 277 (1963).
57. K. Randerath, *Biochim. Biophys. Acta*, **61,** 852 (1962).
58. K. Randerath, *Experientia*, **20,** 406 (1964).
59. J. M. Foster, H. Abbott, and M. L. Terry, *Anal. Biochem.*, **16,** 149 (1966).
60. G. Santini and V. Ulrich, *J. Chromatogr.*, **49,** 560 (1970).
61. E. Koenigk, *J. Chromatogr.*, **37,** 128 (1968).
61a. R. N. Nazar, H. G. Lawford, and J. T.-F. Wong, *Anal. Biochem.*, **35,** 305 (1970).
62. K. Randerath, *Biochim. Biophys. Acta*, **76,** 622 (1963).
63. K. Randerath and E. Randerath, *J. Chromatogr.*, **16,** 111 (1964).
64. L. W. Gonzales and S. E. Geel, *Anal. Biochem.*, **63,** 400 (1975).
65. D. W. Jacobsen and J. Kirchner, *Anal. Biochem.*, **26,** 474 (1968).
66. G. H. Buteau, Jr. and J. E. Simmons, *Anal. Biochem.*, **37,** 461 (1970).
67. K. Randerath, *Angew. Chem. Int. Ed. Engl.*, **1,** 553 (1962).
68. E. Randerath and K. Randerath, *J. Chromatogr.*, **16,** 126 (1964).
69. H. P. Raaen and F. E. Kraus, *J. Chromatogr.*, **35,** 531 (1968).
70. P. D. Fullerton and L. R. Finch, *Anal. Biochem.*, **29,** 544 (1969).
71. M. Cashel, R. A. Lazzarini, and B. Kalbacher, *J. Chromatogr.*, **40,** 103 (1969).
72. G. Pataki and A. Niederwieser, *J. Chromatogr.*, **29,** 133 (1967).
73. O. Schettino and M. I. La Rotonda, *Boll. Soc. Ital. Biol. Sper.*, **44,** 1989, (1968).
74. E. Randerath and K. Randerath, *J. Chromatogr.*, **31,** 485 (1967).
75. *Ibid.*, p. 500.
76. G. Volckaert and W. Fiers, *Anal. Biochem.*, **62,** 573 (1974).
77. G. Volckaert, W. Min Jou, and W. Fiers, *Anal. Biochem.*, **72,** 433 (1976).
78. E. M. Southern, *Anal. Biochem.*, **62,** 317 (1974).
79. M. Litt and V. Hancock, *Biochemistry*, **6,** 1848 (1967).
80. G. Weimann and K. Randerath, *Experientia*, **19,** 49 (1963).

81. K. Randerath and G. Weimann, *Biochim. Biophys. Acta,* **76,** 129 (1963).
82. K. Randerath and E. Randerath, *Angew. Chem. Int. Ed. Engl.,* **3,** 442 (1964).
83. E. Randerath and K. Randerath, *J. Chromatogr.,* **10,** 509 (1963).
84. T. Wieland and H. Determann, *Experientia,* **18,** 431 (1962).
85. T. Hashizume and Y. Sasaki, *Agr. Biol. Chem. (Tokyo),* **27,** 881 (1963); through *Chem. Abstr.,* **60,** 12347 (1964).
86. G. Tortolani and M. E. Colosi, *J. Chromatogr.,* **70,** 182 (1972).
87. J. Tomasz, *J. Chromatogr.,* **84,** 208 (1973).
88. *Ibid.,* **101,** 198 (1974).
89. J. Tomasz, *Anal. Biochem.,* **68,** 226 (1975).
90. J. Tomasz, *Acta Biochim. Biophys. Acad. Sci. Hung.,* **9,** 87 (1974).
91. J. Tomasz and T. Farkas, *J. Chromatogr.,* **107,** 396 (1975).
92. K. Dose and S. Risi, *Z. Anal. Chem.,* **205,** 394 (1964).
93. L. M. Weiner and B. Zak, *Clin. Chim. Acta,* **9,** 407 (1964).
94. T. Borkowski, J. Wojcierowski, and S. Kulesza, *Anal. Biochem.,* **27,** 58 (1969).
95. R. Tsanev, D. Staynov, L. Kokileva, and I. Mladenova, *Anal. Biochem.,* **30,** 66 (1969).
96. B. Zak, L. M. Weiner, and E. Baginski, *J. Chromatogr.,* **20,** 157 (1965).
97. L. M. Weiner, B. Welsh, and B. Zak, *J. Chromatogr.,* **27,** 512 (1967).
98. K.-T. Wang and Po.-H. Wu, *J. Chromatogr.,* **38,** 153 (1968).
99. A. C. Peacock and C. W. Dingman, *Biochemistry,* **6,** 1818 (1967).
100. T. Ikemura and J. Dahlberg, *J. Biol. Chem.,* **248,** 5024 (1973).
101. T. Varricchio and H. J. Ernst, *Anal. Biochem.,* **68,** 485 (1975).
102. R. De Wachter and W. Fiers, *Anal. Biochem.,* **49,** 184 (1972).
103. P. G. Righetti and J. W. Drysdale, *Ann. N.Y. Acad. Sci.,* **209,** 163 (1973).
104. J. E. Caton and G. Goldstein, *Anal. Biochem.,* **42,** 14 (1971).
105. P. G. N. Jeppesen, *Anal. Biochem.,* **58,** 195 (1974).
106. R. K. Fujimura, *Anal. Biochem.,* **36,** 62 (1970).
107. A. Schweiger and H. Guenther, *J. Chromatogr.,* **19,** 201 (1965).
108. J. P. Monjardino, *Anal. Biochem.,* **28,** 313 (1969).
109. G. G. Brownlee and F. Sanger, *Europ. J. Biochem.,* **11,** 395 (1969).
110. H. Boernig and C. Reinicke, *Acta Biol. Med. Ger.,* **11,** 600 (1963).
111. C. M. Chang, S. A. Johnson, J. R. Vercellotti, and J. J. Barboriak, *Anal. Biochem.,* **23,** 35 (1968).
112. R. W. Frei, H. Zuercher, and G. Pataki, *J. Chromatogr.,* **43,** 551 (1969).
113. H. P. Koof and R. V. Noronha, *Z. Anal. Chem.,* **250,** 124 (1970).
114. R. W. Frei, H. Zuercher, and G. Pataki, *J. Chromatogr.,* **45,** 284 (1969).
115. V. T. Lieu, M. M. Frodyma, L. S. Higashi, and L. H. Kunimoto, *Anal. Biochem.,* **19,** 454 (1967).
116. G. Pataki and E. Stracky, *Int. Symp. Chromatogr. Electrophoresis, 4th, 1967,* Ann Arbor Science, Ann Arbor, Mich., 1968, p. 317.
117. G. Pataki and A. Kunz, *J. Chromatogr.,* **23,** 465 (1966).
118. T. A. C. Gissel and U. Stoll, *Fleischwirtschaft,* **52,** 771 (1972).
119. K. Randerath and E. Randerath, *Anal. Biochem.,* **28,** 110 (1969).

120. K. Randerath and E. Randerath, *Experientia,* **24,** 1192 (1968).
121. E. Randerath, C.-T. Yu, and K. Randerath, *Anal. Biochem.,* **48,** 172 (1972).
122. L.-L. S. Y. Chia, K. Randerath, and E. Randerath, *Anal. Biochem.,* **55,** 102 (1973).
123. F. N. Chang, *Anal. Biochem.,* **63,** 371 (1975).
124. G. G. Brownlee and F. Sanger, *Eur. J. Biochem.,* **11,** 395 (1969).
125. M. W. Johnson and G. O. Osuji, *Anal. Biochem.,* **70,** 45 (1976).
126. N. Debreceni, M. T. Behme, and K. Ebisuzaki, *Biochem., Biophys. Res. Commun.,* **41,** 115 (1970).
127. T. M. Dougherty and A. I. Schepartz, *J. Chromatogr.,* **40,** 299, (1969).
128. R. C. Van den Bos, V. C. H. F. De Regt, and R. C. Brand, *Anal. Biochem.,* **41,** 293 (1971).
129. M. Caron and H. Dugas, *J. Chromatogr.,* **101,** 228 (1974).
130. M. W. Neal and J. R. Florini, *Anal. Biochem.,* **55,** 328 (1973).
131. R. F. Derr, C. S. Alexander, and H. T. Nagasawa, *J. Chromatogr.,* **21,** 146, (1966).
132. I. A. Muni, C. H. Altschuler, and J. C. Neicheril, *Anal. Biochem.,* **50,** 354 (1972).

Chapter **XXV**

PESTICIDES

The abundance of publications on the application of thin-layer chromatography to the separation and detection of pesticides is not surprising because of the need for rapid, sensitive methods for determining residues in food products and in some cases for toxicological analysis.

Prior to the chromatography of pesticides they must be extracted from their environment, whether it be food, soil, or an animal. If the history of the sample is known, then the solvent extraction procedure recommended for the particular pesticide or combination of pesticides can be used. However, it is not enough to know the history of pesticide spraying of a particular plant, because other pesticides from previous sprays may be present in the soil to be picked up by the plant. Similarly, an animal treated with one type of pesticide may have eaten alfalfa or grain treated with another pesticide. Among the solvents used for extraction are: acetone, methanol, petroleum ether, benzene, propanol, chloroform, methylene chloride, ethyl acetate, acetonitrile, and various combinations of these. Schnorbus and Phillips [1] have shown propylene carbonate to be an efficient solvent for pesticide extraction. Inefficient sample extraction can be a problem in pesticide analysis [2], especially where plant or animal tissue has to be disintegrated [3]. Probably the use of ultrasonics would be of advantage here, although Hill and Stobbe [4] found no advantage in its use in the extraction of soils as far as recovery was concerned. It did have a time advantage.

Many times the extraction efficiency is checked by adding known amounts of pesticide to the plant or animal tissue, and then an extraction and determination is carried out. This does not check the efficiency of the extraction procedure, although it does check for losses in the other steps of the analysis.

Usually some purification of the sample extract is required prior to thin-layer chromatography, for without proper cleanup, sensitivity is reduced, migration rates of the pesticides are affected so that they differ from the reference standards, resolution is decreased, and spot streaking and distortion occur. This is especially true in the case of fatty samples.

One of the commonly used methods (especially where there is a high content of fat or wax in the sample) for cleaning up pesticide residues is by solvent partition. A hydrocarbon such as hexane or petroleum ether is paired off with a polar solvent like methanol, acetonitrile, acetone, or dimethyl sulfoxide that has been saturated with the hydrocarbon solvent. Solvent ratios vary with the sample type. After the solvent partition separation, the pesticides are in the polar phase and the fats and waxes are in the hydrocarbon phase. The pesticides are then transferred back to fresh hydrocarbon solvent by addition of a sodium chloride solution.

Other cleanup methods include *(a)* passage through a column of Florisil PR [5]. This is a synthetic magnesium silicate which has been heated for 3 hr at 1250°C. The pesticide residues dissolved in petroleum ether are placed on the column and eluted with increasing percentages of ethyl ether in petroleum ether; Beckman and Garber [6] used successive portions of benzene, ethyl ether–benzene (1:2), acetone, and methanol to elute organophosphorus pesticides. Other methods are: *(b)* magnesium oxide–Celite columns; *(c)* acidic–Celite columns (for chlordane and other chlorinated hydrocarbons); *(d)* alkaline hydrolysis where the pesticides are alkali stable; *(e)* carbon columns (not all charcoals are effective) [7, 7a]; *(f)* sweep co-distillation [8]; *(g)* use of a simplified gas chromatographic system [9]; *(h)* Sephadex LH-20 packed in acetone for removal of pigments [10]; *(i)* silica gel columns [11]; *(j)* wedge-shaped layers [12] which use the thick part of the layer for retaining large quantities of impurities (wedge-shaped mold available from Kontes), or a similar idea using plates grooved a short distance from one end thus providing for a coating eight times thicker than the remainder of the plate (available as Chromolay grooved plates from May and Baker); *(k)* multiple layers—a dual layer of silica gel–alumina may be developed along the silica gel to remove interfering co-extractives, then the plate is turned through 90° and developed with a more polar solvent across the alumina to separate the pesticides, or a cellulose–charcoal layer alongside of a silica gel layer so that the former can be used to retain the co-extractives while the silica is used to separate the pesticides in a unidimensional development; and *(l)* vacuum sublimation [13]. Thornburg [14] discussed the preparation and extraction of samples prior to analysis, and McKinley et al. [15] examined cleanup procedures.

After the sample is purified it needs to be concentrated. Concentration is usually done in the Kuderna–Danish evaporator down to about 5 ml, and then it can be concentrated further on a micro-Snyder column. It should never be taken to dryness because the maximum losses are apt to occur during evaporation to low volume or to dryness [16].

Walker and Beroża [17] have examined the separation of 62 pesticides in 19 solvent systems; the R_f values in 10 of these are given in Table 25.1. The solvent combinations for the table are chloroform, chloroform–ether (9:1), chloroform–acetone (9:1), chloroform–acetic aicd (9:1), benzene–ethyl acetate (9:1), benzene–acetone (9:1), benzene–acetic acid (9:1), hexane–acetone (8:2), hexane–methanol (9:1), and hexane–acetic acid (9:1).

In order to visualize the compounds, the completed chromatograms were treated with test T-130 which detects compounds that react with bromine giving yellow spots on a pink background of eosin. This was followed by test T-231. More compounds became visible and were marked before exposing to ultraviolet light for 7 min. This final treatment brought out the remaining spots, except for a few spots which disappeared on exposure to the ultraviolet light.

Suzuki et al. [18] using a two-step solvent system on alumina layers divided approximately 100 pesticides into six groups. Group I contained organochlorine compounds, II organochlorine and phosphorus compounds, III organophosphorus pesticides, IV and V carbamate and triazine herbicides, and VI phenoxy herbicides. The best two-step solvent system was hexane followed by hexane–benzene (7:3). Twenty-one pesticides in the second group and 15 pesticides in the third group were subjected to stepwise elution on a column of silica gel, and the eluates were further separated by thin-layer chromatography and analyzed by gas chromatography [19]. By the use of various solvent mixtures in a stepwise elution from a column of silica gel, 100 pesticides were divided into five groups which were subjected to two-dimensional thin-layer chromatography and also to gas chromatographic analysis [20–21]. Thier and Bergner [21a] used paper and thin-layer chromatography for the separation and identification of important pesticides. The pesticides were divided into three groups. Sandroni and Schlitt [22] developed a technique for screening organochlorine and organophosphorus pesticide residues in

Table 25.1 $R_f \times 100$ Values of Pesticides on Silica Gel G with Various Solvent Systems[a]

Pesticide	Chloro-form	Chloro-form–Ether (9:1)	Chloro-form–Acetone (9:1)	Chloro-form–Acetic Acid (9:1)	Ben-zene–Ethyl Acetate (9:1)	Ben-zene–Acetone (9:1)	Ben-zene–Acetic Acid (9:1)	Hex-ane–Acetone (8:2)	Hex-ane–Meth-anol (9:1)	Hex-ane–Acetic Acid (9:1)
Aldrin[b]	80	78	83	94	73	78	80	76	69	56
Aramite	62[b]	68	79	62	59	71	56[b]	52[d]	40[b]	28
				84				56[b]		
Bayer 25141[c]	12	18	30	62	2	9[b]	18	7	8	2
						13[d]				
Binapacryl (Morocide)[b]	55	69	70	85	66	68	60	41	42	29
Captan[b]	44	56	67	73	37	57	45	37	15	13
Carbophenothion										
(Trithion)	73	75	83	92	73	76	78	60	52	38
Chlordane[b]	79	76	83	93	73	78	78	X[e]	65	41
										53
Chlorthion	0[b]	0[b]	0[b]	0[b]	0	0	23[b]	0	7[b]	2[b]
				85[b]	14	28[b]	61[b]	15[a]		22[b]
					59	67[b]				
Co-Ral	46	61	67	78	53	61	46	23	19	14
DDE,p,p′[b]	69	70	74	90	74	72	68	64	64	54
DDT,o,p[b]	68	69	72	90	72	70	68	58	58	48
DDT,p,p′[b]	66	70	72	90	72	70	68	58	58	46
Delnav	0[b]	0[b]	0	4[b]	0	0[b]	0[b]	0	6[b]	0
	64	72	83	66	58	73	44[b]	43	18[b]	26
				78[b]			53		33	
				86[d]						
				90[b]						

Table 25.1 **(Continued)**

Pesticide	*Chloro-form*	*Chloro-form–Ether* (9:1)	*Chloro-form–Acetone* (9:1)	*Chloro-form–Acetic Acid* (9:1)	*Ben-zene–Ethyl Acetate* (9:1)	*Ben-zene–Acetone* (9:1)	*Ben-zene–Acetic Acid* (9:1)	*Hex-ane–Acetone* (8:2)	*Hex-ane–Meth-anol* (9:1)	*Hex-ane–Acetic Acid* (9:1)
Demeton	25	33	54	65	8	30	35[b]	31	14	14
	66[b]	38[b]	81[b]	85[b]	64	73	60	61[b]	52[b]	34
			84[d]							
Demeton, thiol isomer	26	33[d]	54	68	8	31	34[b]	31	16	13
		37[b]								
Demeton, thiol isomer sulfone	8	13[b]	24[b]	49[b]	0	6	X	7	5[b]	0[b]
Demeton, thiol isomer sulfoxide	0[b]	X	5	23[b]	0	0	X	0	7[b]	0
										11[b]
Demeton, thiono isomer	26	38	0[b]	0[b]	0	28	0[b]	32	8	32
			53	67[b]	9	74	33[b]		18	
			81	84	66		60		51	
Diazinon	0[b]	0[d]	0	2[b]	0	0	0	0	6[b]	0
			48	67[b]	5[b]	22[b]	28[b]	28[b]	13[b]	10[b]
			56		10[b]	33[b]	35[b]			
Dicapthon	66	72	77	84	62	70	65	38	24	45
Dieldrin[b]	64	76	75	79	65	73	71	X	52	37
Dimethoate	8	11[b]	22	47	2	8	12	7	7	0
Dimethoate, oxygen analog	X	0[b]	5[b]	17[b]	0	0	X	0	6	0[b]
Di-syston	63	80	80	83	69	76	72	61	54	41
Di-syston, sulfone	29	49	60	64[b]	22	45	40	23	8	5
				68[d]						

Di-syston, sulfoxide	9	13	25	59	0	6	19	5	7	3
				83[b]	69[b]	75[b]	71[b]		54[b]	40[b]
Dylox (Dipterex)[b]	3	4	10	39	0	3	17	4	7	4
Endosulfan (Thiodan)[b]	73	76	82	91	59	69	65	X	32	26
					71	77	76	X	58	40
Endosulfan, low mp[b]	72	75	82	91	69	76	75	X	56	40
Endosulfan, high mp[b]	59	64	76	85	57	67	64	X	32	27
Endrin[b]	66	78	78	83	68	76	73	X	57	39
EPN	60	78	78	81	68	73	70	48	41	29
Eradex	66	71	79	90	63	73	64	59[d]	38	35
								63[b]	56	
Ethion	66	80	80	85	71	77	74	57	50	37
										44[b]
Genite[b]	63	76	76	81	67	73	70	X	40	30
Guthion	33	55	63	65	33	54	45	27	13	7
Heptachlor[b]	76	80	79	86	75	78	80	X	65	57
Heptachlor epoxide[b]	68	78	78	84	69	74	75	X	56	41
Karathane[b]	66	78	79	85	73	76	75	58	49	X
Kelthane[b]	68	74	76	88	68	74	70	51	35	33
Kepone[b]	X	15	25	37	15	X	20	17	X	6
Lindane[b]	73	74	76	89	68	74	73	X	45	36
Malathion	40	63	72	77	46	63	44	35	25	22
Menazon[f]	0	0	0	35	0	0	8	0	4	0
Methoxychlor[b]	68	73	78	88	66	73	68	45	49	33
Methyl parathion	60	70	75	84	59	67	61	34	29	22
Mirex[b]	81	77	82	92	75	79	81	X	70	62
Naled (Dibrom)[b]	33	45	58	66	0	0	39	25	5	0
					18	36			12	11
Parathion	64	73	78	86	65	71	67	44	38	30
									41[b]	

Table 25.1 (Continued)

Pesticide	Chloroform	Chloroform–Ether (9:1)	Chloroform–Acetone (9:1)	Chloroform–Acetic Acid (9:1)	Benzene–Ethyl Acetate (9:1)	Benzene–Acetone (9:1)	Benzene–Acetic Acid (9:1)	Hexane–Acetone (8:2)	Hexane–Methanol (9:1)	Hexane–Acetic Acid (9:1)
Phorate	71	75	79	89	68	75	72	60	53	42
Phosdrin[b]	X	23	31	57	3	9	11	7	6	2
			41	63		16	18	11		
Phosphamidon[b]	4	9	17	48	0	5	7	4	4	0
Phostex	74	77	81	91	71	77	76	57	4	42
									52	
Schradan	0[b]	0[b]	4[b]	X	0[b]	0[b]	X	X	4	X
Sevin	20[b]	36[b]	51[b]	65[b]	19[b]	34[b]	35[b]	20[b]	6	6[b]
Sulfotepp	56	68	72	85	64	66	57	42	44	32
Sulphenone[b]	X	64	71	80	X	X	51	X	X	18
TDE[b]	X	78	82	92	75	79	83	X	X	41
Tetradifon (Tedion)[b]	66	75	80	89	70	77	75	X	45	31
Toxaphene[b]	76	76	83	93	74	78	81	X	59	40
										53
VC-13	63	72	73	90	69	68	65	52	55	38
Zectran	28[b]	43[b]	60[b]	12[b]	23	44[b]	2	38[b]	10	0

[a] From K. C. Walker and M. Beroza [17]; reproduced with permission of the authors and Assoc. of Agr. Chemists.
[b] Spot appears after exposing sprayed plate to ultraviolet.
[c] *O,O*-diethyl *O-p*-methylsulfinylphenyl phosphorothioate.
[d] Spot disappears after exposing sprayed plate to ultraviolet (For details of detecting reagents see text).
[e] X = No visible spot, too weak, or streaking.
[f] *S*-[(4,6-diamino-*s*-triazin-2-yl)methy] *O,O*-dimethyl phosphorodithioate.

vegetables. For chlorine compounds an Alumina DS-5 (CAMAG) layer prepared with 0.4% aqueous silver nitrate solution was used. After the layer was spotted, it was conditioned at 18% relative humidity for 1 hr by exposing it to 60.6% sulfuric acid solution. Development was continuous with cyclohexane for 90 min. The compounds were visualized by exposure to moisture and then to short-wave ultraviolet light. The organophosphorus pesticides were separated on silica gel G layers placed in a Vario-KS-Chamber using a humidity gradient of 72, 64, 58, 47, 13, 3, 3, 3, 3, and 3% relative humidities in order from the origin to the top of the plate. Methylene chloride was used as a solvent and the spots were detected with the cholinesterase inhibition test. The compounds tested were 17 of the more common pesticides. Schechter and Getz [23] commented on the screening and analysis of multiple pesticide residues.

A number of general reviews have been published including those of Kovacs [24] Abbott and Thompson [25], Takeshita [26], Hill [26a], Sherma and Zweig [27], and on quantitative analysis by MacNeil and Frei [27a].

1 CHLORINATED COMPOUNDS

Holmes and Wood [28] have published a technique for removing interfering substances from vegetable extracts prior to the separation of organochlorine pesticide residues by thin-layer or gas chromatography. To a warm solution of 0.75 g of silver nitrate in 0.7 ml of water is slowly added 4 ml of acetone; this is then stirred with 10 g of alumina (containing 7% moisture). The acetone is then removed. One gram of this "prepared" alumina is placed in a 150 × 8 mm i.d. column as a slurry in *n*-hexane. As the hexane is drained from the column and the meniscus reaches the surface of the alumina, the partially cleaned sample is added in 1.0 ml of *n*-hexane. The sample is then eluted with 11 ml of hexane leaving the carotenes as a red and the sulfur compounds as a black band on the column. Preliminary cleanup should be carried out by one of the methods mentioned above. Woods [28a] recommends dimethyl sulfoxide (DMSO) for the extraction of organochlorine pesticides from fatty materials. It dissolves less fat and the extract can be cleaned by absorbing on a column of deactivated Florisil and then eluting the pesticides with hexane.

Suzuki et al. [29] determined the detection limits of 80 organochlorine pesticides by ultraviolet before and after treating with iodine and *o*-toluidine reagents. The pesticides were then divided into three groups according to the detection limit: first group with a detection limit of 1 to 3 μg, second group with 5 to 10 μg, and third group with over 25 μg.

Abbott et al. [30] described a method for screening for organochlorine pesticides in fats and vegetable. After extraction of the samples and concentration of the extract the latter was purified on an alumina column or by DMSO extraction. Separation took place on alumina G layers prepared with 0.4% aqueous silver nitrate. Solvent for the work was hexane.

Petrowitz [31, 32] used silica gel G layers for chromatographing DDT, four isomers of hexachlorocyclohexane, aldrin, isodrin, dieldrin, endrin, chlordane, and pentachlorophenol. Various solvents were used. Petroleum ether (50 to 70°C) served to separate DDT and the four hexachlorocyclohexanes (Table 25.2). Cyclohexane separated aldrin (R_f = 0.58) from its stereoisomer isodrin (R_f = 0.48) and dieldrin (R_f = 0.57) from its stereoisomer endrin (R_f = 0.48). Aldrin and its isomers were detected by spraying with 0.1 *N* potassium permanganate solution, whereas the DDT and hexachlorocyclohexanes were made visible by spraying first with monoethanolamine and then heating for 20 min at 100°C. This was then followed by a spray of 0.1 *N* silver nitrate–nitric acid (d = 1.40) (10:1). After the second spray the plates were exposed to ultraviolet light or to sunlight. Using three different solvent systems, Petrowitz [33] has also compared the effect of various acidified silica gels on the separation of chlorinated contact insecticides (see Table 3.5). Although dieldrin cannot be separated from aldrin in the silica gel G–cyclohexane system, it can be readily separated on aluminum oxide G with hexane or heptane [34–36] and on silica gel with hexane or heptane [35–37], as well as with cyclohexane–acetone mixtures (38) and cyclohexane–liquid paraffin–dioxane mixtures [35].

Deters [39] has applied the method of Petrowitz [32] to the isolation and determination of pentachlorophenol using acidified silica gel layers prepared with 0.05 *N* oxalic acid instead of water. Using chloroform for the development, the pentachlorophenol had an R_f value of approximately 0.50. Semiquantitative determinations could be made by measuring the area of the spot; for quantitative work the spots were eluted and measured by ultraviolet spectroscopy.

Yamamura et al. [40] have examined the separation of aldrin, dieldrin, and endrin on silica gel layers using different solvent systems. Yamamura and Niwaguchi [38] also examined the same group of compounds along with thiodan in six different solvent systems on starch-bound silicic acid layers. The most suitable solvent for the separation of these four pure compounds was cyclohexane–acetone (9:1), but when used for the separation of technical mixtures it caused tailing of the spots. In order to separate these technical samples, cyclohexane–acetone (23:2) was found to be the best solvent.

Table 25.2 $R_f \times 100$ Values of Some Chlorinated Hydrocarbons on Silica Gel in Various Solvents. Development Distance 10 cm[a]

Compound	*Chloroform*	*Benzene–Methanol* (95:5)	*Petroleum Ether* (50 to 70°C)
DDT	91	93	44
α-Hexachlorocyclohexane	92	96	28
β-Hexachlorocyclohexane	—	—	<2
γ-Hexachlorocyclohexane	90	94	19
δ-Hexachlorocyclohexane	87	96	10

[a] From H.-J. Petrowitz [29]; reproduced with permission of the author and Dr. Alfred Heuthig Verlag.

Morley and Chiba [41] examined a group of five insecticides on silica gel using *n*-hexane as the developing solvent with the following R_f values: aldrin 0.69, *p, p'*-DDE 0.60, heptachlor 0.52, *o,p*-DDT 0.46, and *p,p'*-DDT 0.38. These results were obtained without any preliminary cleanup of the extract obtained from wheat grain. This is in contrast to gas chromatography where some cleanup is necessary prior to inserting the sample into the column. Combination thin-layer and gas chromatography can be applied by first chromatographing on thin layers and then eluting the separated compounds for determination by gas chromatography. Taylor and Fishwick [42] have used loose layers to separate organochlorine pesticides into two groups prior to gas chromatography.

Kawashiro and Hosogai [37] examined a group of reagents for the detection of chlorinated pesticides and preferred a 0.5% alcoholic solution of *o*-toluidine or *o*-dianisidine. After spraying these reagents, the plate was irradiated with short-wave ultraviolet light, whereupon the chlorinated pesticide spots showed up as green spots on a white background. The sensitivity of these two reagents ranges from 0.5 to 3 γ, depending on the compounds involved. Using silica gel with *n*-hexane the following R_f values were obtained: aldrin 0.58, chlordane 0.50, 2,4-dichlorophenoxyacetic acid 0.0, DDT 0.45, dieldrin 0.10, endrin 0.10, heptachlor 0.55, lindane 0.15, pentachlorophenol 0.0, and thiodan 0.10. Anthracene was used as a reference standard.

Kovacs [36] compared a gas-chromatographic and a thin-layer method for recovering chlorinated pesticide residues from samples of food. For the thin-layer work he used both aluminum oxide G and silica gel (Adsorbosil-1) with *n*-heptane as the developing solvent. Detection Test T-231 was used. The sensitivity of this reagent is very high, 0.05 μg for perthane and BHC, 0.1 μg for toxaphene and chlordane, and 0.01 μg for the remaining compounds. In order to obtain the maximum sensitivity, the plates should be washed prior to spotting the sample. This is accomplished by developing twice in distilled water, and can be carried out by removing ½ in. of the adsorbent from the lower edge of the plate and then using a filter-paper wick to carry the water to the adsorbent layer. After the final washing the plates are dried at 75°C for 30 min. The thin-layer chromatographic method compared very favorably with gas chromatography in the examination of pesticide residues in food products, and "in many cases compounds were detected on thin-layer plates which were not detected by gas chromatography even though the sample size was approximately one-tenth that used for gas chromatography" [36]. Later Kovacs [43] spotted samples on aluminum oxide layers and then dipped the layers into a 25% solution of *N,N*-dimethylformamide in ethyl ether immediately before developing in isooctane. Prior to spraying the chromogenic reagent the plates were dried in a forced draft oven at 50°C for 10 min. Adams and Schecter [44] have applied the silver nitrate–2-phenoxyethanol detection method by incorporating silver nitrate in the silica gel plate and the 2-phenoxyethanol in the developing solution.

Thomas et al. [45], using three solvent systems developed by Kovacs, determined the relative R_f values (Table 25.3) of 44 chlorinated pesticides.

Table 25.3 R_S Values of Chlorinated Pesticides in Various Systems Referred to *p,p'*-TDE (1,1-Dichloro-2,2-bis(*p*-chlorophenyl)ethane)[a]

Pesticide	*System I*[b]	*System II*[c]	*System III*[d]
Hexachlorobenzene	2.7	1.7	—[e]
Aldrin	2.1	1.4	4.3
p,p'-DDE	2.0	1.4	3.4
Heptachlor	2.0	1.4	3.7
Chlordane	2.0, 1.8, 1.4, 1.2[f,g]	1.4, 1.3, 1.2, 1.1[f,g]	3.7, 3.4, 2.9, 1.6[f,g]
o,p'-DDT	1.9	1.3	3.0
PCNB	1.8	1.4	3.7
Perthane olefin	1.8	1.4	4.4
p,p'-TDE olefin	1.8	1.4	3.2
TCNB	1.7	1.3	3.2
Telodrin	1.7	1.4	3.5
Toxaphene	1.7, 1.2[g]	1.3, 1.2[g]	3.0, 2.3[g]
Strobane	1.7, 1.2[g]	1.3, 1.2[g]	3.0, 2.3[g]
p,p'-DDT	1.6	1.2	2.2
o,p'-TDE olefin	1.6	1.3	2.9, 2.2

Table 25.3 (Continued)

Pesticide	*System I*[b]	*System II*[c]	*System III*[d]
Chlorobenside	1.3 (gray)	1.2 (fuzzy gray)	2.0, 0.0
BHC	1.3, 1.1, 0.27, 0.10[f]	1.1, 0.92, 0.72, 0.25[f]	1.8, 1.3, 0.75, 0.30[f]
α-BHC	1.3	1.1	—[e]
Perthane	1.3	1.2	2.5
Lindane	1.1	0.92	1.3
o,p'-TDE	1.1	0.95	1.1
p,p'-TDE	1.0	1.0	1.00
Endosulfan	0.88, 0.07[f]	0.92, 0.24[f]	3.1, 0.0[f]
Ronnel	0.85	1.1	2.2
Heptachlor epoxide	0.71	1.0	2.4
Endrin	0.71	1.0	2.9
Dieldrin	0.52	0.90	2.8
Carbophenothion	0.42 (yellow)	1.0 (fuzzy yellow)	1.9
Methoxychlor	0.33, 0.27	0.79	0.85
β-BHC	0.27	0.72	—[e]
Ovex	0.18	0.76	0.61
Dichlone	0.16	0.72, 0.00[f]	0.00
Dyrene	0.15 (gray)	0.51	0.20
Tetradifon	0.11	0.82	0.90
δ-BHC	0.10	0.25	—[e]
Endrin ketone (Delta-Keto 153)	0.09	0.23 (very small)	0.56
Kelthane	0.06	0.28	0.00
Sulphenone	0.00 (large and fuzzy)	0.31 (yellow)	0.00
Captan	0.00 (sharp-edged gray)	0.09	0.25
Chlorobenzilate	0.00 (light)	0.05	0.25
Monuron	0.00 (large and dark)	0.00	0.00
Diuron	0.00	0.00 (dark spot)	0.00
Endrin aldehyde	0.00 (very small)	0.00 (very small)	0.23, 0.00[f]
Endrin alcohol	0.00	0.00	0.14, 0.00[f]

[a] From E. J. Thomas et al. [45]; reproduced with permission of the authors and The Elsevier Publishing Co.

[b] Air-dried (72 hr) aluminum oxide G, *n*-heptane in saturated system.

[c] Air-dried (72 hr) aluminum oxide G, acetone–*n*-heptane (2:98) in saturated system.

[d] Aluminum oxide neutral impregnated with 25% dimethylformamide in ether. Solvent: isooctane.

[e] Not determined.

[f] Most intense spot underlined.

[g] Leaves a streak with these major spots.

Table 25.4 $R_f \times 100$ Values of Organochlorine Pesticides in Various Adsorbent–Solvent Systems. Development Distance 15 cm[a]

Pesticide	Silica Gel–Alumina (1:1)		Silica Gel										Kiesel-guhr	Alumina	
	A	*B*	*C*	*D*	*E*	*F*	*G*	*H*	*I*	*J*	*K*	*L*	*K*	*K*	*L*
Aldrin	88	98	73	58	69	67	70	79	64	62	67	70	98	82	95
α-BHC	—	69	—	—	43	37	—	59	28	29	52	—	92	63	87
γ-BHC	—	58	—	—	37	27	—	47	18	19	46	—	94	55	78
p,p'-DDE	87	98	87	74	62	61	68	73	57	56	65	65	98	78	95
o,p'-DDT	71	90	74	50	58	54	62	71	46	48	59	50	97	73	89
p,p'-DDT	72	91	78	52	54	49	60	69	39	40	57	42	98	69	89
de-HCl-*p,p'*-TDE	85	98	88	67	62	61	72	76	53	51	49	53	98	75	93
p,p'-Dichlorobenzophenone	—	—	—	—	48	45	53	67	27	26	59	14	92	55	31
Dieldrin	69	58	53	30	48	41	46	63	48	54	65	12	88	52	37
Endosulfan A	—	—	—	—	52	47	63	61	35	31	58	17	94	64	65
Endosulfan B	—	—	—	—	—	—	—	—	—	—	12	2	86	9	4
Endrin	—	—	—	—	52	42	58	65	26	26	49	13	88	61	51
Heptachlor	82	98	69	48	62	61	65	73	53	52	65	58	88	78	95
Heptachlor epoxide	—	—	—	—	—	—	—	—	—	—	39	17	88	57	49
Methoxychlor	—	—	—	28	36	27	30	45	10	13	—	—	—	—	—
p,p'-TDE	66	77	58	67	46	33	45	59	26	28	52	25	92	57	71

[a] From D. C. Abbott, H. Egan, and J. Thomson [35]; reproduced with permission of the authors and The Elsevier Publishing Co.

[b] Solvent: (*A*) cyclohexane–liquid paraffin (20%), (*B*) cyclohexane–silicone oil (8%), (*C*) cyclohexane–*n*-hexane (1:1), (*D*) cyclohexane–benzene (1:1)–liquid paraffin (10%), (*E*) cyclohexane–liquid paraffin (20%)–dioxane (10%), (*F*) cyclohexane–liquid paraffin (20%)–dioxane (5%), (*G*) cyclohexane–liquid paraffin (10%)–dioxane (3.5%), (*H*) cyclohexane–liquid paraffin (5%)–dioxane (2%), (*I*) petroleum ether (40–60°C)–liquid paraffin (20%), (*J*) petroleum ether–liquid paraffin (10%), (*K*) petroleum ether–liquid paraffin (5%)–dioxane (1%), (*L*) *n*-hexane.

Abbott et al. [35] chromatographed 16 organochlorine pesticides in 15 different adsorbent–solvent combinations (Table 25.4). By examination of this table the best system can be selected for any given pair of compounds or group of compounds. These workers preferred a 0.5% silver nitrate in ethanol solution as a detecting agent using a 10-min exposure to ultraviolet light after the spraying of the reagent. As an alternative they used test T-229. The pesticides appeared as bright yellow spots on a blue background.

Baeumler and Rippstein [34] separated a group of six chlorinated pesticides on aluminum oxide with hexane as the developing solvent. In order to detect the compounds test T-94 was used. Sensitivity of this color detecting agent was less than 5 γ. In dealing with the visualization of this type of compound, Katz [46] used test T-101. Dichlorodibenzoyl and DDE could not be detected with this, so they were visualized by spraying with 0.005% iodine solution in chloroform.

Adamović [47] investigated a group of aromatic amines as spray reagents for organochlorine pesticides. Among the most useful were dimethylaniline, *p*-phenylenediamine, benzidine, *o*-toluidine, diphenylamine, and α-naphthylamine (See Chapter VII).

Moats [48] used a light spray of hydrogen peroxide to suppress interference from traces of fat in the visualization of chlorinated pesticides on aluminum oxide layers impregnated with silver nitrate. Later [49] he found that incorporation of oxidized vegetable oil eliminated the need for the hydrogen peroxide and increased the sensitivity of detection to 0.01 μg. Layers were prepared from 30 g washed and still moist aluminum oxide G mixed with 0.2 g silver nitrate dissolved in a few drops of water, 20 ml of acetone, and 0.1 ml of rancid vegetable oil. Two 10-ml portions of acetone were used to rinse out the mixing vessels and the entire mixture was mixed for 10 sec in a blender. Plates were air-dried. Kadoum [11] prewashed silica gel and cellulose layers with 1% hydrogen peroxide in redistilled acetone then reactivated at 120°C for 60 min. Five minutes after development the plate was sprayed with fluorescein solution and exposed to ultraviolet for 5 min, this treatment was repeated and then the layers were sprayed with silver nitrate solution and exposed to ultraviolet. This treatment provided an almost white background.

Schmit et al. [50] have reported on the use of thin-layer chromatography to aid in the identification of chlorinated pesticide peaks obtained from gas chromatographic analysis. Ludwig and Freimuth [51] in examining insecticide residues in food were able to separate DDT, BHC, and methoxychlor on thin layers of Supergel; Eder et al. [52] separated and identified 21 insecticides consisting of chlorinated hydrocarbons and phosphoric acid esters by means of paper and thin-layer chromatography.

These insecticide residues were isolated from fruits and legumes. Niwaguchi [53] determined BHC spectrophotometrically by converting it to 1,2,4-trichlorobenzene after purification by plate chromatography. The conversion was accomplished by heating with a large excess of alcoholic potassium hydroxide for 30 min.

Ceresia [54] used two-dimensional thin-layer and paper chromatography for separating a group of pesticides. The best solvent combination was pyridine–95% ethanol (4:1) for the first direction and acetone–water (7:3) for the second direction. This solvent combination separated the pairs: toxaphene–lindane, Perthane–methoxychlor, and Rhothane–chlordane. In dealing with troublesome samples that have not been cleaned up sufficiently, Kawatski and Frasch [55] developed first with *n*-heptane on alumina layers and then rotated 90° and developed with acetonitrile saturated with *n*-heptane. Oily animal extracts were notable problem samples that could be solved in this manner. Eliakis et al. [56] developed a two-dimensional technique for the rapid identification of organochlorine pesticides in blood and tissue. A two-dimensional technique on silica gel G was used with *n*-hexane in the first direction and cyclohexane in the second. Szokolay and Madarič [57] used a multiple development in a one-dimensional separation with 0.3% ethanol in heptane of 13 chlorinated pesticides.

Recently the fact that polychlorobiphenyls (PCB's) may appear in samples with chlorinated pesticides has added another possibility of error in the analysis of the latter. Reynolds [58] used Florisil to separate PCB's from the pesticides; however Bevenue and Ogata [59] observed that a preliminary examination of the Florisil is necessary as not all batches of Florisil are satisfactory. Fehringer and Westfall [60] used a two-dimensional technique to separate DDT analogs in the presence of PCB's. Hattula [61] used a step method for separating fat, PCB's, and chlorinated pesticides in fish tissue. The sample was spotted in hexane and developed for a distance of 8.5 cm with chloroform. After it was dried, the plate was developed to the top (20 cm) with heptane. Bush and Lo [62] have developed a method for the quantitative analysis of PCB's in the presence of pesticides, and Fishbein [63] has reviewed the chromatographic and biological aspects of PCB's.

Quantitative analysis of chlorinated pesticides has been accomplished by eluting the compounds and injecting them into a gas chromatograph [64, 65], by measuring them spectrophotometrically [66], by X-ray fluorescence [67, 68], by potentiometric titration [69, 70], by direct densitometric evaluation [71], by spot area measurements [72, 73], and by polarography [74].

Fishbein [75] reviewed the chromatographic and biological aspects of DDT and its metabolites.

2 PHOSPHORUS COMPOUNDS

In toxicological work to detect the presence of phosphorus-containing insecticides in biological material, Fischer and Klingelhoeller [76, 77] subjected the material to alkaline hydrolysis before extracting the phosphorus compounds. This has an advantage in that more than one cleavage product is formed from each pesticide, thus giving groups of R_f values for each individual compound to assist in the identification. This also helps to differentiate the pesticides from other naturally occurring sulfur compounds in the body. The sulfur-containing compounds were detected with a 3% solution of sodium azide (6 to 8% sodium azide is more sensitive) in 0.1 *N* iodine producing white spots on a brown background. Systox, Meta-Systox, and Thiometon also produced volatile thio ethers which were lost during the hydrolysis process. These may be detected during the hydrolysis by checking the vapors with a strip of filter paper soaked in iodine–azide solution. Where nitrophenols are produced during the hydrolysis, they may be detected on the thin-layer plate by exposing the plate to ammonia vapors which intensifies the yellow color of the nitrophenols. Methylumbelliferone from Potasan and benzacimide from Gusathion have a bright blue fluorescence under ultraviolet light, and the diazine from the Diazinon may be detected by means of the Dragendorff reagent. Baeumler and Rippstein [34] separated a group of phosphorus-containing pesticides by using a mixture of hexane–acetone (4:1) with silica gel plates. As a detecting agent, they used a spray of a weakly acidic 0.5% palladium chloride solution.

The $R_f \times 100$ values of a group of organophosphorus pesticides is given in Table 25.5

Parathion, Meta-Systox, and malathion have been separated on silica gel G using toluene as the developing solvent [83]. In this case, the silica gel layers were prepared by mixing 30 g of silica gel G with 60 ml of fluorescein solution, prepared by dissolving 20 mg of fluorescein in 1.2 ml of 0.1 *N* potassium hydroxide and diluting to 60 ml. Drying of the plates was accomplished at 105°C for 30 min. The compounds were made visible after the development by exposing the plates to bromine vapor. The R_{st} values referred to parathion were as follows: parathion or methyl parathion 1.00, Meta-Systox 0.80, and malathion 0.13. The spots were eluted and tested for activity with *D. melanogaster*.

Hengy and Thirion [84] determined the malathion residues in tobacco

Table 25.5 $R_f \times 100$ Values of Some Organophosphorus Pesticides[a]

Pesticide	*Silica Gel G*			*Polyamide*		*Acidic*[b] Al_2O_3		*Florisil*[b]	
	S_1	S_2	S_3	S_4	S_5	S_6	S_7	S_8	S_9
Azinphos-ethyl	33	90		66	16	60	18	69	12
Azinphos-methyl	19	88	66	54	26	48	15	60	9
Bromophos	85	93	95						
Carbophenothion	83	96				98		93	90
Chlorfenvinphos	24	79							
Coumaphos	33	90	81	66	6	86	25	71	12
Crufomate	6	43							
Delnav				79	3	82	35	81	30
Demeton-S	33	93							
Demeton-S-methyl	17	73							
Diazinon	61	95		87	35	79	33	88	47
Dibrom	0–22[c]	0–89[c]	70	66	49		0	0	0
Dichlorfenthion	77	96							
Dichlorvos	22–27	73	56						0
Dimefox	8	44							
Dimethoate	5	37	35	26	74	18	0	30	3
Disulfoton	82	97	100	88	12	11	78	88	85
Dicapthon						94	73	83	45
Ethion	77	97	100	86	4	94	63	85	72
Ethoate-methyl	7	61							
Fenchlorphos	84	93							
Haloxon	4	71							
Malathion	37	95	82			71	26	75	24
Mecarbam	42	95							
Menazon	0	2							
Meta-Systox-R	0	5	33	15	95	4	0	3	0
Mevinphos	10	64	34	57/65	90	24	0	38	0
Morphothion	6	49							
Parathion	57	91	86	77	9	93	90	86	48
Phenkapton	74	97							
Phorate	80	97	100	88	12	97	66	88	86
Phosalone	39	97							
Phosphamidon	4	34				16	0	22	0
Pyrimithate	62	96							
Ronnel			100	83	4	90	72	92	77
Ruelene				56	35	21	0	30	0
Schradan	0	2							
Sulfotep	75	92							
Tepp	0	0–50[c]		7/71	94				
Thionazin	45	92				79	67	78	27

Table 25.5 (Continued)

Pesticide	Silica Gel G S_1	Silica Gel G S_2	Silica Gel G S_3	Polyamide S_4	Polyamide S_5	Acidic[b] Al_2O_3 S_6	Acidic[b] Al_2O_3 S_7	Florisil[b] S_8	Florisil[b] S_9
Trichlorphon	3	18	8	11	80	0	0	17	0
Trithion				80	3	98		93	90
Thiometon				82	19				
Vamidothion	1	16							
Zytron						89	36	91	68

[a] S_1 hexane–acetone (5:1), S_2 chloroform–acetone (9:1) [78]; S_3 benzene–acetone (9:1) [79, 80]; S_4 hexane–acetone (4:1), S_5 methanol–water (1:1) [81]; S_6 and S_8 cyclohexane–acetone–chloroform (70:25:5), S_7 and S_9 2,2,4-trimethylpentane–acetone–chloroform (70:25:5) [82].

[b] Binder free.

[c] Streaking.

and tobacco smoke. Kadoum [85] used a two-dimensional technique for separating malathion and some of its metabolites found in stored grains. They used test T-79 for detecting the compounds.

Geldmacher-Mallinckrodt [86, 87] has also used a degradation method for the determination of Systox and Meta-Systox. In this case, the compounds were refluxed with sodium ethylate or methylate. The R_f values of the resulting compounds on silica gel G using toluene–petroleum ether (2:1) as a solvent are given in Table 25.6, along with the colors developed

Table 25.6 $R_f \times 100$ of Systox and Meta-Systox and Reduction Products on Silica Gel G with Color Reactions[a]

Compound	Toluene–Petroleum Ether (2:1)	Copper Reagent[b]	Cobalt Reagent[b]
Systox	5	Brown	Turquoise
Meta-Systox	5	Brown	Turquoise
2-Ethylthioethanol	7	Green	Green-brown
1-Methylthio-2-ethylthioethane	40	Green	Green-brown
1,2-Bis(ethylthio)ethane	40	Brown-violet	Brown
1-Ethylthioethanethiol	45	Yellow-green	Blue

[a] From M. Geldmacher-Mallinckrodt [87]; reproduced with permission of the authors and Springer-Verlag.

[b] See text for spray reagents.

with the copper chloride and cobalt nitrate spray reagents. The copper reagent consists of 2 g of cuprous chloride in 10 ml of ethanol and 2.5 ml of concentrated hydrochloric acid, and the cobalt reagent consists of 2.5 g of cobalt nitrate and 1.25 g of ammonium thiocyanate in 10 ml of ethanol.

Stammbach et al. [88] investigated a crude preparation mixture of phencapton. The impurities in this preparation included impurities in the starting materials, oxidation products of the thio ether, and hydrolysis products of the thiophosphate. Two separate chromatograms were run in order to effect a separation. In the first separation on silica gel G using cyclohexane as the developing solvent, the following compounds, listed with their R_f values, were isolated: bis(2,5-dichlorophenylthiomethyl)-ether 0.35, bis(2,5-dichlorophenylthio)methane 0.45, 2,5-dichlorophenyl-thiomethyl chloride 0.56, 2,5-dichlorothiophenol 0.65, and 2,2′,5,5′-tetrachlorodiphenyl disulfide 0.76. The technical phencapton which had an R_f value of 0.17 in this solvent was then rechromatographed in carbon tetrachloride–benzene (95:5). In this case, the phencapton with an R_f value of 0.73 was separated from diethyldithiophosphoric acid (R_f value 0.0), phencaptonsulfone 0.06, the sulfoxide 0.12, *O,O,S*-triethyl dithiophosphate 0.52, and unidentified impurities with an R_f value of 0.95. Visualization was accomplished by means of iodine vapor.

Kováč [89] has separated *O,O*-dimethyl *O*-(3-methyl-4-nitrophenyl) thiophosphate and four other related compounds from a technical product on silica gel layers with petroleum ether (60 to 80°C)–acetone (98.6:1.4), so that the first main compound could be determined polarographically. Kovacs [90] has also used aluminum oxide layers for the separation of 19 organothiophosphate insecticides. The plates were prewashed by developing in distilled water using a filter-paper wick to feed the latter to the thin layer. They were dried at 80°C for 45 min. After the compounds were spotted in ethyl acetate solution, the plates were impregnated with either 15 or 20% dimethylformamide in ether and then immediately placed in the developing solvent of methyl cyclohexane. With a developing distance of 10 cm the following R_f values were obtained on 15 and 20% impregnated plates, respectively: Rogor 0.1, 0.1, Guthion 0.9, 0.6, Imidan 0.9 0.7, methyl parathion 0.17, 0.11, Co-Ral 0.23, 0.15, malathion 0.34, 0.22, Delnav 0.37, 0.24, parathion 0.41, 0.27, Systox (thiol) 0.44, 0.32, EPN 0.49, 0.33, Methyl Trithion 0.50, 0.36, sulfotepp 0.69, 0.55, Trithion 0.74, 0.59, ronnel 0.76, 0.62, ethion 0.77, 0.63, Systox (thiono) 0.79, 0.67, Thimet 0.81, 0.71, Di-syston 0.82, 0.72, and Diazinon 0.86, 0.78. A highly sensitive and specific reagent for detecting sulfur-containing phosphate esters was employed (T-251). Abbott et al. [91] have separated a group of 13 organophosphorus pesticides in 14 different separatory systems using silica gel, kieselguhr–silica gel (1:1), and kieselguhr–alumina (1:1) for the adsorbent layers. The mixed layers were useful in making separations that were not possible on silica gel, for example, carbophenothion from fenchlorphos and parathion from thiometon. Quantitative determinations could be made by infrared measurement of the eluates from the thin-layer separations. Tinox, a Systoxtype thiophosphate ester, and its main metabolite, a sulfoxide derivative, were separated by both paper and thin-layer systems [92]. The best solvent was a five-component mixture composed of toluene–acetonitrile–methanol–isopropanol–water (40:20:16:16:9). The metabolite could be detected with a sensitivity of 1 γ by using hydrochloric acid solutions of potassium iodoplatinate or palladous chloride. In further work using the same solvent system, Ackermann and Spranger [93] have chromatographed a group of esters of thiophosphoric acid of the Systox type. The R_f values are listed as well as the color reactions with seven detecting agents. For the more difficult compounds to detect (under 10 μg), an acetone solution of potassium permanganate or cobalt chloride [94] was used.

Bruaux et al. [95] have developed a gel electrophoresis method for the detection and identification of organophosphorus pesticides. The method is based on the fact that esterases of different bovine organs can be separated into five to seven zones by electrophoresis on agar gel. One or more of these zones disappears when organophosphorus pesticides are added to the extracts of the organs before submitting them to electrophoresis. The inhibition patterns are given for 28 compounds.

There is an abundance of detecting agents for organophosphorus pesticides. Grant et al. [96] tabulated the sensitivity of 12 organophosphorus compounds with 11 different detecting agents, and Watts [97] has reviewed the detection agents for this type of compound. The most sensitive test is the cholinesterase inhibition test (T-113) which can detect as small a quantity as 40 pg. This test is also one of the least sensitive to unclean samples. Care must be taken, however, not to overload the chromatograms; Cassidy et al. [98] found that extracts of alfalfa that had not been sprayed with pesticides produced zones of inhibition when excessively large aliquots were used.

Askew et al. [78] modified the technique for hydrolysis of phosphorus pesticides on TLC plates so that the ammonium molybdate test (T-16) would be more generally applicable. Using this technique 1 μg of pesticide could be detected in water containing 0.001 ppm of pesticide. The use of 4-(p-nitrobenzyl)pyridine for the detection of organophosphorus compounds has been reported [99, 100] (T-183).

For a quantitative method for the determination of Cygon insecticide and its oxygen analog, see Chapter XI, Section 1.

Blinn [101] has investigated the isolation, identification, and determination of Thimet residues. The spray residues are metabolized to the oxygen analog and oxidation products. Blinn has shown that in order to determine the residues as the oxygen analog sulfone, the best way to convert Thimet and its metabolic products is to oxidize with *m*-chloroperbenzoic acid. Once converted, the sulfone can then be separated from other oxidation products, not only from those of Thimet, but also from Ethion, Guthion, Trithion, and Di-syston whose oxidation products can interfere in the chromotropic acid colorimetric determination [102]. Separation is achieved on layers prepared from a mixture of 30 g of silica gel G and 30 g of silica gel HF with 120 ml of pH 6 buffer solution. These plates are air-dried and then prewashed twice by development in freshly distilled acetone. Development of the sample is carried out first in a fresh solution of 1.75% methanol in chloroform in a saturated atmosphere, and then to increase the separation, a second development is carried out in chloroform. Location of the compounds can be made with palladium chloride solution, and a quantitative analysis can be made either by eluting the spot and measuring the absorption in carbon disulfide at 1325 cm^{-1} or colorimetrically, by use of the chromotropic acid method.

Gardner [103] used a two-dimensional method for obtaining confirmation evidence in identifying compounds. This was achieved by oxidizing on the layer after the first development and then developing in the second dimension to separate the oxidation products. The enzyne inhibition technique (T-113) and the 4-(*p*-nitrobenzyl)pyridine test (T-183) were used for detection.

Salamé [104] has tabulated a list of R_f values for 10 organophosphorus insecticides on silica gel with 16 different solvent systems. Stanley [105] likewise has studied the separation of 31 organophosphorus pesticides on silica gel G in six different solvents. The latter work was carried out on microchromatoplates using a travel distance of only 5 cm.

3 CARBAMATES

Chiba and Morley [106] have developed a method for the rapid screening of carbaryl (1-naphthyl methylcarbamate) and its breakdown product, 1-naphthol. The separation can be carried out on silica gel plates using benzene–acetone (19:1) as a solvent with R_f values of 0.17 for carbaryl and 0.33 for 1-naphthol. The compounds can be visualized by spraying the dried plates with 1.5 *N* methanolic sodium hydroxide followed by *p*-nitrobenzenediazonium fluoroborate (10 mg in 100 ml of a 1:1 mixture of diethyl ether–methanol). With this reagent the 1-naphthol appears as a purple spot and the carbaryl as a brilliant blue spot, which later changes to the same color as the naphthol spot. Without cleanup these compounds can be detected in apple and lettuce extracts in as small an amount as 0.02 ppm. By using a preliminary cleanup a sensitivity of 5 ppb can be achieved.

Liebmann and Schuhmann [107] used wedge-shaped layers of silica gel to separate and identify isopropyl *N*-(3-chlorophenyl) carbamate (CIPC) and *m*-chloroaniline isolated from potatoes by extracting with dichloromethane. The spots were detected with a silver nitrate reagent.

Nagasawa et al. [108] chromatographed 22 carbamates on polyamide layers. Detection reagents found effective were the bromine–fluorescein test [2], test T-222, and a 0.1% solution of Pinacryptol Yellow in 95% ethanol with subsequent viewing in transmitted ultraviolet light. Finocchiaro and Benson [109] chromatographed 19 carbamate pesticides and three related phenylureas on aluminum oxide G or on silica gel G-HR. They investigated 14 detection methods; test T-274 with vanillin–sulfuric acid reacted with all the compounds.

Some carbamates can be detected by the cholinesterase inhibition test (T-113). Geike [110, 111] found detection limits of 3 to 400 ng using bovine liver esterase, and Mendoza and Shields [112] found pig liver esterase to be even more sensitive with sensitivities in the range of 0.1 to 100 ng. They found that exposure to bromine or ultraviolet light decreased the inhibitory effect of carbamates in general, although the inhibitory properties of some carbamates are increased by exposure to bromine; carbaryl is an example of this.

MacNeil and Hikichi [113] determined benomyl [methyl 1-(butylcarbamoyl)benzimidazol-2-yl carbamate] on cherries by in situ fluorescence, and Frei et al. [114] determined Sevin by the same technique at the nanogram level by first hydrolyzing to naphthol. Other carbamates can be hydrolyzed and then treated with dansyl chloride to form fluorescent compounds.

4 PYRETHRINS AND OTHER PLANT INSECTICIDES

Spickett [115] was the first to separate pyrethrin I and pyrethrin II from one another. He used silica gel layers with 20% plaster of Paris as a binder and a developing solvent of 20% ethyl acetate in *n*-hexane. As a detecting reagent he used the fluorescein–bromine reagent of Kirchner et al. [116]. The R_f values for these compounds are 0.42 and 0.21, respectively. Stahl [117] also separated the pyrethrins on silica gel including the isopyrethrins. He used a two-dimensional separation and the technique of reactions on thin layers introduced by Miller and Kirchner [118]. A mixture of the pyrethrins was separated in one direction on a thin-layer plate using hexane–ethyl acetate (3:1). The plate was then exposed to ultraviolet or sunlight which catalyzed the oxidation of these compounds. On turning the plates at right angles and developing in the second direction with the same solvent, the oxidation products were separated. The development distance in both directions was 8 cm. A number of visualizing agents including antimony trichloride and pentachloride, and 2,4-dinitrophenylhydrazine were used for detecting the compounds. The pyrethrin peroxides may also be detected by the potassium iodide–starch reaction (T-207). In order to test the activity of the various fractions, a biological test was employed using either *Aedes aegyptici* larvae or 4 to 8-day-old *Drosophila melanogaster*. The peroxides and the lumipyrethrins were inactive. Olive [119] adapted a colorimetric reagent as a specific color test (T-250) for detecting pyrethrins and piperonyl butoxide, because these two appear together in many formulations.

Nash et al. [120] have used thin-layer chromatography to isolate rotenone from a technical grade of the product; Doi [121] has analyzed derris root preparations for rotenone content by chromatographing on an aluminum oxide plate with benzene–ethanol–water (4:2:1) as the solvent giving an R_f value of 0.73. Quantitatively it was determined by spectrophotometric measurement at 294 mμ. Kroeller [122] used heptane–cyclohexane–ethyl acetate–water (120:20:80:0.3) to separate rotenone residues in foods. Levels of 0.05 ppm could be determined within $\pm$10 to 15%. Delfel and Tallent [123] separated and determined rotenone and deguelin on silver nitrate impregnated layers using a direct densitometric method.

Tyihák and Vágujfalvi [124] separated compounds having insecticidal activity from *Chrysanthemum cinerariaefolium*. For the detection of the latter compounds a 1% vanillin in concentrated sulfuric acid solution was used as it was more sensitive than the reagents applied by Stahl [117]. Nalbandov et al. [125] have applied thin-layer chromatography on silica gel G in the separation of a new compound with insecticidal properties from *Nicandra physalodes*. The compound was located by means of a phosphomolybdic acid reagent.

5 MISCELLANEOUS PESTICIDES

Blinn and Gunther [126] developed a thin-layer method to separate Aramite from the acaricide OW-9, which is a mixture of two organo sulfites closely related to Aramite. After an extensive cleanup operation to remove interfering materials, the extract can be chromatographed on silica gel with 3.5% ethyl acetate in benzene to obtain an R_f value of 0.58 for Aramite and 0.46 and 0.30 for the components of OW-9. The compounds are detected by spraying with 1% ethanolic potassium hydroxide and heating in an oven for 5 min at 150°C to hydrolyze the compounds, thus forming the inorganic sulfide. The organic materials are then removed from the plate by washing with acetone, and the latter is then removed by heating for 5 min at 150°C. The potassium sulfite spots are detected by spraying with a mixture of 51 parts of water, 45 parts of acetone, 4 parts of Beckmann #3581 pH-7 buffer, and 1 ml of a saturated acetone solution of malachite green oxalate. White zones appear on a blue background. An alternative reagent is 1,3,5-trinitrobenzene which gives pink to red spots on a colorless background.

Wagner et al. [127–129] have investigated the rodenticide warfarin. Plasma concentrations were measured by spectrophotometry after separation on silica gel layers using 1,2-ethylene dichloride–acetone (9:1) [128]. Ramaut and Benoit [130] separated warfarin, coumachlor, and Racumin on silica gel layers prepared with 0.3 *M* sodium acetate solution. The solvent for the separation was toluene–ethyl formate–formic acid (5:4:1). The spots were visualized by spraying with 0.5% hydrogen peroxide, drying at 100°C, and then spraying with 2% ferric chloride. A final drying at 105°C produced brown spots. Russel [131] separated coumarin and indandione rodenticides on silica gel with dichloromethane–acetic acid (9:1) and detected the compounds with test T-84 or Pauly's reagent (T-249). Janicki et al. [132] have separated and determined the various isomers in norbormide. Perry [133] was able to detect as little as 0.1 μg of sodium monofluoroacetate by using silica gel G layers with 4% ammonium hydroxide in ethanol (vol/vol). The compound was detected with test T-231 and determined by comparison of spot area.

6 INSECTICIDE SYNERGISTS

Pyrethrin and allethrin insecticides commonly contain synergists to increase the effectiveness of the biologically active material. Piperonylbutoxide, bucarpolate, and octachlorodipropyl ether have been separated on silica gel G with hexane–ethyl acetate (3:1) in a saturated atmosphere [117]. With a development distance of 8 cm, the R_f values were 0.35, 0.23, and 0.67, respectively. Beroza [134] has separated all of the nine methylenedioxyphenyl synergists that are used commercially. The synergists were applied as 1% solutions in acetone to silica gel G plates which had been dried at 105 to 110°C for $\frac{1}{2}$ hr. The R_f values were determined in 14 different solvents, some of which are given in Table 25.7. Two detecting agents were used, a chromotropic–sulfuric acid reagent (1 vol of 10% sodium 1,8-dihydroxynaphthlene-3,6-disulfonate in 5 vol of sulfuric acid) and furfural–sulfuric acid reagent (1:50). Charac-

Table 25.7 $R_f \times 100$ Values of 3,4-Methylenedioxyphenyl Synergists on Silica Gel[a] [134]

Synergist	*Acetone–Benzene* (25:97.5)	*Propanol–Benzene* (5:95)	*Ethyl Acetate–Chloroform* (4:1)	*Color with Chromotropic–Sulfuric Acid*[b]
Sulfox-Cide	2	34	28	Purple with blue rim
		43	34	
Sesamex	16	63	42	Orange
Bucarpolate	25	71	58	Purple with blue rim
Piperonyl butoxide	30	70	64	Purple with reddish rim
Sesamin	36	76	60	Purple with brown rim
Sesamolin	52	80	71	Purple with brown rim
Asarinin	52	80	72	Purple
Piperonyl cyclonene	39	73	35	Purple
	47	79	72	Yellowish green
	58	85	80	Purple
	70		86	Reddish pink
n-Porpyl isome	0	1	0	Pink
	36		81	Purple
	52	82		Pink
	58			Dark purple
	77			Pink

[a] For R_f values in additional solvents see original paper.
[b] Color after heating for 30 min at 105 to 110°C.

teristic colors were produced with these reagents which aided in the identification of the compounds (Table 25.7). Asarinin and sesamolin which had practically identical R_f values in all solvent systems could be differentiated from one another by the furfural–sulfuric acid. Sesamolin showed a bright red color in the cold with this spray reagent, whereas asarinin showed no color reaction until it was heated at which time it showed a black spot. Sesamex also showed a bright red color with the same reagent but could be readily differentiated from sesamolin by means of the R_f values. Numerous spots were shown by piperonyl cyclonene and *n*-propyl isome because they were commerical products. Gunner [135] determined methylenedioxy compounds by direct spectrophotometry using the chromotropic–sulfuric acid detection agent.

Fishbein et al. [136] have investigated the metabolites of piperonyl butoxide and tropital in rat urine and bile using thin-layer chromatography. Lichtenstein et al. [137] isolated a number of insecticidal synergists from dill plants by chromatographing extracts on silica gel with benzene. These included myristicin, apiol, dill-apiol, and *d*-carvone. Fishbein et al. [138] chromatographed the metabolites of the synergist octachlorodipropyl ether in rat bile and urine.

Fishbein [139] and Fishbein et al. [140] have reviewed the chromatography of synergists.

7 FUNGICIDES

The methods developed by Kirchner et al. [141] for the separation and determination of biphenyl, used as a fungicide in citrus fruit, has already been described in Chapter XI, Sections 1 and 4. Salo and Maekinen [142] have used Shell Sol A–acetic acid (24:1) with silica gel G to separate biphenyl, *o*-phenylphenol, and 2,4-dichlorophenoxyacetic acid, all of which have also been used on citrus fruit. The R_f values are 0.81, 0.34, and 0.10, respectively.

Because of food regulatory laws the need for a method of identifying food preserving agents is recognized and the versatility of thin-layer chromatography readily lends itself to this work. Gaenshirt and Morianz [143] separated the methyl and propyl esters of *p*-hydroxybenzoic acid, using silica gel G layers activated at 160°C for 2 hr. Pentane–glacial acetic acid (22:3) was used to achieve the separation using a developing distance of 12 to 14 cm. Quantitative determinations were made by a spectrophotometric method after elution of the spots from the silica gel. Rangone and Ambrosio [144] separated the methyl, ethyl, propyl, butyl, and benzyl esters of *p*-hydroxybenzoic acid by reverse phase on silanized silica gel. A pH-11 borate buffer–tetrahydrofuran (95:5) or pH-11 borate buffer–dioxane (90:10) solvent gave the best results. For quantitative results the development was repeated a second time, and the compounds were determined spectrophotometrically. Salo [145] separated *p*-hydroxybenzoic acid and 13 of its esters on 10% acetylated cellulose–polyamide (1:1) with Shell Sol A–acetic acid (1:1) as the solvent. Copius-Peereboom and Beekes [146] have separated a group of nine food preservatives in a number of systems. The separation on cellulose layers took to 5 to 6 hr and achieved a separation of benzoic and sorbic acids only slightly better than with paper chromatography, whereas a mixture of silica gel G–kieselguhr G(1:1) gave a better separation of these two compounds when used with hexane–acetic acid (96:4) with chamber saturation. The color reactions with 12 different reagents were investigated in order to have some specific color tests. As a general reagent that would detect all of the compounds, the plates were sprayed with a bromophenol blue solution followed by a 0.5% potassium permanganate solution containing 1%

sodium carbonate. Covello and Schettino [147, 148] investigated a slightly different group of nine food preservative agents. Their separations were carried out on silica gel G layers deposited on chrome-plated brass plates. Spray reagents were not used and the compounds were recovered from the silica gel layers by the sublimation technique of Baehler [149]. Table 25.8 gives the R_f values for a number of these compounds.

Gosselé [150] used a mixed layer of silica gel GF–cellulose MN 300 F_{254} (15:7.5) (wt/wt) for separating a group of food preservatives with the upper layer of petroleum ether (40 to 60°C)–carbon tetrachloride–acetic acid–chloroform–formic acid (25:20:1:10:4) as the solvent. The plates were developed twice (15 cm each time). Nagasawa et al. [151] chromatographed eight preservatives in six solvents on polyamide layers, and used a two-dimensional separation to obtain complete separations. Chiang [152] compared the separation on silica gel, polyamide, and polyamide–silica gel (10:52) in two solvent systems.

Bajaj et al. [153] separated 11 alkyl gallates used in extending the shelf

Table 25.8 $R_f \times 100$ Values of Food Preservatives in Various Systems[a]

	Silica Gel G[b] [147, 148]			*Cellulose MN 300* [146]	*Silica Gel G–Kieselguhr G (1:1) Development Distance 20 cm*[c] [146]	
Preservative	*A*	*B*	*C*	*D*	*E*	*F*
Benzoic acid	24	65	88	50	154	111
Sorbic acid	23	77	74	58	128	91
Salicylic acid	53	79	70	56	100	100
Dehydroacetic acid	27	81	68	9	60	88
p-Chlorobenzoic acid	24	100	100			
p-Hydroxybenzoic acid	12	75	74	9	7	41
Methyl *p*-hydroxybenzoate	66	88	72	75	12	75
Propyl *p*-hydroxybenzoate	67	90	71	90	18	84
Ethyl *p*-hydroxybenzoate				86	16	79
Cinnamic acid	43	72	61			
o-Phenylphenol				95	13	136

[a] Solvent: *A* = butanol saturated with 2 *N* ammonia; *B* = isopropanol–ethanol–concentrated ammonia–water (5:2:0.5:1); *C* = benzene–acetic acid–water (4:9:2); *D* = *n*-butanol–35% ammonia–water (7:2:1); *E* = hexane–acetic acid (24:1); *F* = petroleum ether–ether–acetic acid (80:20:1).

[b] Development distance = 10 cm in saturated chamber.

[c] $R_{st} \times 100$ values referred to salicylic acid.

life of fatty food materials. Use was made of the complexing characteristics with cinchonine and strychnine by using silica gel layers impregnated with 2% solutions of the alkaloids.

There are many different types of fungicides used to protect seeds, foliage, or fruit and vegetables from the attack of fungi in the field. It is beyond the scope of this book to consider the chromatography of all the types, so only a few select examples will be considered.

Benomyl is one of the important systemic fungicides. Both benomyl and thiophanate methyl, another systemic, yield as a main metabolite 2-benzimidazolecarbamic acid methyl ester (MBC). Von Stryk [154] separated these compounds as well as thiophanate and two other possible metabolites by two-dimensional thin-layer chromatography on silica gel using benzene–methanol (9:1) in the first direction and ethyl acetate–chloroform (6:4) in the second. Abdullaev et al. [155] determined benomyl and MBC by spectrophotometry and Sherma [156] by fluorescent quenching on the thin-layer plate.

Baker et al. [157] developed a TLC method for the identification of eight systemic fungicides. The compounds were detected by ultraviolet or by using test T-208. Thin-layer chromatography has been applied to the detection of thiram [158], polybutene [159], captan and captax [160], chloroneb [161], diathianon [162], and thiabendazole [163].

Tatton and Wagstaffe [164] were able to separate and identify most of the organomercury compounds used as fungicides by converting to their dithizonates before chromatographing on silica or alumina. Geike and Schuphan [165] could detect from 50 ng to 1 μg of the fungicides using urease as a test agent; with sodium sulfide or dithizone as chemical spray reagents 0.5 to 20 μg could be detected. Fishbein and Fawkes [166] and Czeglédi-Jankó [167] chromatographed zinc ethylene bis(dithiocarbamate) (Zineb) and manganous ethylenebis(dichiocarbamate) (Maneb) as well as a number of decomposition products of these compounds on silica gel layers.

Sherma [168] has reviewed the chromatographic (all types) analysis of fungicides.

8 HERBICIDES

Bache [169] has developed a method for the isolation and detection of amiben (3-amino-2,5-dichlorobenzoic acid) in tomatoes. This compound is used as an herbicide for tomato plants. The sample material was obtained by extraction from tomatoes after saponification in order to liberate any bound material. The separation was achieved on silica gel G layers with benzene–acetic acid (5:1) as a solvent. With this solvent and a developing distance of 16 cm, amiben has an R_f value of 0.44. Location on the plate was accomplished by spraying first with 1% sodium nitrite in 1 N hydrochloric acid followed by a light spray of 0.2% (N-1-naphthyl)ethylenediamine dihydrochloride in 2 N hydrochloric acid. Using the extract from the equivalent of 2 g of tomato the method is sensitive to 0.1 ppm. Stammbach et al. [170] chromatographed the triazine herbicides, atrazine, atratone, and prometryne and determined their R_f values along with those of related compounds which occur in the commercial products. Henkel and Ebing [171] have also separated a group of six of the same triazines by using a two-step development on air-dried silica gel, using chloroform–diisopropyl ether (3:2) as the developing solvent. Henkel [172] has reported further on the separation of triazine herbicides using a chloroform–nitromethane solvent in ratios of 1:1 and 5:1. Manner [173] has determined the R_f values for eight triazine herbicides in 91 solvent combinations on fluorescent silica gel layers. Delley et al. [174] ran approximately the same group of compounds in two solvents. Of the three detection methods used, treatment with chlorine gas followed by

Table 25.9 Fromulas of Triazine Herbicides

Compound	R_1	R_2	R_3
Ametryn	SCH_3	NH-i-C_2H_7	NH-C_2H_5
Atratone	OCH_3	NH-i-C_3H_7	NH-C_2H_5
Atrazine	Cl	NH-i-C_2H_7	NH-C_2H_5
Chlorazine	Cl	N-$(C_2H_5)_2$	N-$(C_2H_5)_2$
Desmetryne	SCH_3	NH-CH_3	NH-i-C_3H_7
Ipazine	Cl	N-$(C_2H_5)_2$	NH-i-C_3H_7
Prometone	OCH_3	NH-i-C_3H_7	NH-i-C_3H_7
Prometryne	SCH_3	NH-i-C_3H_7	NH-i-C_3H_7
Propazine	Cl	NH-i-C_3H_7	NH-i-C_3H_7
Simazine	Cl	NH-C_2H_5	NH-C_2H_5
Simeton	OCH_3	NH-C_2H_5	NH-C_2H_5
Simetryn	SCH_3	NH-C_2H_5	NH-C_2H_5
Terbutryne	SCH_3	NH-C_2H_5	NH-t-C_4H_9
Trietazine	Cl	N-$(C_2H_5)_2$	NH-C_2H_5
G 34690	OCH_3	NH-C_3H_6-OCH_3	NH-C_3H_6-OCH_3

potassium iodide–starch (T-207) was the most sensitive, 0.02 μg (could not be used with the diethylamine solvents). On fluorescent layers fluorescent quenching at 254 nm can detect 0.2 μg. Reichling and Fischer [175] used polyamide layers to separate these compounds. By hydrolyzing the triazine herbicides with 1 *N* hydrochloric acid, Lawrence and Laver [176] obtained the corresponding amines which could be coupled with DNS chloride (5-dimethylaminonaphthalene-1-sulfonyl chloride) to yeild highly fluorescent compounds separable by thin-layer chromatography. By this means concentrations as low as 0.05 ppm could be analyzed.

Table 25.9 gives the structural formulas of these compounds and Table 25.10 gives the R_f values obtained in various systems. Stammbach et al. [170] used gas chromatography for the quantitative determination of the triazines.

Abbott et al. [177] have chromatographed a group of eight of these

Table 25.10 $R_f \times 100$ Values of Triazine Herbicides in Various Systems[a]

	Ref. 172		*Ref.* 174		*Ref.* 173		*Ref.* 175		
Herbicide	*A*	*B*	*C*	*D*	*E*	*F*	*G*	*H*	*I*
Ametryn	59		71	76	52	76	65	38	30
Atratone	34		36	50					
Atrazine		37	68	57	44	73	56	15	12
Chlorazine		80							
Desmetryne			58	50	40	60	49	48	38
Ipazine	66								
Prometone	42		45	63	41	36	78	67	60
Prometryne	68		81	76	68	88	78	30	21
Propazine		48	75	69	56	83	67	20	5
Simazine		28	58	44	36	55	37	10	0
Simetone	26								
Symetryn	50								
Terbutryne							75	27	16
Trietazine		60	88	69					
G 34690					5	3			

[a] System: (*A*) silica gel (dried room temperature), chloroform–nitromethane (1:1), 10 cm development; (*B*) same (5:1); (*C* silica gel G, toluene–acetone (17:3), 40 min development; (*D*) same but with carbon tetrachloride–absolute diethylamine (19:3); (*E*) silica gel GF_{254}, *n*-hexane–chloroform–acetonitrile (5:4:1); (*F*) same but with carbon disulfide–ethyl acetate (7:3); (*G*) polyamide 6.6, petroleum ether (40 to 60°C)–chloroform (49:1); (*H*) polyamide 11, water–methanol–acetic acid (5:3:1), (*I*)polyamide 6, water–methanol–acetic acid (14:4:1).

triazines in seven solvent systems on silica gel G and on kieselguhr–silica gel (1:1) in a single solvent system. These results were plotted graphically. Quantitatively, the compounds were determined by plotting the square root of the spot area against the logarithm of the weight of the material. For the quantitative work the spots were visualized by spraying with 0.5% brilliant green in acetone followed by exposure to bromine vapor.

Fishbein [178] has reviewed the chromatography (all types) of triazines.

Henkel and Ebing [171] also separated the following group of six chlorinated herbicides: 2-methyl-4-chlorophenoxyacetic acid hexyl ester (MCPA-hexyl) and 2-butoxyethyl ester (MCPA-butoxyethyl), α-(2-methyl-4-chlorophenoxy)propionic acid hexyl ester (MCPP-hexyl), ethyl ester (MCPP-ethyl), 2-butoxyethyl ester (MCPP-butoxyethyl), and 3-hydroxybutyl ester (MCPP-hydroxybutyl). With a two-step development with cyclohexane–diisopropyl ether (5:1), the separation was accomplished on silica gel layers dried at room temperature. Although these compounds can be detected by spraying with a 0.5% solution of Rhodamine B in ethanol with subsequent observation under ultraviolet light, the sensitivity is not very great (20 μg). A much more effective spray reagent (which would detect 0.5 γ of the compounds as brown to violet spots) was found by spraying with antimony pentachloride in carbon tetrachloride (1:4) and then heating the plates to 105°C. Abbott et al. [179] have worked out a procedure for detection and determination of these same types of herbicides in soil and water. Various mixtures of kieselguhr G and silica gel G were examined to obtain the optimum composition for the separation (Fig. 3.8). The optimum separation was achieved on a layer composed of 60% kieselguhr G and 40% silica gel G and developed with paraffin oil–benzene–acetic acid–cyclohexane (1:3:2:20). The compounds separated were: 2,4-dichlorophenoxyacetic acid (2,4-D), 2,4,5-trichlorophenoxyacetic acid (2,4,5-T), 4-chloro-2-methylphenoxyacetic acid (MCPA), 4-(4-chloro-2-methylphenoxy)butyric acid (MCPB), 4-(2,4-dichlorophenoxy)butyric acid (2,4-DB), and 2,2-dichloropropionic acid (dalapon). In this group of chlorine derivatives was also included 2-(1-methyl-*n*-propyl)-4,6-dinitrophenol (dinoseb) and 2-methyl-4,6-dinitrophenol (DNOC). The compounds were extracted with ether from the soil slurry in a sulfuric acid solution and were then put through an alkaline and acid washing procedure before concentrating and applying to the thin-layer plate.

In working with dinoseb extracts, Abbot and Thomson [180, 181] have used wedge-shaped layers of a 1:1 mixture of alumina G and kieselguhr G tapering from 2 ml to 100 μ in thickness. By applying the extract to the thick end of the layer, the colored impurities were readily adsorbed and allowed the thin-layer portion of the plate to be used for the separation of the dinoseb. The extracting solvent which was used in this case was a 1:1 mixture of methyl ethyl ketone and diethyl ether and gave considerably higher recoveries than did other solvents. The developing solvent for the thin-layer plate in this case was the same as previously described [179]. Quantitative determination of the dinoseb was made by eluting with methyl ethyl ketone and measuring the absorbance at 379 mμ. With this combination of cleanup procedure on wedge-shaped layers in the spectrophotometric determination, recoveries of 80 to 90% at the 0.1 to 0.3 ppm level were obtained.

Phenylurea herbicides were separated by Henkel [172] on air-dried layers of silica gel G with chloroform–nitromethane (1:1). With a development distance of 10 cm, the following R_f values were reported: fenuron 0.31, monuron 0.41, diuron 0.53, monochlorlinuron 0.72, linuron 0.79, and neburon 0.77. After separation, the compounds were split by heating to 150°C for 30 min and then identified by spraying with *p*-dimethylaminobenzaldehyde, with a detection limit of 1 γ. A separation may also be achieved by thermally cleaving the compounds after application to the plate but before chromatographing. In this case the resulting aniline derivatives are chromatographed in chloroform–acetic acid (60:1). The *p*-dimethylaminobenzaldehyde reagent may again be used for locating the compounds but in this case the sensitivity is decreased to 0.5 μg. Geissbuehler and Gross [182], after hydrolyzing these compounds, diazotized the amines and then coupled them with *N*-ethyl-1-naphthylamine. The resulting azo dyes were chromatographed on cellulose layers with dimethylformamide–0.05 *N* hydrochloric acid–ethanol (3:1:1). This method increased the sensitivity of detection to 0.03 to 0.04 μg. Of course, urea herbicides containing the same phenyl moieties (i.e., diuron, linuron, and neburon) could not be distinguished. Hance [183] determined the R_f values of 11 urea herbicides in 14 solvent systems on silica gel layers (including one reverse-phase system) as well as the gas chromatographic retention times.

Golab [184] has investigated the thin-layer separation of trifluralin and related compounds by two-dimensional thin-layer chromatography. Trifluralin is a selective preemergent herbicide and is active against a great variety of broadleaf, weeds and annual grasses. Using silica gel GF layers, the plates were developed in the first direction with benzene–1,2-dichloroethylene (1:1) and in the second direction with *n*-hexane–methanol (98:2). Using the natural color of some of the compounds or the blue absorbing spots under ultraviolet radiation, the sensitivity of detection was 0.5 μg. To increase the sensitivity, the material could be eluted and measured by gas chromatography.

Abbott and Wagstaffe [185] have set up a series of solvents for use with silica gel and silica gel–kieselguhr mixtures for the identification of the active ingredients in mixed herbicide formulations. Ebing (186) has proposed 10 TLC systems for the routine separation and identification of 61 pesticides (mostly herbicides).

Sherma has reviewed the chromatography of herbicides [27].

References

1. R. R. Schnorbus and W. F. Phillips, *J. Agric. Food Chem.,* **15,** 661 (1967).
2. J. H. Lawrence and J. A. Burke, *J. Assoc. Off. Anal. Chem.,* **52,** 817 (1969).
3. W. Γ. Wheeler, D. E. H. Frear, R. O. Mumma, R. H. Hamilton, and R. C. Cotner, *J. Agric. Food Chem.,* **15,** 227 (1967).
4. B. D. Hill and E. H. Stobbe, *J. Agric. Food Chem.,* **22,** 1143 (1974).
5. J. A. Burke and B. Malone, *J. Assoc. Off. Anal. Chem.,* **59,** 1003 (1966).
6. H. Beckman and D. J. Garber, *J. Assoc. Off. Anal. Chem.,* **52,** 286 (1969).
7. R. R. Watts, R. W. Storherr, J. R. Pardue, and T. Osgood, *J. Assoc. Off. Anal. Chem.,* **52,** 522 (1969).

7a. H. A. McLeod, C. Mendoza, P. Wales, and W. P. McKinley, *J. Assoc. Off. Anal. Chem.,* **50,** 1216 (1967).

8. R. W. Storherr and R. R. Watts, *J. Assoc. Off. Agr. Chem.,* **48,** 1154 (1965).
9. K. T. Hartman, *J. Assoc. Off. Anal. Chem.,* **50,** 615 (1967).
10. D. M. Bowker and J. E. Casida, *J. Agric. Food Chem.,* **17,** 956 (1969).
11. A. M. Kadoum, *Bull. Environ. Contam.,* **3,** 354 (1968).
12. D. C. Abbott and J. Thomson, *Chem. Ind. (London),* **1964,** 481.
13. R. P. Farrow, E. R. Elkins, Jr., and L. M. Beacham, *J. Assoc. Off. Agric. Chem.,* **48,** 738 (1965).
14. W. W. Thornburg, *J. Assoc. Off. Agr. Chem.,* **48,** 1023 (1965).
15. W. P. McKinley, D. E. Coffin, and K. A. McCully, *J. Assoc. Off. Agric. Chem.,* **47,** 863 (1964).
16. J. A. Burke, P. A. Mills, and D. C. Bostwick, *J. Assoc. Off. Anal. Chem.,* **49,** 999 (1966).
17. K. C. Walker and M. Beroza, *J. Assoc. Off. Agr. Chem.,* **46,** 250 (1963).
18. K. Suzuki, K. Miyashita, and T. Kashiwa, *Bull. Agric. Chem. Insp. Stn.,* **10,** 24 (1970); through *Chem. Abstr.,* **74,** 13895s (1971) (also see Ref. 19).
19. K. Suzuki, H. Nagayoshi, and T. Kashiwa, *Agric. Biol. Chem.,* **38,** 1433 (1974).
20. K. Suzuki, K. Miyashita, H. Nagayoshi, and T. Kashiwa, *Agric. Biol. Chem.,* **37,** 1959 (1973).

20a. K. Suzuki, H. Nagayoshi, and T. Kashiwa, *Agric. Biol. Chem.,* **38,** 279 (1974).

21. *Ibid.,* **40,** 845 (1976).

21a. H.-P. Thier and K. G. Bergner, *Deut. Lebensm. Rundsch.,* **62,** 399 (1966).

22. S. Sandroni and H. Schlitt, *J. Chromatogr.,* **55,** 385 (1971).

23. M. S. Schechter and M. E. Getz, *J. Assoc. Off. Anal. Chem.,* **50,** 1056 (1967).
24. M. F. Kovacs, Jr., *J. Assoc. Off. Agric. Chem.,* **48,** 1018 (1965).
25. D. C. Abbott and J. Thompson, *Residue Rev.,* **11,** 1 (1965).
26. R. Takeshita, *Eisei Kagaku,* **17,** 8 (1971); through *Chem. Abstr.,* **75,** 87128t (1971).
26a. K. R. Hill, *J. Assoc. Off. Anal. Chem.,* **58,** 1256 (1975).
27. J. Sherma, "Thin-Layer Chromatography: Recent Advances," in *Analytical Methods for Pesticides and Plant Growth Regulators,* Vol. VII, J. Sherma and G. Zweig, Eds., Academic Press, New York, 1974, p. 3.
27a. J. D. MacNeil and R. W. Frei, *J. Chromatogr., Sci.,* **13,** 279 (1975).
28. D. C. Holmes and N. F. Wood, *J. Chromatogr.,* **67,** 173 (1972).
28a. N. F. Wood, *Analyst (London),* **94,** 399 (1969).
29. K. Suzuki, K. Miyashita, and T. Kashiwa, *Noyaku Kensasho Hokoku,* **1970,** 24; through *Chem. Abstr.,* **75,** 4461x (1971).
30. D. C. Abbott, J. O'G. Tatton, and N. F. Wood, *J. Chromatogr.,* **42,** 83 (1969).
31. H.-J. Petrowitz, *Mitt. Deut. Ges. Holzforsch.,* **48,** 57 (1961).
32. H.-J. Petrowitz, *Chem.-Ztg.,* **85,** 867 (1961).
33. *Ibid.,* **86,** 815 (1962).
34. J. Baeumler and S. Rippstein, *Helv. Chim. Acta,* **44,** 1162 (1961).
35. D. C. Abbott, H. Egan, and J. Thomson, *J. Chromatogr.,* **16,** 481 (1964).
36. M. F. Kovacs, Jr., *J. Assoc. Off. Agric. Chem.,* **46,** 884 (1963).
37. I. Kawashiro and Y. Hosogai, *Shokuhin Eiseigaku Zasshi,* **5,** 54 (1964); through *Chem. Abstr.,* **61,** 6262 (1964).
38. J. Yamamura and T. Niwaguchi, *Proc. Japan Acad.,* **38,** 129 (1962).
39. R. Deters, *Chem.-Ztg.,* **86,** 388 (1962).
40. J. Yamamura, M. Chiba, S. Obara, and S. Suzuki, *Kagaku Keisatsu Kenyusho Hokoku,* **15,** 321 (1962).
41. H. V. Morley and M. Chiba, *J. Assoc. Off. Agric. Chem.,* **47,** 306 (1964).
42. A. Taylor and B. Fishwick, *Lab. Pract.,* **13,** 525 (1964).
43. M. F. Kovacs, Jr., *J. Assoc. Off. Agric. Chem.,* **48,** 1018 (1965).
44. M. R. Adams and M. S. Schechter, *Abstracts of Reports and Papers at the 77th Annual Meeting, Assoc. Offic. Agric. Chemists,* October 1963, p. 20.
45. E. J. Thomas, J. A. Burke, and J. H. Lawrence, *J. Chromatogr.,* **35,** 119 (1968).
46. D. Katz, *J. Chromatogr.,* **15,** 269 (1964).
47. V. M. Adamović, *J. Chromatogr.,* **23,** 274 (1966).
48. W. A. Moats, *J. Assoc. Off. Anal. Chem.,* **49,** 795 (1966).
49. *Ibid., J. Assoc. Off. Anal. Chem.,* **52,** 871 (1969).
50. J. A. Schmit, M. L. Laskaris, and U. J. Peters, *Abstracts of Reports and Papers at the 77th Annual Meeting, Assoc. Offic. Agric. Chemists,* October 1963, p. 20.
51. E. Ludwig and U. Freimuth, *Nahrung,* **8,** 559 (1964).
52. F. Eder, H. Schoch, and R. Mueller, *Mitt. Geb. Lebensm. Hyg.,* **55,** 98 (1964).

53. T. Niwaguchi, *Kagaku Keisatsu Kenkyusho Hokoku*, **14**, 419 (1961); through *Chem. Abstr.*, **61**, 2418 (1964).
54. G. B. Ceresia, *N.Y. State Dept. Health, Ann. Rept. Div. Lab. Res.*, **1963**, 63; through *Chem. Abstr.*, **61**, 13816 (1964).
55. J. A. Kawatski and D. L. Frasch, *J. Assoc. Off. Anal. Chem.*, **52**, 1108 (1969).
56. C. E. Eliakis, A. S. Coutselinis, and E. C. Eliakis, *Analyst (London)*, **93**, 368 (1968).
57. A. Szokolay and A. Madarič, *J. Chromatogr.*, **42**, 509 (1969).
58. L. M. Reynolds, *Bull. Environ. Contam. Toxicol.*, **4**, 128 (1969).
59. A. Bevenue and J. N. Ogata, *J. Chromatogr.*, **50**, 142 (1970).
60. N. V. Fehringer and J. E. Westfall, *J. Chromatogr.*, **57**, 397 (1971).
61. M. L. Hattula, *Bull. Environ. Contam. Toxicol.*, **12**, 331 (1974).
62. B. Bush and F.-C. Lo, *J. Chromatogr.*, **77**, 377 (1973).
63. L. Fishbein, *J. Chromatogr.*, **68**, 345 (1972).
64. R. M. Prouty and E. Cromartie, *Environ Sci. Technol.*, **4**, 768 (1970).
65. J. L. Radomski and A. Rey, *J. Chromatogr., Sci.*, **8**, 108 (1970).
66. L. S. Feklisova, *Zh. Anal. Khim.*, **26**, 1446 (1971); through *Anal. Abstr.*, **23**, 4220 (1972).
67. J. Alter, D. Dichlmann, R. Beydatsch, P. Kohler, D. Quaas, and W. Spichale, *Z. Anal. Chem.*, **233**, 188 (1968).
68. H.-J. Petrowitz and W. Berghoff, *Mater. Org.*, **7**, 287 (1972).
69. F. K. Kawahra, R. L. Moore, and R. W. Gorman, *J. Gas Chromatogr.*, **6**, 24 (1968).
70. J. Lauckner and H. Fuerst, *Chem. Tech. (Berlin)*, **20**, 236 (1968).
71. H.-J. Petrowitz and S. Wagner, *Chem.-Ztg.*, **95**, 331 (1971).
72. L. I. Bublik and E. S. Kosmatyi, *Zavod. Lab.*, **36**, 1194 (1970); through *Anal. Abstr.*, **21**, 1524 (1971).
73. E. Gwizdek, *Rocz. Panstw. Zakl. Hig.*, **21**, 647 (1970); through *Anal. Abstr.*, **21**, 4411 (1971).
74. E. S. Kosmatyi, *Vopr. Pitan.*, **28**, 89, (1969); through *Chem. Abstr.*, **71**, 2200e (1969).
75. L. Fishbein, *J. Chromatogr.*, **98**, 177 (1974).
76. R. Fischer and W. Klingelhoeller, *Arch. Toxikol.*, **19**, 119 (1961).
77. R. Fischer and W. Klingelhoeller, *Pflanzenschutz berichte*, **27**, 165 (1961).
78. J. Askew, J. H. Ruzicka, and B. B. Wheals, *Analyst (London)*, **94**, 275 (1969).
79. H. Ackermann, *J. Chromatogr.*, **36**, 309 (1968).
80. *Ibid.*, **44**, 414 (1969).
81. O. Antoine and G. Mees, *J. Chromatogr.*, **58**, 247 (1971).
82. M. E. Getz and H. G. Wheeler, *J. Assoc. Off. Anal. Chem.*, **51**, 1101 (1968).
83. T. Salo, K. Salminen, and K. Fiskari, *Z. Lebensm. Unters.-Forsch.*, **117**, 369 (1962).
84. H. Henzy and J. Thirion, *Beitr. Tabakforsch.*, **5**, 175 (1970); through *Anal. Abstr.*, **21**, 411 (1971).

85. A. M. Kadoum, *J. Agric. Food Chem.*, **18,** 542 (1970).
86. M. Geldmacher-Mallinckrodt, *Deut. Z. Ges. Gerichtl. Med.*, **54,** 90 (1963).
87. M. Geldmacher-Mallinckrodt and U. Weigel, *Arch. Toxikol.*, **20,** 114 (1963).
88. K. Stammbach, R. Delley, R. Suter, and G. Székely, *Z. Anal. Chem.*, **196,** 332 (1963).
89. J. Kováč, *J. Chromatogr.*, **11,** 412 (1963).
90. M. F. Kovacs, Jr., *J. Assoc. Off. Agric. Chem.*, **47,** 1097 (1964).
91. D. C. Abbott, N. T. Crosby, and J. Thomson, Lecture to Society of Analytical Chemistry Conference, Nottingham, England, 1965.
92. H. Woggon, D. Spranger, and H. Ackermann, *Nahrung*, **7,** 608 (1963).
94. R. Donner and Kh. Lohs, *J. Chromatogr.*, **17,** 349 (1965).
95. P. Bruaux, S. Dormal, and G. Thomas, Ann. Biol. Clin., **22,** 375 (1964).
96. D. L. Grant, C. R. Sherwood, and K. A. McCully, *J. Chromatogr.*, **44,** 67 (1969).
97. R. R. Watts, *Residue Rev.*, **18,** 105 (1967).
98. J. E. Cassidy, D. P. Ryskiewich, and R. T. Murphy, *J. Agric. Food Chem.*, **17,** 558 (1969).
99. R. R. Watts, *J. Assoc. Off. Agric. Chem.*, **48,** 1161 (1965).
100. T. H. Ragab, *Bull. Environ. Contam. Toxicol.*, **2,** 279 (1967).
101. R. C. Blinn. *J. Assoc. Off. Agric. Chem.*, **47,** 641 (1964).
102. P. A. Giang and M. S. Schecter, *J. Agric. Food Chem.*, **8,** 51 (1960).
103. A. M. Gardner, *J. Assoc. Off. Anal. Chem.*, **54,** 517 (1971).
104. M. Salamé, *J. Chromatogr.*, **16,** 476 (1964).
105. C. W. Stanley, *J. Chromatogr.*, **16,** 467 (1964).
106. M. Chiba and H. V. Morley, *J. Assoc. Off. Agric. Chem.*, **47,** 667 (1964).
107. R. Liebmann and H. Schuhmann, *Chem. Tech. (Berlin)*, **16,** 267 (1964).
108. K. Nagasawa, H. Yoshidome, and F. Kamata, *J. Chromatogr.*, **52,** 453 (1970).
109. J. M. Finocchiaro and W. R. Benson, *J. Assoc. Off. Anal. Chem.*, **50,** 888 (1967).
110. F. Geike, *J. Chromatogr.*, **53,** 269 (1970).
111. *Ibid.*, **58,** 257 (1971).
112. C. E. Mendoza and J. B. Shields, *J. Chromatogr.*, **50,** 92 (1970).
113. J. D. MacNeil and M. Hikichi, *J. Chromatogr.*, **101,** 33 (1974).
114. R. W. Frei, J. F. Lawrence, and P. E. Belliveau, *Z. Anal. Chem.*, **254,** 271 (1971).
115. R. G. W. Spickett, *Chem. Ind. (London)*, **1957,** 561.
116. J. G. Kirchner, J. M. Miller, and G. J. Keller, *Anal. Chem.*, **23,** 420 (1951).
117. E. Stahl, *Arch. Pharm.*, **293/65,** 531 (1960).
118. J. M. Miller and J. G. Kirchner, *Anal. Chem.*, **25,** 1107 (1953).
119. B. M. Olive, *J. Assoc. Off. Anal. Chem.*, **56,** 915 (1973).
120. N. Nash, P. Allen, A. Bevenue, and H. Beckman, *J. Chromatogr.*, **12,** 421 (1963).
121. Y. Doi, *Kagaku Keisatsu Kenkyusho Hokoku*, **16,** 51 (1963).
122. E. Kroeller, *Dt. Lebensm. Rundsch.*, **65,** 41 (1969).

123. N. E. Delfel and W. H. Tallent, *J. Assoc. Off. Anal. Chem.*, **52,** 182 (1969).
124. E. Tyihák and D. Vágujfalvi, *Acta Biol. Acad. Sci. Hung., Suppl.*, **5,** 77 (1963).
125. O. Nalbandov, R. T. Yamamoto, and G. S. Fraenkel, *J. Agric. Food Chem.*, **12,** 55 (1964).
126. R. C. Blinn and F. A. Gunther, *J. Assoc. Off. Agric. Chem.*, **46,** 204 (1963).
127. J. G. Wagner, P. G. Welling, K. P. Lee, and J. E. Walker, *J. Pharm. Sci.*, **60,** 666 (1971).
128. P. G. Welling, K. P. Lee, U. Khanna, and J. G. Wagner, *J. Pharm. Sci.*, **59,** 1621 (1970).
129. J. G. Wagner, *J. Pharm. Sci.*, **60,** 1272 (1971).
130. J. L. Ramaut and A. Benoit, *J. Pharm. Belg.*, **21,** 293 (1966).
131. H. A. Russel, *Z. Anal. Chem.*, **250,** 125 (1970).
132. C. A. Janicki, R. J. Brenner, and B. E. Schwartz, *J. Pharm. Sci.*, **55,** 1077 (1966).
133. V. A. Perry, *J. Assoc. Off. Anal. Chem.*, **53,** 737 (1970).
134. M. Beroza, *J. Agric. Food Chem.*, **11,** 51 (1963).
135. S. W. Gunner, *J. Chromatogr.*, **40,** 85 (1969).
136. L. Fishbein, J. Fawkes, H. L. Falk, and S. Thompson, *J. Chromatogr.*, **31,** 102 (1967).
137. E. P. Lichenstein, T. T. Lang, K. R. Schulz, K. R. Schnoes, and G. T. Carter, *J. Agric. Food Chem.*, **22,** 658 (1974).
138. L. Fishbein, J. Fawkes, H. L. Falk, and S. Jordan, *J. Chromatogr.*, **37,** 256 (1968).
139. L. Fishbein, *J. Chromatogr. Sci.*, **13,** 238 (1975).
140. L. Fishbein, H. L. Falk, and P. Kotin, *Chromatogr. Rev.*, **10,** 175 (1968).
141. J. G. Kirchner, J. M. Miller, and R. G. Rice, *J. Agric. Food Chem.*, **2,** 1031 (1954).
142. T. Salo and R. Maekinen, *Z. Lebensm. Unters.-Forsch.*, **125,** 170 (1964).
143. H. Gaenshirt and K. Morianz, *Arch. Pharm.*, **293/65,** 1065 (1960).
144. R. Rangone and C. Ambrosio, *J. Chromatogr.*, **50,** 436 (1970).
145. T. Salo, *Z. Lebensm. Unters.-Forsch.*, **124,** 448 (1964).
146. J. W. Copius-Peereboom and H. W. Beekes, *J. Chromatogr.*, **14,** 417 (1964).
147. M. Covello and O. Schettino, "The Application of Thin-Layer Chromatography to Investigations of Antifermentatives in Foodstuffs," in *Thin-Layer Chromatography,* G. B. Marini-Bettòlo, Ed., Elsevier, Amsterdam, 1964, p. 215.
148. M. Covello and O. Schettino, *Riv. Ital. Sostanze Grasse,* **41,** 337 (1964); through *Chem. Abstr.*, **61,** 15260 (1964).
149. B. Baehler, *Helv. Chim. Acta,* **45,** 309 (1962).
150. J. A. W. Gosselé, *J. Chromatogr.*, **63,** 433 (1971).
151. K. Nagasawa, H. Yoshidome, and R. Takeshita, *J. Chromatogr.*, **43,** 473 (1969).
152. H.-C. Chiang, *J. Chromatogr.*, **44,** 201 (1969).
153. I. Bajaj, K. K. Verma, O. Prakash, and D. B. Parihar, *J. Chromatogr.*, **46,** 261 (1970).

154. F. G. von Stryk, *J. Chromatogr.*, **72,** 410 (1972).
155. S. Abdullaev, G. L. Genkina, A. A. Atakuziev, T. T. Shakirov, and C. S. Kadyron, *J. Anal. Chem. USSR, Engl. Trans.*, **30,** 302 (1975).
156. J. Sherma, *J. Chromatogr.*, **104,** 476 (1975).
157. P. B. Baker, J. E. Farrow, and R. A. Hoodless, *J. Chromatogr.*, **81,** 174 (1973).
158. N. G. Porter, *J. Chromatogr.*, **28,** 469 (1967).
159. C. J. Briggs and S. B. Challen, *J. Sci. Food Agric.*, **18,** 602 (1967).
160. L. Fishbein, J. Fawkes, and P. Jones, *J. Chromatogr.*, **23,** 476 (1966).
161. R. C. Rhodes, H. L. Pease, and R. K. Brantley, *J. Agric. Food Chem.*, **19,** 745 (1971).
162. H. Sieper and H. Pies, *Z. Anal. Chem.*, **242,** 234 (1968).
163. H. Otteneder and U. Hezel, *J. Chromatogr.*, **109,** 181 (1975).
164. J. O'G. Tatton and P. J. Wagstaffe, *J. Chromatogr.*, **44,** 284 (1969).
165. F. Geike and I. Schuphan, *J. Chromatogr.*, **72,** 153 (1972).
166. L. Fishbein and J. Fawkes, *J. Chromatogr.*, **19,** 364 (1965).
167. G. Czeglédi-Jankó, *J. Chromatogr.*, **31,** 89 (1967).
168. J. Sherma, *J. Chromatogr.*, **113,** 97 (1975).
169. C. A. Bache, *J. Assoc. Off. Agric. Chem.*, **47,** 355 (1964).
170. K. Stammbach, H. Kilchher, K. Friedrich, M. Larsen, and G. Székely, *Weed Res.*, **4,** 64 (1964).
171. H. G. Henkel and W. Ebing, *J. Chromatogr.*, **14,** 283 (1964).
172. H. G. Henkel, *Chimia (Aarau)*, **18,** 252 (1964).
173. L. P. Manner, *J. Chromatogr.*, **21,** 430 (1966).
174. R. Delley, K. Friederick, B. Karlhuber, G. Székely, and K. Stammbach, *Z. Anal. Chem.*, **228,** 23 (1967).
175. J. Reichling and H. Fischer, *J. Chromatogr.*, **115,** 670 (1975).
176. J. F. Lawrence and G. W. Laver, *J. Chromatogr.*, **100,** 175 (1974).
177. D. C. Abbott, J. A. Bunting, and J. Thomson, *Analyst (London)*, **90,** 356 (1965).
178. L. Fishbein, *Chromatogr. Rev.*, **12,** 167 (1970).
179. D. C. Abbott, H. Egan, E. W. Hammond, and J. Thomson, *Analyst (London)*, **89,** 480 (1964).
180. D. C. Abbott and J. Thomson, *Chem. Ind. (London)*, **481,** (1964).
181. D. C. Abbott and J. Thomson, *Analyst (London)*, **89,** 613 (1964).
182. H. Geissbuehler and D. Gross, *J. Chromatogr.*, **27,** 296 (1967).
183. R. J. Hance, *J. Chromatogr.*, **44,** 419 (1969).
184. T. Golab, *J. Chromatogr.*, **18,** 406 (1965).
185. D. C. Abbott and P. J. Wagstaffe, *J. Chromatogr.*, **43,** 361 (1969).
186. W. Ebing, *J. Chromatogr.*, **65,** 533 (1972).

Chapter **XXVI**

PHARMACEUTICAL PRODUCTS

1 SCREENING BIOLOGICAL SAMPLES FOR DRUGS OF ABUSE

The detection and identification of drugs in biological materials is an every-increasing need. Many times the identity of a drug is needed quickly in order to save a life. The speed of thin-layer chromatography can fill this need. The main difficulty with thin-layer chromatography is that it may give false positive results, which can be due to the interference of other drugs or of non-drug substances. Therefore, it must be remembered that a certain R_f value or a given color reaction is only an indication and the results must be verified. This can take many forms: additional separations in different systems, spectrophotometric methods, gas chromatography, melting point, microcrystallography, multiple spraying techniques, immunoassays, etc. For legal work the confirmation should be by a method other than thin-layer chromatography. Thin-layer chromatography is especially effective where large numbers of samples must be run, as in urine drug testing procedures. These can be run very quickly eliminating those with negative results, and giving some clue as to the nature of those with positive results. In contrast, the expense of gas chromatographic and high performance liquid chromatographic equipment precludes running numerous samples at one time.

A number of methods have been presented for concentrating and removing drugs from biological material. In general, urine is the preferred biological sample although in some cases, as with some barbiturates (amylobarbitone and pentobarbitone), it may be impossible to detect unchanged barbiturates in the urine [1]. In this respect it should be remembered that metabolites and conjugates of the drug may appear in the urine [2]. The appearance of specific metabolites can be additional evidence for the ingestion of a particular drug.

Dole et al. [3] used ion-exchange resin-loaded paper to remove drugs from a urine sample, and subsequently obtained the compounds by extraction of the paper with organic solvents at controlled pH values. Mulé [2], Montalvo et al. [4], and Heaton and Blumberg [5] examined this technique. Mulé placed 5 × cm squares of Reeve Angel ion-exchange paper SA-2 with 50 ml of urine (pH 5 to 6) in 4-oz jars and shook these for 30 min at 80 rpm on an International Shaker machine. After the urine was poured off, the papers were washed twice with distilled water. In order to elute the barbiturates, the SA-2 paper was shaken for 10 min with 20 ml of citrate buffer (pH 2.2) and 10 ml of chloroform. The phases were separated and an additional extraction was carried out with 10 ml of chloroform for 1 min. Narcotic analgesics and psychotic drugs were eluted with 20 ml of borate buffer (pH 9.3) and 20 ml of chloroform–isopropanol (3:1), shaking for 10 min. The *d*-amphetamine and analogs were eluted with 20 ml of carbonate buffer (pH 11.0) and 20 ml of chloroform with shaking for 10 min. After separation of the phases, 50 to 100 μl of glacial acetic acid was added to the chloroform phase. All the organic phases were evaporated to dryness and the residues dissolved in 25 to 50 μl of methanol or chloroform. Methadone could not be detected below 5 to 10 μg/ml of urine, because of the difficulty of extraction by this technique. It was also inadequate for ephedrine, phenmetrazine, psilcybin, glutethimide, chlorpromazine, lysergic acid amide, marihuana, and some sample of barbiturates. The extraction of amphetamines from the resin paper was erratic [5]. Kaistha and Jaffe [6] found the resin-loaded paper technique advantageous for pooling samples of the same patient representing different urine specimens in order to cut down the cost of analysis without decreasing the sensitivity of the test. They preferred this method over the direct extraction method if at least 50 ml of urine was available.

Another method of extraction is by means of the direct liquid–liquid separations. Mulé [2] also used this procedure and extracted 15 ml of urine under the same pH conditions and with the same volume of solvent as explained above for the resin paper except that only 10 ml of solvent was used for the *d*-amphetamine and its analogs. The extractions were carried out by shaking in glass-stoppered centrifuge tubes for 15 min at 300 oscillations/min. Almost all the drugs of abuse tested could be detected in levels of 1 to 2 μg/ml using 15-ml samples. Serfontein et al. [7] developed a micro-phase extraction technique employing organic solvent extractions from solutions containing aqueous buffers of various pH values and conversely extractions of drugs from organic phases by extracting with various aqueous buffers. This method of extraction eliminated concentration steps and yielded "cleaner" chromatograms. Broich et al. [8] extracted the acid–neutral and the basic fractions from separate 2-ml portions of urine sample. After the acid–neutral fraction was isolated in chloroform, this extract was washed with saturated sodium bicarbonate solution to eliminate many interfering materials.

Fujimoto and Wang [9] introduced the use of Amberlite XAD-2 resin (Rohm & Haas) for isolating narcotic analgesics from the urine, and Mulé and co-workers (10–12) extended and improved the technique to include other drugs. The final technique consisted in washing the resin by stirring with four bed volumes of acetone, three times with three bed volumes of methanol, and three times with three bed volumes of distilled water. It was then stored for 7 to 14 days under water in the refrigerator, and subsequently washed twice with one bed volume of water before transferring to the column. Twenty-five milliliters of urine adjusted to pH 8 to 9 with 10% sodium hydroxide was run through a 135 × 10 mm column containing the slurry of XAD-2 resin (1.8 g dry weight) in 20 to 25 min. Two separate elutions were made to isolate the acid and basic components from the resin column. Ten milliliters of isopropyl ether was first passed through the column into a 50-ml centrifuge tube containing 1 ml of 0.1 *N* hydrochloric acid. The tube was then shaken thoroughly in a Genie Vortex mixer. After standing the separated water layer was frozen and the isopropyl layer was poured off. Evaporation of the isopropyl ether left the residues of primarily the acid with some neutral drugs. The resin column was then eluted with two 10-ml portions of chloroform–isopropyl ether (3:1) directly into the centrifuge tube containing the aqueous layer from the first extraction. One milliliter of 0.125 *M* borax solution was added to the centrifuge tube to adjust the pH to 8 to 9.5. After being shaken in the Genie Vortex mixer, the upper aqueous phase was removed and 200 μl of 6 *N* hydrochloric acid in methanol was added to the organic phase. Evaporation at 80°C under an air stream yielded a residue containing mainly the basic with some neutral drugs. Weissman et al. [12a] using the resin column extraction technique, found detection limits of 0.4 μg/ml for amphetamines and barbiturates, and 0.8 μg/ml for alkaloids from 5-ml urine specimens.

Broich et al. [13] isolated drugs from urine by lyophilizing the urine after adding 2.0 ml of glacial acetic acid to 5 ml of the urine. The lyophilized residue was mixed with 3 ml of methanol and allowed to stand for 1 hr. Acetone (9 ml) was then added and the mixture was filtered and evaporated to dryness. This residue could then be submitted to thin-layer chromatography for the separation of the acidic, neutral, and basic compounds, or a preliminary separation could be made by mixing 10 ml of ether with the residue. After standing 30 min and filtering, the filtrate contained acid and neutral compounds with possibly small amounts of the acetates of codeine, quinine, or methadone, for example. The residue from this last extraction was dissolved in methanol to obtain the basic drugs. A slight modification of this method has been used to extract drugs of abuse and pharmaceuticals from bile for which direct extraction procedures are not applicable [14].

Meola and Vanko [15] concentrated drugs in urine by adsorbing them on buffer-moistened charcoal after adding carbonate–bicarbonate buffer (pH 11.0) to the urine. Ethyl ether was used to elute barbiturates, glutethimide, and cocaine from the charcoal. Chloroform–isopropanol was then used to remove amphetamines, alkaloids, and other drugs.

Typical solvents for separating these drugs after removal from the urine include: ethanol–dioxane–benzene–ammonia (5:40:50:5) [2], ethyl acetate–methanol–ammonia (85:10:5) and (85:10:1) [2, 8], methanol–*n*-butanol–benzene–water (60:15:10:15) [2] chloroform–acetone (9:1) [2], hexane–ethanol (93:7) [8], ethyl acetate–methanol–water–ammonia (85:10:3:1) [10], and chloroform–methanol–ammonia (90:10:1) [10]. Additional solvents can be found under the individual classes of compounds.

Usually a succession of sprays are used to detect the various drugs of abuse. As an example of this, Broich et al. [8] first examined the acid/neutral compounds under ultraviolet (a brief exposure to ammonia fumes enhanced the ultraviolet absorption). The area above the phenobarbital standard was then covered and the remainder of the plate was treated with *N*,2,6,-trichlorobenzoquinoneimine (T-265) to disclose the barbiturates and structurally similar compounds as purple or violet spots. The entire plate was then sprayed with mercurous nitrate (T-161) intensifying the barbiturate spots and revealing methylprylon, ethinamate, and carbromal. Meprobamate was next visualized with a solution of 1 g of vanillin in 100 ml of methanol + 2 ml of concentrated sulfuric acid, followed by gentle heating to produce a yellow spot which turned black if sprayed with a solution of 0.5 ml of furfural in 50 ml of methanol + 2 ml of concentrated hydrochloric acid. The plate with the basic compounds was sprayed with a 0.1% ninhydrin solution in methanol and gently heated to disclose amphetamine as a yellow spot. Exposure to 360-mμ ultraviolet light for 10 min revealed methamphetamine as a violet spot and amphetamine as red violet. The plate was then sprayed with iodoplatinate reagent (T-149), air-dried, and sprayed with Dragendorff reagent (T-111). The colors produced were listed for 31 compounds. Additional spray reagents can be found under the individual classes of compounds.

Kaistha and Tadrus [16] have compared the costs and speed of urine screening by thin-layer chromatography and other methods, and Kaistha [17] has reviewed drug abuse screening programs. Sohn et al. [18] has reviewed the preparatory procedures in screening for drugs of abuse. Two bibliographies have been published on drugs of abuse [19, 19a].

2 HYPNOTICS

The widespread use of barbituric acid derivatives and other hypnotics has led to a demand for a rapid method of differentiating between the various drugs. For this reason a number of workers have applied themselves to the use of thin-layer chromatography in this field. Machata [20] demonstrated the feasibility of separating barbiturates on silica gel layers using an ether–chloroform (3:17) mixture. Baeumler and Rippstein [21] separated a group of 17 hypnotics on silica gel with a solvent of chloroform–acetone (9:1). For detection they sprayed the plates with a mercurous nitrate solution.

Porges [22] described the separation of five barbiturates on layers of acidic, alkaline, and neutral silica gel. Reisch et al. [23] report the R_f values for seven barbiturates in two different solvents on thin layers made of a mixture of one part ion-exchange resin and nine parts silica gel G. In analyzing tablets containing mixtures, a preliminary separation was carried out on a column of cation exchanger using 40% methanol to wash the acidic materials through the column. The basic materials were then eluted from the column using a methanol–ammonia mixture. For detecting the barbituric acid derivatives, the plates were sprayed with 1% silver nitrate solution followed by a 1% mercurous nitrate solution. Frahm et al. [24, 25] used silica gel G layers in saturated chambers in examining 18 narcotics, 12 barbiturates, and 6 nonbarbiturates used as soporifics. The compounds with their degradation products were extracted from acidified urine with ether. The developing solvent for the thin layers consisted of isopropanol–25% ammonium hydroxide–chloroform (9:2:9). Niwaguchi and Oki [26] used starch-bound silicic acid plates in identifying nonbarbituric sedatives. Acetone–ethylene dichloride (3:17) and benzene–dioxane–28% ammonium hydroxide (15:4:1) were used as solvents. Sahli and Oesch [27] have chromatographed 13 barbiturates and a number of hydantoin derivatives on silica gel G in benzene–dioxane (5:2), chloroform–acetone (9:1), and benzene–ether (1:1). Detection of the compounds was with a 1% solution of mercurous nitrate solution. Shellard and Osisiogu [28] chromatographed 12 barbiturates on silica gel G in two and sometimes three solvent systems for identification. The solvent systems that were used were: chloroform, isopropyl ether–chloroform–cyclohexane (2:2:1), isopropyl ether–chloroform–benzene (13:8:4), isopropyl ether–chloroform (1:1), isopropanol–chloroform–28% ammonium hydroxide (9:9:2), isopropyl ether–benzene–diethyl ether (2:2:1), acetone–benzene (1:1), and isopropyl ether. In order to detect the barbiturates the plates were sprayed with 5% cobalt nitrate in ethanol followed by exposure to ammonia vapors.

Curry and Fox [29] separated a group of barbiturates on cellulose layers impregnated by dipping in 10% sodium orthophosphate using *n*-amyl methyl ketone as a developing solvent, and Dutkiewicz and Kończalik [30] used cellulose layers impregnated with ammonium nitrate combined with a developing solvent of isopropanol–chloroform–20% ammonium hydroxide (6:3:2). Polyamide has also been used for the separation of barbiturates [31]. Among the solvents used were chloroform–diethyl ether–acetic acid (4:1:0.05) and diethyl ether–*n*-hexane–chloroform–acetic acid (2:1:1:0.05). Detection on the polyamide layers was accomplished by spraying with 0.005% fluorescein in 0.5 *N* sodium hydroxide and observing under ultraviolet light. Sensitivity of detection was 0.5 to 1 μg.

Rosenthal et al. [32] used silica gel G layers with an acidic solvent in a tank containing ammonia vapor thus causing a gradual decrease in the acidity as the solvent front advanced resulting in improved separation of barbiturates. The solvent was chloroform–butanol–formic acid (140:80:7).

Baeumler and Rippstein [33] have separated carbromal and two of its metabolic products on thin layers of silica gel G using chloroform–acetone (9:1) as a solvent. The R_f values are reported as follows: carbromal 0.65, 2-ethyl-2-hydroxybutyric acid 0.30, and 2-ethylbutyrylurea 0.25. The mercurous nitrate spray reagent is not very sensitive to carbromal so the authors turned to a reagent developed for insecticides (T-94). This test is sensitive to 5 γ of carbromal. To detect the 2-ethylbutyrylurea the plate is exposed to a chlorine atmosphere for 5 min and then it is sprayed with the reagent just mentioned to give a violet spot. The remaining metabolic products can be detected by spraying with an acidic solution of potassium permanganate giving a green spot on a reddish background. Lindfors and Ruohonen [34] used the same chromatograhic system for separating carbromal and bromisovalum. For the latter an R_f value of 0.37 is reported and for 3-methylbutyrylurea (metabolite of bromisovalum) 0.17. For a detecting agent, these authors exposed the plates to a chlorine atmosphere for 5 min. Then after excess chlorine was removed by heating 10 min at 105°C, the plate was sprayed with a mixture of 100 ml of 0.5% benzidine acetate and 2 ml of 10% potassium iodide solution. This reagent gave gray-violet spots and less than 0.5 γ could be detected. Lindfors [35] gives the following R_f values for a group of nonbarbituric soporifics on silica gel with chloroform–diethyl ether (17:3): acetylcarbromal 0.39, aponal 0.22, bromisovalum 0.23, 3-methylbutyrylurea 0.10, carbromal 0.54, 2-ethylbutyrylurea 0.18, ethinamate 0.50, and glutethimide 0.66.

Vercruysse [36] used silica gel G with petroleum ether–methanol (1:1) in detecting glutethimide in postmortem material. Dragendorff reagent could detect amounts above 10 γ.

Gaenshirt [37] investigated a series of detecting agents for antineuralgic and soporific drugs. In general the latter can be detected by ammoniacal silver nitrate solution, by potassium iodide–benzidine solution after exposing to chlorine vapors, and by the use of fluorescent plates.

Lehmann and Karamustafaoglu [38] set up a procedure for analyzing for barbiturates in blood serum. The barbiturates are extracted with chloroform from the acidified serum by shaking 15 ml of chloroform, 0.1

ml of concentrated hydrochloric acid, 3 ml of serum, and 2 g of anhydrous sodium sulfate. After separating the extract, 10 ml of the chloroform solution is evaporated to dryness on a water bath. The residue dissolved in 0.2 ml of ethanol is spotted on a silica gel G plate and developed in a filter paper-lined chamber using chloroform–*n*-butanol–ammonium hydroxide (70:40:3.5) as a solvent (note: the chloroform contains 1% ethanol as a stabilizer). Development is carried out with a small beaker of concentrated ammonium hydroxide within the developing chamber in order to keep the chamber saturated with ammonia. If the latter is not carried out, the separation is not as sharp, and the entire group of compounds show increased R_f values. Two visualizing agents may be used: a 0.05% potassium permanganate solution for detecting unsaturated compounds, or a combination spray of 0.1% *sym*-diphenylcarbazide in 95% ethanol followed by 0.33% mercurous nitrate in 0.04 *N* nitric acid. In the latter case, the plates are exposed to sunshine or ultraviolet light in order to bleach the background leaving the compound locations as distinct violet spots. The authors point out that in correlation with the observations in paper chromatography [39] the speed of action of the barbiturate is directly related to the R_f value; that is, the faster acting barbiturates have higher R_f values than the slower acting compounds. In applying a similar method, Petzold et al. [40] used a solvent composed of acetone–*n*-butanol–ammonium hydroxide (9:9:2). The samples were run in duplicate spots, one spot of each sample being treated on the plate with 5 μl of 4 *N* sulfuric acid prior to running the chromatogram. After the acid treatment, the plates were heated for 1 hr at 125°C. With the five barbiturates listed, this treatment changed the R_f values profoundly so that good differentiation was observed. In recovering the barbiturates from the blood samples, the chloroform extract of the blood was passed through a Florisil column which was then eluted with 10% methanol in chloroform. The eluate was evaporated to dryness and the residue was taken up in 100 μl of ethanol. This solution was then used to spot the samples on the silica gel plates. Kelleher and Rollason [41] have used a microchromatoplate for the detection of barbiturates in the blood. De Zeeuw and Wijsbeek [42] used a reaction on the layer with bromine prior to chromatography to help identify closely related thiobarbiturates and barbiturates.

Three of the numerous papers on separations of this group of compounds have been selected to present a group of R_f values in Table 26.1. All three have used silica gel as the adsorbent with various combinations of solvents. Cochin and Daly [43] used three solvents, which permitted two-dimensional chromatography to be carried out, and two detecting agents, which assisted in cases where the R_f values were close. Amo-

Table 26.1 R_f Values of Some Hypnotics on Silica Gel G in Various Solvents, and Some Color Reactions

Hypnotic	Solvents[a]					Color Reagent[c,d]	
	A [43]	*B* [43]	*C* [43]	*D* [46]	*A*[b] [44]	*x* [44]	*y* [46]
Barbital	0.50	0.38	0.40	0.25	1.00	V	G
Phenobarbital	0.50	0.36	0.26	0.1	1.00	V	G
Cyclobarbital	0.68	0.55	0.38	0.2	+[e]	B	G
Vinbarbital					1.19	V	
Heptabarbital				0.2	1.32	V	G
Butabarbital	0.62	0.49	0.52		1.28	V	
Amobarbital	0.60	0.42	0.52		1.44	V	
Aprobarbital					1.30	B	
Allobarbital	0.55	0.43	0.48		1.30	B	
Pentobarbital	0.57	0.40	0.49		1.42	V	
Allylbarbituric acid					1.44	B	
Secobarbital	0.64	0.46	0.54		1.67	B	
Hexobarbital	0.77	0.49	0.58		2.06	V	
Metharbital	0.85	0.53	0.66		2.35	V	
Mephabarbital	0.98	0.85	0.60		2.37	V	
Thiopental	0.94	0.73	0.70		+		
Thiamylal	0.95	0.80	0.70		+	B	
Itobarbital	0.67	0.48	0.50				
Cyclopal					1.38	B	
Butethal					1.41	V	
Talbutal					1.41	B	
Butallyonal					1.50	B	
Hexethal					1.56	V	
Sigmodal					1.77	V	
Methyprylon	0.50	0.81	0.45	0.5			R
Glutethimide	0.80	0.54	0.99	0.9			G
Ethinamate	0.81	0.65	0.55				
Ethchlorvinol	0.95	0.32	0.95				
Dihydroprylon				0.4			G
Carbromal				0.75			W
Bromisovalum				0.6			Br
Ethinamate				0.45			R
2-Methyl-3-*o*-tolyl-4(3*H*)-quinazolinone				0.8			W

[a] *A* = Chloroform–acetone (9:1), solvent travel 10 cm. *B* = Benzene–acetic acid (9:1), solvent travel 10 cm [43]. *C* = Dioxane–benzene–ammonium hydroxide (4:15:1) (saturated atmosphere), solvent travel 10 cm [43]. *D* = Piperidine–petroleum ether (50 to 70°C) (1:5) [46].

[b] R_{st} referred to phenobarbital.

barbital and pentobarbital could not be separated even with two-dimensional chromatography. These two compounds may be separated nicely by the procedure of Petzold et al. [40] as described above, whereby the R_f values after acid treatment, are 0.74 and 0.40, respectively. Chromatographic patterns are shown for the excretion of a number of the drugs, thus showing the metabolic products as well as the original drugs. Sunshine et al. [44] used, in addition to the permanganate spray, a combination spray of mercuric sulfate followed by a 0.001% diphenylcarbazone in chloroform (wt/vol) to detect all the compounds listed. An additional spray that could be used, prior to permanganate spray, was a sodium fluorescein spray for detection of compounds under an ultraviolet light. Much more convenient than this fluorescent spray are the fluorescent layers [45]. Eberhardt et al. [46] followed the appearance of metabolic products of 11 drugs by analyzing urine samples after 3, 6, 12, and 24 hr. The R_f values of the metabolic products are given (Table 26.1) along with those of the original substances.

Dijkhuis [47] located barbiturates by the temporary quenching of fluorescence produced by ammonia vapor. This reaction is highly selective and permits detection to 0.2 μg.

3 PSYCHOTROPIC DRUGS

Tranquilizers

This group of compounds has suddenly become of great importance and is widely used. In this group of course are the Rauwolfia alkaloids which are discussed in Chapter XV, Section 10. The largest group used in the therapy of mental disorders are the phenothiazine derivatives.

Baeumler and Rippstein [21] separated a group of these compounds on silica gel layers by using a methanol–acetone–triethanolamine (1:1:0.03) solvent mixture (Table 26.2). The compounds were located by examination under ultraviolet light and by spraying with palladous chloride and with a modified Dragendorff reagent. Paulus et al. [48], in applying silica gel layers to the separation of a similar group of compounds, used benzene–acetone–25% ammonium hydroxide (10:2:1) as the separating sol-

---

[c] x = Color with mercuric sulfate–diphenylcarbazone spray (see text) [44]. y = Color with 1% mercurous nitrate on eosin-containing silica gel [46].

[d] V = violet, B = blue, G = gray, R = red, Br = brown, W = white. Compounds for which no colors are given can be detected with 0.2% potassium permanganate.

[e] + = Runs with solvent front.

Table 26.2 $R_f \times 100$ Values of Phenothiazine and Drugs of Similar Action on Silica Gel G in Various Solvent Systems as well as the Color Reaction with Sulfuric Acid

	Solvent[a]												
Drug	*A*	*B*	*C*	*D*	*E*	*F*	*G*	*H*	*I*	*J*	*K*	*L*	*Color with Sulfuric Acid*
Acetophenazine	12	36	4	52	18	38							Orange-pink[b]
Butyrylperazine	8	31	6	28	10	45							Orange-pink[b]
Chlorpromazine	37	44	28	14	23	66	94	70	30		5	49–50	Reddish pink[c]
Chlorpromazine sulfoxide	5	10	1	9	5	27	26	47	19	86			Reddish pink[c]
Chlorprotixene	34	78		20	36	88	80	73	46		69	55–57	Light orange[c]
N-Demethylchlorpromazine sulfoxide							20	56	10	68			Reddish pink[c]
Diethazine											81	53–55	Red[d]
N,N-Didemethylchlorpromazine sulfoxide							23	68	18	72			Reddish pink[c]
Ethopropazine											94	57–59	Red[d]
Fluphenazine	37	56	9	68	27	57	34	58	68				Orange[c]
Levomepromazine											72	62–64	Blue[d]
Librium												84–86	
Mepazine	29	46	13	13	13	62	88	57	29		45		Orange[c]
Mepazine sulfoxide							16	59	24	74			Reddish pink[c]
Methdilazine							40	61	16				Red[c]
Methoxypromazine	15	26	12	12	9	45	74	58	24				Purple[c]
Omca											12		Yellow[d]
Perazine											17		Yellow-orange[d]
Perphenazine	28	57	7	48	24	53	36	44	48		9		Reddish pink[c]
Phenothiazine												89–91	
Pipamazine	38	71		41	37	79	12	72	56				Reddish pink[c]

Prochlorperazine	10	31	6	24	8	55	70	27	32		26		Pink[c]
Prochlorperazine sulfoxide							13	28	13	88			
Proketazine	19	45	6	53	25	44							Orange-pink[b]
Promazine	16	31	12	11	11	50	62	38	37		37	37–39	Reddish orange[d]
Promazine sulfoxide							17	46	15	70			Reddish pink[c]
Promethazine							70	59	22		52	55–57	Orange[c]
Promethazine sulfoxide							22	57	30	78			Reddish pink[c]
Prothipendyl	16	21	11	11	8	44							Yellow[b]
Thiethylperazine	14	50		23	13	61							Blue[b]
Thiopropazate							97	67	70				Reddish pink[c]
Thiopropazine	64	79	49	65	53	81							
Thioridazine	24	39	14	24	15	64	97	65	20		45	47–49	Greenish blue[c]
Thioridazine sulfoxide	3	13	1	9	3	33							Blue[b]
Trifluoperazine	18	33	9	34	12	51	69	33	40				Orange[c]
Trifluoperazine sulfoxide	3	11	2	17	3	31							Orange[b]
Trifupromazine	36	50	48	22	22	72	95	79	40		62		Orange[c]
Triflupromazine sulfoxide							26	48	24	93			Orange[c]
Trimeprazine							96	64	30				Orange[c]

[a] A = *tert*-Butanol–1 N ammonium hydroxide (9:1) [49]. B = *n*-Propanol–1 N ammonium hydroxide (22:3) [49]. C = Diethyl ether saturated with water, overrun [49]. D = Methanol–water (7:3) [49]. E = *n*-Propanol–water (17:3) [49]. F = *n*-Butanol saturated with 1 N ammonium hydroxide [49]. G = Benzene–dioxane–ammonium hydroxide (12:7:1), development distance 10 cm [50]. H = Ethanol–acetic acid–water (5:3:2), development distance 10 cm [50]. I = Methanol–butanol (3:2), development distance 10 cm [50]. J = Benzene–dioxane–ammonium hydroxide (1:8:1). Development distance 10 cm [50]. K = Benzene–acetone–25% ammonium hydroxide (10:2:1). Development distance 11.5 cm [48]. L = Methanol–acetone–triethanolamine (1:1:0.03) [3].

[b] Color with 40% sulfuric acid [49].

[c] Color with 5.0% sulufric acid–ethanol (4:1) [50].

[d] Color with sulfuric acid–ethanol (10%) [48].

vent (Table 26.2). Additional values obtained in benzene–ethanol–25% ammonium hydroxide (10:2:1) are as follows: perphenazine 0.49, Omca 0.64, perazine 0.81, promazine 0.92, and Librium 0.64. Several reagents were used for the detection of the compounds including a modified Dragendorff reagent which consisted of freshly mixing 2 ml of Dragendorff reagent, 3 ml of acetic acid, and 10 ml of water; with this reagent all the compounds showed as yellow-orange spots. With a 0.05% potassium permanganate solution, all the compounds appeared as orange-red spots on a rose background except imipramine, which showed up as a green spot. A 10% sulfuric acid in ethanol solution gave various colors to the compounds; however, Taractan showed a color only after 1 hr. A 10% solution of hydrogen peroxide and a 10% solution of nitric acid in ethanol both gave colors similar to those developed with 10% sulfuric acid. Mellinger and Keeler [49] examined a group of phenothiazine compounds by paper chromatography, paper electrophoresis, and thin-layer chromatography on silica gel. The latter gave the best separations, and the R_f values were reported in six different solvents (Table 26.2). In order to detect the compounds, the plates were sprayed with a 40% sulfuric acid solution. The compounds could also be located by examining the plates under ultraviolet light, preferably at 263 mμ.

Cochin and Daly [50] examined a group of 26 phenothiazine and related compounds (Table 26.2) on silica gel in a number of different solvents. The method is set up for the isolation and identification of the compounds in body fluids and tissues. Ethylene dichloride containing 10% isoamyl alcohol is used as a solvent for extracting the compounds from urine which has been adjusted to pH 9.0. Transfer to an ethanol solution is accomplished by first evaporating the ethylene dichloride solution in vacuo. For extracting tissue material, the tissue is first homogenized with 2 to 4 vol of isotonic potassium chloride before applying the extraction procedure as just outlined. Proteins do not interfere; therefore it is unnecessary to precipitate them. After chromatography in one of the solvents given in Table 26.2, the spots are detected by spraying either with a 50% sulfuric acid–ethanol (4:1) or with a potassium iodoplatinate reagent (T-209). The sulfuric acid reagent is especially useful for differentiating compounds that have very similar R_f values because of the difference in colors of the various compounds. It also helps to differentiate the sulfoxides from the parent compounds because the former react much more slowly with the sulfuric acid spray. To differentiate the sulfoxides from the phenothiazines, the plate may also be sprayed with a 2% solution of ferric chloride. This reagent does not react with the sulfoxides but forms red to violet colors with the phenothiazines.

Kofoed et al. [51] used a reaction technique to help identify phenothiazines. Forth phenothiazines were chromatographed on silica gel and then after conversion to the sulfoxides by applying 10 to 20% hydrogen peroxide to the spots with subsequent drying in a hot air stream (60°C), the sulfoxides were chromatographed at right angles. The R_f values for both the phenothiazines and the sulfoxides were reported as well as the spectrophotomaxima for the eluted sulfoxides.

Zingales [52] reported the R_f values of 45 psychotropic drugs in five thin-layer systems along with the color reactions in daylight and ultraviolet light with five detection agents. Kraus and Dumont [53] examining seven phenothiazines on pH gradient layers, determined the optimum pH around 8.5. Hulshoff and Perrin [54] characterized 26 phenothiazine drugs by reverse-phase thin-layer chromatography on oleyl alcohol-impregnated kieselguhr G using 30, 40, and 50% (wt/wt) methanol–water solvents at various pH values.

Using a 5% solution of ammonium sulfate saturated with isobutanol, Noirfalise and Grosjean [55] chromatographed a group of phenothiazine derivatives on cellulose layers.

Ferrari and Tóth [56] examined a number of drugs and their urinary metabolites on silica gel layers. For chlorpromazine and chlorprotixene they used an *n*-butanol–acetic acid–water (88:5:7) solvent, and for imipramine and amitriptyline they used a 65:15:20 mixture. Concentrated sulfuric acid was used for detecting the compounds both in visible and in ultraviolet light.

Seno et al. [57] examined the oxidative and cleavage compounds of prochlorperazine obtained when solutions of the latter are exposed to sunlight. As many as 11 products were separated on silica gel plates using ethylene dichloride–methanol–ammonium hydroxide (13:7:1). Prochlorperazine labeled with ^{35}S proved useful, both as a means of detection and as a quantitative agent. The quantitative work was carried out using a scintillation counter. Autoradiograms were more sensitive to the detection of the spots than were the chemical spraying agents which were used. Rusiecki and Henneberg [58] have investigated the degradation products of chlorpromazine. Eberhardt et al. [59, 60] examined a group of phenothiazine derivatives found in urine. They also examined a series of color producing agents used in detecting compounds of this type. One microgram of the compounds could be detected using sulfuric acid–ethanol (1:9), 10% phosphomolybdic acid, perchloric acid (d = 1.67), or 65 to 68% nitric acid. With 10% ferric chloride solution or iodine–potassium iodide solution (0.25 g of iodine, 0.5 g of potassium iodide, 150 ml of water), the sensitivity fell to 5 μg. Table 26.3 gives the R_f values obtained in pyridine–petroleum ether (50 to 70°C)–methanol (1:4.5:0.1), as well as the color reactions obtained with the various locating reagents.

Table 26.3 $R_f \times 100$ Values of Some Phenothiazine Derivatives on Silica Gel in Pyridine–Petroleum Ether (50 to 70°C)–Methanol (1:4.5:0.1) with Their Color Reactions[a]

Compound	$R_f \times 100$	*Sulfuric Acid–Ethanol (1:9)*	*Perchloric Acid*	*Ferric Chloride*	*Phospho-molybdic Acid*	*Iodine–Potassium Iodide*[b]	*Nitric Acid*[b]
Butyrylperazine	10	Yellow	Red-brown	Yellow	Yellow	Yellow	Yellow-brown→yellow
Perazine	7	Red-brown	Brown	Brown	Brown	Yellow	Brown→yellow
Promazine	17	Red-brown	Pink	Brown	Red-brown	Yellow→green	Red-brown→yellow
Promethazine	27	Pink	Red	Weak red	Pink	Yellow→green	Pink→yellow
Chlorperphenazine	12	Red	Red	Red	Red	Yellow→orange	Red→yellow
Mepazine	22	Red-brown	Pink	Brown	Red-brown	Yellow	Red-brown→yellow
Chloropromazine	38	Red	Dark red	Red	Red	Yellow→green	Red→dark brown
Levomepromazine	47	Violet	Violet	Violet	Violet	Yellow	Violet→yellow
Thioridazine	30	Green	Green	Green	Green	Yellow	Green→red-violet
Triflupromazine	54	Brown	Brown	Weak brown	Weak brown	Yellow-brown	Bright brown→yellow

[a] From H. Eberhardt, O. W. Lerbs, and K. J. Freundt [60]; reproduced with permission of the authors and Editio Cantor K. G.

[b] → = First color changes to second color.

Turano and Turner [61] separated and determined chlorpromazine and its metabolites by two-dimensional thin-layer chromatography and direct-scanning microdensitometry. Kaul et al. [62] converted the metabolites to the dansyl derivatives before chromatographing. Detection down to 0.314 ng was possible by fluorimetry.

Tewari [63] separated a group of 23 psychotropic drugs by electrophoresis on silica gel layers at 500 V using a sodium tetraborate buffer.

Eiden and Stachel [64] found that among the various thin-layer mediums that were tried basic alumina gave the best separation; these layers were then tried with benzene–acetone mixtures of various concentrations. Table 26.4 shows a comparison of the R_f values in several systems.

Fiori and Marigo [65] and Marigo [66, 67] have published a method for the isolation and identification of meprobamate from urine. The drug is extracted from an alkaline solution of urine (0.2 ml of 1 *N* sodium hydroxide, 10 ml of ether, 5 ml of urine). After the ether extract is treated with 60 mg of charcoal it is evaporated, the residue is taken up in alcohol

Table 26.4 $R_f \times 100$ Values of Phenothaizines on Aluminum Oxide and on Cellulose Layers[a]

	Woelm Neutral Alumina	*Woelm Basic Alumina*		*Cellulose Layers*
Phenothiazine	*Benzene–Acetone* (95:5)	*Benzene–Acetone* (95:5)	*Benzene–Acetone* (90:10)	*Water–Acetone* (70:30)
Trifluorperazine	20	40	53	35
Padisal	0	0	0	55
Selvigon	0	4	18	92
Chlorpromazine	31	64	85	46
Isothiazine	79	91	95	60
Combelen	13	38	69	49
Randolectil	10	25	55	37
Diethazine	60	93	95	55
Isothipendyl	24	57	78	65
Prothipendyl	13	45	70	58
Aminopromazine	13	42	63	60
Levomepromazine	21	83	90	62
Promazine	20	54	70	57
Promethazine	35	69	79	60

[a] From F. Eiden and H. D. Stachel [64]; reproduced with permission of the authors and Deutsche Apotheker-Verlag.

for application to the thin-layer plate. Starch-bound silicic acid was used with a solvent composed of cyclohexane–absolute ethanol (17:3) as the developing solvent. Quantitatively, the meprobamate which occurs at an R_f value of 0.30 may be determined by transferring the spot to 1 ml of distilled water and adding 1 ml of 0.2% hydroquinone in concentrated sulfuric acid. After heating for 20 min in a boiling water bath the absorption maximum is determined at 420 mμ.

For the detection of Librium, Baeumler and Rippstein [68] have adopted the procedure of heating the sample with 36% hydrochloric acid for 2 hr in order to hydrolyze and free the metabolic products from combination with glucuronic acid. During this hydrolysis, Librium is converted to 2-amino-5-chlorobenzophenone which can be extracted from the neutralized hydrolysate mixture and chromatographed on silica gel. With benzene as a developing agent, it has an R_f value of 0.5 and can be located as a red spot by diazotization and coupling with β-napthol. As small an amount as 1 γ can be detected. The hydrolysis product can be determined quantitatively by eluting from the plate with benzene and measuring the absorption at 383 mμ.

Schuetz et al. [69] chromatographed five tranquilizers of the 5-phenyl-1,4-benzodiazepine class on silica gel with heptane–ethanol–ethyl acetate–25% ammonia (18:2:14:1). The compounds were then converted into benzophenone derivatives by spraying with 37% hydrochloric acid and then to the free bases by exposing to ammonia vapors. A second development at right angles was then made with benzene. The layer was then exposed to nitrous oxide fumes and sprayed with Bratton–Marshall reagent (T-89) yielding purple-red and violet spots. Sensitivity was 0.02 μg.

Several reviews have been published on the detection and determination of phenothiazines [70–72].

Hallucinogenic Drugs

Clarke [73] has separated a group of eight hallucinogenic drugs on silica gel G with methanol–ammonia (100:1.5). Two new drugs of this type, 3,4-methylenedioxyamphetamine (MDA) and 3-methoxy-4,5-methylenedioxyamphetamine (MMDA), have R_f values of 0.50 and 0.55, respectively, in this solvent [74]. Niwaguchi and Inoue [75] isolated lysergic acid amide and isolysergic acid amide from morning glory seeds. Of the three solvents used chloroform–methanol (4:1) gave the best separation. The compounds were detected with Van Urk's reagent (T-89) which is sensitive to 0.05 μg of *N*,*N*-diethyl-D-lysergamide (LSD) [76]. Alliston et al. [77] used morpholine–toluene (1:9) as a solvent for separating 19 erganes and tryptamine derivatives. Using 5% (wt/vol) of 4-dimethylaminobenzaldehyde in methanol–hydrochloric acid (1:1) as a detecting agent, the sensitivity for LSD was 4 ng. Genest [78] used a direct densitometric method for the determination of lysergic and isolysergic acid amides and clavine alkaloids in morning glory seeds. Niwaguchi and Inoue [79] used an in situ fluorimetric method for LSD and later [79a] a dual-length scanner for the direct densitometric determination of LSD and 2,5-dimethoxy-4-methylamphetamine (STP).

Korte and Sieper [80] have investigated cannabidiolic acid, cannabidiol, cannabinol, and several isomers of tetrahydrocannabinol from hashish extracts. Separation of these components was achieved on silica gel G impregnated with a 60% solution of *N,N*-dimethylformamide in carbon tetrachloride (vol/vol). Impregnation was accomplished by a predevelopment with the impregnating fluid in a saturated chamber. Good separation was achieved by multiple development (3×) with cyclohexane. Phenol reagents were used for detecting the location of the compounds. A quantitative determination of the compounds was made by spraying the plate with Fast Blue B in 0.1 *N* sodium hydroxide and then eluting the spots with acetic acid–methanol (1:1). Spectrophotometric measurements were made against a blank extract with an accuracy of ± 5%. Cannabidiol acid was not determined.

Vinson and Hooyman [81] used layers impregnated with triethylamine and developed with benzene. The impregnated layers were stable and could be stored for up to 10 weeks prior to use. Tewari et al. [82] obtained good separation of cannabis constituents on alumina with benzene–chloroform (1:1), thus avoiding the use of impregnated layers. Segelman [83] combined a histochemical test with thin-layer chromatography for identifying marihuana material. Cannabis users can be identified by swabbing fingers with chloroform and teeth with ethanol, in order to obtain material for TLC examination [84, 85]. Silver nitrate layers gave the best sensitivity (1 ng) [84]. Improved sensitivity of detection of tetrahydrocannabinol can be obtained by converting to a derivative prior to thin-layer chromatography. Just et al. [86] prepared the DANS derivative which was chromatographed on silica gel with heptane–ethyl acetate (95:5) (triple development).Quantitation was by in situ fluorimetry. Fluorimetry on polyamide layers gave a detection limit of 1×10^{-12} *M* of the derivative. Vinson et al. [87] used 2-*p*-chlorosulfophenyl-3-phenylindone as a derivatizing agent [88]. Development was on Bakerflex IB2 silica gel sheets with methanol–water (95:5) in a saturated atmosphere. The spots were visualized under long-wave ultraviolet after spraying with a reagent consisting of 8 g of sodium in 100 ml of methanol and 8 ml of dimethyl sulfoxide. Sensitivity was 0.1 ng.

Parker and Fiske [89] and Merkus et al. [90] have reviewed the analysis of cannabis constituents. Sterling [91] has reviewed the analysis of hallucinogenic drugs (see also Chapter XV, Section 10).

Antidepressants and Stimulants

Tofranil, which is used as an antidepressant, has been investigated by Obersteg and Baeumler [92] because of its use as a suicidal agent. In the case of suicide, unchanged Tofranil could be found in the urine; in patients with a normal dose of Tofranil, only the metabolic products could be isolated from the urine. Using methanol–acetone–triethanolamine (1:1:0.03) on silica gel, the following results are reported for Tofranil and its metabolites: Tofranil 0.45, iminodibenzyl 0.9, 2-hydroxyiminodibenzyl 0.9, dimethylaminopropyl-2-hydroxyiminodibenzyl 0.40, and monomethylaminopropyliminodibenzyl 0.23. Additional solvents and R_f values for Tofranil are given by Mellinger and Keller [49].

The antidepressants isocarboxazid, phenelazine, nialamide, isoniazide, iproclozide, imipramine, desipramine, clomipramine, trimeprimine, opipramol, nortriptylin, amitriptyline, iproniazid, and dibenzepin have been chromatographed on silica gel [93–100]. Some of the solvents used included: chloroform–methanol (1:1), butanol–acetic acid–water (3:1:1), anhydrous peroxide free ether–acetone–diethylamine (90:10:1), and methanol. Some of the detection agents were T-149, T-111, and T-131.

Thunell [100a] developed a method for the isolation and detection of basic psychotropic drugs including narcotic alkaloids and amines that exhibit central stimulation action. Two-dimensional thin-layer chromatography was used on silica gel with methanol–ammonia (49:1) and ammonia-saturated chloroform. Twenty-nine compounds were examined.

See under Purine Alkaloids in Chapter XV, Section 2 for data regarding the purine stimulants such as caffeine and theophylline.

4 ANTIHISTAMINES

Cochin and Daly [50] determined the R_f values of a group of antihistamines using both silica gel and alumina layers (Table 26.5). These R_f values were obtained after the substances were extracted from urine to which the pure compounds had been added. The compounds could be located by the potassium iodoplatinate reagent mentioned under tranquilizers. Fike and Sunshine [101] used three systems to chromatograph 25 antihistamines: silica gel with cyclohexane–benzene–diethylamine (75:15:10), silica gel (prepared with 0.1 M sodium hydroxide) with methanol, and silica gel (prepared with 0.1 M potassium bisulfate) with methanol. Multiple development was also employed to separate all but three unresolved pairs: buclizine–meclizine, pheniramine–chlorpheniramine, and diphenhydramine–bromodiphenhydramine. However, the latter two

Table 26.5 $R_f \times 100$ Values of Antihistamines in Various Systems[a,b]

	Silica Gel			*Aluminum Oxide*
Compound	*Benzene–Dioxane–Ammonium Hydroxide* (12:7:1)	*Ethanol–Acetic Acid–Water* (5:3:2)	*Methanol–Butanol* (3:2)	*Butanol–Butyl Ether–Acetic Acid* (4:8:1)
Antazoline (Antistine)	31	72	40	38
Bromdiphenhydramine (Ambodryl)	86	80	33	63
Brompheniramine (Dimetane)	42	47[c]	16	48
Carbinoxamine (Clistin)	33	40[c]	17	·63
Chlorcyclizine (Di-Paralene)	90	68	40	49
Chlorothen (Tagathen)	70	56[c]	29	66
Chlorpheniramine (Chlortrimeton)	40	41[c]	20	52
Diphenhydramine (Benadryl)	91	61	25	53
Doxylamine (Decapryn)	52	38[c]	17	60
Hydroxyzine (Atarax)	27	70	62	84
Methapyrilene (Thenylpyramine)	83	52[c]	30	56
Phenindamine (Theophorin)	90	67	48	65
Pheniramine (Trimeton)	68	38[c]	19	63
Pyrilamine (Neo-Antergan)	82	43[c]	27	56
Pyrrobutamine (Pyronil)	84	82	32	78
Thonzylamine (Neohetramine)	65	57	32	59
Tripelennamine (Pyribenzamine)	92	48[c]	23	62
Tripolidine (Actidil)	40	48[c]	26	81

[a] From J. Cochin and J. W. Daly [50]; reproduced with permission of the authors and Williams and Wilkins Co.
[b] Development distance 10 cm.
[c] Tailing occurred.

pairs can be separated on alumina (see Table 26.5). Dragendorff's (T-112) and Mandelin's reagent (T-21) were used for visualization. Boonen [102] used a two-dimensional technique on silica gel with ethanol–acetic acid–water (5:3:2) and then methanol.

5 ANALGESICS AND ANTIPYRETICS

In general the alkaloid analgesics are discussed under Alkaloids; however, Mulé [103, 104] has given the R_f values for a large group of narcotic

Table 26.6 $R_f \times 100$ Values[a] of Narcotic Analgesics in Various Systems[b]

Analgesic	Silica Gel G				X^c		Silica Gel G		
	Ethanol–Pyridine–Dioxane–Water (10:4:5:1)	*Ethanol–Acetic Acid–Water (6:3:1)*	*Ethanol–Dioxane–Benzene–Ammonium Hydroxide (1:8:10:1)*	*Methanol–n-Butanol–Benzene–Water (12:3:2:3)*		*tert-Amyl Alcohol–n-Butyl Ether–Water (80:7:13)*		*n-Butanol–Acetic Acid–Water (4:1:2)*	*n-Butanol–Conc. HCL (9:1) Saturated with Water*
Iminoethanophenanthrofurans									
Morphine	29	27	11	21	86	85	7	54	34
Normorphine	8	48	4	7	48	25	S[d]	66	62
Codeine	30	29	39	25	86	91	8	53	30
Norcodeine	12	50	13	9	59	56	6	63	49
Heroin	37	35	76	35	90	65	15	61	32
Nalorphine	71	55	35	67	88	96	25	59	41
Methyldihydromorphinone	16	24	25	15	76	92	S	45	26
Dihydromorphinone	11	21	17	13	65	85	S	41	25
Ethylmorphine	33	25	46	27	84	96	8	53	33
Dihydrohydroxymorphinone	46	29	34	24	63	81	10	45	28
Dihydromorphine	15	21	10	10	67	73	S	43	29
Dihydrocodeinone	17	25	41	19	76	94	S	42	23
Dihydroxycodeinone	46	24	87	29			16	32	34
6-Monoacetylmorphine	38	40	64	29			19	37	37
Iminoethanophenanthrenes									
l-3-Hydroxy-*N*-methylmorphinan	11	47	80	10			7	51	60
l-3-Hydroxymorphinan	5	68	19	10			8	72	80
l-3-Methoxy-*N*-methylmorphinan	13	43	91	8			7	55	59

l-3-Methoxymorphinan	7	65	38	S	S	66	81
l-3-Hydroxy-*N*-allyl-morphinan	65	70	98	41	44	64	73
Dairylalkoneamines							
dl-Methadone	34	59	99	17	17	55	62
l-Acetylmethadol	64	60	99	40	38	52	62
d-Propoxyphene	73	68	97	54	56	53	61
Arylpiperidines							
Pethidine	42	41	97	36	20	46	44
Norpethidine	12	65	51	10	11	58	63
Ketobemidone	31	39	47	24	12	42	40
dl-Alphaprodine	39	40	93	34	20	42	40
Piminodine	88	73	99	85	76	69	58
Benzomorphans							
dl-2′-Hydroxy-5,9-dimethyl-2-phenethyl-6,7-benzo-morphan	88	87	97	82	70	76	77
1-2′-Hydroxy-2,5,9-tri-methyl-6,7-benzomorphan	12	36	56	8	5	43	51
2′-Hydroxy-5,9-dimethyl-2-(3,3-dimethylallyl)-6,7-benzomorphan	73	81	96	25	34	65	77
2′-Hydroxy-5,9-dimethyl-2-cyclopropylmethyl-6,7-benzomorphan	45	71	92	15	16	55	67

[a] Development distance 10 cm at 23 ± 2°C.

[b] From S. J. Mulé; reproduced with permission of the author and The American Chemical Society.

[c] X = Cellulose layers prepared from 15 g of cellulose MN 300 G in 90 ml of 0.1 *M* phosphate buffer, pH 8.0.

[d] Compound streaked.

analgesics using both silica gel G and cellulose layers with seven different solvent systems (Table 26.6). Two-dimensional chromatography was applied by using methanol–*n*-butanol–benzene–water (12:3:2:3) in the first direction, and one of the ethanol or *tert*-amyl alcohol mixtures in the second direction. Detection was accomplished by using an iodoplatinate reagent. Cochin and Daly [105] also separated a group 16 analgesic alkaloids using silica gel and aluminum oxide layers. R_f values are listed in six different solvent-layer combinations.

Vidic [106] has separated the analgesic Jetrium (dextromoramide) from several other analgesics and from Ticarda, an antitussive agent, on silica gel G with 0.1 *N* methanolic ammonia. The R_f values are as follows: Jetrium 0.85, Romilar (*d*-3-methoxy-*N*-methylmorphan hydrobromide) 0.26, Polamidon (*dl*-methadone hydrochloride) 0.42, and Ticarda (6-dimethylamino-4, 4-diphenyl-3-hexanone) 0.59.

Šaršúová and Schwarz [107] have determined the R_f values of a group of eight compounds having antipyretic and analgesic action in 10 different solvents on loose layers of aluminum oxide. The drugs which were investigated are: acetylsalicylic acid, aminopyrine, antipyrine, codeine, caffeine, quinine, papaverine, and phenacetin. Baeumler and Rippstein [21] give the range of R_f values on silica gel G with methanol–acetone–triethanolamine (1:1:0.03) for antipyrine 0.70 to 0.72, dipyrine 0.74 to 0.76, and isopropylantipyrine 0.83 to 0.85.

Gaenshirt [108] separated caffeine, amidopyrin, phenacetin, and benzyl mandelate on silica gel. Quantitative determinations were made by eluting and measuring spectrophotometrically. The mandelate could only be determined semiquantitatively. Frodyma et al. [109] have used ultraviolet reflectance spectroscopy for determining aspirin and salicylic acid. Fuwa et al. [110] applied step chromatography to the analysis of drugs of the Japanese Pharmacopeia VII containing antifebrils. Diphenhydramine hydrochloride, quinine hydrochloride, sulpyrin, caffeine, aminopyrine, pyrabital, quinine ethylcarbonate, phenacetin, and acetanilide were isolated and identified.

Emmerson and Anderson [111] have used silica gel layers combined with neutral solvents by placing a beaker of 28% ammonia in the developing chamber. With this system, the analgesics could be applied as the free bases or as their salts. Best results were obtained using benzene or dichloromethane as the mobile phase.

Thompson and Johnson [112] reported the R_f values for 29 analgesics, antipyretics, and anti-inflammatory agents on fluorescent silica gel with cyclohexane–acetone–acetic acid (40:50:1). Detection limits ranged from 1 to 5 μg under short-wave ultraviolet. Radulovič and co-workers [113, 114] used talc as an adsorbent for analgesic–antipyretic mixtures. Solvents included cyclohexane–acetone–chloroform (7:2:1), benzene, toluene, and various mixtures of cyclohexane–chloroform with 1,4-dioxane, tetrahydrofuran, ethyl acetate, or methyl ethyl ketone.

Hsiu et al. [115] chromatographed the following antipyretics: salicylic acid, salicylamide, acetylsalicylic acid, oxyphenbutazone, antipyrin, acetanilide, aminopyrin, phenacetin, and indomethacin. Polyamide was used with four solvents including: chloroform–benzene–90% formic acid (50:10:1) and cyclohexane–chloroform–acetic acid (4:5:1). Chiang and Chiang [116] used polyamide–silica gel (1:5) with chloroform–cyclohexane–acetic acid (40:60:1) and the same solvent with the addition of 10 parts dioxane for the separation of eight antipyretics.

Unterhalt [117] separated the anti-inflammatory drugs Butazolidin, Tanderil, Anturan, Amuno, Parkemed, and Arlef on silica gel with cyclohexane–chloroform–methanol–acetic acid (12:6:1:1) or methanol–acetic acid–diethyl ether–benzene (1:18:60:120). Pellerin and Mancheron [118] separated caffeine, methylacetanilide, amidopyrine, phenacetin, quinine, and antipyrine on silica with benzene–ether–acetic acid–methanol (120:120:18:1), chloroform–ethanol (99:1), and pyridine–chloroform–cyclohexane (1:12:4). Dragendorff's reagent (T-111) and Millon's reagent (T-166) were used for detection.

The antirheumatic drugs phenacetin, phenazone, aminophenazone, noraminophenazone, phenylbutazone, and salicylamide were separated on silica with butyl acetate–chloroform–85% formic acid (6:4:2) [119]. Detection was obtained by spraying with 5% ferric chloride followed by 1% alcoholic *o*-phenanthroline.

6 SYMPATHOMIMETICS

The R_f values of a group of adrenaline derivatives on buffered and unbuffered layers of silica gel G in four different solvents have already been given in Table 3.5. Waldi [120] converts adrenaline to its triacetyl derivative prior to chromatographing it on silica gel G layers using chloroform–methanol (9:1) as a solvent. As an alternative, cyclohexane–chloroform–methanol–acetic acid (3:5:1:1) may be used for the separation. In this solvent the R_f values of the triacetyl derivatives are for adrenaline 0.36, noradrenaline 0.26, and ephedrine 0.51. In order to detect the compounds, the plates may be sprayed with vanillin–sulfuric acid mixture, or with 40% phosphoric acid followed by 5% phosphomolybdic acid. For the latter detecting agent, the plates are dried at 110°C both after the phosphoric acid spray and after the final spray with the phosphomolybdic acid reagent. By comparison of the size of the spot with spots of known concentrations, a quantitative determination of the amount of adrenaline present may be made. Beckett and Choulis [121, 122] have shown that the sympathomimetic amines show two spots when chromatographing on cellulose thin layers in the presence of trichloroacetic acid using butanol–acetic acid–water (4:1:5). This multiple spot formation does not occur with silica gel layers, and multiple spots are not formed on thin layers of cellulose that have been treated with diazomethane, thus esterifying the carboxyl groups. Choulis [123] used phenol–water (4:1) to separate adrenaline, noradrenaline, and dopamine on cellulose layers. Vahidi and Sankar [124] used ethyl acetate–acetic acid–water–butanol (3:1:3:2) with cellulose layers to separate adrenaline, noradrenaline, 3-*O*-methyladrenaline, dopa, dopamine, and tyramine.

Polyamide layers have also been used for the separation of sympathomimetics. In this case isobutanol–acetic acid–cyclohexane (80:7:10) [125] is a good solvent, as is butanol–acetic acid–water (4:1:1) [126]. Visualization on these layers can be effected by test T-114 or T-249.

Choulis and Carey [127] used butanol–acetic acid–water (4:1:5) containing sodium metabisulfite as an antioxidant in examining quantitative methods on cellulose. Dopa, dopamine, and ephedrine were visualized with ninhydrin, amphetamine by diazotized nitroaniline, and the other amines by T-200. Spectrophotometric measurement of the absorption after elution gave a ±3% error and densitometry with spot area measurement a ±5% error.

Roser and Tocci [128] examined nine diazonium reagents as detection agents for catecholamines and determined that *p*-nitrobenzenediazonium tetrafluoroborate (T-181) was the best of these reagents. Sensitivity ranged from 0.01 to 0.05 μg. Van Hoof and Heyndrickx [129] coupled amphetamine and other sympathomimetic drugs with 4-chloro-7-nitrobenzo-2,1,3-oxadiazole prior to chromatographig on silica gel layers with ethyl acetate–cyclohexane (3:2) or (2:3). The in situ fluorescence was measured at 523 nm with excitation at 482 nm. For visual observation long-wave, 350-nm ultraviolet light could be used. Detection limit was 1 ng. Compounds lacking a primary or secondary alkylamine cannot be detected. Loh et al. [130] converted amphetamine and methamphetamine to 1-dimethylamino-5-naphthalenesulfonyl (DANS) derivatives prior to chromatographing on polyamide with formic acid–water (1.5:100) or water–formic acid–*n*-butanol–ethanol (150:3:4:93). Two-dimensional separations with these two solvents gave even better resolution. Quantitatively by in situ fluorescent scanning 1 to 2 ng could be determined and qualitatively as little as 0.3 to 0.5 ng could be detected.

Table 26.7 $R_f \times 100$ Values[a] of Some Local Anesthetics on Loose Layers of Alumina (III) in Various Solvent Systems[b]

Anesthetic	*Benzene*	*Benzene–Ethanol (98:2)*	*Benzene–Ethanol (95:5)*	*Benzene–Ethanol (90:10)*	*Benzene–Ethanol (80:20)*	*Chloroform*	*Chloroform–Ethanol (99:1)*	*Chloroform–Butanol (98:2)*	*Chloroform–Acetone (1:1)*	*Ether*	*Ether–Petroleum Ether (1:1)*
Procaine hydrochloride	0	14	31	52	65	27	34	43	60	54	11
Cinchocaine hydrochloride	10	19	46	65	75	42	51	47	75	61	20
Cocaine hydrochloride	17	34	65	75	82	57	67	60	80	70	30
Tetracaine hydrochloride	8	15	40	63	72	40	43	56	59	58	20
Ethoform (benzocaine)	18	25	48	—	—	53	52	62	70	75	31
Tutocaine	10	16	43	62	71	45	41	57	60	55	22
Dimethocaine (Larocaine)	12	18	37	54	70	40	50	45	79	68	24
Orthoform	0	0	0	0	0	0	0	0	0	0	0
γ-Eucaine	14	16	28	53	72	41	44	45	51	30	21
Diocaine	10	17	55	67	80	35	51	52	81	80	24
Lidocaine (Xylocaine)	10	17	50	56	78	52	62	56	81	65	22
Phenacaine (Holocaine)	6	24	52	60	77	37	37	55	77	70	27
Hostocaine	10	17	26	42	68	27	—	—	—	—	5
Trimecaine (Mesocaine)	5	16	35	57	74	50	64	55	80	62	23
Dextrocaine (Psicaine)	8	18	46	65	78	55	64	57	78	63	25

[a] Average of six determinations.

[b] From M. Šaršúnová [131]; reproduced with permission of the author and VEB Verlag Volk und Gesundheit.

7 LOCAL ANESTHETICS

Šaršúnová [131] investigated the behavior of 15 local anesthetics in 11 solvent systems on loose layers of alumina (activity III) (Table 26.7). The compounds were located by the Munier-modified Dragendorff reagent and by an acidified iodine–potassium iodide solution.

Guven and Hincal [132] separated *p*-aminobenzoic acid, novocaine, Pantocaine, and benzocaine by two-dimensional thin-layer chromatography on silica gel G with chloroform–acetone–water–methanol (2:4:1:4) and (2:2:1:4). Dragendorff's reagent (T-112) was used for detection. The R_f values in the two solvents were: 0.82, 0.77; 0.46, 0.42; 0.58, 0.59; and 0.92, 0.94, respectively. Fresen [133] used reverse-phase chromatography on cellulose MN 300 impregnated with 10% oleyl alcohol and 0.5 *M* monosodium phosphate–disodium phoshate buffers (pH 0.4 to 11) as solvents for separating benzocaine, procaine, Cinchocaine, and parethoxycaine. Roeder et al. [134] used azeotropic solvent mixtures for separating 10 anesthetics and De Zeeuw [135] applied the principle of vapor programing to the separation of seven compounds. Messerschmidt [136] separated the DANS derivatives of *p*-aminobenzoic acid esters used as anesthetics.

8 SULFA DRUGS

A number of papers have appeared dealing with the separation of these important compounds. Reisch et al. [137] separated 10 compounds on silica gel G using two solvents: *n*-butanol–methanol–diethylamine (9:1:1) and *n*-butanol–methanol–acetone–diethylamine (9:1:1:1). Separations were carried out in a saturated chamber, and the compounds were made visible by diazotization or by spraying with Ehrlich's reagent. Wollish et al. [138] separated a group of four sulfanilamides and sulfanilic acid on silica gel using chloroform–heptane–ethanol (1:1:1). Acetone was used as a solvent for applying these samples in 1 μg amounts. To visualize the compounds, the plates were sprayed with 0.1% *p*-dimethylaminobenzaldehyde in ethanol, containing 1% concentrated hydrochloric acid. Bićan-Fišter and Kajganović [139] used ether and a mixture of chloroform–methanol (10:1) for separating a group of 12 sulfanilamides on silica gel G layers in a saturated atmosphere. Detection of the spots with diazo reagent or with *p*-dimethylaminobenzaldehyde gave a sensitivity of 0.25 μg of the compounds. The R_f values are given for the various compounds.

Bićan-Fišter and Kajganović [140] have also developed a quantitative method for the analysis of mixtures in pharmaceutical preparations. Extraction from tablets is accomplished with a mixture of 50 ml of 70%

ethanol and 2 ml of 25% ammonium hydroxide using one sulfa tablet to 50 ml of solution. After a 15-min extraction period and clarification by centrifuging, the extract (9 μl) is applied in three aliquots to the silica gel G plates along with spots of standard solutions. Suspensions are extracted in the same manner using an amount of the suspension containing 500 mg of the sulfonamide for a final extraction mixture volume of 50 ml. For suppositories a weighed amount containing 250 mg of sulfonamide is shaken with 100 ml of ether, 10 ml of water, and 2 ml of 25% ammonium hydroxide. The layers are separated and the ether is washed with water several times, adding the washings to the water layer, which is finally made up to 25 ml. The developing solvent depends on the mixture that is present. For mixtures of sulfathiazole, sulfamerazine, and sulfadiazine the solvent mixture of chloroform–methanol (9:1) gives excellent separations, but chloroform–methanol–25% ammonium hydroxide (90:15:2.4) is better for a mixture of sulfacetamide, sulfamerazine, and sulfamethazine or for sulfadiazine, sulfamerazine, and sulfamethazine. Ether proved to be the best solvent for a mixture composed of sulfathiazole, sulfadiazine, sulfamethazine, and sulfamerazine. The spots are located by spraying with freshly prepared 0.1% sodium nitrite solution followed by 0.1% *N*-(1-naphthyl)ethylenediamine dihydrochloride solution. Once located, the spots are eluted by transferring to a 25-ml flask and shaking for 20 min with 5 ml of 0.1 *N* hydrochloric acid. A 3-ml aliquot of this centrifuged extract is then placed in a 25-ml flask with 1 ml of sodium nitrite reagent. After standing for 3 min, 1 ml of 0.5% ammonium sulfamate is added and after another 2 min, 1 ml of 0.1% *N*-(1-naphthyl)ethylenediamine is added. The color is measured at 545 mμ after standing 15 min. The absolute amounts are determined by comparison with the standards.

Klein and Kho [141] examined a group of pharmaceutical preparations, many of which contained from two to four mixed sulfanilamides. The samples were extracted with acetone so that approximately 10 mg of the sulfonamide present in the greatest amount would be present in 50 ml of acetone. In the case of suspension samples which contained interfering ingredients, the sample was first mixed with 1 ml of distilled water before extracting with acetone. The chromatoplates were developed a distance of 15 cm in a solvent mixture of chloroform–ethanol–heptane (1:1:1) containing various amounts of water (1.0 to 1.8%), depending on the mixture of sulfonamides that were present. For detection the compounds were diazotized and in this case coupled with 0.1% *N*-(1-naphthyl)ethylenediamine dihydrochloride. Kho and Klein [142] also applied a two-step method for the separation of N^4-substituted sulfonamides. The silica gel plates were developed first for a distance of 5 cm using a solvent composed of methanol–ethanol (1:1) and then after drying

Table 26.8 $R_f \times 100$ Values of Sulfonamides in Various Systems

	Aluminum Oxide G [144]	*Silica Gel G* [144]		*Polyamide* [144]		*Silica Gel G* [143]
Sulfonamide	*Ethyl Acetate–Methanol–25% Ammonium Hydroxide* (17:3:3)	*Ethyl Acetate–Methanol–25% Ammonium Hydroxide* (17:6:5)	*Methyl Isobutyl Ketone–Acetone–25% Ammonium Hydroxide* (1:4:1)	*Methyl Isobutyl Ketone–Acetone–25% Ammonium Hydroxide* (5:20:1)	*Ethyl Acetate–Methanol–25% Ammonium Hydroxide* (17:3:1)	*Petroleum Ether* (60–80°C)–*Chloroform–Butanol* (1:1:1)
Sulfacetamide	19	—	38	12	7	31
Sulfanilamide	96	97	79	84	73	43
Sulfathiazole	42	80	57	18	12	41
Sulfapyridine	28	77	55	84	69	37
Sulfathiourea	28	60	49	10	5	34
Sulfaguanidine	78	50	64	47	33	15
N,N-Dimethylacrosylsulfanilamide	38	74	54	22	12	
Sulfadiazine	19	49	38	30	13	39
Sulfamerazine	25	59	45	49	25	44
Sulfamethazine	35	69	51	67	46	52
Cinnamein sulfanilamide	96	97	79	85	74	
Sulfaethylthiodiazole	41	80	61	14	13	
Maleylsulfanilamide	34	81	61	13	8	
Succinylsulfanilamide (sodium salt)	12	6	39	4	6	
N^1-Benzoylsulfanilamide	50	85	59	10	7	
Phthalylsulfacetamide	6	26	33	0	0	

Phthalylsulfathiazole	6	50	44	1	0	
4-Homosulfanilamide salt of 1-sulfanilyl-2-thiourea	28	60	49	10	5	
N'-3,4-Dimethylbenzoyl-sulfanilamide	85	86	61	10	8	40
Sulfisomidine	39	74	48	20	17	19
N^4-Carbethoxysulfoethyl-thiodiazole	63	88	61	24	23	
Sulfisoxazole	34	77	53	8	7	51
Sulfanilamido-2-phenyl-pyrazole	56	86	57	13	9	77
N^1-Isopropoxysulfanilamide	58	92	61	18	14	
Sulfanilamido-4,5-dimethyl-	—	47+ T[a]	58+ T	25	17	
oxazole				6	5	
Sulfathiazole formaldehyde	—	—	—	0	6	
					23	
2-Sulfanilamido-3-methoxy-pyrazine	34	63	50	25	17	70
N^1-Acetyl-2-sulfanilamido-3-methoxypyrazine	T	76	54	97	T	
Sulfa-4,6-dimethoxypyrimidine	46	85	57	58	45	72
2,4-Dimethyl-6-sulfanilamido-1,3-diazine	45	81	57	57	25	
2-Sulfanilamido-5-methyl-pyrimidine	26	59	45	50	26	46
2-Sulfanilamido-5-methoxydiazine	25	60	45	35	35	56
3-Sulfanilamido-6-methoxypyridazine	36	73	50	47	39	61

Table 26.8 (Continued)

Sulfonamide	*Aluminum Oxide G* [144]	*Silica Gel G* [144]		*Polyamide* [144]		*Silica Gel G* [143]
	Ethyl Acetate–Methanol–25% Ammonium Hydroxide (17:3:3)	*Ethyl Acetate–Methanol–25% Ammonium Hydroxide* (17:6:5)	*Methyl Isobutyl Ketone–Acetone–25% Ammonium Hydroxide* (1:4:1)	*Methyl Isobutyl Ketone–Acetone–25% Ammonium Hydroxide* (5:20:1)	*Ethyl Acetate–Methanol–25% Ammonium Hydroxide* (17:3:1)	*Petroleum Ether* (60–80°C)–*Chloroform–Butanol* (1:1:1)
N^1-Acetyl-3-sulfanilamido-6-methoxypyridazine	T	70	54	97	T	
Sulfonamidodimethoxytriazine						24
Sulfa-5-methylthiodiazole						25
Sulfaproxyline						33
4′-Sulfamyl-2,4-diaminoazobenzene·HCL						47
Sulfadimethoxine						67
5-Methyl-3-sulfanilamidoisoxazole						80
Dimethylacrylsulfonilamide						82

[a] T = Tailing.

at 100°C for 5 min they were developed in *n*-propanol–0.05 *N* hydrochloric acid (4:1) for a distance of 10 cm using a tank lined with filter paper.

Karpitschka [143] determined the R_f values of a large number of sulfonamides (Table 26.8) using silica gel layers with chloroform–petroleum ether (60 to 80°C)–*n*-butanol (1:1:1). The compounds were located with a spray reagent consisting of 1% *p*-dimethylaminobenzaldehyde solution in 5% hydrochloric acid. Van der Venne and T'Siobbel [144] also investigated a large number of these compounds on paper and on thin layers of silica gel G, alumina G, and polyamide. The results with the thin-layer work are recorded in Table 26.8.

Biagi et al. [145] chromatographed 20 compounds on silica gel and polyamide layers with a Veronal acetate buffer (pH 7.4) containing various amounts of acetone. De Zeeuw [146] applied vapor programming to the separation of 15 sulfonamides. Lepri et al. [147] has investigated the chromatographic separation of sulfonamides on Dowex 50-X4 (H^+), AG 1-X4 (CH_3COO^-), microcrystalline cellulose, and Cellex D ($C10_4^-$). Electrophoretic separations were also made on Ag1-X4 (CH_3COO^-), silica gel, and a mixture of these two. Walash and Agarwal [148] recommend a saturated solution of cupric acetate in methanol for detecting sulfonamides. For the determination of sulfonamides, Sigel et al. [149] dipped the developed and dried silica gel chromatogram in 0.5% fluorescamine solution in acetone followed by again drying and dipping in 0.5% triethanolamine in chloroform. Using excitation at 290 nm the fluorescence spectrum was recorded above 400 nm with subsequent measurement of the peak areas. Sulfadiazine down to 0.2 ng was determined in muscle.

9 CARDENOLIDES

This is a group of drugs obtained from medicinal herbs; they are used because of their action on the heart. Most widely known of course are the digitalis glycosides. Because of their importance they have been investigated by quite a number of workers. Tschesche et al. [150] first introduced the use of thin-layer chromatography for this important group of compounds; he gave the relative R_f values for a group of 10 compounds. They were separated on silica gel G layers dried at 130°C using ethyl acetate as a developing solvent. A mixture of diisopropyl ether–acetone (3:1) gave similar R_f values. The compounds were detected by spraying with a mixture of chlorosulfonic acid–acetic acid (1:2) and heating the plates to 130°C. This produced green spots which fluoresced brown-violet under ultraviolet light. Reichelt and Pitra [151, 152] investigated a group of 29 cardenolides in a number of systems. In one system silica gel layers deactivated with 25% water were used with benzene–

Table 26.9 *R_{st}* Values (Relative to Digitoxin) of Some Digitalis Glycosides and Aglycones on Silica Gel Layers and Colors with Detecting Agent[a]

Substance	*Ethyl Acetate–Methanol–Water* (16:1:1)	*Chloroform–Pyridine* (6:1)	*Visible*[b]		*UV Fluorescence*[b]	
			Initial Color	*Final Color*	*Initial Color*	*Final Color*
Digitoxigenin	1.41	2.17	Blue	Blue-green	—	Whitish yellow
Gitoxigenin	1.10	1.28	Red	Green	Yellow-brown	Yellow
Digoxigenin	0.99	1.22	Blue	Blue-green	Reddish violet	White
Gitaloxigenin	1.43	2.05	Red	Green	Yellow-brown	Yellow
Digitoxigenin-mono-digitoxose	1.27	1.41	Blue	Blue	—	Whitish yellow
Gitoxigenin-mono-digitoxose	1.02	0.86	Brown	Green	Blue-white	Blue
Lanadoxin	1.30	1.36	Blue	Blue-green	Yellow-brown	Yellow
Digitoxin	1.00	1.00	Blue	Blue	—	—
Gitoxin	0.75	0.59	Red-brown	Gray-blue	Blue-white	Blue
Digoxin	0.68	0.62	Blue-gray	Blue	Blue-violet	White
Gitaloxin	1.01	0.92	Red-violet	Green-blue	Yellow	Yellow

Purpureaglycoside A	0.14	0.05	Blue-black	Blue-gray	—	—
PurpureaglycosideB	0.09	0.02	Brown-red	Gray	Blue-white	Blue-white
Glucogitaloxin	0.13	0.04	Brown-red	Gray	Yellow-Brown	Yellow
Acetyldigitoxin-1 (α?)	1.32	1.95	Blue	Blue	—	—
Acetyldigitoxin-2 (β?)	1.42	2.03	Blue	Blue	—	—
Acetylgitoxin-α	1.04	1.39	Brown-red	Gray-blue	Yellow-brown	Yellow
Acetylgitoxin-β	1.17	1.47	Brown-red	Gray-blue	Yellow-brown	Yellow
Acetyldigoxin-α	1.00	1.38	Gray-violet	Blue	Blue-violet	White
Acetyldigoxin-β	1.13	1.57	Gray-violet	Blue	Blue-violet	White
Lanatoside A	0.26	0.18	Blue-gray	Blue-violet	—	—
Lanatoside B	0.16	0.12	Red-brown	Blue-gray	Yellow-brown	Yellow-white
Lanatoside C	0.15	0.08	Brown-red	Blue-violet	Red-violet	Blue-white
Odoroside H	0.76	1.49	Blue-green	Blue-green	—	Yellow
Stropeside	0.50	0.91	Red-violet	Green	Yellow-brown	Yellow
Verodoxin	0.77	1.41	Red-violet	Yellow-green	Yellow-brown	Yellow
Digitalinum verum	0.06	0.02	Red-brown	Yellow-green	Yellow-brown	Yellow
Digitonin	0.00	0.00		Yellow		

[a] From I. Sjoeholm [157]; reproduced with permission of the author and Svensk Farmaceutisk Tidskrift.

[b] Spray reagent (freshly prepared) = 0.5 ml of *p*-anisaldehyde, 5 ml of 70% perchloric acid, 10 ml of acetone, 40 ml of water. Plate heated 75 to 80°C for 4 to 5 min after spraying.

ethanol (3:1). A 50% deactivated silica gel with 43% acetic acid as the deactivating agent was also used with the same solvent system. The same authors used silica gel impregnated with borax to increase the retention of compounds with *cis*-vicinal glycol arrangements in the molecule. Braeckman et al. [153] separated the glycosides of *Digitalis purpurea* on silica gel using the solvent systems of methylene chloride–methanol–formamide (80:19:1) and methyl ethyl ketone–methylene chloride–butanol–dimethylformamide (40:40:19:1). Zurkowska et al. [154, 155] and Fauconnet and Waldesbuehl [156] have also applied thin-layer chromatography to the investigation of the digitalis glycosides.

Sjoeholm [157] has separated a group of 28 digitalis glycosides and aglycons using two-dimensional separation on silica gel layers dried at 110°C for 30 min. For the first development, a mixture of ethyl acetate–methanol–water (16:1:1) was used followed by chloroform–pyridine (6:1) in the second direction. In order to remove the solvents the developed chromatograms were heated at 90 to 100°C until the smell of pyridine had disappeared. In separating glucosides a solvent mixture of methyl ethyl ketone–chloroform–formamide (5:2:1) was used. After investigating various color producing agents a *p*-anisaldehyde–perchloric acid reagent (T-25) was selected as the best. With this reagent, characteristic colors are obtained and it is sensitive to 0.1 to 0.2 μg in visible light and 0.02 μg in ultraviolet light. Table 26.9 gives the relative R_f values for this group of compounds. Other detecting agents that have been used are T-263 and T-32.

Pekic [158] separated digitalis glycosides on talc impregnated with 0.134 to 0.2 g of formamide/g using a solvent mixture of xylene–methyl ethyl ketone (1:1) saturated with formamide. Boisio [159] examined the separation of cardiotonic aglycons and glycosides on silica gel with 27 solvent systems that have been reported. The R_f values are listed. In examining the relation between structure and chromatographic behavior, Nover et al. [160] reported the R_f and R_M values for 169 cardenolides and bufadenolides as well as numerous peracetates of these compounds in four solvent systems.

Lutz [161] used a two-dimensional separation for digitoxin in blood and used quantitation by fluorimetry with detection limits of 0.05 μg. Storstein [162] found silica gel impregnated with 15% formamide combined with development (2×) in methyl ethyl ketone–xylene (1:1) to be the best separation system for digitoxin and its seven metabolites. Using a modified ^{86}Ru method [163] determinations could be made down to 0.5 ng per spot; however, recovery was only 59%. Faber et al. [164] exposed the developed layers to hydrochloric acid vapor for at least 1 hı and then measured the fluorescence in situ to determine digitoxin at the 1 to 2 ng level. A 99.1% recovery was obtained with a standard deviation of 11.2%. Watson et al. [164a] used a combination of paper and thin-layer chromatography for isolation of digoxin, digitoxin, and their metabolites from urine. The mother compounds were converted to fluorinated derivatives by treatment with heptafluorobutyric anhydride. The derivatives were measured by electron-capture gas chromatography in amounts as low as 25 pg.

Steinegger and van der Walt [165] separated some Scilla (squill) cardiac glycosides on silica gel layers using the upper phase of a mixture of ethyl acetate–pyridine–water (5:1:4), and Steidle [166] has applied a quantitative method to this group of compounds. Separation was carried out using methyl ethyl ketone saturated with water. For the quantitative determination, the compounds were eluted with methanol and measured spectrophotometrically.

Lukas [167] chromatographed *k*-strophanthin on silica gel with butanol–methanol–formamide (17:2:1) and thus showed that it consists mainly of *k*-strophanthoside and *k*-strophanthin-β with a small amount of cymarin and occasionally an unidentified glycoside. The R_f values are given as 2.7, 5.1, and 7.0, respectively. The aglucon has an R_f value of

Table 26.10 R_f Values and Fluorescent Colors[a] of Constituents of Strophanthus Seeds on Silica Gel Using Ethyl Acetate–Pyridine–water (5:1:4)[a]

Compound	*R_f Range*	*UV Fluorescence*[b]
k-Strophanthoside	0.05	Greenish yellow
g-Strophanthin	0.09–0.12	Yellow-orange (after heating)
Erysimoside	0.18–0.22	Greenish yellow
k-Strophanthin-β	0.25–0.28	Greenish yellow
Cymarol	0.60–0.64	Faded brown
Cymarin	0.70–0.74	Greenish yellow
Periplocymarin	0.77–0.80	Yellow-gray
Sarmentoside A	0.30–0.35	Intense yellow
Emicymarin	0.34–0.38	Faded green
Sarmentocymarin	0.82–0.86	Red-brown
Sarveroside	0.89–0.92	Red-brown

[a] From L. Hoerhammer, H. Wagner, and H. Koenig [173[; reproduced with permission of the author and Deutsche Apotheker-Verlag.

[b] UV fluorescence after spraying with sulfuric acid.

8.2. Khorlin et al. [168] have chromatographed these glycosides on aluminum oxide.

Other solvents used with silica gel for *k*-strophanthin glycosides have been chloroform–ethanol (3:1) and (1:1) [169, 170] and chloroform–acetic acid–methanol (85:2:13) [171]. Kartnig and Danhofer [172] separated Strophanthus glycosides on magnesium oxide layers with acetone–water–ethyl acetate (4:0.2:5.8).

Hoerhammer et al. [173] have examined the glycosides of Strophanthus seeds; liquid pharmaceutical preparations and tablets were also examined. Table 26.10 gives a range of R_f values for the compounds and their fluorescent colors in ultraviolet light after spraying with concentrated sulfuric acid. The compounds may also be located by spraying with Kedde's reagent (T-155) which yields purple spots in visible light.

Sun and Lang [174] reported on the investigation of cardiac glycosides using loose layers of neutral alumina. The solvents used with this adsorbent were: ether, chloroform–methanol (99:1 and 95:5), xylene–methyl ethyl ketone (1:1), and chloroform–dioxane–butanol (14:4:1).

10 ANTICOAGULANTS

Reisch et al. [174a] have chromatographed a group of five anticoagulants. These were 4-hydroxycoumarin derivatives. The separations were accomplished on silica gel G using benzene (saturated with 99% formic acid at room temperature)–methyl ethyl ketone (20:1). Karum [175] used dioxane–toluene–isopropanol–25% ammonia (1:2:1:4:2) and butanol–amyl alcohol (1:1) saturated with ammonia for separating coumarin derivatives or related compounds on silica gel. Seven coumarin derivatives have been chromatographed on silica gel 60 F_{254} with benzene–carbon tetrachloride–dioxane–acetic acid (50:40:10:1) [176]. They can be detected on fluorescent layers under ultraviolet light. The constituents of heparin preparations were separated on cellulose MN with 25% ammonia–pyridine–water (12:7:1) or butanol–water–acetic acid (3:1:1) (177).

11 ANTIDIABETICS

Most of these compounds are sulfonylurea or guanidine derivatives. Baeumler and Rippstein [178] used silica gel–cellulose (1:1) layers with benzene–methanol (4:1) for the former and methanol acidified with acetic acid for the latter. Strickland [179] chromatographed acetohexamide, chlorpropamide, phenformin hydrochloride, and tolbutamide on silica gel with five solvent systems including acetone–benzene–water (13:6:1) and dioxane–ammonia (0.88 sp gr)–water (100:3:10). Compounds were detected with reagents T-239, T-274 or ninhydrin reagent. Guven and Ozsari [180] chromatographed insulin preparations on silica gel with 0.1 *N* hydrochloric acid–96% ethanol (1:1) or 0.5 *N* hydrochloric acid–ethanol–acetone (5:3:0.5).

12 ANTICONVULSANT DRUGS

Pippenger et al. [181] used SilicAR 7-GF (Mallinckrodt) layers with chloroform–acetone (9:1), carbon tetrachloride–acetone (7:3), and benzene–acetone (4:1) to separate anticonvulsant drugs extracted from blood and urine. One chromatogram was developed in each of the solvents. Under ultraviolet light the limit of detection was 1 to 2 μg. Wad et al. [182] used a direct reflectance method for the determination of five anticonvulsant extracts from serum. Silica gel layers with chloroform–acetone (17:3) were used for the separations. Breyer [183] determined carbamazepin in serum by reflectance photometry.

13 ANTIMALARIALS

Quinine from cinchona bark has been chromatographed on silica gel with methyl ethyl ketone–methanol–water (6:2:1) and then with chloroform–acetone–diethylamine (5:4:1) in the reverse direction [184]. Quinidine, dihydroquinidine, and their metabolites were separated on silica gel at 4°C with methanol–acetone (4:1), and the eluted compounds determined by fluorescence [185]. Acridine derivatives used as antimalarials have been separated on silica gel G with benzene–acetone–28% ammonia (37:37:1) [186] and propanol–ethyl acetate–ammonia (13:15:12) among others [187]. Berberine from the powdered drug *(Berberis spp.)* was isolated by chromatography on silica gel with butanol–acetic acid–water (7:1:2) [188].

14 ANTITUSSIVES

Neidlein et al. [189] separated a group of antitussives on silica gel GF with chloroform–ethanol (60:15) containing two drops of ammonia. Prost et al. [190] chromatographed a group of propargyl derivatives with antitussive properties on silica gel with hexane–chloroform (9:1) in an atmosphere of ammonia using triple development. Iodoplatinate reagent (T-149) was used as a detecting agent. Brantner and Vamos [191] chromatographed benzonatate, which is not a homogeneous product, on silica with methyl ethyl ketone–water in various proportions.

15 BACTERICIDES

Bravo Ordenes and Hernandez-A. [192] have separated dichlorophene and hexachlorophene which are used as disinfectants in soap. Starch-bound silica gel layers with *n*-heptane saturated with acetic acid as the solvent was used for the separation giving R_f values of 0.32 and 0.63, respectively. For quantitative determination, the spots were eluted and assayed by ultraviolet spectrophotometry at 290 mμ. Other solvents used with silica gel G include: methyl ethyl ketone, isobutanol–25% ammonia-water (21:6:10), and isobutanol–lactic acid–water (5:1:2) [193]. Chlorhexidine has been identified on silica with butanol–acetic acid–water (3:1:1) at an R_f of 0.64 [194]. Koenig [195] separated a group of bactericides on silica gel GF_{254} with benzene–acetone (4:1). Brominated salicylanilides have been separated on DEAE cellulose with methanol–acetic acid (19:1) [196]. A prior development with methanol helped separate lower brominated compounds, and multiple development was also effective in increasing separations where needed. Quaternary ammonium compounds can be separated on silica gel G with ethanol–chloroform–water (36:60:1) [197] or with chloroform–isopropanol–ammonia (5:5:1) [198]. Nitrofuran and acridin antiseptics were separated with acetone–chloroform–ether–butanol (8:4:6:1) [199]. Cinoxacin and related compounds were separated on silica gel with ethyl acetate–chloroform (7:3) [200]. The compounds were visualized under ultraviolet light.

16 CONTRACEPTIVES

Since chemical contraceptives in use are steroids, the TLC systems used for this class of compounds are applicable to their determination and separation. Roeder [201] examined 19 proprietary preparations using silica gel G with ethyl acetate–cyclohexane–acetone (5:15:2) or two-dimensionally with cyclohexane–ethyl acetate (23:27) as the second solvent. Chlormadinone diacetate, megestrol acetate, norethisterone acetate, ethynodiol acetate, lynoestrenol, ethinylestradiol, and mestranol could be separated. Visualization was with 85% phosphoric acid–methanol (1:1) or antimony trichloride–acetic acid (1:1 wt/vol) with subsequent heating at 120 to 130°C. Simard and Lodge [202] used silica gel with two-dimensional systems for separating the 11 components found in Canadian preparations. The systems were chloroform–methanol (9:1) followed by benzene–methanol (19:1), and benzene–acetone (4:1) followed by methylene chloride–methanol–water (150:9:0.5). The compounds were visualized by ultraviolet and by a sulfuric acid spray followed by heating. Shroff and Shaw [203] used in situ densitometry for measuring norethindrone and mestranol.

17 DIURETICS

Thiazide type diuretics have been chromatographed on silica gel with acetone–benzene–water (3:3:1), (4:3:1), and (5:3:2) [204]; toluene–xylene–dioxane–isopropanol–25% ammonia (1:1:3:3:1) [205] and (1:1:3:3:2) [206]; benzene–ethyl acetate (1:3) [206] and (2:8) [207]; ethyl acetate with 1.5% water [208]; ethyl acetate–benzene–ammonia–25% methanol (25:20:1:5), *n*-hexane–acetone–diethylamine (6:3:1) and (2:2:1), and chloroform–methanol–diethylamine (16:3:1) [209]. Smith and Hermann [210] give a procedure for separating the thiazides from other tablet components before using thin-layer chromatography. Some detection agents include Fearon's reagent (T-240), ammoniacal copper sulfate (T-75), fluorescein–bromine reagent (T-130), Bratton–Marshall reagent (T-174), and (T-235) (sensitivity 5 μg). If sulfonamide drugs are present they may be distinguished from thiazides by spraying with Ehrlich's reagent (T-90), giving yellow spots with the former; thiazides do not react [207].

Guven [211] separated four mercurial diuretics on silica gel G with butanol–acetic acid–water (3:1:1) and located the spots with test T-75.

18 LAXATIVES

Bhandari and Walker [212] chromatographed 11 compounds having laxative properties on silica gel GF with six solvent systems including chloroform–acetone (1:1) and chloroform–benzene–methyl ethyl ketone (1:1:1). Datta and Ghosh [213] separated phenolphthalein and cascara anthraquinones in pharmaceutical products on silica gel G with benzene–ethyl acetate–acetic acid (75:24:1). The anthraquinones were visualized under ultraviolet at R_f 0.55, 0.68, and 0.95, and a 5% potassium hydroxide solution located the phenolphthalein at R_f 0.48. (See also anthraquinones under botanicals.)

19 BOTANICALS

Steinegger and Gebistorf [214] have applied thin-layer chromatography to the determination of 5 to 10% of adulterants in the leaves of stramonium, belladonna, and hyoscyamus; Hoerhammer et al. [215] detected adulteration of *Pimpinella saxifraga* root with *Heracleum spondylium* root by means of chromatographing petroleum ether extracts. Steinegger and Gebistorf examined Tilia drugs [216]. Hoerhammer et al. [217] have examined the components of the leaves, flowers, and fruits of *Crataegus oxyacantha* and their pharmaceutical preparations using thin-layer chromatography on silica gel layers with ethyl acetate–methanol–water (10:2:1). Heusser [218] gives the detailed procedure for the quantitative analysis of a number of components of medicinal plants including capsaicin and capsicum, morphine and powdered opium, digitoxin and esculin, and reserpine.

Fassina and co-workers [219–221] have chromatographed the extracts from the rhizomes of *Atractylis gummifera*. The glycoside (potassium atractylate) could be separated from the aglycon (atractyligen) by chromatographing on silica gel with butanol–methanol–water (8:1:1) with R_f values of 0.17 and 0.57, respectively, or with propanol–xylene–water (7:2:1) with R_f values of 0.16 and 0.61. The rhizomes from Sicily contained from 5 to 10 times as much atractyloside as those from Sardinia.

Hoerhammer et al. [222] have determined the R_f values of arbutin, hydroquinone, and methylhydroquinone as 0.24 to 0.30, 0.89 to 0.93, and 0.96 to 0.97, respectively. These values are for a development distance of 15 cm in ethyl acetate–methanol–water (100:16.5:13.5) on thin layers of silica gel (Woelm). These compounds occur in the leaf drug *Arctostaphylos Uva-ursi*.

The taeniacide action of the male fern rhizome is due to the phloroglucides which are present. These components have been examined by Schantz and co-workers [223, 224] and by Stahl and Schorn [225]. The former group used silica gel G layers buffered with 0.1 *M* citric acid–0.2 *M* disodium hydrogen phosphate buffer of pH 6 using 5% ethanol in an equal mixture of chloroform and petroleum ether as a developing solvent. The following R_f values were obtained: filixic acid 0.90, aspidin 0.88, albaspidin 0.87, desaspidin 0.82, aspidinol 0.41, phloraspin 0.35, and flavaspidic acid 0.07. For a detecting agent, the plates were sprayed with a mixture of equal parts of 1% ferric chloride solution and 1% potassium ferricyanide solution with one drop of concentrated nitric acid added for each 1 ml of the mixture. Deep blue spots appear where the compounds are located. As an alternative, the plates were sprayed with a 0.1% solution of Fast Blue Salt B. Schantz et al. [226] applied the method to a quantitative determination of these materials. After the chromatographic separation, the spots were eluted and determined either spectrophotometrically or by a colorimetric method.

Hoerhammer et al. [227] have separated the anthraquinone drugs found in various plants such as *Aloe, Rhamnus frangula, R. purshiana, Rheum,* and *Senna* leaves on silica gel using ethyl acetate–methanol–water (100:16.5:13.5) and propanol–ethyl acetate–water (4:4:3). The compounds were detected by examination under ultraviolet light (366 mμ) or by spraying with 10% potassium hydroxide in methanol, followed by 0.5% Fast Blue Salt solution. Sieper et al. [228] have also examined the polyhydroxy anthraquinones in *Rhamnus frangula*. The anthraquinones were separated as the glycosides on silica gel layers with dichloromethane–methanol (10:3) and (10:0.5) and with benzene–carbon tetrachloride (1:1). Quantitative measurements were made on the eluted extracts with a spectrophotometer. Chrysophanol and physcion, which are normally difficult to separate, can be separated on silica gel G with petroleum ether–ethyl acetate (9:1) [229].

Boehme and Kreutzig [230, 231] have used both paper and thin-layer chromatography for the examination of a number of different aloes. Janiak and Boehmert [232] separated Sennidin A + B from rhein by using benzene–acetic acid (2:1) with silica gel G layers. The method was used for the quantitative estimation of aloin. Gerritsma and van Rheede [233] have used thin-layer chromatography on silica gel to separate and determine the aloin in dried and fresh aloe juice. The separation was accomplished with chloroform–ethanol (3:1). After the aloin spot was located under ultraviolet light, it was extracted with methanol and determined by measuring the absorption at 355 mμ. The accuracy of the analysis was about 2%.

20 MISCELLANEOUS

Kraus and Veprovska [234–236] have reported on the use of thin-layer chromatography in the analysis of dispensed suppositories. Jones et al. [237] determined stilbestrol in suppositories.

Kreis et al. [238] have used several systems of thin-layer chromatography for the isolation and identification of terephthalanilides extracted from blood or other body fluids.

Ragazzi [239] separated some colored derivatives of phenazine, used as antituberculosis agents, on thin layers of silicic acid; this gave better results than paper chromatography. A large number of different solvent combinations were used in determining the R_f values which are reported in tabular form. Guven [240] separated four antituberculosis agents on silica with methanol, ethanol, and chloroform–methanol (17:3). Detection was with ammoniacal copper sulfate (T-75).

Sulfobromphthalein (used as a liver function test) metabolism has been studied by Whelan and Plaa [240].

Metyrapone, used as a pituitary sufficiency test, and its metabolites were separated on silica gel with methylene chloride–ethanol (25:1) [241].

Baeumler et al. [242] have used silica gel layers in investigating the doping of race horses. Various drugs given *per os,* intravenously, intramuscularly, and subcutaneously could be detected in the saliva in rather large quantities after a short time. Karawya et al. [243] used thin-layer chromatography to determine amphetamine, methylamphetamine, and ephedrine in horse urine. The method may also be used for athletes. Debackere and Laruelle [244] found that urine was the most reliable biological fluid for detecting alkaloids in doped race horses by means of thin-layer chromatography.

Thin-layer chromatography has been applied to the identification of preservatives [245, 246] and tinctures [247] in eye drops.

Other classes of pharmaceuticals have been examined by thin-layer chromatography including anthelmintics [248, 249], antialcoholics [250], antiepileptics [251–253], antihypotensives [254–256] (see also under diuretics), antitumor agents [257–259], and antimotion sickness agents [260].

Besides those references mentioned above under specific topics, a large number of reviews have been written on the use of thin-layer chromatography in the field of pharmaceuticals; some of the more general ones are Refs. 261 to 264.

References

1. S. L. Tompsett, *Analyst (London),* **93,** 740 (1968).
2. S. J. Mulé, *J. Chromatogr.,* **39,** 302 (1969).
3. V. P. Dole, W. K. Kim, and T. Englitis, *J. Am. Med. Assoc.,* **198,** 115 (1966).
4. J. G. Montalvo, Jr., E. Klein, D. Eyer, and B. Harper, *J. Chromatogr.,* **47,** 542 (1970).
5. A. M. Heaton and A. G. Blumberg, *J. Chromatogr.,* **41,** 367 (1969).
6. K. K. Kaistha and J. H. Jaffe, *J. Chromatogr.,* **60,** 83 (1971).
7. W. J. Serfontein, D. Botha, and L. S. De Villiers, *J. Chromatogr.,* **115,** 507 (1975).
8. J. R. Broich, D. B. Hoffman, S. Andryauskas, L. Galante, and C. J. Umberger, *J. Chromatogr.,* **60,** 95 (1971).
9. J. M. Fujimoto and R. I. H. Wang, *Toxicol. Appl. Pharmacol.,* **16,** 186 (1970).
10. S. J. Mulé, M. L. Bastos, D. Jukofsky, and E. Saffer, *J. Chromatogr.,* **63,** 289 (1971).
11. M. L. Bastos, D. Jukofsky, E. Saffer, M. Chedekel, and S. J. Mulé, *J. Chromatogr.,* **71,** 549 (1972).
12. M. L. Bastos, D. Jukofsky, and S. J. Mulé, *J. Chroamatogr.,* **81,** 93 (1973).

12a. N. Weissman, M. L. Lowe, J. M. Beattie, and J. A. Demetriou, *Clin. Chem.*, **17,** 875 (1971).
13. J. R. Broich, D. B. Hoffman, S. J. Goldner, S. Andryauskas, and C. J. Umberger, *J. Chromatogr.*, **63,** 309 (1971).
14. D. B. Hoffman, C. J. Umberger, S. Goldner, S. Andryauskas, D. Mulligan, and J. R. Broich, *J. Chromatogr.*, **66,** 63 (1972).
15. J. M. Meola and M. Vanko, *Clin. Chem.*, **20,** 184 (1974).
16. K. K. Kaistha and R. Tadrus, *J. Chromatogr.*, **109,** 149 (1975).
17. K. K. Kaistha, *J. Pharm. Sci.*, **61,** 655 (1972).
18. D. Sohn, J. Simon, M. A. Hanna, and G. Gahli, *J. Chromatogr. Sci.*, **10,** 294 (1972).
19. Annonymous, *J. Chromatogr. Sci.*, **10,** 352 (1972).
19a. Annonymous, *J. Chromatogr. Sci.*, **12,** 328 (1974).
20. G. Machata, *Mikrochim. Acta,* **47,** 79 (1960).
21. J. Baeumler and S. Rippstein, *Pharm. Acta Helv.*, **36,** 382 (1961).
22. E. Porges, *Bratislav. Lek. Listy,* **44-I,** 3 (1964); *Chem. Abstr.*, **60,** 11847 (1964).
23. J. Reisch, H. Bornfleth, and J. Rheinbay, *Pharm. Ztg. Ver. Apoth.-Ztg.*, **108,** 1183 (1963).
24. M. Frahm, A. Gottesleben, and K. Soehring, *Pharm. Acta Helv.*, **38,** 785 (1963).
25. M. Frahm, A. Gottesleben, and K. Soehring, *Arzneim.-Forsch.*, **11,** 1008 (1961).
26. T. Niwaguchi and H. Oki, *Kagaku Keisatsu Kenkyusho Hokoku,* **16,** 41 (1963); through *Chem. Abstr.*, **59,** 15122 (1963).
27. M. Sahli and M. Oesch, *J. Chromatogr.*, **14,** 526 (1964).
28. E. J. Shellard and I. U. Osisiogu, *Lab. Pract.*, **13,** 516 (1964).
29. A. S. Curry and R. H. Fox, *Analyst (London),* **93,** 834 (1968).
30. T. Dutkiewicz and J. Kończalik, *Farm. Polska,* **24,** 475 (1968).
31. J.-T. Huang and K.-T. Wang, *J. Chromatogr.*, **31,** 587 (1967).
32. W. A. Rosenthal, M. Kaser, and K. N. Milewski, *Clin. Chim. Acta,* **33,** 51 (1971).
33. J. Baeumler and S. Rippstein, *Arch. Pharm.*, **296,** 301 (1963).
34. R. Lindfors and A. Ruohonen, *Arch. Toxikol.*, **19,** 402 (1962).
35. R. Lindfors, *Ann. Med. Exp. Biol. Fenn. (Helsinki),* **41,** 355 (1963).
36. A. Vercruysse, *Chromatogr., Symp., 2nd, Brussels,* **1962,** 207.
37. H. Gaenshirt, *Arch. Pharm.*, **296,** 73 (1963).
38. J. Lehmann and V. Karamustafaoglu, *Scand. J. Clin. Lab. Invest.*, **14,** 554 (1962).
39. J. T. Wright, *J. Clin. Pathol,* **7,** 61 (1954).
40. J. A. Petzold, W. Camp, Jr., and E. R. Kirch, *J. Pharm. Sci.*, **52,** 1106 (1963).
41. J. Kelleher and J. G. Rollason, *Clin. Chim. Acta,* **10,** 92 (1964).
42. R. A. De Zeeuw and J. Wijsbeek, *J. Chromatogr.*, **48,** 222 (1970).
43. J. Cochin and J. W. Daly, *J. Pharmacol. Exp. Therap.*, **139,** 154 (1963).

44. I. Sunshine, E. Rose, and J. LeBeau, *Clin. Chem.*, **9,** 312 (1963).
45. J. G. Kirchner, J. M. Miller, and G. J. Keller, *Anal. Chem.*, **23,** 420 (1951).
46. H. Eberhardt, K. J. Freundt, and J. W. Langbein, *Arzneim.-Forsch.*, **12,** 1087 (1962).
47. I. C. Dijkhuis, *Pharm. Weekbl.*, **106,** 745 (1971).
48. W. Paulus, W. Hoch, and R. Keymer, *Arzneim.-Forsch.*, **13,** 609 (1963).
49. T. J. Mellinger and C. E. Keller, *J. Pharm. Sci.*, **51,** 1169 (1962).
50. J. Cochin and J. W. Daly, *J. Pharmacol. Exp. Therap.*, **139,** 160 (1963).
51. J. Kofoed, C. Fabierkiewicz, and G. H. W. Lucas, *J. Chromatogr.*, **23,** 410, (1966).
52. I. Zingales, *J. Chromatogr.*, **31,** 405 (1967).
53. L. Kraus and E. Dumont, *J. Chromatogr.*, **56,** 159 (1971).
54. A. Hulshoff and J. H. Perrin, *J. Chromatogr.*, **129,** 249 (1976).
55. A. Noirfalise and M. H. Grosjean, *J. Chromatogr.*, **16,** 236 (1964).
56. M. Ferrari and C. S. Tóth, *J. Chromatogr.*, **9,** 388 (1962).
57. S. Seno, W. V. Kessler, and J. E. Christian, *J. Pharm. Sci.*, **53,** 1101 (1964).
58. W. Rusiecki and M. Henneberg, *Acta Polon. Pharm.*, **21,** 25 (1964).
59. H. Eberhardt, O. W. Lerbs, and K. J. Freundt, *Arch. Exp. Pathol Pharmakol.*, **245,** 136 (1963).
60. H. Eberhardt, O. W. Lerbs, and K. J. Freundt, *Arzneim.-Forsch.*, **13,** 804 (1963).
61. P. Turano and W. J. Turner, *J. Chromatogr.*, **64,** 347 (1972).
62. P. N. Kaul, M. W. Conway, and M. L. Clark, *Nature*, **226,** 372 (1970).
63. S. N. Tewari, *Z. Anal. Chem.*, **275,** 31 (1975).
64. F. Eiden and H. D. Stachel, *Deut. Apoth.-Ztg.*, **103,** 121 (1963).
65. A. Fiori and M. Marigo, *Nature*, **182,** 943 (1958).
66. M. Marigo, *Minerva Medicoleg.*, **81,** 70 (1961).
67. M. Marigo, *Arch. Kriminol.*, **128,** 99 (1961); *Chem. Abstr.*, **56,** 5068 (1962).
68. J. Baeumler and S. Rippstein, *Helv. Chim. Acta*, **44,** 2208 (1961).
69. C. Schuetz, D. Post, G. Schewe, and H. Schuetz, *Z. Anal. Chem.*, **262,** 282 (1972).
70. A. M. Chodakowski, *Farm. Polska*, **25,** 637 (1969).
71. G. Cimbura, *J. Chromatogr. Sci.*, **10,** 287 (1972).
72. J. Blazek and Z. Stejskal, *Pharmazie*, **27,** 506 (1972).
73. E. G. C. Clarke, *J. Forensic Sci. Soc.*, **7,** 46 (1967).
74. M. A. Shaw and H. W. Peel, *J. Chromatogr.*, **104,** 201 (1975).
75. T. Niwaguchi and T. Inoue, *J. Chromatogr.*, **43,** 510 (1969).
76. L. A. Dal Cortiva, J. R. Brvich, A. Dihberg, and B. Newman, *Anal. Chem.*, **38,** 1959 (1966).
77. G. V. Alliston, M. J. De Faubert Maunder, and G. F. Phillips, *J. Pharm. Pharmacol.*, **23,** 555 (1971).
78. K. Genest, *J. Chromatogr.*, **19,** 531 (1965).
79. T. Niwaguchi and T. Inoue, *J. Chromatogr.*, **59,** 127 (1971).
79a. *Ibid.*, *J. Chromatogr.*, **121,** 165 (1976).
80. F. Korte and H. Sieper, *J. Chromatogr.*, **13,** 90 (1964).

81. J. A. Vinson and J. E. Hooyman, *J. Chromatogr.*, **106**, 196 (1975).
82. S. N. Tewari, S. P. Harpalani, and S. C. Sharma, *Chromatographia*, **7**, 205 (1974).
83. A. B. Segelman, *J. Chromatogr.*, **82**, 151 (1973).
84. L. Grlic, *Acta Pharm. Jugosl.*, **24**, 63 (1974).
85. H. M. Stone and H. M. Stevens, *J. Forensic Sci. Soc.*, **9**, 31 (1969).
86. W. W. Just, G. Werner, and M. Wiechmann, *Naturwissenschaften*, **59**, 222 (1972).
87. J. A. Vinson, D. D. Patel, and A. H. Patel, *Anal. Chem.*, **49**, 163 (1977).
88. T. P. Ivanon, *Monatsh. Chem.*, **97**, 1499 (1966).
89. J. M. Parker and H. L. Fiske, *J. Assoc. Off. Anal. Chem.*, **55**, 876 (1972).
90. F. W. H. M. Merkus, M. G. J. Jaspers-van Wouw, and J. F. C. Roovers-Bollen, *Pharm. Weekbl.*, **107**, 98 (1972).
91. A. Sterling, *J. Chromatogr. Sci.*, **10**, 268 (1972).
92. J. Im. Obersteg and J. Baeumler, *Arch. Toxikol.*, **19**, 339 (1962).
93. J. Drabner, H. Bauer, and W. Schwerd, *Arch. Toxicol.*, **21**, 367 (1966).
94. A. Alessandro, F. Mari, and S. Settecase, *Farmaco, Ed. Prat.*, **22**, 437 (1967).
95. C. R. Henwood, *Nature*, **216**, 1039 (1967).
96. K. Adank and T. Schmidt, *Chimia*, **23**, 299 (1969).
97. K. C. Guven, *Eczacilik Bul.*, **10**, 43 (1968).
98. R. M. Facino and G. L. Corona, *J. Pharm. Sci.*, **58**, 764 (1969).
99. A. Viala, F. Gouezo, and C. Gola, *J. Chromatogr.*, **45**, 94 (1969).
100. V. J. McLinden, *J. Forensic Sci. Soc.*, **10**, 135 (1970).
100a. S. Thunell, *J. Chromatogr.*, **130**, 209 (1977).
101. W. W. Fike and I. Sunshine, *Anal. Chem.*, **37**, 127 (1965).
102. F. Boonen, *J. Pharm. Belg.*, **27**, 233 (1972).
103. S. J. Mulé, *Anal. Chem.*, **36**, 1907 (1964).
104. S. J. Mulé, *J. Chromatogr. Sci.*, **10**, 275 (1972).
105. J. Cochin and J. W. Daly, *Experientia*, **18**, 294 (1962).
106. E. Vidic, *Arch. Toxikol.*, **19**, 254 (1961).
107. M. Šaršúnová and V. Schwarz, *Pharmazie*, **18**, 34 (1963).
108. H. Gaenshirt, *Arch. Pharm.*, **296**, 129 (1963).
109. M. M. Frodyma, V. T. Lieu, and R. W. Frei, *J. Chromatogr.*, **18**, 520 (1965).
110. T. Fuwa, T. Kido, and H. Tanaka, *Yakuzaigaku*, **22**, 269 (1962); through *Chem. Abstr.*, **59**, 7319 (1963).
111. J. L. Emmerson and R. C. Anderson, *J. Chromatogr.*, **17**, 495 (1965).
112. R. D. Thompson and G. L. Johnson, *J. Chromatogr.*, **88**, 361 (1974).
113. D. Radulovič, Z. Blagojevič, and D. Živanov-Stakič, *Acta Pharm. Jugosl.*, **24**, 173 (1974).
114. Z. Blagojevič, D. Radulovič, and D. Živanov-Stakič, *Acta Pharm. Jugosl.*, **25**, 263 (1975).
115. H. C. Hsiu, T. B. Shih, and K.-T. Wang, *J. Chromatogr.*, **41**, 489 (1969).
116. H.-C. Chiang and T.-M. Chiang, *J. Chromatogr.*, **47**, 128 (1970).

117. B. Unterhalt, *Arch. Pharm.,* **303,** 445 (1970).
118. F. Pellerin and D. Mancheron, *Int. Symp. Chromatogr. Electrophor., Lect. Pap. 6th 1970,* Ann Arbor Science, Ann Arbor, Mich., 1971, p. 501.
119. I. Formanek, L. Fulop, and I. Vereph, *Rev. Med. (Targu-Mures),* **12,** 300 (1966).
120. D. Waldi, *Arch. Pharm.,* **295,** 125 (1962).
121. A. H. Beckett and N. H. Choulis, *J. Pharm. Pharmacol.,* **15,** 236T (1963).
122. A. H. Beckett and N. H. Choulis, *23rd Intern. Kongr. Pharmaz. Wissenschaften, Muenster September 9–14, 1963.*
123. N. H. Choulis, *J. Pharm. Sci.,* **56,** 196 (1967).
124. A. Vahidi and D. V. S. Sanakar, *J. Chromatogr.,* **43,** 135 (1969).
125. R. Segura-Cardona and K. Soehring, *Med. Exp.,* **10,** 251 (1964).
126. J. D. Sapira, *J. Chromatogr.,* **42,** 134 (1969).
127. N. H. Choulis and C. E. Carey, *J. Pharm. Sci.,* **57,** 1048 (1968).
128. R. Roser and P. M. Tocci, *J. Chromatogr.,* **72,** 207 (1972).
129. F. Van Hoof and A. Heyndrickx, *Anal. Chem.,* **46,** 286 (1974).
130. H. H. Loh, I. K. Ho, W. R. Lipscomb, T. M. Cho, and C. Selewski, *J. Chromatogr.,* **68,** 289 (1972).
131. M. Šaršúnová, *Pharmazie,* **18,** 748 (1963).
132. K. C. Guven and A. Hincal, *Eczacilik Bul.,* **8,** 202 (1966).
133. J. A. Fresen, *Pharm. Weekbl.,* **102,** 659 (1967).
134. E. Roeder, E. Mutschler, and H. Rochelmeyer, *Pharm. Acta Helv.,* **44,** 644 (1969).
135. R. A. De Zeeuw, *J. Pharm. Pharmacol. (Suppl.)* **20,** 54 (1968).
136. W. Messerschmidt, *Deut. Apoth.-Ztg.,* **111,** 597 (1971).
137. J. Reisch, H. Bornfleth, and J. Rheinbay, *Pharm. Ztg..Ver. Apoth.-Ztg.,* **107,** 920 (1962).
138. E. G. Wollish, M. Schmall, and M. Hawrylyshyn, *Anal. Chem.,* **33,** 1138 (1961).
139. T. Bićan-Fišter and V. Kajganović, *J. Chromatogr.,* **11,** 492 (1963).
140. *Ibid.,* **16,** 503 (1964).
141. S. Klein and B. T. Kho, *J. Pharm. Sci.,* **51,** 966 (1962).
142. B. T. Kho and S. Klein, *J. Pharm. Sci.,* **52,** 404 (1963).
143. N. Karpitschka, *Mikrochim. Acta,* **1963,** 157.
144. M. Th. Van der Venne and J. B. T'Siobbel, *Chromatogr., Symp., 2nd, Brussels,* **1962,** 196.
145. G. L. Biagi, A. M. Barbaro, M. C. Guerra, C. Cantelli-Forti, and O. Gandolfi, *J. Chromatogr.,* **106,** 349 (1975).
146. R. A. De Zeeuw, *J. Chromatogr.,* **48,** 27 (1970).
147. L. Lepri, P. G. Desideri, and G. Tanturli, *J. Chromatogr.,* **93,** 201 (1974).
148. M. I. Walash and S. P. Agarwal, *J. Pharm. Sci.,* **61,** 277 (1972).
149. C. W. Sigel, J. L. Woolley, Jr., and C. A. Nichol, *J. Pharm. Sci.,* **64,** 973 (1975).
150. R. Tschesche, W. Freytag, and G. Sntazke, *Chem. Ber.,* **92,** 3053 (1959).
151. J. Reichelt and J. Pitra, *Česk. Farm.,* **12,** 416 (1963).
152. J. Reichelt and J. Pitra, *Collect. Czech. Chem. Commun.,* **27,** 1709 (1962).

153. P. Braeckman, R. van Severen, and F. Haerinck, *Pharm. Tijdschr. Belg.*, **40,** 129 (1963); through *Chem. Abstr.*, **60,** 11849 (1964).
154. J. Zurkowska, W. Lukaszewski, and A. Ozarowski, *Acta Pol. Pharm.*, **20,** 115 (1963).
155. J. Zurkowska and A. Ozarowski, *Acta Pol. Pharm.*, **22,** 83 (1965).
156. L. Fauconnet and M. Waldesbuehl, *Pharm. Acta Helv.*, **38,** 423 (1963).
157. I. Sjoeholm, *Svensk Farm. Tidskr.*, **66,** 321 (1962).
158. B. Pekic, *Acta Pharm. Jugosl.*, **18,** 141 (1968).
159. M. L. Boisio, *J. Chromatogr.*, **73,** 279 (1972).
160. L. Nover, G. Juettner, S. Noack, G. Baumgarten, and M. Luckner, *J. Chromatogr.*, **39,** 419 (1969).
161. U. Lutz, Oesterr. *Apoth.-Ztg.*, **25,** 250 (1971).
162. L. Storstein, *J. Chromatogr.*, **117,** 87 (1976).
163. K. Gjerdrum, *Acta Med. Scand.*, **187,** 371 (1970).
164. D. B. Faber, A. De Kok, and U. A. T. Brinkman, *J. Chromatogr.*, **143,** 95 (1977).
164a. E. Watson, P. Tramell, and S. M. Kalman, *J. Chromatogr.*, **69,** 157 (1972).
165. E. Steinegger and J. H. van der Walt, *Pharm. Acta Helv.*, **36,** 599 (1961).
166. W. Steidle, *Planta Med.*, **9,** 435 (1961).
167. G. Lukas, *Sci. Pharm.*, **30,** 47 (1962).
168. A. Y. Khorlin and A. F. Bochkov, *Izv. Akad. Nauk, SSSR, Otd, Khim. Nauk,* **1962,** 1120; through *Chem. Abstr.*, **57,** 12812 (1962).
169. J. Lutomski, Z. Kowalewski, and M. Kortus, *Diss. Pharm. Pharmacol.*, **18,** 409 (1966).
170. *Ibid.*, **21,** 241 (1969).
171. G. L. Corona and M. Raiteri, *J. Chromatogr.*, **19,** 435 (1965).
172. T. Kartnig and R. Danhofer, *J. Chromatogr.*, **52,** 313 (1970).
173. L. Hoerhammer, H. Wagner, and H. Koenig, *Deut. Apoth.-Ztg.*, **103,** 502 (1963).
174. N.-C. Sun and H.-Y. Lang, *Yao Hsueh Hsueh Pao,* **11,** 101 (1964).
174a. J. Reisch, H. Bornfleth, and J. Rheinbay, *Pharm. Ztg. Ver. Apoth.-Ztg.*, **108,** 1183 (1963).
175. G. Karum, *Arzneim.-Forsch.*, **18,** 1336 (1968).
176. R. Vanhaelen-Fastré and M. Vanhaelen, *J. Chromatogr.*, **129,** 397 (1976).
177. K. C. Guven and F. Arabacioglu, *Eczacilik Bul.*, **12,** 188 (1970).
178. J. Baeumler and S. Rippstein, *Deut. Apoth.-Ztg.*, **107,** 1647 (1967).
179. R. D. Strickland, *J. Chromatogr.*, **24,** 455 (1966).
180. K. C. Guven and G. Ozsari, *Eczacilik Bul.*, **8,** 206 (1966).
181. C. E. Pippenger, J. E. Scott, and H. W. Gillen, *Clin. Chem.*, **15,** 255 (1969).
182. N. Wad, E. Hanifl, and H. Rosenmund, *J. Chromatogr.*, **143,** 89 (1977).
183. U. Breyer, *J. Chromatogr.*, **108,** 370 (1975).
184. H. Boehme and R. Bitsch., *Arch. Pharm. (Weinheim),* **303,** 418 (1970).
185. G. Haertel and A. Korhonen, *J. Chromatogr,* **37,** 70 (1968).
186. A. Deleenheer, J. E. Sinsheimer, and J. H. Burckhalter, *J. Pharm. Sci.*, **61,** 1659 (1972).
187. O. N. Yalcindag, *J. Pharm. Belg.*, **26,** 337 (1971).

188. W. Messerschmidt, *J. Chromatogr.*, **39,** 90 (1969).
189. R. Neidlein, E. Hahne, and U. Hahne, *Deut. Apoth.-Ztg.*, **106,** 987 (1966).
190. M. Prost, M. Urbain, and R. Charlier, *Helv. Chim. Acta,* **52,** 1134 (1969).
191. A. Brantner and J. Vamos, *Int. Symp. Chromatogr. Electrophor., Lect. Pap., 6th 1970,* Ann Arbor Science, Ann Arbor, Mich., 1971, p. 401.
192. R. Bravo Ordenes and F. Hernandez-A., *J. Chromatogr.*, **7,** 60 (1962).
193. S. Gecgil, *Eczacilik Bul.*, **11,** 11 (1969).
194. K. C. Guven, *Eczacilik Bul.*, **12,** 111 (1970).
195. H. Koenig, *Z. Anal. Chem.*, **246,** 247 (1969).
196. N. E. Skelly and K. A. Kamke, *Anal. Chem.*, **39,** 1009 (1967).
197. S. M. Henry, G. Jacobs, and A. Achmeteli, *Appl. Microbiol.*, **15** 1489 (1967).
198. J. L. Kiger and J. G. Kiger, *Ann. Pharm. Fr.*, **25,** 601 (1967).
199. R. Neidlein, E. Hohndorf, and J. Rosenblath, *Pharm. Ztg.*, **111,** 874 (1966).
200. R. H. Bishara and I. M. Jakovljevic, *J. Chromatogr.*, **116,** 485 (1976).
201. E. Roeder, *Deut. Apoth.-Ztg.*, **107,** 1007 (1967).
202. M. B. Simard and B. A. Lodge, *J. Chromatogr.*, **51,** 517 (1970).
203. A. P. Shroff and C. J. Shaw, *J. Chromatogr. Sci.*, **10,** 509 (1972).
204. R.-T. Wang and M.-C. Tung, *Hua Hsueh,* **1970,** 25; through *Anal. Abstr.*, **20,** 3349 (1971).
205. K. C. Guven and S. Cobanlar, *Eczacilik Bul.*, **9,** 98 (1967).
206. Z. Gawrych and T. Pomazańska, *Acta Pol. Pharm.*, **25,** 39 (1968); through *Anal. Abstr.*, **16,** 3233 (1969).
207. B. G. Osborne, *J. Chromatogr.*, **70,** 190 (1972).
208. C. L. Lapiere, *J. Pharm. Belg.*, **20,** 275 (1965).
209. R. Maes, M. Gijbels, and L. Laruelle, *J. Chromatogr.*, **53,** 408 (1970).
210. P. J. Smith and T. S. Hermann, *Anal. Biochem.*, **22,** 134 (1968).
211. K. C. Guven, *Eczacilik Bul.*, **10,** 13 (1968); through *Anal. Abstr.*, **16,** 2701 (1969).
212. P. R. Bhandari and H. Walker, *J. Chromatogr.*, **45,** 324 (1969).
213. D. D. Datta and D. Ghosh, *Indian J. Pharm.*, **29,** 179 (1967).
214. E. Steinegger and J. Gebistorf, *Pharm. Acta Helv.*, **37,** 343 (1962).
215. L. Hoerhammer, H. Wagner, and B. Lay, *Pharmazie,* **15,** 645 (1960).
216. E. Steinegger and J. Gebistorf, *Sci. Pharm.*, **31,** 298 (1963).
217. L. Hoerhammer, H. Wagner, and M. Seitz, *Deut. Apoth.-Ztg.*, **103,** 1302 (1963).
218. D. Heusser, *Planta Med.*, **12,** 237 (1964).
219. G. Fassina, *Boll. Soc. Ital. Biol. Sper.*, **36,** 1417 (1960); *Chem Abstr.*, **60,** 20416 (1964).
220. G. Fassina, A. R. Contessa, and C. E. Toth, *Boll. Soc. Ital. Biol. Sper.*, **38,** 260 (1962); *Chem. Abstr.*, **59,** 15598 (1963).
221. *Ibid.*, **34,** 346 (1963); *Chem. Abstr.*, **60,** 9100 (1964).
222. L. Hoerhammer, H. Wagner, and H. Koenig, *Deut. Apoth.-Ztg.*, **103,** 1 (1963).
223. M. von Schantz, L. Ivars, I. Lindgren, L. Laitinen, E. Kukkonen, H. Wallenius, and C. J. Widen, *Planta Med.*, **12,** 112 (1964).

224. M. von Schantz and S. Nikula, *Planta Med.*, **10,** 22 (1962).
225. E. Stahl and P. J. Schorn, *Naturwissenschaften*, **49,** 14 (1962).
225. E. Stahl and P. J. Schorn, *Naturwissenschaften*, **49,** 14 (1962).
226. M. von Schantz, L. Ivars, I. Kukkoven, and A. Ruuskanen, *Planta Med.*, **10,** 98 (1962).
227. L. Hoerhammer, H. Wagner, and G. Bittner, *Pharm. Ztg., Ver. Apoth.-Ztg.*, **108,** 259 (1963).
228. H. Sieper, R. Longo, and F. Korte, *Arch. Pharm.*, **296,** 403 (1963).
229. M. Danilović and O. Naumović-Stevanoić, *J. Chromatogr.*, **19,** 613 (1965).
230. H. Boehme and L. Kreutzig, *Apoth.-Ztg.*, **103,** 505 (1963).
231. H. Boehme and L. Kreutzig, *Arch. Pharm.*, **297,** 681 (1964).
232. B. Janiak and H. Boehmert, *Arzneim.-Forsch.*, **12,** 431 (1962).
233. K. W. Gerritsma and M. C. B. van Rheede, *Pharm. Weekbl.*, **97,** 765 (1962); *Chem. Abstr.*, **58,** 6646 (1963).
234. L. Kraus and E. Veprovska, *Cesk. Farm.*, **12,** 515 (1963).
235. F. Schmidt, *Deut. Apoth.-Ztg.*, **108,** 1326 (1968).
236. F. Schmidt, *Pharm. Ztg.*, **114,** 1523 (1969).
237. L. N. Jones, M. Seidman, and B. C. Southworth, *J. Pharm. Sci.*, **57,** 646 (1968).
238. W. Kreis and D. L. Warkentin, *Cancer Chemother. Rep.*, **32,** 7 (1963).
239. E. Ragazzi, *Boll. Chim. Farm.*, **100,** 402 (1961); through *Chem. Abstr.*, **56,** 10283 (1962).
240. K. C. Guven, *Eczacilik Bul.*, **9,** 186 (1967).
241. F. J. Whelan and G. L. Plaa, *Toxicol. Appl. Pharmacol.*, **5,** 457 (1963).
242. J. Baeumler, A. L. Brault, and J. I. Obersteg, *Schweiz. Arch. Tierheilk.*, **106,** 346 (1964).
243. M. S. Karawya, M. A. El-Keiy, S. K. Wahba, and A. R. Kozman, *J. Pharm. Pharmacol.*, **20,** 650 (1968).
244. M. Debackere and L. Laruelle, *J. Chromatogr.*, **35,** 234 (1968).
245. H. Siedlanowska-Krowczynska, and B. Janik, *Acta Pol. Pharm.*, **30,** 179 (1973).
246. K. Ludwikowska, A. Fiebig, and E. Szcygiel, *Farm. Pol.*, **31,** 37 (1975).
247. B. Pruska-Wysocka, *Farm. Pol.*, **31,** 775 (1975).
248. T. Inoue and K. Juniper, Jr., *J. Chromatogr.*, **42,** 548 (1969).
249. H. D. Beckstead and S. J. Smith, *J. Pharm. Sci.*, **56,** 390 (1967).
250. D. H. Neiderhiser, R. K. Fuller, L. J. Hejduk, and H. P. Roth, *J. Chromatogr.*, **117,** 187 (1976).
251. M. Šaršunová and T. T. Hoang Ba, *Cesk. Farm.*, **15,** 522 (1966).
252. O. V. Olesen, *Acta Pharmacol. Toxicol.*, **25,** 123 (1967).
253. *Ibid.*, **26,** 222 (1968).
254. Z. Oberman, R. Chayen, and M. Herzberg, *Clin. Chim. Acta*, **29,** 391 (1970).
255. A. V. Barooshian, M. J. Lautenschleger, and W. G. Harris, *Anal. Biochem.*, **49,** 569, (1972).
256. S. D. Harrison, Jr., P. Chiu, and R. P. Maickel, *J. Chromatogr.*, **85,** 151 (1973).
257. L. T. Mulligan, Jr., and L. B. Mellett, *J. Chromatogr.*, **43,** 376 (1969).

258. S. Klosowski, *Diss. Pharm.*, **20**, 335 (1968).
259. J. E. Bakke, V. J. Feil, C. E. Fjelstul, and E. J. Thacker, *J. Agric. Food Chem.*, **20**, 384 (1972).
260. S. L. Kidman, *J. Assoc. Publ. Anal.*, **9**, 24 (1971).
261. E. G. C. Clarke, Ed., *Isolation and Identification of Drugs in Pharmaceuticals, Body Fluids and Post-Mortem Material,* Pharmaceutical Press, London, 1969.
262. E. Tyjihák and G. Held, "Thin-Layer Chromatography in Pharmacognosy," in *Progress in Thin-Layer Chromatography and Related Methods,* Vol. II, A. Niederwieser and G. Pataki, Ed., Ann Arbor Science, Ann Arbor, Mich., 1971, p. 183.
263. K. Macek, Ed., *Pharmaceutical Applications of Thin-Layer and Paper Chromatography,* Elsevier, London, 1972.
264. B. Borkowski, Ed., *Chromatografia cienkowarstwowa w analizie farmaceutyznei (Thin-Layer Chromatography in Pharmaceutical Analysis),* PZWL, Warsaw, 1973.

Chapter **XXVII**

PHENOLS

1 DIRECT SEPARATION ON BOUND SILICA GEL LAYERS

As early as 1953, Labat and Montes [1] separated a group of seven phenolic compounds using two different solvent systems with a silica gel-bentonite (6:1) layer. Wagner [2] used a 1:1 mixture of kieselguhr-silicic acid with five different solvent systems of xylene, chloroform, and mixtures of the two. Pastuska and Petrowitz [3, 4] have examined a large number of phenols (Table 27.1) on silica gel G using benzene, benzene-methanol, benzene-dioxane-acetic acid, or benzene-methanol-acetic acid mixtures as solvents. The phenols were detected by spraying with a diazonium salt. Quantitative determinations were investigated and the relationship between spot area and log of the concentration was determined. By using 0.3-mm-thick layers, the density of the spots was measured directly and related to the quantity of material. Gaenshirt and Morianz [5] have separated the methyl and propyl *p*-hydroxybenzoates on silica gel with pentane-acetic acid (22:3). Quantitative analysis was accomplished by eluting the spots with methanol and measuring the ultraviolet absorption between 220 and 300 mμ. The accuracy of the method was between $\pm$2.5 and 3.0%. Petrowitz [6] chromatographed the isomeric xylenols and compared these with the corresponding anisoles. Table 27.1

Table 27.1 $R_f \times 100$ Values of Phenolic Compounds on Silica Gel G in Various Solvents

Phenol	*Benzene–Dioxane–Acetic Acid* (90:25:4)[a] [3, 4]	*Benzene–Methanol–Acetic Acid* (45:8:4)[a] [3, 4]	*Benzene*[a] [4, 6]	*Benzene–Methanol* (95:5)[a] [4, 6]	*Benzene–Acetic Acid* (5:1)[a] [9]	*Color with Diazotized Benzidine* [4]
Caffeic acid	24	43				Yellowish brown
Catechol	58	54			42	Gray-green
o-Cresol			26	52		Yellow
m-Cresol			19	45	57	Yellow
p-Cresol			20	49		Orange-yellow
2,3-Dimethylphenol			27	59		Orange
2,4-Dimethylphenol			24	51		Light yellow
2,5-Dimethylphenol			28	58		Yellow
2,6-Dimethylphenol			39	67		Light Yellow
3,4-Dimethylphenol			15	44		Brownish yellow
3,5-Dimethylphenol			17	44		Orange-yellow
1,3-Dihydroxynaphthalene	70	80	0	28		
o,o'-Dihydroxybiphenyl	71	66	5	35		
Ethyl gallate	36	38				Gray-brown
Ethyl *m*-hydroxybenzoate	76	61				Yellow
Ethyl protocatechuate	50	62				Borwn
Ferulic acid	50	58				Grayish brown
Gallic acid	18	23				Light brown
Gentisic acid	30	40				Orange
Guaicol	83	72			63	Reddish brown
Hydroquinone	54	46			19	Yellow

p-Hydroxybenzaldehyde	63	56		17		Brown
m-Hydroxybenzoic acid	49	51				Light yellow
p-Hydroxybenzoic acid	55	60				Yellow
Isovanillin	56	47		23		Yellowish brown
Kojic acid	6	6				Light red
Methyl gentisate	71	61				Gray
α-Naphthol	81	67	28	60	63	Blue-violet
β-Naphthol	79	63	18	56	58	Blue-violet
o-Hydroxybiphenyl	87	71	46	71		
p-Hydroxybiphenyl	76	62	21	54		
Phenol	76	60	17	43	70	Yellow
Phloroglucinol	34	32			3	Violet
Protocatechuic acid	32	39				Brown
Protocatechuic aldehyde	32	45		6		Gray-brown
Pyrogallol	32	45			15	Dark brown
Resorcinol	56	52			22	Reddish brown
γ-Resorcyclic acid	54	52				Rust red
Salicyclic acid	64	68				V[c]
Salicylaldehyde	82	83		91		V[c]
Syringaldehyde	60	57		14		Orange
Syringic acid	48	60				Orange
Umbelliferon	55	58				No color
Vanillic acid	54	61				Reddish brown
o-Vanillin	74	59		53		Brown
Vanillin	70	64		27		Yellow

[a] Development distance 10 cm.
[b] Development distance 12 cm.
[c] V = Visualized with alkaline permanganate.

gives the R_f values of a group of phenols and the colors obtained by reacting with diazotized benzidine.

Petrowitz [7, 8] has examined the components of coal tar oils and Seeboth [9, 10] has chromatographed phenols occurring in coke-oven effluents. The latter used silica gel and aluminum oxide as well as mixtures of the two. The best separation for the phenols from the low-temperature carbonization distillation process was obtained with a mixture of acidic aluminum oxide and silica gel in a ratio of 1:1, with 20% calcium sulfate as a binder. With these layers and a solvent consisting of chloroform–acetone–diethylamine (4:2:0.2), the following R_f values were obtained: catechol 0.0, resorcinol 0.34, hydroquinone 0.55, and phenol 0.74. These may be contrasted to a separation on silica gel with benzene–acetic acid (5:1) which yielded the R_f values of 0.42, 0.22, 0.19, and 0.70, respectively. For simple phenols the best solvents were petroleum ether (60 to 80°C)–carbon tetrachloride–acetic acid (4:6:1) and chloroform–acetone–diethylamine (4:2:0.2). Polyhydric phenols were best separated using chloroform–acetic acid (5:1), chloroform–acetone–acetic acid (10:2:1), and benzene–acetic acid (5:1). These spots were made visible by spraying with an acetone solution of *p*-nitrobenzenediazonium fluoroborate. Seeboth and Goersch [11] have used this method for the quantitative determination of phenols. This was accomplished by eluting the spots with 4 ml of methanol and then reacting with 2 ml of a 0.1% water solution of *p*-nitrobenzenediazonium fluoroborate containing 2% of a 35% solution of fluoboric acid as a stabilizer. After the addition of 2 ml of 10% sodium acetate solution, the mixture was allowed to stand for 5 min; 10 ml of 1 *N* potassium hydroxide was then added and the mixture was diluted to 50 ml for measuring in a colorimeter. Ragazzi and Veronese [12] measured spectrophotometrically the color produced by reaction with Folin–Ciocalteu reagent (T-131) without prior elution from the adsorbent. Good results were obtained with silica gel or cellulose layers, but polyamide appeared to inhibit the color reaction even though the spots could be detected with the reagent. Ibrahim [13] used an in situ fluorometric method on silica gel–cellulose (1:1) layers for quantities of 0.05 to 10 μg depending on the compounds.

Haub and Kaemmerer [14] give the R_f values for 26 polynuclear compounds obtained from the condensation of *p*-cresol and formaldehyde. Separations were achieved on silica gel layers dried at 110°C; the solvents used for the separation were benzene–methanol–acetic acid (95:2.5:2.5), benzene–methanol (3:1), chloroform–methanol (24:1), and chloroform–methanol–water (95:4:1). The members of a homologous series could be separated by two-dimensional chromatography using different solvents in each direction. Location of the compounds was established by spraying with a 20% solution of antimony pentachloride in carbon tetrachloride.

Halmekoski and Hannikainen [15] examined the relationship between the structure of homologous phenols and their R_f values, paying particular attention to the effect of the size of the para substituent. For most of the solvent systems, plots of log R_f versus n (number of carbon atoms) were linear. The linearity was best in those cases where the para substituent contained only carbon and hydrogen atoms. R_f values were obtained on both silica gel and polyamide layers. Pastuska and Petrowitz [4] also examined a relationship between the constitution of phenols and their R_f values.

Lipina [16] separated mono-, di-, and trihydric phenols on acid-washed silica gel with a starch binder. Development was made with benzene–acetic acid–water (2:2:1) for a distance of 15 cm. Phenol, guaiacol, and catechol were determined quantitatively by eluting with ethanol and measuring the color produced with *p*-nitroaniline.

Naff and Naff [17] have used microchromatoplates in teaching thin-layer chromatography and have used a group of seven phenols as known compounds to be separated on silica gel layers. With a solvent consisting of toluene–dioxane (25:4), the R_f values obtained were *o*-nitrophenol 0.73, *o*-phenylphenol 0.60, *m*-cresol 0.53, *m*-nitrophenol 0.45, *o*-hydroxybenzaldehyde 0.30, resorcinol 0.23, and phloroglucinol 0.07.

Schulz e al. [18] compared the separation of phenols and organic peroxides on silica gels of different pore diameter. The best results were obtained on a 20 to 50 Å material with benzene–acetic acid (5:1).

Waksmundzki and Różylo [19] found that silica gel with a specific surface area of 565 m^2/g gave the best separation of dihydroxynaphthalenes.

For the qualitative and quantitative analysis of pentachlorophenol, Deters [20] has used an acidic silica gel with chloroform as a developing agent; the silica gel layers were prepared with 0.05 N oxalic acid. Quantitative evaluation was achieved by eluting the spot and measuring the absorption in the ultraviolet. Furukawa [21] has also separated a group of phenols on acidic silica gel chromatostrips. Husain [22] has chromatographed a group of 23 chlorinated cresols and xylenols on thin layers of silica gel G. Several solvent systems were investigated and the best solvent appeared to be xylene saturated with formamide. The *p*-chlorocresols could not be separated from the parent cresols, nor could 6-chloro-2-methylphenol be separated from 4,6-dichloro-2-methylphenol. The method was applied to following the progress of chlorination of 2,5-dimethylphenol in carbon tetrachloride with gaseous chlorine. As a detecting agent the chromatograms were sprayed with phosphotungstomolybdic acid solution and then exposed to ammonia. The chlorinated phenols appeared as blue spots on a white background. Quantitatively these compounds were determined by direct densitometry [23].

The lichen acids have been investigated on silica gel plates acidified with oxalic acid [24]. Benzene–chloroform (1:1) was used with tank saturation for the development of the compounds. With this solvent, the following R_f values were reported: usnic acid 0.65, vulpinic acid 0.80, and evernic acid 0.11. Bachmann [25] has used silica gel with a benzene–dioxane–acetic acid (90:25:4) solvent for the separation of lichen acids of the β-orcinol group from *Parmelia*. The following R_f values were reported: atranorin 0.71, psoromic 0.45, thamnolic 0.42, cetraric 0.25, fumaroprotocetraric 0.10, norstictic 0.47, salazinic 0.12, stictic 0.34, and α-methoxysalazinic 0.18. A number of visualizing agents were used for detecting the spots including 10% sodium dichromate solution, 1% ferric chloride solution, and 3.3% potassium permanganate in sulfuric acid. Ramaut [26, 27] has used the same separating system for these compounds and has found that protocetraric acid has the same R_f value as fumaroprotocetraric; although norstictic has the same R_f value as thamnolic acid, the latter can be distinguished by its strong yellow-greenish fluorescence.

Konishi and Kano [28] have reported on the qualitative and quantitative analysis of nonyl phenol isomers. Sopkina and Ryabov [29] have used thin-layer chromatography for the analysis of diphenols and Wenkert et al. [29a] have reported on the wheat-bran phenols. Stambouli and Paris [29b] have chromatographed the hydrolysis products of the three heterosides: scoparoside, cytisoside, and aphloioside.

Aurenge et al. [30] chromatographed a group of synthetic tannins using *n*-butanol–acetic acid–water (4:1:5) in the first direction and *n*-butanol–ethanol–ammonia–water (75:10:15:10) in the second direction. The spots were visible under ultraviolet light. Murko and Jankovič [31] separated industrial tannins with ethyl acetate–formic acid (4:5:1). Fish and Kirk [32] also used a two-dimensional technique for the polyphenols in *Dryopteris* ferns. Optimum separation was achieved with a silica gel G plate prepared with a citric acid–phosphate buffer (pH 6.0) to which 150 mg of ascorbic acid was added to prevent oxidation of the labile polyphenols. Solvents for the separation were petroleum ether (40 to 60°C)–ethanol (95:5) for the first dimension and cyclohexane–ethyl acetate (1:1) for the second. The compounds were visible under ultraviolet light, but were also visualized with Fast Blue B (T-115). Van Sumere et al. [33] reported the R_f values of 93 phenols and coumarins on silica gel and silica gel–cellulose (1:1) layers with toluene–ethyl formate–formic acid

(5:4:1) and chloroform–acetic acid–water (4:1:1). Layers that were steamed prior to use gave better separations of some of the compounds. The colors produced with diazotized *p*-nitroaniline (T-180) were also reported; the mixed layers gave the best color response. Grant and Whetter [34] examined the secondary phenolic compounds in *Lotus* on silica gel, polyamide, and cellulose layers with 30 solvent systems.

Cellulose and polyamide gave unsatisfactory separations; the best results were obtained with silica gel in a combination multiple and two-step development, cyclohexane–ethyl acetate (1:1) developed 2× a distance of 15 cm followed by methanol–chloroform (3:7) for 7.5 cm. Waksmundzki and Mańko [35] separated 10 phenolic compounds on formamide-impregnated silica with benzene–methanol (95:5). Thielemann [36–38] has chromatographed phenols on silver nitrate-impregnated silica gel with benzene, benzene–ethyl acetate (3:2), dichloromethane–ethyl acetate (3:2), dichloromethane–ethyl acetate–diethyl amine (92:5:3), and benzene–methanol–acetic acid (8:1:1). The silver nitrate oxidized *o*- and *p*-dihydroxybenzenes to the corresponding quinones. Gessner and Acara [39] separated isomeric mononitrophenols on silica layers prepared with 0.1 *M* sodium carbonate using diethyl ether–petroleum ether (5:3) as the solvent. Recoveries from the treated gel were good; 93 to 110% in contrast to 80 to 90% for untreated silica gel. Parihar et al. [40] separated a group of polynitrophenols on silica and on buffered silica with four solvents. Clifford et al. [41] chromatographed a large number of compounds in a 2- and a 4-alkyl-substituted homologous series of 2,6-dinitrophenols on silica gel G with petroleum ether (40 to 60°C)–ether–formic acid (45:5:1). Structural relationships were examined. Nouwt and Weingarten [42] give the R_f values of 20 nitrophenols on silica gel with benzene and dibutyl ether as solvents.

For the separation of the phenolic components of hashish, see Chapter XXVI, Section 3.

2 DIRECT SEPARATION ON LOOSE LAYERS

Loose layers of silica gel have also been used for the separation of phenolic compounds by Janák [43]. Single-component solvents such as *n*-hexane, benzene, and chloroform were used in the development. For the location of the compounds, Janák used a saturated solution of tetracyanoethylene in benzene. After application of the coloring agent (which must be done with care in order not to disturb the layer), the plate was heated at 100°C to develop the colors.

Heřmánk et al. [44, 45] used loose layers of aluminum oxide. Table 27.2 gives the R_f values for some substituted phenols (see also Chapter

Table 27.2 $R_f \times 100$ Values of Some Substituted Phenols on Loose Layers of Aluminum Oxide[a] in Various Solvents [44, 45]

Phenol	*Dichloro-ethane*	*Dichloro-methane*	*Chloro-form*	*Diethyl Ether*	*Benzene–Pyridine (9:1)*	*Benzene–Pyridine (4:1)*	*Isopropyl Ether*	*Tetra-hydrofuran*
o-Nitrophenol	9	20	30	5	6	9	21	63
o-Nitro-*p*-bromophenol	2	5	9	0	1	1	3	3
o-Nitro-*p*-hydroxyphenol	1	2	4	5	7	17	38	80
o-Nitro-*p*-methoxyphenol	11	24	30	16	12	18	32	77
o-Nitro-*p*-methylphenol	10	27	32	20	12	19	35	78
o,p-Dinitrophenol	0	0	0	0	0	0	0	0
m-Nitrophenol	—	—	5	10	—	—	44	—
p-Nitrophenol	—	—	3	3	—	—	15	—
o-Aminophenol	—	—	7	4	—	—	—	—
m-Aminophenol	—	—	28	27	—	—	—	—
p-Aminophenol	—	—	37	33	—	—	—	—

[a] Brockmann activity = III.

XXX, Section 3 for phenolic aldehydes). Šaršúnová and Schwarz [46] have also applied loose layers of an acidic aluminum oxide of activity grade V for the separation and identification of medicinally used phenols.

3 DIRECT SEPARATION ON POLYAMIDE LAYERS

Polyamide forms an especially useful adsorbent for the separation of phenols and depends for this usefulness on the formation of hydrogen bonds between the phenolic compounds and the amide groups of the polymer [47]. Wang and co-workers [48–50] have used polyamide layers prepared from a solution of polyamide in formic acid. (For the preparation of these layers see Chapter III, Section 4). Table 27.3 gives the R_f values for separation of some phenols on these polyamide layers. Wang and Lin [49] and Lin et al. [50] have applied this method to the separation of some naturally occurring phenolic substances. Halmekoski and Hanni-

Table 27.3 $R_f \times 100$ Values of Phenols on Polyamide Layers Deposited from Solution [48]

Phenol	*Benzene*	*Chloroform*	*Ethyl Acetate*
Phenol	16	18	75
o-Cresol	30	37	81
m-Cresol	21	30	78
p-Cresol	21	32	78
o-Isopropylphenol	43	47	85
m-Isopropylphenol	36	42	81
p-Isopropylphenol	33	39	78
Thymol	64	57	87
α-Naphthol	14	12	71
β-Naphthol	12	10	68
Hydroquinone	0	2	50
Resorcinol	0	0	29
Catechol	0	0	32
Pyrogallol	0	0	20
Phloroglucinol	0	0	4
Orcinol	0	0	32
m-Chlorophenol	12	9	72
p-Chlorophenol	13	10	74
m-Bromophenol	12	9	7
p-Bromophenol	13	12	70
p-Iodophenol	13	11	70
o-Nitrophenol	100	100	100
p-Nitrophenol	2	3	51

kainen [15] have used polyamide layers in studying the relationship between the R_f values and the number of carbon atoms in the para substituents of phenols of five homologous series. Twelve different solvent systems were employed. Stadler and Endres [51] have investigated the phenols of vegetable tanning extracts. (See also Chapter XXXII, Section 3 for the separation of phenolic antioxidants by means of polyamide layers.) Gstirner and Flach [52] examined the tannins of *Rhizoma rhei* using methanol and acetone with varying amounts of water as solvents. Mosel and Herrmann [53] chromatographed 11 plant phenolics using the same type of solvents.

Grau and Endres [54] have chromatographed a group of quinones, hydroxyquinones, and phenols on acetylated polyamide using methanol–water and acetone–water mixtures for development. This adsorbent was used because of the irreversible binding of some quinones when chromatographed on straight polyamide layers. Acetylation of the polyamide was carried out according to the procedure of Grassmann and co-workers [55]. The acetylation had little effect on the adsorption of the phenols. Wagner et al. [56] also showed this lack of effect.

Soczewiński and Szumilo [57, 58] investigating the effect of structure on adsorption on polyamide chromatographed 33 phenolic compounds in numerous solvents. Bark and Graham [59] and Graham [60] chromatographed a large number of nitrophenols in examining structural effects. Graham et al. [61] found that 100% recovery of *m* and *p*-nitrophenol from polyamide layer was not possible. Bark and Graham [62] used cyclohexane–acetic acid (93:7) and 10% acetic acid as solvents in examining the effect of alkyl, aryl, and alkoxy groups on the chromatographic behavior of phenols. In this case cellulose impregnated with polyamide was used. The layers were prepared from 11 ml of polyamide solution (0.13 g nylon/ml) in 90% formic acid mixed with 13.5 g of cellulose and 74 ml of formic acid. The R_f values for 76 compounds are listed. The same solvents were used with the nitrophenols [59]. Butanol–5 N ammonia (100:33), propanol–water–27% ammonia (8:1:1), and butanol–ethanol–27% ammonia (5:1:1) have been used for phenols containing nitro groups [63]. A mixture of 12 g of polyamide with 2.4 g of cellulose gave the same R_f values as the straight polyamide [56]. Diamond [63a] separated *p*-alkylphenols of chain length C_1 to C_{12} on polyamide with water–dimethylformamide–formamide (6:4:1) and 1 *N* sodium hydroxide–methanol (7:3) as solvents.

4 SEPARATION ON ION-EXCHANGE LAYERS

Sherma and Hood [64] have chromatographed a group of six phenols on layers of ion exchanger containing a starch binder. These layers were prepared in 0.3-mm thickness because thicker layers had a tendency to crack and thinner layers were difficult to prepare uniformly. After preparation the plates were allowed to stand for a period of 15 min before being placed in a storage rack. The resins used were Dowex 50W-X8(H^+), Dowex 50W-X8(NH_4^+), and Dowex 1-X4(Cl^-) with various concentrations of methanol, ranging from 0 to 8 *M* as the solvents. The best resolution was obtained at the higher concentrations on Dowex 50W-X8(H^+) layers, although reproducibility was better at lower concentrations. A number of detecting reagents were tried and a diazotized benzidine solution proved to be the most satisfactory. Lepri et al. [65] reported the R_f values of 58 phenols on Dowex 50-X4(H^+) with water, water–ethanol (4:1 and 1:2) as solvents; and on Rexyn 102(H^+) (Fisher Scientific) with 0.5 *M* hydrochloric acid–ethanol (4:1). The R_f values are also given for Dowex 50-X4(Na^+) in six solvents; 0.1 *M* acetate buffer, 0.1 *M* sodium bicarbonate, 0.05 *M* sodium carbonate, 1 *M* ammonia, 1 *M* ammonia + 0.1 *M* sodium acetate, and 1 *M* ammonia + 1 *M* sodium acetate. The workers [66] also studied this group of compounds on AG 3-X4A (Bio-Rad Labs), PEI cellulose, and DEAE cellulose. Fifty percent methanol, 95% methanol, 0.5 *M* ammonia in 50% ethanol and in 95% ethanol, and cyclohexane–acetic acid (93:7) were used a solvents for the AG 3-X4A in the free base form and 95% methanol for the chloride form of the resin. For the PEI cellulose water, water–methanol (1:4), 0.1 *M* sodium bicarbonate, and acetic acid–cyclohexane (7:93) were used. Multiple and two-dimensional developments were used to obtain improved resolution in some cases. The compounds were also chromatographed on AG 1-X4 and on benzoylated DEAE cellulose (BD-cellulose, Serva) [67]. Methanol was used as a solvent for the AG 1-X4 in the chloride, acetate, and hydroxide forms. Ten molar and 1 *M* acetic acid in methanol were also used as solvents for the acetate form. For the BD-cellulose the solvents were water, water–methanol (4:1, 1:1, and 3:7), methanol, 1 *M* acetic acid, and 3 *M* acetic acid.

5 SEPARATION ON CELLULOSE

Only a limited number of papers have been published on the separation of phenols on thin layers of cellulose. Barton [68] used both silica gel and cellulose and found that the former gave better separations. For the separation on microcrystalline cellulose he used 2% acetic acid as a solvent chromatographing 36 compounds. Diazotized sulfanilic acid (T-249) was used as a detecting agent. Dittmann [69] chromatographed 37 phenols and phenolic acids on cellulose MN 300 layers with ethyl acetate–*n*-propanol–25% ammonia (3:5:2) (developed 2× for 10 cm), amyl alcohol–formic acid–water (20:20:1), *n*-propanol–water (13:7), and *n*-heptane–carbon tetrachloride–methanol (7:4:3). Of these four solvents only the latter was used in a saturated atmosphere. Jangaard [70] separated 71 plant phenolics on MN-polygram CE 300 cellulose sheets with 2% formic acid, 20% potassium chloride, isopropanol–ammonia–water (8:1:1), and 10% acetic acid. The compounds could be detected in the 0.2 to 1.0 μg range with diazotized *p*-nitroaniline (T-180) or sulfanilic acid (T-249). Dass and Weaver [71] also chromatographed 55 plant phenolics on MN-300 and microcrystalline cellulose. The three solvents used were: water–formic acid (98:2), *n*-amyl alcohol–acetic acid–water (10:6:5), and benzene–propionic acid–water (4:9:3). The colors produced with six spray reagents are listed. Haluk et al. [72] used *n*-butanol–pyridine–water (14:3:3), methyl ethyl ketone–water–diethylamine (921:77:2), and 30% acetic acid as solvents for phenolic acids. Two-dimensional separation was also used by chromatographing with 30% acetic acid in the first dimension followed by electrophoresis in the second with a pyridine–acetic acid–water (5:2:493) buffer (pH 5.3) at 1500 V and 30 mA. Joschek and Muller [73] used benzene–methanol–acetic acid (35:4:2) (two developments) and the lower phase of benzene–*n*-butanol–water (9:1:10) as solvents on cellulose 300 GF_{254} in chromatographing 44 phenols.

Phenols have also been separated on impregnated cellulose layers. Formamide, *N*-methylformamide, and *N,N*-dimethylformamide have been used as impregnating agents with hexane as the mobile phase [61, 74–76].

Reverse-phase partition chromatography has been used for alkyl phenols [77, 78] halogenated phenols and halogenated alkyl substituted phenols [79] using ethyl oleate as the stationary phase. The mobile phases included ethanol–water (1:3, 1:9, 3:7, and 3.75:6.25).

6 SEPARATION ON MISCELLANEOUS ADSORBENTS

Bark and Graham have used alumina as an adsorbent with cyclohexane, cyclohexane–dioxane (1:1), dioxane, benzene–methanol (95:5), benzene–ethanol (95:5), and benzene–ethyl acetate (3:7) as solvents for halogenated phenols, halogenated alkyl phenols [80], alkyl phenols [81], and nitrophenols [82]. Srivastava and Dua [83] separated phenol, *o*-, *m*-, and *p*-cresols, *o*-, *m*-, and *p*-chlorophenols, and salicylaldehyde on alumina–calcium hydroxide (2:1) with chloroform or benzene–methanol (9:1). Quinol, *o*-, *m*-, and *p*-nitrophenol, resocinol, catechol, *o*-, *m*-, and *p*-aminophenol, and pyrogallol were separated on zinc carbonate–silica gel (3:2) with benzene–ethyl acetate (9:1) or benzene–methyl ethyl ketone (9:1).

7 SEPARATION BY ELECTROPHORESIS

Pastuska and Trinks [84, 85] have applied electrophoresis to the separation of some naphthols, phenols, and phenolcarboxylic acids. Two types of layers were used for the separation of the phenols. Silica gel layers were prepared with 3% boric acid solution, and the electrophoresis was carried out with an electrolyte consisting of a mixture of 80 ml of ethanol, 30 ml of water, 4 g of boric acid, and 2 g of crystalline sodium acetate. With the pH values adjusted with acetic acid to 4.5 and with a voltage of 20 V/cm at a total potential of 400 V, the separation was completed in approximately 90 min. The kieselguhr G layers, which were also prepared with a 3% boric acid solution, were used with the same elctrolyte solution except that it was adjusted to pH 5.5 with acetic acid. The migration distances were given for 36 phenolic compounds with reference to *m*-hydroxybenzoic acid as a standard.

Haluk et al. [72] used cellulose, silica gel, polyamide, and polyacrylamide gel as substrates for the electrophoresis of 25 phenolic acids. The buffer for the polyacrylamide was pyridine–acetic acid–water (5:2:493) (pH 5.3) used with 1400 V and 40 mA. The silica and polyamide were used two-dimensionally with chromatography in the first direction and then electrophoresis in the second. Ethyl acetate–acetic acid (95:5) was used as a solvent for the polyamide chromatography and Aronsson buffer (pH 8.9) for the electrophoresis at 1800 V and 40 mA in the case of the benzoic series; for the cinnamic series the buffer was pyridine–acetic acid–water (1:6:90) (ph 3.55) with a 20-mA current. (For the cellulose conditions see Section 5 of this chapter.) Walker and Thompson [86] also used cellulose as a support for the electrophoresis of phenols using a pH 5.3 or 9.1 buffer solution and 60 V/cm. Chromatography at right angles gave improved resolution.

8 SEPARATION OF DERIVATIVES

Dhont and de Rooy [87] have used the 3,5-dinitrobenzoates in studying the steam-volatile alcohols and phenols in foods. To prepare the derivatives, it was necessary to extract the distillate with pentane in order to concentrate the smaller amounts of phenols and remove them from major portions of methanol and ethanol which were usually present. The derivatives themselves were chromatographed on silica gel G activated at 110 to 120°C for 15 min using a benzene–light petroleum ether (1:1) solvent. Visualization was obtained by spraying with a 1% solution of 1-naphthylamine in ethanol which produced yellow to orange spots. Homologous members of a series could not be separated.

The azo coupling reaction, which is used to locate phenols on thin-layer plates, may also be used to prepare derivatives prior to their chromatographic separation. Knappe and Rohdewald [88] used the derivatives prepared by coupling the phenolic compounds with Fast Red Salt AL (1-anthraquinone diazonium chloride). Three systems were required to separate phenol, the three cresols, and the six xylenols. For the main separation of these compounds, silica gel G made alkaline with potassium carbonate was used with dichloromethane–ethyl acetate–diethylamine (92:5:3). Separation of phenol, 3,5-xylenol, and *m*-cresol was achieved on the same layer material with chloroform–ethyl acetate–ethanol (93:5:2). An acidic silica gel (30 g silica gel G with 60 ml of 0.5 *N* oxalic acid solution) was needed to separate the 3,4-xylenol, *p*-cresol, and 2,4-xylenol from one another. For this latter separation benzene was used as the solvent. Thielemann [89] separated 1- and 2-naphthol by using this same derivative with a 92:3:3 mixture of the diethylamine solvent. The *o*-, *m*-, and *p*-nitrophenols, the 2,4-dinitro-, and the 2,4,6-trinitrophenol derivatives were separated with benzene as a solvent [90]. The derivatives prepared by coupling phenols with Fast Salt BB have also been used for separating the phenolic compounds [91–93]. For separations on silica gel G–potassium carbonate (1:2) layers dichloromethane–ethyl acetate–diethylamine (92:5:3), dichloromethane–chloroform–benzene–ethyl acetate (88:5:5:2), or dichloromethane–benzene (1:1) was used. Thielemann [94] reported the R_f values for 1- and 2-naphthol after coupling with eight different Fast Salt dyes. Smith and Sullivan [95] have used the *p*-nitrophenylazo derivative in examining the phenols present in cigaret smoke condensate. Separations were carried out on kieselguhr G impregnated with a 5% solution of formamide in acetone. As separating solvents, mixtures of benzene–cyclohexane–dipropylene glycol and benzene–cyclohexane–diethylamine were used. Quantitative measurements were made by eluting the spots in ammonia (sp gr 0.88) and measuring the absorption in a spectrophotometer. Crump [96] has applied two-dimensional chromatography to the separation of 20 alkyl phenols coupled with *p*-nitrophenol azo dyes. A silica gel layer impregnated with 0.5 *N* sodium hydroxide was used for the separations, with chloroform–acetone (9:1) as the solvent in one direction and benzene–dipropylamine (4:1) as the solvent in the other direction. Table 27.4 gives the R_f values for separations of some phenol derivatives.

Churáček et al. [97] chromatographed the *p*-(*N*,*N*-dimethylamino) benzene *p*′-azobenzoates of 22 phenolic compounds. The R_f values are reported for silica gel with chloroform, benzene–chloroform (1:1), hexane–ethyl acetate (4:1), cyclohexane–ethyl acetate (4:1), and cyclohexane–methyl ethyl ketone (7:3) as solvents. Kondo and Kawa-

Table 27.4 $R_f \times 100$ Values of Phenols Coupled with Diazonium Compounds[a,b,c]

	A		*B*	*C*			
Phenol	*t*	*u*	*v*	*w*	*x*	*y*	*z*
Phenol	7	37	Origin	30	23		
3,5-Xylenol	17	44	Origin	56	75		
m-Cresol	18	51	Origin	53	39		
o-Cresol	35	>57	Origin	61	44		
2,3-Xylenol	60	>57	<14	71	70		
2,5-Xylenol	66	>57	<14	74	50		
2,6-Xylenol	78	>57	<14	85	30		
3,4-Xylenol	83	>57	20	100	90–100	23	50
p-Cresol	83	>57	36	100	90–100	15	30
2,4-Xylenol	85	>57	47	100	90–100	53	85
Guaicol				90	100		
o-Ethylphenol				64	64		
m-Ethylphenol				68	54		
p-Ethylphenol				100	90–100	15	50

[a] Adsorbent: *A* = alkaline silica gel G (30 g silica gel + 60 ml 0.5 *N* K_2CO_3) [88]; *B* = acidic silica gel G (30 g silica gel + 60 ml 0.5 *N* oxalic acid) [88]; *C* = kieselguhr G impregnated with 5% formamide in acetone [95].

[b] Solvent: *t* = dichloromethane–ethyl acetate–diethylamine (92:5:3) saturated chamber; *u* = chloroform–ethyl acetate–ethanol (93:5:2) saturated chamber; *v* = benzene; *w* = benzene–cyclohexane–dipropylene glycol (30:70:3); *x* = benzene–cyclohexane–diethylamine (5:5:1); *y* = cyclohexane–diethylamine (9:1); *z* = cyclohexane–benzene–diethylamine (7:3:1).

[c] Derivatives: *A* and *B*; 1-anthraquinoneazo; *C*, *p*-nitroanilineazo.

shiro [98] chromatographed the *p*-nitrobenzenediazonium fluoroborates of some phenols on silica gel with cyclohexane–isopropyl ether–chloroform (7:1:7) and cyclohexane–chloroform (1:1). Renault and Cartron [99] chromatographed the same derivatives of phenols, *o*-, and *m*-cresols, and xylenols not substituted in the 4-position on silica gel layers prepared with 1 *N* sodium hydroxide and using a developing solvent of benzene–methanol (9:1). Derivatives of *p*-cresol and xylenols substituted in the 4-position were separated on silica gel layers prepared with 0.1 *N* sodium hydroxide using a solvent mixture of trichloroethylene–diphenylamine (39:1). Wildenhain and Henseke [100] separated 31 dinitrophenyl ethers of phenolic compounds on magnesium silicate layers in benzene–chloroform (1:1), cyclohexane–cyclohexanol (75:10) and (1:1), cyclohexane–cyclohexanol–acetic acid (50:5:2), benzene–ethyl ether (2:1), benzene–aniline (97:3), and benzene–propylamine (99:1). The 4-acetyl-2-nitrophenyl ethers of 23 of these phenols plus 17 addition compounds have been chromatographed on silica gel G and polyamide layers in cyclohexane–cyclohexanone (3:1) and in dichloromethane–cyclohexane (5:1) [101].

References

1. L. Labat and A. L. Montes, *An. Asoc. Quim. Arg.*, **41,** 166 (1953); *Chem. Abstr.*, **48,** 3637 (1954).
2. G. Wagner, *Pharmazie,* **10,** 302 (1955).
3. G. Pastuska, *Z. Anal. Chem.*, **179,** 355 (1961).
4. G. Pastuska and H.-J. Petrowitz, *Chem.-Ztg.*, **86,** 311 (1962).
5. H. Gaenshirt and K. Morianz, *Arch. Pharm.*, **293/65,** 1065 (1960).
6. H.-J. Petrowitz, *Erdoel Kohle,* **14,** 923 (1961).
7. H.-J. Petrowitz, *Mitt. Dtsch. Ges. Holzforsch.*, **48,** 57 (1961).
8. H.-J. Petrowitz, *Materialpruefung,* **2,** 309 (1960).
9. H. Seeboth, *Chem. Tech. (Berlin),* **15,** 34 (1963).
10. H. Seeboth, *Monatsber. Dtsch. Akad. Wiss. Berlin,* **5,** 693 (1963).
11. H. Seeboth and H. Goersch, *Chem. Tech. (Berlin),* **15,** 294 (1963).
12. E. Ragazzi and G. Veronese, *J. Chromatogr.*, **77,** 369 (1973).
13. R. K. Ibrahim, *J. Chromatogr.*, **42,** 544 (1969).
14. H.-G. Haub and H. Kaemmerer, *J. Chromatogr.*, **11,** 487 (1963).
15. J. Halmekoski and H. Hannikainen, *Suom. Kemistil.*, **36B,** 24 (1963).
16. T. G. Lipina, *Tr. Khim Khim Tekhnol.*, **1962,** 424.
17. M. B. Naff and A. S. Naff, *J. Chem. Educ.*, **40,** 534 (1963).
18. M. Schulz, H. Seeboth, and W. Wieker, *Z. Chem.*, **2,** 279 (1962).
19. A. Waksmundzki and J. Różylo, *Chem. Anal. (Warsaw),* **14,** 1217 (1969).
20. R. Deters, *Chem.-Ztg.*, **86,** 388 (1962).
21. T. Furukawa, *Nippon Kagaku Zasshi,* **80,** 387 (1959); *Chem. Abstr.*, **54,** 13938 (1960).
22. S. Husain, *J. Chromatogr.*, **18,** 419 (1965).
23. S. Husain and P. A. Swaroop, *J. Chromatogr.*, **22,** 180 (1966).
24. E. Stahl and P. J. Schorn, *Z. Physiol. Chem.*, **325,** 263 (1961).
25. O. Bachmann, *Oesterr. Botan. Z.*, **110,** 103 (1963).
26. J. L. Ramaut, *Bull. Soc. Chim. Belg.*, **72,** 97 (1963).
27. *Ibid.*, p. 316.
28. K. Konishi and Y. Kano, *Bunseki Kagaku,* **13,** 1227 (1964).
29. A. K. Sopkina and V. D. Ryabov, *Zh. Anal. Khim.*, **19,** 615 (1964).

29a. E. Wenkert, E. M. Loeser, S. N. Mahapatra, F. Schenker, and E. M. Wilson, *J. Org. Chem.*, **29,** 435 (1964).

29b. A. Stambouli and R. R. Paris, *Ann. Pharm. Fr.*, **19,** 435 (1961).
30. J. Aurenge, G. Barbe-Richaud, and L. Gelpi, *J. Chromatogr.*, **22,** 369 (1966).
31. D. Murko and M. Jankovič, *Kem. Ind. (Zagreb)*, **18,** 730 (1969).
32. F. Fish and W. R. Kirk, *J. Chromatogr.*, **36,** 383 (1968).
33. C. F. Van Sumere, G. Wolf, H. Teuchy, and J. Kint, *J. Chromatogr.*, **20,** 48 (1965).
34. W. F. Grant and J. M. Whetter, *J. Chromatogr.*, **21,** 247 (1966).
35. A. Waksmundzki and R. Mańko, "The Separation of Phenols on Formamide-Impregnated Silica Gel," in *Int. Symp. Chromatog, 2nd,* K. Macek and I. M. Hais, Eds., Elsevier, Amsterdam, 1965, p. 221.
36. H. Thielemann, *Z. Chem.*, **12,** 223 (1972).
37. H. Thielemann, *Mikrochim. Acta,* **1972,** 672.
38. *Ibid.*, p. 718.
39. T. Gessner and M. Acara, *Anal. Biochem.*, **35,** 442 (1970).
40. D. B. Parihar, S. P. Sharma, and K. C. Tewari, *J. Chromatogr.*, **24,** 230 (1966).
41. D. R. Clifford, D. M. Fieldgate, and D. A. M. Watkins, *J. Chromatogr.*, **43,** 110 (1969).
42. J. Nouwt and D. Weingarten, *Int. Symp. Chromatogr., Electrophor., 4th 1966,* Ann Arbor Science, Ann Arbor, Mich., 1968, p. 168.
43. J. Janák, *J. Chromatogr.*, **15,** 15 (1964).
44. S. Heřmánek, V. Schwarz, and Z. Čekan, *Pharmazie,* **16,** 566 (1961).
45. S. Heřmánek, V. Schwarz, and Z. Čekan, *Collect. Czech. Chem. Commun.*, **28,** 2031 (1963).
46. M. Šaršúnová and V. Schwarz, *Pharmazie,* **17,** 527 (1962).
47. L. Hoerhammer and H. Wagner, *Pharm. Ztg.*, **104,** 783 (1959).
48. K.-T. Wang, *J. Chinese Chem. Soc. (Taiwan),* **8,** 241 (1961).
49. K.-T. Wang and Y.-T. Lin, *J. Chinese Chem. Soc. (Taiwan),* **10,** 146 (1963).
50. Y.-T. Lin, K.-T. Wang, and Y.-S. Lin, *J. Chinese Chem. Soc. (Taiwan),* **9,** 68 (1962); *Chem. Abstr.*, **58,** 9412 (1963).
51. P. Stadler and H. Endres, *J. Chromatogr.*, **17,** 587 (1965).
52. F. Gstirner and G. Flach, *Arch. Pharm. (Weinheim),* **303,** 339 (1970).
53. H.-D. Mosel and K. Herrmann, *J. Chromatogr.*, **87,** 280 (1973).
54. W. Grau and H. Endres, *J. Chromatogr.*, **17,** 585 (1965).
55. W. Grassmann, H. Hoermann, and H. von Portatius, *Z. Physiol. Chem.*, **321,** 120 (1960).
56. H. Wagner, L. Hoerhammer, and K. Macek, *J. Chromatogr.*, **31,** 455 (1967).
57. E. Soczewiński and H. Szumilo, *J. Chromatogr.*, **81,** 99 (1973).
58. *Ibid.*, **94,** 229 (1974).
59. L. S. Bark and R. J. T. Graham, *Int. Symp. Chromatogr. Electrophor., 4th 1966,* Ann Arbor Science, Ann Arbor, Mich., 1968, p. 105.
60. R. J. T. Graham, *J. Chromatogr.*, **33,** 118 (1968).
61. R. J. T. Graham, A. E. Nya, and D. A. Tinsley, *Int. Symp. Chromatogr., Electrophor., Lect. Pap. 6th 1970,* Ann Arbor Science, Ann Arbor, Mich., 1971, p. 105.
62. L. S. Bark and R. J. T. Graham, *J. Chromatogr.*, **27,** 116 (1967).

63. M. Trojna and J. Hubaček, *Chem. Prum.*, **22,** 29 (1972).
63a. P. F. Diamond, *J. Chromatogr.*, **32,** 419 (1968).
64. J. Sherma and L. V. S. Hood, *J. Chromatogr.*, **17,** 307 (1965).
65. L. Lepri, P. G. Desideri, M. Landini, and G. Tanturli, *J. Chromatogr.*, **109,** 365 (1975).
66. *Ibid.*, **108,** 169 (1975).
67. *Ibid.*, **129,** 239 (1976).
68. G. M. Barton, *J. Chromatogr.*, **26,** 320 (1967).
69. J. Dittmann, *J. Chromatogr.*, **32,** 764 (1968).
70. N. O. Jangaard, *J. Chromatogr.*, **50,** 146 (1970).
71. H. C. Dass and G. M. Weaver, *J. Chromatogr.*, **67,** 105 (1972).
72. J. P. Haluk, C. Duval, and M. Metche, *Int. Symp. Chromatogr. Electrophor., Lect. Pap., 6th 1970,* Ann Arbor Science, Ann Arbor, Mich., 1971, p. 431.
73. H. I. Joschek and S. I. Muller, *J. Am. Chem. Soc.*, **88,** 3276 (1966).
74. R. J. T . Graham and J. Daly, *J. Chromatogr.*, **48,** 78 (1970).
75. *Ibid.*, p. 67.
76. L. S. Bark, J. Daly, R. J. T. Graham, *Int. Symp. Chromatogr., Electrophor., 4th 1966,* Ann Arbor Sci., Ann Arbor, Mich., 1968, p. 128.
77. L. S. Bark and R. J. T. Graham, *J. Chromatogr.*, **23,** 417 (1966).
78. *Ibid.*, **25,** 357 (1966).
79. L. S. Bark and R. J. T. Graham, *Talanta,* **13,** 1281 (1966).
80. L. S. Bark and R. J. T. Graham, **25,** 347 (1966).
81. *Ibid.*, **23,** 120 (1966).
82. *Ibid., Int. Symp. Chromatogr. Electrophor., 4th 1966,* Ann Arbor Science, Ann Arbor, Mich., 1968, p. 119.
83. S. P. Srivastava and V. K. Dua, *Z. Anal. Chem.*, **275,** 29 (1975).
84. G. Pastuska and H. Trinks, *Chem.-Ztg.*, **86,** 135 (1962).
85. *Ibid.*, **85,** 535 (1961).
86. J. R. L. Walker and J. E. Thompson, *Lab. Pract.*, **18,** 629 (1969).
87. J. H. Dhont and C. de Rooy, *Analyst, (London),* **86,** 527 (1961).
88. E. Knappe and I. Rohdewald, *Z. Anal. Chem.*, **200,** 9 (1964).
89. H. Thielemann, *Z. Chem.*, **9,** 190 (1969).
90. H. Thielemann, *Sci. Pharm.*, **39,** 106 (1971).
91. H. Thielemann, *J. Prakt. Chem.*, **312,** 728 (1970).
92. H. Thielemann, *Pharmazie,* **25,** 365 (1970).
93. H. Thielemann, *Z. Anal. Chem.*, **253,** 38 (1971).
94. H. Thielemann, *Pharmazie,* **25,** 128A (1970).
95. G. A. L. Smith and P. J. Sullivan, *Analyst, (London),* **89,** 312 (1964).
96. G. B. Crump, Anal. Chem., **36,** 2447 (1964).
97. J. Churáček, H. Pechová, and V. Mareš, *J. Chromatogr.*, **67,** 97 (1972).
98. T. Kondo and I. Kawashiro, *Shokuhin Eiseigaku Zasshi,* **6,** 433 (1965).
99. J. Renault and M. F. Çartron, *Ann. Pharm. Fr.*, **25,** 291 (1967).
100. W. Wildenhain and G. Henseke, *J. Chromatogr.*, **19,** 438 (1965).
101. W. Wildenhain, G. Henseke, and G. Bienert, *J. Chromatogr.*, **45,** 158 (1969).

Chapter **XXVIII**

NATURAL PIGMENTS

1 CAROTENOIDS

Thin-layer chromatography has been used to separate the carotenoid pigments. As early as 1952 Mottier and Potterat [1–3] demonstrated the use of circular chromatography on loose layers of alumina for analysis of food coloring materials, including annato which contains bixin. Lagoni and Wortmann [4] also used circular development with petroleum ether on a loose layer of aluminum oxide to isolate β-carotene from butter and margarine. A quantity of 0.05 γ could be detected. The same authors [5] converted this to a quantitative method using a minimum detectable amount procedure. Using starch-bound silicic acid and a solvent mixture of *n*-hexane–ether (3:7 vol/vol), Demole [6] separated carotene, bixin, canthaxanthin, and xeaxanthin. Voelker [7] identified canthaxanthin as the red pigment in the feathers of *Cardinalis cardinalis, Calochaetes coccineus, Pharomachrus moccino, Pyrocephalus rubinus, Guara rubra,* and *Spinus cucullatus*. Thommen and Wackernagel [8], using column and thin-layer chromatography, found this pigment in the feathers and skin of the lesser flamingo *Phoeniconaias minor*. It was also found in fairly high concentration in the liver. Smaller amounts of esterified astaxanthin and traces of β-carotenoid were also found in the lesser flamingo. Thielemann [9] separated canthaxanthin and astaxanthin from *Phoeniconaias minor* using silica with dichloromethane–ether (9:1).

In investigating the occurrence of retinene in nature, Winterstein and Hegedues [10, 11] found that rhodanine could be used as a locating agent. With this agent the limit of detection could be decreased so that 0.3 γ of retinene could be seen on the plate. By virtue of its condensation to form deeply colored products, rhodanine can be used for other faintly colored carotenoid aldehydes. For detection the chromatogram is sprayed first with an alcoholic solution of rhodanine followed by a concentrated aqueous ammonia or sodium hydroxide solution. By means of this reagent on thin-layer plates, the authors were able to detect 8′-apo-β-carotenal in the intestinal mucous membrane.

Thommen [12], using the same technique, isolated 8′-apo-β-carotenal as well as 2′-apo-β-carotenal, 10′-apo-β-carotenal, and 3-hydroxy-8′-apo-β-carotenal from the skin and juice of five species of oranges. Thommen [13], using thin-layer chromatography, also found 8′-apo-β-carotenal along with significant amounts of 8′-apo-β-carotenoic acid in the liver of rats fed the former compound. When chickens were fed 1 mg of labeled β-apo-8′-carotenal, the egg yolks contained 100 to 150 γ of labeled 8′-apo-β-carotenoic acid.

Wildfeuer [14] found 8′-apo-β-carotenoic acid having 30 carbon atoms and 10′-apo-β-carotenoic acid with 27 carbon atoms in the yolks of eggs from hens fed 10 mg/kg of 8′-apo-β-carotenal.

A mixture of 5 g of silica gel G and 20 g of calcium hydroxide was used for the preparation of plates by Winterstein et al. [15] in the investigation of carotenes in various natural products. As a developing agent for separation by adsorption, a 1:1 mixture of petroleum ether (80 to 105°C) and benzene was generally used. This solvent was varied as needed by changing the proportions and also by the addition of 1% methanol to the mixture. In this manner the aldehydes were differentiated into two series: one containing the C_{37}, C_{32}, and C_{27} aldehydes, the other the C_{40}, C_{35}, and C_{25} aldehydes. These same authors also used partition chromatography on thin-layer plates. In this case the dried silica gel plates were immersed in 5% solution of paraffin oil in petroleum ether for 2 min. The petroleum ether was then removed by drying in a horizontal position at 120°C for 10 min. Methanol saturated with paraffin oil was used as a developing solvent. Using the same silica gel–calcium hydroxide mixture Box and Boekenoogen [16] separated the carotenoids and pheophytin in soybean, rape, and linseed oils with benzene–ethyl acetate (77:23) as a solvent.

Bjornland and Aguilar-Martinez [17] separated the carotenoids in red algae using silica–calcium carbonate (1:1) and silica–calcium hydroxide–

magnesium oxide–calcium sulfate (10:4:3:1) with petroleum ether–acetone–isopropanol in various proportions. The acetylenic could be differentiated from the nonacetylenic carotenoids. Sherma [18] used cellulose layers to separate various algal chlorophyll and carotenoid pigments with petroleum ether (20 to 40°C)–benzene–chloroform–acetone–isopropanol (10:7:2:1:0.034). Rai and Lee [19] used a mixed layer of cellulose MN 300–sugar (C & H confectioners)–potato starch (8:2:0.03) for separating planktonic algal pigments with 0.05% *n*-propanol in petroleum ether.

Egger [20] applied reverse-phase partition chromatography using a kieselguhr plate impregnated with vegetable oil. The preferred method of impregnation is to place the dried plate in a 7% solution of vegetable oil (paraffin oil may be substituted for the vegetable oil) in petroleum ether (100 to 140°C) until the solvent is within 3 to 4 cm of the edge of the plate. The solvent may be removed either by allowing the plates to stand for 24 hr at room temperature or by heating at 70°C for 1 hr. The unimpregnated portion of the plate is then used for spotting the sample because this results in the formation of narrow bands instead of round spots, thus giving sharper separations. In contrast, the normally impregnated plate yields round spots, and the sample must be applied in concentrated solutions. For development, a mixture is used of methanol–acetone–water (20:4:3) which is saturated with the impregnating oil. Rhodoxanthin, which had the same R_f value as chlorophyll *b*, could be separated from the latter by using a 3:1 mixture of acetone and water, with R_f values of 0.26 and 0.30, respectively. The impregnating vegetable oil was removed from the layer with petroleum ether and then *N*,*N*-dimethylformamide was used as an eluting solvent for the rhodoxanthine for obtaining the infrared spectrum. Nitsche [21] used a reverse-phase separation on triglyceride-impregnated cellulose. Neoxanthine, deepoxineoxanthin (an allenic carotenoid), violaxanthin, antheraxanthin, zeaxanthin, and cryptoxanthin were separated with methanol–acetone–water (40:10:3). Basix magnesium carbonate was also used to separate cis–trans isomeric xanthophylls using petroleum ether–acetone (5:1) as a solvent.

Isler et al. [22] used a 6:1 ratio of calcium hydroxide and silica gel G with a solvent mixture of petroleum ether and benzene (2:3) in separating various carotenes. They also used the same system for separating the methyl esters of the C_{27} to C_{40} β-apocarotenoic acids.

Grob et al. [23, 24] and Eichenberger and Grob [25–27] have used both silica gel G and aluminum oxide in investigating the various carotenoids in plants. Mixtures of petroleum ether, benzene, and alcohol were used as solvents. These varied in proportion from 50:50:1 and 100:20:7 for the silica gel to 100:100:1 for the aluminum oxide plates. Grob and

Boschetti [28] used cyclohexane with silica gel G to separate lycopersene (as a precursor of carotene) from squalene. In this case location of the spots was made with 10% antimony trichloride in chloroform. Davies, Goodwin, and Mercer [29] used ligroin as the solvent for separating lycopersene from squalene on the same adsorbent. Grob [30] used these thin-layer techniques to illustrate factors in support of his theories on the biogenesis of carotene and carotenoids.

Rispoli and Di Giacomo [31] used solvent mixtures of petroleum ether (50 to 70°C), benzene–acetone (160:40:4), and petroleum ether (50 to 70°C)–ethyl acetate–chloroform (160:40:4) for separating carotenoid pigments used as food colorants in pasta. Separation was achieved on silica gel G. Similarly, Benk et al. [32] separated a series of carotenoids on silica gel as a means of detecting these as added coloring agents in orange juice. In this case, a mixture of petroleum ether–benzene–acetone–acetic acid (80:20:2:1) separated most of the carotenoids now in use. Lycopene and β-carotene were separated with petroleum ether. Some of the slower-moving carotenoids were separated by changing the solvent mixture to 80:20:5:5.

Montag [33] used two-dimensional techniques on silica gel G in working with fat-soluble pigments used in foods. These included the natural pigments bixin, norbixin, carotene, capaxanthins, curcumin, and crocin. Solvents used were benzene, chloroform–acetic anhydride (75:2) and methyl ethyl ketone–acetic acid–methanol (40:5:5). A 20% solution of antimony trichloride in chloroform was used as a detecting agent.

Corbi [34] has used barium sulfate in examining natural food colors.

Duquenois and Meylaender [35] studied the oxidation products of β-carotene and Davies et al. [36, 37] have investigated the formation of lycopersene in *Neurospora crassa*. Egger [38] has reported on the thin-layer partition chromatography of plastid pigments showing the variation in R_f value with varying mixtures of acetone and 95% methanol. Cholnoky et al. [39], in elucidating the structure of kryptocapsin, found that it has an R_f value of 0.47 on silica gel in ligroin (60 to 80°C) containing 2% acetone.

Bolliger et al. [40] separated the carotenes on magnesium oxide (see Table 28.1). Egger and Voigt [41] have separated 31 carotenoids on polyamide containing 15% cellulose with seven solvent systems. Hager and Stransky [42–45] used calcium carbonate–magnesium oxide–calcium hydroxide (15:3:2) to separate a large number of carotenoids with light petroleum ether–acetone–chloroform–methanol (50:50:4:1). Philip and Francis [47] give the following percentages of ethanol–benzene–acetone–petroleum ether for separating various classes of carotenoids on silica gel: polyols and diketo diols (3.5:10:20:66.5); diepoxide diols, monoketo

Table 28.1 $R_f \times 100$ Values of Carotenoids in Various Systems[a]

Carotenoid	*A*	*B*	*C*	*D*	*E*	*F*	*G*	*H*
Alloxanthin					20[b]			
Aphanizophyll					5[c]			
8′-Apo-β-carotenal		34						
Bixin	51	8						
Caloxanthin					24[c]			
Canthaxanthin	38	9			83[d]		80	
Capaxanthin		0						
α-Carotene				75[f]		66		0
β-Carotene	96	97		64[f]		49	80	0
γ-Carotene						11		
δ-Carotene						22		
ϵ-Carotene						70		
Crocetin		16						
Crocoxanthin					75[b]			
Cryptocapsin			47					
Cryptoxanthin					77[d]		74	7
Diadinoxanthin					41[e]			
Diatoxanthin					28[e]			
Equinenone					88[d]		72	
Ethyl 8′-apo-β-carotenoate		54						
3,3′-Hydroxy-equinenone					80[c]			
Heteroxanthin					12[e]			
Isozeaxanthin	63							
Lycopene		96					60	
Lycopersene				30[g]				
Lutein							95	56
Lutein epoxide								72
Methyl 8′-apo-β-carotenoate		47						
Myxoxanthophyll					3[c]			
Neoxanthin					49[d]		96	95
Oscillaxanthin					6[c]			
Phytoene				21[g]				
Phytofluene				12[g]				
Rhodoxanthin							40	26
Vaucheriaxanthin					19[e]			
Violaxanthin						93	84	
Zeaxanthin	17					82		

[a] System: *A* = silica gel, *n*-hexane–ether (3:7), development distance 7.6 cm [6]; *B* = silica gel, petroleum ether–benzene–acetone–acetic acid (80:20:2:1), development distance 10 cm [32]; *C* = silica gel, petroleum ether (60 to 80°C)–acetone (98:2) [39]; *D* = silica gel, light petroleum ether; *E* = calcium carbonate–magnesium oxide–calcium hydroxide (15:3:2), light petroleum–acetone–chloroform–methanol (50:50:4:1); *F* = magnesium oxide, light petroleum

diols, and monoepoxide diols (3.0:10:20:67.0); diols (2.0:10:20:68.0); monols (0.0:10:20:70.0); and hydrocarbons (0.0:0.0:5:95.0). They also list the R_f values of capsanthin and related compounds on silica gel. Singh et al. [48] separated the apocarotenals and related compounds on silica gel using four solvent systems. Knowles and Livingston [49] have used a two-dimensional development on a dual-layer plate in the analysis of carotenoids in feedstuffs. The layer consisted of a strip of magnesium oxide–Celite (analytical filter aid, Johns Manville) (1:4) with the balance of the plate covered with silica gel G–calcium hydroxide (1:6). The first development was made along the strip with hexane–acetone (7:3). The second development across the silica gel–calcium hydroxide layer was made with 2% butanol in benzene. Because of some isomerization of neoxanthin and violaxanthin to their 5,8-epoxide isomers a 3-cm-wide strip of silica gel G–calcium hydroxide was used in conjunction with the balance of the plate covered with magnesium oxide–kieselguhr (1:4). In this case the first development was along the silica gel–lime strip using 2% butanol in benzene, containing ethoxyquin as an antioxidant. The second-dimensional development was made with hexane–acetone (7:3) also containing ethoxyquin.

Rollins [50] has worked out a laboratory demonstration for students using thin-layer chromatography of leaf pigments on silica gel. In this case the solvent was benzene–acetone (7:3).

Fieser [51] has given directions for a demonstration of the thin-layer chromatographic isolation of lycopene from tomato paste and of β-carotene from strained carrots.

Because of the labile character of these pigments, it is advisable to run these chromatograms in the dark. Hager and Bertenrath [52] advise against the use of an acidic adsorbent and also recommend ascorbic acid as an antioxidant. For preparation of the thin layers they used a mixture

Table 28.1 *(Continued)*

ether–benzene (1:1) [40]; *G* = polyamide–cellulose (17:3), methanol–methyl ethyl ketone (1:1) [41]; *H* = kieselguhr impregnated with vegetable oil, methanol–acetone–water (20:4:3) saturated with vegetable oil, development distance 20 cm [20].

[b] Ref. 42.

[c] Ref. 43.

[d] Ref. 44.

[e] Ref. 45.

[f] Ref. 46.

[g] Ref. 29.

consisting of 12 g of kieselguhr G, 3 g of silica gel, 3 g of analytical-grade calcium carbonate, 0.018 g of analytical-grade calcium hydroxide, and 55 ml of a water solution containing 5×10^{-3} mole of ascorbic acid. Since the ascorbic acid begins to decrease after 1 hr it is best to use freshly prepared plates. As a solvent they used a mixture of petroleum ether (100 to 140°C)–isopropanol–water (100:10:0.25). Table 28.1 gives the R_f values of various carotenes. Rahman and Egger [53] chromatographed loniceraxanthin and webbiaxanthin on cellulose with methanol–acetone–water (10:2:1) because these compounds are unstable on silica, magnesium oxide, and zinc carbonate adsorbents.

Foppen [54] has published a group of 232 tables for assisting in the identification of carotenoids. They include data on chromatographic behavior along with other physical data such as absorption spectra, melting points, and infrared spectra.

Quantitative Determination

For determining the amounts of carotenoids present, Winterstein and Hegedues [11] used a semiquantitative determination by comparing the spot size of the sample with a graded series of spots of a standard. The error in this case was ±30%. For more accurate work the pigment may be eluted and measured in a spectrophotometer [15, 23, 25, 55]. Dimethylformamide is especially advantageous as an eluting solvent, because the pigments are stable in this solvent and give spectra like those in ether solution [20]. This solvent is also useful when working with partition chromatography, because the impregnating oil can be removed from the dimethylformamide solution of the pigments by washing with petroleum ether.

Garside and Riley [56] used reflectance densitometry for the determination of phytoplankton pigments after separation on silica gel G with light petroleum ether–ethyl acetate–diethylamine (55:32:13). Chlorophyll C which had remained at the origin was developed with light petroleum–ethyl acetate–dimethylformamide (1:1:2). The precision for most pigments is ±5% or better at the 0.5 μg level.

Preparative Separations

Winterstein et al. [15] used 10 to 20 plates (20 × 20) with 30 spots on each plate in order to obtain enough material for identification work. Each spot contained from 0.1 to 0.3 μg of carotenoid. The colored zones were removed from the developed plate by scraping. Elution was carried out with ether in small centrifuge tubes with the adsorbent being removed by centrifugation.

2 CHLOROPHYLLS

In separating carotenoids added to orange juice, Benk et al. [32] separated the chlorophylls from the carotenoids by using silica gel with petroleum ether–benzene–acetone–acetic acid (80:20:2:1), but this system did not resolve the two chlorophylls. Separation, however, was accomplished by Colman and Vishniac [57] on thin layers of sucrose. These layers were prepared by slurrying confectioner's sugar (10X containing 3% starch) in an equal quantity (wt/vol) of methanol. After the plates were coated, the layers were dried at 40°C for 2 hr. Excess drying should be avoided because this fuses the sucrose to a hard nonabsorbent glassy layer. Petroleum ether (66 to 75°C)–acetone (95:5) separated the chlorophylls, β-carotene, lutein, and neoxanthin. Since the violaxanthin may not be completely separated from chlorophyll A, a second development in petroleum ether (37 to 49°C)–methanol (98:2) can be used to effect a better separation, or the two solvents may be used in a two-dimensional separation. Nutting et al. [58] have applied the use of a powdered sugar layer with chlorophyll compounds.

Hager and Bertenrath [52] have separated chlorophylls A and B from grain plant extracts by using the mixed adsorbent and the solvent mentioned under carotene. Egger [20] has applied reverse-phase separation to the chlorophylls and their degradation products. With kieselguhr plates impregnated with a 7% solution of a triglyceride (vegetable oil) in petroleum ether and a methanol–acetone–water (20:4:3) solvent (saturated with impregnating oil), the following R_f values were obtained: chlorophyll A 0.13, chlorophyll B 0.25, phaeophytin A 0.1, phaeophytin B 0.07, allomeres A′ 0.08, and allomeres B′ 0.19. Since rhodoxanthin has the same R_f value as chlorophyll B in this solvent, the pair can be separated by using a solvent of acetone–water (3:1) in which rhodoxanthin has an R_f value of 0.31 compared to chlorophyll B of 0.26. Jones et al. [59] examined the technique of Egger. Kieselguhr G layers impregnated with 14% peanut oil in isooctane gave the best results; however, some oxidation occurred on that portion of the plate not impregnated which was used for spotting the sample. Egger had shown that sharper spots were formed by applying the sample to an unimpregnated section of the layer rather than to the impregnated section. The later workers found that this deterioration could be prevented by adding an antioxidant to the sample solution (0.1 ml of a solution of 150 mg of BHT in 100 ml of diethyl ether to 1 ml of sample solution). Daley et al. [60] found that the oxidation could be prevented by applying the sample under an atmosphere of nitrogen, but development in a nitrogen atmosphere was not necessary. Because a single-plate system could not separate all of the compounds present in their extracts, they used a series of systems; for "low-range" (pheophytin) compounds a layer impregnated with 8% paraffin oil and solvent mixture of methanol–acetone–isopropanol–water–benzene (35:50:10:10:2), for "mid-range" (chlorophylls) a layer impregnated with 4% triolein and a solvent mixture of 60:20:10:10:2, and for "high-range" (pheophorbides) a layer impregnated with 10% castor oil and a solvent mixture of 80:2.5:2.5:15:2. A number of other combinations were also used in special situations. Petroleum ether (30 to 60°C) was used for impregnating with paraffin and triolein and methanol for impregnating with castor oil. A total of 18 identified and 8 unknown derivatives of chlorophylls A, B, and C were separated.

Shimizu et al. [61], in investigating the metabolism of chlorophyll in higher plants, has applied thin-layer chromatography to the quantitative determination of phytol.

Anwar [62] has used 1 × 8 in.-thin layers in a continuous system for the separation of chloroplast pigments on silicic acid containing a starch binder. Continuous development was carried out by using a paper wick to carry the solvent from the end of the strip out beyond the stopper of the test tube developing chamber. Developing solvents consisted of isooctane–acetone–ether (3:1:1) and isooctane–acetone–carbon tetrachloride (3:1:1).

Bacon [63] has applied cellulose layers to the separation of chlorophylls *a* and *b* and their derivatives using a solvent mixture of petroleum–ether (60 to 80°C)–acetone–*n*-propanol (90:10:45) in a development distance of 15 cm. Schneider [64] separated nine plastid pigments on cellulose MN 300 with methanol–dichloromethane–water (50:9:10).

Loewenschuss and Wakelyn [65] separated chlorophylls A and B from plant extracts on sucrose–starch (97:3) with petroleum ether (30 to 60°C)–chloroform (3:1). The zones were removed from the layer and heated in dilute hydrochloric acid for 4 hr at 90°C to hydrolyze the sucrose. After the addition of a known amount of magnesium, the magnesium in the spot was determined by atomic absorption. Sensitivity was in the range of 1 ng of magnesium.

3 ANTHOCYANINS

These compounds are the water-soluble pigments which occur in plant cell sap, primarily in fruits and flowers, but also to some extent in other portions of some plants. They supply the many shades of blue, purple, violet, mauve, magenta, and most of the reds which appear in the plant kingdom. They are glycosidic in nature and on hydrolysis with dilute mineral acid they yield anthocyanidins and sugar residues. Separations can be achieved using either the anthocyanins or the anthocyanidins.

Nybom [66, 67] has used cellulose layers for the separation of the aglycons and found cellulose MN 300 (Macherey, Nagel) to be a very satisfactory layer. The pigments were extracted from the plants with methanol except when large amounts of pectic substances were present, in which case a 50% isopropanol was used for the extraction. A preliminary cleanup of the extract was made by adjusting to a pH of 9 with ammonium hydroxide and then precipitating the pigments with a saturated solution of lead acetate in methanol. An alternative method was to adsorb the pigments on Dowex 50 W-X 4. After the precipitate or the ion-exchange material was washed, the pigments were recovered with a 2% hydrochloric acid solution in methanol. The aglycons were obtained by hydrolyzing two parts of the purified extracts with one part of concentrated hydrochloric acid for 30 min at 100°C. The anthocyanidins were then transferred to amyl alcohol as a spotting solvent. The actual separation was carried out on a 12 × 16 cm plate using two-dimensional chromatography with a formic acid–hydrochloric acid–water (10:1:3) solvent in the first direction along the short side of the plate, followed by amyl alcohol–acetic acid–water (2:1:1) in the second direction. The method was applied to the examination of the anthocyanidins in 40 different plant species. Paris and Paris [68] have also used cellulose layers in the separation of anthocyanidins and also anthocyanins. For the separation of the anthocyanins, three solvents were used: butanol–2 *N* hydrochloric acid (1:1) with 10% ethyl acetate and with 20% ethyl acetate, and butanol–acetic acid–water (4:1:5). For the anthocyanidins, mixtures of acetic acid–hydrochloric acid–water (10:1:3) and acetic acid–water (3:2) were used. These latter two solvents were also used with polyamide plates.

Gupta [69] tested a number of solvents used in paper chromatography of anthocyanins, but found only *n*-butanol–acetic acid–water (4:1:5) upper phase and water–acetic acid–hydrochloric acid (82:15:3) among those tested to be satisfactory for cellulose layers. The first solvent mentioned and acetic acid–hydrochloric acid–water (30:3:10) were also useful for separating anthocyanidins on cellulose. Mullick [70] used microcrystalline cellulose Avicel SF for two-dimensional chromatography of anthocyanidins. Four nonalcoholic solvents were used in combinations with four alcoholic solvents. These solvents were acetic acid–hydrochloric acid–water (30:3:10), acetic acid–formic acid–hydrochloric acid–water (2:5:1:6), formic acid–4 *N* hydrochloric acid (2:1), formic acid–hydrochloric acid–water (10:1:3), methanol–hydrochloric acid–water (190:1:10), *n*-amyl alcohol–acetic acid–water (2:1:1), *n*-butanol–acetic acid–water (4:1:5) (upper phase), and *tert*-butanol–2 *N* hydrochloric acid–acetic acid–water (6:1:1:2) freshly prepared. The only combination of two alcoholic solvents that was useful was that of the methanolic solvent in the first direction followed by the amyl alcohol solvent in the second. The methanolic solvent proved to be quite useful because it separates the anthocyanidins largely on the basis of the number of hydroxyls on the B ring regardless of their state of methylation. Avicel SF proved to be superior to cellulose MN 300. Barritt and Torre [71] also used microcrystalline cellulose layers for separating anthocyanins, and of two solvent systems evaluated *n*-butanol–hydrochloric acid–water (5:2:1) in the first direction followed by water–hydrochloric acid–formic acid (8:4:1) in the second direction proved to be the best. In isolating the anthocyanins from various fruits the methanol–water (1:1) extract was filtered through a layer of water-insoluble polyvinylpyrrolidone (PVP) which was then washed with water and then methanol. The anthocyanins were eluted from the PVP with 0.1% hydrochloric acid. After taking to dryness under a vacuum the pigments were dissolved in 0.01% hydrochloric acid in methanol. Wrolstad [72] preferred a mixed layer of insoluble PVP (Polycar AT) (General Aniline and Film) 100 to 150 mesh–cellulose (1:9) over straight cellulose layers. The spots were smaller and therefore better for resolution and for quantitative work. Before use, the PVP was purified of traces of hydrogen peroxide by the method of Loomis and Battaile [73]. For anthocyanins acetic acid–water–hydrochloric acid (15:82:3) was used in the first direction and the upper phase of *n*-butanol–acetic acid–water (4:1:5) in the second dimension. After two-dimensional separation on Avicel SF cellulose, Mullick [74] found that exposure to ammonia vapor increased the sensitivity of detection to less than 0.01 μg for anthocyanidins; spraying with fresh 4% ammonium molybdate followed by drying in hot air produced stable reproducible colors at a sensitivity of 0.1 μg. A 1% lead acetate solution in 75% ethanol followed by hot air drying was useful in differentiating pelargonidin from peonidin. Gupta [69] used silica gel sheets with ethyl acetate–formic acid–methyl ethyl ketone–water (5:3:3:1) for the separation of anthocyanins as well as the cellulose mentioned previously. Nybom [75] applied a two-dimensional separation on cellulose for the anthocyanins in black raspberries. Two solvent systems were used: *n*-butanol–hydrochloric acid–water (5:2:1) in the first direction followed by water–hydrochloric acid–formic acid (8:4:1) in the second direction and *n*-butanol–acetic acid–water (6:1:2) in the first direction with water–hydrochloric acid–propionic acid (10:2:3) in the second.

Birkofer et al. [76] have used a layer composed of a mixture of two parts of polyamide with seven parts of polyacrylonitrile which was mixed

Table 28.2 $R_f \times 100$ Values of Anthocyanins and Anthocyanidins in Various Systems[a]

	Polyacrylonitrile–Polyamide (7:2) Mixed with 40 ml of 0.05 M Primary Potassium Phospahate [76]	*Silica Gel D-O (CAMAG)* [77]		*Cellulose MN 300* [55]		*Cellulose* [68]		*Polyamide* [68]	
Compound	*A*	*B*	*C*	*D*	*E*	*F*	*G*	*F*	*G*
Anthocyanidins									
Cyanidin-3-monoglucoside	46								
Cyanidin-3-triglucoside	42								
Cyanidin-3,5-diglucoside		36	52						
Delphinidin-3-monoglucoside	36								
Delphinidin-3,5-diglucoside		35	26						
Malvidin-3-monoglucoside	57	60	61						
Malvidin-3,5-diglucoside	54	44	33						
Paeonidin-3,5-diglucoside		56	37						
Petunidin-3-monoglucoside	38								
Anthocyanins									
Cyanidin	31	83	92	47	53	70	75	57	75
Delphinidin	21			41	34	50	60	50	72
Malvidin	37	95	94	62	61	87	84	84	91
Paeonidin	39			55	65	90	86	83	85
Pelargonidin	41			51	71	95	92	69	78
Petunidin	27			54	46	70	75	72	79

[a] Solvent: *A* = 1-pentanol–1-propanol–acetic acid–water (3:2:2:1), 2 parts 1-heptanol may be added to decrease streaking, development distance 12 cm; *B* = 1-butanol–acetic acid–water (4:1:2); *C* = ethyl formate–methyl ethyl ketone–formic acid–water (3:4:1:2); *D* = formic acid–hydrochloric acid–water (10:1:3); *E* = amyl alcohol–acetic acid–water (2:1:1); *F* = acetic acid–hydrochloric acid–water (10:1:3), development distance 12 cm; *G* = acetic acid–water (3:2), development distance 12 cm.

with 0.05 M primary potassium phosphate solution in place of water. As a solvent for both the anthocyanidins and the anthocyanins, they used pentanol–1-propanol–1-acetic acid–water (3:2:2:1). The tendency to streak could be decreased by adding 1 to 2 parts of 1-heptanol or 1-hexanol.

Silica gel layers have also been used for separating anthocyanins [77]. Solvent mixtures of butanol–acetic acid–water (4:1:2) and ethyl formate–methyl ethyl ketone–formic acid–water (3:4:1:2) were used. Hess and Meyer [51] separated anthocyanins on silica gel G with a mixture of ethyl acetate–formic acid–water (14:3:3) using a development distance of 10 cm, except in certain cases where the R_f values were fairly close together when a distance of 13 cm was used. When the three compounds, pelargonidin-, paeonidin-, and malvidin-3-monoglucosides were present, a mixture of n-butanol–formic acid–water (17:1:2) gave better separation than the ethyl acetate–formic acid–water. For two-dimensional work the ethyl acetate mixture was used in the first direction and the butanol mixture in the second direction. The anthocyanidins were separated using ethyl acetate–acetic acid–water (85:6:9). Although the anthocyanins and the anthocyanidins can be detected by observation under ultraviolet light, the intensity of the spots can be increased by spraying with a 10% solution of oxalic acid in acetone–water (1:1). Table 28.2 gives the R_f values of some anthocyanins and anthocyanidins.

Asen [79] used a mixture of $\frac{2}{3}$ silica gel (Adsorbosil-2) and $\frac{1}{3}$ cellulose powder (MN 300 gypsum-free) in 1-mm-thick layers for the separation of anthocyanins in preparative amounts. These layers gave better separations than straight silica gel. The upper phase of a 24-hr-old mixture of 1-butanol–2 N hydrochloric acid (1:1) was used to separate cyanidin glycosides from pelargonidin glycosides. Individual glycosides were then separated with water–hydrochloric acid formic acid (8:4:1).

4 FLAVONOIDS AND RELATED COMPOUNDS

This group of compounds, which gets its name from the Latin for "yellow," is widely distributed in the plant kingdom. They are found both free and combined as glycosides.

The first application of thin-layer chromatography to this type of compound was in 1957 by Stanley and Vannier [80, 81], who applied it to the separation of some coumarins and furocoumarins found in lemon oil (Table 28.3). The compounds were isolated by chromatographing on a silicic acid column, checking the separation by the method of Miller and Kirchner [82]. The compounds were located in ultraviolet light by the use of fluorescent strips [83]. Stanley and Vannier [84, 85], by using a

Table 28.3 $R_f \times 100$ Values on Starch-Bound Silicic Acid for Some Coumarins and Furocoumarins Found in Lemon Oil[a] [80, 81]

Compound	$R_f \times 100$
5-Geranoxypsoralen	68
7-Methoxy-5-geranoxycoumarin	64
5-(γ,γ-Dimethyl)allyloxypsoralen	57
7-Methoxy-5(γ,γ-dimethyl)allyloxycoumarin	50
8-Geranoxypsoralen	40
5,7-Dimethoxycoumarin	25
Byakangelicin	0

[a] Solvent = Hexane–Ethyl acetate (3:1).

continuous descending method, developed a quantitative method for the determination of coumarin compounds in citrus oils (see Chapter V, Section 3). After the compounds were separated, the spots were eluted with ethanol and then measured with a spectrophotometer. Swift [86] developed a method for the spectrophotometric determination of the flavones in the neutral fraction in orange peel juice. Silica gel layers with 15% n-butanol in hexane as a solvent was used with multiple development to obtain adequate separation which could be followed under ultraviolet light. Hoerhammer and Wagner [87] have also examined citrus flavonoids. Separation was accomplished on thin layers of silica gel (Woelm) using butanol–acetic acid–water (4:1:5) as a developing solvent. Although no R_f values are given, the separation of the following compounds was illustrated: hesperidin, hesperetin, eriodictin, naringin, and naringenin. Although the R_f values of hesperidin and naringin are close, these compounds can be differentiated readily by means of the Fast Blue B reagent which is used for visualizing the compounds. In this case the hesperidin gives a blue color and the naringin a violet color. A quantitative method for determining these compounds was established by first treating them with 2,4-dinitrophenylhydrazine and then with potassium hydroxide to give colored solutions which were measured at 480 mμ. Fisher et al. [88] developed a colorimetric method for determining naringin in grapefruit. Naringin and naringenin-7β-rutinoside were separated on polyamide–silica gel–rice starch (5.5:0.4:0.8) by developing twice with nitromethane–methanol (5:2). The compounds were located by fluorescence after spraying with 1% aluminum chloride in ethanol, and determined by the Davis test [89] with alkaline diethylene glycol after elution. Hoerhammer et al. [90] chromatographed a group of 13 flavonoids on silica gel using benzene–pyridine–formic acid (36:9:5) for the separation. Before spraying the visualizing agents, the plates were dried thoroughly in order to remove the formic acid and pyridine. The two visualizing agents consisted of a 25% solution of basic lead acetate and 10% antimony trichloride solution in chloroform. Observation of the fluorescent colors was made under an ultraviolet lamp with a maximum at 366 mμ. Hoerhammer and Wagner [91] have also investigated isoflavonones in plant extracts.

Duggan [92, 93] separated the flavonol glucosides of pears and other fruits on microcrystalline cellulose using a two-dimensional development with acetic acid–water (15:85) in the first direction and phenol–water (4:1) (wt/wt) in the second. For chromatographing aglycons, acetic acid–hydrochloric acid–water (30:3:10) was used. Jacquin-Dubreuil [94] separated the flavonoids of the yellow weld (*Reseda luteola* L.) on cellulose layers predeveloped with ethanol to remove impurities. A solvent mixture of *n*-butanol–methyl ethyl ketone–water (5:3:3) was used. Quantitative results were obtained by direct densitometry of the ultraviolet absorption. Spiegl [95] used a mixture of cellulose MN 300–polyamide Woelm DC (30:1) for separating flavonoids using 15, 40, and 60% acetic acid as developing solvents. A two-dimensional development with 60 and 15% acetic acid separated aglycons from glucosides. Visualization was achieved with test T-102, T-118, or by fluorescence after spraying with 2% aluminum chloride in 70% ethanol.

Because of the difference in coumarin derivatives, Hoerhammer et al. [96] were able to detect the adulteration of *Radix pimpinellae* with the roots of acanthus (*Heracleum spondylium*). In order to detect the adulteration, a petroleum ether extract of the finely ground root is applied to a silica gel layer and developed with chloroform for a distance of 12 cm. After examination in ultraviolet light, the chromatogram is sprayed with a 20% solution of antimony trichloride in carbon tetrachloride and heated to 110°C. Adulteration is indicated by spots ranging from an R_f value of 0.4 to 0.55. The extract from the *pimpinella* root has only one spot in this region at R_f value 0.45, compared to the four spots for the heracleum root extract.

Tschesche et al. [97] have applied preparative thin-layer chromatography to the isolation of a new coumarin glycoside from *Daphne mezereum*. Vul'fson et al. [98] studied a series of coumarin and furanocoumarin compounds on silica gel with a 1:1 mixture of ether and petroleum ether. Isoimperatorive and peucedanin were found in the roots of *Peucedanum morisonii*. Hsu and Fu [99] chromatographed coumarones on alumina, examining the various factors affecting the separation. Acidic alumina of 100 mesh with petroleum ether–chloroform (1:1), petroleum ether–ethanol (1:1), or petroleum ether–dioxane (5:1) gave satisfactory separations without solvent vapor equilibration of the plate. Weygand et al. [100] investigated the biosynthesis of coumarin. Copenhaver and Carver [101] have separated a group of 13 coumarin compounds in three solvent systems on silica gel layers. Chernobai and Kolesnikov [102] separated a group of coumarins from *Cnidium bubium* by chromatographing on acid-treated aluminum oxide. Enidin, enidicin, and enidilin were isolated from the fruit of this plant.

Goerlich [103] has separated the flavonoids in oleander extract (*Nerium oleander*) by both paper and thin-layer chromatography. The thin-layer separation was carried out on silica gel using ethyl acetate–formic acid–water–methanol (10:2:2:1).

Stambouli and Paris [104] investigated the flavonoids of the plane tree *Platanus occidentalis* on silica gel layers and by paper chromatography.

Ribereau-Gayon [105] identified four flavonoids in the skins of the fruit of red varieties of the genus *Vitis*.

Gabor et al. [106] used *n*-butanol–acetic acid–water (4:1:5) for the separation of bioflavonoids on silica gel, and Lin and Chen [107] reported the R_f values for 18 bioflavonoids and their methylated derivatives on silica gel with benzene–pyridine–formic acid (36:9:5) and (40:10:2).

Daenens and Van Boven [108] give the R_f values of 10 coumarins on silica gel and on phosphate-buffered silica gel with chloroform–methanol (97:3) and chloroform–benzene–formic acid–acetylacetone (49:48:2:1), respectively. Trkovnik et al. [109] separated 14 3-substituted 4-hydroxycoumarins on silica gel in chloroform–methanol–toluene (33:7:10), benzene–light petroleum–acetone–ethanol (61:23:8:8), benzene–acetone (9:1), benzene–methyl ethyl ketone (9:1), benzene–acetic acid–acetone (17:1:2), and benzene–acetic acid–methyl ethyl ketone (8:1:1). R_f values are reported.

Flavonoids have also been separated on polyamide layers. Davídek [110] and Davídek and Davídková [78] demonstrated the usefulness of loose layers of polyamide for the separation of flavonoids. Davídek and Procházka [112] reported on the separation of rutin, quercetin, and naringin. Using polyamide layers with methanol, the following R_f values were reported: rutin 0.45, quercetin 0.20, and naringin 0.79, whereas in a methanol–water mixture (4:1) the R_f values were 0.45, 0.10, and 0.80, respectively. With care to spray from a distance so as not to disturb the loose layers, the compounds could be located with diazotized sulfanilic acid reagent. Davídek [110] determined the rutin and quercitrin content of elder blossoms and buckwheat stalks, by eluting the spots with 30% methanol and determining the quantity, colorimetrically, after reaction with diazotized *p*-aminobenzoic acid. Chia [113] has separated 11 different kinds of flavonoids on polyamide layers using a number of different solvent systems including ethyl acetate saturated with water, *n*-butanol (saturated with water)–acetic acid (100:1 and 100:2), acetone–water (1:1), acetone–95% ethanol–water (2:1:2), 95% ethanol–acetic acid (50:1), and isopropanol–water (3:2). Egger [114] chromatographed a group of 14 flavonoids using both paper and thin layers of polyamide. The R_f values in three different solvents on polyamide are given in Table 28.4. Bhandari [115] has used polyamide for the separation of the flavonoids in hops. Six flavonoids were separated of which three were identified. Astraligin was found for the first time, but the presence of the reported quercetrin could not be confirmed. Beer was found to contain the same spots as hops. Hubacek [116] chromatographed the flavonoids of hops on polyamide layers with methanol–water (4:1) using a 15-cm developing distance. Hoerhammer et al. [117] have chromatographed the flavone glycosides from *Matricaria chamomilla* on polyamide. Nishiura et al. [118] used nitromethane–methanol (5:2) as a solvent for the sep-

Table 28.4 $R_f \times 100$ Values of Some Flavanol Glycosides on Polyamide Layers and on Paper[a]

	Perlon[c]			*Paper*[d]	
Compound[b]	*A*	*B*	*C*	*D*	*E*
K-3-rh	22	22	15	73	64
K-3-arab	22	22	15	72	60
K-3-gluc	22	22	15	62	54
K-3-rhgluc	32	37	27	52	48
K-3-digluc	39	51	32	20	28
K-3,7-digluc	57	63	53	30	31
K-3-rhgal-7-rh	66	65	60	35	51
Q-3-rh	22	22	15	60	48
Q-3-gluc	22	22	15	51	42
Q-3-rhgluc	32	39	29	40	40
My-3rh	22	22	15	48	33
My-3-gluc	22	22	15	35	25

[a] From K. Egger [114]; reproduced with permission of the author and Springer-Verlag.

[b] K = kaempherol, Q = quercetin, My = myricetin, rh = rhamnoside, arab = arabinoside, gluc = glucoside, gal = galactoside.

[c] *A* = Ethanol–water (3:2); *B* = water–ethanol–acetylacetone (4:2:1); *C* = water–ethanol–methyl ethyl ketone acetylacetone (13:3:3:1).

[d] *D* = Partridge mixture; *E* = chloroform–acetic acid (2:3) saturated with water.

aration of isomeric rutinoside and neohesperidoside pairs on polyamide. The compounds were visualized by spraying with 2% sodium borohydride in methanol followed by fuming with hydrochloric acid vapors yielding red to violet colors.

Grisebach and co-workers [119–122] have used thin-layer chromatography in their work on the biogenesis of isoflavones; 5,7-dihydroxy-4′-methoxyisoflavonone and 7-hydroxy-4′-methoxyisoflavone, which could not be separated by paper chromatography, were readily separated on thin layers of silica gel with benzene–ethanol (92:8) giving R_f values of 0.4 and 0.25, respectively. These compounds were isolated from the germ of the chick pea (*Cicer arientinum*). Grisebach and Barz [120] used preparative thin-layer chromatography in studying the biogenesis of the 3-arylcoumarin coumestrol in lucern (*Medicago sativa*).

Bickoff and co-workers [123–125] isolated a group of four isoflavones from alfalfa (*Medicago sativa*), red clover (*Trifolium pratense*), and subterranean clover (*Trifolium subterraneum*). Ladino clover (*Trifolium repens* var. *ladino*) contained three of the isoflavones but did not contain biochanin A. The separations were made on silicic acid chromatostrips [83] using a series of solvents (Table 28.5). The spots were located on the fluorescent strips by examination under ultraviolet light.

Table 28.5 $R_f \times 100$ Values of Four Isoflavones on Silicic Acid Chromatostrips in Various Solvents[a]

Isoflavone[b]	*Daidzein*	*Formo-nonetin*	*Genistein*	*Biochanin A*
Ethyl acetate–Skellysolve B (3:1)	76	88	78	82
Ethyl acetate–Skellysolve B (1:1)	38	56	50	61
Acetone–ethyl acetate–Skellysolve B (4:3:3)	81	95	80	90
Ethyl alcohol–chloroform (1:3)	62	81	—	71
Ethyl alcohol–chloroform (1:1)	84	95	68	82
Ethyl ether–Skellysolve B (7:3)	—	—	36	60

[a] From J. Guggolz, A. L. Livingston, and E. M. Bickoff [123]; reproduced with permission of the authors and the American Chemical Society.

[b] Daidzein = 4′,7-dihydroxyisoflavone. Formononetin = 7-hydroxy-4′-methoxyisoflavone. Genistein = 4′,5,7-Trihydroxyisoflavone. Biochanin A = 5,7-dihydroxy-4′-methoxyflavone.

Larson [126] used Magnesol (magnesium acid silicate) for separating flavonoid compounds. Toluene–ethyl formate–formic acid (5:4:1) was the solvent used in a saturated atmosphere. This system was claimed superior to silica gel, polyamide, or cellulose for separating these compounds. Wang [127] found that separation of daidzin from genistin was better on alumina than on silica gel, and recommends alumina for separating isoflavones with a 5-hydroxyl group from isoflavones lacking this group. Chloroform–methanol–water (65:25:4) was used in the separations.

5 ANTHOCHLOR PIGMENTS

Haensel et al. [128] synthesized a group of aurones and chromatographed them on thin layers of silica gel G buffered with sodium acetate. Solvent mixtures of benzene–ethyl acetate–formic acid (4.5:3.5:2.0), chloroform–ethyl acetate–formic acid (5:4:1) and (6:3:1), and toluene–ethyl formate–formic acid (5:4:1) were used in the separation. A group of chalcones were also chromatographed on silica gel layers prepared with water using a solvent of cyclohexane–ethyl acetate (7:1) saturated with a mixture of 5 ml of water and 10 ml of formamide. Chalcones and aurones can be differentiated according to their behavior under fluorescent light: aurones fluoresce under ultraviolet with a yellow to light green color, whereas chalcones give a dark brown color. Further differentiation can be given by exposure to ammonia vapor which colors the aurones orange and the chalcones a deep red. Kaufmann and El Baya [129] separated the anthochlor pigments of *Dahlia variabilis* and *Cosmos sulphureus* on silica with benzene–ethyl acetate–formic acid (9:7:4). The glycosides were separated with ethyl acetate–methyl ethyl ketone–formic acid–water (5:3:1:1).

6 PORPHYRINS

Using a solvent mixture of benzene–ethyl acetate–ethanol (90:20:7.5) for a development distance of 5.4 cm, Demole [6] has separated some porphyrin esters on thin layers of silicic acid bound with starch. The R_f values obtained are listed as follows: coproporphyrin I ester, 0.68; coproporphyrin III ester, 0.63; uroporphyrin I ester, 0.33; deuteroporphyrin ester, 0.80; and protoporphyrin ester, 0.85. For the free coproporphyrins I and III, Jensen [130] has obtained R_f values of 0.19 and 0.25 on thin layers of silica gel G with a 2,6-lutidine–water–ammonium hydroxide solvent. This solvent was prepared by mixing 10 ml of the 2,6-lutidine with 3 ml of water; the mixture was then placed in the developing chamber and allowed to come to equilibrium with ammonia vapor from a 30% solution of ammonium hydroxide in a separate container in the chamber. Doss and co-workers have published a great deal on the use of thin-layer chromatography of porphyrins and a few of their references can be listed here [131–136]. Silica gel layers were used for the separations with benzene–ethyl acetate–methanol (170:27:3) as a solvent. In most cases the porphyrins were separated as their methyl esters prepared by treating with 5% sulfuric acid in methanol for 6 hr at room temperature. For quantitative work the layers are pre-run with chloroform–methanol (2:1) and then with a 1:2 mixture and then dried at 80°C. After application of the sample a development with light petroleum–ethyl ether (7:3) will remove glycerides, cholesterol esters, and fatty acid methyl esters [132]. Quantitative analysis was carried out by in situ fluorimetric evaluation of the spots [131, 132, 136]. Doss [134] also showed that the stability and molar absorptivity of porphyrins was increased by incorporation of zinc into the ring. The chelate esters were chromatographed in the same system as mentioned above. Mundschenk [137–140] has used 2,6-lutidine–water (62:5:37.5, 66.2:33.8, and 70:30) for the separation of free porphyrins. Ellfolk and Sievers [141] chromatographed free porphyrin acids with benzene–methanol (17:3) and chloroform–methanol (9:1) both solvents 0.3 *M* in formic acid using neutral silica gel layers. For layers prepared with oxalic acid various solvent mixtures of toluene–isopropanol (1:1), benzene–methanol (6:4), as well as benzene with other alcohols were used. Elder [142] used a two-dimensional separation on silica gel for the methyl esters. Carbon tetrachloride–dichloromethane–methyl acetate–methyl propionate (2:2:1:1) was used in the first direction followed by development in the second direction, first with benzene–butanone (40:3) and then with chloroform–kerosene–methanol (200:100:7). Increased resolution of the higher R_f porphyrins could be obtained by repeating the benzene–butanone development and then developing with chloroform–kerosene–methanol (200:100:15).

Cellulose has also been used as an adsorbent for the separation of porphyrins. Solvents that have been used include 2,6-lutidine–water (10:3) [143], 2,6-lutidine–ammonia–water–0.1 *M* EDTA (10:4.2:2.8:0.02) [144], and 2,6-lutidine–water–ammonia (10:3:vapor) [145]. Batlle and Benson [146] separated uroporphyrin octamethyl esters I and III by chromatographing on cellulose containing 10% gypsum using kerosene–dioxane (4:1.5) as a solvent.

Both free and esterified porphyrins have been chromatographed on thin layers of talc. Solvents for these separations include acetone–0.5 *M* hydrochloric acid (3:2) [147], acetone–0.5 *M* hydrochloric acid (7:3)

[148], and ethanol–2,6-lutidine–water (30:3:67) in an ammonia atmosphere [149] for the free porphyrins and acetic acid–pyridine–acetone (1:1:1), chloroform–methanol (1:1), and ethyl acetate–ethanol (2:3) [148] for the esters.

Eriksen [150] has used electrophoresis on thin layers of agar gel for the separation of porphyrins.

Hemoglobins, which are conjugated proteins containing globins and heme (a pigment composed of protoporphyrin and ferrous iron), have been chromatographed on aluminum oxide and diethylaminoethyl cellulose and have been separated by electrophoresis on starch gel [151–153]. Thin-layer electrophoresis on agar [154–157] and on starch [151, 158–163] has been used fairly extensively for the separation of hemoglobins. Schroeder and Nelson [164] used ion-exchange layers of carboxymethyl cellulose CM-52 (Whatman) for the chromatography of hemoglobins.

7 BILE PIGMENTS

Demole [6] has chromatographed biliverdin and bilirubin on silicic acid layers with benzene–ethyl acetate–ethanol (90:20:7.5) using a solvent travel of 5.4 cm to obtain R_f values of 0.06 and 0.92, respectively. O'Carra and Colleran [165] separated IXα, IXβ, IXγ, and IXδ, isomers of protobiliverdin and mesobiliverdin on silica gel G as their dimethyl esters. Best results were obtained with carbon tetrachloride–ethyl acetate–cyclohexane (32:9:4) with multiple development (8×). Propyl or isopropyl esters have been separated with heptane–methyl ethyl ketone (5:2) (4×) or with cyclohexane–methyl ethyl ketone (2:3) [166]. Other solvents that have been used for bilirubin and derivatives on silica gel include: chloroform–ethyl acetate (5:3) [167], chloroform–methanol–formic acid (60:30:1) and methyl acetate–methyl propyl ether–carbon tetrachloride–dichloromethane (1:1:1:1) [168]. Petryka and Watson [169] separated unesterified bile pigments on polyamide with methanol–water (3:1) and methanol–10% ammonia–water (9:1:2). Esterified samples were separated on silica gel G with benzene–ethanol (25:2) and chloroform–ethanol (50:1). Segura-Cardona [170] also used polyamide with propanol–*n*-butanol–pyridine–water (10:3:4:1) for the separation of bile pigments of serum. Petryka [171] used an in situ densitometric method for quantitative evaluation of bile pigments.

Tenhunen [172] coupled bile pigments with diazotized sulfanilic acid and then spotted the azo pigments on silica gel in *n*-butanol. The chromatograms were developed with 50% methanol.

References

1. M. Mottier and M. Potterat, *Mitt. Geb. Lebensm. Hyg.,* **43,** 118 (1952).
2. *Ibid.,* p. 123.
3. M. Mottier, *Mitt. Geb. Lebensm. Hyg.,* **47,** 372 (1956).
4. H. Lagoni and A. Wortmann, *Milchwissenschaft,* **10,** 360 (1955).
5. *Ibid.,* **11,** 206 (1956).
6. E. Demole, *J. Chromatogr.,* **1,** 24 (1958).
7. O. Voelker, *Naturwissenschaften,* **48,** 581 (1961).
8. H. Thommen and H. Wacknernagel, *Biochim. Biophys. Acta,* **69,** 387 (1963).
9. H. Thielemann, *Z. Anal. Chem.,* **271,** 285 (1974).
10. A. Winterstein and B. Hegedues, *Chimia (Aarau),* **14,** 18 (1960).
11. A. Winterstein and B. Hegedues, *Z. Physiol. Chem.,* **321,** 97 (1960).
12. H. Thommen, *Naturwissenschaften,* **49,** 517 (1962).
13. H. Thommen, *Chimia (Aarau),* **15,** 433 (1961).
14. I. Wildfeuer, *Z. Lebensm. Unters. Forsch.,* **140,** 140 (1969).
15. A. Winterstein, A. Studer, and R. Rueegg, *Chem. Ber.,* **93,** 2951 (1960).
16. J. A. Box and H. A. Boekenoogen, *Fette, Seifen, Anstrichm.,* **69,** 724 (1967).
17. T. Bjornland and M. Aguilar-Martinez, *Phytochemistry,* **15,** 291 (1976).
18. J. Sherma, *Anal. Lett.,* **3,** 35 (1970).
19. H. Rai and G. F. Lee, *Anal. Chem.,* **36,** 2208 (1964).
20. K. Egger, *Planta,* **58,** 664 (1962).
21. H. Nitsche, *Z. Naturforsch., Teil C,* **29,** 657 (1974).
22. O. Isler, R. Rueegg, and P. Schudel, *Chimia (Aarau),* **15,** 208 (1961).
23. E. C. Grob, W. Eichenberger, and R. P. Pflugshaupt, *Chimia (Aarau),* **15,** 565 (1961).
24. E. C. Grob and R. P. Pflugshaupt, *Helv. Chim. Acta,* **45,** 1592 (1962).
25. W. Eichenberger and E. C. Grob, *Helv. Chim. Acta,* **45,** 974 (1962).
26. *Ibid.,* p. 1556.
27. *Ibid.,* **46,** 2411 (1963).
28. E. C. Grob and A. Boschetti, *Chimia (Aarau),* **16,** 15 (1962).
29. B. H. Davies, T. W. Goodwin, and E. I. Mercer, *Biochem. J.,* **81,** 40P (1961).
30. E. C. Grob, *Wiss. Veroeff. Dtsch. Ges. Ernaehr.,* **9,** 26 (1963).
31. G. Rispoli and A. Di Giacomo, *Boll. Lab. Chim. Prov. (Bologna),* **13,** 587 (1962); *Chem. Abstr.,* **59,** 8044 (1963).
32. E. Benk, I. Wolff, and H. Treiber, *Deut. Lebensm. Rundsch.,* **59,** 39 (1963).
33. A. Montag, *Z. Lebensm. Untersuch. Forsch.,* **116,** 413 (1962).
34. D. Corbi, *Giorn. Med. Mil.,* **114,** 168 (1964).
35. P. Duquenois and M. Meylaender, *Ann. Fals. Expert. Chim.,* **56,** 371 (1963).
36. B. H. Davies, D. Jones, and T. W. Goodwin, *Biochem. J.,* **87,** 326 (1963).
37. E. I. Mercer, B. H. Davies, and T. W. Goodwin, *Biochem. J.,* **87,** 317 (1963).
38. K. Egger, *Chromatogr. Symp., 2nd, Brussels,* **1962,** 75.

39. L. Cholnoky, J. Szaboles, R. D. G. Cooper, and B. C. L. Weedon, *Tetrahedron Lett.*, **1963**, 1257.
40. H. Bolliger, A. Koenig, and U. Schwieter, *Chimia (Aarau)*, **18**, 136 (1964).
41. K. Egger and H. Voigt, *Z. Pflanzenphysiol.*, **53**, 64 (1965).
42. A. Hager and H. Stransky, *Arch. Mikrobiol.*, **73**, 77 (1970).
43. H. Stransky and A. Hager, *Arch. Mikrobiol.*, **72**, 84 (1970).
44. A. Hager and H. Stransky, *Arch. Mikrobiol.*, **72**, 68 (1970).
45. H. Stransky and A. Hager, *Arch. Mikrobiol.*, **71**, 164 (1970).
46. A. K. Stobart, I. McLaren, and D. R. Thomas, *Phytochemistry*, **6**, 1467 (1967).
47. T. Philip and F. J. Francis, *J. Food Sci.*, **36**, 823 (1971).
48. H. Singh, J. John, and H. R. Cama, *J. Chromatogr.*, **75**, 146 (1973).
49. R. E. Knowles and A. L. Livingston, *J. Chromatogr.*, **61**, 133 (1971).
50. C. Rollins, *J. Chem. Educ.*, **40**, 32 (1963).
51. L. F. Fieser, *Chemistry*, **37**, 23 (1964).
52. A. Hager and T. Bertenrath, *Planta*, **58**, 564 (1962).
53. A. K. Rahman and K. Egger, *Z. Naturforsch., Teil C*, **28**, 434 (1973).
54. F. H. Foppen, *Chromatogr. Rev.*, **14**, 133 (1971).
55. L. A. Vakulova, V. P. Kuznetsova, F. B. Kolot, I. P. Bab'eva, and G. I. Samokhvalov, *Mikrobiologiya*, **33**, 1061 (1964).
56. C. Garside and J. P. Riley, *Anal. Chim. Acta*, **46**, 179 (1969).
57. B. Colman and W. Vishniac, *Biochim. Biophys. Acta*, **82**, 616 (1964).
58. M. D. Nutting, M. Voet, and R. Becker, *Anal. Chem.*, **37**, 445 (1965).
59. I. D. Jones, L. S. Butler, E. Gibbs, and R. C. White, *J. Chromatogr.*, **70**, 87 (1972).
60. R. J. Daley, C. B. J. Gray, and S. R. Brown, *J. Chromatogr.*, **76**, 175 (1973).
61. S. Shimizu, H. Fukushima, and E. Tamaki, *Phytochemistry*, **3**, 641 (1964).
62. M. H. Anwar, *J. Chem. Educ.*, **40**, 29 (1963).
63. M. F. Bacon, *J. Chromatogr.*, **17**, 322 (1965).
64. H. A. W. Schneider, *J. Chromatogr.*, **21**, 448 (1966).
65. H. Loewenschuss and P. J. Wakelyn, *Anal. Chim. Acta*, **63**, 230 (1973).
66. N. Nybom, *Physiol. Planatarum*, **17**, 157 (1964).
67. N. Nybom, *Fruchtsaft-Ind.*, **8**, 205 (1963).
68. R. R. Paris and M. Paris, *Bull. Soc. Chim. Fr.*, **1963**, 1597.
69. S. B. Gupta, *J. Chromatogr.*, **36**, 115 (1968).
70. D. B. Mullick, *J. Chromatogr.*, **39**, 291 (1969).
71. B. H. Barritt and L. C. Torre, *J. Chromatogr.*, **75**, 151 (1973).
72. R. E. Wrolstad, *J. Chromatogr.*, **37**, 542 (1968).
73. W. D. Loomis and J. Battaile, *Phytochemistry*, **5**, 423 (1966).
74. D. B. Mullick, *Phytochemistry*, **8**, 2003 (1969).
75. N. Nybom, *J. Chromatogr.*, **38**, 382 (1968).
76. L. Birkofer, C. Kaiser, H. A. Meyer-Stoll, and F. Suppan, *Z. Naturforsch.*, **17B**, 352 (1962).
77. H. Tanner, H. Rentschler, and G. Senn, *Mitt. (Klosterneuberg) Ser. A., Rebe Wein*, **13**, 156 (1963); through *Chem. Abstr.*, **59**, 13094 (1963).

78. D. Hess and C. Meyer, *Z. Naturforsch.*, **17B**, 853 (1962).
79. S. Asen, *J. Chromatogr.*, **18**, 602 (1965).
80. W. L. Stanley and S. H. Vannier, U.S. Pat. 2,889,337 (June 2, 1959).
81. W. L. Stanley and S. H. Vannier, *J. Am. Chem. Soc.*, **79**, 3488 (1957).
82. J. M. Miller and J. G. Kirchner, *Anal. Chem.*, **24**, 1480 (1952).
83. J. G. Kirchner, J. M. Miller, and G. J. Keller, *Anal. Chem.*, **23**, 420 (1951).
84. W. L. Stanley and S. H. Vannier, *J. Assoc. Off. Agric. Chem.*, **40**, 582 (1957).
85. S. H. Vannier and W. L. Stanley, *J. Assoc. Off. Agric. Chem.*, **41**, 432 (1958).
86. L. J. Swift, *J. Agric. Food Chem.*, **15**, 99 (1967).
87. L. Hoerhammer and H. Wagner, *Deut. Apoth.-Ztg.*, **102**, 759 (1962).
88. J. F. Fisher, H. E. Nordby, and T. J. Kew, *J. Food Sci.*, **31**, 947 (1966).
89. W. B. Davis, *Anal. Chem.*, **19**, 476 (1947).
90. L. Hoerhammer, H. Wagner, and K. Hein, *J. Chromatogr.*, **13**, 235 (1964).
91. L. Hoerhammer and H. Wagner, *Arzneim.-Forsch.*, **12**, 1002 (1962).
92. M. B. Duggan, *J. Agric. Food Chem.*, **17**, 1098 (1969).
93. M. B. Duggan, *J. Assoc. Off. Anal. Chem.*, **52**, 1038 (1969).
94. A. Jacquin-Dubreuil, *J. Chromatogr.*, **71**, 487 (1972).
95. P. Spiegl, *J. Chromatogr.*, **39**, 93 (1969).
96. L. Hoerhammer, H. Wagner, and B. Lay, *Pharmazie*, **15**, 645 (1960).
97. R. Tschesche, U. Schacht, and G. Legler, *Naturwissenschaften*, **50**, 521 (1963).
98. N. S. Vul'fson, V. I. Zaretskii, and L. S. Chetverikova, *Izv. Akad. Nauk SSSR, Ser. Khim.*, **1963**, 1503; through *Chem. Abstr.*, **59**, 15584 (1963).
99. T.-R. Hsu and F.-Y. Fuo, *Yao Hsueh Hsueh Pao*, **11**, 223 (1964); through *Chem. Abstr.*, **61**, 7337 (1964).
100. F. Weygand, H. Simon, H.-G. Floss, and U. Mothes, *Z. Naturforsch.*, **15b**, 765 (1960).
101. J. H. Copenhaver and M. J. Carver, *J. Chromatogr.*, **16**, 229 (1964).
102. V. T. Chernobai and D. G. Kolesnikov, *Dokl. Akad. Nauk SSSR*, **133**, 233 (1960); through *Chem. Abstr.*, **54**, 23188 (1960).
103. B. Goerlich, *Planta Medica*, **9**, 442 (1961).
104. A. Stambouli and R. R. Paris, *Ann. Pharm. Fr.*, **19**, 732 (1961).
105. P. Ribereau-Gayon, *Compt. Rend.*, **258**, 1335 (1964).
106. M. Gabor, B. Matkovics, and G. Gondos, *Pharmazie*, **19**, 785 (1964).
107. Y. M. Lin and F. C. Chen, *J. Chromatogr.*, **104**, D33 (1975).
108. P. Daenens and M. Van Boven, *J. Chromatogr.*, **57**, 319 (1971).
109. M. Trkovnik, M. Kuleš, I. Tabaković, and M. Zečević, *J. Chromatogr.*, **128**, 227 (1976).
110. J. Davídek, *Nahrung*, **4**, 661 (1960).
111. J. Davídek and E. Davídková, *Pharmazie*, **16**, 352 (1961).
112. J. Davídek and Z. Procházka, *Collect. Czech. Chem. Commun.*, **26**, 2947 (1961).
113. Y.-F. Chia, *Yao Hsueh Hsueh Pao*, **11**, 485 (1964).
114. K. Egger, *Z. Anal. Chem.*, **182**, 161 (1961).

115. P. R. Bhandari, *J. Chromatogr.*, **16,** 130 (1964).
116. J. Hubacek, *Collect. Czech. Chem. Commun.*, **35,** 3119 (1970).
117. L. Hoerhammer, H. Wagner, and B. Salfner, *Arzneim.-Forsch.*, **13,** 33 (1963).
118. M. Nishiura, S. Esaki, and S. Kamiya, *Agric. Biol. Chem. (Tokyo)*, **33,** 1109 (1969).
119. H. Grisebach, *Z. Naturforsch.*, **14b,** 802 (1959).
120. H. Grisebach and W. Barz, *Z. Naturforsch.*, **18b,** 466 (1963).
121. H. Grisebach and G. Brandner, *Z. Naturforsch.*, **16b,** 2 (1961).
122. H. Grisebach and L. Patschke, *Chem. Ber.*, **93,** 2326 (1960).
123. J. Guggolz, A. L. Livingston, and E. M. Bickoff, *J. Agric. Food Chem.*, **9,** 330 (1961).
124. A. L. Livingston, E. M. Bickoff, J. Guggolz, and C. R. Thompson, *J. Agric. Food Chem.*, **9,** 135 (1961).
125. R. L. Lyman, E. M. Bickoff, A. N. Booth, and A. L. Livingston, *Arch. Biochem. Biophys.*, **80,** 61 (1959).
126. R. L. Larson, *J. Chromatogr.*, **43,** 287 (1969).
127. L. C. Wang, *Anal. Biochem.*, **42,** 296 (1971).
128. R. Haensel, L. Langhammer, J. Frenzel, and G. Ranft, *J. Chromatogr.*, **11,** 369 (1963).
129. H. P. Kaufmann and A. W. El Baya, *Fette, Seifen, Anstrichm.*, **72,** 372 (1970).
130. J. Jensen, *J. Chromatogr.*, **10,** 236 (1963).
131. M. Doss, B. Ulshoefer, and R. Quast, *J. Chromatogr.*, **41,** 386 (1969).
132. M. Doss, *Z. Anal. Chem.*, **252,** 104 (1970).
133. M. Doss, *Z. Physiol. Chem.*, **350,** 499 (1969).
134. M. Doss, *Z. Klin. Chem. Klin. Biochem.*, **8,** 208 (1970).
135. *Ibid.*, p. 197.
136. M. Doss, B. Ulshoefer, and W. K. Philipp-Dormston, *J. Chromatogr.*, **63,** 113 (1971).
137. H. Mundschenk and J. Fischer, *Z. Klin. Chem. Klin. Biochem.*, **7,** 325 (1969).
138. H. Mundschenk, *J. Chromatogr.*, **25,** 380 (1966).
139. *Ibid.*, **37,** 431 (1968).
140. *Ibid.*, **40,** 393 (1969).
141. N. Ellfolk and G. Sievers, *J. Chromatogr.*, **25,** 373 (1966).
142. G. H. Elder, *J. Chromatogr.*, **59,** 234 (1971).
143. M. Gajdos-Torok, *Bull. Soc. Chim. Biol.*, **50,** 925 (1968).
144. M. Yuan and C. S. Russell, *J. Chromatogr.*, **87,** 562 (1973).
145. C. S. Russell, *J. Chromatogr.*, **25,** 163 (1966).
146. A. M. D. C. Batlle and A. Benson, *J. Chromatogr.*, **25,** 117 (1966).
147. T. K. With, *Clin. Biochem.*, **1,** 30 (1967).
148. T. K. With, *J. Chromatogr.*, **42,** 389 (1969).
149. R. V. Belcher, S. G. Smith, R. Mahler, and J. Campbell, *J. Chromatogr.*, **53,** 279 (1970).
150. L. Eriksen, *Scand. J. Clin. Lab. Invest.*, **10,** 39 (1958).

151. G. Efremov, B. Vaskov, H. Duma, and M. Andrejeva, *Acta Med. Iugosl.*, **17**, 252 (1963); through *Chem. Abstr.*, **61**, 12305 (1964).
152. R. W. Carrell, H. Lehmann, P. A. Lorkin, E. Raik, and E. Hunter, *Nature*, **215**, 626 (1967).
153. T. H. J. Huisman and R. N. Wrightstone, *J. Chromatogr.*, **92**, 391 (1974).
154. R. Dalgelite, L. Juhnjaviciute, and V. Vaiciuvenas, *Lab. Delo*, **9**, 5 (1963).
155. M. van Sande and G. van Ros, *Ann. Soc. Belg. Med. Trop.*, **43**, 537 (1963).
156. V. J. Yakulis, P. Heller, A. M. Josephson, L. Singer, and L. Hall, *Am. J. Clin. Pathol.*, **34**, 28 (1960).
157. J. E. Cradock-Watson, *Nature*, **215**, 630 (1967).
158. E. W. Baur, *J. Lab. Clin. Med.*, **61**, 166 (1963).
159. E. W. Baur, *Clin. Chim. Acta*, **9**, 252 (1964).
160. E. W. Baur, N. M. Rowley, and A. G. Motulsky, *Annual Meeting of the American Society of Human Genetics, Boulder, August 26–28, 1964*.
161. P. Berkeš-Tomašević, J. Rosić, and I. Berkeš, *Acta Pharm. Jugosl.*, **13**, 69 (1963).
162. W. G. Dangerfield, *Nature*, **202**, 520 (1964).
163. C. L. Marsh, C. R. Jolliff, and L. C. Payne, *Tech. Bull. Regist. Med. Technol.*, **34**, 1 (1964).
164. W. A. Schroeder and N. C. Nelson, *J. Chromatogr.*, **115**, 527 (1975).
165. P. O'Carra and E. Colleran, *J. Chromatogr.*, **50**, 458 (1970).
166. *Ibid.*, **108**, 212 (1975).
167. C. C. Kuenzle, J. R. Reuttner, and C. H. Eugster, *Helv. Chim. Acta*, **53**, 1838 (1970).
168. C. S. Berry, J. E. Zarembo, and J. D. Ostrow, *Biochem. Biophys. Res. Commun.*, **49**, 1366 (1972).
169. Z. J. Petryka and C. J. Watson, *J. Chromatogr.*, **37**, 76 (1968).
170. R. Segura-Cardona, *Rev. Espan. Fisiol.*, **22**, 41 (1966).
171. Z. J. Petryka, *J. Chromatogr.*, **50**, 447 (1970).
172. R. Tenhunen, *Acta Chem. Scand.*, **17**, 2127 (1963).

Chapter **XXIX**

STEROIDS

1 STEROLS AND STEROL ESTERS

Separation of Free Sterols

The sterols are saturated and unsaturated alcohols derived from the basic hydrocarbon cholestane. They are widely distributed in nature, occuring both free and combined as esters or glycosides. Because of their close chemical relationship and because more than one sterol is usually present at one time, their separation is often quite difficult.

Avigan et al. [1] separated some sterols on silica gel G prestained with Rhodamine 6G using benzene–ethyl acetate (5:1) in a development distance of 20 cm. The separation of $\Delta^{5,7}$-cholestadienol from the monounsaturated cholestenols and desmosterol, which could not be separated in this system, was effected by using a silver nitrate-impregnated silica gel layer with the same solvent system. To separate cholesterol, desmosterol, 24,25-dihydrolanosterol, and lanosterol it was necessary to acetylate the compounds before chromatographing. Separation was then achieved on 40-cm-long plates using hexane–benzene (5:1) as a solvent. The plates in the latter case were prepared with Mallinckrodt silicic acid which had been screened to pass a 325-mesh sieve because this gave a better resolution than silica gel G. Five percent plaster of Paris was used as a binder. Recently, Claude [2] has used the same system to separate the propionylated products of cholesterol, desmosterol, and 5-dihydrocholesterol on 20 × 20 cm plates. There are two advantages to using this derivative rather than the acetate: (1) the derivative can be prepared immediately by reaction with propionyl chloride in contrast to the 12-hr reaction time for the acetylation in pyridine, and (2) the propionates permit a faster and better separation than the corresponding acetates. Desmosterol and cholesterol have also been separated by continuous horizontal chromatography on silver nitrate-impregnated silica gel using benzene–ethyl acetate (95:5) as a solvent [3].

Copius-Peereboom and Beekes [4] have carried out separations on kieselguhr G using cyclohexane–ethyl acetate (99:5:0.5) as the developing solvent. The compounds were detected by spraying with phosphomolybdic acid. Since the separation of cholesterol from the phytosterols could not be achieved in this system, a reverse-phase separation using kieselguhr impregnated with 10% undecane was applied for this purpose. A satisfactory solvent for the reverse-phase system was acetic acid–water (9:1) saturated with undecane. Acetates of the sterols were separated using the same solvent system in a 92:8 ratio. Since the critical pair, cholesterol–brassicasterol, could not be separated in a reverse-phase system, the brominating technique of Kaufmann et al. [5–7] was employed. By the addition of 0.5% bromine to the solvent, brassicasterol acetate was brominated during the development and showed an R_{st} value of 1.19 referred to cholesterol acetate. Ikan et al. [8] have also used a bromination technique prior to chromatographing to separate sterols from the corresponding stanols. Use of the Kaufmann procedure is much more convenient, however, since it eliminates a separate bromination step. Another way to separate the sterols from the corresponding stanols is to use silver nitrate-impregnated silica gel layers [9]. Separation is achieved with chloroform as a solvent. Table 29.1 shows the R_f values of the stanols and the brominated sterols in six different systems compared to the separation on silver nitrate-treated layers. Idler and Wiseman [10] separated sterol bromides with benzene–ethyl acetate (2:1) on silica gel.

Cargill [11] has used both adsorption and partition thin-layer chromatography on silica gel. For adsorption, mixtures of benzene–ethyl acetate (2:1 and 19:1) were used. The partitioning solvents consisted of heptane with a stationary phase of 2-phenoxyethanol or 2-methoxyethanol (applied as 15% solutions in acetone), methanol with a liquid paraffin stationary phase (0.5% in ether), or undecane (15% in petroleum ether 40 to 60°C). Methanol–ether (49:1) was also used with the undecane stationary phase. Bromination, to separate cholesterol from cholestanol, was accomplished by adding bromine in chloroform directly to the sample spot.

Wientjens et al. [12] applied vapor programming to the separation of sterols. Kartnig and Mikula [13] used a magnesium oxide layer to separate free sterols with cyclohexane–diethyl ether–acetic acid (20:79.5:0.5) using a 15-cm development. By increasing the distance to 25 cm the "critical pairs" of campesterol, cholesterol, and β-sitosterol could be separated. Palmitate, stearate, and acetate esters were also separated by using petroleum ether (40 to 60°C)–acetone (98:2). Nicolaides [14] found that magnesium oxide could be used to separate wax esters from sterol esters; the separation appeared to take place according to the degree of flatness of the molecules. A solvent of 1% acetone in hexane was used.

Barbier et al. [15] have applied silica gel G layers with cyclohexane–ethyl acetate (7:3 and 17:3) in the separation of a group of sterols. Ikan et al. [8] have applied alumina G with benzene–ethanol (19:0.4) to free sterols. Table 29.2 gives a collection of R_f and R_{st} values in various systems for a group of sterols.

Another method for separating nuclear saturated from unsaturated sterols is to oxidize the unsaturated compound to a more polar compound. Azarnoff and Tucker [16] have accomplished this by disolving the sample in 1 to 2 ml of chloroform and epoxidizing with a fivefold molar excess of *m*-chloroperbenzoic acid. After the mixture was allowed to react for 30 min at room temperature, 4 ml of diethyl ether was added and the mixture was washed with 10% sodium bicarbonate solution until the bubbling ceased. The process was completed by washing with aqueous saturated sodium chloride and drying over anhydrous sodium sulfate. Sterols were recovered by evaporating the chloroform and transferring to benzene for application to the thin-layer plate. As an example, cholesterol was separated from β-cholestanol after epoxidation by chromatographing on silica gel layers dried at 110°C for 60 min, using as a

Table 29.1 $R_f \times 100$ Values of Stanols and Brominated Sterols in Numerous Systems[a] Compared to Separation of Stanols and Sterols on Silver Nitrate-Imgregnated Silica Gel [8, 9]

	Benzene–Ethyl Acetate (2:1)	*Benzene–Ethyl Acetate (4:1)*		*Benzene–Ethanol (19:0.2)*	*Benzene–Ethyl (19:0.4)*		*Chloroform*	
Sterol	*Silica Gel G*	*Silica Gel G*	*Alumina G*	*Silica Gel G*	*Silica Gel G*	*Alumina G*	*Silica Gel G*	*$AgNO_3$-Treated Silica Gel G*
Campesterol						69[b]	25	23
Campestanol						51	25	26
Cholesterol	81[b]	78[b]	63[b]	31[b]	44[b]	68[b]	25	23
Allocholesterol	95[b]	92[b]	77[b]	43[b]	58[b]	82[b]	29	28
Desmosterol	82[b]	79[b]			47[b]	69[b]	25	14
Cholestanol	62	60	44	19	27	51	25	26
Coprostanol	76	66				69	35	42
β-Sitosterol	79[b]	80[b]	63[b]	32[b]	43[b]	68[b]	25	23
Stigmasterol	80[b]	76[b]	62[b]		44[b]	68[b]	25	23
β-Sitostanol	60	59	44	18	27	50	25	32
Lanosterol	57[b]	77[b]			59[b]	81[b]	42	41
Dihydrolanosterol	50[b]	55[b]				65[b]	42	45
Agnosterol	100	100				94	92	89
Dihydroagnosterol							92	95

[a] All systems saturated; development distance 12 cm, except for chloroform for which development distance is 15 cm.
[b] As brominated derivatives.

Table 29.2 Some R_{st} and $R_f \times 100$ Values of Sterols[a]

Sterol	Kieselguhr G[b] [4]	Kieselguhr G Impregnated with 10% Decane in Petroleum Ether[b] [4]		Aluminum Oxide G[c] [8]
	A	B	C	D
Cholesterol	100	100	100	50
β-Sitosterol	100	86	83	50
Stigmasterol	100	93	91	52
Dihydrocholesterol	93	90	89	51
Ergosterol	89	116	122	52
7-Dehydrocholesterol	93	112	126	
Zymosterol	102			
Dihydrolanosterol	138			77
Lanosterol	137	84	97	78
Agnosterol	135	76	86	95
Dihydroagnosterol	134			
Brassicasterol		100	100	
Epicholesterol		90	116	
Coprostanol				69
Allocholesterol				53
21-Norcholesterol				50
Desmosterol				50
γ-Sitostanol				49
Campesterol				50
Campestanol				51

[a] Solvent: A = cyclohexane–ethyl acetate (95.5:0.5), development distance 20 cm; B = acetic acid–water (9:1) saturated with undecane at 22 to 24°C, development distance 20 cm; C = acetic acid–water (92:8) saturated with undecane at 22 to 24°C, development distance 20 cm; D = benzene–ethanol (19:0.4), development distance 12 cm.

[b] R_{st} values referred to cholesterol.

[c] R_f values × 100.

[d] Sterol acetates referred to cholesterol acetate.

separating solvent benzene–butyl acetate–butanone (75:25:10). By the use of radioactive cholesterol, it was demonstrated that the epoxidation was nearly quantitative. In oxidizing desmosterol the resulting epoxide had an R_f value similar to cholesterol epoxide indicating that the 24,25-double bond was not oxidized under the conditions that were used. On the other hand, the peroxidation of lanosterol yielded two new spots on chromatographing the mixture, thus indicating the possibility that both the 24,25-double bond and the 8,9-ring double bond had been subjected to oxidation. Smith et al. [17] have examined the autoxidation products of cholesterol. Ethyl acetate–heptane (1:1) was used as a solvent on silica gel layers. Multiple development was used to increase the resolution of 25-hydroxycholesterol and 7-oxocholesterol. Two-dimensional separation with the first development with the above-mentioned solvent and with acetone–heptane (1:1) in the second dimension resolved over 30 components from the oxidation of cholesterol samples. Seven of the compounds were identified by their R_f values and the color test with sulfuric acid in comparison with reference sterols. The sterol hydroperoxides have also been recognized by their in situ reduction on the silica gel layer with sodium borohydride and a subsequent development [18]. The reduced sterol alcohol was more polar than the corresponding hydroperoxide. Neuwald and Fetting [19] examined the oxidation products of cholesterol by chromatographing on silica gel G plates with a chloroform solvent. The peroxides were visualized with test T-207. The peroxides appeared as blue spots with an R_f value less than 0.32 which is the R_f value for peroxide-free cholesterol in this solvent. The cholesterol could be detected as a dark red spot by spraying with 50% antimony trichloride in acetic acid and drying for 10 min at 100°C. Acker and Greve [20] studied the radiation-induced oxidation of cholesterol in egg-dough products. Horvath [21] separated 22 autoxidation products of cholesterol on silica gel H using chloroform–acetone (4:1) using a wedge-layer technique.

Bennett and Heftmann [22] have studied the effect of structural differences remote from polar groups on the separability of sterols. Various combinations of solvents were tried; for example, it was found that cholestane and Δ^{16}-cholestene could be separated with isooctane–carbon tetrachloride (19:1) but not with more polar solvents. Compounds differing in degree or position of unsaturation in ring B were separable in more polar systems. With cyclohexane–ethyl acetate–water (600:400:1) cholesterol could be separated from 7-dehydrocholesterol.

Cholesterol and its esters in serum lipids have been the object of a number of investigations [23-28]. Zoellner et al. [25] have determined the cholesterol esters in as little as 0.4 ml of serum. Five cholesterol esters could be differentiated, with the free cholesterol remaining at the starting point, by using a multiple development on silicic acid with carbon tetrachloride. Beukers et al. [27] used petroleum ether (60 to 80°C)–ethyl ether (4:1) for the separation and determination of free and esterified cholesterol in serum. The lipid areas arranged according to increasing R_f values on silica gel were phospholipids, free cholesterol, triglycerides, and cholesterol esters. Claude and Beaumont [28] combined fractionation on silica gel columns with thin-layer chromatography to study the serum sterols. The fractions containing free and esterified sterols were combined, hydrolyzed, and then converted to the acetates which were separated on silica gel layers impregnated with 13.3% silver nitrate. Hexane–benzene (5:1) was used as a solvent. Progesterone, squalene, 7α-hydroxycholesterol, and cholesterol and its esters were separated on silica gel GF with benzene–ethyl acetate (9:1).

Herz and González [29] separated a number of monohydroxylated cholestanols of the 5α and 5β series on silica gel G with benzene as a solvent. The cholestanones were separated using a double development.

Johnson et al. [30] have demonstrated the presence of cholesterol in higher plants (*Solanum tuberosum* and *Dioscorea spiculiflora*). Purification was effected by a combination of gas chromatography and thin-layer chromatography.

Copius-Peereboom [31] has examined the use of silica gel G, kieselguhr G, and aluminum oxide with various solvent mixtures for the separation in various oils and fats, including pumpkin seed oil. Gad et al. [32] isolated β-sitosterol from Egyptian *Nigella sativa* oil. Kiribuchi et al. [33] analyzed the sterols in soybeans, peanuts, sesame seeds, rapeseed, cotton seeds, and safflower seeds. Silica gel with chloroform–methanol–acetic acid–water (90:8:1:1) was used to separate the sterols after a preliminary fractionation on Florisil columns. Nagasampagi et al. [34] identified the sterols of coffee and Gawienowski and Gibbs [35] isolated cholesterol and progesterone from apple seeds. Wolff et al. [36] used thin-layer chromatography on alumina with benzene–acetone (19:1) as a solvent to identify the oils in paint media by separating the sterols in the unsaponifiable matter.

Silica gel impregnated with silver nitrate has been used to separate a variety of cholestanol isomers and cholesterol using hexane–benzene (97:3) as a solvent [37]. The same adsorbent has been used to separate the free sterols of yeast using chloroform–acetone (95:5) as a solvent. The same solvent gives similar results on neutral alumina impregnated with silver nitrate [39]. In the latter case desmosterol was separated from 7-dehydrocholesterol using chloroform–acetone (13:7).

Using acetic acid–acetonitrile (1:1) that was 70% saturated with undecane, Wolfman and Sachs [40] separated desmosterol, cholesterol, and other sterols by reverse-phase chromatography on silica gel impregnated with a 15% solution of undecane in petroleum ether.

Seher and Homberg [41] improved the separation of sterols by using the 3,5-dinitrobenzoates. Wedge-strip layers of magnesium oxide–alumina–calcium sulfate (15:5:1) were used with hexane–ethyl acetate (9:1) and (23:2), and separations were made according to carbon atom number as well as the number and position of the double bonds.

The sterol content of soft wheat was determined by thin-layer chromatography [42]. No qualitative difference was found among the seven species examined.

Schreiber and Osske [43] isolated α-sitosterol from the leaves of the potato plant *Solanum tuberosum*. β-Sitosterol was isolated from both potato leaves and from the potato beetle *Leptinotarsa decemlineata* [44]. Lavie and Kaye [45] isolated β-sitosterol and β-sitostenone from *Quassia amara* wood and also from the wood of a tree from the genus *Cabralia*. Ikan and Kashman [46] have found both β-sitosterol and β-sitostanol in Hula peat. Wenzel et al. [47] used the Wilzbach technique [48, 49] to label cholesterol with tritium. Purification was accomplished by chromatographing on silica gel with cyclohexane–ethyl acetate (7:3). The radioactivity was measured on the plate with a gas-flow counter.

Richter [50] recommends the use of naphthoquinone-perchloric acid (T-172) for the detection of sterols. It is a sensitive reagent and can detect 0.03 μg of cholesterol. After the chromatogram is sprayed with this reagent the plates are heated at 70 to 80°C. The sterols appear as pink spots which gradually turn blue. Continued heating will finally convert this to a brown-black, as will a higher temperature whereby the intermediate color stages are omitted.

Separation of Sterol Esters

Sterols occur in nature both in the free form and as esters, and it is important in this respect to have reference compounds of the pure esters. Mahadevan and Lundberg [51] have prepared long-chain fatty acid esters of cholesterol by ester interchange between cholesteryl acetate and the fatty acid methyl esters. In general, they proceeded as follows: a mixture of sodium ethylate (0.05 to 0.1 g), 0.01 mole of cholesteryl acetate, and 0.01 mole of the methyl ester was placed in a 200 ml, round-bottom, ground-neck flask. After the flask was flushed with nitrogen, the contents were heated under a vacuum of 20 to 30 mm at 80 to 90°C for 1 hr with occasional gentle mixing. After cooling, the reaction mixture was washed with petroleum ether and filtered to remove insoluble materials. It was evaporated to dryness and the fatty acid ester was recrystallized from an appropriate solvent. The esters which they prepared were chromatographed on silica gel G using petroleum ether (60 to 80°C)–benzene (3:2). Kaufmann et al. [52] and Bergmann et al. [53] have prepared sterol esters by refluxing the sterol with the fatty acid using *p*-toluenesulfonic acid as a catalyst. As an example, the general procedure of Kaufmann et al. can be given: a mixture of 0.03 mole of cholesterol, 0.04 mole of fatty acid, and 1% of *p*-toluenesulfonic acid in 100 ml of benzene is refluxed for 3 to 4 hr at 80 to 85°C on a water bath. After completion of the reaction,

80% of the benzene is removed under vacuum and the residue is dissolved in an equal quantity of ether. The solution is washed with sodium bicarbonate solution and then with water, and finally dried over sodium sulfate for 12 hr. After removing the ether, the ester is purified by recrystallization from ethanol, acetone, and ether–acetone (1:3). For the preparation of radioactive cholesterol esters, Pelick et al. [54] used the ester-interchange method of Mahadevan and Lundberg [51]; as an example, cholesterol acetate was treated with methyl palmitate-1-^{14}C using sodium methoxide as a catalyst. Thin-layer chromatography on silica gel with petroleum ether–benzene (3:2) was used to purify the product.

Van Dam et al. [55] chromatographed fatty acid esters of cholesterol as well as the weakly polar 3-β-acetoxy steroid ketones on silica gel G in 13 solvent systems using both chromatostrips and plates. The R_f values were tabulated. In working with the serum lipids, Weicker [26] separated the stearic, oleic, and butyric esters of cholesterol on silica gel using carbon tetrachloride to obtain R_f values of 0.5, 0.45, and 0.30, respectively. This solvent allowed them to separate the esters from other lipids which remained at the origin. Zoellner and co-workers [25] used the same separating system but increased the resolution by repeated development with the carbon tetrachloride.

Bergmann et al. [53] obtained the R_f values for 15 β-sitosterol esters in six different solvents on silica gel G (Table 29.3). To separate mixtures, two-dimensional chromatography was used with cyclohexane–benzene (2:1) in the first direction and carbon tetrachloride–chloroform (19:1) in the second direction.

Kaufmann and Makus [52] obtained better separation of cholesterol esters by using two-dimensional chromatography in which the first dimension was carried out by adsorption chromatography and the second dimension by reverse-phase chromatography on treated plates. Separation in the first direction was carried out with a mixture of hexane–Decalin (7.5:2.5), and then the unused portion of the plate was immersed in a 5% solution of paraffin oil in petroleum ether. After removal of the solvent, methyl ethyl ketone–acetonitrile (7:3) was used at right angles to the first development. Michalec et al. [56] also obtained excellent separations of cholesterol esters on silicic acid layers impregnated with 0.5% paraffin oil in ether. The solvent in this case was acetic acid. Copius-Perreboom [57, 58] preferred undecane as an impregnating solvent because of the difficulty in removing paraffin and silicone oils that interfered with the detection of minor components in some sterol acetate mixtures. Kieselguhr G layers were impregnated with a 10% solution of undecane (190 to 220°C) in light petroleum ether, and after evaporation of the solvent the plates were shaped for the wedge-strip technique.

Table 29.3 $R_f \times 100$ Values of β-Sitosterol Esters on Silica Gel G in Various Solvents[a]

β-Sitosterol Ester	*Cyclo-hexane–Benzene* (1:1)	*Cyclo-hexane–Benzene* (2:1)	*Cyclo-hexane–Benzene* (4:1)	*Carbon Tetra-chloride–Chloro-form* (19:1)	*n-Hep-tane–Ethyl Acetate* (19:1)	*Chloro-form–Acetone* (19:1)
Acetate	40	22	19	18	48	83
Propionate	53	33	26	25	53	84
Butyrate	55	36	29	26	56	85
Caproate	70	41	33	31	59	86
Caprylate	75	46	39	37	60	88
Pelargonate	74	48	40	36	59	87
Caprate	76	50	42	38	62	88
10-Undecenoate	70	46	35	34	61	87
Laurate	80	53	43	40	63	90
Myristate	82	54	44	41	64	90
Palmitate	85	54	47	43	65	91
Stearate	88	55	50	44	66	92
Oleate	86	53	45	42	65	91
Linoleate	89	56	52	45	67	92
Arachidonate	92	58	54	49	69	93

[a] From E. A. Bergmann, R. Ikan, and S. Harel [53]; reproduced with permission of the authors and The Elsevier Publishing Co.

Acetic acid-acetonitrile (1:3) saturated with undecane was used as a solvent. The bromination technique of Kaufmann et al. [6] was employed for the separation of critical pairs by adding 0.5% bromine to the solvent. The resulting brominated sterols appeared as bright blue bands when sprayed with 50% antimony trichloride in acetic acid. Under these conditions the $\Delta^{5(6)}$ double bond did not brominate. Compounds with conjugated double bonds in the nucleus were completely decomposed and showed only a bright blue band at the solvent front. Nonconjugated bonds in positions other than the $\Delta^{5(6)}$ were also similarly attacked with a resulting blue spot at the solvent front. For assistance in the separation of unsaturated esters, silver nitrate–impregnated layers were used [58] with chloroform–ether–acetic acid (97:2.3:0.5) and chloroform–light petroleum ether–acetic acid (25:75:0.5) as solvents. This method has been applied to the phytosterols of pumpkin seeds [59]. Capella et al. [60] also used silver nitrate-treated layers for triterpene alcohol and sterol mixtures.

Bennett and Heftmann [61] have separated β-sitosteryl acetate, cholesteryl acetate, and stigmasteryl acetate on Anasil B (Analabs) plates using a continuous development with hexane–ether (97:3) for a period of 120 min. Lees et al. [62] used continuous horizontal thin-layer chromatography to separate the acetates of cholesterol and desmosterol on silica gel G with benzene–hexane (1:3). Seher and Homberg [63] used a wedge-strip mixed layer of magnesium oxide–alumina–calcium sulfate (15:5:1) with hexane–diethyl ether (19:1) for separating sterol acetates isolated from vegetable oils. Good separation was achieved and avoided the formation of "critical pairs." Sterol acetates have also been chromatographed on silica gel H impregnated with silver nitrate using benzene–hexane (3:5) as a solvent [64]. Compounds differing in the number of double bonds in the steroid nucleus or side chain were separated. Since the sterol pairs β-sitosterol and stigmasterol, Δ^7-ergosterol and 5-dihydroergosterol, lanosterol and agnosterol, and 7-dehydrocholesterol and ergosterol are difficult or impossible to separate on silver nitrate-impregnated silica gel in the usual solvents, Copius-Peereboom [65] separated them by converting to the acetates and using chloroform–petroleum ether (60 to 80°C)–acetic acid (25:75:0.5) with these layers.

Morris [66] separated human cholesterol esters into seven components in the order of saturation of the fatty acids on silver nitrate-impregnated layers with solvent systems of ether and ether–hexane (1:4).

Haahti and Nikkari [67] have investigated the sterol esters and the waxes of skin surface fat. The sterol esters were separated from other components on aluminum oxide layers with 1% benzene in hexane. By using silver nitrate-impregnated silica gel, the sterol esters were separated into three fractions: one saturated and two unsaturated [68]. Horlich and Avigan [69] investigated the sterols of the skin of normal and tripanol-treated rats.

Quantitative Determination of Sterols and Sterol Esters

A number of methods have been used for determining cholesterol and its esters. Purdy and Truter [70] applied their spot area methods (see Chapter XI, Section 3) for quantitative analysis to the determination of cholesterol esters, and to the analysis of the cholesterol content of wool-wax alcohols. In order to avoid interference from other aliphatic alcohols present in the extract, the plates were sprayed with Liebermann–Buchard reagent which was prepared by adding 10 ml of ice-cold concentrated sulfuric acid to a cold mixture of 50 ml of chloroform and 50 ml of acetic anhydride. The comparison was very good between the spot area method and both the spectroscopic and the digitonin methods. Zoellner [25, 71, 72] has examined a number of quantitative methods for determining cholesterol and cholesterol esters. In one method the plate was evenly sprayed with a 25% antimony trichloride solution in chloroform and then heated 3 to 5 min at 110°C. The colored spots were then measured directly with a photometer, and the areas under the photometric curve were carefully measured with a planimeter. The zones may also be removed from the thin-layer plates and their cholesterol content determined by the method of Zak et al. [73]. The latter method depends on the quantity of the adsorption material and the speed at which the adsorbent is centrifuged from the solution. It is very useful, however for quick orientation results. Kaufmann and Viswanathan [74] described three methods of quantitatively determining cholesterol and its esters after a separation on silica gel layers. Using the Liebermann–Burchard reaction, the sample is shaken with 1 ml of chloroform, 2ml of acetic anhydride, and 0.25 ml of a freshly prepared mixture of acetic acid and sulfuric acid (9:1). The stoppered mixture is then allowed to stand 13 min at 35 ± 1°C. Afterwards, the extinction is measured at 660 mμ. In using Tschugaeff's reagent, the sample and 3 ml of chloroform are mixed with 3 ml of freshly prepared solution of a 2:1 mixture of 20% zinc chloride (in acetic acid) and acetyl chloride. After standing for 3 hr in the dark, the mixture is diluted to 10 ml with chloroform and measured between 520 and 530 mμ. The method is linear between 0.5 and 100 μg and is sensitive to 0.5 μg. The ferric chloride–hexahydrate method used by Kaufmann and Viswanathan in their work was the method of Zlatkis et al. [75] as modified by Quaife et al. [76]. In this case the sample was dissolved in 3 ml of 0.05% ferric chloride hexahydrate in acetic acid and then mixed with 2 ml of concentrated sulfuric acid. After standing 30 min at room temperature, the absorption was measured at 560 mμ. Kaufmann and Viswanathan [74] applied the method to the determination of cholesterol and its esters in human tissues of patients suffering from various types of diseases. Vahouny et al. [77] also used the ferric chloride method for the determination of cholesterol, and found a good comparison with the values obtained with a radioactive method. Ksenofontova et al. [78] applied the ferric chloride method to the determination of β-sitosterol in phytosterols after precipitation with digitonin and separation on an alumina layer.

Samuel et al. [79] determined fecal cholesterol and coprosterol by radioactive assay using both ^{14}C and tritium. The separations were accomplished on silica gel using toluene–ethyl acetate (9:1) as a solvent with a development distance of 12 cm. The following R_f values were obtained: cholesterol 0.30, coprosterol 0.40, epicoprosterol 0.42, coprostenone 0.41, Δ^7-cholesten-3β-ol 0.26, cholestanol 0.27, 7-dehydrocholesterol 0.28, β-sitosterol 0.30, stigmasterol 0.31, and 4–cholesten-3-one-0.50.

Nambara et al. [80] has determined cholestan-3α-ol in the presence of cholestan-3β-ol and cholesterol by chromatographing on silica gel G with hexane–ethyl acetate (4:1) as the developing solvent. The quantitative determination was made by first converting the cholestan-3α-ol to the 3,5-dinitrobenzoyl ester by treating the eluate with 3,5-dinitrobenzoyl chloride in pyridine. The ester which was extracted from the mixture with hexane was then treated with ethylenediamine and dimethylformamide to obtain a colored solution which was measured at 528 mμ. Ivanov et al. [81] determined phytosterols colorimetrically with sulfosalicylic acid after separation on silica gel with petroleum ether–ethyl ether (1:1).

Bondjers and Bjoerkerud [82] used a fluorometric method after elution for the determination of nanogram quantities of cholesterol and cholesterol esters. A modified Tschugaeff reagent and conditions for the reaction was necessary in order to obtain good fluorescence. Good elution of the compounds from the silica gel could be obtained by repeating the extraction three times.

Gas chromatography has also been used for the quantitative determination of thin-layer spots of sterols and their esters. Brandt and Benvista [83] applied this method to the sterols in bean leaves. Parodi [84] used the gas chromatographic determination of sterols separated by thin-layer chromatography to detect the adulteration of butterfat with vegetable oils, and Thrope et al. [85, 86] applied this same principle to the detection of adulteration of vegetable oils with animal fat. Rohmer and Brandt [87] developed a procedure for the systematic analysis of sterols and triterpenes using thin-layer chromatography, gas chromatography, and mass spectrometry.

2 THE C_{18} STEROIDS

Qualitative Separations

The estrogens are derivatives of the hydrocarbon estrane. They are phenolic in nature and occur in the urine of both sexes.

Lisboa and Diczfalusy [88] have published a method for the separation and characterization of 24 steroid estrogens. The separations were carried out on silica gel G layers which were dried at 105°C for 30 min. The following six solvent systems were employed in the separations: I, ethyl acetate–cyclohexane–ethanol (9:9:2); II, ethyl acetate–(water-saturated *n*-hexane)–ethanol (16:3:1); III, ethyl acetate-cyclohexane (1:1); IV, chloroform–ethanol (9:1); V, ethyl acetate–(washer-saturated *n*-hexane)–ethanol–acetic acid (72:13.5:4.5:10); and VI, water-saturated *n*-butanol. A preliminary characterization was obtained by chromatographing in ethyl acetate–cyclohexane–ethanol (9:9:2). From the results of this separation the estrogens were divided into two groups, the more polar group (lower R_f values) and the less polar group (higher R_f values). Elution from the first separation could be obtained with ethanol and after concentrating the eluate it could then be applied to other plates to complete the separations. The more polar group of compounds could be separated in a two-dimensional system using a solvent mixture consisting of ethyl acetate–(water-saturated *n*-hexane)–ethanol (16:3:1) in both directions. Somewhat better resolution could be obtained, however, by developing first in one direction with this solvent system and then in the second direction with ethyl acetate–(water-saturated *n*-hexane)–ethanol–acetic acid (72:13.5:4.5:10), as illustrated in Fig. 29.1. Nineteen of the less polar estrogens obtained in the initial separation were rechromatographed on a two-dimensional system using ethyl acetate–cyclohexane (1:1). From the results of this separation, the 19 compounds were divided into eight different groups for further separation. The first group of estrone and 2-methoxyestrone were separated by two-dimensional chromatography using solvent systems I and III. A confirmational separation could be carried out by reducing the mixture with potassium fluoborate and then separating the 17β-estradiol and 2-methoxy-17β-estradiol that were formed in the solvent systems II and III. The two original compounds could also be identified by forming the nitroso complex directly on the thin layer using the method of Boute [89]. On chromatographing the resulting complexes in system II, the estrone complex had an R_f value of 0.30 compared to the 0.0 of the 2-methoxy derivative. The second group consisting of 16-oxoestrone and 17β-estradiol was chromatographed in solvent I or II and the reduction products could be readily separated in solvent system III. The third group consisting of 7-oxoestrone and 2-methoxy-17β-estradiol could not be separated directly. However, by forming the Girard complex of the 7-oxoestrone it could be readily separated from the accompanying compound by chromatographing in VI; the R_f values for the two hydrazones, one at 0.07 and the other at 0.10, contrasted with the R_f value of 0.66 for the 2-methoxy-17β-estradiol. The four compounds of the fourth group, which includes 6-oxoestrone, 16α-hydroxyestrone, 16β-hydroxyestrone, and 16-oxo-17β-estradiol could not be differentiated by one single system. Solvent system III gave a separation of 16β-hydroxyestrone from the other three compounds. 16-Oxo-17β-estradiol could be identified by its yellow color in the Kaegi–

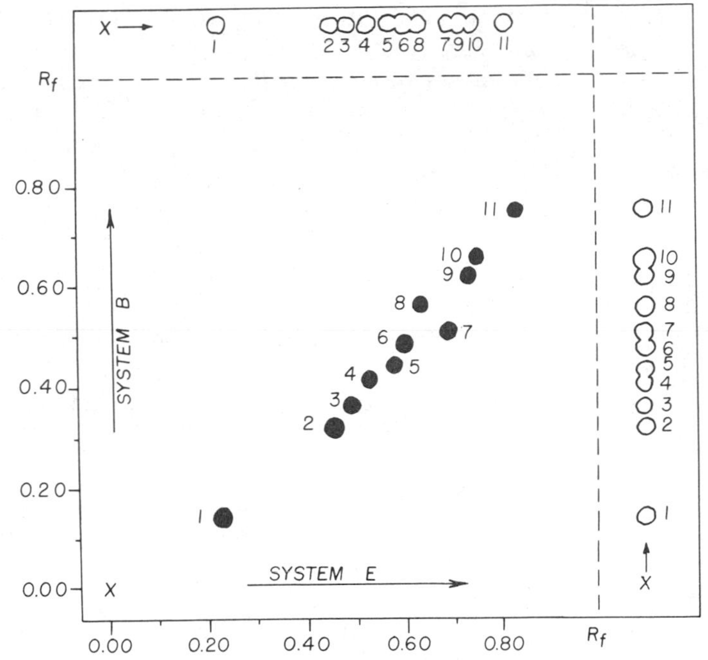

Fig. 29.1 Separation of 4- to 5-μg amounts of polar estrogens by means of two-dimensional thin-layer chromatography in systems *B* (ethyl acetate–water-saturated *n*-hexane–absolute ethanol, 80:15:5) and *E* (ethyl acetate–water-saturated *n*-hexane–absolute ethanol–glacial acetic, 72:13.5:4.5:10). Silica gel G with complete chamber saturation, *1*, 6 ξ-hydroxyestriol; *2*, estriol; *3*, 16,17-epiestriol; *4*, 16-epiestriol; *5*, 17-epiestriol; *6*, 6 α-hydroxy-17β-estradiol; *7*, 2-hydroxy-17β-estradiol; *8*, 16α-hydroxyestrone; *9*, 16-oxo-17β-estradiol; *10*, 17β-estradiol; *11*, estrone. From Lisboa and Diczfalusy [88]; reproduced with permission of the authors and Periodica.

Miescher [90] reaction. Differentiation of the remaining compounds could be obtained by reduction with potassium fluoborate followed by chromatography in solvent III. The 16β-hydroxyestrone and 16-oxo-17β-estradiol both formed the same reduction product (16-epiestriol) and consequently could not be separated by this method. The 6-oxo-17β-estradiol, which by itself forms the fifth group, could be separated from the members of the fourth and sixth groups in solvents I and III. On reduction with potassium fluoborate, 6α-hydroxy-17β-estradiol was formed and could be separated from the parent compound in solvents I, II, and III. The three compounds in group 6, namely, 2-hydroxy-17β-estradiol, 17-epiestriol, and 16-epiestriol, could be separated nicely by two-dimensional separation, first in solvent systems II or IV and then a second-dimensional separation in solvent V. Group 7 contained the four compounds 6α- and 6β-hydroxyestrone and 6α- and 6β-hydroxy-17β-estradiol. A two-dimensional separation starting with solvent II separated the 6-hydroxylated estrones from the corresponding estradiols, and a separation in the second dimension using solvent V gave a separation of the 6α-hydroxyestrone from the corresponding 6β compound. The 6α-hydroxy-17β-estradiol could not be separated from the corresponding β compound in any of the solvent systems used, although they could be separated by paper chromatography using formamide-saturated chloroform–ethyl acetate (5:1) [91]. Finally, the eighth group which contained only one compound, 7α-hydroxy-17β-estradiol, could be separated from the members of group 7 in solvents I, II, and IV. The solvent system has been expanded and R_f values have been obtained for additional compounds (Table 29.4) [92]. Lisboa and Diczfalusy [93] examined 32 estrogens with various color reagents to assist in their identification.

Barbier and Zav'yalov [94] used pentane–ethyl acetate as a solvent for the separation of a group of estrogens on silicic acid, and Diamantstein and Loercher [95] separated a group of six compounds with chloroform as the solvent. Hertelendy and Common [96] report the R_f values for estriol, 16,17-epiestriol, 16-epiestriol, and 17-epiestriol along with various mono-and dimethyl ethers of these compounds in four different solvents. The same authors [97] have separated equol from estrone, 17β-estradiol, and estriol. Chromatography of the mixture on silica gel G with benzene–methanol (85:15) separated the equol and the 17β-estradiol as a single spot from the other compounds. After elution of this spot the mixture was methylated and then rechromatographed in benzene–methanol (95:5). The dimethyl ether of equol was readily separated from the 3-methyl ether of 17β–estradiol.

Fokkens and Polderman [98] have applied thin-layer chromatography in examining the stability of steroid preparations. The sensitivity of the compounds to oxidation could be decreased by the addition of antioxidants such as α-tocopherol or propyl gallate. Kushinsky [99] investigated the stability of estrogens exposed to oxygen on silica gel layers and concluded that oxygen alone was not responsible for the decomposition of estrogens while exposed on a thin layer of silica gel. Coyotupa et al. [100] believed that the decomposition was caused by atmospheric contaminants and could be eliminated or reduced by operating under nitrogen. Doerr [101] was able to operate with picogram quantities of estradiol by adding phenol or 2-mercaptoethanol to the benzene–ethyl acetate (3:1) developing solvent and by evaporating eluates with purified nitrogen. Gelbke and Knuppen [102] were able to recover 2 μg of 2-hydroxy-17β-estradiol quantitatively by chromatographing on ascorbic acid-impregnated silica gel with acetic acid–chloroform–methylcyclohexane (1:2:2) saturated with ascorbic acid as a solvent. Alumina impregnated with ascorbic acid and 2-butanone–acetic acid (1:1) saturated with ascorbic acid were also used. In both cases the sample solution also contained ascorbic acid. In all cases highly purified solvents should be used because impurities can cause degradation of some steroids [103, 104].

Krol et al. [105] used a continuous development at 4°C with cyclohexane–chloroform–triethylamine–collidine (670:280:17:46) in the developing tray; the bottom of the tank contained cyclohexane–chloroform–collidine (67:28:10). Estrone, equilin, equilenin, 17α-estradiol, 17α-dihydroequilin, and 17α-dihydroequilenin were separated on silica gel layers. A multiple development (2×) at room temperature with cyclohexane–chloroform–trimethylamine–collidine (66:29:2:3) was also used to effect a separation. In both cases a silica gel desiccant was placed in the tank to decrease humidity. Crocker and Lodge [106] separated estrone, equilin, equilenin, α-estradiol, α-dihydroequilin, α-dihydroequilenin, β-estradiol, β-dihydroequilin, and β-dihydroequilenin on silver nitrate-impregnated silica gel H with multiple development (4×) using cyclohexane–ethyl acetate (7:3). The R_f values were 0.67, 0.48, 0.59, 0.52, 0.19, 0.36, 0.46, 0.21, 0.34, respectively. The same workers [107] chromatographed these estrogens in the conjugated form. Silver nitrate-impregnated silica gel was used with isooctane–chloroform–ethanol (20:35:9). Sodium β-estradiol sulfate and sodium β-dihydroequilin sulfate were not separated from the corresponding α forms. Abdel-Aziz and Williams [107a] found that 17α-ethynylestradiol and its 3-methyl and 3-cyclopentyl ethers decomposed on silver nitrate-impregnated layers.

Petrović and Petrović [108] obtained a good separation of estrone, estrone 3-methyl ether, 16, 17-seco-17-oxoestrone-16-nitrile 3-methyl ether, 16-oximinoestrone 3-methyl ether, and 17β-hydroxy-16-oximinoestrone 3-methyl ether on silica gel with benzene–ethyl acetate (7:1) giving R_f values of 0.04, 0.26, 0.74, 0.83, and 0.46, respectively. The R_f values of four other solvents are also listed. Touchstone et al. [109] developed a solvent of 20% acetone in isopropyl ether to give a better separation on silica gel of the estrogens present in urine extracts.

Pollow et al. [110] predeveloped silica gel layers with 5% sodium pyrophosphate. These were then dried and used to chromatograph picogram quantities of estradiol with benzene–methanol (19:1) without formation of artifacts.

Since few, if any, of the conventional systems will separate all estrogens in a single system, two-dimensional chromatography can be of advantage. Cavina et al. [111] chromatographed eight estrogens on precoated silica gel 60 F_{254} plates in toluene–95% ethanol (9:1) in the first

Table 29.4 $R_f \times 100$ Values of Steroid Estrogens on Silica Gel G in Different Solvent Systems[a] [92]

Estrogen	*A*	*B*	*C*	*D*	*E*	*F*	*G*	*H*	*I*	*J*	*K*	*L*	*M*	*N*
Estra-1,3,5(10)-trien-3-ol	72	63	67	84	61	56	63	76	44					
Estra-1,3,5(10),16-tetraen-3-ol	65	42	63	78	56	51	44	67	18					
Estra-1,3,5(10)-trien-3-ol 16β,17β-epoxide	69	56	66	80	59	54	47	70	27					
Estra-1,3,5(10)-trien-3-ol 16α,17α-epoxide	67	54	66	79	60	51	47	69	26					
3-Hydroxy-estra-1,3,5(10)-trien-17-one	69	53	65	80	59	51	46	69	20	50	44	60	47	56
3-Hydroxy-estra,1,3,5(10),6-tetraen-17-one	62			75										
3-Hydroxy-estra-1,3,5(10),7-tetraen-17-one	68		66	78								60		55
3-Hydroxy-estra-1,3,5(10),9(11)-tetraen-17-one	65	50	64	79	56	52	45	68	17					
3-Hydroxy-estra-1,3,5(10),6,8-pentaen-17-one	63	47	63	77	58	50	42	65	16					
3-Hydroxy-estra,1,3,5,(10)-trien-6,17-dione	60	33								30	22	44		38
3-Hydroxy-estra-1,3,5(10)-trien-7,17-dione	61	36	55	72	55	44	56							
3-Hydroxy-estra-1,3,5(10)-trien-11,17-dione	59	34	61	73			38	58	5					
3-Hydroxy-estra-1,3,5(10)-trien-16,17-dione	66	43	64				45			36	39	58	40	44
3,4-Dihydroxy-estra-1,3,5(10)-trien-17-one	54	33	39	76			34	61						
3,6α-Dihydroxy-estra-1,3,5(10)-trien-17-one	47	13	35	65										
3,6β-Dihydroxy-estra-1,3,5(10)-trien-17-one	46	12	34	62										
3,7α-Dihydroxyestra-1,3,5(10)-trien-17-one	42	12	40	60	47	30	17	41						
3,7β-Dihydroxyestra-1,3,5(10)-trien-17-one	43			59										
3,11α-Dihydroxyestra-1,3,5(10)-trien-17-one	51													
3,11β-Dihydroxyestra-1,3,5(10)-trien-17-one	56	27	50	72	51	42	26	54	4					
3,15α-Dihydroxyestra-1,3,5(10)-trien-17-one	48	18	46	66	45	30	21	46						
3,16α-Dihydroxyestra-1,3,5(10)-trien-17-one	57	32	58	68			33		6	21	21		21	37
3,16β-Dihydroxyestra-1,3,5(10)-trien-17-one	55	28												
2,3-Dihydroxyestra-1,3,5(10)-trien-17-one 2-methyl ether	68	50	73				66	19	18					
3,4-Dihydroxyestra-1,3,5(10)-trien-17-one 3-methyl ether	66	53	76	79										
Estra-1,3,5(10)-trien-3,17β-diol	61	40	52	74	53	42	30	61	10	30	24	44	25	38
Estra-1,3,5(10)-triene-3,17α-diol	61	43	55	75	52	45	34	62	11	35	27	51	30	43
Estra-1,3,5(10),6-tetraene-3,17β-diol	57	36	52	73	49	41	29	59	8					
Estra-1,3,5(10), 7-tetraene-3,17β-diol	57	36	53	74	50	41	30	59	7					

3,17β-Dihydroxyestra-1,3,5(10)-trien-6-one	52	23												
3,17β-Dihydroxyestra-1,3,5(10)-trien-7-one	44		41			40	22	50						
3,17β-Dihydroxyestra-1,3,5(10)-trien-16-one	59	32	55	69						23	19	40	35	
Estra-1,3,5(10)-triene-2,3,17β-triol	47	21	24	70				47	2					
Estra-1,3,5(10)-triene-3,6α,17β-triol	45	12	23	61			12	44						
Estra-1,3,5(10)-triene-3,6β,17β-triol	44	12	23	61			10	43						
Estra-1,3,5(10)-triene-3,7α,17β-triol	40	9	22	59	37	24	8	40						
Estra-1,3,5(10)-triene-3,11β,17β-triol	34	8	29	54	39	20	9	34	0					
Estra-1,3,5(10)-triene--3,16α,17β-triol	29	6	21	48	38	17	7	29		1	2	7		4
Estra-1,3,5(10)-triene-3,16γ,17β-triol	43	18	26	54			17							
Estra-1,3,5(10)-triene-3,16α,17α-triol	45	20	30	58						8	5	16		15
Estra-1,3,5(10)-triene-3,16β,17α-triol	34	8	18	50										
Estra-1,3,5(10)-triene-3,15α,17β-triol	26		16	38						0	2	5		2
Estra-1,3,5(10)-triene-2,3,17β-triol-2-methyl ether	58	36	58	72	53	45	41	57	8	32	21	37		34
3,16α,17β-Trihydroxy-estra-1,3,5(10)-trien-6-one	21			34				18						
Estra-1,3,5(10)-triene-3,6,7,17β-tetrol	23													
Estra-1,3,5(10)-triene-3,6,16α,17β-tetrol	12	1	3	24			1	10						
Estra-1,3,5(10)-triene-2,3,16α,17β-tetrol	16	5												
3-Hydroxy-16,17-seco-estra-1,3,5(10)-triene-16-methyl-17-oine acid	31		30	78										
Estra,1,3,5(10)-triene-2,3,16α,17β-tetrol 2-methyl ether	29	6		43			13	26						
Estra-1,3,5(10)-triene-3,6α,17β-triol 3-methyl ether							31							
Estra-1,3,5(10)-triene-3,6β,17β-triol 3-methyl ether							28							
Estra-1,3,5(10)-triene-3,16α,17β-triol 3-methyl ether										3	2	8		
Estra-1,3,5(10)-triene-3,16β,17β-triol 3-methyl ether										12	5	18		
Estra-1,3,5(10)-triene-3,16α,17α-triol 3-methyl ether										12	7	20		

[a] *A* = Ethyl acetate–cyclohexane–ethanol (45:45:10); *B* = ethyl acetate–cyclohexane (1:1); *C* = chloroform–ethanol (9:1); *D* = ethyl acetate-*n*-hexane–acetic acid–ethanol (72:13.5:10:4.5); *E* = benzene–ethanol (4:1); *F* = benzene–ethanol (9:1); *G* = chloroform–ethanol (95:5); *H* = ethyl acetate–*n*-hexane–acetic acid (15:4:1); *I* = *n*-hexane–ethylacetate (3:1); *J* = chloroform–ether (3:1); *K* = benzene–ethyl acetate (3:1); *L* = benzene–ethyl acetate (1:1); *M* = chloroform–ethyl acetate (3:1); *N* = benzene–ether (1:1).

direction followed by development at 90° with butyl acetate–acetic acid–petroleum ether (40 to 70°C) (70:1:30). Only estradiol 17β-valerate and 17β-cyclopentylpropionate were not separated. Taylor [112] used a two-dimensional system for separating estrogens and Δ^4-3-oxo steroids. Silica gel was used with development in the first direction with chloroform–methanol–water (94:6:0.5) and cyclohexane–ethyl acetate (1:1) (2×) in the second direction. Thirteen steroids were separated by this method.

Luisi et al. [113] separated four estrogens by horizontal thin-layer chromatography in cyclohexane–ethyl acetate (35:65) using silica gel G plates.

Smith and Foell [114] have chromatographed a group of 11 C_{18} steroids in their work with the separation of steroids on starch-bound silica gel.

Stárka [115] chromatographed six estrogens in 38 different solvent systems on loose layers of alumina (activity IV).

Estrogens have also been separated on thin layers by means of their derivatives. Penzes and Oertel [116] separated the azobenzene-4-sulfonates of estrone, estradiol, and estriol on silica gel with chloroform–dioxane (94:6), chloroform–benzene–ethanol (18:2:1), and cyclohexane–ethyl acetate (3:1). Because of their fluorescence the Dansyl derivatives (1-dimethylamino-5-naphthalene sulfonyl) have been explored, especially as a means for quantitative analysis. Chloroform–benzene–ethanol (18:2:1) [117] and ethanol–chloroform (5:95) [118] have been used as solvents on silica gel for the separation of these derivatives. Touchstone et al. [119] chromatographed the Fast Violet Salt B (6-benzoylamino-4-methoxy-5-toluidine, diazonium salt) derivatives of 11 estrogens on silica gel with 10% methanol in benzene and 30% benzene in butyl acetate. The estradiol 16α and 16β derivatives could not be separated. As little as 0.02 μg could be detected on the chromatogram.

Lisboa [120] reported the R_f values for 55 estrogens on silica gel G in 10 solvent systems. The R_f values of 18 estrogen acetates in three solvents are also given.

Quantitative Determinations

Struck [121] has discussed the various methods that have been used for the quantitative determination of estrogens. For the separation and determination of estrone, estradiol, and estriol, he used silica gel layers to separate the compounds with benzene–ethanol (9:1) as the separating solvent. To get rid of interfering impurities, the silica gel for the layers was allowed to stand overnight in methanol; than after filtering and washing twice with fresh methanol it was dried at 100°C for ½ hr. After the plates were prepared and dried, they were divided into five evenly spaced strips. The center strip was used for the sample for the quantitative determination, a blank strip was left on each side of the sample, and the two outer strips were also spotted with the sample. After developing the plate for a distance of 10 cm, the plates were dried at 100°C for 30 min; then to keep the three center strips from being contaminated by the detecting reagent they were covered by a plexiglass sheet. The two outer strips which were used as the detection zones were sprayed with distilled water, followed by a 12% solution of antimony pentachloride in carbon tetrachloride which was applied with a roller covered with absorbent blotting or filter paper. The solution was applied twice to insure thorough wetting of the layers. The plates were dried at 100°C for 8 to 10 min. Using the outer-colored lanes as guides, the corresponding spot locations in the center strip as well as the two blank strips were scraped into centrifuge tubes where they were extracted for 30 min with 3 ml of a freshly prepared 0.5 *N* solution of sodium hydroxide in 80% ethanol. After cooling in ice, the tubes were centrifuged at 3000 rpm for 4 min after which the clear solution was decanted. The absorbance was measured at 242 mμ against a 0.5 *N* alcoholic sodium hydroxide solution. After deducting the blank value, the quantity of the estrogen was determined from a standard curve.

Schroeder et al. [122] developed a method for the determination of estrone, equilin, equilenin, and estradiol after hydrolysis of their sodium sulfate salts in tablets. After separation of the free estrogens on silica gel with chloroform–cyclohexane–acetone–58% ammonia (30:54:8:8), the equilin, estradol, and estrone were eluted with 1% aqueous sodium hydroxide and measured at 296 nm. Equilenin was eluted with methanol and measured at 231 nm.

Livingston et al. [123] have used a spectrophotometric method for the determination of coumestrol, a naturally occurring plant estrogen, which has been isolated from ladino clover *(Trifolium repens)* and other leguminous plants [124, 125]. After the separation on silicic acid chromatostrips using ether–petroleum ether (7:3) with multiple development (4×), the compound was eluted with methanol and measured at 352 mμ.

Colorimetric methods have also been used in the determination of estrogens separated by thin-layer chromatography. Luisi et al. [126, 127] have applied the method in the determination of estrogens in biological fluids. Separation of estrone, 17β-estradiol, and estriol was accomplished with cyclohexane–ethyl acetate (3:17), although if too many chromogens interfered with the separation, a two-dimensional separation was carried out using isopropanol as the second solvent. Quantitative measurements were made by the method of Marescotti et al. [128]. Jung et al. [129] have also used a colorimetric method for the determination of the same estrogens. Gaenshirt and Polderman [130] applied sulfuric acid or vanillin–sulfuric acid for the determination of Δ^4-17β-hydroxyestrene derivatives. By means of the method, the stability of ethylestrenol tablets was determined.

Wotiz and Chattoraj [131] have applied a gas chromatographic method for the determination of estrogens separated by thin-layer chromatography. For the thin-layer separation, three solvent systems were used. The first system of benzene–ethyl acetate (1:1) was used to separate the estrogens into four different groups; however, when large amounts of androgens were present in the samples, a solvent consisting of petroleum ether–dichloromethane–ethanol (10:9:1) was used as a means of removing these components. The third system of petroleum ether–methanol (9:1) was used whenever it became necessary to purify the steroid acetates prior to injection into the gas chromatographic system. The thin-layer plates were coated with silica gel G and dried at room temperature for 3 hr. They were then given a predevelopment in methanol–concentrated hydrochloric acid (9:1) in order to remove iron and other impurities. Following this they were activated at 105°C. The urine samples were hydrolyzed with acid and then separated into phenolic and nonphenolic fractions by the method of Brown [132]. After separation of the estrogenic fraction on thin-layer plates, the spots were eluted with ethanol and after removal of the eluting solvent "the residue was acetylated by dissolving it in a mixture of five parts of acetic anhydride and one part of pyridine at 68°C for 1 hr. To the acetylated mixture 5 ml of distilled water was added while stirring thoroughly with a glass rod." The acetylated product was thoroughly extracted with petroleum ether, and the latter extract was washed with 8% sodium bicarbonate solution and finally with water. The petroleum ether extract was then taken to dryness and the residue transferred to 2-ml tubes by means of additional petroleum ether and was usually ready for injection into the vapor-phase chromatographic system. If further purification of the acetates was necessary, then it was subjected to thin-layer chromatography in the petroleum ether–methanol solvent. For the gas chromatographic analysis, 2 to 5 μl of solution was injected directly onto a 4 ft by ⅛ in.-diameter stainless steel column packed with 3% SE-30 (General Electric) on 80 to 100 mesh Diatoport S (F and M Scientific Co.). The following conditions were employed: "N_2, 20 psi; H_2, 5 psi; air, 10 psi; column temperature, 228°C; vaporizer temperature, 250°C." A flame-ionization detector was used with the instrument. The precision of the method was not evaluated because of insufficient data; however, the method looks very promising and has a lower limit of detection of 0.02 μg of steroid, which makes the overall technique sensitive to a range of 0.1 to 0.2 μg of individual estrogens per 24 hr collection of urine.

Knuppen et al. [133] separated 25 estrogens by a combination of paper and thin-layer chromatography. The steroids from hydrolyzed urine were first isolated on formamide-impregnated paper and the separated estrogens were then acetylated and separated by thin-layer chromatography on silica gel with cyclohexane–ethyl acetate mixtures. Quantitative determinations were made on the acetates by gas chromatography with a flame-ionization detector. Kushinsky et al. [134] separated estrone, estradiol, and estriol on silica gel layers, then eluted and converted them to the acetates, which were in turn chromatographed on silica gel. The separated acetates were then eluted and converted to the heptafluorobutyrates for determination by gas chromatography. Moretti et al. [135] converted the estrogens, separated as esters by thin-layer chromatography) to the trimethylsilyl esters for determination by gas chromatography. A colorimetric method was also used. Touchstone et al. [136] determined estrogens separated by thin-layer chromatography using gas chromatography without first derivatizing.

In situ determinations on the TLC layer have also come into use. Krol et al. [105] determined six estrogens by spraying the spots with 9 *N* sulfuric acid and charring at 125°C with subsequent scanning by reflectance. Wortmann et al. [137] used a densitometric method for the determination of estriol in pregnancy urine. After isolation of the estriol on silica gel layers with benzene–ethanol (85:15) the spots were detected by dipping the plates in 5% phosphomolybdic acid in ethanol. In situ densitometry at 580 nm was linear between 0.1 and 1 μg. Touchstone et al. [138] determined estriol in pregnancy urine by direct densitometry of the Fast Violet Salt B reaction product after its separation on silica gel layers [119]. The same workers [139] used thin layers of silica gel impregnated with 10% phosphomolybdic acid for the separation of estrogens with a solvent composed of 5% phosphomolybdic acid in ethanol–benzene (1:1) or 15% ethanol in benzene. After separation the color was developed by heating the layers at 110°C for 20 min. The transmission was measured at 560 nm with linearity being observed between 0.05 and 1.0 μg. The azobenzene-4-sulfonates have also been used for densitometry [116] as well as the Dansyl derivatives [117]. The Dansyl derivatives can also be determined by fluorescent measurements; Dvir and Chayen [115] applied this method to the determination of estriol in urine during pregnancy. Penzes and Oertel [140] separated the Dansyl derivatives of estrone, estriol, and estradiol on silica gel with chloroform–benzene–ethanol (18:2:1). After drying, the chromatogram was sprayed with isopropanol–triethylamine (4:1), dried in air, and finally dried over silica gel (indicating) for 24 hr. This treatment increased the sensitivity of the assay to approximately 1 ng. Linearity was found over the range of 1 to 25 ng. Polyamide layers were also used with acetone–methanol–water (2:8:2); however, treatment of these layers with the triethylamine caused a decrease in fluorescence. The silica gel layer was predeveloped in chloroform–ethanol (3:1) and the polyamide layer with methylene chloride–carbon tetrachloride–methanol (1:1:1) with subsequent activation at 110°C.

Jacobsohn [141, 142] has used a photogrammetric procedure for the quantitative determination of estrogens.

3 THE C_{19} STEROIDS

Qualitative Separations

The androgenic hormones, which belong to the C_{19} steroids, are the derivatives of the hydrocarbon etioallocholane (androstane). Dyer et al. [143] investigated the separation of 28 steroids on silica gel G layers. Three different solvent systems were used: ether–chloroform (1:9) for nonpolar diketones and ketol esters, benzene–ethyl acetate (1:1) for ketols, and benzene–ethyl acetate (1:4) for the more polar diols and keto diols. The spots were detected by spraying with concentrated sulfuric acid in ethanol and heating at 90°C. Stárka et al. [144, 145] used loose layers of aluminum oxide with an activity grade II to III. The relationship between the activity of the adsorbent and the composition of the benzene–ethanol solvent was investigated, and it was found that the change in R_f values due to the difference in the adsorbent activity could be compensated for by a change in the solvent system. The R_f values of seven steroids were tabulated in six solvents. The best resolution of urinary steroids was obtained with methylene chloride or methylene chloride containing 7% ethyl acetate. Multiple development increased the separation.

Change [146], on the other hand, has applied partition chromatography in three solvents: 2.5% dichloromethane in methylcyclohexane, benzene–methylcyclohexane (1:1), and benzene–methylcyclohexane–chloroform (1:1:1), all solvents being saturated with ethylene glycol. The R_f values of compounds with hydroxyls at C-17, C-11, and C-3 increased in the following order: C-3 < C-11 < C-17. By introducing ketone or hydroxyl groups into the androstanedione molecule, the polarity was increased in the following sequence: 16-keto < 11-keto < 18-hydroxy < 11β-hydroxy < 6-keto < 6β-hydroxy < 19-hydroxy. The polarity was also increased if the C-10 angular methyl group was removed. The tricomponent solvent was more useful for the polar steroids. Sonanini and Anker [147] used propylene glycol-impregnated kieselguhr layers with cyclohexane–toluene (4:1) for separating a group of 17 compounds. A group of esters were also separated on kieselguhr layers impregnated with 10% paraffin oil in light petroleum using 30, 50, and 70% acetic acid as solvents. James et al. [148] also used a reverse-phase system for separating androgen esters. Silica gel was impregnated with 5% paraffin and used with water–acetone (1:1, 11:9, 3:2, 7:3, and 4:1). The R_M values and the distribution coefficients were calculated for the formates, acetates, propionates, butyrates, and valerates of androstanolone, nandrolone, and testosterone. Vaedtke and Gajewska [149] included a group of five C_{19} steroids in their use of partition chromatography on Celite #545 (Johns Manville). Formamide was used as an impregnating agent and *n*-hexane–benzene (1:1) saturated with formamide was used as the solvent. Watson and Bartosik [150] give some examples of the use of eight Bush-type chromatographic systems for steroids using SilicAR-7 GF (Mallinckrodt) as the adsorbent.

Lisboa [151] has investigated the separation of 29 steroids in 10 different solvent systems (Table 29.5). In addition to the R_f values, which are given in the table, the R_{st} values (st = testosterone), the standard deviation, and the R_M values were calculated. From the results of the separations, a number of generalizations could be made, such as the fact that it is difficult or impossible to separate the isomers of the 5α (androstane) and 5β configuration without a substitution in ring A. This is also true if the ring has only one ketonic group such as the 5α- and 5β-dihydrotestosterone. The introduction of an unsaturated bond changed the R_f values slightly in only a few cases. On the other hand, the introduction of a conjugated double bond changed the mobility so that testosterone could be readily separated from 5α- or 5β-dihydrotestosterone. Two equatorial or axial isomers were difficult to separate, but a compound with an equatorial configuration could be readily separated from an axial one. For the detection of the compounds a number of reactions were used. The anisaldehyde–sulfuric acid reaction was carried out by spraying the plates with a 1% solution (vol/vol) of anisaldehyde in 2% (vol/vol) concentrated sulfuric acid in glacial acetic acid. The colors (Table 29.5) were developed by heating to 95 to 100°C for 12 to 15 min. The Zimmermann reaction was used for the detection of 17- and 3-oxo steroids. To accomplish this the plate was sprayed with a freshly prepared mixture of 2% *m*-dinitrobenzene in ethanol and 1.25 *N* ethanolic potassium hydroxide (1:1). The blue spots of the 3-oxo steroids appeared at once on heating in a stream of hot air and the 17-oxo steroids (with an unsubstituted 16 position) gave a violet color after 3 to 6 min. For the identification of etiocholan-11-one and androstan-11-one, Dragendorff's reagent was employed. This reagent, which must be prepared fresh every second day and kept at low temperatures (4°C), gives an orange color with Δ^4-3-oxo

Table 29.5 $R_f \times 100$ Values of Some C_{19} Steroids on Silica Gel G in Ten Solvent Systems[a] and Their Color Reaction with Anisaldehyde-Sulfuric Acid Reagent (Development Distance 15 cm) [151]

Steroid	A	B	C	D	E	F	G	H	I	J	Color
5α-Androstan-11-one		69		87		72	72		66	64	Colorless
5β-Androstan-11-one		69				73	71		66	64	Colorless
5α-Androstan-17-one	75	73	78		75	78	76	80	66	55	Blue
5α-Androstane-3,17-dione	63	46	74	77	66	63	69	72	53	17	Olive
5β-Androstane-3,17-dione	64	43	74	78	67	60	67	69	49	14	Brown
5α-Androst-1-ene-3,17-dione	61	38	76	75	66	58	67	67	45	13	Violet blue
Androst-4-ene-3,17-dione										8	Salmon
5α-Androst-2-ene-7,17-dione	67	52	77	81	70	61	70	74	55	27	Olive
5β-Androstane-3,11,17-trione	52	25	69	70	66	50	58	56	36	4	Red Brown
5α-Androstan-17-ol	64	54	64	78	62	57	57	73	46	31	Blue
17β-Hydroxy-5α-androstan-3-one	57	35	61	74	57	42	49	62	31		Brown-olive to blackish olive
17β-Hydroxy-5β-androstan-3-one	55	30	61	73	59	42	46	63	28		Bright violet
17β-Hydroxyandrost-4-en-3-one	48	23	58	66	56	49	43	52	24	4	Red-orange to violet purple
3α-Hydroxy-5α-androstan-17-one	55	33	63	74	58	43	50	62	29		Green
3β-Hydroxy-5α-androstan-17-one	52	30	58	72	55	39	44	61	27		Green
3α-Hydroxy-5β-androstan-17-one	49	24	58	71	56	38	43	57	23		Green
3β-Hydroxy-5β-androstan-17-one	57	36	62	73	59	44	50	66	33		Green
11β-Hydroxy-5α-androstane-3,17-dione	57	27	63	75	64	44	46	59	27		Brown
17β-Hydroxy-5α-androstane-3,17-dione	40	12	56	57	49	29	36	42	18		Brown
3α-Hydroxy-5-β-androstane-11,17-dione	38	9	52	61	55	33	31	43	18		Brown-olive
3α-Hydroxy-5α-androstane-7,17-dione	40	12	53	54	47	27	32	40	16		Green-olive
3β-Hydroxy-5α-androstane-7,17-dione	33	6	50	49	45	23	27	34	13		Green-olive
5α-Androstane-3α,17β-diol	50	25	50	70	48	32	32	56	18		Blue
5α-Androstane-3β,17β-diol	48	25	47	68	49	25	29	54	16		Blue
Androst-4-ene-3β,17β-diol	50	25	47	70	50	30	32	58	17		Dark blue
5β-Androstane-3α,17β-diol	44	17	40	65	48	23	23	50	13		Blue
3β,11β-Dihydroxy-5α-androstan-17-one	44	14	43	68	51	28	22	51	13		Green
3α,11β-Dihydroxy-5β-androstan-17-one	43	12	44	66	52	29	23	48	13		Dark gray-blue

[a] *A* = Ethyl acetate–cyclohexane–ethanol (9:9:2); *B* = ethyl acetate–cyclohexane (1:1); *C* = chloroform–ethanol (9:1); *D* = ethyl acetate–(water-saturated *n*-hexane)–ethanol (72:13.5:4.5:10); *E* = benzene–ethanol (4:1); *F* = benzene–ethanol (9:1); *G* = chloroform–ethanol (19:1); *H* = benzene–ethanol (19:1); *I* = ethyl acetate–*n*-hexane–acetic acid (15:4:1); *J* = *n*-hexane–ethyl acetate (3:1).

steroids. Detection of alcoholic steroids could be made by spraying the plates with 0.25% chromic acid anhydride in glacial acetic acid and heating for 15 min at 90 to 95°C. The oxo steroid that was thus formed was then detected by the Zimmermann reaction. A group of 17 steroid acetates were also chromatographed using the ethyl acetate–cyclohexane (1:1) solvent, but in this particular system the results were not as satisfactory as with paper chromatography.

Cohn and Pancake [152] separated six compounds on silica gel G with cyclohexane–ethyl acetate (1:1). Akhrem and Kuznetsova [153] included a group of 10 C_{19} compounds in their steroid separations on silica gel layers, also with cyclohexane–ethyl acetate mixtures. Some additional R_f values for C_{19} steroids are given in Table 29.6 taken from Hara and Takeuchi [154] who used plaster of Paris-bound silica gel layers, and from Smith and Foell [114] who used starch-bound layers. Neher and Wettstein [155] give the R_f values for adrenosterone on silica gel in 10 different solvents and on aluminum oxide in three solvents Menzel and Oertel [156] separated androstenediol acetate, androstenetriol acetate, and estrogen acetate on alumina with cyclohexane–butyl acetate–butanol (10:9:1). Using magnesium silicate without a binder in a number of different solvent systems, Schwarz [157] determined the R_f values of nine C_{19} compounds among a group of 51 steroids. Doenges and Staib [158] used thin-layer silica gel plates for obtaining purified samples for the preparation of infrared spectra. Reisert and Schumacher [159] used 200 × 550 mm plates of silica gel for the separation of urinary C_{19} steroids. The plates were developed in a descending manner with the plates at an angle 10°, because angles greater than 10° were claimed to give uneven solvent development. A chloroform–ethanol (100:1.5) mixture was used for the separation. Chamberlain et al. [160] used thin-layer chromatography to confirm the presence of androstane diones and triones in the oxidized fractions from the urine of pregnant monkeys.

Loose alumina layers have also been used for the separation of these compounds by Černý et al. [161] and Heřmánek et al. [162].

Gower [163] chromatographed the C_{19} 16-dehydro steroids and their acetates on silica gel G in 12 different solvents. Compounds which were difficult to separate by single development were given a multiple development in the same solvent in order to achieve a separation. Among the detecting agents that were used was a solution of 5% uranyl nitrate (wt/vol) in 10% (vol/vol) sulfuric acid. Most of the compounds gave specific colors with this reagent when heated at 110°C.

Wright [164] used repeated development with cyclohexene–cyclohexanone (9:1) or (4:1) to separate closely related pairs of steroids such as 3β-hydroxy-5α- and 3α-hydroxy-5β-androstan-17-one. Starch-bound lay-

Table 29.6 $R_f \times 100$ Values of C_{19} Steroids on Silica Gel Layers in Various Solvents Using a Saturated System

	Silica Gel with 5% Calcium Sulfate Binder [154]					*Silica Gel with 5% Starch Binder* [114]		
Steroid	*Benzene–Acetone* (4:1)	*Benzene–Methanol* (9:1)	*Ether*	*Chloro-form–Acetone* (3:1)	*Chloro-form–Methanol* (97:3)	*Hexane–Ethyl Acetate* (4:1)	*Hexane–Ethyl Acetate* (1:1)	*Ethyl Acetate*
5α-Androstan-3α-ol-17-one	46	32	64	46	77			
5α-Androstan-3β-ol-17-one	38	29	55	41	67			
5α-Androstan-17β-ol-3-one						4	32	63
5α-Androstane-17α-methyl-17β-ol-3-one						3	32	63
5β-Androstane-3α,12α-diol-17-one	37	14	65	38	37			
5α-Androstan-3α-ol-11,17-dione	28	20	33	36	50			
5β-Androstan-3α-ol-12,17-dione	25	16	20	30	41			
5β-Androstan-12α-ol-3,17-dione	20	21	29	25	50			
5α-Androstane,3,17-dione							42	72
5β-Androstane-3,12,7-trione	30	20	14	48	75			
Δ^4-Androsten-17β-ol-3-one	35	23	49	41	63	1	15	47
Δ^4-Androstene-17α-methyl-17β-ol-3-one	41	27	49	48	73	1	17	48
Δ^4-Androsten-17α-ethynyl-17β-ol-3-one						3	28	70
Δ^4-Androstene-3,17-dione	58	53	56	55	80	2	20	55
5α-1-Androstene-3,17-dione							32	69
1,4-Androstadiene-3,17-dione						0	14	44
Δ^4-Androstene-3,11,17-trione	36	40	27	52	82			
3β-Acetoxy-5α-androstane	77	74	87	66	97			

ers of silica gel were used. Remmers et al. [165] separated testosterone and epitestosterone by multiple development (5×) on silica gel with dichloromethane–ethyl acetate (9:1) or by continuous development for 8 hr with the same solvent at 4°C. A two-dimensional development with this solvent as the first solvent and chloroform–cyclohexane–acetic acid (5:4:1) as the second-dimension solvent also provided a good separation.

A large group of androstane steroids have been separated by Lisboa [166–169]. Lisboa and Hoffmann [170] used continuous development for the separation of epimeric $C_{19}O_2$ steroids. Twenty-one steroids were chromatographed in the four solvents: cyclohexane–ethyl acetate (1:1 and 4:1), *n*-hexane–ethyl acetate (3:1), and benzene–ethyl acetate (3:1). Silica gel 60F (Merck) was used as an adsorbent.

The 2,4-dinitrophenylhydrazones have been used for the thin-layer chromatographic separation of the 17-keto steroids [171–175]. Starnes et al. [172] used alumina layers with benzene–hexane–ethyl acetate (3:1:1) for separating these derivatives. Further purification was obtained by rechromatographing in toluene–ethyl acetate (3:1). Penzes et al. [173] give the R_f values for 12 derivatives using polyamide with eight solvent systems including methylene chloride–methanol (2:8), acetone–ethanol–water (5:16:4), and acetone–methanol–acetic acid (20:80:0.5). Hakl [174] has separated seven derivatives on silica gel using chloroform–acetone (98:2, 96:4, 94:6, and 92:8) and also diethyl ether–petroleum ether (7:3) as solvents. A two-dimensional separation with diethyl ether–petroleum ether (7:3) in the first direction (2×) followed by chloroform–acetone (94:6) in the second direction separated all seven of the derivatives. These latter two solvents were used in multiple developments to separate the derivatives of thirteen 17-ketosteroids [175].

Thin-layer chromatography has been used [176–178] for the separation of isomeric *O*-methyloximes of some ketosteroids.

Cerri and Maffi [179] separated a group of five testosterone derivatives on silica gel G with cyclohexane–ethyl acetate (17:3).

The steroids usually appear conjugated in the urine either as glucosiduronates or as sulfates; these can be hydrolyzed before separating the steroids or they may be separated as such. Crépy et al. [180] chromatographed a group of the sulfates of C_{19} steroids by two-dimensional chromatography on plates of a mixture of kieselguhr G–silica gel G (19:1 and 9:1). After the compounds were spotted, the plates were developed for 2 hr in butyl acetate–toluene–4 *N* ammonium hydroxide-methanol (85:35:50:70). Immediately after removal they were placed at right angles in the same solvent composition but in the ratio of 11:9:12:16. The development in the second direction was carried out for 15 cm. Location of the sulfates was accomplished by spraying with methylene blue reagent. Oertel et al. [181] have also chromatographed a group of steroid conjugates of the C_{18}, C_{19}, and C_{21} steroids on DEAE and ECTEOLA cellulose layers. Seven solvent systems were employed and the separations were better than those obtained with the silica gel layers. Segura et al. [182] separated some steroid glucuronides on polyamide layers using ethyl formate–methanol–water–formic acid (10:4:5:1) and methanol–water–formic acid (12:7:1) as solvents. On Woelm polyamide the glucuronides showed a bright fluorescence under ultraviolet light after spraying with 4% aluminum chloride in methanol–water (3:1).

Quantitative Determination

Panicucci et al. [183] used the Zimmermann reaction (see under C_{18} steroids). Vermeulen and Verplancke [184] also used the Zimmermann reaction for the determination of testosterone; in this case, two chromatograms were run with an intermediate oxidation step prior to the evaluation of the spot. The values for normal and unhealthy male and female patients are given.

Cavina et al. [185] used the tetrazolium blue reaction (T-255) for the colorimetric determination of aldosterone. The over-all recovery was 70.7%.

Radioactive-labeled androgens have also been used in determining quantitative amounts. Schulze and Wenzel [186] have designed a special proportional glas-flow counter with an extremely flat aperture plate for measuring the radioactivity of thin-layer plates. The effectiveness was demonstrated with tritium and ^{14}C-labeled steroids. Riondel et al. [187] have applied a scintillation method to the determination of testosterone in human blood. A double isotope procedure was employed in which 1,2-^{3}H-testosterone was used as the indicator and ^{35}S-thiosemicarbazide as the reagent. A combination of thin-layer chromatography and paper chromatography was used to purify the thiosemicarbazone. Actual measurement was carried out on the diacetyl derivative of the thiosemicarbazone. Sparagana [188] also used the double-isotope dilution method for plasma testosterone. In this case the latter was converted to the acetate by using (^{3}H) acetic anhydride. Separation was then achieved by two-dimensional thin-layer chromatography on silica gel with benzene–acetone (3:1) followed by hexane–ethyl acetate (1:1) in the first direction and benzene–ethyl acetate (3:1) followed by dichloromethane–methanol (10:1) in the second direction. Bardin and Lipsett [189] in using a scintillation method for testosterone and androstenedione reduced the latter to testosterone and then after acetylation the acetate was converted to the *O*-methyloxime by treatment at 26°C with methoxylammonium chloride in pyridine.

Gas chromatography has been used for the determination of testosterone in biological samples after separation by thin-layer chromatography

[190–195]. Guerra-Garcia et al. [190] found a sensitivity of approximately 0.1 γ. Average recovery determined by scintillation was 39%. Luisi et al. [191] hydrolyzed urine and ran two separate TLC purifications followed by a TLC separation as the testosterone acetate prior to gas chromatographic determination. Exley [192] obtained a high degree of sensitivity by using electron-capture detection of testosterone diheptafluorobutyrate. The testosterone was purified by thin-layer chromatography and then converted to the derivative. The method could be used to detect nanogram and sub-nanogram amounts of testosterone. It was simpler and quicker, as well as slightly more sensitive, then the double-isotope derivative methods. Fabre et al. [195] used an electron capture method for the determination of aldosterone and tetrahydroaldosterone in blood. Both compounds were converted to the lactones by oxidation with periodic acid which at the same time converted the cortisone and related steroids to etianic acids. The latter could be easily removed by washing with sodium bicarbonate prior to thin-layer chromatography. The limits of analysis for the aldosterone were in the range of 10 to 20 ng. By converting the tetrahydroaldosterone γ-lactone to the monochlorodifluoro derivative this compound could be determined in picogram quantities.

The 2,4-dinitrophenylhydrazone derivatives have been used for the quantitative determination of ketosteroids. Kent and Rawitch [196] eluted the derivatives and measured the optical density at 390 mμ. Knapstein et al. [197, 198] used in situ densitometry for quantitating these derivatives.

Struck et al. [199] used direct reflectance measurements for the determination of testosterone and androst-4-ene-3,17-dione separated by thin-layer chromatography. The coefficient of variation was 10% for 0.2 to 0.5 μg of each compound. Huck [200] separated testosterone on silica gel and then converted it to the trimethylsilyl derivative. The latter was then chromatographed on alumina layers and determined quantitatively by direct fluorometric measurement. The method was sensitive in the parts per billion range.

Graef [201] measured the fluorescence of the dansylhydrazine derivatives of oxosteroids at 511 nm after elution from the silica gel layer. The method was sensitive to 0.02 μg.

4 THE C_{21} STEROIDS

Qualitative Separations

These compounds have as their basic building structure the hydrocarbon pregnane. Quite a number of different adsorbents and solvent systems have been applied in separating these compounds. Schwarz [202] separated a group of eight of these compounds on aluminum oxide which was deactivated by shaking with 2.5% acetic acid for 1 hr. Petroleum ether, benzene, petroleum ether–benzene mixtures, and benzene–ethanol were used for the separations. Schwarz [157] also applied magnesium silicate as an adsorbent without a binder: since the adsorbent had an alkaline reaction it was neutralized by mixing 1.5 ml of acetic acid with 100 g of adsorbent, thus giving a final slurry of pH 6.5. Eight different solvent systems were applied ranging from benzene and chloroform to mixtures of these with more polar solvents such as ethanol, acetic acid, and dioxane. It was found that compounds containing the same polar substituent, but which differed by a few less polar substituents, had R_f values very similar to one another. As a detecting agent the layers were sprayed with concentrated sulfuric acid and then heated to 110 to 130°C after which they were observed with both visible and ultraviolet light. Schwarz and Syhora [203] also applied the acidic alumina to the separation of another group of corticoids. In this case some additional solvents were applied including mixtures of benzene with dioxane, dimethylformamide, and ethanol. Although cortisone and prednisone could not be separated in any of the solvent mixtures on the acidic aluminum oxide, they were well separated on an alkaline aluminum oxide of activity I to II in ethyl ether or mixtures of ether with petroleum ether. In these solvents cortisone remained at the origin whereas prednisone had R_f values ranging from 0.14 to 0.79 depending on the solvent used. Hara and Takeuchi [154] used thin layers of silica gel with mixtures of benzene or chloroform with acetone and methanol as well as straight ethyl ether as solvents. They listed the colors obtained using concentrated sulfuric acid, concentrated sulfuric acid–acetic anhydride mixture, chlorosulfonic acid–acetic acid, and antimony trichloride in chloroform as detecting agents. Smith and Foell [114] used silica gel layers that were bound with rice starch, finding this much more convenient than the rather fragile gypsum-bound layers. For acid-sensitive substances the silica gel–starch mixture was neutralized to a pH of 6.4. These authors applied a number of detecting agents without interference from the starch binder. Ten percent phosphomolybdic acid in ethanol found excellent application and was more sensitive than the corresponding reagent on paper chromatograms. After heating the sprayed plates for approximately 10 min at 100°C, the steroids appeared as blue spots on a lemon yellow background. Sensitivity with this reagent on the acidic plates was 0.125 μg and on the neutral plates 0.25 μg. This reagent was also more sensitive on the starch-bound layers than on the gypsum-bound layers. When using the phosphomolybdic reagent with the neutral layers, 4 ml of concentrated hydrochloric acid was added to each 100 ml of the 10% phosphomolybdic acid reagent in order to obtain good sensitivity. Table 29.7 gives the R_f values of a number of C_{21} steroids in these various systems.

Cavina and Vicari [204] and Waldi [205] have used layers of silica gel G with various solvent systems. R_f values from these sources are given in Table 29.8. Lisboa [206] chromatographed 31 Δ^5-3-hydroxy-C_{21} steroids on silica gel in eight solvent systems. Boric acid-impregnated layers were used to improve the resolution of some compounds. The application of isonicotinic acid hydrazine in dilute acetic acid to the sample spot on silica gel layers permitted the separation of pregnenolone as the hydrazone from 16-dehydroprogesterone, which did not form a hydrazone under the conditions used. Levin et al. [207] found 20% acetone in isopropyl ether to be a useful solvent for chromatography on silica gel. With this solvent cortexolone and corticosterone could be separated.

Stárka and Malíková [208] and Černý et al. [161] have applied loose layers of aluminum oxide in separating corticosteroids. The latter workers have also applied silica gel to a group of six compounds using benzene-ethyl acetate (1:1) as the separating solvent. These workers used an alumina impregnated with morin for the detection of the compounds. In addition to the C_{21} steroids they have also run C_{19} and C_{18} steroids as well as a group of saponins.

Nguen and Chemerisskaya examined the chromatographic behavior of 3-ketosteroids on unbound layers of weakly alkaline alumina [209] and acidic alumina [210]. The R_f values were generally higher on the acidic alumina. Schwarz [211] reported the R_f values of 16α, 17-epoxy-Δ^5-pregnen-3β-ol-20-one and seven related steroids on acidic alumina in five solvents including petroleum ether–benzene (1:1), (2:1), and benzene–ethanol (95:5). Schreiber and Adam [211a] found that alumina could not be used for 3β,16β-diaacetoxy-5-α-pregnan-20-one because it caused a splitting of the 16β-acetoxy group.

Matis et al. [212] found layers of calcium sulfate convenient for separating C_{21} steroids. Using a finely ground adsorbent, they obtained adhesion to the supporting plate without a binder. As a separating solvent a mixture of acetone–benzene (1:4) was used to separate the 17-hydroxy corticoids. In order to get a sharper separation between 5β-pregnane-3α,17α,21-triol-11,20-dione and 5β-pregnane-3α,11β,17α,21-tetrol-20-one, a 3% solution of ethanol in chloroform was used. Development distances of 25 cm were used. These authors have also used silica gel G with methanol-chloroform (5:95) for separating pure corticoids. In this case no binder was used and the plates were developed as loose layers at an angle of about 20°C. Neher and Wettstein [155] determined the R_f values of progesterone and pregn-5-en-3β-ol-20-one on silica gel in eight different

Table 29.7 $R_f \times 100$ Values of C_{21} Steroids in Various Systems[a]

	Silica Gel [154]					*Aluminum Oxide*[b] [203]							*Silica Gel*[c] [114]				*Magnesium Silicate*[d] [157]						
Steroid	*A*	*B*	*C*	*D*	*E*	*F*	*G*	*H*	*I*	*J*	*K*	*L*	*M*	*N*	*O*	*P*	*Q*	*R*	*I*	*S*	*J*	*F*	*T*
Pregnanediol																	32	52	30	45	49	76	93
5α-Pregnane-3,20-dione	67	70	89	63	97																		
5α-Pregnane-3β,20α-diol	30	18	55	38	48																		
5α-Pregnane-3β,20β-diol	32	18	59	38	48																		
Pregn-4-ene-17α,21-diol-3,20-dione	22	13	29	30	31	49	63	75	49	70	40	48		15	51		20	35	10	28	35	47	51
17α-Hydroxyprogesterone																	25	53	20	32	46	60	60
Progesterone	65	57	66	60	97								5	36			37	58	50				
Pregn-4-en-21-ol-3,20-dione	38	37	37	44	84	70			72		60	62		22	56	96	21	43	13	31	40	33	13
Prednisone						26	37	62		52				6	38	69	12	30	4	16	21	24	25
Prednisolone							30	50		29				5	35	46	11	20	2	9	13	29	42
Dexamethasone							37	48		33							13	39	5	10	17	35	52
Triamcinolone							20	16		6							5	20	0	3	5	18	
Cortisone	14	11	13	18	17	32	44	67		57				7	38	75	17	31	5	16	24	34	31
Hydrocortisone	9	10	13	9	8	25	42	53	14	35	18	25		6	38	55	14	33	6	11	16	41	51
Corticosterone	15	12	14	19	25																		
11-Epicortisol						12	21	38		22							6	20	3	4	7	19	15
6α-Methylprednisolone							35	46		30				2	20	33	13	36	3	7	13	32	45
5α-Pregnane-3β,16α-diol-20-one						57			51		41	63					16	44	6	16	24	44	48
Pregnenolone													6	38			27	48	36				
Pregna-5,16-dien-3β-ol-20-one													5	37			27	50	41				
3β-Acetoxy-5,16-pregnadien-20-one													34	75			39	58	64				

5α-Pregn-16-en-3β-ol-20-one													9	44									
Pregn-5-ene-3β,17α-diol-20-one																	25	46	35				
Pregn-5-ene-3β,17α,21-triol-20-one																		40	15			54	83
12α-Acetoxy-3α,7α-dihydroxy- 5β-pregnan-20-one	4	12	4	7	13																		
Pregn-4-en-21-ol-3,20-dione acetate						90			92		79	84					35	60	22	50	57	67	50
Hydrocortisone acetate						67			62		50	73					29	52	12	35	41	62	81
11-Epicrotisol 21-acetate						51			45		38	62											
11-Epicortisol 11,21-diacetate						78	81		76		58	76											
Pregn-4-ene-17α,21-diol-3,20 dione 21-acetate						75	81		73		58	80					34	52	23	54	60	74	
Pregn-4-ene-17α,21-diol-3,20-dione 17,21-diacetate						85																	
17α-Hydroxyprogesterone caproate																	35	61	41	53	60	71	83
Pregn-4-ene-17α,21-diol-3,11,20-trione 21-acetate						68			62		53	70											
Prednisone acetate						64			63		49	67											

[a] A = Benzene–acetone (4:1); B = benzene–methanol (9:1); C = ether; D = chloroform–acetone (3:1); E = chloroform–methanol (97:3); F = benzene–dioxane (2:1); G = benzene–dioxane (1:1); H = benzene–dimethylformamide (9:1); I = chloroform–ethanol (99:1); J = chloroform–ethanol (96:4); K = benzene–ethanol (95:5); L = diisopropyl ether; M = hexane–ethyl acetate (4:1); N = hexane–ethyl acetate (1:1); O = ethyl acetate; P = benzene-2-propanol (4:1); Q = benzene–ethanol (98:2); R = benzene–ethanol (9:1); S = chloroform–ethanol (98:2); T = ether–ethanol (98:2).

[b] Deactivated by shaking with 2.5% acetic acid for 1 hr.

[c] Silica gel with 5% rice starch binder.

[d] Without binder.

Table 29.8 $R_{st} \times 100$ Values (Referred to Cortisone) of C_{21} Steroids in Various Systems in a Saturated Atmosphere on Silica Gel G

	Ref. 204			Ref. 205					
Steroid	*Chloroform–Ethanol* (9:1)	*Chloroform–90% Methanol* (9:1)	*Cyclohexane–Ethyl Acetate* (1:1)	*Chloroform–Acetone* (9:1)	*Chloroform–Acetone* (4:1)	*Cyclohexane–Chloroform–Acetic Acid* (7:2:1)	*Methylene Chloride–Acetone* (4:1)	*Chloroform–Acetic Acid* (9:1)	*Methylene Chloride–Acetic Acid* (9:1)
3α-17α,21-Trihydroxypregnane–11,20-dione	53	38	59						
3α,11β,17α,21-Tetrahydroxypregnan-20-one	33	22	55						
Pregn-4-ene-17α,21-diol-3,20-dione	116	131	230						
17α-Hydroxyprogesterone	145	152		470	300	235	206	<400	<533
Pregn-4-en-21-ol-3,11,20-trione	125	160	100	80	85	78	58		
Pregn-4-en-21-ol-3,20-dione	140	200	300						
3β,11,21-Trihydroxyallopregnan-20-one	73	75	100						
3α,21-Dihydroxypregnane-11,20-dione	79	91	40						

11β,17α,21-Trihydroxypregnane-3,20-dione	70	68	130						
17α,21-Dihydroxypregnane-3,11,20-trione	108	92	140						
17α,21-Dihydroxypregnane-3,20-dione	115	128	360						
3α,17α,21-Trihydroxypregnan-20-one	80	59	170						
Prednisone	94	82	80	80	85	78	58	80	80
Prednisilone	62	45	75		20	22	45	40	53
Dexamethasone	72	51	140	60	55	35	52	50	67
Triamcinolone	0	23	10						
Cortisone	100	100	100	100	100	100	100	100	100
Hydrocortisone	66	50	75	30	30	35	74	40	73
Corticosterone	102	119	100						
Aldosterone	85	85	35		30	109	37	100	146
Pregnenolone				520	290	287	213	<400	<533
Pregn-4-en-3α-ol-20-one				490	300	283	216	<400	<533
Pregnane-3,20-dion-21-ol				530	295	270	219	<400	<533
Allopregnane-3α,20α-diol				390	220	200	200	355	<533
Pregnane-3α,20α-diol				240	170	152	197	325	500
16-Methylene-1-dehydro-11β,17α-dihydroxyprogesterone				180	170	196	136	215	233
16-Methyleneprednisolone					15	52	55	40	73

solvent systems. Johannesen and Sandal [213] used ethyl acetate and a mixture of chloroform–acetone (3:1) for investigating five pharmaceutically important corticosteroids on silica gel. The steroids were extracted from tablet material by using ether. Tschesche and Snatzke [214] used diisopropyl ether and diisopropyl ether–acetic acid (5:2) for a group of pregnane compounds.

Vaedtke and Gajewska [215] used Celite #545 (Johns Manville) impregnated with formamide and applied the Zaffaroni solvent systems of paper chromatography for their separations. One of the big advantages of this system was the rapidity of development, for the chromatograms could be developed for a distance of 10 cm in 4 to 5 min. The solvents used were formamide-saturated benzene–chloroform (1:1) and chloroform saturated with formamide. The latter solvent was also used for the separation of the acetates of the compounds. After chromatographing, the plates were dried in an oven at 90 to 100°C to remove the impregnating agent and the developing solvents; they were then sprayed with the detecting agent. A number of these, among which were concentrated phosphoric acid and triphenyltetrazolium chloride, proved to be quite useful. Yawata and Gold [216] also applied the Zaffaroni systems as well as the Bush systems to the separation of 11 corticosteroids. It was observed that an 11-keto group increased the mobility considerably more than a corresponding C-11 or C-17 OH function. The Bush systems showed somewhat better separation than the Zaffaroni systems. Goeldel et al. [217] also applied the Bush and Zaffaroni type solvents using silica gel layers as a support for the stationary phase. The layers were impregnated with a 40% solution of formamide in acetone by the capillary ascent method. A number of different developing methods were applied. Two-dimensional chromatography was carried out using, for example, chloroform-saturated formamide in one direction followed by development in the second direction either with or without removal of the stationary phase; by removing the second phase, the development in the second direction became an adsorption process and the adsorptive power of the silica gel was apparent. The adsorptive strength of the silica gel could be decreased, both during the partition chromatography and the development after removal of the formamide stationary phase, by exposing the plates to water vapor. Other mixtures of hexane, benzene, chloroform, and butyl acetate, all saturated with formamide, were used as solvents. Continuous descending chromatography was also used to separate the more polar steroids of aldosterone, cortisol, and corticosterone using formamide–butyl acetate–water (1:20:1) in a 6-hr development. As another technique, the step method of development was applied to two-dimensional work (Fig. 29.2). Development in the first case was with

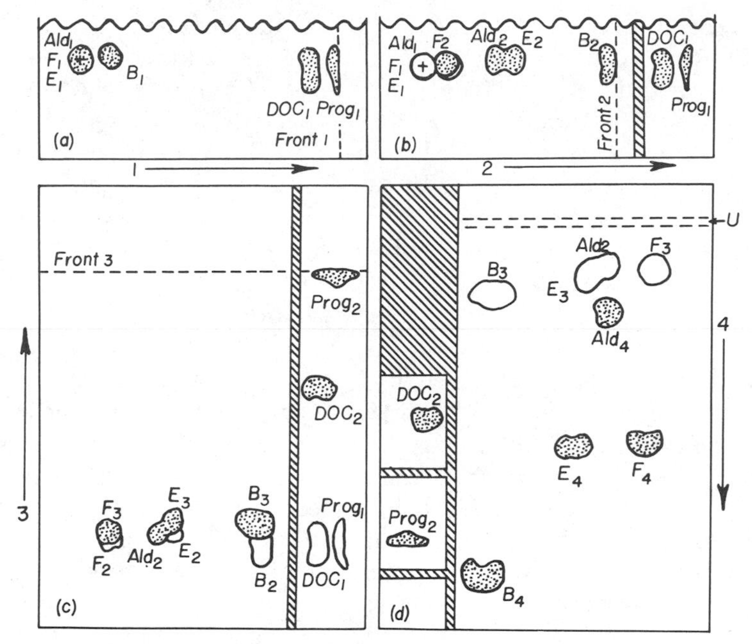

Fig. 29.2 Example of two-dimensional thin-layer chromatography with repeated development for the separation of a corticosteroid mixture. *(a) 1,* Formamide-saturated benzene, ascending; *(b) 2*, run in the same direction with formamide-saturated benzene–chloroform (1:7) (ascending) after scraping out the area designated by shading; *(c) 3*, run at right angles to the second run, ascending in formamide–*n*-butyl acetate–water (1:20:1); *(d) 4*, run in same direction and same solvent as *3*, but descending for 5 to 6 hr after the designated zones are scraped off. Stationary phase: silica gel impregnated with a 40% solution of formamide in acetone. Ald = aldosterone, F = cortisol, E = cortisone, B = corticosterone, Prog = progesterone, DOC = desoxycorticosterone. From Goeldel et al. [217]; reproduced with permission of the authors and Walter de Gruyter and Co.

formamide-saturated benzene to within 1 cm of the upper edge of the plate. After this development the cortisone and progesterone which traveled near the front of the solvent were separated from the remaining steroids by scratching a line through the adsorbent layer to the plate. The plate was then developed in the same direction with a mixture of benzene–chloroform (1:7) (again saturated with formamide) till the solvent front reached a line approximately 1 cm from the separating line. The third development was at right angles to the first and second developments using formamide–*n*-butyl acetate–water (1:20:1) as the developing solvent. The cortexone and the progesterone, which by this time were well separated, were isolated from the remainder of the plate by scratching in separation lines. After this plate was developed for a fourth time in the same direction as the third separation, but by descending chromatography in the same solvent for a period of 5 to 6 hr which resulted in complete separation of all six compounds. Butruk and Vaedtke [218] chromatographed the urinary metabolites of hydrocortisone on kieselguhr layers impregnated with 10% ethanediol in acetone. Ethanediol–dichloromethane was used as a developing solvent.

Quesenberry and Ungar [219] have used thin-layer chromatography for the isolation of aldosterone and the hydroxy steroids from bovine adrenal extracts; Korzun and Bordy [220] have used thin-layer chromatography for the identification of deoxycorticosterone and aldosterone acetates as well as other steroids in sesame oil preparations without doing a preliminary extraction. Lábler and Šorm [221] chromatographed some 18-substituted hydroxycortisones on loose layers of alumina in several solvents.

With a rapid separation technique of this type, it is inevitable that the method would be increasingly used for medical work; an ever-increasing number of papers has appeared on this subject. Scheiffarth and co-workers [222–224] have used thin-layer chromatography in the study of the absorption and metabolism of various cortisone derivatives. Bernauer and Schmidt [225–229] have examined the corticosteroids in the guinea pig under various conditions. Pasqualini and Jayle [230] have isolated and identified 5β-pregnane-11β,17α,21-triol-3,20-dione (dihydrocortisol) in the urine of patients with feminizing cortical cancer.

Auvinen and Favorskaya [231] have used thin-layer chromatography for the separation of 2,4-dinitrophenylhydrazones of steroid ketones. Watson and Romanoff [232] were able to show the presence of two isomeric 2,4-dinitrophenylhydrazones for progesterone-mono-DNPH and progesterone-bis-DNPH by chromatographing the derivatives on silica gel with chloroform–benzene (3:1) or benzene–acetic acid (19:1).

The five *O*-methyloximes of progesterone have been separated on alumina with benzene [178].

Takeuchi [233] has examined 12 pregnane steroids along with a group of C_{18} and C_{19} steroids in 28 solvent systems on silica gel layers.

Using nine different solvent systems, Lisboa [234] investigated the separation and identification of 32 Δ^4-3-keto steroids of the pregnane series on silica gel G. A preliminary run in chloroform–ethanol (9:1) was used to separate the steroids into four major groups. Those of group one, with an R_f value of less than 0.35, were separated by two-dimensional chromatography using the chloroform–ethanol solvent and ethyl acetate–hexane–ethanol–acetic acid (72:13.5:4.5:10). Compounds with R_f values between 0.35 and 0.50 are in group two and could be separated by two-dimensional chromatography in the four-component solvent system and benzene–ethanol (4:1). The third group, with R_f values between 0.50 and 0.67, were chromatographed in the four-component solvent system and in cyclohexane–ethyl acetate–ethanol (9:9:2). The fourth group, which contains compounds with R_f values above 0.67 in the chloroform–ethanol solvent, were separated in the cyclohexane–ethyl acetate–ethanol solvent. Seven different color reactions were given for the identification of the steroids spots. This work was followed [235] by a publication giving 30 color reactions for identifying 37 Δ^4-3-keto-C_{12} steroids on thin-layer chromatograms. The sensitivity, specificity, and optimum conditions are given as well as reaction mechanisms when known.

Lisboa [236] chromatographed 40 21-deoxy-Δ^4-3-oxo steroids on silica gel and silica gel impregnated with silver nitrate with eight solvent systems. Complete separation of all the compounds could not be achieved. Stevens [237] chromatographed a group of corticosteroids and related compounds synthesized from sapogenins. Silver nitrate-impregnated layers were used with chloroform–ethyl acetate (1:1 and 9:1), chloroform–toluene (1:1 and 4:1), and methyl acetate–methylene chloride (4:1).

Frgačić and Kniewald [238] found that ascorbic acid incorporated in the silica gel layer protected corticosteroids from oxidation during the long exposure required for autoradiography. The plates were sprayed with a saturated solution of ascorbic acid in absolute ethanol. The ascorbic acid extinguished the fluorescence of normally fluorescent layers so that ultraviolet light could not be used for locating the compounds.

Taylor [239] developed a technique for separating nanogram amounts of labeled Δ^4-3-oxo steroids from mixtures containing microgram amounts of labeled Δ^5-3β-hydroxy steroids. This was accomplished by using a silica gel layer containing 10% digitonin in the silica gel–calcium sulfate mixture (13% $CaSO_4$). Development was for 4.5 hr in 2-methyl-2-propanol–water–ethyl acetate (5:2:5).

Morena Dalmau et al. [240, 241] have used insoluble polyvinylpyrrolidone layers in separating some C_{21} steroids. The layers were prepared from Polyclar AT (General Aniline and Film) ground and sieved to a 40-μ mesh size; no binder was used. The four anti-inflammatory corticosteroids, hydrocortisone, prednisolone, 9α-fluorohydrocortisone, and 9α-fluoroprednisolone were separated with hydrochloric acid (d = 1.19)–water (1:9) [240]. Betamethasone, dexamethasone, and paramethasone were separated on these layers using dicloromethane–acetic acid (94:6) [241]. Bishara [242] separated six 6-fluoro-16α-hydroxycosteroids on silica gel layers by two-dimensional development. The plate was developed three times in the first direction with benzene–ethyl acetate (1:1) and then twice at right angles with chloroform–acetic acid (20:1).

Whitehouse et al. [243] have stuided the use of iodine for detecting steroids on paper, and Matthews [244] has examined the color reaction with vanillin in sulfuric acid–ethanol solutions on thin-layer plates. In the latter case the 36 steroids examined showed a sensitivity of at least 5 γ/cm^2. Vecsei et al. [245] have increased the sensitivity of the tetrazolium reagent for the detection of corticosteroids on silica gel layers. This was done by incorporating 100 to 200 mg of tetrazolium blue directly in the silica gel slurry prior to preparing the plates. The plates were dried at room temperature for 24 hr. The solvent systems used for development did not dissolve the color reagent from the layers. These were chloroform–ethanol (9:1) and dichloromethane–benzene–acetone–ethanol (75:10:10:5). To develop the color, the chromatograms were sprayed with a methanolic solution of sodium hydroxide (10 g/100 ml 60% methanol). The reaction could be stopped at the desired point by spraying with 2 ml of formic acid in 100 ml of methanol. To prevent the layer from flaking, the finished chromatogram had to be fixed with a plastic spray.

Quantitative Determination

General Quantitative Work

Once the compounds have been separated by thin-layer chromatography, there are a number of methods that can be applied to the determination of the steroids. Stárka and Malíková [208] have employed a colorimetric method by using the color obtained on treatment with concentrated sulfuric aicd [212]. To carry this out, the adsorbent containing the steroid is scraped from the plate and eluted with a small amount of methanol which is made alkaline with concentrated ammonium hydroxide. The solvent is then evaporated to dryness in vacuo and the residue is mixed with 5 ml of concenrated sulfuric acid and allowed to stand for 2 hr at room temperature. The optical density is then determined. Standard and blank runs are made at the same time.

Bernauer [246, 247] has used the tetrazolium blue reaction. With this method the developed chromatogram is dried and sprayed with an alkaline tetrazolium blue solution so that it is throughly moistened. The blue-violet spots are then eluted with ethanol and measured at 546 mμ. A graded series of standard spots must be run on the same plate with the unknown compound in order to reduce sources of error. This color reaction, which depends on the -CO-CH_2-OH group on the G-17 atom, was also used by Cavina and Vicari [204]. They mixed the sample in 2 ml of anhydrous aldehyde-free ethanol with 0.2 ml of a 1% solution of tetramethylammonium hydroxide in ethanol and 0.2 ml of a 0.5% tetrazolium blue solution in anhydrous ethanol. The mixture was allowed to stand for 1 hr in the dark at 25°C after which 1 ml of acetic acid was added and the optical density determined at the absorption maximum. The fluorescein incorporated into the silica gel layer for the detection of the compounds did not interfere with the color reaction. As an alternative these authors also used a spectrophotometric measurement; however, in this case fluorescein could not be used in the layers since this interfered with the quantitative determination. Nishikaze et al. [248, 249] and Luisi and co-workers [250, 251] have also used the tetrazolium blue method for the quantitative determination of urinary corticosteroids. Adamec et al. [252] after locating the compounds with a tetrazolium blue spray determined them colorimetrically by the Porter–Silber reaction [253] with phenylhydrazine.

Bird et al. [254] measured the absorbance at 283 mμ in determining 6-chloro-17α-hydroxypregna-4,6-diene-3,20-dione acetate. Of course, the optimum point for measuring the absorbance will vary from compound to compound. Matthews et al. [255] have used a spectrophotometric method. Feher et al. [256] determined pregnanediol in urine by measuring the absorbance at 390, 425, and 460 nm after TLC separation. Pregnanetriol could be determined simultaneously from the same urine sample.

Hara et al. [257] designed a two-dimensional scanning densitometer with an autorecording integral calculator for the analysis of thin-layer plates. Among other things, this was applied to the measurement of steroids with an accuracy of $\pm 5\%$. A number of sources of error were determined (see Chapter XI, Section 2).

Molina-Andreu et al. [258] determined pregnanediol by direct densitometry after color development by spraying with sodium sulfite–sulfuric acid reagent. Belova et al. [259] separated urinary corticosteroids on silica gel layers containing tetrazolium blue. After separation the plates were sprayed with 10% sodium hydroxide in 60% methanol to develop the blue-violet zones. These were determined by direct densitometry. Uettwiller and Keller [260] used reflectance densitometry for the determination of hydrocortisone, cortisone, corticosterone, and compound S in plasma. Determinations could be made down to 0.1 to 0.2 μg per spot. Hamman and Martin [261] used direct fluorescent measurements for quantitative work.

For the determination of aldosterone, Bruinvels [262] eluted the steroid from the thin layer by means of 1 ml of concentrated sulfuric acid which was mixed with the silica gel and allowed to stand at room temperature for 1 hr; then after centrifugation, the solution was heated in an oil bath at 100°C for an additional hour. The fluorescence of the aldosterone was measured after cooling in ice water. For the determination of aldosterone in urine, Nishikaze and Staudinger [249] first carried out a preliminary purification to remove interfering materials and then separated the aldosterone from other steroids by two-dimensional thin-layer chromatography. Silica gel layers were used with cyclohexane–isopropanol (7:3) in the first direction and chloroform–acetic acid (4:1) in the second direction. Because of the small quantity of aldosterone present, it was important not to spray the spot with the locating reagent. Instead the aldosterone was found by reference to the positions of 11-desoxycortisol and 11-dehydrocortisone which were sprayed with tetrazolium blue reagent. The aldosterone could then be eluted quantitatively and determined colorimetrically with the tetrazolium blue.

By using special precautions Few and Forward [263] were able to determine corticosteroids down to the 0.001 μg level using a fluorescent determination after elution.

Futterweit et al. [264] have used gas chromatography for the analysis of progesterone eluates obtained from the thin-layer chromatography of human pregnancy plasma. Collins and Sommerville [265] also used gas chromatography to quantitatively determine progesterone in human plasma. The extract from the plasma was separated on silica gel G by running in benzene–ethyl acetate (3:2) followed by a second development in ether–dimethylformamide (99:1). The steroid spots were eluted with methanol–chloroform (1:1) and the elutes were then placed on a stainless steel gauze rectangle in the top of the vaporphase chromatographic column. The latter was a 5-ft column of 1% cyclohexane–dimethanol succinate on Chromosorb P, 60 to 80 mesh material, operated at 210°C. The progesterone peak was identified by mixing 0.25 γ of progesterone-4-^{14}C in the original plasma. Using this method, 0.01 γ of progesterone could be determined.

The trimethylsilyl ethers of hydroxylic steroids have been used for the gas chromatographic determination after separation of the ethers on silica gel with cyclohexane–ethyl acetate (9:1) [266].

Determination of Pregnanediol as an Early Pregnancy Test

Because the pregnanediol content of the urine increases to values of 7 to 10 mg/liter in 10 to 14 days after conception as compared to a value of 3 to 5 mg/-liter after ovulation in a normal menstrual cycle, Waldi et al. [267] have proposed a semiquantitative determination of this compound as an early test for pregnancy. This is based on a rapid thin-layer method; further publications have been issued on the method [268–270]. The test is carried out as follows: 2 ml of concentrated hydrochloric acid (sp gr 1.19) is added to 20 ml of a filtered (preferably 24 hr) urine specimen and heated for exactly 10 min at 90°C. After being cooled quickly under running water, the sample is transferred to a 100-ml separatory funnel where it is extracted three times, each time with 25 ml of cyclohexane.

This extract is then washed twice with 20 ml of 1 *N* sodium hydroxide, twice with 30 ml of water, and then dried with 5 to 7 g of anhydrous sodium sulfate. The dried extract is filtered and the sodium sulfate is washed with a few milliliters of cyclohexane to effect a quantitative transfer. The solution is taken to dryness on a water bath at $< 40°C$. After the residue is transferred to a 20-ml thick-walled tube by four additions of chloroform, the latter is removed on a water bath. The residue is taken up in 0.5 ml of chloroform for addition to the silica gel layer.

The silica gel plate is spotted in the following order: (1) a dye control, (2) 2 μg of pregnanediol from a standard solution, (3) 50 μl of the urine extract, (4) 5 μg of pregnanediol, (5) 50μl of urine extract, (6) 10 μg of pregnanediol, (7), 50 μl of urine extract, (8) 15 μg of pregnanediol, (9) 50 μl of urine extract, (10) 20 μg of pregnanediol, (11) 50 μl of urine extract, (12) 30 μg of pregnanediol, (13) the dye control again. The chromatogram is then developed in a saturated atmosphere using a mixture of chloroform–acetone (9:1) as the solvent. If the pregnanediol R_f value occurs near the top of the plate instead of in the lower third, then a solvent mixture of chloroform–acetone (4:1) or methylene chloride–acetone (4:1) is used. The plate is developed for a distance of 10 cm which takes approximately 30 min.

For detection of the steriods the plate is first sprayed with a 40% phosphoric acid solution and heated for 7 to 20 min at 110°C. The pregnanediol appears as a pale greenish gray spot under ultraviolet light. Its location can be confirmed by spraying with 1.5% phosphomolybdic acid in 96% ethanol which gives an immediate blue color with pregnanediol. Evaluation of the quantity is made by comparison with the standard spots.

Other workers [208, 271–275] have used various modifications of this procedure.

5 BILE ACIDS

Qualitative Separations

A number of different systems have been employed for the separation of the bile acids and their derivatives. Frosch and Wagener [276] have used a solvent system of toluene–acetic acid–water (7.5:12.5:1) in separating the acids on silica gel G layers. In this system the taurocholic and taurolithocholic acids were separated from the taurine conjugates of the dihydroxycholanic acids. Taurodesoxycholic and taurochenodesoxycholic acids had the same R_f values. The acids were located by spraying with antimony trichloride in acetic acid (1:1) with observation under untraviolet light. The same authors [277] have also used two-dimensional chromatography on kieselguhr G layers. In this case the solvent for the first direction was a mixture of butanol–acetic acid–water (10:1:1) which was allowed to develop for 5 hr at 18°C in a saturated atmosphere. The plates were then removed, and the butanol evaporated by placing in a cool air stream followed by 12 hr of desiccation over silica gel. Development in the second direction was with toluene–acetic acid–water (7.5:12.5:1) for a period of 2 hr. Gaenshirt et al. [278] have also used toluene–acetic acid–water systems for the separation of cholic acids prior to their quantitative determination. With the silica gel layers they found the upper phase of a 5:5:1 mixture of these solvents to be the most suitable for the separation of individual cholic acids and for the separation of the individual acids from the conjugated bile acids. Resolution of both the free and the conjugated bile acids was accomplished with a 10:10:1 system. Here the detection was accomplished by means of a freshly prepared 5% solution of phosphomolybdic acid in ethanol–ether (1:1), the bile acids appearing as blue spots on a yellow background after heating for 5 min at 100°C. Kritchevsky et al. [279] used solvent mixtures of ether–petroleum ether–methanol–acetic acid (70:30:8:1) and isooctane-isopropyl ether–acetic acid (2:1:1) for separations on silica gel. As a detecting agent they used 0.5 ml of anisaldehyde in 1 ml of concentrated sulfuric acid and 50 ml of acetic acid. The various colors obtained were tabulated. Eneroth [280] has investigated the chromatographic behavior of 40 bile acids in 17 different solvent systems (Table 29.9). The compounds were detected by spraying with concentrated sulfuric acid and heating to 240°C. An examination of the separations achieved permitted a number of conclusions to be drawn as to the effect of various constituents on a separation, as far as the solvents used were concerned. Acids with a C-3 hydroxy group were more polar than the corresponding compounds with a C-7 or a C-12 hydroxy group. For monohydroxy acids the following order of polarity was observed: $3\alpha > 3\beta > 7\beta > 12\beta > 7\alpha \geqq 12\alpha >$ 3-keto > 7-keto $\geqq$ 12-keto. With multiple substituents the effect of individual groupings on the polarity of the compound was more difficult to ascertain; however, the polar effect of the 3α-hydroxyl group was readily apparent. A brief check with two methyl esters indicated that the methyl esters were more easily separated than the free acids.

Schwarz and Syhora [203] have separated a group of 12 bile acids on loose layers of aluminium oxide that had been treated with 2.5% acetic acid. Two solvent systems were used, chloroform and chloroform–ethanol (98:2). Several methyl esters were run in benzene–ethanol (97:3) and petroleum ether–chloroform (1:1).

Barbier et al. [15], using silica gel G layers, have separated a group of methyl esters of cholic, desoxycholic, and etianic acids. Three different solvent mixtures of ethyl acetate–cyclohexane (3:17, 3:7, and 1:3) were used in obtaining an R_f value for each compound. For the separation of cholic acid and deoxycholic acid, starch-bound silicic acid plates were used with acetic acid as the solvent. With this system cholic acid had an R_f value of 0.32 and desoxycholic acid 0.62. Hara and Takeuchi [281–283] have separated a group of 57 bile acids and derivatives on silica gel plates using various mixtures of benzene or hexane with ether and ethyl acetate for the esters and ether–acetic acid for the acids.

Usui [284] separated keto bile acids on silicic acid layers with benzene–acetic acid (9:1 and 7:3). For visualizing these compounds the chromatogram was first treated with 5% sodium borohydride in 80% methanol, and then sprayed with a solution of 5 g phosphomolybdic acid in a mixture of 100 ml of acetic acid and 5 ml of concentrated sulfuric acid followed by heating at 110°C for 5 to 10 min. Koss and Jerchel [285] have separated several bile acids on silica gel layers supported on an aluminum foil. Schwarz [157] has chromatographed a group of seven bile acids on magnesium silicate with chloroform–ethanol (96:4), and mixtures of chloroform–acetic acid (99:1 and 96:4). Hofmann [286, 287] has used silica gel G with acetic acid–carbon tetrachloride–diisopropyl ether–isoamyl acetate–*n*-propanol–benzene (5:20:30:40:10:10) and with propionic acid–isoamyl acetate–water–*n*-propanol (15:20:5:10). With methyl isobutyl ketone as a developing solvent, a group of free and conjugated acids were also separated on hydroxyapatite activated at 160°C. Another paper by Hofmann [288] gives the chromatographic behavior for nine compounds, and the use of thin layers on microscope slides for bile acid derivatives has been reported [289].

Gruette and Gaetner [290] preconditioned their silica gel G layers by equilibrating in an atmosphere exposed to saturated lithium chloride for 30 min. A solvent composed of isooctane–acetic acid–isopropyl ether–isopropanol (10:6:5:1) was used for development. In order to separate bile acids and esters from fatty acids which can interfere in gas chromatography and spectrophotometric analysis, Spears et al. [291] developed the silica gel layers in a saturated atmosphere first with hexane–chloroform–diethyl ether–*n*-butanol–acetic acid (40:10:10:3:0.5). The plates were visualized under short-wave ultraviolet light and a line was scored under the fatty acids. A second development was then made with isooctane–ethyl acetate–*n*-butanol–*n*-propanol–acetic acid (20:10:3:3:3) to separate the bile acids. For methylated bile acids, the second development was made with isooctane–isopropanol–acetic acid (120:40:1). If streaking of deoxycholic acid occurred, the *n*-propanol was omitted and the acetic acid concentration was increased.

Numerous solvent systems have been used with silica gel; Huang and

Table 29.9 $R_{st} \times 100$ Values of Bile Acids on Silica Gel G in Various Solvent Systems[a,b]

Compound[c]	R_D 5.0 *A*	R_L 9.7 *B*	R_D 5.4 *C*	R_C 4.7 *D*	R_C 8.5 *E*	R_C 4.0 *F*	R_C 5.2 *G*	R_C 5.1 *H*	R_C 9.0 *I*	R_C 4.7 *J*	R_D 9.8 *K*	R_L 12.0 *L*	R_D 9.1 *M*	R_D 5.7 *N*	R_L 9.1 *O*	R_L 8.9 *P*	R_L 7.5 *Q*
$3\alpha,7\alpha,12\alpha,23\xi$		13	4	17	12	7	13	28	6	8							
$3\alpha,7\alpha,23\xi$			15	62	43	25	85	43	26	28							
$3\alpha,7\alpha,16\alpha$			67	208	141	250	214	254	162	179	88		72				
$3\alpha,7\alpha,12\alpha$-C_{27}			13	152	128	161	196	188	138	189	85						
$3\alpha,7\alpha,12\alpha$	34	34	17	100	100	100	100	100	100	100	76	50	22	16			
$3\beta,7\alpha,12\alpha$				127	114	144	100	100	87	100							
$3\alpha,7\beta,12\alpha$			30	146	117	147	159	175	122	140							
$3\alpha,6\alpha,7\alpha$			30	135	105	130	147	149	112	136							
$3\alpha,6\beta,7\alpha$				115	98	117	117	114	82	83							
$3\alpha,6\beta,7\beta$				125	89	111	112	86	92	87							
$7\alpha,12\alpha$,3-keto			131	282	155	294	228	302	166	204	87	81	63	126			
$3\alpha,12\alpha$,7-keto			45	195	115	192	183	232	137	163	66	62	44	60			
$3\alpha,7\alpha$,12-keto		66	45	195	115	194	185	232	137	163	66	62	48				
3α,7,12-diketo	116	85	107	256	130	258	194	265	152	139	54	53	52				
$3\alpha,7\alpha$	100		100								91	82	88	100			
$3\beta,7\alpha$			126								85	73	95	133			
$3\alpha,7\beta$			114								77	77	78	100			
$3\alpha,12\alpha$	100	86	100	262	128	259	348	368	167	233	100	83	100	100	21		
$3\beta,12\alpha$			136								96	81	106	140			
$3\alpha,12\beta$			158								100	87	109	153	34		
$3\alpha,6\alpha$		46	63	186	112	180	202	224	124	170	67	71	50	56	10		
$7\alpha,12\alpha$			251									98	162	223	80	84	
3,7,12-Triketo	158	100	213								55	46	91	178	46	41	41

3α,7-Keto	138	90	163	88	70	95	144	45	42	
3α,12-Keto	138		178	103	80	117		55	54	
7α,3-Keto		99	213		93	135	211	74		
12α,3-Keto			185	122	87	145		54		
3,7-Diketo	156	108	265	110	85			93	91	93
3,12-Diketo	156	108	265	117	91			93	91	95
3α	156	100	254	128	100	160	227	100	100	100
3β					104			113	108	115
7α		112	320		108			135	124	140
7β					101			119	110	121
12α					112	171		137		140
12β					112			128		132
3-Keto	162	113	310		112			150	121	135
7-Keto					116			162	127	150
12-Keto					116			162	127	150
Unsubstituted								173	158	189

[a] From P. Eneroth [280]; reproduced with permission of the author and The American Institute of Biological Sciences.

[b] Reference compound: R_C = chloric acid, R_D = desoxycholic acid, R_L = lithocholic acid. The next value is the reference mobility, the distance (in cm) traveled by the reference compound. Solvent distance 17 to 18 cm in unsaturated system. Solvent: *A* = diethyl oxalate–dioxane (40:10); *B* = diethyl oxalate–isopropanol (48:8); *C* = benzene–dioxane–acetic acid (75:20:20); *D* = benzene–dioxane–acetic acid (20:10:2.0); *E* = benzene–dioxane–acetic acid (15:5:2.0); *F* = benzene–dioxane–acetic acid (55:40:2.0); *G* = cyclohexane–ethyl acetate–acetic acid (10:15:4.0); *H* = cyclohexane–ethyl acetate–acetic acid (7:23:3.0); *I* = benzene–isopropanol–acetic acid (30:10:1.0); *J* = cyclohexane–isopropanol–acetic acid (30:10:1.0); *K* = trimethylpentane–isopropanol–acetic acid (30:10:1.0); *L* = trimethylpentane–isopropanol–acetic acid (60:20:0.5); *M* = trimethylpentane–ethyl acetate–acetic acid (10:10:2.0); *N* = trimethylpentane–ethyl acetate–acetic acid (5:25:0.2); *O* = trimethylpentane–ethyl acetate–acetic acid (50:50:0.7); *P* = trimethylpentane–ethyl acetate–acetic acid (10:10:0.25); *Q* = trimethylpentane–ethyl acetate–acetic acid (10:10:0.1).

[c] Hydroxyl groups are indicated by Greek letters. A coprostanic acid is indicated by $-C_{27}$.

Nichols [292] examined 14 of these and compared results with their solvent system of chloroform–methanol–water (70:25:3). The R_f values are listed. Ikawa and Goto [293] list the R_f values for 16 conjugated and unconjugated bile acids in six solvents on silica gel H (type 60). Two-dimensional separation with chloroform–ethanol–28% ammonia (25:35:1) in the first direction and chloroform–ethanol–ethyl acetate–acetic acid–water (8:6:5:4:1) in the second was helpful in obtaining separations of some complex mistures.

In order to remove lipids from serum extracts prior to separating the free and conjugated bile acids, Begemann [294] developed the silica gel layer with chloroform–ethyl ether (6:1) continuously. The bile acids, which remained at the origin, were then chromatographed with butanol–acetic acid–water (100:7:5). Panveliwalla et al. [295] removed lipids by developing in a descending manner with chloroform for 2 hr. The free bile acids and the glycine conjugates were then separated by developing with 2,2,4-trimethylpentane–isopropanol–isopropyl ether–acetic acid (2:1:1:1) for 15 hr in an ascending technique. The taurine conjugates were then separated by "wedge shaping" each lane at the origin and developing with propionic acid–isoamyl acetate–water–*n*-propanol (3:4:1:2).

Polyamide has been used for the separation of six structural isomers of di- and trihydroxycholanic acids by chromatographing with water–2-butanone–25% ammonia (16:4:1) or with 0.05 *N* sodium hydroxide [296]. The acids were visualized with phosphomolybdic acid reagent.

Anthony and Beher [297] examined the colors obtained for 12 bile acids and six conjugated bile acids with four detection reagents. Other detection reagents that can be used for bile acids and their conjugates include: ceric ammonium sulfate–molybdate reagent (T-52) [298], a modified Kagi–Mischer reagent (T-26) [299], and manganous chloride (T-159) [300]. All three of these reagents can differentiate between chenodeoxycholic and deoxycholic acids, which are not always easily distinguished by their R_f values. The manganous chloride reagent is also useful to differentiate cholesterol (pink color which fades after 5 min) from the bile acids.

Quantitative Determination

Gaenshirt et al. [278] has used a spectrophotometric method for the determination of the bile acids. After the components in the systems were separated as described above, the chromatographic plates were dried for 20 min at 100°C. The compounds were located by spraying with water, thus showing the cholic acids as white specks against the translucent background. The spots were scraped into a centrifuge tube and treated with 3 ml of 65% sulfuric acid for 1 hr at 60°C (the wavelength of the absorption maximum varies with the heating time and 60 min was determined as the optimum period). The absorption curves of the different compounds were then determined in the region of 300 to 450 mμ against a corresponding silica gel blank. The quantitative measurements were made at 380 mμ because below 320 mμ a strong background absorption was obtained from treating silica gel with sulfuric acid. Lithocholic acid could not be determined as it had no absorption maximum at 380 mμ. Wollenweber et al. [301] used this same technique for determining the chenodeoxycholic and deoxycholic acid contents in human duodenal contents. Frosch and Wagener [302–304] have also used a spectrophotometric method for the determination of bile acids. In this case the amount of each component was determined by measuring the sum and difference of the absorbencies at 412 and 408 mμ and then comparing the values for known mixtures.

Eastwood et al. [305] determined bile acids and conjugates by using the Pettenkofer chromogen. This is developed by adding 5 ml of 70% sulfuric acid to the scraped-off silica gel spot and heating for 10 min at 42 ± 2°C. Then 1 ml of 0.25% furfural is added and the mixture allowed to stand for 1 hr. The centrifuged supernatant is then measured at 510 mμ, 90 min after the addition of the furfural.

Hara et al. [306] has applied a direct densitometric method for the determination of bile components. Two solvents were used for the separation on silica gel plates; these were isoamyl alcohol–acetic acid–water (18:5:3) and benzene–acetic acid–water (10:10:1). Prior to application of the sample the plates were examined in the densitometer for uniformity of absorption. After development, the chromatograms were heated at 130°C for 1 hr to evaporate the solvents and then they were sprayed uniformly with concentrated sulfuric acid. A uniform heat of 60 to 80°C was applied for 20 min to develop the color. Semiquantitative results were obtained by comparing the values with a standard curve. Bile samples from 21 species of animals were examined and found to give distinctive patterns for each species.

Semenuk and Beher [307] used a direct densitometric method after spraying the layer with a 20% solution of phosphomolybdic acid in ethanol and heating at 110°C for 15 min. Kim and Kritchevsky [308] used the charring technique for direct densitometry.

Cohen et al. [309] converted the bile acids to their trimethylsilyl ethers after thin-layer chromatography for determination by gas chromatography.

6 THE STEROIDAL SAPOGENINS AND SAPONINS

The saponins are glycosides which produce stable foams when their aqueous solutions are shaken. By acid hydrolysis the saponins are split into sugars and the corresponding sapogenins. Silica gel layers can be used for the separation of the steroidal sapogenins. Tschesche and coworkers [310–312] have investigated a number of sapogenins using both paper and thin-layer chromatography. For the thin-layer work ethyl acetate, chloroform–acetone (4:1), and diisopropyl ether–acetone (3:1) were used as solvents. Takeda et al. [313] and Matsumoto [314] investigated a series of 25 different solvent systems for the separation of steroidal sapogenins and selected the most promising of the solvent group. Table 29.10 gives the R_f values of a group of 20 sapogenins in 11 different solvents.

Bennett and Heftmann [315] applied both adsorption on silica gel G and partition chromatography on kieselguhr plates. For adsorption chromatography the following solvent systems were employed: dichloromethane–methanol–formamide (93:6:1), toluene–ethyl acetate–formic acid (57:40:3), cyclohexane–acetone (1:1), cyclohexane–ethyl acetate–water (600:400:1 and 1000:1000:3), and chloroform–methanol–water (485:15:1 and 188:12:1). Two pairs of C-25 isomers, which could not be separated on silica gel G, were separated on a mixture of silica gel G–kieselguhr G (1:1) using a solvent of chloroform–toluene (9:1). This same solvent also gave a good separation of the acetates of the C-25 isomers on silica gel G. For the partition chromatography, kieselguhr G plates were impregnated with water by placing the adsorbent side down over a beaker of boiling water until they were thoroughly wet. The water was then allowed to evaporate from the plate until it began to recede from the corners of the layer at which time it was transferred to the developing chamber. The amount of water on the plate was critical because too much water would cause spreading of the spots with a resultant poor separation, and too little water would cause tailing. For the partition chromatography the chamber was equilibrated by placing a sheet of filter paper dipping into the solvent system on one side of the chamber and a water-saturated sheet of paper, which did not touch the solvent, on the other side. The trifluoroacetates could be separated in the chloroform–toluene solvent mixture. These derivatives could be prepared by applying trifluoroacetic anhydride directly on the sample spot of the hydroxy steroid. The plate was then dried thoroughly to remove the trifluoroacetic acid which was formed as a by-product. As an alternative 2 μl of trifluoroacetic anhydride could be added directly to 0.2 ml of a 0.01 to 0.1% solution of sapogenins in hexane or dichloromethane.

Table 29.10 $R_f \times 100$ Values of Some Steroid Sapogenins on Silica Gel in Various Solvent Systems. Development Distance 15 cm[a]

Sapogenin	*Chloroform–Ethanol* (95:5)	*Chloroform–Acetone* (9:1)	*Benzene–Acetone* (85:15)	*Benzene–Methanol* (92:8)	*n-Hexane–Ethyl Acetate* (1:1)	*n-Hexane–Acetone* (4:1)	*Benzene–Ethanol* (85:15)	*Benzene–Ethanol* (92:8)
Luvigenin	87	76	80	85	80	78	93	72
Neometeogenin	81	52	63	66	64	48	85	48
Meteogenin	81	52	63	66	64	48	85	48
Sarsasapogenin	65	39	46	51	51	42	67	31
Diosgenin	59	35	41	42	46	34	57	25
Tigogenin	59	35	39	39	46	29	53	23
Pennogenin	53	26	30	42	35	22	50	27
Gentrogenin	49	24	25	32	26	16	42	24
Hecogenin	46	22	22	37	21	16	50	25
Convallamarogenin	39	21	24	30	28	22	42	22
Isorhodeasapogenin	39	18	25	38	28	22	47	24
Rhodeasapogenin	37	18	25	32	28	22	47	22
Nogiragenin	27	11	15	20	18	15	37	18
Heloniogenin	26	7	13	16	16	11	30	18
Yonogenin	21	11	11	17	13	9	30	18
Gitogenin	19	7	11	16	9	11	35	15
Tokorogenin	9	2	2	12	3	4	27	10
Metagenin	6	1	1	9	2	1	23	7
Kitigenin	4	1	1	10	0	1	20	0
Kogagenin	3	1	1	8	0	0	23	5

[a] From N. Matsumoto [314]; reproduced with permission of the author and the Pharmaceutical Society of Japan.

This mixture was shaken thoroughly for 1 min and then neutralized with 1 ml of 2 *N* aqueous sodium carbonate solution. The crude mixture in the organic solvent was spotted directly.

Blunden and Hardman [316] used silica gel G for the separation of the crude sapogenin mixtures from *Dioscorea* tubers. Sander and co-workers [317–319] have used thin-layer chromatography extensively in their work with *Solanum* and other sapogenins. Boll [320, 321] has examined the sapogenins from 13 different geographical strains of *Solanum dulcamara*. Kuhn and Loew [322] have worked with the sapogenins of *Solanum chacoense*. Chiarlo [323] isolated the sapogenins from the flowers of *Nerium oleander*. Schreiber et al. [324] investigated 10 steroidal sapogenins.

Smith and Foell [114] have separated a group of 14 sapogenins and their acetates on starch-bound silica gel layers (Table 29.11).

Dawidar and Fayez [325] separated the sapogenins in fenugreek on alumina G impregnated with silver nitrate and on silica gel G with *n*-hexane–ethyl acetate (5:1) and (20:1). The acetates separated with a (100:7) solvent mixture were determined photometrically after treating the layer with phosphomolybdic acid and heating. Held and Vágújfalvi [326] developed a TLC screening method for the analysis of both neutral and basic sapogenins. After hydrolysis to free the conjugated acids, the

Table 29.11 $R_f \times 100$ Values of Steroidal Sapogenins on Starch-Bound Layers of Silica Gel[a]

Sapogenin	*Hexane–Ethyl Acetate* (4:1)	*Hexane–Ethyl Acetate* (1:1)	*Ethyl Acetate*
Diosgenin	18	67	—
Diosgenin 3-acetate	73	—	—
Tigogenin	18	67	—
Tigogenin 3-acetate	74	—	—
Smilagenin	26	72	—
Smilagenin 3-acetate	73	—	—
Hecogenin	2	31	58
Hecogenin 3-acetate	26	89	—
Gentrogenin 3-acetate	25	89	—
Sarsapogenin 3-acetate	68	—	—
Chlorogenin	0	4	18
Kryptogenin	0	22	25
Pennogenin	4	48	68
Tomatidine	0	2	3

[a] From L. L. Smith and T. Foell [114]; reproduced with permission of the authors and The Elsevier Publishing Co.

extract is chromatographed on silica gel with cyclohexane–ethyl acetate–water (60:40:0.1). In this system the six 3-β-mono-ol type neutral sapogenins can not be separated, but additional solvent systems on silica gel layers can be used. The C_{25} epimers can be separated with dichloromethane–ethyl ether (199:1), their acetates with dichloromethane–toluene (7:3). The saturated–unsaturated pairs can be differentiated with benzene–ethyl ether (1:1).

Using a solvent system of butanol saturated with 5% acetic acid, Madaeva and Ryzhkova [327] have used cellulose layers bound with gypsum for the separation of saponins of *Dioscorea tokoro* and *D. gracillima*. Carreras Matas [328] used formamide-impregnated silica gel layers for the partition of steroidal saponins. Van Duuren [329] has separated nonsteroidal sugar beet saponins on silica gel layers with hexane–ethyl acetate (1:1) and butanol–acetic acid–water (4:1:1); Khorlin et al. [330] have chromatographed triterpenoid saponins. Boll [320, 321] has used both silica gel and neutral alumina for separating the saponins of *Solanum dulcamara*. The solvent systems were chloroform–methanol (19:1) and the upper phase of ethyl acetate–pyridine–water (3:1:3). Kawasaki and Miyahara [331] give the R_{st} values (referred to trillin) for a series of 14 steroidal saponins on gypsum-bound silica gel layers using four different solvent systems: butanol saturated with water, chloroform–methanol–water (65:35:10) (lower phase), chloroform–methanol (4:1), and butanol–acetic acid–water (4:1:5) (upper phase). Values are also given for the methylated and acetylated derivatives.

Wolf and Thomas [332] chromatographed the saponins of soybeans on silica gel by multiple development (6×) with chloroform–methanol–water (65:25:4) and the sapogenins by single development with petroleum ether–chloroform–acetic acid (7:2:1).

Pasich [333], after studying seven different methods for the quantitative determination of saponins, introduced the method of separation on thin-layer plates followed by spectrophotometric determination of the isolated saponins. The saponins were separated with *n*-butanol–acetic acid–water (6:1:3) on silica gel layers, and iodine vapor was used to locate the spots. After the location was marked, the iodine was sublimed off and the spots were then removed and treated with the ferric chloride in the acetic acid–sulfuric acid reagent of Zak [334]. After centrifugation, the clarified solution was measured at 445 mμ. The error of the analysis amounted to ± 1.25%.

7 STEROIDAL ALKALOIDS

This group of compounds is related to the sapogenins and occur in nature as glycosides. The best known of these are of course the solanidines which have been isolated from potato sprouts *(Solanum tuberosum)*.

Solanine, chaconine, and solanidine were separated on silica gel G layers [335], whereas solanidine could not be isolated by paper chromatography. Of eight solvent mixtures tabulated, the best was acetic acid–95% ethanol (1:3) which gave R_f values for the three compounds of 0.22, 0.54, and 0.62, respectively. These values were for a development distance of 12 to 15 cm. Schreiber and co-workers [324, 336–340], Boll [320, 321], and Kuhn and Loew [322] have applied thin-layer chromatography in their work with the *Solanum* alkaloids. Thin layers included silica gel, silica gel treated with silver nitrate, and alumina. Bite et al. [341] investigated *Solanum laciniatum* alkaloids. The glycosides were chromatographed on silica gel with butanol saturated with water, and the aglycons were chromatographed on aluminum oxide with chloroform as the developing solvent.

Lábler and Černý [342, 343] and co-workers [344] have investigated the steroidal alkaloids of *Holarrhena antidysenterica*. Fifty-two of the bases were studied [342] on thin layers of silica gel using benzene and ether, both saturated with ammonia. The developing chamber was lined with filter paper and in addition contained a dish filled with concentrated ammonia solution.

Zeitler [345] has chromatographed some of the steroidal veratrum al-

Table 29.12 $R_f \times 100$ Values of Some Bufogenins on Silica Gel G Layers Activated at 100°C[a]

Bufogenin	*Ethyl Acetate*	*Ethyl Acetate–Cyclohexane* (4:1)	*Ethyl Acetate–Acetone* (9:1)	*Ethyl Acetate Saturated with Water*
Resibufogenin	61	61[b]	60	66
Bufalin	62	61	64	62
Bufotalinin	31	22	39	34
Marinobufogenin	43[b]	33[b]	47	50
Gamabufotalin	31	26	37	47
Telocinobufogen	34	23	28	37
Hellebrigenin	28	18	25	30
Hellebrigenol	9	7	17	23

[a] From R. Zelnik and L. M. Ziti [346]; reproduced with permission of the authors and The Elsevier Publishing Co.

[b] Average of two determinations.

kaloids and their derivatives on silica gel HF (Merck) with mixtures of cyclohexane–diethylamine (9:1 and 7:3); separations were increased by multiple development.

8 THE TOAD POISONS

Zelnik and co-workers [346, 347] applied thin-layer chromatography in the isolation and identification of the bufadienolides or bufogenins isolated from the parotid glands of the Brazilian toad. The R_f values of these compounds in four different solvents are given in Table 29.12. Ishikawa and Miyasaka [348] have used a two-step method for these compounds using acetone–chloroform (7:13) followed by hexane–ethyl acetate (7:13).

Komatsu and Okano [349] preferred alumina activated at 180°C for 120 min for the separation of bufosteroids using acetone–chloroform–cyclohexane (3:3:4) or ethyl acetate–ethyl ether (1:1). Both single and double development were used.

9 MISCELLANEOUS

Golab and Layne [350] have examined the chromatographic behavior of 38 19-nor steroids. The R_f values in ethyl acetate–cyclohexane (1:1 and 3:7) have been tabulated, as well as the color reactions with antimony trichloride.

Reviews on the thin-layer chromatography of steroids have been written by Neher [351, 352], Beijleveld [353], Kazuno and Hoshita [354], and Lisboa [355–357] to mention a few.

Rivlin and Wilson [358] and Richardson et al. [359] have applied silica gel layers to the recovery of radioactive-labeled steroids from scintillation mixtures. This enables the radioactive materials to be reused.

Metz [360] used thin-layer chromatography to follow enzymatic steroid transformations. He employed a useful technique for concentrating dilute solutions before applying to thin layers. This involves applying the spot to a short length of filter paper which has been shaped to a point. The base of the paper is then dipped in acetone and the solvent carries the spot of steroid, or other material, to the tip of the paper. It is removed, dried, and processed repeatedly until the desired amount of sample is concentrated in the tip of the paper. This tips is then placed against the thin-layer plate and gradully eluted by allowing acetone to elute the material from the paper.

Dannenberg et al. [361–363] have used thin-layer chromatography in purifying products obtained in the dehydrogenation of steroids.

Stevens [364] has examined the use of three locating reagents that can be used for almost all compounds obtained during the synthesis of corticosteroids from sapogenins and also of synthetic C-21 and C-22 intermediates. These reagents are (1) a freshly prepared mixture of a 1% solution of 2,5-diphenyl-3-(4-styrylphenyl)tetrazolim chloride salt in methanol with 3% solution of sodium hydroxide (1:10), (2) Komarowsky's reagent: 50% (vol/vol) sulfuric acid-2% *p*-hydroxybenzaldehyde in methanol (1:10) (after spraying, the plate is heated to 60°C for 10 min), and (3) a 30% solution of technical zinc chloride in reagent-grade methanol (after spraying the filtered solution, the plate is heated for 60 min at 105°C and then covered immediately with the second plate to prevent contact with atmospheric moisture). The zinc chloride reaction is observed under ultraviolet light at 366 mμ. The styrylphenyl derivative was recommended in preference to the commonly used blue tetrazolium reaction because of the more intense color produced. The effect of the location of various groupings on the use of Komarowsky's reagent and the zinc chloride was surveyed.

Haehnel and Muslim [365] published tables to assist in the identification of steroid conjugates. Biella Souza Valle and Martins Oliveira-Filho [366] applied a modified Hadler histochemical reaction on thin-layer plates to aid in the identification of steroids.

Pinelli and Formento [367] developed a combined thin-layer and gas chromatographic method for the analysis of submicrogram quantities of steroids in small urine samples. A combined enzymatic and chemical hydrolysis was used to free the steroids, after which they were separated on silica gel with diethyl ether–methanol (1:1). The trimethylsilyl ethers were used for the quantitative gas chromatographic determination. Vandenheuvel [368] showed that high concentrations of enzymes were needed in the enzymatic hydrolysis.

Hara and Mibe [369] chromatographed 23 synthetic hormonal steroids on silica gel and alumina with benzene–acetone (4:1). The R_f and ΔR_M values are given as well as the colors with three detection agents.

Lisboa [370] combined thin-layer chromatography, gas chromatography and mass spectrometry to the separation and identification of steroids. A number of examples are given where these techniques complement one another. Curtius and Mueller [371] used a direct coupled gas and thin-layer chromatographic method for steroids. After being separated as the trimethylsilyl ethers by gas chromatography, the esters deposited on the silica gel layers were hydrolyzed in situ by spraying with 1% hydrochloric acid in methanol. Development was then carried out with methylene chloride–acetone (7:1) or chloroform–ethanol (9:1).

Hamman and Martin [327] applied in situ reactions on thin layers to assist in identifying steroids.

References

1. J. Avigan, D. S. Goodman, and D. Steinberg, *J. Lipid Res.*, **4,** 100 (1963).
2. J. R. Claude, *J. Chromatogr.*, **17,** 596 (1965).
3. T. M. Lees, M. J. Lynch, and F. R. Mosher, *J. Chromatogr.*, **18,** 595 (1965).
4. J. W. Copius-Peereboom and H. W. Beekes, *J. Chromatogr.*, **9,** 316 (1962).
5. H. P. Kaufmann and T. H. Khoe, *Fette, Seifen, Anstrichm.* **64,** 81 (1962).
6. H. P. Kaufmann, Z. Makus, and T. H. Khoe, *Fette, Seifen, Anstrichm.* **64,** 1 (1962).
7. H. P. Kaufmann, H. Wessels, and B. Das, *Fette, Seifen, Anstrichm.* **64,** 723 (1962).
8. R. Ikan, S. Harel, J. Kashman, and E. D. Bergmann, *J. Chromatogr.*, **14,** 504 (1964).
9. R. Ikan and M. Cudzinovski, *J. Chromatogr.*, **18,** 422 (1965).
10. D. R. Idler and P. Wiseman, *Int. J. Biochem.*, **2,** 91 (1971).
11. D. I. Cargill, *Analyst,* **87,** 865 (1962).
12. W. H. J. M. Wientjens, R. A. De Zeeuw, and J. Wijsbeek, *J. Lipid Res.*, **11,** 376 (1970).
13. T. Kartnig and G. Mikula, *J. Chromatogr.*, **53,** 537 (1970).
14. N. Nicolaides, *J. Chromatogr., Sci.*, **8,** 717 (1970).
15. M. Barbier, H. Jaeger, H. Tobias, and E. Wyss, *Helv. Chim. Acta,* **42,** 2440 (1959).
16. D. L. Azarnoff and D. R. Tucker, *Biochim. Biophys, Acta,* **70,** 589 (1963).
17. L. L. Smith, W. S. Matthews, J. C. Price, R. C. Bachmann, and B. Reynolds, *J. Chromatogr.*, **27,** 187 (1967).
18. J. E. Van Lier and L. L. Smith, *J. Chromatogr.*, **41,** 37 (1969).
19. F. Neuwald and K. E. Fetting, *Pharm. Ztg. Ver. Apoth.-Ztg.*, **108,** 1490 (1963).
20. L. Acker and H. Greve, *Fette, Seifen, Anstrichm.*, **65,** 1009 (1963).
21. C. Horvath, *J. Chromatogr.*, **22,** 52 (1966).
22. R. D. Bennett and E. Heftmann, *J. Chromatogr.*, **9,** 359 (1962).
23. M. J. D. Van Dam, *Bull. Soc. Chim. Belg.*, **70,** 122 (1961).
24. N. Zoellner, *Z. Klin. Chem.*, **1,** 18 (1963).
25. N. Zoellner, K. Kirsch, and G. Amin, *Verh. Dtsch. Ges. Inn. Med.*, **66,** 677 (1960).
26. H. Weicker, *Klin. Wochenschr.*, **37,** 763 (1959).
27. H. Beukers, W. A. Veltkamp, and G. J. M. Hooghwinkel, *Clin. Chim. Acta,* **25,** 403 (1969).
28. J. R. Claude and J. L. Beaumont, *Ann. Biol. Clin. (Paris),* **22,** 815 (1964).

29. J. E. Herz and E. González, *J. Chromatogr.*, **34,** 251 (1968).
30. D. F. Johnson, R. D. Bennett, and E. Heftmann, *Science,* **140,** 198 (1963).
31. J. W. Copius-Peereboom, *Chromatographic Sterol Analysis,* Pudoc, Wageningen, 1963.
32. A. M. Gad, H. El Dakhakhny, and M. M. Hassan, *Planta Med.,* **11,** 134 (1963).
33. T. Kiribuchi, C. S. Chen, and S. Funahashi, *Agric. Biol. Chem.,* **29,** 265 (1965).
34. B. A. Nagasampagi, J. W. Rowe, R. Simpson, and L. J. Goad, *Phytochemistry,* **10,** 1101 (1971).
35. A. M. Gawienowski and C. C. Gibbs, *Steroids,* **12,** 545 (1968).
36. J. P. Wolff, A. Karleskind, and F. Audiau, *Double-Liaison,* **1966,** 1529.
37. J. D. Gilbert, W. A. Harland, G. Steel, and C. J. W. Brooks, *Biochim. Biophys. Acta,* **187,** 453 (1969).
38. N. W. Ditullio, C. S. Jacobs, Jr., and W. L. Holmes, *J. Chromatogr.,* **20,** 354 (1965).
39. R. Kammereck, W.-H. Lee, A. Paliokas, and G. J. Schroepfer, Jr., *J. Lipid Res.,* **8,** 282 (1967).
40. L. Wolfman and B. A. Sachs, *J. Lipid Res.,* **5,** 127 (1964).
41. A. Seher and E. Homberg, *Fette, Seifen, Anstrichm.,* **73,** 557 (1971).
42. G. Fabriani, *Getreide Mehl,* **12,** 109 (1962).
43. K. Schreiber and G. Osske, *Experientia,* **19,** 69 (1963).
44. K. Schreiber, G. Osske, and G. Sembdner, *Experientia,* **17,** 463 (1961).
45. D. Lavie and I. A. Kaye, *J. Chem. Soc.,* **1963,** 5001.
46. R. Ikan and J. Kashman, *Israel J. Chem.,* **1,** 502 (1963).
47. M. Wenzel, P.-E. Schulze, and H. Wollenberg, *Naturwissenschaften,* **49,** 515 (1962).
48. K. E. Wilzbach, *J. Am. Chem. Soc.* **79,** 1013 (1957).
49. M. Wenzel and P.-E. Schulze, *Tritium-Markierung,* W. de Gruyter, Berlin, 1962.
50. E. Richter, *J. Chromatogr.,* **18,** 164 (1965).
51. V. Mahadevan and W. O. Lundberg, *J. Lipid Res.,* **3,** 106 (1962).
52. H. P. Kaufmann, Z. Makus, and F. Deicke, *Fette, Seifen, Anstrichm.,* **63,** 235 (1961).
53. E. A. Bergmann, R. Ikan, and S. Harel, *J. Chromatogr.,* **15,** 204 (1964).
54. N. Pelick, R. S. Henly, R. F. Sweeny, and M. Miller, *J. Am. Oil Chem. Soc.,* **40,** 419 (1963).
55. M. J. D. Van Dam, G. J. De Kleuver, and J. G. de Heus, *J. Chromatogr.,* **4,** 26 (1960).
56. Č. Michalec, M. Šulc, and J. Měšťan, *Nature,* **193,** 63 (1962).
57. J. W. Copius-Peereboom, "The Analysis of Mixtures of Animal and Vegetable Fats; IV. Separation of Sterol Acetates by Reversed-Phase Thin-Layer Chromatography," in *Thin-layer Chromatography,* G. B. Marini-Bettòlo, Ed., Elsevier, Amsterdam, 1964, p. 197.
58. J. W. Copius-Peereboom and H. W. Beeker, *J. Chromatogr.,* **17,** 99 (1965).

59. J. W. Copius-Peereboom, *Z. Anal. Chem.*, **205,** 325 (1964).
60. P. Capella, E. Fedeli, M. Cirimele, A. Lanzani, and G. Jacini, *Riv. Ital. Sostanze Grasse,* **40,** 645 (1963); through *Chem. Abstr.*, **61,** 4971 (1964).
61. R. D. Bennett and E. Heftmann, *J. Chromatogr.*, **12,** 245 (1963).
62. T. M. Lees, M. J. Lynch, and F. R. Mosher, *J. Chromatogr.*, **18,** 595 (1965).
63. A. Seher and E. Homberg, *Fette, Seifen, Anstrichm.*, **70,** 481 (1968).
64. H. E. Vroman and C. F. Cohen, *J. Lipid Res.*, **8,** 150 (1967).
65. J. W. Copius-Peereboom, "Separation of Sterols on Silver Nitrate Impregnated Adsorbent Layers," in *Stationary Phase in Paper and Thin-Layer Chromatography,* K. Macek and I. M. Hais, Eds., *Proc. Symp. 2nd., 1964,* Elsevier, Amsterdam, 1965 p. 316.
66. L. J. Morris, *J. Lipid Res.*, **4,** 357 (1963).
67. E. Haahti and T. Nikkari, *Acta Chem. Scand.*, **17,** 536 (1963).
68. E. Haahti, T. Nikkari, and K. Juva, *Acta Chem. Scand.*, **17,** 538 (1963).
69. L. Horlich and J. Avigan, *J. Lipid Res.*, **4,** 160 (1963).
70. S. J. Purdy and E. V. Truter, *Analyst (London),* **87,** 802 (1962).
71. N. Zoellner and G. Wolfram, *Klin. Wochenschr.*, **40,** 1098 (1962).
72. N. Zoellner, G. Wolfram, and G. Amin, *Klin. Wochenschr.*, **40,** 273 (1962).
73. B. Zak, R. C. Dickenman, E. G. White, H. Burnett, and P. J. Cherney, *Am. J. Clin. Path.*, **24,** 1307 (1954).
74. H. P. Kaufmann and C. V. Viswanathan, *Fette, Seifen, Anstrichm.*, **65,** 839 (1963).
75. A. Zlatkis, B. Zak, and A. T. Boyle, *J. Lab. Clin. Med.*, **41,** 486 (1953).
76. M. L. Quaife, R. P. Geyer, and H. R. Bolliger, *Anal. Chem.*, **31,** 950 (1959).
77. G. V. Vahouny, C. R. Borja, and S. Weersing, *Anal. Biochem.*, **6,** 555 (1963).
78. E. V. Ksenofontova, M. V. Mukhina, and A. M. Khaletskii, *Zh. Obshch. Khim.*, **39,** 913 (1969).
79. P. Samuel, M. Urivetzky, and G. Kaley, *J. Chromatogr.*, **14,** 508 (1964).
80. T. Nambara, R. Imai, and S. Sakurai, *Yakugaku Zasshi,* **84,** 680 (1964).
81. St. A. Ivanov, P. I. Bicheva, and B. T. Konova, *Rev. Fr. Corps. Gras,* **19,** 177 (1972).
82. G. Bondjers and S. Bjoerkerud, *Anal. Biochem.*, **42,** 363 (1971).
83. R. D. Brandt and P. Benviste, *Biochim. Biophys. Acta,* **282,** 85 (1972).
84. P. W. Parodi, Australian *J. Dairy Technol.*, **22,** 209 (1967).
85. C. W. Thorpe, L. Pohland, and D. Firestone, *J. Assoc. Off. Anal. Chem.*, **52,** 774 (1969).
86. C. W. Thorpe, *J. Assoc. Off. Anal. Chem.*, **52,** 778 (1969).
87. M. Rohmer and R. D. Brandt, *Eur. J. Biochem.*, **36,** 446 (1973).
88. B. P. Lisboa and E. Diczfalusy, *Acta Endocrinol.*, **40,** 60 (1962).
89. J. Boute, *Ann. Endocrinol. (Paris),* **14,** 518 (1953).
90. H. Kaegi and K. Miescher, *Helv. Chim. Acta,* **22,** 683 (1938).
91. H. Breuer, R. Knuppen, and G. Pongels, *Nature,* 190, 720 (1961).
92. B. P. Lisboa, *Clin. Chim. Acta,* **13,** 179 (1966).

93. B. P. Lisboa and E. Diczfalusy, *Acta Endocrinol.*, **43**, 545 (1963).
94. M. Barbier and S. I. Zav'yalov, *Izv. Akad. Nauk SSSR., Otd. Khim, Nauk*, **1960**, 1309; through *Chem. Abstr.*, **54**, 22803 (1960).
95. T. Diamanstein and K. Loercher, *Z. Anal. Chem.* **191**, 429 (1962).
96. F. Hertelendy and R. H. Common, *Steroids*, **2**, 135 (1963).
97. F. Hertelendy and R. H. Common, *J. Chromatogr.*, **13**, 570 (1964).
98. J. Fokkens and J. Polderman, *Pharm. Weekb.*, **96**, 657 (1961); *Chem. Abstr.*, **55**, 26371 (1961).
99. S. Kushinsky, *J. Chromatogr.*, **71**, 161 (1972).
100. J. Coyotupa, K. Kinoshita, R. Y. Ho, C. Chan, W. Paul, M. Foote, and S. Kushinsky, *Anal. Biochem.*, **34**, 71 (1970).
101. P. Doerr, *J. Chromatogr.*, **59**, 452 (1971).
102. H. P. Gelbke and R. Knuppen, *J. Chromatogr.*, **71**, 465 (1972).
103. B. P. Lisboa, *Methods Enzymol.*, **15**, 154 (1969).
104. A. I. Frankel and A. V. Nalbandoy, *Steroids*, **8**, 749 (1966).
105. G. J. Krol, G. R. Boyden, R. H. Moody, J. C. Comeau, and B. T. Kho, *J. Chromatogr.*, **61**, 187 (1971).
106. L. E. Crocker and B. A. Lodge, *J. Chromatogr.*, **62**, 158 (1971).
107. *Ibid.*, **69**, 419 (1972).
107a. M. T. Abdel-Aziz and K. I. H. Williams, *Steroids*, **12**, 167 (1968).
108. J. A. Petrović and S. M. Petrović, *J. Chromatogr.*, **119**, 625 (1976).
109. J. C. Touchstone, T. Murawec, and O. Brual, *J. Chromatogr.*, **37**, 359 (1968).
110. K. Pollow, R. Sinnecker, and B. Pollow, *J. Chromatogr.*, **90**, 402 (1974).
111. G. Cavina, G. Moretti, and M. Petrella, *J. Chromatogr.*, **103**, 368 (1975).
112. T. Taylor, *J. Chromatogr.*, **66**, 177 (1972).
113. M. Luisi, C. Savi, and V. Marescotti, *J. Chromatogr.*, **15**, 428 (1964).
114. L. L. Smith and T. Foell, *J. Chromatogr.*, **9**, 339 (1962).
115. L. Stárka, *J. Chromatogr.*, **17**, 599 (1965).
116. L. P. Penzes and G. W. Oertel, *J. Chromatogr.*, **51**, 322 (1970).
117. G. W. Oertel and L. P. Penzes, *Z. Anal. Chem.*, **252**, 306 (1970).
118. R. Dvir and R. Chayen, *J. Chromatogr.*, **45**, 76 (1969).
119. J. C. Touchstone, A. K. Balin, and P. Knapstein, *Steroids*, **13**, 199 (1969).
120. B. P. Lisboa, *J. Chromatogr.*, **25**, D16 (1966).
121. H. Struck, *Mikrochim. Acta*, **1961**, 634.
122. I. Schroeder, G. López-Sánchez, J. C. Medina-Acevedo, M. del Carmen Espinosa, *J. Chromatogr., Sci.*, **13**, 37 (1975).
123. A. L. Livingston, E. M. Bickoff, J. Guggolz, and C. R. Thompson, *J. Agric. Food Chem.*, **9**, 135 (1961).
124. E. M. Bickoff, R. L. Lyman, A. L. Livingston, and A. N. Booth, *J. Am. Chem. Soc.*, **80**, 3969 (1958).
125. R. L. Lyman, E. M. Bickoff, A. N. Booth, and A. L. Livingston, *Arch. Biochem. Biophys.*, **80**, 61 (1959).
126. M. Luisi, *Ric. Sci. Rend.*, **3**, 369 (1963); *Chem. Abstr.*, **60**, 16154 (1964).
127. M. Luisi, C. Savi, F. Coli, and V. Marescotti, *Folia Endocrinol. (Pisa)*, **15**, 672 (1962).

128. V. Marescotti, M. Luisi, C. Savi, and A. Marcezzi, *Folia Endocrinol. (Pisa),* **14,** 848 (1961).
129. L. Jung, Ch. Bourgoin, J. C. Foussard, P. Audrin, and P. Morand, *Rev. Fr. Etudes Clin. Biol,* **8,** 406 (1963).
130. H. G. Gaenshirt and J. Polderman, *J. Chromatogr.,* **16,** 510 (1964).
131. H. H. Wotiz and S. C. Chattoraj, *Anal. Chem.,* **36,** 1466 (1964).
132. J. B. Brown, *Biochem. J.,* **60,** 185 (1955).
133. R. Knuppen, M. L. Rao, and H. Breuer, *Z. Anal. Chem.,* **243,** 263 (1969).
134. S. Kushinsky, J. Coyotupa, K. Honda, M. Hiroi, K. Kinoshita, M. Foote, C. Chan, R. Y. Ho, W. Paul, and W. J. Dignam, *Mikrochim. Acta,* **1970,** 491.
135. G. Moretti, G. Cavina, and J. Sardi de Valverde, *J. Chromatogr.,* **40,** 410 (1969).
136. J. C. Touchstone, T. Murawec, O. Brual, and M. Breckwoldt, *Steroids,* **17,** 285 (1971).
137. W. Wortmann, B. Wortmann, C. Schnabel, and J. C. Touchstone, *J. Chromatogr., Sci.,* **12,** 377 (1974).
138. J. C. Touchstone, T. Murawec, M. Kasparow, and A. K. Balin, *J. Chromatogr., Sci.,* **8,** 81 (1970).
139. J. C. Touchstone, A. K. Balin, T. Murawec, and M. Kasparow, *J. Chromatogr., Sci.,* **8,** 443 (1970).
140. L. P. Penzes and G. W. Oertel, *J. Chromatogr.,* **74,** 359 (1972).
141. G. M. Jacobsohn, *Anal. Chem.,* **36,** 275 (1964).
142. *Ibid,* p. 2030.
143. W. G. Dyer, J. P. Gould, N. A. Maistrellis, T.-C. Peng, and P. Ofner, *Steroids,* **1,** 271 (1963).
144. L. Stárka and R. Hampl, *J. Chromatogr.,* **12,** 347 (1963).
145. L. Stárka, J. Sulcova, J. Riedlova, and O. Adamec, *Clin. Chim. Acta,* **9,** 168 (1964).
146. E. Change, *Steroids,* **4,** 237 (1964).
147. D. Sonanini and L. Anker, *Pharm. Acta Helv.,* **42,** 54 (1967).
148. K. C. James, G. T. Richards, and T. D. Turner, *J. Chromatogr.,* **69,** 141 (1972).
149. J. Vaedtke and A. Gajewska, *J. Chromatogr.,* **9,** 345 (1962).
150. D. J. Watson and D. Bartosik, *J. Chromatogr.,* **54,** 91 (1971).
151. B. P. Lisboa, *J. Chromatogr.,* **13,** 391 (1964).
152. G. L. Cohn and E. Pancake, *Nature,* **201,** 75 (1964).
153. A. A. Akhrem and A. I. Kuzentsova, *Proc. Acad. Sci. USSR, Chem. Sect. (Engl. Transl.),* **138,** 507 (1961).
154. A. Hara and M. Takeuchi, *Endocrinol. Japan,* **10,** 202 (1963).
155. R. Neher and A. Wettstein, *Helv. Chim. Acta,* **43,** 1628 (1960).
156. P. Menzel and G. W. Oertel, *Arzneim. Forsch.,* **21,** 1034 (1971).
157. V. Schwarz, *Pharmazie,* **18,** 122 (1963).
158. K. Doenges and W. Staib, *J. Chromatogr.,* **8,** 25 (1962).
159. P. M. Reisert and D. Schumacher, *Experientia,* **19,** 84 (1963).

160. J. Chamberlain, B. A. Knights, and G. H. Thomas, *J. Endocrinol.*, **26**, 367 (1963).
161. V. Černý, J. Joska, and L. Lábler, *Collect. Czech. Chem. Commun.*, **26**, 1658 (1961).
162. S. Heřmánek, V. Schwarz, and Z. Čekan, *Collect. Czech. Chem. Commun.*, **26**, 1669 (1961).
163. D. B. Gower, *J. Chromatogr.*, **14, 424** (1964).
164. R. S. Wright, *J. Chromatogr.*, **37**, 363 (1968).
165. V. Remmers, H. Schmitt, H. G. Solbach, W. Staib, and H. Zimmermann, *J. Chromatogr.*, **32**, 760 (1968).
166. B. P. Lisboa, *J. Chromatogr.*, **19**, 333 (1965).
167. *Ibid*, p. 81.
168. *Ibid.*, **13**, 391 (1964).
169. B. P. Lisboa and H. Breuer, *Gen. Comp. Endocrinol.*, **6**, 114 (1966).
170. B. P. Lisboa and U. Hoffmann, *J. Chromatogr.*, **115**, 177 (1975).
171. L. Treiber and G. W. Oertal, *Z. Klin. Chem.*, **5**, 83 (1967).
172. W. R. Starnes, A. H. Rhodes, and R. H. Lindsay, *J. Clin. Endocrinol. Metab.*, **26**, 1245 (1966).
173. L. Penzes, P. Menzel, and G. W. Oertel, *J. Chromatogr.*, **44**, 189 (1969).
174. J. Hakl, *J. Chromatogr.*, **61**, 183 (1971).
175. *Ibid.*, **71**, 319 (1972).
176. G. Gondos, B. Matkovics, and O. Kovacs, *Microchem. J.* **8**, 415 (1964).
177. B. Matkovics, Z. Tegyey, and G. Gondos, *Microchem. J.*, **11**, 548 (1966).
178. F. Dray and I. Weliky, *Anal. Biochem.*, **34**, 387 (1970).
179. O. Cerri and G. Maffi, *Boll. Chim. Farm.*, **100**, 954 (1961).
180. O. Crépy, O. Judas, and B. Lachese, *J. Chromatogr.*, **16**, 340 (1964).
181. G. W. Oertel, M. C. Tornero, and K. Groot, *J. Chromatogr.*, **14**, 509 (1964).
182. R. Segura, J. Oro, and A. Zlatkis, *J. Chromatogr. Sci.*, **8**, 449 (1970).
183. F. Panicucci, C. Savi, F. Coli, and M. Luisi, *Folia Endocrinol. (Pisa)*, **17**, 237 (1964).
184. A. Vermuelen and J. C. M. Verplancke, *Steroids*, **2**, 453 (1963).
185. G. Cavina, G. Giocoli, and D. Sardini, *Steroids*, **14**, 315 (1969).
186. P.-E. Schulze and M. Wenzel, *Angew. Chem. Int. Ed. Engl.*, **1**, 580 (1962).
187. A. Riondel, J. F. Tait, M. Gut, S. A. S. Tait, E. Joachim, and B. Little, *J. Clin. Endocrinol. Metab.*, **23**, 620 (1963).
188. M. Sparagana, *J. Nucl. Med.*, **12**, 253 (1971).
189. C. W. Bardin and M. B. Lipsett, *Steroids*, **9**, 71 (1967).
190. R. Guerra-Garcia, S. C. Chattoraj, L. J. Gabrilvoe, and H. H. Wotiz, *Steroids*, **2**, 605 (1963).
191. M. Luisi, C. Fassorra, C. Levanti, and F. Franchi, *J. Chromatogr.*, **58**, 213 (1971).
192. D. Exley, *Biochem. J.*, **107**, 285 (1968).
193. E. H. Mougey, D. R. Collins, R. M. Rose, and J. W. Mason, *Anal. Biochem.*, **27**, 343 (1969).
194. M. A. Drosdowsky, N.-T.-T. Populu, and M. F. Jayle, *Bull. Soc. Chim. Biol.*, **50**, 1723 (1968).

195. L. F. Fabre, Jr., D. C. Fenimore, R. W. Farmer, H. W. Davis, and G. Farrell, *J. Chromatogr., Sci.,* **7,** 632 (1969).
196. J. R. Kent and A. B. Rawitch, *J. Chromatogr.,* **20,** 614 (1965).
197. P. Knapstein, L. Treiber, and J. C. Touchstone, *Steroids,* **11,** 915 (1968).
198. P. Knapstein and J. C. Touchstone, *J. Chromatogr.,* **37,** 83 (1968).
199. H. Struck, H. Karg, and H. Jork, *J. Chromatogr.,* **36,** 74 (1968).
200. H. Huck, *J. Chromatogr.,* **110,** 125 (1975).
201. V. Graef, *Z. Klin. Chem. Klin. Biochem.,* **8,** 320 (1970).
202. V. Schwarz, *Collect. Czech. Chem. Commun.,* **27,** 2567 (1962).
203. V. Schwarz and K. Syhora, *Collect. Czech. Chem. Commun.,* **28,** 101 (1963).
204. G. Cavina and C. Vicari, "Qualtitative and Quantitative Analysis of Natural and Synthetic Corticosteroids by Thin-Layer Chromatography," in *Thin-Layer Chromatography,* G. B. Marini-Bettòlo, Ed., Elsevier, Amsterdam, 1964, p. 180.
205. D. Waldi, "Steroids," in *Thin-Layer Chromatography,* E. Stahl, Ed., Academic Press, New York, 1965, p. 262.
206. B. P. Lisboa, *J. Chromatogr.,* **39,** 173 (1969).
207. S. S. Levin, J. C. Touchstone, and T. Murawec, *J. Chromatogr.,* **42,** 129 (1969).
208. L. Stárka and J. Malíková, *J. Endocrinol.,* **22,** 215 (1961).
209. B. K. Nguen and A. A. Chemerisskaya, *Zh. Anal. Khim.,* **21,** 1375 (1966); through *Anal. Abstr.,* **15,** 4934 (1968).
210. *Ibid.,* p. 1123; through *Anal. Abstr.,* **15,** 4166 (1968).
211. V. Schwarz, *J. Chromatogr.,* **12,** D6 (1963).
211a. J. Schreiber and G. Adam, *Monatsh. Chem.,* **92,** 1093 (1961).
212. J. Matis, O. Adamec, and M. Galvánek, *Nature,* **194,** 477 (1962).
213. B. Johannesen and A. Sandal, *Medd. Norsk Farm. Selsk.,* **23,** 105 (1961).
214. R. Tschesche and G. Snatzke, *Ann. Chem.,* **636,** 105 (1960).
215. J. Vaedtke and A. Gajewska, *J. Chromatogr.,* **9,** 345 (1962).
216. M. Yawata and E. M. Gold, *Steroids,* **3,** 435 (1964).
217. L. Goeldel, W. Zimmerman, and D. Lommer, *Z. Physiol. Chem.,* **333,** 35 (1963).
218. E. Butruk and J. Vaedtke, *J. Chromatogr.,* **32,** 311 (1968).
219. R. O. Quesenberry and F. Ungar, *Anal. Biochem.,* **8,** 192 (1964).
220. B. P. Korzun and S. Brody, *J. Pharm. Sci.,* **52,** 206 (1963).
221. L. Lábler and F. Šorm, *Collect. Cxech. Chem. Commun.,* **27,** 276 (1962).
222. F. Scheiffarth and L. Zicha, *Acta Endocrinol., Suppl.,* **67,** 93 (1962).
223. F. Scheiffarth, L. Zicha, F.-W. Funck, and M. Engelhardt, *Acta Endocrinol.,* **43,** 227 (1963).
224. L. Zicha, F. Scheiffarth, D. Bergner, and M. Engelhardt, *Acta Endocrinol. Suppl.,* **67,** 94 (1962).
225. W. Bernauer and L. Schmidt, *Arch. Exp. Pathol. Pharmakol.,* **245,** 111 (1963).
226. *Ibid,* **243,** 311 (1962).
227. W. Bernauer, L. Schmidt, and G. Ullman, *Med. Exp.,* **9,** 191 (1963).

228. L. Schmidt and W. Bernauer, *Klin. Wochenschr.*, **40,** 918 (1962).
229. L. Schmidt and W. Bernauer, *Arch. Exp. Pathol. Pharmakol.*, **245,** 112 (1963).
230. J. R. Pasqualini and M. F. Jayle, *Experientia*, **18,** 273 (1962).
231. E. M. Auvinen and I. A. Favorskaya, *Vestn. Leningr. Univ.*, **18,** *Ser. Fiz Khim*, 122 (1963); through *Chem. Abstr.*, **59,** 1187 (1963).
232. D. J. Watson and E. B. Romanoff, *J. Chromatogr.*, **22,** 192 (1966).
233. M. Takeuchi, *Chem. Pharm. Bull. (Tokyo)*, **11,** 1183 (1963); *Chem. Abstr.*, **59,** 15558 (1963).
234. B. P. Lisboa, *Acta Endocrinol.*, **43,** 47 (1963).
235. B. P. Lisboa, *J. Chromatogr.*, **16,** 136 (1964).
236. B. P. Lisboa, *Steroids*, **8,** 319 (1966).
237. P. J. Stevens, *J. Chromatogr.*, **36,** 253 (1968).
238. S. Frgačić and Z. Kniewald, *J. Chromatogr.*, **94,** 291 (1974).
239. T. Taylor, *Anal. Biochem.*, **41,** 435 (1971).
240. J. Moreno Dalmau, J. M. Plá Delfina, and A. del Pozo Ojeda, *J. Chromatogr.*, **48,** 118 (1970).
241. *Ibid.*, **78,** 165 (1973).
242. R. H. Bishara, *J. Chromatogr.*, **43,** 539 (1969).
243. M. W. Whitehouse, A. E. Bresler, and E. Staple, *J. Chromatogr.*, **1,** 385 (1958).
244. J. S. Matthews, *Biochem. Biophys. Acta*, **69,** 163 (1963).
245. P. Vecsei (Weisz), V. Kemény, and A. Goergényi, *J. Chromatogr.*, **14,** 506 (1964).
246. W. Bernauer, *Klin. Wochenschr.*, **41,** 883 (1963).
247. W. Bernauer and L. Schmidt, *Arch. Exp. Pathol. Pharmakol.*, **246,** 68 (1963).
248. O. Nishikaze, R. Abraham, and H. J. Staudinger, *J. Biochem. (Tokyo)*, **54,** 427 (1963).
249. O. Nishikaze and H. J. Staudinger, *Klin. Wochenschr.*, **40,** 1014 (1962).
250. M. Luisi, C. Savi, F. Coli, G. Gambassi, and F. Panicucci, *Boll. Soc. Ital. Biol. Sper.*, **39,** 1267 (1963); *Chem. Abstr.*, **60,** 5838 (1964).
251. M. Luisi, S. Savi, F. Coli, F. Panicucci, and V. Marescotti, *Boll. Soc. Ital. Biol. Sper.*, **39,** 1264 (1963).
252. O. Adamec, J. Matis, and M. Galvanek, *Steroids*, **1,** 495 (1963).
253. C. C. Porter and R. H. Silber, *J. Biol. Chem.*, **185,** 201 (1950).
254. H. L. Bird, H. F. Brickley, J. P. Comer, P. E. Hartsaw, and M. L. Johnson, *Anal. Chem.*, **35,** 346 (1963).
255. J. S. Matthews, A. L. Pereda-V., and A. Aguilera-P., *J. Chromatogr.*, **9,** 331 (1962).
256. T. Feher, G. K. Feher, L. Bodrogi, and K. Vertes, *Kiserl. Orvostud.*, **23,** 327 (1971).
257. S. Hara, H. Tanaka, and M. Takeuchi, *Chem. Pharm. Bull. (Tokyo)*, **12,** 626 (1964).
258. E. Molina-Andreu, S. Masdeu, R. M. Ras, and J. A. Aznar, *Rev. Espan. Fisiol.*, **28,** 129 (1972).

259. T. A. Belova, G. L. Shreiberg, and M. I. Epshtein, *Lab. Delo,* **1968,** 426.
260. A. Uettwiller and M. Keller, *J. Chromatogr.,* **35,** 526 (1968).
261. B. L. Hamman and M. M. Martin, *J. Lab. Clin. Med.,* **73,** 1042 (1969).
262. J. Bruinvels, *Experientia,* **19,** 551 (1963).
263. J. D. Few and T. J. Forward, *J. Chromatogr.,* **36,** 63 (1968).
264. W. Futterweit, N. L. McNiven, and R. I. Dorfman, *Biochim. Biophys. Acta,* **71,** 474 (1963).
265 W. P. Collins and I. F. Sommerville, *Nature,* **203,** 836 (1964).
266. C. J. W. Brooks and J. Watson, *J. Chromatogr.,* **31,** 396 (1967).
267. D. Waldi, F. Munter, and E. Wolpert, *Med. Exp.,* **3,** 45 (1960).
268. D. Waldi, *Klin. Wochenschr.,* **40,** 827 (1962).
269. D. Waldi, *Lab. Sci. (Milan),* **11,** 81 (1963).
270. D. Waldi, *Aerztl. Lab.,* **9,** 221 (1963).
271. L. Stárka and J. Riedlova, *Endokrinologie,* **43,** 201 (1962).
272. H. O. Bang, *J. Chromatogr.,* **14,** 520 (1964).
273. H. L. Lau and G. S. Jones, *Am. J. Obstet. Gynecol.,* **90,** 132 (1964).
274. Z. Kulenda and E. Horáková, *Z. Med. Labortech.,* **4,** 173 (1963).
275. P. Fischer, *Schweiz. Apoth.-Ztg.,* **111,** 703 (1973).
276. B. Frosch and H. Wagener, *Z. Klin. Chem.,* **1,** 187 (1963).
277. H. Wagener and B. Frosch., *Klin. Wochenschr.,* **41,** 1094 (1963).
278. H. Gaenshirt, F. W. Koss, and K. Morianz, *Arzneim.-Forsch.,* **10,** 943 (1960).
279. D. Kritchevsky, D. S. Martak, and G. H. Rothblat, *Anal. Biochem.,* **5,** 388 (1963).
280. P. Eneroth, *J. Lipid Res.,* **4,** 11 (1963).
281. S. Hara, *Bunseki Kagaku,* **12,** 199 (1963).
282. S. Hara and M. Takeuchi, *Tokyo Yakka Daigaku Kenkyu Nempo,* **13,** 75 (1963); through *Chem. Abstr.,* **61,** 16417 (1964).
283. S. Hara and M. Takeuchi, *J. Chromatogr.,* **11,** 565 (1963).
284. T. Usui, *J. Biochem. (Tokyo),* **54,** 283 (1963); *Chem. Abstr.,* **60,** 16191 (1964).
285. F. W. Koss and D. Jerchel, *Naturwissenschaften,* **51,** 382 (1964).
286. A. F. Hofmann, *J. Lipid Res.,* **3,** 127 (1962).
287. A. F. Hofmann, "Thin-Layer Chromatography of Bile Acids and Their Derivatives," *New Biochemical Separations,* A. T. James and L. J. Morris, Eds., Van Nostrand, London, 1964, p. 261.
288. A. F. Hofmann, *Acta Chem. Scand.,* **17,** 173 (1963).
289. A. F. Hofmann, *Anal. Biochem.,* **3,** 145 (1962).
290. F.-K. Gruette and H. Gaertner, *J. Chromatogr.,* **41,** 132 (1969).
291. R. Spears, D. Vukusich, S. Mangat, and B. S. Reddy, *J. Chromatogr.,* **116,** 184 (1976).
292. C. T. L. Huang and B. L. Nichols, *J. Chromatogr.,* **109,** 427 (1975).
293. S. Ikawa and M. Goto, *J. Chromatogr.,* **114,** 237 (1975).
294. F. Begemann, *Z. Klin. Chem. Klin. Biochem.,* **10,** 29 (1972).
295. D. Panveliwalla, B. Lewis, I. D. P. Wootton, and S. Tabaqchali, *J. Clin. Pathol.,* **23,** 309 (1970).

296. U. Freimuth, B. Zawta, and M. Buechner, *J. Chromatogr.,* **30,** 607 (1967).
297. W. L. Anthony and W. T. Beher, *J. Chromatogr.,* **13,** 567 (1964).
298. G. S. Sundaram and H. S. Sodhi, *J. Chromatogr.,* **61,** 370 (1971).
299. D. Kritchevsky, D. S. Martak, and G. H. Rothblat, *Anal. Biochem.,* **5,** 388 (1963).
300. S. K. Goswami and C. F. Frey, *J. Chromatogr.,* **53,** 389 (1970).
301. J. Wollenweber, B. A. Kottke, and C. A. Owen, Jr., *J. Chromatogr.,* **24,** 99 (1966).
302. B. Frosch and H. Wagener, *Z. Klin. Chem.,* **2,** 7 (1964).
303. B. Frosch and H. Wagener, *Klin. Wochenschr.,* **42,** 192 (1964).
304. *Ibid.,* p. 901.
305. M. A. Eastwood, D. Hamilton, and L. Mowbray, *J. Chromatogr.,* **65,** 407 (1972).
306. S. Hara, M. Takeuchi, M. Tachibana, and G. Chihrara, *Chem. Pharm. Bull. (Tokyo),* **12,** 483 (1964).
307. G. Semenuk and W. T. Beher, *J. Chromatogr.,* **21,** 27 (1966).
308. H. K. Kim and D. Kritchevsky, *J. Chromatogr.,* **117,** 222 (1976).
309. B. I. Cohen, R. F. Raicht, G. Salen, and E. H. Mosback, *Anal. Biochem.,* **64,** 567 (1975).
310. R. Tschesche and G. Wulff, *Chem. Ber.,* **94,** 2019 (1961).
311. R. Tschesche, G. Wulff, and G. Balle, *Tetrahedron,* **18,** 959 (1962).
312. R. Tschesche and G. Wulff, *Tetrahedron,* **19,** 621 (1963).
313. K. Takeda, S. Hara, A. Wada, and N. Matsumoto, *J. Chromatogr.,* **11,** 562 (1963).
314. N. Matsumoto, *Chem. Pharm. Bull. (Tokyo),* **11,** 1189 (1963); *Chem. Abstr.,* **59,** 15559 (1963).
315. R. D. Bennett and E. Heftmann, *J. Chromatogr.,* **9,** 353 (1962).
316. G. Blunden and R. Harman, *J. Chromatogr.,* **15,** 273 (1964).
317. H. Sander, H. Hauser, and R. Haensel, *Planta Med.,* **9,** 8 (1961).
318. H. Sander and G. Willuhn, *Flora,* **151,** 150 (1961).
319. H. Sander, M. Alkemeyer, and R. Haensel, *Arch. Pharm.,* **295,** 6 (1962).
320. P. M. Boll, *Acta Chem. Scand.,* **16,** 1819 (1962).
321. P. M. Boll and B. Andersen, *Planta Med.,* **10,** 421 (1962).
322. R. Kuhn and I. Loew, *Chem. Ber.,* **95,** 1748 (1962).
323. I. B. Chiarlo, *Boll. Chim. Farm.,* **103,** 423 (1964).
324. K. Schreiber, O. Aurich, and G. Osske, *J. Chromatogr.,* **12,** 63 (1963).
325. A. M. Dawidar and M. B. E. Fayez, *Z. Anal. Chem.,* **259,** 283 (1972).
326. G. Held and D. Vágújfalvi, *Herba Hung.,* **9,** 127 (1970).
327. O. S. Madaeva and V. K. Ryzhkova, *Med. Prom. SSSR,* **17,** 44 (1963).
328. L. Carreras Matas, *An. Real. Acad. Farm.,* **26,** 371 (1960).
329. A. J. van Duuren, *J. Am. Soc. Sugar Beet Technol.,* **12,** 57 (1962).
330. A. Y. Khorlin, Y. S. Ovodov, and N. K, Kochetkov, *Zh. Obshch. Khim.,* **32,** 782 (1962); through *Chem. Abstr.,* **58,** 4636 (1963).
331. T. Kawasaki and K. Miyahara, *Chem. Pharm. Bull. (Tokyo),* **11,** 1546 (1963); *Chem. Abstr.,* **60,** 11850 (1964).
332. W. J. Wolf and B. W. Thomas, *J. Chromatogr.,* **56,** 281 (1971).

333. B. Pasich, *Planta Med.*, **11**, 16 (1963).
334. B. Zak, *Am. J. Clin. Pathol.*, **27**, 583 (1967).
335. R. Paquin and M. Lepage, *J. Chromatogr.*, **12**, 57 (1963).
336. G. Adam and K. Schreiber, *Tetrahedron Lett.*, **1963**, 943.
337. G. Adam and K. Schreiber, *Z. Chem.*, **3**, 100 (1963).
338. K. Schreiber and G. Adam, *Ann. Chem.*, **666**, 155 (1963).
339. *Ibid*, p. 176.
340. K. Schreiber and H. Roensch, *Tetrahedron Lett.*, **1963**, 329.
341. P. Bite, L. Jókay, and L. Pongŕacz-Sterk, *Acta Chim. Acad. Sci. Hung.*, **34**, 363 (1962).
342. L. Lábler and V. Černý, *Collect. Czech. Chem. Commun.*, **28**, 2932 (1963).
343. L. Lábler and V. Černý, "Thin-Layer Chromatography of Steroidal Bases and Holarrhena Alkaloids," in *Thin-Layer Chromatography*, G. B. Marini-Bettòlo, Ed., Elsevier, Amsterdam, 1964, p. 144.
344. A. Kasal, A. Poláková, A. V. Kamernitzky, L. Lábler, and V. Černý, *Collect. Czech. Chem. Commun.*, **28**, 1189 (1963).
345. H.-J. Zeitler, *J. Chromatogr.*, **18**, 180 (1965).
346. R. Zelnik and L. M. Ziti, *J. Chromatogr.*, **9**, 371 (1962).
347. R. Zelnik, L. M. Ziti, and C. V. Guimarães, *J. Chromatogr.*, **15**, 9 (1964).
348. M. Ishikawa and T. Miyasaka, *Shika Zairyo Kenkyusho Hokoku*, **2**, 397 (1962); through *Chem. Abstr.*, **58**, 11483 (1963).
349. M. Komatsu and S. Okano, *Bunseki Kagaku*, **15**, 1115 (1966).
350. T. Golab and D. S. Layne, *J. Chromatogr.*, **9**, 321 (1962).
351. R. Neher, "Thin-Layer Chromatography of Steroids," in *Thin-Layer Chromatography*, G. B. Marini-Bettòlo, Ed., Elsevier, Amsterdam, 1964, p. 75.
352. R. Neher, "Steroid Separation and Analysis," in *Advances in Chromatography*, Vol. 4, J. C. Giddings and R. A. Keller, Eds., Marcel Dekker, New York, 1967, p. 47.
353. W. M. Beijleveld, *Pharm. Weekbl.*, **97**, 190 (1962).
354. T. Kazuno and T. Hoshita, *Seikagaku*, **34**, 139 (1962); *Chem. Abstr.*, **57**, 4962 (1962).
355. B. P. Lisboa, "Thin-Layer Chromatography of Sterols and Steroids," in *Lipid Chromatographic Analysis*, 2nd Ed., G. V. Marinetti, Ed., Marcel Dekker, New York, 1976, p. 339.
356. B. P. Lisboa, "Steroids," in *Pharmaceutical Applications of Thin-Layer and Paper Chromatography*, K. Macek, Ed., Elsevier, Amsterdam, 1972, p. 275.
357. B. P. Lisboa, "Thin-Layer Chromatography of Steroids, Sterols, and Related Compounds," in *Methods in Enzymology*, Vol. XV, S. P. Colowick and N. O. Kaplan, Eds., Academic Press, New York, 1968, p. 3.
358. R. S. Rivlin and H. Wilson, *Anal. Biochem.*, **5**, 267 (1963).
359. G. S. Richardson, I. Weliky, W. Batchelder, M. Griffith, and L. L. Engel, *J. Chromatogr.*, **12**, 115 (1963).
360. H. Metz, *Naturwissenschaften*, **48**, 569 (1961).
361. H. Dannenberg and K. F. Hebenbrock, *Ann. Chem.*, **662**, 21 (1963).
362. H. Dannenberg and H.-G. Neumann, *Chem. Ber.*, **94**, 3085 (1961).

363. *Ibid,* p. 3094
364. P. J. Stevens, *J. Chromatogr.,* **14,** 269 (1964).
365. R. Haehnel and N. B. Muslim, *Chromatogr., Rev.,* **11,** 215 (1969).
366. L. Biella Souza Valle and R. Martins Oliveira-Filho, *J. Chromatogr.,* **105,** 201 (1975).
367. A. Pinelli and M. L. Formento, *J. Chromatogr.,* **68,** 67 (1972).
368. F. A. Vandenheuvel, *Can. J. Biochem.,* **45,** 191 (1967).
369. S. Hara and K. Mibe, *Anal. Chem.,* **40,** 1605 (1968).
370. B. P. Lisboa, *J. Chromatogr.,* **48,** 364 (1970).
371. H.-C. Curtius and M. Mueller, *J. Chromatogr.,* **32,** 222 (1968).
372. B. L. Hamman and M. M. Martin, *Steroids,* **10,** 169 (1967).

Chapter **XXX**

TERPENES AND ESSENTIAL OILS

1 HYDROCARBONS

Terpene hydrocarbons can be conveniently separated on layers of silicic acid [1]. The layers are dried at 105°C for 15 min, after which they are placed in a desiccator evacuated to 30 mm of mercury and containing potassium hydroxide as a desiccant. (An acidic desiccant such as phosphorus pentoxide should be avoided because enough acidic vapors are adsorbed by the silicic acid to interfere with the fluorescein–bromine test.) Because of the low polar nature of the compounds, solvents of low polarity must be used for the separation. Table 30.1 gives the R_f values of some terpenes on silicic acid chromatostrips in various solvents. Except for *p*-cymene, the compounds were detected by spraying with 0.05% fluorescein in water and then exposing the treated plate to bromine vapor. The compounds appeared as yellow spots on a pink background. The *p*-

Table 30.1 $R_f \times 100$ Values of Some Terpenes and Sesquiterpenes on Silicic Acid with Various Solvents. Development Distance 10 cm[a]

Compound	*Hexane*	2,2-*Dimethylbutane*	*Cyclohexane*	*Methylcyclohexane*	*Isopentane*
Limonene	41	55	54	59	74
Terpinolene	64	67	60	65	89
α-Pinene	83	85	84	90	80
Camphene	84	76	79	80	79
p-Cymene	38	41	62	69	57
Cedrene	82	80	83	85	78
α-Caryophyllene	50	35	47	33	27
β-Caryophyllene	62	60	52	65	75
γ-Caryophyllene	80	82	87	90	83
β-Pinene	80	75	80	85	75

[a] From J. M. Miller and J. G. Kirchner [1]; reproduced with permission of the American Chemical Society.

cymene could be detected by incorporating fluorescent materials into the layer [2] and observing the dark spot under ultraviolet light.

Davies et al. [3], using silica gel G layers and ligroin as a solvent, reported the following R_f values: squalene 0.41, lycopersene 0.30, phytoene 0.21, and phytofluene 0.12. These hydrocarbons were detected by using iodine vapor which was sensitive to 0.05 γ of the compounds. Using petroleum ether with aluminum oxide layers, Grob and Boschetti [4] have isolated lycopersene and squalene from *Neurospora prassa* with R_f values of 0.44 and 0.30, respectively. Vásquez and Janer [5] isolated squalene from olive leaves using column and thin-layer chromatography on silicic acid.

Fukushi and Obata [6] isolated the azulene of camphor blue oil by chromatographing on silicic acid with petroleum ether to give an R_f value of 0.76 (in the same system cadalene had an R_f value of 0.88) and Tétényi et al. [7] investigated the azulene compounds in various species of *Achillea*. Kirby [8] separated a group of azulenes on silica gel impregnated with oxalic acid using toluene as a solvent. Tyihák and Vágújfalvi [9] have investigated proazulenes with silica gel layers and benzene–ethyl acetate (2:1). Tétényi et al. [10] used both paper and thin-layer chromatography in examining the azulenes in *Achillea* species.

Gupta and Dev [11] applied a silver nitrate-impregnated silica gel to the separation of some sesquiterpenes (Table 30.2). (For the preparation of the impregnated gels see Chapter III, Sections 4 and 7.) Development

Table 30.2 R_{Dye} Value of Some Sesquiterpenes on Silica Gel Impregnated with Silver Nitrate[a]

Compound	*Benzene–Acetone* (95:5) R_{st}[b]	*n-Hexane* R_A[c]
Humulene	0.189	—
β-Elemene	0.331	—
Caryophyllene	0.422	—
β-Selinene	0.805	—
β-Bisabolene	0.936	—
Thujopsene	1.149	—
Copaene	1.161	—
ar-Himachalene	1.164	—
α-Gurjunene	1.168	—
β-Himachalene	—	3.155
α-Himachalene	—	5.190
Longifolene	—	5.357
Cuparene	—	5.391
Isolongifolene	—	7.447
Longicyclene	—	8.399

[a] From A. S. Gupta and S. Dev [11]; reproduced with permission of the authors and The Elsevier Publishing Co.
[b] $R_{st} = R_f$ compound/R_f of Sudan III.
[c] $R_A = R_f$ compound/R_f of azobenzene.

was carried out in a saturated atmosphere using *n*-hexane and benzene–acetone (95:5) as solvents. For the detection of the compounds the plates were sprayed with chlorosulfonic acid in acetic acid (1:2), followed by heating for 10 min at 130°C. Westfelt [12] and Chow et al. [13] have also used these layers for the separation of sesquiterpenes. Lawrence [14] investigated the application of silver nitrate-impregnated silica gel to the separation of monoterpenes with benzene as a solvent. Three conclusions were drawn from his studies: *(1)* cyclic terpenes with a single internal double bond did not readily form a π-complex, *(2)* cyclic or acyclic terpenes containing two non-terminal double bonds did not readily form π-complexes unless the bonds were cis conjugated, and *(3)* cyclic and acyclic terpenes with exocyclic or terminal double bonds did form π-complexes. Schantz et al. [15] give the R_f values for 19 monoterpenes separated on this adsorbent using benzene as a solvent.

Attaway et al. [16] give the R_f values for 20 monoterpene, 11 sesquiterpene, and several polyterpene hydrocarbons on silica gel and on alumina using Skellysolve B (hexanes) as a developing solvent. For the separation of sesquiterpene hydrocarbons they preferred alumina with perfluoroalkane (70 to 80°C) as a solvent.

2 ALCOHOLS

Miller and Kirchner [1] investigated a series of terpene alcohols in 12 different solvents. The alcohols were chromatographed on silicic acid containing 5% starch as a binder. The compounds were detected by the fluorescein-bromine reagent and the R_f values are listed in Table 30.3, along with some values from other workers using different separation systems.

(e) H_3C — OH (e) — (e)

1

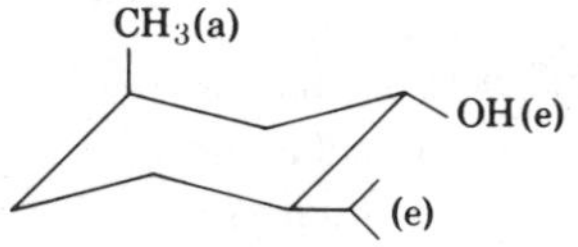

CH_3(a) — OH (e) — (e)

2

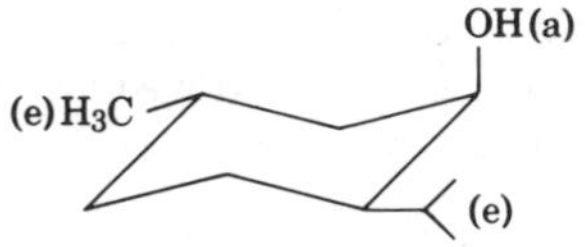

(e) H_3C — OH (a) — (e)

3

CH_3(a) — OH (a) — (e)

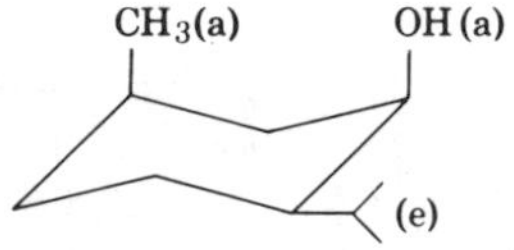

4

Ito [22] found that the steroisomeric menthols with an axial hydroxyl (3 and 4) could be separated from the corresponding equatorial hydroxyl (1 and 2) compounds on silica gel layers using hexane–ethyl acetate (85:15). Thus menthol (1) could be separated from neomenthol (2) and isomenthol (3) could be separated from neoisomenthol (4). Petrowitz [23] separated the same group of compounds in four different solvent systems (Table 30.4). Graf and Hoppe [24] have discussed the stereochemistry of the menthols available on the market. For the detection of neomenthol in menthol, they chromatographed on silica gel with benzene–ethyl acetate (95:5). For the detection of isomenthol in menthol, these authors converted the alcohols to the 3,5-dinitrobenzoates which were then chromatographed for a distance of 15 cm using petroleum ether (105 to 120°C)–isopropyl ether (95:5). By spraying with a 0.04% sodium fluorescein solution, the derivatives could be differentiated by the R_f differential of 0.05. As an alternative method the alcohols could be oxidized to the corresponding methone derivatives, which on chromatographing in petroleum ether (105 to 120°C)–ethyl acetate (95:5) showed an R_f difference

Table 30.3 $R_f \times 100$ Values of Some Terpene Alcohols

	Silic Acid + 5% Starch Binder[a] [1]											*Silica Gel G*[b]		*Al_2O_3 (Grade III) Loose Layer*[c] [20]			
Alcohol	*A*	*B*	*C*	*D*	*E*	*F*	*G*	*H*	*J*	*K*	*L*	*M*	*N*	*O*	*P*	*Q*	*R*
Nerol	26	38	42	30	28	54	38	36	27	0	14		15				
Citronellol	27	39	41	33	36	56	41	41	36	0	19	27	15				
Geraniol	20	40	43	34	29	55	31	30	25	0	12	13	15				
Linalool	36	47	45	31	53	67	45	48	48	0	15	30	20				
α-Terpineol	29	34	36	22	32	56	29	35	25	0	8		15		16	27	43
Nopol	27	48	52	51	59	65	60	43	41	0	12						
Carveol	34	44	49	38	40	60	51	45	29	0	12		18				
Borneol												20		10	17	32	48
Menthol	42[d]										37[d]	25		12	21	28	42

[a] *A* = 15% ethyl acetate in hexane, *B* = 10% ethyl acetate in chloroform (alcohol-free), *C* = 15% ethyl acetate in benzene, *D* =5% ethyl acetate in chloroform (alcohol-free), *E* = 50% 1-nitropropane in hexane, *F* = 30% ethyl acetate in hexane, *G* = 15% ethyl carbonate in chloroform (alcohol-free), *H* = 30% isopropyl formate in hexane, *J* = 50% isopropyl ether in hexane, *K* = hexane, *L* = chloroform (alcohol-free); development distance 10 cm.

[b] *M* = 5% ethyl acetate in benzene, development distance 14 cm [17, 18]; *N* = methylene chloride, development distance 15 cm [19].

[c] *O* = 50% benzene in petroleum ether, *P* = benzene, *Q* = 2% ethanol in benzene, *R* = 5% ethanol in benzene.

[d] Ito et al. [21].

Table 30.4 $R_f \times 100$ Values for Stereoisomers of Menthol[a]

Compound	*Benzene*	*Benzene–Methanol* (95:5)	*Benzene–Methanol* (75:25)	*Methanol*
Methanol (**1**)	16	36	67	90
Neomenthol (**2**)	28	51	73	85
Isomenthol (**3**)	17	37	62	91
Neoisomenthol (**4**)	29	55	76	80

[a] From H.-J. Petrowitz [23]; reproduced with permission of the author and Verlag Chemie.

of 0.1 when detected with 2,4-dinitrophenylhydrazine solution. Using butanol–ethyl acetate (1:4) and acetone–ethyl acetate (3:7) as solvents, Yamamoto and Furukawa [25] separated *cis*- and *trans*-terpene dialcohols on chromatostrips.

As another example of steroisomeric separations, Tyihák et al. [26] were able to separate *trans,trans*-farnesol and *cis,trans*-farnesol on silica gel using 5% ethyl acetate in benzene as a solvent [1]. In this case the solvent path length was 16 cm, giving R_f values of 0.27 and 0.36, respectively. R_f values were also given for a number of ester derivatives. McSweeney [27] used a 4:1 ratio of benzene in ethyl acetate on silica gel to separate the steroisomers of farnesol present in a commercial mixture, as well as other terpene alcohols. Separation was also achieved on kieselguhr layers impregnated by dipping in a 5% solution of paraffin oil in petroleum ether (40 to 60°C). In the latter case, the development was carried out with acetone–water (13:7) saturated with paraffin oil using a solvent travel distance of 15 cm.

Banthorpe and Turnbull [28] chromatographed the isomeric thujols on silica gel G with nine solvent systems.

McSweeney [29] applied reverse-phase thin-layer chromatography to the separation of six terpene alcohols. Kieselguhr impregnated with paraffin was used with acetone–water (65:35) saturated with paraffin oil. The compounds weer also separated on silica gel with benzene–ethyl acetate (95:5). Dunphy et al. [30] also applied partition chromatography to the separation of long-chain isoprenoid alcohols from several sources. Again paraffin-impregnated kieselguhr was used with 90% acetone or pure acetone.

Derivatives of alcohols have been used in the separation of terpene alcohols. The acetates of diterpene alcohols obtained from the *Croton* and *Euphorbia* genera of the family Euphorbiaceae were chromatographed on silica gel and alumina in a number of solvents [31]. Multiple development was employed to increase the separation in some cases. Hedin et al. [32] converted some terpene alcohols to the esters with trichloroacetyl isocyanate. These esters were chromatographed on silica gel with 2-chloropropane. The esters were converted to the corresponding carbamates which were chromatographed in the same system. Ter Heide [33] chromatographed 50 esters (mainly acetates) of monoterpene alcohols on silica gel or silica gel impregnated with silver nitrate using benzene–dichloromethane (2:8) as a solvent. The 3,5-dinitrobenzoates of terpene alcohols have been separated on silica gel with benzene–petroleum ether (38 to 50°C) [34–36] and on polyamide with methanol–water (9:1) [34].

Ikan [37] used silver nitrate-impregnated silica gel to obtain a separation with a group of tetracyclic triterpene alcohols which could not be separated on untreated gel because of the almost identical R_f values. With the treated layers and a developing distance of 15 cm for the chloroform solvent, the following R_f values were obtained: butyrospermol 0.40, cycloartenol 0.33, cyclolaudenol 0.26, euphol 0.30, α-euphorbol 0.27, and parkeol 0.11. Copius-Peereboom [38] has also applied the use of silica gel G–silver nitrate layers to the separation of triterpenoid alcohols.

Cardemil et al. [39] separated some isoprenoid alcohols on silver nitrate-impregnated silica gel with ethyl acetate as a solvent. Pesnelle [40] separated the supposedly sesquiterpene alcohol galbanol into bulnesol, guaiol, and α-eudesmol on silica gel–silver nitrate layers.

Seikel and Rowe [41] chromatographed guaiol and the three isomers of eudesmol on alumina layers with benzene–petroleum ether (1:1) using a descending continuous method in order to obtain sufficient separation of the various compounds.

Budzynska-Topolowska and Rutowski [42] separated five triterpenic alcohols from the unsaponifiable fraction of rapeseed oil using silica gel with cyclohexane–ethyl acetate (4:1). Fedeli et al. [43] investigated the triterpene alcohols in 18 vegetable oils. The free terpene alcohols of *Pelargonium roseum* were separated on silica gel with ethyl acetate–benzene (5:95). [44].

3 CARBONYL COMPOUNDS

Miller and Kirchner [1] separated a group of eight aldehydes and ketones on silicic acid chromatostrips in 10 different solvents (Table 30.5). These compounds appear in various essential oils. The aldehydes were detected by spraying with a solution of *o*-dianisidine in glacial acetic

Table 30.5 $R_f \times 100$ Values of Terpene and Other Essential Oil Carbonyls on Silica Gel in Various Solvents

Compound	*Ethyl Acetate–(n-Hexane)* (3:17)	*Ethyl Acetate–Chloroform* (1:9)[a]	*Ethyl Acetate–Benzene* (3:17)	*Ethyl Acetate–Chloroform* (5:95)[a]	*1-Nitropropane–Hexane* (1:1)	*Ethyl Acetate–Hexane* (3:7)	*Ethyl Acetate–Hexane* (2:3)	*Ethyl Carbonate–Chloroform* (3:17)[a]	*Isopropyl Formate–Hexane* (3:7)	*Isopropyl Ether–Hexane* (1:1)	*Chloroform*[b]	*Benzene*	*Ethyl Acetate–Benzene* (5:95)
Acetaldehyde			27[c]				39[c]				28[c]	20[c]	
Anisaldehyde			22[c]				57[c]				33[c]	22[c]	41[d]
Benzaldehyde			46[c]				70[c]				50[c]	38[c]	
Camphor	56	39	55	43	56	59		78	47	50	28[a]	18[d]	31[d]
Caprylaldehyde												21[d]	42[d]
Carvone	45	72	62	70	76	79	74[c]	75	62	65	37[a]	28[c]	
Cinnamaldehyde	31	70	68	70	68	70	63[c]	78	52	45	9[a]	24[c]	
Citral	45	64	57	47	62	62	46[c]	69	56	51	15[a]	19[c]	
Citronellal	49						81[c]				33[c]	46[c]	
Cuminaldehyde												29[d]	51[d]
Fenchone	51[c]						62[c]				33[c]	30[c]	
Furfural	21	44	41	41	44	41	66	38	37	17	6[a]	14[c]	
Isovanillin	54[c]						32[c]					2[c]	
Lauricaldehyde	58	72	67	62	91	76		84	75	83	50[a]		
Methylheptenone	48	74	62	58	69	70		75	57	60	35[a]		
Perrillaldehyde	41[c]						66[c]				38[c]	26[c]	
Piperonal	30[c]						54[c]				40[c]	22[c]	
Propionaldehyde	58[c]												
Pseudoionone	34[c]						68[c]				20[c]	20[c]	
Pulegone	58	65	76	71	79	83		91	72	79	60[a]		
Vanillin	7[c]						36[c]				9[c]	7[c]	

[a] Alcohol-free.

[b] See remarks in text.

[c] From Katayama [45] on starch-bound silica gel (see text for remarks concerning chloroform values).

[d] From Schantz et al. [46] on silica gel G. Remaining values from Miller and Kirchner [1] on silicic acid bound with starch, development distance 10 cm.

acid. Among the ketones, carvone, methylheptenone, and pulegone could be detected with the fluorescein–bromine spray, but the camphor, which is very unreactive, could be detected only by spraying with concentrated sulfuric acid containing nitric acid as an oxidizing agent, followed by the application of heat to char the compound. Katayama [45] and Schantz et al. [46] have chromatographed some of the same and additional compounds (see Table 30.5). It should be noted, however, that the values of some of the chloroform separations of Katayama are high. Although the carvone value of 0.35 reported by Katayama is comparable to 0.37 by Miller and Kirchner (using alcohol-free chloroform), the citral value of 0.36 is high compared to 0.15 (Miller and Kirchner), which indicates that the lower R_f value compounds of Katayama might have been affected by traces of alcohol preservatives in the chloroform.

Attaway [47] chromatographed carvone and citronellal in various proportions of trifluorotrichloroethane and methylene chloride.

Kheifitis et al. [48] chromatographed 15 alkylcyclohexanones on layers of neutral alumina using benzene–petroleum ether (1:3) as the developing solvent. The compounds were located by exposing to iodine vapors.

Sundt and Saccardi [49] examined the separation of vanilla flavors and other aromatic aldehydes as well as a group of coumarins. Silica gel G layers were used with a number of solvent systems; the most satisfactory are given in Tables 30.6 and 30.7 with the various detecting methods. As a complementary method these authors also reported the R_f values for the same compounds separated by partition chromatography on dimethylformamide-impregnated paper. Kahan and Fitelson [50] have developed an official method based on the work of Sundt and Saccardi for the detection of flavor additives in vanilla extract. Kratzl and Puschmann [51] used thin-layer chromatography on silica gel layers to separate vanillin and related compounds obtained by degradation of lignin. R_f values are tabulated.

Dhont and Dijkman [52] separated α-ionone, β-ionone, α-methylionone, β-methylionone, and pseudoionone on silica gel G layers in an equilibrated system by multiple development (6 $\times$) in benzene. Two to five micrograms of the compounds could be detected with either 2,4-dinitrophenylhydrazine or a vanillin–sulfuric acid spray.

The 2,4-dinitrophenylhydrazones have also been used as a means of separating terpene carbonyl compounds. As early as 1952, Onoe [53] separated the n-aliphatic aldehydes up to C_{10} as well as acrolein, crotonaldehyde, and citral on chromatostrips of silica gel. Vashist and Handa [54] have separated oxoterpene by means of these derivatives using silica gel with chloroform–carbon tetrachloride (1:9, 1:19, and 3:17) as well as petroleum ether–benzene (3:7). The R_f values are given. Lacharme

Table 30.6 $R_f \times 100$ Values from Some Vanillin and Coumarin Compounds on Silica Gel G in Four Solvent Systems[a,b]

Compound	Solvent[c]			
	A	*B*	*C*	*D*
Vanillin (4-hydroxy-3-methoxybenzaldehyde)	30	27	34	29
p-Hydroxybenzaldehyde	32	27	11	20
o-Vanillin (2-hydroxy-3-methoxybenzaldehyde)	39	27	55	45
Methylvanillin (3,4-dimethoxybenzaldehyde)	41	38	62	55
Ethylvanillin (4-hydroxy-3-ethoxybenzaldehyde)	46	42	62	55
m-Hydroxybenzaldehyde	46	37	15	23
2,4-Dimethoxybenzaldehyde	50	44	56	58
Benzylvanillin	61	46	67	64
p-Methoxybenzaldehyde	64	53	70	59
Piperonal (heliotropin)	66	60	72	63
o-Methoxybenzaldehyde	70	62	77	63
Propenylguaethol (vanitrope, 1-ethoxy-2-hydroxy-4-propenylbenzene)	81	74	78	66
Coumarin	50	38	—	—
6-Methylcoumarin	55	43	—	—
Dihydrocoumarin	62	50	—	—
3-Methylcoumarin	68	58	—	—
3-Ethylcoumarin	81	75	—	—

[a] From E. Sundt and A. Saccardi [49]; reproduced with permission of the authors and the Institute of Food Technologists.
[b] Development distance 11.5 cm.
[c] *A* = Petroleum ether (50 to 70°C)–ethyl acetate (5:2.5), *B* = hexane–ethyl acetate (5:2), *C* = chloroform–ethyl acetate (98:2), *D* = Decalin–methylene chloride–methanol (5:4:1).

[55] improved the spectrophotometric determination of sodium camphorsulfonate via the 2,4-dinitrophenylhydrazine by removing excess reagent through chromatography on silica gel layers. Nano and Sancin [56] separated the derivatives of α- and β-thujone as well as other terpene carbonyls on silica gel using benzene–cyclohexane (1:1). R_f values are tabulated. Quantitative evaluation of the thujone can be made by eluting and measuring at 361 mμ. (For the separation, by menas of the 2,4-DNPH derivatives, of some carbonyl compounds found in food products, see Table 20.3.)

Dhont and Mulders-Dijkman [57] chromatographed 22 terpene 2,4-dinitrophenylhydrazones in three systems: silica gel G with benzene–hexane (1:1), silica gel G impregnated with nitromethane vapor using hexane as

Table 30.7 Color Reactions of Some Vanillin and Coumarin Compounds with Various Detection Methods[a]

Compound	1% Hydrazine Sulfate in HCl		Methanolic KOH		Methanolic KOH and Thereafter Diazotized Sulfanilic Acid (Daylight)
	Daylight	Ultraviolet	Daylight	Ultraviolet	
Vanillin (4-hydroxy-3-methoxybenzaldehyde)	Yellow	Orange	—	—	Rose
p-Hydroxybenzaldehyde	Yellow	Yellow	—	—	—
o-Vanillin (2-hydroxy-3-methoxybenzaldehyde)	Yellow	Red	Yellow (weak)	Yellow	Yellow-orange
Methylvanillin (3,4-dimethoxybenzaldehyde)	Yellow	Orange	—	—	—
Ethylvanillin (4-hydroxy-3-ethoxybenzaldehyde)	Yellow	Yellow	—	—	—
m-Hydroxybenzaldehyde	—	—	—	Yellow-green	Yellow (very weak)
2,4-Dimethoxybenzaldehyde	Yellow	Bright yellow	—	—	—
Benzylvanillin	Yellow (weak)	Yellow	—	—	—
p-Methoxybenzaldehyde	Yellow	Yellow-green	—	—	—
Piperonal (heliotropin)	Yellow	Dark spot	—	—	—
o-Methoxybenzaldehyde	Yellow	Yellow	—	—	—
Propenylguaethol (vanitrope, 1-ethoxy-2-hydroxy-4-propenylbenzene)	—	—	—	—	Red-orange
Coumarin	—	—	—	Green	Orange
6-Methylcoumarin	—	—	—	Yellow	Violet
Dihydrocoumarin	—	—	—	—	Orange
3-Methylcoumarin	—	—	—	Blue	Orange-rose
3-Ethylcoumarin	—	—	—	Blue	Orange-rose

[a] From E. Sundt and A. Saccardi [49]; reproduced with permission of the authors and the Institute of Food Technologists.

developing solvent, and silica gel G impregnated with acetonitrile vapor and hexane as the solvent. Vapor impregnation was accomplished by using two layers in a sandwich arrangement with one layer soaked with the impregnating solvent. The R_f values are also given for the derivatives of 24 nonterpene carbonyl compounds. Two-dimensional separations using the unimpregnated layer first, followed in the second direction by one of the impregnated systems, were also very useful in separating components. Martelli [58] chromatographed piperitone, carvone, isomenthone, methone, α- and β-thujone, and some essential oils on silica gel impregnated with 2,4-dinitrophenylhydrazine using benzene, chloroform, and cyclohexane–chloroform (1:1) as solvents. Additional data are also given for a large number of other carbonyl compounds.

Rothbaecher and Suteu [59] chromatographed menthone, dihydrocarvone, pulegone, carvone, piperitone, carvenone, and piperitenone on silica gel G with benzene.

Bell et al. [60] chromatographed sesquiterpene aldehydes on polyamide with chloroform–acetone–formic acid (95:4:1) and ethyl acetate–hexane (1:3).

4 PHENOLS

Klouwen and Ter Heide [61] have examined a large group of phenols and phenol ethers by thin-layer chromatography on silica gel G layers which had been dried at 105°C. For the determination of the R_f values of pure compounds (Table 30.8), the compounds were applied in 1% acetone solution using 10 to 20 μg of compound per spot. Separations were achieved with chloroform, petroleum ether (80 to 100°C)–acetic acid (95:5), and petroleum ether (80 to 110°C)–pyridine (95:5). Seven different color reagents (the results of two of these are given in Table 30.8) were used for the detection of the spots. From the R_f values of a large group of compounds such as this, a number of generalizations concerning the effect of various groups on the R_f value can usually be made (although it must be remembered that reversals occur in some solvents). (1) The polarity of the compound increases with increasing number of OH groups: phenol < pyrocatechol < pyrogallol. (2) Methylation of the hydroxyl group decreases the polarity: dimethylhydroquinone < *p*-hydroxyanisole < hydroquinone and guaiacol < pyrocatechol. (3) Increase in the size of the alkylating group decreases the polarity: propenylguaethol < isochavibetol, benzyleugenol < methyleugenol. (4) Increasing the size of the alkyl substituent on the ring decreases the polarity: dihydroeugenol < cresol (however, note that guaiacol is less than cresol). (5) As might be expected hydrogenation of the side chain had only a slight effect on the polarity; compare dihydroeugenol and eugenol. (6) The linking together of the two adjacent hydroxy groups by means of a methylene group to form a methylenedioxy group decreases the polarity as compared with that of the corresponding compound with two methoxy groups: safrol < methyleugenol and isosafrol < methylisoeugenol. (7) The effect of hydrogen bond formation and the consequent decrease in the polarity of the compound can be seen by a comparison of the two ortho compounds guaiacol and guaethol with *p*-hydroxyanisole.

Kheifitis et al. [48] chromatographed 33 phenols on loose layers of neutral alumina using benzene–methanol (9:1) as the developing solvent. Location of the compounds was accomplished by means of iodine vapor.

Wang [62] and Lin et al. [63] have separated phenols on polyamide layers (see Table 27.3).

Bakshi and Krishnaswamy [64] used thin-layer chromatography to follow the cardanol–formaldehyde reaction under alkaline conditions. Three methylol derivatives were observed along with a fourth spot for the unchanged cardanol.

Thieme [65] separated thymol and carvacrol on unactivated silica gel layers with benzene–carbon tetrachloride–*o*-nitrotoluene (1:1:1) with R_f values of 0.6 and 0.52, respectively.

Fraser and Swan [66] separated 12 phenolic diterpene acetates on silica gel or alumina G with petroleum ether, methylene chloride, or diethyl ether.

5 ACIDS AND ESTERS

Some of the acids that occur in essential oils are discussed in Chapter XIII, and the effect of various complexing agents on the separation of phenol carboxylic acids has been discussed in Chapter III, Section 4.

Tschesche et al. [67] have chromatographed a group of triterpenic acids found in *Bredemeyera floribunda, Alphitonia excelsa,* and *Crataegus oxyacantha*. The separation was carried out on silica gel G layers using for the most part diisopropyl ether–acetone (5:2) (see Table 30.9). Since cochalic, bredemolic, and machaerinic acids streaked in this solvent, they were also chromatographed in the same solvent with the addition of 5% pyridine giving R_f values of 0.15, 0.62, and 0.27, respectively. The compounds were located with chlorosulfonic acid which gave a high sensitivity of detection (0.02 μg of oleanolic acid). Because the oleanolic, ursolic, and petulinic acids could not be separated, the authors applied chromatography on anion-exchange paper with cyclohexane–toluene (4:1) saturated with formic acid, as well as methyl cyclohexane–chloroform (4:1) saturated with formic acid. Methyl esters of some of these

Table 30.8 *R* × 100 (Referred to Thymol) for Phenols and Phenol Ethers Found in Essential Oils[a]

Compound	*Solvent*			*Color with Antimony Pentachloride–carbon Tetrachloride* (1:4)		*Color with 1.5 ml of 3% p-Nitroaniline in 8% Sulfuric Acid + 25 ml of 5% Sodium Nitrite*[b]	
	Chloroform	*Petroleum Ether* (80 to 100°C)–*Acetic Acid* (95:5)	*Petroleum Ether* (80 to 100°C)–Pyridine (95:5)	20°C	110°C	20°C	110°C
Eugenol	124	137	104	Blue-violet	Gray-brown	—	Yellow-brown
Eugenol methyl ether	148	128	235	Blue-biolet	Gray	—	—
Eugenol benzyl ether	153	237	282	Red-brown	Brown	—	—
Myristicin	154	187	393	Brown	Brown	—	—
Chavicol methyl ether	166	500	425	Yellow	Yellow-brown	—	—
Safrol	164	545	573	Gray	Blue-black	—	—
Isoeugenol	120	123	97	Blue-violet	Gray-brown	Bright yellow	Yellow-brown
Isochavibetol	118	133	93	—	Red-brown	—	Yellow
Propenylguaethol	138	146	165	Yellow-brown	Brown	—	Weak pink
Isoeugenol methyl ether	152	130	223	Blue-violet	Gray-violet	Bright yellow	Yellow
Isoeugenol benzyl ether	165	208	244	Red-brown	Brown	—	Weak yellow
Isomyrisiticin	158	191	339	Violet	Violet	—	Yellow
Anethol	166	493	532	Blue-violet	Violet	Bright yellow	Bright yellow
Isosafrol	166	506	550	Blue-violet	Violet	Bright yellow	Bright yellow
α-Naphthol	67	43	50	Gray	Gray	Orange	Orange-brown

β-Naphthol	52	32	41	Yellow-brown	Gray-brown	Pink	Orange-red
β-Naphthol methyl ether	168	341	518	Yellow-brown	Gray	—	Weak pink
β-Naphthol ethyl ether	164	400	637	Gray-brown	Gray	—	Weak pink
β-Naphthol isobutyl ether	176	449	800	Gray-brown	Gray	—	Rose
p-Hydroxyanisole	37	14	35	—	—	—	Violet
3,4-Methylenedioxyphenol	42	18	39	Green	Gray-green	Orange	Orange-brown
Phenol	49	47	60	—	Gray	—	Brown
Guaiacol	113	145	138	—	Gray	—	—
	124	177	177	Gray-blue	Gray-blue	—	Brown
Hydroquinone dimethyl						—	
ether	156	336	396	—	Yellow-brown	—	—
Cresol	107	102	131	Brown	Yellow-brown	Yellow-brown	Rose
Dihydroeugenol	120	198	138	—	Yellow-brown	—	Weak pink
Carvacrol	94	110	110	Brown	Brown	Pink	Brown
Thymol	100	100	100	Red-brown	Red-brown	Pink	Brown
Isothymol	67	63	81	—	—	—	Red-brown
Pyrocatechol	10	0	10	Blue-	Gray-blue	—	Red-brown
Resorcinol	0	0	0	Green	Brown	Orange-yellow	Orange
Hydroquinone	2	0	0	—	Yellow-brown		Bright blue (UV)
Pyrogallol	0	0	0	—	Brown	Yellow	Yellow-brown
Phyloroglucinol	0	0	0	—	Brown	Orange-yellow	Red-brown
Orcin	3	0	0	—	Brown	Orange	Orange

[a] From M. H. Klouwen and R. ter Heide [61]; reproduced with permission of the authors and Dr. Alfred Huethig Verlag.
[b] Most colors can be intensified by spraying with sodium carbonate solution.

Table 30.9 *R_f* Values of Triterpene Acids on Silica Gel G Layers in Diisopropyl Ether–Acetone (5:2)[a]

Acid	*R_f*
Oleanic	0.68
Ursolic	0.68
Betulinic	0.68
Morolic	0.59
Oleanonic	0.68
Masticadienonic	0.47
Isomasticadienonic	0.47
Cochalic	0.18[b]
Bredemolic	0.59[b]
Siaresinolic	0.66
Machaerinic	0.23[b]
Guiiavolic	0.35
Acantholic	0.29
Quinovic	0.55
Medicagenic	0.29[b]
Emmolic	0.59

[a] From R. Tschesche, F. Lampert, and G. Snatzke [67]; reproduced with permission of the authors and The Elsevier Publishing Co.
[b] Tailing.

acids were also separated on silica gel layers in various solvent systems. Triterpene acids of *Liquidambar orientalis* were investigated by Huneck [68]. Brieskorn et al. [69] isolated ursolic acid from the leaves of *Pirus malus*. The triterpenic acids of olive oil were examined by Vioque and Maza [70]. Thomas and Mueller [71] chromatographed the methyl esters of the triterpene acids from *Commiphora glandulosa*. The esters were separated by column chromatography on silica gel, and the fractions were checked by means of thin-layer plates using chloroform–ethyl acetate (4:1) as the developing solvent. Bonati [72] separated 18α- and 18β-glycyrretic acids with ethyl acetate–methanol–diethylamine (14:4:3).

Berosa and Jones [73] separated piperonylic, 5-, 2-, and 6-methoxypiperonylic acids on silica gel layers in a saturated chamber using ethyl acetate–hexane–acetic acid (50:50:0.5) as a solvent. They obtained R_f values of 0.64, 0.54, 0.47, and 0.37, respectively. The compounds were detected with a chromotropic–sulfuric acid reagent.

Norin and Westfelt [74] separated resin acid methyl esters prepared from the resin acids of *Pinus silvestris* on layers of silver nitrate-impreg-

Table 30.10 $R_f \times 100$ Values of Some Terpene Esters on Silic Acid Chromatostrips Using a Starch Binder[a,b]

Ester	*Ethyl Acetate–Hexane* (15:85)	*Ethyl Acetate–Chloroform (Alcohol-free)* (1:9)	*Ethyl Acetate–Benzene* (15:85)	*Ethyl Acetate–Chloroform (Alcohol-free)* (5:95)	*1-Nitropropane–Hexane* (1:1)	*Ethyl Acetate–Hexane* (3:7)	*Ethyl Carbonate–Chloroform (Alcohol-free)* (15:95)	*Isopropyl formate–Hexane* (3:7)	*Isopropyl Ether–Hexane* (1:1)	*Chloroform (Alcohol-free)*
Geranyl acetate	51	69	66	52	81	72	90	71	73	27
Neryl acetate	55	69	55	50	86	69	86	69	73	39
Citronellyl acetate	58	68	66	57	91	81	89	75	84	35
Octyl acetate	72	98	98	82	90	85	92	92	87	50
Terpinyl acetate	58	66	61	55	90	75	86	73	85	42
Methyl anthranilate	42	52	65	53	62	65	72	49	46	26
Ethyl anthranilate	41	58	70	46	79	67	84	58	53	25
N-Methyl methyl anthranilate	58	75	74	71	91	78	90	69	79	56
Carvyl acetate	66	69	79	67	96	84	91	77	84	64

[a] From J. M. Miller and J. G. Kirchner [1]; reproduced with permission of the American Chemical Society.

[b] Development distance 10 cm.

nated silica gel with benzene as the developing solvent. The R_f values for the methyl esters of the acids were as follows: pimaric 0.40, sandaracopimaric 0.27, isopimaric 0.32, levopimaric 0.50, palustric 0.60, dihydroabietic 0.83, abietic 0.75, and neoabietic 0.73.

Miller and Kirchner [1] separated a group of terpene esters in 10 different solvent systems (Table 30.10). The compounds were detected by the fluorescein–bromine test except for the anthranalates which were readily detectable by their fluorescence under ultraviolet light.

Attaway [47] preferred a mixture of trifluorotrichloroethane–methylene chloride (3:2) rather than benzene for the elution chromatography of compounds because of the lower boiling point (36°C compared to 80°C) and the consequent lower loss of compounds through solvent removal. R_f values of a number of esters as well as carvone and citronellal are given for development on silica gel layers with these solvents.

6 OXIDES AND PEROXIDES

Miller and Kirchner [1] separated 1,8-cineol and linalool monoxide in the 10 solvents listed in Table 30.10; they are readily separated by any of this group. For example, in 1-nitropropane–hexane (1:1), they have R_f values of 0.73 and 0.08, respectively, and in ethyl acetate–hexane (15:85), 0.48 and 0.21, respectively. Katayama [45] chromatographed 1,8-cineol and ascaridole in six different solvents obtaining a value for ethyl acetate–*n*-hexane (15:85) of 0.49 and 0.45, respectively, and 0.27 and 0.21, respectively, in chloroform. Jaspersen-Schib [17] has reported an R_f value of 0.82 for menthofuran on silica gel G using 5% ethyl acetate in benzene as the developing solvent. As a detecting agent for this compound he used concentrated nitric acid–acetic acid (1:300). After the plate was dried at 100°C for 5 to 10 min, the menthofuran appeared as a red spot in daylight and as orange-red under ultraviolet light. Felix et al. [75] used aluminum oxide G layers to check the purity of linalool oxide.

Nigam et al. [76] used a coupled gas–liquid thin-layer chromatography system for the isolation and quantitative determination of piperitone oxide as well as piperitone in essential oils. The epoxide was found in *Mentha avensis, M. piperita, Eucalyptus dives,* and *E. numerosa* for the first time.

Mahl et al. [77] used thin-layer chromatography to investigate the epoxidation of terpenoids. Silica gel was used with petroleum ether–ether (5:1). Darcel [78] separated the peroxide components of turpentine.

7 ESSENTIAL OILS

Mint Oils

Ito et al. [21, 79] were the first to apply thin-layer techniques to the examination of mint oils in 1953, and since that time quite a bit of work has been directed toward these oils. The oils were chromatographed on starch-bound silicic acid chromatostrips. The spots were located by spraying first with a saturated aqueous solution of vanillin followed by a spray of concentrated sulfuric acid. Reitsema et al. [80–83] compared the chromatographic patterns of different mint strains. Examination of the leaves from different parts of the plant showed that the amount of constituents varied with the age of the leaf. Increasing amounts of the more reduced forms of the constituents were found in the older leaves. These authors also examined the peppermint oil produced by plants growing in an atmosphere containing radioactive carbon dioxide. Battaile et al. [84, 85] used thin-layer chromatography in investigating the biosynthesis of terpenes in peppermint and related species of mint. From plants grown in radioactive carbon dioxide, it was shown that the terpenes were synthesized in the young leaves. A number of locating agents were used but where the spots were to be eluted and recovered, only a 0.05% Rhodamine B spray was used. Although traces of Rhodamine B were found in the eluate, this could be removed by rechromatographing. Jaspersen-Schib [17] examined different commercial samples of menthol oils and found that adulterations with other species of *Mentha* could be readily recognized in the oils of *Mentha piperita*. Pertsev and Pivnenko [86] and Karawya and Wahba [87] used thin-layer plates for the analysis of peppermint oils. Rothbaecher et al. [88] examined the main components of Rumanian oil of peppermint. Gurvich [89] used a semiquantitative method for the determination of menthol in peppermint oils using radial chromatography on alumino-silicate plates, on which the peppermint oil was compared to standard menthol spots. Nigam and Levi [90] used a coupled vapor-phase thin-layer chromatographic method for the determination of menthofuran in *Mentha arvensis* and other mint species. (See Section 6 for piperitone oxide in mint oils.) Hefendehl [91] investigated the composition of *Mentha aquatica*, and Vlakhov and Ognyanov [92] isolated 23 compounds from the sesquiterpene fraction of Bulgarian peppermint oil. Deryng et al. [93] identified menthol, menthone, menthofuran, and cineol in peppermint oil by using silica gel G with benzene–ethyl acetate (5 : 1) or chloroform.

Hop Oils

Since the quality of the volatile oil in hops affects the aroma and flavor of beer, Rigby and Bethune [94] in 1955 applied thin-layer chromatography to the examination of the aromatic constituents of hops. By a combination of thin-layer and countercurrent distribution, at least 26 constituents were shown to be present. Some of these were identified. Kuroiwa and Hashimoto [95–103] have investigated the lupulones and humulones, the bitter substances in beer and hops. They used starch-bound silica gel chromatostrips and to some extent Zeolite (Dow). Since it is important to know the composition of the bitter substances in hops as a means of controlling the bitterness in beer, these authors developed a chromatographic method for their determination. This consisted in eluting the spots and measuring the density at 325 mμ for humulones and 355 mμ for lupulones. A problem related to this is the off-flavor that occurs in sun-struck beer; this was investigated by the same authors [104, 105] and was found to be due to the 3-methyl-2-butene-1-thiol which appeared to be formed from the reaction of some sulfhydryl compound with the degradation products of isohumulones. Aitken et al. [106, 107] determined the beer-bittering substance by chromatographing extracts on silica gel G with benzene–ethyl ether (16:1). The compounds were then eluted and the acidified solutions measured at 275 nm.

Citrus Oils

The characteristic flavor and aroma of the citrus juices reside in the volatile oil fractions occurring in the juice sacs. As important as they are to the flavor of the juice, some of the constituents are present in exceedingly small amounts and Kirchner and co-workers [108–110] isolated the oils from large quantities of orange and grapefruit juice. Because of the small amount of oil obtainable (35 g of oxygenated constituents from 3000 gal of juice) it was essential that a micromethod be developed for separating the numerous compounds. This was the problem that faced me and my associates when I originated and established the present day system of thin-layer chromatographic analysis [1, 2]. By these techniques the components in the citrus juices were separated and identified. Attaway et al. [19, 111] applied the method along with gas and paper chromatography to the identification of the volatile flavor components obtained from the orange essence recovered from the production of orange concentrate. Peyron [112] applied thin-layer chromatography to the separation and identification of fluorescent components in the oleiferous pockets of citrus fruits. Millet et al. [113] used gas and thin-layer chromatography to compare the oil obtained by direct puncture of the orange secretory cells with that obtained by classical procedures. Landgraf [114] examined the essential oils of lemon, grapefruit, and orange, and Rispoli et al. [115] investigated the oil of Sicilian grapefruit. Stanley and coworkers [116–122] have used the method extensively for the examination of the components of lemon oil including a number of coumarin compounds.

D'Amore and Calapaj [123] isolated the fluorescent substances from lemon, bergamot, tangerine, bitter orange, and sweet orange oils on silica gel using hexane–ethyl acetate (3:7) as a solvent. Cieri [124] isolated the coumarins and furocoumarins from bergamot, lemon, lime, angelica seed, and angelica root oils and identified the compounds by R_f value and their characteristic fluorescence. The compounds were eluted and determined by spectrophotometry. MacLeod and Buigues [125] applied thin-layer chromatography to the analysis of lemon oil. Two-dimensional development on silica gel layers with chloroform–acetic acid–carbon tetrachloride (30:1:69) in the first dimension followed by ethyl acetate–acetic acid–cyclohexane (20:1:79) in the second dimension was used to differentiate cold pressed lime and lemon oils. Distilled lemon and lime oils showed very little fluorescence. Madsen and Latz [126] applied in situ flourimetry to the analysis of lime and lemon oils. Ikeda et al. [127] used a combination of the chromatostrip with gas chromatography to determine the monoterpene composition of various citrus oils. Bernhard [128] applied the chromatostrip technique to the separation and identification of five coumarin compounds in lemon juice. Martinez Nadal [129] has surveyed the application of thin layers to the determination of citral in citrus and other oils. Verderio and Venturini [130] have examined mandarin essential oil. Pozzo-Balbi and Nobile [131] used thin-layer chromatography to separate *Citrus trifoliata* fruit peel oils into classes according to functional groups prior to gas chromatography analysis. Karawya et al. [132] compared the steam-distilled and cold-pressed oils of *Citrus sinensis* and *C. aurantium* by chromatographing on silica gel with hexane–ethyl acetate (17:3). Attaway et al. [133] used thin-layer and gas chromatography as well as mass spectrometry to analyze the leaf oils from 11 citrus varieties. Maier and Grant [134] isolated the bitter principle, limonin, from citrus by development of the extract on silica gel with the upper phase of acetic acid–water–ethanol–benzene (1:15:47:200). Quantitative determination was made by spot densitometry or by visual comparison with known quantities. The method was sensitive to 0.5 ppm.

Seaweed Volatile Constituents

Katayama [135–142] has made an extensive study of the volatile oil constituents of various species of seaweed combining separations on

Table 30.11 Additional Work on Thin-Layer Chromatography as Applied to Essential Oils and Terpenes

Year	*Authors*	*Nature of Work*	*Ref*
1952	Montes	General study of essential oils	150
1953	Gaenshirt	Oils from *Aristolochia clematitus*	151
1954	Gruener and Spaich	Tinctures of *Arnica montana*	152
1955	Coveney et al.	Oil of *Strobilanthopsis linifolia*	153
1955	Kaiser	Oil and resins of *Grindelia* species	154
1955	Garcia de Nadal	Review	155
1955	Wotherspoon and Bedoukian	Oxidized compounds in essential oils	156
1956	Demole	Isophytol in jasmine absolute	157
1956	Onishi et al.	Oil of tobacco leaves	158
1956	Stahl et al.	Chamazulene and derivatives	159
1957	Allentoff and Wright	Reactions products of Grignard reagents with terpenes	160
1957	Frydman et al.	Terpenes	161
1957	Frydman et al.	Numerous essential oils	162
1957	Frydman et al.	Numerous essential oils	163
1957	Garcia de Martinez Nadal	Bay oil deterpenation	164
1957	Gogroef	Patchouli oil	165
1957	Onishi et al.	Oil of tobacco leaves	166
1957	Yamamoto and Furukawa	Terpenes	167
1957	Yamamoto et al.	Terpenes	168
1958	Demole and Lederer	Jasmine oil	169
1958	Klohr-Meinhardt	Action of light on formation of oil in *Pedtoselinum s. and Levisticum o.*	170
1958	Klohr-Meinhardt	Effect of grafting on oil formation	171
1958	Lederer	Review on terpenes	172
1958	Onishi	Oil of tobacco leaves	173
1958	Pryor and Bryant	Oil of eucalyptus	174

1958	Stahl	Camomile oil, artificial nutmeg oil	175
1958	Stahl	Miscellaneous oils, resins, and balsams	176
1958	Suga	Oxidation products of terpenes	177
1959	Katayama and Nagai	Coriander-seed oil	178
1959	Katayama and Nagai	Oil of nutmeg	179
1959	Winkler and Lunau	Oil of *Curcuma xanthorrhiza and C. longa*	180
1960	Akazawa	Ipomeamorone	181
1960	Akazawa and Uritani	Ipomeamorone and coumarins in fungus-injured sweet potatoes	182
1960	Brieskorn and Wenger	Oil of sage	183
1960	Gallardo and Montes	Terpene hydrocarbons	184
1960	Fujita	Oxidation of limonene	185
1960	Lederer	Diterpenes	186
1960	Marbet and Saucy	Pseudionone and analogous compounds	187
1960	Tétényi	Azulene content of chamomile	188
1960	Wulff and Stahl	Oil of *Acorus calamus*	189
1961	Akazawa and Wada	Ipomeamorone, quantitative assay	190
1961	Hoerhammer et al.	Oleanolic acids in *Radix Panax Ginseng*	191
1961	Kratzl	Biogenesis of lignins	192
1961	Neubern de Toledo and Wasicky	Oil of sassafras, citronella, and peppermint	193
1961	Paris and Godon	Miscellaneous essential oils	194
1961	Zanini et al.	Oil of *Pinus pumilio*	195
1962	Akazawa et al.	Biosynthesis of ipomeamorone	196
1962	Brieskorn and Polonius	Germanicol in *Salvia officinalis*	197
1962	Deshusses and Gabbai	Methyl anthranilate in orange blossom honey	198
1962	El-Deeb et al.	Caraway and lemon oils	199
1962	Gabel et al.	Thymol and carvacrol in thyme oil	200
1962	Hoerhammer and Wagner	*Ammi visnaga* and *ammi majus* constitutents	201
1962	Huneck	Triterpenes	202
1962	Huneck and Lehn	Triterpenes	203

Table 30.11 (Continued)

Year	*Authors*	*Nature of Work*	*Ref*
1962	Huneck and Lehn	Triterpenes	204
1962	Jaspersen-Schib and Flueck	Various terpenes	205
1962	Jork	Resins and balsams	206
1962	Lavie et al.	Triterpenes	207
1962	Morgan and Pereira	Grass and corn silage steam distillates	208
1962	Nigam and Kumari	Various essential oils	209
1962	Pertsev and Pevnenko	Oil of coriander, lavender, and nutmeg	210
1962	Scheidegger et al.	Diterpenes	211
1962	Schreiber and Osske	Triterpenes in potato leaves	212
1962	Schulte	Polyynes and terpenes in roots of *Arnica montana* and *A. foliosa*	213
1962	Tyihák and Vágújfalv	Oils of *Matricaria chamomilla, Achillea millefolium,* and *Artemisia absinthium*	214
1963	Bhramaramba and Sidhu	Indian cinnamon leaf oil	215
1963	Borkowski and Pasich	Review of triterpenoids	216
1963	Brud and Daniewski	Quantitative determination of linalool in linalyl acetate	217
1963	Capella et al.	Triterpene alcohols	218
1963	Djerassi et al.	Isomeric cyanohydrocarvones	219
1963	Fu et al.	Triterpenoids in *Oldenlandia pinifolia*	220
1963	Grab	Evaluation of chamomile preparations	221
1963	Graham and McQuillin	Terpene synathesis	222
1963	Hoerhammer et al.	Constituents of fruit of *Angelica silvestris*	223
1963	Hoerhammer et al.	Sesquiterpenes in *Folia farfarae* and *F. petasites*	224
1963	Hoerhammer et al.	Triterpenes in *Crataegus oxyacantha*	225
1963	Ikan and Kashman	Triterpenoids in Hula peat	226
1963	Ikan et al.	Triterpenes	227
1963	Kaufmann and Sen Gupta	Terpenes in coffee bean fat	228

1963	Mangoni and Belardini	Triterpenes in oak galls	229
1963	Paris and Godon	Numerous oils	230
1963	Pasich	Triterpene oils (saponins in *Primula officinalis* and *P. elatioris*)	231
1963	Peyron	Review on radial chromatography, both paper and thin-layer	232
1963	Ramaut	Depsidones of β-orcinol	233
1963	Ramaut	Depsides and depsidones	234
1963	Schantz	Discussion of methods	235
1963	Schulte et al.	Polyacetylenes	236
1963	Smit et al.	Cyclization of isoprenoidal compounds	237
1963	Tschesche et al.	Triterpenes, constitution of escin	238
1963	Tschesche et al.	Triterpenes, structure of bredemolic acid	239
1963	Tyihák et al.	Components in oil of wild and cultivated camomile	240
1963	Vorbrueggen et al.	Terpenoids in Indian medicinal plants	241
1963	Wellendorf	Oil of Danish Pharmacopea	242
1964	Bergstroem and Lagercrantz	Diphenylpicrylhydrazyl as reagent for terpenes	243
1964	Demole	General discussion	244
1964	Demole	Terpene phenols	245
1964	Hoerhammer et al.	Egyptian basil essential oils	246
1964	Hoerhammer et al.	Pharmaceutically used umbelliferon oils	247
1964	Ikan et al.	Tetra and pentacyclic triterpenes	248
1964	Kaufmann and Sen Gupta	Triterpenes in coffee bean	249
1964	Kohen et al.	α- and β-amyrin on $AgNO_3$-alumina layers	250
1964	Kunovits	Identification of turpentines	251
1964	Masse and Paris	Testing of pharmaceuticals containing ethereal oils and resins	252
1964	Moreira and Cecy	Benzoe balsam	253
1964	Nano and Martelli	Separation of thujyl alcohols	254
1964	Schilcher	Evaluation of chamomile	255
1964	Shalaby and Richter	Oil of *Achillea fragrantissima*	256
1964	Stahl	Oil of *Daucus carota*	257

Table 30.11 (Continued)

Year	*Authors*	*Nature of Work*	*Ref*
1964	Takeda et al.	Linderalactone and isolinderalacetone	258
1964	Takeda et al.	Lindestrene and linderene acetate	259
1964	Tschesche et al.	Constituents of *Gratiola officinalis*	260
1964	Vernin	Review	261
1964	Wasicky	Oil of *Peumus boldus*	262
1964	Wollrab	Wax compounds of rose and lavender oils	263
1964	Zinkel and Rowe	Resin methyl esters	264
1965	Betts	Cinnamon	265
1965	Bhatnagar	Concrete essential oils and their clathrates	266
1965	Blanc et al.	Vanillin and ethyl vanillin	267
1965	Elgamal and Fayez	Triterpenoid acids	268
1965	Ikan and Meir	Oxygenated terpenes on $AgNO_3$ silica	269
1965	Kraus and Perenyi	Azulene in oil of *Achillea millefolium*	270
1965	Martin et al.	Components of bourbon whisky	271
1965	Moslé et al.	Analysis of refrigeration machine oils in oleoresins	272
1965	Murakami et al.	Tetra- and pentacyclic triterpenes	273
1965	Nano et al.	Essential oil of *Absenthium gentile*	274
1965	Nigam et al.	Taxonomic applications	275
1965	Paseshnichenko and Guseva	Use of π complexes in the separation and determination of essential oils	276
1965	Schultz and Mohrmann	Components of *Allium sativum*	277
1965	Verderio and Venturini	Mandarin oil	278
1965	Wrolstad and Jennings	Isomerization of terpenes	279
1965	Zacsko-Szasz and Szasz	Oil of anise	280
1966	El-Hamidi and Ahmed	Composition of umbelliferous oils	281

1966	Gogiya	Terpenes and aromatic alcohols in tea essential oil	282
1966	Munoz Teran	Essential oils in pomades, ointments, and solutions	283
1966	Neves et al.	Essential oil of *Eucalyptus saligna*	284
1966	Proenca Da Cunha	Essential oil of *Eucalyptus punctata*	285
1966	Zacsko-Szasz and Szasz	Aniseed oil	286
1967	Barua et al.	Pentacyclic triterpenes on $AgNO_3$ silica	287
1967	Deryng et al.	Volatile oil of *Oleum foeniculi* and *O. Anisi*	288
1967	Domagalina and Zareba	Triterpene fraction in *Sambucus ebulus* roots	289
1967	Foks	Components of rosin	290
1967	Martelli et al.	Essential oil of Sardinian *Eucalyptus rostrata*	291
1967	Pakrashi and Majumdar	Arborinone from *Glycosmis arborea*	292
1967	Tóth	Essential oil of *Foeniculum vulgare*	293
1968	Adcock and Betts	Monoterpene alcohol trimethylsilyl ethers	294
1968	Challen and Kučera	Canada balsam and resin acids	295
1968	Lebez	Archeological amber and Baltic Sea amber	296
1968	Sticher and Flueck	Composition of distilled and extracted oils of *Mentha* species	297
1968	Tyihák et al.	Composition of chamomile essential oil	298
1968	Waisser et al.	Stereochemistry of 1-hydroxytriterpenes	299
1969	Bauer and Brasil e Silva	Essential oil of *Lippia lycioides*	300
1969	Jain	Essential oil of *Heracleum mantegozzianum*	301
1969	Kami et al.	Essential oil of *Geranium* species	302
1969	Pensar	Fatty and resin acids of wood resins	303
1969	Stahl et al.	Differentiation of geographic origin of oregano	304
1970	Babkin et al.	Sesquiterpenes from oleoresin of *Picea ajanensis*	305
1970	El-Hamidi and Sidrak	Essential oil of *Acacia farnesiana*	306
1970	Halim and Collins	Oil of sweet ferm *(Comptonia peregrina)*	307
1970	Jonczyk	Essential oil of *Abies alba* branches	308
1970	Karawya et al.	Essential oil of *Labiate* plants in Egypt	309
1970	Rao and Nigam	Essential oil of *Eugenia bracteata*	310
1970	Sinsheimer et al.	Gymnemic acids in *Gymnema sylvestre* leaves	311

Table 30.11 (Continued)

Year	Authors	Nature of Work	Ref
1970	Vandenburg and Wilder	Constituents of carnauba wax	312
1971	Collins ahd Halim	Essential oil of *Calycanthus floridus*	313
1971	Griffin and Parkin	Constituents of *Parsonsia straminea*	314
1971	Jeannes and Tetau	Quantitative estimation of glycyrrhetic acid	315
1971	Meisinger	Detection of oil of *Flores chamomillae* in pharmaceuticals	316
1971	Poplawski et al.	Sesquiterpene lactones in *Arnica montana* leaves	317
1971	Shah et al.	Composition of *Variyali sowa* and *Ghoda sowa* (dill)	318
1971	Thapa et al.	Essential oil of *Pogostemon plectranthoides*	319
1972	Connell and McLachlan	Oleoresins from ginger and grains of paradise	320
1972	Drozdz et al.	Sesquiterpene lactones from *Eupatorium cannabinum*	321
1972	Herisset et al.	Oils of *Illicium verum, Pimpinella anisum, Foeniculum dulce*	322
1972	Ter Heide	Essential oil of *Cinnamomum cassia*	323
1972	Mancini et al.	Essential oil of leaves of *Coleus barbatus*	324
1972	Popescu and Ciupe	Volatile oil from walnut *(Juglans regia)* leaves	325
1972	Qedan	Essential oil of *Catha edulis*	326
1972	Ruecker	Monocyclic diterpenes in *Commiphora mukul*	327
1972	Rueedi and Eugster	A new diterpenoid from *Coleus barbatus*	328
1972	Schilcher	Oil of *Flores chamomillae*	329
1972	Willuhn	Ingredients of *Arnica* species	330
1973	Ardon and Nakano	Triterpenes from *Ponteria caimito* bark	331
1973	Balbaa et al.	Oil of *Carum copticum* fruit and herb	332
1973	Chauhan et al.	Alkaline hydrolysis products of shellac	333
1973	Habib and Metwally	Sesquiterpene ketolactone from *Senecio*	334
1973	Hoelzl and Demuth	Chromatographic study of chamomile oil	335
1973	Holub and Šamek	Sesquiterpene lactone from *Laserpitium archangelica*	336
1973	Morales and Torres	Resin content of candelilla wax	337

1973	Salehian and Netien	Essential oil of marjoram	338
1973	Sengupta et al.	Differentiation of Apium graveolens and *Carum copticum*	339
1973	Tirimanna	Essential oil of black tea	340
1974	Blechinger	Colophonium and modified resin acids	341
1974	Bungert-Hansing and Jork	Microdehydrogenation of sesquiterpene alcohols	342
1974	Dro and Hefendehl	Analysis of oil of *Ocimum gratissimum*	343
1974	Kaempf and Steinegger	Oil of *Oleum anisi* and *Oleum anisi stellati*	344
1974	Karawya and Hifnawy	Analysis of oil of *Thymus vulgaris*	345
1974	Lichenstein et al.	Insecticidal and synergistic compounds in dill	346
1974	Mesonero et al.	Different essences of Spanish lavender	347
1974	Sen et al.	Differentiation of caraway, coriander, and cumin	348
1974	Thieme and Hartmann	Determination of glycyrrhizic acid in *Radix liquiritiae*	349
1975	Banerji and Mitra	A new tetranortriterpenoid in *Cedrela toona* heartwood	350
1975	Briggs and White	Constituents of essential oil of *Araucaia araucana*	351
1975	Budzikiewicz and Roemer	Sesquiterpene esters in *Euonymus europaeus*	352
1975	Elgamal and El-Tawil	A new triterpenoid from *Glycyrrhiza glabra* roots	353
1975	Haggag et al.	Essential oil of *Achillea millefolium*	354
1975	Hoelzl and Demuth	Ecological factors in *Matricaria chamomilla* oil	355
1975	Hoelzl et al.	Biosynthesis of oil of *Matricaria chamomilla* oil	356
1975	Ivie et al.	Sesquiterpene lactones in *Helenium amarum*	357
1975	Nambudiri et al.	TLC analysis of ginger extract	358
1975	Rustembekova et al.	Oxygen-containing compounds of *Chenopodium botrys* oil	359
1975	Stipanovič et al.	TLC of desocyterpenoids	360
1975	Stipanovič et al.	Antimicrobial terpenoids of *Gossypium*	361
1976	Biering et al.	TLC of *Asarum europaeum* oil	362
1976	Verzar-Petri et al.	Composition of chamomile oil	363

chromatostrips with distillation methods. These consist mainly of various terpenes, although a number of low-molecular-weight aliphatic acids were obtained.

Miscellaneous

Vágújfalvi and Tyihák [143] examined 60 reagents for detecting the constituents in essential oils chromatographed on thin layers. Ten essential oils were used in the tests. Twenty-nine of the reagents were recommended for general use with essential oils. Among these were the concentrated sulfuric acid and fluorescein–bromine tests of Kirchner et al. [2], 70% perchloric acid, concentrated phosphoric acid, and numerous salts dissolved in concentrated sulfuric acid.

Among the reviews on essential oil applications can be mentioned those of Lawrence [144], Petrowitz [145], and Calvarano [146].

There have been many papers on various essential oils where thin-layer chromatography has assisted in separating and identifying constituents. It is impossible to go into detail on all of these; however, Table 30.11 lists some of these along with brief comments to indicate the thin-layer chromatographic application. Additional references may be found in the bibliographies given in the introduction to this edition. Vágújfalvi et al. [147–149] have listed, according to the taxonomical category of the plants investigated, the publications that have utilized thin-layer chromatography for compounds in plants.

References

1. J. M. Miller and J. G. Kirchner, *Anal. Chem.*, **25**, 1107 (1953).
2. J. G. Kirchner, J. M. Miller, and G. J. Keller, *Anal. Chem.*, **23**, 420 (1951).
3. B. H. Davies, T. W. Goodwin, and E. L. Mercer, *Biochem. J.*, **81**, 40P (1961).
4. E. C. Grob and A. Boschetti, *Chimia (Aarau)*, **16**, 15 (1962).
5. A. Vásquez and L. Janer, *Grasas Aceites (Seville, Spain)*, **13**, 242 (1962).
6. S. Fukushi and Y. Obata, *J. Agric. Chem. Soc. Japan*, **27**, 353 (1953); through *Chem. Abstr.*, **50**, 15027 (1956).
7. P. Tétényi, E. Tyihák, I. Máthé, and J. Sváb, *Pharmazie*, **17**, 463 (1962).
8. E. C. Kirby, *J. Chromatogr.*, **80**, 271 (1973).
9. E. Tyihák and D. Vágújfalvi, *Plant Med.*, **15**, 269 (1967).
10. P. Tétényi, E. Tyihák, I. Máthé, and J. Sváb, *Pharmazie*, **17**, 463 (1962).
11. A. S. Gupta and S. Dev, *J. Chromatogr.*, **12**, 189 (1963).
12. L. Westfelt, *Acta Chem. Scand.*, **18**, 572 (1964).
13. P. N. Chow, O. Motl, and V. Lukes, *Collect. Czech. Chem. Commun.*, **30**, 917 (1965).

14. B. M. Lawrence, *J. Chromatogr.*, **38**, 535 (1968).
15. M. von Schantz, S. Juvonen, and R. Hemming, *J. Chromatogr.*, **20**, 618 (1965).
16. J. A. Attaway, L. J. Barabas, and R. W. Wolford, *Anal. Chem.*, **37**, 1289 (1965).
17. R. Jaspersen-Schib, *Pharm. Acta Helv.*, **36**, 141 (1961).
18. R. Jaspersen-Schib and H. Flueck, *Congr. Sci. Farm. Conf. Commun, 21st, Pisa* **1961**, 608 (1962).
19. J. A. Attaway and R. W. Wolford, *25th Intern. Symp. Gas Chromatogr., Brighton, England, September, 1964.*
20. S. Heŕmánek, V. Schwarz, and Z. Čekan, *Pharmazie,* **16**, 566 (1961).
21. M. Ito, S. Wakamatsu, and H. Kawahara, *J. Chem. Soc. Japan, Pure Chem. Sect.*, **75**, 413 (1954); *Chem. Abstr.*, **48**, 13172 (1954).
22. M. Ito, *Nippon Kagaku Zasshi,* **78**, 172 (1957).
23. H.-J. Petrowitz, *Angew. Chem.*, **72**, 921 (1960).
24. E. Graf and W. Hoppe, *Deut. Apoth.-Ztg.*, **102**, 393 (1962).
25. E. Yamamoto and T. Furukawa, *J. Fac. Educ. Hiroshima Univ.*, **4**, 45 (1956).
26. E. Tyihák, D. Vágujfalvi, and P. L. Hágony, *J. Chromatogr.*, **11**, 45 (1963).
27. G. P. McSweeney, *J. Chromatogr.*, **17**, 183 (1965).
28. D. V. Banthorpe and K. W. Turnbull, *J. Chromatogr.*, **37**, 366 (1968).
29. G. P. McSweeney, *J. Chromatogr.*, **17**, 183 (1965).
30. P. J. Dunphy, J. D. Kerr, J. F. Pennock, K. J. Whittle, and J. Feeney, *Biochim. Biophys. Acta,* **136**, 136 (1967).
31. F. J. Evans and A. D. Kinghorn, *J. Chromatogr.*, **87**, 443 (1973).
32. P. A. Hedin, R. C. Gueldner, and A. C. Thompson, *Anal. Chem.*, **42**, 403 (1970).
33. R. ter Heide, *Z. Anal. Chem.*, **236**, 215 (1968).
34. J. P. Minyard, J. H. Tumlinson, A. C. Thompson, and P. A. Hedin, *J. Chromatogr.*, **29**, 88 (1967).
35. J. H. Dhont and C. de Rooy, *Analyst (London),* **86**, 527 (1961).
36. M. von Schantz, S. Juvonen, A. Oksanen, and I. Hakamaa, *J. Chromatogr.*, **38**, 364 (1968).
37. R. Ikan, *J. Chromatogr.*, **17**, 591 (1965).
38. J. W. Copius-Peereboom, *Z. Anal. Chem.*, **205**, 325 (1964).
39. E. Cardemil, J. R. Vicũna, A. M. Jabalquinto, and O. Cori, *Biochem.*, **59**, 636 (1974).
40. P. Pesnelle, *Planta Med.*, **12**, 403 (1964).
41. M. K. Seikel and J. W. Rowe, *Phytochemistry,* **3**, 27 (1964).
42. J. Budzynska-Topolowska and A. Rutkowski, *Rev. Fr. Corps Gras.*, **16**, 695 (1969).
43. E. Fedeli, A. Lanazani, P. Capella, and G. Jacini, *J. Am. Oil Chem. Soc.*, **42**, 254 (1966).
44. H. Wollmann, G. Habicht, I. Lau, and I. Schultz, *Pharmazie,* **28**, 56 (1973).
45. T. Katayama, *Nippon Suisan Gakkaishi,* **26**, 814 (1960).

46. M. von Schantz, A. Lopmeri, E. Stroemer, R. Salonen, and S. Brunni, *Farm. Aikak.*, **71**, 52 (1962).
47. J. A. Attaway, *Anal. Chem.*, **36**, 2224 (1964).
48. L. A. Kheifitis, G. I. Moldovanskaya, and L. M. Shulov, *Zh. Anal. Khim*, **18**, 267 (1963).
49. E. Sundt and A. Saccardi, *Food Technol.*, **16**, 89 (1962).
50. S. Kahan and J. Fitelson, *J. Assoc. Off. Agric. Chem.*, **47**, 551 (1964).
51. K. Kratzl and G. Puschmann, *Holzforschung*, **14**, 1 (1960).
52. J. H. Dhont and G. J. C. Dijkman, *Analyst, (London)* **89**, 681 (1964).
53. K. Onoe, *J. Chem. Soc. Japan, Pure Chem. Sect.*, **73**, 337 (1952).
54. V. N. Vashist and K. L. Handa, *J. Chromatogr.*, **18**, 412 (1965).
55. J. Lacharme, *Bull. Trav. Soc. Pharm. Lyon*, **7**, 55 (1963).
56. G. M. Nano and P. Sancin, *Ann. Chim. (Rome)*, **53**, 677 (1963); *Chem. Abstr.*, **59**, 12189 (1963).
57. J. H. Dhont and G. J. C. Mulders-Dijkman, *Analyst (London)*, **94**, 1090 (1969).
58. A. Martelli, *Riv. Ital. Essenze, Profumi, Piante Off. Aroma, Saponi, Cosmet., Aerosol*, **53**, 607 (1971).
59. A. Rothbaecher and F. Suteu, *J. Chromatogr.*, **100**, 236 (1974).
60. A. A. Bell, R. D. Stipanovic, C. R. Howell, and P. A. Fryxell, *Phytochemistry*, **14**, 225 (1975).
61. M. H. Klouwen and R. ter Heide, *Parfuem. Kosmet.*, **43**, 195 (1962).
62. K.-T. Wang, *J. Chinese Chem. Soc. (Taiwan)*, **8**, 241 (1961).
63. Y. T. Lin, K.-T. Wang, and Y.-S. Lin, *J. Chinese Chem. Soc. (Taiwan)*, **9**, 68 (1962); *Chem. Abstr.*, **58**, 9412 (1963).
64. S. H. Bakshi and N. Krishnaswamy, *J. Chromatogr.*, **9**, 395 (1962).
65. H. Thieme, *Pharmazie*, **22**, 722 (1967).
66. H. S. Fraser and E. P. Swan, *J. Chromatogr.*, **38**, 141 (1968).
67. R. Tschesche, F. Lampert, and G. Snatzke, *J. Chromatogr.*, **5**, 217 (1961).
68. S. Huneck, *Tetrahedron*, **19**, 479 (1963).
69. C. H. Brieskorn, H. Klinger, and W. Polonius, *Arch. Pharm.*, **294**, 389 (1961).
70. E. Vioque and M. P. Maza, *Grasas Aceites (Seville, Spain)*, **14**, 9 (1963); *Chem. Abstr.*, **59**, 8986 (1963).
71. A. F. Thomas and J. M. Mueller, *Experientia*, **16**, 62 (1960).
72. A. Bonati, *Fitoterapia*, **34**, 19 (1963).
73. M. Beroza and W. A. Jones, *Anal. Chem.*, **34**, 1029 (1962).
74. T. Norin and L. Westfelt, *Acta Chem. Scand.*, **17**, 1828 (1963).
75. D. Felix, A. Melera, J. Seible, and E. sz. Kováts, *Helv. Chim. Acta*, **46**, 1513 (1963).
76. I. C. Nigam, M. Sahasrabudhe, and L. Levi, *Can. J. Chem.*, **41**, 1535 (1963).
77. B. S. Mahl, M. S. Wadia, I. S. Bhatia, and P. S. Kalsi, *Perfum. Essent. Oil Rec.*, **59**, 519 (1968).
78. C. le Q. Darcel, *Can. J. Biochem.*, **46**, 509 (1968).

79. M. Ito, S. Wakamatsu, and H. Kawahara, *J. Chem. Soc. Japan, Pure Chem. Sect.*, **74**, 699 (1963); through *Chem. Abstr.*, **48**, 364 (1954).
80. R. H. Reitsema, *Anal. Chem.*, **26**, 960 (1954).
81. R. H. Reitsema, *J. Am. Pharm. Assoc., Sci. Ed.*, **43**, 414 (1954).
82. R. H. Reitsema, F. J. Cramer, N. J. Scully, and W. Chorney, *J. Pharm. Sci.*, **50**, 18 (1961).
83. R. H. Reitsema, F. J. Cramer, and W. E. Fass, *J. Agric. Food Chem.*, **5**, 779 (1957).
84. J. Battaile, R. L. Dunning, and W. D. Loomis, *Biochim. Biophys. Acta*, **51**, 538 (1961).
85. J. Battaile and W. D. Loomis, *Biochim. Biophys. Acta*, **51**, 545 (1961).
86. I. M. Pertsev and G. P. Pivnenko, *Farm. Zh. (Kiev)*, **16**, 28 (1961).
87. M. S. Karawya and S. K. Wahba, *Bull. Fac. Pharm. Cairo Univ.*, **1**, 125 (1961); through *Chem. Abstr.*, **60**, 13092 (1964).
88. H. Rothbaecher, C. Crisan, and E. Bendoe, *Farmacia (Bucharest)*, **12**, 733 (1964).
89. N. L. Gurich, *Vses. Nauchn.-Issled., Inst. Maslichn. Efiromasl. Kul't. Vses. Akad. Sel'skokhoz. Nauk, Kratk. Otchet*, **1956**, 154; through *Chem. Abstr.*, **54**, 25595 (1960).
90. I. C. Nigam and L. Levi, *J. Pharm. Sci.*, **53**, 1008 (1964).
91. F. W. Hefendehl, *Arch. Pharm.*, **300**, 438 (1967).
92. R. Vlakhov and I. Ognyanov, *Riechst., Aromen, Koerperpflegem.*, **17**, 315 (1967).
93. J. Deryng, H. Strzelecka, E. Walewska, and T. Soroczyńska, *Farm. Pol.*, **24**, 187 (1968).
94. F. L. Rigby and J. L. Bethune, *Am. Soc. Brewing Chem. Proc.*, **1955**, 174.
95. Y. Kuroiwa and H. Hashimoto, *Rep. Res. Lab. Kirin Brewery Co., Ltd.*, **3**, 5 (1960).
96. *Ibid.*, p. 11.
97. H. Hashimoto and Y. Kuroiwa, *Hakko Kogaku Zasshi*, **39**, 554 (1961); through *Chem. Abstr.*, **59**, 2132 (1963).
98. *Ibid.*, p. 545.
99. *Ibid.*, p. 541.
100. Y. Kuroiwa and H. Hashimoto, *J. Inst. Brewing*, **67**, 506 (1961).
101. Y. Kuroiwa, E. Kokubo, and H. Hashimoto, *Rep. Res. Lab. Kirin Brewery Co., Ltd.*, **4**, 41 (1961).
102. Y. Kuriowa and H. Hashimoto, *J. Inst. Brewing*, **67**, 352 (1961).
103. *Ibid.*, p. 347.
104. Y. Kuroiwa and N. Hashimoto, *Am. Soc. Brewing Chem. Proc.*, **1961**, 28.
105. *Ibid.*, **1963**, 181.
106. R. A. Aitken, A. Bruce, J. O. Harris, and J. C. Seaton, *J. Inst. Brew. (London)*, **74**, 436 (1968).
107. *Ibid.*, **76**, 29 (1970).
108. J. G. Kirchner, R. G. Rice, J. M. Miller, G. J. Keller, and M. M. Fox, *J. Agric. Food Chem.*, **1**, 510 (1953).

109. J. G. Kirchner and J. M. Miller, *J. Agric. Food. Chem.*, **1,** 512 (1953).
110. *Ibid.*, **5,** 283 (1957).
111. J. A. Attaway, D. V. Hendrick, and R. W. Wolford, *Proc. Florida State Hortic. Soc.*, **77,** 305 (1964).
112. L. Peyron, *Compt. Rend.*, **257,** 235 (1963).
113. F. Millet, M. A. Monghal, M. Rollet, and J. Dorche, *Ann. Pharm. Fr.*, **28,** 63 (1970).
114. H. Landgraf, *Rev. Quim. Ind. (Rio de Janeiro)*, **29,** 24 (1960); *Chem. Abstr.*, **56,** 13028 (1962).
115. G. Rispoli, A. Di Giacomo, and M. L. Tracuzzi, *Riv. Ital. Essenze-Profumi, Piante Off. Oli Veg. Saponi*, **45,** 62 (1963).
116. S. H. Vannier and W. L. Stanley, *J. Assoc. Off. Agric. Chem.*, **41,** 432 (1958).
117. W. L. Stanley and S. H. Vannier, *J. Assoc. Off. Agric. Chem.*, **40,** 582 (1957).
118. W. L. Stanley, R. C. Lindwall, and S. H. Vannier, *J. Agric. Food Chem.*, **6,** 858 (1958).
119. W. L. Stanley, R. M. Ikeda, and S. Cook, *Food Technol.*, **15,** 381 (1961).
120. W. L. Stanley, *J. Assoc. Off. Agric. Chem.*, **42,** 643 (1959).
121. *Ibid.*, **44,** 546 (1961).
122. R. M. Ikeda, W. L. Stanley, S. H. Vannier, and L. A. Rolle, *Food Technol.*, **15,** 379 (1961).
123. G. D'Amore and R. Calapaj, *Rass. Chim.*, **17,** 264 (1965).
124. U. R. Cieri, *J. Assoc. Off. Anal. Chem.*, **52,** 719 (1969).
125. W. D. MacLeod, Jr., and N. M. Buigues, *J. Food Sci.*, **31,** 588 (1966).
126. B. C. Madsen and H. W. Latz, *J. Chromatogr.*, **50,** 288 (1970).
127. R. M. Ikeda, W. L. Stanley, L. A. Rolle, and S. H. Vannier, *J. Food Sci.*, **27,** 593 (1962).
128. R. A. Bernhard, *Nature*, **182,** 1171 (1958).
129. N. G. Martinez Nadal, *Am. Perfum. Cosmet.*, **79,** 43 (1964).
130. E. Verderio and C. Venturini, *Boll. Chim. Farm.*, **104,** 170 (1965).
131. T. Pozzo-Balbi, L. Nobile, *Ann. Chim. (Rome)*, **60,** 171 (1970).
132. M. S. Karawya, S. I. Balbaa, and M. S. Hifnawy, *J. Pharm. Sci.*, **60,** 361 (1971).
133. J. A. Attaway, A. P. Pieringer, and L. J. Barabas, *Phytochemistry*, **5,** 141 (1966).
134. V. P. Maier and E. R. Grant, *J. Agric. Food Chem.*, **18,** 250 (1970).
135. T. Katayama, *Nippon Suisan Gakkaishi*, **24,** 205 (1958).
136. *Ibid.*, p. 346; *Chem. Abstr.*, **53,** 11532 (1959).
137. *Ibid.*, **21,** 412 (1955); *Chem. Abstr.*, **50,** 13184 (1956).
138. *Ibid.*, p. 416.
139. *Ibid.*, p. 412.
140. *Ibid.*, **27,** 75 (1961); *Chem. Abstr.*, **56,** 7710 (1962).
141. *Ibid.*, **24,** 925 (1959); *Chem. Abstr.*, **57,** 15512 (1962).
142. *Ibid.*, **21,** 425 (1955); *Chem. Abstr.*, **50,** 13184 (1956).
143. D. Vágújfalvi and E. Tyihák, *Herba Hung.*, **2,** 361 (1963).

144. B. M. Lawrence, *Perfum. Essent. Oil Rec.*, **59**, 421 (1968).
145. H. Petrowitz, *Riechst., Aromen, Koerperpflegem.*, **16**, 345 (1966).
146. I. Calvarano, *Essenz Deriv. Agrum.*, **35**, 212 (1965).
147. D. Vágújfalvi, E. Tyihák, and G. Held, *Herb Hung.*, **8**, 155 (1969).
148. *Ibid.*, **9**, 79 (1970).
149. *Ibid.*, p. 135.
150. A. L. Montes, *An. Asoc. Quim. Arg.*, **40**, 273 (1952).
151. H. Gaenshirt, *Pharm. Ind.*, **15**, 177 (1953).
152. S. Gruener and W. Spaich, *Arch. Pharm.*, **287/59**, 243 (1954).
153. R. D. Coveney, W. S. A. Matthews, and G. B. Pickering, *Colonial Plant Animal Prod.*, **5**, 150 (1955).
154. H. H. Kaiser, dissertation, Karlsruhe, 1955.
155. N. Garcia de Nadal, *Am. Perfum. Essent. Oil Rev.*, **65**, 17 (1955).
156. P. A. Wotherspoon and P. Z. Bedoukian, *Am. Perfum. Essent. Oil Rev.*, **66**, 17 (1955).
157. E. Demole, *Compt. Rend.*, **243**, 1883 (1956).
158. I. Onishi, H. Tomita, and T. Fukuzumi, *Bull. Agric. Chem. Soc. Japan*, **20**, 61 (1956).
159. E. Stahl, G. Schroeter, G. Kraft, and R. Renz, *Pharmazie*, **11**, 633 (1956).
160. N. Allentoff and F. G. Wright, *Can. J. Chem.*, **35**, 900 (1957).
161. B. J. Frydman, A. L. Montes, and A. Troparevsky, *An. Asoc. Quim. Arg.*, **45**, 248 (1957); *Chem. Abstr.*, **52**, 17622 (1958).
162. *Ibid.*, p. 257.
163. *Ibid.*, p. 261.
164. N. Garcia de Martinez Nadal, *Am. Perfum. Aromat.*, **69**, 27 (1957).
165. G. Gogroef, *Pharmazie*, **12**, 38 (1957).
166. I. Onishi, H. Tomita, and T. Fukuzumi, *Bull. Agric. Chem. Soc. Japan*, **21**, 239 (1957).
167. K. Yamamoto and T. Furukawa, *J. Fac. Educ., Hiroshima Univ.*, **5**, 53 (1957).
168. K. Yamamoto, T. Furukawa, and M. Matsukura, *J. Fac. Educ., Hiroshima Univ.*, **5**, 77 (1957).
169. E. Demole and E. Lederer, *Bull. Soc. Chim. Fr.*, **1958**, 1128.
170. R. Klohr-Meinhardt, *Planta Med.*, **6**, 203 (1958).
171. *Ibid.*, p. 208.
172. E. Lederer, *Accad. Naz. Lincei, Fondazione Donegani, Conso Estivo Chim., 3, Varenna, Italy, September 23–October 7, 1959*, pp. 117–131.
173. I. Onishi, *Nippon Senbai Kosha Kenkyusho Kenkyu Hokoku*, No 163, 19 pp. (1958).
174. L. D. Pryor and L. H. Bryant, *Proc. Linnean Soc. N. S. Wales*, **83**, 55 (1958).
175. E. Stahl, *Chem.-Ztg.*, **82**, 323 (1958).
176. E. Stahl, *Parfuem. Kosmet.*, **39**, 564 (1958).
177. T. Suga, *Chem. Soc. Japan*, **31**, 569 (1958).
178. T. Katayama and I. Nagai, *J. Fac. Fisheries Animal Husb., Hiroshima Univ.*, **2**, 349 (1959).

179. *Ibid.*, p. 355.
180. W. Winkler and E. Lanau, *Pharm. Ztg.*, **104,** 1407 (1959).
181. T. Akazawa, *Arch. Biochem. Biophys.* **90,** 82 (1960).
182. T. Akazawa and I. Uritani, *Arch. Biochem. Biophys.*, **88,** 150 (1960).
183. C.-H. Brieskorn and E. Wenger, *Arch. Pharm.*, **293/65,** 21 (1960).
184. I. Gallardo and A. L. Montes, An. *Asoc. Quim. Arg.*, **48,** 108 (1960); through *Chem. Abstr.*, **55,** 23934 (1961).
185. K. Fujita, *J. Sci. Hiroshima Univ.*, **A24,** 691 (1960); through *Chem. Abstr.*, **56,** 6004 (1962).
186. E. Lederer, *Fr. Parfums,* **3,** 28 (1960).
187. R. Marbet and G. Saucy, *Chimia (Aarau),* **14,** 362 (1960).
188. P. Tétényi, *Pharmazie,* **16,** 273 (1960).
189. H. D. Wulff and E. Stahl, *Naturwissenschaften,* **47,** 114 (1960).
190. T. Akazawa and K. Wada, *Agric. Biol. Chem. (Tokyo),* **25,** 30 (1961).
191. L. Hoerhammer, H. Wagner, and B. Lay, *Pharm. Ztg.* **106,** 1308 (1961).
192. K. Kratzl, *Holz Roh-Werkst.*, **19,** 219 (1961).
193. T. Neubern de Toledo and R. Wasicky, *Tribuna Farm. (Brazil),* **29,** 44 (1961).
194. R. Paris and M. Godon, *Ann. Pharm. Fr.*, **19,** 86 (1961).
195. C. Zanini, A. D. Pozzo, and A. Dansi, *Boll. Chim. Farm.*, **100,** 83 (1961).
196. T. Akazawa, I. Uritani, and Y. Akazawa, *Arch. Biochem. Biophys.*, **99,** 52 (1962).
197. C. H. Brieskorn and W. Polonius, *Pharmazie,* **17,** 705 (1962).
198. J. Deshusses and A. Gabbai, *Mitt. Geb. Lebensm. Hyg.*, **53,** 408 (1962).
199. S. R. El-Deeb, M. S. Karawya, and S. K. Wahba, *J. Pharm. Sci. U. Arab Rep.*, **3,** 81 (1962).
200. E. Gabel, K. H. Mueller, and I. Schoknecht, *Dtsch. Apoth. Ztg.*, **102,** 293 (1962).
201. L. Hoerhammer and H. Wagner, *Dtsch. Apoth.-Ztg.*, **102,** 733 (1962).
202. S. Huneck, *J. Chromatogr.*, **7,** 561 (1962).
203. S. Huneck and J.-M. Lehn, *Bull. Soc. Chim. Fr.*, **1963,** 1702.
204. *Ibid.*, p. 321.
205. R. Jaspersen-Schib and H. Flueck, *Boll. Chim. Farm.*, **101,** 512 (1962).
206. H. Jork, *Chromatogr., Symp., 2nd, Brussels,* **1962, 213.**
207. D. Lavie, E. Glotter, and Y. Shvo, *Tetrahedron,* **19,** 1377 (1963).
208. M. E. Morgan and R. L. Pereira, *J. Dairy Sci.*, **45,** 457 (1962).
209. S. S. Nigam and G. L. Kumari, *Perfum. Essent. Oil Record,* **53,** 529 (1962).
210. I. M. Pertsev and G. P. Pivnenko, *Farm. Zh. (Kiev),* **17,** 35 (1962).
211. U. Scheidegger, K. Schaffner, and O. Jeger, *Helv. Chim. Acta,* **45,** 400 (1962).
212. K. Schreiber and G. Osske, *Kulturpflanze,* **10,** 372 (1962).
213. K. E. Schulte, *Congr. Sci. Farm., Conf. Comun., 21st, Pisa,* **1961,** (Publ. 1962) 798.
214. E. Tyihák and D. Vágúifalv, *Herba Hung.*, **1,** 97 (1962).
215. A. Bhramaramba and G. S. Sidhu, *Perfum. Essent . Oil Rec.*, **54,** 732 (1963).
216. B. Borkowski and B. Pasich, *Farm. Pol.*, **19,** 435 (1963).

217. W. Brud and W. Daniewski, *Chem. Anal. (Warsaw)*, **8,** 753 (1963).
218. P. Capella, E. Fedeli, M. Cirimele, A. Lanzani, and G. Jacini, *Riv. Ital. Sostanze Grasse,* **40,** 645 (1963); through *Chem. Abstr.*, **61,** 4971 (1964).
219. C. Djerassi, R. A. Schneider, H. Vorbrueggen, and N. L. Allinger, *J. Org. Chem.*, **28,** 1632 (1963).
220. F.-Y. Fu, T.-P. Hsu, M.-T. Li, T.-M. Shang, and C.-N. Fang, *Yao Hsueh Hsueh Pao,* **10,** 618 (1963); through *Chem. Abstr.*, **60,** 1485 (1964).
221. R. Grab, *Dtsch. Apoth.-Ztg.*, **103,** 1424 (1963).
222. C. L. Graham and F. J. McQuillin, *J. Chem. Soc.*, **1963,** 4634.
223. L. Hoerhammer, H. Wagner, and W. Eyrich, *Z. Naturforsch.*, **18*b*,** 639 (1963).
224. L. Hoerhammer and H. Wagner, *Dtsch. Apoth.-Ztg.*, **103,** 429 (1963).
225. L. Hoerhammer, H. Wagner, and M. Seitz, *Dtsch. Apoth.-Ztg.*, **103,** 1302 (1963).
226. R. Ikan and J. Kashman, *Israel J. Chem.*, **1,** 502 (1963).
227. R. Ikan, J. Kashman, S. Harel, and E. D. Bergmann, *Israel J. Chem.*, **1,** 248 (1963).
228. H. P. Kaufmann and A. K. Sen Gupta, *Chem. Ber.*, **96,** 2489 (1963).
229. L. Mangoni and M. Belardini, *Ric. Sci. Rend.*, **3,** 528 (1963); *Chem. Abstr.*, **59,** 15330 (1963).
230. R. Paris and M. Godon, *Recherches (Paris),* **13,** 48 (1963).
231. B. Pasich, *Diss. Pharm.*, **15,** 73 (1963); *Chem. Abstr.*, **59,** 13111 (1963).
232. L. Peyron, *Chim. Anal. (Paris),* **45,** 186 (1963).
233. J. L. Ramaut, *Bull. Soc. Chim. Belg.*, **72,** 97 (1963).
234. *Ibid.*, p. 316.
235. M. von Schantz, *Eripainos Farm. Aikak.*, **3,** 95 (1963).
236. K. E. Schulte, F. Ahrens, and E. Sprenger, *Pharm. Ztg. Ver. Apoth.-Ztg.*, **108,** 1165 (1963).
237. V. A. Smit, A. V. Semenovskii, and V. F. Kucherov, *Izv. Akad. Nauk SSSR, Ser. Khim.*, **1963,** 1601; through *Chem. Abstr.*, **59,** 15314 (1963).
238. R. Tschesche, U. Axen, and G. Snatzke, *Ann. Chem.*, **669,** 171 (1963).
239. R. Tschesche, E. Henckel, and G. Snatzke, *Tetrahedron Lett.*, **1963,** 613.
240. E. Tyihák, I. Sárkány-Kiss, and J. Máthe, *Pharm Zentralhalle,* **102,** 128 (1963).
241. H. Vorbrueggen, S. C. Pakrashi, and C. Djerassi, *Ann. Chem.*, **668,** 57 (1963).
242. M. Wellendorf, *Dansk Tidsskr. Farm.*, **37,** 145 (1963).
243. G. Bergstroem and C. Lagercrantz, *Acta Chem. Scand.*, **18,** 560 (1964).
244. E. Demole, "La Chromatographie sur Couches Minces Dans le Domaine des Substances Odorantes Naturelles et Synthetiques," in *Thin-Layer Chromatography,* G. B. Marini-Bettòlo, Ed., Elsevier, Amsterdam, 1964, p. 45.
245. E. Demole, *Helv. Chim. Acta,* **47,** 319 (1964).
246. L. Hoerhammer, E. A. Hamidi, and G. Richter, *J. Pharm. Sci.*, **53,** 1033 (1964).
247. L. Hoerhammer, H. Wagner, G. Richter, H. W. Koenig, and I. Heng, *Dtsch. Apoth.-Ztg.*, **104,** 1398 (1964).

248. R. Ikan, J. Kashman, and E. D. Bergmann, *J. Chromatogr.*, **14,** 275 (1964).
249. H. P. Kaufmann and A. K. Sen Gupta, *Fette, Seifen, Anstrichm.*, **66,** 461 (1964).
250. F. Kohen, B. K. Patnaik, and R. Stevenson, *J. Org. Chem.*, **29,** 2710 (1964).
251. G. Kunovits, *Seifen-Oele-Fette-Wachse,* **90,** 895 (1964).
252. J. Masse and R. Paris, *Ann. Pharm. Fr.,* **22,** 349 (1964).
253. E. A. Moreira and C. Cecy, *Trib. Farm. (Brazil),* **32,** 55 (1964).
254. G. M. Nano and A. Martelli, *Gazz. Chim. Ital.*, **94,** 816 (1964).
255. H. Schilcher, *Dtsch. Apoth.-Ztg.*, **104,** 1019 (1964).
256. A. F. Shalaby and G. Richter, *J. Pharm. Sci.*, **53,** 1502 (1964).
257. E. Stahl, *Arch. Pharm.*, **297,** 500 (1964).
258. K. Takeda, H. Minato, and M. Ishikawa, *J. Chem. Soc.*, **1964,** 4578.
259. K. Takeda, H. Minato, M. Ishikawa, and M. Miyawaki, *Tetrahedron,* **20,** 2655 (1964).
260. R. Tschesche, G. Biernoth, and G. Snatzke, *Ann. Chem.*, **674,** 196 (1964).
261. G. Vernin, *Fr. Parfums,* **7,** 299 (1964).
262. R. Wasicky, *Rev. Fac. Bioquim. (S. Paulo),* **1,** 69 (1963).
263. V. Wollrab, *Riechst. Aromen,* **14,** 321 (1964).
264. D. F. Zinkel and J. W. Rowe, *J. Chromatogr.*, **13,** 74 (1964).
265. T. J. Betts, *J. Pharm. Pharmacol.*, **17,** 520 (1965).
266. V. M. Bhatnagar, *Perfum. Essent. Oil Rec.*, **56,** 374 (1965).
267. P. Blanc, P. Bertrand, G. D. E. Saqui-Sannes, and R. Lescure, *Chim. Anal. (Paris),* **47,** 354 (1965).
268. H. A. Elgamal and M. B. E. Fayez, *Z. Anal. Chem.* **211,** 190 (1965).
269. R. Ikan and R. Meir, *Israel J. Chem..* **3,** 117 (1965).
270. L. Kraus and F. Perenyi, *Cesk, Farm.*, **14,** 423 (1965).
271. G. E. Martin, J. A. Schmit, and R. L. Schoeneman, *J. Assoc. Off. Agric. Chem.*, **48,** 962 (1965).
272. H. G. Moslé, W. Wolfe, and W. Bode, *Z. Anal. Chem.*, **207,** 24 (1965).
273. T. Murakami, H. Itokawa, F. Uzuki, and N. Sawada, *Chem. Pharm. Bull. (Tokyo),* **13,** 1346 (1965).
274. G. M. Nano, G. Biglino, A. Martelli, and P. Sancin, *Atti Accad. Sci. Torino,* **99,** 1 (1965).
275. M. C. Nigam, I. C. Nigam, and L. Levi, *J. Soc. Cosmet. Chem.,* **16,** 155 (1965).
276. V. A. Paseshnichenko and A. R. Guseva, *Prikl. Biokhim. Mikrobiol.,* **1,** 559 (1965).
277. O. E. Schultz and H. L. Mohrmann, *Pharmazie,* **20,** 379 (1965).
278. E. Verderio and D. Venturini, *Riv. Ital. Essenze-Profumi, Piante Off. Oli. Veg. Saponi,* **47,** 430 (1965).
279. R. E. Wrolstad and W. G. Jennings, *J. Chromatogr.*, **18,** 318 (1965).
280. M. Zacsko-Szasz and G. Szasz, *Fette, Seifen, Anstrichm.*, **67,** 332 (1965).
281. A. El-Hamidi and S. S. Ahmed, *Pharmazie,* **21,** 438 (1966).
282. V. T. Gogiya, *Biokh. Progr. Tekhnol. Chai. Proizvod., Akad. Nauk SSSR, Inst. Biokhim.*, **1966,** 57; through *Chem. Abstr.*, **66,** 13995r (1967).

283. O. Munoz Teran, *Rev. Fac. Quim. Farm.*, **5**, 67 (1966).
284. M. T. C. Neves, J. Cardoso Do Vale, and A. C. Neves, *Garcia Orta,* **14**, 431 (1966).
285. A. Proenca Da Cunha, *Garcia Orta,* **14**, 411 (1966).
286. M. Zacsko-Szasz and G. Szasz, *Herba Hung.*, **5**, 91 (1966).
287. A. K. Barua, S. P. Dutta, and S. K. Pal, *J. Chromatogr.*, **31**, 569 (1967).
288. J. Deryng, H. Strzelecka, E. Walewska, and T. Soroczynska, *Farm. Pol.* **23**, 32 (1967).
289. E. Domagalina and S. Zareba, *Diss. Pharm. Pharmacol.*, **19**, 391 (1967).
290. J. Foks, *Zesz. Nauk. Politech. Gdansk., Chem.*, **17**, 35 (1967).
291. A. Martelli, G. M. Nano, G. Biglino, and M. Gallino, *Ann. Chim. (Rome),* **57**, 1027 (1967).
292. S. C. Pakrashi and P. Majumdar, *Indian J. Chem.*, **5**, 129 (1967).
293. L. Tóth, *Planta Med.*, **15**, 371 (1967).
294. J. W. Adcock and T. J. Betts, *J. Chromatogr.*, **34**, 411 (1968).
295. S. B. Challen and M. Kučera, *J. Chromatogr.*, **32**, 53 (1968).
296. D. Lebez, *J. Chromatogr.*, **33**, 544 (1968).
297. O. Sticher and H. Flueck, *Pharm. Acta Helv.*, **48**, 411 (1968).
298. E. Tyihák, S. Behassy, and K. Jahasz, *Mezhdunar. Kongr. Efirnym Maslam, (Mater.), 4th 1968,* **1**, 351 (1971); through *Chem. Abstr.*, **79**, 8335j (1973).
299. K. Waisser, J. Zelinka, and A. Vystrčil, *Collect. Czech. Chem. Commun.*, **33**, 2485 (1968).
300. L. Bauer and G. A. A. Brasil e Silva, *Trib. Farm.*, **37**, 151 (1969).
301. S. R. Jain, *Planta Med.*, **17**, 230 (1969).
302. T. Kami, S. Ōtaishi, S. Hayashi, and T. Matsuura, *Agric. Biol. Chem. (Tokyo),* **33**, 502 (1969).
303. G. Pensar, *Suom. Kemistiseuran Tied.*, **78**, 11 (1969).
304. W. H. Stahl, J. N. Skarzynski, and W. A. Voelker, *J. Assoc. Off. Anal. Chem.*, **52**, 1184 (1969).
305. V. A. Babkin, Z. V. Dubovenko, and V. A. Pentegova, *Izv. Sib. Otd. Akad. Nauk SSSR, Ser. Khim. Nauk,* **1970**, 168; through *Chem. Abstr.*, **73**, 4043w (1970).
306. A. El-Hamidi and I. Sidrak, *Planta Med.*, **18**, 98 (1970).
307. A. F. Halim and R. P. Collins, *Lloydia,* **33**, 7 (1970).
308. J. Joncyzk, *Acta Pol. Pharm.*, **27**, 301 (1970).
309. M. S. Karawya, S. I. Balbaa, and M. S. M. Hifnawy, *Amer. Perfum. Cosmet.*, **85**, 23 (1970).
310. B. G. V. N. Rao and S. Nigam, *Indian Perfum.*, **14**, 4 (1970).
311. J. E. Sinsheimer, G. S. Rao, and H. M. McIlhenny, *J. Pharm. Sci.*, **59**, 622 (1970).
312. L. E. Vandenburg and E. A. Wilder, *J. Am. Oil Chem. Soc.*, **47**, 514 (1970).
313. R. P. Collins and A. F. Halim., *Planta Med.*, **20**, 241 (1971).
314. W. J. Griffin and J. E. Parkin, *Planta Med.*, **20**, 97 (1971).
315. A. Jeannes and M. Tetau, *Plant Med. Phytother.*, **5**, 214 (1971).

316. A. Meisinger, *Apothekerprakt. Pharm. Tech. Assist.*, **17,** 41 (1971).
317. J. Poplawski, M. Holub, Z. Šamek, and V. Herout, *Collect. Czech. Chem. Commun.*, **36,** 2189 (1971).
318. C. S. Shah, J. S. Qadry, and M. G. Chauhan, *J. Pharm. Pharmacol.*, **23,** 448 (1971).
319. R. K. Thapa, V. N. Vashist, C. K. Atal, and R. Gupta, *Planta Med.*, **20,** 67 (1971).
320. D. W. Connell and R. McLachlan, *J. Chromatogr.*, **67,** 29 (1972).
321. B. Drozdz, H. Grabarczyk, Z. Šamek, M. Holub, V. Herout, and F. Šorm, *Collect. Czech. Chem. Commun.*, **37,** 1546 (1972).
322. A. Herisset, J. Jolivet, and P. Rey, *Plant Med. Phytother.*, **6,** 137 (1972).
323. R. ter Heide, *J. Agric. Food Chem.*, **20,** 747 (1972).
324. B. Mancini, C. L. Rubino, G. L. Pozetti, and M. A. D. Mancini, *Rev. Fac. Farm. Odontol. Araraquara,* **6,** 41 (1972).
325. H. Popescu and R. Ciupe, *Chim. Anal. (Bucharest),* **2,** 168 (1972).
326. S. Qedan, *Planta Med.*, **21,** 410 (1972).
327. G. Reucker, *Arch. Pharm. (Weinheim),* **305,** 486 (1972).
328. P. Rueedi and C. H. Eugster, *Helv. Chim. Acta,* **55,** 1994 (1972).
329. H. Schilcher, *Dtsch. Apoth.-Ztg.*, **112,** 1497 (1972).
330. G. Willuhn, *Planta Med.*, **21,** 329 (1972).
331. A. Ardon and T. Nakano, *Planta Med.*, **23,** 348 (1973).
332. S. I. Balbaa, S. H. Hilal, and M. Y. Haggag, *Planta Med.*, **23,** 312 (1973).
333. V. S. Chauhan, N. Sriram, G. B. V. Subramanian, and H. Singh, *J. Chromatogr.*, **84,** 51 (1973).
334. A.-A. Habib and A. M. Metwally, *Planta Med.*, **23,** 88 (1973).
335. J. Hoelzl and G. Demuth, *Dtsch. Apoth.-Ztg.*, **113,** 671 (1973).
336. M. Holub and Z. Šamek, *Collect. Czech. Chem. Commun.*, **38,** 731 (1973).
337. J. C. Morales and G. Torres, E., *Seifen-Oele-Fette-Wachse,* **99,** 17 (1973).
338. A. Salehian and G. Netien, *Trav. Soc. Pharm. Montp.*, **33,** 329 (1973).
339. P. Sengupta, A. R. Sen, A. Bose, and T. V. Mathew, *J. Assoc. Off. Anal. Chem.*, **56,** 1510 (1973).
340. A. S. L. Tirimanna, *Mikrochim. Acta,* **1973,** 9.
341. F. W. Blechinger, *Fette, Seifen, Anstrichm.*, **76,** 275 (1974).
342. I. Bungert-Hansing and H. Jork, *Z. Anal. Chem.*, **268,** 203 (1974).
343. A. S. Dro and F. W. Hefendehl, *Arch. Pharm.*, **307,** 168 (1974).
344. R. Kaempf and E. Steinegger, *Pharm. Acta Helv.*, **49,** 87 (1974).
345. M. S. Karawya and M. S. Hifnawy, *J. Assoc. Off. Anal. Chem.*, **57,** 997 (1974).
346. E. P. Lichtenstein, T. T. Liang, K. R. Schulz, H. K. Schnoes, and G. T. Carter, *J. Agric. Food Chem.*, **22,** 658 (1974).
347. M. Mesonero, J. Cabo Torres, and A. Villar del Fresno, *Boll. Chim. Farm.*, **113,** 131 (1974).
348. A. R. Sen, P. Sengupta, A. Mondal, and B. R. Roy, *J. Assoc. Off. Anal. Chem.*, **57,** 763 (1974).
349. H. Thieme and U. Hartmann, *Pharmazie,* **29,** 50 (1974).

350. R. Banerji and C. R. Mitra, *Planta Med.*, **28**, 52 (1975).
351. L. H. Briggs and G. H. White, *Tetrahedron*, **31**, 1311 (1975).
352. H. Budzikiewicz and A. Roemer, *Tetrahedron*, **31**, 1761 (1975).
353. M. H. A. Elgamal and B. A. H. El-Tawil, *Planta Med.*, **27**, 159 (1975).
354. M. Y. Haggag, A. S. Shalaby, and G. Verzar-Petri, *Planta Med.*, **27**, 361 (1975).
355. J. Hoelzl and G. Demuth, *Planta Med.*, **27**, 37 (1975).
356. J. Hoelzl, C. Franz, D. Fritz, and A. Voemel, *Z. Naturforsch.*, **30**; *C Biosci.*, 853 (1975).
357. G. W. Ivie, D. A. Witzel, and D. D. Rushing, *J. Agric. Food Chem.*, **23**, 845 (1975).
358. E. S. Nambudiri, A. G. Mathew, N. Krishnamurthy, and Y. S. Lewis, *Int. Flavours Food Addit.*, **6**, 135 (1975).
359. G. B. Rustembekova, M. I. Goryaev, G. I. Krotova, and A. D. Dembitskii, *Izv. Akad. Nauk Kaz, SSR, Ser. Khim.*, **25**, 32 (1975).
360. R. D. Stipanovič, A. A. Bell, and C. R. Howell, *Phytochemistry*, **14**, 1809 (1975).
361. R. D. Stipanovič, A. A. Bell, M. E. Mace, and C. R. Howell, *Phytochemistry*, **14**, 1077 (1975).
362. W. E. Biering, I. Bungert-Hansing, and H. Jork, *Planta Med.*, **29**, 133 (1976).
363. G. Verzar-Petri, G. Marczal, and E. Lemberkovics, *Pharmazie*, **31**, 256 (1976).

Chapter **XXXI**

VITAMINS

1 FAT-SOLUBLE VITAMINS

Vitamin A and Related Compounds

Some of the early work on thin layers was on the determination of vitamin A and carotene in edible fats and oils [1, 2]. Planta et al. [3] have chromatographed various isomeric vitamin A and vitamin A_2 compounds on silica gel. Katsui et al. [4] have reported on the separation of vitamin A alcohol, vitamin A acid, vitamin A aldehyde, and various vitamin A esters using both silica gel G and alumina G with either benzene or chloroform as solvents. R_f values for these compounds are given in Table 31.1. Davidek and Blattná [5] used loose layers of alumina to determine the R_f values of a series of fat-soluble vitamins including vitamin A and

Table 31.1 $R_f \times 100$ Values for Vitamin A Compounds with Color Reactions[a]

Compound	Silica Gel G		Aluminum Oxide G		Reagent			Sensitivity
	C_6H_6	$CHCl_3$	C_6H_6	$CHCl_3$	$SbCl_3$	H_2SO_4	$HClO_4$	$SbCl_3$
Vitamin A alcohol	8	22	8	28	Blue	Blue	Red-violet	0.1
Vitamin A acetate	41	69	62	78	Blue	Blue	Violet	0.1
Vitamin A palmitate	75	94	74	82	Blue	Blue	Violet	0.1
Vitamin A benzoate	63	88	66	78	Blue	Blue-violet	Violet	—
Vitamin A trimethoxy-benzoate	15	35	38	76	Blue	Blue-violet	Violet	—
Vitamin A senecionate	57	82	65	78	Blue	Blue-violet	Violet	—
Vitamin A pivalate	61	84	66	78	Blue	Violet	Violet	—
Vitamin A aldehyde	18	38	35	68	Dark blue	Violet	Violet	0.05
Vitamin A acid	0	0	0	0	Red-violet	Red → red violet	Red-violet	0.2
Vitamin A acid methyl ester	50	72	66	81	Violet	Red → red-violet	Red-violet	0.2
β-Carotene	100	100	78	86	Green-blue	Blue	Blue	0.05

[a] Reproduced from G. Katsui, S. Ishikawa, M. Shimizu, and Y. Nishimoto [4] (translation kindly furnished by G. Katsui).

α and β carotene (Table 31.2). Detection of these spots was accomplished by allowing 70% perchloric acid or 98% sulfuric acid to develop across the plate at right angles to the original development of the spots.

Varma et al. [6] used a bound alumina G layer for the separation of vitamin A compounds found in fish liver oils and in liquid multivitamin preparations. The R_f values for a series of solvents are given in Table 31.3. John et al. [7] have chromatographed a similar group of compounds as well as some 5,6-monoepoxy vitamin A compounds on silica gel layers. Development was for a distance of 16 to 18 cm with 6% acetone in petroleum ether (40 to 60°C), 15% diethyl ether in petroleum ether, or 3% acetone in isooctane.

Kuznetsova and Koval'ova [8] separated the alcohol and ester components of vitamin A on layers of aluminum oxide using chloroform as a developing solvent. After locating the spots by fluorescence under ultraviolet light, they were eluted with acetone and determined colorimetrically by reaction with glycerol dichlorohydrin. To prevent oxidation the separation was carried out in an inert atmosphere.

Eremina [9] employed alumina with 8% water content for the separation of vitamin A alcohol, aldehyde, and palmitate, and of β-carotene. Petroleum ether–acetone–ethanol (94:4:2) was used as a solvent. Adam-

Table 31.2 $R_f \times 100$ Values of Various Fat-Soluble Vitamins in Numerous Solvents on Loose-Layer Alumina[a,b]

	Carotene		*Vitamin*					
Solvent	α	β	*A*	D_2	*E*	K_1	K_2	K_3
Methanol	76		78	80	81	78	71	93
Anhydrous ethanol	87		71	91	98	89	84	91
n-Butanol	90	89	90	93	91	92	91	92
Benzyl alcohol	90	89	89	98	92	89	92	93
Hexane	93		90	79	90	85	85	80
Cyclohexane	92		88	98	100	90	94	98
Petroleum ether	75	64	24	5	5	31	21	29
Petrol	50	50	13	0	0	21	11	10
Benzene	87		91	24	93	94	88	74
Toluene	88		91	17	69	90	88	71
Xylene	89		91	12	72	91	87	72
Chloroform	94		93	58	87	94	95	92
Carbon tetrachloride	95	90	63	9	54	74	70	49

[a] From J. Davidek and J. Blattná [5]; reproduced with permission of the authors and The Elsevier Publishing Co.

[b] Development distance 30 cm.

Table 31.3 $R_f \times 100$ Values of Vitamin A and Related Compounds[a]

Compound	R_f Value (±5%)						
	Cyclohexane	*5% Benzene in Cyclo-hexane*	*0.25% Methanol in Cyclohexane*	*1% Methanol in Cyclohexane*	*3% Methanol in Cyclohexane*	*3% Ethanol in Cyclohexane*	*8% Ethanol in Cyclohexane*
Anhydrovitamin A_1	63	90	97				
β-Carotene	06	80					
retro-Vitamin A_1 acetate	0	36	90				
Vitamine A_1 acetate		19	88				
Vitamin A_1 aldehyde		0	66				
Vitamin A_2 aldehyde			59	100			
retro-Vitamin A_1 alcohol			12	36	42		
Vitamin A_1 alcohol			6	16	28	45	
Vitamin A_1 epoxide			3	12		32	
Vitamin A_2 alcohol			0	8	26	28	58
Vitamin A_1 acid				0		0	5

[a] From T. N. R. Varma, T. Panalaks, and T. K. Murray [6]; reproduced with permission of the authors and the American Chemical Society.

ski and Dobrucki [10] determined vitamin A palmitate in ointments by chromatographing on alumina with benzene–cyclohexane (1:2) with the addition of butylated hydroxytoluene to prevent oxidation. For quantitative determination the vitamin was eluted with cyclohexane and determined spectrophotometrically at 325 to 328 mμ.

Ludwig and Freimuth [11] analyzed pharmaceutical preparations for vitamins A, D, and E by saponifying the fat-containing media of the preparations and then separating the extracted material on silica gel layers. Quantitative results were obtained by spectrophotometric measurements of the fractions. Koleva et al. [12] determined vitamins A, E, and D_2 in pharmaceutical preparations by using silica gel DG with cyclohexane–ethyl ether (4:1) in a saturated atmosphere. Direct densitometry was used after visualizing vitamins D and A with 22% antimony trichloride in chloroform and α-tocopherol with ceric sulfate reagent. Baczyk et al. [13] used a multiple development with chloroform on silica gel for the separation of vitamins A, D_3, and E.

Johnson and Vickers [14] have developed a procedure for the identification and semiquantitative assay of vitamin A, its acetate and palmitate, vitamin D_2 (ergocalciferol), vitamin D_3 (cholecalciferol), α-tocopherol, and its acetate, as well as antioxidants incorporated in pharmaceutical products and animal feeds. The separations were made on silica gel with *n*-hexane–methyl ethyl ketone–di-*n*-butyl ether (34:7:6) with the following R_f values: propyl, octyl, and dodecyl gallates 0, hydroquinone 0.03, vitamin A (alcohol) 0.17, vitamin D_2 0.22, vitamin D_3 0.22, pre-vitamin D_2 0.30, pre-vitamin D_3 0.30, 2,6-di-*test*-butyl-4-methylphenol 0.75, 2-*tert*-butyl-4-methoxyphenol 0.32, ethoxyquin 0.42, α-tocopherol 0.51, vitamin A acetate 0.58, α-tocopherol acetate 0.66, anhydrovitamin A 0.81, vitamin A palmitate 0.85, and β-carotene 0.88. For visualization test T-121 was used, which detects all except tocopherol acetate; for the latter, the modified acidic agent was used because it detects α-tocopherol without previous hydrolysis. Since the D_2 and D_3 were not separated chromatographically, they were differentiated by removing the solvent from the vitamins at room temperature under a nitrogen stream and dissolving the residues in 0.1 ml of glacial acetic acid in a test tube. Then 2 ml of 72% perchloric acid (**Caution: HANDLE WITH CARE; AVOID STRONG REDUCING AGENTS**) was added and the mixture was heated on a water bath at 70°C for approximately 1 min before cooling and shaking with 1 ml of added chloroform. Vitamin D_2 gives a red or purple color and vitamin D_3 a greenish yellow color. Vitamin A interferes with this test. The vitamins were protected from light, and in order to avoid oxidation these workers added triethylamine to the cyclohexane spotting solvent in a 1:9 ratio. After development the plates should be sprayed as soon as the solvent has been removed by a stream of cold air.

Hanewald et al. [15] added 0.01% squalene and 0.01% BHT (butylated hydroxytoluene) to the vitamin solution and 0.01% BHT to the developing solvent to reduce decomposition in the determination of vitamin D. Kuznetsova and Koval'ova [16] separated vitamin A components in a chamber filled with inert gas. Zile and DeLuca [17] pointed out that in using silicic acid columns for separating vitamin A compounds it is necessary to incorporate an antioxidant (they used *dl*-α-tocopherol) in order to avoid oxidation. Without this, recovery yields dropped from quantitative to 48 to 60%. Using 10% methanol in benzene and straight benzene on silica layers they separated all-*trans*-retinoic acid, all-*trans*-retinol, all-*trans*-retinene, all-*trans*-retinyl acetate, 9,13-*cis*-methyl retinoate, all-*trans*-methyl retinoate, 13-*cis*-methyl retinoate, and some unidentified isomers absorbing at 362 to 363 mμ. Extinction values are given for all of these compounds. Adamski et al. [18] found the best separation of decomposition products of vitamin A palmitate to be achieved on unactivated silica gel layers with cyclohexane–petroleum ether–ethyl ether (7:3:1). Because of the sensitivity of vitamin A to peroxides the ether must be free of peroxides [19].

Keefer and Johnson [20] separated retinol (vitamin A), retinal, retinyl acetate, and some related carotenoids on magnesium hydroxide using carbon disulfide, carbon tetrachloride, and benzene, as well as other solvents. Perišić-Janjić et al. [21] chromatographed the fat-soluble vitamins retinyl acetate and palmitate, vitamins K_1, K_3, K_4, K_5, D_2, D_3, E, and E acetate by reverse-phase chromatography on layers of starch, cellulose, and talc impregnated with paraffin oil. Relative separations were comparable on the three supports. Vitamins D_2 and D_3 could not be separated and vitamin A acetate was very close to these two; however, it could be distinguished by its color with antimony pentachloride. Ropte and Gu [22] separated vitamin A alcohol, vitamin D, and α-tocopherol on silica gel G impregnated with liquid paraffin or castor oil. Acetone–water (4:1) was used for the separation. A quantitative separation was obtained with the castor oil-impregnated layers, but not with the paraffin layers. Richter and Ropte [23] used polyethylene glycol as an impregnating agent for partition chromatography with petroleum ether (30 to 50°C)–benzene (1:1) as the developing solvent for separating vitamins A, D, and E.

Vitamin D

Davídek and Blattná [5, 24] have investigated the behavior of vitamin D_2 on loose layers of alumina with numerous solvents. The R_f values for these results are listed in Table 31.2. The spots were detected by means of 75% perchloric acid or 98% sulfuric acid.

Kakáč et al. [25] also used loose alumina layers for separating vitamins A, D, and E with toluene, chloroform, or xylene as developing solvents. Quantitative results were obtained by visible or ultraviolet spectrophotometry after elution.

Norman and DeLuca [26] have chromatographed vitamin D and related compounds on silica gel in several solvents. The thin-layer plates were activated at 140°C for at least 16 hr before use. These plates were used for the investigation of the compounds formed during the irradiation of ergosterol or 7-dehydrocholesterol. Development time was approximately 35 min for a solvent travel distance of 10 cm. For detecting the zones, the air-dried plates were sprayed with either 0.2% potassium permanganate in 1% sodium carbonate or with 0.2 *M* sulfuric acid. After spraying, the plates were heated for 15 min with two 250-W heat lamps in order to visualize the spots. The R_f values for these various compounds are shown in Table 31.4.

Parekh and Wasserman [27] have used a combination of column and thin-layer chromatography for the purification of tritium-labeled vitamin D_3.

Janecke and Maass-Goebels [28] have developed a minimum-detectable

Table 31.4 $R_f \times 100$ Values of Vitamin D and Related Compounds Following Chromatography on Silica Gel[a]

	Solvent[b]		
Compound	*10% Acetone in Skellysolve B*	*100% Chloroform*	*5% Acetone in Skellysolve B*
Vitamin D_2	33	44	15
Vitamin D_3	32	44	15
Ergosterol	27	35	12
7-Dehydrocholesterol	27	35	12
Dihydrotachysterol	49	75	24
Cholesterol	30	41	—
Cholesterol acetate	98	99	86
Ergosterol acetate	97	99	79
Vitamin D_2-3,5-dinitrobenzoate	96	—	56

[a] From A. W. Norman and H. F. DeLuca [26]; reproduced with permission of the authors and the American Chemical Society.
[b] Development distance 10 cm.

spot method for the quantitative analysis of vitamin D. For this determination plates are sprayed with phosphotungstic acid after which they are heated in an oven at 70°C. Since the developing time for the chromatograph is approximately 30 min, the method provides a very rapid evaluation of vitamin D. Castren [29] has developed a colorimetric method for the determination of vitamin D by eluting the spots and using antimony pentachloride as a color reagent.

Mitta et al. [29] have used thin layers of starch-bound silica gel to separate vitamin D, lumisterol, phytosterol, cholesterol, and 7-dehydrocholesterol. Separations were accomplished with six solvent systems consisting of: (1) petroleum ether (60 to 70°C)-benzene (1:1), (2) chloroform, (3) benzene-chloroform (4:1), (4) benzene-chloroform (1:1), (5) methyl ethyl ketone-benzene (1:1), and (6) methyl ethyl ketone-benzene-chloroform (1:2:1).

Fuerst [31] separated vitamin D_2 (ergocalciferol) from its decomposition products on weakly activated alumina with petroleum ether–methanol (250:1) as the developing solvent. Only vitamin D_2 and ergosterol move on the layer under these conditions. Nerlo and Palak [32] separated vitamin D_3 from its decomposition products by chromatographing on silica gel G with chloroform. Adamski and Sawicka [33] determined vitamin D_3 in the presence of vitamin A by separating the extract from aqueous solutions on silica gel G layers with ethyl ether–hexane (1:1); for oily preparations the solution was diluted with hexane and then applied to an alumina plate for separation with ether–hexane–chloroform (1:1:1). Determination was made spectrophotometrically using the reaction with antimony trichloride reagent. Hashmi et al. [34] separated vitamins A, D_2, D_3, E, K_1, and K_3 by circular thin-layer chromatography on alumina layers with cyclohexane–methyl isopropyl ketone (9:1) giving R_f values of 0.94, 0.46, 0.40, 0.77, 0.85, and 0.60, respectively. All except D_2 and D_3 could also be separated on silica gel with cyclohexane–ether (4:1). Vitamin D may be separated from cholesterol on silica gel or alumina using dichloromethane [35, 36]. Pinelli et al. [36] preferred ethylene dichloride–methyl isobutyl ketone (9:1) as a solvent because this gave maximum separation for isolating submicrogram amounts of vitamin D from biological materials. For biological extracts they used silanized silica gel HF_{254}. Campion et al [37] have used an isotope dilution method for the measurement of nanogram quantities of vitamin D_3.

Pasalis and Bell [38] chromatographed 13 esters each of vitamins D_2 and D_3 on silica gel with hexane–benzene (1:1 and 1:2).

Vitamin E

The wide distribution of the tocopherols in nature and growing importance of these substances as antioxidants has established a need for isolation and analyzing techniques. Thin-layer chromatography lends itself well to this type of work. The R_f values of α-tocopherol in a series of solvents are given in Table 31.2.

Shone has reported on the tocopherols in tung oil [39] and on a new oxidation product of γ-tocopherol isolated from the oil[40]. This component has an R_f value of 0.50 on silica gel G using a solvent of benzene-petroleum ether (40 to 60°C) (1:1) or a value of 0.73 using a solvent composed of ethyl ether-petroleum ether (1:17). On aluminum oxide (Fluka containing 5% additional calcium sulfate binder), it had an R_f value of 0.57 in ethyl ether-petroleum ether (3:17). Skinner et al. [41] have also investigated the oxidation products of tocopherols. In this case they examined *dl*-α-tocopherol and its oxidation products. For separating these components they used a cyclohexane-chloroform (2:1) solvent with silica gel layers. Four locating agents were used: (1) 5% aqueous potassium ferricyanide followed by 5% aqueous ferric chloride, (2) 60% sulfuric acid with heating to 150°C, (3) 20% antimony pentasulfide in chloroform, and (4) 10% ammonium molybdate in 10% sulfuric acid followed by heating to 150°C. The components were compared with authentic samples. Skinner and Parkhurst [42] have also chromatographed a group of phenolic compounds related to the tocopherols. Dilley [43] and Dilley and Crane [44, 45] have investigated α-tocopherol and the oxidation product α-tocopherylquinone in spinach, applying silica gel G thin-layer chromatography for their separation and identification. The components were extracted from spinach leaves with acetone and separated on the thin layers by using benzene, chloroform, or 1% ethyl ether in chloroform as the developing solvent. The R_f values for the separations are listed in Table 31.5.

Sturm et al. [46] determined the α-, γ-, and δ-tocopherol in peanut oil also using silica gel G with chloroform for separating the compounds. The quantitative results were obtained by eluting and reacting with the Emmerie–Engle reagent. This work must be carried out in dim artificial light. Lovelady [47] tried seven solvent systems and found the best

Table 31.5 $R_f \times 100$ Values of the Tocopherols on Silica Gel G in Two Solvents [45]

Compound	*Benzene*	*Chloroform*
α-Tocopherol	52	26
β-Tocopherol	37	20
γ-Tocopherol	37	18
δ-Tocopherol	24	15

separation of β- and γ-tocopherols on silica gel G was achieved with cyclohexane–n-hexane–isopropyl ether–ammonia (20:20:10:1). By use of a spray reagent composed of 1.6 g of phosphomolybdic acid and 0.092 g of 2,7-dichlorfluorescein in 60 ml of ethanol + 7.6 ml ammonia and then diluted to 100 ml with deionized distilled water, 0.08 μg per μl of the vitamins could be separated and detected. The visualizing agent stabilized the visibility for several months. The method was applied to the determination of individual tocopherols in plasma and red blood cells [48]. The vitamins were quantitated by eluting the bands and preparing trimethylsilyl derivatives for determination by gas chromatography. The detection limit for the hydrogen flame detector was 0.03 μg. Whittle and Pennock [49] used a two-dimensional technique for separating α-, β-, γ-, and δ-tocopherols. Silica gel G was used with chloroform in the first direction and petroleum ether (40 to 60°C)–isopropyl ether (5:1) in the second dimension. The Emmerie–Engel reagent (T-108) was used for quantitative determination after elution of the spots. Recovery was approximately 92%. Rao et al. [50] using the same quantitative technique separated the same compounds one-dimensionally on silica gel with petroleum ether (60 to 80°C)–ether–isopropyl ether–acetone–acetic acid (254:3:32:12:3). Recovery in this case was 97 to 98%. This was the same system used earlier by Stowe (51) for separating β- and γ-tocopherols.

Schmandke [52] has used zinc carbonate-aluminum oxide (1:3) and zinc carbonate-silica gel (1:1) mixtures for the separation of several tocopherols as well as α-tocopherylquinone, tocopheronolactone, and 2,5,7,8-tetramethyl-2-(β-carboxyethyl)-6-hydroxychromane, the latter being a decomposition product of α-tocopherol through β-oxidation. Chloroform was the best solvent for the zinc carbonate-aluminum oxide layer, and benzene–chloroform (1:1) for the zinc carbonate-silica gel layer.

Roughan [53] examined 15 adsorbents for separating α- and β-tocopherol and found aluminum oxide–basic zinc carbonate (3:1) with chloroform and kieselguhr G–basic zinc carbonate (2:1) with benzene–cyclohexane (3:7) gave the best separations. The R_f values for the α-compound were 0.92 and 0.70, and for the β-compound 0.54 and 0.33, respectively.

In checking for the presence of various tocopherols in fish oils and marine organisms, Braekkan et al. [54] employed not only chromatography on silica gel using 10% ethyl ether in hexane, but also a partition method on squalene-impregnated Celite using ethanol-water (85:15). For observing the spots the plates were sprayed with Emmerie–Engel ferric chloride–dipyridyl reagent [55].

Other methods have been published on the quantitative analysis of various tocopherols by thin-layer systems. Seher [56, 57] used a method of comparing spot area to quantity for the analysis of various tocopherols. For observation of the spots they are sprayed with phosphomolybdic acid followed by treatment with ammonia vapor. For differentiation of the β- and γ-tocopherols, the Schultz and Strauss [58] modification of the Sonnenschein ceric sulfate reagent is applied. With this reagent β-tocopherol shows up as a brown spot and γ-tocopherol is observed as a blue spot. Castren [29] also used a planimetric method for the determination of vitamin E in multivitamin preparations. For this determination the aqueous solution of the powder of tablet is extracted with petroleum ether after having been saponified with an alcoholic potassium hydroxide solution. The extract is then transferred to benzene after removing the petroleum ether under vacuum. Development of the silica gel plate is carried out in chloroform or ethyl acetate-benzene (3:7). Visualization can be made with phosphomolybdic acid.

Vuilleumier et al. [59] used a spot comparison method for the analysis of α-tocopherols in food and feeds. For the thin-layer separation, trichloroethylene was used as a solvent on silica gel plates. Lambertsen et al. [60] have used a spectrophotometric analysis after separation of the tocopherols from nuts by using an alumina column. The identity of the tocopherol fractions was ascertained by thin-layer chromatography on silica gel G using 10% ethyl ether in hexane as the developing solvent. Dilley and Crane [61] have also used a spectrophotometric method for the determination of tocopherols as well as naturally occurring quinones and hydroquinones. The specific assay for tocopherols was accomplished by oxidizing the eluted tocopherol with gold chloride to the tocopherylquinone. The spectrophotometric assay was completed by measuring

Table 31.6 $R_f \times 100$ Values[a] of Some Vitamin K's in Various Systems

	Silica Gel G [4]		*Al_2O_3 G* [4]		*Loose-Layer Al_2O_3* [5]		
Vitamin	*C_6H_6*	*$CHCl_3$*	*C_6H_6*	*$CHCl_3$*	*MeOH*	*C_6H_6*	*CCl_4*
K_1	67	81	73	80	78	94	74
K_2		73[b]			71	88	70
K_3	29	49	63	75	93	74	49
K_4 monoacetate	1	3	0	0			
K_4 diacetate	3	18	22	70			

[a] Development distance 10 cm.
[b] From Wagner and Dengler [71] and Dilley [43].

the drop in the absorbance at 262 mμ after reduction of the quinone to the hydroquinone with potassium borohydride. Katsumi et al. [62] separated the tocopherols on silica gel and aluminum oxide layers using benzene as the developing solvent. The α-tocopherol was determined quantitatively by eluting from the plate with ethanol and determining by the Emmerie–Engel method obtaining recoveries of 98%.

Aratani et al. [63] have employed an in situ densitometric method for the determination of tocopherols. Very thin layers (53 μm) of silica gel (particle size about 325 μm) on 76 × 26 × 0.7 mm quartz slides were used. There was only a slight shift in the absorption maximum from that obtained in solution, and the standard deviation of area was only 2%.

Vitamin K and Related Quinones

Davídek and Blattná [5] have given the R_f values for vitamins K_1, K_2, and K_3 in 13 different solvents on loose layers of alumina (Table 31.2).

Katsui et al. [4] have separated vitamin K_1, K_3, K_4 monoacetates, and K_4 diacetate on silica gel G and aluminum oxide G using benzene and chloroform as solvents. As detecting agents after the separation, 95% sulfuric acid, 60% perchloric acid, or 65% nitric acid were used (Table 31.6).

Thielemann [64] separated vitamins K_1, K_2, and K_3 on silica gel G with benzene–acetone (9:1). Vitamin K_3 formed a red spot when sprayed with an aqueous solution of 4-aminophenazone. Perišić-Janjić et al. [21] separated vitamins K_3, K_4, and K_5 on layers of starch or cellulose impregnated with paraffin oil. Water–dioxane–acetone–formaldehyde (17:4:3:5) was used as a solvent. Unimpregnated talc layers could also be used with the same solvent system. Visualization was achieved by incorporating a fluorescent indicator in the layer. Matschiner and Amelotti [65] used silica gel impregnated with silver nitrate to separate the vitamin K fractions from bovine liver. Benzene–heptane (4:1) was the solvent. Reverse-phase thin-layer chromatography was also employed with silica gel G impregnated with paraffin and acetone–water (96:4) as a solvent. The vitamins were mainly menaquinone-10, menaquinone-11, and menaquinone-12. Hammond and White [66] used silica gel G impregnated with hexadecane for the reverse-phase separation of vitamin K_2 isoprenologues with acetone–water (95:5) as the solvent. The compounds could be recovered quantitatively from the layers. If it was necessary to remove the hexadecane this could be accomplished by chromatographing on silica gel with 5% chloroform in hexane or 10% chloroform in methanol.

Manes et al. [67] used in situ reflectance densitometry for the determination of vitamin K_1 in infant formula products. The vitamin extract was separated from lipids by passing through an alumina column and then chromatographed on silica gel first with carbon tetrachloride and then a second development in the same direction with benzene.

Billeter and Martius [68] have investigated the change of vitamin $K_{2(30)}$ and $K_{2(10)}$ into vitamin $K_{2(20)}$ in the organisms of birds and mammals, by means of a two-step chromatographic separation on silica gel G employing heptane-benzene (1:1) followed by benzene. For quantitative work a scintillation method was used.

Wagner et al. [69–71] have worked out a method for the isolation and determination of ubiquinones in crude lipid extracts. A preliminary separation of the ubiquinones is first made by chromatographing the crude mixture on silica gel layers using a benzene-chloroform (1:1) mixture. The ubiquinones are located as a single spot by spraying with 0.25% Rhodamine B solution in ethanol. This is observed as a violet fluorescent spot having an R_f value of 0.86. As an alternative the ubiquinones may be located with an antimony trichloride solution in chloroform. By using a spray reagent, as little at 0.5 γ of ubiquinones may be detected. To separate the individual ubiquinones and to determine the length of the isoprene chain, the spot is extracted several times with a few milliliters of warm acetone. The acetone is removed under a vacuum and the residue is dissolved in a small amount of cyclohexane. Separation is accomplished by reverse-phase chromatography on a silica gel plate impregnated with 5% paraffin oil in ether. The developing solvent to be used is a mixture of nine parts of acetone and one part of paraffin oil-saturated water. For a development distance of 8 cm, the following R_f values are given: $U_{30} = 0.77$, $U_{35} = 0.65$, $U_{40} = 0.55$, $U_{45} = 0.43$, $U_{50} = 0.28$. For the quantitative determination, the area from an unsprayed spot (located by means of an adjacent test sample) is eluted with acetone. After removal of the acetone under vacuum the residue is taken up in cyclohexane and measured with a spectrophotometer at 272 mμ. Threlfall and Goodwin [72] have isolated ubiquinone-50 and plastoquinone-45 from meristematic tissue cultures of Paul's Scarlett Rose.

Eck and Trebst [73] have examined a number of ubiquinones and plastoquinones, as well as dimers of the latter, in working with these compounds which are present in chestnut tree leaves. Separations were made on silica gel and paraffin oil-impregnated silica gel. R_f values were tabulated for the compounds.

Henninger and Crane [74], in examining the functions of various natural quinones, have obtained separation of the mixed quinones on thin layers of silica gel G (Table 31.7). Dilley [43, 44] and Henninger et al. [75] have applied the methods to various investigations.

Drews et al. [76] examined the occurrence and distribution of ubiqui-

Table 31.7 $R_f \times 100$ Values of Various Natural Quinones on Silica Gel G

Quinone	15% *Trichloroethyl Acetate in Benzene* [74]	1% *Ether in Chloroform* [74]	*Chloroform*[a] [43]
Plastoquinone A	81	74	89
Plastoquinone B	88	78	94
Plastoquinone C	0	49	58
Plastoquinone D	0	40	45
α-Tocopherylquinone	21	37	26
β-Tocopherylquinone	19	33	20
γ-Tocopherylquinone	16	25	18
δ-Tocopherylquinone			15

[a] Development distance 10 cm.

nones in barley and malt. The ubiquinones were separated on silica gel with carbon tetrachloride–ethyl acetate (93:7). The individual compounds were reduced with sodium borohydride and the resulting quinols determined spectrophotometrically with the Emmerie–Engel reagent.

Ikan et al [77] investigated the ubiquinones of the oriental hornet queens after a preliminary column separation on Florisil to remove lipids. Further purification of the ubiquinone fraction was achieved on silica gel G with petroleum ether–ether–acetic acid (20:4:1). Final separation of the ubiquinones was accomplished by reverse-phase on silica gel G impregnated with 10% paraffin oil in hexane. The dried plates were developed with acetone–water (96:4).

2 WATER-SOLUBLE VITAMINS

Mixtures of Water-Soluble Vitamins

Gaenshirt and Malzacher [78] used silica gel G layers for the separation of soluble vitamins present in vitamin preparations. Prior to preparation of the layers, the silica gel G was mixed with 2% fluorescent material. A mixed solvent, consisting of acetic acid–acetone–methanol–benzene (5:5:20:70), was allowed to migrate a distance of approximately 19 cm in the dark. For detection of the spots, vitamin B_1, vitamin C, and niacinamide appear as dark spots on the fluorescent plate under ultraviolet light. Biotin appears as a white spot on a pink background, when the plate is sprayed with a potassium iodoplatinate solution. Also with this reagent vitamin B_1 gives a gray color, niacinamide a light yellow, and vitamin C a yellow color. Vitamin B_6 will form a blue colored zone if it is sprayed with 0.1% of dichloroquinonechlorimide in ethanol; the color develops after treating with ammonia vapor. In order to locate the calcium pantothenate, the plate is heated for ½ hr at 160°C, after which it is sprayed with 0.5% ethanolic ninhydrin solution; a purple spot appears after a further short heating period at 160°C. Ishikawa and Katsui [79] used the same solvent system on both silica gel G and alumina oxide G for the separation and identification of water-soluble vitamins and other factors in commercial vitamin preparations. Huettenrauch et al. [80] have chromatographed the main components of the vitamin B complex on thin layers of air-dried ion-exchange resin (Wofatit CP 300). Dittmann [81] separated vitamins B_1, B_2, B_2 phosphate, B_6, and B_{12} on cellulose layers with propanol–water (3:2); nicotinic acid and nicotinamide were separated with propanol–ethyl acetate–ammonia (5:3:2). Thielemann [82, 83] used water as a solvent to separate vitamins B_1, B_2, B_6, B_{12}, C, nicotinic acid, nicotinamide, and panthenol on silica gel layers. All the vitamins except vitamin C could be detected with the chlorine–toluidine test (T-61). Vitamin C was detected under ultraviolet light on luminescent layers. Petrović et al. [84] separated vitamin B_1, B_6, B_{12}, B_2, C, nicotinic acid, and calcium pantothenate on starch layers with *n*-propanol–pyridine–acetic acid–water (15:10:3:13). Calcium pantothenate could not be separated from *p*-aminobenzoic acid. Chiang et al. [85] chromatographed a group of water-soluble vitamins on polyamide–silica gel (66:34) layers with 10% sodium chloride and sodium acetate as solvents. Hashmi et al. [86] used circular chromatography for a semiquantitative determination of water-soluble vitamins in a mixture by means of a minimum detectable amount. Using silica gel D-O (CAMAG) with water–96% ethanol–2 *M* hydrochloric acid (45:48:0.2) the following compounds were separated: vitamins B_1, B_2, B_6, B_{12}, *p*-aminobenzoic acid, and choline chloride with R_f values of 0.48, 0.56, 0.61, 0.31, 0.98, and 0.54, respectively. Nicotinic acid (R_f 0.55) and vitamin C (R_f 0.28) were separated with benzene–acetic acid–acetone (4:1:1). Bićan-Fišter and Dražin [87] quantitatively determined vitamins B_1, B_2, B_6, nicotinamide, and *p*-aminobenzoic acid in tablets or granules after separation on silica gel with acetic acid–acetone–methanol–benzene (1:1:4:14). Vitamin B_2 was determined by fluorimetry after elution, *p*-aminobenzoic acid by colorimetry after diazotizing, and the remainder by ultraviolet spectrophotometry after elution of the compounds. Good reproducibility and precision were achieved. Hashmi et al. [87a] using circular thin-layer chromatography on alumina layers with acetone–acetic acid–methanol–benzene (1:1:4:14) obtained the following R_f values: B_1 0.22, nicotinic acid 0.46, biotin 0.53, choline chloride 0.28, *p*-aminobenzoic acid 0.58, B_{12} 0.28, and B_6 0.30 (note: the R_f values for B_1 and B_2 are just the reverse of those given by Ishikawa and Katsui [79] for the same adsorbent and solvent system; see Table 31.8).

Nuttall and Bush [88] separated the water-soluble vitamins from a multivitamin preparatin by applying aliquots of the aqueous extract to two kieselgel $HF_{254+366}$ layers and to one cellulose layer. One silica gel layer was developed with 2-propanol–*n*-butanol–water (1:3:1) and the region below R_f 0.15 was treated with T-107 to reveal choline tartrate as a deep red spot at R_f 0.04. The balance of the layer was then sprayed with ninhydrin, and after heating at 80°C for 5 min, the methionine was visible as a deep red spot at R_f 0.26; further heating at 140°C revealed the panthenol at R_f 0.56. The second silica gel layer was developed with acetone–acetic acid–benzene–methanol (1:1:14:4). Riboflavin at R_f 0.32 and nicotinamide at 0.64 were visible under ultraviolet light; the region above R_f 0.50 was also visualized with T-61 to detect biotin as a deep blue spot at R_f 0.78; the remainder of the layer was treated with T-84 to visualize pyridoxin as a blue spot at R_f 0.13 (in using the latter two tests, the balance of the plate is covered with another glass plate to protect the balance of the layer from ammonia and chlorine). The cellulose layer was developed with 10% ammonia–methanol (3:7). Vitamin B_1 was detected under ultraviolet light at R_f 0.45, and after irradiation for 30 min with ultraviolet light, test T-174 detected folic acid at R_f 0.30. Sensitivities given were: choline tartrate 3 μg, methionine 0.3 μg, panthenol 1 μg,

Table 31.8 $R_f \times 100$ Values of Some Water-Soluble Vitamins

	HAc–CH₃COCH₃–MeOH–C₆H₆ (5:5:20:70)			10% *EtOH*
	Silica Gel G			*WOFATIT*
Vitamin	*Ref. 78*	*Ref. 79*	*Al₂O₃ G* [79]	CP300 [80]
B_1·HCl or HNO_3	0	0	54	0
B_2	35	29	24	42
B_6·HCl	15	12	26	35
Biotin	80	55	54	
Calcium panthothenate	57	40	0	
Nicotinamide	65	44	62	70
Ascorbic acid	30	25	0	
Folic acid		7	0	
B_{12}		0	23	100
Nicotinic acid				100
Carnitine·HCl		3	20	
Inositol		2	0	

Table 31.9 $R_f \times 100$ Values for the Separation of Thiamine and Its Phosphate Esters on Cellulose and Cellulose Derivatives[a,b]

Adsorbent	*Solvent*	*Thiamine*	*MPT*[c]	*DPT*[d]	*TPT*[e]
Cellulose MN 300 G	*n*-Propanol–acetate buffer pH 5–water (7:2:1)	65	26	14	5
MN 300 G	*n*-Propanol–phosphate buffer pH 4.9–water (3:1:1)	80	47	30	1
Cellulose monophosphate MN 300 P	Hydrochloric acid 0.03 *N*	10	30	73	80
MN 300 P	Glycine–hydrochloric acid, sodium chloride buffer pH 1.4	23	48	75	
MN 300 P	Glycine–hydrochloric acid buffer pH 2.05	15	40	75	
MN 300 P	Acetate buffer pH 3.58	5	15	60	
Carboxymethyl cellulose MN 300 CM	Hydrochloric acid 0.03 *N*	30			
MN 300 CM	Acetate buffer pH 3.58	30			

[a] From S. David and H. Hirshfeld [89]; reproduced with permission of the authors and Société Chimique de France.
[b] Development distance 10 cm.
[c] MPT = Thiamine monophosphate.
[d] DPT = Cocarboxylase (thiamine diphosphate).
[e] TPT = Thiamine triphosphate.

biotin 0.3 μg, pyridoxin 0.1 μg, riboflavin 0.2 μg, nicotinamide 3 μg, and folic acid 0.2 μg. Semiquantitative results could be obtained by comparison with spots from known amounts of the vitamins.

The R_f values of some water-soluble vitamins are given in Table 31.8.

Vitamin B_1

David and Hirshfeld [89] have used thin layers of cellulose and cellulose derivatives for separating thiamine and its phosphate esters. The separations achieved are shown in Table 31.9. On silica gel G layers the same compounds had R_f values of 0.92, 0.50, 0.17, and 0.08, respectively using pyridine–water–ammonia–methanol–acetic acid (6:6:5:1:1) [90]. Chuang et al. [91] chromatographed thiamine and its derivatives on polyamide Amilan CM 1011 (Toyo Rayon) in acetone, acetone–diethyl ether–acetic acid (20:20:1), methyl ethyl ketone, and chloroform–ethyl acetate–acetic acid (20:20:1). Dwivedi et al. [92, 93] used thin-layer chromatography to investigate the thermal degradation products of thiamine. Silica gel was used with acetonitrile–water (adjusted to pH 2.54 with formic acid) (4:1). Ariaey-Nejad et al. [94] used cellulose layers with eight solvent systems to investigate the thiamin metabolic products in man.

Vitamin B_6

Nuernberg [95] has separated pyridoxine hydrochloride from pyridoxal hydrochloride and pyridoxamine hydrochloride by using a step technique on silica gel plates. The first development for a distance of 14 cm was with acetone. This was followed by a second development for the same distance in acetone-dioxane-25% ammonium hydroxide (4.5:4.5:1) (developments were carried out in a saturated chamber). However, in extracting these compounds from the body, as is usually done with methanol, the pyridoxal is partially converted to the corresponding acetal and therefore gives two or three spots. This can be prevented by refluxing the sample with methanol in the dark for 1 hr to convert the pyridoxal to the acetal. This can then be separated from the pyridoxine and the pyridoxamine by the same solvent system. The R_f values are as follows: pyridoxine 0.2, pyridoxal methyl acetal 0.25, and pyridoxamine 0.55. For detection of the spots the plate was sprayed with 0.4% 2,6-dibromoquinonechlorimide in methanol.

Deyl and Rosmus [96] in working with the pyridoxine sulfur derivatives which occur in sterilized condensed milk were able to separate bis-4-pryidoxyl disulfide, 4-pyridoxthiol, and pyridoxol from one another by chromatographing their azo dye derivatives. The obtained R_f values of 0.80, 0.52, and 0.85, respectively, on loose-layer aluminum oxide, using ethanol–amyl alcohol–water (1:1:1) as a developing solvent.

Dement'eva et al. [97] separated nonphosphorylated vitamin B_6 compounds on silica gel micro plates with ethyl acetate–acetone–25% ammonia (40:20:3). The 5'-phosphoric esters were separated with *n*-butanol–ethanol–5% ammonia–acetic acid (10:10:10:1). Ahrens and Korytnyk [98] chromatographed nine vitamin B_6 compounds in seven solvent systems on silica gel and on cellulose. On silica gel HF_{254} with ammonia–water (1:139) and chloroform–methanol (3:1) the $R_f \times 100$ values were as follows: pyridoxol 62, 47; pyridoxal 68, 56; pyridoxamine 12, 5; pyridoxal ethyl acetal 54, 84; 4-pyridoxic acid 91, 49; 4-pyridoxic acid lactone 91, 18; pyridoxol phosphate 95, 0; pyridoxal phosphate 95, 0; pyridoxamine phosphate 86, 0. The three phosphates could be separated with methyl ethyl ketone–ethanol–ammonia–water (15:5:5:5) giving $R_f \times 100$ values of 30, 54, and 41, respectively. Of the visualizing agents used, diazotized *p*-nitroaniline (T-180) produced a somewhat more stable color (orange-red) than the Gibb's reagent (T-264) although both had a sensitivity of 0.1 μg.

Electrophoresis may also be used for the separation of these vitamin B_6 compounds. Pyridoxol, pyridoxal, pyridoxamine, and the corresponding phosphates were separated on silica gel layers impregnated with acetate buffers (pH 3.95 or 4.53) using 500 V, the current increasing from 19 to 30 mA and 28 to 43 mA for the two buffers, respectively [98]. Colombini and McCoy [99] also used thin-layer electrophoresis on cellulose layers impregnated with 0.5 *M* sodium acetate buffer to separate the vitamins obtained from tissue extracts. A 46 × 20 cm layer was required for the separation using 750 V for 3 hr or 600 V for 4 hr. Scintillation counting was used for radioactive material. Unlabeled compounds could be determined spectrofluorimetrically after elution with a sensitivity of 0.1 μg.

Vitamin B_{12}

Cima and Mantovan [100] separated cyanocobalamin and hydroxocobalamin on silica gel plates prepared with 0.066 *M* potassium dihydrogen phosphate buffer, using either a solvent mixture of butanol–acetic acid–water–methanol (20:10:20:5) or a buffered solvent with a silica gel plate prepared with water. After eluting the spots the compounds were determined spectrophotometrically. Covello and Schettino [101] in determining the same compounds converted both of them to dicyanocobalamin with potassium cyanide in 0.1% concentration. The chromatograms were run on either silica gel G or alumina G, using 95% methanol as the developing solvent. The errors in the determination in 11 different pharmaceuticals ranged from 6% at 0.05 γ/ml to 0.4% at 500 γ/ml. Ono and Kawasaki [102] found R_f values of 0.05 and 0.23, respectively, for hydroxocobalamin and cyanocobalamin after separating them on silica gel

G using glacial acetic–water–methanol–chloroform–butanol (9:11:5:10:25) as a solvent. The spots were detected by bioautography using vitamin B_{12} agar medium and *Lactobacillus leichmannii* ATCC 7830 as a test organism. The minimum detectable amounts were 0.005 mγ for hydroxocobalamin and 0.025 mγ for cyanocobalamin.

Popova et al. [103] separated cyanocobalamin and hydroxocobalamin on neutral alumina (Brockmann activity II) with isobutanol–isopropanol–water (1.5:1:1.25) adjusted to pH 8.5 with ammonium hydroxide. The R_f values were 0.30 and 0.46, respectively. Basic alumina (Brockmann activity II) was used with isobutanol–isopropanol–water (1:1:1) to separate cyanocobalamin, pseudovitamin B_{12}, and factors B, B_{12}, V(nB), and A; R_f values of 0.62, 0.25, 0.74, 0.46, 0.12, and 0.37, respectively [104]. The samples were applied in a 5% sodium cyanide solution so as to form the dicyano complexes which were then decomposed to the monocyano form during chromatography. Sasaki [105] used cellulose MN 300 CM layers to separate cyanocobalamin, hydroxocobalamin, adenylcobamide coenzyme, benzimidazolylcobamide coenzyme, and dimethylbenzimidazolylcobamine coenzyme. In this case the sample solution must be free of inorganic ions in order to obtain good resolution. The solvent used was the lower phase of *sec*-butanol–0.1 *M* acetate buffer (pH 3.5)–methanol (4:12:1).

Vitamin B_{12} compounds can be detected in a concentration of 0.3 μg under visible light, because of their red color. Covello and Schettino [106] used in situ densitometry for the determination of cyanocobalamin and hydroxocobalamin.

Vitamin B_2

In studying the photochemical degradation of riboflavin, Smith and Metzler [107] used silica gel G plates with two solvent systems: butanol–ethanol–water (7:2:1) and water saturated with isoamyl alcohol. The compounds were detected by their blue or yellow-green fluorescence under an ultraviolet light. Treadwell and Metzler [108] developed a procedure for detecting as small an amount as 0.2 ng of flavin in plant tissue. After extraction of the tissue with ammonium sulfate solution and clarifying by centrifuging, the extract was adsorbed on a resorcinol–formaldehyde resin column to separate the flavins from salts, residual proteins, and some pigments by washing with water; flavin nucleotides pass through the resin during application and washing [109]. The flavins eluted with acetone–water were then applied in a water solution to an unwashed talc column to remove most of the nonflavin pigments with acetone–water (1:9). The flavins were eluted from the talc with acetone–water (1:1) and applied to silica gel layers for separation. Ten solvents were listed including: the lower phase of benzyl alcohol–acetic acid–water

(3:1:3) (lumichrome 0.91, riboflavin 0.46, riboflavinyl glucoside 0.24) and acetic acid–2-butanone–methanol–benzene (5:5:20:70). The compounds were detected under ultraviolet illumination; however, for very small amounts of lumichrome it was necessary to predevelop the plate repeatedly prior to sample application in order to remove fluorescent impurities in the gel. Haworth et al. [109] used silica gel layers with toluene–methanol–acetic acid (10:9:1) to separate riboflavin and its metabolites from urine. A preliminary purification was made by precipitation and column chromatography of interfering constituents. Riboflavin at R_f 0.44 was determined by fluorescent densitometry. Ismaiel and Yassa [110] isolated riboflavin from pharmaceutical preparations on silica gel with chloroform–98% ethanol–water (50:25:1) and eluted with pH 4 acetate buffer for adsorption measurement at 444 nm.

Nicotinic Acid and Nicotinamide

Nuernberg [111] separated nicotinic acid and nicotinamide on air-dried silica gel G layers with a freshly prepared solution of *n*-propanol-10% ammonia (95:5), without chamber saturation. With a developing distance of 8 cm, the resulting R_f values are 0.35 and 0.6 to 0.7, respectively. For detecting the compounds, the plate was first sprayed with a 5% solution of *p*-aminobenzoic acid in methanol and then placed in a chamber with cyanogen chloride vapors (**Danger: poison!**), prepared by mixing 20 ml of a 28% chloramine suspension, 20 ml of 1 *N* hydrochloric acid, and 10 ml of 10% potassium cyanide solution. Nicotinic acid shows up as a red spot and nicotinamide as an orange-red color. The method can be used for a semiquantitative determination by comparing the spots with a series of known concentrations. It is highly specific and can be used for the determination of these compounds in multivitamin preparations without a preliminary separation. Sensitivity of this detection is 0.1 μg. The same separation system gives an R_f value of 0.69 for nicotinic acid benzyl ester [112]. Washuettl [113] used a spot area method for quantitative work and Ismaiel and Yassa [114] used spectrophotometry at 261 nm after elution from the layer. Brunink and Wessels [115] determined the nicotinic acid in meat by in situ fluorimetry. Mlodecka and Sekowska [116] chromatographed nicotinic acid, amide, and nitrile as well as isonicotinic acid, amide, and nitrile on silica gel G with 10 solvent systems. In water and ethanol–chloroform–25% ammonia–water (35:20:10:1) the R_f values were 0.64, 0.53; 0.41, 0.71; 0.45, 0.84; 0.76, 0.53; 0.49, 0.71; and 0.47, 0.84, respectively; the R_f values for the other solvents were listed.

Vitamin C

For the determination of ascorbic acid in potatoes, Hasselquist and Jaarma [117] applied the potato extract to silica gel G layers and used a solvent consisting of 2 g of oxalic acid dissolved in 20 ml of methanol and mixed with 60 ml of chloroform containing a trace of potassium cyanide. The 10 cm the solvent traveled was accomplished in 35 min, after which the plate was dried for 2 min at room temperature. The vitamin C was obtained as a blue spot after spraying with a solution of phosphomolybdic acid containing 5 g per 100 ml of 96% ethanol. The minimum detectable amount was used as a measure for determining the amount of ascorbic acid present in the potato. Although semiquantitative in nature, the method gave more reliable results than the spectrophotometric method, because the chromatographing method removed interfering reducing substances. Strohecker and Pies [118] used the 2,4-dinitro-phenylhydrazone derivative of vitamin C in determining the amount present in foods containing soluble carbohydrates. The separation was accomplished on air-dried silica gel G plates using chloroform-ethyl acetate (1:1) as the developing solvent. With a running space of 15 cm the vitamin C derivative

Table 31.10 Vitamin C Content of a Few Food Substances Determined by the Modified Dinitrophenylhydrazine Method (TLC)[a]

Material	*TLC Method*	*Titration with Dichlorophenol–Indophenol*	*Polarographic*
Model solution A (500 mg ascorbic acid + 50 g glucose/liter)	497 mg/liter	498 mg/liter	
Model solution B (500 mg ascorbic acid + 50 g sucrose/liter)	492 mg/liter	497 mg/liter	
Orange lemonade A	325 mg/liter 328 mg/liter	340 mg/liter	335 mg/liter
Orange lemonade B	442 mg/liter	455 mg/liter	460 mg/liter
Orange lemonade C	463 mg/liter	404 mg/liter	475 mg/liter
Chocolate A (vitaminized)	31.9 mg/15 g	35 mg/15 g	
Chocolate B (vitaminized)	56.3 mg/ 100 g	60.6 mg/100 g	
Currant juice, black	550 mg/liter	Not titrated	
Dietetic fluid with added vitamins in a malt base	53.5 mg/5 ml	54.2 mg/5 ml titrated with 0.1 *N* chloramine	49.5 mg/5 ml photometrically

[a] From R. Strohecker, Jr. and H. Pies [116]; reproduced with permission of the authors and J. F. Bergmann.

had an R_f value of about 0.25. For the quantitative determination, the spot was eluted with 85% sulfuric acid and measured in a photometer. A comparison of the results obtained with other methods is shown in Table 31.10.

Guven and Alpar [119] separated ascorbic (R_f 0.8) and dehydroascorbic acid (R_f 0.5) on silica gel G with acetic acid–acetone–methanol–benzene (3:3:10:24). The plate was sprayed with 1% phenylenediamine in acetic acid and subsequently heated to 100°C for 2 min to visualize the compounds. Huettenrauch and Keiner [120] used polyamide (with 2% fluorescent indicator) to separate ascorbic acid (R_f 0.10) and dehydroascrobic acid (R_f 0.72) with ethanol–water (13:7) as a solvent.

Folic Acid

Folic acid has an R_f value of 0.07 on silica gel G and 0.0 on alumina with acetic acid–acetone–methanol–benzene (1:1:4:14) [79, 87a]. With acetic acid–butanol–water (1:4:5) the value is 0.23 on alumina G [79]. For the determination of folic acid in vitamin preparations, Iyer and Apte [121] used silica gel G with butanol–acetic acid–ethanol–water (250:1:100:125) to isolate the compound and then eluted with 3% disodium phosphate. The folic acid was then transferred to isobutanol after the addition of sodium chloride and measured at 550 nm with a recovery of 98%. Spinelli and Ciuffi [122] used *n*-butanol–acetic acid–water (6:1:1) on silica gel HF to isolate the acid after a preliminary separation on a combination alumina–talc column. For quantitative evaluation the folic acid was eluted with 30% ammonia and then determined in a spectrophotofluorimeter after the addition of Sonneschein reagent. Popova and Kovacheva [123, 124] oxidized the folic acid with potassium permanganate in an acid medium to 2-amino-4-hydroxy-6-pteridinecarboxylic acid which was separated on an unbound layer of neutral alumina using 0.1 *M* sodium citrate for development. This was then eluted with 0.1 *M* monosodium phosphate and determined fluorimetrically, with a relative error of 0.53%. Guven and Pekin [125] chromatographed folic acid with *n*-propanol–10% ammonia–glycerol (4:1:1) on silica gel G obtaining an R_f value of 0.41. Kreuzig [126] used a dual layer composed of a narrow strip of cellulose MN 300–Dowex 50-X8 resin (4:1) and the balance of the layer of cellulose. The crude extract of fungal mycelium was spotted on the ion-exchange layer and desalted by developing with ammonia–methanol–water (3:10:7) just onto the cellulose. The ion-exchange layer was removed and then development of the folic acids was completed with 5% citric acid adjusted to pH 9 with ammonia. Sensitivity was approximately 50 pg for fresh inocula. Copenhaver and O'Brien [127] used a cellulose–AG 50-W-X4 (H^+) (200 to 400 mesh) (17.5:2.5) layer with 15% disodium phosphate buffer (pH 8.5) containing 0.1 *M* mercaptoethanol as an antioxidant for a group of pteroylglutamates and related compounds. Brown et al. [128] applied a partition system of cellulose MN 300 layers with 3% ammonium chloride (wt/vol) containing 0.5% 2-mercapto-ethanol for the chromatography of folates in the blood. The R_f values of a group of naturally occurring folates as well as a number of related compounds are given in both the latter works. Sensitivity by fluorescent quenching was usually less than 3.0 μg and fluorescent compounds could be detected in subnanogram amounts.

Pantothenic Acid

Pantothenic acid has an R_f value of 0.57 on silica gel G with acetic acid–acetone–methanol–benzene (1:1:4:14) [79], 0.31 with butanol–acetic acid–water (20:4:3) [125], and 0.41 with propanol–10% ammonia–glycerol (4:1:1) [125]. Zivanov-Stakič et al. [129], among other water-soluble vitamins, chromatographed calcium pantothenate and panthenol on silica gel with *tert*-butanol–dioxane–isopropylether–water (1:1:2.5:0.5), chloroform–butanol (3:2), and chloroform–dioxane–ethyl ether (2:3:1). Panthenol was visualized with ferric chloride containing iodine and calcium pantothenate with Ehrlich's reagent (T-90). Pantothenic acid gives a blue color with 10% copper sulfate in 2% ammonia [125]. Frank et al. [130] chromatographed calcium pantothenate on cellulose MN 300 with butanol–ethanol–water (4:3:3) to obtain an R_f value of 0.65 with a second spot at 0.40 due to β-alanine (a hydrolysis product of pantothenic acid). Visualization was with ninhydrin with subsequent heating at 105°C for 30 min.

Miscellaneous Water-Soluble Vitamins

On alumina G and silica gel G biotin has R_f values of 0.54 and 0.50, respectively, when using acetic acid–acetone–methanol–benzene (1:1:4:14) [79].

Silica gel G with acetic acid–acetone–methanol–benzene (1:1:4:14) gives and R_f value of 0.64 for *p*-aminobenzoic acid [79]; with this solvent Bićan-Fišter and Dražin [87] separated *p*-aminobenzoic acid from other vitamins and determined it quantitatively by colorimetry after elution and diazotization. On silica gel D-O (CAMAG) with water–96% ethanol–2 *M* hydrochloric acid (45:48:0.2) the R_f value was 0.98 [86]. Sherma and Dorflinger [131] have set up a student experiment to determine this vitamin in urine using silica gel with benzene–acetic acid (5:1) and direct densitometry after diazotization on the layer.

Hashmi et al. [87a] obtained an R_f value of 0.28 for choline chloride using alumina G with acetic acid–acetone–methanol–benzene (1:1:4:14) and a value of 0.54 on silica gel with water–96% ethanol–2 *M* hydrochloric acid (45:48:0.2) [86]. Sullivan and Brady [132] separated betaine (0.57), choline (0.77), and muscarine (0.87) on aluminum oxide G with methanol–carbon tetrachloride–acetic acid (28:12:1). Skidmore and Entenman [133] obtained an R_f value of 0.09 on silic gel G with methanol–water–7 *N* ammonia (6:3:1). Radecka et al. [134] analyzed pharmaceutical preparations. Alumina with chloroform–methanol–water (75:23:3) was used for choline chloride and chloroform–methanol–1 *N* hydrochloric acid (40:60:3) for other salts of choline. The compounds were detected with a spray of 5% aqueous cobalt chloride–ethanol (1:1) followed by 1% aqueous potassium ferrocyanide–ethanol (1:2). A quantitative determination was made by in situ densitometry.

References

1. H. Lagoni and A. Wortmann, *Intern. Dairy Congr., 14th, Rome,* **1956.**
2. H. Lagoni and A. Wortmann, *Milchwissenschaft,* **11,** 206 (1956).
3. C. V. Planta, U. Schwieter, L. Chopard-dit-Jean, R. Rueegg, M. Kofler, and O. Isler, *Helv. Chim. Acta,* **45,** 548 (1962).
4. G. Katsui, S. Ishikawa, M. Shimizu, and Y. Nishimoto, *Bitamin,* **28,** 41 (1963); *Chem. Abstr.,* **60,** 9577 (1964).
5. J. Davídek and J. Blattná, *J. Chromatogr.,* **7,** 204 (1962).
6. T. N. R. Varma, T. Panalaks, and T. K. Murray, *Anal. Chem.,* **36,** 1864 (1964).
7. K. V. John, M. R. Lakshmanan, F. B. Jungalwala, and H. R. Cama, *J. Chromatogr.,* **18,** 53 (1965).
8. L. M. Kuznetsova and V. M. Koval'ova, *Ukr. Biokhim. Zh.,* **36,** 302 (1964).
9. G. V. Eremina, *Lab. Delo,* **1968,** 81.
10. R. Adamski and R. Dobrucki, *Acta Pol. Pharm.,* **25,** 307 (1968).
11. E. Ludwig and U. Freimuth, *Nahrung,* **8,** 563 (1964).
12. M. Koleva, M. Dzhoneidi, and O. Budevski, *Pharmazie,* **30,** 168 (1975).
13. S. Baczyk, K. Baranowska, W. Jakubowska, and I. Sobisz, *Z. Anal. Chem.,* **255,** 132 (1971).
14. G. W. Johnson and C. Vickers, *Analyst (London),* **98,** 257 (1973).
15. K. H. Hanewald, F. J. Mulder, and K. J. Keuning, *J. Pharm. Sci.,* **57,** 1308 (1968).
16. L. M. Kuznetsova and V. M. Koval'ova, *Ukr. Biokhim. Zh.,* **36,** 302 (1964).
17. M. Zile and H. F. DeLuca, *Anal. Biochem.,* **25,** 307 (1968).
18. R. Adamski, R. Dobrucki, and B. Marchlewska, *Farm. Pol.,* **30,** 1023 (1974).
19. B. D. Drujan, R. Castillon, and E. Guerrero, *Anal. Biochem.,* **23,** 44 (1968).
20. L. K. Keefer and D. E. Johnson, *J. Chromatogr.,* **69,** 215 (1972).
21. N. Perišić-Janjić, S. Petrović, and P. Hadzić, *Chromatographia,* **9,** 130 (1976).

22. D. Ropte and J. U. Gu, *Pharmazie,* **27,** 544 (1972).
23. J. Richter and D. Ropte, *Pharmazie,* **24,** 601 (1969).
24. J. Blattná and J. Davídek, *Experientia*, **17,** 474 (1961).
25. B. Kakáč, M. Šaršúnová, Tran Thi Hoang Ba, and J. Vachek, *Pharmazie,* **22,** 202 (1967).
26. A. W. Norman and H. F. DeLuca, *Anal. Chem.,* **35,** 1247 (1963).
27. C. K. Parekh and R. H. Wasserman, *J. Chromatogr.,* **17,** 261 (1965).
28. H. Janecke and I. Maass-Goebels, *Z. Anal. Chem.,* **178,** 161 (1960).
29. E. Castren, *Farm. Aikak.,* **71,** 351 (1962).
30. A. E. A. Mitta, A. Troparevsky, and M. L. P. de Troparevsky, *Arg., Rep., Com. Nacl. Energia At., Informe,* **123,** 7 pp. (1964); through *Chem. Abstr.,* **62,** 2667 (1965).
31. W. Fuerst, *Arch. Pharm. (Weinheim),* **300,** 359 (1967).
32. H. Nerlo and W. Palak, *Acta Pol. Pharm.,* **24,** 399 (1967).
33. R. Adamski and J. Sawicka, *Farm. Pol.,* **26,** 131 (1970).
34. M. H. Hashmi, F. R. Chughtai, and M. I. D. Chughtai, *Mikrochim. Acta,* **1969,** 53.
35. A. L. Fisher, A. M. Parfitt, and H. M. Lloyd, *J. Chromatogr.,* **65,** 571 (1972).
36. A. Pinelli, F. Witzke, and P. P. Nair, *J. Chromatogr.,* **42,** 271 (1969).
37. T. H. Campion and S. Dilley, *Anal. Lett.,* **6,** 139 (1973).
38. J. Pasalis and N. H. Bell, *J. Chromatogr.,* **20,** 407 (1956).
39. G. Shone, *J. Sci. Food. Agric.,* **13,** 315 (1962).
40. G. Shone, *Chem. Ind. (London),* **1963,** 335.
41. W. A. Skinner, R. M. Parkhurst, and P. Alaupovic, *J. Chromatogr.,* **13,** 240 (1964).
42. W. A. Skinner and R. M. Parkhurst, *J. Chromatogr.,* **13,** 69 (1964).
43. R. A. Dilley, *Anal. Biochem.,* **7,** 240 (1964).
44. R. A. Dilley and F. L. Crane, *Biochim. Biophys. Acta,* **75,** 142 (1963).
45. R. A. Dilley and F. L. Crane, *Plant Physiol.,* **38,** 452 (1963).
46. P. A. Sturm, R. M. Parkhurst, and W. A. Skinner, *Anal. Chem.,* **38,** 1244 (1966).
47. H. G. Lovelady, *J. Chromatogr.,* **78,** 449 (1973).
48. *Ibid.,* **85,** 81 (1973).
49. K. J. Whittle and J. F. Pennock, *Analyst (London),* **92,** 423 (1967).
50. M. K. Govind Rao, S. Venkob Rao, and K. T. Achaya, *J. Sci. Food Agric.,* **16,** 121 (1965).
51. H. D. Stowe, *Arch. Biochem. Biophys.,* **103,** 42 (1963).
52. H. Schmandke, *J. Chromatogr.,* **14,** 123 (1964).
53. P. G. Roughan, *J. Chromatogr.,* **29,** 293 (1967).
54. O. R. Braekkan, G. Lambertsen, and H. Myklestad, *Fisk. Skr., Ser. Teknol. Unders.,* **4,** 3 (1963).
55. A. Emmerie and C. Engel, *Rec. Trav. Chim.,* **57,** 1371 (1938).
56. A. Seher, *Mikrochim. Acta.,* **1961,** 308.
57. A. Seher, *Nahrung,* **4,** 466 (1960).
58. O. E. Schultz and D. Strauss, *Arzneim.-Forsch.,* **5,** 342 (1955).

59. J. P. Vuilleumier, G. Brubacher, and M. Kalivoda, *Helv. Chim. Acta.* **46,** 2983 (1963).
60. G. Lambersten, H. Myklestad, and O. R. Braekkan, *J. Sci. Food. Agric.,* **13,** 617 (1962).
61. R. A. Dilley and F. L. Crane, *Anal. Biochem.,* **5,** 531 (1963).
62. G. Katsumi, Y. Ichimura, and Y. Nishimoto, *Yakuzaigaku,* **23,** 299 (1963); through *Chem. Abstr.,* **61,** 2168 (1964).
63. T. Aratani, K. Mita, and F. Mizui, *J. Chromatogr.,* **79,** 179 (1973).
64. H. Thielemann, *Mikrochim. Acta,* **1972,** 227.
65. J. T. Matschiner and J. M. Amelotti, *J. Lipid Res.,* **9,** 176 (1968).
66. R. K. Hammond and D. C. White, *J. Chromatogr.,* **45,** 446 (1969).
67. J. D. Manes, H. B. Fluckiger, and D. L. Schneider, *J. Agric. Food Chem.,* **20,** 1130 (1972).
68. M. Billeter and C. Martius, *Biochem. Z.,* **334,** 304 (1961).
69. H. Wagner, L. Hoerhammer, and D. Dengler, *J. Chromatogr.,* **7,** 211 (1962).
70. H. Wagner, *Chromatogr., Symp., 2nd, Brussels,* **1962,** 243.
71. H. Wagner and B. Dengler, *Biochem. Z.,* **336,** 380 (1962).
72. D. R. Threlfall and T. W. Goodwin, *Biochim. Biophys. Acta,* **78,** 532 (1963).
73. H. Eck and A. Trebst, *Z. Naturforsch.,* **18b,** 446 (1963).
74. M. D. Henninger and F. L. Crane, *Biochemistry,* **2,** 1168 (1963).
75. M. D. Henninger, R. A. Dilley, and F. L. Crane, *Biochem. Biophys. Res. Commun.,* **10,** 237 (1963).
76. B. Drews, H. Specht, and H.-J. Hinze, *Monatsschr. Brau.,* **20,** 7 (1967).
77. R. Ikan, R. Gottlieb, E. D. Bergmann, and J. Ishay, *J. Insect Physiol.,* **14,** 1215 (1968).
78. H. Gaenshirt and A. Malzacher, *Naturwissenschaften,* **47,** 279 (1960).
79. S. Ishikawa and G. Katsui, *Bitamin,* **29,** 203 (1964).
80. R. Huettenrauch, L. Klotz, and W. Mueller, *Z. Chem.,* **3,** 193 (1963).
81. J. Dittmann, *Dtsch. Gesundheitsw.,* **22,** 1217 (1967).
82. H. Thielemann, *Z. Chem.,* **13,** 15 (1973).
83. H. Thielemann, *Sci. Pharm.,* **42,** 94 (1974).
84. S. E. Petrović, B. E. Belia, and D. B. Vukajlović, *Anal. Chem.,* **40,** 1007 (1968).
85. H.-C. Chiang, Y. Lin, and Y.-C. Wu, *J. Chromatogr.,* **45,** 161 (1969).
86. Manzur-ul-Haque Hashmi, F. R. Chugtai, and M. I. D. Chughtai, *Mikrochim Acta,* **1969,** 951.
87. T. Bićan-Fišter and V. Dražin, *J. Chromatogr.,* **77,** 389 (1973).
87a. M. H. Hashmi, F. R. Chughtai, A. S. Adil, and T. Qureshi, *Mikrochim. Acta,* **1967,** 1111.
88. R. T. Nuttall and B. Bush, *Analyst (London),* **96,** 875 (1971).
89. S. David and H. Hirshfeld, *Bull. Soc. Chim. Fr.,* **1963,** 1011.
90. T. Ono and M. Hara, *Bitamin,* **33,** 512 (1966).
91. H.-P. Chuang, H.-C. Chiang, and K.-T. Wang, *J. Chromatogr.,* **41,** 487 (1969).
92. B. K. Dwivedi, R. G. Arnold, and L. M. Libbey, *J. Food Sci.,* **37,** 689 (1972).
93. B. K. Dwivedi and R. G. Arnold, *J. Food Sci.,* **37,** 886 (1972).

94. M. R. Ariaey-Nejad, M. Balaghi, E. M. Baker, and H. E. Sauberlich, *Am. J. Clin. Nutr.*, **23,** 764 (1970).
95. E. Nuernberg, *Deut. Apoth.-Ztg.*, **101,** 268 (1961).
96. Z. Deyl and J. Rosmus, *J. Chromatogr.*, **8,** 537 (1962).
97. E. N. Dement'eva, N. A. Drobinskaya, L. V. Ionova, M. Y. Karpeiskii, and V. L. Florent'ev, *Biokhimiya*, **33,** 350 (1968).
98. H. Ahrens and W. Korytnyk, *Anal. Biochem.*, **30,** 413 (1969).
99. C. E. Colombini and E. E. McCoy, *Anal. Biochem.*, **34,** 451 (1970).
100. L. Cima and R. Mantovan, *Farmaco Ed. Prat.*, **17,** 473 (1962); through *Chem. Abstr.*, **57,** 16986 (1962).
101. M. Covello and O. Schettino, *Farmaco Ed. Prat.*, **19,** 38 (1964).
102. T. Ono and M. Kawasaki, *Bitamin*, **30,** 280 (1964); through *Chem. Abstr.*, **62,** 1957 (1965).
103. Ya. Popova, Ch. Popov, and M. Ilieva, *J. Chromatogr.*, **24,** 263 (1966).
104. *Ibid.*, *J. Chromatogr.*, **21,** 164 (1966).
105. T. Sasaki, *J. Chromatogr.*, **24,** 452 (1966).
106. M. Covello and O. Schettino, *Farmaco Ed. Prat.*, **20,** 581 (1965).
107. E. C. Smith and D. E. Metzler, *J. Am. Chem. Soc.*, **85,** 3285 (1963).
108. G. E. Treadwell, Jr., and D. E. Metzler, *Anal. Biochem.*, **46,** 261 (1972).
109. C. Haworth, R. W. A. Oliver, and R. A. Swaile, *Analyst (London)*, **96,** 432 (1971).
110. S. A. Ismaiel and D. A. Yassa, *Analyst (London)*, **98,** 1 (1973).
111. E. Nuernberg, *Dtsch. Apoth.-Ztg.*, **101,** 142 (1961).
112. H. Thielemann and M. Papke, *Sci. Pharm.*, **40,** 210 (1972).
113. J. Washuettl, *Mikrochim. Acta*, **1970,** 621.
114. S. A. Ismaiel and D. A. Yassa, *Analyst (London)*, **98,** 816 (1973).
115. H. Brunink and E. J. Wessels, *Analyst (London)*, **97,** 258 (1972).
116. J. Mlodecka and B. Sekowska, *Chem. Anal. (Warsaw)*, **13,** 159 (1968).
117. H. Hasselquist and M. Jaarma, *Acta Chem. Scand.*, **17,** 529 (1963).
118. R. Strohecker, Jr., and H. Pies, *Z. Lebensm.-Untersuch.-Forsch.*, **118,** 394 (1962).
119. K. C. Guven and O. Alpar, *Eczacilik Bul.*, **9,** 160 (1971); through *Chem. Abstr.*, **69,** 24366z (1968).
120. R. Huettenrauch and I. Keiner, *Pharmazie*, **23,** 157 (1968).
121. H. R. S. Iyer and B. K. Apte, *Indian J. Pharm.*, **31,** 58 (1969).
122. P. Spinelli and M. Ciuffi, *Boll. Chim. Farm.*, **105,** 849 (1966).
123. Y. Popova and E. Kovacheva, *Nauch. Tr., Vissh Inst. Khranit. Vkusova Prom., Plovdiv*, **16,** 323 (1969); through *Chem. Abstr.*, **77,** 168685r (1972).
124. *Ibid.*, *Nauch. Tr., Vissh Inst. Khranit. Vkusova Prom. Plovdiv*, **17,** 351 (1970); through *Chem. Abstr.*, **77,** 52391n (1972).
125. K. C. Guven and O. Pekin, *Eczacilik Bul.*, **8,** 113 (1966); through *Anal. Abstr.*, **14,** 5689 (1967).
126. F. Kreuzig, *Z. Anal. Chem.*, **255,** 126 (1971).
127. J. H. Copenhaver and K. L. O'Brien, *Anal. Biochem.*, **31,** 454 (1969).
128. J. P. Brown, G. E. Davidson, and J. M. Scott, *J. Chromatogr.*, **79,** 195 (1973).

129. D. Zivanov-Stakič, D. Radulovič, and G. Ilič, *Arhiv Farm. (Belgrade)*, **25** 85 (1975); through *Chem. Abstr.*, **84,** 79758u (1976).
130. M. J. Frank, P. G. Trager, J. B. Johnson, and S. H. Rubin, *J. Pharm. Sci.*, **62,** 108 (1973).
131. J. Sherma and L. Dorflinger, *Am. Lab.*, **8,** 63 (1976).
132. G. Sullivan and L. R. Brady, *Lloydia*, **28,** 68 (1965).
133. W. D. Skidmore and C. Entenman, *J. Lipid Res.*, **3,** 471 (1962).
134. C. Radecka, K. Genest, and D. W. Hughes, *Arzneim.-Forsch.*, **21,** 548 (1971).

Chapter **XXXII**

MISCELLANEOUS

1 MYCOTOXINS

Mycotoxins are toxic substances produced by molds. Meyer and Leistner [1] stated that there are about 240 toxicogenic mold species known and listed 75 of these. Fishbein and Falk [2] have reviewed the chromatography of mycotoxins, Stoloff [3] has reviewed the analytical methods for these compounds, and Schuller et al. [4] reviewed aflatoxin methodology.

The best known of the toxins are the aflatoxins produced by *Aspergillus flavus* and *A. parasiticus,* as well as *A. ochraceus* and a *Rhizopus* species [5]. At present the known aflatoxins are B_1, B_2, B_{2a}, B_3 (parasiticol), R_0 (aflatoxol), G_1, G_2, G_{2a}, GM_1, GM_{2a}, M_1, M_2, M_{2a}, RB_1, RB_2, and P_1. Aflatoxin P_1, a phenol, is a metabolite of aflatoxin B_1 and has not been identified from mold cultures; B_{2a} is a hemiacetal of B_1, M_1 is a metabolite of B_1, and M_2 is a metabolite of B_2. Because of their known carcinogenic properties most of the work has been concerned with the detection and determination of these compounds in foods and feed products. Thin-layer chromatography has been widely used for this purpose.

The aflatoxins may be extracted with water–chloroform (1:10), chloroform, chloroform–methanol–water (8:1:103) [6], methanol [7], dimethyl sulfoxide–water (1:1) and dimethylformamide–water (1:1) [8], acetone [9], 70% acetone [10], and chloroform–methanol (9:1) [11]. Samples containing large amounts of fat or other interfering materials will require a preliminary cleanup before the extracts are applied to the thin layers [10, 12–15]. Yin [16] recommends the use of acetonitrile–water (9:1) for extraction as giving an extract fairly free of interfering materials including oils, although if the extract contains material precipitable by lead an appropiate amount of lead acetate may be used. Brown et al. [17] tested a number of methods for extracting aflatoxin from animal tissue and found the latter method did not give the desired sensitivity. They preferred the methanol extraction of Jacobson et al. [7], and then used suitable cleanup procedures to obtain the sample for thin-layer chromatography. Altenkirk [18] used the TLC layer to clean up the sample; after the sample was applied to a 6 × 14 cm silica gel layer, the plate was developed in the short direction with dry ether. The solvent was removed by drying in an oven at 80°C for 15 min and then development was made at right angles in the long direction with chloroform–trichloroethylene–*tert*-butanol–formic acid (85:10:4:1). If the region of the aflatoxin was not completely clear of interfering background, it was developed again in ether in the short dimension. Leistner [19] removed interfering substances by applying the samples (benzene extracts of feed materials) on a line in the middle of a 20 × 20 cm silica gel layer on an aluminum sheet and then developing twice with benzene and twice with ether, leaving the aflatoxins at the origin. The upper part of the layer was then cut off 2 cm above the origin and turned through 180°C for development with chloroform–trichloroethylene–amyl alcohol–formic acid (80:15:4:1). This development was repeated for samples containing low amounts of aflatoxins. Crowther [20] found that a metabolite of *Macrophomina phaseoli* (Maubl) Ashby, exhibited the same behavior as aflatoxin B on silica gel with chloroform–methanol (19:1) and had a bright-blue fluorescence under 363-nm radiation. Another metabolite exhibited a green fluorescence and appeared at R_f 0.6 to 0.75. Since neither metabolite was toxic to mice or ducklings, he recommended that chromatographs apparently positive to aflatoxin B and having the green fluorescence at R_f 0.6 to 0.75 should be rechromatographed in ether which places the metabolite from *M. phaseoli* at R_f 0.9 and the green fluorescent spot at 1.0; aflatoxin B remains at R_f 0,5. Roberts and Patterson [20a] added a dialyzing step to the purification procedure for mycotoxins.

Chloroform has been the main spotting solvent for aflatoxins; however, Stoloff et al. [21] recommended the use of benzene as a spotting solvent for a number of reasons: (*a*) the higher boiling point of benzene causes less problems in transferring and in sample applications, (*b*) the spots are smaller and more compact, and (*c*) the alcohol preservative in chloroform can affect the R_f values; the addition of 2% acetonitrile improved the solvent properties of the benzene and still retained the ability to produce small spots [22].

The preferred adsorbent for the thin-layer chromatography of aflatoxins is silica gel. Heathcote and Hibbert [23] examined several silica gels for the best resolution; kieselgel G (Merck), the adsorbent most used in this work in the past, was inferior to the other silica gels tested. Best results were obtained with SilicAR TLC-7G (Mallinckrodt).

A number of solvents have been used for the separation of aflatoxins: chloroform–methanol (97:3), chloroform–methanol–acetic acid (94.5:5:0.5), acetone–chloroform (9:1) and (3:17), benzene–acetic acid–methanol (90:5:5), and chloroform–acetone–2-propanol (33:6:1 and 34:5:1). Chloroform–methanol solvents were found to be susceptible to variations in humidity and temperature, and a number of investigators searched for more reliable solvent systems. Benzene–ethanol–water (46:35:19) was found to be a good solvent, but only under optimum conditions of humidity and temperature [23]. Stubblefield et al. [24] found their best system to be isopropanol–acetone–chloroform (5:10:85) for the separation of B_1, B_2, G_1, G_2, M_1, and M_2. The best resolution of M_1 and M_2 was obtained with *n*-amyl alcohol–acetone–chloroform (1:1:8), but in this case the other aflatoxins ran with the solvent front. Toluene–ethyl acetate–acetone–acetic acid (50:35:15:2) gives a good separation of G_{2a} and the more polar derivatives M_{2a} and GM_{2a} [23]; small quantities of acetyl derivatives are formed, but these can be removed by using chloroform–methanol (97:3). Reddy et al. [25] found that toluene–isoamyl alcohol–methanol (90:32:3) gave a clear separation of aflatoxin B_1

(0.56), B_2 (0.48), G_1 (0.42), and G_2 (0.34); still further improvement in resolution could be obtained by developing a second time in the same solvent. Engstrom [26] reported three solvent systems that gave improved resolution over that obtained with the usual solvents. These were: methylene chloride–trichloroethylene–*n*-amyl alcohol–formic acid (80:15:4:1) (development 2×) (this solvent reversed the usual order of B_2 and G_1), chloroform–trichloroethylene–*n*-amyl alcohol–formic acid (80:15:4:1) (2× development), and a partition method on silica gel H–silica gel G-HR (1:1) impregnated with *tert*-butanol–formic acid–water (10:1:25) and air-dried for 30 min; in this case the developing solvent was xylene–*tert*-butanol–formic acid (94:5:1).

Two-dimensional techniques have been used to improve the resolution. These include a combination of the solvents chloroform–acetone (9:1) and ethyl ether–methanol–water (188:9:3) [27], acetone–chloroform (1:9) and toluene–ethyl acetate–90% formic acid (5:4:1) [28], benzene–acetic acid–methanol (90:5:5) and acetone–chloroform (1:9) [29], acetone–chloroform (1:9) and ethyl acetate–isopropanol–water (10:2:1) [30], and ether–methanol–water (188:9:3) and chloroform–acetone–isopropanol (85:10:5) [31]. De Zeeuw and Lillard [32] applied vapor programming to increase the resolution of aflatoxins B_1, B_2, G_1, and G_2.

Ethoxyquin is used as an antioxidant in some foods, forage crops, and animal feeds and may show up in aflatoxin extracts. Some of the solvents used in development give fluorescent spots with commercial ethoxyquin, one of which is similar in R_f values to those for B_1. Stefaniak [33] and Issaq et al. [34] have investigated this and proposed a number of methods for avoiding this problem.

Kwon and Ayres [35] in purifying G_1 found that silica gel layers impregnated with EDTA and solvents containing BHT as an antioxidant decreased deterioration on the layer. This does not eliminate the need to work in subdued artificial light when working with aflatoxins.

Andrellos and Reid [36] have worked out three confirmatory tests for aflatoxin B_1 and G_1. After a preparatory separation by thin-layer chromatography the spots are eluted with methanol; the solvent is removed and then the residue is reacted with (1) 0.2 ml of glacial acetic acid in one drop of colorless thionyl chloride, (2) 0.2 ml of glacial acetic acid in one drop of 90% reagent-grade formic acid, or (3) three drops of anhydrous trifluoroacetic acid. After the reaction [5 min for (1) and (2) and 60 sec for (3)], the reaction products are evaporated to dryness by gentle heat in a stream of nitrogen. The chloroform solution of the reaction material is then spotted on a silica gel plate for development in an equilibrated atmosphere in methanol–chloroform (5:95). Under long-wave ultraviolet light, the reaction products give typical fluorescent spots. Stoloff [37] modified these tests by adding an improved silica gel column cleanup. Przybylski [38] used the reaction with trifluoroacetic acid as an in situ thin-layer test with good results. Pohland et al. [39] obtained the same confirmatory products as Andrellos and Reid by heating with concentrated hydrochloric acid and water to form the water adducts and with hydrochloric acid and acetic anhydride to form the acetate. Ashoor and Chu [40, 41] have developed an additional confirmatory test for B_1 and B_2 by reducing the aflatoxins with sodium borohydride to form the corresponding aflatoxins RB_1 and RB_2. The R_f values of these in chloroform–ethyl acetate (1:3) are: B_1 0.54, RB_1 0.71, B_2 0.46, and RB_2 0.65. Stack et al. [42] obtained a confirmatory test for M_1 by treating this aflatoxin with pyridine–acetic anhydride (1:1) for 15 min at room temperature and chromatographing the derivative. Haddon et al. [43] used mass spectrometry to confirm the identification of aflatoxins. The toxicity of aflatoxin to the chick embryo is also used as confirmatory information [44], and Clements [45] has investigated the use of growth inhibition of *Bacillus megaterium* as confirmatory evidence for aflatoxin B_1.

Quantitatively the aflatoxins are best determined by in situ fluorimetry [46–49]. A sensitivity of 0.1 to 0.2 ng can be achieved by this method [50–51]. Berman and Zare [52] have used laser fluorescent analysis for the detection of subnanogram quantities. Kiermeier and Groll [53] have determined aflatoxin B_1 by spectrophotometric measurement at 363 nm after elution from the plate with methanol.

Ďuračková et al. [54] reported that there are more than 100 known mycotoxins, and it is beyond the scope of this book to report all the TLC work that has been carried out on these compounds.

After the aflatoxins, the ochratoxins are the most widely investigated mycotoxins. Ochratoxins A, B, and C have been chromatographed on silica gel with benzene–acetic acid (3:1) with R_f values of 0.5, 0.35, and 0.65, respectively [55]. They exhibit a blue fluorescence under ultraviolet light after spraying with 0.1 N sodium hydroxide. In methanol–acetic acid–benzene (5:5:90) the R_f values of A and B are 0.65 and 0.5, respectively [56]. Confirmation of the identity of A and B was obtained by forming the ethyl esters which have R_f values of 0.8 and 0.7, respectively, in the same solvent system. Confirmation may also be obtained by microbiological assay [57]. Stoloff et al. [58] used benzene–methanol–acetic acid (90:5:5), hexane–acetone–acetic acid (18:2:1), chloroform–acetic acid–ethyl ether (17:1:3), and chloroform–acetone (9:1) to separate and detect mixtures of aflatoxins, ochratoxins, zearalenone, sterigmatocystin, and patulin. Four silica gel layers were used, two for the first two solvents monodimensionally and two for two-dimensional separations using the

Table 32.1 $R_f \times 100$ of Some Mycotoxins on Silica Gel [54][a]

	Solvent[b]							
Mycotoxin	*A*	*B*	*C*	*D*	*E*	*F*	*G*	*H*
4-Acetamido-4-hydroxy-2-butenoic acid-γ-lactone	12.0	11.5	2.5	57.0	5.0	8.0	7.0	55.0
Aflatoxin B_1	28.0[c]	31.5[c]	19.5[c]	80.5[c]	22.5[c]	32.5[c]	31.5[c]	61.5[c]
Aflatoxin B_2	30.0	18.0	14.0	80.0	9.0	33.0	20.0	61.0
Aflatoxin G_1	19.0[c]	19.0[c]	13.5[c]	84.5[c]	12.0[c]	25.0[c]	20.0[c]	50.0[c]
Aflatoxin G_2	22.5	8.0	8.0	78.0	4.5	17.0	11.0	51.5
Aflatoxin M_1	21.5	15.5	10.0	78.5	0.0	7.5	6.5	64.0
Altenuene	24.0	33.0	7.0	66.0	3.5	6.5	12.0	83.0
Alternariol	30.5	65.5	17.0	83.0	32.0	28.5	58.0	95.0
Alternariol monomethyl ether	57.5	72.5	33.0	88.0	67.0	56.5	73.5	93.0
Aspergillic acid	67–80[d]	62–82[d]	35–57[d]	74–94[d]	4–21[d]	33–71[d]	65–86[d]	92.0[d]
Citrinin	2–25[d]	0–48[d]	0–5[d]	0–24[d]	0–13[d]	0–20[d]	37.0[c]	72.0[c]
Citreoviridin	25.5[c]	30.0[c]	9.5[c]	80.0[c]	1.0[c]	9.5[c]	8.0[c]	86.0[c]
Cyanein	18.0	27.0	5.0	64.5	0.0	0.0	2.5	90.0
Diacetoxyscirpenol	39.0	30.5	19.5	90.0	14.5	34.5	27.5	82.0
N,N'-dibenzylethylenediamine	31.0	27–44[d]	2.0	31.0	0.0	0.0	21.0	69.0
Fusaric acid	23.5	11.5	0.11[d]	70.0	0–12[d]	0–14[d]	12.5	58.0
Gliotoxin	50.0	42.5	31.5	85.0	18.5	32.0	30.5	81.0
Griseofulvin	40.5	46.0	19.0	91.0	24.0	38.0	43.0	66.5
Helvolic acid	52.5	53.0	28.0	89.0	15.0	29.0	50.0	93.5

	A	B	C	D	E	F	G	H
Kojic acid	12.5	15.0	2.5	52.0	0.0	2.0	6.0	65.0
Luteoskyrin	0–45[d]	60.0[c]	0–12[d]	0–46[d]	0–17[d]	0–15[d]	0–38[d]	0–100[d]
Mycophenolic acid	57.0	65.5	29.5	81.5	29.0	42.5	71.5	87.0
Nivalenol	7.0	5.0	0.0	50.0	0.0	0.0	0.0	67.5
Ochratoxin A	52.0	59.5	34.5[c]	79.0	0–11[d]	0–23[d]	56.5	95.0
Ochratoxin B	41.0	46.0	12.0[c]	65.0	0.0	2.0	33.5	79.0
Ochratoxin C	80.0	72.0	75.0	91.0	53.0[c]	73.5	86.0	87.0
Patulin	34.0	37.0	22.0	72.0	20.5	25.0	30.0	78.0
Penicillic acid	32.0	43.0	18.5	76.0	18.5	22.5	31.0	88.5
Rubratoxin B	0.0[c]	0.0[c]	0.0[c]	34.0[c]	0.0[c]	0.0[c]	0.0[c]	84.0[c]
Rugulosin	0–34[d]	51.5[c]	1.5	0–38[d]	0–4[d]	0–5[d]	27.5[c]	92.0[c]
Secalonic acid	45.0[c]	55.5	0.0	54.0[c]	0.0	5.0	20–37[d]	72.0[c]
Sterigmatocystin	67.5	71.0	59.0	93.0	71.5	74.0	78.0	84.5
Terreic acid	55.0	66.0	55.0	79.0[c]	51.0	49.0	63.0	90.5
T-2 toxin	43.5	32.0	16.5	92.0	16.0	38.5	38.5	81.0
Trichothecin	68.5	55.5	43.0	92.0	63.0	73.5	69.5	81.5
Viridicatum toxin	0–12[d]	0–41[d]	0.0	0–12[d]	0.0	0.0	0–17[d]	57.5[c]
Zearalenone	56.5	58.5	40.0	88.5	61.5	61.5	64.0	84.0

[a] Original also contains R_f values with Silufol sheets and the color reactions.

[b] Solvent: *A* = benzene–methanol–acetic acid (24:2:1); *B* = toluene–ethyl acetate–90% formic acid (6:3:1); *C* = benzene–ethanol (95:5); *D* = chloroform–methanol (4:1); *E* = chloroform–methyl isobutyl ketone (4:1); *F* = chloroform–acetone (9:1); *G* = chloroform–acetic acid–diethyl ether (17:1:3); *H* = *n*-butanol–acetic acid–water (4:1:4).

[c] Tailing occurs.

[d] Elongated spot, values show top and bottom of spot.

third and fourth solvents in the first and second dimensions, respectively. Both nonfluorescent and fluorescent layers were used in detecting the various compounds.

Steigmatocystin has been separated from grain samples by chromatographing with benzene–acetic acid–methanol (90:5:5) and confirmed by preparation of the acetate derivatives and an acid hydrolysis derivative with subsequent separation of the derivatives by thin-layer chromatography [59].

Other examples of the use of thin-layer chromatography for detection of mycotoxins include: byssochlamic acid in food stuffs [60], tricothecenes [61], wedeliatoxin [62], grayanotoxins in honey [63], tremortins in agricultural products [64], rubratoxin B in corn [65], citrinin [66], patulin [67–71], citreoviridin [72], β-nitropropionic acid [73], hoplonemertine toxin [74], sterigmatocystin and its isomers [75], and altenuene and its derivatives [76].

Steyn [77] determined the R_f values of 11 mycotoxins on silica gel layers impregnated with 0.4 *N* oxalic acid. The solvent used was chloroform–methyl isobutyl ketone (4:1). Concentrated sulfuric acid with subsequent heating at 110°C for 10 min was used as a visualizing agent; 1% ceric sulfate in 6 *N* sulfuric acid was also recommended, as well as several specific color reactions. Roberts and Patterson [20a] separated 12 mycotoxins on silica gel G-HR plates; the R_f values in nine solvent systems are given.

Ďuračková et al. [54] reported the R_f values for 37 mycotoxins on silica gel G and Silufol sheets in eight solvent systems. Five detection systems are listed; spraying with *p*-anisaldehyde (0.5 ml + 70 ml methanol + 10 ml acetic acid + 5 ml concentrated sulfuric acid) and then heating at 130°C (60°C for Silufol sheets) for 8 to 20 min revealed all the compounds under ultraviolet light; however, the other reagents were useful for confirmatory tests. Some of the R_f values are given in Table 32.1. The same workers [78] also reported a bioautographic method for mycotoxins using *Artemia salina* larvae. Gedek [79] describes the biological methods for detection of mycotoxins including the use of chick embryos and cell cultures.

2 ADHESIVES

Dietl [80] chromatographed adhesives by applying a 0.1 to 5% solution in a nonpolar solvent on silica gel G or aluminum oxide G layers and developing with 90% butanol. The spots were visualized by spraying with sulfuric acid and heating to 120°C. The method was sensitive to 0.01 γ of adhesive.

3 ANTIOXIDANTS

Antioxidants are used not only in foods to prevent or delay the development of rancidity in fats and other components, but also in many industrial products such as rubber and plastics to prevent deterioration of these materials. Anderson et al. [81] have used thin-layer chromatography to investigate the disappearance of BHT (3,5-di-*tert*-butyl-4-hydroxy-toluene) and BHA (3-*tert*-butyl-4-hydroxyanisole) from breakfast cereals in relation to an increase in the peroxide content of the products. Hexane-ether (90:10) with silica gel layers was used in the analysis to see if oxidation products of the two antioxidants could be isolated.

For the detection of synthetic antioxidants in edible oils Seher [82, 83] has studied a group of 24 of these materials. Separations were obtained by the use of one-dimensional chromatography on silica gel G using chloroform as a solvent, or by means of two-dimensional chromatography using chloroform in the first direction and benzene in the second direction. For detecting the zones a number of agents were used. The dry chromatoplates can be sprayed with a 20% solution of phosphomolybdic acid in methanol. After 1 or 2 min the antioxidants appear as blue spots, and on additional treatment of the plates with ammonia vapor the background becomes pure white and the antioxidants appear as blue-violet or green spots. Antioxidants with low reducing power appear only after the plates are heated for 10 min at 120°C. As an alternative method for identification, the plates may be sprayed with a 1% solution of 2,6-dichlorobenzoquinone chloroimide in ethanol. The plates are then exposed to a neutral atmosphere for 15 min for the colored zones to develop, In some cases these are converted to characteristic colors if the plates are then sprayed with a 2% borax solution in 40% ethanol. The colors for these various reactions are given in Table 32.2. Meyer [84] in working with antioxidants for fat and fat-containing foods has used thin layers of a mixture of silica gel G and kieselguhr (25:5). For a solvent he used a mixture of hexane–acetic acid (2:0.5). Ishikawa and Katsui [85] have used both silica gel and polyamide layers for the separation of some antioxidants used in vegetable oils. Jonas [86] has given the R_f values of some common antioxidants in a series of solvents with the best results obtained on paraffin oil-impregnated silica gel using 75% methanol as the solvent. Zentz [87] has also separated some natural and synthetic antioxidants. Table 32.3 presents R_f values for some commonly used antioxidants, and Table 32.4 lists the R_f values of gallates separated on acetylated cellulose [88] and polyamide [89]. Davídek and Pokorný [90, 91] have used unbound layers of polyamide powder as the adsorbent for separating a group of phenolic antioxidants [Table 32.5].

Table 32.2 Detection of a Group of Antioxidants with Various Reagents[a]

Compound	Abbreviated Designation	Phosphomolybdic Acid	2,6-Dichloroquinone Chlorimide	
			Neutral	Borax Spray
α-Tocopherol		+	Yellow-brown	Yellow-brown
α-Tocopherol acetate		–	Rose	(Rose)
2,2,5,6,7-Pentamethyl-6-hydroxychromane	PMHC	+	Brown-yellow	Yellow-brown
Propyl gallate	PG	+	Brown	Gray-brown
Octyl gallate	OG	+	Brown	Gray-brown
Dodecyl gallate	DG	+	Brown	Gray-brown
2-*tert*-Butyl-4-hydroxyanisole	BHA	+	Rust-brown	Violet
3-*tert*-Butyl-4-hydroxyanisole	BHA	+	Rust-brown	Violet
2,5-Di-*tert*-butyl-4-hydroxyanisole	DBHA	+	Purple	Violet
3-*tert*-Butyl-4-hydroxytoluene	BHT	+	Orange	Orange
3,5-Di-*tert*-butyl-4-hydroxytoluene	JONOL	+	Yellow	(Bright yellow)
Nordihydroguaiaretic acid	NDGA	+	Violet	Brown-violet
Guaiacum resin	GH	+	Olive green	Olive green
Ascorbyl palmitate	AP	+	Red	Bright violet
Hydroquinone, monomethyl ether	HA	+	Red-violet	Blue-violet
4-*tert*-Butoxyanisole	BOA	+	Red-violet	Blue-violet
Monoglyceride citrate	MGC	+[b]	Rose	Bright violet
N-Lauroyl-*p*-phenetidine	Suconox 12	+	Bright rose	Bright rose
N-Stearoyl-*p*-phenetidine	Suconox 18	+	Bright rose	Bright rose

Table 32.2 (Continued)

Compound	Abbreviated Designation	Phosphomolybdic Acid	2,6-Dichloroquinone Chlorimide: Neutral	2,6-Dichloroquinone Chlorimide: Borax Spray
N,N-Diphenyl-*p*-diphenylenediamine	DPPD	+	Gray-brown	Gray-brown
Tetraethylthiuram disulfide	TETD	+	Rust-brown	Brown
β,β′-Thiodipropionic acid	TDP	+	Bright brown	Orange
4(*n*-Butylmercapto)butanone	BMB	+	Canary yellow	Bright brown
2,4,6-Tri-*tert*-butylphenol	TBPh	+	Orange	Purple

[a] From A. Seher [82]; reproduced with permission of the author and Industrieverlag von Hernhaussen KG.
[b] Reaction after warming.

Wang and Chou [92] separated BHA, BHT; steryl, cetyl, lauryl, amyl, and propyl gallates; and ethyl protocatechuate on polyamide sheets with petroleum ether (30 to 70°C)–benzene–acetone(8:2:5) and acetone–ethanol–water (4:1:2). The compounds were detected with 0.25% sodium fluorescein in dimethylformamide–ethanol (5:100) and with bromine–carbon tetrachloride (1:20). The R_f values and sensitivity of detection were: 0.98, 0.33, 4; 0.96, 0.46, 1; 0.64, 0.08, 1; 0.48, 0.20, 2.5; 0.37, 0.30, 1; 0.17, 0.49, 0.5; 0.09, 0.53, 0.05; and 0.31, 0.57, 1 μg, respectively. Lee [93] used a polyamide–silica gel (8:15) layer with a 1:4 mixture of acetic acid with benzene, chloroform, or carbon tetrachloride. Ammoniacal silver nitrate solution was used as a detecting agent. Eight antioxidants were separated.

Daniels et al. [94] have investigated the natural phenolic antioxidants in oats.

Quantitative methods have been applied to the thin-layer separation of antioxidants [95–97] for the determination of BHA, BHT, propyl gallate (PG), and nordihydroguaiaretic acid (NDGA) in lard. Sahasrabudhe [95] extracted the antioxidants from a hexane solution of the fat by means of 80% ethanol and by acetonitrile. Separations of mixtures of antioxidants were achieved by using two-dimensional chromatography with benzene

Table 32.3 $R_f \times 100$ Values of Some Common Antioxidants

	Polyamide			*Silica Gel*		
Antioxidant	$MeOH:Me_2CO:H_2O$ (6:1:3) [91]	$CHCl_3$[a] [85]	$MeOH$[a] [85]	$CHCl_3$[a] [85]	C_6H_6[a] [85]	*Impregnated with paraffin oil* $MeOH:H_2O$ (3:1)[a] [86]
Propyl gallate (PG)	56	00	61	00	00	82
Butylhydroxyanisole (BHA)	67	52	66	31	31	65
Norhydroguaiaretic acid (NDGA)	27	00	45			
Butylhydroxytoluene (BHT)		89	70	79	75	12
Isoamyl gallate		00	61	00	00	
Octyl gallate						50
Lauryl gallate						20

[a] Development distance 10 cm.

Table 32.4 $R_f \times 100$ Values of Some Gallates Used as Antioxidants on 10% Acetylated Cellulose and on Polyamide

	A-C [88]	*Polyamide* [89]					
Gallate	*Shell Sol A–Propanol–Acetic Acid–Formic acid* (15:2:1:2)	*Butanol–Acetic Acid–Water* (4:1:5)	*Carbon Tetra-chloride–Ethanol* (7:3)	*Carbon Tetra-chloride–Ethanol* (3:2)	*Ethanol*	*Carbon Tetra-chloride–Methanol* (7:3)	*Diethyl Ether*
Gallic acid	3	28	5	8	31	6	4
Methyl gallate	10	60	19	33	53	29	7
Ethyl gallate	19	70	26	44	67	39	13
Propyl gallate	29	68	44	47	55	46	28
Butyl gallate	40						
Octyl gallate	67	89	62	73	80	63	45
Dodecyl gallate	83						
Lauryl gallate		89	80	85	80	80	67

Table 32.5 $R_f \times 100$ Values of Some Phenolic Antioxidants on Loose-Layer Polyamide[a]

Antioxidants	*MeOH–CCl$_4$* (1:9)	*MeOH–H$_2$O* (6:4)
Phenol	67	37
4-Methylphenol	57	10
4-*tert*-Butylphenol	67	59
2-*tert*-Octylphenol	54	25
4-*tert*-Octylphenol	73	35
4-Phenylphenol	46	36
2-Phenylphenol	76	37
4-Cyclohexylphenol	70	33
Pyrocatechol	29	73
3-Methylpyrocatechol	24	70
4-Methylpyrocatechol	20	69
4-*tert*-Butylpyrocatechol	25	53
4-*tert*-Octylpyrocatechol	56	56
3,4-Di-*tert*-butylpyrocatechol	73	63
4-*tert*-Octylhomopyrocatechol	33	36
Resorcinol	50	62
Hydroquinone	72	5
Pyrogallol	67	4
Gallic acid	1	32
4-*tert*-Octylpyrogallol	43	29
4-*tert*-Octylhydroxybenzoquinone	46	23
Phloroglucinol	54	8
Hydroxyhydroquinone	18	95
α-Naphthol	38	10
β-Naphthol	42	37

[a] From J. Davídek and J. Pokorný [90]; reproduced with the permission of the authors and publishers.

and acetonitrile as the solvents. The spots were eluted and then determined colorimetrically with recoveries ranging from 82 to 101%. Amano et al. [96] used a spot area measurement method for the determination of BHT in edible oils. In this case the oils were subjected to stream distillation and the antioxidant was recovered from the distillate by extraction with carbon tetrachloride. Chromatography was carried out on silica gel plates using hexane-carbon tetrachloride (3:1) as the developing solvent. After the plate was visualized with 10% phosphomolybdic acid in ethanol, the square root of the spot area was found to be proportional to the log of the quantity of antioxidant. Rutkowski et al. [97] have also used a colorimetric method for the determination of propyl gallate and BHA. A number of gallate esters, NDGA, BHA, and BHT were separated by paper chromatography and by thin-layer chromatography.

4 NON-FOOD ANTIOXIDANTS

Slonaker and Sievers [98] and Heide and Wouters [99] have used silica gel for the identification of trace amounts of antioxidants in polyethylene. As solvents, either 4% methanol in cyclohexane or 10% ethyl acetate in petroleum ether can be used. For the extraction of the antioxidants from the polyethylene, 1 kg of granulated polyethylene is extracted at 50°C for 4 hr with 1.5 liters of hexane. The extract is then cooled to 0°C and filtered; the filtrate is concentrated (below 50°C) to 35 ml on a steam bath. Cooling again to 0°C removes additional low-molecular-weight polyethylene. The extract is then concentrated to 5 ml and the antioxidants transferred to ethanol by extracting twice with 5-ml portions. The alcohol solution is then ready to be applied to the thin-layer plate. As detecting agents, either a 3% solution of phosphomolybdic acid in ethanol followed by exposure to ammonia vapor, or a 2% solution of 2,6-dichloroquinone chloroimide in alcohol followed in 15 min by a 2% borax solution can be used. Separations of the antioxidants can be achieved on silica gel layers using 4% methanol in cyclohexane or 10% ethyl acetate in petroleum ether. The R_f values of a group of these agents are listed in Table 32.6.

Dobies [100] used silica gel G with cyclohexane–methanol (150:3) and

Table 32.6 Range of $R_f \times 100$ Values on Silica Gel for Antioxidants Used in Polyethylene[a]

Antioxidant	*Petroleum Ether–Ethyl Acetate (9:1)*
Santonox R [4,4′-Thiobis(6-*tert*-butyl-*m*-cresol)]	19–25
Nonox DPPD (Diphenyl-*p*-phenyldiamine)	24–30
Neozone A or ASM-PAN (*N*-phenyl-α-naphthylamine)	58–65
Agerite or ASM-PBN (*N*-phenyl-β-naphthylamine)	45–50
Stabilizer 2246 [2,2-Methylenebis(4-methyl-6-*tert*-butylphenol)]	60–70
Ionol (2,6-Di-*tert*-butyl-*p*-cresol)	78–85
Agerite white (Di-*N*-β-naphthyl-*p*-phenylenediamine)	17–25 (streak)

[a] From R. F. van der Heide and O. Wouters [99]; reproduced with permission of the authors and J. F. Bergmann.

silica gel G impregnated with 5% silicone with ethanol–water (3:1) for the separation of phenolic type antioxidants in polyethylene and polypropylene films. The compounds were visualized with phosphomolybdic acid and then determined by in situ densitometry. Using a 5-g sample 0.02% could be determined in contrast to the method of Slonaker and Sievers [98], by which 1 ppm could be detected with the larger sample. Heide et al. [101] used thin-layer chromatography to determine the presence of non-permitted antioxidants in plastics used for food packaging. Miles [102] used silica gel–aluminum oxide (1:1) layers for separating antioxidants found in polyolefinic polymers. Chloroform was used for developing the plates. Simpson and Currell [103] used silica gel GF_{254} to separate 21 antioxidants used in plastics. The R_f values are reported for benzene–ethyl acetate–acetone (100:5:2) and chloroform–hexane (2:1).

Thin-layer chromatography has been used for the separation of antioxidants and other additives in lubricating oils [104–108]. Delves [104] found that silica gel layers prepared with 2 *N* sodium hydroxide gave better separations of the nitrogen-containing antioxidants than the neutral layers. The R_f values for seven compounds using benzene–*n*-hexane (2:3) as a solvent are reported. A 20% solution (vol/vol) of antimony pentachloride in carbon tetrachloride could be used to detect 1 μg. Cox [105] used silica gel with a fluorescent indicator to separate dialkyl and diaryl dithiophosphates used as antioxidants in lubricating oils. Carbon tetrachloride–toluene (4:1) was used as a solvent. For detection T-189 was used and after heating at 100°C for 5 min brown spots appeared on a white background. Densitometry was used for quantitative determination. Coates [106] extracted antioxidants from lubricating oils with methanol. Separations were made on silica gel using isopropyl ether–isooctane (3:97), carbon tetrachloride, and ethyl acetate–isooctane (3:97) for the separation of phenolic antioxidants. The latter two solvents were used for the separation of nitrogen-containing antioxidants. Coates also used thin-layer chromatography to analyze the other numerous additives used in lubricating oils. Amos [107] determined the antioxidant 6-*tert*-butyl-2,4-xylenol used in aviation turbine fuel. The compound was first removed with alumina and then eluted and transferred to a fluorescent silica gel where it was chromatographed with carbon tetrachloride (R_f 0.50). The quantity was determined visually or by densitometry. Crump [108] used a two-dimensional technique for separating phenols from kerosine. The phenols were first coupled with diazotized *p*-nitroaniline and then applied to silica gel prepared with 0.5 *N* sodium hydroxide. Development in the first direction was with chloroform–acetone (9:1) and in the second with benzene–di-*n*-propylamine (4:1). Thin-layer chromatography has also been used for the separation and identification of antioxidants, antiozonants, and accelerators in rubber [109–114]. Kreiner [115] has reviewed the literature published from 1958 to mid-1970 on the thin-layer chromatography of rubber-compounding ingredients.

5 EXPLOSIVES

Quite a number of publications have appeared on the use of thin-layer chromatography for the examination of explosives and residues from their combustion. Harthon [116] investigated the separation of hexogen (hexahydro-1,3,5,-trinitro-*s*-triazine) and octogen (octahydro-1,3,5,7-tetranitro-*s*-tetrazine) as well as various compounds associated with the manufacturing process. Fauth and Roecker [117] have extended the work of Harthon to a quantitative densitometric method suitable for quality control. Rao et al. [118] have investigated the analysis of blasting explosives containing nitrate esters of glycol, glycerol, diethylene glycol, and diglycerol. Hansson and Alm [119] have examined diphenylamine occurring as a stabilizer in powders and various other explosive compositions. It stabilizes by reacting with the oxides of nitrogen formed during the slow decomposition of the explosives, thus forming nitro and *N*-nitroso derivatives. These compounds can be separated on silica gel layers which have been activated at 110°C. Samples may be applied as an acetone solution and development is carried out with benzene, chloroform, or toluene. Hansson and Alm used a 0.2% solution of sodium nitrite in alcohol and 1 *N* sulfuric acid for detecting the diphenylamine derivatives. Yasuda [120] has used a two-dimensional method to separate and identify 19 *N*-nitroso- and nitrodiphenylamines. Tetranitro- and pentanitrodiphenylamines could not be separated. In order to reduce the nitro compounds for the reaction with the *p*-diethylaminobenzaldehyde used as the detecting reagent, zinc dust was incorporated directly into the thin layer. For this purpose 3 g of zinc dust was blended with 30 g of silica gel and 65 ml of water. After the layers were spread they were activated at 110°C for 1 to 2 hr. The two-dimensional separation was carried out by developing first in acetone-benzene-petroleum ether (2:99:99) and then in ethyl acetate-petroleum ether (1:4). R_f values for the compounds were plotted graphically. These methods were applied to the identification of the reaction products of ethyl centralite (a stabilizer) with nitrogen tetroxide [121] using 1,2-dichloroethane as the first-dimension solvent and ethyl acetate–petroleum ether (1:3) as the second solvent, the identification of tetryl and related compounds [122], and the identification of the impurities in 1,3,5-triamino-2,4,6-trinitrobenzene [123].

Hansson [124] has chromatographed a number of explosives using benzene, chloroform, and petroleum ether-acetone (5:3) as the developing solvents. These explosives were detected by spraying with diphenylamine and then examining under ultraviolet light. Table 32.7 gives the R_f values of some of these compounds.

Chandler et al. [125] separated all of the major oxidation products from the various steps in the TNT continuous nitration process. Starch-bound silica gel layers were used with benzene–cyclohexane–ethyl acetate (10:9:1) and benzene–ethyl ether–ethanol (5:3:2) (several drops of ammonia were added to the solvents to prevent tailing, caused by the acidity of the nitration mixture). Visualization was by ultraviolet light and T-114. Midkiff and Washington [126, 127] have used, among other techniques,

Table 32.7 $R_f \times 100$ Values[a] of the Components of Various Explosive Materials on Silica Gel G Activated at 110°C [119, 124]

	Solvent		
Substance	*Benzene*	*Chloroform*	*Petroleum Ether–Acetone (5:3)*
Ammonium nitrate	0	0	0
Dipicrylamine	0	0	6
Picric acid	0	0	9
Octogen[b]	4	0	23
Hexogen[c]	5	10	39
DINA[d]	16	42	56
Tetryl[e]	26	46	62
Trinitrobenzene	40	62	71
Penthrit[f]	41	61	74
Trotyl[g]	48	68	73
Diphenylamine	62	86	
N-Nitrosodiphenylamine	46	81	
4-Nitrosodiphenylamine	7	28	
2-Nitrodiphenylamine	58	86	
2,4-Dinitrodiphenylamine	43	81	
2,4′-Dinitrodiphenylamine	37	77	
sym-Hexanitrodiphenylamine	5	5	
Triphenylamine	75	91	

[a] Development distance 10 cm.
[b] Octogen = 1,3,5,7-Tetranitrotetramethylenetetramine.
[c] Hexogen = 1,3,5-Trinitrotrimethylenetriamine.
[d] DINA = Dinitroxydiethylnitramine.
[e] Tetryl = *N*-Methyl-*N*,2,4,6-tetranitroaniline.
[f] Penthrit = Pentaerythritol tetranitrate.
[g] Trotyl = Trinitrotoluene.

thin-layer chromatography for the detection of explosive residues from various explosives. The R_f values are given for 2,4,6-trinitrotoluene, cyclotrimethylenetrinitramine, and the components (nitroglycerin and ethylene glycol dinitrate) of some dynamites with numerous solvents. Hoffsommer and Glover [128] used a capillary decomposition method to obtain products for analysis by thin-layer chromatography in combination with spectrophotometry or by gas chromatography, and Hoffsommer [129] used gas chromatography with the ^{63}Ni electron capture detector in combination with thin-layer chromatography for the quantitative analysis of picogram amounts of nitro compounds. Archer [130] applied thin-layer chromatography to the separation and identification of additives and stabilizers in smokeless powder. Parihar et al. [131] found alumina and magnesium silicate gave better separations than silica gel for diethylene glycol dinitrate, nitroglycerine, ethylene glycol dinitrate, pentaerythritol tetranitrate, mannitol hexanitrate, and sorbitol hexanitrate. Solvents included: xylene, toluene, monochlorobenzene, and petroleum ether–ethylene dichloride (4:1). Parihar and co-workers [132, 133] have used the π-complexes of explosives with amines for separating and identifying the former. This permitted the separation of compounds not separable directly.

6 ORGANOMETALLIC COMPOUNDS

Organotin Stabilizers

The organotin compounds are used in stabilizing polyvinyl chloride which is used in the preparation of plastics. Because of the new regulations on trace contaminants in foods and on the composition of plastics which come in contact with foods, it is important to have a method of detecting and analyzing for the stabilizing compounds. Tuerler and Hoegl [134] chromatographed a group of these compounds on layers of silica gel G incorporating into the layer 0.1 to 0.2 g of disodium ethylenediaminetetraacetate (for 30 g of silica gel G) in order to mask the metal salts in the layer material. Using *n*-butanol-acetic acid (60:1) saturated with water, it was found that the dibutyltin compounds were clearly separated from the tributyltin compounds but that the dibutyl compounds, namely, dibutyltin dilaurate, dichloride, dioleate, and dimaleate, all had the same R_f values. On the other hand, in the same solvent system dibutyltin compounds could be separated from dioctyltin compounds. The dibenzyl compounds had the same R_f values as the dioctyl compounds. Dibutyl-, dioctyl-; and dibenzyltin salts could, however, be separated with water–butanol–ethanol–acetic acid (10:5:5:0.15). Detection of the compounds was accomplished by spraying with a solution of 10 mg of dithizone in 100 ml of chloroform. Diphenylcarbazone could also be used for the detection of the dialkyltin salts but not for the trialkyl compounds. Buerger [135] investigated a series of solvents for the separation of organotin compounds on layers of silica gel. Table 32.8 gives the R_f values for the separations that were achieved. The compounds were located by spraying with 0.1% alcoholic pyrocatechol violet and then examining under an ultraviolet lamp. Neubert [136], Heide [137,

Table 32.8 $R_f \times 100$ Values of Organo-Tin Compounds on Silica Gel in Various Solvents[a,b]

	Solvent[c]					
Compound	*A*	*B*	*C*	*D*	*E*	*F*
Dimethyltin chloride	0	0	0	0	2	0
Diethyltin chloride	0	0	0	0	9	0
Triethyltin chloride	2	7	48	17	—	—
Tetraethyltin	100	100	100	100	—	100
Tripropyltin acetate	9	34	71	29	—	—
Dibutyltin dichloride	0	0	0	0	38	—
Tributyltin chloride	21	45	85	41	—	83
Tetrabutyltin	100	100	100	100	—	100
Dihexyltin dichloride	0	0	0	0	57	—
Trihexyltin chloride	45	54	92	60	—	—
Tetrahexyltin	100	100	100	100	—	100
Di-2-ethylhexyltin dichloride	0	0	0	0	—	—
Dioctyltin dichloride	0	0	0	0	68	—
Tri-2-ethylhexyltin chloride	78	69	100	82	—	—
Tetra-2-ethylhexyltin	100	100	100	100	—	—
Diphenyltin dichloride	0	0	0	0	29	—
Triphenyltin acetate	55	55	84	60	—	—
Tetraphenyltin	100	100	100	100	—	—
Butyltin trichloride	—	—	—	—	—	0
Butylthiostannous acid	0	0	0	0	—	0

[a] From K. Buerger [135]; reproduced by permission of the authors and Springer-Verlag.

[b] Development distance 10 cm.

[c] *A* = Butanol–pyridine (15:7) saturated with water; *B* = butanol–ethanol (3:1) saturated with water; *C* = butanol saturated with 25% ammonia; *D* = upper phase of butanol + 2.5% ammonia; *E* = Isopropanol–(1 vol 1 *N* sodium acetate + 1 vol 1 *N* acetic acid) (2:1); *F* = isopropanol–(2 vol 10% ammonium carbonate + 1 vol 5 *N* ammonia) (2:1).

138], Koch and Figge [139], and Simpson and Currell [103] also published on the analysis of organotin stabilizers.

Ferrocene Derivatives

Schloegl and co-workers [140–148] have prepared and examined a large number of ferrocene derivatives. Thin-layer chromatography on silica gel G was used both in the purification of many of the compounds as well as for their characterization. The less polar compounds such as ferrocene and the alkylferrocenes were chromatographed with hexane, and in some cases with propylene glycol–methanol (1:1) and chlorobenzene–propylene glycol–methanol (1:1:1). The more polar glycols, alcohols, and carbonyl compounds were chromatographed in benzene and benzene–ethanol mixtures in proportions of 15:1 and 30:1. The R_f values for 85 of these compounds have been given graphically [140, 145, 146]. In most cases the colors of the compounds were sufficient for locating the spots; however, with weakly colored components the colors could be intensified by treating with an oxidizing agent such as bromine or 1% sodium periodate.

Miscellaneous Organometallic Compounds

Vobecky et al. [149] chromatographed triphenylarsine, triphenylstilbine, triphenylbismuthine, triphenylphosphine, and di-*o*-methylphenyl telluride on aluminum oxide. Vobecky et al. [150] separated organotellurium compounds on alumina with petroleum ether.

Johnson and Vickers [151] separated organomercury compounds as their chlorides on silica gel with cyclohexane–acetone (4:1) using continuous development. Tatton and Wagstaffe [152] chromatographed the dithizonates of organomercury compounds on silica gel and alumina with several mixtures of hexane–acetone and light petroleum–acetone; the R_f values of seven compounds are given. Takeshita et al. [153] used a reverse-phase system to separate the dithizonates of alkylmercury compounds. Best results were obtained with Avicel SF (Avicel Sales Division, FMC) impregnated with 20% liquid paraffin in combination with methyl Cellosolve–water (3:1) as a solvent. Fishbein [154] reviewed the chromatographic and biological aspects of organomercurials.

Druding and Shupack [155] have chromatographed some platinum and palladium complexes on silica gel with dimethyl sulfoxide containing three drops of reagent chloroform per 100 ml to prevent tailing. The detection agent was 5% aqueous stannous chloride.

7 ORGANIC PHOSPHORUS AND SULFUR COMPOUNDS

Klement and Wild [156] have chromatographed a group of phosphorus compounds including tertiary alkyl phosphates, triphenyl phosphate, amidophosphoric acid esters, thiophosphoric acid esters, and phosphorous acid esters, as well as ammonium salts of dialkyl and diaryl phosphoric acid esters. A group of eight different solvent mixtures (consisting of various mixtures of hexane, benzene, methanol, chloroform, dimethylformamide, ethanol, acetic acid, and methylene chloride) was used for the separations depending on the compounds to be separated. The compounds were detected by an ammonium molybdate–perchloric acid spray reagent. The thiophosphoric esters were detected by means of a 1% silver nitrate solution containing a few drops of concentrated sulfuric acid. Donner and Lohs [157] reported the R_f values for a large number of esters of phosphoric and phosphorous acid in two solvent systems: the hexane–benzene–methanol (2:1:1) solvent of Klement and Wild [156], and hexane–methanol–ether (6:1:1). The separations were achieved on silica gel. As a more sensitive reagent for the detection of these compounds, they used a 1% acetone solution of cobalt chloride (water-free). The spots appeared even in the cold, except for small quantities of esters which appeared on warming to 40 to 50°C. Reuter and Hanke [158] have chromatographed the ethyl esters of phosphoric acid.

Schindlbauer and Mitterhofer [159] reported that phosphines and phosphine sulfides can be separated on silica gel layers with weakly or moderately polar solvents with R_f values increasing in the order > tertiary > secondary > primary phosphines. The positional isomers such as the three tritolylphosphines were separated best on basic alumina with hexane–benzene (5:1). Phosphine oxides were best separated with ether–ethyl acetate. Phosphines could be oxidized to the oxides directly on the layer by spotting with iodine in carbon tetrachloride prior to development. Lamotte and coworkers [160–163] have investigated the TLC separation of phosphorus compounds. With silica gel, *tert*-butanol–acetone–water–ammonia (5:4:1:1) was used for simple phosphates. Neutral organophosphorus compounds were also well separated on silica with 3:1 mixtures of hexane–acetone or hexane–methyl isobutyl ketone. Other solvents included acetone–water–methanol (75:25:10) containing 100 mg of trichloracetic acid, *tert*-butanol–acetonitrile–ammonia (5:4:1), and acetone–*tert*-butanol–water–ammonia (5:4:1:1). Phosphates were separated with hexane–acetone–ethyl acetate (3:1:1), phosphonates with an 8:3:9 mixture, and phosphinates with *tert*-butanol–hexane (1:3) [162]. These compounds can be detected with ammonium molybdate–perchloric acid (T-15). Other workers separating organophosphorus compounds on thin layers include: Bloom [164], Ivanova et al. [165], Petschik and Steger [166], de Licastro and Rúveda [167], and Neubert [168].

Using solvent mixtures of hexane–acetone (10:1 and 4:1), Mastryukova et al. [169] chromatographed the esters and amides of thiophosphoric acids as well as the pyrophosphoric analogs on thin layers of alumina and on silica gel layers containing 6% water. Location of the spots for the 24 compounds was obtained with potassium permanganate or iodine vapor. Petschik and Steger [170] separated aliphatic thiophosphoric acid esters on thin layers of aluminum oxide using a starch binder with *n*-heptane–acetone (10:1) as a solvent. They used a visualizing agent composed of 10% paraperiodic acid in 70% perchloric acid and containing a few milligrams of vanadium pentoxide per 100 ml. The same reagent was found to be more sensitive for sulfur- and selenium-containing organic and inorganic compounds than the iodine–sodium azide reagent [171]. Ertel and Horner [172] chromatographed some phosphinoxides along with phenylbenzyl sulfide and its oxidation products on silica gel in several solvents. Dichromate–sulfuric acid and permanganate–sulfuric acid reagents were used to detect these compounds. (**Caution:** the latter should not be mixed in large quantities because manganous heptoxide is explosive.) Stephan and Erdman [173] recommend the following method for detecting divalent sulphur compounds such as *dl*-methionine, aliphatic and aromatic thiols, sulfides, and thioketones. The plate is first sprayed with 0.1% sodium metaperiodate, followed after 4 min with 0.5% benzidine in butanol–acetic acid (4:1). This procedure gives white spots on a dark blue background of the oxidized benzidine. The sensitivity of the reaction with methionine is 5 to 10 γ and with aromatic sulphur compounds, 20 to 30 γ.

Curtis and Phillips [174] have used silica gel G and alumina G for chromatographing 26 thiophene derivatives. Nonpolar thiophenes were separated on alumina with petroleum ether (40 to 60°C) as the developing agent, and moderately polar thiophenes were separated on silica gel with benzene–chloroform (9:1). The very polar thiophenes, such as those containing a carboxylic group, were separated on silica gel with methanol as the solvent. The compounds were visualized by examination under ultraviolet light and by spraying with a 0.4% solution of isatin in sulfuric acid. With this latter reagent the colors were observed both at room temperature and after heating at 120°C. Mayer et al. [175] chromatographed 53 trithiones and 16 1,2-thiazoline-5-thione compounds on silica gel G layers. The nonpolar trithiones were chromatographed with a mixture of petroleum ether–benzene (1:1) or with carbon disulfide. More polar trithiones and the thiazolinethiones were chromatographed with benzene–ethyl acetate (3:1); pure acetone was used for very polar compounds containing carboxyl or hydroxyl groups. A combination of thin-layer chromatography and absorption spectra were used as identifying characteristics for these compounds. The compounds were made visible on the plates with tetracyanoethylene.

Fishbein and Fawkes [176] separates sulfides, sulfones, and sulfoxides on silica gel DF-5 with toluene–ethyl acetate (1:1) and 2.5% acetone in benzene. A number of detecting agents were examined; T-252 and T-80 were the best of the five tested for helping to detect and differentiate these compounds. These compounds can also be separated on alumina [177–182]. Karaulova et al. [177] used alumina layers without a binder and acetone–carbon tetrachloride (1:4) as a solvent for dialkyl, alkyl aryl, diaryl, heterocyclic, and alicyclic sulfoxides. Iodine vapor as a detection agent gave a sensitivity of 3.5 to 10 μg. Snegotskii and Snegotskaya [178] using alumina (activity grade II without binder) separated thiols, sulfides, and disulfides with hexane. Ethyl ether was used for sulfoxides and sulfones. The 2-(alkylthio)ethanols were chromatographed with hexane–ether (1:2) and their sulfoxides and sulfones with hexane–acetone (1:1). Ether was used for chlorosulfolans and hexane–ether (1:3) for alkyl vinyl sulfones. Using alumina (loose layer, activity III or IV) Novitskaya et al. [179] give the R_f values of some organic sulfides, sulfoxides, and sulfones with seven solvents. Prinzler and coworkers (180) have used 96% acetic acid–acetonitrile (1:3) for alkyl sulfides and methanol–chloroform–water (5:15:1) for aryl sulfides. These workers have also used two-dimensional thin-layer chromatography for separating sulfoxides [181] and for mercaptans and thiophenols;[182] on impregnated layers. Hiley and Cameron [183] separated polysulfides on silica gel with carbon disulfide.

Thin-layer chromatography on silica gel layers has also been used for the isolation and purification of thio sugars using ethyl acetate–acetone (4:1) as the developing solvent [184].

8 PEROXIDES, EPOXY COMPOUNDS, AND OZONIDES

Thin-layer chromatography has been used in lipid chemistry to separate the epoxy fatty acids [185–187]; it has also been used with various other lipid oxidation products [188–191]. Silicic acid can be used as an adsorbent in these cases with 3 to 10% diethyl ether in hexane or petroleum ether as the solvent, the concentration depending on the components present. As a detecting agent the chromatograms may be sprayed with 50% sulfuric acid and subsequently heated for 15 min at 105 to 110°C. Table 32.9 gives the R_f values of some epoxy esters and acids. Kaufmann and Makus [185] have separated some epoxy acids on silica gel G im-

Table 32.9 R_f × 100 Values of Some Epoxy Fatty Esters, Acids, and Alcohols on Silica Gel

Compound	*30% Ether in Petroleum Ether* (40 to 60°C)[a] [191]	*10% Ether in Petroleum Ether* (40 to 60°C) + 1% *Acetic Acid*[a,b] [191]	*30% Ether in Petroleum Ether* (40 to 60°C) + 1% *Acetic Acid*[a,b] [191]	*5% Ether in Petroleum Ether* (35 to 45°C) [187]
Methyl *cis*-9,10 epoxystearate	86			
cis-9,10-Epoxystearyl alcohol	32			
cis-9,10-Epoxystearic acid			61	
cis-13,14-Epoxydocosanoic acid		56	69	
trans-13,14-Epoxydocosanoic acid			79	
Methyl 9,10-epoxystearate				43
Methyl 12,13-epoxyoleate				51

[a] Development distance 15 cm.
[b] Added to prevent streaking.

pregnated with a 15% solution of undecane in petroleum ether. The developing solvent in this case was 96% acetic acid.

The separation of ozonides of various lipids has been carried out by Privett and Nickell [192] and Privett and Blank [193]. Separations were achieved on silica gel G using various ratios of diethyl ether in petroleum ether, the ratios ranging all the way from 0.6 to 25% concentration. The ozonides can be separated into classes based on the number of ozonide groups in the molecule. Triglyceride ozonides that differ by only one ozonide group can be separated by this procedure, and cis–trans ozonides can be separated from one another [194].

If the ozonides are to be recovered or used for quantitative analysis, then the silica gel plates must be washed thoroughly with diethyl ether prior to the chromatographic run in order to remove organic contaminants. Quantitative analysis of these materials is made by spraying with chromic–sulfuric acid and heating the plates at 180°C for 20 min in order to char the spots for densitometry [195].

Neuwald and Fetting [189] separated cholesterol peroxides by chromatographing samples on silica gel with chloroform. Peroxides were visible as blue spots with an R_f of < 0.32 when the plates were sprayed with 5 ml of 5% potassium iodide in 20 ml of acetic acid and then after 5 min with a starch solution. Cholesterol itself appeared as a red spot at R_f 0.32 when treated with 50% antimony trichloride in acetic acid followed by heating at 100°C for 10 min.

Knappe and Peteri [196] have separated a series of 14 organic peroxides on silica gel (Table 32.10). The ketone peroxides exhibited more than one component. For detection the chromatograms were sprayed with reagent T-94 or with freshly prepared reagent T-127.

Kavčič and Plesničar [197] compared seven polyamide and cellulose adsorbents for separating mono- and disubstituted peroxybenzoic acids

Table 32.10 $R_f \times 100$ Values of Various Organic Peroxides on Silica Gel G[a]

Peroxide	*Toluene–Carbon Tetrachloride* (2:1)	*Toluene–Acetic Acid* (19:1)	*Petroleum Ether* (50 to 70°C)–*Ethyl Acetate* (49:1)
Lauroyl peroxide	85	95	
2,4-Dichlorobenzoyl peroxide	81	88	
4-Chlorobenzoyl peroxide	74	94	
Benzoyl peroxide	55	70	
tert-Butyl peroctoate	28	55	
Methyl isobutyl ketone peroxide			
Component A	25	55	
Component B	00	12	
tert-Butyl perbenzoate	24	47	
Cyclohexanone peroxide			
Component A	21	38	
Component B	00	12	
Component C	00	10	
Methyl ethyl ketone peroxide			
Component A	16	42	
Component B	10	10	
tert-Butylperacetate	12	32	18
Cumene hydroperoxide	11	33	9
2,2-Bis(*tert*-butylperoxy)butane	10	35	
tert-Butyl hydroperoxide	5	30	
Di-*tert*-butyl peroxide	00	39	
Hydrogen peroxide	00	00	00

[a] From E. Knappe and D. Peteri [196]; reproduced with the permission of the authors and Springer-Verlag.

using carbon tetrachloride–acetic acid (10:1) and chlorobenzene–acetic acid (10:1) as solvents. Silica gel G caused partial or complete decomposition of these compounds.

Sorokina et al. [198] chromatographed 32 organic peroxides on alumina with hexane–ether (3:2), acetone–toluene–heptane (1:1:2), acetone–tetrachloromethane (1:2), and tetrachloromethane. Buzlanova et al. [199] obtained the best separation of dialkyl and diacyl peroxides with heptane–ethyl ether (15:1) and of hydroperoxides with toluene–methanol (20:3). Both alumina and silica gel were used. Some peroxides partially decomposed on alumina.

See Chapter XXX, Section 6 for some terpene peroxides.

9 PLANT HORMONES

Gibberellins

Mandava [200] reported that as of 1973, 38 gibberellins had been characterized. A number of workers have investigated their separation on thin-layer plates. Kutáček et al. [201] used loose layers of alumina, developing in a continuous manner with a mixture of (thiophene-free) benzene–acetic acid (100:23). With the passage of 60 ml of solvent, which took about 6 hr, gibberellin A_1 could be separated from gibberellin A_3. Much better and faster separations can be achieved on silica gel or on kieselguhr. Sembdner et al. [202] used silica gel layers with varying proportions of chloroform–ethyl acetate–acetic acid as well as mixtures of *n*-butanol or *n*-propanol with 3 *N* ammonia. MacMillan and Suter [203] used both silica gel and kieselguhr layers with various solvents (Table 32.11). The benzene–propionic acid–water solvent was the slowest system and required 70 min for a run of 15 cm compared to only 25 min for the diisopropyl ether–acetic acid mixture. With the benzene–acid solvent systems, the plates were equilibrated overnight with the lower phase and then developed with the upper phase. Except for A_4 and A_7, the methyl esters were also separated and could be resolved in two solvent systems on silica gel G (Table 32.12). Ikekawa et al. [204] and Kagawa et al. [205] separated the gibberellins with a group of five solvents on both silica gel and kieselguhr layers (Table 32.11). In using the carbon tetrachloride solvents, these authors equilibrated the plates overnight with the upper phase and then developed with the lower phase or with the lower phase with added ethyl acetate as required. In addition to separating the methyl esters on thin-layer plates they applied gas chromatography to the latter. All the gibberellins can be separated by using a minimum of two systems from Table 32.11, and all the methyl esters can be separated from one

Table 32.11 $R_f \times 100$ Values of Gibberellins in Various Systems[a,b]

	Silica Gel [204, 205]			*Kieselguhr* [204, 205]			*Silica Gel* [203]		*Kieselguhr* [203]	
Gibberellin	*A*	*B*	*C*	*D*	*C*	*E*	*F*	*G*	*H*	*G*
A_1	20	49	0	0	28	49	11	0	54	26
A_2	17	40	0	0	23	37	4	0	64	30
A_3	19	54	0	0	18	40	11	0	42	18
A_4	63	95	67	67	90	100	37	82	100	100
A_5	53	87	27	45	85	90	31	35	100	88
A_6	59	87	11	33	86	84	25	21	95	76
A_7	60	90	57	45	85	91	37	70	100	100
A_8	4	30	0	0	6	10	4	0	28	6
A_9	87	95	100	100	100	100	75	100	100	100

[a] Development distance 15 cm.

[b] Solvent: *A* = benzene–*n*-butanol–acetic acid (16:3:1); *B* = benzene–*n*-butanol–acetic acid (14:5:1); *C* = carbon tetrachloride–acetic acid–water (8:3:5) lower phase; *D* = lower phase of *C* + 10% ethyl acetate; *E* = lower phase of *C* + 20% ethyl acetate; *F* = diisopropyl ether–acetic acid (95:5); *G* = benzene–acetic acid–water (8:3:5) upper phase (see text); *H* = benzene–propionic acid–water (8:3:5) upper phase (see text).

another (Table 32.12), except for A_4 and A_7. The compounds can be visualized by spraying with a water–concentrated sulfuric acid (3:7) solution which is very sensitive, ranging from 0.00025 μg of gibberellin A_3 to 0.01 μg for gibberellin A_6 [203]. After the plates are sprayed with the reagent, they are heated for 10 min at 120°C and then examined under

Table 32.12 $R_f \times 100$ of the Methyl Esters of Gibberellins in Various Systems on Silica Gel

	Methyl Ester of									
Solvent	A_1	A_2	A_3	A_4	A_5	A_6	A_7 A_8	A_9	*Ref.*	
Benzene–acetic acid–water (8:3:5) upper phase (see text)	33	44	26	100	100	100	100	10	100	203
Diisopropyl ether–acetic acid (98:2)	18	8	16	50	38	31	48	8	80	203
Ethyl ether–benzene (4:1)	31	23	35	73	60	66	71	17	98	204
Ethyl ether–petroleum ether (4:1)	29	13	32	75	69	67	72	12	96	204

[a] Development distance 15 cm.

ultraviolet light. Antimony trichloride solution in chloroform may also be used, but it is less sensitive than the sulfuric acid spray.

Cavell et al. [206] separated 17 gibberellins by thin-layer chromatography using ethyl acetate–chloroform–acetic acid with silica gel. Musgrave [207] used a dual band composed of 5.5 cm of kieselguhr G and 10.5 cm of silica gel G with benzene–acetic acid–propionic acid–water (8:2:1:5) to separate gibberellins A_1 through A_9 plus A_{13}. The plate with samples applied was equilibrated with the vapor from the lower phase overnight in an unlined tank prior to chromatographing in the upper phase.

Indoleacetic Acid and Other Factors

Stahl and Kaldewey [208] and Ballin [209] have chromatographed auxin (indoleacetic acid) as well as other related compounds on silica gel layers (see Table 16.9). Kaldewey and Stahl [210] applied a modification of the well-known *Avena* test for the quantitative evaluation of auxins isolated by thin-layer chromatography.

Kaldewey [211] also used the bioassay method on thin-layer-separated compounds from the flower stalks of *Fritillaria meleagris*. The flower stalks contained only indoleacetic acid, two or three precursors, and two or three inhibitors. Dubouchet and Pilet [212] have found a synergistic effect of silica gel on the growth effect of indoleacetic acid with *Triticum coleoptiles*, and Collet [213] confirming this has also shown an effect with calcium sulfate.

Railton [214] used thin layers of polyamide to clean up plant extracts. With methyl ethyl ketone–methanol (99:1) indole-3-acetic acid migrated with the solvent front, and fluorescing phenolic substances were located between R_f 0 and 0.6. Traces of pigments and phenolics that migrated with the solvent front were removed by rechromatographing in the reverse direction with benzene–hexane (1:1). It is important that these phenolic compounds be removed before running a bioassay. Raj and Hutzinger [215] have chromatographed eight indole acids on cellulose layers in five solvents; benzene–acetic acid–water (8:3:5) was one of the better solvents.

10 PLASTICIZERS

Plasticizers are used in the formulation of the many plastics used today in industry. Not all plasticizers are suitable for incorporation into packaging material for food materials, because of the toxic nature of the compounds. It is necessary therefore to have available a method for detecting unsuitable plasticizers. Silica gel provides a satisfactory thin layer for their separation. Peereboom [216] incorporated 0.005% of the

Table 32.13 R_f and R_{st} × 100 Values of Plasticizers on Silica Gel G

			R_{st}[b] [216]		
Solvent	R_f[a] [218] *Methylene Chloride*	R_f[a] [217] *Ethyl Acetate–Benzene; for Phthalates* (5:95), *Citrates* (1:19)	*Ethyl Acetate–Isooctane* (1:9)	*Ethyl Acetate–Benzene; for Phthalates* (5:95), *Citrates* (1:19)	*Diethyl Ether–Hexane* (4:1)
Citric acid esters					
Acetyl tributyl citrate	32	62	53	85	70
Acetyl triethyl citrate	19	29	26	51	29
Acetyl tri-2-ethyl hexyl citrate	52	90			
		96			
Tributyl citrate	20	35			
Triethyl citrate	15	17			
Adipic acid esters					
Benzyl octyl adipate	50				
Dinonyl adipate	50				
Dioctyl adipate	49				
Diisobutyl adipate			83	86	85
2-Ethylhexyl adipate	50				
Polyester adipate	0				
Phosphoric acid esters					
Diphenylcresyl phosphate	43				
Diphenyloctyl phosphate	38				
2-Ethylhexyl diphenyl phosphate			46	77	58
Trichlorethyl phosphate	13				
Tricresyl phosphate	49		42	86	69
Trioctyl phosphate	24				
Triphenyl phosphate	42		33	80	50
Phthalic acid esters					
Benzylbutyl phthalate	53				

Dibutyl phthalate	52		74	103	84
Dicyclohexyl phthalate	53	73			
Didecyl phthalate	63				
Diethyl phthalate		58	51	79	60
Di-(2-ethylhexyl) phthalate	58	89	114	116	115
Dihexyl phthalate	57				
Diisobutyl phthalate		74			
Diisodecyl phthalate	57				
Diisononyl phthalate	56				
Dimethoxy ethyl phthalate	10	19			
Dimethyl phthalate	38	48			
Dimethylcyclohexyl phthalate	48				
Dinonyl phthalate	60		101	118	114
Dioctyl phthalate	59	88			
Sebacic acid esters					
Dibutyl sebacate	35		100	100	100
Dioctyl sebacate	47				
Ethylhexyl sebacate	47				
Sebacic acid polyester	0				
Miscellaneous					
Methyl acetoricinoleate	33				
Butyl acetylricinoleate	40				
N-Butylbenzenesulfonamide	38				
Butyl phthalyl butyl glycolate			43	90	65
Butyl stearate			161	123	128
Di-2-ethylhexyl thiobutyrate	52				
Ethyl phthalyl ethyl glycolate			22	66	30
2-Ethylhexyl *p*-hydroxybenzoate	16				
Triacetin	18		18	34	17

[a] Development distance 10 cm.

[b] R_{st} = R_f value referred to dibutyl sebacate. Values obtained on silica gel containing 0.005% of the fluorescent indicator, Ultraphor (Badische Anilin und Soda Fabrik).

fluorescent indicator Ultraphor, because all the plasticizers he investigated either fluoresced or appeared as dark spots on the fluorescent background under ultraviolet light. Using three solvents (Table 32.13), all the compounds could be separated except for three (critical) pairs as follows: tricresyl phosphate and butyl phthalyl butyl glycolate, tricresyl phosphate and 2-ethylhexyl diphenyl phosphate, and 2-ethylhexyl diphenyl phosphate and acetyl tributyl citrate. These pairs, however, could be differentiated by means of one or more of the nine color reagents which were listed. Paraflex G2 (expoxidized natural glycerides), which is not listed in the R_f values in Table 32.13, gave numerous spots in all three solvents.

Jaminet [217] separated citrate and phthalate esters used as plasticizers. With silica gel G the citrates were separated with 5% ethyl acetate in benzene. For the phthalate esters, three systems were used: petroleum ether (40 to 60°C)-ethyl acetate (9:1), isooctane-ethyl acetate (9:1), and benzene–ethyl acetate (19:1). Acetylated citrates could be detected by spraying with 2 N alcoholic potassium hydroxide and heating to 80°C. Unacetylated citrates were first acetylated on the plate by spraying with a mixture of 5 ml of acetic anhydride, 0.5 ml concentrated phosphoric acid, and 5 ml of dioxane. The acetylation was carried out by heating for 30 min at 100°C. Detection could then be carried out by using the potassium hydroxide spray. The phthalate esters were located by spraying with a mixture of 4 N sulfuric acid–20% resorcinol in alcohol (1:1) and heating the sprayed plates in an oven at 120°C for 10 min. The brown spots which appeared could be converted to orange by exposing to ammonia vapor.

Braun [218] chromatographed a large number of plasticizers using methylene chloride as a solvent (Table 32.13).

Samples of the plasticizers can be applied to the thin layer in ether or benzene solutions. Braum extracted plasticizers from thin plastic sheets by means of methylene chloride, and Jaminet macerated 1 g of the plastic in 25 ml of ether for 10 to 15 hr.

Bloom [219] examined phthalic esters by means of thin-layer and gas chromatography. Twenty-seven compounds were chromatographed on silica gel F with isooctane–isoamyl acetate (85:15); R_f values for 10 other solvents are reported for some of the compounds. Groebel [220] examined the plasticizers and stabilizers in poly(vinylidene chloride) copolymer film used for packaging foods. Aliphatic plasticizers (tributyl citrate acetate and dibutyl sebacate) were separated on silica gel with dichloromethane. Haase [221] reported the R_f values for 30 plasticizers using the same solvent. Kreiner [222] chromatographed 14 plasticizers and stabilizers used in epoxy materials. The R_f values on silica gel G are reported with 1,1,1-trichloroethane–dichloromethane (3:1) and 1,1,1-trichloroethane–dichloromethane–methyl ethyl ketone (75:25:2) as solvents. Only 1,1,1-trichloroethane inhibited with 1,4-dioxane was used in the solvent mixtures.

Braun and Geenen [223] used thin-layer chromatography for identifying the acids which are present in the esters of the plasticizers (for the dibasic acids in this work see Chapter XIII, Section 4). In addition, the following R_f values were obtained on silica gel G with 96% ethanol–water–25% ammonium hydroxide (100:12:16): phthalic acid 0.26, terephthalic acid 0.73, benzoic acid 0.76, *p*-toluic acid 0.76, and phosphoric acid 0.0.

11 SURFACE-ACTIVE AGENTS

Obruba [224] has used thin-layer chromatography on silica gel for the determination of free polyethylene glycols in nonionic adducts of ethylene oxide. Three solvent systems were used: ethanol–methanol–ammonium hydroxide (12:3:2) and (12:4:2), and ethanol–methanol–water (12:4:2). The spots were detected with Dragendorff reagent, with the exception of ethylene glycol, which was detected with a silver nitrate spray. Thoma et al. [225] have used a combination two-dimensional and continuous method for the separation of surface-active esters and ethers of polyethylene glycol. Ascending chromatography in the first direction with *n*-butanol–ethanol–25% ammonia (14:3:5) was followed by continuous chromatography in a BN-chamber (Desaga) in the second dimension. The solvent for the second dimension was either water-saturated methyl ethyl ketone or chloroform–methanol–water (3:25:12). As an example of the separating characteristics of this method, a sample of polyethylene glycol-900 stearate was separated into 17 individual spots by using the *n*-butanol mixture in the second direction. In the separation of mixtures of different polyethylene glycol stearates using the chloroform mixture as a solvent for the second direction, polyethylene glycol stearates of the designations 400, 900, 2000, and 4700 were separated. All these separations were carried out on silica gel layers for a distance of 15 cm for the ascending chromatography, and for a period of 3 hr for the continuous separation. For detection of the compounds, a modified Dragendorff reagent was used as well as a 0.005 *N* iodine solution. In the latter case, with higher concentrations of the polyethylene glycol derivatives, a second spray of 0.2% starch solution yielded violet to brown spots. Free fatty acids were visualized by spraying with 0.2% Rhodamine B solution in ethanol followed by 10 *N* potassium hydroxide in 50% methanol. The acids appeared as dark red spots which showed a yellow fluorescence under ultraviolet light at 366 mμ [226].

Seher [227] also used silica gel layers for the analysis of nonionic surface-active agents. Various polyglycerols could be separated from each other and from glycerin using a solvent mixture of ethyl acetate–isopropanol–water (65:22.7:12.3) in a saturated chamber. For detecting the compounds, the chromatograms were sprayed with reagent T-242. A second detecting agent that was used was an ammoniacal silver nitrate solution which gave brown flecks on a bright background after heating the plate at 100°C for 10 to 20 min. Using these procedures, 1% of diglycerin could be detected in the presence of 99% glycerin. Seher and Janssen [228] analyzed a commercial mixture of glycerol monoesters of diacetoxysuccinic acid on silica gel G with petroleum ether (50 to 70°C)–ethyl ether–acetic acid (60:40:1); molybdophosphoric acid was used as a visualizing agent. Koenig [229] separated nonionic detergents from 16 commercial products consisting of fatty acid esters of polyvalent alcohols, fatty acid alkanolamides, and ethylene oxide adducts on cellulose with heptane–acetic acid–butanol (65:20:15); on alumina with ethyl acetate–pyridine (2:3); and on silica gel with chloroform–methanol (9:1 or 4:1). Stancher et al. [230] combined thin-layer gas chromatography to analyze polyoxyethylene nonionic surfactants.

Mangold and Kammereck [231] have discussed the separation, on silica gel G layers containing 10% ammonium sulfate, of some surface-active agents prepared from aliphatic lipids. Mixtures of *N*-acylated sarcosine, oleic acid ester of hydroxysulfonic acid, and *N*-acylated short-chain amino acids were separated with a mixture of 3% acidic methanol (containing 5% 0.1 *N* sulfuric acid) in chloroform. Alkyl sulfates, sulfonates, phosphates, and phosphonates were also separated on the same layers using a 20% concentration of the acidic methanol in chloroform. Hofmann [232] has used a solvent mixture of isoamyl acetate–propionic acid–*n*–propanol–water (4:3:2:1) for the separation of some anionic detergents such as sodium oleyl taurate, sodium lauryl sulfate, etc. Alkyl sulfates could be separated from alkyl sulfonates on hydroxyapatite by using a solvent system of *n*–butyl ether–methanol–acetic acid (5:5:1). Takagi and Fukuzumi [233] have also separated synthetic surfactants on silica gel plates.

Desmond and Borden [234] have chromatographed a group of surface-active agents on thin layers of aluminum oxide using isopropanol as the developing solvent. These included alkylarylsulfonates, soaps, xylenesulfonates, toluenesulfonates, sulfated alcohol ethoxylates, sulfated alkylphenolethoxylates, amine oxides, alkanolamides, and ethoxylates. For visualizing the spots the chromatograms were sprayed with a 0.05% solution of pinacryptol yellow solution in ethanol. Examination under ultraviolet light revealed variously colored spots, except for the alkanolamides which do not fluoresce under ultraviolet light. These were detected by exposing to iodine vapor. Allen and Martin [235] separated alkene and hydroxyalkane sulfonates on ammonium sulfate-impregnated silica gel with chloroform–methanol–0.1 *N* sulfuric acid (35:16:3). They were visualized by charring using sulfur trioxide and then quantitated by densitometry. Hordyńska and Legatowa [236] determined anionic and cationic detergent residues by separation on silica gel with ethyl acetate–methanol–ammonia (12:3:1). Takeshita et al. [237] used polyamide bound with cellulose for chromatographing 16 sodium alkanesulfonates and eight sodium alkylbenzenesulfonates. The solvent systems consisted of 0 to 1.0 *N* ammonia–pyridine (15:1) and 0.1 *N* ammonia–pyridine–methanol (15:1:0 to 15) (various ammonia normalities and methanol concentrations were used).

Groves and Mustafa [238] observed that mixtures of surfactants sometimes interacted so that additional spots appeared on chromatograms.

12 SYNTHETIC SWEETENERS

Waldi [239] chromatographed saccharin and dulcin on layers of silica gel G using chloroform–acetic acid (9:1) as the developing solvent. A preliminary separation was made by extracting an acidified aqueous solution with ethyl acetate which removed the saccharin. Dulcin was also extracted with ethyl acetate after making the solution alkaline. The compounds were detected on the chromatograms by spraying with a 0.5% ethanolic solution of Rhodamine B followed by ammoniacal silver nitrate. Salo and co-workers [108, 109] chromatographed saccharin, dulcin, and cyclamate on layers prepared from a mixture of 60% acetylated cellulose (MN cellulose powder 300 Ac) and 40% polyamide (Woelm). The saccharin and dulcin were spotted in 0.1% methanol solution, and the cyclamate in 0.1% water–methanol solution (1:1). Development was carried out with a mixture of Shell Sol A–*n*-propanol–acetic acid–formic acid (45:6:7:2) giving the following R_f values: saccharin 0.47, dulcin 0.66, and cyclamate 0.28. The compounds were visualized by spraying with a 0.2% ethanolic dichlorofluorescein solution with observation under 254-mμ ultraviolet light.

Korbelak and Bartlett [242] separated calcium cyclamate, sodium saccharin, dulcin, and P4000 (5-nitro-2-*n*-propoxyaniline) obtaining R_f values of 0.30, 0.41, 0.75, and 0.84, respectively. Silica gel was used with butanol–95% ethanol–28% ammonia–water (40:4:1:9). Under 254-mμ ultraviolet, saccharin appeared as a blue fluorescent spot and P4000 as a dark spot. A heavy spray of 1% chloranil in benzene followed by heating at 100°C for 5 min showed these two as white spots on a lavender background.

Following a spray of 1% *p*-dimethylaminobenzaldehyde in 10% hydrochloric acid dulcin was revealed as bright yellow and P4000 remained lavender. Sensitivities were: 5, 2, 1, and 1 μg, respectively. Woidich et al. [243] separated dulcin, suosan [*N*-(*p*-nitrophenylcarbamoyl)-β-alanine], *p*-methoxy-*o*-benzoylbenzoic acid, and Ultra-Sweet (5-nitro-2-*n*-propoxyaniline) by a combination extraction–thin-layer chromatography method.

Dihydrochalcone sweeteners have been chromatographed on polyamide sheets K541 V (Eastman) with nitromethane–methanol (3:2) [244].

13 ULTRAVIOLET ABSORBERS

Because of the widespread use of ultraviolet absorbers in the plastics and varnish industries, there is a need for an analytical method to isolate and identify these compounds. The 2-hydroxybenzophenone compounds are extensively used for this purpose; Knappe et al. [245] have examined the behavior of these and several other ultraviolet absorbers with thin-layer chromatography. These separations were carried out on silica gel G, kieselguhr G, aluminum oxide G, and cellulose powder G impregnated with 80 to 82% adipic acid triethylene glycol polyester in methylglycol (Glasurit-Werke, Polyester IK 123). The solvent was a mixture of *m*-xylene-formic acid (98:2) saturated with the polyester. The R_f values for eight 2-hydroxy-benzophenone compounds, phenylsalicylate, *p*-*tert*-butylphenylsalicylate, and 2,4-dibenzoylresorcinol were tabulated together with the colors observed under ultraviolet light and also by spraying with Fast Red AL. The separation was mainly a partition separation, for the support appeared to have only a minor influence.

Dobies [246] worked out a procedure for extracting ultraviolet absorbers from paraffin wax (used in coating papers) and then separated a group of four of the absorbers on silica gel impregnated with 5% silicone. A number of methanol–water and ethanol–water mixtures were used as solvents varying from (3:1 to 9:1).

14 MISCELLANEOUS QUINONES

Barbier [247] obtained excellent separations of *p*-benzoquinones isolated from natural sources on starch-bound silicic acid layers using hexane-ethyl acetate (17:3) as the developing solvent. Since plates dried at 105°C adsorbed the quinones too strongly, they were allowed to stand in the air for 48 hr before use. Pettersson [248] separated benzoquinones on silica gel G layers and then obtained quantitative measurements by eluting and measuring at 270 mμ. The recoveries were in the range of 95 to 100%.

Two-dimensional separations were also employed in the work. Pettersson [249] has isolated some toluquinones from *Aspergillus fumigatus*.

Grau and Endres [250] chromatographed a group of quinones on an acetylated polyamide, because some of these compounds are irreversibly adsorbed on polyamide. Methanol-water (1:1) and acetone-water (3:1) were used as developing solvents.

15 DIVERSE COMPOUNDS

Nealey [251] separated a group of polyphenyl ethers on silica gel G by using a multiple development with benzene-cyclohexane (5:95). Two developments were required to give a clean separation and four of course gave an even further separation. In this case prior to each redevelopment, it was necessary to reactivate the plates at 110°C for $\frac{1}{2}$ hr. This could be done without danger of decomposition since the polyphenyl ethers used are base fluids for high-temperature lubricants and are extremely thermal stable. Location of the spots was by means of iodine vapor.

Nine pairs of cycloalkane and threoerythro isomers including alcohols, methyl esters, acids, and nitrile were separated by using the continuous flow method of Brenner and Niederwieser [83] on silica gel and aluminum oxide plates [252].

The methanol–chloroform (1:2) extract of the larval foods of bees has been examined on thin-layer plates by Patel et al. [253]. It is interesting to note that the constituents of the royal jelly fed to the queen larvae remain fairly constant, whereas the food fed to the drones and workers varies with the age of the larvae and is different from that of the royal jelly.

Gehrmann and Schneider [254] checked a number of reagents which are used for the photometric determinations of various substances. Of the four compounds examined, only one was shown to consist of a single component. As an example, a sample of dithizone gave three spots on a silica gel plate, each of which gave a different color with a basic Pb^{2+} test solution. It was suggested that reagents for photometric analysis should be checked for purity in order to avoid erroneous absorption results.

Korte and Vogel [255] chromatographed a group of lactones, lactams, and thiolactones on layers of silica gel G. The R_f values are tabulated for the solvents used. These were: isopropyl ether, isopropyl ether-ethyl acetate (4:1 and 1:4), and isopropyl ether-octane (3:2). The lactams could be detected by first converting to the hydroxamic acids by spraying with 12.5% sodium hydroxide in methanol and 5% hydroxylamine hydrochloride in methanol.

Wusteman et al. [256] chromatographed a group of alkyl, aryl, and steroid sulfuric acid esters on silica gel layers. The ester sulfates of weakly polar compounds could be resolved in general using a solvent mixture of benzene–methyl ethyl ketone–ethanol–water (3:3:3:1), while the more polar compounds required a more polar solvent such as 1-butanol-acetic acid-water (3:1:1). The R_f values for a group of representative compounds were tabulated.

References

1. H. Meyer and L. Leistner, *Fleischwirtschaft,* **50,** 1414 (1970).
2. L. Fishbein and H. L. Falk, *Chromatogr. Rev.,* **12,** 42 (1970).
3. L. Stoloff, *Clin. Toxicol.,* **5,** 465 (1972).
4. P. L. Schuller, W. Horwitz, and L. Stoloff, *J. Assoc. Off. Anal. Chem.,* **59,** 1315 (1976).
5. W. Vanwalbeck, P. M. Scott, and F. S. Thatcher, *Can. J. Microbiol.,* **14,** 131 (1968).
6. G. Krug and B. Teichmann, *Nahrung,* **19,** 255 (1975).
7. W. C. Jacobson, W. C. Harmeyer, and H. G. Wiseman, *J. Dairy Sci.,* **54,** 21 (1971).
8. M. S. Masri, *J. Am. Oil Chem. Soc.,* **47,** 61 (1970).
9. I. F. H. Purchase and M. Steyn, *Brit. J. Cancer,* **23,** 800 (1969).
10. K. Ranfft, *Z. Lebensm.-Untersuch.-Forsch.,* **150,** 130 (1972).
11. K. Lemieszek-Chodorowska, *Rocz. Panstw. Zakl. Hig.,* **18,** 563 (1967).
12. A. F. Cucullu, L. S. Lee, W. A. Pons, Jr., and L. A. Goldblatt, *J. Am. Oil Chem. Soc.,* **47,** 226 (1970).
13. J. Velasco, *J. Assoc. Off. Anal. Chem.,* **53,** 611 (1970).
14. M. S. Masri, J. R. Page, and V. C. Garcia, *J. Assoc. Off. Anal. Chem.,* **52,** 641 (1969).
15. S. N. Hagan and W. H. Tietjen, *J. Assoc. Off. Anal. Chem.,* **58,** 620 (1975).
16. L. Yin, *J. Assoc. Off. Anal. Chem.,* **52,** 880 (1969).
17. N. L. Brown, S. Nesheim, M. E. Stack, and G. M. Ware, *J. Assoc. Off. Anal. Chem.,* **56,** 1437 (1973).
18. B. Altenkirk, *J. Chromatogr.,* **65,** 456 (1972).
19. L. Leistner, *Fleischwirtschaft,* **55,** 985 (1975).
20. P. C. Crowther, *Analyst (London),* **93,** 623 (1968).

20a. B. A. Roberts and D. S. P. Patterson, *J. Assoc. Off. Anal. Chem.,* **58,** 1178 (1975).

21. L. Stoloff, A. C. Beckwith, and M. E. Cushmac, *J. Assoc. Off. Anal. Chem.,* **51,** 65 (1968).
22. L. Stoloff, A. D. Campbell, A. C. Beckwith, S. Nesheim, J. S. Winbush, Jr., and O. M. Fordham, Jr., *J. Am. Oil Chem. Soc.,* **46,** 678 (1969).
23. J. G. Heathcote and J. R. Hibbert, *J. Chromatogr.,* **108,** 131 (1975).
24. R. D. Stubblefield, O. L. Shotwell, and G. M. Shannon, *J. Assoc. Off. Anal. Chem.,* **55,** 762 (1972).

25. T. V. Reddy, L. Viswanathan, and T. A. Venkitasurbramanian, *Anal. Biochem.*, **38**, 568 (1970).
26. G. W. Engstrom, *J. Chromatogr.*, **44**, 128 (1969).
27. P. L. Schuller, C. A. H. Verhuelsdonk, and W. E. Paulsch, *Arzneim.-Forsch.*, **20**, 1517 (1970).
28. P. M. Scott and B. P. C. Kennedy, *J. Assoc. Off. Anal. Chem.*, **56**, 1452 (1973).
29. L. Yin, A. D. Campbell, and L. Stoloff, *J. Assoc. Off. Anal. Chem.*, **54**, 102 (1971).
30. R. E. Peterson and A. Ciegler, *J. Chromatogr.*, **31**, 250 (1967)
31. L. G. M. T. Tuinstra and J. M. Bronsgeest, *J. Chromatogr.*, **111**, 448 (1975).
32. R. A. de Zeeuw and H. S. Lillard, *J. Assoc. Off. Anal. Chem.*, **54**, 98 (1971).
33. B. Stefaniak, *J. Chromatogr.*, **44**, 403 (1969).
34. H. J. Issaq, E. W. Barr, and W. L. Zielinski, Jr., *J. Chromatogr.*, **132**, 115 (1977).
35. T.-W. Kwon and J. C. Ayres, *J. Chromatogr.*, **31**, 420 (1967).
36. P. J. Andrellos and G. R. Reid, *J. Assoc. Off. Agric. Chem.*, **47**, 801 (1964).
37. L. Stoloff, *J. Assoc. Off. Anal. Chem.*, **50**, 354 (1967).
38. W. Przybylski, *J. Assoc. Off. Anal. Chem.*, **58**, 163 (1975).
39. A. E. Pohland, L. Yin, and J. G. Dantzman, *J. Assoc. Off. Anal. Chem.*, **53**, 101 (1970).
40. S. H. Ashoor and F. S. Chu, *J. Assoc. Off. Anal. Chem.*, **58**, 492 (1975).
41. *Ibid.*, p. 617.
42. M. E. Stack, A. E. Pohland, J. G. Dantzman, and S. Nesheim, *J. Assoc. Off. Anal. Chem.*, **55**, 313 (1972).
43. W. F. Haddon, M. Wiley, and A. C. Waiss, Jr., *Anal. Chem.*, **43**, 268 (1971).
44. M. J. Verrett and J. P. Marliac, *J. Assoc. Off. Agric. Chem.*, **47**, 1003 (1964).
45. M. L. Clements, *J. Assoc. Off. Anal. Chem.*, **51**, 611 (1968).
46. P. R. Beljaars and F. H. M. Fabry, *J. Assoc. Off. Anal. Chem.*, **55**, 775 (1972).
47. P. R. Beljaars, "A Contribution to the Determination of Aflatoxin B_1, Quinine Hydrochloride, and L (+)-ascorbic Acid in Foodstuffs by Quantitative *in Situ* Thin-Layer Chromatographic Analysis," Doctoral Dissertation, Wageningen, 1974.
48. L. Tóth, F. Tauchmann, and L. Leistner, *Fleischwirtschaft*, **50**, 1235 (1970).
49. W. A. Pons, Jr., *J. Assoc. Off. Anal. Chem.*, **54**, 870 (1971).
50. J. L. Ayers and R. O. Sinnhuber, *J. Am. Oil Chem. Soc.*, **43**, 423 (1966).
51. W. A. Pons, Jr., A. F. Cucullu, A. O. Franz, Jr., and L. A. Goldblatt, *J. Am. Oil Chem. Soc.*, **45**, 694 (1968).
52. M. R. Berman and R. N. Zare, *Anal. Chem.*, **47**, 1200 (1975).
53. F. Kiermeier and D. Groll, *Z. Lebensm.-Untersuch.-Forsch.*, **142**, 120 (1970).
54. Z. Ďuračková, V. Betina, and P. Nemec, *J. Chromatogr.*, **116**, 141 (1976).

55. P. S. Steyn and K. J. Merwe, *Nature,* **211,** 418 (1966).
56. S. Nesheim, N. F. Hardin, O. J. Francis, Jr., and W. S. Langham, *J. Assoc. Off. Anal. Chem.,* **56,** 817 (1973).
57. D. Broce, R. M. Grodner, R. L. Killebrew, and F. L. Bonner, *J. Assoc. Off. Anal. Chem.,* **53,** 616 (1970).
58. L. Stoloff, S. Nesheim, L. Yin, J. V. Rodricks, M. Stack, and A. D. Campbell, *J. Assoc. Off. Anal. Chem.,* **54,** 91 (1971).
59. M. Stack and J. V. Rodricks, *J. Assoc. Off. Anal. Chem.,* **54,** 86 (1971).
60. I. Schmidt and H.-J. Rehm, *Z. Lebensm.-Unters.-Forsch.,* **139,** 20 (1968).
61. R. M. Eppley, *J. Assoc. Off. Anal. Chem.,* **58,** 906 (1975).
62. P. B. Oelrichs and W. A. Muller, *Toxicon,* **10,** 63 (1972).
63. P. M. Scott, B. B. Coldwell, and G. S. Wieberg, *Food Cosmet. Toxicol.,* **9,** 179 (1971).
64. C. T. Hou, A. Ciegler, and C. W. Hesseltine, *J. Assoc. Off. Anal. Chem.,* **54,** 1035 (1971).
65. A. W. Hayes and H. W. McCain, *Food Cosmet. Toxicol.,* **13,** 221 (1975).
66. C. Damodaran, C. S. Ramadoss, and E. R. B. Shanmugasundaram, *Anal. Biochem.,* **52,** 482 (1973).
67. P. M. Scott, W. F. Miles, P. Toft, and J. G. Dubé, *J. Agric. Food Chem.,* **20,** 450 (1972).
68. A. E. Pohland and R. Allen, *J. Assoc. Off. Anal. Chem.,* **53,** 686 (1970).
69. J. Reiss, *Naturwissenschaften,* **59,** 37 (1972).
70. J. Reiss, *Mikrochim. Acta, 1975,* 473.
71. I. Alperden, H.-J. Mintzlaff, F. Tauchmann, and L. Leistner, *Fleischwirtschaft,* **53,** 566 (1973).
72. G. Engel, *J. Chromatogr.,* **130,** 293 (1977).
73. T. Iwasaki and F. V. Kosikowski, *J. Food Sci.,* **38,** 1162 (1973).
74. W. R. Kem, B. C. Abbott, and R. M. Coates, *Toxicon,* **9,** 15 (1971).
75. G. Sullivan, D. D. Maness, G. J. Yakatan, and J. Scholler, *J. Chromatogr.,* **116,** 490 (1976).
76. R. W. Pero, R. G. Owens, and D. Harvan, *Anal. Biochem.,* **43,** 80 (1971).
77. P. S. Steyn, *J. Chromatogr.,* **45,** 473 (1969).
78. Z. Ďuračková, V. Betina, and P. Nemec, *J. Chromatogr.,* **116,** 155 (1976).
79. B. Gedek, *Zentralbl. Veterinaermed., Reihe B,* **19,** 15 (1972).
80. A. Dietl., *Allgem. Papierrundsch.,* **1962,** 1262.
81. R. H. Anderson, T. E. Huntley, W. M. Schwecke, and J. H. Nelson, *J. Am. Oil Chem. Soc.,* **40,** 349 (1963).
82. A. Seher, *Fette, Seifen, Anstrichm.,* **61,** 345 (1959).
83. A. Seher, *J. Soc. Cosmet. Chem.,* **13,** 385 (1962).
84. H. Meyer, *Dtsch. Lebsensm-Rundsch.,* **57,** 170 (1961).
85. S. Ishikawa and G. Katsui, *Bitamin,* **30,** 203 (1964); *Chem. Abstr.,* **62,** 806 (1965).
86. J. Jonas, *J. Pharm. Belg.,* **17,** 103 (1962).
87. C. Zentz, *Sonderh. Z. Landwirtsch. Forsch.,* **18,** 152 (1964); through *Chem. Abstr.,* **62,** 2241 (1965).
88. T. Salo and K. Salminen, *Z. Lebensm. Untersuch-Forsch.,* **125,** 167 (1964).

89. J. Davídek, *J. Chromatogr.*, **9,** 363 (1962).
90. J. Davídek and J. Pokorný, *Rev. Univ. Ind. Santander (Columbia),* **4,** 111 (1962).
91. J. Davídek and J. Pokorný, *Z. Lebensm-Untersuch.-Forsch.*, **115,** 113 (1961).
92. R. T. Wang and S. S. Chou, *J. Chromatogr.*, **43,** 522 (1969).
93. S. C. Lee, *Hua Hsueh,* **1968,** 155; through *Anal. Abstr.*, **18,** 2780 (1970).
94. D. G. H. Daniels, H. G. C. King, and H. F. Martin, *J. Sci. Food Agr.*, **14,** 385 (1963).
95. M. R. Sahasrabudhe, *J. Assoc. Off. Agric. Chem.*, **47,** 888 (1964).
96. R. Amano, K. Kawada, and I. Kawashiro, *Shokuhin Eiseigaku Zasshi,* **5,** 333 (1964); through *Chem. Abstr.*, **61,** 15266 (1964).
97. A. Rutkowski, H. Kozlowska, and J. Szerszynski, *Rocz. Panstw. Zakl. Hig.*, **14,** 361 (1963).
98. D. F. Slonaker and D. C. Sievers, *Anal. Chem.*, **36,** 1130 (1964).
99. R. F. van der Heide and O. Wouters, *Z. Lebensm.-Untersuch-Forsch.*, **117,** 129 (1962).
100. R. S. Dobies, *J. Chromatogr.*, **40,** 110 (1969).
101. R. F. van der Heide, A. C. Maagdenberg, and J. H. van der Neut, *Chem. Weekbl.*, **61,** 440 (1965).
102. D. T. Miles, *Analyst (London),* **99,** 724 (1974).
103. D. Simpson and B. R. Currell, *Analyst (London),* **96,** 515 (1971).
104. R. B. Delves, *J. Chromatogr.*, **26,** 296 (1967).
105. R. Cox, *J. Chromatogr.*, **105,** 57 (1975).
106. J. P. Coates, *J. Inst. Petrol.*, **57,** 209 (1971).
107. R. Amos, *J. Inst. Petrol.*, **54,** 9 (1968).
108. G. B. Crump, *Anal. Chem.*, **36,** 2447 (1964).
109. G. M. C. Higgins and G. P. McSweeney, *Rubber Chem. Technol.*, **47,** 1206 (1974).
110. L. Ibarta Rueda and L. Gonzales Hernandez, *Rev. Plast. Mod.*, **24,** 82 (1973).
111. L. Sluzewska, *Rocz. Panstw. Zakl. Hig.*, **25,** 495 (1974); through *Anal. Abstr.*, **28,** 6081 (1975).
112. G. Ivan and R. Ciutacu, *J. Chromatogr.*, **88,** 391 (1974).
113. J. G. Kreiner and W. C. Warner, *J. Chromatogr.*, **44,** 315 (1969).
114. K. Nagasawa and K. Ohta, *Bunseki Kagaku,* **16,** 1285 (1967; also *J. Chromatogr.*, **37,** D86 (1968).
115. J. G. Kreiner, *Rubber Chem. Technol.*, **44,** 381 (1971).
116. J. G. L. Harthon, *Acta Chem. Scand.*, **15,** 1401 (1961).
117. M. I. Fauth and G. W. Roecker, *J. Chromatogr.*, **18,** 608 (1965).
118. K. R. K. Rao, A. K. Bhalla, and S. K. Sinha, *Current Sci. (India),* **33,** 12 (1964).
119. J. Hansson and A. Alm, *J. Chromatogr.*, **9,** 385 (1962).
120. S. K. Yasuda, *J. Chromatogr.*, **14,** 65 (1964).
121. *Ibid.*, **16,** 488 (1964).
122. *Ibid.*, **50,** 453 (1970).

123. *Ibid.*, **71**, 481 (1972).
124. J. Hansson, *Explosivstoffe*, **10**, 73 (1963).
125. C. D. Chandler, J. A. Kohlbeck, and W. T. Bolleter, *J. Chromatogr.*, **64**, 123 (1972).
126. C. R. Midkiff, Jr., and W. D. Washington, *J. Assoc. Off. Anal. Chem.*, **57**, 1092 (1974).
127. *Ibid.*, **59**, 1357 (1976).
128. J. C. Hoffsommer and D. J. Glover, *J. Chromatogr.*, **62**, 417 (1971).
129. J. C. Hoffsommer, *J. Chromatogr.*, **51**, 243 (1970).
130. A. W. Archer, *J. Chromatogr.*, **108**, 401 (1975).
131. D. B. Parihar, S. P. Sharma, and K. K. Verma, *J. Chromatogr.*, **31**, 551 (1967).
132. D. B. Parihar, S. P. Sharma and K. K. Verma, *J. Chromatogr.*, **31**, 120 (1967).
133. D. B. Parihar, O. Prakash, I. Bajaj, R. P. Tripathi, and K. K. Verma, *Mikrochim. Acta*, **1971**, 393.
134. M. Tuerler and D. Hoegl, *Mitt. Geb. Lebensm. Hyg.*, **52**, 123 (1961).
135. K. Buerger, *Z. Anal. Chem.* **192**, 280 (1962).
136. G. Neubert, *Z. Anal. Chem.*, **203**, 265 (1964).
137. R. F. van der Heide, *Z. Lebensm.-Unters.-Forsch.*, **124**, 198 (1964).
138. *Ibid.*, p. 348.
139. J. Koch and K. Figge, *J. Chromatogr.*, **109**, 89 (1975).
140. K. Schloegl, H. Pelousek, and A. Mohar, *Monatsh. Chem.*, **92**, 533 (1961).
141. K. Schloegl, A. Mohar, and M. Peterlik, *Monatsh. Chem.* **92**, 921 (1961).
142. K. Schloegl and H. Pelousek, *Ann. Chem.* **651**, 1 (1962).
143. K. Schloegl and M. Peterlik, *Tetrahedron Lett.*, **1962**, 573.
144. K. Schloegl, M. Peterlik, and H. Seiler, *Monatsh. Chem.*, **93**, 1309 (1962).
145. K. Schloegl and M. Peterlik, *Monatsh. Chem.*, **93**, 1328 (1962).
146. K. Schloegl and H. Egger, *Monatsh. Chem.*, **94**, 376 (1963).
147. K. Schloegl and M. Fried, *Monatsh. Chem.* **94**, 537 (1963).
148. K. Schloegl and M. Fried, *Tetrahedron Letters*, **1963**, 1473.
149. M. Vobecky, V. D. Nefedov, and E. N. Sinotova, *Zh. Obshch. Khim.*, **33**, 4023 (1963); through *Chem. Abstr.*, **60**, 8672 (1964).
150. M. Vobecky, V. D. Needov, and E. N. Sinotova, *J. Chromatogr.*, **30**, D1 (1967).
151. G. W. Johnson and C. Vickers, *Analyst (London)*, **95**, 356 (1970).
152. J. O'G. Tatton and P. J. Wagstaffe, *J. Chromatogr.*, **44**, 284 (1969).
153. R. Takeshita, H. Akagi, M. Fujita, and Y. Sakagami, *J. Chromatogr.*, **51**, 283 (1970).
154. L. Fishbein, *Chromatogr., Rev.*, **13**, 83 (1970).
155. L. F. Druding and S. I. Shupack, *J. Chromatogr.*, **24**, 491 (1966).
156. R. Klement and A. Wild, *Z. Anal. Chem.*, **195**, 180 (1963).
157. R. Donner and Kh. Lohs, *J. Chromatogr.*, **17**, 349 (1965).
158. H. Reuter and H. Hanke, *Pharm. Zentralhalle*, **104**, 323 (1965).
159. H. Schindlbauer and F. Mitterhofer, *Angew. Chem. Int. Eng. Ed.*, **5**, 680 (1966).

160. A. Lamotte, M. Porthault, and J.-C.-Merlin, *Bull. Soc. Chim. Fr.*, **1965,** 919.
161. A. Lamotte and J.-C. Merlin, *J. Chromatogr.*, **38,** 296 (1968).
162. *Ibid.*, **45,** 432 (1969).
163. A. Lamotte, A. Francina, and J.-C. Merlin, *J. Chromatogr.*, **44,** 75 (1969).
164. P. J. Bloom, *J. Chromatogr.*, **75,** 261 (1973).
165. N. L. Ivanova, A. I. Zavalishima, I. V. Fursenko, I. S. Konyaeva, I. P. Komlev, and E. E. Nifant'ev, *Zh. Obshch. Khim.*, **42,** 91 (1972).
166. H. Petschik and E. Steger, *J. Chromatogr.*, **31,** 369 (1967).
167. S. A. de Licastro and M. A. Rúveda, *J. Chromatogr.*, **92,** 207 (1974).
168. G. Neubert, *J. Chromatogr.*, **20,** 342 (1965).
169. T. A. Mastryukova, T. B. Sakharova, and M. I. Kabachnik, *Izv. Akad. Nauk SSSR, Ser. Khim.*, **1963,** 2211; through *Chem. Abstr.*, **60,** 9882 (1964).
170. H. Petschik and E. Steger, *J. Chromatogr.*, **9,** 307 (1962).
171. *Ibid.*, **7,** 135 (1962).
172. H. Ertel and L. Horner, *J. Chromatogr.*, **7,** 268 (1962).
173. R. Stephan and J. G. Erdman, *Nature,* **203,** 749 (1964).
174. R. F. Curtis and G. T. Phillips, *J. Chromatogr.*, **9,** 366 (1962).
175. R. Mayer, P. Rosmus, and J. Fabian, *J. Chromatogr.*, **15,** 153 (1964).
176. L. Fishbein and J. Fawkes, *J. Chromatogr.*, **22,** 323 (1966).
177. E. N. Karaulova, T. S. Bobruiskaya, and G. D. Gal'pern, *Zh. Anal. Khim.*, 21 893 (1966).
178. V. I. Snegotskii and V. A. Snegotskaya, *Zavodsk. Lab.*, **35,** 429 (1969).
179. N. N. Novitskaya, L. E. Zhuraleva, and L. P. Ivanova, *J. Chromatogr.*, **43,** D45 (1969).
180. H. W. Prinzler, D. Pape, and M. Teppke, *J. Chromatogr.*, **19,** 375 (1965).
181. H. W. Prinzler, H. Tauchmann, and C. Tzscharnke, *J. Chromatogr.*, **29,** 151 (1967).
182. H. W. Prinzler, D. Pape, H. Tauchmann, M. Teppe, and C. Tzscharnke, *Ropa Uhlie,* **8,** 13 (1966).
183. R. W. Hiley and A. Cameron, *J. Chromatogr.*, **107,** 393 (1975).
184. M. L. Wolfrom, D. Horton, and D. H. Hutson, *J. Org. Chem.*, **28,** 845 (1963).
185. H. P. Kaufmann and Z. Makus, *Fette, Seifen, Anstrichm.*, **62,** 1014 (1960).
186. H. Kaunitz, D. C. Malins, and D. G. McKay, *J. Exp. Med.*, **115,** 1127 (1962).
187. L. J. Morris, R. T. Holman, and K. Fontell, *J. Lipid Res.*, **2,** 68 (1961).
188. L. Acker and H. Greve, *Fette Seifen, Anstrichm.*, **65,** 1009 (1963).
189. F. Neuwald and K. E. Fetting, *Pharm. Ztg. Ver. Apoth.-Ztg.*, **108,** 1490 (1963).
190. E. Schauenstein and H. Esterbauer, *Monatsh. Chem.*, **94,** 164 (1963).
191. R. Subbarao, M. W. Roomi, M. R. Subbaram, and K. T. Achaya, *J. Chromatogr.*, **9,** 295 (1962).
192. O. S. Privett and E. C. Nickell, *J. Am. Oil Chem. Soc.*, **41,** 72 (1964).
193. O. S. Privett and M. L. Blank, *J. Am. Oil Chem. Soc.*, **40,** 70 (1963).
194. O. S. Privett and E. C. Nickell, *J. Lipid Res.*, **4,** 208 (1963).

195. O. S. Privett and M. L. Blank, *J. Am. Oil Chem. Soc.*, **39**, 520 (1962).
196. E. Knappe and D. Peteri, *Z. Anal. Chem.*, **190**, 386 (1962).
197. R. Kavčič and B. Plesničar, *J. Chromatogr.*, **38**, 515 (1968).
198. A. N. Sorokina, A. E. Batog, and M. K. Romantsevich, *Zh. Obshch. Khim.*, **37**, 766 (1967).
199. M. M. Buzlanova, V. F. Stepanovskaya, and V. L. Antonovsky, *Zh. Anal. Khim.*, **21**, 1491 (1966).
200. N. Mandava, *Am. Lab.*, **6**, 27 (1973).
201. M. Kutáček, J. Rosmus, and Z. Deyl, *Biol. Plant. Acad. Sci. Bohemoslov.*, **4**, 226 (1962).
202. G. Sembdner, R. Gross, and K. Schreiber, *Experientia*, **18**, 584 (1962).
203. J. MacMillan and P. J. Suter, *Nature*, **197**, 790 (1963).
204. N. Ikekawa, T. Kagawa, and Y. Sumiki, *Proc. Japan Acad.* **39**, 507 (1963).
205. T. Kagawa, T. Fukinbara, and Y. Sumiki, *Agric. Biol. Chem. (Tokyo)*, **27**, 598 (1963).
206. B. D. Cavell, J. MacMillan, R. J. Pryce, and A. C. Sheppard, *Phytochemistry*, **6**, 867 (1967).
207. A. Musgrave, *J. Chromatogr.*, **36**, 388 (1968).
208. E. Stahl and H. Kaldewey, *Z. Physiol. Chem.*, **323**, 182 (1961).
209. G. Ballin, *J. Chromatogr.*, **16**, 152 (1964).
210. H. Kaldewey and E. Stahl, *Planta*, **62**, 22 (1964).
211. H. Kaldewey, *Colloq. Int. Cent. Natl. Rech. Sci. (Paris)*, **123**, 421 (1963 pub. 1964).
212. J. Dubouchet and P.-E. Pilet, *Ann. Physiol. Veg.*, **5**, 175 (1963).
213. G. Collet, *Compt. Rend.*, **259**, 871 (1964).
214. I. D. Railton, *J. Chromatogr.*, **70**, 202 (1972).
215. R. K. Raj and O. Hutzinger, *Anal. Biochem.*, **33**, 471 (1970).
216. J. W. Copius-Peereboom, *J. Chromatogr.*, **4**, 323 (1960).
217. F. Jaminet, *Farmaco (Pavia), Ed. Prat.*, **18**, 633 (1963).
218. D. Braun, *Kunstoffe-Plastics*, **52**, 2 (1962).
219. P. J. Bloom, *J. Chromatogr.*, **72**, 35 (1972).
220. W. Groebel, *Z. Lebensm.-Unters.-Forsch.*, **137**, 7 (1968).
221. H. Haase, *Kautsch. Gummi Kunst.*, **21**, 9 (1968).
222. J. G. Kreiner, *J. Chromatogr.*, **75**, 271 (1973).
223. D. Braun and H. Geenen, *J. Chromatogr.*, **7**, 56 (1962).
224. K. Obruba, *Collect. Czech. Chem. Commun.*, **27**, 2968 (1962); through *Chem. Abstr.*, **58**, 9337 (1963).
225. K. Thoma, R. Rombach, and E. Ullmann, *Arch. Pharm.*, **298**, 19 (1965).
226. L. Anker and D. Sonanini, *Pharm. Acta Helv.*, **37**, 360 (1962).
227. A. Seher, *Fette, Seifen, Anstrichm.*, **66**, 371 (1964).
228. A. Seher and J. Janssen, *Fette, Seifen, Anstrichm.*, **72**, 773 (1970).
229. H. Koenig, *Z. Anal. Chem.*, **251**, 167 (1970).
230. B. Stancher, L. F. Gabrielli, and L. Favretto, *J. Chromatogr.*, **111**, 459 (1975).
231. H. K. Mangold and R. Kammereck, *J. Am. Oil Chem. Soc.*, **39**, 201 (1962).

232. A. F. Hofmann, "Thin-Layer Adsorption Chromatography of Lipids," in *Biochemical Problems of Lipids,* A. C. Frazer, Ed., Elsevier, Amsterdam, 1963, p. 1.
233. T. Takagi and K. Fukuzumi, *Yukagaku,* **13,** 520 (1964).
234. C. T. Desmond and W. T. Borden, *J. Am. Oil Chem. Soc.,* **41,** 552 (1964).
235. M. A. Allen and T. T. Martin, *J. Am. Oil Chem. Soc.,* **48,** 790 (1971).
236. S. Hordyńska and B. Legatowa, *Rocz. Panstw. Zakl. Hig.,* **18,** 189 (1967).
237. R. Takeshita, N. Jinnai, and H. Yoshida, *J. Chromatogr.,* **123,** 301 (1976).
238. M. J. Groves and R. M. A. Mustafa, *J. Chromatogr.,* **97,** 297 (1974).
239. D. Waldi, "Synthetic Organic Materials," in *Thin-Layer Chromatography,* E. Stahl, Ed., Academic Press, New York, 1965, p. 365.
240. T. Salo, E. Airo, and K. Salminen, *Z. Lebensm.-Untersuch.-Forsch,* **124,** 20 (1964).
241. T. Salo and K. Salminen, *Suom. Kemistil.,* **A37,** 161 (1964).
242. T. Korbelak and J. N. Bartlett, *J. Chromatogr.,* **41,** 124 (1969).
243. H. Woidich, H. Gnauer, and J. Tunka, *Z. Lebensm-Unters.-Forsch.,* **147,** 284 (1971).
244. B. Gentili and R. M. Horowitz, *J. Chromatogr.,* **63,** 467 (1971).
245. E. Knappe, D. Peteri, and I. Rohdewald, *Z. Anal. Chem.,* **197,** 364 (1963).
246. R. S. Dobies, *J. Chromatogr.,* **35,** 370 (1968).
247. M. Barbier, *J. Chromatogr.,* **2,** 649 (1959).
248. G. Pettersson, *J. Chromatogr.,* **12,** 352 (1963).
249. G. Pettersson, *Acta Chem. Scand.,* **17,** 1771 (1963).
250. W. Grau and H. Endres, *J. Chromatogr.,* **17,** 585 (1965).
251. R. H. Nealey, *J. Chromatogr.,* **14,** 120 (1964).
252. M. Maugras, Ch. Robin, and R. Gay, *Bull. Soc. Chim. Biol.,* **44,** 887 (1962).
253. N. G. Patel, M. H. Haydak, and R. Lovell, *Nature,* **191,** 362 (1961).
254. J. Gehrmann and F. L. Schneider, *Microchem. J.,* **6,** 561 (1962).
255. F. Korte and J. Vogel, *J. Chromatogr.,* **9,** 381 (1962).
256. F. S. Wusteman, K. S. Dodgson, A. G. Lloyd, F. A. Rose, and N. Tudball, *J. Chromatogr.,* **16,** 334 (1964).

Chapter **XXXIII**

INORGANIC IONS

1 CATIONS

Separation on Silica Gel and Cellulose Layers

The first inorganic work on thin layers was done by Meinhard and Hall in 1949 [1] on layers of a mixture of aluminum oxide and Celite bound with starch. In this case zinc and iron were separated by radial chromatography. The separation is mainly of historical interest. Separations can be achieved, however, on silica gel. Since the silica gel contains impurities in the form of sodium, magnesium, calcium, and iron, these must first be removed by washing with acid and water [2]. (For a description of this procedure see Chapter III, Section 2.) Because this treatment also removes the calcium sulfate binder, it must be replaced where gypsum-bound layers are desired or with starch when gypsum-bound layers cannot be used. Specially purified gel is available commercially as MN silica gel HR. Seiler [3] examined the separation of cations and found that their separation depends on the ion-exchange properties of the adsorbent and the coordination tendencies of the solvent.

Among the reviews are those of Merkus [4], Lederer [5], Lesigang-Buchtela [6], Volynets and Ermakov [7], Garel [8], and Takitani and Kawanabe [9]. Merkus [10] has reviewed the separations on cellulose and Fishbein [11] the chromatography of mercury. Brinkman and de Vries [11a] collected TLC data.

The Hydrogen Sulfide Group

Seiler and Seiler [2] used layers of purified silica gel bound with gypsum for the separation of this group. Copper, mercury, bismuth, cadmium, and lead were separated in a saturated chamber using a solvent composed of *n*-butanol–1.5 *N* hydrochloric acid–acetonylacetone (100:20:0.5). The compounds were detected by spraying with a 2% solution of potassium iodide and then exposing to ammonia vapor after drying the plates. Finally, they were inserted into a container filled with hydrogen sulfide to obtain the following colors in increasing order of R_f value: Cu^{2+} dark brown, Pb^{2+} brown, Cd^{2+} yellow, Bi^{3+} brown-black, and Hg^{2+} brown-black. Kuenzi [12] and Kuenzi et al. [13] investigated the separation of heavy metals of forensic importance including thallium, copper, lead, arsenic, cadmium, antimony, bismuth, and mercury. Using the same adsorbent as Seiler, these authors investigated the use of various complexing agents with organic solvents and found a mixture of 100 ml acetone–benzene (3:1) saturated with tartaric acid and 6 ml nitric acid (10%) to be the best solvent. However, in this solvent the mercury tended to overlap the bismuth spot and the lead overlapped the copper; cadmium showed three spots. A mixture of methanol–acetonylacetone–nitric acid (proportions not given) gave a selective separation of thallium, with an R_f value of 0.72, from the remaining ions which traveled at or near the solvent front. Both Seiler and Kuenzi observed that individual R_f values were in general of little value since mixtures of ions affected the absolute value of the individual ions in the mixture. Of value then, is the relative order of the ions and their color with various spray reagents. The order with the Kuenzi solvent was as follows: Hg > Bi > Sb > Cd > As > Pb > Cu > T1. Table 33.1 gives some color reactions for the various ions of this group. The Kuenzi workers have applied their method to the quantitative determination of a number of metals in practical cases such as the determination of arsenic in flour, thallium in blood, mercury in urine, and arsenic and cadmium in tea. The evaluation was by means of comparison of spot size with standard solutions. For the determination of arsenic and cadmium in tea, the calculated error was 10%, but for the determination of mercury in urine a value of 0.5 mg % was obtained in 3 hr compared to a value of 0.4 to 0.5 mg % obtained in 12 hr by an electrolytic method.

Johri and Kaushik [14] separated Cu^{2+}, Pb^{2+}, Bi^{2+}, Cd^{2+}, and Hg^{2+} on silica gel with *n*–butanol–1.5 *N* hydrochloric acid–methyl ethyl ketone (75:15:2). For the separation of As^{2+}, Sb^{2+}, and Sn^{2+}, *n*–butanol–1 *M* tartaric acid–3 *N* hydrochloric acid (10:1:1) was used. Detection was accomplished with potassium thiocarbonate (T-215); sensitivites varied from 1.27 to 11.85 μg. Johri and Mehra [15] separated Hg^{+} and Hg^{2+} with *n*-butanol-3 *N* hydrochloric acid–2% tartaric acid (4:1:1); Sb^{3+} and Sb^{5+} with *tert*-butanol–2 *N* hydrochloric acid (4:1); and $T1^{+}$ and $T1^{3+}$ with *tert*-butanol–acetic acid (2:1). Baffi et al. [16] examined the behavior of 60 inorganic ions on cellulose with tartaric acid–water–ethanol and tartaric acid–water–ethanol–ammonia solutions. Separations of As^{3+}, Sb^{3+}, Sn^{2+}, and Pb^{2+} can be accomplished on silica gel G with amyl acetate–hydrochloric acid (sp gr 1.16) [17]. As^{3+} and As^{5+} were separated on silica gel containing 5% calcium sulfate using acetone–15 *M* orthophosphoric acid (50:1) as a solvent [18].

Rai and Kukreja [19] separated the diethyl dithiocarbamate complexes

Table 33.1 Color Reactions of Some of the Hydrogen Sulfide Group

	Dithizone [12, 13]				
Reagent	*Acid*	*Alkaline (NH_4OH)*	*$(NH_4)_2S$* [12, 13]	*KI* [2]	*H_2S* [2]
Hg^{2+}	Pink	Orange-red	Black	Red	Brown-black
Bi^{3+}	Purple	Red-orange	Brown	Red-yellow	Brown-black
Sb^{3+}	Reddish	Light brown	Orange		
Cd^{2+}	Lilac	Orange	Yellow		Yellow
As^{5+}	Yellow		Yellow		
Pb^{2+}		Pink	Brown	Yellow-brown	Brown
Cu^{2+}	Yellow-green	Gray-brown	Brown	Brown	Dark brown
Tl^{+}		Pink	Black		

of Ag, Hg^{2+}, Cu^{2+}, Bi^{3+}, Tl^{3+}, Pb^{2+}, and Pd^{2+} using benzene, toluene, xylene, and chloroform–carbon tetrachloride (1:1) as solvents on silica gel. Senf [20] used *n*-hexane–chloroform–diethylamine (20:2:1) to separate this complex of Pb^{2+} (R_f 0.00), Bi^{2+} (0.27), Cd^{2+} (0.34), Cu^{2+} (0.44), and Hg^{2+} (0.56) on silica. Again on silica gel, Schwedt and Lippmann [21] applied this complex to the separation of Zn, Cu, Ni, Pb, Hg, and Cd with benzene–hexane (5:1).

Hranisavljević-Jakovljević et al. [22] have chromatographed the dithizonates of mercury, lead, copper, bismuth, cadmium, and zinc on silica gel G with benzene–methylene chloride (5:1) as a developing solvent giving the following R_f values: Cd^{2+} 0.13, Bi^{3+} 0.37, Pb^{2+} 0.34, Cu^{2+} 0.48, Zn^{2+} 0.50, and Hg^{2+} 0.58. Other solvents used for the separation of dithizone complexes on silica gel include benzene, toluene, xylene, and carbon tetrachloride–chloroform (5:2) [23]. Gregorowicz et al. [24] used silica gel G–kieselguhr G (7:3) layers with benzene–dichloromethane–heptane (25:27:10) in separating Zn (R_f 0.41), Cu (0.37), Ni (0.34), Co (0.29), Pb (0.24), Bi (0.20), and Cd (0.05) as the dithizone complexes. Baudot et al [25] have used a combination dithizone–carbon tetrachloride extraction and TLC separation on silica gel 60 (Merck) for isolating and identifying silver, cadmium, cobalt, copper, mercury, nickel, lead, and zinc in toxicological cases. The extractions were carried out at four different pH values, and the developing solvent was benzene for a distance of 6 to 7 cm.

Tsunoda et al. [26–28] have chromatographed metallic chelates of acetylacetone and ethylenediaminetetraacetic acid with cobalt, manganese, nickel, copper, chromium, and iron. The trivalent 6-coordinated chelates had higher R_f values than the bivalent 4-coordinated chelates. A number of separations were accomplished on silica gel and on alumina. Masoomi and Haworth [29] examined the separation of the EDTA complexes of chromium, manganese, iron, cobalt, copper, nickel, zinc, cadmium, and mercury on microcrystalline cellulose in 15 solvent systems, and Vanderdeelen [30] chromatographed EDTA complexes of cobalt, copper, nickel, manganese, chromium, and iron on silica gel without gypsum binder in three solvents.

Verma and Rai [31] separated zinc, copper, cadmium, nickel, and cobalt on silica gel G as their thiocyanate complexes. The ammonium thiocyanate was incorporated directly in the solvents. A good separation was achieved with ammonium thiocyanate–*n*–butanol–ammonia (sp g 0.910)–pyridine (4:80:10:10) (wt/vol/vol/vol).

The Ammonium Sulfide Group

Seiler and Seiler [2] used a solvent composed of acetone-concentrated hydrochloric acid–acetonylacetone (100:1:0.5) for separating iron, zinc, cobalt, manganese, chromium, nickel, and aluminum on their purified silica gel layers. For visualizing the resulting spots the plates were exposed to ammonia gas and then sprayed with a solution of 0.5 g of 8-hydroxyquinoline in 100 ml of 60% alcohol. All the spots could be seen under ultraviolet light. Here again, the location of spots in a mixture were affected by the components of that mixture (Fig. 33.1). These authors [32] also separated UO_2^{2+} from a mixture of Fe^{3+}, Cu^{2+}, Co^{2+}, Ni^{2+} Cr^{3+}, Al^{3+}, and Th^{4+} ions by using a mixture of 50 ml of ethyl acetate, 50 ml of water-saturated ether, and 2 ml of tri-*n*-butylphosphate. To obtain this separation, the spotting solution was adjusted to 4.7 *N* with nitric acid. This resulted in the formation of a uranyl nitrate-(tri-*n*-butylphosphate) complex which traveled quite readily in the developing solvent, whereas the other cations remained near or at the starting point. As little as 1 μg of uranium could be detected by spraying with 0.25% pyridylazo-naphthol in ethanol. Gallium (Ga^{3+}) was separated from a hundredfold excess of aluminum ion by chromatographing in a mixture of 100 ml of acetone with 0.5 ml of concentrated hydrochloric acid. The gallium was detected by spraying with a 0.5% solution of 8-hydroxyquinoline in 60% ethanol. After spraying, the plate was exposed to concentrated ammonia and then observed under ultraviolet light. Lesigang-Buchtela and Buchtela [33] used isobutyl methyl ketone–tributyl phos-

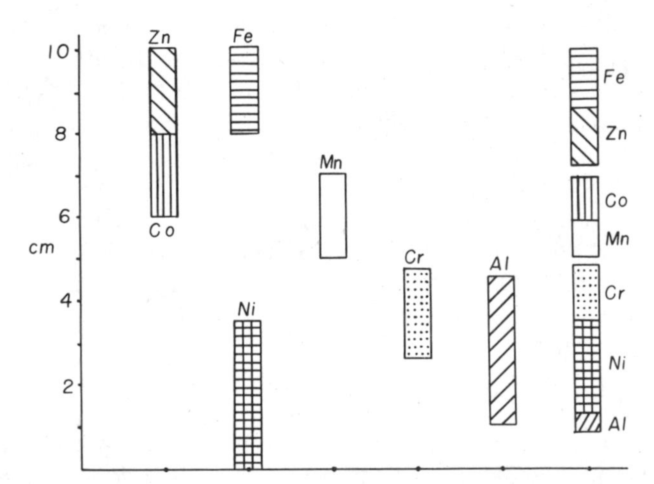

Fig. 33.1 Separation of the ammonium sulfide group on purified silica gel G with acetone–concentrated hydrochloric acid–acetonylacetone (100:1:0.5) as a solvent. Development distance 10 cm. Detection by exposure to ammonium vapor and then spraying with a solution of 0.5 g of 8-hydroxyquinoline in 100 ml of 60% alcohol. From Seiler and Seiler [2]; reproduced with permission of the authors and Verlag Helvetica Chimica Acta.

phate–4.7 *M* nitric acid (100:1:10) for separating uranium from other ions. Hu [34] has used tributyl phosphate saturated with various concentrations of nitric acid as a developing solvent for separating titanium, zirconium, thorium, scandium, and uranium on silica gel layers. The nitric acid solutions of higher concentrations gave the best separations. Butanol–ethoxyacetone–acetic acid was used to separate uranium from thorium, zirconium, and the rare earth elements. Phipps [35] separated perrhenate, molybdate, and selenite ions on Eastman "Chromagram" silica gel sheets with 0.6 *M* hydrochloric acid in methanol and containing 10% water. The ions could not be separated on silica gel layers on glass plates. Gaibakyan and Babayan [36] separated these ions and also tungsten on alumina with methanol–3 *N* sodium hydroxide (3:1).

Cellulose layers can also be used to advantage in separating many metals. Šoljić and Marjanović [37] separated Fe, Al, Ga, and Ti with butanol saturated with 1 *N* hydrochloric acid; small amounts of Ga can be separated from these and also from In with *n*-butanol–water–hydrochloric acid (8:1:1) [38]. Gagliardi and Brodar [39] investigated the separation of Al, Ga, I, Tl^{3+}, and Tl^{+} on cellulose with various alcohols and concentrations of the halogen acids. Zaye et al. [40] chromatographed Al^{3+}, Bi^{3+}, Cd^{3+}, Co^{2+}, Cu^{2+}, Fe^{3+}, Pb^{2+}, Mn^{2+}, Hg^{2+}, Ni^{2+}, Ag, Sn^{2+}, and Zn^{2+} on cellulose with hydrochloric acid–butanol (1:3). Miketuková and Frei [41] chromatographed Bi^{3+}, Cd^{2+}, Co^{2+}, Cu^{2+}, Fe^{3+}, Mn^{2+}, Ni^{2+}, Pb^{2+}, U^{6+}, V, and Zn^{2+} on silica gel and on cellulose with isopropanol–acetic acid–6 *N* hydrochloric acid–water (8:1:1:1); silica gel gave a better separation of Mn and Ni than cellulose. Buchbauer [42] separated cations that might be associated with organic acids used in pharmaceuticals. Pb^{2+}, Cu^{2+}, Cd^{2+}, Bi^{3+}, and Hg^{2+} were separated on cellulose with butanol–1.5 *N* hydrochloric acid–acetonylacetone (200:40:1) and Al^{3+} Cr^{3+}, Ni^{2+}, Mn^{2+}, Co^{2+}, Zn^{2+}, and Fe^{3+} with acetic acid–pyridine–hydrochloric acid (40:3:10). Merkus [43] discussed the separation of various groups of ions on cellulose. Lesigang-Buchtela [44] separated a large number of cations on cellulose or silica gel using various solvent combinations. Lesigang-Buchtela and Buchtela [45] separated U^{6+} from 52 other metals by chromatographing on silica gel HR or cellulose with the upper phase of isobutyl methyl ketone–tributyl phosphate–4.7 *N* nitric acid (100:1:10).

Markl and Hecht [46] have applied the principle of using a complexing agent to assist in the separation of ions using thin layers of purified silica gel under saturated chamber conditions. The developing solvent was a mixture of 100 ml of ethyl ether–ethyl acetate (1:1) and 8 ml of triisooctylamine. This mixture was equilibrated with an equal volume of the acid which was used for applying the samples to the plate. With 1 *M* sulfuric acid as the equilibrating acid, both uranium and molybdenum could be separated from iron or from a mixture of nickel, zinc, manganese, and cobalt; however, in separating uranium from molybdenum, the molybdenum spot was obtained with a slight uranium impurity in it, but the major quantity of the uranium remained behind at the starting point. Using this separation procedure for the chloride complexes and using 3 *N* hydrochloric acid as the acidifying acid Fe^{3+} formed a complex and moved with the solvent a distance of 4.5 to 5.0 cm with a solvent travel distance of 7 cm and could therefore be separated from Ni^{2+} and Mn^{2+} which did not form complexes and remained at the origin. The same was true with a mixture of Zn^{2+}, Mn^{2+}, and Ni^{2+} where again zinc could be separated because of the formation of a zinc complex. Uranium, which forms a complex, could be separated from the noncomplex-forming titanium. Using 8 *N* hydrochloric acid, uranium could be separated from thorium which did not form a complex. A somewhat different application by these workers [47] followed the work of Cerrai and Testa [48] and Testa [49], who used paper impregnated with tri-*n*-octylamines. The silica gel was impregnated prior to the preparation of the thin layers. To accomplish this, 2 ml of tri-*n*-octylamine dissolved in 25 ml of ether was equilibrated by shaking for 2 min with 50 ml of the acid that was to be used. The ether solution was then added to 16.8 g of purified silica gel; after removal of the ether, the gel was mixed with 2.4 g of plaster of Paris and the required amount of water to prepare the thin-layer plate. The plates were dried at 100°C for 2 hr. By this procedure the silica gel was given the properties of an ion-exchange resin, and separations could be achieved by using various concentrations of acid as solvents. Using 3 *N* hydrochloric acid the following separations were possible: Fe/Co/Ni, Zn/Co/Ni, Mo/Co/Ni, U/Co/Ni, U/Zr/Th, Zn/Co/Mn, Fe/Zr, and Fe/Th. The separation of U, Mo, Fe, Zn, and Ti was not possible in this solvent. Substituting 1 *M* sulfuric acid as the separating acid, and of course equilibrating the plate with the same acid, gave a good separation of Mo/U and also Mo/Fe. By using 8 *N* hydrochloric acid, a good separation could be made of U/Zr/Th as well as U/Ti/Th. If 5 *N* nitric acid was used as the solvent, then molybdenum could be separated from most of the other cations. Molybdenum or uranium could also be separated from other ions by using tri-*n*-butyl phosphate as the impregnating agent. For detecting the various ions several reagents were used. A 10% solution of 8-hydroxyquinoline in ammoniacal alcohol was used for molybdenum, zinc, manganese, and cobalt, and a 1% solution of potassium ferrocyanide was applied for uranium and iron. After exposure of the plates with ammonia vapor, titanium was detected with hydrogen peroxide and zirconium with quercetin. Graham and co-workers [50–54] and Brinkman and co-workers [55–61] have investigated the use of silica gel and cellulose layers impregnated with various liquid anion exchangers using a variety of solvents for development.

Rai and Kukreja [62] chromatographed gallium, indium, thallium, zinc, and cadmium as their diethyldithiocarbamate complexes. Silica gel G was used with benzene, toluene, or chloroform–carbon tetrachloride (1:3). With 0.5% ethanolic 8-hydroxyquinoline the sensitivity was 3, 2, 4, 20, and 10 μg, respectively. These authors also separated Mn^{2+}, Ni^{2+}, Co^{2+}, Cr^{6+}, and Va^{5+} in the same systems [17]. Cassidy et al. [63] were able to detect 0.1 to 1 ng of Co^{2+}, Ni^{2+} Cu^{2+}, Zn^{2+}, Ag^{+}, Cd^{2+}, Hg^{2+}, Pb^{2+}, and Bi^{3+} by separating the diethylthiocarbamates on activated silica gel G layers. The layers were sprayed with Pd^{2+}–calcein reagent (T-188) and observed under ultraviolet light for fluorescence. For the TLC separation, cyclohexane–chloroform (3:7) was the best solvent. Olsina et al. [64] separated Zr^{4+} and Hf^{4+} on silica gel G with hydrochloric acid–phosphoric acid–water (10:1:9) as the solvent. The R_f values of 41 other ions in this system are reported.

Radionuclides and the Rare Earths

Because of the rapid separation, thin-layer chromatography is especially suited to the separation of nuclides with short half-lives. Breccia and Spalletti [65, 66] separated ^{95}Zr from ^{95}Nb on silica gel G layers. The greatest separation was obtained using a solvent which was 2 N with respect to hydrochloric acid and 0.2 N with respect to hydrogen fluoride; the R_f values were 0.88 and 0.20, respectively. However, with this solvent the adsorbent beneath the surface of the solvent slid from the plate. Good separation could also be achieved with methanol-10 N hydrogen fluoride (25:1), which gave R_f values of 0.03 and 0.05, respectively, for a 45-min run. For the separation of ^{90}Sr from ^{90}Y the silica gel was purified by the method of Seiler and Rothweiler [67], and the plates were then prepared without a binder. The best eluant for this pair was methanol containing 0.2% EDTA and 10% water; this gave R_f values of 0.16 and 0.85, respectively, in a 40-min run. The compounds were detected with a Geiger–Müller counter with the window shielded by a copper plate having a 1 mm by 1 cm slit. Kuroda and Oguma [68] separated ^{90} Sr and ^{90}Y by isotopic-exchange thin-layer chromatography. Thin layers of strontium sulfate with 1 N sulfuric acid as well as other combinations were used for the work. With this solvent system and adsorbent ^{90}Sr is retained by the isotopic exchange, and ^{90}Y has an R_f of approximately 0.90.

Bottura et al. [69] have carried out further investigations on the use of complexing agents for the separation of ^{235}U fission products. One application of the complexing agents was in treating the silica gel layer in order to remove interfering impurities instead of the acid treatment according to Seiler and Rothweiler [67]. In comparing silica gel purified by treatment with hydrochloric acid and silica gel treated with EDTA, using the same solvent system (0.3 *M* hydrochloric acid in methanol), ^{140}Ba and ^{140}La were separated with R_f values of 0.00 and 0.85, respectively, for the hydrochloric acid-treated gel and 0.20 and 0.50 for the EDTA. ^{90}Sr and ^{90}Y were effectively separated using a solution of 0.1 *M* TTA (4,4,4-trifluoro-1(2)-thienylbutane-1,3-dione) in methanol as the developing solvent to obtain R_f values of approximately 0.25 and 0.8, respectively, for a development distance of 17.5 cm. ^{113}Sn and ^{113}In were separated on acid purified-silica gel by using a 0.1 *M* ammonium thiocyanate solution in methanol as the developing solvent. This gave R_f values of approximately 0.05 and 0.7 with a solvent travel distance of 10 cm. A four-component system of ^{140}Ba, ^{140}La, ^{113}In was run on acid-washed silica using 0.4 *M* hydrochloric acid in methanol for a development distance of 10.5 cm to yield R_f values of Ba 0.00, La 0.74, and Sn 0.82. Although the R_f values of La and Sn were close, the radioactive count showed a sharp separation. For the separation of ^{141}Ce, ^{113}Sn, and ^{115}Cd, a thin-layer plate of EDTA-treated gel was used with a solvent composed of 0.0025 *M* EDTA solution containing 8% methanol. With a development distance of 14.3 cm, the R_f values were 0.34, 0.69, and 0.85, respectively. Practically identical separations could be obtained by substituting ammonium thiocyanate for the EDTA. Spot locations were identified by running radioactive counts and by chromatograms of individual components. The combination of two-dimensional and stepwise development was applied to the separation of a seven-component mixture of ^{140}Va, ^{140}La, ^{95}Zr, ^{181}Hf, ^{95}Nb, ^{113}Sn, and ^{113}In. The first development was accomplished with 0.3 *M* hydrochloric acid in methanol and then the plates were turned at right angles and developed in 0.1 *M* ammonium thiocyanate in methanol; for the third development (which was made in the same direction as the first development) 2 *M* hydrochloric acid containing 0.2 *M* hydrofluoric acid was used.

Bottura and Breccia [70] applied this technique to the separation of a mixture of elements from irradiated reactor fuel. Complete separation was achieved by two-dimensional thin-layer chromatography. Acetate buffer solution–water–methanol (4:5:41) containing 0.01 *M* EDTA or water–methanol (1:10) with the same amount of EDTA was used in the first direction and acetone–acetonylacetone–hydrochloric acid (200:1:2) in the second direction.

Moghissi [71] has separated a group of nuclides on silica gel G layers with various solvent mixtures (Table 33.2). Inactive ions were located

Table 33.2 $R_f \times 100$ Values for Separation of Some Inorganic Radionuclides on Silica Gel [71]

Radionuclide	$R_f \times 100$	*Solvent*
^{198}Au (colloidal), Au^{3+}	0,90	2 *M* HCl in acetone–water (7:3)
Ba, La, Cs	0,100,80	Butanol–6 *N* HCl (3:7)
^{140}Ba, ^{140}La	0,100	Butanol–6 *N* HCl (3:7)
^{133}Ba, ^{133}Cs	0,80	Butanol–6 *N* HCl (3:7)
^{47}Sc, ^{47}Ca	10,80	0.8 *M* NH_4SCN in water–ethanol (5:3)
Sr, Y	0,100	4% HNO_3 (d = 1.52) in ether–ethanol (1:1)
^{72}Ga, ^{27}Zn	10,60	Butanol (saturated with 1 *N* HCl)
^{95}Nb, ^{182}Ta	10,80	0.8 *M* HCl and 0.1 *M* $H_2C_2O_4$ in acetone–water (4:1)
^{95}Zr, ^{95}Nb	0,90	0.25 *M* $H_2C_2O_4$ and 0.1 *M* HCl in methyl ethyl ketone–dioxane–water (1:1:1)

with typical reagents and the active ions were located by a modified counter.

Volynets and Guseva [72] separated uranium from plutonium and the transplutonium elements by chromatographing on silica gel with tributyl phosphate–benzene (1:1); americium, curium, and plutonium remained at the origin while U^{6+} traveled with the solvent front. In this case the samples were applied in 1 *M* perchloric acid. By application of the samples in 1 *M* nitric acid and development with tributyl phosphate–benzene (1:10) the transplutonium elements could be separated from plutonium. Seiler and Seiler [73] used silica gel Mn S-HR with methanol–2 *M* sodium chloride–1 *M* hydrochloric acid (90:10:1) as a solvent to separate ^{113}Sn and ^{113}In.

Babayan and Varshal [74] separated rare earth metals on silica gel containing 10 mg of starch and 100 to 150 mg ammonium nitrate per gram of silica gel. Development was carried out with 0.11 *M* thiocyanic acid in methyl ethyl ketone. The R_f values were La 0.08, Ce 0.24, Pr 0.31, Nd 0.38, Sm 0.52, Eu 0.56, Gd 0.60, Y 0.65, Er 0.68, and Yb 0.72. Detection was achieved with 0.1% arsenazo I in a 35% solution of urotropine in aqueous ethanol. Vagina and Volynets [75] separated the rare earth metals with tributyl phosphate–benzene (1:1 and 1:10) and with water. Volynets et al. [76] used the tributyl phosphate-benzene solvent in a 1:19 mixture to separate the rare earths from uranyl nitrate of high purity. They were determined spectrophotometrically with arsenoazo III after elution from the gel. Rare earth metals from 0.1 to 10 ppm could be determined. Ce^{3+} can be separated from Ce^{4+} by this technique [77]. Oguma [78] separated Th and U^{6+} on cellulose with dioxane–12 *M* hydrochloric acid or 14 *M* nitric acid in ratios of 7:3, 1:1, or 3:7.

Pierce and Flint [79] have separated rare earth mixtures on layers of Corvic (vinyl chloride–vinyl acetate copolymer) impregnated with bis(2-ethylhexyl) phosphate with various concentrations of hydrochloric acid as the developing agent. Daneels et al. [80] separated yttrium, ytterbium, and gadolinium on starch-bound layers of silica gel H which were impregnated with 0.4 *M* perchloric acid. Bis(2-ethylhexyl)phosphoric acid (equilibrated with the stationary phase) in carbon tetrachloride was used as the developing solvent. The following separations were also achieved: Zr–Gd, Eu–Gd, and Eu–Gd–Sm using various modifications of the stationary phase.

Holzapfel et al. [81] used silica gel impregnated with bis(2-ethylhexyl) phosphate with various concentrations of hydrochloric acid and nitric acid as solvents. All pairs of the elements La, Ce, Pr, Nd, Sm, Eu, Gd, Tb, Dy, Ho, Er, Y, and Yb could be separated except for Pr–Nd and Er–Y. The separations were improved by increasing the concentration of the impregnant and decreasing the thickness of the layer [82]; two-dimensional runs separated nearly all the rare earths.

The Alkaline Earth Group

Seiler [83] has separated this group by using a starch-bound silica gel to obtain the following order of decreasing R_f values: Ca > Sr > Ba. The sample was applied in the form of the acetates, and in order to prevent streaking 0.001 ml of acetic acid was added to the sample spot. Development was carried out with a solvent composed of ethanol–*n*–propanolacetic acid–acetylacetone–water (37.5:37.5:5:1:20). By spraying the developed chromatogram with 1.5% violuric acid in water and heating for 20 min at 100°C, calcium appeared as a yellow-orange spot, strontium as a pink spot, and barium as a red-violet spot. Merkus [43] separated barium, strontium, calcium, and magnesium on cellulose MN 300 with methanol–hydrochloric acid–water (7:1:2). In order to separate Be from this group, acetone–hydrochloric acid–water (10.3:10.5:79.2) (mole %) can be used [4]. Detection was made with T-145.

Hammerschmidt and Mueller [84] have also used thin layers of cellulose for the identification of paper fillers and the friction compositions for striking matches by separating the inorganic cations. These authors purified the cellulose (MN 300) by refluxing with 1.5% nitric acid at 50°C for 2 hr. For the preparation of the sample, the paper or striking material was ashed and then fused in a platinum crucible with alkali and finally treated with 1% hydrochloric acid solution. A development distance of

10 cm was used in a sandwich-type chamber. For the separation, a solvent mixture of ethanol–water–hydrochloric acid (17.5:2.6:3.0) was used giving the following R_f values: barium 0.03, calcium 0.14, magnesium 0.34, aluminum 0.41, titanium 0.49, iron 0.86, and zinc 0.97. The spots were located by spraying first with a saturated solution of alizarin in alcohol and then with a 25% solution of ammonium.

The Alkali Group

This group can be separated as their acetates on silica gel layers, but the silica gel must be purified and should not contain gypsum as a binder because it interferes with the detection of the ions [67]. A starch binder is therefore used with the gel and the acetates are separated by using ethanol–acetic acid (100:1). If the sample is in the form of the sulfates, they can be converted directly to the acetates by spotting the plate with an equivalent quantity of barium acetate. The sample containing the alkali sulfates is then spotted on top of the barium spot. Using this procedure, the alkalies are converted to the acetate salts and can be separated. In this latter case the development is carried out for a distance of 15 cm rather than the 10 cm used for the direct acetate application. The spray reagent of violuric acid gives a red spot at the origin where it reacts with the barium ion. The order of development and the color reaction with 1.5% solution of violuric acid is as follows: Li^+ light red > Mg^{2+} yellow-orange > Na^+ red-violet > K^+ blue-violet. Seiler [85] has also determined sodium, potassium, and magnesium quantitatively after separation on thin layers. After separation of the ions by the method just described, the individual components were determined by several methods. Determination on the basis of the size of the spot gave an error of ±10%; direct photometric measurement with a densitometer provided measurements with an error of ±4%, and the determination of sodium and potassium by means of radioactive tracers gave an error of ±1%. Purdy and Truter [86] determined potassium and magnesium by determining the square root of the area of the spot. Separation of the potassium and magnesium (as acetates) was accomplished with ethanol–methanol (1:1) containing 1% acetic acid, and the spots were located by spraying the plate with a 1.5% solution of Acid Violet 6BN.

Janauer et al. [87] separated the alkali metals as their polyiodide complexes on air-dried silica gel with 0.1 *M* iodine in nitromethane–benzene (2:3) as the solvent. Lesigang-Buchtela [88] separated sodium, potassium, rubidium, and cesium as well as barium, lanthanum, strontium, and yttrium on silica gel containing 5% ammonium phosphododecamolybdate. With 0.01 *N* hydrochloric acid the R_f values were 0.85, 0.52, 0.21, 0.05, 0.55, 0.12, 0.58, and 0.03, respectively.

Handa and Johri [89] obtained a separation of sodium, potassium, rubidium, and cesium on alumina G freed of sodium ions by predevelopment with 0.05 *M* hydrochloric acid. After predevelopment the plates are dried and activated. Phenol saturated with 6 *N* hydrochloric acid gave R_f values of 0.0, 12, 23, and 43, respectively.

Markus [43] separated lithium, sodium, and potassium on cellulose MN 300 using methanol–10 *N* hydrochloric acid–water (20:3:2). Detection was made with T-276. Since this test is not very sensitive for potassium, Markus [10] recommends Pollard's method [90]; after detection of sodium and lithium as above, the plate is sprayed with a saturated solution of barium nitrate to precipitate any sulfate ions. Then T-70 reveals potassium as a grayish-black precipitate of Pb–Co–K triple nitrite; however T-253 appears to be somewhat more sensitive for potassium.

The Noble Metals

Johri et al. [91] separated Pd^{2+}, Rh^{3+}, and Ru^{3+} on silica gel with *tert*-butanol–acetic acid (6:1) and Os^{8+}, Ir^{4+}, and Pt^{4+} with *tert*-butanol–acetic acid–hydrochloric acid and (20:3:1). The R_f values were 0.58, 0.33, 0.77, 0.08, 0.68, and 0.93, respectively. In the first case T-215 was used as a detection agent and benzidine in the second; iridium was fumed with hydrochloric acid. Verma and Rai [92] used amyl acetate–hydrochloric acid (18:1) to separate platinum, rhodium, gold, palladium, and mercury applied as chlorides on silica gel. Yamamoto and Uno [93] separated Au (R_f 0.94), Pt (0.68), Pd (0.81), and Cu (0.0) on silica gel with acetone as a solvent. Rai and Kukreja [19] chromatographed the diethyldithiocarbamate complexes of gold, palladium, and platinum as well as a number of other metals on silica gel with benzene, toluene, xylene, and chloroform–carbon tetrachloride (1:1).

Hashmi and Adil [94] used circular thin-layer chromatography to separate rhodium, gold, ruthenium, platinum, osmium, palladium, and copper with acetone–acetylacetone–2 *M* hydrochloric acid (100:10:3). Acetone–acetonylacetone–water–2 *M* hydrochloric acid (20:2:1:1) was used for palladium. A 1% rubeanic acid solution in 96% ethanol detected all the compounds except rhodium and osmium with a sensitivity ranging from 0.1 to 2.0 μg [95]. A 4% solution of thiourea in 2 *M* hydrochloric acid detects osmium at the 0.3-μg leveland 2% stannous chloride in 4 *M* hydrochloric acid containing 0.5% potassium chloride detects rhodium at 1.3 μg. Au^{3+}, Se^{4+}, and Te^{4+} can be separated on aluminum oxide with 3 *M* Sodium hydroxide–ammonia–water (2:5:13) with R_f values of 0.0, 0.52, and 0.34, respectively [96].

Gagliardi and Shambri [97] separated gold, platinum, palladium, ruthenium, rhodium, and silver on cellulose by using a 40-cm layer and using a development, first with isobutyl methyl ketone–hydrochloric acid–*n*-butanol–methanol–acetylacetone (11:6:5:2:0.5) for 17.5 cm and then with acetone–hydrochloric acid–water–*n*-hexanol (14:2:4:2.5) for another 10 cm. Osmium was separated from ruthenium, palladium, platinum, and iridium by a two-dimensional technique.

Total Analysis

Takitani et al. [98] have used thin layers of purified silica gel with a starch binder and three developing solvents for the analysis of 20 common metal ions. Separation of Ni, Co, Cu, Fe, Pb, Mn, Cr, and (As) was first accomplished by developing in a solvent composed of acetone–3 *N* hydrochloric acid (99:1). Methanol–butanol–35% hydrochloric acid (8:1:1) was used to give a separation for Ba, Sr, Ca, Mg, Al, NH_4, Na, K, and Li. The third solvent, butanol–benzene–1 *N* nitric acid–1 *N* hydrochloric acid (50:46:2.6:1.4), was used for the separation of Sb, As, (Cu), Cd, Sn, Bi, Zn, and Hg. The separation of the same element with different valence states was also examined. Takitani et al. [99], using a combination ammonium sulfide method and thin-layer chromatography, examined the effect of various anions on the identification of the cations. The identification limits for 24 cations was found to be 10 to 100 times more sensitive than those with paper chromatography.

Using purified silica gel layers without a binder, Druding [100] applied two solvent systems for the separation of 18 ions. Acetic acid–95% ethanol (5:95) was used for the separation of the alkali metals and *tert*-butanol–2 *N* hydrochloric acid (95:5) for the remainder of the ions. Table 33.3 shows the R_f values of the ions in both solvents along with the color reactions; however, it should be remembered that when the ions are run in mixtures, the R_f values are displaced as illustrated in Table 33.4 for the alkali metals. Two detecting agents were used, a violuric acid agent (T-1) and T-253.

Merkus [101] has applied thin layers of cellulose (MN 300) in the toxicological analysis of metals. The two solvents used for the separation were acetone–4 *N* hydrochloric acid (7:3) and acetone–25% nitric acid (7:3). The R_f values obtained with these solvents are given in Table 33.5. For the location of the spots, seven general reagents were used; the author lists the colors with these reagents as well as with 13 specific reagents for various ions.

Zetlmeisl and Haworth [102] tabulate the separation of various combinations of 21 cations on cellulose, and Kaushik and Johri [103] give the separations of 27 ions on silica gel G including R_f values, color, and limits of identification with T-215.

Hashmi et al. [104] separated 39 cations into five groups by means of

Table 33.3 $R_f \times 100$ Values of Some Inorganic Ions on Purified Silica Gel Along with Color Reactions[a,b]

Ion	*Acetic Acid–Ethanol* (5:95)	2 *N* *Hydrochloric Acid–tert-Butanol* (5:95)	*Violuric Acid*	*Li(TCNQ)*[c]	*Sensitivity of TCNQ*$^-$ (μg)
Li	98		Yellow		0.5
Na	89	39	Orange	Bright blue	0.5
K	60	20	Red	Bright blue	2
Rb	53	17	Red-violet	Blue	2
Cs	38	10	Red-violet	Gray	20
Be	99	100[d]	Yellow-green	Yellow	20
Mg	95	96[d]	Yellow	Olive green	20
Ca	94	93[d]	Orange	Gray-green	50
Sr	93	15	Red-pink	Pale green	50
Ba	90	0	Pink	Blue	25
Cu^{2+}	80[e]	95		Green	0.1
Ag	89[e]	38[e]		Bright blue	10
Zn	90	100[e]		Blue	10
Cd	100	100		Blue	10
Hg^+	98	0		Bright yellow	10
Hg^{2+}	100	97		Bright yellow	40
Tl^+	30[e]	8	Red	Gray	75
Pb^{2+}	90	2	Orange	Blue	

[a] From Druding [100]; reproduced with permission of the author and The American Chemical Society.
[b] Development distance 10 cm.
[c] 0.5% lithium tetracyanoquinodimethanide in ethanol–water (1:1).
[d] Very narrow band.
[e] Tailing.

solvent extractions. These were then chromatographed by circular thin-layer chromatography on silica gel except for those of the fifth group which were separated on alumina. Group I containing Fe^{3+}, Au^{3+}, Mo^{6+}, V^{5+}, Ga^{3+}, Sb^{5+}, As^{3+}, Te^{4+}, and Ge^{4+} and group II consisting of Pd^{2+}, Pt^{4+}, Co^{2+}, Zn^{2+}, and Sn^{4+} were chromatographed with acetone–4 *M* hydrochloric acid–acetylacetone (45:3:2); group III (Cu^{2+}, Ru^{3+}, U^{6+}, Ti^{4+}, Zr^{4+}, Cr^{3+}, La^{3+}, and Al^{3+}) was separated with acetone–4 *M* hydrochloric acid (47:3); group IV (Hg^{2+}, Cd^{2+}, Bi^{3+}, Pb^{2+}, Se^{4+}, Rh^{2+}, Ni^{2+}, and Mn^{2+}) was separated with acetone–4 *M* hydrochloric acid–acetylac-

Table 33.4 Comparison of $R_f \times 100$ Values of Alkali Metal Ions Run Individually and as a Mixture in 5% Acetic Acid in Methanol[a,b]

Ion	*Individually*	*As a Mixture*
Li	98	98
Na	89	89
K	60	60
Rb	5	27
Cs	38	15

[a] From Druding [100]; reproduced with permission of the author and The American Chemical Society.

[b] Development distance 10 cm.

Table 33.5 $R_f \times 100$ Values of Some Toxic Inorganic Ions on Cellulose Layers[a,b]

	Solvent			*Solvent*	
Ion	*Acetone–4 N Hydrochloric Acid* (7:3)	*Acetone–25% Nitric Acid* (7:3)	*Ion*	*Acetone–4 N Hydrochloric Acid* (7:3)	*Acetone–25% Nitric Acid* (7:3)
Ag^{+}	50–90	30	Hg^{2+}	90	90
As^{3+}	80	40	Mn^{2+}	30	50
Ba^{2+}	0–10	10	Ni^{2+}	20	30
Be^{2+}	70	60	Pb^{2+}	70	30
Bi^{3+}	100	90	SeO_3^{2-}	100	50
Cd^{2+}	100	40	Sb^{3+}	100	90
Ce^{4+}	10	0	TeO_3^{-}	30	20
Co^{2+}	40	30	Tl^{+}	0	20
Cu^{2+}	80	50	UO_2^{2+}	90	90
			Zn^{2+}	100	30

[a] From F. W. H. M. Merkus [101]; reproduced with permission of the author and de Koninklijke Nederlandsche Maatschappij ter Bevordering der Pharmacie.

[b] Development distance 10 cm.

etone (48 : 1.5 : 0.5), and group V (Sr^{2+}, Ba^{2+}, Th^{4+}, Ce^{3+}, Mg^{2+}, K^+, Rb^+, Cs^+, and Li^+) was chromatographed on alumina with acetone–4 *M* hydrochloric acid (46:4). Group V was also run on silica with acetone–concentrated hydrochloric acid (47 : 3). The R_f values are given as well as the colors obtained with various detection agents.

Husain and Eivazi [105] tabulate the separations in various combinations of 44 cations on the inorganic ion exchanger stannic arsenate.

Separation on Ion-Exchange Layers

The Hydrogen Sulfide Group

Zabin and Rollins [106] have investigated the use of inorganic compounds as ion exchangers for the separation of cations. Zirconium phosphate and hydrous zirconium oxide in the ammonium form, with 3% cornstarch as a binder, were used for making the thin-layer plates. With the ammonium form of zirconium oxide Hg (R_f 0.9) could be separated from Cd (R_f 0.3) and from the group Cu, Ag, Fe, Pb (R_f's 0.0), as well as from Ni and Co (R_f's 0.5), by using 2.0 *M* ammonium nitrate solution as the developing solvent over a distance of 10 cm. With the hydrogen form of zirconium phosphate and with 0.1 *M* hydrochloric acid as a developing agent, the following R_f values were obtained: Pb 0.0, Ag 0.0, Cu 0.1, Cd 0.4, and Hg 0.85. In addition, in this system iron had an R_f value of 0 to 0.1 and nickel and cobalt tailed badly. With both these inorganic exchangers the spots were detected by spraying with an ammoniacal solution of ammonium sulfide.

Oguma [107] chromatographed 39 cations on ECTEOLA cellulose and on cellulose with various concentrations of hydrochloric acid in methanol, hydrochloric acid, and acetic acid–hydrochloric acid. Cd^{2+} was separated from 38 other ions in various combinations. Kuroda et al. [108] chromatographed Cd^{2+}, Cu^{2+}, and Hg^{2+} on DEAE cellulose in aqueous sodium azide–hydrochloric acid and aqueous azide–hydrochloric acid–methanol mixtures. The research also included the transition metals. Frache and Dadone [109] examined the behavior of Pb^{2+}, Bi^{3+}, Sn^{2+}, Sb^{3+}, Cd^{2+}, Cu^{2+}, Cr^{3+}, and Hg^{2+} on the anionic resin Amberlite CG 400 and the cationic resin Amberlite CG 120 with various concentrations of hydrochloric acid as developing solvent and with various concentrations of nitric acid [110]. Shimizu et al. [111] studied this same group of ions except with Sn at a valency state of IV on carboxymethyl cellulose layers with sulfuric acid and acidic ammonium sulfate solutions. The R_f values are listed for a number of multicomponent mixtures. Lepri and Desideri [112] determined the R_f values of divalent mercury, lead, copper, cadmium, and monovalent mercury on sodium carboxymethylcellulose and on Dowex 50-X4 (Na^+) with various concentrations of acetate, lactate, and oxalate buffer solutions as well as in neutral solutions of univalent and divalent ions and alkaline solutions of univalent ions [113].

The Ammonium Sulfide Group

Berger et al. [114] have applied a double layer plate for the separation of Fe^{3+}, Ni^{2+}, and Co^{2+}. A narrow strip of the plate was covered with a mixture of cellulose MN 300 (3 g) and 0.1 g of dimethylglyoxime in 35 ml of water. The balance of the plate was covered with Dowex 50 WX 2 in the ammonium form (30 g) mixed with cellulose (5 g) and 60 ml of water. After drying the layer, the sample was spotted on the dimethylglyoxime portion of the layer and developed with water–ethanol (70:20) containing 5 g of tartaric acid and neutralized with ammonia to a definite ammonia odor. With this system the nickel forms the dioxime complex and is retained at the point of application. Part of the cobalt stops at the line of the separation of the two layers and the remainder distributes itself successively in several spots. The spots are probably complexes of the cobaltammine type for the colors may be accentuated by spraying with an aqueous solution of sodium sulfide. The iron travels with the solvent front probably as a tartrate complex.

Various members of this group have been chromatographed on EC-TEOLA cellulose [107], Amberlite CG 400, Amberlite CG 120 [109, 110], carboxymethyl cellulose [111], sodium carboxymethyl cellulose and Dowex 50-X4 (Na^+) [112, 113] in the solvents mentioned above under the hydrogen sulfide group.

The Rare Earths

The lanthanides have been chromatographed on sodium carboxymethyl cellulose and Dowex 50-X4 (Na^+) with lactate buffer solutions [115]. Shimizu and Muto [116] chromatographed the trivalent rare earths and zirconium, hafnium, thorium, and uranium on DEAE cellulose with 0.1 *M* sulfuric acid–0.05 *M* ammonium sulfate; these cations can be separated completely by a two-dimensional separation using this solvent in the first dimension and 0.1 *M* sulfuric acid–1 *M* ammonium sulfate in the second dimension. Ishida [117] examined the behavior of the rare earths on DEAE cellulose with methanol–nitric acid solvents. The cations from lanthanum to neodymium were best separated with methanol–8 *N* nitric acid (5:1) and methanol–1 *N* nitric acid (20:1). The cations samarium, europium, and gadolinium were separated with methanol–14 *N* nitric acid (20:1).

The Alkaline Earth Elements

Berger et al. [114, 118] also applied the use of Dowex 50 WX 2 (H^+) to the separation of a mixture of barium, calcium, and cesium. The layers were bound with cellulose as described under the ammonium sulfide group and the separation was carried out with 0.75 *M* ammonium lactate. By applying this method to the separation of the radioactive isotopes, ^{45}Ca, ^{89}Sr, and ^{131}Ba could be separated, as well as the transformation product ^{131}Cs which travels with the solvent front.

Cations of this group have been chromatographed on ECTEOLA cellulose [107], Amberlite CG 400 and Amberlite CG 120 [109, 110], carboxymethyl cellulose [111], and sodium carboxymethyl cellulose and Dowex 50-X4 (Na^+) [112, 113] in the solvents as mentioned under the hydrogen sulfide group. Berger et al. [119] used a dual layer of the anionic resin Dowex 1-X10 and the cationic Chelax 100 (Bio-Rad) in order to separate Ca^{2+}, Sr^{2+}, and Ba^{2+} as well as the anions I^-, Br^-, and Cl^- at one time with 2 *M* ammonium nitrate.

The Alkali Group

For the separation of the alkali metals, Berger et al. [114] used the Dowex 50 (H^+)–cellulose (30:5) mixture. With 1 *M* lithium chloride as the development solvent, barium, cesium, and sodium could be separated from one another and from the pair of rubidium and potassium which could not be separated. The radioactivity of the spots was measured by a Geiger–Müller counter.

Potassium and rubidium, however, can be separated very nicely on an inorganic ion exchanger. Lesigang [120] has examined the use of inorganic ion-exchange materials for the separation of the isotopic alkali metals. The compounds examined were: ammonium phosphododecamolybdate (APM), ammonium arsenododecamolybdate (AAM), ammonium germanododecamolybdate (AGM), and oxinegermanododecamolybdate (OGM). The layers were prepared without a binder by preparing a slurry of the crystals with acetone or acetone–water mixtures as required; spreading on the plate, and allowing to dry in the air for 24 hr. As an example, the APM prepared according to the method of Smit et al. [121] was slurried in an acetone–water (9:1) mixture and spread on the supporting plate so as to obtain approximately 10 to 12 mg/cm². The ^{22}Na, ^{42}K, ^{86}Rb, and ^{137}Cs were applied to the plate and developed with various concentrations of ammonium nitrate solution (Table 33.6). The best overall separation was given with APM with 5.0 *M* ammonium nitrate; however, the separation of rubidium and cesium was improved by using a 10.0 *M* solution of the same solvent. An excellent separation of cesium from the other three ions was possible by using 0.1 *M* ammonium nitrate solution with OGM layers. In this system cesium did not move, whereas the other three had high R_f values. Location of the compounds was made by means of a Geiger–Müller counter with the

Table 33.6 $R_f \times 100$ Values of Alkali Metals on Inorganic Ion-Exchange Materials in Various Concentrations of Ammonium Nitrate[a]

Thin-layer[b]	*Molarity of NH_4NO_3 Solvent*	$R_f \times 100$[c]			
		^{22}Na	^{24}K	^{86}Rb	^{137}Cs
APM	0.1	80	21	14	14
	1.0	90	42	20	16
	5.0	96	72	24	20
	10.0	100	100	40	20
AAM	0.1	60	18	14	16
	1.0	80	25	20	20
AGM	0.1	50	20	0	0
	1.0	70	30	14	14
OGM	0.1	94	90	92	0
	1.0	100	100	100	96

[a] From M. Lesigang [120]; reproduced with permission of the author and Springer-Verlag.

[b] APM = Ammonium phosphododecamolybate; AAM = ammonium arsenododecamolybdate; AGM = ammonium germanododecamolybdate; OGM = oxinegermanododecamolybdate.

[c] Development distance 10 cm.

window covered by aluminum foil having a 5-mm slit. Lesigang and Hecht [122] also examined other heteropoly acids.

Kawamura et al. [123] obtained a good separation of sodium, potassium, rubidium, and cesium on zinc ferrocyanide (an inorganic ion exchanger)–cellulose (1:10) layer with 0.2 *M* ammonium nitrate as the developing solvent. Berger [119] used a dual layer composed of Dowex 1-X10 (OH^-) and Dowex 50-WX2 (H^+) with 1 *M* lithium nitrate as a solvent to separate both the anions and cations from a mixture of sodium iodide, potassium bromide, and cesium chloride.

Miscellaneous

The weakly basic anion exchanger DEAE cellulose has also been investigated for separating cations. Kuroda and co-workers have examined the behavior of 48 cations on this exchanger with binary solvent mixtures containing various concentrations of sulfuric acid and an organic solvent such as acetone, dioxane, and acetic acid [124]; with thiocyanic acid–organic solvent mixtures [125]; and with various concentrations of phosphoric acid [126]. Kuroda et al. [127] measured the R_f values of 52 cations as the function of hydrochloric acid concentration in hdyrochloric acid–organic solvents (1:20).

Electrophoresis

Pfrunder et al. [128] has investigated the separation of inorganic ions on thin layers of agar-agar at a voltage of 120 V. The separation of cobalt, nickel, copper, and iron took place in 10 to 12 min, with decreasing mobility in the order named. For the detection of these compounds, the thin agar layer was washed with distilled water and then exposed to ammonia vapor for 20 min, after which it was immersed in a 1% solution of dithiooxamide in alcohol for 5 min. By this method Pb^{2+}, Hg^{2+}, and Cd^{2+} were separated with a voltage of 110 V for 10 min. After the initial washing and treatment with ammonia gas, the visualizing solution was a 1% solution of diphenylcarbazide in ethanol. This showed the mercury at the origin with the lead and cadmium well separated, the latter having the highest mobility. In separating chromium, manganese, and nickel in the presence of a large amount of ferric ion, it was found that the large excess of iron diffused over the layer and interfered with the other ions. This difficulty was solved by preparation of the agar layers with a buffer solution consisting of 2 *M* ammonium hydroxide-2 *M* ammonium chloride–water (1:9:190). With this buffered layer the sample was spotted in the center, and after application of 80 V for 20 min, the nickel and iron were found to have traveled toward the cathode, with the nickel having the higher mobility, while the chromium and manganese moved toward the anode with the chromium moving faster than the manganese. In this case the color-developing agent for the nickel and the iron was a 1% solution of dimethylglyoxime in ethanol giving brown and red spots, respectively; then after a further treatment with ammonia gas and development with a 0.05% benzidine solution in 10% acetic acid, manganese appeared at once as a blue spot and chromium behaved similarly after a period of time.

Moghissi [129] has applied high-voltage separations at 45 V/cm for the separation of cations. Two types of equipment were used, one having a water-cooled chamber and the other carried out in a water-vapor atmosphere at 90°C. With 0.5 *M* lactic acid as the electrolyte, the following migrations, in millimeters, were observed: Fe^{3+} 45, Zr^{4+} 0, Nb^{5+} 40, Pt^{4+} 0 and 15, Hf^{4+} 50 (tailing), Co^{2+} 75, Ni^{2+} 70, Ba^{2+} 65, Pb^{2+} 0, Ga^{2+} 55, Sr^{2+} 20, Ce^{3+} 60, La^{3+} 50, Y^{3+} 60 (tailing), Pd^{2+} 35, W^{6+} 55 (tailing), Cd^{2+} 75, Sc^{3+} 40, Bi 55 (tailing), Zn^{2+} 85 (tailing), Rh^{3+} 45, Ir^{4+} 50, Ru^{3+} 62, Tl^{+} 58, Ag^{+} 90, Sn^{2+} 75, Ti^{4+} 25, Cu^{2+} 65, Be^{2+} 58, Sb^{3+} 60, Al^{3+} 50, Li^{+} 65, and Mg^{2+} 70 mm. These separations were achieved in 5 min.

Takitani et al. [130] used 1200 V/30cm for 20 min to separate Na^+, Li^+, Ca^{2+}, Mg^{2+}, Ba^{2+}, Sr^{2+}, Cu^{2+}, Ni^{2+}, Co^{2+}, Fe^{3+}, Pb^{2+}, Zn^{2+}, Cr^{2+}, Mn^{2+}, Al^{3+}, Cd^{2+}, Bi^{3+}, Sn^{4+}, Sb^{3+}, As^{3+}, Ag^+, and Hg^{2+} on starch-bound silica gel layers with citric acid solution as the electrolyte. Buchtela and Lesigang-Buchtela [131] give the migration distances for 32 metal ions on cellulose at 50 V/cm and 5 to 15 mA for 20 min in 0.1 *M* 2-hydroxy-2-methylpropionic acid at pH 2.72, 4.0, 6.0, and 8.8.

EDTA solutions have been used as the electrolyte for the electrophoresis of some rare earths [132, 133] as well as for cations of the hydrogen sulfide group [134].

Van Ooij and Houtman [135] applied a two-dimensional technique of electrophoresis in the first dimension followed by thin-layer chromatography in separating radioactive iridium compounds.

2 ANIONS

Phosphates

Because of the reaction of calcium sulfate with the phosphates to form insoluble calcium phosphates, Seiler [136] used layers of purified silica gel bound with starch for the separation of secondary pyrophosphate, primary orthophosphate, primary orthophosphite, and primary hypophosphite. After numerous solvents, both basic and acidic were tried, a mixture of methanol–ammonium hydroxide–10% trichloroacetic acid–water (10:3:1:6) was selected as giving the best results using a development distance of 10 cm. After the developed plates were dried, they were sprayed with 1% ammonium molybdate, followed by 1% stannous chloride in 10% hydrochloric acid. With this procedure the ions appeared as blue spots (the hypophosphite after a period of time). The separation of these ions is in the following order of decreasing R_f values: $H_2PO_2 > H_2PO_3 > H_2PO_4 > H_2P_2O_7$.

Pilson and Fragala [137] separated silicate and phosphate on ECTEOLA cellulose with isopropanol–water–acetic acid (20:5:2). Phosphate was detected with T-19 and silicate with T-18.

Roessel [138] investigated a number of different adsorbents for the separation of a group of condensed phosphates and selected double acid-washed cellulose powders, 140 and 142 dg (Schleicher & Schuell). He used cornstarch as a binding agent since these cellulose powders will not adhere sufficiently to the plate by themselves. This survey of adsorbents included not only silica gel G and silica gel Woelm without a binder, but also six other cellulose powders of various manufacturers. Five different solvent mixtures were used in the separations; the R_f values for the condensed phosphates are given in Table 33.7. Two-dimensional separations were also applied for the separation of mono-, di-, tri-, trimeta-, tetrameta-, and long-chain phosphates. For the development in the first direction, solvent A (see Table 33.7) was used and for the second direction the following solvent mixture was used: 67.5 ml of methanol; 22.6 ml of a solution of 700 ml of isopropanol and 100 ml of distilled water; 50 ml of a solution of 75 g of trichloroacetic acid and 80 ml of 25% ammonia diluted to 1000 ml with distilled water; 6 ml of a solution of 96% acetic acid and 800 ml of distilled water. The same two-dimensional

Table 33.7 $R_f \times 100$ Values of Condensed Phosphates on Cellulose Layers in Various Solvents[a]

	Cellulose 140 dg (Schleicher & Schuell) [138][b]		*Cellulose 142 dg (Schleicher & Schuell)* [138][b]		*Purified Cellulose MN 300* [139][c]	
Phosphate	*A*	*B*	*B*	*C*	*D*	*E*
Mono-	84	84	83	55	55	34
Di-	75	64	67	72	40	15
Tri-	60	46	45	58	31	15
Tetra-	46	30	30	45	22	
Penta-	33	18	20	30	16	
Hexa-	23		12	19	11	
Hepta-	16		7	11	7	
Octa-	10		4	6	5	
Trimeta-	45	19	17	34		6
Tetrameta-	24	10	19	18		47
Pentameta-			5	12		
Hexameta-				8		
Heptameta-				5		

[a] *A* = 60 ml of methanol, 30 ml of dioxane, 30 ml of isopropanol–water (7:1), 8 ml of 96% acetic acid–water (1:4), 40 ml of a solution of 125 g of trichloroacetic acid and 35 ml of 25% ammonia diluted to 1000 ml. *B* = 70 ml of dioxane, 30 ml of a solution of 160 g of trichloroacetic acid and 8 ml of 25% ammonia diluted to 1000 ml. *C* = 75 ml of methanol, 20 ml of isopropanol–water (7:1), 25 ml of a solution of 125 g of trichloroacetic acid and 32 ml of 25% ammonia diluted to 1000 ml, 6 ml of 96% acetic acid–water (1:4). *D* = 30 ml of water, 35 ml of ethanol, 15 ml of isobutanol, 20 ml of isopropanol, 0.4 ml of ammonia 22° Bé, 5 g of trichloroacetic acid. *E* = 9 ml of ammonia 22° Bé, 50 ml of methanol, 10 ml of isobutanol, 31 ml of water, 0.3 ml of formic acid.

[b] Running time 60 min; development distance about 16 cm.

[c] See text for purification procedure. Development distance 14 cm.

technique was also applied to the separation of a mixture of linear condensed phosphates and cyclic condensed phosphates. In the latter case spots were obtained for trimeta-, tetrameta-, pentameta-, hexameta-, and heptametaphosphates.

For the detection of the various compounds, the dried plates were sprayed with a molybdate solution consisting of 40 g of sodium molybdate dihydrate and 50 g of ammonium nitrate dissolved in water and diluted to 1000 ml. This solution was then poured into 100 ml of concentrated nitric acid (d_{20} = 1.40). This was followed by drying the plates and then spraying with a reducing solution composed of 300 g of sodium pryosulfite, 10 g of sodium sulfite, and 2 g of Metol dissolved in 1000 ml of water and filtered [138].

Aurenge et al. [139] have also applied thin layers of cellulose for the separation of condensed phosphates. In this case the authors purified the cellulose for acid solvents by washing cellulose MN 300 successively with hydrochloric acid and 8-hydroxyquinoline followed by rinsing with ethanol and distilled water to a pH of 7. After preparation of the plates, they were given a predevelopment with trichloroacetic acid in order to move the remaining trace impurities to the top of the plate. After the treated cellulose was rinsed with alcohol for the alkaline solvents, a final supplementary wash of trichloroacetic acid was used followed by washing with distilled water to a pH of 7. With a multicomponent solvent system, the effect of varying the components on the separation was investigated. The R_f values for the best alkaline and acidic solvents are given in Table 33.7. Two-dimensional chromatography could be carried out by developing in one direction with the acidic solvent and then in the second direction with the basic solvent.

Clesceri and Lee [140] have separated orthophosphate and pyrophosphate on cellulose layers using dioxane–water–ammonium hydroxide–trichloroacetic acid (65 ml : 27.5 ml : 0.25 ml : 5 g) to obtain R_f values of 0.83 and 0.61, respectively. Baudler and Stuhlmann [141] and Baudler and Mengel [142, 143] have separated mono-, di-, and triphosphoric acids on layers of cellulose MN 2300 using various mixtures of methanol, ammonium hydroxide, acetic acid, and water with acetone or ethanol. Covello and Schettino [144] have applied a photodensitometric determination to polyphosphates in foods after separating them by thin-layer chromatography.

Halides

Seiler and Kaffenberger [145] applied solutions of the alkaline salts for separating the anions in acetone–*n*-butanol–ammonium hydroxide–water (13 : 4 : 2 : 1) on purified silica gel G layers. Although applied as the alkali salts, the anions moved as the ammonium salts with the fluoride remaining at the origin. In the order of increasing R_f values the remaining ions were: chloride, bromide, and iodide. These were located with a 0.1% solution of bromocresol purple in ethanol which was adjusted to the turning point with dilute ammonia. The spots were also detected by a combination of 1% ammoniacal silver nitrate and 0.1% fluorescein in ethanol. The fluoride ion, which could not be detected by either of these reagents, was made visible with a 0.1% solution of zirconium–alizarin in strong hydrochloric acid.

Tustanowski [146] separated Cl^-, Br^-, and I^- on alumina with 0.2 *M* potassium nitrate. Petrović and Canić [147] used maize starch layers to separate Cl^- (R_f 0.40), Br^- (0.60), I^- (0.78), and F^- (0.04) with acetone–3 *N* ammonia (7:3). These four ions may also be separated on microcrystalline cellulose with acetone–water (4:1) or acetone–ethyl acetate–water (3:1:1) [148].

Berger et al. [114, 118, 149] have applied the Dowex 1X2 anion exchanger to the separation of chloride, bromide, and iodide. The layers were prepared by mixing 30 g of 100 to 200 mesh resin with 5 g of cellulose MN 300 and 60 ml of water. The cellulose acted as a binder to retain the resin on the thin-layer plate. The resin was used in either the OH^- or Cl^- form with 1 *M* sodium nitrate solution as the developing solvent. The radioactive forms of the ions were separated and were detected by means of a scintillation counter. On the ion-exchange resin the order was $I^- < Br^- < Cl^-$; this was the reverse of that on silica gel.

Muto [150, 151] has applied thin-layer precipitation chromatography to the separation of calcium chloride, calcium bromide, calcium iodide, and also calcium phosphate. The silica gel layer was prepared with 3% aqueous silver nitrate solution and development was carried out with water saturated with isobutanol–40% ammonium acetate (4:1) as the solvent.

Other Anions

Seiler and Erlenmeyer [152] used thin layers of silica gel bound with starch (MN-KieselgelS-HR) for the separation and identification of some oxo acids of sulfur and some polythionic acids as their alkali salts. The oxo acids were best separated by a mixture of methanol–1-propanol–ammonium hydroxide–water (10:10:1:2), and the polythionates by a mixture of methanol–dioxane–ammonium hydroxide–water (3:6:1:1). Ammoniacal silver nitrate solution and bromocresol green were used as locating reagents. Handa and Johri [153] obtained a separation of S^{2-}, $SO_3{}^{2-}$, $SO_4{}^{2-}$, and $S_2O_3{}^{2-}$ on microcrystalline cellulose. The sodium salts were applied to spots of 1% zinc acetate (previously dried) and then developed with *n*-propanol–1 *M* ammonia–acetone (15:10:1). The R_f values were 0.0, 0.70, 0.28, and 0.51, respectively. Quantitative results were obtained by ring colorimetry.

Chlorate (R_f 0.82), perchlorate (0.97), bromate (0.53), iodate (0.06), and periodate (0.06) were chromatographed on maize starch [147] with acetone–3 *N* ammonia (7:3). Iodate, periodate, and also iodide may be separated on silica gel with methanol–25% ammonia–water–10% acetic acid (18:2:2:1) [154]. Lederer and Sinibaldi [155] obtained R_f values of 0.67, 0.27, 0.10, and 0.07 for perbromate, bromate, periodate, and iodate, respectively. Cellulose layers on aluminum foil were used with butanol–pyridine–1 *N* ammonia (2:1:2). Peschke [156] separated perchlorate, chlorate, chlorite, bromate, bromite, and iodate on alumina–silica gel (1:1) with *n*-butanol–acetone–ammonia–water (8:10:2:1).

Phipps [157] separated perrhenate, molybdate, and selenite ions on Chromagram silica gel sheets with nitric acid–methanol–water in various concentrations. Silica gel layers on glass plates did not give as good a separation because the molybdate ion was very close to the perrhenate ion. Johri et al. [158] separated molybdate, selenite, tellurite, and vanadate ions on silica gel with *n*-butyl acetate–hydrochloric acid (40:0.6). A better separation was obtained between selenite and molybdate using diethyl oxalate–hydrochloric acid (60:1).

Table 33.8 $R_f \times 100$ Values of Various Anions on Layers of Starch-Bound Lanthanum Oxide[a]

	Solvent			*Solvent*	
Ion	1 *N* *Ammonium Hydroxide*	1 *N* *Ammonium Hydroxide–Acetone* (1:1)	*Ion*	1 *N* *Ammonium Hydroxide*	1 *N* *Ammonium Hydroxide–Acetone* (1:1)
SO_4^{2-}	100	60[b]	Br^-		90
PO_4^{3-}	0	0	Cl^-		100
$Fe(CN)_6^{4-}$	100	45	SO_3^{2-}	100	80
$Fe(CN)_6^{3-}$	100	90	$S_2O_3^{2-}$	100	60
MnO_4^-	0	0	SeO_3^{2-}	75	30
SCN^-	100[b]	90	TeO_3^{2-}	0	0
$Cr_2O_7^{2-}$	80	35[b]	MoO_4^{2-}	90	60
IO_3^-	85	45	NO_2^-	90	85
I^-	95	90	NO_3^-	90	90

[a] From A. Moghissi [71]; reproduced with permission of the author and The Elsevier Publishing Co.

[b] Tailing.

Moghissi [71] has determined the R_f values of a series of anions on layers of lanthanum oxide with a starch binder. These R_f values are given in Table 33.8 for two solvents. Iodide, iodate, and tellurite were separated on silica gel layers purified according to Seiler and Seiler [2]. The best solvent for this latter separation was a mixture of acetone and 6 *N* ammonia (1:1) which gave R_f values of 0.9, 0.4, and 0.1, respectively.

Siechowski [159] applied thin-layer chromatography on silica gel layers to the quantitative determination of chromic acid using a spot area method to give results with an average error of ±6%. Developing agent for the method was methanol–water (8:2). With low concentrations of chromic acid, the sensitivity of detection could be increased tenfold by using a spray of 1% of diphenylcarbazide in acetone which gave violet-colored spots

Separation by Electrophoresis

Dobici and Grassini [160] separated periodate and iodate on thin layers of plaster of Paris. The plaster of Paris layer was saturated with 0.05 *M* ammonium carbonate solution as an electrolyte and the separation was carried out at a potential of 300 to 400 V for 1.5 to 2 hr. The separation was superior to electrophoresis on paper. Moghissi [129] used silica gel or kieselguhr layers for the electrophoretic separation of anions. Both low-and high-voltate equipment was used, although the results reported are from the high-voltage apparatus. For the separation of the anions an electrolyte of 0.1 *N* sodium hydroxide was used and the migration distances, in millimeters, for a period of 2 min at 45 V/cm were as follows: SCN^{-35}, SeO_3^{2-} 30 (tailing), TeO_3^{2-} 22 (tailing), I^- 60, IO_3^- 51, Cl^- 55, ClO_3^- 53, Br^- 60, BrO_3^- 45, NO_3^- 58, No_2^- 55, SO_4^{2-} 56, and PO_4^{3-} 0. Radionuclides were used for detection.

Total Analysis of Anions

Kawanabe et al. [161, 162] have applied thin-layer chromatography on purified silica gel with 5% starch binder to the total analysis of anions. Group A, containing SCN^-, I^-, Cl^-, $Fe(CN)_6^{3-}$, $Fe(CN_6)^{4-}$, ClO_3^-, BrO_3^-, IO_3^-, and NO_3^-, were separated with acetone–water (10:1). Group B, containing F^-, NO_2^-, $S_2O_3^{2-}$, SO_4^{2-}, CrO_4^{2-}, PO_4^{3-}, AsO_4^{3-}, and AsO_3^{3-}, were chromatographed in methanol–butanol–water (3:1:1), and group C, containing $C_2O_4^{2-}$ and BO_2^{2-}, was developed with butanol saturated with 2 *N* nitric acid.

Hashmi et al. [104] give the R_f values, color reactions, and sensitivities for 19 anions separated by circular thin-layer chromatography mainly on aluminum oxide layers with the upper phase of *n*-butanol–pyridine–water–ammonia (8:4:8:1). Hashmi and Chughtai [163] chromatographed chromate, chloride, bromide, iodide, bromate, chlorate, ferricyanide, ferrocyanide, thiocyanate, arsenite, and sulfite on silica and on alumina using circular thin-layer chromatography with various solvents, including butanol–water–pyridine (2:2:1) and acetone. Canić et al. [164] separated NO_2^-, $S_2O_3^{2-}$, CrO_4^{2-}, N_3^-, Cn^-, SCN^-, BO_3^{3-}, S^{2-}, AsO_4^{4-}, NO_3^-, SO_4^{2-}, and PO_4^{3-} on maize starch with acetone–3 *M* ammonia (1:1).

References

1. J. E. Meinhard and N. F. Hall, *Anal. Chem.*, **21,** 185 (1949).
2. H. Seiler and M. Seiler, *Helv. Chim. Acta,* **43,** 1939 (1960).
3. H. Seiler, *Helv. Chim. Acta,* **45,** 381 (1962).
4. F. W. H. M. Merkus, "Kwalitatieve Analyse van Kationen met Behulp van Dunnelaagchromatografie," Thesis, Amsterdam, 1966.
5. M. Lederer, *Chromatogr., Rev.,* **9,** 115 (1967).
6. M. Lesigang-Buchtela, *Oesterr. Chem. Ztg.,* **67** 115 (1966).
7. M. P. Volynets and A. N. Ermakov, *Usp. Khim.,* **39,** 934 (1970).
8. J.-P. Garel, *Bull. Soc. Chim. Fr.,* **1965,** 1899
9. S. Takitani and K. Kawanabe, *Kagaku No Ryoiki, Zokan* No. **64,** 221 (1964); through *Chem. Abstr.,* **62,** 10800c (1965).
10. F. W. H. M. Merkus, "Progress in Inorganic Thin-Layer Chromatography," in *Progress in Separation and Purification,* Vol. 3, E. S. Perry and C. J. Van Oss, Eds., Wiley-Interscience, 1970, p. 234.
11. L. Fishbein, *Chromatogr., Rev.,* **15,** 195 (1971).

11.a U. A. T. Brinkman and G. de Vries, *J. Chromatogr.,* **85,** 187 (1973).

12. P. Kuenzi, dissertation, Basel University, 1962.
13. P. Kuenzi, J. Baeumler, and J. I. Obersteg, Dtsch. *Z. Ges. Gerichtl. Med.,* **52,** 605 (1962).
14. K. N. Johri and N. K. Kaushik, *Indian J. Appl. Chem.,* **33,** 173 (1970).
15. K. N. Johri and H. C. Mehra, *Chromatographia,* **4,** 80 (1971).
16. F. Baffi, A. Dadone, and R. Frache, *Chromatographia,* **9,** 280 (1976).
17. J. Rai and V. P. Kukreja, *Chromatographia,* **2,** 404 (1969).
18. K. Oguma, *Talenta,* **14,** 685 (1967).
19. J. Rai and V. P. Kukreja, *Int. Symp. Chromatogr. Electrophor., Lect. Pap., 6th 1970,* Ann Arbor Science, Ann Arbor, Mich., 1971, p. 453.
20. H.-J. Senf, *J. Chromatogr.,* **21,** 363 (1966).
21. G. Schwedt and C. Lippmann, *Dtsch. Lebensm.-Rundsch.,* **70,** 204 (1974).
22. M. Hranisavljević-Jakovljević, I. Pejković-Tadić, and K. Jakovljević, "Thin-Layer Chromatography of Inorganic Ions," in *Thin-Layer Chromatography,* G. B. Marini-Bettólo, Ed., Elsevier, Amsterdam, 1964, p. 221.
23. S. N. Tewari and N. Bhatt, *Chromatographia,* **5,** 624 (1972).
24. Z. Gregorowicz, J. Kulicka, and T. Suwinska, *Chem. Anal. (Warsaw),* **16,** 169 (1971).

25. P. Baudot, J. L. Monal, M. H. Livertoux, and R. Truhaut, *J. Chromatogr.*, **128,** 141 (1976).
26. Y. Tsunoda, T. Takeuchi, and Y. Yoshino, *Nippon Kagaku Zasshi,* **85,** 275 (1964).
27. *Ibid.*, p. 103.
28. Y. Tsunoda, T. Takeuchi, and Y. Yoshino, *Sci. Papers Coll. Gen. Educ., Univ. Tokyo,* **14,** 63 (1964); through *Chem. Abstr.*, **61,** 15325 (1964).
29. Z. Masoomi and D. T. Haworth, *J. Chromatogr.*, **48,** 581 (1970).
30. J. Vanderdeelen, *J. Chromatogr.*, **39,** 521 (1969).
31. M. R. Verma and J. Rai, *Int. Symp. Chromatogr., Electrophoresis, 4th, 1966,* Ann Arbor Science, Ann Arbor, Mich., 1968, p. 544.
32. H. Seiler and M. Seiler, *Helv. Chim. Acta,* **44,** 939 (1961).
33. M. Leisigang-Buchtela and K. Buchtela, *Mikrochim. Acta.*, **1967,** 670.
34. C.-T. Hu, *K'o Hsueh T'ung Pao,* **1963,** 63; through *Chem. Abstr.*, **60,** 4757 (1964).
35. A. M. Phipps, *Anal. Chem.*, **43,** 467 (1971).
36. D. S. Gaibakyan and R. S. Babayan, *Zavodsk. Lab.*, **37,** 9 (1971).
37. Z. Šoljić and V. Marjanović, *Z. Anal. Chem.*, **242,** 245 (1968).
38. Z. Šoljić, S. Turina and V. Marjanović, *Mikrochim. Acta,* **1969,** 894.
39. E. Gagliardi and B. Brodar, *Chromatographia,* **2,** 267 (1969).
40. D. F. Zaye, R. W. Frei, and M. M. Frodyma, *Anal. Chim. Acta,* **39,** 13 (1967).
41. V. Miketuková and R. W. Frei, *J. Chromatogr.*, **47,** 427 (1970).
42. G. Buchbauer, *Sci. Pharm.*, **40,** 190 (1972).
43. F. W. H. M. Merkus, *Int. Symp. Chromatogr., Electrophor., 5th 1968,* Ann-Arbor-Humphrye Science, Ann Arbor, Mich., 1969, p. 72.
44. M. Lesigang-Buchtela, *Mikrochim. Acta,* **1966,** 408.
45. M. Lesigang-Buchtela and K. Buchtela, *Mikrochim. Acta,* **1967,** 570.
46. P. Markl and F. Hecht, *Mikrochim. Acta,* **1963,** 889.
47. *Ibid.*, p. 970.
48. E. Cerrai and C. Testa, *J. Chromatogr.*, **5,** 442 (1961).
49. C. Testa, *J. Chromatogr.*, **5,** 236 (1961).
50. R. J. T. Graham, L. S. Bark, and D. A. Tinsley, *J. Chromatogr.*, **35,** 416 (1968).
51. D. McCormick, R. J. T. Graham, and L. S. Bark, *Int. Symp. Chromatogr., Electrophor., 4th, 1966,* Ann-Arbor Science, Ann Arbor, Mich., 1968, p. 199.
52. R. J. T. Graham, L. S. Bark, and D. A. Tinsley, *J. Chromatogr.*, **39,** 200 (1969).
53. R. J. T. Graham and A. Carr, *J. Chromatogr.*, **46,** 293 (1970).
54. *Ibid.*, p. 301.
55. U. A. T. Brinkman and G. de Vries, *J. Chromatogr.*, **18,** 142 (1965).
56. U. A. T. Brinkman, G. de Vries, and E. Van Dalen, *J. Chromatogr.*, **22,** 407 (1966).
57. *Ibid.*, **23,** 287 (1966).
58. *Ibid.*, **25,** 447 (1966).

59. U. A. T. Brinkman, P. J. J. Steerenburg, and G. de Vries, *J. Chromatogr.*, **54,** 449 (1971).
60. U. A. T. Brinkman and G. de Vries, *J. Chromatogr.*, **56,** 103 (1971).
61. H. R. Leene, G. de Vries, and U. A. T. Brinkman, *J. Chromatogr.*, **80,** 221 (1973).
62. J. Rai and V. P. Kukreja, *Chromatographia,* **3,** 499 (1970).
63. R. M. Cassidy, V. Miketuková, and R. W. Frei, *Anal. Lett.*, **5,** 115 (1972).
64. R. Olsina, R. Dapas, and C. Marone, *J. Chromatogr.*, **75,** 93 (1973).
65. A. Breccia and F. Spalletti, *Nature,* **198,** 756 (1963).
66. A. Breccia, *Metodi Sep. Nella Chim. Inorg.*, **2,** 137 (1963).
67. H. Seiler and W. Rothweiler, *Helv. Chim. Acta,* **44,** 941 (1961).
68. R. Kuroda and K. Oguma, *Anal. Chem.*, **39,** 1003 (1967).
69. G. Bottura, A. Breccia, F. Marchetti, and F. Spalletti, *Ric. Sci., Rend., Ser. A,* **6,** 373 (1964).
70. G. Bottura and A. Breccia, *Ric. Sci.*, **37,** 295 (1967).
71. A. Moghissi, *J. Chromatogr.*, **13,** 542 (1964).
72. M. P. Volynets and L. I. Guseva, *Zh. Anal. Khim.*, **23,** 947 (1968).
73. H. Seiler and M. Seiler, *Helv. Chim. Acta,* **53,** 601 (1970).
74. K. S. Babayan and G. M. Varshal, *Zh. Anal. Khim.*, **28,** 921 (1973).
75. N. S. Vagina and M. P. Volynets, *Zh. Anal. Khim.*, **23,** 521 (1968).
76. M. P. Volynets, N. S. Vagina, T. V. Fomina, and L. K. Fakina, *Zh. Anal. Khim.*, **24,** 1477 (1969).
77. N. S. Vagina and L. K. Fakina, *Zavodsk. Lab.*, **34,** 928 (1968).
78. K. Oguma, *Talanta,* **15,** 860 (1968).
79. T. B. Pierce and R. F. Flint, *Anal. Chim. Acta,* **31,** 595 (1964); trhough *Chem. Abstr.*, **62,** 4592 (1965).
80. A. Daneels, D. L. Massert, and J. Hoste, *J. Chromatogr.*, **18,** 144 (1965).
81. H. Holzapfel, Le-Viet-Lan, and G. Werner, *J. Chromatogr.*, **20,** 580 (1965).
82. *Ibid.*, **24,** 153 (1966).
83. H. Seiler, "Thin-layer Chromatography of Inorganic Ions," in *Thin-Layer Chromatography,* E. Stahl, Ed., Academic Press, New York, 1965, p. 478.
84. H. Hammerschmidt and M. Mueller, *Papier,* **17,** 448 (1963).
85. H. Seiler, *Helv. Chim. Acta,* **46,** 2629 (1963).
86. S. J. Purdy and E. V. Truter, *Analyst (London),* **87,** 802 (1962).
87. G. E. Janauer, J. D. Carrano, and R. C. Johnston, *Mikrochim. Acta,* **1968,** 61.
88. M. Lesigang-Buchtela, *Mikrochim. Acta,* **1969,** 1027.
89. A. C. Handa and K. N. Johri, *Chromatographia,* **4,** 530 (1971).
90. F. H. Pollard, J. F. W. McOmie, and H. M. Stevens, *J. Chem. Soc.*, **1951,** 771.
91. K. N. Johri, B. S. Saxena, and H. C. Mehra, *Chromatographia,* **4,** 351 (1971).
92. M. R. Verma and J. Rai, J. *Less-Common Metals,* **15,** 237 (1968).
93. D. Yamamoto and H. Uno, *J. Chromatogr.*, **51,** 348 (1970).
94. M. H. Hashmi and A. S. Adil, *Mikrochim. Acta,* **1968,** 947.
95. M. H. Hashmi and A. S. Adil, *Z. Anal. Chem.*, **277,** 170 (1967).

96. D. S. Gaibakyan and R. T. Egikyan, *Arm. Khim. Azh.*, **23,** 16 (1970).
97. E. Gagliardi and H. Shambri, *Mikrochim. Acta,* **1969,** 155.
98. S. Takitani, M. Fukazawa, and H. Hasegawa, *Bunseki Kagaku,* **12,** 1156 (1963) through *Chem. Abstr.,* **60,** 4767 (1964).
99. S. Takitani, N. Fukuoka, Y. Iwasaki, and H. Hasegawa, *Bunseki Kagaku,* **13,** 469 (1964).
100. L. F. Druding, *Anal. Chem.* **35,** 1582 (1963).
101. F. W. H. M. Merkus, *Pharm. Weekbl.,* **98,** 947 (1963).
102. M. J. Zetlmeisl and D. T. Haworth, *J. Chromatogr.,* **30,** 637 (1967).
103. N. K. Kaushik and K. N. Johri, *Chromatographia,* **9,** 233 (1976).
104. M. H. Hashmi, M. A. Shahid, A. A. Ayaz, F. R. Chugtai, N. Hassan, and A. S. Adil, *Anal. Chem.,* **38,** 1554 (1966).
105. S. W. Husain and F. Eivazi, *Chromatographia,* **8,** 277 (1975).
106. B. A. Zabin and C. B. Rollins, *J. Chromatogr.,* **14,** 534 (1964).
107. K. Oguma, *Chromatographia,* **8,** 669 (1975).
108. R. Kuroda, N. Kojima, and K. Oguma, *J. Chromatogr.,* **69,** 223 (1972).
109. R. Frache and A. Dadone, *Chromatographia,* **4,** 156 (1971).
110. *Ibid.,* **6,** 475 (1973).
111. T. Shimizu, Y. Kogure, H. Aria, and T. Suda, *Chromatographia,* **9,** 85 (1976).
112. L. Lepri and P. G. Desideri, *J. Chromatogr.,* **84,** 155 (1973).
113. L. Lepri, P. G. Desideri, V. Coas, and D. Cozzi, *J. Chromatogr.,* **47,** 442 (1970).
114. J. A. Berger, C. Meyniel, J. Petit, and P. Blanquet, *Bull. Soc. Chim. Fr.,* **1963,** 2662.
115. L. Lepri, P. G. Desideri, and R. Mascherini, *J. Chromatogr.,* **70,** 212 (1972).
116. T. Shimizu and A. Muto, *J. Chromatogr.,* **88,** 351 (1974).
117. K. Ishida, *Bunseki Kagaku,* **19,** 1250 (1970); through *Chem. Abstr.,* **74,** 68148n (1971).
118. J. A. Berger, C. Meyniel, and J. Petit, *Compt. Rend.,* **259,** 2231 (1964).
119. J. A. Berger, G. Meyniel, and J. Petit, *J. Chromatogr.,* **29,** 190 (1967).
120. M. Lesigang, *Mikrochim. Acta,* **1964,** 34.
121. J. van R. Smit, J. J. Jacobs, and W. Robb, *J. Inorg. Nucl. Chem.,* **12,** 95, 104 (1959).
122. M. Lesigang and F. Hecht, *Mikrochim. Acta,* **1964,** 508.
123. S. Kawamura, K. Kurotaki, H. Kuraku, and M. Izawa *J. Chromatogr.,* **26,** 557 (1967).
124. K. Oguma and R. Kuroda, *J. Chromatogr.,* **61,** 307 (1971).
125. *Ibid.,* **52,** 339 (1970).
126. R. Kuroda and T. Kondo, *J. Chromatogr.,* **80,** 241 (1973).
127. R. Kuroda, N. Yoshikuni, and K. Kawabuchi, *J. Chromatogr.,* **47,** 453 (1970).
128. B. Pfrunder, R. Zurflueh, H. Seiler and H. Erlenmeyer, *Helv. Chim. Acta,* **45,** 1153 (1962).
129. A. Moghissi, *Anal. Chim. Acta,* **30,** 91 (1964).

130. S. Takitani, M. Suzuki, N. Fujita, and K. Hozumi, *Bunseki Kagaku,* **14,** 597 (1965).
131. K. Buchtela and M. Lesigang-Buchtela, *Mikrochim. Acta,* **1967,** 380.
132. T. P. Makarova and A. V. Stepanov, *Zh. Anal. Khim.,* **26,** 1823 (1971).
133. R. Frache and A. Dadone, *Chromatographia,* **6,** 430 (1973).
134. *Ibid.,* p. 266.
135. W. J. Van Ooij and J. P. W. Houtman, *Z. Anal. Chem.,* **236,** 407 (1969).
136. H. Seiler, *Helv. Chim. Acta,* **44,** 1753 (1961).
137. M. E. Q. Pilson and R. J. Fragala, *Anal. Chim. Acta,* **52,** 553 (1970).
138. T. Roessel, *Z. Anal. Chem.,* **197,** 333 (1963).
139. J. Aurenge, M. Degeorges, and J. Normand, *Bull. Soc. Chim. Fr.,* **1964,** 508.
140. N. L. Clesceri and G. F. Lee, *Anal. Chem.,* **36,** 2207 (1964).
141. M. Baudler and F. Stuhlmann, *Naturwissenschaften,* **51,** 57 (1964).
142. M. Baudler and M. Mengel, *Z. Anal. Chem.,* **206,** 8 (1964).
143. *Ibid.,* **211,** 42 (1965).
144. M. Covello and O. Schettino, *Farmaco (Pavia), Ed. Prat.,* **20,** 396 (1965).
145. H. Seiler and T. Kaffenberger, *Helv. Chim. Acta,* **44,** 1282 (1961).
146. S. Tustanowski, *J. Chromatogr.,* **31,** 270 (1967).
147. S. M. Petrović and V. D. Canić, *Z. Anal. Chem.,* **228,** 339 (1967).
148. D. T. Haworth and R. M. Ziegert, *J. Chromatogr.,* **38,** 544 (1968).
149. J. A. Berger, G. Meyniel, and J. Petit, *Compt. Rend,* **255,** 1116 (1962).
150. M. Muto, *Nippon Kagaku Zasshi,* **85,** 147 (1964); through *Chem. Abstr.,* **61,** 15326 (1964).
151. *Ibid.,* **86,** 91 (1965).
152. H. Seiler and H. Erlenmeyer, *Helv. Chim. Acta,* **47,** 264 (1964).
153. A. C. Handa and K. N. Johri, *Talanta,* **20,** 219 (1973).
154. S. Palagyi and M. Zaduban, *Chem. Zvesti,* **23,** 876 (1969).
155. M. Lederer and M. Sinibaldi, *J. Chromatogr.,* **60,** 275 (1971).
156. W. Peschke, *J. Chromatogr.,* **20,** 572 (1965).
157. A. M. Phipps, *Anal. Chem.,* **43,** 467 (1971).
158. K. N. Johri, N. K, Kaushik, and K. Singh, *Mikrochim. Acta,* **1969,** 737.
159. J. Siechowski, *Chem. Anal. (Warsaw),* **9,** 391 (1964).
160. F. Dobici and G. Grassini, *J. Chromatogr.,* **10,** 98 (1963).
161. K. Kawanabe, S. Takitani, M. Miyazaki, and Z. Tamura, *Bunseki Kagaku,* **14,** 354 (1965).
162. *Ibid.,* **13,** 976 (1964).
163. M. H. Hashmi and N. A. Chughtai, *Mikrochim. Acta,* **1968,** 1040.
164. V. D. Canić, M. N. Turčić, M. B. Bugarski-Vojinović, and N. U. Perišić, *Z. Anal. Chem.,* **229,** 93 (1967).

Appendix **A**

REAGENTS FOR THE DETECTION OF COMPOUNDS OR TYPES OF COMPOUNDS*

	Test Number
Acetyoxymercuric-methoxy derivatives	104
Acetylated citrate esters	204
Acetylene derivatives	86, 185
Acids	42, 71, 82
Acyl sugar derivatives	225
Adenine	111, 273
Adenyl compounds	273
Adrenaline and derivatives	85, 200
Alcohols, higher	274
Aldehydes	77, 96, 140, 128, 137
Aldoses	35, 97
Alkali metals	1
Alkaline earth metals	1, 145
Alkaloids	22, 51, 53, 55, 90, 111, 112, 148, 149, 190, 209
Alkaloids, pyrrolizidine	244
Alkylated purine and pyrimidine bases	61
Aloe compounds	83
Aliphatic amines (secondary)	238
Amides	48
Amines	88, 177
Amines, aromatic	78, 129, 139, 181, 201, 249, 252
Amines, secondary	58, 128
Amines, aliphatic	238, 239
Amino acids	152, 153, 176, 177, 178, 198, 205, 267
Amino acids, primary	128
Amino acids, sulfur containing	147, 237
Amino sugars	11, 177
Amphetamines	264

* See also specific chapters or sections for additional tests.

	Test Number
Anesthetics	264
Anions	24
Anthocyanidins	8
Anthraquinones	116
Antihistamines	21, 264
Antioxidants	85, 121, 179, 195
Arbutin	166
Argenine	239, 271
Aromatic amines	68, 79, 120, 129, 139, 184, 201, 206, 236, 249, 252
Aromatic ethers	87
Aromatic nitro compounds	204, 247
Aromatic sulfur compounds	241
Arsenate and arsenite	226
Arylhydroxyamines	62
Aryl-*N*-methyl carbamates	181
Barbiturates	67, 161, 166, 224, 228, 264
Barbiturates (with double bonds in side chain at C-5)	46
Basic compounds with primary amine or reactive methylene groups	235
Bile acids	26, 52, 160, 216
Bile salts	216
Biotin	61, 84, 209
Bufadienolides	57
Caffein	56, 59, 61, 166
Cannabidiol	34, 115
Cannabiodolic acid	34
Carbamates	91, 274
Carbazoles	79, 80
Carbobenzoxy amino acids	61
Carbobenzocyamino acid esters	60
Carbohydrates	9, 7, 10, 23, 24, 27, 157, 171, 193, 259
Carbonyls	95
Carboxin	229
Carbromal	161
Cardenolides	57, 155, 262, 263
Carotenoid aldehydes	223
Carotenoids	31
Catechol amines	85, 114, 180, 181, 200, 249
Cations, inorganic	2, 103, 109, 110, 142,

	Test Number
	145, 188, 203, 215, 219, 220, 253
Cations, H_2S group	206
Cations, heavy metals	103, 110, 203
Cerebrosides	29
Chalcones	141
Chlorinated hydrocarbons (see next line)	
Chlorinated pesticides	35, 94, 99, 101, 173, 226, 229, 231, 260
Chlorophenols	252
Chloroquin	180
Cholesterol	160, 197
Choline	107
Citrate esters	204
Cobalt	2, 109, 145, 215, 256
Colchicine	55
Copper	2, 203, 215, 256
Coumarines	102, 204
Creatine and creatinine	239
Cyanamide derivatives	239
Cyanocobalamin	61
Cyclohexylamines	112
Decongestants	264
2-Deoxysugars	76
Dicaboxylic acids	43
Diethanolamine	238
N,*N*-Diethyl-D-lysergamide (LSD)	89
Digitalis glycosides	25, 57, 63, 155, 262
4-Dimethylaminoazo benzene	141
3,5-Dinitrobenzamides	173
3,5-Dinitrobenzoates	173
2,4-Dinitrophenylamino acids	184
2,4-Dinitrophenylhydrazones	201, 233
1,2-Diols	157
Disulfides	237
Diuretics	75, 235, 240, 264
Dipeptides	176
Diphenylamine	49
Diphosphoric acids	16
Drugs (many classes)	264
Erythromycins	139
Essential oils (see under terpenes)	
Esters	144
Esters (differentiate saturated and unsaturated)	123

	Test Number
Imidazoles	68, 69, 181
Imides	48
Imino acids	152
Indoles	50, 65, 132, 181
Insecticide synergists	64
Iodides, organic	55
Iodoamino acids	122
Keto acids	82
Ketones	77, 96
Ketoses	29, 96, 97
Δ^4-3-Keto-C_{21}-steroids	6, 33
Lactams	112
Lactones	144
Lipids	47, 66, 81, 98, 112, 138, 186, 195, 197, 209, 222
Lipids, phospho	125, 167
Lipoproteins	248a
Lithium	2, 276
Menthofuran	262
Mercaptans	237
Mercury ions	2, 203
Meta-Systox	74
Methanol (in ethanol)	246
Methionine	241
Methoxycinnamic acids	8
Methylated hydrazines	131
3,4-Methylenedioxyphenyl compounds	64, 136
Methyprylon	161
Monoglycerides, unsaturated	192
Monophosphoric acids	16
Morpholine	238
Mucopolysaccharides	261
Naphthaquinones	175
Narcotics	264
Narcotine	64
Neomycines	239
Nickel	93, 256
Nicotinic acid	5
Nicotinic acid amide	5, 61
Nitrate esters	99
Nitro derivatives	88
Nitro compounds, aromatic	204
Nitrogen compounds, organic	111, 112, 148
Nitrophenols	45

	Test Number
Nitrosamines	100
Nitrose diphenylamines	88
Organic sulfite pesticides	266
Organic tin salts	110, 217
Organomercury fungicides	110
Organophosphorus compounds	17, 41, 68
Organophosphorus pesticides	16, 19, 44, 113, 124, 165, 183
Organothiophosphates	251
Oximes	73
Penicillins	147, 150
Peptides, cyclic	48, 198
Peroxides	20, 94, 207
Peroxides, stable	128
Persulfates	35
Pesticides, carbamate	274
Pesticides, chlorinated	94, 99, 101, 173, 230, 260
Pesticides, chlorophenolic esters	252
Pesticides, organic sulfites	159, 265
Pesticides, organophosphorus	113, 165, 183
Pesticides	101, 124, 189, 230
Phenol carboxylic acids	131
Phenolic steroids	120
Phenols	22, 36, 40, 50, 79, 84, 108, 115, 116, 117, 118, 120, 124, 131, 166, 180, 181, 245, 249, 252, 274
Phenols (except *p*-substituted)	264
Phenothiazines	89, 118, 119, 133, 189, 190, 195, 209
Phenylthiohydantoins	147
Phenylurea derivatives	91
Phosphate ions	19, 226
Phosphate groups (in lipids)	125
Phosphite ions	19, 226
Phospholipids	14, 167, 196, 234, 243
Phosphoric mono- and diesters	79
Phosphorthioic mono- and diesters	79
Piperonyl butoxide	250
Plasticizers	180, 221, 258, 274
Polyacetylene compounds	86
Polyalcohols	12, 254, 257
Polyeneacids	192

	Test Number
Polyene aldehydes	223
Polyethylene glycols	112, 148
Polyphenyls	54
Polysaccharides (acid)	66
Potassium	70
Potassium (in high concentrations)	276
Primary aromatic amines	58, 78, 62
Prostaglandins	27
Proteins	3, 128, 169, 170, 181, 199, 261
Psychotropic drug	65, 264
Purines	56, 61
Pyrazoles	237
Pyrazolones	118
Pyrethrins	250
Pyridine compounds	5
Pyridoxine	61
α- and γ-Pyrones	102
Pyrrolizidine alkaloids	244
Quinidine	135
Quinine	135
Quinones	175
Quinones (not naphthaquinones)	164
Reducing compounds	24, 108, 197, 201, 212, 226, 255, 267
Resins	30
Riboflavin	61
Sapogenins	63, 276
Sedatives	59
Silicates	18
Sodium	276
Sphingolipids	37, 66
Steroids	22, 26, 32, 63, 106, 143, 154, 158, 163, 172, 186, 195, 197, 202, 211, 262, 267, 268, 269, 272, 274, 275, 276
Steroids, reducing	255
Steroids, α, β-unsaturated	112
Sterols, unsaturated	158
Steroid glycosides	31, 155
Steroid sapogenins	30, 31, 190, 276
Steroid sulfates	163
Steroids, keto-	95, 154

	Test Number
Sterols	63, 172, 197
Streptomycins	239
Sucrose esters	270
Sugar acetates	144
Sugar alcohols	157, 242
Sugar phenylhydrazones	162
Sugar phenylosazones	249
Sugars	7, 23, 26, 157, 162, 171, 225, 242, 259
Sugars, amino-	11, 177
Sugars, 2-deoxy-	76
Sugars, reducing	10, 24, 28, 226, 267
Sulfa drugs	264
Sulfhydryl groups	79
Sulfides	147, 252
Sulfonamides	78, 90, 92, 174, 236
Sulfones	252
Sulfoxides	252
Sulfur compounds	80, 241
Sulfur containing anions	226
Synergists, 3,4-methylenedioxyphenyl-	136
Systemic fungicides	112, 208
Systox	74
Terpene alcohols, carbonyls, ester, ethers, hydrocarbons, and oxides	105
Terpene phenols	118, 226
Terpenes	26, 30, 31, 63, 158, 195, 248, 274
Testosterone derivatives	96
Thiamine	61
Thiazide diuretics	235
Thiocyanate ions	118
Thiols	237
Thiophene derivatives	151
Thiophosphates	44, 124, 189
Thiophosphoric esters	147, 191
Thioureas	79, 239
Tocopherols	108, 121
Trialkylphosphates	68
Triazine herbicides	41
Triosephosphates	15, 194, 247
Trithiofluorobenzaldehydes	72
Tryptamine	120
Unsaturated compounds	130, 188, 186

Appendix **B**

ADDRESSES OF COMMERCIAL FIRMS CITED IN THE TEXT

Aldrich Chemical Co., Inc., 940 W. Saint Paul Ave., Milwaukee, Wis. 53233.

Allied Chemical Corporation, General Chemical Division, 40 Rector St., New York, N.Y. 10006

American Cyanamide Co., Industrial Chemicals and Plastics Division, Berdan Ave., Wayne, N.J. 07470.

Analabs, Analytical Engineering Laboratories, Inc., P.O. Box 5215, Hamden, Conn. 06514.

Analtech, Inc., 100 South, Justison St., Wilmington, Del. 19801.

Anton Paar, AG, Graz, Austria.

Applied Science Laboratories, Inc., P.O. Box 140, State College, Pa.

Arthur H. Thomas Co., Inc., Vine Street at Third, P.O. Box 779, Philadelphia, Pa. 19105.

Avicel Sales Division, American Viscose Co., FMC Corp., Marcus Hook, Pa.

Badische Anilin und Soda Fabrik, Ludwigshafen am Rhein, Germany.

Baird Atomic, 33 University Road, Cambridge, Mass. 02139.

Baird and Tatlock, Ltd., Chadwell Heath, Essex, England.

J. T. Baker Chemical Co., Phillipsburg, N.J. 08865.

Best Foods Division, CPC International, Inc., 25 Continental Dr., Wayne, N.J. 07674.

Biomed Instruments, Inc., Suite 2000, 6 N. Michigan Ave., Chicago, Ill., 60602.

Bio-Rad Laboratories, 32nd and Griffin Ave., Richmond, Calif. 94804.

Brinkmann Instruments, Inc., Cantiague Road, Westbury, N.Y. 11590.

Buchard Scientific, Ltd., Rickmansworth, Great Britain.

Calbiochem, 10933 N. Terrey Pines Rd., La Jolla, Calif. 92037.

CAMAG A.G., Muttenz B. L., Homburger, Str., 24 Switzerland.

CAMAG. Inc., 16229 W. Ryerson Rd., New Berlin, Wis., 53151.

Chemetron, Via Gustavo Modena, 24, Milano, Italy.

Chemical Research & Development Center, FMC Corp., Box 8, Princeton, N.J. 08540.

Chemirad Corp. P.O. Box 187, East Brunswick, N.J. 08816 (U.S. Rep. of Badische Anilin und Soda Fabrik).

Chemplast, Inc., 150 Dey Rd., Wayne, N.J. 07470.

Chen-Chin Trading Co., Ltd., #75, Section 1, Hankow St., Taipei, Taiwan, Republic of China.

Chinoin-Nagytétény, Budapest, Hungary.

Clarkson Chemical Co., Inc., 213 Main, S. W., Williamsport, Pa., 17707.

Clinton Corn Processing Co., Clinton, Iowa.

CoLab, Gleenwood, Ill.

Colab Laboratories, Inc., Chicago Heights, Ill.

Columbia Chemicals Co., Columbia, S.C.

Connaught Medical Laboratories, University of Toronto, Toronto, Canada.

Consolidated Laboratories, Inc., P.O. Box 234, Chicago Heights, Ill. 60412 (U.S. Rep. of Shandon Scientific Co., Ltd.).

Corning Glass Works, 1964 Crystal St., Corning, N.Y. 14830.

Crystal X Corp., Darky, Pa.

C. Desaga, G.m.b.H., Heidelberg, Haupstr. 60, Germany.

Distillation Products Industries, Ridge Road West, Rochester, N.Y. 14603.

The Dow Chemical Corp., 1000 Miami, Midland, Mich. 48640.

Drummond Scientific Co., 500 Parkway, Broomall, Pa. 19008.

E. I. du Pont de Nemours & Co., Photo Products Department, Wilmington, Del.

Dylon International, London, Great Britain.

Eastman Kodak Co., 343 State St., Rochester, N.Y. 14608.

E-C Apparatus Corp., 3831 Tyrone Blvd., N., St. Petersburg, Fla. 33709.

Fabwerke Hoechst, A.G., Frankfurt, am Main, Germany.

Fisher Scientific, Fair Lawn, N.J.

Floridin Co., 2 Gateway Center, Pittsburgh, Pa.

Fluka, A. G., Buchs S. G., Switzerland.

FMC Corporation, 2000 Market St., Philadelphia, Pa. 19103.

Gallard-Schlesinger Chemical Mfg. Corp., 580 Mineola Ave., Carle Place, Long Island, N.Y. 11514 (Rep. of Serva-Entwicklungslabor Co.).

Gelman Instrument Co., P.O. Box 1148, Ann Arbor, Mich. 48107.

General Aniline and Film Corp., Dyestuff and Chemical Division, 435 Hudson St., New York, N.Y. 10014.

General Electric Co., Insulating Materials Department, 1 Campbell Road, Schenectady, N.Y. 12306.

Glasurit-Werke M. Winkelmann A.G., Analytisches Laboritorium, 4403 Hiltrup, Westf., Germany.

M. Grumbacher, Inc., 466 W. 34th St., New York, N.Y. 10001.

Hamilton Co., Inc., P.O. Box 307, Whittier, Calif. 90608.

Harshaw Chemical Co., 1945 East 97th St., Cleveland, Ohio 44106.

Hercules Powder Co., 900 Market St., Wilmington, Del. 19801.

Hoffmann-La Roche, Inc., Kingsland Rd. and Bloomfield Ave., Nutley, N.J. 07110.

Ingold, A.G., Zurich, Switzerland.

Instrumentation Laboratory, Inc., Watertown, Mass.

Johns-Manville Corp., Celite Division, 22 E. 40th St., New York, N.Y. 10016.

Kensington Scientific Corp., 1717 Fifth Street, Berkeley, Calif. 94710.

Kontes, Vineland, N.J. 08360.

Leuchtstoffwerk, G.m.b.H. & Co., Vertriebsgesellschaft, Heidelberg, Germany.

Lightner Instrument Co., West Chester, Pa.

LKB Producter-AB, Stockholm, Sweden.

Macherey, Nagel & Co., Dueren, Rhineland, Germany.

Mallinckrodt Chemical Works, St. Louis, Mo. 63160.

K. Marggraf Co., Berlin, Germany.

Matheson, Coleman & Bell, 333 Paterson Plank Road, East Rutherford, N.J. 07073.

May and Baker, Ltd., Dagenheim, Essex RM107XS, England.

Mercer Chemical Corp., 216 Lake Ave., Yonkers, N.Y. 10701.

E. Merck, A.G., Chemische Fabrik, Darmstadt, Germany.

E. Merck, Ltd., Winchester Rd., Four Marks, Alton, Hampshire, Great Britain.

Microchemical Specialties Co., 1825 Eastshore Highway, Berkeley, Calif. 94710.

3M Company, Atlanta Branch, Industrial Chemical Division, 2860 Bankers Industrial Dr., Atlanta, Ga. 30360.

Packard Instrument Co., P.O. Box 428, La Grange, Ill. 60526.

Pallflex Products Corp., Glen Cove, N.Y. 11542.

Peninsular Chemical Research, Inc., Gainesville, Fla.

Pharmacia, Uppsala, Sweden.

Pharmacia Fine Chemicals, Inc., 501 Fifth Avenue, New York, N.Y. 10017 (U.S. Rep. of Pharmacia).

Philip A. Hunt Co., Palisade Park, N.J. 07650.

Pittsburgh Plate Glass Co., Chemical Division, One Gateway Center, Pittsburgh 22, Pa.

R.S.A. Corp., Ardsley, N.Y. 10502.

H. Reeve Angel & Co., Inc., 9 Bridewell Place, London, E.C. 4, Great Britain.

H. Reeve Angel & Co., Inc., 52 Duane St., New York, N.Y. 10007.

Regis Chemical Co., Morton Grove, Ill. 60053.

Research Specialties, Co., 200 S. Garrard Blvd., Richmond, Calif. 94804.

Rohm & Haas, Independence Hall W., Philadelphia, Pa. 19105.

C. Schleicher & Schuell Co., Dassel, Krs. Einbeck, Germany.

C. Schleicher & Schuell Co., 543 Washington St., Keene, N.H. 03431.

Serometrics, Inc., P.O. Box 66, Chicago Heights, Ill. 60412 (U.S. Rep. of Chemetron).

Serva-Entwicklungslabor Co., Heidelberg, Germany.

Shandon Scientific Co., Ltd., 65 Pound Lane, Willesden, London NW 10, Great Britain.

Spolana N.E., Neratovice, CSSR (Czechoslovakia).

Stein-Hall Co., Inc., 285 Madison Ave., New York, N.Y. 10017.

Technicon Corp., Tarrytown, N.Y. 10591.

Toyo Rayon Co., Tokyo, Japan.

G. K. Turner Associates, 2524 Pulgas Ave., Palo Alto, Calif. 94303.

Ultra-Violet Products, Inc., 5114 Walnut Grove Ave., San Gabriel, Calif. 91776.

Union Carbide Plastics Division, Clifton, N.J. 07012.

Union Carbide Corp., Film Packaging Division, 6733 West 65 St., Chicago Ill. 60638.

U.S. Radium Corp., 537 Pearl St., New York, N.Y. 10007.

Wako Pure Chemicals, Ltd., Osaka, Japan.

Warner-Chilcott Laboratories Instrument Division, Warner-Lambert Pharmaceutical Co., 20 South Garrard Blvd., Richmond, Calif. 94801.

Waters Associates, Inc. (U.S. distributer for Woelm), 61 Fountain St., Framingham, Mass. 10701.

Waverly Chemical Co., Mamaroneck, N.Y. 10543.

Westvaco Chloralkali Division, Food, Machinery & Chemical Corp., 161 East 42nd St., New York, N.Y. 10017.

West Virginia Pulp & Paper Co., Industrial Chemical Sales Division, 230 Park Ave., New York, N.Y. 10017.

Whatman, Inc., 9 Bridewell Place, Clifton, N.J. 07014.

M. Woelm Co., Eschwege, Germany.

COMPOUND INDEX

A

B

C

D

E

F

G

H

I

J

K

L

M

N

O

P

Q

R

T

U

V

W

SUBJECT INDEX

A

B

C

D

E

I

K

L

M

N

O

P

Q

R

S

T

U

V

W

X

Z